Emerging Human Viral Diseases, Volume II

Prudhvi Lal Bhukya • Anbazhagan Subbaiyan
Editors

Emerging Human Viral Diseases, Volume II

Encephalitic, Gastroenteric, and Immunodeficiency Viral Infections

Springer

Editors
Prudhvi Lal Bhukya
Preclinical Vaccine Testing Laboratory,
Rodent Experimentation Facility
ICMR-National Animal Facility Resource
Facility for Biomedical Research
Hyderabad, Telangana, India

Anbazhagan Subbaiyan
Preclinical Vaccine Testing Laboratory,
Rodent Experimentation Facility
ICMR-National Animal Facility Resource
Facility for Biomedical Research
Hyderabad, Telangana, India

ISBN 978-981-97-4479-4 ISBN 978-981-97-4480-0 (eBook)
https://doi.org/10.1007/978-981-97-4480-0

This Springer imprint is published by the registered company Springer Nature Singapore Pte Ltd.
The registered company address is: 152 Beach Road, #21-01/04 Gateway East, Singapore 189721, Singapore

I devote this little book to my parents Bharathi Bhukya, Balaji Bhukya, my mentor Dr. Kavita S. Lole, my better half Lavanya and to my little daughter Nievi. This book is also dedicated to the families of COVID-19 warriors.

Dr. Prudhvi Lal Bhukya

Preface

This book illustrates the importance and significance of the emerging human viral infectious diseases and their impact on global public health. It provides information on certain infectious human and animal viruses which cause life-threatening encephalitic, gastroenteric, immunodeficiency, and oncogenic viral diseases. The book also reflects the thorough compilation of the factors contributing to the emergence and re-emergence of viruses, rapid diagnostics, therapeutic and computational bioinformatics tools to combat and clearly understand transmission and emergence of viral pathogens. *Emerging Human Viral Infectious Diseases-Volume II* book clearly summarizes the host and viral factors favouring the emergence of novel encephalitic, gastroenteric, hepatic, oncogenic, and immunodeficiency viral infections.

The previous book, *Emerging Human Viral Diseases, Volume-I,* is particularly devoted to three diverse modules and they are characterized as (1) emerging respiratory viral diseases, (2) haemorrhagic fever viral infections, and (3) laboratory diagnosis of viral infections. These three modules concentrate on modules-specific determinants that relate to viral pathogenesis, pathobiology, mode of transmission, reservoir, high-risk populations and deal with many societal and environmental factors that can be instrumental in disease progression and emergence in detail.

This book *Emerging Human Viral Diseases, Volume II: Encephalitic, Gastroenteric, and Immunodeficiency Viral Infections* covers the in-depth recent updates on potentiality of emergence, host pathogenesis, diagnosis and treatment of emerging viral infections causing encephalitic, gastroenteric, immunodeficiency, and oncogenic diseases. As such, a comprehensive book on this subject is timely and needed to disseminate knowledge and insights to researchers. It would be relevant to healthcare practitioners, researchers, pharmaceutical and medical device industry professionals.

The book is organized into four parts offering the comprehensive updates on the emerging encephalitic, gastroenteric, immunodeficiency, and oncogenic viral diseases. *Part I* explicates the details about the familiar encephalitic viral infections like Nipah, Japanese encephalitis, chikungunya, Lymphocytic Choriomeningitis Virus (LCMV), West Nile virus, herpes viruses, Zika enteroviruses associated with acute flaccid paralysis, rabies, and Kunjin viral infection. Along with this viral infection, this also highlights the importance of genetic, molecular, and functional nature of prions, and also of the diseases caused by them. Over all this section addresses challenges to public health systems and ongoing efforts in prevention and control of these viral diseases. For researchers, healthcare professionals, and students, this is a resourceful impact on human and animal health. Covering historical context, transmission details, clinical manifestations, viral structure, replication processes, pathogenesis, immune response, virulence factors, diagnostics, therapeutics, and prevention, this section serves as a comprehensive guide to understand emerging encephalitis pathogens.

Part II gives a detailed insight into emerging gastroenteric viral diseases like Astro virus, Noro virus, Sapo virus, and Rota virus. This section aims to provide insights into different aspects of gastroenteric viral biology with a focus on the viral life cycle, immunopathogenesis, viral evolution, and genetic diversity. Further, these chapters discuss viral epidemiology and disease burden, clinical features of these emerging gastroenteric viral infections, as well as current strategies and challenges for diagnosis, prevention, and control.

Part III focuses on the emerging viral hepatitis which is primarily transmitted through the faecal-oral: hepatitis A virus, hepatitis E and blood-borne pathogens like hepatitis b and hepatitis. This section highlights the emerging and re-emerging concerns, immunopathogenesis, and challenges faced in designing the hepatitis E and hepatitis C vaccines.

Finally, *Part IV* emphasizes on the oncogenic and immunodeficiency viral diseases. This section highlights on the emerging concerns of HIV biology, human cytomegalovirus, Epstein-Barr virus, human papillomavirus, and human adult T-cell leukaemia virus Type 1. This section at the end also highlights on the CAR T technology targeting the viral diseases like Human Immunodeficiency Virus (HIV), Hepatitis B Virus (HBV), Hepatitis C Virus (HCV), CytoMegalo Virus (CMV), and Epstein-Barr Virus (EBV).

Over all this book is a valuable guide covering most recent scientific progress in the field of emerging viral diseases management and will serve as the best resource for undergraduates, graduates, medical professionals, researchers, public health physicians, and national and international health authorities. The information in this book can help them apply their expertise to more advanced study that could have positive societal and economic effects.

All together the present editorial work provides an in-depth understanding of emerging encephalitic, gastroenteric, and immunodeficiency viral infections. Our authors have explored all the current advancements in the treatment of emerging encephalitic, gastroenteric, and immunodeficiency viral infections. It gives us immense pleasure to present this comprehensive summary to the scientific community for the benefit of health care and society at large.

This book offers a piece of valuable information and updates to undergraduates, graduates, medical professionals, clinical researchers, social and basic scientists about preventing and controlling viral diseases and biosafety measures in combating emerging novel viral pathogens.

This book creates an awareness of the emerging virus and the global impact of emerging viral outbreaks. Lastly, the book also elucidates a comprehensive yet a representative description of a large number of challenges associated with the emerging viral diseases.

Hyderabad, Telangana, India Prudhvi Lal Bhukya
Hyderabad, Telangana, India Anbazhagan Subbaiyan

Contents

Editors and Contributors

About the Editors

Prudhvi Lal Bhukya is a Virologist currently serving as a Scientist in ICMR-National Animal Resource Facility for Biomedical Research (NARFBR), Hyderabad. Presently, he is involved in preclinical vaccine testing facility activities at NARFBR. His area of working includes preclinical vaccinology studies, host–viral interaction studies of emerging human viral infections, prevention and control of diseases, and development of tools to prevent outbreaks. He has done his Ph.D. from ICMR-National Institute of Virology-Pune on human hepatitis B virus. He proved his credentials by working on HBV and HEV with his publication record. He has presented his work in numerous national and international conferences. He has proved his accolades with several publications in international journals. He is a well-experienced manuscript writer, editor, and has contributed to several books and chapters as an author for publication bodies. His contribution to these books as an Editor will act as catalyst as his comments are really critical and it is an achievement for the book.

Anbazhagan Subbaiyan completed his undergraduate studies at TANUVAS, Tamil Nadu, and obtained his postgraduate degree from ICAR-IVRI, Uttar Pradesh. He has carried out his doctoral research in the field of veterinary microbiology at ICAR-IVRI, UP. During his doctoral studies, he obtained expertise working with risk group 2 and 3-level zoonotic pathogens. He had experience in the development of genome-edited vaccines for high-risk pathogens. He has expertise in small and large animal experiments for preclinical analysis of viral and bacterial candidate vaccines. He also has proficiency working with next-generation sequencing for emerging and re-emerging microbial genome characterization. Along with this, he possesses expertise in isolation, characterization, and analysis of several emerging antimicrobial resistance pathogens. Additionally, he conducted molecular docking studies for SARS-CoV-2, identifying therapeutic molecules and exploring the interaction of SARS-CoV-2 with ACE2 receptors in domestic and wild animals. His multidisciplinary approach displays a comprehensive understanding of veterinary microbiology and one health, contributing significantly to the understanding and development of strategies against infectious agents.

Contributors

Deeksha Agrawal IIS (Deemed to be University), Jaipur, Rajasthan, India

P. S. Akshay Hepatitis Division, ICMR-National Institute of Virology, Pune, Maharashtra, India

Ambika ICAR-Indian Veterinary Research Institute, Bareilly, Uttar Pradesh, India

Anbazhagan Subbaiyan Preclinical Vaccine Testing Laboratory, Rodent Experimentation Facility, ICMR-National Animal Resource Facility for Biomedical Research, Hyderabad, Telangana, India

Ripu Daman Arora Department of ENT & Head and Neck Surgery, AIIMS, Raipur, Chhattisgarh, India

Sampan Attri Viral Testing Facility, Forensic Science Laboratory, Mohali, Punjab, India

Vivekanand Bahuguna School of Biological Sciences, Doon University, Dehradun, Uttarakhand, India

Shruti Banswal Department of Biomedical Science, Shaheed Rajguru College of Applied Sciences for Women, University of Delhi, Delhi, India

Sridivya Bhagavatula Hepatitis Division, ICMR-National Institute of Virology, Pune, Maharashtra, India

Bacteriology Division, ICMR-National Institute of Virology, Pune, India

Mahalakshmi Bhat Integrative Virology, Center for Integrative Infectious Diseases Research (CIID), uniklinik Heidelberg, Germany

Aishwarya Bhatta Warwick Medical School, University of Warwick, Coventry, United Kingdom

Neha Bhise Hepatitis Division, ICMR-National Institute of Virology, Pune, Maharashtra, India

Prudhvi Lal Bhukya Preclinical Vaccine Testing Laboratory, Rodent Experimentation Facility, ICMR-National Animal Facility Resource Facility for Biomedical Research, Hyderabad, Telangana, India

Muneesh Kumar Burman Dr. Reddy's Institute of Life sciences, Hyderabad, India

Kailash Chand Dr. Reddy's Institute of Life Sciences, Hyderabad, Telangana, India

Sarah Cherian Bioinformatics Group, ICMR-NIV, Pune, Maharashtra, India

Shobha D. Chitambar Enteric Viruses Group, ICMR-NIV, Pune, Maharashtra, India

Srijita Chowdhury Department of Biotechnology, National Institute of Technology, Rourkela, Odisha, India

Prasad Deepak Dandekar MIT School of Bioengineering Sciences & Research, MIT Art, Design and Technology University, Pune, Maharashtra, India

Shalini Das ICMR-National Institute of Virology, Pune, Maharashtra, India

Manjusha Dharavath Department of Molecular Biology, Institute of Genetics and Hospital for Genetic Diseases, Osmania University, Hyderabad, Telangana, India

Abhranil Gangopadhayya Hepatitis Division, National Institute of Virology, Pune, Maharashtra, India

Santosh L. Gaonkar Department of Chemistry, Manipal Institute of Technology, Manipal Academy of Higher Education, Manipal, India

Vivek Kumar Garg Department of Medical Lab Technology, University Institute of Allied Health Sciences, Chandigarh University, Chandigarh, Punjab, India

Garima Department of Biotechnology, Delhi Technological University, Delhi, India

Shailender Gugalavath CAR T Cell Therapy Centre, The Advanced Centre for Treatment, Research and Education in Cancer, Tata Memorial Centre, Navi Mumbai, India

Department of Medical Oncology, Tata Memorial Hospital, Tata Memorial Center, Mumbai, India

Navita Gupta Department of Allied Health Sciences, Chitkara School of Health Sciences, Chitkara University, Rajpura, Punjab, India

Bharath Raja Guru Department of Biotechnology, Manipal Institute of Technology, Manipal Academy of Higher Education, Manipal, India

Sushruta S. Hakkimane Department of Biotechnology, Manipal Institute of Technology Bengaluru, Manipal Academy of Higher Education, Manipal, India

Sunita Jetly Department of Biomedical Science, Acharya Narendra Dev College, University of Delhi, New Delhi, India

Manisha Khatri Department of Biomedical Science, Shaheed Rajguru College of Applied Sciences for Women, University of Delhi, Delhi, India

Bidhan Chandra Koner Multidisciplinary Research Unit, Maulana Azad Medical College, and Associated Hospitals, New Delhi, India

Ruta Kulkarni Department of Communicable Diseases, Interactive Research School for Health Affairs, Bharati Vidyapeeth (Deemed to be University), Pune, Maharashtra, India

Neelesh Kumar Department of Biotechnology, Delhi Technological University, Delhi, India

Umesh Kumar Department of Biosciences, IMS Ghaziabad (University Courses Campus), Ghaziabad, Uttar Pradesh, India

Mansi Kumari Dr. D. Y. Patil Biotechnology and Bioinformatics Institute, Dr. D. Y. Patil Vidyapeeth, Pune, Maharashtra, India

Rongala Laxmivandana ICMR-National Institute of Virology (ICMR-NIV), Pune, Maharashtra, India

Varsha Mahadik Biochemical Sciences Division, National Chemical Laboratory, Pune, Maharashtra, India

Urmi Majumdar Interactive Research School for Health Affairs (IRSHA), Bharati Vidyapeeth (Deemed to be University), Pune, India

Shivani Manglic Department of Biotechnology, Jaypee Institute of Information and Technology, Noida, Uttar Pradesh, India

Salonee Martins ICMR-National Institute of Virology, Pune, Maharashtra, India

Anil Kumar Mavi Department of Pulmonary Medicine, Vallabhbhai Patel Chest Institute, University of Delhi, New Delhi, India

Department of Botany and Life Sciences, Sri Aurobindo College, University of Delhi, Delhi, India

G. K. Megha Division of Veterinary Public Health and Epidemiology, Indian Veterinary Research Institute, Bareilly, Uttar Pradesh, India

Suhas T. Mhaske Preclinical Vaccine Testing Laboratory, Rodent Experimentation Facility, ICMR-National Animal Facility Resource Facility for Biomedical Research, Hyderabad, Telangana, India

Piya Paul Mudgal Manipal Institute of Virology, Manipal Academy of Higher Education, Manipal, India

Sonal Mundhra Encephalitis Group, ICMR-National Institute of Virology, VRDC, Venky's India Ltd., Pune, Maharashtra, India

Nitin M. Nagarkar AIIMS, Raipur, Chhattisgarh, India

Gaurav Narula CAR T Cell Therapy Centre, The Advanced Centre for Treatment, Research and Education in Cancer, Tata Memorial Centre, Navi Mumbai, India

Department of Medical Oncology, Tata Memorial Hospital, Tata Memorial Center, Mumbai, India

Ambika Nayak ICAR-Indian Veterinary Research Institute, Bareilly, Uttar Pradesh, India

Sanjay Singh Negi State VRDL, Department of Microbiology, AIIMS, Raipur, Chhattisgarh, India

Albeena Nisar CAR T Cell Therapy Centre, The Advanced Centre for Treatment, Research and Education in Cancer, Tata Memorial Centre, Navi Mumbai, India

Tripti Pande Department of Veterinary Microbiology, ICAR-Indian Veterinary Research Institute, Bareilly, Uttar Pradesh, India

Archana Pandey Department of Biomedical Science, Acharya Narendra Dev College, University of Delhi, New Delhi, India

Deepali Pandit Dr. Bhubaneswar Borooah Cancer Institute, Guwahati, India

Khushali Pandit CAR T Cell Therapy Centre, The Advanced Centre for Treatment, Research and Education in Cancer, Tata Memorial Centre, Navi Mumbai, India

Department of Medical Oncology, Tata Memorial Hospital, Tata Memorial Center, Mumbai, India

Aishwarya Pathare ICMR-National Institute of Virology, Pune, Maharashtra, India

L. Preethi Public Health, London School of Hygiene and Tropical Medicine, London, UK

Zeenal Punamiya Department of Allied Health Sciences, Chitkara School of Health Sciences, Chitkara University, Rajpura, Punjab, India

Deepika Raina School of Pharmacy, Graphic Era Hill University, Dehradun, Uttarakhand, India

Mudasir M. Rather Department of Veterinary Microbiology, ICAR-Indian Veterinary Research Institute, Bareilly, Uttar Pradesh, India

Sakshi School of Biotechnology, Gautam Buddha University, Greater Noida, Uttar Pradesh, India

Shalini Sakthivel Bar-Ilan University, Azrieli Faculty of Medicine, Safed, Israel

Afnan Saleem Division of Animal Biotechnology, FVSc and AH, Shuhama, Sher-E-Kashmir University of Agriculture Sciences and Technology, Jammu, Kashmir, India

Purva Salvi ICMR-National Institute of Virology, Pune, Maharashtra, India

Sanjana Division of Biological Products, ICAR-Indian Veterinary Research Institute, Bareilly, Uttar Pradesh, India

Bunty Sharma Department of Bio-Sciences and Technology, Maharishi Markandeshwar Engineering College, Maharishi Markandeshwar (Deemed to be University), Ambala, Haryana, India

Ritu Sharma Department of Biochemistry, Government Medical College, Amritsar, Punjab, India

Ujjawal Sharma Department of Human Genetics and Molecular Medicine, School of Health Sciences, Central University of Punjab, Bathinda, India

Devendra Kumar Singh Department of Respiratory Medicine, School of Medical Sciences and Research, Sharda University, Noida, Uttar Pradesh, India

Pushpendra Singh State VRDL, Department of Microbiology, AIIMS, Raipur, Chhattisgarh, India

Satendra Singh Department of Biomedical Science, Acharya Narendra Dev College, University of Delhi, New Delhi, India

Madhav Singla Chitkara College of Pharmacy, Chitkara University, Rajpura, Punjab, India

Subhash Sonkar Multidisciplinary Research Unit, Maulana Azad Medical College, and Associated Hospitals, New Delhi, India

Korra Bhanu Teja Hepatitis Division, ICMR-National Institute of Virology, Pune, Maharashtra, India

Subrat Thanapati Department of Research and Development, Techinvention Lifecare Pvt Ltd., Mumbai, Maharashtra, India

Shilpa J. Tomar Hepatitis Division, ICMR-National Institute of Virology, Pune, Maharashtra, India

Koppu Vasavi Department of Veterinary Microbiology, Indian Veterinary Research Institute, Bareilly, Uttar Pradesh, India

Manisha Yadav Multidisciplinary Research Unit, Maulana Azad Medical College, and Associated Hospitals, New Delhi, India

Asima Zehra Centre for One Health, GADVASU, Ludhiana, Punjab, India

Part I

Viral Encephalitis: Familiar Infections and Emerging Pathogens

1 Nipah Virus: Through a Pandemic Lens

Purva Salvi, Salonee Martins, Shalini Das,
Mahalakshmi Bhat, and Aishwarya Pathare

Abstract

Nipah virus is an emerging, highly pathogenic virus belonging to the genus *Henipavirus* of the family *Paramyxoviridae*. It is a zoonotic virus; fruit bats are its natural reservoir. It was first discovered when it caused the 1998–1999 outbreak in pigs—which were the intermediate hosts, and humans, in Malaysia and Singapore. However, in Bangladesh and India where subsequent outbreaks have predominantly occurred, there has been no involvement of an intermediate host, and human-to-human transmission has been reported as well. In humans, the infection may cause acute respiratory disease and fatal encephalitis. Currently, no licensed vaccines or treatments are available. This chapter describes various aspects of the Nipah virus, emphasizes its significance as a public health threat and the importance of the one health approach in combating the disease.

Keywords

Nipah virus · Outbreaks · Zoonoses · Public health

P. Salvi (✉) · S. Martins (✉)
ICMR-National Institute of Virology, Pune, Maharashtra, India
e-mail: purvasalvi97@gmail.com; salonee.martins@gmail.com

S. Das · A. Pathare
Human Influenza Group, ICMR-National Institute of Virology, Pune, Maharashtra, India

M. Bhat

Integrative Virology, Center for Integrative Infectious Diseases Research (CIID), uniklinik Heidelberg, Germany

P. L. Bhukya, A. S (eds.), *Emerging Human Viral Diseases, Volume II*,
https://doi.org/10.1007/978-981-97-4480-0_1

1.1 Introduction

Nipah virus (NiV) was first confirmed in the countries of Malaysia and Singapore in the year 1998–1999 (CDC 1999). The Nipah virus was named after the village of Sungai Nipah, which is located by the Nipah River in Malaysia's Negeri Sembilan state. This village is the site at which the presence of the virus, when tested for IgM Abs against Hendra viral antigenic proteins in cerebrospinal fluid (CSF) samples, was authenticated in 1999, through sera specimens obtained from patients with symptomology of encephalitis (Chua et al. 1999; Parashar et al. 2000).

Subsequently, in the suburban area of Ipoh, part of Perak on the Malaysian peninsula, an outbreak of an acute febrile illness accompanied by encephalitis broke out among humans on a pig farm in late September 1998 and it displayed a high fatality rate with 15 deaths, 9 among which were confirmed for NiV infection following an autopsy. Preceding the outbreak, the pigs on this particular farm were plagued by a sudden insurgence of respiratory and encephalitis-related illness. The outbreak was initially allocated as Japanese encephalitis (JE) as 4 of the 28 patients' serum samples tested positive for IgM specific to JE, later confirmed by Centre for Tropical Disease in Japan in collaboration with the WHO, and with the help of RT-PCR testing conducted at the University of Malaya, nucleic acids belonging to JE were detected in some samples (Looi and Chua 2007).

Alas, the pig farms that had been affected by this outbreak had already sold pigs to farms throughout the country. Come February of the year 1999, outbreaks had occurred in Sikimat, Sungai Nipah and Bukit Pelandok in the region Negri Sembilan. This was some 300 km away from the original epicentre in Ipoh and this outbreak being more severe than the previous, contributed to around 180 cases of which 89 deaths occurred (Wong et al. 2002). However, following clandestine transactions and further movement of these infected pigs, cases were reported in Sungai Buloh and Sepang in Selangor. Subsequently, in March of 1999, slaughterhouse employees in Singapore who had direct contact with pigs imported from Malaysia reported 11 cases along with one death. By this time, the news of the outbreak had spread nationwide panic and almost led to the collapse of the pig-rearing industry of Malaysia (Looi and Chua 2007).

Healthcare workers involved with patients of this outbreak were suspicious that the causative agent was not JE, fairly early on in the outbreak (Tan and Wong 2003). This was due to the fact that many infected individuals had past immunizations to JE and the agent seemed to affect adults rather than children, which was unlike JE. Moreover, the autopsy results were inconsistent with those usually seen in a JE infection. A huge chunk of the victim pool was infected due to their direct contact with pigs, which was unlike any mosquito-borne disease and infected pigs manifested a severe bark-like cough and many succumbed to the illness. Unfortunately, the conflicting reports issued by the government got in the way of implementing suitable outbreak prevention and control strategies. The incident that finally acknowledged the emergence of a new agent was the isolation of NiV from CSF of an infected individual belonging to Sungai Nipah province by a virologist at the University of Malaya in March of 1999. When clubbed with HeV, the newly discovered genus was called Henipavirus, a combination of the words Hendra and Nipah, and they were designated to the Paramyxoviridae family (Chua 2003). Reports showed that HeV and NiV shared a high level of epitope similarity and; thus, HeV antigens were used in a prototype to detect NiV antibodies in patient samples via serological testing (Chua et al. 2000b). This was extremely helpful in the screening and diagnosis of NiV outbreaks that followed (Looi and Chua 2007).

The end of the outbreak in Singapore was brought about by placing a prohibition on the import of pigs reared in Malaysia and the closure of slaughterhouses; while in Malaysia, it was brought to a close by the culling of a million pigs following widespread surveillance of swine populations. The last human fatality was recorded on May 27th, 1999; however by this time, cases recorded having acute NiV-related encephalitis were 265 and the deaths recorded were 105, this brought the rate of mortality in Malaysia to up to 40% (Chua 2003).

1.2 Epidemiology

1.2.1 Age, Mortality and Morbidity

One of the major reasons to implement pertinent control measures in order to avoid an epidemic is the high case fatality rate which is between 40% and 75% for the disease caused by NiV (Narayanan 2020; World Health Organization 2021a). Reports made as per Kenmoe et al. encephalitis caused due to Nipah virus diseases has an overall 61% case fatality rate (CFR) (Kenmoe et al. 2019; Aborode et al. 2022). According to the successive suggestion made by Arunkumar et al., the CFR associated with Nipah virus disease (NVD) was estimated to be about 50–75%, where the patients lost the battle against the disease (Arunkumar et al. 2019a; Khan et al. 2022).

From 1998 to 2021, reports of more than 634 cases in people have been made (Skowron et al. 2022; Chaiyes et al. 2022). Case fatality rates during outbreaks have greatly varied from 40% to 90%, with lower CFR in the initial outbreaks in Malaysia. However, a high CFR was observed in the latest outbreaks that occurred in India and Bangladesh (Hossain et al. 2008; Goh et al. 2000; Arunkumar et al. 2019a; McEntire et al. 2021).

1.2.2 Origin and Transmission of Virus

NiV belongs to a group of zoonotic pathogens that have sources and transmission vectors that are wild as well as domesticated animals. During some outbreaks of the Nipah virus, the manifestation of the disease in humans has been correlated to the presence of certain animal species (Skowron et al. 2022). The main route of NiV spread from bats to humans, in South Asian countries such as Bangladesh, where the bat-contaminated unprocessed date palm sap is subsequently consumed by humans. However, in other Asian countries such as India and Malaysia, the intermediate hosts are domesticated animals that are likely to get infected by consuming foods contaminated with the saliva or urine of *Pteropus* spp. These livestock then transmit the disease to surrounding humans (Khan et al. 2022) (Fig. 1.1).

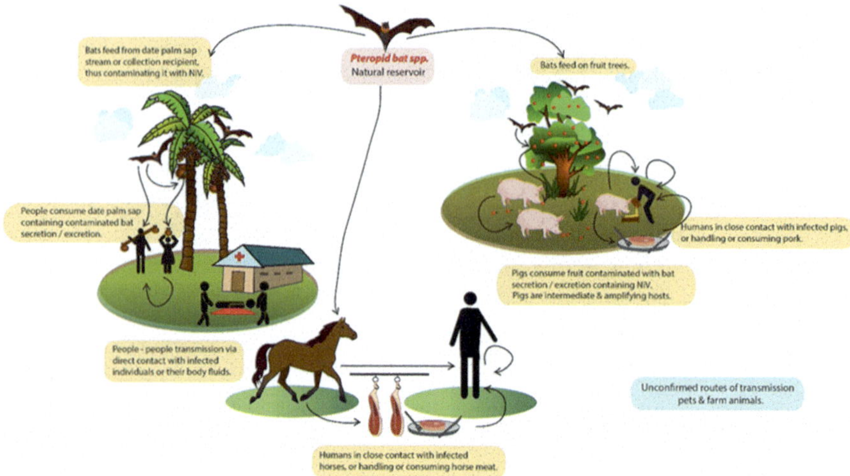

Fig. 1.1 Possible transmission routes of Nipah

1. **Livestock:** According to reports, an important route of the spread of the Nipah virus is direct personal contact between infected animals like pigs, goats and cattle, and humans (Luby et al. 2009). In Malaysia, pigs were associated to be amplifying hosts in the emergence of Nipah (CDC 1999; Chua et al. 1999; Mohd Nor et al. 2000). Some data obtained from Bangladesh suggest that domesticated species such as goats and cattle have also shown to be infected with NiV, while horses were found to be the intermediate reservoir for human infection with NiV in the Philippines (Chowdhury et al. 2014; Ching et al. 2015).
2. **Foodborne sources:** The potential of a NiV particle to infect and survive in fruit pulp and juices depends on the type and pH of the fruit, at that point in time (Fogarty et al. 2008). The likelihood of human infections occurring as a result of the consumption of contaminated fruit, which has been contaminated by bat saliva, is considered plausible (Luby et al. 2006; Chattu et al. 2018).
3. **Humans:** The risk of human spill-over depends on population density and the location of the village (Hahn et al. 2014).

1.2.2.1 Bats

During the early investigations of natural reservoirs for NiV, the focal point was held on to bats, which were known reservoirs of Hendra virus, a close relative of NiV. (Chua 2003). Subsequently, the virus was confirmed to be present in species such as *Pteropus vampyrus*, *Pteropus hypomelanus* and *Pteropus lylei* in Malaysia (Sharma et al. 2019). Meanwhile in India, the search for the identification of the NiV vector started with insectivorous bats called *Megaderma spasma*; the presence of the virus was substantiated in fruit bats that also belonged to the *Pteropus* spp., e.g. *Pteropus giganteus* (Plowright et al. 2019; Thakur and Bailey 2019). The first case of Nipah, in India, was detected in 2019, where the virus and IgG antibodies were proven in the bat species of *P. medius* (Sudeep et al. 2021). The Nipah viral genes obtained from infected bats and those taken from patient samples in India showed huge similarity in their sequences. In comparison, the degree of similarity between the genes obtained from Bangladesh, Cambodia and Malaysia was not that high. This inferred that the major source of the epidemic was bats. The epidemic could have been caused due to the inhalation of aerosols or the ingestion of fruits that have been contaminated by bat saliva or urine (Yadav et al. 2019). The coinciding dates of the palm fruit harvesting and juicing period, and that of the NiV epidemic in Bangladesh, confirm that unprocessed date palm juice could also be a significant source of NiV infection (Aditi and Shariff 2019).

When serologically tested by using ELISA, the bat samples from Bangladesh, Cambodia, Malaysia and Thailand indicated the presence of 9–25% of NiV antibodies (Sharma et al. 2019). The suggestion that the virus may have been present in the organs of bats for years and that it has been causing sporadic infections in intermediate hosts and humans was presented based on the fact that when tested, neither the wild NiV-positive bats nor those vaccinated in the laboratory with NiV showed a lack of any characteristic clinical features. This led to the conclusion that the only way to monitor and identify this virus is to develop appropriate diagnostic methods, which include the N gene-specific qRT-PCR for NiV (Thakur and Bailey 2019). At this point, the identification and confirmation of the presence of the virus are investigated by examining the body fluids of bats, and also other animals, such as faeces, saliva or urine (Looi and Chua 2007; Sharma et al. 2019; Sudeep et al. 2021). Detection of NiV or other viruses related to it, by surveillance, be it serological or genomic, has been successful in various bat species such as *P. hypomelanus* (Chua et al. 2002; Singh et al. 2019), *P. lylei* in Cambodia (Reynes et al. 2005) and *P. vampyrus* in Malaysia (Rahman et al. 2010).

Between September 1998 and April 1999, an outbreak was reported in pigs and humans in Malaysia. The number of infected individuals was 265, out of which 105 deaths were reported, and one million pigs were slaughtered. According to initial reports, the cause of the outbreak was considered to be Japanese encephalitis but the in vitro culturing of samples in Vero cells and an electron micrograph confirmed it to be Nipah virus (Chua et al. 2000a). It is reported that the virus is transmitted from bats to an intermediate host (mostly pigs) and then from the intermediate host to human beings.

In the strains of NiV isolated in Bangladesh (2001), heterogeneity was observed in their nucleotide sequences. This suggested that the virus was repeatedly introduced into the people of Bangladesh from the host (Gurley et al. 2007a; Montgomery et al. 2008; Chua et al. 2002; Hong-Wu et al. 2016; Khanna et al. 2009; Debnath et al. 2021). After the Bangladesh outbreak of 2001, two outbreaks were reported in India, in 2007 in West Bengal and in May 2018 in the state of Kerala. The reservoir for the outbreaks in Bangladesh and India was found to be *P. medius* (Hsu et al. 2004; Yadav et al. 2012). As reported by Chua et al., the presence of neutralizing antibodies in the wildlife species against the Nipah virus was discovered in Bangladesh, during an outbreak of *P. hypomelanus* (Chua et al. 2002). When the swabs of eaten fruits and urine samples, from these Island Flying Foxes, were cultured, a cytopathic effect of syncytia formation in Vero cells was reported (Gupta et al. 2021).

1.2.2.2 Pigs

During the Malaysian NiV epidemic, even though bats were considered the primary natural reservoir, the source of human infection was found to be pigs (Looi and Chua 2007). Here, the transmission of the Nipah virus happened due to close contact of humans with infected pigs. The virus isolated from the pigs of southern regions of Malaysia showed identical genetic markers to those of the virus found in humans (Abu Bakar et al. 2004). Direct association between pig-rearing activities, such as insemination, application of drugs or disposal of dead animals and infection in humans, has been confirmed (Parashar et al. 2000; Thakur and Bailey 2019). The case fatality rate in pigs infected by NiV is relatively low, even as the incidence is 100% (Mohd Nor et al. 2000). Similarly, NiV cases in Singapore have been confirmed to be a result of the direct exposure of pigs, their faeces and other bodily fluids, to those working at slaughterhouses (Paton et al. 1999; Chan et al. 2002; Aditi and Shariff 2019).

In the course of the Malaysian outbreak of NiV in 1998–1999, pigs suffered dearly; however, they were diagnosed as a fraction of an investigation into human infections. The degree of severity of disease manifestations in pigs was different in different age groups. Suckling pigs manifested a fever of 39.9 °C with respiratory symptoms like laboured respiration, a loud bark-like cough, breathing with an open mouth and epistaxis (www.oie. Int/en/animal-health-in-the-world/animal-diseases/Nipah-Virus). A few of these cases also manifested with neurological symptoms such as trembling, twitching, fasciculation of the muscle, tetanic spasms, poor coordination, weakness of limbs or partial body paralysis. Pigs at this stage of life showed low mortality (<5%) but high morbidity (OIE Nipah Virus 2018; Aziz et al. 1999; Mohd Nor et al. 2000). Few animals recorded to be over 6 months succumbed within 24 h without displaying any symptoms of clinical illness. In adult pigs, respiratory manifestations included difficulty in breathing, bloody discharge from the nose and an increase in salivation but were less obvious as compared to the young ones, while the neurological manifestations included pressing of heads, biting the surrounding rails, tetanic spasms and convulsions. Abortions during the first trimester were also reported (OIE Nipah Virus 2018; Aziz et al. 1999; Mohd Nor et al. 2000).

The various names for the disease caused in pigs by NiV were 'porcine respiratory and neurological syndrome,' 'porcine respiratory and encephalitis syndrome' or 'barking pig syndrome' (Malaysia) (Mohd Nor et al. 2000). NiV infection was included as the sixth pig-related disease that was notifiable to OIE, a World Organization for Animal Health (OIE 2018; McLean and Graham 2019).

During the Bangladesh and Indian epidemics, there was an implication of pigs as vectors of NiV but their impact with respect to infection was not observed (CDC 1999; Thakur and Bailey 2019). On another note, the people infected with NiV who declared zero contact with pigs acted as a stimulus to further search for hosts that could act as viral hosts and pose a threat to the human population. Various species, such as birds, bats, pigs and dogs, which live in close proximity with humans were assessed after the epidemic in Bangladesh, but no virus was found in them; however, antibody presence was detected in bats, which confirmed they could be accidentally infected (Hsu et al. 2004). Though Nipah virus was shown to infect domestic animals like cats and dogs, by direct contact with pig secretions, no definitive proof has been found referring to their role in causing further human infection. As the epidemic in Malaysia came to an end after the mass culling of pig herds, it was proven that the participation of other animals in the spread of the disease was negligible (Parashar et al. 2000; Chua 2003; Mills et al. 2009).

1.2.2.3 Cats, Dogs and Horses

Throughout the Malaysian outbreaks, several dogs residing on pig farms were seropositive for NiV; however, only two manifested the disease. These cases were, in all likelihood, due to them having been in direct contact with pigs or consuming pig meat or its products. When autopsies were conducted for the two dogs with a fatal disease, the results showed indications of meningitis, interstitial pneumonia and pulmonary oedema. One of the cases showed signs that were analogous to those observed in animals having canine distemper infection (Field et al. 2001; Hooper et al. 2001; Mills et al. 2009; Parashar et al. 2000; Lawrence and Escudero-Pérez 2022).

Similarly, during the same outbreak, many horses showed seroconversion and one developed meningitis. This infection in horses can be most likely attributed to the consumption of bat-contaminated fruits. Incidentally, in the Philippines, during an outbreak, horses were identified to be intermediate reservoirs. In the duration of this outbreak, ten fatal cases, in horses, were identified and many manifested neurological symptoms. However, it was impossible to detect NiV via histopathological analysis as no samples were available. A dog and four cats are suspected to have ingested infected horse meat and were found deceased (Ching et al. 2015; Lawrence and Escudero-Pérez 2022). There, however, have not been any reports of our canine and feline friends being identified as intermediate reservoirs nor have there been any reports of them working as transmission agents of NiV (Sizemore 2021).

1.2.3 Serosurveillance in Bats in India

Surveillance conducted in the Northeast in 2015 in India, disclosed the presence of the Nipah virus among bats of the species *P. medius*, these were collected from the Cooch district in West Bengal and the Dhubri district of Assam (Yadav et al. 2018). Similarly, surveillance was carried out among pigs in eight districts of the state of Mizoram in the Northeast of India; however, the samples collected from these pig populations tested negative for anti-NiV IgG Abs (Mourya et al. 2019; Gokhale et al. 2022).

During the period of 2018–2019, the most recent emergence of the Nipah virus was reported in the Western Ghats of the state of Kerala (Arunkumar et al. 2019a; Sahay et al. 2020). The subsequent investigation that followed the 2018 outbreak revealed that the index case was exposed to bats and then followed the human-to-human transmission (Paul 2018). Similarly, the principal source of infection in the single case found in 2019 in Kerala was also linked to bats; but there was no confirmed cause for the spill-over found. The presence of anti-NiV IgG antibodies and Nipah virus RNA was detected when bat populations, of the locations where index cases were found, were studied (Sudeep et al. 2021). The strain found in Kerala, upon sequence analysis, showed a deviation from those found in Bangladesh and the Northeastern states of India. The fact that the Nipah virus emerged in Kerala in 2018–2019, a southern state in India, which was located on the opposite side of the country from the previously considered hotspots, was an astonishment (Gokhale et al. 2022).

Gokhale et al. undertook a study in the southern states and union territories of India, in order to identify the presence of Nipah virus activity in bat populations present in those states. These states are geographically close to Kerala, the new hotspot of NiV. The study also took place in the state of Odisha, which is located to the southeast of West Bengal. In 2019, 255 bats of species *P. medius* were captured and their serum samples were collected. Of these, 20% of bats were captured from the states of Karnataka, Kerala, Puducherry and Tamil Nadu and they were reported to possess anti-NiV IgG antibodies; although no virus presence was detected in any of the specimens (Gokhale et al. 2022).

A report earlier to this one showed positive anti-Nipah IgG specimens among the fruit bats related to *Pteropus* spp. that had been recaptured in Madagascar (Brook et al. 2019). The loss of antibodies over time or the return of infected bats to a specific geographical unit can lead to variations in Ab level in these bat populations. This occurrence supports the findings of Gokhale et al., that there was the presence of anti-Nipah IgG in only a few of the specimens sampled in a specific geographical area. Similarly, in Bangladesh, varied seroprevalence of NiV was in various locations over the 6 years of duration of the study. Along with this, NiV RNA was detected among a few bat specimens as well (Epstein et al. 2020; Fleming 2019). In tropical and sub-tropical regions, the migratory behaviour of bats is less common as compared to those in the temperate regions and is not associated with hibernation. The main drive behind migration is food resources (Fleming 2019). The scarcity of knowledge on the migration patterns of bats in India is the reason for limited information on migratory bat populations (Gokhale et al. 2022).

In India, during the NiV outbreaks, both NiV RNA and IgG antibodies were found in samples of *Pteropus medius* and; therefore, they are considered the main reservoir of the Nipah virus (Yadav et al. 2012; Sudeep et al. 2021). Very few studies on other species of bats as potential reservoirs are available in India. Thus, in order to study the prevalence of Nipah virus in bats, a cross-sectional survey was instituted. A random sampling of bats, that are widely prevalent in India, was done. The bat populations included in this survey were *Pipistrellus pipistrellus, P. medius* and *Rousettus leschenaultii.* As widely known, *Pteropus* spp. consists of large fruit-eating bats. In comparison, *Rousettus* spp. bats are medium in size and are also fruit eating. *Pipistrellus* spp. bats are tiny and are insectivorous bats, majorly found in similar habitats as *Rousettus* spp. bats (Gokhale et al. 2021). Out of the bat specimens captured in 2020, 01/04 *Pipistrellus* and 33/56 *Rousettus* spp. bat sera samples were found to have the presence of anti-NiV IgG antibodies, respectively. NiV RNA was detected in various organs, of both these species as well as in the throat and rectal swabs, using N-gene-based qRT-PCR. One bat from each species group was reported to be positive for both anti-NiV IgG antibodies and NiV RNA (Gokhale et al. 2021). Although the aforementioned study signals the possibility of these bats being infected by the Nipah virus, it is necessary to systematically sample a larger number of bats from *Rousettus* and *Pipistrellus* spp. to determine and confirm their role in Nipah virus transmission (Gokhale et al. 2021).

The findings obtained in the most recent outbreak of NiV in 2021, amidst the COVID-19 pandemic, suggest that the spill-over of Nipah infections is sporadic in humans and the seroprevalence in *P. medius* suggests that the virus is prevalent in this bat population (Yadav et al. 2022).

1.2.4 Human-to-Human Transmission

Certain studies conducted during the Bangladesh epidemic, on the spread of Nipah virus, concluded that the primary cause of the disease spread was human-to-human transmission via direct contact with infected individuals or their bodily secretions. It was challenging to identify the source of primary infection, however, it is probable that they were secretions from *Pteropus* spp. (Montgomery et al. 2008). Accordingly, the study conducted in 2003 detected the virus presence in a shared household, indicating that human-to-human transmission via the respiratory route had occurred; however, this was not enough evidence to exclude the chances of external infection. The basis of these uncertainties was that the healthcare workers who came in contact with infected individuals did not contract the disease (Hsu et al. 2004). However, other research from Bangladesh between 2001 to 2014 suggested that among 248 patients, 33.06% of patients could have contracted the disease via human-to-human transmission.

Contrastingly, a study performed by Gurley et al. (2007a) stated that 92% of the infected individuals were known to have primary contact with infected patients before showing symptoms, and 5 cases were associated with secondary or tertiary transmission. Similarly, this way of transmission was confirmed in the Indian epidemics, viz., West Bengal (2001) and Kerala (2018), as well. NiV infections had been detected among hospital workers and in an outbreak in Siliguri, India, Nipah virus nosocomial transmission was reported (Chadha et al. 2006). The infections were associated with the hospital environment as the healthcare staff was repeatedly affected and the visitors of patients in hospitals, where NiV cases were confirmed, were affected as well (Chadha et al. 2006; Thomas et al. 2019).

In order to protect the hospital staff, it is essential to detect whether the infected patients are transmitting the disease to the surrounding individuals, by testing the hospital surfaces for NiV RNA. Gurley et al. (2007b) using RT-PCR demonstrated that samples obtained from hospital surroundings showed contamination by NiV RNA. The exact lifespan of NiV on surface objects is yet unknown. The mechanism by which NiV is transmitted among the human population has not been thoroughly established, however, exposure to saliva, blood, urine and other bodily fluids has been shown to elevate the risk of transmission as opposed to being in the immediate surrounding of the infected individual (Kumar et al. 2019). Studies in non-human primates have shown the infection route by exposure to aerosols (small- to medium-sized) produced by the infected (Cong et al. 2017; Hammoud et al. 2018; Sizemore 2021).

1.2.5 Spread of Disease—Outbreaks

1.2.5.1 Malaysia

In the year 1998, the month of September saw the reports of the early human cases of Nipah virus infection among pig farmers, during a Malaysian epidemic. Though primarily misinterpreted for Japanese encephalitis (JE) cases, the infections occurring in adults as well as transmission via direct contact with infected pigs did not fit the pattern of the JE virus (JEV) (CDC 1999). The virus was officially designated as the Nipah virus in May 1999, when it was isolated from the CSF of an infected person from the area of Sungai Nipah (Chua 2003). The demographic that was majorly hit was men who were in constant contact with infected pigs while very few young children were infected (Lam and Chua 2002). Symptoms reported during this epidemic were very archetypal and included reduced consciousness, headaches and fever. Based on reports received by certain sources, the number of cases ranged from 238 to 256 in Malaysia. The death toll was 105–109, which indicated a high CFR and this was confirmed during the later outbreaks (Looi and Chua 2007; Aditi and Shariff 2019; Ambat et al. 2019). The accurate detection of CFR in humans is dependent on the capacity of epidemiological surveillance and clinical management systems (Lam and Chua 2002). The death toll in Malaysia was 105 out of 265 cases and this led to the culling of a million pigs with the aim of restraining the further transmission of the virus (Ang et al. 2018; Khan et al. 2022).

1.2.5.2 Singapore

Singapore witnessed a Nipah virus outbreak in March of 1999. During this epidemic, infections involving the neurological and respiratory systems were detected in workers from slaughterhouses that received pigs from Malaysia. Eleven cases were detected, out of which one fatality occurred. The median age of the infected was 44 years, with the range of infection being between 26 and 66 years of age. In Singapore, all the infected individuals were male, 10 of Chinese and 1 of Indian descent (Paton et al. 1999). The end of the epidemic was brought upon by declaring a ban on pigs imported out of Malaysia (Chua 2003; Aditi and Shariff 2019).

1.2.5.3 Bangladesh

Following the Malaysia and Singapore epidemics, an outbreak with encephalitis as the main symptom broke out in Meherpur, Bangladesh, in 2001 (Rahman and Chakraborty. 2012). The source of this outbreak was discovered to be the ingestion of raw date palm sap or the scaling of trees that showed the presence of bat excreta (Rahman et al. 2012; Yadav et al. 2022).

Thereafter, Nipah outbreaks arose almost yearly from 2001 to 2013, chiefly in the winter season. Cases were detected throughout the country with certain areas being plagued frequently. In 2001, the reported mortality was 69%, however, it increased to 83% in 2013. From April 2001 to March 2012, the number of NiV cases was 209, of which 77% (161) proved fatal (Ambat et al. 2019; Anderson et al. 2019). Gurley et al. (2007a, showcased NiV as an anthropological factor in the Faridpur district outbreak, in April–May of 2004. Lo et al. (2012a) presented the genomic sequence of the Bangladesh epidemic strain. According to their report, these genomic sequences, which were taken from two patients in 2008 and three in 2010, comprised a new genetic group (Lo et al. 2012a) from the earlier identified strain of NiV. According to their division, the genomic sequences found in Malaysia and Cambodia were designated as genotype M, while those from Bangladesh and India were defined as genotype B. Researchers and physicians were chiefly interested in the way the virus was spread in Bangladesh, as a result of the ingestion of unprocessed date palm sap and the human-to-human transmission due to direct contact with various body fluids, including aerosols (Aditi and Shariff 2019; Skowron et al. 2022).

1.2.5.4 India

Nipah outbreaks in India have occurred often but not as frequently as those in Bangladesh. The first recorded NiV outbreak occurred in 2001 in the state of West Bengal, a region bordering Bangladesh's NiV belt (Aditi and Shariff 2019). Out of 66 confirmed infections, 45 confirmed fatal cases were recorded; which was the highest of the time. The following outbreak of 2007, arose from the same region, with five cases followed by five confirmed deaths (Arankalle et al. 2011). The month of May 2018 saw the rise of Nipah in the southern Indian state of Kerala. The cases here were mainly characterized by encephalitis and acute respiratory syndrome. There were at least 17 confirmed deaths, characterizing this NiV outbreak by the highest mortality, which exceeded 90% (Ambat et al. 2019; Plowright et al. 2019). Arunkumar et al. (2019a) carried out a phylogenetic analysis showing a 97.37–98.64% identity of genotype B of NiV (AY988601—Bangladesh 2004) with the strain responsible for the Kerala outbreak of 2018. Although they were clustered with genotype B, the Kerala strains created variations within the clade (Arunkumar et al. 2019a, b). Whereas the phylogenetic analysis carried out by Yadav et al. (2019), which was based on three fruit bats (*Pteropus medius*) and four human samples from the Kerala outbreak of 2018, reported 96.15% similarity with the Bangladesh strain. However, it showed 99.7–100% similarity to the bat virus, leading to the conclusion that bats were the original source of the epidemic. A recent incidence of NiV infection in Kerala was in 2019, where only one case was diagnosed in Ernakulam district and the patient showed a complete recovery (Sahay et al. 2020). However, in 2021, amid the COVID-19 pandemic, an NiV-infected 12-year-old boy died on the fifth of September in the Kozhikode district of Kerala (Thiagarajan. 2021). After contact tracing, samples were taken from healthcare workers, friends and family but all tested negative (World Health Organization [WHO] 2021).

The 2001 outbreak of Siliguri was the foremost outbreak of NiV which presented with neurological manifestations as well as febrile illness with a CFR of 74% (M.S. Chadha et al. 2006). Following this, in the Nadia district of West Bengal in 2007, an outbreak affected five patients and they all succumbed to the disease taking the CFR to 100% (Arankalle et al. 2011; Gokhale et al. 2022). Similarly, in the 2018 Kerala outbreaks, of the 23 reported cases, 21 confirmed deaths were reported with CFR of 91% (WHO-Nipah virus outbreak 2019; Soman Pillai et al. 2020; Foster et al. 2022).

In 2010, from Maynaguri, West Bengal, and in 2015 from Cooch Behar, West Bengal, and Dhubri in Assam, all situated close to the border of Bangladesh, the Nipah virus was found in *Pteropus medius* (Yadav et al. 2012, 2018). A decade later, an outbreak broke out in Kozhikode district in Kerala in 2018 with a resulting CFR of 89%, however, the state and national health systems were effective in containing it (Arunkumar et al. 2019a, Sahay et al. 2020). In the following year, however, another outbreak materialized in the Ernakulam district in Kerala. This outbreak reported only a single case, and the early detection curtailed the further spread of the virus. Upon investigation, it was found that the surrounding bats *P. medius* showed the presence of Nipah virus and anti-Nipah antibodies (Sudeep et al. 2021). In the most recent outbreak in September 2021, a 12-year-old male child presented with acute encephalitis and tested positive for Nipah disease, he succumbed to the disease (Yadav et al. 2022).

Currently, the state of Kerala is facing a new Nipah virus disease outbreak in Kozhikode district. A 47-year-old male who succumbed to the infection on August 30th, 2023, has been identified as the index case. The state has taken precautionary measures to restrict the spread of the virus. As of September 17th, 2023, two fatalities and four active cases have been reported (source: Times of India (TOI); 2023, Sept. 17).

1.2.5.5 Philippines

In 2014, a NiV epidemic broke out in the southern Philippines. It involved 17 cases and was characterized by a high mortality rate, which exceeded 80%. The main transmitter of this outbreak was horse meat, its exposure or consumption. The strain responsible for it was found to be genetically close to the Malaysian strain (Ching et al. 2015; Aditi and Shariff 2019; Ambat et al. 2019). The suspected culprits behind horse infections were fruit bats (Ching et al. 2015).

1.2.6 Genetic Diversity

The impact of the virus on the host organism may vary depending on the strain that caused the infection (Skowron et al. 2022). The genetic analyses of all the outbreak strains have resulted in the identification of two main lineages of the Nipah virus, viz., the Malaysian (NiV-My) and Bangladesh (NiV-Bd) lineage (Ang et al. 2018; Devnath and Masud 2021). Further analyses found that the strains found in Bangladesh and India have a 18,252 nucleotide-long genome, which is about 6 nucleotides longer than the Malaysian strain (Ang et al. 2018; Aditi and Shariff

2019). Similarly, the nucleotide sequences of strains found in the Indian regions of Kerala and Siliguri were observed to have 97% and 99% sequence similarity, respectively, to the strains of Bangladesh (Pallivalappil et al. 2020; Skowron et al. 2022). Subsequently, the strains found in new NiV hotspots in Kerala, when phylogenetically analysed for N gene sequence, showed grouping into a distinct cluster from those found in Bangladesh and Malaysia. This suggested the possibility of the presence of a new genotype evolving independently in the southern part of India (Paul 2018; Gokhale et al. 2022).

1.2.6.1 Case Fatality Rate (CFR)

The culprit of the initial NiV 1998–1999 epidemics in Malaysia and Singapore was the Malaysian strain (NiV-My), which infected over 270 patients and had a CFR of ~40%, due to an outbreak in pigs and humans (Sharma et al. 2019). In Malaysia, the virus manifested in the form of encephalitis and the CFR was 39% among humans (Chua 2003; Chaiyes et al. 2022). It also caused an outbreak in 2014 in the Philippines with ~52% CFR (Ching et al. 2015; Foster et al. 2022).

Since 2001, regions of India bordered by Bangladesh and regions of northern and western Bangladesh have frequently reported cases of bat-to-human transmission of NiV-Bd; few clusters of subsequent human-to-human transmission have also been reported (Luby et al. 2009). During the 2018 outbreak of Kerala, a CFR that was significantly higher than that in Siliguri was observed (Pallivalappil et al. 2020). As compared to NiV-My epidemics, the CFR of the NiV-Bd outbreaks is significantly higher (43–100%) (Hsu et al. 2004; Chadha et al. 2006; Gurley et al. 2007a).

Demonstrations in experimental animal models have highlighted distinctions in symptomology and the course of disease among Malaysia and Bangladesh strains and suggested the hand of a genetic component:

1. Hamsters—Ones subjected to NiV-My infection demonstrated an accelerated course of disease inclusive of replication, pathology and death in comparison to those subjected to NiV-Bd (DeBuysscher et al. 2013). These results were opposite to the data obtained from human outbreaks in Bangladesh and Malaysia.
2. Ferrets—Comparability in histopathology of CNS and lungs and progression of disease were seen in ferrets infected with NiV-My and NiV-Bd. However, the group infected with NiV-Bd showed high levels of shedding of the virus in oral discharge compared to animals infected with NiV-My (Clayton et al. 2012). Consistently, humans infected with NiV during the Bangladesh epidemic demonstrated increased shedding (Hossain et al. 2008).
3. African Green Monkey—AGMs demonstrated 50% lethality when infected with NiV-My while at an equivalent dose of NiV-Bd displayed 100% fatality. The Bangladesh strain displays severe histopathology of the lungs and has a short window of treatment with m102.4 MAb (Mire et al. 2016).

Based on the above finding, it is concluded that AGMs mimic human infection of NiV more accurately than hamsters and ferrets and the results obtained suggest that the differences in clinical symptomatology and epidemiology of Malaysian and

Bangladesh epidemics are partly due to genetic variations and also due to differences in transmission, variation in population health and quality of healthcare available (Sizemore 2021).

1.2.6.2 Clinical Features

The three main clinical distinctions among the Malaysia and Bangladesh epidemics are:

1. The time from onset to death was 6–7 days and 16 days while the CFR was 74% and 38% in Bangladesh and Malaysia, respectively (Ahmad and Tan 2014; Ang et al. 2018; Hossain et al. 2008; Lo and Rota 2008; Ang et al. 2018; Pallivalappil et al. 2020).
2. A high incidence (50–75%) of respiratory manifestations was reported in Bangladesh. This along with the increased level of NiV-Bd RNA found in oral discharge pointed towards transmission via aerosols (Ahmad and Tan 2014).
3. Neurological manifestations observed in Bangladesh were comparable to those found in Malaysia.

Although there have been reports of subclinical infections during the epidemics in Malaysia and Singapore, there have been variations in the percentage of confirmed cases. In Singapore, the percentage was more than 45%, whereas in Malaysia it ranged from 8% to 17% (Tan et al. 1999; Parashar et al. 2000; Chan et al. 2002; Chua 2003; Sharma et al. 2019). However, in the Bangladesh epidemic, subclinical infections were either sporadic or did not occur (Hsu et al. 2004). Asymptomatic cases were rare in India as well. This verified that individuals who had interaction with diseased people but not their bodily fluids showed a lower risk of developing subclinical disease (Kumar et al. 2019).

In context to neurological manifestations, some studies in Malaysia showcased more than 50% of the diseased exhibited brain stem dysfunction, low level of consciousness (Chua 2003) and muscle twitching with convulsions and hyporeflexia; while in the Bangladesh epidemics, the percentage of patients exhibiting these symptoms was above 90% (Sharma et al. 2019; Ang et al. 2018; Hossain et al. 2008). Though these records have no definite rationale behind them, they could point to the involvement of distinct CNS areas with different tropisms, distinct infection routes or decelerated disease courses which enable infection of neural tissues of different types (Sizemore 2021).

In addition to neurological manifestations, the disease also shows severe respiratory illness. During the Bangladesh and India epidemics, 50–75% of cases were diagnosed with respiratory symptoms like cough, cold and shortness of breath, while in severe cases acute respiratory distress syndrome was also diagnosed. In the Malaysia and Singapore epidemics, a low number of individuals reported respiratory symptoms, the main reason for this was the difference between the circulating strains responsible for infection and the high percentage of human-to-human transmission reported in Bangladesh and India (Goh et al. 2000; Chua 2003; Sharma et al. 2019; Thomas et al. 2019). The 2018 Kerala outbreak also saw the occurrence of a new, previously unreported symptom, namely the dysfunction of the heart muscle in patients (Pallivalappil et al. 2020).

1.2.6.3 Epidemiology

The epidemiological effects caused by the different strains of Nipah virus vary due to the genetic differences between them. In Malaysia, the infections were less severe, coupled with a low mortality rate and a higher number of subclinical cases (Kumar et al. 2019). Contrastingly, in Bangladesh and India, the strains circulating were genetically similar and caused a high fatality rate, though this could be partially attributed to the low quality of healthcare and the occurrence of new virus outbreaks (Ochani et al. 2019; Sharma et al. 2019; Thakur and Bailey 2019). In the 2001 epidemic of Siliguri in India, the clinical symptoms observed in NiV cases were analogous to those reported in the Bangladesh outbreaks (Chadha et al. 2006). The NiV-Bd strain showed a high infection rate in Bangladesh with respect to human-to-human transmission, this indicates the chances of instantaneous mutational capacity (Devnath and Masud 2021).

1.2.6.4 Pathogenicity and Virulence

Though the two strains show 91.8% sequence homology, the reports that they both show such contrasting distinctions in human-to-human transmission rate and CFRs are fascinating (Harcourt et al. 2005; Foster et al. 2022). Moreover, the NHP models developed for both strains, NiV-My (Geisbert et al. 2010) and NiV-Bd (Mire et al. 2016), showcased that the latter is more pathogenic in African Green Monkeys (AGMs) as compared to NiV-My, even under identical experimental conditions (Mire et al. 2016). It was also reported that therapeutics that showed protection against NiV-My had a shorter therapeutic window against NiV-Bd. This showcased the significance of a medical countermeasure to protect against NiV-Bd (Mire et al. 2016; Foster et al. 2022).

1.3 Potential Risk of Emergence and Re-Emergence

The ease of transmission, high mortality and the absence of effective countermeasures have led to NiV being classified as a BSL-4 viral agent. Additionally, it has been classified by many US government agencies, like the National Institute of Allergy and Infectious Diseases (NIAID) and the CDC, as a 'Category C priority pathogen and transboundary select agent'; among the agents classified as posing a risk of terrorism (Ochani et al. 2019; Gómez Román et al. 2022). The priority lists on which NiV and henipavirus caused diseases are included are WHO's R&D Blueprint List of Priority Pathogens and the former is also included on the Coalition for Epidemic Preparedness Innovation's (CEPI) priority list due to its nature of zoonosis, high CFR, its possible ability of human-to-human transmission and the lack of reliable vaccines and therapeutics (Anderson et al. 2019; WHO 2021b; Gómez Román et al. 2022). As a result of the ongoing pandemic of COVID-19, the security concerns with respect to respiratory pathogens have been heightened. The

US CDC in 2020 advocated the addition of NiV to a list Tier 1 select Agents (CDC/USDA 2021; Foster et al. 2022).

1.3.1 Potential of Emergence/Re-Emergence

About 60% of human infectious diseases and those emerging have infectious agents with zoonotic origins (Karesh et al. 2012; Jones et al. 2008). The Anthropocene is characterized by the often occurring epidemics and pandemics caused by emerging and re-emerging zoonotic pathogens (Lancet Zoonoses 2020). In the past few centuries, pathogenic zoonotic agents have caused public health emergencies worldwide, such as bird flu, Ebola, Hendra, Nipah, Middle-East Respiratory Syndrome (MERS), Rift Valley Fever (RVF), Swine flu, Severe Acute Respiratory Syndrome (SARS) and Zika. These pathogens have, over the years, shown a capability to return, spread like wildfire and cause high mortality rates killing millions worldwide (Zumla and Hui 2019; Singh et al. 2019; Devnath and Masud 2021).

As these diseases have zoonotic origins, it is essential to understand the animal–human interface in order to understand their emergence, possible preventive measures and therapeutic measures to be taken. Dissimilar species existing in close quarters, tight-knit small mammal densities and human encroachment, and in turn, disturbance of the surviving ecosystems have all affected the dynamics of the emergence of zoonotic diseases (Gupta 2005; Gibb et al. 2020a, b).

Though zoonotic diseases are a global threat they are especially germane to developing and under-developed countries, where a long range of environmental, socio-economic and other corresponding factors may lead to increasing the risk of disease outbreak. Among these factors are poor healthcare resources, rapidly increasing population, resulting urbanization, variable literacy rate, unstable economic conditions, consumption of game-meat, and the infringement of wildlife ecosystems and landscapes for human utilization (Karesh et al. 2012; Gibb et al. 2020a, b; Allen et al. 2017; Wolfe et al. 2005, 2007). Additionally, due to the lack of appropriate resources, certain fractions of the population may not have access to appropriate healthcare and diagnostic settings, may be unaware of government-recommended health guidelines and have little to no surveillance systems, spill-over events are more often than not under-reported (Johnson et al. 2020a, b; Li et al. 2019; Durrance-Bagale et al. 2021). A recently conducted study has reported that tropical countries, like India that are rich in wildlife flora and fauna, are at high risk of disease spread by an agent with zoonotic origins, due to the changes they are undergoing with landscaping with respect to agriculture (Allen et al. 2017; Durrance-Bagale et al. 2021).

The impressive rate at which the Nipah virus mutates gives it certain advantages such as adapting to divergent environments, jumping hosts easily, generating new virulence factors, bypassing the host immune system and/or accelerating the rate of transmission (Kulkarni et al. 2009). Diversity in hosts and changing surroundings may lead to a further increase in the rate of mutation of RNA viruses, this explains the new, emerging cases during the COVID-19 pandemic. When compared with

SARS-CoV-2, NiV has a low prevalence. However, seeing as the nation's current healthcare resources and policies are directed towards restraining the spread of SARS-CoV-2, and the focus of the entire scientific community on COVID-19, reports of NiV cases amidst this are extremely concerning. The matter warrants further attention from public health authorities and the institution of policies to curb the spread (Aborode et al. 2022).

Deka and Morshed in 2018 reported that countless Southeast Asian countries have been identified as potential hotspots for Nipah virus emergence and disease spread (Deka and Morshed 2018; Gokhale et al. 2021). In countries of the Asia-Pacific belt like Malaysia, Singapore, Bangladesh, India and the Philippines where the virus has been causing sporadic outbreaks for decades, Nipah virus associated encephalitis has become a significant public health concern (Ang et al. 2018; Ambat et al. 2019; Sharma et al. 2019; Hauser et al. 2021). The most recent reports confirm the re-emergence of the Nipah virus in Kerala, India, during September of 2022 and 2023 (World Health Organization [WHO] 2021; Skowron et al. 2022).

1.3.2 Pandemic Potential

Factors promoting NiV's pandemic potential are the susceptibility of humans, the capability of human-to-human transmission, a high mutation rate of RNA viruses; and the infection of communities of low-income countries with high population density and increased international travel due to globalization may lead to rapid spread of the virus (Luby 2013; Foster et al. 2022).

1.3.2.1 Bat as Primary Reservoir of NiV Outbreak

In the case of NiV, the main reservoir is fruit bats of the species *Pteropus giganteus* (Luby et al. 2006). Bangladesh, home to no less than 33 species of bats commonly the species known as flying foxes or Indian fruit bats, was the prime area where NiV outbreaks took place. It is reported to have the highest number of NiV cases, with sporadic outbreaks from 2001 to 2014 (Nikolay et al. 2019). Since the Asia-Pacific region is known to be rich in bat fauna, it is highly plausible that NiV outbreaks may occur frequently, as bats are the primary reservoir (Devnath and Masud 2021). The existence of viruses from the genus *Henipavirus* in bats from West Africa and China suggests that the diseases associated with them could also occur on continents other than Asia and Australia (Halpin and Rota 2014; Kasloff et al. 2019; Skowron et al. 2022).

1.3.2.2 Species Tropism

The key role in regulating the selection of the virus host, pathogenesis and tissue tropism is played by the interaction between the virus and its receptor on the host cell surface (Maginnis 2018). It is essential for the virus-receptor binding site to be highly specific as it plays a significant role in initiating the entrance of the virus into the host, thus causing infection (Devnath and Masud 2021).

The binding receptors of NiV, ephrinB2 and B3 receptors, are common among epithelial cells of mammals and; thus, the virus has a wider host range, as compared to other members of the family Paramyxoviridae (Luby 2014). Additional opportunity for transmission to humans is created by its ability to infect domestic as well as farm animals (Ang et al. 2018). Pigs were eliminated as a source of infection during the Bangladesh epidemic, however, they were found to be susceptible to the NiV-Bd strain without manifesting a clinical disease by recent studies. The virus was detected after a nasal wash of the pigs and these pigs are suggested to have spread the virus to other susceptible animals (Kasloff et al. 2019; Skowron et al. 2022).

1.3.2.3 Cell and Tissue Tropism

The ability of NiV to bind to the ephrin (B2/B3) receptor, which is present in almost all tissues in mammals, widens its host pool (Xu et al. 2012; Taylor et al. 2017). The function of this receptor is to disseminate the virus, an essential step necessary for the infection of cells (de Wit and Munster 2015; Maisner et al. 2009). During a natural infection, the ephrin (B2/B3) receptor present in leukocytes aids NiV in binding to the white blood cells and thus aids their dispersion into the host bloodstreams (Mathieu et al. 2011). Since the receptors mentioned above are densely expressed in the central nervous system (CNS), NiV is considered to have the highest potential of causing neurological disease, which is often fatal (Navaratnarajah et al. 2020; Devnath and Masud 2021).

1.3.2.4 Zoonotic Spill-Over

When there is successful transmission coupled with the development of an infection and its following stages in humans, only then is the virus said to be capable of jumping hosts (Plowright et al. 2017). However, the spill-over phenomenon is dependent on factors such as the density of the population it was transmitted to and the reservoir-host-disease dynamics (Johnson et al. 2020a, b). An analysis carried out by Luby et al., with respect to spill-over of NiV and the resulting few cases in Bangladesh between 2001 and 2007, led to the conclusion that patients suffering from NiV disease with severe respiratory syndrome presumably had a higher potential to spread the Nipah virus than the others (Luby et al. 2009). Their hypothesis was that a bat of the *Pteropus* spp., carrying NiV may infect one or more individuals and they, in turn, could infect others and initiate an epidemic chain via human-to-human transmission. Since the study had many limitations, the need for additional research with a larger sample size and using a well-ordered means of surveillance is essential to confirm the hypothesis (Nikolay et al. 2019; Devnath and Masud 2021).

1.3.2.5 Virus Mutations

Due to the low fidelity of RNA polymerase enzyme, the rate of mutation of RNA viruses is exponential as compared to their hosts (Duffy 2018; Devnath and Masud 2021). Among the verified and tested strains of the Nipah virus, dissimilarities have been reported in methods of transmission, the type of symptoms, their severity and the size of genomes when confirmed by molecular analysis. Incidentally, it is necessary to take into account the risk of mutations, e.g.: facilitation of person-to-person transmission during the assessment of the pandemic potential NiV holds. This suggestion arose when the genomic sequence of a Filipino patient (2014) displayed a great range of similarities with the Malaysian strain. Although during the Malaysian epidemic, there was no evidence of human-to-human transmission reported, the likelihood was confirmed in the Philippines (Ching et al. 2015). Owing to their high nucleotide substitution rate, inability to correct the said substitutions and the concluding ability to adapt to new hosts, RNA viruses are considered severely hazardous and; thus, are required to be under constant surveillance (Jones et al. 2008; Skowron et al. 2022).

1.3.2.6 Human-to-Human Transmission

During the outbreaks that took place initially, there was little to no evidence of human-to-human transmission of NiV, with the exception of a nurse who had no direct contact with pigs but developed a subclinical NiV infection and displayed seropositivity and MRI findings similar to those of an infected individual (Tan et al. 2000; Tan and Tan 2001; McEntire et al. 2021).

The quick and effective transmission of the pathogen among people via people, especially in immunocompromised individuals or those with lower immunity, is one of the key features that indicate NiV's pandemic potential. During the Bangladesh and Indian epidemics, this route of transmission was found to be very significant. As is well known, infection may occur due to the exposure of a healthy individual to the infected body fluids, while an additional risk reported is the likelihood of spread due to aerosols created by patients with respiratory distress accompanied by coughing (Luby et al. 2009). There happens to be plenty of research revolving around the indication that aerosols created by respiratory secretions are the main source of infection and substantiate the transmission of the virus via this route and the degree of development of symptoms of advanced changes in the respiratory system (Luby 2013; Skowron et al. 2022).

The clustering of human cases in recent outbreaks suggests a concerning new attribute of human-to-human transmission (Singh et al. 2019; Gurley et al. 2007a, Sazzad et al. 2013; Ching et al. 2015). In the 2018 Kerala outbreak, during which 16 out of the 18 confirmed cases died, the overwhelming number of case fatalities was due to person-to-person transmissions that occurred after a few primary animal–human transmissions (Gurley et al. 2007a, Thomas et al. 2019; Wilson et al. 2020). These occurrences raise valid concerns regarding the risk of a lethal pandemic being caused by NiV and; thus, the WHO has designated the virus as high-risk 'priority disease' (Gómez Román et al. 2020; McEntire et al. 2021).

Due to NiV's ability to cause a high rate of mortality, this factor can limit the spread as it does not infect another person who comes in contact with the dead host. However, it is essential to take into account the fact that the NiV outbreaks were detected in areas that were thinly populated (Devnath and Masud 2021). In the event of an epidemic, it can be assumed that if it occurs in a densely populated area, the rate of infections would be higher, typically since the WHO suggested that the incubation period for NiV in extreme cases may last for 45 days. Transmission via human-to-human contact is facilitated by the rapidly growing population, trade, globalization or voguish travel patterns (Carrasco-Hernandez et al. 2017). The standing of NiV as a potential agent responsible for a global pandemic increases when the risk associated with the emergence of a particular percentage of asymptomatic cases is taken into account (Skowron et al. 2022).

1.3.2.7 Speed of Virus Spread

The estimated speed at which a malady can transmit within a population is known as the basic reproduction number or R0. The number is crucial when researching the rate of transmission in a given population. The R0 value is essential in providing crucial information regarding an outbreak which permits the construction of national policies with respect to preparedness for an epidemic (Ridenhour 2014). Although there are exceptions in existence, an R0 value of >1 is an indication of an increase in the transmission of disease while an R0 value of <1 indicates a reduction in the transmission ability of the pathogen (Biggerstaff et al. 2014; Devnath and Masud 2021).

The R0 value of NiV is 0.48 (Luby 2013), however similar to MERS, NiV has a low R0 value coupled with an elevated rate of mortality. It is a well-established fact that if the R0 value is below 1, the cases gradually decrease and disappear as a result of a drop in the virus's ability to be transmitted. Several pathogens have failed to attain endemicity due to the limitation of susceptible hosts, therefore, the R0 value is correlated with host density. For instance, villages in Malaysia, Bangladesh and India where the respective outbreaks were reported, the population density was low and this was suggested to be a reason for no large outbreaks being reported (Ang et al. 2018). One more reason for the low value of R0 could be the infrequency of spill-over due to disease-related death. Thus, diseases caused by zoonotic pathogens that have high mortality rates usually reveal low rates of incidence (Belay and Monroe 2014; Devnath and Masud 2021).

The change in the R0 value depends on contrasting variables like population, geographical location, intervals between outbreaks and accuracy of data collection. In accordance with this, the R0 value could be used to implement many interventions; however, extensive study is necessary because a single mutational change in the virus may cause the prediction to be obsolete, as was starkly evident during the COVID-19 pandemic (Ridenhour 2014; Devnath and Masud 2021).

1.3.3 Risk Factors

1.3.3.1 Bats

1.3.3.1.1 Reservoir of Zoonotic Viruses

A total of 60 zoonotic pathogenic viruses have been estimated to be suitably hosted by bats. Over the past few decades, these mammalian reservoirs have been considered to be at the heart of the spill-over (wildlife to humans) or spill-back (human to wildlife) phenomena of zoonotic viruses (Naicker 2011; Khan et al. 2022). Presently, more than 12,000 bat-related viral sequences have been detected in 30/168 viral families (see database of bat-associated viruses, http://www.mgc.ac.cn/DBatVir/) (Chen et al. 2014; Tian et al. 2022). Several zoonotic viruses, emerging and re-emerging, possess bats as their reservoirs, the current pandemic of SARS-CoV-2 has a high possibility of being included in these (Mackenzie and Smith 2020). The jumping of a viral species from its bat reservoir to a new host (animals) occurs when said species gains the capability to infect and proliferate among the new host population (Flanagan et al. 2011). A crucial role, in the spread of virus species from bats to humans, is played by the virus–host–environment factor. Viruses associated with bats usually emerge when there is an interaction between humans and wildlife, possibly due to the changing ecological conditions; the primary reason being human activity that contributes to landscape expansion. This increases the chances of human–bat interactions and opportunities for disease transmission (McMichael 2004). A spill-over event can be possible directly via bat–human interaction or through a suitable intermediate reservoir, like domesticated pets, livestock or other wildlife (Allocati et al. 2016; Khan et al. 2022).

Pteropodidae family fruit bats are NiV's natural host reservoirs (Mougari et al. 2022). Bats of the Pteropodidae family were detected to be harbouring viruses of families such as Paramyxoviridae (*Henipaviruses*—pteropid bats), Filoviridae (*Marburg, Ebola*) and Reoviridae (*Rotavirus*). These bats consume fruits and are capable of flying lengthy distances; thus, they are found to reside on all six continents (Simsek et al. 2021; Tian et al. 2022). From the family Pteropodidae, genus *Pteropus*, fruit bats also known as flying foxes, act as the primary reservoir of NiV and the infection produced by NiV is asymptomatic in these reservoirs. When experimentally inoculated with extremely high NiV doses, the infection was seen to be subclinical in nature and no viraemia was observed, however, the organism underwent seroconversion and was detected to be shedding virus particles, rarely, while urinating (Geisbert et al. 2012; Middleton et al. 2007).

Tropical environments are where these reservoirs thrive best and; therefore, regions of Asia and Africa are considered to have a rich host biodiversity coupled with large pathogen pools (Khan et al. 2022). Bats belonging to *Eidolon* genus are native to the continent of Africa and are considered a reservoir for NiV as they demonstrated NiV seropositivity (Enchéry and Horvat 2017). In countries like China, Indonesia, Madagascar, New Caledonia, Papua New Guinea and Vietnam, bats have been detected to be seropositive for the Nipah virus (Sun et al. 2018). In Bangladesh, a 6-year study carried out on *Pteropus medius* indicated the widespread distribution of NiV with similar infection patterns throughout the entire year throughout the country and not only the 'Nipah belt' (Price et al. 2021).

In the past few decades, there have been four bat-borne, emergent, zoonotic viruses that have grabbed the attention of researchers, Nipah [1998–1999], severe acute respiratory syndrome coronavirus (SARS-CoV) [2002–2003], Middle-East respiratory syndrome coronavirus (MERS-CoV) [2012] and Ebola [2014]. These have caused major epidemics leading to high fatality rates globally (Goh et al. 2020; Clayton et al. 2013; Nikolay et al. 2019; Khan et al. 2022).

1.3.3.1.2 Ecology and Physiology

Though *Pteropid* bats are found on all six continents, their chief habitats are tropical and sub-tropical regions, inclusive of cities, farmlands, deserts, grasslands and tropical rainforests. Their widespread habitation is an indicator of their success as a radiation in the evolution of mammals (Gunnell et al. 2017). Unlike other flying animals belonging to the kingdom Animalia, bats lead an extremely sundry lifestyle, including flying long distances, elevated metabolic activity, being extremely sociable and prolonged life span (Hayman et al. 2013; Khan et al. 2022). Like humans, bats display social tendencies, live in large colonies composed of multiple species (Serra-Cobo and López-Roig 2017) and have a long lifespan (Wilkinson and South 2002), which was associated with telomerase maintenance (Power et al. 2021; Mougari et al. 2022) and change in DNA methylation with respect to genes of innate immunity and tumorigenesis (Wilkinson et al. 2021) though they are small mammals. Long lifespan permits prolonged cartage of pathogens among colonies and facilitates vertical transmission, thus ensuring persistence throughout generations (Mougari et al. 2022) or via urine used for grooming through the oronasal route (Middleton et al. 2007; Lawrence and Escudero-Pérez 2022). Viral infection in bats often produces asymptomatic disease; this may be attributed to a strong innate immune response (Gorbunova et al. 2020; Jebb et al. 2020). The positive selection of various immunity-associated genes was observed in bats and this may bestow upon them the boon of tolerance in the face of viral pathogens (Jebb et al. 2020; Wang et al. 2021). Bats infected with NiV are known to shed the virus through oronasal secretions, urine, faeces and blood (Edson et al. 2015). Bats display territorial behaviour; however, the encroachment of humans on their habitats leads them to venture close to human settlements and; thus, cause transmission of viral diseases (Frick et al. 2020; Tian et al. 2022). Sizes of bats vary (≤ 1 m length) (Teeling et al. 2018), they are active at night, tend to hang upside down while roosting and several species fly using echolocation (Springer et al. 2001). Notably, during echolocation, aerosols are generated and this could assist virus dispersion into the environment (Calisher et al. 2006; Constantine et al. 1972; Tian et al. 2022). The variations in bat populations and their spatial distribution have proven to be an essential factor in providing viruses opportunities to cause acute as well as persistent infections (Hayman et al. 2013; Khan et al. 2022).

1.3.3.1.3 Cell Receptors and Biochemical Pathways

The only mammals capable of powered flight are bats (O'Shea et al. 2014; Brook and Dobson 2015; Mougari et al. 2022) and this is considered an advantage not only because it helps them find new roosts but also because the action requires a high RBC concentration and; thus, consume four times higher amount of oxygen in comparison to when resting (Thomas and Suthers 1972; Lawrence and Escudero-Pérez 2022). Extremely demanding metabolic processes are required for flight and, therefore, bats display increased possession of OXPHOS genes in the mitochondria and nuclei of cells and these genes are actively involved in metabolic mechanisms (Shen et al. 2010). Previous studies have shown the impact of oxidative stress on hosts as well as the pathogens they harbour (Beck et al. 2000); in keeping with this report, it has been observed that when oxidative stress is upregulated, there is a decrease in the persistence and spread of the pathogens (Ivanov et al. 2017). There is an increase in their body temperature when flying and this is proposed to imitate an intermittent febrile state which activates the immune system daily (O'Shea et al. 2014; Brook and Dobson 2015). Elevated body temperature and upregulated oxidative stress have resulted in the increased expression of heat-shock protein (HSP) in bats (Chionh et al. 2019) and HSP90 is suggested to chaperone polymerase activity of NiV and other members of the order Mononegavirales, permitting maturation (Bloyet et al. 2016) and facilitation of early replication phase of viruses (Mougari et al. 2022). Additionally, genomic analysis of *P. alecto* made evident the positive selection of DNA damage repair genes such as DNA-PKc, KU80, ATM, MDM2 and RAD50 (Zhang et al. 2013; Mougari et al. 2022). These adaptations brought about by evolution in bats may have affected their metabolism, immunity and their capability of harbouring pathogens without undergoing pathogenesis (Lawrence and Escudero-Pérez 2022).

The evolution of viruses within bat species has enabled them to utilize cellular receptors and biochemical pathways that are highly conserved among mammals (Schountz 2014). This evolution gives these emerging viruses an advantage and, therefore, enhances their chances of being transmitted to other mammals (Khan et al. 2022).

1.3.3.1.4 NiV Spill-Over from Bats to Humans

Belonging to the family Paramyxoviridae, genus *Henipavirus*, NiV was first isolated during the Malaysian outbreak in 1999 from pigs (Eaton et al. 2006). Encompassing 13 Megachiroptera species and 60 Microchiroptera species, Southeast Asian countries are considered to be a hub of bat population diversity (Calisher et al. 2006; Islam et al. 2015). Clinical signs such as fever and respiratory illness coupled with encephalitis were first observed in a Chinese pig rearer, who was in constant direct contact with pigs. During post-outbreak surveillance conducted by the Malaysian authorities, it was observed that *Pteropus vampyrus,* flying foxes of the large variety and *Pteropus hypomelanus,* those of the smaller variety, were the natural reservoirs for NiV (Kulkarni et al. 2013). In Bangladesh, since 2001, the Northeastern region has seen sporadic outbreaks of Nipah virus disease (Anderson et al. 2019; Epstein et al. 2020; Rahman et al. 2021). Though these outbreaks seem to be similar on many accounts to the Malaysian outbreaks, no intermediate host was identified in them; the evidence suggested human-to-human transmission. Serological surveillance was performed using wild and domesticated animals, in areas where the outbreaks occurred. The findings of this survey suggested the presence of NiV antibodies in the Indian variety of flying foxes (*Pteropus medius*) (Epstein et al. 2020; Singh et al. 2019). In India in 2001, widespread infections were caused by a virus that was detected in a range of *Pteropus medius,* it was thought to be NiV, however, later it was suggested that it could be a closely related virus to Nipah virus, mostly Hendra (Kulkarni et al. 2013). The evolution of Nipah and Hendra viruses within the bat populations for decades, the altered behavioural and foraging patterns of the natural reservoirs and the expansive viral diversity within the bat populations have led to an increase in their vicinity to livestock, domesticated animals and humans; and thus, increased risk of transmission (Islam et al. 2016; Khan et al. 2022).

When comparative studies for NiV were conducted in hamsters and fruit bats, the results obtained demonstrated a lower rate of multiplication and less shedding from ECs of bats accompanied by poor response of antibodies when presented with virus challenge (Wong et al. 2003; Lo et al. 2010; de Wit et al. 2011; Singh et al. 2019). Except for humans and rodents, bats are considered to be the most abundant animals among mammals. They are extensively distributed among mainland animals of Africa and Asia (Calisher et al. 2006). They may dwell as metapopulations or be panmictic; which in turn, gives them the potential to cause seasonal, yearly or sporadic outbreaks (Calisher et al. 2006). The high population density in bat nests and their crowding behaviour during roosting increase the intra-species as well as inter-species transmissibility of viral infections. Additionally, the migratory and non-migratory communities increase the potential for spill-over and spillback phenomena (Khan et al. 2022). Several countries like Cambodia, Ghana, Indonesia, Madagascar, Philippines and Thailand are considered to be at future risk of infection as NiV has been detected in *Pteropus* spp. populations of these areas (Olival et al. 2020; Khan et al. 2022).

1.3.3.2 Pigs

In order to assess the risk of transmission from pigs, it is essential to take into account the widely used pig-rearing method in the particular region. In Malaysia, it was found that pigs are reared in large crowded pigsties. However, in Bangladesh, which has a largely Islamic population, pigs are bred in small herds, owned by individual families. This has been suggested to decrease the chances of transmission between animals and subsequently to humans. The assumption is that due to these practices, pigs were not found to be participating in the Bangladesh outbreaks of NiV (Looi and Chua 2007; Aditi and Shariff 2019). Similarly, in India, pig rearing is done at a lower rate as compared to Malaysia (Ang et al. 2018; Skowron et al. 2022) and the correlation of the disease with pigs is few and far between.

1.3.3.3 Age and Sex

The basic factors when considering the susceptibility to NiV infection are the age and sex of the individual. During the Malaysian epidemic, studies conducted reported that 75–81.6% of male adults in the age range of 37–44 years were mainly affected by NiV (Goh et al. 2000; Parashar et al. 2000; Chua 2003). In early 2004, during the Bangladesh outbreak, most affected patients were young boys below the age of 15 years. The reason for this low average age of infected individuals was suspected to be attributed to high-risk activities such as climbing trees with residing bat populations carrying NiV. These activities, more than the low average age, were considered to be relevant to the outbreak stats (Montgomery et al. 2008). Between 2001 to 2004, an analysis of Nipah virus disease cases reported in Bangladesh showed the median age of infected individuals to be 24 years and 64% of them were male (Nikolay et al. 2020). A detailed analysis, however, indicated varying reports from different regions of the country. The cases with confirmed or probable Nipah virus infection showed a median age of 12 years in Naogaon, while it was reported to be 38 years in Meherpur districts of Bangladesh. The time span between the manifestation of the first symptoms and death of the patients was observed to be shorter in the younger patients found in the Naogaon region. These reports showcased that both the younger as well as the older populous are susceptible to NiV infections (Hsu et al. 2004). The cases reported in the Siliguri district of West Bengal in India were observed to be above 15 years of age and the male-to-female ratio was 1:1.4 (Chadha et al. 2006). However, in the following epidemic that took place in Kerala, the affected were predominantly male and the mean age was reported to be 41 years (Thomas et al. 2019; Skowron et al. 2022).

As most animal husbandry-related occupations are related to men, the lofty percentage of NiV infections in men could be attributed to the direct contact they have with infected animals and their secretions. Pigs have been confirmed to be an intermediate reservoir for NiV in Malaysia and the contact between the pigs and humans has been established as the predominant mode of transmission, 70% of cases of NiV disease consisted of pig farmers or farmhands on pig farms (Chua 2003). The studies that followed this supposition reported that 90% of infected individuals worked or lived on pig farms, were exposed to pigs or had direct contact with their saliva, urine or faeces (Parashar et al. 2000). Beyond 40% of the animals that the infected had come in contact with were reported to have died (Goh et al. 2000).

1.3.3.4 Healthcare Workers

Along with people associated with animals, the group at severe risk of infection are healthcare workers. During the various epidemics and outbreaks that happened over the years, they were frequently diagnosed with NiV infection due to the human-to-human mode of transmission. The risk of exposure while working and living in a healthcare facility is significant as the virus may enter the environment due to the coughing and sneezing of patients and be suspended in the air in the form of droplets, this is extremely dangerous when the room is not properly ventilated. The presence of the pathogen on hospital surfaces was proven in Bangladesh, however, the studies could not conclusively report it as a possible route of infection (Gurley et al. 2017). Another factor that has been shown to increase the risk of infection to hospital staff is the contact with patients' body fluids (Sayed et al. 2019). More than 600 healthcare workers were examined in Malaysia, who were both exposed and not exposed to patients infected with NiV. Though some of the workers confirmed exposure to bodily fluids as well as needle stick injuries, no cases were observed (Mounts et al. 2001). Three infections among healthcare workers were reported in the duration of the Bangladesh epidemics from 2001 to 2014 (Nikolay et al. 2020). The primary cause of infection in Siliguri in 2001 was established to be direct contact with infected people residing in healthcare facilities. About 75% of the cases were of patients residing in a hospital along with the healthcare professionals working at the said facility (Chadha et al. 2006). In the duration of the Kerala epidemic of 2018, nosocomial transmission of NiV was observed and two healthcare professionals were reported to be infected (Thomas et al. 2019). In accordance with the potential risk of NiV transmission in hospital environments witnessed during the Bangladesh and Indian outbreaks, the emphasis was on the importance of basic hygienic practices like cleaning hands with soap or the proper and regular use of personal protective gear, which is confirmed to be neglected often (Chadha et al. 2006; Sayed et al. 2019).

1.3.3.5 Cultural Attributes

Caring for patients infected with NiV also increases the risk of infection through contact. The cultural practices of the countries where outbreaks have taken place support this model of assessing the risk of infection (Thakur and Bailey 2019). Family members of infected patients who acted as caregivers and contracted the disease were grouped together to form clusters of infection in India (Thomas et al. 2019). In some countries, religious activities and practices are suspected to have influenced the transmission of NiV. For example, in Bangladesh, the Muslims of the country do not consume pork and hence the transmission through an intermediate host is bypassed (Montgomery et al. 2008). However, the practice of bathing corpses before burial is an important part of Islamic funeral rites. This ritual, when performed on the corpse of an infected individual, could promote the spread of the virus to family members (Ambat et al. 2019; Skowron et al. 2022).

1.3.3.6 Anthropogenic Habitat Destruction

Zoonotic infections and the diseases caused as a result of them comprise 60% of all infectious diseases, whereas 75% of all emerging diseases are zoonotic in nature (Nulkar et al. 2021). A report submitted by the United Nations Environment Programme (UNEP) in 2016 put forth the increase in the risk of emerging diseases and their 'amplification increases with the intensification of human activities surrounding and encroaching into natural habitats.' This enables the spill-over of pathogens and aids them in crossing the species barrier to infect livestock and subsequently, humans. The main triggers of spill-over events have been observed to be the encroachment of diverse ecosystems (UNEP Frontiers Reports 2016), the expansion of agricultural land and the altering of landscapes for human settlement (Nulkar et al. 2021). The progress in the process of urbanization and the resulting forest area reduction lead to a decrease in the habitats of the bats which in turn leads to their movement close to human civilization in search of sustenance. High stress induced by starvation and low immunity lead to increased viraemia and high virus titres in the bodily secretions (saliva, urine and semen) of bats (Ambat et al. 2019; Skowron et al. 2022).

The emergence of NiV in Malaysia is hypothesized to be due to two major factors,

1. Deforestation for the planting of orchards, increasing bat–fruit interaction.
2. Industrialization of pig farming, increasing availability of intermediate hosts in congregation.

This enabled the virus to remain persistent in the surroundings of the human populace (Nulkar et al. 2021).

Evidence of the transmission of NiV from bats to humans through the intake of date palm sap was found, according to a study conducted in Bangladesh (Luby and Gurley, 2012). Harvesting the sap is done by making an incision in the tree trunk and the sap accumulates in an earthen container. Upon conduction of infrared studies, it was reported that *Pteropus gigantus* bats often sojourn at these tree incision sites and lick the sap while it is being collected (Khan et al. 2010). The virus-laden saliva is hence introduced to the collected sap. The virus is known to survive in glucose-rich medium, such as the date palm sap, for days (Fogarty et al. 2008). With the consumption of this infected sap, a door to enter the human organ systems is opened for the virus (Nulkar et al. 2021).

1.3.3.7 Non-Anthropogenic Climate Events

Severe climatic events that are independent of human intervention are an important factor in determining the intensity and scale of epidemics. Catastrophes combined with progressive human-influenced degradation of the natural ecosystems can lead to a modification in the lifestyle of natural reservoirs of NiV (Skowron et al. 2022).

The El Niño anomaly of 1997–1998 and the resultant famine that occurred followed by the human-induced fires of Indonesia in 1997 are considered as an inducing factor of NiV outbreaks, in many hypotheses. The dense fog that was seen in the aftermath of these disasters stunted the flowering and fruiting of trees in the forest. Due to this situation, the orchards situated close to nearby human settlements saw visits from a flurry of forest fruit bats, which put them in the proximity of farm animals and humans (Looi and Chua 2007). During the Malaysian epidemic, the primary spread of the virus from bats to pigs was established to begin from a pig ranch which was surrounded by fruit trees. Thought to be an additional source of income, the trees provided an enriching habitat for the natural reservoirs of the virus, leading to the contamination of the water and pig feed (Chua 2003; Daszak et al. 2013). El Niño Southern Oscillation (ENSO) brings about warm (El Niño), cool (La Niña) and neutral phases leading to modifications in atmospheric circulation leading to altered patterns in temperature and precipitation, globally (Hanley et al. 2003) and causing noteworthy cascade-like repercussions on agriculture (Iizumi et al. 2014) and ecosystems (Kogan 2000; Rifai et al. 2018; McPhaden et al. 2020). Mechanisms using which ENSO influences and eases the spread of zoonotic infections are alteration in replication and transmission patterns, change in the distribution of hosts and vectors, modifying cycle of seasons and population dynamics (McIntyre et al. 2017; Anyamba et al. 2019; Morand et al. 2013). The 2014 emergence of NiV in the Philippines could not be linked to ENSO phases but was found to have occurred due to the major volcanic activity of Mt. Mayon which began a year prior to NiV emergence (Wunderman 2013; Latinne and Morand 2022).

The major climatic component associated with spill-over events is the drop in temperature during winter (McKee et al. 2021). The study by Latinne and Morand (2022) confirmed that temperature influenced spill-overs; however, their logistic regression model proposed spill-over events and rainfall to be indirectly proportional. The findings of the Environmental Compliance Audit (ECA), however, proposed a link between Bangladesh spil-lover of NiV and ENSO; whose warm and cool phases in Bangladesh are categorized by rainfall and temperature fluctuations in many large areas of Bangladesh (Wahiduzzaman and Luo 2021). Climate change has been proposed to bring warmer winters in Bangladesh (Chowdhury and Ndiaye 2017) and; therefore, lead to a reduction in spillover events (McKee et al. 2021). An expected increase in droughts around some regions of Bangladesh due to global warming has been noticed (Chowdhury and Ndiaye 2017; Islam et al. 2021), which is bound to negatively impact bat behaviour and movement and lead to more spill-over events (Latinne and Morand 2022). Modelling of climate led to a strong suggestion of EL Niño intensification in the foreseeable period (Cai et al. 2022; Hu et al. 2021) which may lead to a potential increase in outbreak occurrence and emergence of zoonotic bat-borne infectious disease. Climate change is bound to bring about the migration of bats and change their distribution globally which in turn increases the threat of the emergence of new bat-borne viruses (Beyer et al. 2021; Latinne and Morand 2022).

1.3.3.8 Risk-Associated Odds Ratios (OR)

Activities that were closely associated with proximity and contact between humans and bats were described as related to the risk of incurring NiV (Arunkumar 2019a, Chakraborty et al. 2016; Hegde et al. 2016; Rahman et al. 2012; Luby et al. 2006; Montgomery et al. 2008; Gurley et al. 2017; Khan et al. 2012; Br et al. 2019). Every source indicated that bat contact with risk factors for infection such as climbing known bat roosting trees, proximity to them during breeding season and consumption of processed or raw date palm sap (Chakraborty et al. 2016; Hegde et al. 2016; Rahman et al. 2012; Luby et al. 2006; Montgomery et al. 2008; Gurley et al. 2017; Nahar et al. 2015; Homaira et al. 2010; Islam et al. 2016). During one such outbreak in Bangladesh, it was seen that the infected showed five times higher odds (adjusted odds ratio 4.91, 95% CI 3.16–7.65) of having ingested raw date palm sap as compared to the controls (Hegde et al. 2016). Another similar study found the odds ratio to be 7.9 (95% CI 1.6–38) (Luby et al. 2006). One of the deaths out of the 11 that occurred was of the son of a date palm sap collector. He reported hearing the sounds of bats, at night, in the trees and the presence of bat excreta in the pots in which the sap is collected. The collector presented fresh sap to a few relatives, two of whom developed encephalitis thought to be associated with NiV and died, though the infection could not be confirmed (Luby et al. 2006). In two studies conducted in Bangladesh, consumption of contaminated feed or fruit by livestock and hunting of bats were reported to be frequent behaviours (Openshaw et al. 2016, 2017). Almost 94% of households in these outbreak areas reported the consumption of fruits by bats, from trees surrounding their properties (Openshaw et al. 2016); while in some rural communities, the use of bats as food or medicine was reported (Openshaw et al. 2017; Durrance-Bagale et al. 2021).

A report from the Nipah virus endemic region in Bangladesh suggested that lack of awareness could be considered as a driver of risk associated with infection. This report highlighted that out of the 50% population that was observed to drink raw sap, only 5% were aware of the risk of incurring NiV; however, those participants who were aware of the risk of infection were just as likely to drink the raw sap as the ones that were unaware (i.e. 69% versus 67%) (Nahar et al. 2015; Durrance-Bagale et al. 2021). From Bangladesh, 12 sources discussing ecological factors influencing NiV infections have been identified. It was found during a case–control study that cases of infection were linked to having bats nearby human settlements during the day (OR 6.5, 95% CI 1.1–37.5) or having settlements close to trees in which bats roost (OR 40.1, 95% CI 3.9–416.7) (Chakraborty et al. 2016). The main potential driver cited in many papers was residing near where fruit bats were often found (Chakraborty et al. 2016; Hegde et al. 2016; Rahman et al. 2012; Luby et al. 2006; Montgomery et al. 2008; Gurley et al. 2017; Nahar et al. 2015; Homaira et al. 2010; Islam et al. 2016). Upon an outbreak investigation, it was reported that the areas where NiV cases were found, often contained a suitable habitat for *P. giganteus* (OR 2.6, 95% CI 1.2–5.8), which is regarded as a chief transmitter of the Nipah virus, as compared to the areas that remained unaffected (Hahn et al. 2014). During the northwest and central Bangladesh outbreaks, it was observed that the cases were mainly found among areas with high population density and scattered forested areas. This suggested that the increase in human population and encroachment of wildlife ecology affect the spill-over of NiV into humans (Hahn et al. 2014). According to Hegde et al., after taking into consideration the controls for age, sex and raw date palm sap consumption; the most significant association of Nipah virus cases was with the presence of bats around houses, for a month preceding the manifestation of illness (Hegde et al. 2016). Data from 57 Nipah virus spill-over events, from 2007 to 2013, were analysed by Cortes et al. In their analysis, they reported that these events were connected to a lack of rainfall and low temperatures, which accounted for 36% variation in the cumulative number of spill-over events that took place each winter (Cortes et al. 2018; Durrance-Bagale et al. 2021).

1.4 Structural and Molecular Organization

1.4.1 Classification and Phylogeny

The Nipah virus is classified under the order Mononegavirales, family Paramyxoviridae subfamily Paramyxovirinae and the genus Henipavirus (Goldsmith et al. 2003). The ultrastructural features, molecular, antigenic and serological made it an appropriate fit for the Paramyxoviridae family. The most striking unusual biological features of *Henipaviruses* are their zoonotic potential and the widely ranged tropism including pigs, dogs, cats, horses, hamsters, guinea pigs, bats and humans (Majee et al. 2021; Diederich and Maisner 2007). Apart from NiV and HeV, *Henipaviruses* include a trio of viruses known to be harmless to humans, viz., Mojiang virus (MojV), Cedar virus (CedV) and Kumasi virus (KV). The potential for pathogenesis of KV and MojV is undetermined due to the inability to isolate viruses but they are identified using sequencing. HeV and NiV display 80% similarity with regard to the homology of nucleotides (Skowron et al. 2022).

So far two different NiV strains were identified. Based on the differences in the viral properties, the strains were named NiV Bangladesh (NiV-Bd) and NiV Malaysia (NiV-My) (Devnath and Masud 2021). Phylogenetic analysis showed similarities between Malaysian NiV strain with the *Henipavirus* genus. The sequence similarity among NiV-Bd strain and NiV-My strain was found to be 91.8% and the variations were distributed unevenly along the genome (Sun et al. 2018). Phylogenetic analyses of samples from early Indian outbreaks have shown that the isolates from India are closer to the Bangladesh strain and they diverged away from the Malaysian strain (Arankalle et al. 2011). The strain found to be circulating in South India displayed a distinction from NiV-Bd and was designated to belong to a separate cluster. The net evolutionary divergence in NiV nucleotides among strains from Malaysia-Bangladesh and Kerala was 8.24% and 1.96%, respectively. Sequences obtained from West Bengal and Bangladesh demonstrated 0.75% divergence, whereas Bangladesh and Kerala displayed 2.2% divergence. A separate genotype for the Kerala strains was described as 'I' (India) genotype (Sudeep et al. 2021). Isolates from a very recent outbreak that occurred in Kerala, India, amidst the COVID-19 clustered into the previously described 'I' genotype. Nucleotide similarity, among sequences acquired from *P. medius* specimen obtained in 2019 and human samples in 2018, was 99.51% and 99.62%, respectively (Yadav et al. 2022) (Figs. 1.2 and 1.3).

As Nipah outbreaks in India have been restricted to West Bengal (2001, 2007) in the Northeast, and Kerala (2018, 2019, 2021) in the South, the following should be noted: NiV sequences from the West Bengal cluster with those from Bangladesh are segregated from Malaysian sequences (Arankalle et al. 2011), while NiV sequences from Kerala are divergent from Bangladesh sequences and cluster into a separate Indian (I) genotype (Yadav et al. 2022).

1.4.2 Morphology and Virion Structure

The enveloped virus particle appears to be pleomorphic but roughly spherical, with an average diameter of 500 nm (Chua et al. 2000a). Thin-section electron microscope studies of infected cells showed filamentous nucleocapsids forming thread-like appearance ranging up to 40–1900 nm in length (Skowron et al. 2022). Nucleocapsids revealed a typical herringbone pattern with an average diameter of 21 nm, a characteristic of *Paramyxoviruses* (Goldsmith et al. 2003). The envelope consists of two transmembrane glycoproteins, a protein that binds to cellular receptors, glycoprotein (G) and a fusion (F) protein (Chua et al. 2000a) (Fig. 1.4).

1.4.3 Genome Structure and Organization

With a high replication and mutation rate, RNA viruses show a high proclivity to spread to a new host by transcending the species barrier, almost 25–44% of novel emerging agents belong to this category of viruses (Skowron et al. 2022). The Nipah

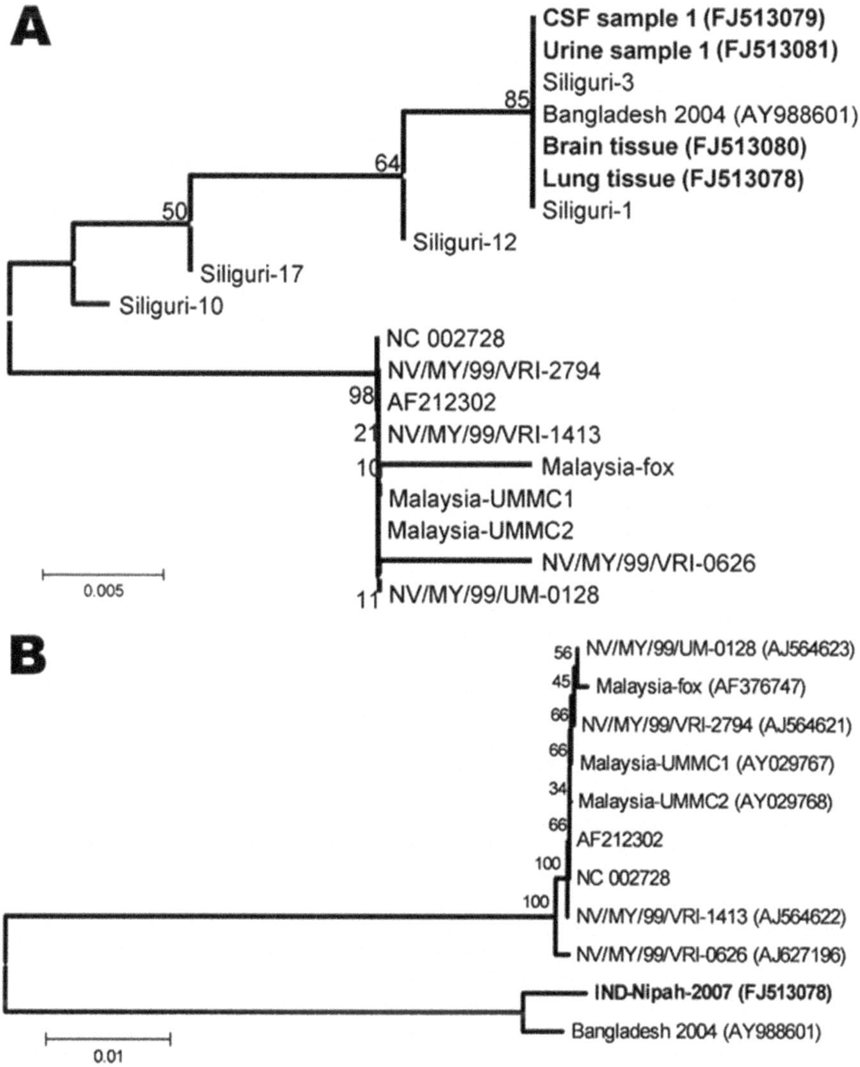

Fig. 1.2 Phylogenetic analysis of Nipah virus (NiV) sequences from the 2007 Nipah outbreak in Nadia district, West Bengal, Northeast India. (**a**) Based on partial NiV nucleocapsid (N) gene nucleotide sequences, using 4 sequences from the Nadia outbreak (**bold**), a Bangladesh 2004 sequence, 5 sequences from the 2001 outbreak in Siliguri, West Bengal, and representative sequences (GenBank accession numbers indicated). (**b**) Based on full NiV genome sequences, using a single sequence from the Nadia outbreak (**bold**) and representative sequences (GenBank accession numbers indicated). From, Arankalle, V. A., Bandyopadhyay, B. T., Ramdasi, A. Y., Jadi, R., Patil, D. R., Rahman, M., Majumdar, M., Banerjee, P. S., Hati, A. K., Goswami, R. P., Neogi, D. K., and Mishra, A. C. (2011). Genomic characterization of Nipah virus, West Bengal, India. *Emerging infectious diseases*, *17* (5), 907–909. https://doi.org/10.3201/eid1705.100968 (this publication is in the public domain)

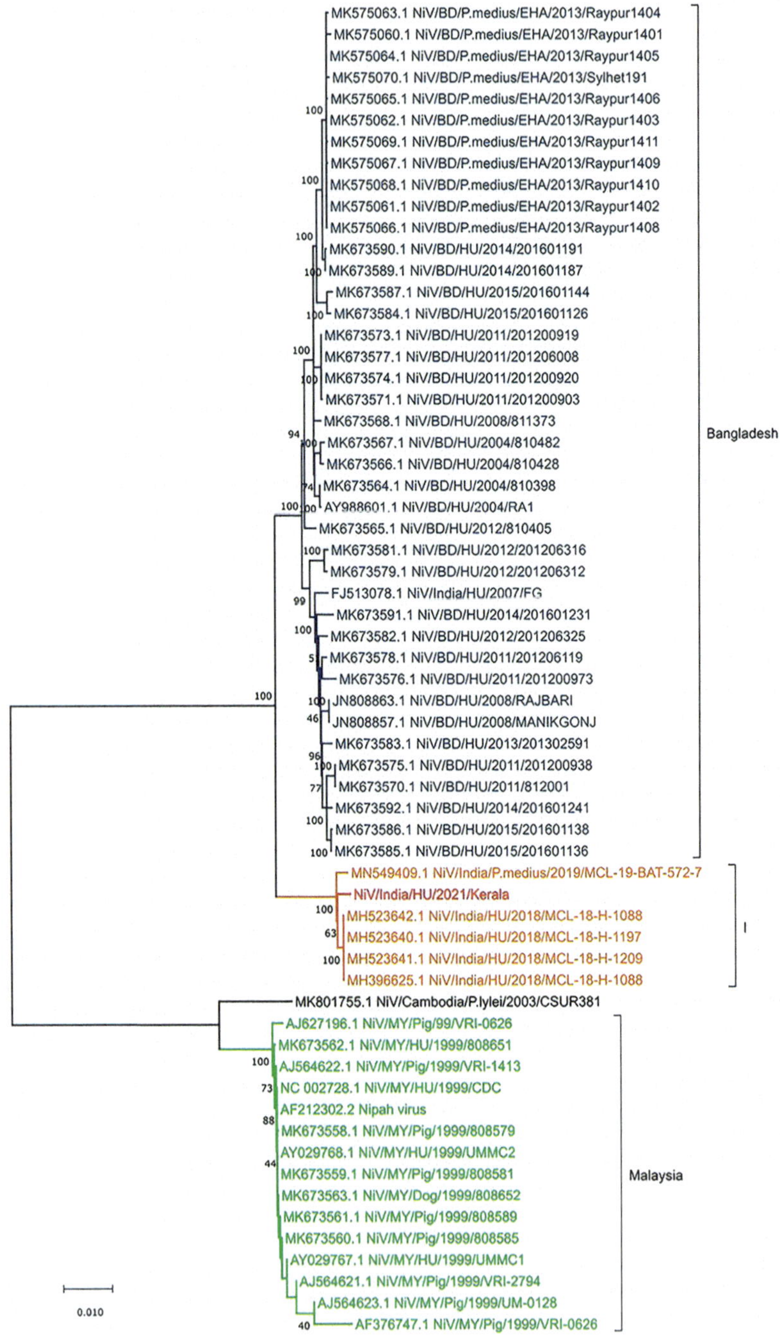
MK575063.1 NiV/BD/P.medius/EHA/2013/Raypur1404
MK575060.1 NiV/BD/P.medius/EHA/2013/Raypur1401
MK575064.1 NiV/BD/P.medius/EHA/2013/Raypur1405
MK575070.1 NiV/BD/P.medius/EHA/2013/Sylhet191
MK575065.1 NiV/BD/P.medius/EHA/2013/Raypur1406
MK575062.1 NiV/BD/P.medius/EHA/2013/Raypur1403
MK575069.1 NiV/BD/P.medius/EHA/2013/Raypur1411
MK575067.1 NiV/BD/P.medius/EHA/2013/Raypur1409
MK575068.1 NiV/BD/P.medius/EHA/2013/Raypur1410
MK575061.1 NiV/BD/P.medius/EHA/2013/Raypur1402
MK575066.1 NiV/BD/P.medius/EHA/2013/Raypur1408
MK673590.1 NiV/BD/HU/2014/201601191
MK673589.1 NiV/BD/HU/2014/201601187
MK673587.1 NiV/BD/HU/2015/201601144
MK673584.1 NiV/BD/HU/2015/201601126
MK673573.1 NiV/BD/HU/2011/201200919
MK673577.1 NiV/BD/HU/2011/201206008
MK673574.1 NiV/BD/HU/2011/201200920
MK673571.1 NiV/BD/HU/2011/201200903
MK673568.1 NiV/BD/HU/2008/811373
MK673567.1 NiV/BD/HU/2004/810482
MK673566.1 NiV/BD/HU/2004/810428
MK673564.1 NiV/BD/HU/2004/810398
AY988601.1 NiV/BD/HU/2004/RA1
MK673565.1 NiV/BD/HU/2012/810405
MK673581.1 NiV/BD/HU/2012/201206316
MK673579.1 NiV/BD/HU/2012/201206312
FJ513078.1 NiV/India/HU/2007/FG
MK673591.1 NiV/BD/HU/2014/201601231
MK673582.1 NiV/BD/HU/2012/201206325
MK673578.1 NiV/BD/HU/2011/201206119
MK673576.1 NiV/BD/HU/2011/201200973
JN808863.1 NiV/BD/HU/2008/RAJBARI
JN808857.1 NiV/BD/HU/2008/MANIKGONJ
MK673583.1 NiV/BD/HU/2013/201302591
MK673575.1 NiV/BD/HU/2011/201200938
MK673570.1 NiV/BD/HU/2011/812001
MK673592.1 NiV/BD/HU/2014/201601241
MK673586.1 NiV/BD/HU/2015/201601138
MK673585.1 NiV/BD/HU/2015/201601136
Bangladesh
MN549409.1 NiV/India/P.medius/2019/MCL-19-BAT-572-7
NiV/India/HU/2021/Kerala
MH523642.1 NiV/India/HU/2018/MCL-18-H-1088
MH523640.1 NiV/India/HU/2018/MCL-18-H-1197
MH523641.1 NiV/India/HU/2018/MCL-18-H-1209
MH396625.1 NiV/India/HU/2018/MCL-18-H-1088
I
MK801755.1 NiV/Cambodia/P.lylei/2003/CSUR381
AJ627196.1 NiV/MY/Pig/99/VRI-0626
MK673562.1 NiV/MY/HU/1999/808651
AJ564622.1 NiV/MY/Pig/1999/VRI-1413
NC 002728.1 NiV/MY/HU/1999/CDC
AF212302.2 Nipah virus
MK673558.1 NiV/MY/Pig/1999/808579
AY029768.1 NiV/MY/HU/1999/UMMC2
MK673559.1 NiV/MY/Pig/1999/808581
MK673563.1 NiV/MY/Dog/1999/808652
MK673561.1 NiV/MY/Pig/1999/808589
MK673560.1 NiV/MY/Pig/1999/808585
AY029767.1 NiV/MY/HU/1999/UMMC1
AJ564621.1 NiV/MY/Pig/1999/VRI-2794
AJ564623.1 NiV/MY/Pig/1999/UM-0128
AF376747.1 NiV/MY/Pig/1999/VRI-0626
Malaysia
0.010

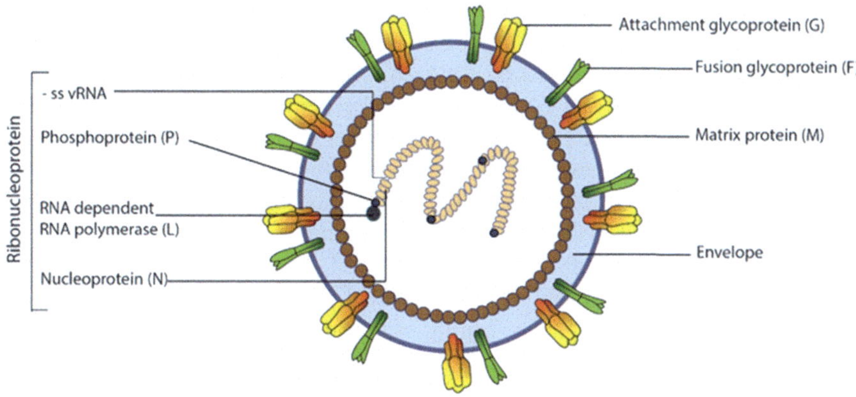

Fig. 1.4 Structure of the Nipah virus

virus contains non-segmented ssRNA with genome length ranging from 18.23 kb to 18.25 kb (Kojom and Singh 2021). The length of *Henipavirus* genomes is longer and is in the form of long untranslated regions. Negative-stranded RNA genome of the Nipah virus consists of six essential genes: the fusion protein (F) gene, the glycoprotein (G) gene, the large protein aka RNA polymerase protein (L) gene, the matrix protein (M) gene, nucleocapsid (N) gene and phosphoprotein (P) gene (Diederich and Maisner 2007).

1.5 Viral Proteins and Life Cycle

The RNA is encapsidated by the nucleocapsid protein in a ratio of one N for every six nucleotides. The NiV RNA polymerase replicates only viral genomes counting $6n + 0$ nucleotides. An exact nucleotide-N match at the RNA 3′-end is required for

Fig. 1.3 Phylogenetic analysis based on Nipah virus (NiV) full genome sequences, using NiV sequence from the index case (red) of the 2021 outbreak in Kerala, South India, and representative sequences from previous outbreaks (GenBank accession numbers indicated). From, Yadav, P. D., Sahay, R. R., Balakrishnan, A., Mohandas, S., Radhakrishnan, C., Gokhale, M. D., Balasubramanian, R., Abraham, P., Gupta, N., Sugunan, A. P., Khobragade, R., George, K., Shete, A., Patil, S., Thankappan, U. P., Dighe, H., Koshy, J., Vijay, V., Gayathri, R., Kumar, P. J., … Keerthi, K. V. (2022). Nipah Virus Outbreak in Kerala State, India Amidst of COVID-19 Pandemic. *Frontiers in public health*, *10*, 818,545. https://doi.org/10.3389/fpubh.2022.818545.

the formation of an active replication promoter (Diederich and Maisner 2007). The P gene codes for the phosphoprotein and three other proteins, viz., W, C and V. These additional proteins are generated by RNA editing (V and W protein) or are transcribed from an overlapping reading frame (C protein) (Diederich and Maisner 2007). C and P proteins are produced using distinct ORFs on the P gene; an alternate translational site present on the P gene is used for the translation of the C protein. The original transcript of the P gene is used to code for the P protein; however, the W and V proteins are the products of the addition of one/two G nucleotides during mRNA editing (Kolakofsky et al. 2005; Lo et al. 2009; Lawrence and Escudero-Pérez 2022). The editing site is present in the region that codes for the N-terminal of the phosphoprotein. During transcription the polymerase stutters at this position and incorporates one, two or more pseudo-templated Gs, resulting in shifts of the reading frame. V and W are non-structural proteins (NSPs) and are generated through different mRNAs generated in the process. Additionally, a second gene coding for the C protein is present at the start of the P gene in the same reading frame as the V protein. This is expressed from three different mRNAs by a leaky ribosome scanning mechanism (Jensen et al. 2020). Proteins C and V are responsible for the antagonization of the host innate IR, while the N-terminal domain of proteins W, P and V display capability of binding to STAT1 and interfere with the JAK/STAT pathway by hindering the phosphorylation of STAT1. The M gene encodes for the matrix protein, the G gene encodes for glycoproteins that aid in receptor recognition and binding activity and the L gene encodes for viral

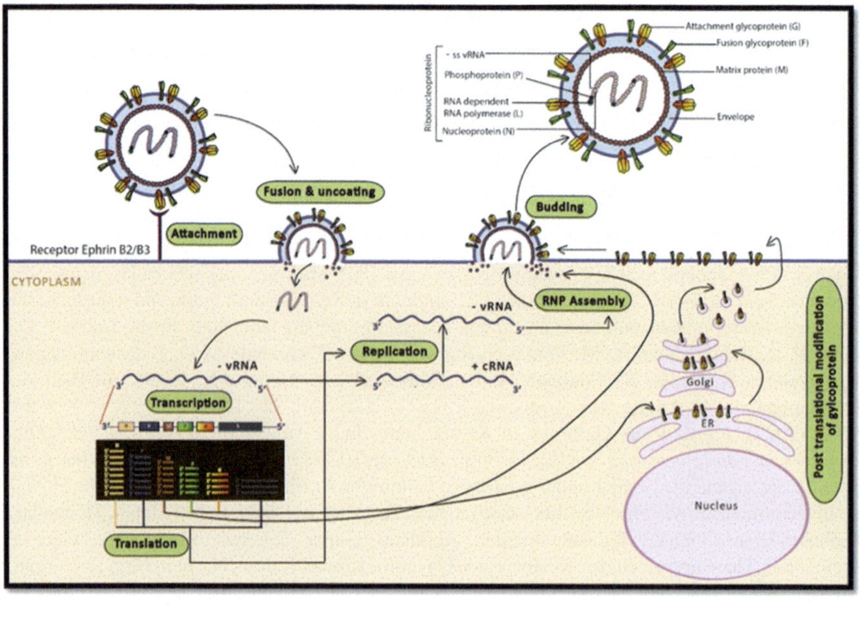

Fig. 1.5 Life cycle of the Nipah virus

RNA-dependent RNA polymerase (RdRp) that is required for genome replication (Majee et al. 2021) (Fig. 1.5).

The surface glycoprotein (G) along with the fusion protein (F) play an important role in the beginning phase of the life cycle of NiV. NiV G protein lacks haemagglutinatin and neuraminidase activities and does not bind to carbohydrate moieties (Diederich and Maisner 2007). The G protein binds to the cellular receptor EphrinB2/B3, as soon as the binding takes place, protein conformational changes begin in the F protein. The fusion peptide is pushed into the cell membrane of the host upon the change in conformation of the F protein from pre-fusion (PF) to a hair-pin intermediate (PHI) conformation. Following this the F protein conforms to a thermodynamically stable beam of six helices (6HB). The intermediary steps that take place between these events are external flake fusion aka hemi-fusion and fusion pore formation followed by widening. The fusion of the host and viral membrane is pH dependent and the cascade of events in the fusion process is important for the initiation of the infection (Skowron et al. 2022). The F protein is glycosylated at multiple sites and the N-glycans might play a role in protecting the protein from neutralizing antibodies (Hauser et al. 2021). Following attachment and entry into the host cell, the released genome undergoes replication using host machinery (Skowron et al. 2022). In genome replication, two important steps occur (1) transcription of the single-stranded RNA genome into monocistronic, capped, and polyadenylated messenger RNAs (positive-sense RNA) that can be used by the cellular ribosomes to generate new viral proteins and (2) replication of the negative-sense RNA genome into positive-sense antigenome copies, which are then used to generate a progeny of negative-sense genomes (Jensen et al. 2020). An unusual cytoplasmic inclusion, formed by open and closed ring-like structures that looked like a network of membrane-like reticular structures, was found during an examination by Goldsmith et al. (2003). These structures were different from the typical nucleocapsid inclusions (NCIs) and were called reticular inclusion (RI). It was found that genomic RNA, anti-genomic RNA template and mRNA were localized exclusively over the RIs (Goldsmith et al. 2003).

The N protein encapsidates the viral genome in a helical and long nucleocapsid. This is responsible for protecting the viral RNA and also serves as a template for mRNA transcription and aids the replication of the released viral genome with RdRp. In the first phase of the life cycle, there is predomination of the transcription of mRNA and viral proteins are produced using the host translational machinery. Genome replication is green lighted only after enough quantity of N protein is synthesized. The novel RNA produced is then encapsidated into the N protein such that it forms a helical assembly where the viral RNA is housed between the C- and N-terminal domains of the N protein. These domains bind, in an exposed and buried pattern, to the nucleotide triplets in alternation. The helical nucleocapsid assembly is stabilized due to the subunit-to-subunit interplay among the adjoining N proteins and the protein to RNA interactions. Highest conserved sequences are found at the RNA-binding surfaces and also at subunit–subunit interfaces, whereas the outer exposed surfaces of the nucleocapsid display the most sequence variation. Besides the typical spiral assembly, particles that resemble a clam-shaped assembly were also observed (Ker et al. 2021).

The matrix protein (M) is located at the inner surface of the lipid bilayer and mediates the contact between the RNP and the surface glycoproteins (Diederich and Maisner 2007). During virion formation, the nucleocapsid and RdRp complex are transported to the plasma membrane for assembly and budding. The M protein interactions with the G protein and nucleocapsid assembly facilitate the budding of the virus particle (Ker et al. 2021).

The phosphoprotein aids in viral genome transcription. P gene products V, W and C are not incorporated into the virus particle but help in host immunomodulation (Diederich and Maisner 2007). Although the length of the P protein varies among the families of order Mononegavirales, the architecture of the phosphoprotein is conserved in terms of functional and structural modules. NiV phosphoprotein behaves as a nanometer-size octopus-like protein that uses its long flexible arms to expose binding sites for multiple viral and cellular partners and to orchestrate their participation in virus replication, hijacking cellular machines and interfering with the interferon response (Jensen et al. 2020).

A common cytopathic effect observed in NiV infection is syncytia formations wherein the infected cells fuse together to form a multinucleated large cell mass. The surface G protein facilitates the viral entry and syncytia formation during infection (Ortega et al. 2022). The formation of multinucleated giant cells that comprise infected ECs is suspected to be involved in the occurrence of vasculitis (DeBuysscher et al. 2021).

NiV has shown survival for up to 7 days in artificially produced raw palm sap, composed of 0.2% BSA and 13% sucrose in water and pH 7.0, and at least 3 days in fruit like a mango or a few juices. The half-life of the virus is designated to be 18 h in the urine of reservoir bats. Nipah virus shows relative stability in the environment and is found to be viable after exposure to 70 °C for 1 hr. with a slight reduction in virus concentration. After heat inactivation at 100 °C for 15 min, the virus is found to be completely inactivated and incapable of causing infection (Singh et al. 2019). Although the Nipah virus' viability may vary depending on changes in its natural environment, the virus is promptly inactivated by detergents, soap and commercial disinfectants such as bleach (Hassan et al. 2018; Singh et al. 2019).

The availability of data on effectively and securely inactivating samples of NiV in BSL-4 settings is very limited. For instance, a reference showed that the use of icy absolute methanol is successful in the inactivation of cells infected with NiV, however, no details have been provided pertaining to the time required for inactivation (Aljofan et al. 2008). Additionally, two chemical methods of inactivation were proposed, viz., 10% neutral-buffered formalin (NBF), for inactivation of infected organs of ferrets, this displayed thorough virus inactivation, similarly, 4% PFA for infected cell monolayers for 15 min showed complete inactivation (Edwards et al. 2021). Other successful procedures for inactivation are UV inactivation, heat inactivation for NiV-spiked sera (Watanabe et al. 2020), pH-dependent inactivation and desiccation of samples (Fogarty et al. 2008). Under adequate conditions, however, the virus can survive in culture medium, fruit flesh, urine and blood for several days (Fogarty et al. 2008; Smither et al. 2019). Though essential to thoroughly inactivate the virus, in some cases it is also prudent to preserve the viral nucleic acids and/or proteins for research or diagnostic purposes. Of the methods so far used for NiV inactivation, most do not preserve proteins of the virus or are not included in the common procedures followed in laboratories, leading to the allowance for conducting further processing for research or diagnostics (Widerspick et al. 2022).

A study with an aim to '*compare some of the most common inactivation methods in order to evaluate their efficacy at inactivating NiV in infected cells, supernatants and organs*' was conducted by Widerspick et al. (2022). Accordingly, many chemical and physical methods of inactivation and their combinations were performed and the results analysed. Replication of NiV was kept track of for 3 weeks and its presence was detected using indirect immunofluorescence, plaque assay and RT-qPCR. Upon analyses, 19 processes demonstrated the reduction of infectious NiV particles in organs, cells and supernatants to a level undetectable even by RT-qPCR. The authors of this study have provided information on the methods that were most effective and secure (Widerspick et al. 2022). The article describing the study posted results pertaining to the evaluation of the efficacy of two physical methods of virus inactivation, viz., heat and UV inactivation, and 11 methods of chemical inactivation, viz., acetone, acetone/methanol, methanol, RLT + EtOH (ethanol), AVL + EtOH, PFA, NBF, Cytoperm/Cytofix (C/C), SDS, Triton-X100 and Trizol. The study used organs for 2 methods, cell cultures for 10 methods and supernatants of cell cultures for 7 methods, concluding in the testing of a total 19 methods. The authors of the study envisioned that the many procedures tested in this study would aid in the improvement of SOPs for virus inactivation processes for NiV as well as other viruses, especially, enveloped negative sense single-stranded viruses and facilitate further processing following inactivation. This study provided a list of procedures for the inactivation of NiV samples, safely and effectively, outside BSL-4 settings (Widerspick et al. 2022).

1.6 Pathogenesis and Pathology

Fig. 1.6.

1.6.1 Host-Virus Interaction

1.6.1.1 Cell Tropism

In vitro susceptibility to the Nipah virus has been known to depend on ephrin B2/B3 cellular receptor expression. For example, endothelial cells (ECs) that lack the ephrin B2/B3 receptors have been shown to be resistant to NiV infection, whereas human brain ECs aka HBMECs and porcine microvascular endothelial cells (PBMECs), which are known to highly express ephrin B2, were shown to be infected by NiV readily (Erbar et al. 2008). Except dendritic cells (DC) and macrophages, other blood cell populations, which have shown low affinity for NiV infection, are able to surface bind the virus and transport it to new sites of inflammation during cell recruitment (Pernet et al. 2012; Maisner et al. 2009; Mathieu et al. 2011) and

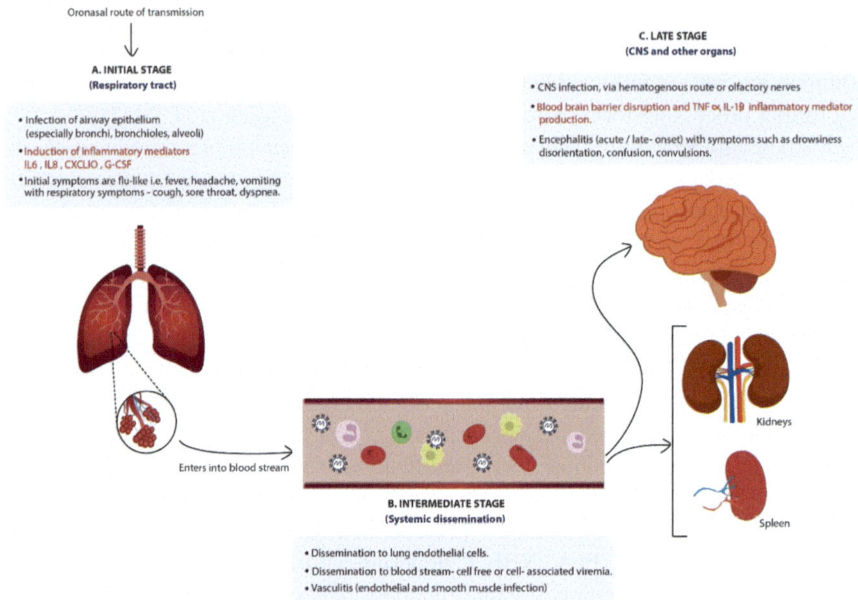

Fig. 1.6 Pathogenesis of the Nipah

may lead to HNVs possessing ligands that help virus attachment to certain cells but do not allow infection. The proof of this with respect to NiV was shown with the binding of the virus to EC lectin via the ligand LSECtin (Bowden et al. 2008; Lawrence and Escudero-Pérez 2022).

The ephrin receptors belong to the subclass of tyrosine kinases called receptor tyrosine kinases (RTKs), which are known to be greatly conserved within species. Namely, upon comparison, ephrin B2 of humans to that found in mice, pigs and fruit bats, the per cent difference was found to be 3%, 4% and 5%, respectively. In some instances, the HNV glycoproteins have been seen to bind to a variety of ephrin B2 homologues found in mammals, with indistinguishable affinity (Bossart et al. 2008). Ephrin B2, in humans, is absent on venous ECs, however, is present on arterial ECs, neuroglia, neurons and epithelial cells belonging to the upper respiratory system and alveolar sacs, smooth muscle cells (SMCs), phagocytes from the spleen as well as lymph glands and in pigs on macrophages (Maisner et al. 2009; Gale et al. 2001). As mentioned above, the distinction in the expression of ephrin receptors in arteries and veins can be shown through the Syrian golden hamster model, wherein when infected with NiV, the virus was detected in small- and medium-sized arteries, however, no presence of the virus was displayed in veins (Baseler et al. 2015; Lawrence and Escudero-Pérez 2022).

The part of the human organ system that is extremely rich in ephrin B2 expression is the human brain. Particularly, the cells of the prefrontal cortex and neural epithelial cells. The receptor is also found in lower levels such as the amygdala and olfactory bulb. Along with the brain, the ephrin B2 receptors are also expressed in organs such as the lungs, placenta and prostate (Liebl et al. 2003). As opposed to this, ephrin B3 is seen to be majorly expressed along the central nervous system (CNS), the cardiac muscles and the cells of the prostate. The maximum level of ephrin B3 along the CNS is in areas of the occipital lobe, the prefrontal cortex and the amygdala, whereas it is found in areas such as the pons, temporal lobe, subthalamic nucleus, globus pallidus, hippocampus, corpus callosum and the hypothalamus, in low levels (Benson et al. 2005). Fascinatingly, it is ephrin B3, not B2, that is expressed in the cells of the brainstem (Negrete et al. 2007; Lawrence and Escudero-Pérez 2022). In conclusion, the ubiquitous distribution of these receptors on distinct tissues and in a variety of host species could help in shedding light on the nature of systemic infection and the mystery of this group of viruses having a wide host range in comparison to other members of Paramyxoviridae (Lawrence and Escudero-Pérez 2022).

The primary targets of infection are vascular endothelial cells (EC). The induced cytopathic effect in ECs by NiV is displayed as the formation of syncytia of the infected ECs and is directly correlated with the instigation of vasculitis (Chua et al. 1999, 2000a, Wong et al. 2002; Wong and Tan 2012). Syncytium formation is mainly observed in alveolar cavities which leads to the development of a pulmonary oedema, while antigens were detected in tunica media cells, the endothelium and in alveolar cavities (DeBuysscher et al. 2013, 2021).

The infection of smooth muscle cells (SMC) from the tunica media of the small blood vessel lining is seen in animal models and humans. However, no histological or pathological changes that can be associated with SMCs that show the presence of the viral antigen have been reported. In order to control vascular stability and function, these cells are tightly packed, during and after the formation of a blood vessel. There is a developmental link between SMCs and ECs such that they both originate from the lateral mesodermal plate. Similar to ECs, SMCs are also known to express ephrins and Eph receptors, and the interaction between these receptors and the membrane-bound ligands of the two cell types plays a major role in cell motility and adhesion during the process of angiogenesis (Foo et al. 2006; DeBuysscher et al. 2021). A study with the aim to 'gain an understanding of the possible contributions that SMCs might have in the development of NiV disease' was conducted. They carried out an investigation to confirm the in vitro susceptibility and cytopathogenic effects occurring in SMCs of humans. SMCs showed permissibility to NiV entry and infection, and produced high viral titres for prolonged periods, despite no visible cytopathogenic effects and a total absence of any detectable ephrin-B2/B3. These results led to the conclusion that human SMCs harbour NiV, allowing it to proliferate without suffering any visible cytopathic injury (DeBuysscher et al. 2021).

1.6.1.2 Pathogenesis

1.6.1.2.1 Entry and Initial Infection

In order to infect human and animal hosts, the virus enters the host body using the oronasal route. Accordingly, the initial replication is seen in epithelial cells alongside the respiratory tract and, subsequently, a large quantity of viral particles were detected among the cells of the respiratory and lymphoid tissues (Clayton et al. 2016; Liew et al. 2022). During the acute phase, the virus can be detected in bronchiolar epithelial cells (Chua et al. 2001). The primary targets for the detection of viral particles in experimental models of disease are epithelial bronchiolar cells and alveolar type-2 pneumocytes (Rockx et al. 2011). With the infection of the epithelium along the respiratory tract comes the induction of inflammatory mediators {cytokines, IL-1a, IL-6, IL-8, CXCL-10, G-CSF, etc.} and recruitment of immune cells, which leads to the development of disease manifestations similar to acute respiratory disease syndrome (ARDS) (Rockx et al. 2011; Escaffre et al. 2013).

1.6.1.2.2 Viremia and Multi-Organ Dissemination

The inflammatory response generated in the acute phase of the infection due to replication paves the way for a systemic infection characterized by the production of generalized symptoms. Viral dissemination into cells takes place in the following ways:

1. Viraemia facilitated by replication in ECs followed by dispersion through blood.
2. Neuronal replication leads to spread in the CNS.
3. Replication in the ECs leading to the disruption of the blood–brain barrier (BBB).
4. Formation of syncytia.
5. Attachment to blood cells without infection, for transport.
6. Replicating in the nasal cavity and disseminating via olfactory neurons (de Wit and Munster 2015; Weingartl et al. 2005; Munster et al. 2012; Lawrence and Escudero-Pérez 2022).

Following primary viraemia, secondary replication is said to occur in the endothelium. The beginning of the infection is marked by the attachment of the viral G protein to ephrin B2/B3 receptors (Bonaparte et al. 2005; Xu et al. 2008; Negrete et al. 2006). In the span of a week, the virus quickly disseminates to organs such as the heart, liver, kidneys and spleen (Mire et al. 2016; Geisbert et al. 2010; Guillaume et al. 2009; Liew et al. 2022). As mentioned earlier in this section, ephrin B2/B3 receptors are widespread in all cell types inclusive of epithelial cells, ECs and neuronal cells (Bossart et al. 2008). It has also been reported that NiV can spread through the CNS with the help of monocytes and immature DCs (Tiong et al. 2018). Even though these immune cells were found to be permissive to NiV, no virus replication was reported. However, they did assist in the migration of the virus, across an in vitro BBB and displayed infection similar to that which was observed in human neurons and the development of foci-like lesions on the brain belonging to humans and animals infected with NiV (Liu et al. 2019; Wong et al. 2009; Liew et al. 2022).

In the later stage of the disease, the virus replicating in the respiratory epithelium is disseminated to the ECs of the lungs. Following this, the virus dispersion in the blood occurs freely or via attachment to host leukocytes. Besides the lungs, the spleen, kidneys and the brain are essential targets and dissemination to them may lead to multiple organ failure (Rockx et al. 2011; Escaffre et al. 2013). In the hamster model of infection, when NiV-attached leukocytes are passively transferred, a lethal infection is developed (Mathieu et al. 2011). In pigs, monocytes, $CD6^+$ $CD8^+$ T lymphocytes and Natural killer (NK) cells are seen to display infection followed by viral production (Stachowiak and Weingartl, 2012).

The two distinctly involved pathways in the entry of the virus into the CNS are:

1. Hematogenous route (via choroid plexus or cerebral blood vessels).
2. Anterograde entry via the olfactory nerves (Weingartl et al. 2005).

The development of neurological signs is a result of the CNS infection due to the disruption of the BBB and the expression of IL-1b with the tumour necrosis factor a (TNF-a) (Rockx et al. 2011). In humans, there may be inclusion bodies present in cases of CNS infection; the evidence of plaques with necrosis may be present in grey and white matter (Escaffre et al. 2013). In various experimental models, direct viral entry into the CNS via the olfactory route is evident. In these animal models, the NiV infection is seen in the epithelium of the nasal turbinate and following this, the infection goes through the cribriform plate to the olfactory bulb. In the end, the dissemination of the virus occurs throughout the ventral cortex alongside the olfactory tubercle (Munster et al. 2012; Weingartl et al. 2005; Escaffre et al. 2013).

1.6.2 Immune Response

1.6.2.1 Innate Immune Response

1.6.2.1.1 In vitro Discoveries

Endothelial cells infected in vitro were shown to instigate secretion of IFN-1, various host antiviral proteins, inflammatory cytokines and chemokines. After primary attachment followed by fusion of the virus with the host cell membrane, the RNA helicase present in the host cytoplasm recognizes the NiV RNA and triggers a sturdy response with the release of IFN-I and varied IFN-instigated antiviral genes like OAS1, IP-10 and ISG56 (Habjan et al. 2008; Leon et al. 2018; Liew et al. 2022).

1.6.2.1.2 Findings in *Pteropid* Bats

The *Pteropid* spp. displays an absence of upregulated response of IFN-I following exposure to the virus. These bats serve as reservoirs for various virus species and tend to have an IFN system that is highly active which leads to the instigation of specified IFN-induced genes. This response in bats is suspected to confer the ability to regulate the replication of viruses and co-exist with various species of zoonotic viruses and remain asymptomatic. Furthermore, they induce a response of IFN type-III and keep a certain level of IFN-I constant for the duration of the infection. In bats, the activation of immune response takes place after infection and it is regulated such that it is just enough to restrict replication but not clear the virus (Zhou et al. 2016; Zhou et al. 2011a, b; Liew et al. 2022). Fruit bats demonstrate the presence of tethrin, which is capable of inhibiting replication and functioning as an antiviral to aid the host immune system (Hoffmann et al. 2019; Singh et al. 2019).

1.6.2.1.3 In vivo Discoveries

In tandem with the results of in vitro experiments, in vivo confirmation of host IR stimulation and signalling of IFN-I was conducted. Hamster models demonstrated the signalling of cytokine, chemokine and IFN-I, first in the lungs and then in the brain and this coincided with previous findings in regard to disease development and progression (Rockx et al. 2011); similar findings were reproduced in a ferret model of disease (Leon et al. 2018). In IFNAR-KO mice, the deficiency of IFN response was demonstrated and the results displayed a complete lack of resistance to infection that was present in wild-type mice; this highlighted the significance of innate IR during a NiV infection (Dhondt et al. 2013; Liew et al. 2022). Along with the antiviral response of IFN-I, many pro-inflammatory cytokines like IL-1β and TNF-α were produced (Rockx et al. 2011). Inordinate production of these factors could be disorderly and cause unrestricted inflammation which could in turn advance the process of pathogenesis and this was apparent in SARS-CoV-2 infections (Darif et al. 2021). With respect to NiV, the production of IL-1β and TNF-α is associated with cerebral infection (Rockx et al. 2011). It was reported that the pro-inflammatory nature of these factors led to the disruption of the BBB and aided in the occurrence of neurological manifestations typically seen in patients suffering from NiV disease (Ang et al. 2018; Sejvar et al. 2007; Liew et al. 2022).

1.6.2.1.4 Viral Protein Modulation and Virus Evasion

NiV is capable of modulating the inflammatory responses and the stages involved in the signalling pathway of this process. It expresses many structural as well as non-structural proteins in order to bring about this modulation (Keiffer et al. 2020; Sugai et al. 2017). For example, the V protein expressed by NiV interacts with STAT-1 and 2 in order to sequester them from ISGF3 (a complex of transcription factors) (Rodriguez et al. 2003), a downstream effector of IFN modulation seen in the late phase of host immune antiviral response (Honda and Taniguchi 2006). Additionally, the host innate antiviral response is said to be modulated via the W, V and P proteins of NiV via interaction with STAT-5 and 4 proteins and effectively altering STAT activity (Keiffer et al. 2020). The interaction between NiV V protein and MDA-5 leads to the suppression of activation of IFN via dephosphorylation of Melanoma differentiation associated protein-5 (MDA-5) (Davis et al. 2014; Rodriguez and Horvath 2014; Parisien et al. 2009). The ability of NiV proteins P, V and W was demonstrated by their construction into a recombinant Newcastle disease virus (NDV) and infecting a primary human dendritic monocyte-derived cell line (Seto et al. 2010). When compared to infection with original NDV, which led to a sturdy activation of IFN-α/β, the expression of Nipah genes on NDV overturned the results and led to the subversion of IFN-α/β response accompanied by a low cytokine response. On another note, the M protein of NiV displayed the ability to degrade host TRIM6 protein. This, in turn, led to the blockage of IFN-I signalling pathways (Rajsbaum et al. 2014; Bharaj et al. 2016). The data available so far have indicated the ability of NiV to evade the host immune system using a multifaceted approach to interference (Liew et al. 2022).

Antagonization of the IFN signalling pathways is a virulence mechanism found in genus *Henipavirus* attributed to C, V and W viral proteins (Basler 2012). During in vivo studies, recombinant NiV (rNiV) with deletions of C or V protein, but not W, demonstrated attenuation in Syrian hamster models leading to no disease manifestation and 100% survival. In addition to that, there was a low replication rate and seroconversion found in models that lacked C and V, respectively, that were infected with rNiV (Yoneda et al. 2010). In a different study, two rNiV strains with deletion of C proteins were used and they showed high levels of attenuation and the survival rate ranged from 10% to 80% depending on the strain used and the dose administered (Mathieu et al. 2012). These results combined with results of the in vitro demonstration of the infection in ECs by rNiV lacking C protein, which displayed the expression of high levels of pro-inflammatory cytokines when compared to wild-type strain infection, suggested that the C proteins are helpful in preventing the control of infection via inhibition of host pro-inflammatory response (Mathieu et al. 2012; Lawrence and Escudero-Pérez 2022).

The P gene on the NiV genome encodes four viral proteins, viz., P, V, W and C (Lawrence and Escudero-Pérez 2022). These four proteins are all known to antagonize the signalling of the IFN pathway; V in the cytoplasm binds to STAT-1 and STAT-2 and; therefore, prevents dimerization and inhibits their delivery to the nucleus for ISG activation. Similarly, the W protein performs the sequestration of STAT-1 in the nucleus which in turn inhibits ISG activation. The P protein is also capable of performing this sequestration, however, it is done in lower levels as compared to V and W proteins (Ciancanelli et al. 2009; Shaw et al. 2004). In the cytoplasm, protein C performs antagonistic activity against IFN but the mechanism is not very well understood yet. It is however known that C antagonizes IFN-α recruitment dependent on TLR-7 or 9 by impeding the RNA synthesis and binding to the inhibitor of κB kinase α aka IKKα (Horie et al. 2016; Lo et al. 2012b, 2014b; Yamaguchi et al. 2014; Lawrence and Escudero-Pérez 2022).

Other IFN antagonizing mechanisms seen in the course of a NiV infection are:

1. Protein W blocks signalling of IFN via interaction with TBK-1 and IKKε (Shaw et al. 2005).
2. Protein W can be located in the nucleus and interferes with phosphorylated IRF-3 forms, hence, interferes with IFN-β promoter activation and thus, the regulation of IFN-I synthesis (Yoneda et al. 2010).
3. Protein V prevents downstream signalling via the inhibition of LGP-2, RIG-I (Childs et al. 2012), MDA-5 (Childs et al. 2007) and STAT-2 (Rodriguez et al. 2002; Lawrence and Escudero-Pérez 2022).

Not simply the P gene but also the NiV matrix protein (M) is guilty of inhibiting IFN responses. Though the main function of M protein is assembly and enabling budding, the protein also displays the ability to inhibit IFN-I production. During the process of replication, the M protein is seen to interact with TRIM6, promoting its degradation leading to reduced IKKε poly-ubiquitination and thus, subsequently suppressing IFN-associated response (Bharaj et al. 2016) The nucleoprotein (N) also exhibits the capability to inhibit IFN type I and II responses. It demonstrates the interference with STAT-complex and thus reduces nuclear assimilation of STAT and subsequent activation of ISG. Additionally, it is also capable of preventing the direct import of STAT into the nucleus (Sugai et al. 2017; Lawrence and Escudero-Pérez 2022) (Fig. 1.7).

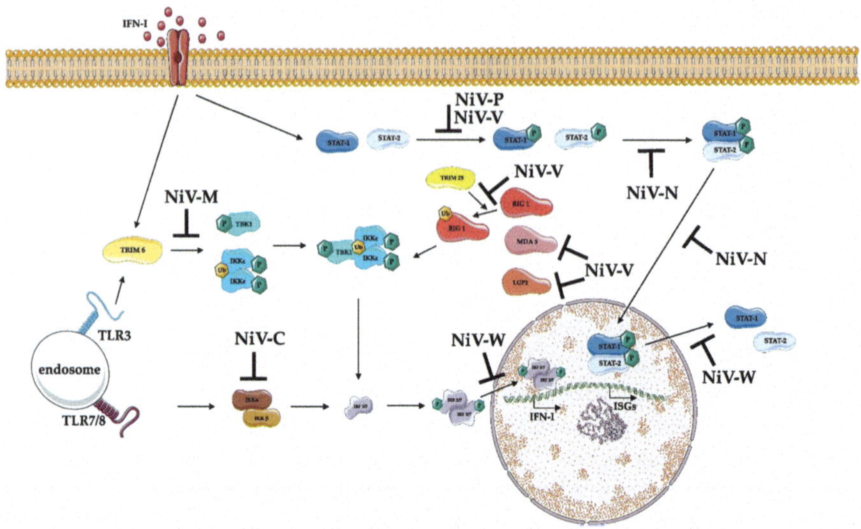

Fig. 1.7 Mechanisms of innate immunomodulation by Nipah virus proteins. From, Lawrence, P., and Escudero-Pérez, B. (2022). Henipavirus Immune Evasion and Pathogenesis Mechanisms: Lessons Learnt from Natural Infection and Animal Models. *Viruses*, *14* (5), 936. https://doi.org/10.3390/v14050936. This is an open-access article distributed under a Creative Commons Attribution (CC BY) license (https://creativecommons.org/licenses/by/4.0/). Copyright © 2022 Philip Lawrence and Beatriz Escudero-Pérez

1.6.2.2 Humoral Immune Response

1.6.2.2.1 Findings in Humans

The presence of inadequate clinical samples spanning the course of the disease and the inability to obtain samples from fatal cases have resulted in limited resources to investigate the humoral response of human hosts to NiV infection. However, from two survivors of the 2018 outbreak of NiV in Kerala, India, a descriptive account of the adaptive IR during the acute phase followed by the convalescent phase was obtained (Arunkumar et al. 2019b). Upon testing, the sera of one of the survivors was detected to consist of IgG and IgM Ab that showed NiV specificity, and the viraemic clearance was an indication of pathogen-specific response of IgG Abs; both survivors demonstrated an increase in B lymphocyte populations, which compares to the production of IgM and IgG Ab specific to NiV (Liew et al. 2022).

1.6.2.2.2 Findings in Bats

As a natural source of NiV, bats have evolved to control the immune modulation brought about by proteins of the virus. Bats demonstrated increased levels of highly specific naïve Ig, and this enables the control of replication through direct B lymphocyte clonal development which does not necessitate affinity maturation for increased production of Abs. Furthermore, in the absence of affinity maturation and divested hypermutation phases, these reservoirs develop weak IR and restricted Ab generation (Schountz et al. 2017). These factors may aid in viral persistence and thereby delay clearance and thus, this could explain the pattern of seroconversion that was demonstrated during in vivo infection of bats with HNV, wherein 50% of the infected population displayed seroconversion with weak NAb response (Halpin et al. 2011; Liew et al. 2022).

1.6.2.2.3 In vivo Discoveries

The elicitation of Abs is analogous to protection and is associated with clearance of the virus followed by recovery of the organism. With the limitation of adequate human samples, animal models such as hamsters, ferrets and AGMs were established and are being used to accumulate a better understanding of humoral IR. Pig models that were infected experimentally, demonstrated a development of NAb titres within a week PID which was accompanied by a spike in their titres following 2 weeks PID (Berhane et al. 2008); even though NAbs were present in the sera of infected pigs, viral genome was detected for up to a month PID, demonstrating slow rate of clearance of virus. AGMs displayed an exhaustive B lymphocyte population at 12 days PID (Lara et al. 2019), i.e. in the acute phase and this corresponded with the quick progression of the disease. In a study conducted by Riddell et al. (2007) involving six AGM models, only one survived upon incurring infection. The surviving AGM demonstrated the development of NiV-specific IgM and decreased NAb responses following 12 days of PID. The titre of IgM Abs showed an incline at 14 days PID and parallelly, the detection of an IgG response occurred. This incline in the amount of NiV-specific Ab response corresponds to the expansion of the B lymphocyte population (Riddell et al. 2007). This indicated the activation of a host humoral IR upon infection with the Nipah virus; it also suggested that the temporary immune suppression of the host and slow clearance of infection could be a characteristic of Nipah virus disease (Liew et al. 2022). Following this, in 2022, a study with the aim to 'Analyse the polyclonal serum antibody responses elicited by vaccination of macaques with NiV G' indicated the receptor binding head domain to be immunodominant (Wang et al. 2022). This discovery that the NiV G head domain is a neutralization target in Rhesus monkeys vaccinated with tetrameric NiV G ectodomains has helped inspire the development of vaccine candidates focused on producing a robust Ab response against this domain (Wang et al. 2022).

1.6.2.3 Cell-Mediated Immune Response

1.6.2.3.1 Findings in Humans

Relating to human infection, as mentioned before, there is limited information on adaptive IR due to the inadequacy of samples. The information gained from animal models reports the presence of T lymphocyte responses. Two survivors of NiV infection were studied to understand the changes in T lymphocyte populations during the acute phase as well as convalescence (Arunkumar et al. 2019b). The study showed that the definitive number of T cells in blood remained constant; however, there was a peak in the CD8 T-cell population that expressed PD-1, Ki67 and granzyme-B. These results demonstrated the importance of increased cytotoxic cell populations with respect to the elimination of NiV-infected cell populations (Liew et al. 2022).

1.6.2.3.2 In vivo Discoveries

Akin to the study of B lymphocytes, T lymphocyte-specific animal models have been established to study and analyse the pathogenesis of NiV, this circumvents the limitation of samples from infected patients. For example, studies in pigs and AGMs correlated insights similar to those found in a human study (Lara et al. 2019; Arunkumar et al. 2019b; Cong et al. 2017; Pickering et al. 2016). Upon analysing the peripheral IR of AGM models of NiV infection, an increase in the $CD8^+$ and $CD4^+$ memory effector cells was reported and this was correlated to a spike in chemokines (Ki67) and cytokines. When infected with NiV, a group of pigs demonstrated an increase in the CD25 upregulation on T_H memory cells along with T_C cells (Pickering et al. 2016); while further analysis revealed a decrease in viral RNA levels and viral clearance from tissues. When compiled together they highlighted the significant role played by CMI in the survival of patients suffering from NiV disease (Liew et al. 2022).

As a member of the Paramyxoviridae family, NiV is grouped together with viruses like MuV, MeV, HPIV and RSV, but is most closely related to HeV (Chua et al. 2000a). The shared features among genomes of Paramyxoviruses make it possible for cross-reactivity of T cells to occur among the viruses belonging to the same family. In order to support this hypothesis, a study was conducted and it was found that a former infection by a human paramyxovirus resulted in the instigation of cellular cross-reactivity (de Vries et al. 2020). Two T-cell clones (TCCs) specific to the F protein, viz., $CD8X^{react}1$ and $CD8X^{react}2$, were obtained from donors previously infected by HPIV and MeV; while the epitope F129–37 was recognized to be highly conserved among the Paramyxoviridae members. Subsequently, when NiV-infected cells were co-cultured with TCCs, the infected cells were thoroughly eliminated and this confirmed the functionality of TCCs, suggesting their potential to provide efficacious protection against NiV (Liew et al. 2022).

1.6.2.4 Immunomodulation

NiV is capable of exhibiting an association of two types with hosts, i.e. remaining persistent in the environment through reservoirs and afflicting humans as well as domestic animals like pigs, with a lethal clinical disease (Chua et al. 2000a; Negrete et al. 2005; Kulkarni et al. 2013; Singh et al. 2019). In order to ensure successful NiV infection, replication and avoid premature mortality of the host, a critical balance has to be maintained between the host's innate IR and the ability of the virus to neutralize said response (Liew et al. 2022).

Subsequently, the elevation of vascular permeability due to the release of some inflammatory cytokines in spite of suppression by viral machinery at an early phase leads to the creation of favourable conditions for viral spread within the system (Schountz 2014; Singh et al. 2019). In the brain, the CXCL10 over-expression leads to neuronal cell apoptosis as a result of indirect caspase-3 activation and Cl downregulation where the latter leads to subsequent cell demise (Sui et al. 2004, 2006).

The NiV replication was closely followed by the expression of CXCL10 mRNA, this phenomenon was observed in various organs, especially, the brain of the hamster model; thus, CXCL10 was suspected to play a significant role in the manifestation of NiV disease-related encephalitis. These reports corroborate the findings of epithelial cells in patients, whose cause of death was recorded as NiV-related encephalitis during the Malaysian epidemic. When stained, the cerebral epithelial cells were demonstrated to be deeply stained by CXCL10. This supported the suggestion of a correlation between brain injury and fatality to the presence of chemokines. CXCL10 was found to be highly expressed in the lungs and spleen of bats infected with HeV, though the infection of NiV/HeV is non-lethal in bats. Therefore, here the increased level showcased a protective role (Woon et al. 2020). This leads to the acceptance of CXCL10 as a potential marker of fatal encephalitis associated with the disease and an excellent target for the development of therapeutics against NiV (Mathieu et al. 2012; Liew et al. 2022).

Additionally, in response to NiV, the expression of pro-inflammatory cytokines was observed; like, IL-6 instigates maturation of DCs which in turn plays a key role in the maturation of T cells (Gupta et al. 2013); IL-8 induced chemotaxis of granulocytes to the site of infection (Prasad et al. 2020b) and MCP-1 regulates the permeability of the blood–brain barrier (BBB) (Stamatovic et al. 2005). The expression of all these cytokines has the ability to promote the induction of T-cell movement and that of functional monocytes to the site of infection (Lo et al. 2010; Escaffre et al. 2016; Valbuena et al. 2014). This heightened activity of cytokines resulted in systemic vasculitis which was also seen in human patients (Liew et al. 2022).

Though the dysregulation of inflammatory cytokines may cause harm to the host, when correctly regulated and balanced, they are capable of benefiting the host organism. For instance, along with the initiation of maturation of DCs, IL-6 is capable of amplifying the function of antigen presentation on DCs and thus, enhancing the efficiency of initiation of Ag-specific IR (Stamatovic et al. 2005; Levroney et al. 2005). This can be further enhanced by an innate effector protein Galectin-1, which displays a high expression level at the site of inflammation. The high exposure of DCs to Galectin-1 results in the upregulation of IL-6 in DCs derived from monocytes (Levroney et al. 2005) and this further implies that DCs and Galectin-1 act as an intermediary of a protective response of innate inflammatory factors against infection of the Nipah virus (Liew et al. 2022).

Evidence suggesting that the viral proteins of NiV are capable of suppressing the expression of MHC-I in the host has been discovered. This leads to the quelling of both, the antigen presentation by APCs and the induction of adaptive response which consequently leads to spread and viral persistence in target organs (Dasgupta et al. 2007; Seto et al. 2010). Aside from this, it has been computationally predicted that the virus also interacts with other genes such as TJP-1, GRIA-3 and FHL-1 associated with BBB and encephalitis, TLR genes of defence in host, and notch genes involved in neurogenesis. miRNAs present along the NiV genome are reported to play a crucial role in inhibiting host defence genes and, thus help in causing increased pathogenesis (Saini et al. 2018; Singh et al. 2019).

In concern with the adaptive IR, there is very little that is understood and no information is currently known as to whether it is efficient at clearing the virus or if it aids in the process of pathogenesis (Lawrence and Escudero-Pérez 2022). The detection of inclusion bodies that are positive for NiV antigens in patients with NiV-associated encephalitis, however, shows the deficiency of the innate and adaptive IR in the prevention of viral dissemination. This suggested the inadequacy of DCs present at the target for entry, the lungs and respiratory tract, to capture antigen and restrict its entry into tissues (Chua et al. 1999; Chong and Tan 2003; Singh et al. 2019).

1.7 Clinical Features

1.7.1 Case Definitions

Case definitions about Nipah virus disease according to the National Centre for Disease Control, India are as follows:

1. **Suspected Case:** 'Person from a community affected by a Nipah outbreak who has:
   - Fever with new onset altered mental status or seizure and/or.
   - Fever with headache and/or.
   - Fever with cough or shortness of breath.'
2. **Probable Case:** 'Suspect case-patients/s who resided in the same village where confirmed case-patient/s were living during the outbreak period and who died before diagnostic specimens could be collected.

   OR

   Suspect case-patients who came in direct contact with confirmed case-patients in a hospital setting during the outbreak period and who died before complete diagnostic specimens could be collected.'
3. **Confirmed Case:** 'Suspected case who has laboratory confirmation of Nipah virus infection either by:
   - Nipah virus RNA identified by PCR from respiratory secretions, urine or cerebrospinal fluid.
   - Isolation of the Nipah virus from respiratory secretions, urine or cerebrospinal fluid.'

1.7.2 Clinical Presentation

NiV disease may manifest itself as apparent or subclinical in nature (Skowron et al. 2022) and a relapse of the asymptomatic infection has been reported to occur months or years following the acute phase (Thakur and Bailey 2019; Skowron et al. 2022). Patients with previous non-encephalitic or asymptomatic Nipah infection can develop late-onset encephalitis or relapsed encephalitis. The Nipah virus may be unique in its ability to cause late-onset encephalitis as well as relapsing and remitting encephalitis by direct viral attack in a significant number of survivors of acute Nipah encephalitis. Most patients with relapsed and late-onset encephalitis had an acute illness with the usual features of encephalitis, such as fever, headache, drowsiness, coma and CSF pleocytosis (Tan et al. 2002). Brainstem involvement and meningism were not observed in most of the relapsed encephalitis cases. Although the mortality rate was low in relapsed encephalitis and late-onset encephalitis cases as compared with acute encephalitis, worse neurological outcomes were observed with irreversible neuronal damage observed in the former cases. Pathological, clinical and radiological results interpret that the relapsing encephalitis is distinct from NiV-associated acute-onset encephalitis but is similar to late-onset encephalitis associated with NiV disease (Wong et al. 2012).

1.7.3 Phases of Disease

The Nipah virus demonstrates an incubation period of 4–45 days (WHO 2021a), wherein the associated symptoms are inapparent until the clinical quiescence period of 4 days to 2 weeks is not over (Ochani et al. 2019; Thakur and Bailey 2019; Skowron et al. 2022). After the quiescent phase mild, symptomatic or severe encephalitis may occur. Relapsing or 'late-onset' infections are also been identified; one such relapsing infection occurred 11 years after an asymptomatic infection (Ong and Wong 2015).

1.7.4 Clinical Signs and Symptoms

Symptoms in the acute phase resemble common flu-like infections, namely headache, fever, vomiting and dizziness (Thakur and Bailey 2019). These symptoms display the ability to rapidly progress into encephalitis which is categorized by severe neurological manifestations along with headache and fever. Patients displayed a decline in consciousness along with convulsions and apparent cerebellar manifestation accompanied by tachycardia (Sharma et al. 2019) and this severe mental disorientation might result in a comatose state for 1–2 days (Singh et al. 2019). Following an incubation period of 5–21 days, an infected individual usually develops a headache, high fever and meningism. Though it was seldom detected in the Malaysian outbreaks initially, cases of viral pneumonia have been noted in ≥50% of cases from the Indian and Bangladesh outbreaks. Following the onset of neurological manifestations, there is a rapid decline in the mental state and within 24–48 h the patient progresses from severe encephalopathy to a comatose state; seizures occur frequently and chances of mortality are increased (McEntire et al. 2021). NiV infection usually consists of a single phase; however, there have been reports of cases that suffered fatal encephalitic syndrome years following the initial infections, including patients who manifested only mild symptoms (McEntire et al. 2021).

The main clinical manifestation of NiV disease is both respiratory and neurologic. Fatal encephalitis is considered a primary representation of many NiV-infected cases (Ong and Wong 2015; Singh et al. 2019; Chua et al. 1999). Fever, headache, dizziness, vomiting, reduced level of consciousness, non-productive cough and myalgia are some of the main features presented in acute Nipah infection. Reduced level of consciousness absent or reduced reflexes, abnormal pupils, tachycardia (heart rate > 120/min), hypertension (blood pressure > 160/90 mm hg), abnormal doll's-eye reflex, segmental myoclonus, meningism, seizures, nystagmus cerebellar signs, bilateral ptosis, bilateral postural tremor, dysarthria and dysphasia are some of the neurological features presented in the Malaysian pig farmers. Profuse and segmental sweating were also seen in a few cases. Disseminated micro-infarction in the brain secondary to vasculitis-induced thrombosis was observed pathologically (Chong et al. 2002). MRI of the brain showed multifocal lesions (subcortical and deep white matter lesions, probably corresponding to areas of microinfarction) that appeared to be specific for acute Nipah encephalitis (Goh et al. 2000). The chemical profile of the CSF of an infected patient resembles those suffering from non-haemorrhagic meningitis of viral origins. MRIs display ≤ 1 cm diameter, punctate, patchy and diffused, subcortical as well as deep white matter, scarcely involving brainstem and cortex, lesions. When studied closely, pathology shows small vessel necrotizing vasculitis present throughout the CNS and other organs such as the lung, heart and kidney (McEntire et al. 2021).

Both encephalitis and respiratory symptoms were presented in patients of Bangladesh outbreaks. Infections occurred in all age groups and fever, altered mental status, cough and respiratory symptoms were the most common symptoms. Severe neurological manifestations were also seen and chronic sequelae were observed in some of the survivors of acute Nipah infection (Hossain et al. 2008). Depressive symptoms, personality change and chronic fatigue syndrome were also observed in certain cases. Depression following Nipah virus encephalitis is indistinguishable from an idiopathic depressive disorder. Sadness, anhedonia, loss of energy, weight loss, insomnia, psychomotor agitation or retardation, feelings of worthlessness or hopelessness, diminished ability to concentrate, and thoughts of death or suicide are the characteristic features. Depression occurs most frequently in patients with lesions affecting the frontal and temporal lobes and the basal ganglia, particularly the caudate nuclei (Ng et al. 2004). Dysautonomia or signs of brain stem involvement were highly associated with a fatal outcome (McCarthy 2003).

A-reflexia/hypo-reflexia was seen in cases of acute encephalitis (Ong and Wong 2015). Abdominal pain, anorexia, diarrhoea, gastritis and constipation were also reported in a few cases (International Centre for Diarrheal Diseases Research Bangladesh, 2007; Sharma et al. 2019). Additionally, in Indian and Bangladesh outbreaks, the cases presented with respiratory manifestations like cold, shortness of breath along with encephalitis, sepsis and acute respiratory distress syndrome (ARDS) (Pallivalappil et al. 2020). Lymphopenia, thrombocytopaenia and hyperglycaemia were observed simultaneously as a requirement for ventilation in a few cases (Chong et al. 2002).

Studies involving the two NiV isolates suggest that NiV-My is closely aligned with neurological disease manifestations while the NiV-Bd is highly associated with respiratory syndrome in human patients. Nevertheless, NiV-My infection demonstrated respiratory symptomology followed by a rapid course of the disease. When exposed to a low dose of large aerosolized particles, it was observed that the disease progression on slow and lasted longer with manifestation of neurological disease in all animal models (Lee et al. 2020).

1.7.5 Nipah Virus Disease Models

Nipah virus is capable of infecting a wide range of animal hosts among both wild and domesticated species inclusive of horses, dogs, cats, pigs and fruit bats. The symptoms in each species are seen to be varied and most often do not manifest similarly to human disease. For example, in *Pteropid* bats, the infection is subclinical and no clinical manifestation of the disease is observed (Geisbert et al. 2012). Clinical disease manifestations are seen only in laboratory mice that have been intracranially inoculated (Mungall et al. 2006), or when inoculated in older mice such as BALB/c or C57B6 or in IFNAR KO—the transgenic species of mice that have had IFN-I knocked out (Dhondt et al. 2013). There have been no reports of laboratory-adapted strains of NiV in mice or other animal models. The models that have been shown to accurately reproduce pathogenesis and symptoms of NiV disease in humans are Ferrets, Syrian Golden Hamsters and African Green Monkeys (AGMs) (Price et al. 2021).

1.7.5.1 Hamster

The most commonly used NiV disease animal challenge model is the Syrian Golden Hamster (*Mesocricetus auratus*) due to its size it is economical to house. It has displayed the capability of developing pathologies and symptoms that are similar to NiV-related disease in humans, inclusive of neurological and respiratory manifestations (Guillaume et al. 2006). Inoculation of the model is carried out via intraperitoneal (IP) and intranasal (IN) routes and the pathological manifestations induced after inoculation via both routes are similar; however, the course of the disease is variable depending on the administered dose and route of administration (Rockx et al. 2011; Wong et al. 2003); where inoculation via the IP route shows a rapid progression of the disease when compared to the IN route of inoculation. In hamsters, the respiratory symptoms manifest in the form of difficulty in respiration, blood in the nasal discharge, pulmonary oedema and the presence of haemorrhagic abrasion on the lungs (Rockx et al. 2011), whereas the neurological symptoms comprise lethargy, improper co-ordination, twitching of muscles, behavioural decline and partial limb paralysis (Dhondt and Horvat 2013). The level of dose administered is vital when considering the symptomology of NiV in hamsters. When high doses are supplied intranasally, the response manifests primarily in the form of a respiratory syndrome. Low doses induce respiratory as well as neurological signs (Rockx et al. 2011). Thrombosis and vasculitis lead to the development of lesions on many organs, involving the brain and lungs, and encephalitis in hamsters (Torres-Velez et al. 2008). Variance in pathology among hamsters in the context of the disease pathology of NiV has been reported in different studies conducted using this model. The NiV-My inoculation in hamsters by Wong et al. (2003) was reportedly characterized by neurological manifestations (Wong et al. 2003). Contrastingly, Rockx et al. (2011) recorded that inoculation with the same strain, using the same route and administration of a similar dose level displayed primarily a respiratory syndrome (Rockx et al. 2011). The reason behind the discordance in both these studies is a mystery and hence, this suggests the need for further studies for understanding the infection of NiV in hamsters (Hassell et al. 2021).

Tests comprising both the NiV strains, viz., NiV-My and NiV-Bd, have been conducted in hamster models; however, most have studied the NiV-My strain as the challenge agent. One study displayed the Malaysian strain when IP inoculated demonstrated rapid replication (Lawrence and Escudero-Pérez 2022) and hence, an accelerated disease course in hamsters. This was contrary to what has been observed during outbreaks in humans. Although as a whole, the symptoms and distribution of the virus of both strains were similar (DeBuysscher et al. 2013; Price et al. 2021), a second study observed indistinguishable pathology after inoculation via intranasal route (Lawrence and Escudero-Pérez 2022).

Cellular pathogenicity of the Nipah virus in hamsters is seen to be influenced majorly by receptor dissemination, as the detection of antigens occurred in arteries, small as well as medium-sized, but not in veins (Baseler et al. 2015). Few animals that were inoculated using the IN route displayed evidence of the transport of the virus by infecting the sensory neurons (Munster et al. 2012). Similarly, hamsters that were infected via the consumption of contaminated, artificially prepared date palm sap demonstrated the ability of the virus to infiltrate the sensory neurons thereby causing an invasion of the CNS (de Wit et al. 2014; Lawrence and Escudero-Pérez 2022).

Certain studies have shed light on the increase in signalling of inflammatory mediators and cells of the immune systems in many organs of animals infected with NiV, especially the brain and lungs (Mathieu et al. 2012), while other experiments have used the hamster model to communicate the role played by NiV NSPs during infection (Yoneda et al. 2010; Lawrence and Escudero-Pérez 2022). The significance of IFN functionality in a NiV infection was reproduced in hamsters by stimulating IFN response using analogous RNA and it demonstrated protection against fatal NiV challenge (Georges-Courbot et al. 2006; Lawrence and Escudero-Pérez 2022). A study confirming the capability of NiV to repress the expression of IFN innate response genes in ECs, such as STAT-1 and 2, IFN-α7, CXCL-10 and GTPase-1 induced by IFN, was conducted (Schountz et al. 2019). Reports of host IR genes including CXCL-10, IL-4, IL-6, IFN-γ and TNF-α showing increased activity during NiV-My infection in comparison to NiV-Bd infections have been published (DeBuysscher et al. 2013; Lawrence and Escudero-Pérez 2022).

1.7.5.2 Ferret

Mustela putorius furo, a species of ferret that is domesticated, is a carnivore that is small in size and displays susceptibility to respiratory viruses such as *Influenza*, *Paramyxoviruses* (HeV, NiV), *Morbilliviruses* (CDV) and *Pneumoviruses* (RSV, HMPV). In comparison to the hamster model, ferrets are simpler to monitor for pathological signs and display easily interpretable behaviours (Enkirch and von Messling 2015). The snout of the ferret is larger than that of hamsters and hence, this makes it easier to perform an IN challenge (R. Gomez-Roman, personal communication; Price et al. 2021).

Severe respiratory and neurological manifestations are observed in ferrets infected with NiV. The virus is commonly introduced into the animals via IN or oronasal routes. Development of clinical manifestations is observed within 6 days post infection, symptoms frequently seen are depression, fever, spasms in muscles and partial paralysis (Clayton et al. 2012). Symptoms of the respiratory kind commonly seen are laboured breathing, bloody nasal secretion, cough and grievous respiratory distress. Pathological manifestations include cerebral subcutaneous oedema, lymphadenopathy (haemorrhagic), petechial haemorrhage on the kidney and lung, and vasculitis (Bossart et al. 2009). Commonly present pathologies are inflammation in alveolar sacs and vasculitis, meningitis may appear but encephalitis was not demonstrated (de Wit and Munster 2015). Syncytium formation leads to the development of lesions (Torres-Velez et al. 2008). Most studies in ferrets have been performed using the NiV-My strain (Price et al. 2021). Ferrets inoculated with Bangladesh and Malaysian strains of NiV display comparable clinical manifestations and course of disease; however, the ones infected with NiV-My demonstrated an increasingly prominent haemorrhagic condition, and those infected with NiV-Bd induced higher oronasal discharge (Clayton et al. 2012; Leon et al. 2018).

Progression of disease course and symptoms demonstrate comparability among studies with most models displaying similar neurological and respiratory symptoms and a humane endpoint is reached 7–10 days post-infection date (PID) (Price et al. 2021). Vaccination studies have shown affirmative results with various candidates against a NiV challenge in animal models by providing long-lasting and efficacious protection. An experimental study demonstrated continued efficacious protection against the fatal challenge of NiV, 12 months post vaccine administration with limited localized virus replication detectable (Pallister et al. 2013). Similarly, the effectiveness of passive immunization against disease is observed in animals upon administration hours to days PID (Mire et al. 2020; Bossart et al. 2009; Lawrence and Escudero-Pérez 2022).

1.7.5.3 African Green Monkeys (AGMs)

In order to study the complete pathogenesis of NiV and the effects of therapeutics and prophylactics, various animal models have been investigated such as AGM, cats, ferrets, hamster, guinea pigs, pigs and squirrel monkeys. Among the aforementioned animal models, AGMs (*Chlorocebus sabaeus*) are the most relevant NHP model of NiV-related disease (Price et al. 2021), as it is known to show pathological characteristics closest to what NiV disease is in humans (Geisbert et al. 2012; Lee et al. 2020; Johnston et al. 2015), such as lethal ARDS, dyspnoea along with vasculitis and neurological manifestations (Hammoud et al. 2018; Liu et al. 2019; Lee et al. 2020; Geisbert et al. 2010; Cong et al. 2017; Marianneau et al. 2010). As this experimental model reiterates the neurological, vascular as well as respiratory aspects of the disease, it is essential in a study of the pathogenesis of NiV as well as therapeutic efficacy. In addition to this, the magnetic resonance imaging (MRI) results in AGMs closely resemble that of disease manifestations in humans (Lee et al. 2020; Cline et al. 2022). Similar to human infection and other animal infection models, in AGMs the virus antigens can be detected in tissues of the brain, lungs, liver, kidneys and spleen, pointing to systemic dissemination of the virus (Mire et al. 2016; Prasad et al. 2020a; Lawrence and Escudero-Pérez 2022).

Inoculation strategies currently used for inoculation of AGMs with NiV comprise aerosol or intratracheal exposure (Johnston et al. 2015; Hammoud et al. 2018; Cong et al. 2017), intranasal mucosal atomization (Geisbert et al. 2020) or a combination of intratracheal and intranasal exposure (Prasad et al. 2020a; Mire et al. 2016). When it comes to replicating natural infections, the above routes of inoculations are considered clinically pertinent, in which they imitate the exposure to infectious aerosols, although these are not well characterized due to the limited research that has been performed (Cline et al. 2022). Inoculation of NiV via the intratracheal (IT) route in AGMs may prove fatal with mortality occurring within 9–12 days of administration. Higher doses have been seen to accelerate the course of the disease and reduce the time of survival (Geisbert et al. 2012) thus, providing less time for studying pathology. The animals primarily display signs of fatigue, depression, febrile illness and appetite loss. NiV is capable of manifesting ARDS in AGMs categorized by shortness of breath and serous nasal exudate. Enlargement of the lungs is due to pulmonary oedema, congestion and haemorrhagic lesions (Geisbert et al. 2010). Histopathology indicates haemorrhage in alveolar sacs, syncytium in ECs of lungs and general vasculitis. Neurologically, the signs are less congruous in AGMs than in ferrets or hamsters. Some demonstrate muscular twitching and paresis (Geisbert et al. 2014), tremors and histopathological indications of encephalitis (Johnston et al. 2015), however, no brain anomalies are overtly present. One study led to the identification of lesions in the brain via neuroimaging (Hammoud et al. 2018) whereas another did not (Cong et al. 2017). The incongruous results of these studies could be due to the different levels of dosage or administration routes (Hassell et al. 2021).

Various isolates of NiV-My have been used in AGM studies, but only one isolate of NiV-Bd has been used. Upon comparison of infection of the AGM model with NiV-My and NiV-Bd with human infection, it was observed that mortality and morbidity were acceded by ~3 days in the NiV-Bd infection model compared to the NiV-My infection model. This accelerated progression of disease was inconsistent with what was demonstrated in Ferret and hamster models but was consistent with NiV disease in humans. Other than that, no particular distinction was observed among the two AGM groups (Mire et al. 2016; Price et al. 2021). Curtis et al., conducted a study in AGMs, with a goal to 'further characterize pathologic changes associated with the NiV-MY AGM intratracheal model.' The results of the study further substantiated the claim that the AGM model of NiV disease in humans is relevant since it demonstrates two major components of the disease, viz., neurological and respiratory. In addition to this, they also demonstrated the presence of a chronic stage of the disease which could be important to study the yet unknown late-onset encephalitis as well as a relapse of it, as it has been described in humans (Cline et al. 2022).

1.8 Diagnostics and Therapeutics

1.8.1 Laboratory Assays

Taking into account the similarity in symptoms of NiV with those of some other diseases, it is essential to make an early diagnosis to provide good patient care and control the outbreak. Laboratory diagnosis of NiV is done by two types of procedures, Ab detection via ELISA and PRNT, or Ag detection via isolation, PCR, immunohistochemistry, etc. (Daniels et al. 2001; Kulkarni et al. 2013; Wang et al. 2012). Other tests, such as electron microscopy, are used for the visualization and confirmation of the agent, and sequencing is used for the characterization of the virus; although they are not easily available and do not aid in initial diagnosis (Aditi and Shariff 2019).

The use of cell culture in NiV is not affected by variations in virus and its use aids in getting a detailed description of viral activities (Hyatt et al. 2001). Isolation and neutralization assays are conducted for confirmation and can only be conducted in a BSL-4 facility under strict biosafety regulations (Mazzola and Kelly-Cirino 2019). The standard method used widely for detecting the presence of NiV is the serum neutralization test (Daniels et al. 2001); though, both the above-mentioned methods require BSL-4 laboratory facilities (Ma et al. 2019). Due to their high specificity, sensitivity and ability to issue rapid results RT-PCR and RT-qPCR are considered to be the gold standard in diagnosis (Guillaume et al. 2004; Ma et al. 2019). NiV IgM ELISA is an alternative approach for conducting serological diagnosis when PCR is unavailable. However, though ELISA is a reliable method for the detection of NiV (Yadav et al. 2012), it has lower sensitivity and specificity when compared to molecular methods of detection (Ma et al. 2019). Post mortem, in order to confirm the presence of NiV in lethal cases, immunohistochemistry is utilized. Electron microscopy is seldom used, due to its inability to distinguish between NiV, HeV and other members of Paramyxoviridae (Chow et al. 2000).

Sample specimens usually collected from animals are blood, urine, respiratory secretions, faeces and organ biopsy, this is due to previous evidence suggesting the course of NiV is the initial entry through the respiratory tract following which there is large-scale dissemination through the CNS (Aditi and Shariff 2019). Samples such as nasal and throat swabs are transported in a viral transport medium (VTM), while blood is collected in different colour-coded vacutainers based on the component of blood that is to be tested, and urine and faeces are collected in leakproof containers; for organ biopsies, it is essential for the tissue to be fixed immediately and this fixation is carried out using 10% NBF. Samples to be used for virus detection can be collected from infected patients displaying symptoms or during postmortem examination. For serological diagnosis, the samples are to be collected in the late phase of infection, roughly 10–14 days after disease onset. The National Centre for Disease Control (NCDC) of India recommends nasal and throat swabs (in VTM), blood, urine and CSF be used for diagnosis. Sample collection must be done while following proper biosafety measures and the samples should be transported in triple-layer packaging using a cold chain of 2–8 °C. If the sample is not to be processed immediately, storage at −20 °C is recommended after 48 h of collection. Sample processing is to be done in BSL-4 settings using strict biosafety measures (Chadha et al. 2006; Aditi and Shariff 2019).

1.8.1.1 Molecular Methods

Nucleic Acid Amplification Tests (NAATs) like PCR are the preferred tests for the diagnosis of an active infection due to their high sensitivity and specificity (detect up to as few as 20 genome particles), but the equipment and infrastructure required for these tests may not be available or economical in all endemic areas. Additionally, the tests may become less sensitive in the event of genomic changes occurring due to mutations or the probe design is incompatible with the strain circulating. In order to ensure the inclusion of genetic variants, probes are designed to be compatible with highly conserved regions of the viral genome, which are resistant to mutations and are generally present in all circulating strains of the agent being detected (Feldman et al. 2009). The two NiV strains majorly circulating, namely NiV-My and NiV-Bd, including subvariants and the Philippines and Cambodian variants, have been subjected to whole genome sequencing (Chua et al. 2000a; Harcourt et al. 2000; Lo et al. 2010; Lo et al. 2012a, b; Halpin et al. 2007; Chadha et al. 2006; Harcourt et al. 2005; Reynes et al. 2005). The targets for RT-PCR of NiV are the highly conserved regions of P, N or M genomic segments. Conventional PCR, RT-PCR, RT-qPCR and nested duplex RT-PCR (nRT-PCR) are utilized in the detection of NiV followed by nucleotide sequencing for confirmation. The US CDC developed a conventional PCR that targets the N gene (Chua et al. 2000a; Aditi and Shariff 2019); however, RT-PCR is a thousand times more sensitive than conventional PCR and, therefore, is used to near exclusivity for diagnosis (Wang et al. 2012; Mazzola and Kelly-Cirino 2019).

The RT-qPCR assay based on a TaqMan probe developed in 2004 is designed to detect the N protein gene of NiV and has shown extremely high sensitivity (~1 pfu). The assay displayed high specificity such that it was capable of detecting NiV RNA from blood samples of infected hamsters but failed to detect viral RNA of HeV (Guillaume et al. 2004). Commercially, a kit based on the aforementioned technology was developed by a company named Zoologix in order to detect Nipah virus in samples such as CSF, blood, serum plasma, infected tissues collected from NiV animal models or other secretions (Mazzolla and Kenny 2019; Thakur and Bailey 2019). A nRT-PCR developed has aided in the detection of two distinct strains of virus from *Pteropus lylei* (Wacharapluesadee and Hemachudha 2007), and has proved to be a vital tool for the surveillance of NiV in Thailand (Wacharapluesadee and Hemachudha 2007; Liu et al. 2016; Thakur and Bailey 2019). Similarly, during outbreaks in Malaysia and Singapore, nested primer encoding the highly conserved regions of genes P, M and N was most frequently used (Harcourt et al. 2000; Chua 2003; Sharma et al. 2019). For the detection of *Henipaviruses*, specific RT-qPCR SOPs have been drawn up and these have been useful in the detection of NiV during infection (Wang and Daniels 2012; Kulkarni et al. 2013). An assay based on SYBR Green-I has been developed and it targets the N gene of NiV, however, it has low sensitivity (~100 pfu) when compared to TaqMan it also is capable of detecting HeV (Chang et al. 2006). Similarly, a TaqMan assay targeting the P gene was developed which was envisaged as an investigative tool against known NiV strains using higher sensitivity and specificity (Wacharapluesadee and Hemachudha 2007; Liu et al. 2016; Thakur and Bailey 2019). For the quantitative detection of NiV RNA, a novel RT-qPCR assay has been developed. It is a one-step assay and targets the region separating the G and F genes, it is designed in such a way that it avoids the amplification of viral mRNA and is a representation of a highly precise assay in comparison to the conventional real-time RT-PCR (Jensen et al. 2018; Singh et al. 2019). PCR assays that target the N protein gene of NiV have proved to be a significant help in the generation of data that was eventually used to perform phylogenetic studies (Ong and Wong 2015; Daniels et al. 2001; Thakur and Bailey 2019).

1.8.1.2 Tissue Culture Methods

1.8.1.2.1 Immunohistochemistry

For detection using immunohistochemistry, infected tissue that has been fixed using formalin is used (Daniels et al. 2001; Aditi and Shariff 2019). Tissue section analysis is a significant step towards understanding and identifying pathogenesis related to NiV, including lesions of inflammation, vasculitis and necrosis (Ong and Wong 2015; Thakur and Bailey 2019). In pregnant animals, the products of conception, the placenta and the uterus are also analysed (Daniels et al. 2001). For example, in a ferret model of disease, upon experimentally inducing infection, viral lesions were observed in tissues of the respiratory tract, kidneys, liver and lymphoid tissue, while viral antigens were detected in the vascular ECs and neurons (Clayton et al. 2012; Thakur and Bailey 2019). Formerly, the use of convalescent serum of humans was utilized in immunohistochemistry, but now it has been replaced by anti-NiV rabbit antisera (Daniels et al. 2001; Aditi and Shariff 2019). For the detection of antigens, anti-NiV Abs have been utilized successfully to stain infected tissues fixed using formalin (Ong and Wong 2015; Daniels et al. 2001; Thakur and Bailey 2019).

1.8.1.2.2 Virus Isolation and Characterization

In the event of outbreaks or sporadic cases being discovered, it is essential to isolate the aetiological agent and this is usually attempted using samples taken from the human brain, lung, spleen, kidney (Kulkarni et al. 2013; Daniels et al. 2001; Thakur and Bailey 2019) and biopsy samples obtained during the acute phase of infection (Chua 2003; Goh et al. 2000; Chua et al. 2001). Among animals, NiV can be isolated using tissue samples obtained from the serum, brain, spleen, lung and kidneys

(Middleton et al. 2002; Wong et al. 2003; Chang et al. 2006; Sharma et al. 2019) and all isolation-related procedures, be it human or animal sample processing, must be carried out in a BSL-4 facility (Daniels et al. 2001). For the isolation of NiV, the establishment of cell lines obtained from *Pteropid* bats has been brought about (Crameri et al. 2009). The Nipah virus proliferates fluently in Vero cells and manifests CPE after 72 h of infection of the cell line, which is in the form of a syncytium along with characteristic Nipah virus-produced plaques on the monolayers (Kulkarni et al. 2013; Daniels et al. 2001; Thakur and Bailey 2019). The syncytium formation upon NiV infection is larger than that forms upon HeV infection and, therefore, can be used to distinguish between the two viruses (Hyatt et al. 2011). Confirmatory diagnostic methods used in companionship with virus isolation are serum neutralization, PCR using supernatant of culture and immunostaining. Immuno-electron microscopy and electron microscopy can be vital tools to detect the interactions between viruses and Abs and for virus identification by viewing a distinct structure, respectively (Kulkarni et al. 2013; Daniels et al. 2001; Thakur and Bailey 2019)

Yadav et al. (2019) mentioned cell fusion and syncytia as CPE in NiV-infected Vero cells. The authors reported the progression of CPE as follows: 'As the infection advanced, the number of multinucleated giant cells increased. Cells with dendritic extensions were observed, 48 hrs post-infection. 64 hrs post-infection, considerable cellular damage and detachment was seen. Morphological changes indicative of apoptosis were also observed.' (Fig. 1.8).

1.8.1.3 Serological Methods

Serological testing is done to detect viral antigens directly or indirectly as well as for the detection of IgM and IgG Abs that are raised against NiV. Generally, tests of the serological type are broadly cross-reactive among strains and subtypes, and possess low sensitivity with regard to slight genetic variations (Lo et al. 2012a, b). In order to detect an active infection, IgM and antigen detection tests can be conducted; detection of IgM Abs raised against NiV in patient sera is possible 9 days after onset and may persist for 3 months. As IgG Abs persist for longer, their peak is seen 17 days post onset and last up to 8 months; thus, these tests are mainly used for surveillance and epidemiological studies (Ramasundpum et al. 2000; Mazzola and Kelly-Cirino 2019). Fifty percent of patients were shown to have a detectable IgM titre on day 1 of illness while detectable IgG titre was seen in 100% of patients after the 18th day of illness (Ramasundaram et al. 2000; Aditi and Shariff 2019). During vaccine development, it is essential to be able to detect virus-specific serum IgG levels in order to characterize the magnitude and specificity of the humoral IR induced by the vaccine (Price et al. 2021).

1.8.1.3.1 ELISA

In serological diagnosis, ELISA is the most frequently utilized, as it is a highly sensitive and rapid assay method while being simple and safe to use. Additionally, it is well-established, comparatively low-tech and affordable, with a broad range of available reagents and varied applications (Price et al. 2021; Aditi and Shariff 2019).

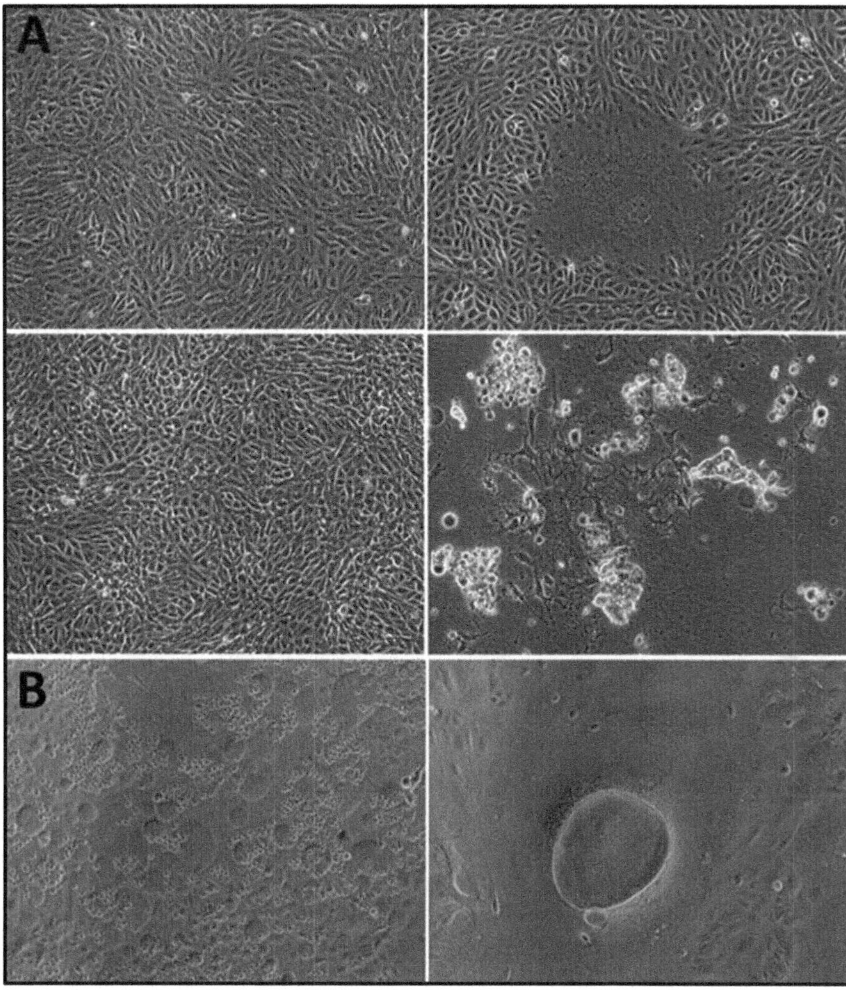

Fig. 1.8 Cytopathic effect (CPE) displayed by the Nipah virus (NiV) from a throat swab sample of a patient from the 2018 outbreak in Kerala, South India, inoculated into Vero CCL81 cells. (**a**) Upper and lower, right—controls. Upper and lower, left—CPE at days post infection-1 and 2, respectively. (**b**) NiV-infected cells (original magnification x10). From, Yadav, P. D., Shete, A. M., Kumar, G. A., Sarkale, P., Sahay, R. R., Radhakrishnan, C., Lakra, R., Pardeshi, P., Gupta, N., Gangakhedkar, R. R., Rajendran, V. R., Sadanandan, R., and Mourya, D. T. (2019). Nipah Virus Sequences from Humans and Bats during Nipah Outbreak, Kerala, India, 2018. *Emerging infectious diseases*, *25* (5), 1003–1006. https://doi.org/10.3201/eid2505.181076 (this publication is in the public domain)

The CDC, USA formulated the first ELISAs for NiV-specific IgG and IgM detection in serum samples, which were employed for diagnosis during the initial Nipah outbreak in Malaysia. A major disadvantage of these early ELISAs was the utilization of crude NiV-infected cell lysates, which meant non-specific binding in the assay, and the requirement of a BSL-4 facility for this reagent preparation (Daniels et al. 2001; Price et al. 2021).

The next ELISAs began to use rNiV proteins generated in a wide range of expression systems; and along with higher specificity, bypassed the need for a BSL-4. *E. coli* expressed rNiV N, F and G proteins exhibited reactivity with NiV-infected human sera; additionally, the rNiV G was shown to be reactive against serum derived from a patient with late-onset relapsing Nipah disease. Soluble NiV F and G proteins expressed via *E. coli*, insect cell and mammalian cell systems have been used in various ELISAs (Eshaghi et al. 2004, 2005; Keshwara et al. 2019, Kong et al. 2012; Tiong et al. 2017; Thakur and Bailey 2019; Mazzola and Kelly-Cirino 2019). In the context of NiV serodiagnosis, IgM ELISA is generally the first-line test, after which SNT or PCR may be carried out for confirmation (Sherrini and Chong 2014; Daniels et al. 2001; Wang et al. 2012). Yu et al. (2006) described an rNiV N protein-based indirect IgG ELISA and IgM capture ELISA by the Institute of Tropical Medicine, Japan; utilized for testing serum samples of both humans and pigs, and only humans, respectively (Singh et al. 2019). The ICAR-National Institute of High Security Animal Diseases developed an indirect IgG ELISA using rNiV N protein as the target Ag, for the purpose of screening porcine sera (Kulkarni et al. 2013; Singh et al. 2019). Fischer et al. (2018) gave an account of indirect IgG ELISAs made using recombinant proteins as follows—full-length NiV N expressed in *E. coli* and truncated soluble HeV/NiV G expressed in *L. tarentolae*. The N protein-based ELISA performed preliminary pig sera screening for *Henipavirus* disease, whereas the G protein-based-ELISAs permitted distinction between HeV and NiV diseases (Singh et al. 2019).

A mAb-based virus Ag-capture ELISA that targets NiV/HeV N and HeV P proteins was formulated by the CDC, USA. Along with detecting NiV, it also differentiates it from HeV (Chiang et al. 2010; Singh et al. 2019). Kaku et al. (2012) report a virus Ag-capture sandwich ELISA employing rabbit anti-NiV G protein polyclonal Abs created by the National Institute for Infectious Diseases, Japan, which targets the NiV and HeV G proteins (Singh et al. 2019).

1.8.1.3.2 Virus Neutralization Test

The gold-standard assay employed for serological confirmation of Nipah disease is the virus neutralization test. Generally, the classical PRNT as well as the SNT are the formats used for NiV neutralization. However, the chief drawback is the live virus needed for the assay, which necessitates the presence of a BSL-4 facility (Crameri et al. 2002; Daniels et al. 2001).

In order to avoid the requirement for a high containment facility, surrogate neutralization assays have been developed which make use of pseudotype viruses, i.e. pseudoviruses, which are non-infectious in nature, thereby enabling the assay to be carried out in a BSL-2 facility. Construction basically involves a parental virus, e.g. VSV, whose envelope displays foreign virus proteins that play roles in cell binding and/or fusion—which is why they are neutralized by virus-specific sera. Also, to act as a marker of infection, most contain a reporter gene like GFP. The first devised pseudovirus assay for NiV had VSV pseudotyped with NiV G and F proteins, with a luciferase reporter. It exhibited a good correlation of titres upon comparison with standard PRNT, albeit with decreased sensitivity for very low serum Ab levels. A similar assay with a GFP reporter was developed, which showed good specificity and greater sensitivity when evaluated against a conventional SNT. Another assay which demonstrated in general higher sensitivity than an SNT is a modification of the aforementioned test; it allows the usage of a regular ELISA plate reader for the assessment of the SEAP reporter, from the culture supernatant (Wang and Daniels 2012; Tamin et al. 2009; Kaku et al. 2009, 2012).

1.8.1.3.3 Cell Fusion Assay

Utilizing the activation of the reporter gene in a quantifiable cell fusion assay has been established. Target cells coding for an *E. coli* lacZ cassette that is driven by a T7 promoter are cultured together with effector cells that express NiV glycoproteins G and F as well as an RNA polymerase for T7. Upon the merger of the cytoplasm of the fused cells, it is possible to measure the activity of β-galactosidase, permitting the quantification of cell fusion mediated by NiV. Following this step, a demonstration of cell–cell fusion inhibition was done using serially diluted sera that were incubated, prior to the addition of the target cell, with effector cells (DeFang et al. 2010). This test is envisioned to pair with VNTs and examine the impediment of cell–cell fusion and virus–cell fusion, if it displays adaptability for higher throughput analyses (Thakur and Bailey 2019).

1.8.1.3.4 Luminex® Assay

The Luminex®, a liquid protein array system based on the use of beads, is a relatively recent assay based on the binding Ab assay principle, utilized to detect NiV G protein-specific IgG in serum (Vignali 2000). This assay consists of fluorescent microspheres covalently bonded to purified soluble NiV G (sGNiV) protein (Bossart et al. 2007). Following the step of binding to analyte (NiV G-specific serum IgG), Protein A (biotinylated) is introduced, after which streptavidin-conjugated phycoerythrin (PE), a fluorescent signal molecule that emits at a wavelength distinct from the microspheres, is added. This bead mixture is then analysed through an instrument that is based on dual laser flow detection. Of the two lasers, one detects the bead and ascertains the presence of the analyte; while the other laser detects the signal produced by PE, which is directly proportional to the amount of bound anti-NiV G protein (Source: R&D Systems website). The assays that are based on Luminex® show high specificity and sensitivity and the programme for running the experiment is as simple as running an assay based on ELISA. The second advantage is that it is possible to develop a multiplex assay for distinguishing between distinct types of Abs and their specificities. However, the biggest limitation is the specialized equipment required and the presence of unfamiliar technology in comparison to ELISA (Price et al. 2021).

A secondary neutralization assay for NiV that utilizes a variation of Luminex® method was used to detect the binding of serum IgG. Similar to the binding Ab assay, this Luminex® variation (Bossart et al. 2007) utilizes sGNiV bound covalently to fluorescent microspheres. The cell receptor of NiV, ephrin-B2 is biotinylated and added while the binding is detected via streptavidin-PE. NAbs specific to the G protein of NiV disrupt the Ephrin-B2–sGNiV interaction and thus the PE signal is diminished. The specificity and sensitivity of this assay are comparable to the SNT assay for NiV utilizing controls and immune sera from different sources (Price et al. 2021).

1.8.1.3.5 Rapid Test—RT-LAMP

LAMP stands for loop-mediated isothermal amplification and was developed in 2000 by Notomi et al. (2000); it is used for the amplification of DNA and does so effectively under isothermal conditions with high specificity (Notomi et al. 2000). The DNA synthesis is carried out by large fragment DNA polymerase of *B. stearothermophilus* (*Bst*), which possesses high activity of chain displacement. The isothermal conditions under which the reaction is performed range from 60 °C to 65 °C and, therefore, there is no loss of time for transition of temperature. A set of target-specific primers identifies as low as six distinct fragments; thus, the reaction has high specificity for the sequence targeted. The time of reaction can be reduced and the sensitivity can be potentially increased if loop primers are added to the LAMP (Nagamine et al. 2002). Qualitative detection of positive samples is possible and can be visualized as an increase in turbidity or the turbidity measurement is done to quantify the reaction. Moreover, calcein dye addition aids in the reading of positive results based on changes in colour with the naked eye (Tomita et al. 2008). Thus, due to its rapidity of testing and high specificity, the LAMP assay is considered the most advantageous in comparison to other NAATs (Ma et al. 2019).

A study carried out by Ma et al. (2019) aimed to develop '*isothermal (65°C) reverse transcription-loop-mediated isothermal amplification (RT-LAMP) method, targeting the nucleocapsid protein (N) gene, for the rapid detection of NiV, and was compared with conventional RT-PCR.*' Construction of three pseudoviruses of the N gene of NiV as a representative of known strains was done to replace the use of live NiV. Few primers of RT-LAMP were designed such that they targeted the N gene's highly conserved regions and; therefore, is able to identify all known strains of the Nipah virus. When sensitivity tests were conducted, the results indicated the limit of detection (LoD) to be ~100 pg of total pseudoviral RNA, i.e. tenfold higher than conventional RT-PCR. Results of specificity tests concluded zero cross-reactivity with N protein belonging to HeV, NDV, JEV and influenza A virus. This assay issues results within 45 min and does not require any kind of specialized instruments. Clinical sample analysis results concluded that the assay was stable. Finally, it was concluded that the RT-LAMP assay developed by Ma et al. (2019) detects all three pseudoviruses representing NiV strains, with great stability, sensitivity and specificity (Ma et al. 2019). The application of this assay can be done for routine surveillance and outbreak detection (Ma et al. 2019).

1.8.1.3.6 Multiplex Assays

The understanding of virus structure, divergence and other aspects of the life cycle has been aided by sequence characterization of NiV following each outbreak in league with tools such as electron microscopy. The use of recombinant proteins as a replacement for virulent pathogens has led to the development of diagnostic tools that can be used in low-level containment and detect neutralizing and non-neutralizing Abs accurately and rapidly in the field. They provide an insight into molecular mechanisms involved in virus transmission and their targeting in order to control outbreaks in the future (Thakur and Bailey 2019).

As there is a similarity between NiV symptoms and those of some other diseases, cases with severe infection are at risk of mortality due to misdiagnosis and delay in treatment. The application of the syndromic approach is done, where testing for agents based on the syndrome manifested such as fever or encephalitis is done; individual tests are shifted to multiplex testing in order to rapidly identify the pathogen using a single specimen. The cost of added reagents and specialized machinery must be taken into account; however, an efficient and quick diagnosis is more important and may end up saving costs in an overall sense (Demby et al. 1994; Perkins and Kessel 2015). NiV outbreaks being seasonal as well as regional, the preparation of panels that are region specific could be more economical and would make it easier to deploy when required for risk factor surveillance and detection of new cases (Mazzola and Kelly-Cirino 2019).

The development of multiplex PCR assays for differential diagnosis of multiple agents of infection utilize microspheres, micro-card and micro-array technologies. Multiplex bead-based RT-qPCR (Luminex®) assay was developed for detection and differentiation between NiV and HeV (Foord et al. 2013), this microsphere utilizing assay differentiates between NiV and HeV with sensitivity comparable to individual real-time PCR. The Luminex bead-based assays have been utilized for the detection of HeV Abs in pigs and fruit bats (Sendow et al. 2013; Field et al. 2007; Breed et al. 2010; Hayman et al. 2011; Peel et al. 2012, 2013). TaqMan developed a multiplex RT-qPCR based on micro-array card technology and it utilizes a micro-fluidic card comprising 384 wells for running simultaneous PCR reactions and differentially detects acute febrile illnesses, inclusive of NiV, from RT-PCR panels made up of 21 and 26 members (Liu et al. 2016; Onyango et al. 2017). The assay exhibited an overall 97–99% specificity and 86–88% sensitivity when compared to qPCR individual assays. The panel composed of 26 members included 3 protozoa, 8 bacteria and 15 viruses; while the 21-member CNS multi-pathogen assay included 13 viruses, 6 bacteria and 2 parasites (Mazzola and Kelly-Cirino 2019).

The Luminex bead-based immuno-assay is a platform for conducting differential diagnosis by using soluble NiV G and HeV protein-coated microspheres (Bossart et al. 2007). During detection, two separate test formats were examined, viz., a binding assay for differentiating NiV and HeV Abs and a blocking assay for detection followed by differentiation between NiV and HeV NAbs. Comparatively, ELISA is incapable of differentiating between antibodies (Mazzola and Kelly-Cirino 2019). The Luminex microsphere assay has also been used for detecting NiV-specific antibodies raised against G protein in pig, goat and cattle sera (Chowdhury et al. 2014; Singh et al. 2019).

1.8.1.3.7 POC Test—Truenat ™

Some major hurdles in the containment of NiV outbreaks are the absence of containment facilities that cater to quick diagnosis, delay in sample diagnosis with respect to the time taken for the sample to get to the reference laboratory and time to positivity (TTP). Quick detection of NiV can grant valuable time for patient management and help contain the spread. The three main diagnostic methods via which NiV infections are diagnosed are molecular detection, serological assays and/or virus isolation (Aditi and Shariff 2019). Real-time RT-PCR is the most accurate and rapid method for diagnosis of NiV (Gurley et al. 2007a, b; Guillaume et al. 2006); however, the technique requires good infrastructure, highly priced reagents and expensive equipment; which cannot be made available in areas with poor resources and remote locations. Considering the issues that arise with respect to biosafety during the collection and transport of samples to the reference laboratories, it is imperative to develop specific Point-of-Care (PoC) diagnostic tests for a quick diagnosis of the Nipah virus, which can be used effectively in the field or remote areas where NiV outbreaks occur (Yadav et al. 2021). As the locations of NiV outbreaks are remote and the bat surveillance is mainly carried out in areas surrounded by forests, a user-friendly PoC system would be considered a blessing with respect to the detection of NiV at the source. This would be extremely beneficial to developing countries such as India and Bangladesh, where the resources are not at par and the access to containment facilities is limited (Yadav et al. 2021).

In keeping with the growing need for PoCs, Molbio Diagnostics PvT. Ltd., Goa, teamed up with ICMR-National Institute of Virology (NIV), Pune, and together developed a NiV PoC test which was validated by ICMR-NIV, Pune. The system developed is a lightweight, portable, battery-operated PCR analyser named Truenat™ Nipah PoC system, it also includes a Trueprep® AUTO universal cartridge-based sample prep kit and a device to aid sample processing at the PoC. The kit contains a reagent combination that has the ability to inactivate the clinical samples during the extraction of nucleic acid (Molbio Diagnostics Pvt. 2020; Vijayalakshmi et al. 2019). This method also facilitates the extraction of the built-in NiV internal positive control (PC) along with sample extraction (Yadav et al. 2021). When studied, the performance of the Trueprep® AUTO system and Truenat™ Nipah PoC test was compared to standardized and validated methods like MagMAX machine-based RNA extraction (Applied Biosystems, USA) and TaqMan NiV real-time RT-PCR (Applied Biosystems, USA) (Guillaume et al. 2006; Yadav et al. 2021). During the study, the sensitivity and specificity of the PoC were revealed to be 97% (95% CI, 90–100%) and 100% (95% CI, 98–100%), respectively (Yadav et al. 2021). There was no cross-reactivity reported with the PC panels obtained from ICMR-NIV, Pune, and the Truenat™ Nipah PoC test was found to be specific for Nipah virus detection. The PC panel comprised 35 human specimens that were positive for viruses from the genus such as Alphaviridae—*Chikungunya* virus (Hsu et al. 2004), Flaviviridae—*Dengue* virus (Hsu et al. 2004), Orthomyxoviridae—*Influenza* virus (Hsu et al. 2004), Paramyxoviridae—*Measles* virus (Mazzola et al. 2019) and Togaviridae—*Rubella* virus (Mazzola et al. 2019; Yadav et al. 2021). The device was designed in such a way that it would aid in the detection of the Nipah virus in both human and bat specimens in field settings (Yadav et al. 2021).

In view of the high CFR associated with a NiV infection, it is essential to process the clinical samples only in containment facilities. The fact that the current Truenat™ Nipah PoC system contains a lysis buffer that was found to thoroughly inactivate the Nipah virus is noteworthy. That is why this assay can be performed in a field setting while donning standard personal protective equipment such as Tyvek suits with N95 or respirators. The TTP associated with Truenat™ Nipah PoC system takes an hour compared to the two and a half hours taken by the TaqMan real-time RT-PCR. The product is predicted to bring a remarkable change in the worldly battle against control and management of the Nipah virus (Yadav et al. 2021).

1.8.2 Therapeutics

As is with every disease with a high mortality rate, many attempts have been made to test different active molecules that have the ability to inhibit the proliferation of NiV. During the Malaysia and Singapore epidemics, ribavirin and acyclovir were administered, respectively, to treat the symptoms; however, their effectiveness has been considered debatable (Ambat et al. 2019; Sharma et al. 2019; Devnath and Al Masud 2021). When tested in vitro, chloroquine, a drug used for treating malaria, showed promising results with respect to inhibiting the Nipah virus; however, this was not confirmed in animal models (Ochani et al. 2019). The application of favipiravir and the monoclonal antibody m102.4 produced promising results in animal models (Sayed et al. 2019; Singh et al. 2019; Devnath and Al Masud 2021) and; therefore, m102.4 is under human clinical trials. No deaths were reported (Playford et al. 2020). Lo et al. (2020a, b) evaluated in vitro the antiviral effects of GRFT (Griffithsin) and 3mG, its trimeric tandemer, against NiV and other viruses from four different families, and reported that a preliminary challenge of GRFT that is oxidation resistant (aka QGRFT) given in vivo in golden Syrian hamsters showed protection against a lethal dose of NiV. A varied spectrum of in vivo activity is shown by GRFT, against various viruses, which is a lectin that binds to oligosaccharides and contains high levels of mannose (Lo et al. 2020a, b).

1.8.2.1 mAbs—Human and Humanized Cross-Reactive nAbs

As mentioned earlier, the most promise was seen in the trials conducted with m102.4 monoclonal antibody (mAb) as a mAb-based therapeutic treatment against a NiV infection. The antibody was affinity matured in such a way that it strongly neutralizes both HeV and NiV attachment glycoprotein G such that it blocks the interaction of the glycoprotein with Ephrin B2 and B3, the host cellular receptors that facilitate entry (Xu et al. 2013; Zhu 2008). In both, the ferret model and the non-human primate (NHP) model of infection, m102.4 has been seen to be effective in protecting the host against HNV infections. In the ferret model of disease, 10 h after intranasal infection of NiV, a single intravenous (IV) dose of the mAb was said to have afforded complete protection (Bossart et al. 2009). There were even more promising results seen in the post exposure African Green Monkey (AGM) model; where when treatment was initiated after 3 days post infection date (PID) with HeV, m102.4 showed full protection and similar results were obtained at 5 days PID with NiV-My, even though clinical symptoms and viraemia had already been established (Bossart et al. 2011; Geisbert et al. 2014). In both studies, a follow-up second dose was administered 2 days later. Fascinatingly, a study that compared the pathogenicity of NiV-My and NiV-Bd in AGM models of disease determined the treatment window of NiV-Bd to be shorter than that of NiV-My. In keeping with the accelerated disease onset timeline of NiV-Bd, it was seen that m102.4 only showed protection when administered up to 3 days PID with NiV-Bd. Animals that were infected at 5 days PID succumbed to the disease (Mire et al. 2016; Johnson et al. 2021).

The results obtained from the above studies made it possible to justifiably use m102.4 in humans for compassionate as well as phase 1 clinical trial use. Following high-risk exposure, to date, m102.4 has been used for compassionate therapy about 14 times (Playford et al. 2020; Broder et al. 2013). The assessment of tolerance, safety and immunogenicity in healthy adults for phase 1 clinical trials was confirmed following two important stages: the use of m102.4 as compassionate therapy without any display of adverse effects and promising data in preclinical studies conducted in animal models (Playford et al. 2020). This phase 1 trial comprising 30 participants concluded that the dose administered was safe and well tolerable and no adverse effects that could result in the withdrawal of participants were reported. Pharmacokinetic studies reported that the mAb remained active at levels suitable for virus neutralization for at least 8 days post administration. Immunogenicity tests revealed that there was no generation of anti-m102.4 antibodies. Although the trial had a limited number of participants and; thus, was unable to determine the protective efficacy of m102.4, the reports made for safety and tolerance make m102.4 the most promising therapeutic candidate for the treatment of patients exposed to HNV (Johnson et al. 2021).

A second potential contender for antibody therapy under investigation is h5B3.1. It is a cross-reactive, humanized, neutralizing mAb whose target is fusion glycoprotein F of HeV and NiV, this blocks the conformational change that facilitates the membrane fusion and virus infection (Mire et al. 2020; Dang et al. 2019). The development of h5B3.1 humanized mAb as a potential candidate was supported by some previous studies that demonstrated efficacious protection of anti-F mouse polyclonal or mAbs against HeV or NiV challenge in hamsters (Guillaume et al. 2006, 2009). In ferrets, intraperitoneal administrations of h5B3.1 at 1 and 3 days as well as 3 and 5 days PID with a lethal dose of HeV or NiV showed efficacious protection. The researchers of the study have also proposed a treatment that consists of a combination of h5B3.1 and m102.4 antibodies that target viral surface glycoproteins, in order to minimize the opportunity for the emergence of escape mutants. Although as the in vivo characterization studies regarding h5B3.1 are limited, further studies are required before introduction in the human population alone or in combination with m102.4 (Johnson et al. 2021).

1.8.2.2 Antivirals

1.8.2.2.1 In humans

Ribavirin, a broad-spectrum nucleoside analogue, is a therapeutic option that has been clinically utilized in patients suffering from NiV disease (Goh et al. 2000; Chong et al. 2001; Banerjee et al. 2019). The drug is licensed to be used as a therapy against Hepatitis C virus (HCV), viral haemorrhagic fevers and Respiratory syncytial virus (RSV) and is part of WHO's list of essential medicines (Bausch et al. 2010; Fisher-Hoch et al. 1995; Snell et al. 2001; WHO 2019). Consisting of 140 patients, an open-label trial of ribavirin was conducted during the 1998–1999 outbreaks in Malaysia; with a control group ($n = 54$) consisting of patients who were treated prior to the trial and those who refused treatment (Chong et al. 2001). Following this study it was concluded that the treatment with ribavirin led to a 36% reduction in mortality and the neurological deficits seen were fewer among survivors. An earlier description of that outbreak, however, stated that no significant difference was observed in the clinical features and in the outcome after ribavirin treatment (Goh et al. 2000). The 2018 Kerala outbreak, comprising of a total 23 infected cases, saw the use of ribavirin (Arunkumar et al. 2019a; Banerjee et al. 2019). The two survivors received oral administration of ribavirin out of a subset of six infected individuals who received the treatment (Arunkumar et al. 2019a; Banerjee et al. 2019; Chandni et al. 2020). A corresponding reduction of 20% was seen in mortality, as compared with the 100% seen in the group of 6 patients who did not receive ribavirin. However, the sample size was too compact to draw a conclusion about the efficacy of the drug against NiV. As a post exposure preventive measure, ribavirin was administered to eight healthcare workers who were exposed to patients with insufficient PPE during the 2018 outbreak (Banerjee et al. 2019). None of the healthcare workers manifested NiV disease; however, most of them reported to have suffered mild to adverse effects from the treatments. The overall consensus was that the efficacy of ribavirin in Kerala in 2018 was inconclusive. In animal models, the studies were not found to be promising. When ribavirin was administered either alone or in combination with chloroquine, it was not found to be protective in the hamster model (Freiberg et al. 2010; Georges-Courbot et al. 2006; Johnson et al. 2021).

A second antiviral, acyclovir (Zovirax), was used in combination with ceftriaxone in order to treat nine infected individuals who worked in a slaughterhouse during the NiV outbreak in Singapore in 1999. Out of the nine, eight are said to have survived (Paton et al. 1999). No data with respect to the in vitro studies of acyclovir against NiV are available (Liew et al. 2022). Subsequently, an immunomodulator by the name of Rintatolimid (Ampligen) was concluded to be effective in inhibiting the replication of NiV and in inducing IFN- α and IFN-β to protect against a viral challenge in hamsters (Georges et al. 2006; Liew et al. 2022).

1.8.2.2.2 In vivo

Only a few existing therapeutics have been utilized in animal models against NiV infections. A nucleotide analogue, Remdesivir (GS-5734; Veklury®), a broad-spectrum antiviral which has demonstrated antiviral activity against *Coronaviruses*, *Paramyxoviruses* as well *Filoviruses*, is one of the few (Lo et al. 2017, 2019; Sheahan et al. 2017). In an AGM model with a lethal challenge with NiV-Bd, when remdesivir was administered, intravenously, daily starting after 24 h after infection and continued for 12 days, 100% survival rate was observed (Lo et al. 2019). Out of the four NHPs involved in this study, two developed mild respiratory symptoms that were resolved by day 14 post infection. At 92 days post infection, the termination point of the study, one animal showed the presence of viral RNA in the brain. Though the results were highly encouraging, more studies are essential to determine the antiviral efficacy of remdesivir (Mulangu et al. 2019). During the COVID-19 pandemic, remdesivir has been used as a compassionate form of therapy for treating patients with SARS-CoV-2 infection, although the efficacy is unclear; it has also been previously proven to be safe in trials against Ebola in the Democratic Republic of Congo (Mulangu et al. 2019; Grein et al. 2020; Veklury 2021; Okoli et al. 2021). Multiple trials involving remdesivir are ongoing globally (Gilead n.d.; Johnson et al. 2021).

In small animal models, certain additional antivirals demonstrating promising efficacy are being evaluated as treatment options against NiV infection. In Japan, a small molecule purine analogue antiviral by the name of Favipiravir (T-705; Avigan®) has been licensed to manage the emergence of Influenza virus infections (Furuta et al. 2013). When tested in the Syrian golden hamster model, favipiravir displayed complete protection against infection with lethal NiV-My when administered immediately after infection and continued for 14 days PID (Dawes et al. 2018). No animals treated with the drug showed any clinical signs of the disease throughout the study and no viral RNA or any pathological changes in the tissues were observed. In future studies, it is important to evaluate the post-exposure antiviral efficacy of favipiravir. GRFT (Griffithsin), a homodimeric high mannose oligosaccharide binding lectin, is being evaluated as a topical microbicide in clinical trials against human immunodeficiency virus 1 (HIV-1) (Mori et al. 2005; clinicaltrial.gov). In in vitro studies, GRFT along with 3mG and Q-GRFT have demonstrated antiviral activity against NiV in the nanomolar range (Lo et al. 2020a, b). 3mG and Q-GRFT were tested for their prophylactic potential in Syrian golden hamsters against NiV-Bd and the results demonstrated an overall of 15% and 35% of survival rates, respectively. Further studies are needed to develop these products in treatments against NiV infections (Johnson et al. 2021).

1.8.2.2.3 In vitro

A cytidine analogue with broad spectrum antiviral activity called 4′azidocytidine (R1479) has been known to show antiviral activity against flaviviruses and a *Pneumovirus*, exhibited a similar effect against NiV, in vitro, in low micromolar range (Klumpp et al. 2006; Nguyen et al. 2013; Wang et al. 2015; Hotard et al. 2017). Although a prodrug of R1479, named balapiravir did not demonstrate good results in trials against *Flaviruses*; the trials, however, were discontinued due to poor efficacy and certain negative side effects (Nguyen et al. 2013; Roberts et al. 2008; Nelson et al. 2012). Other derivatives of R1479 have proved to be promising as compared to R1479 itself. In vitro study of 2′-monofluoro- and 2′difluoro-modified derivatives of R1479 (2′F-4′N3-C and 2′diF-4′N3-C) have demonstrated 20-fold increased antiviral activity against NiV and shown less cytotoxicity (Lo et al. 2018). The high efficacy and low toxicity of these derivatives of R1479 make them promising candidates for therapeutics; with the need for future evaluations (Johnson et al. 2021).

Another cytidine analogue is 4′-chloromethyl-2′-deoxy-2′-fluorocytidine (ALS-8112) and its parent nucleoside called lumicitabine have undergone phase I and phase II clinical trials for the treatment of RSV infections (Lo et al. 2020a, b; DeVincenzo et al. 2015; Patel et al. 2019). In the low micromolar range, ALS-8112 demonstrated strong antiviral effects against NiV in vitro, with minimum cytotoxicity in several cell lines except lymphoblastoid cells and peripheral blood mononuclear cells (PBMCs) (Lo et al. 2020a, b). However, caution must be observed when optimizing this molecule's dosage in order to prevent the occurrence of lymphopenia and neutropenia, if it is to be considered a potential candidate for the treatment of NiV infections (Johnson et al. 2021).

Evaluation of peptide fusion inhibitors that aim to prevent viral fusion with host cellular membrane has been done in addition to those that target viral replication machinery. After prophylactic administration, certain optimized lipopeptide inhibitors (having tocopherol or cholesterol conjugated to polypeptide using dPEG) demonstrated protection in AGMs (33% survival) and Syrian golden hamsters (50% survival) against a challenge with lethal NiV (Mathieu et al. 2017, 2018). A lipopeptide inhibitor called Enfuvirtide (Fuzeon™), an FDA-approved analogue for HIV-1 therapy, has the potential to become an effective antiviral if developed further (Lalezari et al. 2003; Kilby et al. 1998). Development of more lipopeptide inhibitors against NiV infections is necessary for the future (Johnson et al. 2021).

1.8.2.3 Defective-interfering and Therapeutic-infectious Particles

A natural by-product of viral replication is defective interfering particles (DIPs) (Welch et al. 2022). A strategy that has exhibited potential is the use of defective interfering particles (DIPs) against NiV. These contain genomes that are defective and can be used to alter the dynamic of the viral populations such that they inhibit the replication of NiV. Assays conducted in vitro have demonstrated that in silico developed as well as naturally occurring DIPs decreased the viral titre by 100-fold and limited cytopathic effects in Vero cells (Welch et al. 2020). Studies that reported the use of DIPs against influenza A virus show promising results. Though in the case of NiV, DIPs for treatment are still in the first stage of research, it would help us take a step further if they are tested in animal models (Dimmock et al. 2012a, b; Scott et al. 2011a, b; Johnson et al. 2021).

Welch et al. (2022) investigated 'the efficacy of virus-like particles containing DI genomes (therapeutic infectious particles [TIPs]) in the Syrian hamster model of NiVD.' They found that co-administration of high doses of TIPs intraperitoneally along with virus challenge showed to improve the course of clinical disease and lower lethality (Welch et al. 2022). They demonstrated that TIPs are able to provide sturdy protection in hamster models with lethal disease by NiV. One dose of TIP was shown to reduce mortality and the hamsters that were treated with TIPs showed lesser chances of exhibiting clinical signs of infection. If at all they did present with clinical signs, they were shorter in duration (Welch et al. 2022).

The study provided support for the continuation of the clinical usage of TIP treatment. Since NiV is a disease with high mortality rates, is lethal and has a probability of causing a pandemic. The development of a candidate that can be administered easily while also achieving a high level of protection is required to prevent this disease and to facilitate intervention during outbreaks. The authors of the study have provided primary evidence that TIPs are able to confer protection against Nipah disease and its lethality in sturdy animal models. This supports further investigations of the therapeutic use of DIPs for NiV and other similar pathogens (Welch et al. 2022).

1.8.2.4 Computer Identified Multitarget Inhibitors

While conventional approaches to drug discovery adhering to the principle of 'one drug, one target,' the next step in this field is the discovery of drugs that affect multiple targets, and, hence, the idea of multitarget drugs has attracted a high amount of interest the past decade (Csermely et al. 2005; Ramsay et al. 2018; Viana et al. 2018; Makhoba et al. 2020).

Computational methods for drug discovery (Viana et al. 2018; Zhang et al. 2017) such as molecular docking, molecular modelling, machine learning, in silico ADME, pharmacophore, QSAR and virtual screening have been developed and implemented in drug discovery and show the ability to screen for new candidates that possess multi-targeting capabilities (Ma et al. 2010; Randhawa et al. 2022).

The significant roles played by NiV proteins G, F and N in replication and transmission make them attractive as targets for the discovery of anti-NiV drugs. Therefore, a study was conducted '*to prospect for potential multitarget chemical/phytochemical inhibitor(s) against NiV, using a sequential molecular docking and molecular-dynamics-based approach and was implemented by simultaneously targeting NiV-G, NiV-F, and NiV-N.*' Molecules seen to bind to G, F and N proteins were considered to be potential multitarget inhibitors. The study identified phytochemical molecules CARS0358 (NA) and 17-O-Acetyl-nortetraphyllicine (RASE0125), which are indole alkaloids, as distinctive multitarget inhibitors of G, F and N viral proteins and also molecule ND_nw_193 (RSV604) as a chemical inhibitor of N and G proteins of NiV. These molecules need to be further evaluated in vitro, in vivo and subsequently in clinical trials (Randhawa et al. 2022).

1.8.2.5 N Protein-Specific Universal siRNAs

The protein coding the nucleocapsid of NiV is the N gene which is vital in the replication of the virus as it aids in the encapsidation of the genome and, hence, protects it from host cellular defence mechanisms; without encapsidation, the genome will not be recognized by RdRp (Ogino and Green 2019) and no replication will occur (Ray et al. 2016; Mahfuz et al. 2022).

Small interfering RNAs also known as siRNAs play a crucial role in the inhibition of viral protein expression via hybridization and consequent inactivation of viral mRNA using the RNA interference (RNAi) pathway. Therefore, a study aimed at 'designing siRNAs against gene encoding N protein and their use as molecular therapeutics' was conducted. Accordingly, ten prospective candidates were identified using conserved N gene sequences from all NiV strains available globally, then in order to precisely predict the suitability GC content, off-target binding, target affinity binding, binding capacity to human argonaute protein-2, secondary structure and melting temperature of these candidates were evaluated. The results obtained following the study concluded that the siRNA candidates demonstrated great promise in repressing the N gene and, therefore, may aid in alleviating the severity of this disease (Mahfuz et al. 2022).

Following the promising results of the previous study, a new study to design siRNAs, that are universal, against NiV using computational tools was conducted. Oany et al. (2015) designed siRNAs against 5 NiV-Bd strains' N gene. However, the previously mentioned study used all current strains and the results obtained are complete and not restricted due to geographical distribution. The siRNAs have been cross-checked against two databases for the prediction of efficacy and their binding affinity to human argonaute protein-2 was checked using molecular dynamic modelling and molecular docking (Mahfuz et al. 2022).

The above-mentioned siRNAs are universal with the ability to target highly conserved regions of distinct NiV strains from various countries and may not show ineffectiveness against certain strains like other molecular therapeutics (Mahfuz et al. 2022). Utilizing nanovesicles for the delivery of siRNAs to targets and delivering them in a desired dose is possible (Gupta et al. 2019). A major hurdle in NiV-related infection is crossing over the BBB and causing encephalitis. However, siRNAs conjugated to various nanoparticles were successfully shown to cross the BBB in mice models (Eyford et al. 2021; Zou et al. 2020; Zhou et al. 2020).

1.9 Prevention and Control

1.9.1 Management of Disease

The current method of management of NiV involves symptomatic, supportive and prophylactic treatment as there is a lack of therapeutics for the Nipah virus (Chakraborty et al. 2019; Sharma et al. 2019; Aborode et al. 2022; Skowron et al. 2022). Prophylaxis of venous thrombosis, maintaining airway patency along with a balance of fluids and electrolytes (Aditi and Shariff 2019; Ambat et al. 2019). The use of broad-spectrum antivirals and mechanical ventilation during severe respiratory conditions are also some of the ways the disease can be managed (Ambat et al. 2019; Skowron et al. 2022).

1.9.2 Vaccines

Safety issues with production (use of BSL-4 containment) and those associated with the administration of an inactivated or live-attenuated vaccine, and the necessity to elicit a neutralizing antibody response, most attempts at the development of a NiV vaccine have been targeted at recombinant viral vectors and sub-unit vaccines. In every case, the target antigens have been F and G glycoproteins (Price et al. 2021).

Owing to the lack of new outbreaks, clinical trials of the NiV vaccine being conducted are limited. Therefore, the testing of potential candidates is mainly done in animal models. To date, over ten potential vaccine preparations have been researched based on mRNA, recombinant proteins and their subunits, virus-like particles as well as viral vectors (Broder et al. 2016; Amaya and Broder 2020). The candidate that is majorly studied is a subunit type of vaccine that comprises the soluble Hendra glycoprotein HEV-sG, it induces a cross-reactive immune response to the virus. Testing it with NiV-My, NiV-Bd and HeV, it has been proven to confer complete protection in cats, ferrets, horses and NHPs except for pigs (Mungall et al. 2006; McEachern et al. 2008; Bossart et al. 2012; Pallister et al. 2013; Middleton et al. 2014; Mire et al. 2013). In one study, a single dose of a lipid nanoparticle nucleoside-modified messenger RNA vaccine encoding the soluble HeV glycoprotein protected up to 70% of Syrian hamsters from lethal NiV challenge, indicating the promise of this particular platform for future prevention of NiV disease (CEPI 2020; Lo et al. 2020a, b; Johnson et al. 2021).

The vaccine produced by Zoetis Inc. known as Equivac is the only vaccine that has been officially registered and approved by the Australian Pesticides and Veterinary Medicines Authority which is used as a preventive treatment for equines (Aditi and Shariff 2019; Ambat et al. 2019; Ochani et al. 2019; Geisbert et al. 2021). Two studies have been conducted on this vaccine by Geisbert et al. (2021). The first reports from the first study concluded that one dose of the vaccine is efficacious, while another reported that 7 days after administration of the first dose, the vaccine was still effective in conferring protection in AGMs against debris form of NiV, which is enough proof to approve its potential usage in case of an outbreak of Nipah (Geisbert et al. 2021; Skowron et al. 2022).

The efficacy of the rVSV candidate has been confirmed in animal models of rVSV and the rMV candidate against animal models of rMV (Yoneda et al. 2013; Debuysscher et al. 2014; Ang et al. 2018). Yoneda et al. (2013) carried out thorough research to determine the efficacy of rMV in hamsters and AGMs. They demonstrated a spike in anti-NiV antibodies in the hamster serum of the models, 3 weeks following inoculation. No disease signs or mortality were observed in the hamsters. AGMs immunized with two doses of vaccination displayed no signs of clinical disease and there was an induction of the anti-NiV G antibodies. In 2014, Debuysscher et al., assessed candidate rMV that expressed NiV G or F glycoproteins or the N protein in a Syrian Golden hamster model and the results showed the production of elevated humoral IR after the first dose (Debuysscher et al. 2014; Skowron et al. 2022).

The search for new candidates is constantly being carried on. One such candidate created by Soltan et al. (2021) is a multi-epitope candidate on the basis of the highest ranked epitopes HTL, BCL and CTL from a selection of Nipah proteins. They were then tested for tolerance, ability to be detected as an allergen and toxicity, antigenicity, physico-chemical features and solubility via computer analysis. The results of these tests concluded that the vaccine was non-allergenic and non-toxic (Soltan et al. 2021). However, more experimental investigation is needed to prove efficacy in practicality. A second such attempt by Loomis et al. in 2020, examined the structural integrity with antigenicity of the protein subunits to be turned into vaccine candidates (stabilized pre-F and G to form a chimeric protein) using electron microscopy, kinetic binding assays and certain other biophysical properties. The efficacy of this vaccine candidate was determined in mouse models. Mice vaccinated by pre-F demonstrated indistinguishable levels of binding for pre-F and post-F. In contrast, those vaccinated with multimers of G were seen to induce only responses that were specific to G, and those vaccinated by chimeras (pre-F + G or pre-F/G) showed an Ab response that was built to oppose F and G monomers. The researchers confirmed that glycoprotein G of NiV is an excellent vaccine immunogen and it is possible to construct a single vaccine composed of pre-F and G that has the ability to confer a diverse antibody response (Loomis et al. 2020). They also concluded, that pre-fusion F glycoprotein induced a stronger neutralizing activity as compared to post fusion F, therefore confirming that stability in the conformation of pre-fusion F is one of the detrimental attributes for increasing efficacy. In a follow-up study, based on protein structure, they attempted to stabilize the pre-fusion F in its trimeric form and obtain increased immunogenicity and expression. The resultant mRNA vaccines of Pre-F and G were shown to induce effective neutralizing Abs responses (Loomis et al. 2021; Skowron et al. 2022). With the recent success of mRNA vaccines in the SARS-CoV-2 pandemic, they are considered to be the most promising potential candidates in case of the emergence of NiV infections.

1.9.2.1 rVSV Vaccines

In hamsters, an rVSV-based vaccine that was replication competent, coding for NiV-My G showed high efficacy (DeBuysscher et al. 2014) and protected the organism against NiV-My even if it was administered 1 day prior to exposure (DeBuysscher et al. 2016). In AGMs, this vaccine conferred protection from NiV-My disease even 1 month after a single dose was administered intramuscularly (Prescott et al. 2015). Although this vaccine is reliant on the inculcation of a replicative glycoprotein from the Zaire Ebolavirus (EBOV) and has not been tested against NiV-Bd. When an rVSV that expresses either NiV-My F or G was inoculated in mice, high neutralizing titres were generated and this led to the consideration that single-cycle replication rVSVs when developed confer a strong immunogenicity (Chattopadhyay and Rose 2011). These single cycle replication vectors provided homogenous protection throughout in the hamster model of infection for the Malaysian NiV strain (Lo et al. 2014a, b). An alternative single-cycle rVSV vaccine vector was developed to encode NiV-Bd F or G proteins with a Green fluorescent protein (GFP) (Mire et al. 2019). This candidate was assessed 28 days following a single dose administration in ferrets as well as in AGM models, where they exhibited complete protection against a lethal challenge of NiV-My (Mire et al. 2019), while AGMs showed protection even against NiV-Bd lethal challenge (Mire et al. 2019; Foster et al. 2022).

Most of the candidates that have been under pre-clinical vaccine studies have assessed protective efficacy approximately 4–5 weeks prior to NiV exposure. However, the timeline seen in previous epidemics makes it crucial to produce a vaccine that confers effective protection within days of exposure in order to control outbreaks (Foster et al. 2022).

A study was conducted in order to 'assess the ability of the rVSV-ΔG-NiV_BG vaccine to induce protective immunity in the uniformly lethal AGM model of NiV-Bd at 3 or 7 d after vaccination.' The authors of this study constructed an rVSV single-cycle replication-based vector that encoded NiV-Bd G protein and this was coded as rVSV-ΔG-NiV_BG. The study exhibited that vaccines that were rVSV bases may show utility in conferring rapid protection on humans against an NiV infection. The candidate was found to be an immunogenically safe and efficacious vaccine that effectively protects AGMs against a lethal dosage of NiV-Bd that was inoculated following a 7-day interval following vaccine administration. This study highlighted the fact that this vaccine candidate completely protected AGM NiV models against a high dose of NiV-Bd when vaccinated 7 days prior and showed 67% protection when vaccinated 3 days prior to exposure to NiV-Bd (Foster et al. 2022).

1.9.2.2 Computer Constructed Single v/s Multiple Epitope Vaccines

The interface between immunological data and computational methods employed to analyse this data is known as the immunoinformatics field (Oli et al. 2020; Soltan et al. 2021). The use of immunoinformatics for designing a new NiV vaccine via epitope prediction has been done in plenty of studies. In a study in 2014, epitopes of B and T cells of F and G protein were predicted and their potential as a vaccine candidate was analysed (Sakib et al. 2014). In order to design peptide-based vaccines, in a similar approach, T-cell epitopes were predicted (Kamthania et al. 2015). In addition, the analysis of B- and T-cell epitopes of membrane proteins and RNA-dependent RNA polymerase (RdRp) proteins was carried out by Saha et al. (2017) and Ravichandran et al. (2019), respectively (Soltan et al. 2021). The common ground between the studies was that the candidates were single epitopes and that there was no clear rationale for their selection for investigation. As some of these candidates were single viral protein vaccines, a major disadvantage was in store, i.e. since RNA viruses have a high mutation rate and continuous mutation leads to increased viral pathogenicity and, hence, early loss of effectivity of vaccine (Sanjuán et al. 2016; Soltan et al. 2021).

A study for a proteome-based computational approach for the design and analysis of a multitope vaccine against NiV was performed. In it, all the proteins in the proteome of NiV were analysed to determine their antigenicity score and role in virulence, where they zeroed in on the fusion glycoprotein (F), protein (V), protein (W) and glycoprotein (G) as epitope prediction candidates. Following these decisions, the candidate, a multi-epitope vaccine, was designed by using high-ranking

HTL, CTL and BCL epitopes from the selected proteins. Suitable linkers, adjuvants and PADRE peptides were utilized to finalize the construction of the vaccine. This was then analysed for antigenicity, allergenicity, toxicity, solubility and physiochemical features and passed all the assessments through computational analysis and when a docking search was run between the candidate and TLR-3. The docked complex was validated through molecular dynamics solutions which predicted a strong bond, and this supported the candidate nomination as a vaccine for the supposed solution for NiV infections (Soltan et al. 2021).

Taking into account the above-mentioned disadvantage of single viral protein vaccines, the protein selected for this study was chosen after whole proteome analysis for the virulence role of every protein and antigenicity score. A major advantage of this candidate was that it was a multi-epitope construction which has higher efficacy protection against infectious agents as compared to single epitope-based vaccines. Though the vaccine candidate passed all assessments of physicochemical, structural and immunological characteristics, further experimental trials are necessary to prove efficacy (Soltan et al. 2021).

The next study was conducted by Rahman et al. (2022) and was aimed at 'predicting a dual-antigen multi-epitope subunit chimeric vaccine against surface-glycoproteins G and F of NiV.' Epitope prediction of T- and B-cell antigens was performed using immuno-informatic analysis on target proteins. Based on conservancy, codon adaptation, molecular docking, population to be covered, 2° mRNA structures, in silico cloning and simulations of immune response, vaccine designs were proposed and implemented. The constructed vaccines displayed high levels of Ig and cytokine responses in humans and using the pADTrack-CMV expression vector can be expressed effectively. Further investigations would be required and the conduction of clinical trials would be necessary for the validation of safety and efficacy (Rahman et al. 2022). Of the COVID-19 vaccines available in the market, most target the Spike (S) protein (Salvatori et al. 2020; Malik et al. 2021) which is constantly mutating and leading to the emergence of new variants, this may create concerns regarding the efficacy of available vaccines (Mittal et al. 2022). Additionally, the increased selection pressure of the vaccine that targets the S protein of SARS-CoV-2 may lead to a rise of escape mutants via changes in the structure of the said protein (Moore and Oft 2021). Taking into consideration these issues during the process of designing a vaccine for NiV, candidates that are dual-antigenic multitope vaccines are considered well suited even in the event of natural mutations occurring in the targeted genes of NiV. Thus, the study used a combination of F and G antigenic proteins to design multitope candidate vaccines. The selection pressure on the proposed candidate vaccine may be low as it targets multiple epitopes; this is presuming no mutations would occur in F and G proteins, simultaneously (Rahman et al. 2022). The resulting vaccine candidate predicted using immuno-informatics was an 'effective dual antigenic multitope chimeric subunit vaccine' and it was capable of eliciting a robust CMI and humoral IR. The predicted designs met all criteria such as allergenicity, antigenicity, physicochemical properties, immunogenicity, toxicity and the induction of IR without disturbing housekeeping functions of host cells. Therefore, the proposed constructs could be promising potential candidates for protection against the Nipah virus (Rahman et al. 2022).

1.9.2.3 CEPI

The Coalition for Epidemic Preparedness Innovations (CEPI) was created with the mission of accelerating vaccine development through safety, proof-of-concept and immunogenicity studies, in order to enable efficient testing of efficacy and licensing in the case of a public-health emergency, and innovative initiatives to develop vaccines for emerging infectious diseases where there could be no path towards a phase III clinical trial (Røttingen et al. 2017). In keeping with its mission, the coalition announced financial assistance for the development of candidate vaccines against seven priority emerging pathogens by WHO, including Nipah virus (NiV) (Nikolay et al. 2021). CEPI has funded four potential candidates that exhibited efficacy in animal models which were disease relevant, to be further evaluated in phase I and phase II clinical trials. They include three recombinant viral vector vaccines undergoing preclinical development (CEPI 2019; Christodoulou 2018, 2019) and a HeV glycoprotein subunit vaccine which is in phase I clinical trials (ClinicalTrials.Gov n.d.; Johnson et al. 2021).

1.9.3 Transmission Prevention and Control Strategies

1.9.3.1 Zoonotic Transmission Prevention

Preventive measures in order to avoid new cases and the expansion of an initiated outbreak are majorly related to the avoidance of first-hand contact with the source (fruit bats and pigs), their secretory fluids and the consumption of contaminated eatables. It is recommended to check fruits for injury and wash them thoroughly when picked from trees where bats roost and to put in place procedures (bamboo skirt method) to limit their contact with areas as well as storage apparatus in which the raw palm sap is collected and stored (Ambat et al. 2019; Sharma et al. 2019). Recommendations are to avoid the planting of fruit trees, that attract bats, close to pig farms (Ang et al. 2018). In occupations that require first-hand contact with farm animals, especially in slaughterhouses, appropriate, well-maintained protection gear is essential (Aditi and Shariff 2019; Skowron et al. 2022).

1.9.3.1.1 Vaccination in Pigs

On a global scale, pig rearing allocated 25% of energy, while the protein obtained from them makes up 9% of the total protein obtained from animals. Historically, it is observed that an increase in agricultural production is directly proportional to the increase in pork consumption. The rearing of pigs is seen as an ever-expanding sector of animal husbandry and this is most observed in developed and developing countries of Asia. While in developed parts, a large portion of pork is generated with the help of meticulously methodical technology, in developing parts, farmers still own small herds of livestock in their backyard inclusive of pigs. The production of pigs is restricted due to several factors, one of which is emerging infectious diseases like NiV and their zoonotic characteristics. In such cases, the administration of vaccines would prevent not only losses in the industry but also curb the spread of infection between pigs and humans. This may result in alleviating poverty by increasing the income of households leading to the availability of income for food, education or healthcare. Thus, novel effective vaccine candidates that target agents hampering farming activities in developing countries will bestow great socio-economic benefits on them (Yoneda et al. 2010; Pedrera et al. 2020).

Between September 1998 and May 1999, the initial and most calamitous outbreak of NiV occurred on the Malaysian peninsula (Paton et al. 1999; Chua et al. 2003). The association of pigs with this epidemic was established early as almost 93% of infected individuals had been in contact with infected pigs (Parashar et al. 2000). In the event that an NiV strain adapts to humans and is transmitted among communities with large human and swine populations and where pigs are exported from, the disease will spread like wildfire and the world will face the most disastrous pandemic (Luby et al. 2013; Donaldson et al. 2018; Huynh et al. 2007). Reports have shown that NiV proteins F or G expressing ALVAC, designated ALVAC-F or ALVAC-G respectively, were seen to protect pigs against a challenge with NiV, 2 weeks after administration of the second dose (Weingartl et al. 2006). Increased titres of neutralizing Abs (NAbs) against NiV were induced by ALVAC-G, however, in spite of low NAbs titres generated after immunization with ALVAC-F, vaccinated animals were protected from a lethal challenge of NiV. An attenuated recombinant NDV expressing glycoproteins of Nipah virus induced lasting NAbs specific to NiV in pigs. The vector that expressed the G protein performed better in comparison to the one that expressed the F protein (Kong et al. 2012).

McLean and Graham (2022) assessed the immunogenicity of the ALVAC-G vaccine against recombinant bovine herpesvirus-4 (rBoHV-4) vectors that express F or G proteins of NiV in pigs. Both vectors of BoHV-4 seemed to induce potent T-cell and Ab responses when compared to ALVAC-G; however, no virus challenge was issued and hence, the efficacy of protection could not be established (Thakur et al. 2019; McLean and Graham 2022).

Other candidates being evaluated for efficacy of protection and immunogenicity against NiV are:

1. An NiV adjuvanted soluble G (sG-NiV) protein (orthologous with HeV Equivac R vaccine).
2. A non-replicative simian adenovirus vector (ChAdOx1-NiV-G) expressing the G protein of NiV.
3. An adjuvanted F protein of NiV that is stabilized using a molecular clamp (mcsF protein) (McLean and Graham 2019).

1.9.3.2 Human Transmission Prevention

The most important pre-emptive measure that can be effective in reducing the spread of the disease is the prevention of human-to-human transmission. Following proper hygiene protocols like handwashing, disinfection using 70% IPA and donning gloves along with other PPE while also avoiding direct contact with the bodily secretions of the infected are standard protocols to be followed in the case of contact with a suspected or a confirmed case (Sharma et al. 2019). Within the hospital, no two patients should be allowed to share food or beds; this is considered absolutely unacceptable. The transmissibility of the virus to healthcare workers could be reduced if there is proper accessibility to hygienic facilities and well-maintained PPE is provided. Another particularly important rule to obey is to follow officially laid down procedures during a funeral like donning gloves and masks, and cleaning up efficiently, immediately after being exposed to the corpse (Ambat et al. 2019; Skowron et al. 2022).

1.9.3.3 Disinfectants

The evaluation of the disinfection efficiency of both physical and chemical disinfectants against NiV is important. Like many *Paramyxoviruses*, the Nipah virus is readily deactivated with the treatment of detergents as well as many other disinfectants. In Malaysia, pig farms have been recommended to be disinfected by the use of sodium hypochlorite (bleach) (Spiclera 2016). An effective decrease in infectivity of NiV was demonstrated by Eickmann et al. (2020) in plasma concentrates and platelets utilizing UV-C as well as increasing doses of visible light. Treating NiV with 0.2 J/cm^2 UV-C lowered the infectiousness ($\geq$4.3 log) to the limit of detection (LOD) in platelet samples, while MB (methylene blue) treatment (120 J/cm^2) lessened NiV ($\geq$2.7 log) to LOD in the plasma (Skowron et al. 2022).

1.9.3.4 Public Health Awareness

Raising public cognizance of the outbreak-associated risks as well as the importance of following the officially established preventive measures is a major attribute that may help reduce the threat of a new outbreak. Regional media such as TV news, radio along with print media have been crucial in spreading awareness; for instance, the perils in the consumption of raw date palm juice and how it should be restricted (Aditi and Shariff 2019). In contradiction with the reigning social and cultural practices, attempts to abolish the practice of consuming fresh juices were unsuccessful (Ang et al. 2018).

1.9.3.5 NiV Surveillance

Surveillance of the NiV outbreak belt is done to enable strain analysis, premature detection of outbreaks and monitoring the interaction of environmental factors and developmental dynamics of an epidemic. In order to achieve these purposes, specialized teams of researchers are established and their job is to identify suspected cases and analyse the probability of the emergence of potential new reservoirs accountable for NiV transmission to humans (Aditi and Shariff 2019; Ambat et al. 2019).

1.9.3.5.1 NCDC-Integrated Disease Surveillance Programme (IDSP) for Emerging Infectious Diseases (EID)-NiV — India

Two districts of Kerala state in India saw NiV outbreaks in 2018 (Baggchi et al. 2018; Sahay et al. 2020). The recorded initial case was at the Baby Memorial

Hospital in Kozhikode district and the doctors in charge suspected the presence of a rare virus infection and immediately sent samples to the referral centre, namely Manipal Centre for Virus Research, Karnataka, followed by ICMR-NIV, Pune, Maharashtra, who then confirmed the agent to be Nipah virus (Baggchi et al. 2018; Sahay et al. 2020). Rapidly, the Govt. of Kerala with the help of GOI and other national agencies, supported by WHO through a set of measures, including tracking and isolation of contact cases, conducted widespread surveillance and mass awareness campaigns (Baggchi et al. 2018; Sahay et al. 2020). The resultant 'No Nipah' campaign has been efficient in reaching the masses with simple messages such as covering the date palm sap container, etc., on how to prevent the spread of NiV infection. After 2018, no NiV cases were observed in Kerala other than the 2021 outbreak amidst a global pandemic (Kojom and Singh 2021) and the recent ongoing outbreak of September 2023. Involvement of various agencies like the Ministry of Health and Family Welfare, Directorate of Health Research, Indian Council of Agricultural Research, State Health Department, State Animal Husbandry, District Administration and a resultant multidisciplinary response mechanism was seen in the 2018 and 2019 NiV outbreaks (Singhai et al. 2021).

1.9.3.5.2 One Health Surveillance of NiV—Thailand

In Thailand, One Health surveillance has been conducted for NiV, including a triangulated surveillance of bats, pigs and humans since 2001 in areas considered at a high risk of NiV emergence, in collaboration with countless laboratories and universities, the Ministry of Public Health, Ministry of Agriculture and Co-operatives and Ministry of Natural Resources and Environment (Wacharapluesadee et al. 2021). From Wat Luang village, Chonburi province, considered the home to Thailand's largest *P. lylei* bat roost and certain other provinces, periodically samples were collected from pigs, bats and human volunteers and tested for NiV since 2001. Cerebrospinal fluid (CSF) specimens from encephalitic cases between 2001 and 2012, which had been archived, were also tested for NiV (Wacharapluesadee et al. 2021). In fruit bats, NiV RNA was detected using RT-PCR from 2002 to 2020. Results of the whole genome sequencing sampled from bat urine in 2017 suggest that the sample shared 99.17% identity to the NiV sequence obtained from a Bangladesh patient in 2004. No anti-NiV IgG antibodies or RNA have been found in the healthy volunteers that were tested or in encephalitis cases or in pigs, to date. On sample collection trips, at least a hundred community members were trained on how to live safely with bats (Wacharapluesadee et al. 2021).

The outbreak potential of NiV in Thailand was highlighted by the high per cent identity of the NiV genome obtained from Thai bats and patients from Bangladesh. As a result of these cross-sectoral surveillances carried out, the findings were shared with national authorities and the residents of the villages which in turn led to the implementation of preventive measures, the increase in surveillance of pigs and humans residing in close vicinity of NiV-infected roosts and reduced perilous behaviour at the community level (Wacharapluesadee et al. 2021). As Thailand initiated outbreak preparedness using the One Health approach, prior to human or domestic animal outbreaks of NiV, this approach of encouraging cross-sectoral co-operation between various health sectors and raising awareness from villages to national government in Thailand is an excellent example of a successful transboundary model of One Health for Asia-Pacific region in order to successfully monitor NiV outbreak, develop preventive measures and; therefore, improve preparedness for the next pandemic (Wacharapluesadee et al. 2021).

1.9.3.5.3 Ecological Niche Mapping Approach—Reservoir Ecology and Zoonoses Transmission Prediction

Valuable tools in the prediction of zoonotic disease transmission and determination of the distribution of suitable habitats for reservoirs are spatial modelling and risk mapping (Ostfeld et al. 2005, Hahn et al. 2014). Ecological Niche Mapping (ENM) establishes a relationship between environmental variables and occurrence records and can be utilized to predict habitat suitability for target vector or species (Peterson et al. 2007; Peterson. 2014; Peterson et al. 2011). This is a theoretical framework that takes into consideration environmental conditions (A), mobility (M) and biotic interactions (B) to understand species distribution (Soberòn and Peterson 2005; Chaiyes et al. 2022).

A study aimed at the identification of the current and future range of *P. lylei,* a key bat reservoir of NiV in Cambodia, Thailand and Vietnam using ecological niche modelling, was done. Based on three different transmission pathways, they overlaid habitat suitability maps with high-res in-country data from Thailand in order to identify potential sites for spill-over of NiV. This data helped improve the understanding of the ecological influences on NiV and its transmission. The risk map provided a baseline for guiding active surveillance of NiV and can help in shaping surveillance policy decisions depending on the primary transmission pathway that is prominent in the region. This approach can help improve awareness and prevention aimed at reducing outbreaks and the economic losses that go hand-in-hand (Chaiyes et al. 2022).

1.10 Current Challenges and Future Perspective

Though many candidates are available for use as treatments in the course of NiV disease, the limitation of information and research available is the biggest challenge we face in curtailing this potential mammoth. Among a range of tested antivirals against NiV, two stand out and demonstrate efficacy in therapy in the NHP model: the nucleotide analogue Remdesivir and the monoclonal antibody m102.4. m102.4 is the strongest contender however, the intravenous route of administration and requirement of a cold chain may prove it to be impractical for field use (Johnson et al. 2021). Though its efficacy is unclear, the use of Ribavirin has been carried out for compassionate treatment for NiV-related encephalitis (Snell 2004; Randhawa et al. 2022) and post-exposure prophylaxis (Johnson et al. 2021); however, various side effects to its consumption are observed such as convulsions, nausea and vomiting, in course of treatment (Chong et al. 2001; Randhawa et al. 2022). Nucleoside analogues and other small molecule antivirals have exhibited proficiency in blocking NiV infection and need further evaluation for therapeutic potential. Combinatorial therapy of small molecule antivirals and mAbs could be an effective strategy for treatment (Johnson et al. 2021). Targeting NiV proteins that are involved in fusion and infection of host cells might lead to the development of therapeutics that are NiV-specific (Aborode et al. 2022). The biggest hurdle in the evaluation of the efficacy of NiV-specific antivirals is the occurrence of infrequent cases and hence, alternative routes for licensing are necessary. Timely diagnosis and initiation of treatment are essential as the developed therapeutic candidates work within a short treatment efficacy window (Johnson et al. 2021).

A study was conducted in which due to the alterations in the monocytic, Th1/Th2 and cytokines and chemokine balances, COVID-19, Ebola, MERS, NiV and SARS were suggested to be considered to be reclassified into a new immunological category, namely acute onset immune dysrhythmia syndrome (n-AIDS) (Kelleni et al. 2021c). NiV reclassification may be due to its ability to modulate inflammatory and immune responses as well as interferon homeostasis (Pelissier et al. 2019), it is also known to influence leucocyte attracting and proinflammatory cytokines that determine disease course (Satterfield et al. 2015) and to dysregulate immune enlistment that causes acute vasculitis amidst other dysregulations (Prescott et al. 2012; Kelleni et al. 2021c). The authors of this study suggested that the novel reclassification as n-AIDS may help guide us through the quest for the search of a potential cure, i.e. if we understand the cause, we can effectively employ it towards therapeutic research; and focus on immunomodulation of effective therapy of COVID-19 (Gupta et al. 2020; Kelleni et al. 2021b, c), NiV (Pelissier et al. 2019) and EBOV (Bixler et al. 2017). They also hypothesized that their immunomodulatory COVID-19 management protocol (Kelleni et al. 2021a, b), which proved effective during the pandemic, may be beneficial in the initial management of different RNA viruses that have been re-classified into n-AIDS and other potential diseases caused by the alteration of the Th1/Th2 balance (Kelleni et al. 2021c).

NiV disease-related data on infection of humans are limited due to the sporadicity of outbreaks in endemic areas and this brings the necessity of animal models capable of mimicking human disease into the spotlight. NiV has a wide host range due to the highly conserved cellular receptors that are present throughout many species; however, not all NiV hosts show the same degree of infection (Lawrence and Escudero-Pérez 2022). Therefore, several NiV-specific animal models of infection have been generated successfully (Liew et al. 2022) and are used to study pathogenesis, shedding, clinical features, transmission and antiviral and vaccine efficacy (Lawrence and Escudero-Pérez 2022). The models of infection that mimic NiV disease in humans most closely are AGMs, hamsters and ferrets (Lawrence and Escudero-Pérez 2022). Though the cell culture and animal models of NiV have highlighted certain comparabilities among immunopathogenesis and immune responses that are invoked in hosts, the mechanisms in humans are yet not completely known and understood, and no evidence has been found to confirm the results seen in animal models are accurate and would be reproducible in humans. Thus, it is important to conduct further studies with respect to NiV pathogenesis in humans in order to provide solid ground for developing therapeutic and control strategies against NiV (Liew et al. 2022).

Moreover, though a few experimental studies have been conducted on bats, reservoirs and the source of NiV transmission, it is necessary to understand the mechanism by which bats control and regulate NiV infection to produce asymptomatic disease and yet shed and transmit the virus; this may help provide helpful insights into the prevention and treatment of infection in intermediate hosts and humans and reduce mortality in both populations (Lawrence and Escudero-Pérez 2022).

In the past few decades, the universal agreement that neutralizing antibodies (NAbs) play a crucial role in conferring protection against a pathogen has led to their elicitation becoming the focus of vaccine development procedures (Broder et al. 2012; Prescott et al. 2012; Satterfield et al. 2016). However, no correlate of protection on the basis of NAb response has been defined and in animal models where protection was elicited, all were protected from disease, however, this made it difficult to establish a threshold. Contrastingly, the role played by CMI in protection is an avenue few have ventured into. A study conducted on pigs reported that CMI may play a vital role in conferring complete protection; however, no mechanism was described and conclusions were distorted as, unlike other models, the Nipah virus is known to infect a wide range of immune cells in pigs (Pickering et al. 2016; Price et al. 2021).

Out of the majority of vaccine candidates that exhibited a sturdy immune response in animal models, most required multiple dose administration to confer efficacious protection. Few have showcased complete protection in the NHP model which is the gold standard model for NiV. When considering NiV, which is native to the Asia-Pacific region and is considered a potential bioterrorism candidate, a solitary dose immunization is most suitable. When trying to prevent natural infections, multi-dose vaccines are uneconomical as well as impractical in low-income countries; while in case of deliberate release, it would prove impractical due to the deployment of multiple doses and hence prove redundant. Thus, a single immunization conferring quick protection is the necessity of the hour (Foster et al. 2022). The question of the hour is whether to take vaccine candidates that have been concluded to be safe and immunogenic to test for efficacy in phase III trials or whether to find an alternate route of licensure. This dilemma is due to the sporadicity of human cases and since the outbreaks are small, they are detected late (Nikolay et al. 2019). Bangladesh is most suited as the epidemiology is better understood and it reports the most outbreaks. However, even if there are only seven spill-overs from bats to humans per year on average are reported (Nikolay et al. 2020); outbreaks do not extend over 34 cases or five generations and ~ 10% of cases transmit the virus from human-to-human (Nikolay et al. 2019).

A study was carried out with the aim to 'assess the feasibility of a phase III vaccine trial for NiV considering the current epidemiological characteristics.' Simulations based on previous NiV cases from Bangladesh, an assumed efficacy of 90% and other characteristics of vaccine targets, were run. They compared three study designs:

I. Cluster randomized ring vaccination.
II. Cluster randomized mass vaccination.
III. Observational case–control study (Nikolay et al. 2021).

The findings suggest that ring or mass vaccination trials in Bangladesh are unlikely to be fulfilled within a reasonable period with the epidemiological characteristics of the country. Ring vaccination trials are unsuitable due to short chains of interhuman transmission. In light of this, globally the following options can be considered:

- No conduction of phase III trials until epidemiology changes—potentially leading to further delays in the case of a large outbreak.
- Identifying alternate routes of licensure of vaccines—no clinical trials, observational or controlled animal studies as in the US FDA (Nikolay et al. 2021).

An observational case–control design is a viable strategy which we estimate to be completed within 7 years, without investments to improve case detection (Nikolay et al. 2021). Four vaccine candidates are being funded by CEPI to undergo safety and immunogenicity testing. However, the simultaneous testing of multiple candidates in phase III may take long and hence it may be efficacious to focus efforts to evaluate the protection of one single vaccine candidate (Nikolay et al. 2021). Another method could be the improvement of the identification of NiV cases via enhanced surveillance such as increasing the target area of surveillance and the number of hospitals or healthcare facilities in frequently affected areas (Nikolay et al. 2021).

The use of Animal Rule could be another alternative to licensing, where a thoroughly designed animal study is considered sufficient to prove efficacy (Burns et al. 2012). These results can be bridged to humans by the use of well-established immunity correlates (Nikolay et al. 2021). When it comes to NiV-related diseases, the standardization and refinement of animal models are necessary when considering licensure of vaccines as NiV outbreaks are sporadic and the cases of infection are a relatively small number hence, getting permission for human trials is a challenge. An alternative path of licensure is the 'Animal rule' as described by the US FDA wherein phase-I/II immunogenicity and safety tests are conducted in humans but efficacy trials are conducted and approved in an established animal model Similarly, the European Medical Agency (EMA) does not have a model analogous to FDA's 'Animal Rule,' existing guidelines contain a loophole using which it is possible to demonstrate the efficacy of vaccines in animal models (Price et al. 2021). The variety of stocks used for vaccine studies in animal models throughout published studies has made it difficult to establish a standardized, well-characterized and common virus stock used for challenges in order to produce comparable results in animal models (Price et al. 2021). Additionally, the use of the Malaysian strain for frequent challenges is unclear as most outbreaks globally are a result of infection with the Bangladesh strain. Besides the strain used for virus challenge, dose and route of administration are significant variables to be considered when evaluating the performance of vaccine models (Price et al. 2021).

The application of Bioinformatics tools for the prediction and validation of vaccine designs against varied pathogens is a promising technique that is relatively inexpensive and enables the shortening of periods for a few experimental steps when moved to the wet lab for validation (Soltan et al. 2021).

The expanding diversity of HNVs leads to questions of preparedness for potential spill-overs and the emergence of new pathogens from the group, which is essential to consider with respect to therapeutics and vaccines. Although m102.4 and h5B3.1 exhibit excellent cross-protection among NiV and HeV, it has been reported that Abs elicited by G proteins of these viruses have limited cross-reactivity and zero cross-protection against Ghana virus (GhV) and Mojiang virus (MojV) (Li et al. 2020). A study proposed the development of a multitude fusion protein broad-spectrum vaccine inclusive of epitopes from MojV, GhV, HeV, along with NiV that would elicit protection against all four viruses. This signifies the importance of developing a pan-HNV vaccine in the event of the emergence of a new member from the group (Johnson et al. 2021).

1.10.1 Transmission Prevention and Control Strategies—India

Strict surveillance measures and awareness campaigns should be organized to limit interactions with bats as the dynamics of NiV transmission depend on multiple host factors including human behaviour and human–bat interface. In order to achieve early detection and containment of outbreaks, it is crucial to have strong human surveillance for Severe Acute Respiratory Infection and Acute Encephalitis Syndrome inclusive testing in areas susceptible to NiV (Yadav et al. 2022). Implementation and adherence to policies such as regular testing for healthcare professionals, regular screening for individuals presenting with possible symptoms of the Nipah virus as part of history taking and strict quarantine policies in regions of the outbreak can help in early detection and treatment (Aborode et al. 2022). In order to break the transmission cycle, contact tracing and quarantine for close contacts who are possibly exposed could be helpful. Increased resources directed towards public health policy in order to generate advertisements and radio broadcasts can help spread awareness about risks and their impact on public health (Aborode et al. 2022).

1.10.2 Major Gaps in Outbreak Preparedness and Response Strategy

With past experiences, it is easily apparent that each instance of a major disease that has posed a challenge to the sector of public health is owing to:

(i) poor preparedness to deal with disastrous health events,
(ii) restricted resources, mostly in developing countries,
(iii) little to no surveillance,
(iv) lack of cognizance with regard to basic practices during outbreaks,
(v) lack of interdisciplinary one-health approach,
(vi) ineffective management in public and bureaucratic capacities (Debnath et al. 2021).

Due to repetitions of these events, the first-hand health impact is disastrous while the indirect impact leads to a rise in mortality along with morbidity due to re-routing and subsequent depletion of reserves used in daily care and limited access to these due to travel limitations and fear, this in turn may lead to depletion in the number of frontline health-care staff during an epidemic or pandemic. The impact of epidemics and pandemics is felt in society as well as on a national level leaving the infected to face social stigma, which in turn may reduce the capability of the institution to manage the precariousness of the situation (Debnath et al. 2021).

1.10.3 Zoonoses and One Health Approach

The need to approach the public health issue globally in a holistic way has been highlighted due to the several emerging and re-emerging zoonotic disease outbreaks. This need is felt urgently in countries like India, which have a high diversity of life forms, with communities living in areas that are biodiverse, with a high degree of constant wildlife–livestock–human interface (Nulkar et al. 2021). The concept of One Health is a holistic approach to tackling emerging global health concerns with interdisciplinary attempts, chiefly with respect to zoonotic diseases. In the One Health approach, health is considered to be at three levels, viz., individual level, population level and ecosystem level (WHO One Health 2017). The approach takes into consideration the rapidly evolving environment which aids the emergence and transmission of an infectious disease affecting a living system able to harbour the pathogen, while allowing an opportunity to use data in order to visualize predictive models for the spread and control of disease (Nulkar et al. 2021).

1.10.4 India and One Health

Constant epidemiological evolution, involvement of intermediate hosts and sources of infection in humans of different areas call for constant surveillance of NiV-infected bats, animals and humans. This would help understand the dynamics of NiV in that geographical region and help make new effective policies for prevention and control (Gokhale et al. 2022). The heightened need for in-depth proactive nationwide surveillance for NiV with the One Health approach throughout the year in bats and other animals is warranted by three consecutive incidences of NiV in Kerala, far away from the Nipah belt, with no known intermediate host or confirmed route of entry in the human population (Gokhale et al. 2022).

The key to preventing the emergence of future zoonotic outbreaks and protecting community health are national strategies. However, the design of such strategies is a complex issue due to the lack of effective surveillance and several socio-ecological factors that influence disease spread. The One Health approach is crucial for implementing effective surveillance, prevention and curbing strategies and involves a collaboration between human health, animal health and environmental sectors at every level of governance. It aims to establish inter-sectoral coordination at the national, state and district levels, and utilize the existing surveillance systems of these stakeholders to detect early warning signs of impending outbreaks for timely and effective response; it also facilitates sharing of relevant information among the stakeholders for taking necessary action (Singhai et al. 2021).

An approach like this must involve veterinarians, environmental specialists and medical professionals working with members of the community to foster deep roots to conduct effective research and practices. Communities lead to pinpointing what works in their context and this could influence surveillance, reporting and enforcement of regulations. Policy and legislation need to be put into place, although these processes take much political will and effort and are predicated on consistent governance and co-production with communities to design effective and workable strategies. Some encouragement for these initiatives may be possible following the recent pandemic, as policymakers and the citizens have become aware of the importance of emerging zoonotic pathogens and the potential ramifications of their transmission. Further research should focus on encouraging a One Health response, working within communities to identify their priorities, barriers, requirements and enable effective addressing of risk factors around zoonotic outbreaks, and how behaviourial change initiatives could be supported by governmental and multilateral bodies (Durrance-Bagale et al. 2021).

The interdisciplinary response during the 2018 outbreak in Kerala helped further the understanding of the significance of the One Health approach for utilizing the existing surveillance systems. This also helped in the subsequent 2019 outbreak that occurred, where there was no documented spread from the index case and no mortality was observed. This reiterates the importance of the One Health approach during public health emergencies, especially zoonotic diseases (Singhai et al. 2021).

1.10.5 Lessons Learned Post-Intervention for Future Activities—India

- Strengthening the healthcare system.
- Inter-sectoral coordination.
- Modelling transmission dynamics.
- Public–private partnership.
- IEC activities (Singhai et al. 2021).

1.11 Conclusion

Crossing of the species barrier by bat-borne viruses is a recurring event, owing to their continual evolution and relationship with humans. The ongoing COVID-19 pandemic has attested to the fact that human–human transmissible zoonotic viruses are highly capable of causing a pandemic. The possibility of global dissemination of NiV is evidenced by the latest outbreaks of 2018, 2019 and 2021 in Kerala, South India; which occurred thousands of kilometres away from the original outbreak areas of Malaysia and the subsequent outbreak regions of Bangladesh and East India. Consequently, the at-risk zone for NiVD encompasses approximately two billion people.

The absence of antivirals and vaccines with proven effectiveness for NiV, along with the inadequate healthcare structure of developing countries in particular, leads to the inevitable question of whether re-emergence and a worldwide spread of NiVD would lead once again to global catastrophe. As a zoonotic disease, NiVD would certainly benefit from the 'One Health' approach. Additionally, the current pandemic has conveyed the significance of a comprehensive surveillance system to supervise emerging and re-emerging bat-associated viruses, along with environmental conservation policies to decrease human–wildlife interactions—both of which are pertinent in the context of NiV.

Therefore, concrete efforts made in the above-mentioned areas would certainly aid in the prevention and control of future NiV outbreaks.

References

Aborode AT, Wireko AA, Mehta A, Abdul-Rahman T, Nansubuga EP, Kundu M, Pustake M, Mehmood Q, Tillewein H (2022) Concern over Nipah virus cases amidst the COVID-19 pandemic in India. J Med Virol 3–5:3488–3490. https://doi.org/10.1002/jmv.27745

Abu Bakar S, Chang LY, Ali AR, Sharifah SH, Yusoff K, Zamrod Z (2004) Isolation and molecular identification of Nipah virus from pigs. Emerg Infect Dis 10:2228–2230. https://doi.org/10.3201/eid1012.040452

Aditi, Shariff M (2019) Nipah virus infection: a review. Epidemiol Infect 147:1–6. https://doi.org/10.1017/S0950268819000086

Ahmad B, Sherrini & Tan, Chong. (2014) Nipah Encephalitis – An Update. Med J Malaysia 69:2014

Aljofan M, Porotto M, Moscona A, Mungall BA (2008) Development and validation of a chemiluminescent immunodetection assay amenable to high throughput screening of antiviral drugs for Nipah and Hendra virus. J Virol Methods 149(1):12–19

Allen T, Murray KA, Zambrana-Torrelio C, Morse SS, Rondinini C, Di Marco M, Breit N, Olival KJ, Daszak P (2017) Global hotspots and correlates of emerging zoonotic diseases. Nat Commun 8(1):1124. https://doi.org/10.1038/s41467-017-00923-8

Allocati N, Petrucci AG, Di Giovanni P, Masulli M, Di Ilio C, De Laurenzi V (2016) Bat-man disease transmission: zoonotic pathogens. Cell Death Discov 2:16048. https://doi.org/10.1038/cddiscovery.2016.48. eCollection 2016

Amaya M, Broder CC (2020) Vaccines to Emerging Viruses: Nipah and Hendra. Annual review of virology 7(1):447–473. https://doi.org/10.1146/annurev-virology-021920-113833

Ambat AS, Zubair SM, Prasad N, Pundir P, Rajwar E, Patil DS et al (2019) Nipah virus: a review on epidemiological characteristics and outbreaks to inform public health decision making. J Infect Public Health 12:634–639. https://doi.org/10.1016/j.jiph.2019.02.013

Anderson DE, Islam A, Crameri G, Todd S, Islam A, Khan SU, Foord A, Rahman MZ, Mendenhall IH, Luby SP, Gurley ES, Daszak P, Epstein JH, Wang LF (2019) Isolation and full-genome characterization of Nipah viruses from bats, Bangladesh. Emerg Infect Dis 25:166–170. https://doi.org/10.3201/eid2501.180267

Ang B, Lim T, Wang L (2018) Nipah virus infection. J Clin Microbiol 56(6):e01875–e01817. https://doi.org/10.1128/JCM.01875-17

Anyamba A, Chretien JP, Britch SC, Soebiyanto RP, Small JL, Jepsen R et al (2019) Global disease outbreaks associated with the 2015–2016 El Niño event. Sci Rep 9(1):1–14

Arankalle VA, Bandyopadhyay BT, Ramdasi AY, Jadi R, Patil DR, Rahman M, Majumdar M, Banerjee PS, Hati AK, Goswami RP, Neogi DK, Mishra AC (2011) Genomic characterization of Nipah virus, West Bengal, India. Emerg Infect Dis 17(5):907–909. https://doi.org/10.3201/eid1705.100968

Arunkumar G, Chandni R, Mourya DT, Singh SK, Sadanandan R, Sudan P, Bhargava B, Nipah Investigators People and Health Study Group (2019a) Outbreak investigation of Nipah virus disease in Kerala, India, 2018. J Infect Dis 219:1867–1878

Arunkumar G, Devadiga S, McElroy AK, Prabhu S, Sheik S, Abdulmajeed J, Robin S, Sushama A, Jayaram A, Nittur S, Shakir M, Kumar K, Radhakrishnan C, Sakeena K, Vasudevan J, Reena KJ, Sarita RL, Klena JD, Spiropoulou CF, Laserson KF et al (2019b) Adaptive immune responses in humans during Nipah virus acute and convalescent phases of infection. Clin Infect Dis 69(10):1752–1756. https://doi.org/10.1093/cid/ciz010

Aziz AJ, Mahendran R, Daniels P, Shahiruddin S, Narasiman M, Azizah D, et al (1999) The status, public response and challenges in overcoming emerging and exotic diseases—Nipah virus disease experience. In: National Congress on Animal Health and Production: Environmental Care in Animal Production. (Alor Gajah)

Baggchi S (2018) Bye Bye Nipah. Lancet Infect Dis 18(11):1189. https://doi.org/10.1016/S1473-3099(18)30621-2

Banerjee S, Niyas VKM, Soneja M et al (2019) First experience of ribavirin postexposure prophylaxis for Nipah virus, tried during the 2018 outbreak in Kerala, India. J Infect 78(6):491–503

Baseler L, de Wit E, Scott DP, Munster VJ, Feldmann H (2015) Syrian hamsters (Mesocricetus auratus) oronasally inoculated with a Nipah virus isolate from Bangladesh or Malaysia develop similar respiratory tract lesions. Vet Pathol 52:38–45. https://doi.org/10.1177/0300985814556189

Basler CF (2012) Nipah and hendra virus interactions with the innate immune system. Curr Top Microbiol Immunol 359:123–152. https://doi.org/10.1007/82_2012_209

Bausch DG, Hadi CM, Khan SH et al (2010) Review of the literature and proposed guidelines for the use of oral ribavirin as postexposure prophylaxis for Lassa fever. Clin Infect Dis 51(12):1435–1441

Beck MA, Handy J, Levander OA (2000) The role of oxidative stress in viral infections. Ann N Y Acad Sci 917:906–912. https://doi.org/10.1111/j.1749-6632.2000.tb05456.x

Belay ED, Monroe SS (2014) Low-incidence, high-consequence pathogens. Emerg Infect Dis 20:319–321. https://doi.org/10.3201/eid2002.131748

Benson MD, Romero MI, Lush ME, Lu QR, Henkemeyer M, Parada LF (2005) Ephrin-B3 is a myelin-based inhibitor of neurite outgrowth. Proc Natl Acad Sci USA 102:10694–10699. https://doi.org/10.1073/pnas.0504021102

Berhane Y, Weingartl HM, Lopez J, Neufeld J, Czub S, Embury-Hyatt C, Goolia M, Copps J, Czub M (2008) Bacterial infections in pigs experimentally infected with Nipah virus. Transbound Emerg Dis 55:165–174

Beyer RM, Manica A, Mora C (2021) Shifts in global bat diversity suggest a possible role of climate change in the emergence of SARS-CoV-1 and SARS-CoV-2. Sci Total Environ 767:145413. https://doi.org/10.1016/j.scitotenv.2021.145413

Bharaj P, Wang YE, Dawes BE, Yun TE, Park A, Yen B, Basler CF, Freiberg AN, Lee B, Rajsbaum R (2016) The matrix protein of Nipah virus targets the E3-ubiquitin ligase TRIM6 to inhibit the IKKepsilon kinase-mediated type-I IFN antiviral response. PLoS Pathog 12:e1005880. https://doi.org/10.1371/journal.ppat.1005880

Biggerstaff M, Cauchemez S, Reed C, Gambhir M, Finelli L (2014) Estimates of the reproduction number for seasonal, pandemic, and zoonotic influenza: a systematic review of the literature. BMC Infect Dis 14:480. https://doi.org/10.1186/1471-2334-14-480

Bixler SL, Duplantier AJ, Bavari S (2017) Discovering drugs for the treatment of Ebola virus. Curr Treat Options Infect Dis 9(3):299–317. https://doi.org/10.1007/s40506-017-0130-z

Bloyet LM, Welsch J, Enchery F, Mathieu C, de Breyne S, Horvat B, Grigorov B, Gerlier D (2016) HSP90 chaperoning in addition to phosphoprotein required for folding but not for supporting enzymatic activities of measles and Nipah virus L polymerases. J Virol 90(15):6642–6656. https://doi.org/10.1128/JVI.00602-16

Bonaparte MI, Dimitrov AS, Bossart KN, Crameri G, Mungall BA, Bishop KA, Choudhry V, Dimitrov DS, Wang LF, Eaton BT, Broder CC (2005) Ephrin-B2 ligand is a functional receptor for Hendra virus and Nipah virus. Proc Natl Acad Sci USA 102(30):10652–10657. https://doi.org/10.1073/pnas.0504887102

Bossart KN, McEachern JA, Hickey AC et al (2007) Neutralization assays for differential henipavirus serology using bio-plex protein array systems. J Virol Methods 142:29–40. https://doi.org/10.1016/j.jviromet.2007.01.003

Bossart KN, Tachedjian M, McEachern JA, Crameri G, Zhu Z, Dimitrov DS, Broder CC, Wang LF (2008) Functional studies of host-specific ephrin-B ligands as Henipavirus receptors. Virology 372:357–371. https://doi.org/10.1016/j.virol.2007.11.011

Bossart KN, Zhu Z, Middleton D, Klippel J, Crameri G, Bingham J, McEachern JA, Green D, Hancock TJ, Chan YP et al (2009) A neutralizing human monoclonal antibody protects against lethal disease in a new ferret model of acute nipah virus infection. PLoS Pathog 5:e1000642. https://doi.org/10.1371/journal.ppat.1000642

Bossart KN, Geisbert TW, Feldmann H et al (2011) A neutralizing human monoclonal antibody protects African green monkeys from hendra virus challenge. Sci Transl Med 3(105):105ra103

Bossart KN, Rockx B, Feldmann F, Brining D, Scott D, LaCasse R, Geisbert JB, Feng YR, Chan YP, Hickey AC, Broder CC, Feldmann H, Geisbert TW (2012) A Hendra virus G glycoprotein subunit vaccine protects African green monkeys from Nipah virus challenge. Sci Transl Med 4(146):146ra107. https://doi.org/10.1126/scitranslmed.3004241

Bowden TA, Aricescu AR, Gilbert RJ, Grimes JM, Jones EY, Stuart DI (2008) Structural basis of Nipah and Hendra virus attachment to their cell-surface receptor ephrin-B2. Nat Struct Mol Biol 15:567–572. https://doi.org/10.1038/nsmb.1435

Br A, Dovih P, Ramakrishnan U, Liang E, Mendenhall I, Hong DLW, Smith GJD (2019) Evidence of filovirus and henipavirus in bats and bat harvesters, India. Int J Infect Dis 79:60. https://doi.org/10.1016/j.ijid.2018.11.156

Breed AC, Yu M, Barr JA et al (2010) Prevalence of henipavirus and rubulavirus antibodies in pteropid bats, Papua New Guinea. Emerg Infect Dis 16:1997–1999. https://doi.org/10.3201/eid1612.100879

Broder CC, Geisbert TW, Xu K, Nikolov DB, Wang LF, Middleton D, Pallister J, Bossart KN (2012) Immunization strategies against henipaviruses. Curr Top Microbiol Immunol 359:197–223

Broder CC, Xu K, Nikolov DB et al (2013) A treatment for and vaccine against the deadly Hendra and Nipah viruses. Antivir Res 100(1):8–13

Broder CC, Weir DL, Reid PA (2016) Hendra virus and Nipah virus animal vaccines. Vaccine 34(30):3525–3534. https://doi.org/10.1016/j.vaccine.2016.03.075

Brook CE, Dobson AP (2015) Bats as 'special' reservoirs for emerging zoonotic pathogens. Trends Microbiol 23:172–180

Brook CE, Ranaivoson HC, Broder CC, Cunningham AA, H'eraud JM, Peel AJ, Gibson L, Wood JL, Metcalf CJ, Dobson AP (2019) Disentangling serology to elucidate henipa-and filovirus transmission in Madagascar fruit bats. J Anim Ecol 88(7):1001–1016

Burns DL (2012) Licensure of vaccines using the animal rule. Current Opin Virol 2(3):353–356. https://doi.org/10.1016/j.coviro.2012.01.004

Cai W, Ng B, Wang G, Santoso A, Wu L, Yang K (2022) Increased ENSO sea surface temperature variability under four IPCC emission scenarios. Nat Clim Chang 12(3):228–231

Calisher CH, Childs JE, Field HE, Holmes KV, Schountz T (2006) Bats: important reservoir hosts of emerging viruses. Clin Microbiol Rev 19(3):531–545

Carrasco-Hernandez R, Jácome R, López Vidal Y, Ponce de León S (2017) Are RNA viruses candidate agents for the next global pandemic? A review. ILAR J 58:343–358. https://doi.org/10.1093/ilar/ilx026

CDC (1999) Outbreak of Hendra-like virus—Malaysia and Singapore, 1998–1999. Morb Mortal Wkly Rep 48(13):265–269

CDC/USDA (2021) Federal Select Agent Program, HHS and USDA Select Agents and Toxins 7CFR Part 331, 9 CFR Part 121, and 42 CFR Part 73. https://www.selectagents.gov/sat/list.htm?CDC_AA_refVal=https%3A%2F%2Fwww.selectagents.gov%2FSelectAgentsandToxinsList.html. Accessed 9 Dec 2021

Centers for Disease Control and Prevention [CDC] (1999) Update: outbreak of Nipah virus–Malaysia, and Singapore, 1999. MMWR Morb Mortal Wkly Rep 48:335–337

CEPI (2019) CEPI awards up to US$43.6 million to Public Health Vaccines, LLC. for development of a single-dose Nipah virus vaccine candidate

CEPI: Our portfolio. Reference Source 78, Lo MK, Spengler JR, Welch SR et al (2020) Evaluation of a single-dose nucleoside-modified messenger RNA vaccine encoding Hendra virus-soluble glycoprotein against lethal Nipah virus challenge in Syrian Hamsters. J Infect Dis 221(Suppl 4):S493–S498

Chadha MS, Comer JA, Lowe L, Rota PA, Rollin PE, Bellini WJ, Ksiazek TG, Mishra A (2006) Nipah virus-associated encephalitis outbreak, Siliguri, India. Emerg Infect Dis 12(2):235–240. https://doi.org/10.3201/eid1202.051247

Chaiyes A, Duengkae P, Suksavate W, Pongpattananurak N, Wacharapluesadee S, Olival KJ, Srikulnath K, Pattanakiat S, Hemachudha T (2022) Mapping risk of Nipah virus transmission from bats to humans in Thailand. EcoHealth 19:175. https://doi.org/10.1007/s10393-022-01588-6

Chakraborty A, Sazzad HM, Hossain MJ, Islam MS, Parveen S, Husain M, Banu SS, Podder G, Afroj S, Rollin PE, Daszak P, Luby SP, Rahman M, Gurley ES (2016) Evolving epidemiology of Nipah virus infection in Bangladesh: evidence from outbreaks during 2010-2011. Epidemiol Infect 144(2):371–380. https://doi.org/10.1017/S0950268815001314

Chakraborty S, Deb B, Barbhuiya PA, Uddin A (2019) Analysis of codon usage patterns and influencing factors in Nipah virus. Virus Res 263:129–138. https://doi.org/10.1016/j.virusres.2019.01.011

Chan KP, Rollin PE, Ksiazek TG, Leo YS, Goh KT, Paton NI et al (2002) A survey of Nipah virus infection among various risk groups in Singapore. Epidemiol Infect 128:93–98. https://doi.org/10.1017/s0950268801006422

Chandni R, Renjith TP, Fazal A, Yoosef N, Ashhar C, Thulaseedharan NK, Suraj KP, Sreejith MK, Sajeeth Kumar KG, Rajendran VR, Remla Beevi A, Sarita RL, Sugunan AP, Arunkumar G, Mourya, D T, Murhekar M (2020). Clinical Manifestations of Nipah Virus-Infected Patients Who Presented to the Emergency Department During an Outbreak in Kerala State in India, May 2018. Clinical infectious diseases : an official publication of the Infectious Diseases Society of America, 71(1), 152–157. https://doi.org/10.1093/cid/ciz789

Chong HT, Tan CT (2003) Relapsed and late-onset Nipah encephalitis, a report of three cases. Neurol J Southeast Asia 8:109–112

Chang LY, Mohd Ali AR, Hassan SS, AbuBakar S (2006) Quantitative estimation of Nipah virus replication kinetics in vitro. Virol J 3(1):1–7

Chattopadhyay A, Rose JK (2011) Complementing defective viruses that express separate paramyxovirus glycoproteins provide a new vaccine vector approach. J Virol 85(5):2004–2011. https://doi.org/10.1128/JVI.01852-10

Chattu VK, Kumar R, Kumary S, Kajal F, David JK (2018) Nipah virus epidemic in southern India and emphasizing "one health" approach to ensure global health security. J Family Med Prim Care 7(2):275–283

Chen L, Liu B, Yang J, Jin Q (2014) DBatVir: the database of bat-associated viruses. Database 2014:bau021. https://doi.org/10.1093/database/bau021

Chiang CF, Lo MK, Rota PA, Spiropoulou CF, Rollin PE (2010) Use of monoclonal antibodies against Hendra and Nipah viruses in an antigen capture-ELISA. Virol J 7(1):115. https://doi.org/10.1186/1743-422X-7-115

Childs K, Stock N, Ross C, Andrejeva J, Hilton L, Skinner M, Randall R, Goodbourn S (2007) mda-5, but not RIG-I, is a common target for paramyxovirus V proteins. Virology 359:190–200. https://doi.org/10.1016/j.virol.2006.09.023

Childs K, Randall R, Goodbourn S (2012) Paramyxovirus V proteins interact with the RNA helicase LGP2 to inhibit RIG-I-dependent interferon induction. J Virol 86:3411–3421. https://doi.org/10.1128/JVI.06405-11

Ching PK, de los Reyes VC, Sucaldito MN, Tayag E, Columna-Vingno AB, Malbas FF Jr, Bolo GC Jr, Sejvar JJ, Eagles D, Playford G, Dueger E, Kaku Y, Morikawa S, Kuroda M, Marsh

GA, McCullough S, Foxwell AR (2015) Outbreak of henipavirus infection, Philippines, 2014. Emerg Infect Dis 21(2):328–331. https://doi.org/10.3201/eid2102.141433
Chionh YT, Cui J, Koh J, Mendenhall IH, Ng J, Low D, Itahana K, Irving AT, Wang LF (2019) High basal heat-shock protein expression in bats confers resistance to cellular heat/oxidative stress. Cell Stress Chaperones 24(4):835–849. https://doi.org/10.1007/s12192-019-01013-y
Chong HT, Kamarulzaman A, Tan CT, Goh KJ, Thayaparan T, Kunjapan SR, Chew NK, Chua KB, Lam SK (2001) Treatment of acute Nipah encephalitis with ribavirin. Ann Neurol 49(6):810–813. https://doi.org/10.1002/ana.1062
Chong HT, Kunjapan SR, Thayaparan T, Tong JM, Petharunam V, Jusoh MR, Tan CT (2002) Nipah encephalitis outbreak in Malaysia, clinical features in patients from Seremban. Can J Neurol Sci 29(1):83–87. http://ovidsp.ovid.com/ovidweb.cgi
Chow VT, Tambyah PA, Yeo WM, Phoon MC, Howe J (2000) Diagnosis of nipah virus encephalitis by electron microscopy of cerebrospinal fluid. J Clin Virol 19(3):143–147. https://doi.org/10.1016/s1386-6532(00)00094-9
Chowdhury MR, Ndiaye O (2017) Climate change and variability impacts on the forests of Bangladesh–a diagnostic discussion based on CMIP5 GCMs and ENSO. Int J Climatol 37(14):4768–4782
Chowdhury S, Khan SU, Crameri G, Epstein JH, Broder CC, Islam A, Peel AJ, Barr J, Daszak P, Wang LF, Luby SP (2014) Serological evidence of henipavirus exposure in cattle, goats and pigs in Bangladesh. PLoS Negl Trop Dis 8(11):e3302. https://doi.org/10.1371/journal.pntd.0003302
Christodoulou M (2018) CEPI awards contract worth up to USD$19 million to Oxford university and Janssen vaccines to develop MERS, Lassa, and Nipah vaccines
Christodoulou M (2019) CEPI awards a contract worth up to US$ 31 million to the University of Tokyo to develop a vaccine against Nipah virus
Chua KB (2003) Nipah virus outbreak in Malaysia. J Clin Virol 26(3):265–275. https://doi.org/10.1016/s1386-6532(02)00268-8
Chua KB, Goh KJ, Wong KT, Kamarulzaman A, Tan PS, Ksiazek TG, Zaki SR, Paul G, Lam SK, Tan CT (1999) Fatal encephalitis due to Nipah virus among pig-farmers in Malaysia. Lancet (London, England) 354(9186):1257–1259. https://doi.org/10.1016/S0140-6736(99)04299-3
Chua KB, Bellini WJ, Rota PA, Harcourt BH, Tamin A, Lam SK, Ksiazek TG, Rollin PE, Zaki SR, Shieh WJ, Goldsmith CS (2000a) Nipah virus: a recently emergent deadly paramyxovirus. Science 288(5470):1432–1435. https://doi.org/10.1126/science.288.5470.1432
Chua KB, Lam SK, Tan CT, Hooi PS, Goh KJ, Chew NK, Tan KS, Kamarulzaman A, Wong KT (2000b) High mortality in Nipah encephalitis is associated with presence of virus in cerebrospinal fluid. Ann Neurol 48(5):802–805
Chua KB, Lam SK, Goh KJ, Hooi PS, Ksiazek TG, Kamarulzaman A, Olson J, Tan CT (2001) The presence of Nipah virus in respiratory secretions and urine of patients during an outbreak of Nipah virus encephalitis in Malaysia. J Infect 42(1):40–43. https://doi.org/10.1053/jinf.2000.0782
Chua KB, Koh CL, Hooi PS, Wee KF, Khong JH, Chua BH, Chan YP, Lim ME, Lam SK (2002) Isolation of Nipah virus from Malaysian Island flying-foxes. Microbes Infect 4(2):145–151. https://doi.org/10.1016/s1286-4579(01)01522-2
Ciancanelli MJ, Volchkova VA, Shaw ML, Volchkov VE, Basler CF (2009) Nipah virus sequesters inactive STAT1 in the nucleus via a P gene-encoded mechanism. J Virol 83(16):7828–7841. https://doi.org/10.1128/JVI.02610-08
Clayton BA, Middleton D, Bergfeld J, Haining J, Arkinstall R, Wang L, Marsh GA (2012) Transmission routes for nipah virus from Malaysia and Bangladesh. Emerg Infect Dis 18(12):1983–1993. https://doi.org/10.3201/eid1812.120875
Clayton BA, Wang LF, Marsh GA (2013) Henipaviruses: an updated review focusing on the pteropid reservoir and features of transmission. Zoonoses Public Health 60(1):69–83. https://doi.org/10.1111/j.1863-2378.2012.01501.x

Clayton BA, Middleton D, Arkinstall R, Frazer L, Wang LF, Marsh GA (2016) The nature of exposure drives transmission of Nipah viruses from Malaysia and Bangladesh in ferrets. PLoS Negl Trop Dis 10(6):e0004775. https://doi.org/10.1371/journal.pntd.0004775

Cline C, Bell TM, Facemire P, Zeng X, Briese T, Ian Lipkin W, Shamblin JD, Esham HL, Donnelly GC, Johnson JC, Hensley LE, Honko AN, Johnston SC (2022) Detailed analysis of the pathologic hallmarks of Nipah virus (Malaysia) disease in the African green monkey infected by the intratracheal route. PLoS One 17(2 February):1–26. https://doi.org/10.1371/journal.pone.0263834

ClinicalTrials.gov: Safety and Immunogenicity of a Nipah Virus Vaccine. https://clinicaltrials.gov/study/NCT04199169

Cong Y, Lentz MR, Lara A, Alexander I, Bartos C, Bohannon JK, Hammoud D, Huzella L, Jahrling PB, Janosko K, Jett C, Kollins E, Lackemeyer M, Mollura D, Ragland D, Rojas O, Solomon J, Xu Z, Munster V, Holbrook MR (2017) Loss in lung volume and changes in the immune response demonstrate disease progression in African green monkeys infected by small-particle aerosol and intratracheal exposure to Nipah virus. PLoS Negl Trop Dis 11(4):e0005532. https://doi.org/10.1371/journal.pntd.0005532

Constantine DG, Emmons RW, Woodie JD (1972) Rabies virus in nasal mucosa of naturally infected bats. Science (New York, NY) 175(4027):1255–1256. https://doi.org/10.1126/science.175.4027.1255

Cortes MC, Cauchemez S, Lefrancq N, Luby SP, Jahangir Hossain M, Sazzad H, Rahman M, Daszak P, Salje H, Gurley ES (2018) Characterization of the spatial and temporal distribution of Nipah virus spillover events in Bangladesh, 2007-2013. J Infect Dis 217(9):1390–1394. https://doi.org/10.1093/infdis/jiy015

Crameri G, Wang LF, Morrissy C, White J, Eaton BT (2002) A rapid immune plaque assay for the detection of Hendra and Nipah viruses and anti-virus antibodies. J Virol Methods 99(1–2):41–51. https://doi.org/10.1016/s0166-0934(01)00377-9

Crameri G, Todd S, Grimley S, McEachern JA, Marsh GA, Smith C, Tachedjian M, De Jong C, Virtue ER, Yu M, Bulach D, Liu JP, Michalski WP, Middleton D, Field HE, Wang LF (2009) Establishment, immortalisation and characterisation of pteropid bat cell lines. PLoS One 4(12):e8266. https://doi.org/10.1371/journal.pone.0008266

Csermely P, Agoston V, Pongor S (2005) The efficiency of multi-target drugs: the network approach might help drug design. Trends Pharmacol Sci 26(4):178–182. https://doi.org/10.1016/j.tips.2005.02.007

Dang HV, Chan YP, Park YJ, Snijder J, Da Silva SC, Vu B, Yan L, Feng YR, Rockx B, Geisbert TW, Mire CE, Broder CC, Veesler D (2019) An antibody against the F glycoprotein inhibits Nipah and Hendra virus infections. Nat Struct Mol Biol 26(10):980–987. https://doi.org/10.1038/s41594-019-0308-9

Daniels P, Ksiazek T, Eaton BT (2001) Laboratory diagnosis of Nipah and Hendra virus infections. Microbes Infect 3(4):289–295. https://doi.org/10.1016/s1286-4579(01)01382-x

Darif D, Hammi I, Kihel A, El Idrissi Saik I, Guessous F, Akarid K (2021) The pro-inflammatory cytokines in COVID-19 pathogenesis: what goes wrong? Microb Pathog 153:104799. https://doi.org/10.1016/j.micpath.2021.104799

Dasgupta A, Hammarlund E, Slifka MK, Früh K (2007) Cowpox virus evades CTL recognition and inhibits the intracellular transport of MHC class I molecules. Journal of immunology (Baltimore, Md. : 1950), 178(3), 1654–1661. https://doi.org/10.4049/jimmunol.178.3.1654

Daszak P, Zambrana-Torrelio C, Bogich TL, Fernandez M, Epstein JH, Murray KA, Hamilton H (2013) Interdisciplinary approaches to understanding disease emergence: the past, present, and future drivers of Nipah virus emergence. Proc Natl Acad Sci U S Am 110 Suppl 1(Suppl 1):3681–3688. https://doi.org/10.1073/pnas.1201243109

Davis ME, Wang MK, Rennick LJ, Full F, Gableske S, Mesman AW, Gringhuis SI, Geijtenbeek TB, Duprex WP, Gack MU (2014) Antagonism of the phosphatase PP1 by the measles virus V protein is required for innate immune escape of MDA5. Cell Host Microbe 16(1):19–30. https://doi.org/10.1016/j.chom.2014.06.007

Dawes BE, Kalveram B, Ikegami T et al (2018) Favipiravir (T-705) protects against Nipah virus infection in the hamster model. Sci Rep 8(1):7604
de Vries RD, de Jong A, Verburgh RJ, Sauerhering L, van Nierop GP, van Binnendijk RS, Osterhaus A, Maisner A, Koopmans M, de Swart RL (2020) Human paramyxovirus infections induce T cells that cross-react with zoonotic henipaviruses. mBio 11(4):e00972–e00920. https://doi.org/10.1128/mBio.00972-20
de Wit E, Munster VJ (2015) Animal models of disease shed light on Nipah virus pathogenesis and transmission. J Pathol 235(2):196–205. https://doi.org/10.1002/path.4444
de Wit E, Bushmaker T, Scott D, Feldmann H, Munster VJ (2011) Nipah virus transmission in a hamster model. PLoS Negl Trop Dis 5(12):e1432. https://doi.org/10.1371/journal.pntd.0001432
Debnath F, Chakraborty D, Deb AK, Saha MK, Dutta S (2021) Increased human-animal interface & emerging zoonotic diseases: an enigma requiring multi-sectoral efforts to address. Indian J Med Res 153(5&6):577–584. https://doi.org/10.4103/ijmr.IJMR_2971_20
DeBuysscher BL, de Wit E, Munster VJ, Scott D, Feldmann H, Prescott J (2013) Comparison of the pathogenicity of Nipah virus isolates from Bangladesh and Malaysia in the Syrian hamster. PLoS Negl Trop Dis 7(1):e2024. https://doi.org/10.1371/journal.pntd.0002024
DeBuysscher BL, Scott D, Marzi A, Prescott J, Feldmann H (2014) Single-dose live-attenuated Nipah virus vaccines confer complete protection by eliciting antibodies directed against surface glycoproteins. Vaccine 32:2637–2644
DeBuysscher BL, Scott D, Thomas T, Feldmann H, Prescott J (2016) Peri-exposure protection against Nipah virus disease using a single-dose recombinant vesicular stomatitis virus-based vaccine. NPJ Vaccines 1:16002
DeBuysscher BL, Scott DP, Rosenke R, Wahl V, Feldmann H, Prescott J (2021) Nipah virus efficiently replicates in human smooth muscle cells without cytopathic effect. Cells 10(6):1319. https://doi.org/10.3390/cells10061319
Defang GN, Khetawat D, Broder CC, Quinnan GV Jr (2010) Induction of neutralizing antibodies to Hendra and Nipah glycoproteins using a Venezuelan equine encephalitis virus in vivo expression system. Vaccine 29(2):212–220. https://doi.org/10.1016/j.vaccine.2010.10.053
Deka MA, Morshed N (2018) Mapping disease transmission risk of Nipah virus in South and Southeast Asia. Trop Med Infect Dis 3(2):57. https://doi.org/10.3390/tropicalmed3020057
Demby AH, Chamberlain J, Brown DW, Clegg CS (1994) Early diagnosis of Lassa fever by reverse transcription-PCR. J Clin Microbiol 32(12):2898–2903. https://doi.org/10.1128/jcm.32.12.2898-2903
DeVincenzo JP, McClure MW, Symons JA, Fathi H, Westland C, Chanda S, Lambkin-Williams R, Smith P, Zhang Q, Beigelman L, Blatt LM, Fry J (2015) Activity of Oral ALS-008176 in a respiratory syncytial virus challenge study. N Engl J Med 373(21):2048–2058. https://doi.org/10.1056/NEJMoa1413275
Devnath P, Masud H (2021) Nipah virus: a potential pandemic agent in the context of the current severe acute respiratory syndrome coronavirus 2 pandemic. New Microbes New Infect 41:100873. https://doi.org/10.1016/j.nmni.2021.100873
Dhondt KP, Mathieu C, Chalons M, Reynaud JM, Vallve A, Raoul H, Horvat B (2013) Type I interferon signaling protects mice from lethal henipavirus infection. J Infect Dis 207(1):142–151. https://doi.org/10.1093/infdis/jis653
Diederich S, Maisner A (2007) Molecular characteristics of the Nipah virus glycoproteins. Ann N Y Acad Sci 1102:39–50. https://doi.org/10.1196/annals.1408.003
Dimmock NJ, Dove BK, Meng B, Scott PD, Taylor I, Cheung L, Hallis B, Marriott AC, Carroll MW, Easton AJ (2012a) Comparison of the protection of ferrets against pandemic 2009 influenza A virus (H1N1) by 244 DI influenza virus and oseltamivir. Antivir Res 96(3):376–385. https://doi.org/10.1016/j.antiviral.2012.09.017
Dimmock NJ, Dove BK, Scott PD, Meng B, Taylor I, Cheung L, Hallis B, Marriott AC, Carroll MW, Easton AJ (2012b) Cloned defective interfering influenza virus protects ferrets from pandemic 2009 influenza A virus and allows protective immunity to be established. PLoS One 7(12):e49394. https://doi.org/10.1371/journal.pone.0049394

Donaldson H, Lucey D (2018) Enhancing preparation for large Nipah outbreaks beyond Bangladesh: preventing a tragedy like Ebola in West Africa. Int J Infect Dis 72:69–72. https://doi.org/10.1016/j.ijid.2018.05.015

Duffy S (2018) Why are RNA virus mutation rates so damn high? PLoS Biol 16:e3000003. https://doi.org/10.1371/journal.pbio.3000003

Durrance-Bagale A, Rudge JW, Singh NB, Belmain SR, Howard N (2021) Drivers of zoonotic disease risk in the Indian subcontinent: a scoping review. One Health (Amsterdam, Netherlands) 13:100310. https://doi.org/10.1016/j.onehlt.2021.100310

Eaton BT, Broder CC, Middleton D, Wang LF (2006) Hendra and Nipah viruses: different and dangerous. Nat Rev Microbiol 4(1):23–35. https://doi.org/10.1038/nrmicro1323

Edson D, Field H, McMichael L, Vidgen M, Goldspink L, Broos A, Melville D, Kristoffersen J, de Jong C, McLaughlin A, Davis R, Kung N, Jordan D, Kirkland P, Smith C (2015) Routes of Hendra virus excretion in naturally-infected flying-foxes: implications for viral transmission and spillover risk. PLoS One 10(10):e0140670. https://doi.org/10.1371/journal.pone.0140670

Edwards SJ, Caruso S, Suen WW, Jackson S, Rowe B, Marsh GA (2021) Evaluation of henipavirus chemical inactivation methods for the safe removal of samples from the high-containment PC4 laboratory. J Virol Methods 298:114287. https://doi.org/10.1016/j.jviromet.2021.114287

Eickmann M, Gravemann U, Handke W, Tolksdorf F, Reichenberg S, Müller TH, Seltsam A (2020) Inactivation of three emerging viruses - severe acute respiratory syndrome coronavirus, Crimean-Congo haemorrhagic fever virus and Nipah virus - in platelet concentrates by ultraviolet C light and in plasma by methylene blue plus visible light. Vox Sang 115(3):146–151. https://doi.org/10.1111/vox.12888

Enchéry F, Horvat B (2017) Understanding the interaction between henipaviruses and their natural host, fruit bats: paving the way toward control of highly lethal infection in humans. Int Rev Immunol 36(2):108–121. https://doi.org/10.1080/08830185.2016.1255883

Enkirch T, von Messling V (2015) Ferret models of viral pathogenesis. Virology 479-480:259–270. https://doi.org/10.1016/j.virol.2015.03.017

Epstein JH, Anthony SJ, Islam A, Kilpatrick AM, Ali Khan S, Balkey MD, Ross N, Smith I, Zambrana-Torrelio C, Tao Y, Islam A, Quan PL, Olival KJ, Khan M, Gurley ES, Hossein MJ, Field HE, Fielder MD, Briese T, Rahman M et al (2020) Nipah virus dynamics in bats and implications for spillover to humans. Proc Natl Acad Sci USA 117(46):29190–29201. https://doi.org/10.1073/pnas.2000429117

Erbar S, Diederich S, Maisner A (2008) Selective receptor expression restricts Nipah virus infection of endothelial cells Abstract. Virol J 5(1):10.1186/1743-422X-5-142

Escaffre O, Borisevich V, Rockx B (2013) Pathogenesis of Hendra and Nipah virus infection in humans. J Infect Dev Ctries 7(4):308–311. https://doi.org/10.3855/jidc.3648

Escaffre O, Borisevich V, Vergara LA, Wen JW, Long D, Rockx B (2016) Characterization of Nipah virus infection in a model of human airway epithelial cells cultured at an air-liquid interface. J Gen Virol 97(5):1077–1086. https://doi.org/10.1099/jgv.0.000441

Eshaghi M, Tan WS, Mohidin TB, Yusoff K (2004) Nipah virus glycoprotein: production in baculovirus and application in diagnosis. Virus Res 106(1):71–76. https://doi.org/10.1016/j.virusres.2004.06.011

Eshaghi M, Tan WS, Chin WK, Yusoff K (2005) Purification of the extra-cellular domain of Nipah virus glycoprotein produced in Escherichia coli and possible application in diagnosis. J Biotechnol 116(3):221–226. https://doi.org/10.1016/j.jbiotec.2004.10.020

Eyford BA, Singh C, Abraham T, Munro L, Choi KB, Hill T, Hildebrandt R, Welch I, Vitalis TZ, Gabathuler R, Gordon JA, Adomat H, Guns E, Lu CJ, Pfeifer CG, Tian MM, Jefferies WA (2021) A nanomule peptide carrier delivers siRNA across the intact blood-brain barrier to attenuate ischemic stroke. Front Mol Biosci 8:611367. https://doi.org/10.3389/fmolb.2021.611367

Feldman KS, Foord A, Heine HG, Smith IL, Boyd V, Marsh GA, Wood JL, Cunningham AA, Wang LF (2009) Design and evaluation of consensus PCR assays for henipaviruses. J Virol Methods 161(1):52–57. https://doi.org/10.1016/j.jviromet.2009.05.014

Field H, Young P, Yob JM, Mills J, Hall L, Mackenzie J (2001) The natural history of Hendra and Nipah viruses. Microbes Infect 3(4):307–314. https://doi.org/10.1016/s1286-4579(01)01384-3

Field HE, Mackenzie JS, Daszak P (2007) Henipaviruses: emerging paramyxoviruses associated with fruit bats. Curr Top Microbiol Immunol 315:133–159. https://doi.org/10.1007/978-3-540-70962-6_7

Fisher-Hoch SP, Khan JA, Rehman S, Mirza S, Khurshid M, McCormick JB (1995) Crimean Congo-haemorrhagic fever treated with oral ribavirin. Lancet (London, England) 346(8973):472–475. https://doi.org/10.1016/s0140-6736(95)91323-8

Fischer K, Diederich S, Smith G, Reiche S, Pinho Dos Reis V, Stroh E, Groschup MH, Weingartl HM, Balkema-Buschmann A (2018) Indirect ELISA based on Hendra and Nipah virus proteins for the detection of henipavirus specific antibodies in pigs. PLoS One 13(4):e0194385. https://doi.org/10.1371/journal.pone.0194385

Flanagan M, Leighton T, Dudley J (2011) Anticipating viral species jumps: bioinformatics and data needs. https://doi.org/10.21236/ada555216

Fleming TH (2019) Bat migration. Encyclopedia of Animal Behavior, pp 605–610. https://doi.org/10.1016/B978-0-12-809633-8.20764-4

Fogarty R, Halpin K, Hyatt AD, Daszak P, Mungall BA (2008) Henipavirus susceptibility to environmental variables. Virus Res 132(1–2):140–144. https://doi.org/10.1016/j.virusres.2007.11.010

Foo SS, Turner CJ, Adams S, Compagni A, Aubyn D, Kogata N, Lindblom P, Shani M, Zicha D, Adams RH (2006) Ephrin-B2 controls cell motility and adhesion during blood-vessel-wall assembly. Cell 124(1):161–173. https://doi.org/10.1016/j.cell.2005.10.034

Foord AJ, White JR, Colling A, Heine HG (2013) Microsphere suspension array assays for detection and differentiation of Hendra and Nipah viruses. Biomed Res Int 2013:289295. https://doi.org/10.1155/2013/289295

Foster SL, Woolsey C, Borisevich V, Agans KN, Prasad AN, Deer DJ, Geisbert JB, Dobias NS, Fenton KA, Cross RW, Geisbert TW (2022) A recombinant VSV-vectored vaccine rapidly protects nonhuman primates against lethal Nipah virus disease. Proc Natl Acad Sci USA 119(12):e2200065119. https://doi.org/10.1073/pnas.2200065119

Freiberg AN, Worthy MN, Lee B, Holbrook MR (2010) Combined chloroquine and ribavirin treatment does not prevent death in a hamster model of Nipah and Hendra virus infection. J Gen Virol 91(Pt 3):765–772. https://doi.org/10.1099/vir.0.017269-0

Frick WF, Kingston T, Flanders J (2020) A review of the major threats and challenges to global bat conservation. Ann N Y Acad Sci 1469(1):5–25. https://doi.org/10.1111/nyas.14045

Furuta Y, Gowen BB, Takahashi K, Shiraki K, Smee DF, Barnard DL (2013) Favipiravir (T-705), a novel viral RNA polymerase inhibitor. Antivir Res 100(2):446–454. https://doi.org/10.1016/j.antiviral.2013.09.015

Gale NW, Baluk P, Pan L, Kwan M, Holash J, DeChiara TM, McDonald DM, Yancopoulos GD (2001) Ephrin-B2 selectively marks arterial vessels and neovascularization sites in the adult, with expression in both endothelial and smooth-muscle cells. Dev Biol 230(2):151–160. https://doi.org/10.1006/dbio.2000.0112

Geisbert TW, Daddario-DiCaprio KM, Hickey AC, Smith MA, Chan YP, Wang LF, Mattapallil JJ, Geisbert JB, Bossart KN, Broder CC (2010) Development of an acute and highly pathogenic nonhuman primate model of Nipah virus infection. PLoS One 5(5):e10690. https://doi.org/10.1371/journal.pone.0010690

Geisbert TW, Feldmann H, Broder CC (2012) Animal challenge models of henipavirus infection and pathogenesis. Curr Top Microbiol Immunol 359:153–177. Epub 2012/04/06. https://doi.org/10.1007/82_2012_208

Geisbert TW, Mire CE, Geisbert JB, Chan YP, Agans KN, Feldmann F, Fenton KA, Zhu Z, Dimitrov DS, Scott DP, Bossart KN, Feldmann H, Broder CC (2014) Therapeutic treatment of Nipah virus infection in nonhuman primates with a neutralizing human monoclonal antibody. Sci Transl Med 6(242):242ra82. https://doi.org/10.1126/scitranslmed.3008929

Geisbert JB, Borisevich V, Prasad AN, Agans KN, Foster SL, Deer DJ, Cross RW, Mire CE, Geisbert TW, Fenton KA (2020) An intranasal exposure model of lethal Nipah virus infection in African Green monkeys. J Infect Dis 221(Suppl 4):S414–S418. https://doi.org/10.1093/infdis/jiz391

Geisbert TW, Bobb K, Borisevich V, Geisbert JB, Agans KN, Cross RW, Prasad AN, Fenton KA, Yu H, Fouts TR, Broder CC, Dimitrov AS (2021) A single dose investigational subunit vaccine for human use against Nipah virus and Hendra virus. NPJ Vaccines 6(1):23. https://doi.org/10.1038/s41541-021-00284-w

Georges-Courbot MC, Contamin H, Faure C, Loth P, Baize S, Leyssen P, Neyts J, Deubel V (2006) Poly(I)-poly(C12U) but not ribavirin prevents death in a hamster model of Nipah virus infection. Antimicrob Agents Chemother 50(5):1768–1772. https://doi.org/10.1128/AAC.50.5.1768-1772.2006

Gibb R, Franklinos L, Redding DW, Jones KE (2020a) Ecosystem perspectives are needed to manage zoonotic risks in a changing climate. BMJ 371:m3389. https://doi.org/10.1136/bmj.m3389

Gibb R, Redding DW, Chin KQ, Donnelly CA, Blackburn TM, Newbold T, Jones KE (2020b) Zoonotic host diversity increases in human-dominated ecosystems. Nature 584(7821):398–402. https://doi.org/10.1038/s41586-020-2562-8

Gilead (n.d.) Remdesivir Clinical Trials

Goh KJ, Tan CT, Chew NK, Tan PSK, Kamarulzaman A, Sarji SA, Wong KT, Abdullah BJJ, Chua KB, Lam SK (2000) Clinical features of Nipah virus encephalitis among pig farmers in Malaysia. N Engl J Med 342(17):1229–1235. https://doi.org/10.1056/nejm200004273421701

Gokhale MD, Sreelekshmy M, Sudeep AB, Shete A, Jain R, Yadav PD, Mathapati B, Mourya DT (2021) Detection of possible Nipah virus infection in Rousettus leschenaultii and Pipistrellus pipistrellus bats in Maharashtra, India. J Infect Public Health 14(8):1010–1012. https://doi.org/10.1016/j.jiph.2021.05.001

Gokhale M, Sudeep AB, Mathapati B, Balasubramanian R, Ullas PT, Mohandas S, Patil DR, Shete AM, Gopale S, Sawant P, Jain R, Holeppanavar M, Suryawanshi AT, Chopade G, Dhaigude S, Patil DY, Mourya DT, Yadav PD (2022) Serosurvey for Nipah virus in bat population of southern part of India. Comp Immunol Microbiol Infect Dis 85(March):22–25. https://doi.org/10.1016/j.cimid.2022.101800

Goldsmith CS, Whistler T, Rollin PE, Ksiazek TG, Rota PA, Bellini WJ, Daszak P, Wong KT, Shieh WJ, Zaki SR (2003) Elucidation of Nipah virus morphogenesis and replication using ultrastructural and molecular approaches. Virus Res 92(1):89–98. https://doi.org/10.1016/S0168-1702(02)00323-4

Gómez Román R, Wang LF, Lee B, Halpin K, de Wit E, Broder CC, Rahman M, Kristiansen P, Saville M (2020) Nipah@20: lessons learned from another virus with pandemic potential. mSphere 5(4):e00602–e00620. https://doi.org/10.1128/mSphere.00602-205

Gómez Román R, Tornieporth N, Cherian NG, Shurtleff AC, L'Azou Jackson M, Yeskey D, Hacker A, Mungai E, Le TT (2022) Medical countermeasures against henipaviruses: a review and public health perspective. Lancet Infect Dis 22(1):e13–e27. https://doi.org/10.1016/S1473-3099(21)00400-X

Gorbunova V, Seluanov A, Kennedy BK (2020) The world goes bats: living longer and tolerating viruses. Cell Metab 32(1):31–43. https://doi.org/10.1016/j.cmet.2020.06.013

Grein J, Ohmagari N, Shin D et al (2020) Compassionate use of Remdesivir for patients with severe Covid-19. N Engl J Med 382(24):2327–2336

Guillaume V, Lefeuvre A, Faure C, Marianneau P, Buckland R, Lam SK, Wild TF, Deubel V (2004) Specific detection of Nipah virus using real-time RT-PCR (TaqMan). J Virol Methods 120(2):229–237. https://doi.org/10.1016/j.jviromet.2004.05.018

Guillaume V, Contamin H, Loth P, Grosjean I, Courbot MC, Deubel V, Buckland R, Wild TF (2006) Antibody prophylaxis and therapy against Nipah virus infection in hamsters. J Virol 80(4):1972–1978. https://doi.org/10.1128/JVI.80.4.1972-1978.2006

Guillaume V, Wong KT, Looi RY, Georges-Courbot MC, Barrot L, Buckland R, Wild TF, Horvat B (2009) Acute Hendra virus infection: analysis of the pathogenesis and passive antibody protection in the hamster model. Virology 387(2):459–465. https://doi.org/10.1016/j.virol.2009.03.001

Gunnell GF, Smith R, Smith T (2017) 33 million year old Myotis (Chiroptera, Vespertilionidae) and the rapid global radiation of modern bats. PLoS One 12(3):e0172621. https://doi.org/10.1371/journal.pone.0172621

Gupta R (2005) Recent outbreak of rabies infections in Brazil transmitted by vampire bats. Euro Surveill 10(11):E051110.3. https://doi.org/10.2807/esw.10.45.02831-en
Gupta A (2020) Is Immuno-modulation the key to COVID-19 pandemic? Indian J Orthop 54(3):394–397. https://doi.org/10.1007/s43465-020-00121-7
Gupta M, Lo MK, Spiropoulou CF (2013) Activation and cell death in human dendritic cells infected with Nipah virus. Virology 441(1):49–56. https://doi.org/10.1016/j.virol.2013.03.004
Gupta N, Rai DB, Jangid AK, Pooja D, Kulhari H (2019) Nanomaterials-based siRNA delivery: routes of administration, hurdles and role of nanocarriers. Nanotechnology in Modern Animal Biotechnology: Recent Trends and Future Perspectives, pp 67–114. https://doi.org/10.1007/978-981-13-6004-6_3
Gupta P, Singh MP, Goyal K, Tripti P, Ansari MI, Obli Rajendran V, Dhama K, Malik YS (2021) Bats and viruses: a death-defying friendship. Virusdisease 32(3):467–479. https://doi.org/10.1007/s13337-021-00716-0
Gurley ES, Montgomery JM, Hossain MJ, Bell M, Azad AK, Islam MR, Molla MA, Carroll DS, Ksiazek TG, Rota PA, Lowe L, Comer JA, Rollin P, Czub M, Grolla A, Feldmann H, Luby SP, Woodward JL, Breiman RF (2007a) Person-to-person transmission of Nipah virus in a Bangladeshi community. Emerg Infect Dis 13(7):1031–1037. https://doi.org/10.3201/eid1307.061128
Gurley ES, Montgomery JM, Hossain MJ, Islam MR, Molla MA, Shamsuzzaman SM, Akram K, Zaman K, Asgari N, Comer JA, Azad AK, Rollin PE, Ksiazek TG, Breiman RF (2007b) Risk of nosocomial transmission of Nipah virus in a Bangladesh hospital. Infect Control Hosp Epidemiol 28(6):740–742. https://doi.org/10.1086/516665
Gurley ES, Hegde ST, Hossain K, Sazzad H, Hossain MJ, Rahman M, Sharker M, Salje H, Islam MS, Epstein JH, Khan SU, Kilpatrick AM, Daszak P, Luby SP (2017) Convergence of humans, bats, trees, and culture in Nipah virus transmission, Bangladesh. Emerg Infect Dis 23(9):1446–1453. https://doi.org/10.3201/eid2309.161922
Habjan M, Andersson I, Klingström J, Schümann M, Martin A, Zimmermann P, Wagner V, Pichlmair A, Schneider U, Mühlberger E, Mirazimi A, Weber F (2008) Processing of genome 5′ termini as a strategy of negative-strand RNA viruses to avoid RIG-I-dependent interferon induction. PLoS One 3(4):e2032. https://doi.org/10.1371/journal.pone.0002032
Hahn MB, Gurley ES, Epstein JH, Islam MS, Patz JA, Daszak P, Luby SP (2014) The role of landscape composition and configuration on Pteropus giganteus roosting ecology and Nipah virus spillover risk in Bangladesh. Am J Trop Med Hyg 90(2):247–255. https://doi.org/10.4269/ajtmh.13-0256
Halpin K, Mungall BA (2007) Recent progress in henipavirus research. Comp Immunol Microbiol Infect Dis 30(5-6):287–307. https://doi.org/10.1016/j.cimid.2007.05.008
Halpin K, Rota P (2014) A review of Hendra virus and Nipah virus infections in man and other animals. Zoonoses—Infections Affecting Humans and Animals: Focus on Public Health Aspects. pp 997–1012. https://doi.org/10.1007/978-94-017-9457-2_40
Halpin K, Hyatt AD, Fogarty R, Middleton D, Bingham J, Epstein JH, Rahman SA, Hughes T, Smith C, Field HE, Daszak P, Henipavirus Ecology Research Group (2011) Pteropid bats are confirmed as the reservoir hosts of henipaviruses: a comprehensive experimental study of virus transmission. Am J Trop Med Hyg 85(5):946–951. https://doi.org/10.4269/ajtmh.2011.10-0567
Hammoud DA, Lentz MR, Lara A, Bohannon JK, Feuerstein I, Huzella L, Jahrling PB, Lackemeyer M, Laux J, Rojas O et al (2018) Aerosol exposure to intermediate size Nipah virus particles induces neurological disease in African green monkeys. PLoS Negl Trop Dis 12:e0006978. https://doi.org/10.1371/journal.pntd.0006978
Hanley DE, Bourassa MA, O'Brien JJ, Smith SR, Spade ER (2003) A quantitative evaluation of ENSO indices. J Clim 16(8):1249–1258
Harcourt BH, Tamin A, Ksiazek TG, Rollin PE, Anderson LJ, Bellini WJ, Rota PA (2000) Molecular characterization of Nipah virus, a newly emergent paramyxovirus. Virology 271(2):334–349. https://doi.org/10.1006/viro.2000.0340
Harcourt BH, Lowe L, Tamin A, Liu X, Bankamp B, Bowden N, Rollin PE, Comer JA, Ksiazek TG, Hossain MJ, Gurley ES, Breiman RF, Bellini WJ, Rota PA (2005) Genetic characteri-

zation of Nipah virus, Bangladesh, 2004. Emerg Infect Dis 11(10):1594–1597. https://doi.org/10.3201/eid1110.050513

Hassan MZ, Sazzad H, Luby SP, Sturm-Ramirez K, Bhuiyan MU, Rahman MZ, Islam MM, Ströher U, Sultana S, Kafi M, Daszak P, Rahman M, Gurley ES (2018) Nipah virus contamination of hospital surfaces during outbreaks, Bangladesh, 2013–2014. Emerg Infect Dis 24(1):15–21. https://doi.org/10.3201/eid2401.161758

Hauser N, Gushiken AC, Narayanan S, Kottilil S, Chua JV (2021) Evolution of nipah virus infection: past, present, and future considerations. Trop Med Infect Dis 6(1):24. https://doi.org/10.3390/tropicalmed6010024

Hayman DT, Wang LF, Barr J, Baker KS, Suu-Ire R, Broder CC, Cunningham AA, Wood JL (2011) Antibodies to henipavirus or henipa-like viruses in domestic pigs in Ghana, West Africa. PLoS One 6(9):e25256. https://doi.org/10.1371/journal.pone.0025256

Hayman DT, Bowen RA, Cryan PM, McCracken GF, O'Shea TJ, Peel AJ, Gilbert A, Webb CT, Wood JL (2013) Ecology of zoonotic infectious diseases in bats: current knowledge and future directions. Zoonoses Public Health 60(1):2–21. https://doi.org/10.1111/zph.12000

Hegde ST, Sazzad HM, Hossain MJ, Alam MU, Kenah E, Daszak P, Rollin P, Rahman M, Luby SP, Gurley ES (2016) Investigating rare risk factors for Nipah virus in Bangladesh: 2001–2012. EcoHealth 13(4):720–728. https://doi.org/10.1007/s10393-016-1166-0

Hoffmann M, Nehlmeier I, Brinkmann C, Krähling V, Behner L, Moldenhauer AS, Krüger N, Nehls J, Schindler M, Hoenen T, Maisner A, Becker S, Pöhlmann S (2019) Tetherin inhibits Nipah virus but not Ebola virus replication in fruit bat cells. J Virol 93(3):e01821–e01818. https://doi.org/10.1128/JVI.01821-18

Homaira N, Rahman M, Hossain MJ, Epstein JH, Sultana R, Khan MS, Podder G, Nahar K, Ahmed B, Gurley ES, Daszak P, Lipkin WI, Rollin PE, Comer JA, Ksiazek TG, Luby SP (2010) Nipah virus outbreak with person-to-person transmission in a district of Bangladesh, 2007. Epidemiol Infect 138(11):1630–1636. https://doi.org/10.1017/S0950268810000695

Honda K, Taniguchi T (2006) Toll-like receptor signaling and IRF transcription factors. IUBMB Life 58(5–6):290–295. https://doi.org/10.1080/15216540600702206

Hong-Wu Z, Jing Y, Wen-Xian H, Hong-Jun L, Lu PX, Xiao-Hua L et al (2016) Influenza A (H1N1). In: Lu PX, Zhou BP (eds) Diagnostic imaging of emerging infectious diseases. Springer, The Netherlands, pp 57–76

Hooper P, Zaki S, Daniels P, Middleton D (2001) Comparative pathology of the diseases caused by Hendra and Nipah viruses. Microbes Infect 3(4):315–322. https://doi.org/10.1016/s1286-4579(01)01385-5

Horie R, Yoneda M, Uchida S, Sato H, Kai C (2016) Region of Nipah virus C protein responsible for shuttling between the cytoplasm and nucleus. Virology 497:294–304. https://doi.org/10.1016/j.virol.2016.07.013

Hossain MJ, Gurley ES, Montgomery JM, Bell M, Carroll DS, Hsu VP, Formenty P, Croisier A, Bertherat E, Faiz MA, Azad AK, Islam R, Molla MAR, Ksiazek TG, Rota PA, Comer JA, Rollin PE, Luby SP, Breiman RF (2008) Clinical presentation of Nipah virus infection in Bangladesh. Clin Infect Dis 46(7):977–984. https://doi.org/10.1086/529147

Hotard AL, He B, Nichol ST, Spiropoulou CF, Lo MK (2017) 4′-Azidocytidine (R1479) inhibits henipaviruses and other paramyxoviruses with high potency. Antivir Res 144:147–152. https://doi.org/10.1016/j.antiviral.2017.06.011

Hsu VP, Hossain MJ, Parashar UD, Ali MM, Ksiazek TG, Kuzmin I, Niezgoda M, Rupprecht C, Bresee J, Breiman RF (2004) Nipah virus encephalitis reemergence, Bangladesh. Emerg Infect Dis 10(12):2082–2087. https://doi.org/10.3201/eid1012.040701

Hu K, Huang G, Huang P, Kosaka Y, Xie SP (2021) Intensification of El Niño-induced atmospheric anomalies under greenhouse warming. Nat Geosci 14(6):377–382

Huynh TT, Aarnink AJ, Drucker A, Verstegen MW (2007) Pig production in Cambodia, Laos, Philippines and Vietnam: a review. Asian J Agric Dev 1:323–339

Hyatt AD, Zaki SR, Goldsmith CS, Wise TG, Hengstberger SG (2001) Ultrastructure of Hendra virus and Nipah virus within cultured cells and host animals. Microbes Infect 3(4):297–306. https://doi.org/10.1016/s1286-4579(01)01383-1

Iizumi T, Luo JJ, Challinor AJ, Sakurai G, Yokozawa M, Sakuma H, Brown ME, Yamagata T (2014) Impacts of El Niño southern oscillation on the global yields of major crops. Nat Commun 5:3712. https://doi.org/10.1038/ncomms4712

International Centre for Diarrheal Diseases Research Bangladesh (2007) Person-to-person transmission of Nipah infection in Bangladesh. Health Sci Bull 5:1–6

Islam MN, Shaikat AH, Islam KMF, Shil SK, Akter S, Rahman MM, Hassan MM, Islam A, Khan SA, Furey N (2015) First record of Ratanaworabhans' s Fruit Bat Megaerops niphanae Yenbutra and Felten, 1983 (Chiroptera: Pteropodidae) from Bangladesh. J Threatened Taxa 7:7821–7824

Islam MS, Sazzad HM, Satter SM, Sultana S, Hossain MJ, Hasan M, Rahman M, Campbell S, Cannon DL, Ströher U, Daszak P, Luby SP, Gurley ES (2016) Nipah virus transmission from bats to humans associated with drinking traditional liquor made from date palm sap, Bangladesh, 2011–2014. Emerg Infect Dis 22(4):664–670. https://doi.org/10.3201/eid2204.151747

Islam ARM, Salam R, Yeasmin N, Kamruzzaman M, Shahid S, Fattah M et al (2021) Spatiotemporal distribution of drought and its possible associations with ENSO indices in Bangladesh. Arab J Geosci 14(23):1–19

Ivanov AV, Bartosch B, Isaguliants MG (2017) Oxidative Stress in Infection and Consequent Disease. Oxidative Med Cell Longev 2017:3496043. https://doi.org/10.1155/2017/3496043

Jebb D, Huang Z, Pippel M, Hughes GM, Lavrichenko K, Devanna P, Winkler S, Jermiin LS, Skirmuntt EC, Katzourakis A, Burkitt-Gray L, Ray DA, Sullivan K, Roscito JG, Kirilenko BM, Dávalos LM, Corthals AP, Power ML, Jones G, Ransome RD et al (2020) Six reference-quality genomes reveal evolution of bat adaptations. Nature 583(7817):578–584. https://doi.org/10.1038/s41586-020-2486-3

Jensen KS, Adams R, Bennett RS, Bernbaum J, Jahrling PB, Holbrook MR (2018) Development of a novel real-time polymerase chain reaction assay for the quantitative detection of Nipah virus replicative viral RNA. PLoS One 13(6):e0199534. https://doi.org/10.1371/journal.pone.0199534

Jensen MR, Yabukarski F, Communie G, Condamine E, Mas C, Volchkova V, Tarbouriech N, Bourhis JM, Volchkov V, Blackledge M, Jamin M (2020) Structural description of the Nipah virus phosphoprotein and its interaction with STAT1. Biophys J 118(10):2470–2488. https://doi.org/10.1016/j.bpj.2020.04.010

Johnson CK, Hitchens PL, Pandit PS, Rushmore J, Evans TS, Young CCW et al (2020a) Global shifts in mammalian population trends reveal key predictors of virus spillover risk. Proc R Soc B Biol Sci 287(1924):20192736

Johnson CK, Hitchens PL, Pandit PS, Rushmore J, Evans TS, Young C, Doyle MM (2020b) Global shifts in mammalian population trends reveal key predictors of virus spillover risk. Proc Biol Sci 287(1924):20192736. https://doi.org/10.1098/rspb.2019.2736

Johnson K, Vu M, Freiberg AN (2021) Recent advances in combating Nipah virus. Fac Rev 10:74. https://doi.org/10.12703/r/10-74

Johnston SC, Briese T, Bell TM, Pratt WD, Shamblin JD, Esham HL, Donnelly GC, Johnson JC, Hensley LE, Lipkin WI, Honko AN (2015) Detailed analysis of the African green monkey model of Nipah virus disease. PLoS One 10(2):e0117817. https://doi.org/10.1371/journal.pone.0117817

Jones KE, Patel NG, Levy MA, Storeygard A, Balk D, Gittleman JL, Daszak P (2008) Global trends in emerging infectious diseases. Nature 451(7181):990–993. https://doi.org/10.1038/nature06536

Kaku Y, Noguchi A, Marsh GA, McEachern JA, Okutani A, Hotta K, Bazartseren B, Fukushi S, Broder CC, Yamada A, Inoue S, Wang LF (2009) A neutralization test for specific detection of Nipah virus antibodies using pseudotyped vesicular stomatitis virus expressing green fluorescent protein. J Virol Methods 160(1–2):7–13. https://doi.org/10.1016/j.jviromet.2009.04.037

Kaku Y, Noguchi A, Marsh GA, Barr JA, Okutani A, Hotta K, Bazartseren B, Fukushi S, Broder CC, Yamada A, Inoue S, Wang LF (2012) Second generation of pseudotype-based serum neutralization assay for Nipah virus antibodies: sensitive and high-throughput analysis utilizing secreted alkaline phosphatase. J Virol Methods 179(1):226–232. https://doi.org/10.1016/j.jviromet.2011.11.003

Kamthania M, Sharma DK (2015) Screening and structure-based modeling of T-cell epitopes of Nipah virus proteome: an immunoinformatic approach for designing peptide-based vaccine. *3*. Biotech 5(6):877–882. https://doi.org/10.1007/s13205-015-0303-8

Karesh WB, Dobson A, Lloyd-Smith JO, Lubroth J, Dixon MA, Bennett M, Aldrich S, Harrington T, Formenty P, Loh EH, Machalaba CC, Thomas MJ, Heymann DL (2012) Ecology of zoonoses: natural and unnatural histories. Lancet (London, England) 380(9857):1936–1945. https://doi.org/10.1016/S0140-6736(12)61678-X

Kasloff SB, Leung A, Pickering BS, Smith G, Moffat E, Collignon B, Embury-Hyatt C, Kobasa D, Weingartl HM (2019) Pathogenicity of Nipah henipavirus Bangladesh in a swine host. Sci Rep 9(1):5230. https://doi.org/10.1038/s41598-019-40476-y

Keiffer TR, Ciancanelli MJ, Edwards MR, Basler CF (2020) Interactions of the Nipah virus P, V, and W proteins across the STAT family of transcription factors. mSphere 5(6):e00449–e00420. https://doi.org/10.1128/mSphere.00449-20

Kelleni M (2021a) NSAIDs/nitazoxanide/azithromycin immunomodulatory protocol used in adults, geriatric, pediatric, pregnant, and immunocompromised COVID-19 patients: a prospective observational study and case-series. Authorea (Preprint). https://doi.org/10.22541/au.162126601.15715282/v5

Kelleni MT (2021b) NSAIDs/nitazoxanide/azithromycin repurposed for COVID-19: potential mitigation of the cytokine storm interleukin-6 amplifer via immunomodulatory effects. Expert Rev Anti-Infect Ther. https://doi.org/10.1080/14787210.2021.1939683

Kelleni MT (2021c) COVID-19, Ebola virus disease, and Nipah virus infection reclassification as novel acute immune dysrhythmia syndrome (n-AIDS): potential crucial role for immunomodulators. Immunol Res 69(5):457–460. https://doi.org/10.1007/s12026-021-09219

Kenmoe S, Demanou M, Bigna JJ, Nde Kengne C, Fatawou Modiyinji A, Simo FBN, Eyangoh S, Sadeuh-Mba SA, Njouom R (2019) Case fatality rate and risk factors for Nipah virus encephalitis: a systematic review and meta-analysis. J Clin Virol 117:19–26. https://doi.org/10.1016/j.jcv.2019.05.009

Ker DS, Jenkins HT, Greive SJ, Antson AA (2021) CryoEM structure of the Nipah virus nucleocapsid assembly. PLoS Pathog 17(7):e1009740. https://doi.org/10.1371/journal.ppat.1009740

Keshwara R, Shiels T, Postnikova E, Kurup D, Wirblich C, Johnson RF, Schnell MJ (2019) Rabies-based vaccine induces potent immune responses against Nipah virus. NPJ Vaccines 4:15. https://doi.org/10.1038/s41541-019-0109-5

Khan MS, Hossain J, Gurley ES, Nahar N, Sultana R, Luby SP (2010) Use of infrared camera to understand bats' access to date palm sap: implications for preventing Nipah virus transmission. EcoHealth 7(4):517–525. https://doi.org/10.1007/s10393-010-0366-2

Khan SU, Gurley ES, Hossain MJ, Nahar N, Sharker MA, Luby SP (2012) A randomized controlled trial of interventions to impede date palm sap contamination by bats to prevent nipah virus transmission in Bangladesh. PLoS One 7(8):e42689. https://doi.org/10.1371/journal.pone.0042689

Khan SA, Imtiaz MA, Islam MM, Tanzin AZ, Islam A, Hassan MM (2022) Major bat-borne zoonotic viral epidemics in Asia and Africa: a systematic review and meta-analysis. Vet Med Sci 8(4):1787–1801. https://doi.org/10.1002/vms3.835

Khanna M, Kumar B, Gupta N, Kumar P, Gupta A, Vijayan VK, Kaur H (2009) Pandemic swine influenza virus (H1N1): A threatening evolution. Indian J Microbiol 49(4):365–369. https://doi.org/10.1007/s12088-009-0064-3

Kilby JM, Hopkins S, Venetta TM, DiMassimo B, Cloud GA, Lee JY, Alldredge L, Hunter E, Lambert D, Bolognesi D, Matthews T, Johnson MR, Nowak MA, Shaw GM, Saag MS (1998) Potent suppression of HIV-1 replication in humans by T-20, a peptide inhibitor of gp41-mediated virus entry. Nat Med 4(11):1302–1307. https://doi.org/10.1038/3293

Klumpp K, Lévêque V, Le Pogam S, Ma H, Jiang WR, Kang H, Granycome C, Singer M, Laxton C, Hang JQ, Sarma K, Smith DB, Heindl D, Hobbs CJ, Merrett JH, Symons J, Cammack N, Martin JA, Devos R, Nájera I (2006) The novel nucleoside analog R1479 (4′-azidocytidine) is a potent inhibitor of NS5B-dependent RNA synthesis and hepatitis C virus replication in cell culture. J Biol Chem 281(7):3793–3799. https://doi.org/10.1074/jbc.M510195200

Kogan FN (2000) Satellite-observed sensitivity of world land ecosystems to El Nino/La Nina. Remote Sens Environ 74(3):445–462

Kojom LP, Singh V (2021) A review on emerging infectious diseases prioritized under the 2018 WHO research and development blueprint: lessons from the Indian context. Vector Borne Zoonotic Dis 21(3):149–159. https://doi.org/10.1089/vbz.2020.2661

Kolakofsky D, Roux L, Garcin D, Ruigrok RWH (2005) Paramyxovirus mRNA editing, the "rule of six" and error catastrophe: a hypothesis. J Gen Virol 86:1869–1877. https://doi.org/10.1099/vir.0.80986-0

Kong D, Wen Z, Su H, Ge J, Chen W, Wang X, Wu C, Yang C, Chen H, Bu Z (2012) Newcastle disease virus-vectored Nipah encephalitis vaccines induce B and T cell responses in mice and long-lasting neutralizing antibodies in pigs. Virology 432(2):327–335. https://doi.org/10.1016/j.virol.2012.06.001

Kulkarni S, Volchkova V, Basler CF, Palese P, Volchkov VE, Shaw ML (2009) Nipah virus edits its P gene at high frequency to express the V and W proteins. J Virol 83(8):3982–3987. https://doi.org/10.1128/JVI.02599-08

Kulkarni DD, Tosh C, Venkatesh G, Senthil Kumar D (2013) Nipah virus infection: current scenario. Indian J Virol 24(3):398–408. https://doi.org/10.1007/s13337-013-0171-y

Kumar C, Sugunan AP, Yadav P, Kurup KK, Aarathee R, Manickam P, Bhatnagar T, Radhakrishnan C, Thomas B, Kumar A 5th, Jayasree J, Philomina B, Kumar K, Thulaseedharan NK, Gupta N 5th, Rajendran R, Saritha RL, Mourya DT, Gangakhedkar RR, Murhekar MV (2019) Infections among contacts of patients with Nipah virus, India. Emerg Infect Dis 25(5):1007–1010. https://doi.org/10.3201/eid2505.181352

Lalezari JP, Henry K, O'Hearn M, Montaner JS, Piliero PJ, Trottier B, Walmsley S, Cohen C, Kuritzkes DR, Eron JJ Jr, Chung J, DeMasi R, Donatacci L, Drobnes C, Delehanty J, Salgo M, TORO 1 Study Group (2003) Enfuvirtide, an HIV-1 fusion inhibitor, for drug-resistant HIV infection in North and South America. N Engl J Med 348(22):2175–2185. https://doi.org/10.1056/NEJMoa035026

Lam SK, Chua KB (2002) Nipah virus encephalitis outbreak in Malaysia. Clin Infect Dis 34(Suppl 2):S48–S51. https://doi.org/10.1086/338818

Lara A, Cong Y, Jahrling PB, Mednikov M, Postnikova E, Yu S, Munster V, Holbrook MR (2019) Peripheral immune response in the African green monkey model following Nipah-Malaysia virus exposure by intermediate-size particle aerosol. PLoS Negl Trop Dis 13(6):e0007454. https://doi.org/10.1371/journal.pntd.0007454

Latinne A, Morand S (2022) Climate anomalies and spillover of bat-borne viral diseases in the Asia-Pacific Region and the Arabian Peninsula. Viruses 14(5):1100. https://doi.org/10.3390/v14051100

Lawrence P, Escudero-Pérez B (2022) Henipavirus immune evasion and pathogenesis mechanisms: lessons learnt from natural infection and animal models. Viruses 14(5):936. https://doi.org/10.3390/v14050936

Lee JH, Hammoud DA, Cong Y, Huzella LM, Castro MA, Solomon J, Laux J, Lackemeyer M, Bohannon JK, Rojas O, Byrum R, Adams R, Ragland D, St Claire M, Munster V, Holbrook MR (2020) The use of large-particle aerosol exposure to Nipah virus to mimic human neurological disease manifestations in the African Green monkey. J Infect Dis 221(Suppl 4):S419–S430. https://doi.org/10.1093/infdis/jiz502

Leon AJ, Borisevich V, Boroumand N, Seymour R, Nusbaum R, Escaffre O, Xu L, Kelvin DJ, Rockx B (2018) Host gene expression profiles in ferrets infected with genetically distinct henipavirus strains. PLoS Negl Trop Dis 12(3):e0006343. https://doi.org/10.1371/journal.pntd.0006343

Levroney EL, Aguilar HC, Fulcher JA, Kohatsu L, Pace KE, Pang M, Gurney KB, Baum LG, Lee B (2005) Novel innate immune functions for galectin-1: galectin-1 inhibits cell fusion by Nipah virus envelope glycoproteins and augments dendritic cell secretion of proinflammatory cytokines. J Immunol 175(1):413–420. https://doi.org/10.4049/jimmunol.175.1.413

Li H, Mendelsohn E, Zong C, Zhang W, Hagan E, Wang N, Li S, Yan H, Huang H, Zhu G, Ross N, Chmura A, Terry P, Fielder M, Miller M, Shi Z, Daszak P (2019) Human-animal interactions and bat coronavirus spillover potential among rural residents in southern China. Biosaf Health 1(2):84–90. https://doi.org/10.1016/j.bsheal.2019.10.004

Li Y, Li R, Wang M, Liu Y, Yin Y, Zai X, Song X, Chen Y, Xu J, Chen W (2020) Fc-based recombinant henipavirus Vaccines elicit broad neutralizing antibody responses in mice. Viruses 12(4):480. https://doi.org/10.3390/v12040480

Liebl DJ, Morris CJ, Henkemeyer M, Parada LF (2003) mRNA expression of ephrins and Eph receptor tyrosine kinases in the neonatal and adult mouse central nervous system. J Neurosci Res 71(1):7–22. https://doi.org/10.1002/jnr.10457

Liew YJM, Ibrahim PAS, Ong HM, Chong CN, Tan CT, Schee JP, Gómez Román R, Cherian NG, Wong WF, Chang L-Y (2022) The immunobiology of Nipah virus. Microorganisms 10(6):1162. https://doi.org/10.3390/microorganisms10061162

Liu J, Ochieng C, Wiersma S, Ströher U, Towner JS, Whitmer S, Nichol ST, Moore CC, Kersh GJ, Kato C, Sexton C, Petersen J, Massung R, Hercik C, Crump JA, Kibiki G, Maro A, Mujaga B, Gratz J, Jacob ST et al (2016) Development of a TaqMan array card for acute-febrile-illness outbreak investigation and surveillance of emerging pathogens, including Ebola virus. J Clin Microbiol 54(1):49–58. https://doi.org/10.1128/JCM.02257-15

Liu J, Coffin KM, Johnston SC, Babka AM, Bell TM, Long SY, Honko AN, Kuhn JH, Zeng X (2019) Nipah virus persists in the brains of nonhuman primate survivors. JCI Insight 4(14):e129629. https://doi.org/10.1172/jci.insight.129629

Lo MK, Rota PA (2008) The emergence of Nipah virus, a highly pathogenic paramyxovirus. J Clin Virol 43(4):396–400. https://doi.org/10.1016/j.jcv.2008.08.007

Lo MK, Harcourt BH, Mungall BA, Tamin A, Peeples ME, Bellini WJ, Rota PA (2009) Determination of the henipavirus phosphoprotein gene mRNA editing frequencies and detection of the C, V and W proteins of Nipah virus in virus-infected cells. J Gen Virol 90(Pt 2):398–404. https://doi.org/10.1099/vir.0.007294-0

Lo MK, Miller D, Aljofan M, Mungall BA, Rollin PE, Bellini WJ, Rota PA (2010) Characterization of the antiviral and inflammatory responses against Nipah virus in endothelial cells and neurons. Virology 404(1):78–88. https://doi.org/10.1016/j.virol.2010.05.005

Lo MK, Lowe L, Hummel KB, Sazzad HM, Gurley ES, Hossain MJ, Luby SP, Miller DM, Comer JA, Rollin PE, Bellini WJ, Rota PA (2012a) Characterization of Nipah virus from outbreaks in Bangladesh, 2008–2010. Emerg Infect Dis 18(2):248–255. https://doi.org/10.3201/eid1802.111492

Lo MK, Peeples ME, Bellini WJ, Nichol ST, Rota PA, Spiropoulou CF (2012b) Distinct and overlapping roles of Nipah virus P gene products in modulating the human endothelial cell antiviral response. PLoS One 7(10):e47790. https://doi.org/10.1371/journal.pone.0047790

Lo MK, Bird BH, Chattopadhyay A, Drew CP, Martin BE, Coleman JD, Rose JK, Nichol ST, Spiropoulou CF (2014a) Single-dose replication-defective VSV-based Nipah virus vaccines provide protection from lethal challenge in Syrian hamsters. Antivir Res 101:26–29. https://doi.org/10.1016/j.antiviral.2013.10.012

Lo MK, Søgaard TM, Karlin DG (2014b) Evolution and structural organization of the C proteins of paramyxovirinae. PLoS One 9(2):e90003. https://doi.org/10.1371/journal.pone.0090003

Lo MK, Jordan R, Arvey A, Sudhamsu J, Shrivastava-Ranjan P, Hotard AL, Flint M, McMullan LK, Siegel D, Clarke MO, Mackman RL, Hui HC, Perron M, Ray AS, Cihlar T, Nichol ST, Spiropoulou CF (2017) GS-5734 and its parent nucleoside analog inhibit filo-, pneumo-, and paramyxoviruses. Sci Rep 7:43395. https://doi.org/10.1038/srep43395

Lo MK, Jordan PC, Stevens S, Tam Y, Deval J, Nichol ST, Spiropoulou CF (2018) Susceptibility of paramyxoviruses and filoviruses to inhibition by 2′-monofluoro- and 2′-difluoro-4′-azidocytidine analogs. Antivir Res 153:101–113. https://doi.org/10.1016/j.antiviral.2018.03.009

Lo MK, Feldmann F, Gary JM, Jordan R, Bannister R, Cronin J, Patel NR, Klena JD, Nichol ST, Cihlar T, Zaki SR, Feldmann H, Spiropoulou CF, de Wit E (2019) Remdesivir (GS-5734) protects African green monkeys from Nipah virus challenge. Sci Transl Med 11(494):eaau9242. https://doi.org/10.1126/scitranslmed.aau9242

Lo MK, Amblard F, Flint M, Chatterjee P, Kasthuri M, Li C, Russell O, Verma K, Bassit L, Schinazi RF, Nichol ST, Spiropoulou CF (2020a) Potent in vitro activity of β-D-4′-chloromethyl-2′-deoxy-2′-fluorocytidine against Nipah virus. Antivir Res 175:104712. https://doi.org/10.1016/j.antiviral.2020.104712

Lo MK, Spengler JR, Krumpe L, Welch SR, Chattopadhyay A, Harmon JR, Coleman-McCray JD, Scholte F, Hotard AL, Fuqua JL, Rose JK, Nichol ST, Palmer KE, O'Keefe BR, Spiropoulou CF (2020b) Griffithsin inhibits Nipah virus entry and fusion and can protect Syrian Golden hamsters from lethal Nipah virus challenge. J Infect Dis 221(Supplement_4):S480–S492. https://doi.org/10.1093/infdis/jiz630

Looi LM, Chua KB (2007) Lessons from the Nipah virus outbreak in Malaysia. Malays J Pathol 29(2):63–67

Loomis RJ, Stewart-Jones G, Tsybovsky Y, Caringal RT, Morabito KM, McLellan JS, Chamberlain AL, Nugent ST, Hutchinson GB, Kueltzo LA, Mascola JR, Graham BS (2020) Structure-based design of Nipah virus vaccines: a generalizable approach to paramyxovirus immunogen development. Front Immunol 11:842. https://doi.org/10.3389/fimmu.2020.00842

Loomis RJ, DiPiazza AT, Falcone S, Ruckwardt TJ, Morabito KM, Abiona OM, Chang LA, Caringal RT, Presnyak V, Narayanan E, Tsybovsky Y, Nair D, Hutchinson GB, Stewart-Jones G, Kueltzo LA, Himansu S, Mascola JR, Carfi A, Graham BS (2021) Chimeric fusion (F) and attachment (G) glycoprotein antigen delivery by mRNA as a candidate Nipah vaccine. Front Immunol 12:772864. https://doi.org/10.3389/fimmu.2021.772864

Luby SP (2013) The pandemic potential of Nipah virus. Antivir Res 100(1):38–43. https://doi.org/10.1016/j.antiviral.2013.07.011

Luby SP (2014) Viruses: Nipah virus. Encyclopaedia Food Saf 2:214–217

Luby SP, Rahman M, Hossain MJ, Blum LS, Husain MM, Gurley E et al (2006) Foodborne transmission of Nipah virus, Bangladesh. Emerg Infect Dis 12:1888–1894. https://doi.org/10.3201/eid1212.060732

Luby SP, Gurley ES, Hossain MJ (2009) Transmission of human infection with Nipah virus. Clin Infect Dis 49(11):1743–1748. https://doi.org/10.1086/647951

Luby SP, Gurley ES (2012) Epidemiology of henipavirus disease in humans. Curr Top Microbiol Immunol 359:25–40. https://doi.org/10.1007/82_2012_207

Ma XH, Shi Z, Tan C, Jiang Y, Go ML, Low BC, Chen YZ (2010) In-silico approaches to multi-target drug discovery : computer aided multi-target drug design, multi-target virtual screening. Pharm Res 27(5):739–749. https://doi.org/10.1007/s11095-010-0065-2

Ma L, Chen Z, Guan W, Chen Q, Liu D (2019) Rapid and specific detection of all known Nipah virus strains' sequences with reverse transcription-loop-mediated isothermal amplification. Front Microbiol 10:1–10. https://doi.org/10.3389/fmicb.2019.00418

Mackenzie JS, Smith DW (2020) COVID-19: a novel zoonotic disease caused by a coronavirus from China: what we know and what we don't. Microbiol Aust:MA20013. https://doi.org/10.1071/MA20013

Maginnis MS (2018) Virus–receptor interactions: the key to cellular invasion. J Mol Biol 430:2590–2611. https://doi.org/10.1016/j.jmb.2018.06.024

Mahfuz A, Khan MA, Sajib EH, Deb A, Mahmud S, Hasan M, Saha O, Islam A, Rahaman MM (2022) Designing potential siRNA molecules for silencing the gene of the nucleocapsid protein of Nipah virus: a computational investigation. Infect Genet Evol 102:105310. https://doi.org/10.1016/j.meegid.2022.105310

Maisner A, Neufeld J, Weingartl H (2009) Organ- and endotheliotropism of Nipah virus infections in vivo and in vitro. Thromb Haemost 102(6):1014–1023. https://doi.org/10.1160/TH09-05-0310

Majee P, Jain N, Kumar A (2021) Designing of a multi-epitope vaccine candidate against Nipah virus by in silico approach: a putative prophylactic solution for the deadly virus. J Biomol Struct Dyn 39(4):1461–1480. https://doi.org/10.1080/07391102.2020.1734088

Makhoba XH, Viegas C Jr, Mosa RA, Viegas F, Pooe OJ (2020) Potential impact of the multi-target drug approach in the treatment of some complex diseases. Drug Des Devel Ther 14:3235–3249. https://doi.org/10.2147/DDDT.S257494

Malik JA, Mulla AH, Farooqi T, Pottoo FH, Anwar S, Rengasamy K (2021) Targets and strategies for vaccine development against SARS-CoV-2. Biomed Pharmacother 137:111254. https://doi.org/10.1016/j.biopha.2021.111254

Marianneau P, Guillaume V, Wong T, Badmanathan M, Looi RY, Murri S, Loth P, Tordo N, Wild F, Horvat B, Contamin H (2010) Experimental infection of squirrel monkeys with Nipah virus. Emerg Infect Dis 16(3):507–510. https://doi.org/10.3201/eid1603.091346

Mathieu C, Pohl C, Szecsi J, Trajkovic-Bodennec S, Devergnas S, Raoul H, Cosset FL, Gerlier D, Wild TF, Horvat B (2011) Nipah virus uses leukocytes for efficient dissemination within a host. J Virol 85(15):7863–7871. https://doi.org/10.1128/JVI.00549-11

Mathieu C, Guillaume V, Sabine A, Ong KC, Wong KT, Legras-Lachuer C, Horvat B (2012) Lethal Nipah virus infection induces rapid overexpression of CXCL10. PLoS One 7(2):e32157. https://doi.org/10.1371/journal.pone.0032157

Mathieu C, Augusto MT, Niewiesk S, Horvat B, Palermo LM, Sanna G, Madeddu S, Huey D, Castanho MA, Porotto M, Santos NC, Moscona A (2017) Broad spectrum antiviral activity for paramyxoviruses is modulated by biophysical properties of fusion inhibitory peptides. Sci Rep 7:43610. https://doi.org/10.1038/srep43610

Mathieu C, Porotto M, Figueira TN, Horvat B, Moscona A (2018) Fusion inhibitory lipopeptides engineered for prophylaxis of Nipah virus in primates. J Infect Dis 218(2):218–227. https://doi.org/10.1093/infdis/jiy152

Mazzola LT, Kelly-Cirino C (2019) Diagnostics for Nipah virus: a zoonotic pathogen endemic to Southeast Asia. BMJ Glob Health 4(Suppl 2):e001118. https://doi.org/10.1136/bmjgh-2018-001118

McCarthy M (2003) Newer viral encephalitides. Neurologist 9(4):189–199. https://doi.org/10.1097/01.nrl.0000080957.78533.56

McEachern JA, Bingham J, Crameri G, Green DJ, Hancock TJ, Middleton D, Feng YR, Broder CC, Wang LF, Bossart KN (2008) A recombinant subunit vaccine formulation protects against lethal Nipah virus challenge in cats. Vaccine 26(31):3842–3852. https://doi.org/10.1016/j.vaccine.2008.05.016

McEntire CRS, Song KW, McInnis RP, Rhee JY, Young M, Williams E, Wibecan LL, Nolan N, Nagy AM, Gluckstein J, Mukerji SS, Mateen FJ (2021) Neurologic manifestations of the World Health Organization's list of pandemic and epidemic diseases. Front Neurol 12(February). https://doi.org/10.3389/fneur.2021.634827

McIntyre KM, Setzkorn C, Hepworth PJ, Morand S, Morse AP, Baylis M (2017) Systematic assessment of the climate sensitivity of important human and domestic animals pathogens in Europe. Sci Rep 7(1):1–10

McKee CD, Islam A, Luby SP, Salje H, Hudson PJ, Plowright RK, Gurley ES (2021) The ecology of Nipah virus in Bangladesh: a nexus of land-use change and opportunistic feeding behavior in bats. Viruses 13(2):169. https://doi.org/10.3390/v13020169

McLean RK, Graham SP (2019) Vaccine development for Nipah virus infection in pigs. Front Vet Sci 6:16. https://doi.org/10.3389/fvets.2019.00016

McLean RK, Graham SP (2022) The pig as an amplifying host for new and emerging zoonotic viruses. One Health 14:100384. https://doi.org/10.1016/j.onehlt.2022.100384

McMichael AJ (2004) Environmental and social influences on emerging infectious diseases: past, present and future. Philos Trans R Soc Lond Ser B Biol Sci 359(1447):1049–1058. https://doi.org/10.1098/rstb.2004.1480

McPhaden MJ, Santoso A, Cai W (eds) (2020) El Niño southern oscillation in a changing climate, vol 253. John Wiley & Sons

Middleton DJ, Westbury HA, Morrissy CJ, van der Heide BM, Russell GM, Braun MA, Hyatt AD (2002) Experimental Nipah virus infection in pigs and cats. J Comp Pathol 126(2–3):124–136. https://doi.org/10.1053/jcpa.2001.0532

Middleton DJ, Morrissy CJ, van der Heide BM, Russell GM, Braun MA, Westbury HA, Halpin K, Daniels PW (2007) Experimental Nipah virus infection in pteropid bats (Pteropus poliocephalus). J Comp Pathol 136(4):266–272. https://doi.org/10.1016/j.jcpa.2007.03.002

Middleton D, Pallister J, Klein R, Feng YR, Haining J, Arkinstall R, Frazer L, Huang JA, Edwards N, Wareing M, Elhay M, Hashmi Z, Bingham J, Yamada M, Johnson D, White J, Foord A, Heine HG, Marsh GA, Broder CC et al (2014) Hendra virus vaccine, a one health approach to protecting horse, human, and environmental health. Emerg Infect Dis 20(3):372–379. https://doi.org/10.3201/eid2003.131159

Mills JN, Alim AN, Bunning ML, Lee OB, Wagoner KD, Amman BR, Stockton PC, Ksiazek TG (2009) Nipah virus infection in dogs, Malaysia, 1999. Emerg Infect Dis 15(6):950–952. https://doi.org/10.3201/eid1506.080453

Mire CE et al (2013) Single injection recombinant vesicular stomatitis virus vaccines protect ferrets against lethal Nipah virus disease. Virol J 10:353

Mire CE, Satterfield BA, Geisbert JB, Agans KN, Borisevich V, Yan L, Chan YP, Cross RW, Fenton KA, Broder CC, Geisbert TW (2016) Pathogenic differences between Nipah virus Bangladesh and Malaysia strains in primates: implications for antibody therapy. Sci Rep 6:30916. https://doi.org/10.1038/srep30916

Mire CE et al (2019) Use of single-injection recombinant vesicular stomatitis virus vaccine to protect nonhuman primates against lethal Nipah virus disease. Emerg Infect Dis 25:1144–1152

Mire CE, Chan YP, Borisevich V, Cross RW, Yan L, Agans KN, Dang HV, Veesler D, Fenton KA, Geisbert TW, Broder CC (2020) A cross-reactive humanized monoclonal antibody targeting fusion glycoprotein function protects ferrets against lethal Nipah virus and Hendra virus infection. J Infect Dis 221(Suppl 4):S471–S479. https://doi.org/10.1093/infdis/jiz515

Mittal A, Khattri A, Verma V (2022) Structural and antigenic variations in the spike protein of emerging SARS-CoV-2 variants. PLoS Pathog 18(2):e1010260. https://doi.org/10.1371/journal.ppat.1010260

Mohd Nor MN, Gan CH, Ong BL (2000) Nipah virus infection of pigs in peninsular Malaysia. Rev Sci Tech 19(1):160–165. https://doi.org/10.20506/rst.19.1.1202

Molbio Diagnostics Pvt. Ltd (2020) Trueprep® AUTO Universal Sample Pre-treatment Pack. http://www.molbiodiagnostics.com/uploads/product_download/20200901.175931~Trueprep-AUTO-Universal-Sample-pretreatment-Pack-V-04.pdf. Accessed 14 Jan 2020

Montgomery JM, Hossain MJ, Gurley E, Carroll GD, Croisier A, Bertherat E, Asgari N, Formenty P, Keeler N, Comer J, Bell MR, Akram K, Molla AR, Zaman K, Islam MR, Wagoner K, Mills JN, Rollin PE, Ksiazek TG, Breiman RF (2008) Risk factors for Nipah virus encephalitis in Bangladesh. Emerg Infect Dis 14(10):1526–1532. https://doi.org/10.3201/eid1410.060507

Moore JP, Offit PA (2021) SARS-CoV-2 Vaccines and the growing threat of viral variants. JAMA 325(9):821–822. https://doi.org/10.1001/jama.2021.1114

Morand S, Owers KA, Waret-Szkuta A, McIntyre KM, Baylis M (2013) Climate variability and outbreaks of infectious diseases in Europe. Sci Rep 3:1774. https://doi.org/10.1038/srep01774

Mori T, O'Keefe BR, Sowder RC 2nd, Bringans S, Gardella R, Berg S, Cochran P, Turpin JA, Buckheit RW Jr, McMahon JB, Boyd MR (2005) Isolation and characterization of griffithsin, a novel HIV-inactivating protein, from the red alga Griffithsia sp. J Biol Chem 280(10):9345–9353. https://doi.org/10.1074/jbc.M411122200

Mungall BA, Middleton D, Crameri G, Bingham J, Halpin K, Russell G, Green D, McEachern J, Pritchard LI, Eaton BT, Wang LF, Bossart KN, Broder CC (2006) Feline model of acute nipah virus infection and protection with a soluble glycoprotein-based subunit vaccine. J Virol 80(24):12293–12302. https://doi.org/10.1128/JVI.01619-06

Mougari S, Gonzalez C, Reynard O, Horvat B (2022) Fruit bats as natural reservoir of highly pathogenic henipaviruses: balance between antiviral defense and viral toleranceInteractions between Henipaviruses and their natural host, fruit bats. Curr Opin Virol 54:101228. https://doi.org/10.1016/j.coviro.2022.101228

Mounts AW, Kaur H, Parashar UD, Ksiazek TG, Cannon D, Arokiasamy JT, Anderson LJ, Lye MS, Nipah Virus Nosocomial Study Group (2001) A cohort study of health care workers to assess nosocomial transmissibility of Nipah virus, Malaysia, 1999. J Infect Dis 183(5):810–813. https://doi.org/10.1086/318822

Mourya D, Yadav P, Rout M, Pattnaik B, Shete A, Patil D (2019) Absence of Nipah virus antibodies in pigs in Mizoram State, North East India. Indian J Med Res 149(5):677–679. https://doi.org/10.4103/ijmr.IJMR_1086_18

Mulangu S, Dodd LE, Davey RT Jr, Tshiani Mbaya O, Proschan M, Mukadi D, Lusakibanza Manzo M, Nzolo D, Tshomba Oloma A, Ibanda A, Ali R, Coulibaly S, Levine AC, Grais R, Diaz J, Lane HC, Muyembe-Tamfum JJ, PALM Writing Group, Sivahera B, Camara M et al (2019) A randomized, controlled trial of Ebola virus disease therapeutics. N Engl J Med 381(24):2293–2303. https://doi.org/10.1056/NEJMoa1910993

Munster VJ, Prescott JB, Bushmaker T, Long D, Rosenke R, Thomas T, Scott D, Fischer ER, Feldmann H, de Wit E (2012) Rapid Nipah virus entry into the central nervous system of hamsters via the olfactory route. Sci Rep 2:736. https://doi.org/10.1038/srep00736

Nagamine K, Hase T, Notomi T (2002) Accelerated reaction by loop-mediated isothermal amplification using loop primers. Mol Cell Probes 16(3):223–229. https://doi.org/10.1006/mcpr.2002.0415

Nahar N, Paul RC, Sultana R, Gurley ES, Garcia F, Abedin J, Sumon SA, Banik KC, Asaduzzaman M, Rimi NA, Rahman M, Luby SP (2015) Raw sap consumption habits and its association with knowledge of Nipah virus in two endemic districts in Bangladesh. PLoS One 10(11):e0142292. https://doi.org/10.1371/journal.pone.0142292

Naicker PR (2011) The impact of climate change and other factors on zoonotic diseases. Arch Clin Microbiol 2

Narayanan CS (2020) A novel cohort analysis approach to determining the case fatality rate of COVID-19 and other infectious diseases. PLoS One 15(6):e0233146. https://doi.org/10.1371/journal.pone.0233146

Navaratnarajah CK, Generous AR, Yousaf I, Cattaneo R (2020) Receptor-mediated cell entry of paramyxoviruses: mechanisms, and consequences for tropism and pathogenesis. J Biol Chem 295:2771–2786. https://doi.org/10.1074/jbc.REV119.009961

Negrete OA, Levroney EL, Aguilar HC, Bertolotti-Ciarlet A, Nazarian R, Tajyar S, Lee B (2005) EphrinB2 is the entry receptor for Nipah virus, an emergent deadly paramyxovirus. Nature 436(7049):401–405. https://doi.org/10.1038/nature03838

Negrete OA, Wolf MC, Aguilar HC, Enterlein S, Wang W, Mühlberger E, Su SV, Bertolotti-Ciarlet A, Flick R, Lee B (2006) Two key residues in ephrinB3 are critical for its use as an alternative receptor for Nipah virus. PLoS Pathog 2(2):e7. https://doi.org/10.1371/journal.ppat.0020007

Negrete OA, Chu D, Aguilar HC, Lee B (2007) Single amino acid changes in the Nipah and Hendra virus attachment glycoproteins distinguish ephrinB2 from ephrinB3 usage. J Virol 81(19):10804–10814. https://doi.org/10.1128/JVI.00999-07

Nelson DR, Zeuzem S, Andreone P, Ferenci P, Herring R, Jensen DM, Marcellin P, Pockros PJ, Rodríguez-Torres M, Rossaro L, Rustgi VK, Sepe T, Sulkowski M, Thomason IR, Yoshida EM, Chan A, Hill G (2012) Balapiravir plus peginterferon alfa-2a (40KD)/ribavirin in a randomized trial of hepatitis C genotype 1 patients. Ann Hepatol 11(1):15–31

Ng BY, Lim CCT, Yeoh A, Lee WL (2004) Neuropsychiatrie sequelae of Nipah virus encephalitis. J Neuropsychiatry Clin Neurosci 16(4):500–504. https://doi.org/10.1176/jnp.16.4.500

Nguyen NM, Tran CN, Phung LK, Duong KT, Huynh H, Farrar J, Nguyen QT, Tran HT, Nguyen CV, Merson L, Hoang LT, Hibberd ML, Aw PP, Wilm A, Nagarajan N, Nguyen DT, Pham MP, Nguyen TT, Javanbakht H, Klumpp K et al (2013) A randomized, double-blind placebo controlled trial of balapiravir, a polymerase inhibitor, in adult dengue patients. J Infect Dis 207(9):1442–1450. https://doi.org/10.1093/infdis/jis470

Nikolay B, Salje H, Hossain MJ, Khan AKMD, Sazzad HMS, Rahman M, Daszak P, Ströher U, Pulliam JRC, Kilpatrick AM, Nichol ST, Klena JD, Sultana S, Afroj S, Luby SP, Cauchemez S, Gurley ES (2019) Transmission of Nipah virus—14 years of investigations in Bangladesh. N Engl J Med 380:1804–1814

Nikolay B, Salje H, Khan A, Sazzad H, Satter SM, Rahman M, Doan S, Knust B, Flora MS, Luby SP, Cauchemez S, Gurley ES (2020) A framework to monitor changes in transmission and epidemiology of emerging pathogens: lessons from Nipah virus. J Infect Dis 221(Suppl 4):S363–S369. https://doi.org/10.1093/infdis/jiaa074

Nikolay B, Ribeiro dos Santos G, Lipsitch M, Rahman M, Luby SP, Salje H, Gurley ES, Cauchemez S (2021) Assessing the feasibility of Nipah vaccine efficacy trials based on previous outbreaks in Bangladesh. Vaccine 39(39):5600–5606. https://doi.org/10.1016/j.vaccine.2021.08.027

Notomi T, Okayama H, Masubuchi H, Yonekawa T, Watanabe K, Amini N et al (2000) Loop-mediated isothermal amplification of DNA. Nucleic Acids Res 28:E63. https://doi.org/10.1093/nar/28.12.e63

Nulkar G, Bedarkar M, Ghate K, Nulkar S (2021) Hitchhiking microbes: declining biodiversity and emerging zoonoses. Indian J Med Res 153(3):367–374. https://doi.org/10.4103/ijmr.IJMR_620_21

Oany AR, Hossain MU, Ahmad SAI (2015) Computational approach to design a potential SiRNA molecule to silence the Nucleocapsid gene of different Nipah virus strains of Bangladesh. Bioresearch Communications-(BRC) 1(1):40–44. Retrieved from https://www.bioresearch-communications.com/index.php/brc/article/view/171

Okoli GN, Rabbani R, Copstein L, Al-Juboori A, Askin N, Abou-Setta AM (2021) Remdesivir for coronavirus disease 2019 (COVID-19): a systematic review with meta-analysis and trial sequential analysis of randomized controlled trials. Infectious diseases (London, England) 53(9):691–699. https://doi.org/10.1080/23744235.2021.1923799

O'Shea TJ, Cryan PM, Cunningham AA, Fooks AR, Hayman DT, Luis AD, Peel AJ, Plowright RK, Wood JL (2014) Bat flight and zoonotic viruses. Emerg Infect Dis 20(5):741–745. https://doi.org/10.3201/eid2005.130539

Ochani RK, Batra S, Shaikh A, Asad A (2019) Nipah virus—the rising epidemic: a review. Infez Med 27(2):117–127

Ogino T, Green TJ (2019) RNA synthesis and capping by non-segmented negative Strand RNA viral polymerases: lessons from a prototypic virus. Front Microbiol 10:1490. https://doi.org/10.3389/fmicb.2019.01490

OIE (2018) OIE-Listed Diseases, Infections and Infestations in Force in 2018 [Online]. http://www.oie.int/animal-health-in-the-world/ oie-listed-diseases-2018/. Accessed 27 Sept 2018

OIE Nipah Virus [Online] (2018). http://www.oie.int/en/animal-health-in-the-world/animal-diseases/Nipah-Virus/. Accessed 27 Sept 2018

Oli AN, Obialor WO, Ifeanyichukwu MO, Odimegwu DC, Okoyeh JN, Emechebe GO, Adejumo SA, Ibeanu GC (2020) Immunoinformatics and vaccine development: an overview. Immunotargets Ther 9:13–30. https://doi.org/10.2147/ITT.S241064

Olival KJ, Latinne A, Islam A, Epstein JH, Hersch R, Engstrand RC, Gurley ES, Amato G, Luby SP, Daszak P (2020) Population genetics of fruit bat reservoir informs the dynamics, distribution and diversity of Nipah virus. Mol Ecol 29(5):970–985. https://doi.org/10.1111/mec.15288

Ong KC, Wong KT (2015) Henipavirus encephalitis: recent developments and advances. Brain Pathol 25(5):605–613. https://doi.org/10.1111/bpa.12278

Onyango CO, Loparev V, Lidechi S, Bhullar V, Schmid DS, Radford K, Lo MK, Rota P, Johnson BW, Munoz J, Oneko M, Burton D, Black CM, Neatherlin J, Montgomery JM, Fields B (2017) Evaluation of a TaqMan array card for detection of central nervous system infections. J Clin Microbiol 55(7):2035–2044. https://doi.org/10.1128/JCM.02469-16

Openshaw JJ, Hegde S, Sazzad HM, Khan SU, Hossain MJ, Epstein JH et al (2016) Increased morbidity and mortality in domestic animals eating dropped and bitten fruit in Bangladeshi villages: implications for zoonotic disease transmission. EcoHealth 13(1):39–48

Openshaw JJ, Hegde S, Sazzad H, Khan SU, Hossain MJ, Epstein JH, Daszak P, Gurley ES, Luby SP (2017) Bat hunting and bat-human interactions in Bangladeshi villages: implications for zoonotic disease transmission and bat conservation. Transbound Emerg Dis 64(4):1287–1293. https://doi.org/10.1111/tbed.12505

Ortega V, Zamora JLR, Monreal IA, Hoffman DT, Ezzatpour S, Johnston GP, Contreras EM, Vilchez-Delgado FJ, Aguilar HC (2022) Novel roles of the Nipah virus attachment glycoprotein and its mobility in early and late membrane fusion steps. MBio 13(3):e0322221. https://doi.org/10.1128/mbio.03222-21

Ostfeld RS, Glass GE, Keesing F (2005) Spatial epidemiology: an emerging (or re-emerging) discipline. Trends Ecol Evol 20(6):328–336. https://doi.org/10.1016/j.tree.2005.03.009

Pallister JA, Klein R, Arkinstall R, Haining J, Long F, White JR, Payne J, Feng YR, Wang LF, Broder CC, Middleton D (2013) Vaccination of ferrets with a recombinant G glycoprotein subunit vaccine provides protection against Nipah virus disease for over 12 months. Virol J 10:237. https://doi.org/10.1186/1743-422X-10-237

Pallivalappil B, Ali A, Thulaseedharan NK, Karadan U, Chellenton J, Dipu KP, Anoop Kumar AS, Sajeeth Kumar KG, Rajagopal TP, Suraj KP, Santosh Kumar GR, Supreeth RN, Yelisetti M, Muthyala PR, Aryasree KS, Apurva Rao KS (2020) Dissecting an outbreak: a clinico-epidemiological study of Nipah virus infection in Kerala, India, 2018. J Global Infect Dis 12(1):21–27. https://doi.org/10.4103/jgid.jgid_4_19

Parashar UD, Sunn LM, Ong F, Mounts AW, Arif MT, Ksiazek TG, Kamaluddin MA, Mustafa AN, Kaur H, Ding LM, Othman G, Radzi HM, Kitsutani PT, Stockton PC, Arokiasamy J, Gary HE Jr, Anderson LJ (2000) Case-control study of risk factors for human infection with a new zoonotic paramyxovirus, Nipah virus, during a 1998-1999 outbreak of severe encephalitis in Malaysia. J Infect Dis 181(5):1755–1759. https://doi.org/10.1086/315457

Parisien JP, Bamming D, Komuro A, Ramachandran A, Rodriguez JJ, Barber G, Wojahn RD, Horvath CM (2009) A shared interface mediates paramyxovirus interference with antiviral RNA helicases MDA5 and LGP2. J Virol 83:7252–7260

Patel K, Kirkpatrick CM, Nieforth KA, Chanda S, Zhang Q, McClure M, Fry J, Symons JA, Blatt LM, Beigelman L, DeVincenzo JP, Huntjens DR, Smith PF (2019) Respiratory syncytial virus-A dynamics and the effects of lumicitabine, a nucleoside viral replication inhibitor, in experimentally infected humans. J Antimicrob Chemother 74(2):442–452. https://doi.org/10.1093/jac/dky415

Paton NI, Leo YS, Zaki SR, Auchus AP, Lee KE, Ling AE, Chew SK, Ang B, Rollin PE, Umapathi T, Sng I, Lee CC, Lim E, Ksiazek TG (1999) Outbreak of Nipah-virus infection among abattoir workers in Singapore. Lancet 354(9186):1253–1256. https://doi.org/10.1016/S0140-6736(99)04379-2

Peterson AT (2007) Ecological niche modelling and understanding the geography of disease transmission. Vet Ital 43(3):393–400

Paul L (2018) Nipah virus in Kerala: a deadly zoonosis. Clin Microbiol Infect 24(10):1113–1114. https://doi.org/10.1016/j.cmi.2018.06.017

Pedrera M, Macchi F, McLean RK, Franceschi V, Thakur N, Russo L, Medfai L, Todd S, Tchilian EZ, Audonnet JC, Chappell K, Isaacs A, Watterson D, Young PR, Marsh GA, Bailey D, Graham SP, Donofrio G (2020) Bovine herpesvirus-4-vectored delivery of Nipah virus glycoproteins enhances T cell immunogenicity in pigs. Vaccine 8(1):115. https://doi.org/10.3390/vaccines8010115

Peel AJ, Baker KS, Crameri G, Barr JA, Hayman DT, Wright E, Broder CC, Fernández-Loras A, Fooks AR, Wang LF, Cunningham AA, Wood JL (2012) Henipavirus neutralising antibodies in an isolated Island population of African fruit bats. PLoS One 7(1):e30346. https://doi.org/10.1371/journal.pone.0030346

Peel AJ, McKinley TJ, Baker KS, Barr JA, Crameri G, Hayman DT, Feng YR, Broder CC, Wang LF, Cunningham AA, Wood JL (2013) Use of cross-reactive serological assays for detecting novel pathogens in wildlife: assessing an appropriate cutoff for henipavirus assays in African bats. J Virol Methods 193(2):295–303. https://doi.org/10.1016/j.jviromet.2013.06.030

Pelissier R, Iampietro M, Horvat B (2019) Recent advances in the understanding of Nipah virus immunopathogenesis and anti-viral approaches. F1000Research 8:F1000 Faculty Rev-1763. https://doi.org/10.12688/f1000research.19975.1

Perkins MD, Kessel M (2015) What Ebola tells us about outbreak diagnostic readiness. Nat Biotechnol 33(5):464–469. https://doi.org/10.1038/nbt.3215

Pernet O, Wang YE, Lee B (2012) Henipavirus receptor usage and tropism. Curr Top Microbiol Immunol 359:59–78. https://doi.org/10.1007/82_2012_222

Peterson AT (2014) Mapping disease transmission risk. Johns Hopkins University Press, Baltimore, MD

Peterson AT, Soberón J, Pearson RG, Anderson RP, Martínez-Meyer E, Nakamura M, Araújo MB (2011) Ecological niches and geographic distributions. Princeton University Press, Princeton, NJ

Pickering BS, Hardham JM, Smith G, Weingartl ET, Dominowski PJ, Foss DL, Mwangi D, Broder CC, Roth JA, Weingartl HM (2016) Protection against henipaviruses in swine requires both, cell-mediated and humoral immune response. Vaccine 34(40):4777–4786. https://doi.org/10.1016/j.vaccine.2016.08.028

Playford EG, Munro T, Mahler SM, Elliott S, Gerometta M, Hoger KL, Jones ML, Griffin P, Lynch KD, Carroll H, El Saadi D, Gilmour ME, Hughes B, Hughes K, Huang E, de Bakker C, Klein R, Scher MG, Smith IL, Wang LF et al (2020) Safety, tolerability, pharmacokinetics, and immunogenicity of a human monoclonal antibody targeting the G glycoprotein of henipaviruses in healthy adults: a first-in-human, randomised, controlled, phase 1 study. Lancet Infect Dis 20(4):445–454. https://doi.org/10.1016/S1473-3099(19)30634-6

Plowright RK, Parrish CR, McCallum H, Hudson PJ, Ko AI, Graham AL, Lloyd-Smith JO (2017) Pathways to zoonotic spillover. Nat Rev Microbiol 15(8):502–510. https://doi.org/10.1038/nrmicro.2017.45

Plowright RK, Becker DJ, Crowley DE, Washburne AD, Huang T, Nameer PO, Gurley ES, Han BA (2019) Prioritizing surveillance of Nipah virus in India. PLoS Negl Trop Dis 13(6):e0007393. https://doi.org/10.1371/journal.pntd.0007393

Power ML, Foley NM, Jones G, Teeling EC (2021) Taking flight: an ecological, evolutionary and genomic perspective on bat telomeres. Mol Ecol 31:6053–6068. https://doi.org/10.1111/mec.16117

Prasad AN, Agans KN, Sivasubramani SK, Geisbert JB, Borisevich V, Mire CE, Lawrence WS, Fenton KA, Geisbert TW (2020a) A lethal aerosol exposure model of Nipah virus strain Bangladesh in African Green monkeys. J Infect Dis 221(Suppl 4):S431–S435. https://doi.org/10.1093/infdis/jiz469

Prasad AN, Woolsey C, Geisbert JB, Agans KN, Borisevich V, Deer DJ, Mire CE, Cross RW, Fenton KA, Broder CC, Geisbert TW (2020b) Resistance of cynomolgus monkeys to Nipah and Hendra virus disease is associated with cell-mediated and humoral immunity. J Infect Dis 221(Suppl 4):S436–S447. https://doi.org/10.1093/infdis/jiz613

Prescott J, de Wit E, Feldmann H, Munster VJ (2012) The immune response to Nipah virus infection. Arch Virol 157(9):1635–1641. https://doi.org/10.1007/s00705-012-1352-5

Prescott J, DeBuysscher BL, Feldmann F, Gardner DJ, Haddock E, Martellaro C, Scott D, Feldmann H (2015) Single-dose live-attenuated vesicular stomatitis virus-based vaccine protects African green monkeys from Nipah virus disease. Vaccine 33(24):2823–2829. https://doi.org/10.1016/j.vaccine.2015.03.089

Price A, Sizemore D, Hassell T, Gómez Román R, Holst J, Kristiansen P (2021) Nipah virus assays and animal models for vaccine development. Landscape Analysis:1–6. https://media.tghn.org/medialibrary/2021/02/Nipah_Virus_Assays_and_Animal_Models_for_Vaccine_Development_final.pdf

Rahman M, Chakraborty A (2012) Nipah virus outbreaks in Bangladesh: a deadly infectious disease. WHO South East Asia J Public Health 1(2):208–212. https://doi.org/10.4103/2224-3151.206933

Rahman SA, Hassan SS, Olival KJ, Mohamed M, Chang LY, Hassan L, Saad NM, Shohaimi SA, Mamat ZC, Naim MS, Epstein JH, Suri AS, Field HE, Daszak P, Henipavirus Ecology Research Group (2010) Characterization of Nipah virus from naturally infected Pteropus vampyrus bats, Malaysia. Emerg Infect Dis 16(12):1990–1993. https://doi.org/10.3201/eid1612.091790

Rahman MA, Hossain MJ, Sultana S, Homaira N, Khan SU, Rahman M, Gurley ES, Rollin PE, Lo MK, Comer JA, Lowe L, Rota PA, Ksiazek TG, Kenah E, Sharker Y, Luby SP (2012) Date palm sap linked to Nipah virus outbreak in Bangladesh, 2008. Vector Borne Zoonotic Dis 12(1):65–72. https://doi.org/10.1089/vbz.2011.0656

Rahman MZ, Islam MM, Hossain ME, Rahman MM, Islam A, Siddika A, Hossain M, Sultana S, Islam A, Rahman M, Rahman M, Klena JD, Flora MS, Daszak P, Epstein JH, Luby SP, Gurley ES (2021) Genetic diversity of Nipah virus in Bangladesh. Int J Infect Dis 102:144–151. https://doi.org/10.1016/j.ijid.2020.10.041

Rahman MM, Puspo JA, Adib AA, Hossain ME, Alam MM, Sultana S, Islam A, Klena JD, Montgomery JM, Satter SM, Shirin T, Rahman MZ (2022) An Immunoinformatics prediction of novel multi-epitope Vaccines candidate against surface antigens of Nipah virus. Int J Pept Res Ther 28(4):123. https://doi.org/10.1007/s10989-022-10431-z

Rajsbaum R, Versteeg GA, Schmid S, Maestre AM, Belicha-Villanueva A, Martínez-Romero C, Patel JR, Morrison J, Pisanelli G, Miorin L, Laurent-Rolle M, Moulton HM, Stein DA, Fernandez-Sesma A, tenOever BR, García-Sastre A (2014) Unanchored K48-linked polyubiquitin synthesized by the E3-ubiquitin ligase TRIM6 stimulates the interferon-IKKε kinase-mediated antiviral response. Immunity 40(6):880–895. https://doi.org/10.1016/j.immuni.2014.04.018

Ramasundram V et al (2000) Kinetics of IgM and IgG seroconversion in Nipah virus infection. Neurol J Southeast Asia 5:23–28

Ramasundpum, V., Chong Tin, T. A. N., Kaw Bing, C. H. U. A., CHONG, H. T., & Jin, K. (2000). Kinetics of IgM and IgCî seroconversion in Nipah virus infection. Malay, 6, 0.

Ramsay RR, Popovic-Nikolic MR, Nikolic K, Uliassi E, Bolognesi ML (2018) A perspective on multi-target drug discovery and design for complex diseases. Clin Transl Med 7(1):3. https://doi.org/10.1186/s40169-017-0181-2

Randhawa V, Pathania S, Kumar M (2022) Computational identification of potential multitarget inhibitors of Nipah virus by molecular docking and molecular dynamics. Microorganisms 10(6):1181. https://doi.org/10.3390/microorganisms10061181

Ravichandran L, Venkatesan A, Dass FP, J. (2019) Epitope-based immunoinformatics approach on RNAdependent RNA polymerase (RdRp) protein complex of Nipah virus (NiV). J Cell Biochem 120(5):7082–7095. https://doi.org/10.1002/jcb.27979

Ray G, Schmitt PT, Schmitt AP (2016) C-terminal DxD-containing sequences within paramyxovirus Nucleocapsid proteins determine matrix protein compatibility and can direct foreign proteins into budding particles. J Virol 90(7):3650–3660. https://doi.org/10.1128/JVI.02673-15

Reynes JM, Counor D, Ong S, Faure C, Seng V, Molia S, Walston J, Georges-Courbot MC, Deubel V, Sarthou JL (2005) Nipah virus in Lyle's flying foxes, Cambodia. Emerg Infect Dis 11(7):1042–1047. https://doi.org/10.3201/eid1107.041350

Riddell MA, Moss WJ, Hauer D, Monze M, Griffin DE (2007) Slow clearance of measles virus RNA after acute infection. J Clin Virol 39(4):312–317. https://doi.org/10.1016/j.jcv.2007.05.006

Ridenhour B, Kowalik JM, Shay DK (2014) Unraveling R0: considerations for public health applications. Am J Public Health 104(2):e32–e41. https://doi.org/10.2105/AJPH.2013.301704

Rifai SW, Girardin CAJ, Berenguer E, Del Aguila-Pasquel J, Dahlsjö CAL, Doughty CE, Jeffery KJ, Moore S, Oliveras I, Riutta T, Rowland LM, Murakami AA, Addo-Danso SD, Brando P, Burton C, Ondo FE, Duah-Gyamfi A, Amézquita FF, Freitag R, Pacha FH et al (2018) ENSO Drives interannual variation of forest woody growth across the tropics. Philos Trans R Soc Lond Ser B Biol Sci 373(1760):20170410. https://doi.org/10.1098/rstb.2017.0410

Roberts SK, Cooksley G, Dore GJ, Robson R, Shaw D, Berns H, Hill G, Klumpp K, Najera I, Washington C (2008) Robust antiviral activity of R1626, a novel nucleoside analog: a randomized, placebo-controlled study in patients with chronic hepatitis C. Hepatology 48(2):398–406. https://doi.org/10.1002/hep.22321

Rockx B, Brining D, Kramer J, Callison J, Ebihara H, Mansfield K, Feldmann H (2011) Clinical outcome of henipavirus infection in hamsters is determined by the route and dose of infection. J Virol 85(15):7658–7671. https://doi.org/10.1128/JVI.00473-11

Rodriguez KR, Horvath CM (2014) Paramyxovirus V protein interaction with the antiviral sensor LGP2 disrupts MDA5 signaling enhancement but is not relevant to LGP2-mediated RLR signaling inhibition. J Virol 88(14):8180–8188. https://doi.org/10.1128/JVI.00737-14

Rodriguez JJ, Parisien JP, Horvath CM (2002) Nipah virus V protein evades alpha and gamma interferons by preventing STAT1 and STAT2 activation and nuclear accumulation. J Virol 76(22):11476–11483. https://doi.org/10.1128/jvi.76.22.11476-11483.2002

Rodriguez JJ, Wang LF, Horvath CM (2003) Hendra virus V protein inhibits interferon signaling by preventing STAT1 and STAT2 nuclear accumulation. J Virol 77(21):11842–11845. https://doi.org/10.1128/jvi.77.21.11842-11845.2003

Røttingen JA, Gouglas D, Feinberg M, Plotkin S, Raghavan KV, Witty A, Draghia-Akli R, Stoffels P, Piot P (2017) New vaccines against epidemic infectious diseases. N Engl J Med 376(7):610–613. https://doi.org/10.1056/NEJMp1613577

Sahay RR, Yadav PD, Gupta N, Shete AM, Radhakrishnan C, Mohan G, Menon N, Bhatnagar T, Suma K, Kadam AV, Ullas PT, Anu Kumar B, Sugunan AP, Sreekala VK, Khobragade R, Gangakhedkar RR, Mourya DT (2020) Experiential learnings from the Nipah virus outbreaks in Kerala towards containment of infectious public health emergencies in India. Epidemiol Infect 148:e90. https://doi.org/10.1017/S0950268820000825

Saini S, Thakur CJ, Kumar V, Tandon S, Bhardwaj V, Maggar S, Namgyal S, Kaur G (2018) Computational prediction of miRNAs in Nipah virus genome reveals possible interaction with human genes involved in encephalitis. Mol Biol Res Commun 7(3):107–118. https://doi.org/10.22099/mbrc.2018.29577.1322

Sakib MS, Islam MR, Hasan AK, Nabi AH (2014) Prediction of epitope-based peptides for the utility of vaccine development from fusion and glycoprotein of nipah virus using in silico approach. Adv Bioinforma 2014:402492. https://doi.org/10.1155/2014/402492

Saha CK, Mahbub Hasan M, Saddam Hossain M, Asraful Jahan M, Azad AK (2017) In silico identification and characterization of common epitope-based peptide vaccine for Nipah and Hendra viruses. Asian Pac J Trop Med 10(6):529–538. https://doi.org/10.1016/j.apjtm.2017.06.016

Salvatori G, Luberto L, Maffei M, Aurisicchio L, Roscilli G, Palombo F, Marra E (2020) SARS-CoV-2 SPIKE PROTEIN: an optimal immunological target for vaccines. J Transl Med 18(1):222. https://doi.org/10.1186/s12967-020-02392-y

Sanjuán R, Domingo-Calap P (2016) Mechanisms of viral mutation. Cell Mol Life Sci 73(23):4433–4448. https://doi.org/10.1007/s00018-016-2299-6

Satterfield BA, Cross RW, Fenton KA, Agans KN, Basler CF, Geisbert TW, Mire CE (2015) The immunomodulating V and W proteins of Nipah virus determine disease course. Nat Commun 6:7483. https://doi.org/10.1038/ncomms8483

Satterfield BA, Cross RW, Fenton KA, Borisevich V, Agans KN, Deer DJ, Graber J, Basler CF, Geisbert TW, Mire CE (2016) Nipah Virus C and W Proteins Contribute to Respiratory Disease in Ferrets. J Virol 90(14):6326–6343. https://doi.org/10.1128/JVI.00215-16

Sayed A, Bottu A, Qaisar M, Mane MP, Acharya Y (2019) Nipah virus: a narrative review of viral characteristics and epidemiological determinants. Public Health 173:97–104. https://doi.org/10.1016/j.puhe.2019.05.019

Sazzad HM, Hossain MJ, Gurley ES, Ameen KM, Parveen S, Islam MS, Faruque LI, Podder G, Banu SS, Lo MK, Rollin PE, Rota PA, Daszak P, Rahman M, Luby SP (2013) Nipah virus infection outbreak with nosocomial and corpse-to-human transmission, Bangladesh. Emerg Infect Dis 19(2):210–217. https://doi.org/10.3201/eid1902.120971

Schountz T (2014) Immunology of bats and their viruses: challenges and opportunities. Viruses 6(12):4880–4901. https://doi.org/10.3390/v6124880

Schountz T, Baker ML, Butler J, Munster V (2017) Immunological control of viral infections in bats and the emergence of viruses highly pathogenic to humans. Front Immunol 8:1098. https://doi.org/10.3389/fimmu.2017.01098

Schountz T, Campbell C, Wagner K, Rovnak J, Martellaro C, DeBuysscher BL, Feldmann H, Prescott J (2019) Differential Innate Immune Responses Elicited by Nipah Virus and Cedar Virus Correlate with Disparate In Vivo Pathogenesis in Hamsters. Viruses 11(3):291. https://doi.org/10.3390/v11030291

Scott PD, Meng B, Marriott AC, Easton AJ, Dimmock NJ (2011a) Defective interfering influenza A virus protects in vivo against disease caused by a heterologous influenza B virus. J Gen Virol 92(Pt 9):2122–2132. https://doi.org/10.1099/vir.0.034132-0

Scott PD, Meng B, Marriott AC, Easton AJ, Dimmock NJ (2011b) Defective interfering influenza virus confers only short-lived protection against influenza virus disease: evidence for a role for adaptive immunity in DI virus-mediated protection in vivo. Vaccine 29(38):6584–6591. https://doi.org/10.1016/j.vaccine.2011.06.114

Sejvar JJ, Hossain J, Saha SK, Gurley ES, Banu S, Hamadani JD, Faiz MA, Siddiqui FM, Mohammad QD, Mollah AH, Uddin R, Alam R, Rahman R, Tan CT, Bellini W, Rota P,

Breiman RF, Luby SP (2007) Long-term neurological and functional outcome in Nipah virus infection. Ann Neurol 62(3):235–242. https://doi.org/10.1002/ana.21178

Sendow I, Ratnawati A, Taylor T, Adjid RM, Saepulloh M, Barr J, Wong F, Daniels P, Field H (2013) Nipah virus in the fruit bat Pteropus vampyrus in Sumatera, Indonesia. PLoS One 8(7):e69544. https://doi.org/10.1371/journal.pone.0069544

Serra-Cobo J, López-Roig M (2017) Bats and emerging infections: an ecological and virological puzzle. Adv Exp Med Biol 972:35–48. https://doi.org/10.1007/5584_2016_131

Seto J, Qiao L, Guenzel CA, Xiao S, Shaw ML, Hayot F, Sealfon SC (2010) Novel Nipah virus immune-antagonism strategy revealed by experimental and computational study. J Virol 84(21):10965–10973. https://doi.org/10.1128/JVI.01335-10

Sharma V, Kaushik S, Kumar R, Yadav JP, Kaushik S (2019) Emerging trends of Nipah virus: a review. Rev Med Virol 29(1):e2010. https://doi.org/10.1002/rmv.2010

Shaw ML, García-Sastre A, Palese P, Basler CF (2004) Nipah virus V and W proteins have a common STAT1-binding domain yet inhibit STAT1 activation from the cytoplasmic and nuclear compartments, respectively. J Virol 78(11):5633–5641. https://doi.org/10.1128/JVI.78.11.5633-5641.2004

Shaw ML, Cardenas WB, Zamarin D, Palese P, Basler CF (2005) Nuclear localization of the Nipah virus W protein allows for inhibition of both virus- and toll-like receptor 3-triggered signaling pathways. J Virol 79(10):6078–6088. https://doi.org/10.1128/JVI.79.10.6078-6088.2005

Sheahan TP, Sims AC, Graham RL, Menachery VD, Gralinski LE, Case JB, Leist SR, Pyrc K, Feng JY, Trantcheva I, Bannister R, Park Y, Babusis D, Clarke MO, Mackman RL, Spahn JE, Palmiotti CA, Siegel D, Ray AS, Cihlar T et al (2017) Broad-spectrum antiviral GS-5734 inhibits both epidemic and zoonotic coronaviruses. Sci Transl Med 9(396):eaal3653. https://doi.org/10.1126/scitranslmed.aal3653

Shen YY, Liang L, Zhu ZH, Zhou WP, Irwin DM, Zhang YP (2010) Adaptive evolution of energy metabolism genes and the origin of flight in bats. Proc Natl Acad Sci USA 107(19):8666–8671. https://doi.org/10.1073/pnas.0912613107

Sherrini BA, Chong TT (2014) Nipah encephalitis—an update. Med J Malaysia 69 Suppl A:103–111

Simsek C, Corman VM, Everling HU, Lukashev AN, Rasche A, Maganga GD, Binger T, Jansen D, Beller L, Deboutte W, Gloza-Rausch F, Seebens-Hoyer A, Yordanov S, Sylverken A, Oppong S, Sarkodie YA, Vallo P, Leroy EM, Bourgarel M, Yinda KC et al (2021) At least seven distinct rotavirus genotype constellations in bats with evidence of reassortment and zoonotic transmissions. mBio 12(1):e02755–e02720. https://doi.org/10.1128/mBio.02755-20

Singh RK, Dhama K, Chakraborty S, Tiwari R, Natesan S, Khandia R, Munjal A, Vora KS, Latheef SK, Karthik K, Singh Malik Y, Singh R, Chaicumpa W, Mourya DT (2019) Nipah virus: epidemiology, pathology, immunobiology and advances in diagnosis, vaccine designing and control strategies–a comprehensive review. Vet Q 39(1):26–55. https://doi.org/10.1080/01652176.2019.1580827

Singhai M, Jain R, Jain S, Bala M, Singh S, Goyal R (2021) Nipah virus disease: recent perspective and one health approach. Ann Glob Health 87(1):102. https://doi.org/10.5334/aogh.3431

Sizemore, D.R. (2021). Nipah Virus Assays and Animal Models for Vaccine Development.

Skowron K, Bauza-Kaszewska J, Grudlewska-Buda K, Wiktorczyk-Kapischke N, Zacharski M, Bernaciak Z, Gospodarek-Komkowska E (2022) Nipah virus–another threat from the world of zoonotic viruses. Front Microbiol 12:811157. https://doi.org/10.3389/fmicb.2021.811157

Smither SJ, Eastaugh LS, Findlay JS, O'Brien LM, Thom R, Lever MS (2019) Survival and persistence of Nipah virus in blood and tissue culture media. Emerg Microbes Infect 8(1):1760–1762. https://doi.org/10.1080/22221751.2019.1698272

Snell NJ (2001) Ribavirin—current status of a broad spectrum antiviral agent. Expert Opin Pharmacother 2(8):1317–1324. https://doi.org/10.1517/14656566.2.8.1317

Snell NJ (2004) Ribavirin therapy for Nipah virus infection. J Virol 78(18):10211. https://doi.org/10.1128/JVI.78.18.10211.2004

Soberon J, Peterson AT (2005) Interpretation of models of fundamental ecological niches and species' distributional areas. Biodivers Inform 2:1–10

Soltan MA, Eldeen MA, Elbassiouny N, Mohamed I, El-Damasy DA, Fayad E, Ali OAA, Raafat N, Eid RA, Al-Karmalawy AA (2021) Proteome based approach defines candidates for designing a multitope vaccine against the Nipah virus. Int J Mol Sci 22(17):9330. https://doi.org/10.3390/ijms22179330

Soman Pillai V, Krishna G, Valiya Veettil M (2020) Nipah virus: past outbreaks and future containment. Viruses 12(4):465. https://doi.org/10.3390/v12040465

Spiclera A. R. (2016). Nipah Virus Infection.. Available online at: https://www.cfsph.iastate.edu/Factsheets/pdfs/nipah.pdf

Springer MS, Teeling EC, Madsen O, Stanhope MJ, de Jong WW (2001) Integrated fossil and molecular data reconstruct bat echolocation. Proc Natl Acad Sci USA 98(11):6241–6246. https://doi.org/10.1073/pnas.111551998

Stachowiak B, Weingartl HM (2012) Nipah virus infects specific subsets of porcine peripheral blood mononuclear cells. PLoS One 7(1):e30855. https://doi.org/10.1371/journal.pone.0030855

Stamatovic SM, Shakui P, Keep RF, Moore BB, Kunkel SL, Van Rooijen N, Andjelkovic AV (2005) Monocyte chemoattractant protein-1 regulation of blood-brain barrier permeability. J Cereb Blood Flow Metab 25(5):593–606. https://doi.org/10.1038/sj.jcbfm.9600055

Sudeep AB, Yadav PD, Gokhale MD, Balasubramanian R, Gupta N, Shete A, Jain R, Patil S, Sahay RR, Nyayanit DA, Gopale S, Pardeshi PG, Majumdar TD, Patil DR, Sugunan AP, Mourya DT (2021) Detection of Nipah virus in Pteropus medius in 2019 outbreak from Ernakulam district, Kerala, India. BMC Infect Dis 21(1):162. https://doi.org/10.1186/s12879-021-05865-7

Sugai A, Sato H, Takayama I, Yoneda M, Kai C (2017) Nipah and Hendra virus nucleoproteins inhibit nuclear accumulation of signal transducer and activator of transcription 1 (STAT1) and STAT2 by interfering with their complex formation. J Virol 91(21):e01136–e01117. https://doi.org/10.1128/JVI.01136-17

Sui Y, Potula R, Dhillon N, Pinson D, Li S, Nath A, Anderson C, Turchan J, Kolson D, Narayan O, Buch S (2004) Neuronal apoptosis is mediated by CXCL10 overexpression in simian human immunodeficiency virus encephalitis. Am J Pathol 164(5):1557–1566. https://doi.org/10.1016/S0002-9440(10)63714-5

Sui Y, Stehno-Bittel L, Li S, Loganathan R, Dhillon NK, Pinson D, Nath A, Kolson D, Narayan O, Buch S (2006) CXCL10-induced cell death in neurons: role of calcium dysregulation. Eur J Neurosci 23(4):957–964. https://doi.org/10.1111/j.1460-9568.2006.04631.x

Sun B, Jia L, Liang B, Chen Q, Liu D (2018) Phylogeography, transmission, and viral proteins of Nipah virus. Virol Sin 33(5):385–393. https://doi.org/10.1007/s12250-018-0050-1

Tamin A, Harcourt BH, Lo MK, Roth JA, Wolf MC, Lee B, Weingartl H, Audonnet JC, Bellini WJ, Rota PA (2009) Development of a neutralization assay for Nipah virus using pseudotype particles. J Virol Methods 160(1–2):1–6. https://doi.org/10.1016/j.jviromet.2009.02.025

Tan CT, Tan KS (2001) Nosocomial transmissibility of Nipah virus. J Infect Dis 184(10):1367. https://doi.org/10.1086/323996

Tan CT, Wong KT (2003) Nipah encephalitis outbreak in Malaysia. Ann Acad Med Singap 32(1):112–117

Tan KS, Tan CT, Goh KJ (1999) Epidemiological aspects of Nipah virus infection. Neurol J Southeast Asia 4:77–81

Tan K, Sarji SA, Tan C-T, Abdullah B, Chong H, Thayaparan T et al (2000) Patients with asymptomatic Nipah virus infection may have abnormal cerebral MR imaging. Neurol J Southeast Asia 5:69–73

Tan CT, Goh KJ, Wong KT, Sarji SA, Chua KB, Chew NK, Murugasu P, Loh YL, Chong HT, Tan KS, Thayaparan T, Kumar S, Jusoh MR (2002) Relapsed and late-onset Nipah encephalitis. Ann Neurol 51(6):703–708. https://doi.org/10.1002/ana.10212

Taylor H, Campbell J, Nobes CD (2017) Ephs and ephrins. Curr Biol 27:R90–R95. https://doi.org/10.1016/j.cub.2017.01.003

Teeling EC, Vernes SC, Dávalos LM, Ray DA, Gilbert MTP, Myers E, Bat1K Consortium (2018) Bat biology, genomes, and the Bat1K project: to generate chromosome-level genomes for all living bat species. Ann Rev Anim Biosci 6:23–46

Thakur N, Bailey D (2019) Advances in diagnostics, vaccines and therapeutics for Nipah virus. Microbes Infect 21(7):278–286. https://doi.org/10.1016/j.micinf.2019.02.002

The Lancet (2020) Zoonoses: beyond the human-animal-environment interface. Lancet 396(10243):1. https://doi.org/10.1016/S0140-6736(20)31486-0

Thiagarajan K. December 11,2021;12:2021. Why the world should be more than a bit worried about India's Nipah virus outbreak. NPR, September .https://www.npr.org/sections/goatsandsoda/2021/09/12/1035571714/why-the-world-should-be-more-than-a-bit-worriedabout-indias-nipah-virus-outbreak

Thomas SP, Suthers RA (1972) The physiology and energetics of bat flight. J Exp Biol 57:317–335. https://doi.org/10.1242/jeb.57.2.317

Thomas B, Chandran P, Lilabi MP, George B, Sivakumar CP, Jayadev VK, Bindu V, Rajasi RS, Vijayan B, Mohandas A, Hafeez N (2019) Nipah virus infection in Kozhikode, Kerala, South India, in 2018: epidemiology of an outbreak of an emerging disease. Indian J Community Med 44(4):383–387. https://doi.org/10.4103/ijcm.IJCM_198_19

Tian J, Sun J, Li D, Wang N, Wang L, Zhang C, Meng X, Ji X, Suchard MA, Zhang X, Lai A, Su S, Veit M (2022) Emerging viruses: cross-species transmission of coronaviruses, filoviruses, henipaviruses, and rotaviruses from bats. Cell Rep 39(11):110969. https://doi.org/10.1016/j.celrep.2022.110969

Tiong V, Lam CW, Phoon WH, AbuBakar S, Chang LY (2017) Serum from Nipah virus patients recognises recombinant viral proteins produced in Escherichia coli. Jpn J Infect Dis 70(1):26–31. https://doi.org/10.7883/yoken.JJID.2015.501

Tiong V, Shu MH, Wong WF, AbuBakar S, Chang LY (2018) Nipah virus infection of immature dendritic cells increases its transendothelial migration across human brain microvascular endothelial cells. Front Microbiol 9:2747. https://doi.org/10.3389/fmicb.2018.02747

Tomita N, Mori Y, Kanda H, Notomi T (2008) Loop-mediated isothermal amplification (LAMP) of gene sequences and simple visual detection of products. Nat Protoc 3(5):877–882. https://doi.org/10.1038/nprot.2008.57

Torres-Velez FJ, Shieh W-J, Rollin PE et al (2008) Histopathologic and Immunohistochemical Characterization of Nipah Virus Infection in the Guinea Pig. Vet Pathol 45(4):576–585. https://doi.org/10.1354/vp.45-4-576

United Nations Environment Programme (2016) UNEP Frontiers 2016 Report: Emerging Issues of Environmental Concern. https://wedocs.unep.org/20.500.11822/7664

Valbuena G, Halliday H, Borisevich V, Goez Y, Rockx B (2014) A human lung xenograft mouse model of Nipah virus infection. PLoS Pathog 10(4):e1004063. https://doi.org/10.1371/journal.ppat.1004063

Veklury (remdesivir) EUA Letter of Approval (10/22/2020). FDA. 2020. Reference Source 45, Okoli GN, Rabbani R, Copstein L et al (2021) Remdesivir for coronavirus disease 2019 (COVID-19): A systematic review with meta-analysis and trial sequential analysis of randomized controlled trials. Infect Dis (Lond) 53(9):691–699

Viana JD et al (2018) Drug discovery and computational strategies in the multitarget drugs era. Brazil J Pharmac Sci 54:e01010. https://doi.org/10.1590/s2175-97902018000001010

Vijayalakshmi J, Surekha A, Devi AR, Devi SU (2019) Truenat—A novel diagnostic tool for rapid detection of mycobacterium tuberculosis and rifampicin resistance in pulmonary samples. Int J Curr Microbiol App Sci 8:1260–1267

Vignali DA (2000) Multiplexed particle-based flow cytometric assays. J Immunol Methods 243(1-2):243–255. https://doi.org/10.1016/s0022-1759(00)00238-6

Wacharapluesadee S, Hemachudha T (2007) Duplex nested RT-PCR for detection of Nipah virus RNA from urine specimens of bats. J Virol Methods 141(1):97–101. https://doi.org/10.1016/j.jviromet.2006.11.023

Wacharapluesadee S, Ghai S, Duengkae P, Manee-Orn P, Thanapongtharm W, Saraya AW, Yingsakmongkon S, Joyjinda Y, Suradhat S, Ampoot W, Nuansrichay B, Kaewpom T, Tantilertcharoen R, Rodpan A, Wongsathapornchai K, Ponpinit T, Buathong R, Bunprakob S, Damrongwatanapokin S et al (2021) Two decades of one health surveillance of Nipah virus in Thailand. One Health Outlook 3(1):12. https://doi.org/10.1186/s42522-021-00044-9

Wahiduzzaman M, Luo JJ (2021) A statistical analysis on the contribution of El Niño–Southern Oscillation to the rainfall and temperature over Bangladesh. Meteorog Atmos Phys 133(1):55–68
Wang LF, Daniels P (2012) Diagnosis of henipavirus infection: current capabilities and future directions. Curr Top Microbiol Immunol 359:179–196. https://doi.org/10.1007/82_2012_215
Wang G, Deval J, Hong J, Dyatkina N, Prhavc M, Taylor J, Fung A, Jin Z, Stevens SK, Serebryany V, Liu J, Zhang Q, Tam Y, Chanda SM, Smith DB, Symons JA, Blatt LM, Beigelman L (2015) Discovery of 4′-chloromethyl-2′-deoxy-3′,5′-di-O-isobutyryl-2′-fluorocytidine (ALS-8176), a first-in-class RSV polymerase inhibitor for treatment of human respiratory syncytial virus infection. J Med Chem 58(4):1862–1878. https://doi.org/10.1021/jm5017279
Wang LF, Gamage AM, Chan WOY, Hiller M, E.C. (2021) Teeling decoding bat immunity: the need for a coordinated research approach. Nat Rev Immunol 21:269–271. https://doi.org/10.1038/s41577-021-00523-0
Wang Z, Amaya M, Addetia A, Dang HV, Reggiano G, Yan L, Hickey AC, DiMaio F, Broder CC, Veesler D (2022) Architecture and antigenicity of the Nipah virus attachment glycoprotein. Science 375(6587):1373–1378. https://doi.org/10.1126/science.abm5561
Watanabe S, Fukushi S, Harada T, Shimojima M, Yoshikawa T, Kurosu T, Kaku Y, Morikawa S, Saijo M (2020) Effective inactivation of Nipah virus in serum samples for safe processing in low-containment laboratories. Virol J 17(1):151. https://doi.org/10.1186/s12985-020-01425-8
Weingartl H, Czub S, Copps J, Berhane Y, Middleton D, Marszal P, Gren J, Smith G, Ganske S, Manning L, Czub M (2005) Invasion of the central nervous system in a porcine host by nipah virus. J Virol 79(12):7528–7534. https://doi.org/10.1128/JVI.79.12.7528-7534.2005
Weingartl HM, Berhane Y, Caswell JL, Loosmore S, Audonnet JC, Roth JA, Czub M (2006) Recombinant nipah virus vaccines protect pigs against challenge. J Virol 80(16):7929–7938. https://doi.org/10.1128/JVI.00263-06
Welch SR, Tilston NL, Lo MK, Whitmer S, Harmon JR, Scholte F, Spengler JR, Duprex WP, Nichol ST, Spiropoulou CF (2020) Inhibition of Nipah virus by defective interfering particles. J Infect Dis 221(Suppl 4):S460–S470. https://doi.org/10.1093/infdis/jiz564
Welch SR, Spengler JR, Harmon JR, Coleman-McCray JAD, Scholte FEM, Genzer SC, Lo MK, Montgomery JM, Nichol ST, Spiropoulou CF (2022) Defective interfering viral particle treatment reduces clinical signs and protects hamsters from lethal Nipah virus disease. MBio 13(2):1–15. https://doi.org/10.1128/mbio.03294-21
WHO (2019). - Nipah Virus Outbreak in Kerala https://www.who.int/southeastasia/outbreaks-and-emergencies/health-emergencyinformation-risk-assessment/surveillance-and-risk-assessment/nipah-virus-outbreak-in-kerala
WHO (2021a) Nipah virus disease—India (17 Sep 2021). https://www.who.int/emergencies/diseaseoutbreak-news/item/nipah-virus-disease-india. Accessed 10 December 2021
WHO (2021b) Outbreak of Nipah virus encephalitis in the Kerala state of India (01 Jun 2018). https://www.who.int/southeastasia/outbreaks-and-emergencies/health-emergency-informationrisk-assessment/surveillance-and-risk-assessment/nipah-virus-outbreak-in-kerala. Accessed 10 Dec 2021
Widerspick L, Vázquez CA, Niemetz L, Heung M, Olal C, Bencsik A, Henkel C, Pfister A, Brunetti JE, Kucinskaite-Kodze I, Lawrence P, Muñoz Fontela C, Diederich S, Escudero-Pérez B (2022) Inactivation methods for experimental Nipah virus infection. Viruses 14(5):1052. https://doi.org/10.3390/v14051052
Wilkinson GS, South JM (2002) Life history, ecology and longevity in bats. Aging Cell 1(2):124–131. https://doi.org/10.1046/j.1474-9728.2002.00020.x
Wilkinson GS, Adams DM, Haghani A, Lu AT, Zoller J, Breeze CE, Arnold BD, Ball HC, Carter GG, Cooper LN, Dechmann D, Devanna P, Fasel NJ, Galazyuk AV, Günther L, Hurme E, Jones G, Knörnschild M, Lattenkamp EZ, Li CZ et al (2021) DNA methylation predicts age and provides insight into exceptional longevity of bats. Nat Commun 12(1):1615. https://doi.org/10.1038/s41467-021-21900-2
Wilson A, Warrier A, Rathish B (2020) Contact tracing: a lesson from the Nipah virus in the time of COVID-19. Trop Dr 50(3):174–175. https://doi.org/10.1177/0049475520928217

Wolfe ND, Daszak P, Kilpatrick AM, Burke DS (2005) Bushmeat hunting, deforestation, and prediction of zoonoses emergence. Emerg Infect Dis 11(12):1822–1827. https://doi.org/10.3201/eid1112.040789

Wolfe ND, Dunavan CP, Diamond J (2007) Origins of major human infectious diseases. Nature 447(7142):279–283. https://doi.org/10.1038/nature05775

Wong KT, Tan CT (2012) Clinical and pathological manifestations of human henipavirus infection. Curr Top Microbiol Immunol 359:95–104. https://doi.org/10.1007/82_2012_205

Wong KT, Shieh WJ, Zaki SR, Tan CT (2002) Nipah virus infection, an emerging paramyxoviral zoonosis. Springer Semin Immunopathol 24(2):215–228. https://doi.org/10.1007/s00281-002-0106-y

Wong KT, Grosjean I, Brisson C, Blanquier B, Fevre-Montange M, Bernard A, Loth P, Georges-Courbot MC, Chevallier M, Akaoka H, Marianneau P, Lam SK, Wild TF, Deubel V (2003) A golden hamster model for human acute Nipah virus infection. Am J Pathol 163(5):2127–2137. https://doi.org/10.1016/S0002-9440(10)63569-9

Wong K, Robertson T, Ong B, Chong J, Yaiw K, Wang L, Ansford A, Tannenberg A (2009) Human Hendra virus infection causes acute and relapsing encephalitis. Neuropathol Appl Neurobiol 35:296–305

Woon AP, Boyd V, Todd S, Smith I, Klein R, Woodhouse IB, Riddell S, Crameri G, Bingham J, Wang LF, Purcell AW, Middleton D, Baker ML (2020) Acute experimental infection of bats and ferrets with Hendra virus: insights into the early host response of the reservoir host and susceptible model species. PLoS Pathog 16(3):e1008412. https://doi.org/10.1371/journal.ppat.1008412

World Health Organization [WHO] (2021) R&D Blueprint and Nipah Virus. https://www.who.int/teams/blueprint/nipah. Accessed 1 Sept 2021

World Health Organization. One Health (2017). https://www.who.int/news-room/q-a-detail/one--health. Accessed 20 Feb 2021

Wunderman R (2013) (ed.) Global Volcanism Program, 2013. Report on Mayon (Philippines). Bulletin of the Global Volcanism Network, 38:4. Smithsonian Institution. https://doi.org/10.5479/si.GVP.BGVN201304-273030

Xu K, Rajashankar KR, Chan YP, Himanen JP, Broder CC, Nikolov DB (2008) Host cell recognition by the henipaviruses: crystal structures of the Nipah G attachment glycoprotein and its complex with ephrin-B3. Proc Natl Acad Sci USA 105(29):9953–9958. https://doi.org/10.1073/pnas.0804797105

Xu K, Broder CC, Nikolov DB (2012) Ephrin-B2 and ephrin-B3 as functional henipavirus receptors. Semin Cell Dev Biol 23(1):116–123. https://doi.org/10.1016/j.semcdb.2011.12.005

Xu K, Rockx B, Xie Y, DeBuysscher BL, Fusco DL, Zhu Z, Chan YP, Xu Y, Luu T, Cer RZ, Feldmann H, Mokashi V, Dimitrov DS, Bishop-Lilly KA, Broder CC, Nikolov DB (2013) Crystal structure of the Hendra virus attachment G glycoprotein bound to a potent cross-reactive neutralizing human monoclonal antibody. PLoS Pathog 9(10):e1003684. https://doi.org/10.1371/journal.ppat.1003684

Yadav PD, Raut CG, Shete AM, Mishra AC, Towner JS, Nichol ST, Mourya DT (2012) Detection of Nipah virus RNA in fruit bat (Pteropus giganteus) from India. Am J Trop Med Hyg 87(3):576–578. https://doi.org/10.4269/ajtmh.2012.11-0416

Yadav P, Sudeep A, Gokhale M, Pawar S, Shete A, Patil D, Kumar V, Lakra R, Sarkale P, Nichol S, Mourya D (2018) Circulation of Nipah virus in *Pteropus giganteus* bats in northeast region of India, 2015. Indian J Med Res 147(3):318–320. https://doi.org/10.4103/ijmr.IJMR_1488_16

Yadav PD, Shete AM, Kumar GA, Sarkale P, Sahay RR, Radhakrishnan C, Lakra R, Pardeshi P, Gupta N, Gangakhedkar RR, Rajendran VR, Sadanandan R, Mourya DT (2019) Nipah virus sequences from humans and bats during Nipah outbreak, Kerala, India, 2018. Emerg Infect Dis 25(5):1003–1006. https://doi.org/10.3201/eid2505.181076

Yadav PD, Majumdar T, Gupta N, Kumar MA, Shete A, Pardeshi P, Sultana S, Sahay RR, Manoj MN, Patil S, Floura S, Gangakhedkar R, Mourya DT (2021) Standardization & validation of Truenat™ point-of-care test for rapid diagnosis of Nipah. Indian J Med Res 154(4):645–649. https://doi.org/10.4103/ijmr.IJMR_4717_20

Yadav PD, Sahay RR, Balakrishnan A, Mohandas S, Radhakrishnan C, Gokhale MD, Balasubramanian R, Abraham P, Gupta N, Sugunan AP, Khobragade R, George K, Shete A, Patil S, Thankappan UP, Dighe H, Koshy J, Vijay V, Gayathri R et al (2022) Nipah virus outbreak in Kerala state, India amidst of COVID-19 pandemic. Front Public Health 10. https://doi.org/10.3389/fpubh.2022.818545

Yamaguchi M, Kitagawa Y, Zhou M, Itoh M, Gotoh B (2014) An anti-interferon activity shared by paramyxovirus C proteins: inhibition of toll-like receptor 7/9-dependent alpha interferon induction. FEBS Lett 588:28–34. https://doi.org/10.1016/j.febslet.2013.11.015

Yoneda M, Guillaume V, Sato H, Fujita K, Georges-Courbot MC, Ikeda F, Omi M, Muto-Terao Y, Wild TF, Kai C (2010) The nonstructural proteins of Nipah virus play a key role in pathogenicity in experimentally infected animals. PLoS One 5(9):e12709. https://doi.org/10.1371/journal.pone.0012709

Yoneda M, Georges-Courbot MC, Ikeda F, Ishii M, Nagata N, Jacquot F, Raoul H, Sato H, Kai C (2013) Recombinant measles virus vaccine expressing the Nipah virus glycoprotein protects against lethal Nipah virus challenge. PLoS One 8(3):e58414. https://doi.org/10.1371/journal.pone.0058414

Yu F, Khairullah NS, Inoue S, Balasubramaniam V, Berendam SJ, Teh LK, Ibrahim NS, Abdul Rahman S, Hassan SS, Hasebe F, Sinniah M, Morita K (2006) Serodiagnosis using recombinant nipah virus nucleocapsid protein expressed in Escherichia coli. J Clin Microbiol 44(9):3134–3138. https://doi.org/10.1128/JCM.00693-06

Zhang, G., Cowled, C., Shi, Z., Huang, Z., Bishop-Lilly, K. A., Fang, X., Wynne, J. W., Xiong, Z., Baker, M. L., Zhao, W., Tachedjian, M., Zhu, Y., Zhou, P., Jiang, X., Ng, J., Yang, L., Wu, L., Xiao, J., Feng, Y., Chen, Y., … Wang, J. (2013). Comparative analysis of bat genomes provides insight into the evolution of flight and immunity. Science (New York, N.Y.), 339(6118), 456–460. https://doi.org/10.1126/science.1230835

Zhang W, Pei J, Lai L (2017) Computational multitarget drug design. J Chem Inf Model 57(3):403–412. https://doi.org/10.1021/acs.jcim.6b00491

Zhou P, Cowled C, Marsh GA, Shi Z, Wang LF, Baker ML (2011a) Type III IFN receptor expression and functional characterisation in the pteropid bat, Pteropus alecto. PLoS One 6(9):e25385. https://doi.org/10.1371/journal.pone.0025385

Zhou P, Cowled C, Todd S, Crameri G, Virtue ER, Marsh GA, Klein R, Shi Z, Wang LF, Baker ML (2011b) Type III IFNs in pteropid bats: differential expression patterns provide evidence for distinct roles in antiviral immunity. J Immunol 186(5):3138–3147. https://doi.org/10.4049/jimmunol.1003115

Zhou P, Tachedjian M, Wynne JW, Boyd V, Cui J, Smith I, Cowled C, Ng JH, Mok L, Michalski WP, Mendenhall IH, Tachedjian G, Wang LF, Baker ML (2016) Contraction of the type I IFN locus and unusual constitutive expression of IFN-α in bats. Proc Natl Acad Sci USA 113(10):2696–2701. https://doi.org/10.1073/pnas.1518240113

Zhou Y, Zhu F, Liu Y, Zheng M, Wang Y, Zhang D, Anraku Y, Zou Y, Li J, Wu H, Pang X, Tao W, Shimoni O, Bush AI, Xue X, Shi B (2020) Blood-brain barrier-penetrating siRNA nanomedicine for Alzheimer's disease therapy. Sci Adv 6(41):eabc7031. https://doi.org/10.1126/sciadv.abc7031

Zhu Z, Bossart KN, Bishop KA, Crameri G, Dimitrov AS, McEachern JA, Feng Y, Middleton D, Wang LF, Broder CC, Dimitrov DS (2008) Exceptionally potent cross-reactive neutralization of Nipah and Hendra viruses by a human monoclonal antibody. J Infect Dis 197(6):846–853. https://doi.org/10.1086/528801

Zou Y, Sun X, Wang Y, Yan C, Liu Y, Li J, Zhang D, Zheng M, Chung RS, Shi B (2020) Single siRNA nanocapsules for effective siRNA brain delivery and glioblastoma treatment. Adv Mater 32(24):e2000416. https://doi.org/10.1002/adma.202000416

Zumla A, Hui D (2019) Emerging and reemerging infectious diseases: global overview. Infect Dis Clin N Am 33(4):xiii–xix. https://doi.org/10.1016/j.idc.2019.09.001

Is JEV the New Emerging Encephalitis?

2

L. Preethi and Prudhvi Lal Bhukya

Abstract

Arboviruses are at the forefront of societal, governmental and public health concerns and have attracted national attention and so are mosquito-borne diseases. Both epidemic arboviral infections in rural areas and urban epidemics are increasingly more likely to occur. Japanese encephalitis (JE) has been linked to approximately 12,000 cases in India in recent years. Japanese encephalitis virus (JEV) infection is deadly, and its effects on the nervous system are of concern. The abundance of vector population due to suitable agro-climatic conditions has contributed significantly to the rise in JEV cases. In order to lower the danger, vector control programmes should be adequate. For people with JE, there is currently no antiviral medicine available. Symptom relief and patient stabilisation are the main objectives of supportive treatment. Expanded JEV research should focus on the creation of powerful antivirals and an improved immunisation approach. The distribution of various JEV genotypes has significantly changed in recent years. The virus's genetic change might be a serious danger. Priority should be given to improving vector surveillance and control activities and developing fast diagnostic techniques. We have access to highly cutting-edge scientific advancements; therefore, there is hope for the future creation of high-coverage vaccination schemes and effective JE treatments.

L. Preethi
MSc Public Health, London School of Hygiene and Tropical Medicine, London, UK

P. L. Bhukya (✉)
Preclinical Vaccine Testing Laboratory, Rodent Experimentation Facility, ICMR-National Animal Facility Resource Facility for Biomedical Research, Hyderabad, Telangana, India
e-mail: saiprudhvi21@gmail.com; prudhvi.bhukya@icmr.gov.in

P. L. Bhukya, A. S (eds.), *Emerging Human Viral Diseases, Volume II*,
https://doi.org/10.1007/978-981-97-4480-0_2

Keywords

Japanese encephalitis · Vector · Genotypes · Vaccine · Policy

2.1 Introduction

Arboviruses have gained national attention and are at the forefront of societal, governmental and public health concerns. And so are mosquito-borne diseases. There is now a greater risk of urban epidemics as well as epidemic arboviral illnesses in rural areas. Over 80% of the world's population is at risk of acquiring a vector-borne disease, with mosquito-borne diseases accounting for the majority of the incidence (Franklinos et al. 2019). Mosquito-borne infections are gaining a greater worldwide health concern, posing a hazard to more than 40% of the population (Jones et al. 2021), dengue, chikungunya and Japanese encephalitis (JE) continue to strike tropical nations around the world.

Japanese encephalitis, a mosquito-borne disease, is caused by the Japanese encephalitis virus (JEV), a flavivirus belonging to the family of Flaviviridae. Other viruses belonging to this family include West Nile Virus (WNV), Murrey Valley Encephalitis Virus (MVEV), Kunjin Virus (KUNV) and St. Louis Encephalitis Virus (SLEV). While many flaviviruses are responsible for encephalitis, JEV often causes severe neurological consequences (Turtle and Solomon 2018). Transmission of the virus between the vertebral hosts is maintained by mosquitoes, especially by *Culex tritaeniorhynchus*. In 1938, the role of *C. tritaeniorhynchus* as a vector was demonstrated, with pigs and ardeids being the reservoir hosts (van den Hurk et al. 2008).

Over the recent years (2014–2021), 23 states have reported around 12,944 JEV cases. Mohan et al. (2019) conducted a retrospective study, where the overall seroprevalence of JEV in Northeast India was found to be 23.7%. In Southern and Eastern Asia, JE has now been linked to a mortality rate of 30–50%. Convulsions, tremors, paralysis, ataxia and other neurological signs affect a substantial percentage of survivors (30–60% in some circumstances) (Sarkar et al. 2012). Yet another study conducted by Balakrishnan et al. (2017) in 2015 showed a positivity rate of 15.9% in the Alappuzha district of Kerala. Birendra Gupta in his book titled *Seroprevalence of JEV Infection in GKP and Basti Districts, UP, INDIA* has reported 64.9% seroconversion 1 month following vaccination with the SA-14-14-2 vaccine, while in Basti, it was 35.29%. In light of the increasing number of cases across the country, the Ministry of Health and Family Welfare released a notice on Japanese encephalitis on September 21, 2016, in order to address its challenges in specific areas of the country and thus declare it a notifiable disease.

Additionally, the abundance of vector population is due to suitable agro-climatic conditions. The key vector *Culex tritaeniorhynchus* has a very high distribution in the Indian subcontinent. Due to the sheer massive range of favourable environments for these mosquitos, JEV vectors are found all across the nation. As a result, determining the likelihood of a JEV outbreak in India necessitates taking into account a

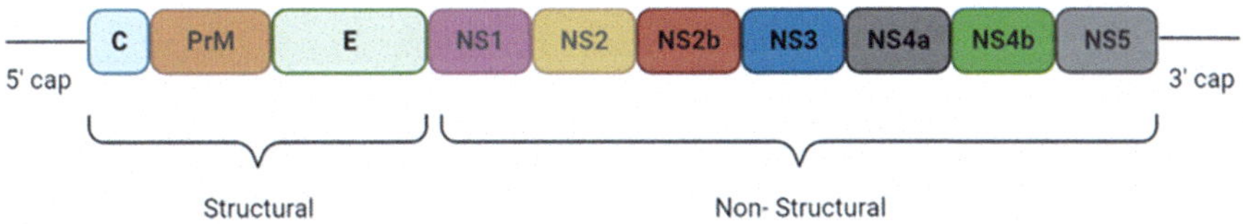

Fig. 2.1 Represents the Japanese encephalitis virus structure

wider range of geographies, which include panoramas of wetland areas, rain-fed agriculture and wildlife hosts (Walsh et al. 2021).

JE appears to be linked to a lack of health knowledge, a lack of preventive actions at home and the consumption of dirty water. Ansari and Nagar (2020) reported that the majority of parents whose kids were afflicted with JE had no knowledge of what the condition was or how to prevent it.

2.2 Genome Organisation and Life Cycle

JEV is a positive sense, single-stranded RNA virus with a 9–12 kb genome. It codes for three structural proteins, namely Capsid (C), envelope protein (E), membrane (prM/M) and a few non-structural proteins which include NS1-NS5 (NS1, NS2a, NS2b, NS3, NS4a, NS4b and NS5) (Fig. 2.1).

2.3 Dispersion of Genotype

JEV is divided into five genetic variants: GI (clades GI-a and GI-b), GII, GIII, GIV and GV. JEV GI is mostly found in tropical regions, whereas GIII is prevalent in temperate zones (Gao et al. 2018). Figure 2.2 depicts the genotype distribution among various countries. It only shows the major genotype wherein few countries tend to have more than one genotype. GIII genotype that was previously dominant in the Indian subcontinent started getting replaced by GI during the early 2000s (Sarkar et al. 2012). GI strain has been associated with recent breakouts in India (Fulmali et al. 2011).

2.4 Epidemiology

2.4.1 Global Scenario

The very first recognised clinical occurrence of JE was recorded in Japan in 1871 (Halstead 2010). A large outbreak later occurred in the following decades. The first known strain of the Nakayama virus was isolated in 1935.

Around 24 countries, most of them belonging to Western-Pacific and South Eastern Asia, have been identified as JE endemic regions. It is the most vital,

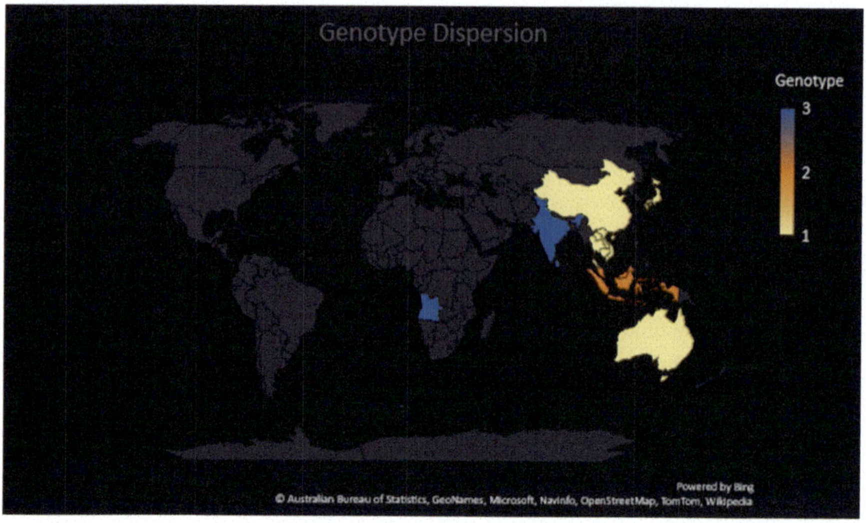

Fig. 2.2 Highlights the major genotypes present in various countries with a high risk of JEV (some countries have more than one genotype)

prevalent and scattered form of encephalitis; however, associated neurotropic viruses are unfolded throughout the globe (Bonville and Domachowske 2021).

Of 24, 12 countries of the Western Pacific Region (WPR) are endemic to JE transmission. The total number of cases in WPR has substantially reduced by almost 73% from 3428 in 2008 to 899 in 2019. Whereas a gradual increase in cases has been observed in South East Asia.

2.4.2 Indian Scenario

Unexplained infections infiltrate India almost every year, costing thousands of innocent lives; viral encephalitis is one of them. The first case of JE in India was diagnosed in 1955 in Tamil Nadu's Vellore and Puducherry districts. The virus was not isolated from human brain tissue until 1958, when three isolates were discovered from instances of encephalitis (Dhillon and Raina 2008). Since then, many major outbreaks have been reported in various parts of the country (Reuben and Gajanana 1997). Few remain undiagnosed too (Saxena et al. 2009). In 1973, a severe epidemic in the Bankura District of West Bengal resulted in a 42.6% mortality rate. Following that, the disease spread to neighbouring states, resulting in a series of outbreaks across the country. There were instances recorded from 21 states and union territories in 1978 (Dhillon and Raina 2008). JE instances have been pronounced particularly from Assam, Bihar, Jharkhand, Karnataka, Manipur, Meghalaya, West Bengal, Odisha, Tamil Nadu and Uttar Pradesh.

Since 2008, a number of cases have shown a gradual surge wherein 2017 experienced the worst hit by AES with a total of 12,578 cases throughout the country. Of 12,578 cases, 2043 were associated with the Japanese encephalitis virus.

Keeping in view the rising number of cases in the country, on September 21, 2016, the Ministry of Health and Family Welfare published a notification on Japanese encephalitis in order to address its issue in specific parts of the country and thus announced it as a notifiable disease.

2.5 Ecological Determinants of JEV

The epidemiology of the Japanese encephalitis virus is heavily reliant on the ecology of both amplifying hosts and mosquito vectors, both of which are influenced by both natural and human influences. The dynamics of arthropod-borne diseases, particularly JE, is complicated by the fact that they are extremely dependent on a variety of biotic and abiotic environmental conditions, as well as on the geographic range at which they are studied.

2.5.1 Hosts

Cattle, goats, fowl, ducks, dogs, horses and pigs may be infected in the JEV rural domestic loop, although very few of them have been effective virus amplifiers.

Pigs are the prime amplification hosts for JEV, particularly in epidemic areas, and they also serve as maintenance hosts in endemic areas. Buescher and Scherer (1959) demonstrated the role of pigs in the transmission of viruses attributed to their high JEV seroprevalence rate and sustained viraemia after natural infection lasting up to 4 days. They had also established that the pigs were capable of infecting *Culex tritaeniorhynchus* mosquitoes. High viraemia in pigs with no outright clinical manifestations, such as miscarriage, and stillbirth in pregnant sows, all contribute to the disease's persistence.

2.5.1.1 Host Abundance

The country's overall livestock population as per the 20th livestock census is 535.78 million, gaining 4.6% from the previous census in 2012. The total pig population in the country was found to be 9.06 million in the recent census with Uttar Pradesh accounting for nearly 4.54% (0.41 million) pig population of the total. Pig farmers practically live harmoniously with their animals. As a result, the odds of infecting people increase considering pigs are the JEV's amplifying vertebrate hosts.

2.5.2 Birds (Vertebrate Reservoir)

The importance of birds as reservoir hosts for JEV has been acknowledged since 1958, but their position as possible amplifying hosts has received less attention.

Many arboviruses rely heavily on birds for their maintenance and spread. Furthermore, they are prevalent in urban and suburban areas, where they coexist with humans and mosquitoes. Although this hypothesis is based on relatively limited data that is difficult to assess collectively, members of the bird family Ardeidae (herons, egrets, black-crowned night herons and bitterns) have been identified as key JEV amplifying hosts. Several avian species have developed JEV viraemia after natural and experimental infections, and spontaneously acquired anti-JEV antibodies have been found in a range of free-ranging birds in India, Japan, Korea and West Bengal (Nemeth et al. 2012).

2.5.3 Vectors

About 14 mosquito species have been recognised as JEV vectors, with yet another 11 species demonstrating experimental vector competence. *An. subpictus*, on the other hand, had the maximum susceptibility (as judged by the minimal infection rate). Overall, *Cx. annulirostris*, *Cx. sitiens*, *Cx. fuscocephala* and *Ae. japonicus* had the greatest JEV infection rates. According to a recent study, *Cx. tritaeniorhynchus*, *Cx. gelidus*, *Cx. sitiens* and *Cx. fuscocephala* are important JEV vector species, and all these species are classified as verified mosquito vectors (Auerswald et al. 2021). The most efficient JEV vectors are Culex species. *Culex tritaeniorhynchus* is often considered to be the most influential source throughout most of Asia, however, other there are other species such as *Cx. gelidus, Cx. vishnui, Cx. Fuscocephala* and *Cx. pseudovishnui* (Mackenzie et al. 2006). Females of *C. tritaeniorhynchus* seek huge animals for blood, such as cattle and pigs, and are profoundly anthropophilic (Faizah et al. 2021; Longbottom et al. 2017). *Culex tritaeniorhynchus* is found throughout Southeast Asia and surrounding tropical areas, as well as the Middle East and Africa. It was subsequently discovered in Europe.

2.5.3.1 Vector Abundance

Vector abundance is largely influenced by two factors: one is the agro-climatic factor and the other is the insecticide resistance.

2.5.3.2 Agro-Climatic Factor

The principal vector, *Culex tritaeniorhynchus*, loves to breed in irrigated rice fields and feed on pigs. All of these epidemiological factors create an excellent habitat for mosquito population density amplification and JEV transmission among people in pig-associated rural home cycles, as well as mosquitoes. JEV transmission is influenced by environmental conditions because it is a mosquito-borne disease. The intricate combination of climate trends, agricultural practices, mosquito population dynamics and the existence of susceptible amplifying hosts result in significant geographical and temporal variation in mosquito infection rates.

2.5.3.3 Agriculture

JE threat and occurrence have been influenced by changes in land use and land cover. Since the early 1960s, rice cultivation surfaces in Asia have been steadily expanding. Because paddy fields offer protracted *Culex* sp. nesting sites and attract a large number of wading birds for feeding and nesting, they help mosquito and wading bird stocks circulate and expand. As a result, in places where paddy farming is common, infection transmission could become less reliant on rain (Le Flohic et al. 2013).

2.5.3.4 Climatic Factor

Changes in climate, which include a global temperature rise as well as fluctuations in rainfall and wind dynamics, could have a significant impact on viral vector dispersion, reservoir species, amplifying hosts and perhaps even JEV genotypes. In line with rising temperatures, possible JEV hotspots may even see prolonged and more severe droughts, along with intervals of higher precipitation and potentially an upsurge in the frequency and severity of severe weather patterns including cyclones and hurricanes.

Temperature and precipitation have been reported to alter the mosquito life cycle, including the growth time of immature mosquito stages and mosquito density, driving JE transmission (Mansfield et al. 2017). Variability in the climate has a direct impact on the epidemiology of vector-borne diseases. Rainfall, temperature, relative humidity, virus infection from reservoirs and the consequences of social and economic situations are only a few of the risk factors involved in disease transmission. It is critical to investigate the role of weather in the transmission of Japanese encephalitis, as global warming may alter temperature and rainfall patterns, affecting mosquito density directly or indirectly (Luo et al. 1995). The life cycle pattern of mosquitoes is heavily influenced by relative humidity. For a long period of time, JEV aetiology has been discussed only on a global level and principally in terms of meteorological factors (i.e. rainfall and temperature) that do have a significant impact on vector abundance. It has been discovered that a temperature of 28 °C with a relative humidity of 50–55% is more conducive to an increase in mosquito density than a temperature of 22 °C with a relative humidity of 80–85% (Murty et al. 2010).

2.6 Transmission Cycle

The virus can be transmitted by domestic pigs and wild birds. and is spread by mosquitoes, pigs and/or water birds in a cycle (enzootic cycle). Infected mosquitoes, especially *Culex* spp., bite and spread the disease (*Culex tritaeniorhynchus*). Humans serve as unintentional dead-end hosts. The JEV cannot be passed from one person to another. Once infected, humans do not acquire enough JEV concentration in their circulation to infect feeding mosquitoes (Barrett 2008; Chapter 29—Flaviviridae 2017; Karthikeyan et al. 2017; Singh et al. 2012).

2.7 Clinical Manifestation

Most people infected with JEV are asymptomatic. On average, only one in 300 individuals have symptoms. Symptoms begin to appear between 6 and 14 days after infection. A fever above 38 °C, chills, muscle pain and meningitis-type headaches usually precede vomiting. Children usually present with gastrointestinal symptoms: nausea, vomiting and abdominal pain (Tiwari et al. 2012). Neurological manifestations include paralysis, confusion, disorientation, Parkinsonian movement disorder, dystonia, seizures and coma (Misra and Kalita 2010).

Recovery usually leaves a variety of serious sequelae, such as severe mental retardation and the persistent effects of sensorium deprivation. The duration of coma is associated with a variety of neurodevelopmental disorders (Tiwari et al. 2012) (Fig. 2.3).

2.7.1 Stages of Disease

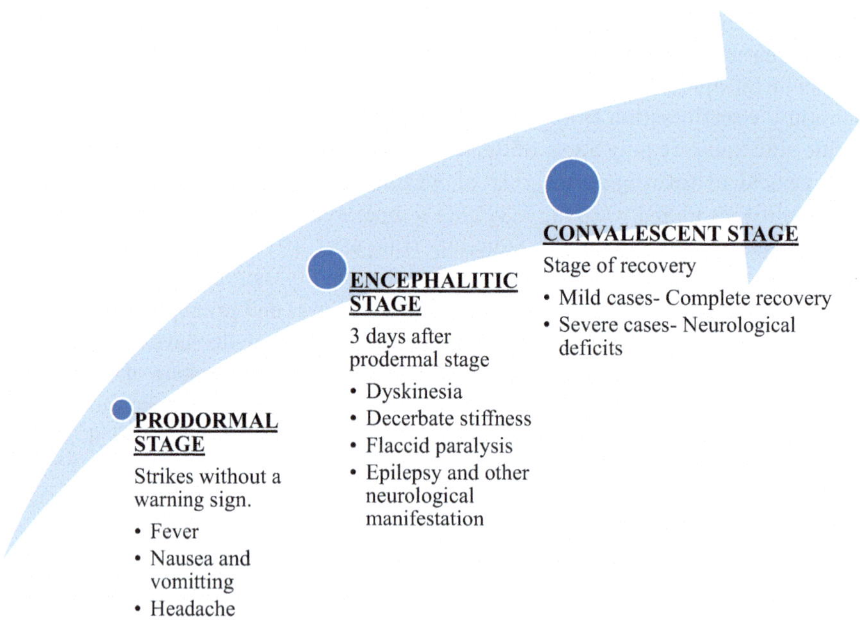

Fig. 2.3 Stages of the disease

2.8 Pathogenesis

It is thought that the virus replicates in Langerhans' dendritic cells (as in the case of dengue virus) and/or in keratinocytes (West Nile Virus) after inoculation into the skin. It is then transported to a local lymph node, where the Japanese encephalitis virus (JEV) continues to replicate. Viraemia develops and the virus crosses the blood–brain barrier (BBB) and invades the central nervous system (CNS) via endocytosis (Turtle and Solomon 2018; Li et al. 2015).

2.8.1 Immunopathogenesis

When humans are infected with JEV for the first time, they usually have a strong physiologic immune response, including adaptive humoral and cell-mediated responses.

The major site of infection for JEV is the human dermis layer, which is infected by an infected mosquito bite. JEV replication takes place in the peripheral network, especially PBMCs, infecting macrophages, dendritic cells (DCs) and monocytes. When JEV enters the bloodstream, it infects monocytes, causing viral replication to peak without causing cell death and resulting in the generation of TNF-alpha, which causes monocytes to become activated and differentiate into monocyte-derived dendritic cells (MDDCs) and monocyte-derived macrophages (MDMs) (Sooryanarain et al. 2012). JEV prefers to modulate DCs in order to undermine the immune system and develop a successful infection. DCs play an important role in the production of T cells and proinflammatory cytokines. The JEV-infected DCs produce a significant number of cytokines such as IL-6, TNF-alpha and IL-8 (Gupta et al. 2014). The upregulation of co-stimulatory molecules is the conventional mechanism for regulating macrophages. As a result of this stimulation of DCs and macrophages, an inflammatory milieu is induced, which aids in the permeability of the BBB, allowing the virus to migrate into the CNS. Damage to dendritic cells leads to loss of $CD4^+$ and $CD8^+$ T-cell response.

2.8.2 Entry into CNS

Viral entry into CNS can occur through many mechanisms, JEV tends to enter via paracellular transit across the BBB through the cleaved tight junctions (Hsieh and St. John 2020). This BBB compromise may be attributed to mast cell (MC) degranulation which occurs during the infection. The vasoactive protease MC-derived chymase has a functionally important role in the breakdown of TJ proteins such as ZO-1, ZO-2, claudin-5 and occludin, resulting in enhanced BBB cleavage (Hsieh and St. John 2020). Although neuronal cells are the most significant target cells in JEV, as the infection spreads to the CNS, astrocytes, which are a component of the BBB and an important portion of the CNS, also become infected. Astrocytes are also thought to aid in the transport of JEV from peripheral tissues to the cerebrospinal fluid. JEV infects the microglial cell, which is thought to be the CNS's resident immune cells/macrophage. Microglial cells are important in the CNS during JEV infection because they act as a virus reservoir (Chen et al. 2010). TNF-alpha and IL-6 are proinflammatory cytokines produced by activated microglia, which trigger neuronal cell death. JEV manages to diffuse into the CNS, eluding all immune responses, and it is this damage to the CNS that makes JEV infection more deadly.

2.8.3 Adaptive Immune Response

Both the humoral and cell-mediated arms of adaptive immunity have been associated with JEV infection defence and recuperation.

2.8.4 Cell-Mediated Immunity

Cell-mediated immunity (CMI), unlike its humoral analogue, is poorly described and poorly defined during JEV infection. Research in mice employing the leukocyte migratory inhibition experiment revealed the relevance of CMI, and it was discovered that the passive transfer of immunological T cells confers lifetime cell-mediated immunity.

T suppressor cells, T helper cells and B memory cells have all been found in JEV experiments involving an animal model.

2.8.5 Humoral Immunity

The function of the humoral immune response in the control and spread of JEV infection has been widely studied and documented. In cases where JEV is the first flavivirus to infect an individual, the patient's blood and CSF will show a fast and accurate monotypic IgM response. Within 7 days of infection, IgM production hits a peak, and attempts to effectively avoid the virus are usually ineffective. Failure to mount a significant IgM response leads to positive viral isolation and death. Antibodies are intended to safeguard the host by limiting JEV replication during the viraemic phase, which occurs before the virus passes the blood–brain barrier and enters the central nervous system. Asymptomatic JEV infection is linked with higher IgM levels in the blood, while antibodies in the CSF are linked to a better prognosis in individuals recovering from JE. Antibodies transition from IgM to IgG in surviving individuals, and most patients have IgG in serum and CSF by 30 days of infection (Kumar et al. 2019).

As per Burke et al. (1985), IgM for JEV detected in CSF were 68% on the first day, 100% on day seven, 96% on day 30 and 72% on 180th day. The frequencies of IgG in CSF were 47% on day 1, 89% on day 7, 100% on day 30 as well as day 180.

2.9 Diagnosis

Japanese encephalitis (JE) should be examined in an individual with signs of a neurologic infection (e.g. meningitis, encephalitis or acute flaccid paralysis) who has recently gone to or stayed in an endemic region in Asia or the western Pacific. Since JEV is irreversible, early identification is crucial for preventing an epidemic breakout, primarily because the first indications are often misdiagnosed as dengue or malaria.

Due to the obvious brief duration of viraemia and the asymptomatic aspect of JEV, diagnosis is difficult. These characteristics have made it difficult to establish a reliable diagnosis of JEV infection or determine prevalence levels. Another key consideration when selecting a diagnostic method is the test's specificity, which is hampered by flavivirus cross-reactivity. To confirm JEV infection and distinguish it from other flavivirus infections, and several other sources of bacterial and viral encephalitis, a definitive diagnosis is required. Consequently, laboratory testing is frequently performed later in the course of disease, after encephalitis has indeed been established as well as the aetiology determined. Anti-JEV IgM antibodies can be identified earlier in the phase of the illness, and IgG antibodies in the later phase of illness in the serum samples or cerebrospinal fluid, suggesting that an immune reaction is developing (Filgueira and Lannes 2019). To accurately identify JEV, a combination of clinical, serological and pathological studies is required. This covers older, time-consuming procedures like virus isolation, hemagglutination inhibition, Plaque Reduction Neutralisation Test (PRNT), complement fixation, molecular and immunofluorescence assays, as well as newer, faster detection techniques like ELISA and immunosensors.

2.9.1 Molecular Tests

The peripheral blood picture indicates moderate peripheral leukocytosis, mild anaemia and neutrophilia. The analysis of CSF (cerebrospinal fluid): In early CSF samples, neutrophils may predominate, but lymphocytic pleocytosis is common. In nearly half of the cases, CSF protein levels are moderately high.

2.9.2 Serological Test: PRNT, IgM CAPTURE ELISA, IgG ELISA

Neutralising assays, such as the **P**laque **R**eduction **N**eutralisation **T**est (PRNT), are more specific and are rated the gold standard in detecting previous flavivirus exposure since they quantify the dilution of blood necessary to neutralize live viruses. For **JEV**, a PRNT titre of 1:10 dilution by PRNT50 is considered cut-off against infection; however, for epidemiological studies of previous exposure, a more rigorous threshold, PRNT90, may be chosen, minimising the potential of background serum cross-reactivity (Cha et al. 2014).

JE is diagnosed in the laboratory by looking for virus-specific IgM antibodies in the serum or spinal fluid (CSF). IgM antibodies to the JEV are normally detected 3–8 days after the onset of the disease and last 30–90 days, while prolonged latency has been reported. As a result, positive IgM antibodies may potentially indicate a previous attack or immunisation. If IgM is not really evident in serum taken within 10 days of the commencement of the infection, the test should be performed on a convalescent serum. A capture ELISA approach is used for detecting **IgM** antibodies to the JEV. Microplates were sensitised with rabbit anti-human IgM at a dilution of 1:1000 overnight at 4 °C and blocked with BSA. Plates were then incubated for 1 h at 37 °C with test samples at a dilution of 1:1OO, recombinant and control antigens at a dilution of 1:500, horseradish peroxidase-conjugated human anti-flavivirus at a dilution of 1:200, and 1.8 mg/mL *o*-phenylene diamine in the presence of 0.03% H_2O_2 at 37 °C for 20 min. The difference between recombinant and control antigen absorbance values was compared to the difference between infected and normal mouse brain antigen absorbance values (Ravi et al. 2006). Using the commercial JE Detect IgG ELISA kit, the presence of JEV-specific IgG antibodies was determined. The microtitre plates will be inoculated with recombinant antigens made up of peptides derived from various JEV genes. The reaction well shall be filled with 50 μL of 1:300 diluted serum samples. Based on the absorbance at 450 nm, the immunological status ratio was computed for each sample. Positive samples show an immune status ratio of greater than 5, whereas negative samples shall give an immune status ratio of less than 2. Samples having a value of 2–5 will be deemed ambiguous (Cha et al. 2014).

2.9.3 Hemagglutination Inhibition

Hemagglutination inhibition assays have been widely utilised for viruses that hemagglutinate red blood cells of one or more species, such as many arthropod-borne viruses, influenza viruses and parainfluenza viruses. The methods for detecting and quantifying antibodies in animal serum are sensitive, specific, simple, reliable and very inexpensive. Hemagglutination inhibition assays are still the gold standard for assessing antibody responses to specific influenza A viruses, despite all of the technical improvements. The assay works on the simple premise that the virus interacts with red blood cells via receptors on their surface. Hemagglutination is blocked by antiviral antibodies that attach to these receptors. A constant amount of virus, usually four or eight hemagglutinating units, is introduced to each well of the microtitre plate after the serum is diluted serially in the wells of the microtitre plate, usually in twofold stages. The hemagglutination-inhibition titre of the serum is the reciprocal of the maximum dilution of serum that inhibits the agglutination of red blood cells by the specified amount of virus. Numerous previous sero-surveys based on results of hemagglutination inhibition tests, notably for paramyxoviruses, should be interpreted with caution, as non-specific agglutination inhibitors produced many false-positive test results in some of those studies.

2.9.4 CSF Findings

After an overnight fast, CSF will be collected via lumbar puncture. CSF was taken from participants who were going to receive spinal anaesthesia for operational procedures as a control group. After discarding the initial 1 mL of CSF, a total of 5 mL shall be recovered. CSF will be stored frozen at 200 degree celcius until examined for TNF-a and IL-2 quantification by ELISA after centrifugation at 3500 rev/min for 10 min at 4 degree celcius (Babu et al. 2006).

2.9.5 EEG

In JE, the EEG indicates diffuse delta slowing, with epileptiform discharges occurring rarely, especially in youngsters. 'Alpha pattern' coma in JE may be more common than initially assumed. The outcome is unrelated to EEG alterations, including alpha pattern coma. The EEG changes in JE, on the other hand, could help rule out herpes simplex encephalitis (Kalita and Misra 1998).

2.10 Therapeutic Approach

There is no antiviral medication available for JE patients. Treatment is supportive in nature, with the goal of relieving symptoms and stabilising the patient. To avert future outbreaks, JE prevention is critical.

S. no.	Author	Drugs	Discussion
1.	Ajibowo et al. (2021), Hoke Jr. et al. (1992), Rathi et al. (1993)	Dexamethasone	Within 5 days of admission, patients with Japanese encephalitis (JE) passed either from respiratory arrest or coma. There were no changes in mortality, hospital stays, days spent on a ventilator, or usage of mannitol between the treatment and placebo groups in a clinical trial involving dexamethasone. For cerebral oedema, steroids were investigated, but no discernible advantages were found
2.	Ajibowo et al. (2021), Kumar et al. (2009)	Ribavirin	With the exception of hyperventilation being more prevalent in placebo patients, there were no appreciable differences in the ribavirin and placebo groups' clinical, laboratory or demographic characteristics. There were no discernible variations in death rates when hyperventilation was taken into consideration
3.	Ajibowo et al. (2021), Rayamajhi et al. (2015)	Intravenous immunoglobulin	Trial with 22 children diagnosed with JE; no significant differences in recovery between IVIG and placebo groups. The IVIG group showed higher neutralising antibody titres and anti-JEV IgM titres. Author suggests exploring higher IVIG dosages for improved outcomes in future investigations

S. no.	Author	Drugs	Discussion
4.	Ajibowo et al. (2021), Ayukawa et al. (2004)	Acyclovir	Acyclovir was given to six JE patients in Japan during an outbreak. Four with neurological problems, one without, and one fatality, which indicates that more placebo-controlled studies are necessary to confirm that acyclovir is ineffective in treating juvenile encephalitis
5.	Li et al. (2018)	Erythrosin B	Erythrosin B strongly inhibited flaviviruses in cell culture, including DENV2, ZIKV, WNV, YFV and JEV. Indicates potential for orthosteric inhibitors. Erythrosin B is identified as a potent inhibitor with a good safety profile
6.	Fang et al. (2013) Fang et al. (2013))	Cilnidipine and niclosamide	Efficient inhibitors of JEV, targeting NS3 protease and helicase. No impact on methyltransferase or RNA-dependent RNA-polymerase. Potential as anti-JEV medication
7.	Ye et al. (2014) Ye et al. (2014)	Etanercept	By lowering proinflammatory cytokine release, glial activation and neuronal damage, Etanercept injection shielded animals against JEV-induced death
8.	Shi et al. (2014) Rossignol (2014)	Nitazoxanide	The anti-JEV effect of Nitazoxanide was evaluated in mice and cell models. It significantly reduced the in vitro replication of JEV. Ninety per cent of the treated in vivo mice made it through a lethal JEV challenge, which indicates the possibility of treating JE by suggesting anti-JEV action

2.11 Prevention

In India, the main method of preventing JE is to reduce the density of the vector population by using various pesticides. As a result, a significant necessity of the national control effort is to monitor the susceptibility status of JE vectors to commonly used insecticides at regular intervals in order to ensure that an appropriate pesticide is selected and administered adequately (Karunaratne and Hemingway 2000). Collateral approaches for the control of JE vectors include IRS with DDT and the use of insecticide-treated mosquito nets, both of which are used to fight malaria vectors. In addition, in JE-affected areas of India, thermal fogging with malathion is commonly used to minimise the density of JE vectors. Organophosphorus and synthetic pyrethroids are also commonly utilised in agricultural areas where JE vectors breed. As a result, the resistance of JE vectors to three regularly used pesticides was determined (Sahu et al. 2019). Malathion fogging is thought to be an effective method for controlling disease vectors. The goal of fogging is to stop the spread of infections by killing the vectors that carry them. In India, a fogging operation was initiated in 1970 as a rapid fix. Because of its unpredictable and uncommon occurrence, thermal fogging with malathion is particularly efficient in malaria and dengue epidemics, but it has very little influence on Japanese encephalitis (JE) outbreaks. Malathion thermal fogging has become an essential aspect of vector control in the JE outbreak due to the lack of any effective control technique for JE vectors. During JE epidemics, technical fogging with Malathion is therefore advised (Srivastava et al. 2008). The presence of high levels of insensitive AChE enzyme (RS and RR) in *Culex tritaeniorhynchus* and *Culex quinquefasciatus* suggests that insensitive AChE is responsible for malathion resistance in these mosquito populations. Finally, it is concluded that two significant JE vectors, *Culex vishnui* and *Culex tritaeniorhynchus*, were shown to be resistant to DDT, malathion and deltamethrin in this investigation. The study's findings will inform the NVBDCP's approach by giving updated information on the resistance status of JE vectors. This will be very helpful in determining which pesticides to use and at what rates for JE vector control.

In regions where JE is a public health hazard, the World Health Organization (WHO) has long recommended JE vaccine programmes (Vannice et al. 2021).

JE vaccination began in India in 2006, following large outbreaks of the virus in several parts of Bihar and Eastern Uttar Pradesh. Children aged 1–15 years received a single dose of the Chinese live attenuated SA-14-14-2 JE vaccine. Eleven of the country's most endangered zones were vaccinated in 2006, followed by 27, 22 and 30 districts over the next 3 years. In 2011, the Universal Immunisation Program (UIP) introduced the same JE vaccine into routine immunisation in the 181 endemic locations. In the 16th to 24th month, it was started as a single dosage with DPT.

In 2013, a two-dose JE immunisation schedule was introduced, with the first dose (JE-1) given at 9 and 12 months with measles vaccine. Over the next 10 years and beyond, the Indian government skilfully executed a resilient and sustainable JE vaccination campaign, initially vaccinating India's most susceptible children and then boosting regular JE immunisation in dozens of endemic areas across the country. The Ministry of Health and Family Welfare (MoHFW) of the Indian Government initiated Mission Indradhanush in 2014, with the goal of vaccinating 90% of kids under the age of 2 and pregnant women by 2020. As of mid-2018, approximately 156 million children had been vaccinated, according to the Ministry of Health and Family Welfare.

The production of neutralising antibodies against the envelop (E) protein, which is capable of binding to and eventual penetration into the specific receptor, is critical for a protective humoral immune response. Animal tests have shown that neutralising antibodies against JEV E protein can be employed effectively for passive immunisation in the early treatment of JEV-infected individuals. So far, five genotypes have been found with E protein variants ranging from 1% to 5%, associated epitope changes and resultant antibodies (Filgueira and Lannes 2019).

2.11.1 Types of Vaccines

2.11.1.1 Inactivated Vaccine

Since 1967, China has used an inactivated vaccine made from the Beijing-3 strain of the JEV-2 virus, utilizing primary hamster kidney cell culture (Ajibowo et al.

2021). In the last decade, various JEV-based vaccines have been developed. One of these is a Vero cell-cultured vaccine that uses various local JEV isolates. Using an attenuated SA14-14-2 strain, researchers have been developing a Vero cell-culture-derived formalin-inactivated vaccine, which has elicited high titres of neutralising antibodies in mice after two injections. Recently, a formaldehyde-inactivated JE vaccine developed from Vero cell culture utilising P20778 (Indian isolate) was created, and it produced significant titres of anti-JEV antibodies in mice; sera from vaccinated animals neutralised several JEV strains with varying efficacies (Appaiahgari and Vrati 2004).

2.11.1.2 Cell Culture-Derived Live Attenuated JE Vaccine

Since 2001, the Chimer Ivax-JE has been introduced. Because less virus is required to elicit a sufficient immune response, live attenuated vaccines seem to have promising future vaccine development prospects, as they require fewer doses and are easier to administer.

2.11.1.3 Mouse Brain-Derived JE Vaccine

Inactivated vaccines developed from mouse brain are based on the Nakayama and Beijing-1 strains (seroconversion incidences of 80–90%). The vaccine developed from the original Nakayama strain is made in Japan and was first approved in 1954. Under the Biken label, it is available all over the world. China, India, Thailand and Taiwan all develop this vaccine on their own. In India, the manufacturer is the Central Research Institute in Kasauli. It comes in a lyophilised form with stabilisers including gelatine and sodium glutamate, as well as a preservative called thimerosal. The initial immunisation is given subcutaneously between the ages of one and three in doses ranging from 0.5 mL to 1 mL (0.25–0.5 mL with children under the age of three). One injection is given on days 0, 7 and 30, with a booster at 1 year and then every 3 years until age 10. The protective effectiveness is greater than 90%. This vaccine is not practicable to deliver in poor rural areas, where it is vitally required, due to its high production cost, lack of long-term immunity and unpleasant allergic reactions. As a result of these challenges, better vaccinations have been developed.

2.12 Complexity in Policy and Vulnerability in JEV Immunisation

Despite significant success in JE vaccination and containment over the previous 15 years, problems remain that threaten these gains and may hamper continued development. These difficulties cover the following: Focusing on increasing JE vaccine coverage, a regular flow of JE vaccine, providing vigilance and immunisation data.

2.13 Future Perspective

Concerns exist regarding fatal JEV infection and its effects on the nervous system. Programmes for vector control ought to be appropriate to reduce the risk. The development of efficient antivirals and a better vaccination strategy should be a part of the expanded JEV research. Strengthening vector surveillance and control programmes and creating quick diagnostic tools should be prioritised. With incredibly cutting-edge biomedical advances at our disposal, there is optimism for the development of high-coverage immunisation programmes and efficient JE treatments in the future.

2.14 Emerging Aspects

JEV, or the Japanese encephalitis virus, is a flavivirus with distinct traits that have influenced its evolution. Because JEV lacks these processes, it is extremely vulnerable to evolutionary changes, in contrast to those organisms that have strong repair systems during genome replication. Its capacity for genetic diversity over time and its adaptability are both influenced by this trait.

JEV is still a serious public health issue in India that is not getting enough attention. It is especially well known that the virus is the main cause of meningoencephalitis in several Asian countries. The significance of comprehending the dynamics and evolution of the virus is highlighted by the severity of Japanese encephalitis in these areas.

The constant shift in the genotype distribution of JEV is one noteworthy feature of the virus. Current research indicates a notable genetic change in JEV, as demonstrated by a study conducted in 2023 (Pegu et al. 2023). In the usual epidemic hotspots across Asia, JEV G1 has emerged as the prevalent genotype, displacing the historical dominance of JEV genotype G3.

In addition to being an interesting feature of the virus's evolutionary biology, this shift in genotype distribution also has useful ramifications for public health. The epidemiological landscape of Japanese encephalitis may be influenced by changes in the incidence of particular genotypes. It prompts worries about possible effects on the disease's severity and rate of spread.

In recent years, JEV's regional distribution has also changed. In addition to being a genetic change, the transition from G3 to G1 indicates a shift in the virus's geographic distribution. This shift in location is critical to comprehending the dynamics of Japanese encephalitis outbreaks and to developing efficient control and prevention strategies.

Geographic and genetic variations in JEV present a serious risk to public health. The virus's modified genetic composition may have an impact on its virulence, transmissibility and possible interactions with host immune systems. Furthermore, if the virus's geographic distribution shifts, new populations may get infected and outbreaks may occur in previously unaffected areas.

To effectively treat Japanese encephalitis, a thorough comprehension of these evolutionary dynamics is necessary. To effectively stop the virus from spreading, surveillance, research and public health initiatives must adjust to these new realities. To lessen the effects of this changing public health concern, tailored treatments including vaccination campaigns and vector control measures must be developed. This requires careful monitoring of the genotypic and geographic alterations in JEV.

References

Ajibowo AO, Ortiz JF, Alli A, Halan T, Kolawole OA (2021) Management of Japanese encephalitis: a current update. Cureus 13(4):e14579–e14579. https://doi.org/10.7759/cureus.14579

Ansari K, Nagar V (2020) Awareness among the families of patients affected by Japanese encephalitis and acute encephalitis syndrome: a case study of Gorakhpur region of Uttar Pradesh

Appaiahgari MB, Vrati S (2004) Immunogenicity and protective efficacy in mice of a formaldehyde-inactivated Indian strain of Japanese encephalitis virus grown in Vero cells. Vaccine 22(27):3669–3675. https://doi.org/10.1016/j.vaccine.2004.03.024

Auerswald H, Maquart P-O, Chevalier V, Boyer S (2021) Mosquito vector competence for Japanese encephalitis virus. Viruses 13(6):1154. https://doi.org/10.3390/v13061154

Ayukawa R, Fujimoto H, Ayabe M, Shoji H, Matsui R, Iwata Y et al (2004) An unexpected outbreak of Japanese encephalitis in the Chugoku district of Japan, 2002. Jpn J Infect Dis 57(2):63–66

Babu GN, Kalita J, Misra UK (2006) Inflammatory markers in the patients of Japanese encephalitis. Neurol Res 28(2):190–192. https://doi.org/10.1179/016164106X98062

Balakrishnan A, Thekkekare RJ, Sapkal G, Tandale BV (2017) Seroprevalence of Japanese encephalitis virus & West Nile virus in Alappuzha district, Kerala Indian. J Med Res 146(Supplement):S70–s75. https://doi.org/10.4103/ijmr.IJMR_1638_15

Barrett ADT (2008) Japanese encephalitis virus. In: Mahy BWJ, Van Regenmortel MHV (eds) Encyclopedia of virology, 3rd edn. Academic Press, Oxford, pp 182–188

Bonville C, Domachowske J (2021) Japanese encephalitis. In: Domachowske J, Suryadevara M (eds) Vaccines: a clinical overview and practical guide. Springer, Cham, pp 211–218

Buescher EL, Scherer WF (1959) Ecologic studies of Japanese encephalitis virus in Japan. IX. Epidemiologic correlations and conclusions. Am J Trop Med Hyg 8:719–722. https://doi.org/10.4269/ajtmh.1959.8.719

Burke DS, Nisalak A, Ussery MA, Laorakpongse T, Chantavibul S (1985) Kinetics of IgM and IgG responses to Japanese encephalitis virus in human serum and cerebrospinal fluid. J Infect Dis 151(6):1093–1099. https://doi.org/10.1093/infdis/151.6.1093

Cha GW, Cho JE, Ju YR, Hong Y-J, Han MG, Lee W-J et al (2014) Comparison of four serological tests for detecting antibodies to Japanese encephalitis virus after vaccination in children. Osong Public Health Res Perspect 5(5):286–291. https://doi.org/10.1016/j.phrp.2014.08.003

Chapter 29—Flaviviridae (2017) In: MacLachlan NJ, Dubovi EJ (eds) Fenner's veterinary virology, 5th edn. Academic Press, Boston, MA, pp 525–545

Chen CJ, Ou YC, Lin SY, Raung SL, Liao SL, Lai CY et al (2010) Glial activation involvement in neuronal death by Japanese encephalitis virus infection. J Gen Virol 91(Pt 4):1028–1037. https://doi.org/10.1099/vir.0.013565-0

Dhillon GP, Raina VK (2008) Epidemiology of Japanese encephalitis in context with Indian scenario. J Indian Med Assoc 106(10):660–663

Faizah AN, Kobayashi D, Amoa-Bosompem M, Higa Y, Tsuda Y, Itokawa K et al (2021) Evaluating the competence of the primary vector, Culex tritaeniorhynchus, and the invasive mosquito species, Aedes japonicus japonicus, in transmitting three Japanese encephalitis virus genotypes. PLoS Negl Trop Dis 14(12):e0008986. https://doi.org/10.1371/journal.pntd.0008986

Fang JE, Sun L, Peng G, Xu J, Zhou R, Cao S et al (2013) Identification of three antiviral inhibitors against Japanese encephalitis virus from library of pharmacologically active compounds 1280. PLoS One 8(11):e78425–e78425. https://doi.org/10.1371/journal.pone.0078425

Filgueira L, Lannes N (2019) Review of emerging Japanese encephalitis virus: new aspects and concepts about entry into the brain and inter-cellular spreading. Pathogens (Basel, Switzerland) 8(3):111. https://doi.org/10.3390/pathogens8030111

Franklinos LHV, Jones KE, Redding DW, Abubakar I (2019) The effect of global change on mosquito-borne disease. Lancet Infect Dis 19(9):e302–e312. https://doi.org/10.1016/S1473-3099(19)30161-6

Fulmali PV, Sapkal GN, Athawale S, Gore MM, Mishra AC, Bondre VP (2011) Introduction of Japanese encephalitis virus genotype I, India. Emerg Infect Dis 17(2):319–321. https://doi.org/10.3201/eid1702.100815

Gao X, Liu H, Li X, Fu S, Cao L, Shao N et al (2018) Changing geographic distribution of Japanese encephalitis virus genotypes, 1935–2017. Vector Borne Zoonotic Dis 19(1):35–44. https://doi.org/10.1089/vbz.2018.2291

Gupta N, Hegde P, Lecerf M, Nain M, Kaur M, Kalia M et al (2014) Japanese encephalitis virus expands regulatory T cells by increasing the expression of PD-L1 on dendritic cells. Eur J Immunol 44(5):1363–1374. https://doi.org/10.1002/eji.201343701

Halstead SB (2010) Japanese encephalitis. In: Artenstein AW (ed) Vaccines: a biography. Springer, New York, NY, pp 317–333

Hoke CH Jr, Vaughn DW, Nisalak A, Intralawan P, Poolsuppasit S, Jongsawas V et al (1992) Effect of high-dose dexamethasone on the outcome of acute encephalitis due to Japanese encephalitis virus. J Infect Dis 165(4):631–637. https://doi.org/10.1093/infdis/165.4.631

Hsieh JT, St. John AL (2020) Japanese encephalitis virus and its mechanisms of neuroinvasion. PLoS Pathog 16(4):e1008260. https://doi.org/10.1371/journal.ppat.1008260

Jones RT, Ant TH, Cameron MM, Logan JG (2021) Novel control strategies for mosquito-borne diseases. Philos Trans R Soc Lond B Biol Sci 376(1818):20190802. https://doi.org/10.1098/rstb.2019.0802

Kalita J, Misra UK (1998) EEG in Japanese encephalitis: a clinico-radiological correlation. Electroencephalogr Clin Neurophysiol 106(3):238–243. https://doi.org/10.1016/S0013-4694(97)00123-5

Karthikeyan A, Shanmuganathan S, Pavulraj S, Prabakar G, Pavithra S, Porteen K et al (2017) Japanese encephalitis, recent perspectives on virus genome, transmission, epidemiology, diagnosis and prophylactic interventions. J Exp Biol Agric Sci 5:730–748

Karunaratne SHPP, Hemingway J (2000) Insecticide resistance spectra and resistance mechanisms in populations of Japanese encephalitis vector mosquitoes, Culex tritaeniorhynchus and Cx. gelidus, in Sri Lanka. Med Vet Entomol 14(4):430–436. https://doi.org/10.1046/j.1365-2915.2000.00252.x

Kumar R, Tripathi P, Baranwal M, Singh S, Tripathi S, Banerjee G (2009) Randomized, controlled trial of oral ribavirin for Japanese encephalitis in children in Uttar Pradesh, India. Clin Infect Dis 48(4):400–406. https://doi.org/10.1086/596309

Kumar A, Sharma P, Shukla KK, Misra S, Nyati KK (2019) Japanese encephalitis virus: associated immune response and recent progress in vaccine development. Microb Pathog 136:103678. https://doi.org/10.1016/j.micpath.2019.103678

Le Flohic G, Porphyre V, Barbazan P, Gonzalez J-P (2013) Review of climate, landscape, and viral genetics as drivers of the Japanese encephalitis virus ecology. PLoS Negl Trop Dis 7(9):e2208. https://doi.org/10.1371/journal.pntd.0002208

Li F, Wang Y, Yu L, Cao S, Wang K, Yuan J et al (2015) Viral infection of the central nervous system and neuroinflammation precede blood-brain barrier disruption during Japanese encephalitis virus infection. J Virol 89(10):5602–5614. https://doi.org/10.1128/jvi.00143-15

Li Z, Sakamuru S, Huang R, Brecher M, Koetzner CA, Zhang J et al (2018) Erythrosin B is a potent and broad-spectrum orthosteric inhibitor of the flavivirus NS2B-NS3 protease. Antivir Res 150:217–225. https://doi.org/10.1016/j.antiviral.2017.12.018

Longbottom J, Browne AJ, Pigott DM, Sinka ME, Golding N, Hay SI et al (2017) Mapping the spatial distribution of the Japanese encephalitis vector, Culex tritaeniorhynchus Giles, 1901 (Diptera: Culicidae) within areas of Japanese encephalitis risk. Parasit Vectors 10(1):148. https://doi.org/10.1186/s13071-017-2086-8

Luo D, Ying H, Yao R, Song J, Wang Z (1995) Socio-economic status and micro-environmental factors in relation to the risk of Japanese encephalitis: a case-control study. Southeast Asian J Trop Med Public Health 26(2):276–279

Mackenzie JS, Williams DT, Smith DW (2006) Japanese encephalitis virus: the geographic distribution, incidence, and spread of a virus with a propensity to emerge in new areas. In: Tabor E (ed) Perspectives in medical virology, vol 16. Elsevier, pp 201–268

Mansfield KL, Hernández-Triana LM, Banyard AC, Fooks AR, Johnson N (2017) Japanese encephalitis virus infection, diagnosis and control in domestic animals. Vet Microbiol 201:85–92. https://doi.org/10.1016/j.vetmic.2017.01.014

Misra UK, Kalita J (2010) Spectrum of movement disorders in encephalitis. J Neurol 257(12):2052–2058. https://doi.org/10.1007/s00415-010-5659-4

Mohan DG, Gogoi M, Sharma A (2019) Seroprevalence of Japanese encephalitis amongst cases of acute encephalitis syndrome in a tertiary care centre of north east India: a four year retrospective study. J Int J Res Med Sci 7(9):3490–3349. https://doi.org/10.18203/2320-6012.ijrms20193934

Murty US, Rao MS, Arunachalam N (2010) The effects of climatic factors on the distribution and abundance of Japanese encephalitis vectors in Kurnool district of Andhra Pradesh, India. J Vector Borne Dis 47(1):26–32

Nemeth N, Bosco-Lauth A, Oesterle P, Kohler D, Bowen R (2012) North American birds as potential amplifying hosts of Japanese encephalitis virus. Am J Trop Med Hyg 87(4):760–767. https://doi.org/10.4269/ajtmh.2012.12-0141

Pegu SR, Das PJ, Sonowal J, Sengar GS, Deb R, Yadav AK et al (2023) Japanese encephalitis virus genotype III strains detection and genome sequencing from Indian pig and mosquito vector. Vaccine 11(1). https://doi.org/10.3390/vaccines11010150

Rathi AK, Kushwaha KP, Singh YD, Singh J, Sirohi R, Singh RK, Singh UK (1993) JE virus encephalitis: 1988 epidemic at Gorakhpur. Indian Pediatr 30(3):325–333

Ravi V, Desai A, Balaji M, Apte MP, Lakshman L, Subbakrishna DK et al (2006) Development and evaluation of a rapid IgM capture ELISA (JEV-Chex) for the diagnosis of Japanese encephalitis. J Clin Virol 35(4):429–434. https://doi.org/10.1016/j.jcv.2005.11.004

Rayamajhi A, Nightingale S, Bhatta N, Singh R, Ledger E, Bista K et al (2015) Correction: a preliminary randomized double blind placebo-controlled trial of intravenous immunoglobulin for Japanese encephalitis in Nepal. PLoS One 10:e0122608. https://doi.org/10.1371/journal.pone.0136008

Reuben R, Gajanana A (1997) Japanese encephalitis in India. Indian J Pediatr 64(2):243–251. https://doi.org/10.1007/bf02752458

Rossignol J-F (2014) Nitazoxanide: a first-in-class broad-spectrum antiviral agent. Antivir Res 110:94–103. https://doi.org/10.1016/j.antiviral.2014.07.014

Sarkar A, Taraphdar D, Mukhopadhyay SK, Chakrabarti S, Chatterjee S (2012) Molecular evidence for the occurrence of Japanese encephalitis virus genotype I and III infection associated with acute encephalitis in patients of West Bengal, India, 2010. Virol J 9(1):271. https://doi.org/10.1186/1743-422X-9-271

Sahu SS, Sonia T, Dash S, Gunasekaran K, Jambulingam P (2019) Insecticide resistance status of three vectors of Japanese encephalitis in east Central India abstract. Med Vet Entomol 33(2):213–219. https://doi.org/10.1111/mve.2019.33.issue-2 https://doi.org/10.1111/mve.12352

Saxena SK, Mishra N, Saxena R, Singh M, Mathur A (2009) Trend of Japanese encephalitis in North India: evidence from thirty-eight acute encephalitis cases and appraisal of niceties. J Infect Dev Ctries 3(7):517–530. https://doi.org/10.3855/jidc.470

Shi et al (2014) Nitazoxanide inhibits the replication of Japanese encephalitis virus in cultured cells and in a mouse model abstract Virol J 11(1) https://doi.org/10.1186/1743-422X-11-10

Singh A, Saxena SK, Srivastava AK, Mathur A (2012) Japanese encephalitis: a persistent threat. Proc Natl Acad Sci India Sect B Biol Sci 82(1):55–68. https://doi.org/10.1007/s40011-011-0005-x

Sooryanarain H, Ayachit V, Gore M (2012) Activated CD56(+) lymphocytes (NK+NKT) mediate immunomodulatory and anti-viral effects during Japanese encephalitis virus infection of dendritic cells in-vitro. Virology 432(2):250–260. https://doi.org/10.1016/j.virol.2012.05.013

Srivastava VK, Singh A, Thapar BR (2008) Field evaluation of malathion fogging against Japanese encephalitis vector, Culex tritaeniorhynchus. J Vector Borne Dis 45:249–250

Tiwari S, Singh RK, Tiwari R, Dhole TN (2012) Japanese encephalitis: a review of the Indian perspective. Braz J Infect Dis 16(6):564–573. https://doi.org/10.1016/j.bjid.2012.10.004

Turtle L, Solomon T (2018) Japanese encephalitis—the prospects for new treatments. Nat Rev Neurol 14(5):298–313. https://doi.org/10.1038/nrneurol.2018.30

van den Hurk AF, Ritchie SA, Mackenzie JS (2008) Ecology and geographical expansion of Japanese encephalitis virus. Annu Rev Entomol 54(1):17–35. https://doi.org/10.1146/annurev.ento.54.110807.090510

Vannice KS, Hills SL, Schwartz LM, Barrett AD, Heffelfinger J, Hombach J et al (2021) The future of Japanese encephalitis vaccination: expert recommendations for achieving and maintaining optimal JE control. npj Vaccines 6(1):82. https://doi.org/10.1038/s41541-021-00338-z

Walsh MG, Pattanaik A, Vyas N, Saxena D, Webb C, Sawleshwarkar S, Mukhopadhyay C (2021) High risk landscapes of Japanese encephalitis virus outbreaks in India converge on wetlands, rainfed agriculture, wild Ardeidae, and domestic pigs. medRxiv. https://doi.org/10.1101/2021.09.07.21263238

Ye J, Jiang R, Cui M, Zhu B, Sun L, Wang Y et al (2014) Etanercept reduces neuroinflammation and lethality in mouse model of Japanese encephalitis. J Infect Dis 210(6):875–889. https://doi.org/10.1093/infdis/jiu179

3 Conquering Chikungunya Virus: Global Threats and Promising Solutions

L. Preethi, Anbazhagan Subbaiyan, Suhas T. Mhaske, Prudhvi Lal Bhukya, and Subrat Thanapati

Abstract

The Chikungunya virus (CHIKV), identified in Tanzania during 1952–1953, has triggered numerous outbreaks worldwide, causing fever, headache, rash, and debilitating joints pain in humans. Its transmission involves two distinct cycles: sylvatic and urban, with *Aedes* group of mosquitos, *Aedes aegypti* and *Aedes albopictus.* Considering broad range climatic adaptation of *Aedes albopictus,* raises concern about the virus's potential global dissemination. Given the lack of specific treatment, Chikungunya has evolved into a significant global health issue, necessitating reliance on symptomatic therapies such as analgesics and antipyretics. The imperative for combating this virus lies in the development of a vaccine. Studies underscore the pivotal role of CHIKV-specific humoral responses in effectively countering the virus. Notably, vaccines based on virus-like particles (VLPs) have demonstrated efficacy in inducing protective neutralizing antibodies. Despite these advancements, further research is essential in unraveling optimal vaccine strategies and understanding the adaptive immune responses potential in controlling viremia and rheumatic clinical manifestations associated with CHIKV infection. The quest for a comprehensive understanding of these aspects is crucial in advancing our ability to tackle and mitigate the impact of this global health challenge.

L. Preethi
MSc Public Health, London School of Hygiene and Tropical Medicine, London, UK

A. Subbaiyan · S. T. Mhaske · P. L. Bhukya
Preclinical Vaccine Testing Laboratory, Rodent Experimentation Facility, ICMR-National Animal Facility Resource Facility for Biomedical Research, Hyderabad, Telangana, India

S. Thanapati (✉)
Department of Research and Development, Techinvention Lifecare Pvt Ltd., Mumbai, Maharashtra, India
e-mail: ssubrat123@gmail.com

P. L. Bhukya, A. S (eds.), *Emerging Human Viral Diseases, Volume II*,
https://doi.org/10.1007/978-981-97-4480-0_3

Keywords

Chikungunya · Burden · Vector · Transmission · Vaccine

3.1 Introduction

3.1.1 Alphaviruses

Alphaviruses belong to the family *Togaviridae* and are mostly arthropod-borne viruses. The genus alphavirus consists of nearly 30–40 viruses which are categorized as Old World and New World alphavirus based on their location of discovery (Strauss and Strauss 1994). Old World alphaviruses, such as CHIKV, o'nyong-nyong virus (ONNV), Ross River virus (RRV), Semliki Forest virus (SFV), and Sindbis virus (SINV), usually cause fever, rash, and polyarthritis in humans and other mammals. New World alphaviruses, such as Eastern equine encephalitis virus (EEEV), Venezuelan equine encephalitis virus (VEEV), and Western equine encephalitis virus (WEEV), are usually associated with encephalitis (Ryman and Klimstra 2008).

3.1.2 Chikungunya Virus (CHIKV)

CHIKV is mosquito-borne viral disease in humans. The word chikungunya is derived from Swahili or Makonde word "Kun qunwala," which means "to become contorted" or "that which bends up" (Burt et al. 2012). The distinctive features of chikungunya are headache, myalgia, rash self-limiting fever, nausea, arthritis, and persistent arthralgia (Charrel et al. 2007). The incubation period of Chikungunya virus lasts for 4–7 days following CHIKV infection and mostly resolves within the acute phase, while persistent joints pain can last for several months to years (Kelvin et al. 2011; Dupuis-Maguiraga et al. 2012). After a gap of 32 years, CHIKV's re-emergence in Indian Ocean islands during 2004–2007 has shown an adverse form disease associated with altered clinical manifestations associated with morbidity and mortality (Economopoulou et al. 2009; Jarahian et al. 2009). Mother-to-fetus transmission, severe neurological complications, and fulminant hepatitis cases have also been reported in outbreaks of La Re'union islands, India, and Sri Lanka outbreaks during the year 2005–2006 (Robillard et al. 2006; Ramful et al. 2007; Gerardin et al. 2008), (Rao et al. 2008; Senanayake et al. 2009). In this epidemic, one-third of the total population of Reunion Island got infected with CHIKV with a fatality rate of 0.1%, while in India it was estimated to be 1.4–6.5 million cases (Mavalankar et al. 2008; Pialoux et al. 2006; Schwartz and Albert 2010; Manimunda et al. 2011). Invasion of CHIKV to new world has been reported in the recent years. Since November 2013, chikungunya outbreaks in the Caribbean and the Americas have been recorded (Cassadou et al. 2013; Leparc-Goffart et al. 2014; Van Bortel et al. 2014). As of August 29, 2015, Pan American Health Organization and World Health Organization have reported 967,760 suspected chikungunya cases from different territories of

America (Cases of Chikungunya Fever in the Americas, 2015 Cases reported by IHR NFPs to PAHO/WHO and/or through Member States websites or official news publication). Over three million suspected cases of chikungunya have been recorded worldwide till now (Seppa 2015). As per National Institute of Allergy and Infectious Diseases CHIKV has been listed as category B priority pathogen (https://www.niaid.nih.gov/research/emerging-infectious-diseases-pathogens). In the absence of a specific treatment, recent epidemics in previously naïve parts of the world have elevated chikungunya to a global health problem. Currently, symptomatic therapies such as analgesics, antipyretics, and/or nonsteroidal anti-inflammatory drugs (NSAIDS) are used to treat the patients. Chloroquine phosphate has been used to manage prolonged arthralgias and aspirin is strictly avoided (Brighton 1984; Chikungunya—An overview—CiteSeerX 2008; Jaffar-Bandjee et al. 2010).

3.1.3 Epidemiology and Geographical Range of CHIKV

The virus was first isolated from the serum of a febrile patient during a dengue such as outbreak on the Makonde Plateau of Tanzania (formerly Tanganyika) in 1952–1953 (Robinson 1955). However, the first chikungunya outbreak occurred long ago in 1779 at Jakarta, when it was misidentified as a dengue outbreak (Carey 1971). Since discovery, CHIKV has caused numerous outbreaks in different parts of the world. The major outbreaks were recorded in Thailand in the late1950s to early 1960s and in India from late 1960s to early 1970s (Myers and Carey 1967; Halstead et al. 1969a, b). In 1962 outbreak in Thailand, 31% of the total population of Bangkok was reported to be infected with CHIKV. The first chikungunya case in India was recorded in 1963 in West Bengal (Shah et al. 1964) that was followed by epidemics in Chennai, Pondicherry, Vellore, Visakhapatnam, Rajahmundry, Kakinada, Nagpur, and Barsi between 1964 and 1973. In 1963, more than 400,000 individuals were estimated to be infected in Chennai alone (Arankalle et al. 2007) (Fig. 3.1).

Subsequently, epidemics have been recorded in Congo during 1999–2000 (50,000 cases) (Pastorino et al. 2004) and Indonesia in 2001–2003 (6000 cases) (Buonaguro et al. 2013).

In 2004, the largest ever recorded chikungunya epidemic originated in Kenya (estimated 13,500 cases) that had spread to Comoros (estimated 215,000 cases) and subsequently to Reunion Island (estimated 300,000 cases in 2005–2006) (Schuffenecker et al. 2006). In 2006, the chikungunya epidemic spread to India where 150 districts of 8 states were affected with more than one million reported cases (Lahariya and Pradhan 2006). From India, the epidemic had spread to Southeast Asia including Malaysia (3870 reported cases in 2006–2009) (Azami et al. 2013), Thailand (46,069 reported cases in 2008–2009) (Chusri et al. 2011), and subsequently northern part of Italy (300 cases in 2007) (Moro et al. 2010; Rezza et al. 2007). Due to the huge number of suspected cases, this epidemic caught global attention. Studies of viral isolates suggested that the epidemic started in Reunion Island was due to ECSA genotype (one of the three genotypes of CHIKV) of CHIKV containing E1: A226V mutation (Schuffenecker et al. 2006; Pialoux et al.

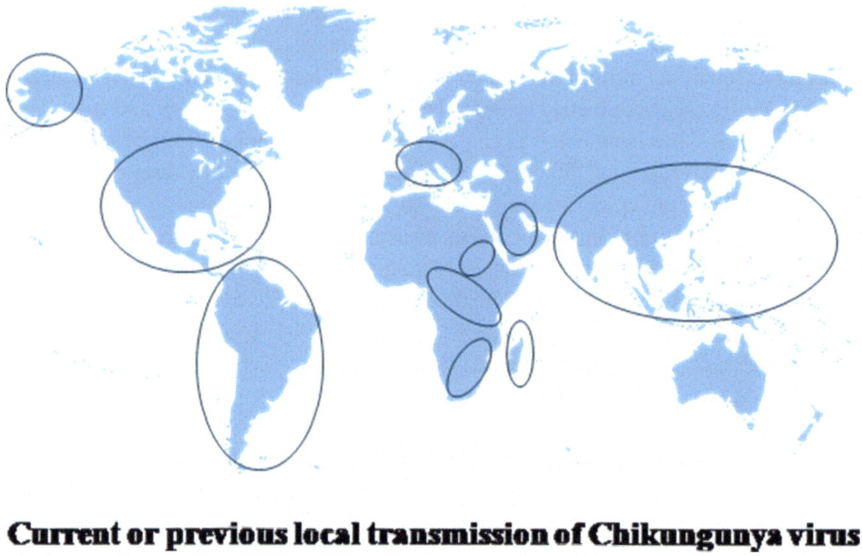

Fig. 3.1 Geographical distribution of chikungunya. Map is showing the current and previous local transmission of CHIKV in different parts of the world. The map was updated on April 22, 2016

2007). This new clade (also known as Indian Ocean lineage) of CHIKV can be effectively transmitted by *Aedes albopictus* (Charrel et al. 2007; Delatte et al. 2008).

Several outbreaks have been reported in the recent years in: Cambodia (1500 cases in 2011) (Bigongiari J., 2012 "Cambodia sees 1500 cases of Chikungunya". Vaccine News. Retrieved 14 June 2012), Papua New Guinea (1590 cases in 2012) (Horwood et al. 2013), Colombia (83,588 confirmed cases with 7 deaths in 2013–2014), United States of America (484 suspected cases, in 2013–2014), Venezuela (117,000 suspected cases in 2013–2014), and Caribbean (800,000 suspected cases in 2013–2014). As on May 13, 2016, Pan American Health Organization reported 8396 confirmed chikungunya cases from North America, Central America Isthmus, Latin Caribbean, Andean Area, and Southern Cone (www.chikungunyavirusnet.com/epidemiology.html; www.paho.org). On the basis of environmental suitability predictions, 240 million, 260 million, and 270 million people residing in the areas of Africa, Americas, and Asia, respectively, are at the risk of CHIKV transmission. Due to the micro-evolution of Indian Ocean lineage of CHIKV, Europe is at a high risk of chikungunya outbreaks (Nsoesie et al. 2016; Fischer et al. 2013).

There are three lineages of CHIKV, with distinct genotypic and antigenic properties: (i) the east, central, and south African (ECSA) genotype, (ii) the west African genotype, and (iii) the Asian genotype (Volk et al. 2010).

3.1.4 Vectors and Transmission Cycles

CHIKV follows two distinct transmission cycles, namely (Aaronson and Horvath 2002) sylvatic cycle and (Aaskov et al. 1981) urban cycle. In Africa, the virus follows the sylvatic cycle where several species of *Aedes* mosquitoes (*Aedes furcifer, Aedes africanus, Aedes luteocephalus, Aedes dalzieli, Aedes vigilax, Aedes camptorhynchites, Aedes vittatus, Aedes fulgens, and Aedes aegypti*) serve as vector and nonhuman primates as reservoir/amplifying host (Diallo et al. 1999). These enzootic vectors may infect humans residing nearby the forests and lead to outbreaks. *Aedes furcifer* is recognized as the principal enzootic vector that possibly transmits the virus from monkeys to humans (Peyrefitte et al. 2007, 2008). Other than Africa, sylvatic transmission cycle was reported from Southeast Asia, where antibodies against CHIKV and infectious CHIKV were detected in some primates (*Macaca fascicularis* and *Macaca nemestrina leonine*) (McIntosh et al. 1964; Marchette et al. 1978; Apandi et al. 2010; Nakaya et al. 2012).

In Asia, CHIKV follows the urban cycle for transmission where virus circulates between mosquitoes and humans (amplifying host). *Aedes aegypti* and *Aedes albopictus* are the major vectors of this cycle because they live in close contact with the people. *Aedes aegypti* is recognized as the principal vector in 2005–2006 epidemics in India (Yergolkar et al. 2006). *Aedes aegypti* is found in tropical and subtropical regions, whereas *Aedes albopictus* has a broader range of distribution and possibly disseminates the virus to new geographical regions (Burt et al. 2012). Although, there is evidence for the existence of sylvatic cycle in Asia, still urban cycle is the major cause for chikungunya epidemics outside Africa (Powers and Logue 2007).

3.1.5 Genomic Organization and Structure of CHIKV

As of now, several CHIKV isolates have been sequenced, such as Ross strain and the S27 CHIKV African prototype from 1952 Tanzania outbreak (Khan et al. 2002), strain from 1983 Senegal outbreak (Vanlandingham et al. 2005), isolate from Reunion Island (Charrel et al. 2007), and six isolates from Reunion and Seychelles islands (Schuffenecker et al. 2006). CHIKV is a positive-sense, single-stranded RNA virus which has a genome of approximately 11.8 kb nucleotides. The viral genome contains two open reading frames (ORFs)—ORF1 and ORF2–which encode the nonstructural proteins (nsP) and structural proteins [capsid (C) and envelope (E)], respectively (Solignat et al. 2009). The genome of CHIKV is organized as 5′ cap-nsP1-nsP2-nsP3-nsP4-(junction region)-C-E3-E2-6k-E1-poly (A) 3′ (49S RNA). The genome of this virus resembles eukaryotic mRNA as it contains a 5' 7-methylguanosine cap and a 3' poly (A) tail. The nontranslatable regions (NTRs) in the 5' and 3' proximal sequences are of 76 and 526 nucleotides long, respectively (Solignat et al. 2009). The viral mRNA translated to form two nonstructural polypeptide precursors, which cleaved and synthesized nsP1-nsP4. nsP1, has RNA capping properties and helps in the synthesis of negative strand of CHIKV RNA. nsP2 plays a major role in host's cell transcription shut-off and has RNA helicase, RNA tri-phosphatase, and proteinase activities. nsP3 is a part of replicase unit, and nsP4 is the viral RNA polymerase (Salonen et al. 2005; Garmashova et al. 2007; Fields et al. 2013). These nonstructural proteins of the virus assemble together to make the virus replication complex, which synthesizes negative strand from the viral mRNA. The negative-strand RNA serves as the template for the synthesis of subgenomic (26S) and whole-genome (49S) of CHIKV. The structural proteins are synthesized from a polyprotein precursor, which is translated from the subgenomic RNA. The structural protein pE1 is involved in virus-host cell membrane fusion, pE2 helps in receptor binding, and pE3 mediates proper folding of pE2 and further associates with pE1 (Schwartz and Albert 2010). The 6K protein helps the viral morphogenesis but is not a part of the virion (Gaedigk-Nitschko and Schlesinger 1990, 1991) (Fig. 3.2).

The mature chikungunya virion is spherical with a diameter of 70 nm and has 240 heterodimers of E1/E2 proteins assembled to form 80 glycoprotein spikes on the surface (Weaver et al. 2012). The glycoprotein spikes arrange with triangulation number (T) = 4 quasi-symmetry on the surface to make an icosahedral structure. The viral genome gets encapsulated in 240 capsid (called nucleocapsid) proteins, and then, heterodimeric protein spikes are formed on the surface of virion (Akahata et al. 2010). The spikes on the virion get inserted into the plasma membrane of the

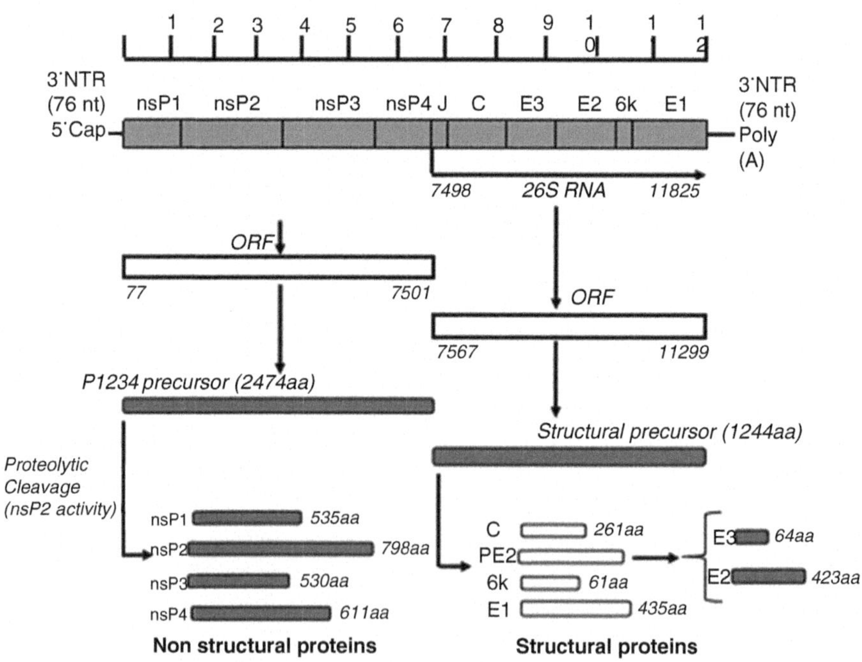

Fig. 3.2 Genome organization of CHIKV and their gene products. The CHIKV genome resembles eukaryotic mRNA which contains 5′ cap and 3′ poly (A) tail region. The genome encodes two open reading frames (ORFs). The ORF toward 5′ end transcribes to form nonstructural polypeptide that translates into nonstructural proteins (nsP1–4). The ORF toward 3′ end transcribes to form 26S RNA which serves as mRNA for the synthesis of structural proteins (C, capsid; E1, E2, E3, envelope; 6 K). Both 5′ and 3′ ends of the CHIKV genome possess nontranslated region (NTR) of 76 nucleotides length

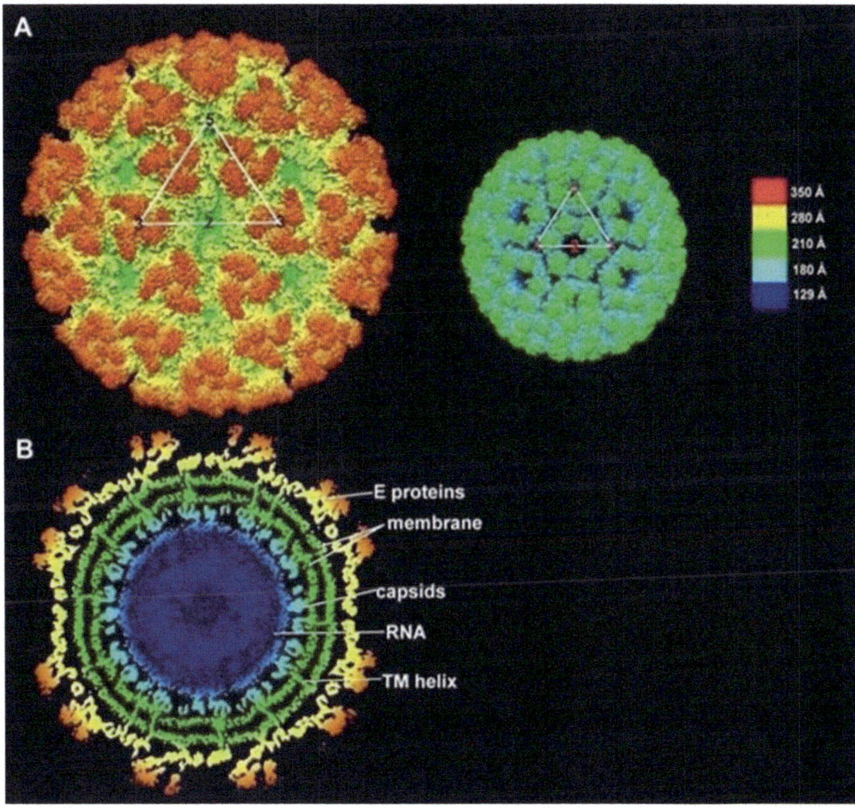

Fig. 3.3 Chikungunya virus-like particle (VLP). (**a**) Left figure, the ectodomain, and right figure, the nucleocapsid of the CHIKV surface shaded with colors. Colors are designating the radial distance from the center of the virus. White triangles are representing one icosahedral unit. (**b**) Cross section of the virus showing density above 1.5 σ also shaded with colors that designate the radial distance from the center of the virus

infected cells and acquire host's lipid on its surface after passing through secretory pathway (Weaver et al. 2012) (Fig. 3.3).

3.1.6 CHIKV Replication Cycle

3.1.6.1 Entry of Virus into Human Cells

CHIKV enters into the susceptible cells by an unknown receptor present on the cell surface. It follows clathrin-dependent endocytic pathway with low endosomal pH for entry into human cells (Solignat et al. 2009). The cholesterol-enriched domains on the plasma membrane of human cells play a major role in the entry of CHIKV (Solignat et al. 2009). The C-type lectins such as dendritic cell-specific ICAM-3-grabbing non-integrin 1 (DC-SIGN) and liver/lymph node-specific ICAM-3-grabbing non-integrin (L-SIGN)/DC-SIGN-related (DC-SIGNR), which recognizes a mannose-rich region on the viral glycoproteins, mediate CHIKV attachment and increase its infectivity (Salvador et al. 2009). The delivery of viral particle from clathrin vesicle into the endosomal compartment mediates irreversible conformational changes in the viral glycoproteins (Wahlberg et al. 1989). These changes lead to dissociation of the E1-E2 heterodimer and expose E1 fusion domain which carries out viral membrane fusion with the endosomal membrane (Kielian and Jungerwirth 1990; Wahlberg et al. 1992). This fusion encapsulates nucleocapsid of the virus and leads to the release of the viral genome (49S) into the cytoplasm (Powers et al. 2001).

3.1.6.2 CHIKV Replication and Translation

Upon release of 49S RNA into the host's cell cytoplasm, the translation of viral genome starts. The 5′ ORF of the virus translates into nonstructural polyprotein (nsP1234), which participates in virus replication. Once delivered into host's cell cytoplasm, the capsid protein binds to the large ribosomal subunit which is likely responsible for disassembly of nucleocapsid in *Alphaviruses* (Singh and Helenius 1992). The 5′ ORF translates as nsP1234 polyprotein in CHIKV, whereas nsP123 polyprotein is formed in some *Alphaviruses* due to the presence of a leaky termination codon in between nsP3 and nsP4 (Li and Rice 1993). The protease activity of nsP2 helps in the cleavage of nsP1234 polypeptide. At first, the polypeptide is cleaved into nsP123 and nsP4; afterward the nsP123 is cleaved to form nsP1 and nsP23. The nsP1, nsP23, and nsP4 along with the host proteins form a replication complex for the synthesis of negative strand of the virus. Finally, nsP23 is cleaved into nsP2 and nsP3 (Lulla et al. 2012). All the mature nonstructural proteins assemble to form a replication complex that synthesizes the subgenomic (26S) and the whole-genome (49S) of the virus (Fayzulin and Frolov 2004; Frolov et al. 2001).

The 26S RNA serves as a template for the synthesis of structural polyproteins' precursor C-pE2-6K-E1. Subsequently, the precursor undergoes co-translational and posttranslational modifications to form C (capsid), pE2 and E1, and 6K (may act as a signal sequence for the translocation of E1) proteins. The C-terminal region of the capsid protein contains a conserved auto-protease domain (such as serine protease). Thus, the capsid protein is released from the nascent polypeptide by auto-protease activity. The N-terminal signal sequence present in the nascent polypeptide directs it to the lumen of endoplasmic reticulum, where signalase catalyzes the cleavage of pE2 from 6K. pE2 and E1 are glycosylated in the endoplasmic reticulum (Garoff et al. 1990). The pE2-E1 heterodimers are formed in the Golgi apparatus and transported to the cell surface where the heterodimers trimerize to form the spike. During the process of translocation, pE2 is cleaved by furin (host protease) and the PC5A convertase to form E2 and E3 (de Curtis and Simons 1988; Ozden et al. 2008). E2 protein is involved in receptor binding while E3 in initial folding of the virus (Lobigs and Garoff 1990). The capsid protein associates with 49S RNA to form nucleocapsid complex. The nucleocapsids then combine with the viral spike to form mature structural proteins and subsequently with host-lipid bilayer to form the mature virus particles. The newly formed virus particles then bud out from the host's cell membrane which is controlled by interaction between E2 cytoplasmic domains with the nucleocapsids (Lopez et al. 1994) (Fig. 3.4).

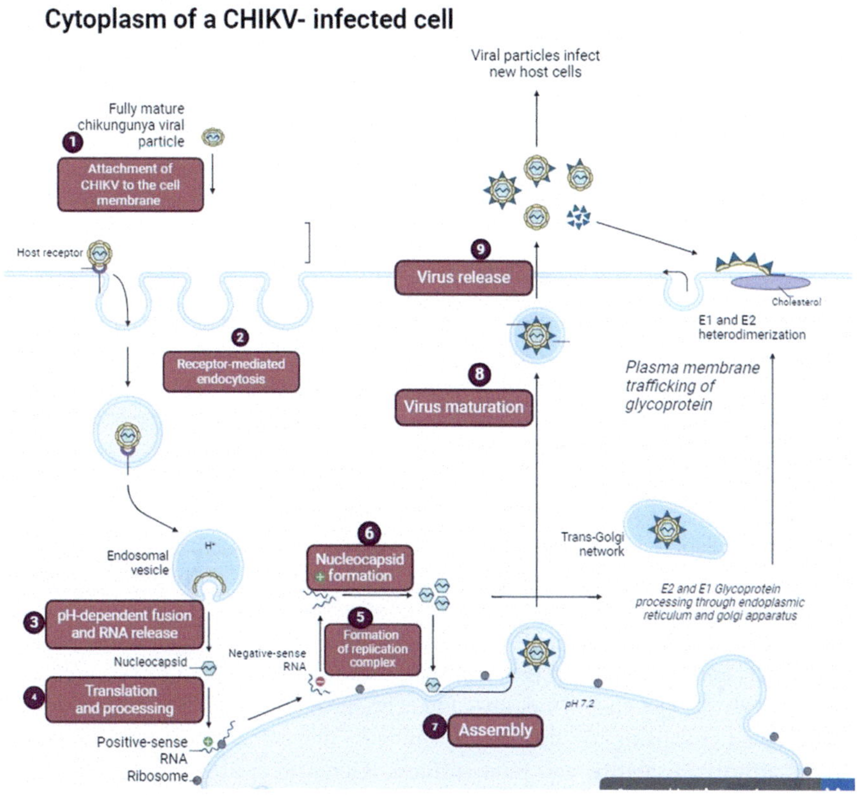

Fig. 3.4 Replication cycle of CHIKV. (**a**) CHIKV enters into the susceptible cells by an unknown receptor present on the cell surface. (**b**) Following endocytosis, the acidic environment of the endosomal compartment triggers the conformational changes in viral envelopes that expose E1 structural protein resulting in the fusion of host and virus membranes. This allows release of nucleocapsids into the cytoplasm. (**c**) First, the viral genome translates into the 4 nsPs that assemble to form the replication complex. (**d**) In the next step, synthesis of negative strand takes place which is used as the template for the 49S viral RNA and 26S subgenomic mRNA synthesis in the subsequent phase. (**e**) Then, the 26S subgenomic mRNA drives the expression of structural proteins (C-pE2-6 K-E1) that are subsequently processed through autoproteolytic cleavage by serine proteases and the capsid is released into the cytoplasm. (**f**) Further, the remaining structural proteins undergo posttranslational modifications in the endoplasmic reticulum and subsequently in the Golgi apparatus. E1 and E2 assemble to form dimers and are transported to the host plasma membrane, where they get incorporated onto the virion surface as trimeric spikes. (**g**) Capsid proteins form the icosahedral nucleocapsids that encapsulate viral genomic RNA transcribed from 49S subgenome before the formation of mature virions. (**h**) The virions then bud out from the host's plasma membrane and acquire a membrane bilayer from the host

3.1.7 Clinical Manifestations

3.1.7.1 Acute Phase

The acute phase of chikungunya lasts for approximately 2 weeks. The symptoms of this disease appear after 4–7 days of initial infection with CHIKV (Dupuis-Maguiraga et al. 2012) and, however, resolve within the acute phase. During CHIKV transmission, the virus replicates in the skin and then disseminates into liver and joints, likely through the blood (Weaver et al. 2012). The acute phase symptoms include high fever (>38.9°C) (Erin Staples et al. 2009), joints pain (mostly symmetrical) usually in wrists, elbows, fingers, knees, and ankles but can also affect more-proximal joints (Simon et al. 2007), rash (Taubitz et al. 2007), headache, fatigue, nausea, vomiting, conjunctivitis, and myalgias (does not occur in febrile illness) (Erin Staples et al. 2009). High levels of viremia (10^5–10^{12} RNA copies per milliliter of blood as quantitated by real-time PCR) have also been detected in patients (Das et al. 2010). The case fatality rate is estimated to be 1:1000, with most of the deaths observed in neonates, in adults with abovementioned conditions, and in older people (Lemant et al. 2008). Asymptomatic CHIKV infection cases are also reported and are about 3–25% (Queyriaux et al. 2008; Sissoko et al. 2008). CHIKV can infect individuals of all age groups and any gender (Jaffar-Bandjee et al. 2010).

3.1.7.2 Chronic Phase

The acute phase lasts for 2 weeks, when most of the symptoms resolve, but in some individuals, the joint pain (polyarthralgia and/or polyarthritis) can persist for several months to years. The joint pain is considered the hallmark of chronic phase of chikungunya (Kelvin et al. 2011; Dupuis-Maguiraga et al. 2012). The chronic polyarthritis is mostly symmetrical and at times erosive. It involves small and large joints of the hands and feet, mimicking rheumatoid arthritis (RA) (Ganu and Ganu 2011; Mathew et al. 2011). Other symptoms of the chronic phase include arthralgia, myalgia, asthenia, relapsing or lingering rheumatic musculoskeletal pain (RMSP), chronic arthritic disability, and joint stiffness (Gerardin et al. 2013; Schilte et al. 2013; Sissoko et al. 2009; Borgherini et al. 2008; de Andrade et al. 2010; Couturier et al. 2012; Larrieu et al. 2010; Moro et al. 2012; Chopra et al. 2012; Kularatne et al. 2012; Brighton et al. 1983). The nature of joint pain during and after CHIKV infection is presumably inflammatory (Brighton et al. 1983; Manimunda et al. 2010; Rodriguez-Morales et al. 2016). Although chronic phase of this disease resembles RA, there are no classic symptoms of RA such as cartilage and bone erosion in chikungunya (Weaver et al. 2012). Manimunda et al. reported the presence of anticyclic-citrullinated peptide (anti-CCP), while absence of rheumatoid factor (RF) in CHIKV-infected patients with chronic arthritis (Manimunda et al. 2010). Reports suggested that 40–80% of the patients infected with CHIKV develop chronic rheumatic symptoms (Chopra et al. 2012; Win et al. 2010; Sissoko et al. 2009; Schilte et al. 2013; Borgherini et al. 2008; Essackjee et al. 2013; Larrieu et al. 2010). Prospective studies from India and La Réunion showed that 46 and 69% of patients experienced persistent joint pain after 10 and 36 months of acute chikungunya, respectively (Manimunda et al. 2010; Schilte et al. 2013).

3.1.7.3 Other Severe Manifestations Associated with Chikungunya

During 2005–2006 outbreaks in La Réunion Islands, many unusual clinical presentations came into picture. The manifestations included hepatitis, autoimmune neurologic pathologies (Guillain–Barré syndrome), cardiologic manifestations, and deaths (Borgherini et al. 2007; Economopoulou et al. 2009). Mother-to-child transmission of CHIKV with high rates of morbidity was reported for the first time from this outbreak. Vertical transmission of virus from mother to fetus associated with congenital illness and fetal death was also reported (Gerardin et al. 2008). Studies reported the ophthalmologic involvement manifested as optic neuritis, retinitis, episcleritis, and iridocyclitis in chikungunya (Mahendradas et al. 2008; Rose et al. 2011). Prevalence of mixed cryoglobulinemia in patients infected with CHIKV was documented (Oliver et al. 2009). Clinical abnormalities such as leukopenia (38%), thrombocytopenia (37%), increased aspartate amino-transferase and alanine amino-transferase levels in blood (32 and 7%, respectively), and hypocalcemia (39%) were observed in a retrospective patients' study including 157 individuals. Lymphopenia ($<1000/mm^3$ of blood) associated with viremic patients in chikungunya has also been documented (Borgherini et al. 2007).

3.1.8 Route of Infection and Virus Dissemination to Target Cells

CHIKV infects the host through mosquitoe bite at the skin, where it can infect the resident fibroblasts and dermal macrophages (Sourisseau et al. 2007; Couderc et al. 2008; Kam et al. 2009). It is the place where the first round of virus replication and counteraction of host's immune system against invading pathogen starts (Kupper and Fuhlbrigge 2004; Schilte et al. 2010). However, CHIKV rapidly disseminates into the lymph nodes in spite of the host's immune response and subsequently into other tissues (e.g., muscles) through the circulatory system (Kam et al. 2009). The replication of the virus in these peripheral tissues is called the viremic phase of chikungunya. During this phase, CHIKV can be transmitted from human to mosquito through blood meal. The viremic phase lasts for 2–10 days post-infection (Panning et al. 2008; Kam et al. 2009).

3.1.9 CHIKV Susceptible Cells

For the understanding of virus pathogenesis, it is important to know the cells susceptible to CHIKV. CHIKV can infect a broad range of cell types including primary and secondary cell lines. The virus can infect and replicate in epithelial, endothelial, and primary fibroblast cells (Sourisseau et al. 2007; Salvador et al. 2009; Wikan et al. 2012). The human muscle satellite cells are also susceptible to this virus, while it cannot infect the differentiated myotubes (Ozden et al. 2007). Furthermore, Rohatgi et al. have demonstrated that infection of muscle connective tissue fibroblasts and myofibers was associated with severity of CHIKV infection in murine model (Rohatgi et al. 2014). Issac et al. have shown the susceptibility of human rhabdomyosarcoma cell line (SJCRH30) to CHIKV infection (Issac et al. 2014). There are contrasting reports on the susceptibility of peripheral blood mononuclear cells (PBMCs) to CHIKV (Sourisseau et al. 2007; Her et al. 2010; Teng et al. 2012).

3.1.10 Animal Models of CHIKV

Several animal models have been developed to study CHIKV-induced pathophysiology and therapies against chikungunya.

3.1.10.1 Mouse Models

CHIKV pathogenesis has been studied using mouse as a model since many years. It has been seen that age, genetic background, and route of virus inoculation are the key factors for virus pathology outcomes (especially the rheumatic manifestations) in these mouse models. People have implemented neonates/young mice (Ziegler et al. 2008) or immune-compromised mice (Couderc et al. 2008; White et al. 2011; Werneke et al. 2011; Gardner et al. 2010) as CHIKV infection models. However, these studied mouse models are not suitable for vaccine response study/pathologic outcomes in CHIKV infection, because they die before appearance of the symptoms due to immature immune system.

Studies on adult C57BL/6 wild-type mice have shown some pathophysiology resembling the pathological outcomes in humans after CHIKV infection. CHIKV inoculations on the ventral side of the footpad of adult C57BL/6 mice resulted in the development of rheumatic manifestations as seen in humans (Gardner et al. 2010; Lum et al. 2013; Teo et al. 2013). Viremia has also been detected in the spleen, lymph nodes, liver, and muscles of these mice which are in line with the data obtained from in vitro studies. The systemic viremia in C57BL/6 mice after infection appears such as CHIKV-infected patients (Gardner et al. 2010; Lum et al. 2013; Teo et al. 2013). A study on C57BL/6 young mouse model inoculated with CHIKV by subcutaneous route has shown foot swelling, arthritis, tenosynovitis, and myositis as similar to CHIKV-induced disease features in human (Morrison et al. 2011). However, the younger mice demonstrated severe form of the disease than the adult C57BL/6 mice which suggests role of age in pathological outcome. The adult C57BL/6 wild-type mice have been considered suitable model for CHIKV infection and vaccine studies (Wang et al. 2011; Prow et al. 2010; Metz et al. 2013b; Garcia-Arriaza et al. 2014; Kam et al. 2012b; Teo et al. 2012; Rudd et al. 2012; Nakaya et al. 2012).

3.1.10.2 Nonhuman Primate Models

Nonhuman primates are the natural reservoir in sylvatic cycle of CHIKV transmission in Africa. In addition, their physiology and immune system resemble those of human beings. Thus, they are suitable models for the study of CHIKV pathogenesis and assessment of therapies. Labadie et al. have used the intravenous/intradermal route of CHIKV inoculation to infect the cynomolgus macaques (*Macaca fascicularis*). CHIKV-infected macaques have shown infection of lymphoid tissues, liver, central nervous system, joints, and muscles during the acute phase of infection. The macrophages, dendritic cells, and some endothelial cells are found to be mainly infected. At the later stage of infection, they have found the persistence of CHIKV in lymphoid organs, liver, joints, and muscles. The authors have also observed the presence of CHIKV in macrophages up to 3 months after the initial virus inoculation. Thus, macrophages are considered to be the main cellular reservoir of CHIKV that may be responsible for the long-lasting symptoms as observed in humans (Labadie et al. 2010). The resemblance of symptoms in CHIKV-infected macaques and humans has made this model ideal for new intervention studies against CHIKV. Later, this primate model has been used for the investigation of anti-CHIKV antibody's specificity against "E2EP3" (an epitope)-based vaccine candidate (Kam et al. 2012a, b, c). Furthermore, vaccine efficacies have been assessed in cynomolgus macaques using a single dose of live-attenuated vaccine which showed protection against CHIKV (Roy et al. 2014).

Akahata et al. have used rhesus macaques (*Macaca mulatta*) to evaluate humoral responses induced by CHIKV virus-like particles (VLPs) which have conferred protection against CHIKV infection (Akahata et al. 2010). Rhesus macaques immunized with live-attenuated vaccine have been used for challenge study against CHIKV (Levitt et al. 1986). *Cercopithecus* monkeys, baboons, and bonnet monkeys (*Macaca radiata*) are also being used for CHIKV infection studies (McIntosh et al. 1963; Paul and Singh 1968).

3.2 Host Cellular Response Against CHIKV

In vivo and in vitro studies have shown that CHIKV infection induces cytopathic events with rapid onset of apoptosis at the cellular level (Sourisseau et al. 2007; Dhanwani et al. 2011, 2012; Wikan et al. 2012; Puttamallesh et al. 2013). Apoptosis is thought to be a host protective mechanism for the containment of the pathogen, but different viruses including alphaviruses can hijack this machinery for their survival in the host (Griffin and Hardwick 1997; Chen et al. 2006). CHIKV uses apoptotic blebs from the dyeing cells to infect neighboring cells and macrophages (Krejbich-Trotot et al. 2011a, b). This is a clever evasion strategy adopted by CHIKV to conceal from host's immune system (Long and Heise 2015).

Autophagy is another cellular process that has been investigated in the context of alphavirus infection. It has been reported that autophagy in human cells increases CHIKV replication while decreases replication in mouse embryonic fibroblasts (Krejbich-Trotot et al. 2011a, b). Later, it has been demonstrated that the human autophagy receptor NDP52 interacts with viral nsP2 that promotes virus replication which is not in the case of mouse ortholog of NDP52 (Judith et al. 2013)

Heat shock proteins (HSPs) are known to be involved in cellular stress response where they mediate protein folding/unfolding and protein transport and provide protection against apoptosis. Furthermore, HSPs can stimulate professional antigen-presenting cells (Long and Heise 2015). Interaction between HSP90 and CHIKV nsP3/nsP4 with higher viral replication has been demonstrated in vitro and in mouse model (Rathore et al. 2014). In an in vitro study, HSP90 is shown to stabilize viral nsP2 and that promotes CHIKV replication (Das et al. 2014). In addition, HSP70 is shown to bind CHIKV in both HEK-293 and Vero-E6 cells (Paingankar and Arankalle 2014). Recently, Scholte et al. have suggested proviral role of GTPase-activating protein (SH3 domain)-binding protein 1 and 2 (key markers of stress granule) in CHIKV replication (Scholte et al. 2015).

3.2.1 Host Immune Response Against CHIKV

Chikungunya infection is known to be self-limiting, and most of the acute phase symptoms resolve within 2 weeks. However, the chronic rheumatic symptoms last for months to years in a very small percentage of patients (Suhrbier et al. 2012). CHIKV viremia has been reported to disappear from the peripheral circulation within 2 to 10 days post-infection (Panning et al. 2008). Therefore, the virus elimination probably starts before induction of IgG antibodies and T-cell responses in the host that indicates a significant role of the innate immune system in controlling disease at the initial phase of infection.

3.2.1.1 Innate Responses to CHIKV

Host's innate immune response plays prime role in providing protection before induction of a specific response against viral infections. The innate immune system can sense the presence of viruses, limit viral replication, and ultimately assist in the induction of effective adaptive immune response. Therefore, immunological research is focused on how the innate immune system protects the host from CHIKV infection (Wauquier et al. 2011; Long et al. 2013; Thanapati et al. 2015; Van Den Bossche et al. 2015). However, in addition to protective response, some components of the innate system are also involved in the development of CHIKV-induced pathologies in the mammals (Long et al. 2015).

3.2.1.1.1 Type I Interferon

Type I interferon (IFN) pathway has a direct antiviral role in vertebrates and may also regulate the production of pro-inflammatory cytokines that induce a potent host immune response. IFNs and IFN-stimulated genes (ISGs) act through various mechanisms against virus invasions (Akira and Takeda 2004; Stetson and Medzhitov 2006). IFN was first described in 1957 by Isaacs and Lindenmann as an antiviral protein for influenza virus (Isaacs and Lindenmann 1957). In 1963, Gifford and Heller have shown IFN production in chicken fibroblasts 3 h post-CHIKV infection (Gifford and Heller 1963). It has been reported that CHIKV can elicit type I IFN responses along with the production of other pro-inflammatory cytokines in humans, nonhuman primates, and mice (Gardner et al. 2010; Labadie et al. 2010; Chow et al. 2011; Wauquier et al. 2011; Teng et al. 2012). A study in zebrafish model revealed that type I response is critical for survival upon CHIKV infection (Palha et al. 2013). IFN-α and IFN-β, in particular, play a crucial antiviral role as being reported in humans and mice and in cell lines of both the species infected with CHIKV (Sourisseau et al. 2007; Couderc et al. 2008; Gardner et al. 2010; Schilte et al. 2012; Teng et al. 2012). Mice deficient of both IFN-α and IFN-β receptor (IFNAR$^{-/-}$) develop severe form of disease with higher viral titers and central nervous system involvement after CHIKV infection (Couderc et al. 2008; Schilte et al. 2010). Fros et al. have reported that nsP2 protein of CHIKV has the ability to block type I IFN pathway leading to lower expression of antiviral mediators (Fros et al. 2010). Type I IFN production upon CHIKV infection is mediated through cellular pattern recognition receptors (PRRs), such as toll-like receptors (TLRs), retinoic acid-inducible gene I (RIG-I)-like receptor, the melanoma differentiation-associated protein 5 (MDA5) (Schilte et al. 2010; Teng et al. 2011; Her et al. 2014; Priya et al. 2014). The RIG-I and MDA5 sense the presence of double-stranded RNA (dsRNA) intermediates during CHIKV replication upon infection. Interaction of viral intermediates with innate receptors activates the downstream signaling cascade that involves the CARD adaptor inducing IFN-β adaptor protein (CARDIF) and crucial interferon regulatory factors (IRF) 3 and 7 (Kawai et al. 2005; Honda et al. 2006; Stetson and Medzhitov 2006). The activation of signaling cascade leads to the production of type I IFNs which mediate the ISGs production through Janus kinase (JAK) and signal transducers and activators of transcription (STAT) signaling pathway (Decker et al. 2002; Stetson and Medzhitov 2006). Anti-CHIKV properties of ISGs such as viperin, GADD44, tetherin (BST-2), ISG15, Mx proteins, protein kinase R (PKR), and 2',5'-oligoadenylate synthase 3 (OAS3) have been reported (Brehin et al. 2009; Schoggins et al. 2011; Werneke et al. 2011; White et al. 2011; Clavarino et al. 2012; Teng et al. 2012; Jones et al. 2013).

Induction of antiviral ISGs through TLR signaling is also described (Kawai and Akira 2010). Recently, Her et al. have demonstrated the importance of TLR3 in controlling CHIKV infection in hematopoietic cells (Her et al. 2014). It is known that TLR3 can sense dsRNA. A study in TLR3 deficient mice has shown pronounced disease with profound viremia (Rudd et al. 2012). Higher levels of IFN-α, IFN-β, and TLR3 transcripts in the PBMCs of CHIKV-infected patients than in controls have been detected (Teng et al. 2012; Her et al. 2014). In a human fibroblast culture study, ISG viperin is shown to be upregulated following CHIKV infection (White et al. 2011). Further, mice deficient of viperin suffer with higher viremia along with severe joint inflammation compared with wild-type mice after CHIKV infection (Kam et al. 2012a, b, c). Bone marrow stromal antigen 2 (BST-2) or tetherin possibly limits CHIKV spread by restraining tethering of budding virus-like particles to the host cell membrane (Jones et al. 2013). In vivo study showed BST-2 deficiency results in increased CHIKV titers and tropism for lymphoid tissues (Mahauad-Fernandez et al. 2014). These observations further corroborate that the progression of CHIKV infection could be controlled through the combined effects of two or more sensor pathways which are derived from both hematopoietic and non-hematopoietic cells (Schilte et al. 2010; Her et al. 2014). In toto, these observations suggest that type I IFNs play a critical role in controlling CHIKV replication at the early phase of chikungunya.

3.2.1.1.2 Cytokines and Chemokines

Cytokines and chemokines are the major regulators of the immune system and their critical contribution toward viral immune pathogenesis is well known. Pro-inflammatory cytokines such as interleukin (IL)-1β and IL-6 are found to be elevated in the patients with chikungunya fever (Ng et al. 2009). These cytokines return to their normal range after fever and viremia disappeared from the patients (Wauquier et al. 2011; Chow et al. 2011; Kelvin et al. 2011). Several patient cohort studies have shown higher levels of IL-6, monocyte chemotactic protein 1 (MCP-1), and IFN-ϒ during acute phase of chikungunya (Ng et al. 2009; Hoarau et al. 2010; Chow et al. 2011; Kelvin et al. 2011; Wauquier et al. 2011). Positive correlation of IL-6 or MCP-1 with high viral load is observed in CHIKV-infected patients (Chow et al. 2011). Increased production of IL-1β, IL-6, and TNF-α has been reported in chikungunya patients having severe joint pathology associated with inflammation and tissue destruction (Ng et al. 2009; Hoarau et al. 2010; Chow et al. 2011). Elevated MCP-1, IL-6, IL-8, macrophage inflammatory protein (MIP)-1α, and MIP-1β levels in the chronic chikungunya patients have been corroborated with arthritides (Chaaitanya et al. 2011). The aforementioned cytokines and chemokines have been described in relation to arthritides, including epidemic polyarthritis (EPA) caused by RRV and rheumatoid arthritis (Chaaitanya et al. 2011).

Recently, a meta-analysis study by Teng et al. has identified pro-inflammatory cytokines such as IFN-α, IFN-ϒ, IL-2, IL-2R, IL-6, IL-7, IL-12, IL-15, IL-17, and IL-18; anti-inflammatory cytokines such as IL-1Rα, IL-4 and IL-10; chemokines: granulocyte colony-stimulating factor (GM-CSF), IP-10, MCP-1, monokine induced by gamma interferon (MIG), MIP-1α, and MIP-1β; and growth factor: basic fibroblast growth factor (FGF) that are common in chikungunya patient cohorts in spite of their differences in geographical origin and ethnicity (Teng et al. 2015). A study in Chikungunya patients has reported increased levels of IL-6, granulocyte colony-stimulating factor (G-CSF), GM-CSF, MCP-1, and TNF-α in acute patients' sera compared with the control sera while these cytokines levels were lower in convalescent patients' sera (Chirathaworn et al. 2013). Role of regulatory T cells and IL-10 with recovery from disease has suggested multifaceted character of IL-10 with disease outcome in Chikungunya (Kulkarni et al. 2017).

The production of IFN-α, IL-6, and CCL2 (MCP-1), followed by IFN-ϒ, TNF-α, CCL3 (MIP-1α), and CCL4 (MIP-1β), has been observed in the early stage of CHIKV infection in a nonhuman primate model (Labadie et al. 2010).

3.2.1.1.3 Innate Immune Cells in CHIKV infection

Monocytes/Macrophages

Monocytes originate from the myeloid progenitor cells in the bone marrow by the process of hematopoiesis. The monocytes are released into the bloodstream and then migrate to tissues where they differentiate into macrophages and dendritic cells (Auffray et al. 2009). Studies on humans and mouse models have demonstrated higher levels of MCP-1, a monocyte/macrophage chemo-attractant in the acute phase of chikungunya (Chen et al. 2010, 2015; Gardner et al. 2010; Hoarau et al. 2010; Teng et al. 2015). Elevated MCP-1 has been shown to be accompanied by increased infiltration of monocytes into the site of inflammation in CHIKV-infected animal models (Gardner et al. 2010; Labadie et al. 2010; Poo et al. 2014a, b). Her et al. have shown that monocytes are the main site of replication during acute phase of CHIKV infection (Her et al. 2010). Primary monocytes from humans, mice, and macaques have also been shown to be infected by CHIKV (Labadie et al. 2010; Her et al. 2010; Gardner et al. 2010).

The above observations suggest a critical role of monocytes/macrophages in CHIKV-induced disease pathology. Activated macrophages are the source of inflammatory mediators (including prostaglandins, TNF-α, and IL-6), which are known to be involved in tissue damage, inflammation, and arthritis (Suhrbier and Mahalingam 2009). A study in CHIKV-infected mouse model revealed that liposome-containing clodronate (known to deplete macrophages) treatment alleviates CHIKV-induced pathology (Gardner et al. 2010). Recently, Chen et al. have reported higher levels of MCP-1, MCP-2, and MCP-3 in the joints of CHIKV-infected mice (Chen et al. 2015). In addition, several studies demonstrated that treatment of bindarit, a MCP inhibitor (Bhatia et al. 2005), entirely abolishes CHIKV-induced pathology (Rulli et al. 2011; Chen et al. 2015). Data from human and animal studies have suggested that monocytes could also be the cellular vehicle for virus dissemination along with the main reservoir for persistent CHIKV infection (Her et al. 2010; Hoarau et al. 2010; Labadie et al. 2010; Stoermer et al. 2012). Further, CHIKV-infected CCR2-deficient mice showed pronounced arthritic manifestations with no effects on viral replication (Poo et al. 2014a, b). In toto, the observations suggest monocytes/macrophages are the main cellular reservoirs and are associated with virus persistence.

Dendritic Cells (DCs)

DCs are the professional antigen-presenting cells identified in 1868 by Paul Langerhans (Langerhans 1868) who thought they are the part of the nervous system. In 1973, Ralph Steinman and Zanvil Cohn showed that DCs are white blood cells with distinctive characteristics (Steinman and Cohn 1973). Study from acute chikungunya patients has demonstrated the activation of plasmacytoid DCs (pDCs) (Hoarau et al. 2010). pDCs are the known producers of type I IFNs and are believed to initiate and activate natural killer (NK) cell-mediated cytolysis (Siegal et al. 1999; Colonna et al. 2004). However, Schilte et al. have shown that pDCs from acute chikungunya patients cannot secret type I IFNs (Schilte et al. 2010).

Study on dendritic cell immune receptor (DCIR)-deficient mouse has revealed their protective role during CHIKV infection (Long et al. 2013). In toto, these observations suggest that DCs impart protection during CHIKV infection. However, the literature on DCs involvement in CHIKV pathogenesis is least documented and more research is needed.

NK Cells and NK-like T cells

The literature defining the role of on NK cells in alphavirus including CHIKV infections is less documented. A study from chikungunya patients has shown transient expansion of NK cells expressing CD94/NKG2C during the acute phase of the disease that positively correlated with the viral load (Petitdemange et al. 2011). Higher activating receptors and perforin on NK cells with increased NK cell cytotoxicity activity have been observed in acute Chikungunya patients. Increased frequency of IFN-γ expressing NK-like T cells in the convalescent patients has suggested critical role of NK/NK-like T cells immune response in the early stages of Chikungunya (Thanapati et al. 2015). NK-like T cells are the non-classical NKT cells that recognize antigens presented by classical MHC. These lymphocytes are a subset of αβ T cells that express NK-activating receptors and also exhibit a specific effector memory phenotype (Peralbo et al. 2007; Tang et al. 2013). Upregulation of NK cell inhibitory receptors with lower perforin and CD107a expression on NK cells in chronic patients have abrogated NK cell functions and could be one of the reason for CHIKV-induced chronicity (Thanapati et al. 2017b). A study in chronic Chikungunya and rheumatoid patients has revealed pathogenic role of NK-like T cells in chronic arthritic conditions. TNF-α+ NK-like T and IFN-γ+ NK-like T cells have been emerged as the markers for chronic arthritic diseases (Thanapati et al. 2017a). Study on nonhuman primates has demonstrated increased NK cells number in the early stages of CHIKV infection and suggested their participation in the early control of virus replication (Labadie et al. 2010). In human RRV infection, rapid expansion of NK cells has been reported (Aaskov et al. 1981, 1987). Both protective and pathogenic roles of NK cells have been described in alphavirus infections. In a mouse model study, infiltration of NK cells in the arthritic feet of mice following CHIKV infection has been observed (Gardner et al. 2010). This observation is in line with the higher IL-12 levels in the feet of CHIKV-infected mice (Stoermer et al. 2012). Furthermore, Teo et al. have demonstrated pathogenic role of NK cells toward the development of joint pathology in mice infected with CHIKV from Indian ocean isolates (LR2006 OPY1) (Teo et al. 2015).

Due to the conflicting reports on NK cells, their contribution toward CHIKV-induced arthritis is not clear. However, disease-promoting role of NK cells has been suspected in rheumatoid arthritis as they produce broad range of pro-inflammatory cytokines such as IFN-γ, receptor activator of nuclear factor kappa-B ligand, and macrophage colony-stimulating factor (Ahern and Brennan 2011).

3.2.1.2 Fine-Tuning of Innate Immune Response by Cytokine Signaling Inhibitors

The reports available on CHIKV pathogenesis suggest a potential role of innate immune response to control chikungunya progression. Activation of innate immune responses in chikungunya includes the involvement of innate immune cells such as monocytes, DCs, and NK cells and also the production of pro-inflammatory cytokines. Induction of pro-inflammatory responses has the capability to limit the pathogen progression and the ability of harming the host. Thus, activation of innate immune response needs to be tightly controlled to avoid unwanted immune-pathologic side effects. Although a lot of studies are available describing the activation process within innate immunity, the questions of termination and inhibition are still largely unsolved. However, in clinical situations, it is very clear that failures in negative-regulatory circuits can have detrimental outcomes (Dalpke et al. 2008).

A large number of cytokines are reported to be produced during chikungunya. In mammals, Janus kinase/signal transducers and activators of transcription (JAK/STAT) pathway is the major signaling mechanism that is used by many cytokines and growth factors (Rawlings et al. 2004) (Fig. 3.5a). After binding to the receptors, cytokines induce receptor dimerization on the cell surface and recruit JAK tyrosine kinases in the cytoplasmic domain which juxtaposes JAKs and allows their trans-phosphorylation. The trans-phosphorylation of JAKs subsequently phosphorylates

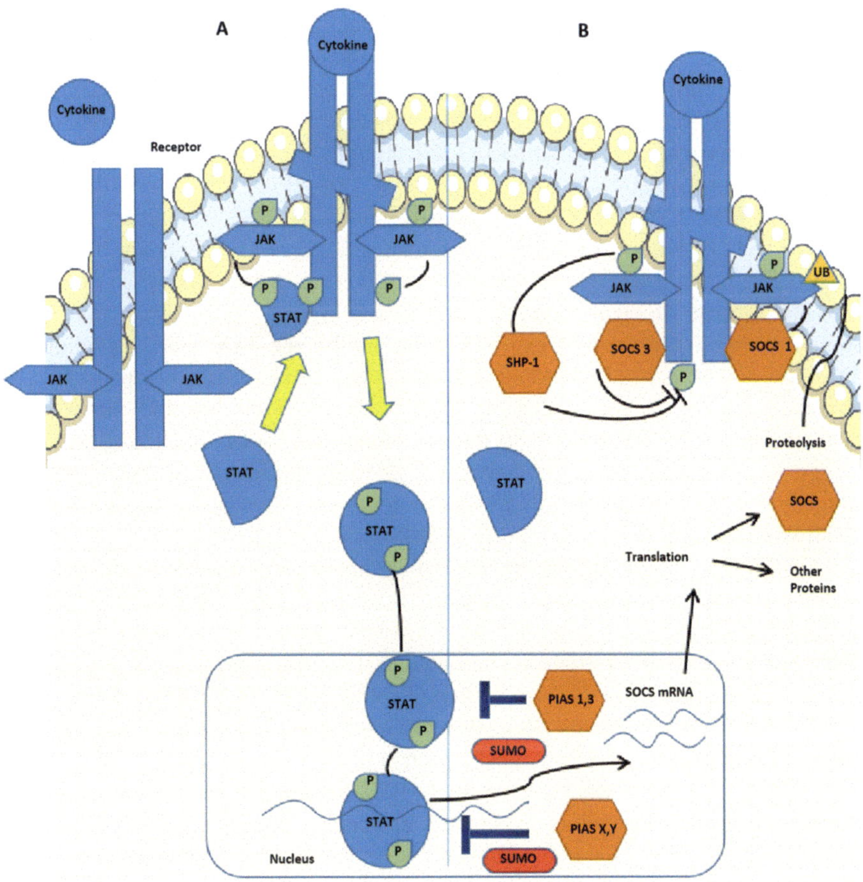

Fig. 3.5 JAK/STAT signaling pathway and the role of cytokine signaling inhibitors. (**a**) Binding of cytokine to its receptor leads to receptor dimerization and activates receptor-associated JAKs. Activated JAK phosphorylates the receptor and facilitates binding of STATs. JAK also phosphorylates STAT. Upon phosphorylation, STAT proteins can dimerize and move into the nucleus where it induces transcription of cytokine-regulated genes. (**b**) Regulation of JAK/STAT pathway by cytokine signaling inhibitors. SHP-1 can repress the pathway by dephosphorylating the activated JAKs or receptors. PIAS proteins can sumoylate STATs and inhibit their transcriptional activation. SOCS proteins can bind to the JAKs or receptor or can compete with STATs for binding to the receptor. (Adapted from Wormald and Hilton 2004)

STATs which are present in the cytoplasm. After phosphorylation, the STATs become active and migrate to nucleus where they regulate transcription of target genes (Xu et al. 2016). However, this signaling pathway is finely tuned by some negative regulators (Greenhalgh and Hilton 2001).

There are three major classes of negative regulators that contribute to cytokine signaling inhibition: suppressors of cytokine signaling (SOCS), protein inhibitors of activated STATs (PIAS), and protein tyrosine phosphatases (PTPs) (Krebs and Hilton 2000; Greenhalgh and Hilton 2001; Wormald and Hilton 2004). PTPs are the simplest negative regulators among these three that reverse the activity of JAKs. Src homology domain 2 containing protein tyrosine phosphatase1 (SHP-1)/protein tyrosine phosphatase non-receptor6 (PTPN6) is the best characterized PTP that can bind to either phosphorylated JAKs or phosphorylated receptors to facilitate dephosphorylation of these activated signaling molecules. CD45 is another PTP that appears to have a role in regulating JAK/STAT signaling through a subset of receptors (Rawlings et al. 2004; Wormald and Hilton 2004) (Fig. 3.5b). The PIAS family consists of PIAS1, PIAS2, PIASx, and PIASy proteins. They can inhibit the function of STATs by binding directly to the activated STAT and prevent their binding to DNA (Liu et al. 2004; Wormald and Hilton 2004) (Fig. 3.5b). The SOCS family of proteins consists of eight members: SOCS 1 to 7 and cytokine-inducible SH2 domain-containing protein (CIS). SOCS can inhibit the JAK/STAT signaling by three different ways: (1) By binding to the phosphotyrosines on the receptors, SOCS physically blocks the recruitment of STATs; (2) SOCS inhibits JAK activity by binding directly to the JAKs or to the receptors; and (3) by interacting with the elongin BC complex and cullin 2 which leads to ubiquitination of JAKs and presumably the receptors and finally toward proteasomal degradation (Rawlings et al. 2004; Wormald and Hilton 2004) (Fig. 3.5b).

JAK/STAT signaling pathway is one of the primary pathways that regulate innate immune responses against viral infections in a broad range of species including drosophila to vertebrates (Aaronson and Horvath 2002; Randall and Goodbourn 2008; Fros et al. 2010; Myllymaki and Ramet 2014). Viruses prevent the production of antiviral compounds by blocking different components of JAK/STAT pathway by various evolved mechanisms (Randall and Goodbourn 2008; Hong and Laimins 2013). Fros et al. have shown that nsP2 protein of CHIKV can inhibit IFN-induced JAK/STAT pathway. In the same study, CHIKV replication blocks STAT1 phosphorylation in the mammalian cells that is induced by either type I IFN or type II IFN (Fros et al. 2010). Recently, Xu et al. have demonstrated modulation of JAK/STAT signaling pathway by salmonid alphavirus subtype 3 (Xu et al. 2016). Furthermore, PTP-mediated inhibition of IFN-induced JAK/STAT pathway has been demonstrated in Japanese encephalitis virus (JEV) infection (Lin et al. 2006). A pathological role of SHP-1/PTPN6 has been identified in a virus-infected skeletal muscle which promotes inflammatory macrophage differentiation and myofiber damage suggesting its crucial role in virus-induced myopathy (Watson et al. 2015). Increased PTPN6 and TNF-α expression with significant upregulation of TNF-α upon PTPN6 siRNA treatment in CHIKV infected human muscle cell line has been reported. The observation indicating the ability of CHIKV to induce PTPN6 or

SHP-1 to lessen host's pro-inflammatory immune response which ultimately helps CHIKV persistence in the host (Thanapati et al. 2018). Induction of SOCS family members has been reported in a number viral infections including HIV-1, HCV, HBV, herpes simplex virus type 1, respiratory syncytial virus, Ebola virus, influenza A virus, and coxsackievirus that contributes to virus replication and pathogenesis (Akhtar and Benveniste 2011). PIASy has been reported to inhibit virus-induced, IFN-stimulated transcription, and ISGs expressions (Kubota et al. 2011). However, in the context of alphavirus infections, involvement of cytokine signaling inhibitors has not yet been reported.

3.2.1.3 Adaptive Immune Responses in CHIKV Infection

Adaptive immune responses include humoral and cell-mediated responses which are demonstrated by B and T cells, respectively. These immune responses are specific against viral infection and are also important for the development of vaccine. The reported higher vulnerability of old individuals to CHIKV infections and higher disease severity could be due to a weaker adaptive immune responses (Borgherini et al. 2008; Sissoko et al. 2009; Dorshkind et al. 2009; Hoarau et al. 2010; Gerardin et al. 2013). Although a lot of studies have deciphered the importance of adaptive immune responses in chikungunya, detailed studies are needed to find out the vaccine strategies, potential role of adaptive immune responses on viremia control, and rheumatic manifestations.

3.2.1.3.1 B-Cell Responses

CHIKV-specific humoral responses have been studied in last few years (Kam et al. 2012a, b, c; Lum et al. 2013; Poo et al. 2014b). There are reports demonstrating association of antibody production by B cells and CHIKV infection control (Eckels et al. 1970; Lum et al. 2013). A study from CHIKV-infected patients has revealed IgG3 to be the dominant antibody isotype (Kam et al. 2012a). Furthermore, CHIKV load is thought to be the driving force behind the production of these neutralizing antibodies (Kam et al. 2012a). It has been shown that virus-like particle (VLP)-based vaccine can induce protective neutralizing antibodies against CHIKV in non-human primates (Akahata et al. 2010). Similarly, Brandler et al. have demonstrated that live-attenuated measles vaccine which expresses chikungunya VLPs can protect mice against lethal CHIKV challenge (Brandler et al. 2013). Further, this vaccine has been tested in healthy individuals with no adverse effects (Chang et al. 2014; Ramsauer et al. 2015). The CHIKV-specific antibodies have been demonstrated to clear viremia in μMT mice where it has been previously reported to persist for more than a year (Lum et al. 2013; Poo et al. 2014b). The virus-resolving capacity of antibodies has been further proven by the passive transfer of human anti-CHIKV antibodies into the IFNAR$^{-/-}$ or neonatal mice (Couderc et al. 2009). The CHIKV-specific antibodies are reported to target E2EP3-dominant epitope that is present in CHIKV E2 glycoprotein (Kam et al. 2012b, c; Lum et al. 2013). In addition, several CHIKV-specific antibodies of either human or mice origin have been designed targeting CHIKV E2, E1, capsid, and nsP2 which are shown to be protective in mice against CHIKV infection (Lum and Ng 2015). Anti-CHIKV antibodies against E2 glycoprotein, Capsid, nsP1, nsP3 and nsP4 proteins are neutralizing and are the targets of the anti-CHIKV antibody response in macaques. This study has provided critical information that has enhanced understanding of CHIKV infection and immunity, vaccine design, and pre-clinical efficacy studies (Kam et al. 2014). Hawman et al. have demonstrated that pathogenic CHIKV strains evade E2 domain-B-neutralizing antibodies to establish persistence. Thus, the murine and human antibodies targeting E2 epitope of B cells cannot neutralize pathogenic CHIKV strains efficiently (Hawman et al. 2016). There is a report suggesting transfer of CHIKV-specific antibodies from infected mother to offspring which protects the child from the illness until 9 months of age (Watanaveeradej et al. 2006). CHIKV infection is believed to confer lifelong immunity (CDC n.d.; Sam and AbuBakar n.d.).

The contributions of B cells toward the development of alphavirus-induced arthritic manifestations are not clear. In RRV infection, it is reported that B cells have no role in the development of arthritis (Morrison et al. 2006, 2007), while deficiency of B cells resulted in severe arthritic disease in CHIKV-infected mice (Lum et al. 2013). In addition, studies in patients with high levels of CHIKV-specific IgG antibodies have shown to be associated with long-term rheumatic manifestations; however, the mechanism behind this pathology is unclear (Moro et al. 2012; Gerardin et al. 2013). Further studies are needed to ascertain the role of B cells in viremia control, rheumatic manifestations and for designing vaccine strategies.

3.2.1.3.2 T-Cell Responses

A study from CHIKV-infected patients has demonstrated the rise in $CD8^+$ T cells in the acute phase followed by $CD4^+$ T cells in the later phase of chikungunya (Wauquier et al. 2011). However, Hoarau et al. have detected $CD4^+$ T cells, but not $CD8^+$ T cells in the synovial fluid of the patients suffering from chronic inflammation following CHIKV infection (Hoarau et al. 2010). Further, infiltration of $CD4^+$ and $CD8^+$ T cells into the CHIKV-infected tissues has been demonstrated in mice model. In addition, a pathogenic role of $CD4^+$ T cells independent of viral titer and IFN-ϒ production has been reported (Teo et al. 2013). Vaccination of B-cell-deficient μMT mice with whole inactivated virus-induced $CD4^+$ T cells to the levels of wild-type mice (Poo et al. 2014b).

Further challenge with CHIKV in these mice has shown severe joint pathology supporting the pathogenic role of $CD4^+$ T cells (Poo et al. 2014b). A defective T-cell immune response in chronic patients following CHIKV infections is reported (Hoarau et al. 2013). Cytokine profiling has indicated the involvement of Th1-related pathway in CHIKV infection (Gardner et al. 2010; Nakaya et al. 2012; Teo et al. 2013, 2015). In other alphaviruses such as RRV and VEEV infections, T cells are shown to be protective (Soden et al. 2000; Yun et al. 2009).

CHIKV vaccines are reported to induce $CD8^+$ T cells targeting E1 and E2 of CHIKV that suggest the potential role of $CD8^+$ T cells against CHIKV infection (Chattopadhyay et al. 2013; Garcia-Arriaza et al. 2014). However, subsequent study revealed that adoptively transferred vaccine-induced CHIKV-specific $CD8^+$ T cells cannot prevent CHIKV infection (Chu et al. 2013). This observation is in concordance with the study showing no effect of $CD8^+$ T deficiency on the disease outcome in CHIKV-infected mice. Also, $CD8^+$ T cells deficiency has nothing to do with the arthritis level (foot swelling) suggesting that $CD8^+$ T cells are not involved in CHIKV-induced arthritis (Teo et al. 2012).

IL-17, produced by Th17 cells, has been detected in several cohorts of chikungunya patients (Ng et al. 2009; Chow et al. 2011; Teng et al. 2015). Th17 cells have been reported to be pathogenic in other alphavirus infections and in patients suffering from rheumatoid arthritis (RA) (Kotake et al. 1999; Chabaud et al. 2000; Kulcsar et al. 2014). As the disease outcome of CHIKV infection is quite similar to RA, it would be interesting to unravel the role of Th17 in chikungunya. Recently, regulatory T cells that alleviate joint pathology in mice through inhibition of CHIKV-specific $CD4^+$ T effector cells have been demonstrated (Lee et al. 2015). A study in mouse model has revealed the protective role of ϒδ T cells limiting inflammatory outcome and joint pathology following CHIKV infection (Long et al. 2015). However, further investigations are needed because of the conflicting reports on T cells to ascertain their role in chikungunya.

3.2.2 Therapeutic Antibody and Vaccine Development Against Chikungunya

In CHIKV infection, humans are thought to be the amplifying host in urban cycle of transmission. Thus, vaccination is the most promising way of controlling CHIKV infection. However, till date, no specific vaccine is available. Several strategies have been employed to develop vaccine against this virus including inactivated viral vaccines, live-attenuated viruses, alphavirus chimeras, recombinant viral vaccines, consensus-based DNA vaccines, recombinant subunit vaccines, and virus-like particle (VLP)-based vaccine (Weaver et al. 2012). The development of immunotherapeutics against this disease has also been tried (Couderc et al. 2009; Warter et al. 2011; Pal et al. 2013; Fric et al. 2013).

3.2.2.1 Chikungunya Vaccines

The CHIKV infection has been shown to induce long-lasting immunity. Cross-protection conferred by different isolates of CHIKV against other alphaviruses has also been demonstrated in animal models (Petitdemange et al. 2015). These observations suggest effective vaccination would be the best way to control chikungunya.

The first vaccine candidate developed against CHIKV was formalin-inactivated virus cultured in African green monkey's kidney cells (Harrison et al. 1971). This vaccine was trialed in 16 US army personnel and had induced hemagglutination inhibition, complement fixation, and neutralizing antibodies production (Harrison et al. 1971). Another CHIKV vaccine trial was also conducted in US army using live-attenuated CHIKV of Asian isolates derived from the MRC5 cells (Edelman et al. 2000). This vaccine showed promising results with 58 of the vaccinated army recruits developing neutralizing antibodies that persisted for 1 year in 85% of the recipients. However, this vaccine trial was discontinued because 5 of the recipients developed arthralgia such as symptoms. Recently, VLPs have been developed as vaccine candidates. VLPs are immunogenic but not infectious, thus, considered safer than attenuated vaccines. Promising results have been obtained from VLP vaccine derived from the CHIKV strain 37997 (VRC-CHKVLP059-00-VP) which has induced high-tittered neutralizing antibodies production in monkeys and protected the animals against virus after challenge (Akahata et al. 2010). This vaccine has been trialed in 25 volunteers and shown to be safe, well-tolerated, and immunogenic (Chang et al. 2014). Recently, Goo et al. have demonstrated that VLP vaccine from a single CHIKV strain can elicit cross-reactive neutralizing antibodies against three genotypes of CHIKV (Goo et al. 2016).

Along with the above discussed vaccines, a number of other vaccines have also been developed against CHIKV including recombinant protein-based vaccine from our laboratory (Kumar et al. 2012), live-attenuated CHIKV vaccine (Levitt et al. 1986), VLP vaccines (Metz et al. 2013a, b), recombinant CHIKV vaccines with an encephalomyocarditis virus internal ribosome entry site (Plante et al. 2011; Partidos et al. 2012; Chu et al. 2013; Roy et al. 2014), chimeric alphavirus (VEEV/EEV/CHIKV) vector-based vaccines (Chattopadhyay et al. 2013), other virus vector-based vaccines (Wang et al. 2011; Brandler et al. 2013; Garcia-Arriaza et al. 2014), and DNA vaccines (Muthumani et al. 2008; Mallilankaraman et al. 2011).

3.2.2.2 Therapeutic Antibody Against Chikungunya

Antibody based therapeutic can be a successful treatment strategy in Chikungunya as CHIKV is less divergent unlike Human immunodeficiency virus, Influenza, and Hepatitis. Passive immunotherapy has been used for the treatment of several viral infections since long time. Couderc et al. have demonstrated that purified polyvalent immunoglobulins from convalescent plasma of chikungunya patients provide protection to neonatal mice against CHIKV in a dose-dependent manner (Couderc et al. 2009). Recently, to CHIKV-exposed neonates, administration of anti-CHIKV hyperimmune human immunoglobulins is in Phase 1 clinical trial carried at Pointe-a-Pitre hospital (Guadeloupe, France). (ClinicalTrials.gov Identifier: NCT02230163) (Petitdemange et al. 2015). Furthermore, human recombinant mAbs have been developed which have potent neutralizing ability against several isolates of CHIKV strains tested in vitro (Warter et al. 2011; Pal et al. 2013). In mice model, these antibodies are shown to delay CHIKV-induced lethality (Fric et al. 2013). Antibodies targeting E2 domain of CHIKV have been isolated and tested in animal models to evaluate their therapeutic efficiencies (Jin and Simmons 2019).

References

Aaronson DS, Horvath CM (2002) A road map for those who don't know JAK-STAT. Science 296:1653–1655

Aaskov JG, Fraser JR, Dalglish DA (1981) Specific and non-specific immunological changes in epidemic polyarthritis patients. Aust J Exp Biol Med Sci 59:599–608

Aaskov JG, Dalglish DA, Harper JJ, Douglas JF, Donaldson MD, Hertzog PJ (1987) Natural killer cells in viral arthritis. Clin Exp Immunol 68:23–32

Ahern DJ, Brennan FM (2011) The role of natural killer cells in the pathogenesis of rheumatoid arthritis: major contributors or essential homeostatic modulators? Immunol Lett 136:115–121
Akahata W, Yang ZY, Andersen H, Sun S, Holdaway HA, Kong WP, Lewis MG, Higgs S, Rossmann MG, Rao S, Nabel GJ (2010) A virus-like particle vaccine for epidemic Chikungunya virus protects nonhuman primates against infection. Nat Med 16:334–338
Akhtar LN, Benveniste EN (2011) Viral exploitation of host SOCS protein functions. J Virol 85:1912–1921
Akira S, Takeda K (2004) Toll-like receptor signalling. Nat Rev Immunol 4:499–511
Apandi Y, Lau SK, Izmawati N, Amal NM, Faudzi Y, Mansor W, Hani MH, Zainah S (2010) Identification of Chikungunya virus strains circulating in Kelantan, Malaysia in 2009. Southeast Asian J Trop Med Public Health 41:1374–1380
Arankalle VA, Shrivastava S, Cherian S, Gunjikar RS, Walimbe AM, Jadhav SM, Sudeep AB, Mishra AC (2007) Genetic divergence of Chikungunya viruses in India (1963–2006) with special reference to the 2005–2006 explosive epidemic. J Gen Virol 88:1967–1976
Auffray C, Sieweke MH, Geissmann F (2009) Blood monocytes: development, heterogeneity, and relationship with dendritic cells. Annu Rev Immunol 27:669–692
Azami NA, Salleh SA, Shah SA, Neoh HM, Othman Z, Zakaria SZ, Jamal R (2013) Emergence of chikungunya seropositivity in healthy Malaysian adults residing in outbreak-free locations: chikungunya seroprevalence results from the Malaysian cohort. BMC Infect Dis 13:67
Bhatia M, Ramnath RD, Chevali L, Guglielmotti A (2005) Treatment with bindarit, a blocker of MCP-1 synthesis, protects mice against acute pancreatitis. Am J Physiol Gastrointest Liver Physiol 288:G1259–G1265
Borgherini G, Poubeau P, Staikowsky F, Lory M, Le Moullec N, Becquart JP, Wengling C, Michault A, Paganin F (2007) Outbreak of chikungunya on Reunion Island: early clinical and laboratory features in 157 adult patients. Clin Infect Dis 44:1401–1407
Borgherini G, Poubeau P, Jossaume A, Gouix A, Cotte L, Michault A, Arvin-Berod C, Paganin F (2008) Persistent arthralgia associated with chikungunya virus: a study of 88 adult patients on Reunion Island. Clin Infect Dis 47:469–475
Brandler S, Ruffie C, Combredet C, Brault JB, Najburg V, Prevost MC, Habel A, Tauber E, Despres P, Tangy F (2013) A recombinant measles vaccine expressing chikungunya virus-like particles is strongly immunogenic and protects mice from lethal challenge with chikungunya virus. Vaccine 31:3718–3725
Brehin AC, Casademont I, Frenkiel MP, Julier C, Sakuntabhai A, Despres P (2009) The large form of human 2',5'-Oligoadenylate Synthetase (OAS3) exerts antiviral effect against Chikungunya virus. Virology 384:216–222
Brighton SW, Prozesky OW, de la Harpe AL (1983) Chikungunya virus infection. A retrospective study of 107 cases. S Afr Med J 63:313–315
Brighton SW (1984) Chloroquine phosphate treatment of chronic Chikungunya arthritis. An open pilot study. S Afr Med J 66:217–218
Buonaguro L, Tagliamonte M, Visciano ML, Tornesello ML, Buonaguro FM (2013) Developments in virus-like particle-based vaccines for HIV. Expert Rev Vaccines 12:119–127
Burt FJ, Rolph MS, Rulli NE, Mahalingam S, Heise MT (2012) Chikungunya: a re-emerging virus. Lancet 379:662–671
Carey DE (1971) Chikungunya and dengue: a case of mistaken identity? J Hist Med Allied Sci 26:243–262
Cassadou S, Boucau S, Petit-Sinturel M, Huc P, Leparc-Goffart I, Ledrans M (2013) Emergence of chikungunya fever on the French side of Saint Martin Island, October to December 2013. Euro Surveill 19
CDC (n.d.) Chikungunya fever fact sheet available on www.cdc.gov/ncidod/dvbid/chikungunya/chikvfact. Accessed 19 Aug 2007 V17.
Sam IC, AbuBakar S (n.d.) Chikungunya virus infection. www.mma.org.my/info/2_chikungunya_06. Accessed 29 Aug 2007
Chaaitanya IK, Muruganandam N, Sundaram SG, Kawalekar O, Sugunan AP, Manimunda SP, Ghosal SR, Muthumani K, Vijayachari P (2011) Role of proinflammatory cytokines and chemokines in chronic arthropathy in CHIKV infection. Viral Immunol 24:265–271

Chabaud M, Garnero P, Dayer JM, Guerne PA, Fossiez F, Miossec P (2000) Contribution of interleukin 17 to synovium matrix destruction in rheumatoid arthritis. Cytokine 12:1092–1099
Chang LJ, Dowd KA, Mendoza FH, Saunders JG, Sitar S, Plummer SH, Yamshchikov G, Sarwar UN, Hu Z, Enama ME, Bailer RT, Koup RA, Schwartz RM, Akahata W, Nabel GJ, Mascola JR, Pierson TC, Graham BS, Ledgerwood JE (2014) Safety and tolerability of chikungunya virus-like particle vaccine in healthy adults: a phase 1 dose-escalation trial. Lancet 384:2046–2052
Charrel RN, de Lamballerie X, Raoult D (2007) Chikungunya outbreaks—the globalization of vectorborne diseases. N Engl J Med 356:769–771
Chattopadhyay A, Wang E, Seymour R, Weaver SC, Rose JK (2013) A chimeric vesiculo/alphavirus is an effective alphavirus vaccine. J Virol 87:395–402
Chen CI, Clark DC, Pesavento P, Lerche NW, Luciw PA, Reisen WK, Brault AC (2010) Comparative pathogenesis of epidemic and enzootic Chikungunya viruses in a pregnant Rhesus macaque model. Am J Trop Med Hyg 83:1249–1258
Chen W, Foo SS, Taylor A, Lulla A, Merits A, Hueston L, Forwood MR, Walsh NC, Sims NA, Herrero LJ, Mahalingam S (2015) Bindarit, an inhibitor of monocyte chemotactic protein synthesis, protects against bone loss induced by chikungunya virus infection. J Virol 89:581–593
Chen YB, Seo SY, Kirsch DG, Sheu TT, Cheng WC, Hardwick JM (2006) Alternate functions of viral regulators of cell death. Cell Death Differ 13:1318–1324
Chikungunya-An overview—Cite Seer X (2008). citeseerx.ist.psu.edu/viewdoc/download?doi=10.1.1.690.8904&rep=rep1
Chirathaworn, C, Poovorawan Y, Lertmaharit S, Wuttirattanakowit N (2013) Cytokine levels in patients with chikungunya virus infection Asian Pacific Journal of Tropical Medicine 6(8):631–634 https://doi.org/10.1016/S1995-7645(13)60108-X
Chopra A, Anuradha V, Ghorpade R, Saluja M (2012) Acute Chikungunya and persistent musculoskeletal pain following the 2006 Indian epidemic: a 2-year prospective rural community study. Epidemiol Infect 140:842–850
Chow A, Her Z, Ong EK, Chen JM, Dimatatac F, Kwek DJ, Barkham T, Yang H, Renia L, Leo YS, Ng LF (2011) Persistent arthralgia induced by Chikungunya virus infection is associated with interleukin-6 and granulocyte macrophage colony-stimulating factor. J Infect Dis 203:149–157
Chu H, Das SC, Fuchs JF, Suresh M, Weaver SC, Stinchcomb DT, Partidos CD, Osorio JE (2013) Deciphering the protective role of adaptive immunity to CHIKV/IRES a novel candidate vaccine against Chikungunya in the A129 mouse model. Vaccine 31:3353–3360
Chusri S, Siripaitoon P, Hirunpat S, Silpapojakul K (2011) Case reports of neuro- Chikungunya in southern Thailand. Am J Trop Med Hyg 85:386–389
Clavarino G, Claudio N, Couderc T, Dalet A, Judith D, Camosseto V, Schmidt EK, Wenger T, Lecuit M, Gatti E, Pierre P (2012) Induction of GADD34 is necessary for dsRNA-dependent interferon-beta production and participates in the control of Chikungunya virus infection. PLoS Pathog 8:e1002708
Colonna M, Trinchieri G, Liu YJ (2004) Plasmacytoid dendritic cells in immunity. Nat Immunol 5:1219–1226
Couderc T, Chretien F, Schilte C, Disson O, Brigitte M, Guivel-Benhassine F, Touret Y, Barau G, Cayet N, Schuffenecker I, Despres P, Arenzana-Seisdedos F, Michault A, Albert ML, Lecuit M (2008) A mouse model for Chikungunya: young age and inefficient type-I interferon signaling are risk factors for severe disease. PLoS Pathog 4:e29
Couderc T, Khandoudi N, Grandadam M, Visse C, Gangneux N, Bagot S, Prost JF, Lecuit M (2009) Prophylaxis and therapy for Chikungunya virus infection. J Infect Dis 200:516–523
Couturier E, Guillemin F, Mura M, Leon L, Virion JM, Letort MJ, De Valk H, Simon F, Vaillant V (2012) Impaired quality of life after chikungunya virus infection: a 2-year follow-up study. Rheumatology (Oxford) 51:1315–1322
Van Den Bossche D, Cnops L, Meersman K, Domingo C, Vang A, Vane M (2015) Chikungunya virus and West Nile virus infections imported into Belgium, 2007–2012. Epidemiol Infect 143:2227–2236
Dalpke A, Heeg K, Bartz H, Baetz A (2008) Regulation of innate immunity by suppressor of cytokine signaling (SOCS) proteins. Immunobiology 213:225–235

Das I, Basantray I, Mamidi P, Nayak TK, B MP, Chattopadhyay S, Chattopadhyay S (2014) Heat shock protein 90 positively regulates Chikungunya virus replication by stabilizing viral nonstructural protein nsP2 during infection. PLoS One 9: e100531
Das T, Jaffar-Bandjee MC, Hoarau JJ, Krejbich Trotot P, Denizot M, Lee-Pat-Yuen G, Sahoo R, Guiraud P, Ramful D, Robin S, Alessandri JL, Gauzere BA, Gasque P (2010) Chikungunya fever: CNS infection and pathologies of a re-emerging arbovirus. Prog Neurobiol 91:121–129
de Andrade DC, Jean S, Clavelou P, Dallel R, Bouhassira D (2010) Chronic pain associated with the Chikungunya fever: long lasting burden of an acute illness. BMC Infect Dis 10:31
de Curtis I, Simons K (1988) Dissection of Semliki Forest virus glycoprotein delivery from the trans-Golgi network to the cell surface in permeabilized BHK cells. Proc Natl Acad Sci USA 85:8052–8056
Decker T, Stockinger S, Karaghiosoff M, Muller M, Kovarik P (2002) IFNs and STATs in innate immunity to microorganisms. J Clin Invest 109:1271–1277
Delatte H, Dehecq JS, Thiria J, Domerg C, Paupy C, Fontenille D (2008) Geographic distribution and developmental sites of Aedes albopictus (Diptera: Culicidae) during a Chikungunya epidemic event. Vector Borne Zoonotic Dis 8:25–34
Dhanwani R, Khan M, Alam SI, Rao PV, Parida M (2011) Differential proteome analysis of Chikungunya virus-infected new-born mice tissues reveal implication of stress, inflammatory and apoptotic pathways in disease pathogenesis. Proteomics 11:1936–1951
Dhanwani R, Khan M, Bhaskar AS, Singh R, Patro IK, Rao PV, Parida MM (2012) Characterization of Chikungunya virus infection in human neuroblastoma SH-SY5Y cells: role of apoptosis in neuronal cell death. Virus Res 163:563–572
Diallo M, Thonnon J, Traore-Lamizana M, Fontenille D (1999) Vectors of Chikungunya virus in Senegal: current data and transmission cycles. Am J Trop Med Hyg 60:281–286
Dorshkind K, Montecino-Rodriguez E, Signer RA (2009) The ageing immune system: is it ever too old to become young again? Nat Rev Immunol 9:57–62
Dupuis-Maguiraga L, Noret M, Brun S, Le Grand R, Gras G, Roques P (2012) Chikungunya disease: infection-associated markers from the acute to the chronic phase of arbovirus-induced arthralgia. PLoS Negl Trop Dis 6:e1446
Eckels KH, Harrison VR, Hetrick FM (1970) Chikungunya virus vaccine prepared by tween-ether extraction. Appl Microbiol 19:321–325
Economopoulou A, Dominguez M, Helynck B, Sissoko D, Wichmann O, Quenel P, Germonneau P, Quatresous I (2009) Atypical Chikungunya virus infections: clinical manifestations, mortality and risk factors for severe disease during the 2005-2006 outbreak on Reunion. Epidemiol Infect 137:534–541
Edelman R, Tacket CO, Wasserman SS, Bodison SA, Perry JG, Mangiafico JA (2000) Phase II safety and immunogenicity study of live Chikungunya virus vaccine TSI-GSD-218. Am J Trop Med Hyg 62:681–685
Teng TS, Kam YW, Tan JJ, Ng LF (2011) Host response to Chikungunya virus and perspectives for immune-based therapies. Future Virol 6(8):975–984
Essackjee K, Goorah S, Ramchurn SK, Cheeneebash J, Walker-Bone K (2013) Prevalence of and risk factors for chronic arthralgia and rheumatoid-like polyarthritis more than 2 years after infection with chikungunya virus. Postgrad Med J 89:440–447
Fayzulin R, Frolov I (2004) Changes of the secondary structure of the 5′ end of the Sindbis virus genome inhibit virus growth in mosquito cells and lead to accumulation of adaptive mutations. J Virol 78:4953–4964
Fields BN, Knipe DM, Howley PM (2013) 1950; Fields virology, 6th edn. Wolters Kluwer Health/ Lippincott Williams & Wilkins, Philadelphia, PA
Fischer D, Thomas SM, Suk JE, Sudre B, Hess A, Tjaden NB, Beierkuhnlein C, Semenza JC (2013) Climate change effects on Chikungunya transmission in Europe: geospatial analysis of vector's climatic suitability and virus' temperature requirements. Int J Health Geogr 12:51
Fric J, Bertin-Maghit S, Wang CI, Nardin A, Warter L (2013) Use of human monoclonal antibodies to treat Chikungunya virus infection. J Infect Dis 207:319–322
Frolov I, Hardy R, Rice CM (2001) Cis-acting RNA elements at the 5′ end of Sindbis virus genome RNA regulate minus- and plus-strand RNA synthesis. RNA 7:1638–1651

Fros JJ, Liu WJ, Prow NA, Geertsema C, Ligtenberg M, Vanlandingham DL, Schnettler E, Vlak JM, Suhrbier A, Khromykh AA, Pijlman GP (2010) Chikungunya virus nonstructural protein 2 inhibits type I/II interferon-stimulated JAK-STAT signaling. J Virol 84:10877–10887

Gaedigk-Nitschko K, Schlesinger MJ (1990) The Sindbis virus 6K protein can be detected in virions and is acylated with fatty acids. Virology 175:274–281

Gaedigk-Nitschko K, Schlesinger MJ (1991) Site-directed mutations in Sindbis virus E2 glycoprotein's cytoplasmic domain and the 6K protein lead to similar defects in virus assembly and budding. Virology 183:206–214

Ganu MA, Ganu AS (2011) Post-chikungunya chronic arthritis—our experience with DMARDs over two year follow up. J Assoc Physicians India 59:83–86

Garcia-Arriaza J, Cepeda V, Hallengard D, Sorzano CO, Kummerer BM, Liljestrom P, Esteban M (2014) A novel poxvirus-based vaccine, MVA-CHIKV, is highly immunogenic and protects mice against chikungunya infection. J Virol 88:3527–3547

Gardner J, Anraku I, Le TT, Larcher T, Major L, Roques P, Schroder WA, Higgs S, Suhrbier A (2010) Chikungunya virus arthritis in adult wild-type mice. J Virol 84:8021–8032

Garmashova N, Gorchakov R, Volkova E, Paessler S, Frolova E, Frolov I (2007) The Old World and New World alphaviruses use different virus-specific proteins for induction of transcriptional shutoff. J Virol 81:2472–2484

Garoff H, Huylebroeck D, Robinson A, Tillman U, Liljestrom P (1990) The signal sequence of the p62 protein of Semliki Forest virus is involved in initiation but not in completing chain translocation. J Cell Biol 111:867–876

Gerardin P, Barau G, Michault A, Bintner M, Randrianaivo H, Choker G, Lenglet Y, Touret Y, Bouveret A, Grivard P, Le Roux K, Blanc S, Schuffenecker I, Couderc T, Arenzana-Seisdedos F, Lecuit M, Robillard PY (2008) Multidisciplinary prospective study of mother-to-child chikungunya virus infections on the Island of La Reunion. PLoS Med 5:e60

Gerardin P, Fianu A, Michault A, Mussard C, Boussaid K, Rollot O, Grivard P, Kassab S, Bouquillard E, Borgherini G, Gauzere BA, Malvy D, Breart G, Favier F (2013) Predictors of Chikungunya rheumatism: a prognostic survey ancillary to the TELECHIK cohort study. Arthritis Res Ther 15:R9

Gifford GE, Heller E (1963) Effect of actinomycin D on interferon production by 'active' and 'inactive' Chikungunya virus in chick cells. Nature 200:50–51

Goo L, Dowd KA, Lin TY, Mascola JR, Graham BS, Ledgerwood JE, Pierson TC (2016) A virus-like particle vaccine elicits broad neutralizing antibody responses in humans to all Chikungunya virus genotypes. J Infect Dis 214(10):1487–1491

Greenhalgh CJ, Hilton DJ (2001) Negative regulation of cytokine signaling. J Leukoc Biol 70:348–356

Griffin DE, Hardwick JM (1997) Regulators of apoptosis on the road to persistent alphavirus infection. Ann Rev Microbiol 51:565–592

Halstead SB, Udomsakdi S, Scanlon JE, Rohitayodhin S (1969a) Dengue and chikungunya virus infection in man in Thailand, 1962–1964. V. Epidemiologic observations outside Bangkok. Am J Trop Med Hyg 18:1022–1033

Halstead SB, Scanlon JE, Umpaivit P, Udomsakdi S (1969b) Dengue and chikungunya virus infection in man in Thailand, 1962–1964. IV. Epidemiologic studies in the Bangkok metropolitan area. Am J Trop Med Hyg 18:997–1021

Harrison VR, Eckels KH, Bartelloni PJ, Hampton C (1971) Production and evaluation of a formalin-killed Chikungunya vaccine. J Immunol 107:643–647

Hawman DW, Fox JM, Ashbrook AW, May NA, Schroeder KM, Torres RM, Crowe JE Jr, Dermody TS, Diamond MS, Morrison TE (2016) Pathogenic Chikungunya virus evades B cell responses to establish persistence. Cell Rep 16:1326–1338

Her Z, Malleret B, Chan M, Ong EK, Wong SC, Kwek DJ, Tolou H, Lin RT, Tambyah PA, Renia L, Ng LF (2010) Active infection of human blood monocytes by Chikungunya virus triggers an innate immune response. J Immunol 184:5903–5913

Her Z, Teng TS, Tan JJ, Teo TH, Kam YW, Lum FM, Lee WW, Gabriel C, Melchiotti R, Andiappan AK, Lulla V, Lulla A, Win MK, Chow A, Biswas SK, Leo YS, Lecuit M, Merits A, Renia L,

Ng LF (2014) Loss of TLR3 aggravates CHIKV replication and pathology due to an altered virus-specific neutralizing antibody response. EMBO Mol Med 7:24–41
Hoarau JJ, Jaffar Bandjee MC, Krejbich Trotot P, Das T, Li-Pat-Yuen G, Dassa B, Denizot M, Guichard E, Ribera A, Henni T, Tallet F, Moiton MP, Gauzere BA, Bruniquet S, Jaffar Bandjee Z, Morbidelli P, Martigny G, Jolivet M, Gay F, Grandadam M, Tolou H, Vieillard V, Debre P, Autran B, Gasque P (2010) Persistent chronic inflammation and infection by Chikungunya arthritogenic alphavirus in spite of a robust host immune response. J Immunol 184:5914–5927
Hoarau JJ, Gay F, Pelle O, Samri A, Jaffar-Bandjee MC, Gasque P, Autran B (2013) Identical strength of the T cell responses against E2, nsP1 and capsid CHIKV proteins in recovered and chronic patients after the epidemics of 2005–2006 in La Reunion Island. PLoS One 8:e84695
Honda K, Takaoka A, Taniguchi T (2006) Type I interferon [corrected] gene induction by the interferon regulatory factor family of transcription factors. Immunity 25:349–360
Hong S, Laimins LA (2013) The JAK-STAT transcriptional regulator, STAT-5, activates the ATM DNA damage pathway to induce HPV 31 genome amplification upon epithelial differentiation. PLoS Pathog 9:e1003295
Horwood PF, Reimer LJ, Dagina R, Susapu M, Bande G, Katusele M, Koimbu G, Jimmy S, Ropa B, Siba PM, Pavlin BI (2013) Outbreak of chikungunya virus infection, Vanimo, Papua New Guinea. Emerg Infect Dis 19:1535–1538
https://www.niaid.nih.gov/research/emerging-infectious-diseases-pathogens
Isaacs A, Lindenmann J (1957) Virus interference. I. The interferon. Proc R Soc Lond B Biol Sci 147:258–267
Issac TH, Tan EL, Chu JJ (2014) Proteomic profiling of chikungunya virus-infected human muscle cells: reveal the role of cytoskeleton network in CHIKV replication. J Proteome 108:445–464
Jaffar-Bandjee MC, Ramful D, Gauzere BA, Hoarau JJ, Krejbich-Trotot P, Robin S, Ribera A, Selambarom J, Gasque P (2010) Emergence and clinical insights into the pathology of Chikungunya virus infection. Expert Rev Anti-Infect Ther 8:987–996
Jarahian M, Watzl C, Fournier P, Arnold A, Djandji D, Zahedi S, Cerwenka A, Paschen A, Schirrmacher V, Momburg F (2009) Activation of natural killer cells by Newcastle disease virus hemagglutinin-neuraminidase. J Virol 83:8108–8121
Jin J, Simmons G (2019) Antiviral functions of monoclonal antibodies against chikungunya virus. Viruses 11(4):305
Jones PH, Maric M, Madison MN, Maury W, Roller RJ, Okeoma CM (2013) BST-2/tetherin-mediated restriction of chikungunya (CHIKV) VLP budding is counteracted by CHIKV non-structural protein 1 (nsP1). Virology 438:37–49
Judith D, Mostowy S, Bourai M, Gangneux N, Lelek M, Lucas-Hourani M, Cayet N, Jacob Y, Prevost MC, Pierre P, Tangy F, Zimmer C, Vidalain PO, Couderc T, Lecuit M (2013) Species-specific impact of the autophagy machinery on chikungunya virus infection. EMBO Rep 14:534–544
Kam YW, Ong EK, Renia L, Tong JC, Ng LF (2009) Immuno-biology of Chikungunya and implications for disease intervention. Microbes Infect 11:1186–1196
Kam YW, Simarmata D, Chow A, Her Z, Teng TS, Ong EK, Renia L, Leo YS, Ng LF (2012a) Early appearance of neutralizing immunoglobulin G3 antibodies is associated with chikungunya virus clearance and long-term clinical protection. J Infect Dis 205:1147–1154
Kam YW, Lum FM, Teo TH, Lee WW, Simarmata D, Harjanto S, Chua CL, Chan YF, Wee JK, Chow A, Lin RT, Leo YS, Le Grand R, Sam IC, Tong JC, Roques P, Wiesmuller KH, Renia L, Rotzschke O, Ng LF (2012b) Early neutralizing IgG response to Chikungunya virus in infected patients targets a dominant linear epitope on the E2 glycoprotein. EMBO Mol Med 4:330–343
Kam YW, Lee WW, Simarmata D, Harjanto S, Teng TS, Tolou H, Chow A, Lin RT, Leo YS, Renia L, Ng LF (2012c) Longitudinal analysis of the human antibody response to Chikungunya virus infection: implications for serodiagnosis and vaccine development. J Virol 86:13005–13015
Kam YW, Lee WW, Simarmata D, Le Grand R, Tolou H, Merits A, Roques P, Ng LF (2014) Unique epitopes recognized by antibodies induced in Chikungunya virus-infected non-human primates: implications for the study of immunopathology and vaccine development. PLoS One 9:e95647

Kawai T, Takahashi K, Sato S, Coban C, Kumar H, Kato H, Ishii KJ, Takeuchi O, Akira S (2005) IPS-1, an adaptor triggering RIG-I- and Mda5-mediated type I interferon induction. Nat Immunol 6:981–988

Kawai T, Akira S (2010) The role of pattern-recognition receptors in innate immunity: update on toll-like receptors. Nat Immunol 11:373–384

Kelvin AA, Banner D, Silvi G, Moro ML, Spataro N, Gaibani P, Cavrini F, Pierro A, Rossini G, Cameron MJ, Bermejo-Martin JF, Paquette SG, Xu L, Danesh A, Farooqui A, Borghetto I, Kelvin DJ, Sambri V, Rubino S (2011) Inflammatory cytokine expression is associated with chikungunya virus resolution and symptom severity. PLoS Negl Trop Dis 5:e1279

Khan AH, Morita K, Parquet Md Mdel C, Hasebe F, Mathenge EG, Igarashi A (2002) Complete nucleotide sequence of chikungunya virus and evidence for an internal polyadenylation site. J Gen Virol 83:3075–3084

Kielian M, Jungerwirth S (1990) Mechanisms of enveloped virus entry into cells. Mol Biol Med 7:17–31

Kotake S, Udagawa N, Takahashi N, Matsuzaki K, Itoh K, Ishiyama S, Saito S, Inoue K, Kamatani N, Gillespie MT, Martin TJ, Suda T (1999) IL-17 in synovial fluids from patients with rheumatoid arthritis is a potent stimulator of osteoclastogenesis. J Clin Invest 103:1345–1352

Krebs DL, Hilton DJ (2000) SOCS: physiological suppressors of cytokine signaling. J Cell Sci 113(Pt 16):2813–2819

Krejbich-Trotot P, Denizot M, Hoarau JJ, Jaffar-Bandjee MC, Das T, Gasque P (2011a) Chikungunya virus mobilizes the apoptotic machinery to invade host cell defenses. FASEB J 25:314–325

Krejbich-Trotot P, Gay B, Li-Pat-Yuen G, Hoarau JJ, Jaffar-Bandjee MC, Briant L, Gasque P, Denizot M (2011b) Chikungunya triggers an autophagic process which promotes viral replication. Virol J 8:432

Kubota T, Matsuoka M, Xu S, Otsuki N, Takeda M, Kato A, Ozato K (2011) PIASy inhibits virus-induced and interferon-stimulated transcription through distinct mechanisms. J Biol Chem 286:8165–8175

Kularatne SA, Weerasinghe SC, Gihan C, Wickramasinghe S, Dharmarathne S, Abeyrathna A, Jayalath T (2012) Epidemiology, clinical manifestations, and long-term outcomes of a major outbreak of chikungunya in a hamlet in Sri Lanka, in 2007: a longitudinal cohort study. J Trop Med 2012:639178

Kulcsar KA, Baxter VK, Greene IP, Griffin DE (2014) Interleukin 10 modulation of pathogenic Th17 cells during fatal alphavirus encephalomyelitis. Proc Natl Acad Sci USA 111:16053–16058

Kulkarni SP, Ganu M, Jayawant P, Thanapati S, Ganu A, Tripathy AS (2017) Regulatory T cells and IL-10 as modulators of chikungunya disease outcome: a preliminary study. Eur J Clin Microbiol Infect Dis. 36:2475–2481

Kumar M, Sudeep AB, Arankalle VA (2012) Evaluation of recombinant E2 protein-based and whole-virus inactivated candidate vaccines against chikungunya virus. Vaccine 30:6142–6149

Kupper TS, Fuhlbrigge RC (2004) Immune surveillance in the skin: mechanisms and clinical consequences. Nat Rev Immunol 4:211–222

Labadie K, Larcher T, Joubert C, Mannioui A, Delache B, Brochard P, Guigand L, Dubreil L, Lebon P, Verrier B, de Lamballerie X, Suhrbier A, Cherel Y, Le Grand R, Roques P (2010) Chikungunya disease in nonhuman primates involves long-term viral persistence in macrophages. J Clin Invest 120:894–906

Lahariya C, Pradhan SK (2006) Emergence of chikungunya virus in Indian subcontinent after 32 years: a review. J Vector Borne Dis 43:151–160

Langerhans P (1868) Uber die nerven der menschlichen haut. Arch Pathol Anat 44:325–337

Larrieu S, Pouderoux N, Pistone T, Filleul L, Receveur MC, Sissoko D, Ezzedine K, Malvy D (2010) Factors associated with persistence of arthralgia among Chikungunya virus-infected travellers: report of 42 French cases. J Clin Virol 47:85–88

Lee WW, Teo TH, Her Z, Lum FM, Kam YW, Haase D, Renia L, Rotzschke O, Ng LF (2015) Expanding regulatory T cells alleviates chikungunya virus-induced pathology in mice. J Virol 89(15):7893–7904

Lemant J, Boisson V, Winer A, Thibault L, Andre H, Tixier F, Lemercier M, Antok E, Cresta MP, Grivard P, Besnard M, Rollot O, Favier F, Huerre M, Campinos JL, Michault A (2008) Serious

acute chikungunya virus infection requiring intensive care during the Reunion Island outbreak in 2005–2006. Crit Care Med 36:2536–2541
Leparc-Goffart I, Nougairede A, Cassadou S, Prat C, de Lamballerie X (2014) Chikungunya in the Americas. Lancet 383:514
Levitt NH, Ramsburg HH, Hasty SE, Repik PM, Cole FE Jr, Lupton HW (1986) Development of an attenuated strain of chikungunya virus for use in vaccine production. Vaccine 4:157–162
Li G, Rice CM (1993) The signal for translational readthrough of a UGA codon in Sindbis virus RNA involves a single cytidine residue immediately downstream of the termination codon. J Virol 67:5062–5067
Lin RJ, Chang BL, Yu HP, Liao CL, Lin YL (2006) Blocking of interferon-induced Jak-Stat signaling by Japanese encephalitis virus NS5 through a protein tyrosine phosphatase-mediated mechanism. J Virol 80:5908–5918
Liu B, Mink S, Wong KA, Stein N, Getman C, Dempsey PW, Wu H, Shuai K (2004) PIAS1 selectively inhibits interferon-inducible genes and is important in innate immunity. Nat Immunol 5:891–898
Lobigs M, Garoff H (1990) Fusion function of the Semliki Forest virus spike is activated by proteolytic cleavage of the envelope glycoprotein precursor p62. J Virol 64:1233–1240
Long KM, Whitmore AC, Ferris MT, Sempowski GD, McGee C, Trollinger B, Gunn B, Heise MT (2013) Dendritic cell immunoreceptor regulates Chikungunya virus pathogenesis in mice. J Virol 87:5697–5706
Long KM, Heise MT (2015) Protective and pathogenic responses to Chikungunya virus infection. Curr Trop Med Rep 2:13–21
Long KM, Ferris MT, Whitmore AC, Montgomery SA, Thurlow LR, McGee CE, Rodriguez CA, Lim JK, Heise MT (2015) Gammadelta T cells play a protective role in Chikungunya virus-induced disease. J Virol 90:433–443
Lopez S, Yao JS, Kuhn RJ, Strauss EG, Strauss JH (1994) Nucleocapsid-glycoprotein interactions required for assembly of alphaviruses. J Virol 68:1316–1323
Lulla A, Lulla V, Merits A (2012) Macromolecular assembly-driven processing of the 2/3 cleavage site in the alphavirus replicase polyprotein. J Virol 86:553–565
Lum FM, Teo TH, Lee WW, Kam YW, Renia L, Ng LF (2013) An essential role of antibodies in the control of Chikungunya virus infection. J Immunol 190:6295–6302
Lum FM, Ng LF (2015) Cellular and molecular mechanisms of Chikungunya pathogenesis. Antivir Res 120:165–174
Mahauad-Fernandez WD, Jones PH, Okeoma CM (2014) Critical role for bone marrow stromal antigen 2 in acute Chikungunya virus infection. J Gen Virol 95:2450–2461
Mahendradas P, Ranganna SK, Shetty R, Balu R, Narayana KM, Babu RB, Shetty BK (2008) Ocular manifestations associated with chikungunya. Ophthalmology 115:287–291
Mallilankaraman K, Shedlock DJ, Bao H, Kawalekar OU, Fagone P, Ramanathan AA, Ferraro B, Stabenow J, Vijayachari P, Sundaram SG, Muruganandam N, Sarangan G, Srikanth P, Khan AS, Lewis MG, Kim JJ, Sardesai NY, Muthumani K, Weiner DB (2011) A DNA vaccine against chikungunya virus is protective in mice and induces neutralizing antibodies in mice and nonhuman primates. PLoS Negl Trop Dis 5:e928
Manimunda SP, Vijayachari P, Uppoor R, Sugunan AP, Singh SS, Rai SK, Sudeep AB, Muruganandam N, Chaitanya IK, Guruprasad DR (2010) Clinical progression of chikungunya fever during acute and chronic arthritic stages and the changes in joint morphology as revealed by imaging. Trans R Soc Trop Med Hyg 104:392–399
Manimunda SP, Mavalankar D, Bandyopadhyay T, Sugunan AP (2011) Chikungunya epidemic-related mortality. Epidemiol Infect 139:1410–1412
Marchette NJ, Rudnick A, Garcia R, MacVean DW (1978) Alphaviruses in Peninsular Malaysia: I. Virus isolations and animal serology. Southeast Asian J Trop Med Public Health 9:317–329
Mathew AJ, Goyal V, George E, Thekkemuriyil DV, Jayakumar B, Chopra A (2011) Rheumatic-musculoskeletal pain and disorders in a naive group of individuals 15 months following a Chikungunya viral epidemic in South India: a population based observational study. Int J Clin Pract 65:1306–1312
Mavalankar D, Shastri P, Bandyopadhyay T, Parmar J, Ramani KV (2008) Increased mortality rate associated with chikungunya epidemic, Ahmedabad, India. Emerg Infect Dis 14:412–415

McIntosh BM, Harwin RM, Paterson HE, Westwater ML (1963) An epidemic of Chikungunya in South-Eastern Southern Rhodesia. Cent Afr J Med 9:351–359
McIntosh BM, Paterson HE, McGillivray G, Desousa J (1964) Further studies on the Chikungunya outbreak in southern Rhodesia in 1962. I. Mosquitoes, wild primates and birds in relation to the epidemic. Ann Trop Med Parasitol 58:45–51
Metz SW, Gardner J, Geertsema C, Le TT, Goh L, Vlak JM, Suhrbier A, Pijlman GP (2013a) Effective chikungunya virus-like particle vaccine produced in insect cells. PLoS Negl Trop Dis 7:e2124
Metz SW, Martina BE, van den Doel P, Geertsema C, Osterhaus AD, Vlak JM, Pijlman GP (2013b) Chikungunya virus-like particles are more immunogenic in a lethal AG129 mouse model compared to glycoprotein E1 or E2 subunits. Vaccine 31:6092–6096
Moro ML, Gagliotti C, Silvi G, Angelini R, Sambri V, Rezza G, Massimiliani E, Mattivi A, Grilli E, Finarelli AC, Spataro N, Pierro AM, Seyler T, Macini P (2010) Chikungunya virus in North-Eastern Italy: a seroprevalence survey. Am J Trop Med Hyg 82:508–511
Moro ML, Grilli E, Corvetta A, Silvi G, Angelini R, Mascella F, Miserocchi F, Sambo P, Finarelli AC, Sambri V, Gagliotti C, Massimiliani E, Mattivi A, Pierro AM, Macini P (2012) Long-term chikungunya infection clinical manifestations after an outbreak in Italy: a prognostic cohort study. J Infect 65:165–172
Morrison TE, Whitmore AC, Shabman RS, Lidbury BA, Mahalingam S, Heise MT (2006) Characterization of Ross River virus tropism and virus-induced inflammation in a mouse model of viral arthritis and myositis. J Virol 80:737–749
Morrison TE, Fraser RJ, Smith PN, Mahalingam S, Heise MT (2007) Complement contributes to inflammatory tissue destruction in a mouse model of Ross River virus-induced disease. J Virol 81:5132–5143
Morrison TE, Oko L, Montgomery SA, Whitmore AC, Lotstein AR, Gunn BM, Elmore SA, Heise MT (2011) A mouse model of chikungunya virus-induced musculoskeletal inflammatory disease: evidence of arthritis, tenosynovitis, myositis, and persistence. Am J Pathol 178:32–40
Muthumani K, Lankaraman KM, Laddy DJ, Sundaram SG, Chung CW, Sako E, Wu L, Khan A, Sardesai N, Kim JJ, Vijayachari P, Weiner DB (2008) Immunogenicity of novel consensus-based DNA vaccines against Chikungunya virus. Vaccine 26:5128–5134
Myers RM, Carey DE (1967) Concurrent isolation from patient of two arboviruses, Chikungunya and dengue type 2. Science 157:1307–1308
Myllymaki H, Ramet M (2014) JAK/STAT pathway in Drosophila immunity. Scand J Immunol 79:377–385
Nakaya HI, Gardner J, Poo YS, Major L, Pulendran B, Suhrbier A (2012) Gene profiling of Chikungunya virus arthritis in a mouse model reveals significant overlap with rheumatoid arthritis. Arthritis Rheum 64:3553–3563
Ng LF, Chow A, Sun YJ, Kwek DJ, Lim PL, Dimatatac F, Ng LC, Ooi EE, Choo KH, Her Z, Kourilsky P, Leo YS (2009) IL-1beta, IL-6, and RANTES as biomarkers of Chikungunya severity. PLoS One 4:e4261
Nsoesie EO, Kraemer MU, Golding N, Pigott DM, Brady OJ, Moyes CL, Johansson MA, Gething PW, Velayudhan R, Khan K, Hay SI, Brownstein JS (2016) Global distribution and environmental suitability for chikungunya virus, 1952 to 2015. Euro Surveill 21
Oliver M, Grandadam M, Marimoutou C, Rogier C, Botelho-Nevers E, Tolou H, Moalic JL, Kraemer P, Morillon M, Morand JJ, Jeandel P, Parola P, Simon F (2009) Persisting mixed cryoglobulinemia in Chikungunya infection. PLoS Negl Trop Dis 3:e374
Ozden S, Huerre M, Riviere JP, Coffey LL, Afonso PV, Mouly V, de Monredon J, Roger JC, El Amrani M, Yvin JL, Jaffar MC, Frenkiel MP, Sourisseau M, Schwartz O, Butler-Browne G, Despres P, Gessain A, Ceccaldi PE (2007) Human muscle satellite cells as targets of chikungunya virus infection. PLoS One 2:e527
Ozden S, Lucas-Hourani M, Ceccaldi PE, Basak A, Valentine M, Benjannet S, Hamelin J, Jacob Y, Mamchaoui K, Mouly V, Despres P, Gessain A, Butler-Browne G, Chretien M, Tangy F, Vidalain PO, Seidah NG (2008) Inhibition of Chikungunya virus infection in cultured human muscle cells by furin inhibitors: impairment of the maturation of the E2 surface glycoprotein. J Biol Chem 283:21899–21908

Paingankar MS, Arankalle VA (2014) Identification of chikungunya virus interacting proteins in mammalian cells. J Biosci 39:389–399

Pal P, Dowd KA, Brien JD, Edeling MA, Gorlatov S, Johnson S, Lee I, Akahata W, Nabel GJ, Richter MK, Smit JM, Fremont DH, Pierson TC, Heise MT, Diamond MS (2013) Development of a highly protective combination monoclonal antibody therapy against Chikungunya virus. PLoS Pathog 9:e1003312

Palha N, Guivel-Benhassine F, Briolat V, Lutfalla G, Sourisseau M, Ellett F, Wang CH, Lieschke GJ, Herbomel P, Schwartz O, Levraud JP (2013) Real-time whole-body visualization of Chikungunya virus infection and host interferon response in zebrafish. PLoS Pathog 9:e1003619

Panning M, Grywna K, van Esbroeck M, Emmerich P, Drosten C (2008) Chikungunya fever in travelers returning to Europe from the Indian Ocean region, 2006. Emerg Infect Dis 14:416–422

Partidos CD, Paykel J, Weger J, Borland EM, Powers AM, Seymour R, Weaver SC, Stinchcomb DT, Osorio JE (2012) Cross-protective immunity against o'nyong-nyong virus afforded by a novel recombinant chikungunya vaccine. Vaccine 30:4638–4643

Pastorino B, Muyembe-Tamfum JJ, Bessaud M, Tock F, Tolou H, Durand JP, Peyrefitte CN (2004) Epidemic resurgence of Chikungunya virus in Democratic Republic of the Congo: identification of a new central African strain. J Med Virol 74:277–282

Paul SD, Singh KR (1968) Experimental infection of Macaca radiata with Chikungunya virus and transmission of virus by mosquitoes. Indian J Med Res 56:802–811

Peralbo E, Alonso C, Solana R (2007) Invariant NKT and NKT-like lymphocytes: two different T cell subsets that are differentially affected by ageing. Exp Gerontol 42:703–708

Petitdemange C, Becquart P, Wauquier N, Beziat V, Debre P, Leroy EM, Vieillard V (2011) Unconventional repertoire profile is imprinted during acute chikungunya infection for natural killer cells polarization toward cytotoxicity. PLoS Pathog 7:e1002268

Petitdemange C, Wauquier N, Vieillard V (2015) Control of immunopathology during chikungunya virus infection. J Allergy Clin Immunol 135:846–855

Peyrefitte CN, Rousset D, Pastorino BA, Pouillot R, Bessaud M, Tock F, Mansaray H, Merle OL, Pascual AM, Paupy C, Vessiere A, Imbert P, Tchendjou P, Durand JP, Tolou HJ, Grandadam M (2007) Chikungunya virus, Cameroon, 2006. Emerg Infect Dis 13:768–771

Peyrefitte CN, Bessaud M, Pastorino BA, Gravier P, Plumet S, Merle OL, Moltini I, Coppin E, Tock F, Daries W, Ollivier L, Pages F, Martin R, Boniface F, Tolou HJ, Grandadam M (2008) Circulation of Chikungunya virus in Gabon, 2006–2007. J Med Virol 80:430–433

Pialoux G, Gauzere BA, Strobel M (2006) Chikungunya virus infection: review through an epidemic. Med Mal Infect 36:253–263

Pialoux G, Gauzere BA, Jaureguiberry S, Strobel M (2007) Chikungunya, an epidemic arbovirosis. Lancet Infect Dis 7:319–327

Plante K, Wang E, Partidos CD, Weger J, Gorchakov R, Tsetsarkin K, Borland EM, Powers AM, Seymour R, Stinchcomb DT, Osorio JE, Frolov I, Weaver SC (2011) Novel chikungunya vaccine candidate with an IRES-based attenuation and host range alteration mechanism. PLoS Pathog 7:e1002142

Poo YS, Nakaya H, Gardner J, Larcher T, Schroder WA, Le TT, Major LD, Suhrbier A (2014a) CCR2 deficiency promotes exacerbated chronic erosive neutrophil-dominated chikungunya virus arthritis. J Virol 88:6862–6872

Poo YS, Rudd PA, Gardner J, Wilson JA, Larcher T, Colle MA, Le TT, Nakaya HI, Warrilow D, Allcock R, Bielefeldt-Ohmann H, Schroder WA, Khromykh AA, Lopez JA, Suhrbier A (2014b) Multiple immune factors are involved in controlling acute and chronic chikungunya virus infection. PLoS Negl Trop Dis 8:e3354

Powers AM, Brault AC, Shirako Y, Strauss EG, Kang W, Strauss JH, Weaver SC (2001) Evolutionary relationships and systematics of the alphaviruses. J Virol 75:10118–10131

Powers AM, Logue CH (2007) Changing patterns of chikungunya virus: re-emergence of a zoonotic arbovirus. J Gen Virol 88:2363–2377

Priya R, Patro IK, Parida MM (2014) TLR3 mediated innate immune response in mice brain following infection with Chikungunya virus. Virus Res 189:194–205

Prow TW, Chen X, Prow NA, Fernando GJ, Tan CS, Raphael AP, Chang D, Ruutu MP, Jenkins DW, Pyke A, Crichton ML, Raphaelli K, Goh LY, Frazer IH, Roberts MS, Gardner J, Khromykh AA, Suhrbier A, Hall RA, Kendall MA (2010) Nanopatch-targeted skin vaccination against West Nile virus and Chikungunya virus in mice. Small 6:1776–1784

Puttamallesh VN, Sreenivasamurthy SK, Singh PK, Harsha HC, Ganjiwale A, Broor S, Pandey A, Narayana J, Prasad TS (2013) Proteomic profiling of serum samples from chikungunya-infected patients provides insights into host response. Clin Proteomics 10:14

Queyriaux B, Simon F, Grandadam M, Michel R, Tolou H, Boutin JP (2008) Clinical burden of chikungunya virus infection. Lancet Infect Dis 8:2–3

Ramful D, Carbonnier M, Pasquet M, Bouhmani B, Ghazouani J, Noormahomed T, Beullier G, Attali T, Samperiz S, Fourmaintraux A, Alessandri JL (2007) Mother-to-child transmission of Chikungunya virus infection. Pediatr Infect Dis J 26:811–815

Ramsauer K, Schwameis M, Firbas C, Mullner M, Putnak RJ, Thomas SJ, Despres P, Tauber E, Jilma B, Tangy F (2015) Immunogenicity, safety, and tolerability of a recombinant measles-virus-based chikungunya vaccine: a randomised, double-blind, placebo-controlled, active-comparator, first-in-man trial. Lancet Infect Dis 15:519–527

Randall RE, Goodbourn S (2008) Interferons and viruses: an interplay between induction, signalling, antiviral responses and virus countermeasures. J Gen Virol 89:1–47

Rao G, Khan YZ, Chitnis DS (2008) Chikungunya infection in neonates. Indian Pediatr 45:240–242

Rathore AP, Haystead T, Das PK, Merits A, Ng ML, Vasudevan SG (2014) Chikungunya virus nsP3 & nsP4 interacts with HSP-90 to promote virus replication: HSP-90 inhibitors reduce CHIKV infection and inflammation in vivo. Antivir Res 103:7–16

Rawlings JS, Rosler KM, Harrison DA (2004) The JAK/STAT signaling pathway. J Cell Sci 117:1281–1283

Rezza G, Nicoletti L, Angelini R, Romi R, Finarelli AC, Panning M, Cordioli P, Fortuna C, Boros S, Magurano F, Silvi G, Angelini P, Dottori M, Ciufolini MG, Majori GC, Cassone A (2007) Infection with chikungunya virus in Italy: an outbreak in a temperate region. Lancet 370:1840–1846

Robillard PY, Boumahni B, Gerardin P, Michault A, Fourmaintraux A, Schuffenecker I, Carbonnier M, Djemili S, Choker G, Roge-Wolter M, Barau G (2006) Vertical maternal fetal transmission of the chikungunya virus. Ten cases among 84 pregnant women. Presse Med 35:785–788

Robinson MC (1955) An epidemic of virus disease in Southern Province, Tanganyika Territory, in 1952–1953. I. Clinical features. Trans R Soc Trop Med Hyg 49:28–32

Rodriguez-Morales AJ, Gil-Restrepo AF, Ramirez-Jaramillo V, Montoya-Arias CP, Acevedo-Mendoza WF, Bedoya-Arias JE, Chica-Quintero LA, Murillo-Garcia DR, Garcia-Robledo JE, Castrillon-Spitia JD, Londono JJ, Bedoya-Rendon HD, Cardenas-Perez Jde J, Cardona-Ospina JA, Lagos-Grisales GJ (2016) Post-chikungunya chronic inflammatory rheumatism: results from a retrospective follow-up study of 283 adult and child cases in La Virginia, Risaralda, Colombia. F1000Res 5:360

Rohatgi A, Corbo JC, Monte K, Higgs S, Vanlandingham DL, Kardon G, Lenschow DJ (2014) Infection of myofibers contributes to increased pathogenicity during infection with an epidemic strain of Chikungunya virus. J Virol 88:2414–2425

Rose N, Anoop TM, John AP, Jabbar PK, George KC (2011) Acute optic neuritis following infection with chikungunya virus in southern rural India. Int J Infect Dis 15:e147–e150

Roy CJ, Adams AP, Wang E, Plante K, Gorchakov R, Seymour RL, Vinet-Oliphant H, Weaver SC (2014) Chikungunya vaccine candidate is highly attenuated and protects nonhuman primates against telemetrically monitored disease following a single dose. J Infect Dis 209:1891–1899

Rudd PA, Wilson J, Gardner J, Larcher T, Babarit C, Le TT, Anraku I, Kumagai Y, Loo YM, Gale M Jr, Akira S, Khromykh AA, Suhrbier A (2012) Interferon response factors 3 and 7 protect against Chikungunya virus hemorrhagic fever and shock. J Virol 86:9888–9898

Rulli NE, Rolph MS, Srikiatkhachorn A, Anantapreecha S, Guglielmotti A, Mahalingam S (2011) Protection from arthritis and myositis in a mouse model of acute chikungunya virus disease by bindarit, an inhibitor of monocyte chemotactic protein-1 synthesis. J Infect Dis 204:1026–1030

Ryman KD, Klimstra WB (2008) Host responses to alphavirus infection. Immunol Rev 225:27–45

Salonen A, Ahola T, Kaariainen L (2005) Viral RNA replication in association with cellular membranes. Curr Top Microbiol Immunol 285:139–173

Salvador B, Zhou Y, Michault A, Muench MO, Simmons G (2009) Characterization of Chikungunya pseudotyped viruses: identification of refractory cell lines and demonstration of cellular tropism differences mediated by mutations in E1 glycoprotein. Virology 393:33–41

Schilte C, Couderc T, Chretien F, Sourisseau M, Gangneux N, Guivel-Benhassine F, Kraxner A, Tschopp J, Higgs S, Michault A, Arenzana-Seisdedos F, Colonna M, Peduto L, Schwartz O, Lecuit M, Albert ML (2010) Type I IFN controls chikungunya virus via its action on nonhematopoietic cells. J Exp Med 207:429–442

Schilte C, Buckwalter MR, Laird ME, Diamond MS, Schwartz O, Albert ML (2012) Cutting edge: independent roles for IRF-3 and IRF-7 in hematopoietic and nonhematopoietic cells during host response to Chikungunya infection. J Immunol 188:2967–2971

Schilte C, Staikowsky F, Couderc T, Madec Y, Carpentier F, Kassab S, Albert ML, Lecuit M, Michault A (2013) Chikungunya virus-associated long-term arthralgia: a 36-month prospective longitudinal study. PLoS Negl Trop Dis 7:e2137

Schoggins JW, Wilson SJ, Panis M, Murphy MY, Jones CT, Bieniasz P, Rice CM (2011) A diverse range of gene products are effectors of the type I interferon antiviral response. Nature 472:481–485

Scholte FE, Tas A, Albulescu IC, Zusinaite E, Merits A, Snijder EJ, van Hemert MJ (2015) Stress granule components G3BP1 and G3BP2 play a proviral role early in Chikungunya virus replication. J Virol 89:4457–4469

Schuffenecker I, Iteman I, Michault A, Murri S, Frangeul L, Vaney MC, Lavenir R, Pardigon N, Reynes JM, Pettinelli F, Biscornet L, Diancourt L, Michel S, Duquerroy S, Guigon G, Frenkiel MP, Brehin AC, Cubito N, Despres P, Kunst F, Rey FA, Zeller H, Brisse S (2006) Genome microevolution of chikungunya viruses causing the Indian Ocean outbreak. PLoS Med 3:e263

Schwartz O, Albert ML (2010) Biology and pathogenesis of chikungunya virus. Nat Rev Microbiol 8:491–500

Senanayake MP, Senanayake SM, Vidanage KK, Gunasena S, Lamabadusuriya SP (2009) Vertical transmission in chikungunya infection. Ceylon Med J 54:47–50

Seppa N (2015) https://www.sciencenews.org/article/chikungunyamove

Shah KV, Gilotra SK, Gibbs CJ Jr, Rozeboom LE (1964) Laboratory studies of transmission of Chikungunya virus by mosquitoes: a preliminary report. Indian J Med Res 52:703–709

Siegal FP, Kadowaki N, Shodell M, Fitzgerald-Bocarsly PA, Shah K, Ho S, Antonenko S, Liu YJ (1999) The nature of the principal type 1 interferon-producing cells in human blood. Science 284:1835–1837

Simon F, Parola P, Grandadam M, Fourcade S, Oliver M, Brouqui P, Hance P, Kraemer P, Ali Mohamed A, de Lamballerie X, Charrel R, Tolou H (2007) Chikungunya infection: an emerging rheumatism among travelers returned from Indian Ocean islands. Report of 47 cases. Medicine (Baltimore) 86:123–137

Singh I, Helenius A (1992) Role of ribosomes in Semliki Forest virus nucleocapsid uncoating. J Virol 66:7049–7058

Sissoko D, Moendandze A, Malvy D, Giry C, Ezzedine K, Solet JL, Pierre V (2008) Seroprevalence and risk factors of chikungunya virus infection in Mayotte, Indian Ocean, 2005–2006: a population-based survey. PLoS One 3:e3066

Sissoko D, Malvy D, Ezzedine K, Renault P, Moscetti F, Ledrans M, Pierre V (2009) Post-epidemic Chikungunya disease on Reunion Island: course of rheumatic manifestations and associated factors over a 15-month period. PLoS Negl Trop Dis 3:e389

Soden M, Vasudevan H, Roberts B, Coelen R, Hamlin G, Vasudevan S, La Brooy J (2000) Detection of viral ribonucleic acid and histologic analysis of inflamed synovium in Ross River virus infection. Arthritis Rheum 43:365–369

Solignat M, Gay B, Higgs S, Briant L, Devaux C (2009) Replication cycle of chikungunya: a re-emerging arbovirus. Virology 393:183–197

Sourisseau M, Schilte C, Casartelli N, Trouillet C, Guivel-Benhassine F, Rudnicka D, Sol-Foulon N, Le Roux K, Prevost MC, Fsihi H, Frenkiel MP, Blanchet F, Afonso PV, Ceccaldi PE, Ozden S,

Gessain A, Schuffenecker I, Verhasselt B, Zamborlini A, Saib A, Rey FA, Arenzana-Seisdedos F, Despres P, Michault A, Albert ML, Schwartz O (2007) Characterization of reemerging chikungunya virus. PLoS Pathog 3:e89
Staples JE, Breiman RF, Powers AM (2009) Chikungunya fever: an epidemiological review of a re-emerging infectious disease. Clin Infect Dis 49:942–948
Steinman RM, Cohn ZA (1973) Identification of a novel cell type in peripheral lymphoid organs of mice. I. Morphology, quantitation, tissue distribution. J Exp Med 137:1142–1162
Stetson DB, Medzhitov R (2006) Type I interferons in host defense. Immunity 25:373–381
Stoermer KA, Burrack A, Oko L, Montgomery SA, Borst LB, Gill RG, Morrison TE (2012) Genetic ablation of arginase 1 in macrophages and neutrophils enhances clearance of an arthritogenic alphavirus. J Immunol 189:4047–4059
Strauss JH, Strauss EG (1994) The alphaviruses: gene expression, replication, and evolution. Microbiol Rev 58:491–562
Suhrbier A, Mahalingam S (2009) The immunobiology of viral arthritides. Pharmacol Ther 124:301–308
Suhrbier A, Jaffar-Bandjee MC, Gasque P (2012) Arthritogenic alphaviruses—an overview. Nat Rev Rheumatol 8:420–429
Sun S, Xiang Y, Akahata W, Holdaway H, Pal P, Zhang X, Diamond MS, Nabel GJ, Rossmann MG (2011) Structural analyses at pseudo atomic resolution of Chikungunya virus and antibodies show mechanisms of neutralization. elife 2:e00435
Tang Y, Li X, Wang M, Zou Q, Zhao S, Sun B, Xu L, Jiang Y (2013) Increased numbers of NK cells, NKT-like cells, and NK inhibitory receptors in peripheral blood of patients with chronic obstructive pulmonary disease. Clin Dev Immunol 2013:721782
Taubitz W, Cramer JP, Kapaun A, Pfeffer M, Drosten C, Dobler G, Burchard GD, Loscher T (2007) Chikungunya fever in travelers: clinical presentation and course. Clin Infect Dis 45:e1–e4
Teng TS, Foo SS, Simamarta D, Lum FM, Teo TH, Lulla A, Yeo NK, Koh EG, Chow A, Leo YS, Merits A, Chin KC, Ng LF (2012) Viperin restricts chikungunya virus replication and pathology. J Clin Invest 122:4447–4460
Teng TS, Kam YW, Lee B, Hapuarachchi HC, Wimal A, Ng LC, Ng LFA (2015) Systematic meta-analysis of immune signatures in patients with acute Chikungunya virus infection. J Infect Dis 211:1925–1935
Teo TH, Lum FM, Lee WW, Ng LF (2012) Mouse models for Chikungunya virus: deciphering immune mechanisms responsible for disease and pathology. Immunol Res 53:136–147
Teo TH, Lum FM, Claser C, Lulla V, Lulla A, Merits A, Renia L, Ng LF (2013) A pathogenic role for CD4+ T cells during Chikungunya virus infection in mice. J Immunol 190:259–269
Teo TH, Her Z, Tan JJ, Lum FM, Lee WW, Chan YH, Ong RY, Kam YW, Leparc-Goffart I, Gallian P, Renia L, de Lamballerie X, Ng LF (2015) Caribbean and La Reunion chikungunya virus isolates differ in their capacity to induce proinflammatory Th1 and NK cell responses and acute joint pathology. J Virol 89:7955–7969
Thanapati S, Das R, Tripathy AS (2015) Phenotypic and functional analyses of NK and NKT-like populations during the early stages of chikungunya infection. Front Microbiol 6:895
Thanapati S, Ganu M, Giri P, Kulkarni S, Sharma M, Babar P, Ganu A, Tripathy AS (2017a) Impaired NK cell functionality and increased TNF-α production as biomarkers of chronic chikungunya arthritis and rheumatoid arthritis Human Immunology 78(4):370–374 https://doi.org/10.1016/j.humimm.2017.02.006
Thanapati S, Ganu MA, Tripathy AS (2017b) Differential inhibitory and activating NK cell receptor levels and NK/NKT-like cell functionality in chronic and recovered stages of chikungunya PLOS ONE 12(11) e0188342 https://doi.org/10.1371/journal.pone.0188342
Van Bortel W, Dorleans F, Rosine J, Blateau A, Rousset D, Matheus S, Leparc-Goffart I, Flusin O, Prat C, Cesaire R, Najioullah F, Ardillon V, Balleydier E, Carvalho L, Lemaitre A, Noel H, Servas V, Six C, Zurbaran M, Leon L, Guinard A, van den Kerkhof J, Henry M, Fanoy E, Braks M, Reimerink J, Swaan C, Georges R, Brooks L, Freedman J, Sudre B, Zeller H (2014) Chikungunya outbreak in the Caribbean region, December 2013 to March 2014, and the significance for Europe. Euro Surveill 19

Vanlandingham DL, Hong C, Klingler K, Tsetsarkin K, McElroy KL, Powers AM, Lehane MJ, Higgs S (2005) Differential infectivities of o'nyong-nyong and chikungunya virus isolates in Anopheles gambiae and Aedes aegypti mosquitoes. Am J Trop Med Hyg 72:616–621

Volk SM, Chen R, Tsetsarkin KA, Adams AP, Garcia TI, Sall AA, Nasar F, Schuh AJ, Holmes EC, Higgs S, Maharaj PD, Brault AC, Weaver SC (2010) Genome-scale phylogenetic analyses of chikungunya virus reveal independent emergences of recent epidemics and various evolutionary rates. J Virol 84:6497–6504

Wahlberg JM, Boere WA, Garoff H (1989) The heterodimeric association between the membrane proteins of Semliki Forest virus changes its sensitivity to low pH during virus maturation. J Virol 63:4991–4997

Wahlberg JM, Bron R, Wilschut J, Garoff H (1992) Membrane fusion of Semliki Forest virus involves homotrimers of the fusion protein. J Virol 66:7309–7318

Wang D, Suhrbier A, Penn-Nicholson A, Woraratanadharm J, Gardner J, Luo M, Le TT, Anraku I, Sakalian M, Einfeld D, Dong JY (2011) A complex adenovirus vaccine against chikungunya virus provides complete protection against viraemia and arthritis. Vaccine 29:2803–2809

Warter L, Lee CY, Thiagarajan R, Grandadam M, Lebecque S, Lin RT, Bertin-Maghit S, Ng LF, Abastado JP, Despres P, Wang CI, Nardin A (2011) Chikungunya virus envelope-specific human monoclonal antibodies with broad neutralization potency. J Immunol 186:3258–3264

Watanaveeradej V, Endy TP, Simasathien S, Kerdpanich A, Polprasert N, Aree C, Vaughn DW, Nisalak A (2006) The study transplacental chikungunya virus antibody kinetics, Thailand. Emerg Infect Dis 12:1770–1772

Watson NB, Schneider KM, Massa PT (2015) SHP-1-dependent macrophage differentiation exacerbates virus-induced myositis. J Immunol 194:2796–2809

Wauquier N, Becquart P, Nkoghe D, Padilla C, Ndjoyi-Mbiguino A, Leroy EM (2011) The acute phase of chikungunya virus infection in humans is associated with strong innate immunity and T CD8 cell activation. J Infect Dis 204:115–123

Weaver SC, Osorio JE, Livengood JA, Chen R, Stinchcomb DT (2012) Chikungunya virus and prospects for a vaccine. Expert Rev Vaccines 11:1087–1101

Werneke SW, Schilte C, Rohatgi A, Monte KJ, Michault A, Arenzana-Seisdedos F, Vanlandingham DL, Higgs S, Fontanet A, Albert ML, Lenschow DJ (2011) ISG15 is critical in the control of chikungunya virus infection independent of UbE1L mediated conjugation. PLoS Pathog 7:e1002322

White LK, Sali T, Alvarado D, Gatti E, Pierre P, Streblow D, Defilippis VR (2011) Chikungunya virus induces IPS-1-dependent innate immune activation and protein kinase R- independent translational shutoff. J Virol 85:606–620

Wikan N, Sakoonwatanyoo P, Ubol S, Yoksan S, Smith DR (2012) Chikungunya virus infection of cell lines: analysis of the East, Central and South African lineage. PLoS One 7:e31102

Win MK, Chow A, Dimatatac F, Go CJ, Leo YS (2010) Chikungunya fever in Singapore: acute clinical and laboratory features, and factors associated with persistent arthralgia. J Clin Virol 49:111–114

Wormald S, Hilton DJ (2004) Inhibitors of cytokine signal transduction. J Biol Chem 279:821–824. www.chikungunyavirusnet.com/epidemiology.html

Xu C, Evensen O, Munang'andu HM (2016) A de novo transcriptome analysis shows that modulation of the JAK-STAT signaling pathway by salmonid alphavirus subtype 3 favors virus replication in macrophage/dendritic-like TO-cells. BMC Genomics 17:390

Yergolkar PN, Tandale BV, Arankalle VA, Sathe PS, Sudeep AB, Gandhe SS, Gokhle MD, Jacob GP, Hundekar SL, Mishra AC (2006) Chikungunya outbreaks caused by African genotype, India. Emerg Infect Dis 12:1580–1583

Yun NE, Peng BH, Bertke AS, Borisevich V, Smith JK, Smith JN, Poussard AL, Salazar M, Judy BM, Zacks MA, Estes DM, Paessler S (2009) CD4+ T cells provide protection against acute lethal encephalitis caused by Venezuelan equine encephalitis virus. Vaccine 27:4064–4073

Ziegler SA, Lu L, da Rosa AP, Xiao SY, Tesh RB (2008) An animal model for studying the pathogenesis of chikungunya virus infection. Am J Trop Med Hyg 79:133–139

Lymphocytic Choriomeningitis Virus (LCMV): Agent for Neurological Diseases and Model for Studying Human Viruses

4

Navita Gupta, Zeenal Punamiya, Sakshi, Anil Kumar Mavi, Vivekanand Bahuguna, and Srijita Chowdhury

Abstract

Lymphocytic choriomeningitis virus (LCMV) is an arenavirus—single-stranded RNA virus. It causes asymptomatic infection in rodents and was identified more than a decade ago in the USA. Though rodents are the main reservoir for LCMV, it can also be transmitted to humans directly or indirectly (through congenital disease or organ donation). Human viral encephalitis caused by LCMV infection can spread through mice, usually starting in the late autumn. LCMV is not a lytic virus and is capable of producing robust lymphocyte cytotoxic responses and triggering inflammation. It enters via aerosol and settles in the lungs causing edema. It then travels via blood to the "meninges," "choroid plexus," and "ventricular ependymal linings," where the virus replicates. Currently, no antiviral therapy, other than "favipiravir," is promising. A number of important concepts have arisen with the use of LCMV as a model pathogen such as "major histocompatibility complex (MHC) restriction," "T-cell memory," chronic infections, exhaustion of T cells, and a major role in the identification of immune-dominant

N. Gupta (✉) · Z. Punamiya
Department of Allied Health Sciences, Chitkara School of Health Sciences, Chitkara University, Rajpura, Punjab, India
e-mail: navita.gupta@chitkara.edu.in

Sakshi
School of Biotechnology, Gautam Buddha University, Greater Noida, Uttar Pradesh, India

A. K. Mavi
Department of Pulmonary Medicine, Vallabhbhai Patel Chest Institute, University of Delhi, New Delhi, Delhi, India

V. Bahuguna
School of Biological Sciences, Doon University, Dehradun, Uttarakhand, India

S. Chowdhury
Department of Biotechnology, National Institute of Technology, Rourkela, Odisha, India

P. L. Bhukya, A. S (eds.), *Emerging Human Viral Diseases, Volume II*,
https://doi.org/10.1007/978-981-97-4480-0_4

and subdominant T-cell epitopes. This viral system has also been used to identify tolerance, autoimmune tissue damage, and the efficacy of treatment interventions and vaccine preparations. The knowledge and progress of immunology have greatly benefitted from studying this virus.

Keywords

LCMV · Arenaviridae · Viral encephalitis · Inflammation · Autoimmune diseases

4.1 Introduction

In the late eighteenth century, a scientific investigation of the smallpox vaccine was conducted by Edward Jenner. He routinely immunized people with a mild form of the cowpox virus to protect them from the deadly virus (Riedel 2005). Nearly a century later, Louis Pasteur tackled rabies (another virus) with similar efforts. By deliberately causing the virus to become attenuated and safe for use as a vaccination, Pasteur made major and vital advances to our understanding of pathogens, human defense mechanisms, and Jenner's vaccination method (Smith 2012). Roads of virology and immunology frequently cross one another. An analogous scientist from the nineteenth century was Pasteur. Contemporary virology and immunology both emerged at the end of the twentieth century. Due to the viral kingdom's enormous diversity and wide range of viral–host interactions in distinct host species, there are many viruses that target different organs. This offered the ideal tool for researching numerous mammalian immune mechanisms. The distinction between the hosts targets the capacity to make a number of decisions about immunological events that occur across species and in many environments (Pircher et al. 1990; Wherry 2011). Most of these viruses have low protein content and a straightforward genetic structure Oldstone 2016). A significant advantage of these pathogens is the fact that they have fewer genes and encoded proteins than any other of their kind. As a result, it makes it simpler to examine the immune responses of these few proteins and identify linkages between host and viral proteins. Furthermore, given a limited number of pathogenicity components compared to other types of pathogens, a causal relationship between viral infections, disease, and proteins is established (Abdel-Hakeem 2019).

A large number of people are infected due to the common human pathogen "lymphocytic choriomeningitis virus" (LCMV). Despite the possibility of major neurological problems (due to LCMV) such as meningitis, encephalitis, and abnormal neurological births, neurologists typically lack knowledge about them. This article aimed to close the gap in knowledge for pediatric neurologists and define the basis and causes of LCMV infection.

LCMV is a "single-stranded RNA" virus known as arenavirus. The name was derived from the Latin word arenosus, which means "sandy," as tiny granularities can be seen inside the virion when seen under an electron microscope (Buchmeier and Zajac 1999).

Like all arenaviruses, LCMV also primarily relies on rodents as a reservoir. The mouse, *Mus musculus,* is the "natural host" and "reservoir" of the virus that is vertically transmitted from parents to offspring's population via infection inside the womb. Hamsters are additional capable reservoirs (Biggar et al. 1975). Because the virus does not break down in the cell and rodents who acquire the infection genetically have immunological tolerance for the pathogen, rodents with LCMV infection may be severely affected, although most instances are asymptomatic. LCMV-infected mice and hamsters secrete large amounts of virus in their nasal secretions, urine, feces, and saliva throughout their lives.

LCMV is typically transmitted to postnatal humans through direct contact with contaminated fomites or by breathing in a viral aerosol. It can also infect through organ transplantation (Fischer et al. 2006). A pregnant woman, who already has a primary LCMV infection, acquires a congenital LCMV infection. The fetus can possibly contract the virus during labor and delivery. With the exception of the transplantation of diseased tissues, there is no evidence of LCMV horizontal transmission from person to person (Bonthius 2012).

4.2 Structure of LCMV

The "lymphocytic choriomeningitis virus" (LCMV) is the original arenavirus group enclosed by an ambisense RNA virus. It has two segments of RNA: one large fragment (L), which encodes a "zinc finger-binding Z protein" and an "RNA-dependent RNA polymerase"; and one small (S) segment encodes a nucleoprotein (NP) and a glycoprotein precursor (GP0) that is cleaved into two subunits, GP1 and GP2. Despite having a mouse as its native host, this virus can infect a broad variety of animals, including humans.

Numerous strains of LCMV have been isolated from the wild. The "Armstrong strain" was discovered by isolating from a monkey experiencing lymphocytic choriomeningitis. Another one is the "Traub strain," which was isolated from a colony of mice that were repeatedly infected in a laboratory, and the "WE strain," which was isolated from a human after contact with those repeatedly infected mice. These have been the subject of the majority of studies in scientific laboratories. While the "Traub" and "WE" strains are sometimes referred to as "viscerotropic," the Armstrong strain is sometimes referred to as "neurotropic." Each strain can grow easily and multiply within the brain system; hence, the term "neurotropic" is a confusing misnomer (Fig. 4.1).

Virus replicates very fast in the viscera, and this leads to the redirection of T cells, causing an immune response that would otherwise prevent T-cell-dependent encephalitis and meningitis. T-cell-dependent disease may be studied using the LCMV-mediated infection model. There are also other variations of these strains that have been studied in animals. Notably, the WE strain—a derivative of Armstrong clone 13—and its "docile" offspring appear to replicate more successfully in mice than the parent strains' high antigen loads do. Using escape mechanisms, mutated T-cell or antibody epitopes are produced (Welsh and Seedhom 2008).

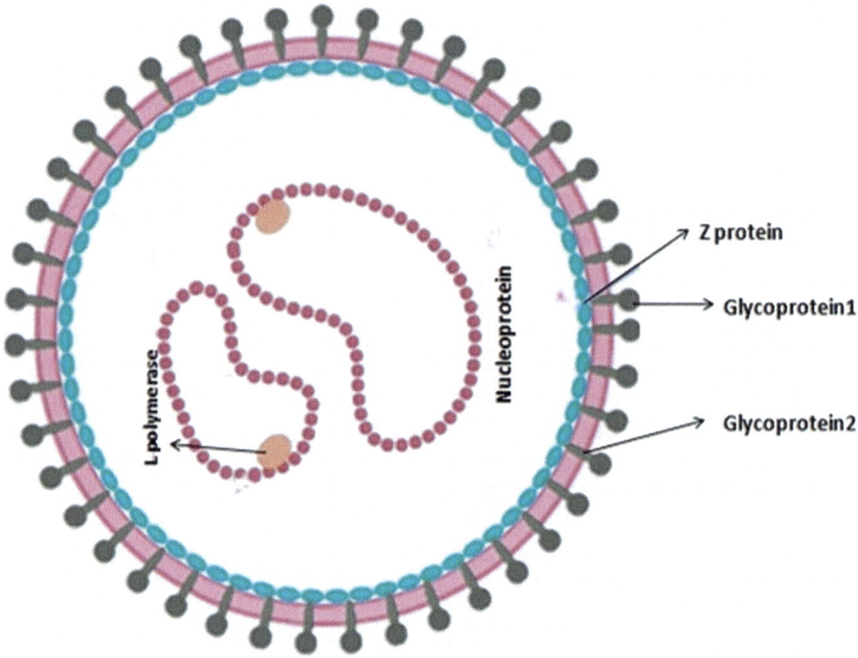

Fig. 4.1 Structure of LCMV. LCMV virions are enveloped with a dense lipid envelope. Surface is covered by glycoproteins, which are aligned with Z protein. Viral ribonucleoproteins are packed into two circular structures and are associated with RNA-dependent RNA polymerase (L polymerase)

4.3 Life Cycle of LCMV

LCMV is an enclosed negative-strand RNA virus with a bi-segmented genome and a cell-cytoplasm-specific life cycle, similar to other arenaviruses. The lipid envelope surrounding the virion's surface layer is dense with distinctive, uniformly distributed spike-like surface structures called GP, which stands for the viral glycoprotein. The virion's surface layer is spherical to pleomorphic in shape and ranges in size from 40 nm to 200 nm (mean: 90–110 nm). Cryo-EM tests with high resolution showed that the arenavirus particles have a highly ordered structure. The glycoproteins on the surface are in alignment with the nearby Z protein packed into a two-dimensional structure and viral ribonucleoproteins. Viral membrane's inner surface has a lattice (Neuman et al. 2005).

The genome is made up of portions of small (S, about 3.5 kb) and large (L, about 7.2 kb) RNA. A noncoding intergenic region (IGR) made up of a sequence anticipated to create a stable hairpin structure is located between each of the genomic segments, which directs the synthesis of two polypeptides in the opposite direction. The large fragment (L) instructs the viral "RNA-dependent RNA polymerase" (RdRp or L protein) (ca 200 kDa) and a tiny RING finger-containing Z protein, whereas the small fragment (S) RNA instructs the viral nucleoprotein (NP, ca 63 kDa) and glycoprotein precursors (GPC, ca 75 kDa) (ca 11 kDa) (Fig. 4.2).

The virus envelope glycoprotein triggers LCMV binding to a cellular receptor and its entrance into the cell. "Stable signal peptide" (SSP, 58aa), GP1 (40–46 kDa), and GP2 are the three substances that make up the GP complex after post-translational proteolytic processing of GPC (35 kDa) (Fig. 4.3).

After the cleavage, GPC is further processed into the GP1 and GP2 subunits by the "cellular proprotein convertase" SKI-1/S1P ("subtilisin kexin isozyme-1/site-1 protease") in the endoplasmic reticulum (Beyer et al. 2003; Kunz et al. 2003). Because it regularly interacts with the GP complex and plays important functions not only in the GP-mediated cell fusion and formation and secretory pathway trafficking of GP, GP1 and GP2 interact ionically, and the GP1/GP2 homotrimer produces the infectious. The viral surface is decorated with spikes. GP1 is the protein that attaches to virion and mediates virus contact situated at the top of the host cell surface receptor spike. The transmembrane protein known as GP2 anchors GP1 to the virus' exterior. The highly conserved and widely expressed alpha-dystroglycan (-DG) serves as the cellular receptor for LCMV cell surface receptor for extracellular matrix proteins. After attaching to a receptor, LCMV virions are transported to acidified endosomes in smooth-walled vesicles that are not connected to clathrin where the host is first invaded by the viral ribonucleoprotein (RNP) cytoplasm of the cell by a pH-dependent membrane fusion process.

It is believed that the late endosomes' acidic environment causes conformational changes in the GP2 to expose a peptide that can mediate fusing of the host cell membrane and the virion. According to a thorough analysis of LCMV entry, the virus primarily enters cells through an unusual endocytotic pathway that is clathrin- and caveolin-independent mechanism, does not need the actin, ARF6, flotillin, or dynamin proteins, and confirms some dependency on membrane cholesterol (Rojek et al. 2008). For Successful infection phosphatidylinositol 3-kinase is necessary for LCMV to function. In addition, the rare phospholipid lysophosphatidic acid leads to the creation of intra-luminal vesicles of the late endosome's multivillus body. Moreover, experimental evidence indicates that endosomal sorting has a function.

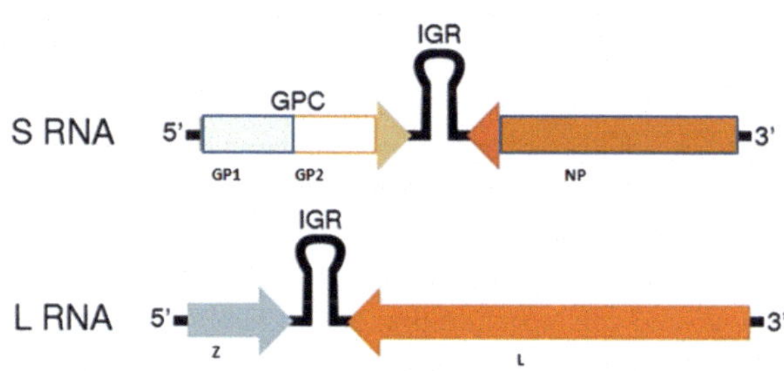

Fig. 4.2 LCMV encodes two single-stranded RNA segments S and L. Each encodes two proteins in opposite orientation separated by an intergenic region. Short segment S encodes glycoprotein precursor post-translationally cleaved into GP1 and GP2 and nucleoprotein. Large segment L encodes Z protein and L polymerase

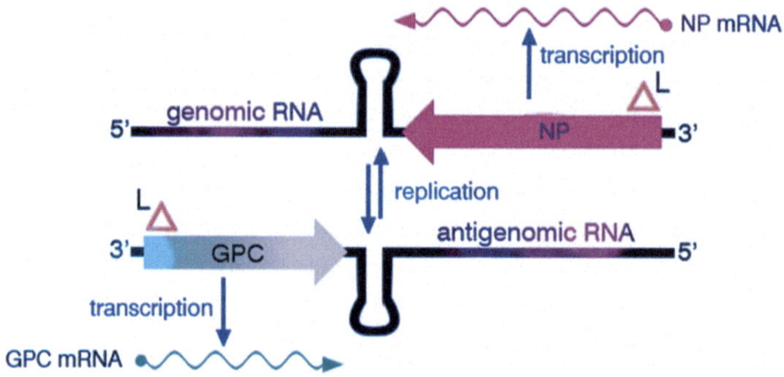

Fig. 4.3 Replication and transcription of S segment ambisense coding strategy of replication means that while NP is transcribed by L polymerase into a genomic complementary RNA, GPC region is transcribed from full-length antigenomic RNA. This antigenomic RNA also serves as a template for replication of viral genome. The GPC and Z coding regions are translated from the genomic sense mRNAs, which are transcribed using the appropriate antigenome RNA (agRNA) species as templates and also serve as replicative intermediates, according to the ambisense organization, while the NP and L coding regions are transcribed into a genomic complementary mRNA

LCMV cell entrance requires a complex for transport (ESCRT), Hrs, Tsg101, Vps22, and other ESCRT components, in particular Alix, Vps4, and the ESCRT-associated ATPase Vps24. As soon as viral RNP enters the cytoplasm of the host cell, the LCMV polymerase uses the infected cell as a template for both replication and transcription.

As a result of genomic promoter's activity, virus recognizes both sequences specifically inside the polymerase, the integrity of highly conserved "3-terminal genomes" of the sequence-formed anticipated panhandle structure complementarity between the viral 5′ and 3′ termini DNA-based RNAs (Perez et al. 2003). The harmful NPL proteins are both required and sufficient for effective transcriptional and replication RNA synthesis (Lee et al.). Starting point for primary transcription is the genome. The production of NP and L mRNA is caused by promoters that are present at the 3′ ends of the RNA segments, respectively. After the disease, polymerase might switch to a replicase mode and make progress along the IGR to produce a duplicate of the uncapped complete agRNA.

The NP encapsulates this agRNA, which acts as a model for the synthesis of the GPC and Z mRNA and for amplification of the associated species of genome RNA (Perez et al. 2003). Termination of non-polyadenylated subgenomic transcripts viral mRNAs has been identified in numerous places (Tortorici et al. 2001), who contend that IGR serves as a signal to stop transcription for the virus polymerase. The mRNAs of LCMV include 1–5 heterogeneous, non-templated nucleotide(s) sequences and the 5′-end cap structure that are probably acquired from mRNAs in the cell by means of cap-snatching processes.

The precise method of grabbing the cap structure is not made clear. A newly discovered special cap-binding property of LASV NP and/or the N-terminal endonuclease activity of the arenavirus L polymerase domain (Morin et al. 2010) could be important. In this process, the generation and spread of LCMV's contagious offspring viral RNPs must be assembled from infected cells and must be related, having membranes that are abundant with the cell surface maturing GP virus. The primary element in the blossoming process is the Z protein with a RING finger that serves as a matrix protein in assembly of the arenavirus particle (Strecker et al. 2003).

The Z proline-rich late domain motifs PTAP and PPPY mediate the budding process. Strictly dependent on Z myristoylation (Perez et al. 2004), Tsg101 is recruited by the LCMV Z protein to the plasma membrane. This belongs to the ESCRT-I. Focusing on Tsg101 via RNA interference results in a significant decrease in Z-mediated blossoming, indicating that Tsg101 is essential for arenavirus development. There are probably more cellular proteins to aid in the emergence of the arenavirus. The virion particles' ability to incorporate GPs depends on the interaction between GP and Z (Capul et al. 2007). Despite the fact that a GPC that has not been digested can go to the cell surface, the GPC must undergo proteolytic processing in order to be incorporated into virion and produce infectious viral particles (Burri et al. 2012; Kunz et al. 2003).

4.4 Immunogenicity of LCMV

LCMV has been widely used for the study and interpretation of the immune response of the host against viral infections. It is characterized by nonlytic nature, simple genome, and presence of two strains that distinguish acute and viral infections (Pascutti et al. 2016). The species *Mus musculus* is the actual primary host and can be used for the study of human viral infections. The mouse infected for 30 days with the virus showed restrained development of hematopoietic stem cells (HSCs) and dendritic cells, and this was related to changes in inflammatory mediators (Yoo et al. 2023).

4.5 LCMV Infection in Host

The common house mouse species *Mus musculus* and *Mus domesticus*, which are found all over the world, serve as the LCMV's "natural host and reservoir." The virus is vertically passed from one mouse group to another through "intrauterine infections." Mice infected during pregnancy fail to generate a defense mechanism using immune cells and develop a chronic but asymptomatic infection that lasts their entire lives. They continuously release significant amounts of the virus into their saliva, milk, semen, urine, and feces throughout their lifetimes.

Humans might be infected if the infected rodent bites or if they have direct contact with rodents, or contact with objects contaminated with rat excrement, or mucosal exposure to aerosols. As additional sources of infection, in addition to mice, pet mice, and hamsters, experimentally infected rodents used in research and mice and hamsters used as pets have also been mentioned (Emonet et al. 2007). There are several exceptions to vertical transmission from mother to fetus, solid-organ donation, and person-to-person infection that have not been documented now (Fischer et al. 2006; MacNeil et al. 2012; Palacios et al. 2008).

Evidence shows the highly focused character of the LCMV and the potential for human illness as a result of virus exposure (Foster et al. 2006). Any clinically serious LCMV infection in people is currently not known. According to serologic research conducted in the USA, the prevalence of LCMV-specific antibodies in people is around 5% with a trend to decline. In Nova Scotia, samples of human serum had a 4% seropositivity rate for LCMV. Between 1998 and 2003, the incidence of LCMV antibodies in an urban setting in Argentina ranged from 1% to 3.6% in humans and 12.9% in house mice (Riera et al. 2005).

In Germany, the presence of mice was less widespread and a prevalence of 9% was reported in rural people. The prevalence of LCMV infection was reported to be 1.7% in humans and 9% in wild rats in Spain in 2003 (Lledó et al. 2003). 2.5% of the blood samples from forestry workers and 5.6% of the tested rats in Italy contained anti-LCMV antibodies (Kallio-Kokko et al. 2006). Contrarily, in our earlier research, we discovered a prevalence of 37.5% of anti-LCMV antibodies in human sera from Bratislava, Slovakia. Additionally, 36% prevalence was reported for Croatia by Dobec et al. (2006).

Despite serosurveys being done on populations of wild mice, nothing is known regarding the frequency of LCMV in domestic pets and laboratory rodents. Pet hamsters with LCMV infection can continue virus transmission even in the absence of overt symptoms. Exposure to persistently infected hamsters or LCMV-contaminated tumor lines has been connected to several outbreaks of LCMV infection in humans. There were 181 human cases of LCMV during the greatest outbreak, which affected 12 states between 1973 and 1974. A single distributor's supply of pet hamsters was linked to this outbreak. It was also discovered that a pet hamster was the potential source of LCMV in a transplant-related outbreak in 2005. Although pet rodents may contract the virus after contact with mice harboring the infection, wild house mice remain the virus' primary reservoir and are a significant cause of human exposure.

In immune-competent people, LCMV infections frequently go unnoticed or cause a minor, self-limiting sickness. Symptoms include fever, chills, nausea, vomiting, myalgia, and headaches, and they start to occur 1–3 weeks after exposure. In the majority of cases, the illness goes away on its own in a few days. A small percentage of patients may develop aseptic meningitis or meningoencephalitis as a result of their disease. Meningitis rarely develops without a prodromal symptom. "Transverse myelitis, Guillain-Barre syndrome, hydrocephalus, and sensory-neural hearing loss" are other neurological side effects linked to adult LCMV infections that have been documented. "Pancreatitis, orchitis, arthritis, parotitis, and pericarditis" are uncommon non-neurological symptoms of disease (Jamieson et al. 2006). In most cases of adult human LCMV infection, individuals make a full recovery, and fatalities are uncommon. Even though recovery is usually complete, it could take months for the same.

However, in a few cases, the illness develops very differently and resembles an infection caused by Lassa virus (Bonthius and Perlman 2007). Three lymphoma patients who received LCMV inoculations in an effort to cause the regression of their chemotherapy-resistant malignancies also experienced the same pathogenetic process. These incidents strikingly resembled fatal LCMV infections that have recently been identified in transplant patients. There have been five clusters of reports of LCMV and LCMV-like arenavirus transmission after solid-organ donation (Fischer et al. 2006; MacNeil et al. 2012; Palacios et al. 2008). There were 17 receivers in these clusters, and 14 of them died from multisystem organ failure, with LCMV-associated hepatitis being a key symptom.

Ribavirin (an antiviral medication with in vitro action against LCMV) and reduced immunosuppression medication were used to treat the survivor in one cluster (Fischer et al. 2006). Without getting therapy intended to treat LCMV infection, two additional infected patients lived (MacNeil et al. 2012). In three additional cases, the outcomes of laboratory testing suggested that the donors may be the source of LCMV infection because there were no clinical indications of LCMV infection or evidence of infection by PCR or serologic examination in the donors in two other cases. One of the donors' infections was positively traced to a pet hamster, while the other sources are still unknown. Viral culture, electron microscopy, RT-PCR, and specific immunohistochemical and serologic testing were used to confirm LCMV infections in four clusters (Fischer et al. 2006; MacNeil et al. 2012).

A novel arenavirus was discovered in one cluster for the first time using objective sequencing at high throughput. After that, further methods were used to confirm the infection (Palacios et al. 2008). These CNS abnormalities were present in a number of patients, although the majority of them were systemic illnesses that resembled VHF. Therefore, these pathogenesis symptoms seemed to depend on persistent viremia versus the T-cell immune response to a virus in tissues. In humans and mice, meningeal syndromes are also seen. Indeed, immunosuppression in these situations is not protective, but rather makes people susceptible to serious diseases. Prenatal LCMV infection is linked to a serious detrimental influence on the health of the fetus, in contrast to adult infection in humans, where severe sickness is uncommon.

First-trimester pregnancy infection with LCMV is linked to a higher incidence of spontaneous abortion (Barton et al. 2002). Congenital intrauterine infection, which manifests as hydrocephalus, macrocephaly or microcephaly, psychomotor retardation, "peri-ventricular calcifications," "gyral dysplasia," "cerebellar hypoplasia," "focal cerebral destruction," visual loss, and chorioretinitis, has been linked to infection in the second and third trimesters (Barton et al. 2002; Bonthius and Perlman 2007; Jamieson et al. 2006).

It is believed that mid- to late-pregnancy maternal viremia causes transplacental infection of the fetus. According to Komrower et al. (1955), this was the first incident of congenital infection to be documented in England. Congenital LCMV infection was documented in numerous cases all over the world in the decades that followed (Bonthius) (Jamieson et al. 2006). Not even half of the cases were linked to the mother's clinical sickness. Pregnant woman's rodent exposure was detected in more than one-third of the instances. Chorioretinitis and hydrocephalus were identified as congenital in the majority of children with LCMV infection.

Additional ophthalmologic abnormalities include cataracts, optic atrophy, nystagmus, esotropia, microphthalmia, and scarring of the chorio-retinal tissue (Jamieson et al. 2006). A new strain known as LCMV-LE was discovered as a result of viral genome analysis (Meritet et al. 2009). Congenital LCMV problems claim the lives of about 35% of babies. More than 66% of the survivors showed neurological conditions such as "microcephaly, mental retardation, seizures, and visual impairment."

The real frequency of congenital LCMV infection is unknown since it resembles congenital toxoplasmosis or cytomegalovirus infection. Evidence suggests that the frequency of congenital LCMV infection may be far more than previously thought. Therefore, LCMV should be considered in the differential diagnosis for any prenatal human infection in which the classic TORCH pathogens (toxoplasmosis, rubella, cytomegalovirus, and herpes virus) have been ruled out. The cornerstone of intervention is prevention because there is no treatment for this virus. Pregnant women should be cautioned to avoid rats at all costs.

4.6 Conclusion

The complications that arise in children are non-specific, and the medical problems that arise due to infections are related to neurological disturbances such as encephalopathy. These complications include shunt failure or infection, aspiration pneumonia, injuries from falls, and joint contractures. A vaccine for the prevention of LCMV infection is not available, nor is there an availability of an antiviral medicine for it, but measures can be taken to reduce the occurrence of LCMV infection. This infection is particularly dangerous when acquired during pregnancy as it may lead to congenital defects. So, proper precautions are required, especially for pregnant women like those working in laboratories and animal care facilities in research institutions. They should wear gloves, face masks, and other barriers for the prevention of infection (Bonthius 2012).

References

Abdel-Hakeem MS (2019) Viruses teaching immunology: role of LCMV model and human viral infections in immunological discoveries. Viruses 11(2):106

Barton LL, Mets MB, Beauchamp CL (2002) Lymphocytic choriomeningitis virus: emerging fetal teratogen. Am J Obstet Gynecol 187(6):1715–1716

Beyer WR, Pöpplau D, Garten W, von Laer D, Lenz O (2003) Endoproteolytic processing of the lymphocytic choriomeningitis virus glycoprotein by the subtilase SKI-1/S1P. J Virol 77(5):2866–2872

Biggar RJ, Woodall JP, Walter PD, Haughie GE (1975) Lymphocytic choriomeningitis outbreak associated with pet hamsters: fifty-seven cases from New York State. JAMA 232(5):494–500

Bonthius DJ (2012) Lymphocytic choriomeningitis virus: an underrecognized cause of neurologic disease in the fetus, child, and adult. Sem Pediatr Neurol 19(3):89–95

Bonthius DJ, Perlman S (2007) Congenital viral infections of the brain: lessons learned from lymphocytic choriomeningitis virus in the neonatal rat. PLoS Pathog 3(11):e149

Buchmeier MJ, Zajac AJ (1999). Lymphocytic choriomeningitis virus. Persistent viral infections, p 575
Burri DJ, Da Palma JR, Kunz S, Pasquato A (2012) Envelope glycoprotein of arenaviruses. Viruses 4(10):2162–2181
Capul AA, Perez M, Burke E, Kunz S, Buchmeier MJ, de la Torre JC (2007) Arenavirus Z-glycoprotein association requires Z myristoylation but not functional RING or late domains. J Virol 81(17):9451–9460
Dobec M, Dzelalija B, Punda-Polic V, Zoric I (2006) High prevalence of antibodies to lymphocytic choriomeningitis virus in a murine typhus endemic region in Croatia. J Med Virol 78(12):1643–1647
Emonet S, Retornaz K, Gonzalez JP, De Lamballerie X, Charrel RN (2007) Mouse-to-human transmission of variant lymphocytic choriomeningitis virus. Emerg Infect Dis 13(3):472
Fischer SA, Graham MB, Kuehnert MJ, Kotton CN, Srinivasan A, Marty FM et al (2006) Transmission of lymphocytic choriomeningitis virus by organ transplantation. N Engl J Med 354(21):2235–2249
Foster ES, Signs KA, Marks DR, Kapoor H, Casey M, Stobierski MG, Walker ED (2006) Lymphocytic choriomeningitis in Michigan. Emerg Infect Dis 12(5):851
Jamieson DJ, Kourtis AP, Bell M, Rasmussen SA (2006) Lymphocytic choriomeningitis virus: an emerging obstetric pathogen? Am J Obstet Gynecol 194:1532–1536. https://doi.org/10.1016/j.ajog.2005.11.040
Kallio-Kokko H, Laakkonen J, Rizzoli A, Tagliapietra V, Cattadori I, Perkins SE et al (2006) Hantavirus and arenavirus antibody prevalence in rodents and humans in Trentino, Northern Italy. Epidemiol Infect 134(4):830–836
Komrower GM, Williams BL, Stones PB (1955) Lymphocytic choriomeningitis in the newborn. Probable transpiacental infection. Lancet:697–698
Kunz S, Edelmann KH, de la Torre JC, Gorney R, Oldstone MB (2003) Mechanisms for lymphocytic choriomeningitis virus glycoprotein cleavage, transport, and incorporation into virions. Virology 314(1):168–178
Lledó L, Gegúndez MI, Saz JV, Bahamontes N, Beltrán M (2003) Lymphocytic choriomeningitis virus infection in a province of Spain: analysis of sera from the general population and wild rodents. J Med Virol 70(2):273–275
MacNeil A, Ströher U, Farnon E, Campbell S, Cannon D, Paddock CD et al (2012) Solid organ transplant–associated lymphocytic choriomeningitis, United States, 2011. Emerg Infect Dis 18(8):1256
Meritet JF, Krivine A, Lewin F, Poissonnier MH, Poizat R, Loget P et al (2009) A case of congenital lymphocytic choriomeningitis virus (LCMV) infection revealed by hydrops fetalis. Prenat Diagn 29(6):626–627
Morin B, Coutard B, Lelke M, Ferron F, Kerber R, Jamal S et al (2010) The N-terminal domain of the arenavirus L protein is an RNA endonuclease essential in mRNA transcription. PLoS Pathog 6(9):e1001038
Neuman BW, Adair BD, Burns JW, Milligan RA, Buchmeier MJ, Yeager M (2005) Complementarity in the supramolecular design of arenaviruses and retroviruses revealed by electron cryomicroscopy and image analysis. J Virol 79(6):3822–3830
Oldstone MB (2016) An odyssey to viral pathogenesis. Ann Rev Pathol 11:1–19
Palacios G, Druce J, Du L, Tran T, Birch C, Briese T et al (2008) A new arenavirus in a cluster of fatal transplant-associated diseases. N Engl J Med 358(10):991–998
Pascutti MF, Erkelens MN, Nolte MA (2016) Impact of viral infections on hematopoiesis: from beneficial to detrimental effects on bone marrow output. Front Immunol 7:364
Perez M, Craven RC, de la Torre JC (2003) The small RING finger protein Z drives arenavirus budding: implications for antiviral strategies. Proc Natl Acad Sci 100(22):12978–12983
Perez M, Greenwald DL, de La Torre JC (2004) Myristoylation of the RING finger Z protein is essential for arenavirus budding. J Virol 78(20):11443–11448
Pircher H, Moskophidis D, Rohrer U, Bürki K, Hengartner H, Zinkernagel RM (1990) Viral escape by selection of cytotoxic T cell-resistant virus variants in vivo. Nature 346(6285):629–633

Riedel S (2005) Edward Jenner and the history of smallpox and vaccination. Proc (Bayl Univ Med Cent) 18(1):21–25

Riera L, Castillo E, Saavedra MDC, Priotto J, Sottosanti J, Polop J, Ambrosio AM (2005) Serological study of the lymphochoriomeningitis virus (LCMV) in an inner city of Argentina. J Med Virol 76(2):285–289

Rojek JM, Perez M, Kunz S (2008) Cellular entry of lymphocytic choriomeningitis virus. J Virol 82(3):1505–1517

Smith KA (2012) Louis Pasteur, the father of immunology? Front Immunol 3:68

Strecker T, Eichler R, Meulen JT, Weissenhorn W, Dieter Klenk H, Garten W, Lenz O (2003) Lassa virus Z protein is a matrix protein sufficient for the release of virus-like particles. J Virol 77(19):10700–10705

Tortorici MA, Albarino CG, Posik DM, Ghiringhelli PD, Lozano ME, Pomar RR, Romanowski V (2001) Arenavirus nucleocapsid protein displays a transcriptional antitermination activity in vivo. Virus Res 73(1):41–55

Welsh RM, Seedhom MO (2008) Lymphocytic choriomeningitis virus (LCMV): propagation, quantitation, and storage. Curr Protoc Microbiol 8(1):15A.1

Wherry EJ (2011) T cell exhaustion. Nat Immunol 12(6):492–499

Yoo S, Jeong YH, Choi HH, Chae S, Hwang D, Shin SJ, Ha SJ (2023) Chronic LCMV infection regulates the effector T cell response by inducing the generation of less immunogenic dendritic cells. Exp Mol Med 55(5):999–1012

West Nile Virus: A Comprehensive Overview of Epidemiology and Pathology

5

Sonal Mundhra

Abstract

West Nile Virus (WNV), a Flavivirus, has become a global public health concern since its discovery in 1937. This arthropod-borne virus exhibits diverse clinical manifestations in humans, ranging from asymptomatic infections to severe neurological complications, transmitted primarily by mosquitoes and affecting both humans and other mammals. This book chapter provides a concise yet thorough exploration of WNV, covering molecular biology, host interactions, transmission dynamics, clinical outcomes, diagnostics, and therapeutics. Navigating through viral replication pathways and host interactions, it delves into factors influencing WNV epidemiology. The chapter also addresses challenges to public health systems and ongoing efforts in prevention and control. For researchers, healthcare professionals, and students, this resource offers insights into WNV's impact on human and animal health. Covering historical context, transmission details, clinical manifestations, viral structure, replication processes, pathogenesis, immune response, virulence factors, diagnostics, therapeutics, and prevention, this chapter serves as a comprehensive guide to understanding the epidemiology and pathology of WNV.

Keywords

West Nile virus · Flavivirus · Arthropod-borne · Mosquitoes · Neurotropic Encephalitis · Epidemiology · Global concern

S. Mundhra (✉)
Encephalitis Group, ICMR-National Institute of Virology, VRDC, Venky's India Ltd., Pune, Maharashtra, India
e-mail: sonal.mundhra@gmail.com

P. L. Bhukya, A. S (eds.), *Emerging Human Viral Diseases, Volume II*,
https://doi.org/10.1007/978-981-97-4480-0_5

5.1 Introduction

West Nile virus (WNV) is an arbovirus belonging to the *Flavivirus* genus and *Flaviviridae* family, transmitted primarily through mosquito vectors. First identified in 1937, the virus gained global attention in recent decades due to its rapid spread across different continents. From the mid-1990s onward, WNV outbreaks witnessed an increase in frequency, severity, and geographic range (Murgue et al. 2002). In 1999, WNV crossed the Atlantic and reached the Western Hemisphere when a cluster of encephalitis cases emerged in the metropolitan area of New York City, United States, after which the virus spread across most of the states in the USA and the neighboring Canada and Mexico within a span of 3 years (Hayes and Gubler 2006). Over the course of 86 years since its discovery, WNV has disseminated across a vast region of the world (Fig. 5.1) and is now recognized as the primary cause of viral encephalitis on a global scale due to its potential for large-scale outbreaks and its ability to cause substantial morbidity and mortality. In the wild, WNV is maintained in an enzootic cycle of transmission between birds and mosquito vectors, while humans and other mammals are incidental hosts. The clinical spectrum of WNV infection can vary widely, with the majority of individuals being asymptomatic or experiencing mild flu-like symptoms. However, a small proportion of infected individuals develop severe neurological manifestations, including meningitis, encephalitis, and acute flaccid paralysis (Sejvar et al. 2003). This chapter aims to provide a comprehensive overview of the current understanding of WNV, including its epidemiology, clinical manifestations, virulence factors, immune response, diagnostic methods, and prevention strategies. By analyzing the available literature and highlighting key research findings, the authors seek to contribute to the existing

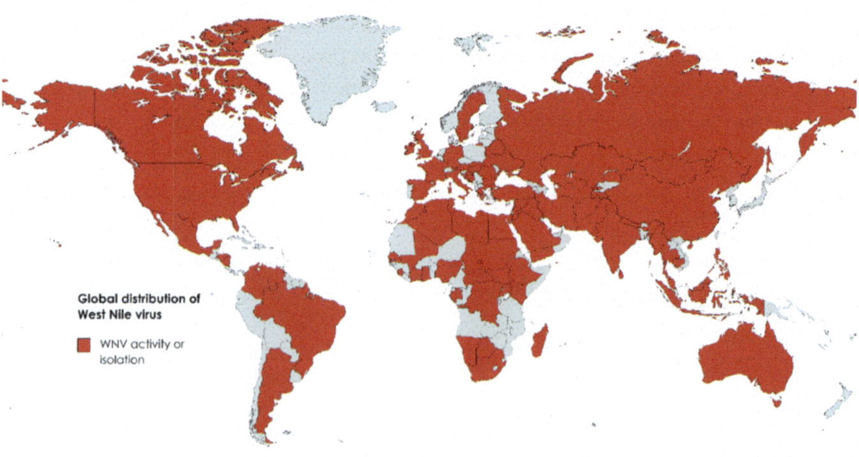

Fig. 5.1 Global distribution of WNV. The map was generated using a free online tool Mapchart with data from published literature (Caron et al. 2017; Chowdhury et al. 2014; Eybpoosh et al. 2019; Lorenz and Chiaravalloti-Neto 2022; Mencattelli et al. 2022; Saiz et al. 2021; Sule et al. 2018; Tinto et al. 2022; Zakhia et al. 2021; ECDC Report 2023)

knowledge and assist in guiding future efforts aimed at combating the challenges posed by its global expansion.

5.1.1 Brief History

WNV was first isolated in the West Nile region of Uganda in 1937 from a febrile adult woman during a survey on yellow fever (Smithburn et al. 1940). Later cases of WN fever were documented in the Middle East, France, and South Africa (1950 onwards) (Murgue et al. 2002). Until the 1950s, sporadic cases and infrequent outbreaks of WNV with mild febrile illness have been reported in India, Egypt, France, and South Africa (Chancey et al. 2015). WNV became recognized as a cause of severe human meningoencephalitis after an outbreak of WN neuroinvasive disease (WNND) was reported in Israel (Weinberger et al. 2001). WNV has been reported in human serosurveillance studies in the Middle East, Africa, Europe, Asia, Australia (subtype Kunjin), Central America and the Caribbeans, South America, and North America (Chancey et al. 2015; Murgue et al. 2002). Starting from the 1990s, the frequency, severity, and geographical range of WNV infection increased tremendously with cases in almost every continent except Antarctica (Campbell et al. 2002). The most striking introduction of WNV is its appearance in North America in 1999 in New York City when 62 confirmed cases, including 7 deaths were reported (Nash et al. 2001). Later, the virus rapidly spread to most US states, neighboring Canada, and Mexico and to date is responsible for >7 million human infections with more than 28,000 neuroinvasive cases and 2700 deaths documented in the USA alone (Ronca et al. 2019; https://www.cdc.gov/westnile/statsmaps/historic-data.html).

5.1.2 Classification

WNV belongs to the family Flaviviridae, composed of enveloped positive-sense single-stranded RNA viruses divided into three main genera (flavivirus, hepacivirus, and pestivirus). Genus flavivirus is composed of more than 70 neurotropic and hepatotropic virus species, most of which are transmitted by arthropods (e.g., dengue, Japanese encephalitis, yellow fever, and tick-borne encephalitis) (Chen et al. 2019). WNV belongs to the Japanese encephalitis serocomplex, which includes Japanese Encephalitis virus (JEV), Murray Valley encephalitis virus (MVEV), St. Louis encephalitis virus (SLEV), Cacipacore virus, Koutango virus, Usutu virus, and Yaoundé virus (Mackenzie et al. 2006). Genetic analysis of globally distributed WNV strains shows 20–26% nucleotide divergence (Bondre et al. 2007). WNV is biologically diverse; up to nine lineages have been proposed (Fall et al. 2017). Most human outbreaks have been linked to lineages 1 and 2. Lineage 1 (L1) has three distinct clades. Clade 1a comprises strains isolated from Europe, Africa, Asia, and the Americas. Clade 1b (Kunjin virus) is limited to Oceania and clade 1c, comprising of strains endemic to India, is now proposed as a distinct lineage 5 (Bondre et al.

2007). Clade 1a is the cause of massive outbreaks affecting humans, horses, and birds while clade 1b is fairly restricted to horses and neurological disease is rarely reported (Hall et al. 2002). Lineage 2 (L2) was exclusively reported in Africa up until 2004, when it was isolated from humans and birds in Hungary, Greece, and Italy (Bagnarelli et al. 2011; Bakonyi et al. 2006; Botha et al. 2008; Papa et al. 2011). Besides L1 and L2, other lineages are less widespread. Lineage 3 (Rabensburg virus) was reported in the Czech Republic (Bakonyi et al. 2005). Lineage 4 has been isolated from Russia (Lvov et al. 2004). A supposed sixth lineage, identified from a small gene fragment, has been described from Spain (Vázquez et al. 2010). Koutango virus forms a distinct lineage (lineage 7) of WNV (Fall et al. 2014). Additionally, two new lineages (lineage 8 and 9) were isolated from mosquitoes in Senegal and Austria, respectively (Fall et al. 2014; Pachler et al. 2014).

5.2 Epidemiology

The first occurrence of neuroinvasive disease caused by WNV was documented in Israel in the 1950s (Weinberger et al. 2001). Since then, there were some sporadic reports of WNV circulation in Europe until the 1990s, after which the virus reemerged with higher frequency and severity of clinical cases (Calistri et al. 2010; Hubálek and Halouzka 1999). The geographical distribution of WNV has expanded significantly in recent years, encompassing regions in Africa, Europe, Asia, Australia, and the Americas. WNV encephalitic outbreaks have been seen in North America, Europe, Africa, Australia, and the Middle East affecting humans, horses, and birds (Chancey et al. 2015). Over the span of three decades since the 1990s, several outbreaks have been documented in the Americas, Mediterranean basin, and southern Europe, including in Italy (1990s and 2000s), Algeria (1994), Morocco (1996, 2003), Tunisia (1997), Romania (1996, 2009), the Czech Republic (1997), Russia (1999–2001, 2004), Israel (1999–2000, 2005–2010, 2012), France (2003), Hungary (2003–2013), Portugal (2004), Spain (2004, 2010, 2010), and in the United States of America (1999 onwards) (Calistri et al. 2010; Ronca et al. 2021). Most of the above outbreaks and WNV circulation in the above regions have been seen to be associated with lineage 1 (Ciota and Kramer 2013). The WNV strains originally isolated from Africa (Sub-Saharan and Madagascar regions) are classified as lineage 2 (L2). Although L2 strains were initially considered to be of low virulence, they have caused deaths in horses and increased human cases involving neuroinvasive disease since 2007 in South Africa (Venter and Swanepoel 2010) and epidemics in Europe and Russia since 2008 (Bakonyi et al. 2013; Danis et al. 2011; Platonov et al. 2011). Lineage 2 strains are now responsible for major outbreaks in Italy, Spain, and Russia and have now become predominant in Central and Southern Europe with an expanding geographical range (Platonov et al. 2014). Bird serosurveillance and mortality have closely been associated with increased human and equine outbreaks which were extensively conducted in the Americas and Europe (Hernández-Triana et al. 2014; Ronca et al. 2021). The Australian Kunjin strain (lineage 1c) has also reported increased virulence in horses with rare human infections (Frost et al. 2012). The lineage 5 strains (earlier 1b) are endemic to India and though they haven't caused major outbreaks since their isolation, they are responsible for sporadic cases and introductions in new areas (Bondre et al. 2007; Chowdhury et al. 2014). Besides lineages 1, 2, and 5, other lineages have not been associated with human cases.

5.2.1 WNV Transmission Cycle

WNV has a complex life cycle involving birds acting as reservoirs and amplifying hosts and mosquitoes as transmission vectors. Humans and other vertebrates act as incidental hosts as seen in (Fig. 5.2).

5.2.2 Vector Transmission

Mosquitoes belonging to the *Culex* spp. (family Culicidae) are the primary amplification vectors and also act as bridge vectors. In North America, Europe, Australia, Asia, and Africa, several mosquito species have been recognized as capable vectors, with *Culex* species being the most common and effective (Brault 2009; Hall et al. 2002; Hayes et al. 2005; Zeller and Schuffenecker 2004). Besides mosquitoes, WNV has been sporadically isolated in other arthropods such as ticks, mites, and hippoboscid flies although it seems unlikely that they play a substantial role in the viral transmission cycle (Farajollahi et al. 2005; Lwande et al. 2014; Mumcuoglu et al. 2005).

5.2.3 Reservoir Hosts

Birds are recognized as amplifying hosts and are the WNV's natural reservoirs. Since WNV's first introduction in 1999, hundreds of avian species in North America have been identified as capable of being infected by the virus (Beasley 2005). The Passeriformes, primarily those of the Corvidae family, have been the most severely impacted. When infected, several avian species secrete considerable amounts of the

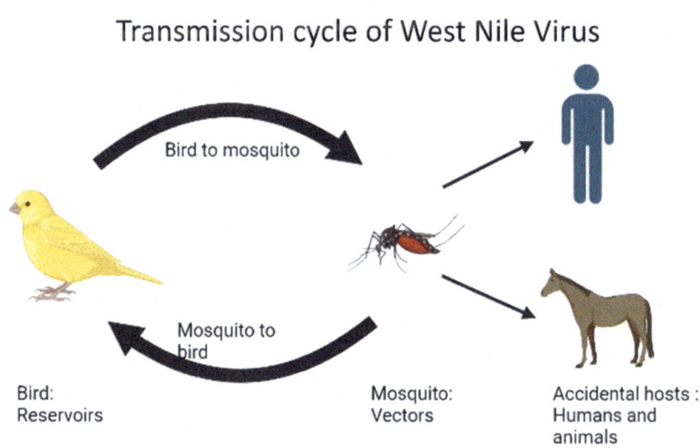

Fig. 5.2 Transmission cycle of West Nile virus (created on BioRender.com)

virus in their feces or oral secretions (Komar et al. 2003), facilitating direct transmission between birds and even between birds and humans. The introduction, reintroduction, and spread of WNV throughout Europe and the Americas have mostly been attributed to migrating birds (Calistri et al. 2010; Dusek et al. 2009). In the southern, eastern, and central regions of India, migrating wild birds are shown to have serological evidence of WNV infection (Mishra et al. 2012).

5.2.4 Humans, Horses, and Other Animals

WNV infection in humans, horses, and other animals is typically regarded as a "dead-end" infection since viral replication in these hosts does not result in enough viremia to allow transmission from vertebrate hosts to feeding mosquitoes (Bunning et al. 2002). Mammals such as dogs, sheep, pigs, deer, raccoons, bears, wolves, squirrels, bats, etc., are susceptible to WNV infection (Beasley 2005; Blitvich 2008). Besides mammals, snakes, crocodiles, frogs, etc., have also been found to harbor WNV with sometimes high viremia. However, the role of hosts other than birds in maintaining the virus in nature is unknown (Byas and Ebel 2020).

5.2.5 Nonvector Transmission

In addition to transmission between birds and mosquitoes, West Nile Virus (WNV) can also be maintained through nonvector mechanisms. In humans, blood transfusion, organ transplants, and mother-to-child transmission through placenta or breast milk have been documented (Byas and Ebel 2020). WNV RNA has been detected in carcasses and feces of certain birds like crows and oral and cloacal sheddings of geese suggesting bird-to-bird and possibly even bird-to-human transmission. Vertical transmission from adult female mosquitoes to their offspring is also possible but inefficient (Nelms et al. 2013). Overwintering in mosquitoes and birds coupled with vertical transmission allows the pathogen's reappearance each spring.

5.3 Clinical Manifestations

Most human infections are subclinical (almost 80%), and most persons developing symptoms (about 20%) have a febrile illness called West Nile fever (WNF), accompanied by myalgia, malaise, headache, and gastrointestinal distress (Petersen and Marfin 2002). Only a tiny percentage of infected people (<1%) develop encephalitis, and when this condition progresses to serious neurological illness, it can cause acute flaccid paralysis after meningitis or encephalitis, with quickly worsening symptoms that may affect all four limbs. Symptoms usually develop between 2 and 15 days after infection and last for 2–5 days (Campbell et al. 2002). Extremely poor prognosis can result from a severe poliomyelitis-like condition due to the development of respiratory problems. Elderly people, infants, immunocompromised patients, transplant patients, and those with a history of cardiovascular disease, hypertension, or hepatitis C virus infection are more prone to neurological involvement that may result in death (Murray et al. 2006). As many as 50% of older adults who survive may experience significant post-illness morbidity for at least a year following infection (Campbell et al. 2002). Children with WNV have better neurological outcomes, shorter hospital stays, less neurological sequelae, and decreased deaths when compared to adults (LaBeaud et al. 2006).

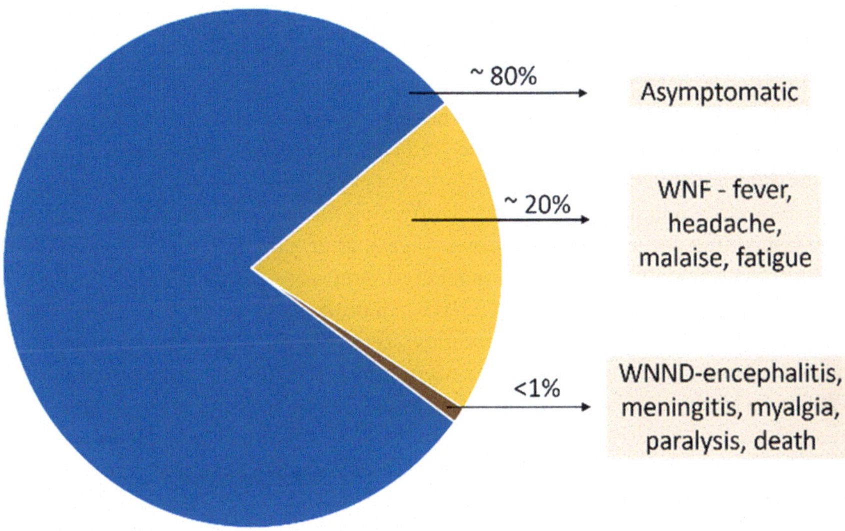

Compared to humans, horses develop symptoms in only 10% of infections and are characterized by fever and muscle weakness (Bunning et al. 2002). Mortality has drastically reduced due to the availability of vaccines for horses (Ward 2006). Infections in birds are characterized by non-neurological signs like ruffled feathers, anorexia, dehydration, and neurological symptoms like ataxia and tremors (Gamino and Höfle 2013).

5.3.1 Morbidity and Mortality

WNE is associated with considerable short-term and long-term morbidity and mortality, especially in older and immunosuppressed individuals (Campbell et al. 2002). A case fatality rate of more than 10% has been estimated in hospitalized cases (Petersen and Marfin 2002). Neuromuscular symptoms are recognized as crucial indicators of increased morbidity and mortality in WNV neuroinvasive illness (Leis and Stokic 2012). As many as 50% of older adults who survive WNND may experience significant postillness morbidity for at least a year following infection (Campbell et al. 2002). Recent studies indicate that WNF may be linked to more serious ongoing issues than previously believed especially in older adults (Sejvar 2014). Although the underlying causes of chronic morbidity after WNF are still unknown, accumulating evidence points to the possibility that even "mild" WNV infections might cause severe short- and long-term morbidity (Watson et al. 2004).

5.3.2 Viral Persistence

While there has been a prevailing assumption that successful survival post-WNV infection, with or without signs of neuroinvasive disease, implies effective control and likely elimination of the virus, emerging evidence challenges this conventional understanding. Findings from investigations involving laboratory-infected animals consistently demonstrate that WNV can establish an asymptomatic persistent infection in vertebrates with a competent immune system. Virus was isolated from slain rhesus monkeys that did not show any indication of WND but presented WNV-specific antibodies in the serum (Pogodina et al. 1983). Later, persistent viral infection was described in the hamster model, affecting the brain and kidneys along with urinary shedding for as long as 8 months (Tesh et al. 2005). The wild-type WNV persisted preferentially in the skin, spinal cord, brain, and lymphoid organs of mice after systemic infection (Appler et al. 2010). All of these results from animal experiments imply that persistent infections may arise in people and should be closely watched. Clearly enough, a recent investigation has found WNV-RNA in the urine of 20% of a study population 7 years after infection, without recovery of infectious virus (Murray et al. 2010). Similar to this, 99 days after the onset of symptoms, WNV-RNA and antigens were also found in the CNS of an immunocompromised patient (Penn et al. 2006).

5.3.3 Long-Term Sequelae

Long-term outcomes of WN infection indicate events that occur for months to years following acute illness. WN Neuroinvasive disease (WNND) can result in long-term sequelae, including cognitive impairments, movement disorders, muscle weakness, fatigue, memory loss, parkinsonism, and functional difficulties (Sejvar 2014). This is especially seen in those patients who had WNE or WN meningitis (WNM). The prognosis of those with WN poliomyelitis (WNP) is far poorer than those with WNE or WNM and its long-term outcomes include muscle weakness in one to all four limbs, respiratory failure, and the need for ongoing rehabilitation (Sejvar 2007).

5.4 Structure of the Virus

5.4.1 Morphology and Virion Structure

WNV is a small enveloped virus with icosahedral symmetry and a diameter of 50 nm. The protein capsid (C) forms a nucleocapsid or covering around the genomic RNA, which is then encased in a lipid bilayer made of host cell lipids. The

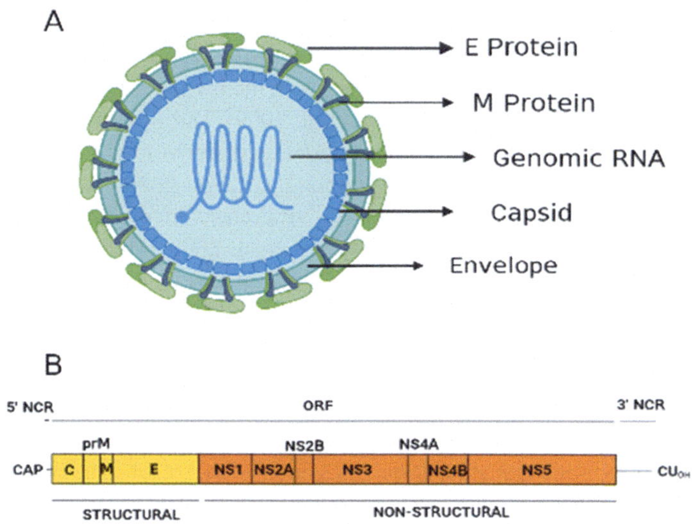

Fig. 5.3 (**A**) Schematic representation of WN viral structure. (**B**) Schematic representation of WN viral genome

membrane (M) protein and the Envelope (E) glycoprotein, 180 copies of each, are embedded in the lipid bilayer surrounding the nucleocapsid. The E protein is arranged in 90 antiparallel homodimers in three different symmetry environments (Fig. 5.3A). The exterior surface of mature virions is smooth and devoid of protrusion or spikes (Mukhopadhyay et al. 2003).

5.4.2 Genome Structure and Organization

WNV genome is composed of a single-stranded positive-sense RNA molecule flanked by two noncoding regions (NCRs) at the 5′ and 3′ ends of the genome that are 100 and 400–700 nucleotides in length, respectively (Fig. 5.3B). The WNV genome has a cap structure at the 5′ end and terminates with CU_{OH} but no 3′ poly(A) tract. The RNA consists of a single open reading frame of ~11,000 kb which encodes a single polyprotein (Brinton 2002). Ten primary viral proteins result from the proteolytic processing of the polyprotein by viral and cellular proteases: three structural proteins, the capsid (C), premembrane (prM), and envelope (E), and seven nonstructural proteins (NS1, 2A, 2B, 3, 4A, 4B, and 5).

5.4.2.1 Structural and Nonstructural Proteins

The structural proteins play important roles in the assembly, dissemination, binding, fusion, and internalization of a mature virion among other functions. The C protein associates with the viral genome to form a nucleocapsid making it essential for viral assembly and replication (Brinton 2013). It has also been linked to the activation of apoptosis (Yang et al. 2008) and breakdown of the blood–brain barrier

(BBB) (Medigeshi et al. 2009). The prM/M is a short transmembrane glycosylated protein with chaperone-like function. It is involved in correct folding, maturation, and assembly of the E protein and interferes with the premature fusion of E protein during virion assembly and release (Tan et al. 2009). During virion maturation, a furin-like protease in the trans-Golgi network cleaves prM to produce the M protein (Brinton 2002). The E is the most immunogenic transmembrane glycosylated protein and is involved in receptor binding, viral entry, membrane fusion, vector transmission, and neuroinvasiveness (Moudy et al. 2009; Mukhopadhyay et al. 2003). Of the three domains of WNV E protein, domain (D) I and DII are involved in viral fusion with host endosomal membranes where DII has the fusion loop, DII and DIII take part in E protein homodimerization while DIII is involved in binding to neutralizing antibodies by its several immunogenic epitopes (Kyle Austin and Dowd 2014). E proteins undergo a pH-dependent structural change to facilitate the fusion of viral and host membranes inside the endosome. WNV E has a single N-linked glycosylation site at position 154 which is important for viral neuroinvasiveness and pathogenesis (Carbaugh and Lazear 2020). The NS proteins facilitate the formation of host membrane-derived replication complexes (RCs) that are important for viral replication and assembly. The NS1 protein performs several functions in WNV replication, virulence, immunomodulation, and pathogenicity (Brinton 2013). The NS2A plays a role in the assembly of virion and the reorganization of intracellular membranes for the formation of RCs and seems to have an immunomodulatory role by inhibiting α/β interferon (IFN) production (Leung et al. 2008; Liu et al. 2006). The NS2B protein acts as a cofactor for the viral trypsin-like serine protease NS3 (Chappell et al. 2008). The NS3/NS2B complex and NS3 alone have been associated with host cell apoptosis (Ramanathan et al. 2006). The NS4A is involved in the establishment of the viral RC along with NS4B while the latter also has a significant role in WNV's suppression of IFN signaling (Evans and Seeger 2007). The NS5 has two distinct enzymatic activities and colocalizes with dsRNA in the viral RC (Mackenzie and Westaway 2001). The C-terminal encodes the viral RNA-dependent RNA polymerase (RdRp) required for replication (Malet et al. 2008) and is a potent antagonist of IFN signaling (Laurent-Rolle et al. 2010). The N-terminal encodes the methyltransferase necessary for capping of viral RNA at the 5′ end (Zhou et al. 2007).

5.4.3 Viral Replication

The replication cycle of flaviviruses, including WNV, involves several stages: binding and entry, translation and replication, assembly and release. During the binding and entry stage, WNV attaches to the cell surface and the E glycoprotein binds to a cellular receptor, initiating internalization via endocytosis. Various molecules have been suggested as potential cellular receptors for WNV, including glycosaminoglycans, c-type lectins like DCSIGNR, TIM phosphatidylserine-binding protein, integrin αvβ3, and the ubiquitin ligase CBLL1 (Saiz et al. 2021). Additionally, certain G-coupled receptor kinases (GRK) act as cofactors that aid in viral entry. Endocytosis occurs primarily in clathrin-coated pits while viral uncoating takes place in the mildly acidic environment of endosomes, leading to fusion of the viral envelope with the endosomal membrane (Chu and Ng 2004). The positive-strand RNA genome is released into the cytoplasm. The viral polyprotein is synthesized and then cleaved by host and viral proteases into structural and nonstructural proteins. The synthesis of the polyprotein occurs at the endoplasmic reticulum (ER). The rearrangement of the endoplasmic reticulum (ER) is orchestrated by viral nonstructural proteins in collaboration with host proteins to form replication organelles (RO). These ROs, characterized by convoluted membranes and invaginations, serve as sites where NS5 uses viral RNA to generate negative-sense intermediates which serve as templates for the production of positive-sense RNAs. The progeny viral RNA is utilized either for the assembly of new virus particles or for translation into viral proteins. The newly synthesized capsid protein is directed to vesicle packets along the rough ER to encapsulate viral RNA into the nucleocapsid. Concurrently, structural prM and E proteins are recruited to vesicle packets forming immature virions in the ER lumen. Viral NS proteins interact with the host endosomal sorting complexes for the transport of virions. Prior to release into the extracellular space, the prM protein is cleaved releasing a short pr fragment, forming mature virions.

5.5 Pathogenesis in Humans

5.5.1 Viral Entry in the CNS

Upon a mosquito bite, WNV initially infects skin cells. The virus then begins to replicate locally at the site of entry. Following this initial replication, the virus may spread to regional lymph nodes, where it can encounter immune cells. From the lymph nodes, WNV can further disseminate through the bloodstream, reaching various organs, including spleen and the central nervous system (CNS) (Fig. 5.4). WNV can get access to the CNS by multiple possible mechanisms, including the direct infection of the vascular endothelium, crossing the BBB following its compromise, carried by immune infected cells via a "Trojan horse" mechanism, infection of the olfactory neurons, and direct axonal retrograde transport from infected peripheral neurons (Bai et al. 2019). The main CNS residential cells, including neurons, and glial cells are presumably the target cells for WNV (Shrestha et al. 2003).

5.5.2 WNV CNS Pathogenesis

The survival and pathogenicity of WNV in the host are contingent on its ability to early invade the CNS, replicate in high titers, and evade host immune responses (Winkelmann et al. 2016). A key aspect of WNV neuropathogenesis is CNS invasion followed by neuronal cell death (Garcia-Tapia et al. 2007). WNV produces lesions in various brain regions, causing neuroinvasive disease (Guarner et al. 2004; Omalu et al. 2003). Neuroinflammatory reactions, while a

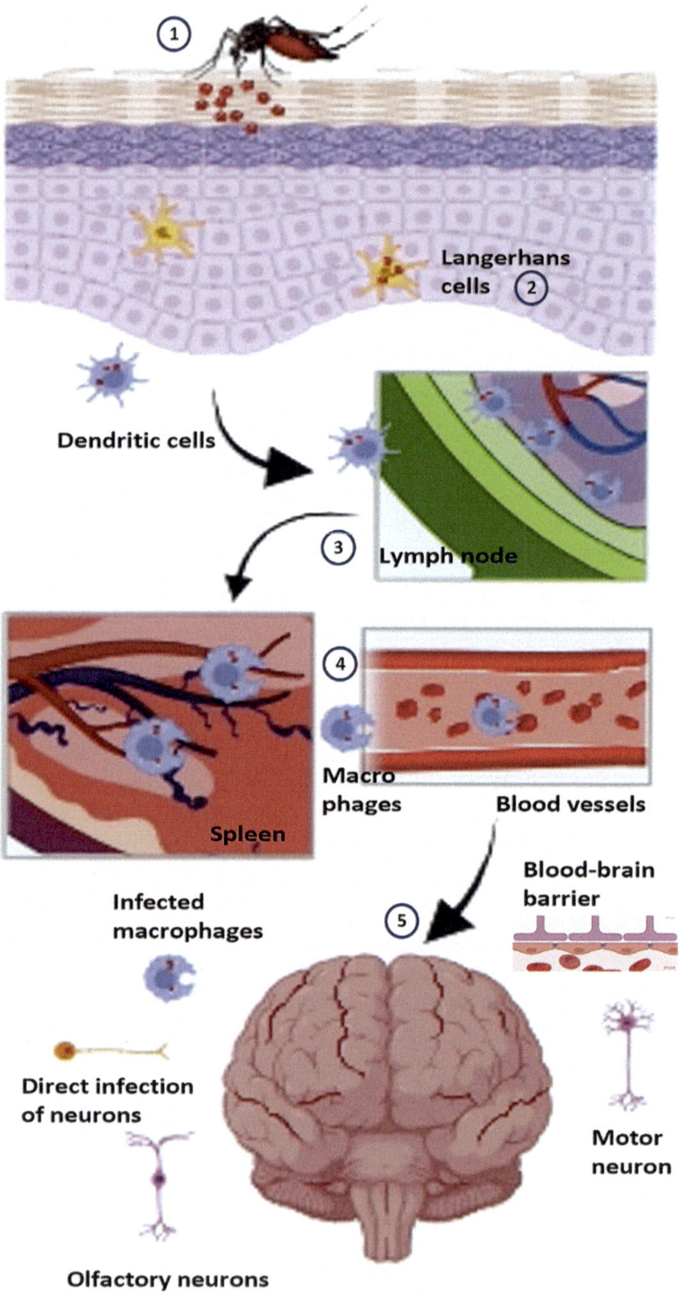

Fig. 5.4 Entry of WNV and migration to the CNS. WNV first replicates in keratinocytes (**1**) at the site of infection proceeding to Langerhans cells and dendritic cells (**2**). The virus subsequently spreads to regional lymph nodes and the spleen (**3**), infecting macrophages and neutrophils that enter the bloodstream along with circulating virions (**4**). WNV may enter the CNS via multiple routes (**5**) like direct infection through brain endothelium, infected peripheral motor or olfactory neurons, transport across the BBB, or within infected macrophages or neutrophils. Created on BioRender.com

defense mechanism, are also implicated in neuropathogenesis (Angeles Muñoz-Fernández and Fresno 1998; Das et al. 2008). Histological examinations reveal brain inflammation, microglial nodules, plasma cell and T-cell accumulation, and activation of resident microglia and invading macrophages (Eldadah and Nathanson 1967; Omalu et al. 2003). Abnormalities in brain areas are observed in WNV-infected patients through magnetic resonance scans (Petropoulou et al. 2005). WNV infection induces encephalitis in humans, marked by inflammatory lesions, degenerating neurons, and the presence of T lymphocytes (Swanson and McGavern 2015). The virus triggers apoptotic pathways, including caspase-3-dependent apoptosis, as well as autophagy, pyroptosis, and necroptosis (Kleinschmidt et al. 2007; Lim et al. 2017; Medigeshi et al. 2007; Mundhra and Bondre 2023; Ramanathan et al. 2006; Yang et al. 2008). Overall, WNV's migration to the CNS, infection, and neuronal damage define the course of infection, disrupting normal neuronal function and contributing to observed neurological symptoms (Sampson et al. 2000).

5.5.3 Host Immune Response (Innate and Adaptive)

Innate antiviral defenses are essential for viral control and include pro-inflammatory cytokines, type I IFNs, and the expression of antiviral genes (Suthar et al. 2010). WNV is injected intradermally by a mosquito bite and most likely initially replicates in keratinocytes and antigen-presenting cells (APCs) like Langerhans cells (LCs). The infected Langerhans cells migrate to draining lymph nodes from which the virus enters the bloodstream (Johnston et al. 2000). Circulating and migratory innate cells like neutrophils and mast cells are attracted to the inoculation site due to the chemokines released by injured cells and the APCs after being stimulated by both damage- and pathogen-associated molecular patterns (Suthar et al. 2013). The cell senses the presence of the virus through its pattern recognition receptors (PRRs), such as toll-like receptors (TLRs) and retinoic acid-inducible gene (RIG)-I-like receptors, leading to the release of soluble mediators, including chemokines and type I IFNs (Barton and Medzhitov 2003). Neutrophils arrive promptly at the designated location and enhance the initial stages of infection by aiding the replication of the WNV virus. However, in later stages, they play a role in controlling the infection, as demonstrated in mice where, when neutrophils are depleted, susceptibility to the virus increases (Bai et al. 2010). WNV is thought to first multiply in myeloid APCs like myeloid dendritic cells (mDCs) and monocytes/macrophages before spreading to the brain (Diamond et al. 2003). Activated macrophages and dendritic cells (DCs) lead to the release of pro-inflammatory cytokines and chemokines like type I IFN, tumor necrosis factor (TNF), interleukin (IL)-1, CCL2, CCL3, and IL-8 in culture (Martina et al. 2008; Shrestha et al. 2008; Silva et al. 2007). The release of TNF-α and IFN-α either allows the virus to breach the blood–brain barrier or infected macrophages potentially acting as carriers, transport the virus into the brain (Samuel and Diamond 2006). Complement activation can induce several antiviral effector functions of the immune response, including pathogen opsonization and/or lysis, priming of B cells, and enhancement of T cell killing of infected target cells (Stoermer and Morrison 2011). Adaptive immunity is crucial for the removal of the virus and protection from infection. Antibodies produced by B cells help neutralize the virus or stop it from binding to the host cells. $CD4^+$ T cells assist B cells in producing antibodies while $CD8^+$ T lymphocytes induce apoptosis in infected cells through perforin- or Fas-mediated pathways (Cho and Diamond 2012). According to research, regulatory $CD4^+$ T cells called Tregs, inhibit the replication and spread of WNV in both mice and human models (Lanteri et al. 2009). Linking of both types of immunity is crucial for a successful antiviral immune response, for which DCs, NK cells, and macrophages are essential. Mice deficient in innate immune recognition molecules or adaptors like TLR3, MyD88, etc., had reduced humoral response (Xia et al. 2013).

5.6 Factors Affecting Virulence

The virulence of a viral strain depends on multiple factors, including viral load, viral genetics, vector competence, the host's immune response, and genetic makeup. Consequently, the virulence of WNV strains may waver based on the interplay between virus and host factors.

5.6.1 Host Factors

Clinically, host age has been deemed the most important factor for the development of severe WNND and higher case fatality. Besides old age, other risk factors include immunosuppression due to organ transplantation or human immunodeficiency virus infection, or comorbidities like hypertension and diabetes (Campbell et al. 2002). A study by Bigham et al. (2011) identified variants in three loci (IRF3, MX1, and OAS1) associated with WNV disease. Notably, IRF3 and MX1 had not been previously associated with WNV disease in humans. IRF3 is involved in the upregulation of type 1 IFN genes and has been reported to protect mice from WNV-induced disease through interferon-dependent and -independent mechanisms (Daffis et al. 2007). MX1, belonging to the MX family of interferon-induced proteins, exhibits antiviral functions by diminishing viral replication. OAS1, along with its downstream effector RNASEL, blocks viral replication. In neuronal cell lines, WNV infection was less efficient in cells that overexpress wild-type Oas1b as compared to those that express the inactive form (Lucas et al. 2003). A variant in OAS1 (SNP rs34137742) was identified as a risk factor for human WNV disease progression (Bigham et al. 2011). Overall, the findings suggest that the interferon pathway may play a crucial role in modulating human WNV disease, impacting both the risk of symptomatic infection and disease progression.

5.6.2 Viral Factors

Several virally encoded determinants of pathogenesis exist in natural populations. One well-studied factor is the glycosylation state of the envelope (E) protein. Most US strains and those responsible for severe WN disease outbreaks or neuroinvasiveness in mice have glycosylation at position 154 of the E protein (Beasley 2005; Shirato et al. 2004). Glycosylation affects virus attachment to host cells, progeny release, and vector transmission rates (Diamond 2009b). Another mutation in the NS3 helicase has been implicated in crow pathogenesis and is found to be subject to positive selection (Brault et al. 2007). Variable nucleotide changes have been detected in different strains attenuated for neuroinvasiveness, suggesting that several genetic factors may be responsible for attenuation.

5.6.3 Vector Factors

Vector competence, referring to the ability of a mosquito species to transmit a particular virus, significantly influences WNV virulence. Notably, a study explored the variation in vector competence among different mosquito species and its impact on the amplification and transmission of WNV in North America (Goddard et al. 2002). Furthermore, insights into viral dose, lineage, and environmental factors on vector competence have also been reported (Ciota 2017). Mosquito saliva plays a dual role in WNV infection by both promoting and inhibiting immune responses, increasing vasodilatation and blood vessel permeability, and affecting mast cell degranulation and T-cell recruitment (Saiz et al. 2021). The salivary factors enhance viral infection by suppressing antiviral responses and promoting the infiltration of WNV-target cells, leading to systemic infection.

5.7 Diagnostics and Therapeutic Approaches

Primary WNV infections can be diagnosed using a number of tests that detect infectious particles, nucleic acid, and specific IgM and/or IgG antibodies. In most cases, a laboratory diagnosis is made by checking serum or cerebrospinal fluid (CSF) for WNV-specific IgM antibodies IgM antibody capture enzyme-linked immunosorbent assay (MAC-ELISA). Commercially available MAC-ELISA kits for WNV-specific IgM are offered by private laboratories in the country and performed by certified government and private diagnostic laboratories including the National Institute of Virology, Pune. WNV-specific IgM in blood or CSF is considered a reliable indicator of recent infection since it declines to undetectable levels around 2–3 months postinfection with most viral diseases (Murray et al. 2013). The absence of detectable virus-specific IgM does not necessarily rule out the diagnosis of WNV infection if serum is taken within the first 8 days of the commencement of the symptoms. So, it might be necessary to perform the test on a subsequent sample. IgG antibody response seems to occur 4–5 days after the onset of illness and has a prolonged presence years after infection (Prince et al. 2005). WNV IgA appears at about the same time as WNV IgM and before WNV IgG. All positive results from commercially available WNV IgM assays should be verified by neutralizing antibody testing of acute- and convalescent-phase serum specimens. For this, plaque-reduction neutralization tests (PRNTs) can be done. Other tests include viral cultures/isolation and identification of viral RNA (by reverse transcriptase-polymerase chain reaction [RT-PCR] method) which can be conducted on samples of serum, cerebrospinal fluid (CSF), and tissue gathered early in the disease's progression. If these tests yield positive outcomes, they can confirm the presence of an infection. However, the probability of identifying a WNV infection using molecular testing is relatively limited. Immunohistochemistry (IHC) is capable of recognizing WNV antigens in tissue that has been fixed with formalin. It is important to note that negative results from these tests do not definitively rule out a WNV infection. The option to perform viral culture, RT-PCR, and IHC is available through state public health or certified private laboratories.

5.8 Prevention and Control

Effective surveillance systems, prompt diagnosis, and vector control measures are crucial for managing and controlling WNV infections. In the absence of vaccines for the general public, preventing mosquito bites becomes the most effective way of combating WNV spread and infection. Preventive measures include wearing loose clothing, installing window and door screens, using insect repellents and mosquito nets, reducing skin surface exposure to bites, avoiding outdoor activities during mosquito season, and updating the local authorities with information about dead birds in the area (https://www.cdc.gov/westnile/prevention/). Additionally, public health education and community awareness play a vital role in reducing the risk of exposure to mosquito bites and preventing WNV transmission. Reducing mosquito populations in a community is also vital. Key measures to achieve this include source reduction by eliminating stagnant water sources, trimming vegetation and removing leaf litter, cleanup campaigns and education about mosquito control, using larvicides and insecticides, deploying mosquito traps, as well as encouraging natural mosquito predators, such as birds and bats, is also beneficial.

5.9 Management of Disease

Although recent years have seen significant progress in our understanding of the molecular biology of WNV, no vaccine or targeted therapy has yet been licensed for humans, and clinical care is only supportive (Diamond 2009b). However, potential strategies are emerging, which can be summed up as the quest for antivirals, the development of therapeutic antibodies, and vaccine development.

5.9.1 Antiviral Compounds

Finding antiviral substances that are effective against WNV infection involves identifying those that target various stages of the viral replication cycle. The candidate may be created to target the viral particle directly, preventing entry and/or infection, or it may prevent WNV from multiplying inside infected cells. The availability of numerous methods based on engineered viruses that encode reporter genes and sub-genomic replicons has helped in the initial identification of novel putative antiviral substances. Some of those prevent membrane fusion and viral entry, some act as polymerase or methyltransferase inhibitors and target NS5 enzymatic activities while some others inhibit NS2B/NS3 protease complex (Chappell et al. 2008; Dong et al. 2008; Malet et al. 2008). Candidates for antiviral compounds come in a variety of forms, such as peptide inhibitors, short-interfering RNAs, antisense oligonucleotides, iminosugars, or other small molecules (Diamond 2009b). Besides creation, repurposing of approved drugs has also demonstrated antiviral effects against WNV. Ribavirin, mycophenolic acid, and sodium valproate are some of the few that have shown promise in in vitro and in vivo assays (Diamond 2009a; Vázquez-Calvo et al. 2011). The use of medications that boost the immune system's ability to combat infection is another promising antiviral option for the treatment of WNV. These immunomodulators include substances that promote the expression of interferon or are derivatives of interferon (Beasley 2011). A small number of WNV-infected patients can benefit from interferon treatment since the virus is very sensitive to it (Kalil et al. 2005; Lewis and Amsden 2007). Thus, advancement in WNV treatments should come from a mix of pharmacological techniques that aim to inhibit viral replication, stimulate protective immune responses, reduce neuronal damage, and prevent the emergence of resistant variations.

5.9.2 Therapeutic Antibodies

Therapeutic antibodies that have been tested so far either directly target WNV particles (mostly directed against E glycoprotein and a small fraction against prM/M) or NS1 protein (Diamond et al. 2008). The epitopes on Domain III of the E protein are recognized by the majority of neutralizing antibodies produced in mice (Beasley and Barrett 2002; Choi et al. 2007). To reduce the risk of WNV neuroinvasive illness in the elderly and immunocompromised individuals, therapeutic antibody delivery should be beneficial. Intravenous immunoglobulin therapy derived from pooled human sera aids in the recovery of sick patients (Nathan et al. 2010). Humanized monoclonal antibodies were also tested for the treatment of WNV CNS infections, however, the study was terminated due to a lack of subjects (Beigel et al. 2010, WNV therapeutics review of literature, CDC, USA).

5.9.3 Vaccines

Although significant progress has been made in the development of WNV vaccines, no vaccines have been authorized for use in humans (Dauphin and Zientara 2007). On the other hand, effective, approved horse vaccines have significantly reduced the incidence of equine deaths in the US (De Filette et al. 2012). They include the formalin-inactivated, whole virus West Nile-Innovator® (Pfizer), the killed virus Vetera® (Boehringer Ingelheim Vetmedica), and the chimeric recombinant canary-pox virus vaccine Recombitek® (Sanofi Aventis), which contains the prM and E genes derived from a 1999 New York isolate. Some novel vaccine strategies for human use have undergone clinical trials and have been explored in preclinical animals (Ledgerwood et al. 2011). The strategies employed in vaccine development include vaccines based on live attenuated or chemically inactivated whole viruses, recombinant subunit vaccines containing viral proteins or fragments of them, live attenuated chimeric, and DNA-based. Several factors have hindered WNV vaccines from moving forward into later stages of clinical development, including challenges with designing and implementing efficacy studies, potential vaccine safety concerns, and anticipated costs of WNV vaccine programs.

5.10 Future Perspective

An "emerging" infectious disease's ability to quickly and effectively establish itself in a new environment was demonstrated by the advent of WNV in North America and its subsequent spread across the continent. Since 1996, there have been more frequent and severe outbreaks of WNV worldwide, however, the cause is still unknown. The virus's spread into places with large immunologically naive populations belonging to an age structure that includes many old and immunocompromised people, and the emergence of more virulent strains of the virus have all been hypothesized (Deubel et al. 2001). The virus has become established in a broad range of ecologic settings and is transmitted by a relatively large number of mosquito species, making it likely that WNV will continue to be a public health concern for the foreseeable future. The future epidemiology of WNV in North America and Europe is unknown, and the virus's historical pattern doesn't offer much insight into its likely future. It remains to be seen whether continual infection in a population will cause a decrease in the number of susceptible avian and human hosts, and result in a decrease in the number of cases. Although the future trend cannot be foreseen, the fact that adults and the elderly appear to be more susceptible to WNV sickness than younger people may indicate that a sizable immune population will develop over time (Sejvar 2014). The cornerstones of the public health response to WNV are expected to remain to be efforts in managing and reducing vector populations, reinforcement of public health messages of protection from mosquito bites, attentive surveillance, and public awareness campaigns. However, the complex epidemiology of the illness continues to present challenges for surveillance and prevention measures. Besides age, immune status, and certain host genetic factors, the risk factors for the development of severe neuroinvasive WNV disease have yet to be unraveled. Through continued research, it is hoped that the remaining questions regarding the pathogenesis and clinical outcomes of WNV may be answered. Although West Nile virus (WNV) continues to inflict significant morbidity and mortality at great expense, both funding and research dedicated to this issue have experienced a decline in recent years. In contrast, other flaviviruses such as yellow fever and Japanese encephalitis have successfully obtained licensed vaccines that have been in use for several years. Unfortunately, no approved vaccines for WNV have been developed for human use thus far. Furthermore, climate change is widely recognized as a crucial factor that could elevate the risk and spread of vector-borne diseases, including WNV (Paz 2015). The diminishing research efforts and funding could profoundly hinder the future prospects of creating vaccines and treatments for WNV. While extensive research has been conducted on antivirals against WNV, there is still no specific treatment available for West Nile neuroinvasive disease (WNND). The poor prognosis, long-term consequences, and decline in quality of life following recovery underscore the urgent need for further research and the development of effective treatments. Studies on the genetic variations, evolutionary patterns, and transmission dynamics of the virus have added immensely to the epidemiological and molecular knowledge about WNV. Available data indicate the immense genetic diversity of WNV strains within and between hosts. However, the significance of this diversity and its effect on replication fitness, transmission efficiency, and clinical outcome is not yet clear. Hence addressing these issues, however complicated, is important given the massive success and increasing spread of the virus in the northern hemisphere. WNV will probably continue to evolve, as evidenced by the emergence of lineage 2 along with the potential for the appearance of more neuroinvasive strains in regions that previously did not view WNV as a significant public health concern. Unless we are adequately prepared to confront this emerging and reemerging infectious threat, we may find ourselves in a dire situation with the expanding prevalence of WNV and other vector-borne pathogens.

References

Angeles Muñoz-Fernández M, Fresno M (1998) The role of tumour necrosis factor, interleukin 6, interferon-γ, and inducible nitric oxide synthase in the development and pathology of the nervous system. Prog Neurobiol 56(3):307–340. https://doi.org/10.1016/S0301-0082(98)00045-8

Appler KK, Brown AN, Stewart BS, Behr MJ, Demarest VL, Wong SJ, Bernard KA (2010) Persistence of West Nile virus in the central nervous system and periphery of mice. PLoS One 5(5):e10649. https://doi.org/10.1371/journal.pone.0010649

Bagnarelli P, Marinelli K, Trotta D, Monachetti A, Tavio M, Del Gobbo R, Capobianchi MR, Menzo S, Nicoletti L, Magurano F, Varaldo PE (2011) Human case of autochthonous West Nile virus lineage 2 infection in Italy, September 2011. Eur Secur 16(43):20002. https://doi.org/10.2807/ese.16.43.20002-en

Bai F, Kong K-F, Dai J, Qian F, Zhang L, Brown CR, Fikrig E, Montgomery RR (2010) A paradoxical role for neutrophils in the pathogenesis of West Nile virus. J Infect Dis 202(12):1804–1812. https://doi.org/10.1086/657416

Bai F, Ashley Thompson E, Vig PJS, Arturo Leis A (2019) Current understanding of West Nile virus clinical manifestations, immune responses, neuroinvasion, and immunotherapeutic implications. Pathogens 8(4):193. https://doi.org/10.3390/pathogens8040193

Bakonyi T, Hubálek Z, Rudolf I, Nowotny N (2005) Novel flavivirus or new lineage of West Nile virus, Central Europe. Emerg Infect Dis 11(2):225–231. https://doi.org/10.3201/eid1102.041028

Bakonyi T, Ivanics É, Erdélyi K, Ursu K, Ferenczi E, Weissenböck H, Nowotny N (2006) Lineage 1 and 2 strains of encephalitic West Nile virus, Central Europe. Emerg Infect Dis 12(4):618–623. https://doi.org/10.3201/eid1204.051379

Bakonyi T, Ferenczi E, Erdélyi K, Kutasi O, Csörgő T, Seidel B, Weissenböck H, Brugger K, Bán E, Nowotny N (2013) Explosive spread of a neuroinvasive lineage 2 West Nile virus in Central Europe, 2008/2009. Vet Microbiol 165(1–2):61–70. https://doi.org/10.1016/j.vetmic.2013.03.005

Barton GM, Medzhitov R (2003) Linking toll-like receptors to IFN-α/β expression. Nat Immunol 4(5):432–433. https://doi.org/10.1038/ni0503-432

Beasley D (2005) Recent advances in the molecular biology of West Nile virus. Curr Mol Med 5(8):835–850. https://doi.org/10.2174/156652405774962272

Beasley DW (2011) Vaccines and immunotherapeutics for the prevention and treatment of infections with West Nile virus. Immunotherapy 3(2):269–285. https://doi.org/10.2217/imt.10.93

Beasley DWC, Barrett ADT (2002) Identification of neutralizing epitopes within structural domain III of the West Nile virus envelope protein identification of neutralizing epitopes within structural domain III of the West Nile virus envelope protein. J Virol 76(24):13097–13100. https://doi.org/10.1128/JVI.76.24.13097

Beigel JH, Nordstrom JL, Pillemer SR, Roncal C, Goldwater DR, Li H, Holland PC, Johnson S, Stein K, Koenig S (2010) Safety and pharmacokinetics of single intravenous dose of MGAWN1, a novel monoclonal antibody to West Nile virus. Antimicrob Agents Chemother 54(6):2431–2436. https://doi.org/10.1128/AAC.01178-09

Bigham AW, Buckingham KJ, Husain S, Emond MJ, Bofferding KM, Gildersleeve H, Rutherford A, Astakhova NM, Perelygin AA, Busch MP, Murray KO, Sejvar JJ, Green S, Kriesel J, Brinton MA, Bamshad M (2011) Host genetic risk factors for West Nile virus infection and disease progression. PLoS One 6(9):e24745. https://doi.org/10.1371/journal.pone.0024745

Blitvich BJ (2008) Transmission dynamics and changing epidemiology of West Nile virus. Anim Health Res Rev 9(1):71–86. https://doi.org/10.1017/S1466252307001430

Bondre VP, Jadi RS, Mishra AC, Yergolkar PN, Arankalle VA (2007) West Nile virus isolates from India: evidence for a distinct genetic lineage. J Gen Virol 88(3):875–884. https://doi.org/10.1099/vir.0.82403-0

Botha EM, Markotter W, Wolfaardt M, Paweska JT, Swanepoel R, Palacios G, Nel LH, Venter M (2008) Genetic determinants of virulence in pathogenic lineage 2 West Nile virus strains. Emerg Infect Dis 14(2):222–230. https://doi.org/10.3201/eid1402.070457

Brault AC (2009) Changing patterns of West Nile virus transmission: altered vector competence and host susceptibility. Vet Res 40(2):43. https://doi.org/10.1051/vetres/2009026

Brault AC, Huang CY-H, Langevin SA, Kinney RM, Bowen RA, Ramey WN, Panella NA, Holmes EC, Powers AM, Miller BR (2007) A single positively selected West Nile viral mutation confers increased virogenesis in American crows. Nat Genet 39(9):1162–1166. https://doi.org/10.1038/ng2097

Brinton MA (2002) The molecular biology of West Nile virus: A new invader of the Western hemisphere. Ann Rev Microbiol 56(1):371–402. https://doi.org/10.1146/annurev.micro.56.012302.160654

Brinton MA (2013) Replication cycle and molecular biology of the West Nile virus. Viruses 6(1):13–53. https://doi.org/10.3390/v6010013

Bunning ML, Bowen RA, Cropp BC, Sullivan KG, Davis BS, Komar N, Godsey M, Baker D, Hettler DL, Holmes DA, Biggerstaff BJ, Mitchell CJ (2002) Experimental infection of horses with West Nile virus. Emerg Infect Dis 8(4):380–386. https://doi.org/10.3201/eid0804.010239

Byas AD, Ebel GD (2020) Comparative pathology of West Nile virus in humans and non-human animals. Pathogens 9(1):48. https://doi.org/10.3390/pathogens9010048

Calistri P, Giovannini A, Hubalek Z, Ionescu A, Monaco F, Savini G, Lelli R (2010) Epidemiology of West Nile in Europe and the Mediterranean Basin. Open Virol J 4:29

Campbell GL, Marfin AA, Lanciotti RS, Gubler DJ (2002) West Nile virus. Lancet Infect Dis 2(9):519–529. https://doi.org/10.1016/S1473-3099(02)00368-7

Carbaugh DL, Lazear HM (2020) Flavivirus envelope protein glycosylation: impacts on viral infection and pathogenesis. J Virol 94(11):10–128. https://doi.org/10.1128/jvi.00104-20

Caron A, Chiweshe N, Mundava J, Abolnik C, Capobianco Dondona A, Scacchia M, Gaidet N (2017) Avian viral pathogens in swallows, Zimbabwe. Ecohealth 14(4):805–809. https://doi.org/10.1007/s10393-017-1282-5

CDC (n.d.) West Nile virus disease therapeutics Review of the literature for healthcare providers. http://clinicaltrials.gov/ct2/results?term=west+nile+virus&Search=Search

Chancey C, Grinev A, Volkova E, Rios M (2015) The global ecology and epidemiology of West Nile virus. BioMed Res Int 2015(Figure 2):376230. https://doi.org/10.1155/2015/376230

Chappell K, Stoermer M, Fairlie D, Young P (2008) West Nile virus NS2B/NS3 protease as an antiviral target. Curr Med Chem 15(27):2771–2784. https://doi.org/10.2174/092986708786242804

Chen T, He X, Zhang P, Yuan Y, Lang X, Yu J, Qin Z, Li X, Zhang Q, Zhu L, Zhang B, Wu Q, Zhao W (2019) Research advancements in the neurological presentation of flaviviruses. Rev Med Virol 29(1):e2021. https://doi.org/10.1002/rmv.2021

Cho H, Diamond MS (2012) Immune responses to West Nile virus infection in the central nervous system. Viruses 4(12):3812–3830. https://doi.org/10.3390/v4123812

Choi KS, Nah JJ, Ko YJ, Kim YJ, Joo YS (2007) The DE loop of the domain III of the envelope protein appears to be associated with West Nile virus neutralization. Virus Res 123(2):216–218. https://doi.org/10.1016/j.virusres.2006.09.002

Chowdhury P, Khan SA, Dutta P, Topno R, Mahanta J (2014) Characterization of West Nile virus (WNV) isolates from Assam, India: insights into the circulating WNV in northeastern India. Comp Immunol Microbiol Infect Dis 37(1):39–47. https://doi.org/10.1016/j.cimid.2013.10.006

Chu JJH, Ng ML (2004) Infectious entry of West Nile virus occurs through a Clathrin-mediated endocytic pathway. J Virol 78(19):10543–10555. https://doi.org/10.1128/jvi.78.19.10543-10555.2004

Ciota AT (2017) West Nile virus and its vectors. In: Current opinion in insect science, vol 22. Elsevier Inc., pp 28–36. https://doi.org/10.1016/j.cois.2017.05.002

Ciota AT, Kramer LD (2013) Vector-virus interactions and transmission dynamics of West Nile virus. Viruses 5(12):3021–3047. https://doi.org/10.3390/v5123021

Daffis S, Samuel MA, Keller BC, Gale M, Diamond MS (2007) Cell-specific IRF-3 responses protect against West Nile virus infection by interferon-dependent and -independent mechanisms. PLoS Pathog 3(7):e106. https://doi.org/10.1371/journal.ppat.0030106

Danis K, Papa A, Theocharopoulos G, Dougas G, Athanasiou M, Detsis M, Baka A, Lytras T, Mellou K, Bonovas S, Panagiotopoulos T (2011) Outbreak of West Nile virus infection in Greece, 2010. Emerg Infect Dis 17(10):1868–1872. https://doi.org/10.3201/eid1710.110525

Das S, Mishra MK, Ghosh J, Basu A (2008) Japanese encephalitis virus infection induces IL-18 and IL-1β in microglia and astrocytes: correlation with in vitro cytokine responsiveness of glial cells and subsequent neuronal death. J Neuroimmunol 195(1–2):60–72. https://doi.org/10.1016/j.jneuroim.2008.01.009

Dauphin G, Zientara S (2007) West Nile virus: Recent trends in diagnosis and vaccine development. Vaccine 25(30):5563–5576. https://doi.org/10.1016/j.vaccine.2006.12.005

De Filette M, Ulbert S, Diamond MS, Sanders NN (2012) Recent progress in West Nile virus diagnosis and vaccination. Vet Res 43(1):16. https://doi.org/10.1186/1297-9716-43-16

Deubel V, Fiette L, Gounon P, Drouet MT, Khun H, Huerre M, Banet C, Malkinson M, Desprès P (2001) Variations in biological features of West Nile viruses. Ann N Y Acad Sci 951:195–206. http://www.ncbi.nlm.nih.gov/pubmed/11797777

Diamond MS (2009a) Progress on the development of therapeutics against West Nile virus. Antiviral Res 83(3):214–227. https://doi.org/10.1016/j.antiviral.2009.05.006

Diamond MS (2009b) West Nile encephalitis virus infection. In: West Nile Encephalitis Virus Infection. Springer, New York, NY. https://doi.org/10.1007/978-0-387-79840-0

Diamond MS, Shrestha B, Mehlhop E, Sitati E, Engle M (2003) Review innate and adaptive immune responses determine protection against disseminated infection by West Nile encephalitis virus. Viral Immunol 16(3):259–278

Diamond MS, Pierson TC, Fremont DH (2008) The structural immunology of antibody protection against West Nile virus. Immunol Rev 225(1):212–225. https://doi.org/10.1111/j.1600-065X.2008.00676.x

Dong H, Zhang B, Shi P-Y (2008) Flavivirus methyltransferase: A novel antiviral target. Antivir Res 80(1):1–10. https://doi.org/10.1016/j.antiviral.2008.05.003

Dusek RJ, McLean RG, Kramer LD, Ubico SR, Dupuis AP, Ebel GD, Guptill SC (2009) Prevalence of West Nile virus in migratory birds during spring and fall migration. Am J Trop Med Hyg 81(6):1151–1158. https://doi.org/10.4269/ajtmh.2009.09-0106

ECDC Report (2023) Weekly updates: 2023 West Nile virus transmission season. Retrieved October 25, 2023, from https://www.ecdc.europa.eu/en/west-nile-fever/surveillance-and-disease-data/disease-data-ecdc

Eldadah AH, Nathanson N (1967) Pathogenesis of West Nile virus encephalitis in mice and rats. II. Virus multiplication, evolution of immunofluorescence, and development of histological lesions in the brain. Am J Epidemiol 86(3):776–790. https://doi.org/10.1093/oxfordjournals.aje.a120786

Evans JD, Seeger C (2007) Differential effects of mutations in NS4B on West Nile virus replication and inhibition of interferon signaling. J Virol 81(21):11809–11816. https://doi.org/10.1128/jvi.00791-07

Eybpoosh S, Fazlalipour M, Baniasadi V, Pouriayevali MH, Sadeghi F, Ahmadi Vasmehjani A, Karbalaie Niya MH, Hewson R, Salehi-Vaziri M (2019) Epidemiology of West Nile virus in the Eastern Mediterranean region: a systematic review. PLoS Negl Trop Dis 13(1):e0007081. https://doi.org/10.1371/journal.pntd.0007081

Fall G, Loucoubar C, Diallo M, Sall AA, Faye O (2014) Vector competence of Culex neavei and Culex quinquefasciatus (Diptera: Culicidae) from Senegal for lineages 1, 2, Koutango and a putative new lineage of West Nile virus. Am J Trop Med Hyg 90(4):747–754. https://doi.org/10.4269/ajtmh.13-0405

Fall G, Di Paola N, Faye M, Dia M, Freire CC, Loucoubar C, Zanotto PM, Faye O, Sall AA (2017) Biological and phylogenetic characteristics of West African lineages of West Nile virus. PLoS Negl Trop Dis 11(11):1–23. https://doi.org/10.1371/journal.pntd.0006078

Farajollahi A, Crans WJ, Nickerson D, Bryant P, Wolf B, Glaser A, Andreadis TG (2005) Detection of West Nile virus RNA from the louse fly Icosta americana (Diptera: Hippoboscidae). J Am Mosq Control Assoc 21(4):474–476. https://doi.org/10.2987/8756-971X(2006)21[474:DOWNVR]2.0.CO;2

Frost MJ, Zhang J, Edmonds JH, Prow NA, Gu X, Davis R, Hornitzky C, Arzey KE, Finlaison D, Hick P, Read A, Hobson-Peters J, May FJ, Doggett SL, Haniotis J, Russell RC, Hall RA, Khromykh AA, Kirkland PD (2012) Characterization of virulent West Nile virus Kunjin strain, Australia, 2011. Emerg Infect Dis 18(5):792–800. https://doi.org/10.3201/eid1805.111720

Gamino V, Höfle U (2013) Pathology and tissue tropism of natural West Nile virus infection in birds: a review. Vet Res 44(1):39. https://doi.org/10.1186/1297-9716-44-39

Garcia-Tapia D, Hassett DE, Mitchell WJ, Johnson GC, Kleiboeker SB (2007) West Nile virus encephalitis: sequential histopathological and immunological events in a murine model of infection. J Neurovirol 13(2):130–138. https://doi.org/10.1080/13550280601187185

Goddard LB, Roth AE, Reisen WK, Scott TW (2002) Vector competence of California mosquitoes for West Nile virus. Emerg Infect Dis 8(12):1385–1391. https://doi.org/10.3201/eid0812.020536

Guarner J, Shieh WJ, Hunter S, Paddock CD, Morken T, Campbell GL, Marfin AA, Zaki SR (2004) Clinicopathologic study and laboratory diagnosis of 23 cases with West Nile virus encephalomyelitis. Hum Pathol 35(8):983–990. https://doi.org/10.1016/j.humpath.2004.04.008

Hall RA, Broom AK, Smith DW, Mackenzie JS (2002) The ecology and epidemiology of Kunjin virus, pp 253–269. https://doi.org/10.1007/978-3-642-59403-8_13

Hayes EB, Gubler DJ (2006) West Nile virus: epidemiology and clinical features of an emerging epidemic in the United States. Annu Rev Med 57:181–194. https://doi.org/10.1146/annurev.med.57.121304.131418

Hayes EB, Komar N, Nasci RS, Montgomery SP, O'Leary DR, Campbell GL (2005) Epidemiology and transmission dynamics of West Nile virus disease. Emerg Infect Dis 11(8):1167–1173. https://doi.org/10.3201/eid1108.050289a

Hernández-Triana LM, Jeffries CL, Mansfield KL, Carnell G, Fooks AR, Johnson N (2014) Emergence of West Nile virus lineage 2 in Europe: a review on the introduction and spread of a mosquito-borne disease. Front Public Health 2:271. https://doi.org/10.3389/fpubh.2014.00271

Hubálek Z, Halouzka J (1999) West Nile fever—a reemerging mosquito-borne viral disease in Europe. Emerg Infect Dis 5(5):643–650. https://doi.org/10.3201/eid0505.990505

Johnston LJ, King NJC, Halliday GM (2000) Langerhans cells migrate to local lymph nodes following cutaneous infection with an arbovirus. J Invest Dermatol 114(3):560–568. https://doi.org/10.1046/j.1523-1747.2000.00904.x

Kalil AC, Devetten MP, Singh S, Lesiak B, Poage DP, Bargenquast K, Fayad P, Freifeld AG (2005) Use of interferon-alpha in patients with West Nile encephalitis: report of 2 cases. Clin Infect Dis 40(5):764–766. https://doi.org/10.1086/427945

Kleinschmidt MC, Michaelis M, Ogbomo H, Doerr HW, Cinatl J (2007) Inhibition of apoptosis prevents West Nile virus induced cell death. BMC Microbiol 7:1–8. https://doi.org/10.1186/1471-2180-7-49

Komar N, Langevin S, Hinten S, Nemeth N, Edwards E, Hettler D, Davis B, Bowen R, Bunning M (2003) Experimental infection of North American birds with the New York 1999 strain of West Nile virus. Emerg Infect Dis 9(3):311–322. https://doi.org/10.3201/eid0903.020628

Kyle Austin S, Dowd KA (2014) B cell response and mechanisms of antibody protection to west nile virus. Viruses 6(3):1015–1036. https://doi.org/10.3390/v6031015

LaBeaud AD, Lisgaris MV, King CH, Mandalakas AM (2006) Pediatric West Nile virus infection: neurologic disease presentations during the 2002 epidemic in Cuyahoga County, Ohio. Pediatr Infect Dis J 25(8):751–753. https://doi.org/10.1097/01.inf.0000227830.73271.65

Lanteri MC, O'Brien KM, Purtha WE, Cameron MJ, Lund JM, Owen RE, Heitman JW, Custer B, Hirschkorn DF, Tobler LH, Kiely N, Prince HE, Ndhlovu LC, Nixon DF, Kamel HT, Kelvin DJ, Busch MP, Rudensky AY, Diamond MS, Norris PJ (2009) Tregs control the development of symptomatic West Nile virus infection in humans and mice. J Clin Invest 119:3266. https://doi.org/10.1172/JCI39387

Laurent-Rolle M, Boer EF, Lubick KJ, Wolfinbarger JB, Carmody AB, Rockx B, Liu W, Ashour J, Shupert WL, Holbrook MR, Barrett AD, Mason PW, Bloom ME, García-Sastre A, Khromykh AA, Best SM (2010) The NS5 protein of the virulent West Nile virus NY99 strain is a potent antagonist of type I interferon-mediated JAK-STAT signaling. J Virol 84(7):3503–3515. https://doi.org/10.1128/JVI.01161-09

Ledgerwood JE, Pierson TC, Hubka SA, Desai N, Rucker S, Gordon IJ, Enama ME, Nelson S, Nason M, Gu W, Bundrant N, Koup RA, Bailer RT, Mascola JR, Nabel GJ, Graham BS (2011) A West Nile virus DNA vaccine utilizing a modified promoter induces neutralizing antibody in younger and older healthy adults in a phase I clinical trial. J Infect Dis 203(10):1396–1404. https://doi.org/10.1093/infdis/jir054

Leis AA, Stokic DS (2012) Neuromuscular manifestations of West Nile virus infection. Front Neurol 3:37. https://doi.org/10.3389/fneur.2012.00037

Leung JY, Pijlman GP, Kondratieva N, Hyde J, Mackenzie JM, Khromykh AA (2008) Role of nonstructural protein NS2A in flavivirus assembly. J Virol 82(10):4731–4741. https://doi.org/10.1128/jvi.00002-08

Lewis M, Amsden JR (2007) Successful treatment of West Nile virus infection after approximately 3 weeks into the disease course. Pharmacotherapy 27(3):455–458. https://doi.org/10.1592/phco.27.3.455

Lim SM, van den Ham HJ, Oduber M, Martina E, Zaaraoui-Boutahar F, Roose JM, van IJcken WFJ, Osterhaus ADME, Andeweg AC, Koraka P, Martina BEE (2017) Transcriptomic analyses reveal differential gene expression of immune and cell death pathways in the brains of mice infected with West Nile virus and chikungunya virus. Front Microbiol 8:1556. https://doi.org/10.3389/fmicb.2017.01556

Liu WJ, Wang XJ, Clark DC, Lobigs M, Hall RA, Khromykh AA (2006) A single amino acid substitution in the West Nile virus nonstructural protein NS2A disables its ability to inhibit alpha/beta interferon induction and attenuates virus virulence in mice. J Virol 80(5):2396–2404. https://doi.org/10.1128/jvi.80.5.2396-2404.2006

Lorenz C, Chiaravalloti-Neto F (2022) Why are there no human West Nile virus outbreaks in South America? Lancet Reg Health Am 12:100276. https://doi.org/10.1016/j.lana.2022.100276

Lucas M, Mashimo T, Frenkiel MP, Simon-Chazottes D, Montagutelli X, Ceccaldi PE, Guénet JL, Desprès P (2003) Infection of mouse neurones by West Nile virus is modulated by the interferon-inducible 2′-5′ oligoadenylate synthetase 1b protein. Immunol Cell Biol 81(3):230–236. https://doi.org/10.1046/j.1440-1711.2003.01166.x

Lvov DK, Butenko AM, Gromashevsky VL, Kovtunov AI, Prilipov AG, Kinney R, Aristova VA, Dzharkenov AF, Samokhvalov EI, Savage HM, Shchelkanov MY, Galkina IV, Deryabin PG, Gubler DJ, Kulikova LN, Alkhovsky SK, Moskvina TM, Zlobina LV, Sadykova GK et al (2004) West Nile virus and other zoonotic viruses in Russia: examples of emerging-reemerging situations. In: Emergence and control of zoonotic viral encephalitides. Springer Vienna, pp 85–96. https://doi.org/10.1007/978-3-7091-0572-6_7

Lwande OW, Venter M, Lutomiah J, Michuki G, Rumberia C, Gakuya F, Obanda V, Tigoi C, Odhiambo C, Nindo F, Symekher S, Sang R (2014) Whole genome phylogenetic investigation of a West Nile virus strain isolated from a tick sampled from livestock in north eastern Kenya. Parasit Vectors 7(1):542. https://doi.org/10.1186/s13071-014-0542-2

Mackenzie JM, Westaway EG (2001) Assembly and maturation of the Flavivirus Kunjin virus appear to occur in the rough endoplasmic reticulum and along the secretory pathway, respectively. J Virol 75(22):10787–10799. https://doi.org/10.1128/JVI.75.22.10787-10799.2001

Mackenzie JS, Williams DT, Smith DW (2006) Japanese encephalitis virus: the geographic distribution, incidence, and spread of a virus with a propensity to emerge in new areas. Perspect Med Virol 16:201–268. https://doi.org/10.1016/S0168-7069(06)16010-3

Malet H, Massé N, Selisko B, Romette J-L, Alvarez K, Guillemot JC, Tolou H, Yap TL, Vasudevan S, Lescar J, Canard B (2008) The flavivirus polymerase as a target for drug discovery. Antivir Res 80(1):23–35. https://doi.org/10.1016/j.antiviral.2008.06.007

Martina BEE, Koraka P, van den Doel P, Rimmelzwaan GF, Haagmans BL, Osterhaus ADME (2008) DC-SIGN enhances infection of cells with glycosylated West Nile virus in vitro and virus replication in human dendritic cells induces production of IFN-α and TNF-α. Virus Res 135(1):64–71. https://doi.org/10.1016/j.virusres.2008.02.008

Medigeshi GR, Lancaster AM, Hirsch AJ, Briese T, Lipkin WI, DeFilippis V, Früh K, Mason PW, Nikolich-Zugich J, Nelson JA (2007) West Nile virus infection activates the unfolded protein response, leading to CHOP induction and apoptosis. J Virol 81(20):10849–10860. https://doi.org/10.1128/jvi.01151-07

Medigeshi GR, Hirsch AJ, Brien JD, Uhrlaub JL, Mason PW, Wiley C, Nikolich-Zugich J, Nelson JA (2009) West Nile virus capsid degradation of Claudin proteins disrupts epithelial barrier function. J Virol 83(12):6125–6134. https://doi.org/10.1128/JVI.02617-08

Mencattelli G, Ndione MHD, Rosà R, Marini G, Diagne CT, Diagne MM, Fall G, Faye O, Diallo M, Faye O, Savini G, Rizzoli A (2022) Epidemiology of West Nile virus in Africa: an underestimated threat. PLoS Negl Trop Dis 16(1):e0010075. https://doi.org/10.1371/JOURNAL.PNTD.0010075

Mishra N, Kalaiyarasu S, Nagarajan S, Rao MVS, George A, Sridevi R, Behera SP, Dubey SC, McCracken T, Newman SH (2012) Serological evidence of West Nile virus infection in wild migratory and resident water birds in Eastern and Northern India. Comp Immunol Microbiol Infect Dis 35(6):591–598. https://doi.org/10.1016/j.cimid.2012.08.002

Moudy RM, Zhang B, Shi PY, Kramer LD (2009) West Nile virus envelope protein glycosylation is required for efficient viral transmission by Culex vectors. Virology 387(1):222–228. https://doi.org/10.1016/j.virol.2009.01.038

Mukhopadhyay S, Kim B-S, Chipman PR, Rossmann MG, Kuhn RJ (2003) Structure of West Nile virus. Science 302(5643):248–248. https://doi.org/10.1126/science.1089316

Mumcuoglu KY, Banet-Noach C, Malkinson M, Shalom U, Galun R (2005) Argasid ticks as possible vectors of West Nile virus in Israel. Vector Borne Zoonotic Dis 5(1):65–71. https://doi.org/10.1089/vbz.2005.5.65

Mundhra S, Bondre VP (2023) Higher replication potential of West Nile virus governs apoptosis induction in human neuroblastoma cells. Apoptosis 28(7–8):1113–1127. https://doi.org/10.1007/s10495-023-01844-2

Murgue B, Zeller H, Deubel V (2002) The ecology and epidemiology of West Nile virus in Africa, Europe and Asia, pp. 195–221. https://doi.org/10.1007/978-3-642-59403-8_10

Murray KO, Baraniuk S, Resnick M, Arafat R, Kilborn C, Cain K, Shallenberger R, York TL, Martinez D, Hellums JS, Hellums D, Malkoff M, Elgawley N, McNeely W, Khuwaja SA, Tesh RB (2006) Risk factors for encephalitis and death from West Nile virus infection. Epidemiol Infect 134(6):1325–1332. https://doi.org/10.1017/S0950268806006339

Murray K, Walker C, Herrington E, Lewis JA, McCormick J, Beasley DWC, Tesh RB, Fisher-Hoch S (2010) Persistent infection with West Nile virus years after initial infection. J Infect Dis 201(1):2–4. https://doi.org/10.1086/648731

Murray KO, Garcia MN, Yan C, Gorchakov R (2013) Persistence of detectable immunoglobulin M antibodies up to 8 years after infection with West Nile virus. Am J Trop Med Hyg 89(5):996–1000. https://doi.org/10.4269/ajtmh.13-0232

Nash D, Mostashari F, Fine A, Miller J, O'Leary D, Murray K, Huang A, Rosenberg A, Greenberg A, Sherman M, Wong S, Campbell GL, Roehrig JT, Gubler DJ, Shieh W-J, Zaki S, Smith P, Layton M (2001) The outbreak of West Nile virus infection in the new York City area in 1999. N Engl J Med 344(24):1807–1814. https://doi.org/10.1056/NEJM200106143442401

Nathan DB, Samina I, Orr N (2010) High titer human immunoglobulin as a specific therapy against West Nile virus encephalitis. Hum Vaccin 6(3):279–281. https://doi.org/10.4161/hv.6.3.10298

Nelms BM, Fechter-Leggett E, Carroll BD, Macedo P, Kluh S, Reisen WK (2013) Experimental and natural vertical transmission of West Nile virus by California Culex (Diptera: Culicidae) mosquitoes. J Med Entomol 50(2):371–378. https://doi.org/10.1603/ME12264

Omalu BI, Shakir AA, Wang G, Lipkin WI, Wiley CA (2003) Fatal fulminant pan-meningo-polioencephalitis due to West Nile virus. Brain Pathol 13(4):465–472. https://doi.org/10.1111/j.1750-3639.2003.tb00477.x

Pachler K, Lebl K, Berer D, Rudolf I, Hubalek Z, Nowotny N (2014) Putative New West Nile virus lineage in *Uranotaenia unguiculata* mosquitoes, Austria, 2013. Emerg Infect Dis 20(12):2119–2122. https://doi.org/10.3201/eid2012.140921

Papa A, Bakonyi T, Xanthopoulou K, Vázquez A, Tenorio A, Nowotny N (2011) Genetic characterization of west nile virus lineage 2, Greece, 2010. Emerg Infect Dis 17(5):920–922. https://doi.org/10.3201/eid1705.101759

Paz S (2015) Climate change impacts on West Nile virus transmission in a global context. Philos Trans R Soc B Biol Sci 370(1665):1–11. https://doi.org/10.1098/rstb.2013.0561

Penn RG, Guarner J, Sejvar JJ, Hartman H, McComb RD, Nevins DL, Bhatnagar J, Zaki SR (2006) Persistent neuroinvasive West Nile virus infection in an immunocompromised patient. Clin Infect Dis 42(5):680–683. https://doi.org/10.1086/500216

Petersen LR, Marfin AA (2002) Review West Nile virus: a primer for the clinician. Ann Intern Med 137(3):173–179. https://doi.org/10.7326/0003-4819-137-3-200208060-00009

Petropoulou KA, Gordon SM, Prayson RA, Ruggierri PM (2005) West Nile virus meningoencephalitis: MR imaging findings. AJNR Am J Neuroradiol 26(8):1986–1995

Platonov AE, Karan' LS, Shopenskaia TA, Fedorova MV, Koliasnikova NM, Rusakova NM, Shishkina LV, Arshba TE, Zhuravlev VI, Govorukhina MV, Valentseva AA, Shipulin GA (2011) Genotyping of West Nile fever virus strains circulating in southern Russia as an epidemiological investigation method: principles and results. Zh Mikrobiol Epidemiol Immunobiol 2:29–37

Platonov AE, Tolpin VA, Gridneva KA, Titkov AV, Platonova OV, Kolyasnikova NM, Busani L, Rezza G (2014) The incidence of West Nile disease in Russia in relation to climatic and environmental factors. Int J Environ Res Public Health 11(2):1211–1232. https://doi.org/10.3390/ijerph110201211

Pogodina VV, Frolova MP, Malenko GV, Fokina GI, Koreshkova GV, Kiseleva LL, Bochkova NG, Ralph NM (1983) Study on West Nile virus persistence in monkeys. Arch Virol 75(1–2):71–86. https://doi.org/10.1007/BF01314128

Prince HE, Tobler LH, Lapé-Nixon M, Foster GA, Stramer SL, Busch MP (2005) Development and persistence of West Nile virus-specific immunoglobulin M (IgM), IgA, and IgG in viremic blood donors. J Clin Microbiol 43(9):4316–4320. https://doi.org/10.1128/JCM.43.9.4316-4320.2005

Ramanathan MP, Chambers JA, Pankhong P, Chattergoon M, Attatippaholkun W, Dang K, Shah N, Weiner DB (2006) Host cell killing by the West Nile virus NS2B-NS3 proteolytic complex: NS3 alone is sufficient to recruit caspase-8-based apoptotic pathway. Virology 345(1):56–72. https://doi.org/10.1016/j.virol.2005.08.043

Ronca SE, Murray KO, Nolan MS (2019) Cumulative incidence of West Nile virus infection, continental United States, 1999–2016. Emerg Infect Dis 25(2):325–327. https://doi.org/10.3201/eid2502.180765

Ronca SE, Ruff JC, Murray KO (2021) A 20-year historical review of west Nile virus since its initial emergence in North America: has west Nile virus become a neglected tropical disease? PLoS Negl Trop Dis 15(5):1–20. https://doi.org/10.1371/journal.pntd.0009190

Saiz JC, Martín-Acebes MA, Blázquez AB, Escribano-Romero E, Poderoso T, Jiménez de Oya N (2021) Pathogenicity and virulence of West Nile virus revisited eight decades after its first isolation. Virulence 12(1):1145–1173. https://doi.org/10.1080/21505594.2021.1908740

Sampson BA, Ambrosi C, Charlot A, Reiber K, Veress JF, Armbrustmacher V (2000) The pathology of human West Nile virus infection. Hum Pathol 31(5):527–531. https://doi.org/10.1053/hp.2000.8047

Samuel MA, Diamond MS (2006) Pathogenesis of West Nile virus infection: a balance between virulence, innate and adaptive immunity, and viral evasion. J Virol 80(19):9349–9360. https://doi.org/10.1128/JVI.01122-06

Sejvar JJ (2007) The long-term outcomes of human West Nile virus infection. Clin Infect Dis 44(12):1617–1624. https://doi.org/10.1086/518281

Sejvar JJ (2014) Clinical manifestations and outcomes of West Nile virus infection. Viruses 6(2):606–623. https://doi.org/10.3390/v6020606

Sejvar JJ, Haddad MB, Tierney BC, Campbell GL, Marfin AA, Van Gerpen JA, Fleischauer A, Leis AA, Stokic DS, Petersen LR (2003) Neurologic manifestations and outcome of West Nile virus infection. JAMA 290(4):511–515. www.jama.com

Shirato K, Miyoshi H, Goto A, Ako Y, Ueki T, Kariwa H, Takashima I (2004) Viral envelope protein glycosylation is a molecular determinant of the neuroinvasiveness of the New York strain of West Nile virus. J Gen Virol 85(Pt 12):3637–3645. https://doi.org/10.1099/vir.0.80247-0

Shrestha B, Gottlieb D, Diamond MS (2003) Infection and injury of neurons by West Nile encephalitis virus. J Virol 77(24):13203–13213. https://doi.org/10.1128/jvi.77.24.13203-13213.2003

Shrestha B, Zhang B, Purtha WE, Klein RS, Diamond MS (2008) Tumor necrosis factor alpha protects against lethal West Nile virus infection by promoting trafficking of mononuclear leukocytes into the central nervous system. J Virol 82(18):8956–8964. https://doi.org/10.1128/JVI.01118-08

Silva MC, Guerrero-Plata A, Gilfoy FD, Garofalo RP, Mason PW (2007) Differential activation of human monocyte-derived and plasmacytoid dendritic cells by West Nile virus generated in different host cells. J Virol 81(24):13640–13648. https://doi.org/10.1128/JVI.00857-07

Smithburn KC, Hughes TP, Burke AW, Paul JH (1940) A neurotropic virus isolated from the blood of a native of Uganda 1. Am J Trop Med Hyg s1-20(4):471–492. https://doi.org/10.4269/ajtmh.1940.s1-20.471

Stoermer KA, Morrison TE (2011) Complement and viral pathogenesis. Virology 411(2):362–373. https://doi.org/10.1016/j.virol.2010.12.045

Sule WF, Oluwayelu DO, Hernández-Triana LM, Fooks AR, Venter M, Johnson N (2018) Epidemiology and ecology of West Nile virus in sub-Saharan Africa. Parasit Vectors 11(1):414. https://doi.org/10.1186/s13071-018-2998-y

Suthar MS, Ma DY, Thomas S, Lund JM, Zhang N, Daffis S, Rudensky AY, Bevan MJ, Clark EA, Kaja M-K, Diamond MS, Gale M (2010) IPS-1 is essential for the control of West Nile

virus infection and immunity. PLoS Pathog 6(2):e1000757. https://doi.org/10.1371/journal.ppat.1000757
Suthar MS, Diamond MS, Gale M (2013) West Nile virus infection and immunity. Nat Rev Microbiol 11(2):115–128. https://doi.org/10.1038/nrmicro2950
Swanson PA, McGavern DB (2015) Viral diseases of the central nervous system. In: Current opinion in virology, vol 11. Elsevier B.V, pp 44–54. https://doi.org/10.1016/j.coviro.2014.12.009
Tan TTT, Bhuvanakantham R, Li J, Howe J, Ng ML (2009) Tyrosine 78 of premembrane protein is essential for assembly of West Nile virus. J Gen Virol 90(5):1081–1092. https://doi.org/10.1099/vir.0.007872-0
Tesh RB, Siirin M, Guzman H, Travassos da Rosa APA, Wu X, Duan T, Lei H, Nunes MR, Xiao S (2005) Persistent West Nile virus infection in the Golden hamster: studies on its mechanism and possible implications for other Flavivirus infections. J Infect Dis 192(2):287–295. https://doi.org/10.1086/431153
Tinto B, Kaboré DPA, Kagoné TS, Constant O, Barthelemy J, Kiba-Koumaré A, Van de Perre P, Dabiré RK, Baldet T, Gutierrez S, Gil P, Kania D, Simonin Y (2022) Screening of circulation of Usutu and West Nile viruses: A one health approach in humans, domestic animals and mosquitoes in Burkina Faso, West Africa. Microorganisms 10(10):2016. https://doi.org/10.3390/microorganisms10102016
Vázquez A, Sánchez-Seco MP, Ruiz S, Molero F, Hernández L, Moreno J, Magallanes A, Tejedor CG, Tenorio A (2010) Putative new lineage of West Nile virus, Spain. Emerg Infect Dis 16(3):549–552. https://doi.org/10.3201/eid1603.091033
Vázquez-Calvo A, Saiz J-C, Sobrino F, Martín-Acebes MA (2011) Inhibition of enveloped virus infection of cultured cells by valproic acid. J Virol 85(3):1267–1274. https://doi.org/10.1128/JVI.01717-10
Venter M, Swanepoel R (2010) West Nile virus lineage 2 as a cause of zoonotic neurological disease in humans and horses in Southern Africa. Vector Borne Zoonotic Dis 10(7):659–664. https://doi.org/10.1089/vbz.2009.0230
Ward MP (2006) Spread of equine West Nile virus encephalomyelitis during the 2002 Texas epidemic. Am J Trop Med Hyg 74(6):1090–1095
Watson JT, Pertel PE, Jones RC, Siston AM, Paul WS, Austin CC, Gerber SI (2004) Clinical characteristics and functional outcomes of West Nile fever. Ann Intern Med 141(5):360. https://doi.org/10.7326/0003-4819-141-5-200409070-00010
Weinberger M, Pitlik SD, Gandacu D, Lang R, Nassar F, Ben David D, Rubinstein E, Izthaki A, Mishal J, Kitzes R, Siegman-Igra Y, Giladi M, Pick N, Mendelson E, Bin H, Shohat T, Chowers MY (2001) West Nile fever outbreak, Israel, 2000: epidemiologic aspects. Emerg Infect Dis 7(4):686–691. https://doi.org/10.3201/eid0704.017416
Winkelmann ER, Luo H, Wang T (2016) West Nile virus infection in the central nervous system. F1000Research 5:105. https://doi.org/10.12688/f1000research.7404.1
Xia J, Winkelmann ER, Gorder SR, Mason PW, Milligan GN (2013) TLR3- and MyD88-dependent signaling differentially influences the development of West Nile virus-specific B cell responses in mice following immunization with RepliVAX WN, a single-cycle Flavivirus vaccine candidate. J Virol 87(22):12090–12101. https://doi.org/10.1128/JVI.01469-13
Yang M-R, Lee SR, Oh W, Lee E-W, Yeh J-Y, Nah J-J, Joo Y-S, Shin J, Lee H-W, Pyo S, Song J (2008) West Nile virus capsid protein induces p53-mediated apoptosis via the sequestration of HDM2 to the nucleolus. Cell Microbiol 10(1):165–176. https://doi.org/10.1111/j.1462-5822.2007.01027.x
Zakhia R, Dupuis AP, Khodr F, Fadel M, Kramer LD, Haddad N (2021) Evidence of West Nile virus circulation in Lebanon. Viruses 13(6):994. https://doi.org/10.3390/v13060994
Zeller HG, Schuffenecker I (2004) West Nile virus: an overview of its spread in Europe and the Mediterranean basin in contrast to its spread in the Americas. Eur J Clin Microbiol Infect Dis 23(3):147–156. https://doi.org/10.1007/s10096-003-1085-1
Zhou Y, Ray D, Zhao Y, Dong H, Ren S, Li Z, Guo Y, Bernard KA, Shi P-Y, Li H (2007) Structure and function of flavivirus NS5 methyltransferase. J Virol 81(8):3891–3903. https://doi.org/10.1128/JVI.02704-06

Herpes Simplex Virus: Insights into the Emerging Ocular Viral Infection

6

Sushruta S. Hakkimane, Santosh L. Gaonkar, Piya Paul Mudgal, and Bharath Raja Guru

Abstract

The article is focused on the overview of Herpes Simplex Virus (HSV), the leading cause of blindness due to corneal disease called Herpetic keratitis. The first part narrates epidemiology, the virus HSV, its structure, life cycle, and pathogenesis. There is an estimate of around 1.5 million people affected globally by HSV keratitis with an addition of about 40,000 cases every year. It usually infects in the early age of life, and impacts the vision thereby affecting the quality of life of young and productive adults. The chapter shall discuss on diagnosis and respective treatment emphasizing on the role of two essential drugs, namely, Acyclovir and Ganciclovir against HSV and their effective delivery. The diagnosis approaches were established, and upcoming treatment methods are based on the latest research techniques, especially the topical and oral antiviral drug

S. S. Hakkimane (✉)
Department of Biotechnology, Manipal Institute of Technology Bengaluru, Manipal Academy of Higher Education, Manipal, India
e-mail: sushruta.s@manipal.edu

S. L. Gaonkar
Department of Chemistry, Manipal Institute of Technology, Manipal Academy of Higher Education, Manipal, India

P. P. Mudgal
Manipal Institute of Virology, Manipal Academy of Higher Education, Manipal, India

B. R. Guru (✉)
Department of Biotechnology, Manipal Institute of Technology, Manipal Academy of Higher Education, Manipal, India
e-mail: bharath.guru@manipal.edu

P. L. Bhukya, A. S (eds.), *Emerging Human Viral Diseases, Volume II*,
https://doi.org/10.1007/978-981-97-4480-0_6

administration to control HSV keratitis in particular. Herpes Simplex Keratitis (HSK), is a frequently observed condition which is described as recurrent corneal infection with the possibility of blindness. HSV type-1 (HSV-1) is observed to be the predominant cause of infection compared to HSV type-2 (HSV-2), which is infected in lesser extent.

6.1 Introduction

6.1.1 Brief History

Herpes virus infections of the eyes have been a major issue across the globe. Herpes Simplex Virus type-1 (HSV-1) and Herpes Simplex Virus type-2 (HSV-2) are the major causal organisms for ocular infections. Among the 25+ viruses present in the Herpesviridae family, only 8 can cause infections in humans. HSV type-1 and HSV type-2 (HSV-1 and HSV-2) are the two double-stranded DNA viruses that cause ocular infections, especially HSV-1, which leads to corneal inflammation known as Herpes Simplex Keratitis (HSK) is among the significant causes of blindness that is characterized by recurrent infections of the cornea (Karsten et al. 2012; Hunt 2016).

6.1.2 Discovery

HSV and their human hosts have coevolved over millions of years. The name herpes, comes from Greek, which means "to creep." It has a history of more than 2000 years. "Blisters over ladies' lips" are referred in *Romeo and Juliet* in the sixteenth century, and in the eighteenth century, due to the occurrence of cold sores among prostitutes on their lips that was referred to as "a vocational sickness of women." In the Richard Boulton's book "A System of Rational and Practical Surgery" of 1713, the names "herpes miliaris" and "herpes exedens," were mentioned and the word "herpes simplex" first emerged. Jean Astruc identified the illness as a sexually transmitted one in 1736 and noticed its high frequency among homosexuals (Leo 1982). Six clinical entities, including genital and facial herpes, were documented in the nineteenth century, but the illness was not thought to be contagious. Herpes was proven to be contagious in 1921 when rabbit infection was spread by lesion material that had been filtered through membrane filters (Nahmias and Dowdle 1968; Misra (n.d.)).

6.2 Epidemiology

Geographical Distribution/Demographics Age, Mortality, and Morbidity.

HSK infection and disease development is a significant cause of visual morbidity, with 1.5 million people affected per year globally and a rough addition of 40,000 cases every year (Farooq et al. 2012). Almost two million people were impacted with vision impairment among the ten million individuals who have had HSK (Wilhelmus 2010).

There is a wide range of factors that influence the infection rates of HSV like age, sex, geographical location, social class, and race. In a survey with the population aged between 0—49 years around 67% and 11.3% of a population were affected by HSV-1 and HSV-2 respectively (Looker et al. 2015a, b). As per a study in 2002, the adult population of countries like the United States, Tanzania, and Germany had a HSV-1 seroprevalence of >50, >90, and > 75%, respectively, while the global seroprevalence is estimated to be between 80 and 90% (Rabenau et al. 2002; Smith and Robinson 2002). Remarkably, in many industrialized nations, there has been a general reduction in HSV-1 prevalence in the past 30 years, which could be due to improved hygiene and healthier living conditions (Bradley et al. 2014; Steiner et al. 2007; Xu et al. 2006; Farooq et al. 2012).

6.2.1 Spread of Disease and Biosafety Measures

HSV will be transmitted through ingestion, accidental parenteral inoculation, droplet exposure of the mucous membranes of nose, mouth, or eyes and inhalation of concentrated aerosolized material. The HSV-1 (enveloped virus) is sensitive and is quickly inactivated by physical inactivation and disinfectants. Outside of the host, the virus does not exhibit any metabolic activity. According to the Department of Biotechnology's List of Infectious Microorganisms corresponding to different Risk Group's (2021), HSVs are categorized as Risk Group 2 pathogens. Facilities, tools, and operational procedures are followed at Containment Level 2. Hand gloves and lab coats are used to avoid direct skin contact with infected materials or animals. In case of known or possible exposure risk to splashes, eye protection must be worn. Needles, syringes, and other sharp instruments should only be used in certain situations (Jerome and Morrow 2015).

6.3 Potential Risk of Emergence and Re-Emergence

Although the ocular disease is caused by the primary infection, the recurrent infections tend to give rise to more destructive effects, which can cause permanent visual impairment, a substandard life, more frequent medical visits, and a significant financial burden. The rate of recurrence is estimated to be about 9.6%, 22.9%, and 63.2% in the first, second, and 20 years after the first episode of HSV keratitis, respectively. Environmental and physiological stress factors including fever, hormonal variation, immunosuppression, and ultraviolet exposure triggers reactivation of the virus in the tissues innervated trigeminal ganglion, leading to HSK, which may be in different forms like epithelial, stromal, or endothelial. (Liesegang et al. 1989; Borderie et al. 2012; Souza et al. 2003).

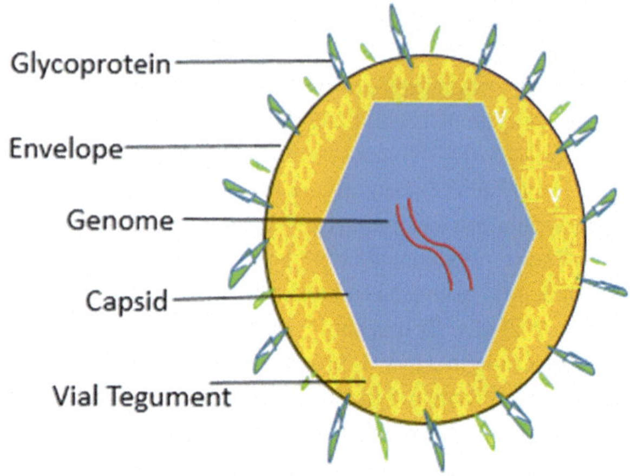

Fig. 6.1 Structure of HSV

6.4 Organization of Infectious Agents (Structural and Molecular)

HSV is the member of *Herpesviridae* family and the Alpha-herpesvirinae is one of the three subfamilies of Herpesviridae (the others being beta and gamma). Herpes virus family members are composed of a unique four-layered structure. As shown in Fig. 6.1, first is a lipid bilayer envelope that secretes many glycoproteins required for viral entry; the second is a tegument surrounding the capsid, which mainly comprises amorphous protein coats, having transcriptional regulators necessary for the early infection. Finally, an icosahedral capsid composed of capsomeres, a linear-copy surrounded by DNA genome with double-stranded, in a circular pattern (Karasneh and Shukla 2011). The virion is a spherical structure of about 200 nm which has an icosahedral capsid of approximate diameter of 125 nm with projection of glycoprotein spikes from the viral envelop as per electron micrograph observations (Yuan et al. 2018; Dai and Zhou 2018;) (Shukla and Spear 2001).

6.5 Genome Structure and Organization

The tegument being innermost space of virus, an icosahedral nucleocapsid encloses the linear double-stranded viral genome. The relatively large DNA genome has about 150Kbp, encoding at least 80 genes or more (Whitley and Roizman 2001). It has high GC (Guanine-cytosine) content that is about 67% and the genome codes differentially expressed proteins. The early proteins are fist synthesized which later activates the transcription of middle and late genes. DNA replication is handled by the middle genes while structural proteins are handled by late genes (McGeoch et al. 1988).

6.5.1 Viral Proteins and Life Cycle

The virus enters the host by fusing or undergoing endocytosis which is dependent on the type of cell being infected. Virion attachment and entry into the cell are both necessary steps in the life cycle of the virus. For HSV to enter a host cell, numerous outer envelope glycoproteins on its surface must connect to the cell's trans-membrane receptors to facilitate the penetration. It then goes through a number of processes, including transport to the nucleus, immediate, early, and late viral gene expression, and viral DNA replication. The viral capsid enters the cell cytoplasm and is then carried to the cell nucleus. Late proteins contribute to the formation of the capsid and the surface receptors that help with arrangement of capsid and viral DNA packing, both of which take place in the nucleus. The nucleocapsid exits nucleus either through nuclear membrane budding or through the nuclear pore (Liu et al. 2006; Weller and Coen 2012).

6.6 Pathogenesis in Humans

Pathogenesis includes replication of virus, latency state, and reactivation. Through interactions with particular viral and host receptors, HSV gains access to corneal epithelial cells. Primary HSV infection depends on the immunological status, but it must dodge several barriers formed by the host to show symptoms of infection in the cornea. Initial HSV infection of the eyes is known as primary ocular herpes. This arises mainly due to the inoculation on the eye's surface. Direct inoculation happens due to the presence of active HSV elsewhere in the body, in those who are immune compromised (Gimenez et al. 2013). Any part of the eye can be affected by HSV with symptoms like blurry vision, ocular discomfort, and redness of eye. Virus enters the latency state by moving to the trigeminal ganglion from the corneal nerve to cause recurrent infection after reactivation.

6.6.1 Host Immune Response (Innate and Immune)

Infection is usually controlled by immune mechanisms through innate and adaptive which inhibits the viral multiplication. The innate immune mechanisms help to inhibit infection, and also produces type 1 interferons. (Ma and He 2014). Cytokines control the lytic and latent HSV infections, and functions of IFN-ά or IFN-γ drive the virus to latency and preventing recurrence (De Regge et al. 2010). The protection of the corneal tissue against HSV-1 infection is taken up by innate immune cells. Dendritic cells (DCs), which produces type I IFNs, and NK cells' antiviral cytokines, both help in defending against infected cells (Bigley 2014). Local and infiltrating macrophages helps in defending against infections (Mott et al. 2009). Whenever eye is infected with HSV-1 infection, most stood out innate immune response is accumulation of neutrophils on the cornea. T-cell subtype CD4 known as regulatory T-cells (Treg) are put up in herpes keratitis development, though its immunomodulatory aspect is not distinct (Varanasi et al. 2017). HSV-1 also employs a number of techniques to limit host innate immune responses such that it promotes host cells infection. The virus can remain dormant inside the cells for a long time, evading the particular immune response.

6.7 Clinical Manifestations

People with frequent infections, especially repeated stromal keratitis, lose their vision in HSK. Clinical indications rely on the anatomic site, host immune status, age, and antigenic type () of the virus. Initial HSV infections are presented as conjunctivitis with eyelid vesicles which can appear or infect in children and young adults if they come in direct proximity via mucous membrane. In most of the cases, the reactivation is triggered by latent HSV. When HSV is reactivated, it moves into ocular division of trigeminal ganglion, then basal epithelium is infiltrated and exits through corneal surface, causing viral shedding. Stress, fever, exposure to UV rays, and other environmental and iatrogenic factors like laser therapy or topical corticosteroids can cause HSV reactivation. HSVK is typically a unilateral disease, but 2–19% of cases, especially in immunosuppressed individuals, have turned to be bilateral diseases (Lobo et al. 2019; Rowe et al. 2013; Tasman and Jaeger 2013).

6.7.1 Serology and Molecular Approaches

The value of any laboratory test for the diagnosis of HSV infection depends on the kind of test, the specimen's quality, the lab's ability to execute the test reliably, and how the clinician interprets the test results. Recent improvements in HSV infection diagnostic procedures include novel serological assays and viral detection techniques. Confirmation of herpes clinical diagnosis to be considered only based on Laboratory test. Herpes definitive diagnosis depends on representing the presence of HSV in the afflicted area, either through virus isolation or antigen detection. Viral culture and antigen detection are being replaced in certain labs by the HSV DNA detection utilizing molecular diagnostic techniques. When direct approaches give negative results in symptomatic patients or in asymptomatic individuals to determine previous or present infection, serological testing may be helpful.

6.7.2 Phase and Disease

Primary infection is usually asymptomatic. Symptoms are generally not specific and vary based on the individual affected and the phase of the disease. Recurrent infection signs and symptoms include conjunctival hyperaemia, lacrimation, photophobia, and foreign body sensation. Recurrent keratitis patients typically are aware of their symptoms, allowing for quick referral and treatment. HSV infections in children might be asymptomatic or manifest as conjunctivitis or blepharoconjunctivitis on the ocular surface of the host that may include ulcers, unilateral lid vesicles, and eyelid inflammation (blepharitis) (Rowe et al. 2013; Darougar et al. 1985). HSV can also cause corneal lesions that stay on the epithelium layer and are known as infectious epithelial keratitis or move into deeper layer of stroma referred to as stromal keratitis. The dendritic ulcer, which has small branches and terminal bulbs, is the most typical form. Other signs of epithelial keratitis include sub epithelial infiltration of inflammatory cells, bulbar conjunctival hyperaemia, limbal hyperaemia, and sub epithelial stromal oedema at the site of the ulcer. Stromal recurrent keratitis may also include epithelial keratitis in the form of a dendritic ulcer. The neurotropic ulcer, as opposed to the dendritic ulcer, is a more likely clinical type if there are repeated recurrences.

Reduced corneal transparency and chronic scarring caused by stromal inflammation can result in permanent tissue disease such as stromal necrosis and vascularization. A mild anterior chamber reaction and stromal localized or widespread corneal oedema frequently accompany the inflammatory process. In addition, the disease's severity could worsen with each episode as the inflammatory response intensifies despite the absence of detectable viral activity.

Once the ocular tissue is damaged, they hardly possess the capacity to regenerate. Every structure in our eye has evolved with a range of overlaying cellular and molecular mechanisms to swamp potential destructive T-cells reciprocations along with maintaining tissue integrity (Hori 2008; Stein-Streilein 2008). The local tissue destruction is lessened by these mechanisms and help preserve vision and making eye susceptible to infections arising a powerful inflammatory reaction for suppression (Dartt et al. 2011).

6.8 Diagnostic and Therapeutic Approaches

Clinical signs are the primary basis for the diagnosis of ocular HSV. Through slit lamp inspection, HSV keratitis-related unilateral corneal lesions were identified. Diagnostic testing is rarely required because of the defining clinical symptoms. For a lab diagnosis of HSV keratitis, the viral isolation with culture is considered as gold standard. For reliability, however, other modern approaches can also be employed, like faster diagnostic analysis. A few are immunological tests based on enzymes and fluorescence that detect viral antigens as well as polymerase chain reaction (PCR) techniques that identify viral DNA. Using cutting-edge methods, an immunochromatographic assay or multiplex dot hybridization (MDH) may help with diagnosis. (Farhatullah et al. 2004). Majority of the time, primary ocular HSV typically resolves on its own, effective therapy with right care can shorten the disease's course, lessen its severity, and minimize damage, which lessens the risk of long-term problems. Over the past century, a variety of physical, pharmacological, and antiviral treatments have been employed to treat HSVK, includes iodinization, and carbonization in the first half of the twentieth century, curettage and cauterization in the 1890s (Wilhelmus 2010).

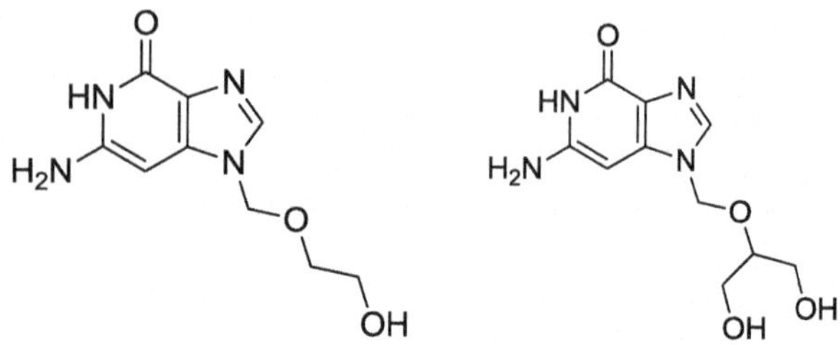

A B

Fig. 6.2 Structure of acyclovir (**a**) and ganciclovir (**b**)

6.8.1 Role of Acyclovir, Ganciclovir, and Their Effective Delivery

Decreased viral load helps to regenerate the corneal epithelium that encourages quick recovery. The researchers had initiated screening of anti-infective agents towards prospective therapies after understanding the association of HSV infection with epithelial keratitis. A pyrimidine analogue Idoxuridine was the first antiviral drug and was developed in 1962 (Kaufman 1962). Due to its success, several synthetic nucleoside analogues that are currently utilized to treat HSV were created. Antiviral medications are now the cornerstone of HSV therapy.

One of the most well-known and preferred antiviral medications available currently in the market is acyclovir (Zovirax), Fig. 6.2a. Acyclovir can be taken orally, applied as an ointment, or given intravenously. Nausea and diarrhoea are frequent adverse effects of this medication, while other severe side effects include kidney ailments and platelets reduction. As first-line medications, ganciclovir ophthalmic gel 0.15% with acyclovir ophthalmic ointment 3% is used to treat acute superficial herpetic keratitis outside of the United States. The ganciclovir is a synthetic nucleoside analogue of 2′-deoxyguanosine, Fig. 6.2b. It has a molecular weight of 255.23 and the following structure: 9-(1,3-dihydroxy-2-propoxymethyl)-guanine. Since 1996, ganciclovir ophthalmic gel 0.15% (GCV 0.15%; European tradename: Virgan®), was commercially accessible outside the United States, has established itself as the gold standard of care towards the treatment of acute herpetic keratitis (Guess et al. 2007).

6.8.2 Mode of Action

Acyclovir and ganciclovir hinder HSV infection and replication through similar mechanisms. Acyclovir is a guanosine analogue with an aliphatic functional group attached to its side chain. Acyclovir is phosphorylated into acyclovir monophosphate by the viral-selective binding of the enzyme thymidine kinase with the aid of the enzyme guanylate kinase. Acyclovir monophosphate is changed into a diphosphate. Acyclovir diphosphate is then being transformed to its active form acyclovir-triphosphate (ACV-TP) by the human enzymes as shown in Fig. 6.3. The affinity of ACV-TP for viral DNA polymerase was greater than cellular DNA polymerase. It limits HSV replication through total inhibition of viral DNA synthesis while transcription of viral genome. This substantially decreases the severity of symptoms and the spread of illness (Hung et al. 1984). Due to its poor water solubility, acyclovir contributes only 20–30% of the over-all free drug supply in the body. Acyclovir quickly eliminates from the body, whereas the drug's half-life is said to be 3 h (Elion 1983; Elion 1993; Piret and Boivin 2011; Wilhelmus 2010). The delivery of anti-viral medication via ocular route is considered to be a challenging task, which affects drug permeability and achieves therapeutic concentration. In fact, delivery systems based on upcoming technology is indeed a great subject for research, especially biodegradable Drug Delivery Systems (DDS) have attracted significant global attention in recent years which uses biocompatible materials and are non-toxic to cells (Kondel et al. 2019); (Pinto et al. 2021; Hakkimane et al. 2018). Nano carriers play crucial aspect in extending the drug residence time in order to attain the steady and controlled flow of active compounds to the infection site. This results in better bioavailability with less systemic adverse effects. Since the last decade, acyclovir has been a focus as one of the drugs with limited bioavailability that can be loaded into different kinds of recently developed drug delivery systems (Monavari et al. 2014). Different types of nano-carriers have been designed, developed to entrap and deliver antiviral drugs via various routes of administration (Shirazi et al. 2015; Chaudhari et al. 2016; Oh et al. 2016). Many strategies were used to improve the therapy and reduce the drug dosage to combat side effects of antiviral drugs. Nano drug delivery which increases bioavailability in the targeted tissue is the current requirement to address the limitations of antiviral drug delivery. Liposomes and niosomes,

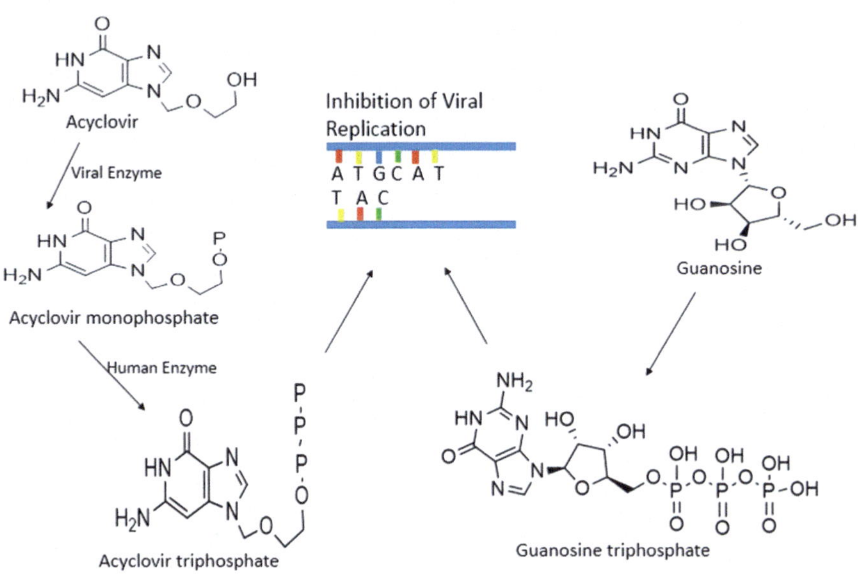

Fig. 6.3 Acyclovir's mode of action to inhibit viral replication

nanostructured lipid carriers, nanoemulsions, polymeric-based nanocarriers, nanomicelles, polymeric nanoparticles and nanosuspensions, prodrug approach, and peptide delivery approaches are the different carriers for Ocular Drug Delivery (Rabanel et al. 2012; Vedula and Chopra 2016). In the drug delivery system, the carrier is likely to possess these qualities: repeatability, biocompatibility, biodegradation duration, non-toxicity, and target specific active compound. The use of nanoparticles in medicine has expanded the possibilities for creating new drugs and for upgrading the pharmacologically constrained formulations that are now on the market. Synthetic and natural ingredients used in this drug delivery mechanism are constructed of medically suitable and biologically inert materials. This innovative technique is regarded as the next development in antiviral ocular therapy (De Koker et al. 2012).

Furthermore, in depth studies are required to test the choice of acyclovir delivery method. As a requirement for clinical trials, its optimum therapeutic efficacy and toxicity data can be obtained, along with its stability test. The recent data from several studies that were carried out using diverse methodologies, technologies, and resources have drawn worldwide attention and, as a result, have provided significant knowledge to this field of research, even though more research and studies are still needed (Pandey et al. 2020).

6.9 Prevention and Control

Direct contact with the lesions and secretions of a patient with active HSV should be avoided in order to prevent herpetic infection. Antiviral medications are the most effective medications for people infected with HSV. Although they cannot cure the disease completely, these can help to lessen the severity and frequency of symptoms. Acyclovir is an effective antiviral agent for treatment and prophylaxis. Other HSV-1 susceptible antivirals include valaciclovir, famciclovir, and cidofovir. Acyclovir resistant HSV-1 strains have been documented (WHO 2022). The patient should be continuously observed and if no improvement has been seen, it is important to rule out other diagnoses and investigate the possibilities of antiviral resistance to therapy, antiviral toxicity, neurotropic disease, poor compliance with medication, and/or a different diagnosis should be considered. Foscarnet antiviral agent is used for acyclovir resistant HSV cases.

6.10 Future Perspective: Anything Outstanding with Respect to Virus

HSVs (Herpes Simplex Virus) are the root cause of numerous deadly diseases. Patients who are immunosuppressed have an increased risk of infection. WHO is aiming to enhance access to antiviral drugs and raise awareness of HSV infection and associated symptoms. Additionally, research is being supported by WHO and its partners to create fresh approaches towards prevention and management of such HSV infections, vaccinations, and topical microbicides. Clinical studies and research into targeted drug delivery could offer fresh hope for the treatment of herpes infections.

References

Bigley NJ (2014) Complexity of interferon-gamma interactions with HSV-1. Front Immunol 5:77777

Borderie VM, Gineys R, Goldschmidt P, Batellier L, Laroche L, Chaumeil C (2012) Association of anti-herpes simplex virus IgG in tears and serum with clinical presentation in patients with presumed herpetic simplex keratitis. Cornea 31(11):1251–1256

Bradley H, Markowitz LE, Gibson T, McQuillan GM (2014) Seroprevalence of herpes simplex virus types 1 and 2—United States, 1999-2010. J Infect Dis 209:325–333

Chaudhari MB, Desai PP, Patel PA, Patravale VB (2016) Solid lipid nanoparticles of amphotericin B (AmbiOnp): in vitro and in vivo assessment towards safe and effective oral treatment module. Drug Deliv Transl Res 6:354–364

Dai X, Zhou ZH (2018) Structure of the herpes simplex virus 1 capsid with associated tegument protein complexes. Science 360:eaao7298. https://doi.org/10.1126/science.aao7298

Darougar S, Wishart MS, Viswalingam ND (1985) Epidemiological and clinical features of primary herpes simplex virus ocular infection. Br J Ophthalmol 69:2–6

Dartt DA, Dana R, D'Amore p, Niederkorn J. (2011) Immunology, inflammation and diseases of the eye. Academic Pres, San Diego

De Koker R, Hoogenboom B, De Geest G (2012) Polymeric multilayer capsules for drug delivery. Chem Soc Rev 41:2867–2884

De Regge N, Van Opdenbosch N, Nauwynck HJ, Efstathiou S, Favoreel HW (2010) Interferon alpha induces establishment of alphaherpesvirus latency in sensory neurons in vitro. PLoS One 5:e13076

Elion GB (1983) The biochemistry and mechanism of action of acyclovir. J Antimicrob Chemother 12:9–17

Elion GB (1993) Acyclovir: discovery, mechanism of action, and selectivity. J Med Virol 1(1):2–6

Farhatullah S, Kaza S, Athmanathan S, Garg P, Reddy SB, Sharma S (2004) Diagnosis of herpes simplex virus-1 keratitis using Giemsa stain, immunofluorescence assay, and polymerase chain reaction assay on corneal scrapings. Br J Ophthalmol 88:142–144

Farooq AV, Shukla D (2012) Herpes simplex epithelial and stromal keratitis: an epidemiologic update. Survey of ophthalmology 57(5):448–62.

Gimenez F, Suryawanshi A, Rouse BT (2013) Pathogenesis of herpes stromal keratitis: a focus on corneal neovascularization. Prog Retin Eye Res 33:1–9

Guess S, Stone DU, Chodosh J (2007) Evidence-based treatment of herpes simplex virus keratitis: a systematic review. Ocul Surf 5:240–250

Hakkimane SS, Shenoy VP, Gaonkar SL, Bairy I, Guru BR (2018) Antimycobacterial susceptibility evaluation of rifampicin and isoniazid benz-hydrazone in biodegradable polymeric nanoparticles against *mycobacterium tuberculosis* H37Rv strain. Int J Nanomedicine 13:4303–4318

Hori J (2008) Mechanisms of immune privilege in the anterior segment of the eye: what we learn from corneal transplantation. J Ocul Biol Dis Infor 1:94–100

Hung SO, Patterson A, Rees PJ (1984) Pharmacokinetics of oral acyclovir (Zovirax) in the eye. Br J Ophthalmol 68:192–119

Richard Hunt (2016). Virology Chapter 11 Herpes Viruses Microbiology and Immunology. https://www.microbiologybook.org/virol/herpes.htm

Jerome KR, Morrow RA (2015) Herpes simplex viruses and herpes B virus. Manual Clin Microbiol 11(2):1687–1703

Karasneh GA, Shukla D (2011) Herpes simplex virus infects most cell types in vitro: clues to its success. Virol J 8:481. https://doi.org/10.1186/1743-422X-8-481

Karsten E, Watson SL, Foster LJ (2012) Diversity of microbial species implicated in keratitis: a review. Open Ophthalmol J 6:110–124

Kaufman HE (1962) Clinical cure of herpes simplex keratitis by 5-iodo-2-deoxyuridine. Proc Soc Exp Biol Med 109:251–252

Kondel R, Shafiq N, Kaur IP, Singh MP, Pandey AK, Ratho RK, Malhotra S (2019) Effect of acyclovir solid lipid nanoparticles for the treatment of herpes simplex virus (HSV) infection in an animal model of HSV-1 infection. Pharm Nanotechnol 7(5):389–403

Leo J (1982) The New Scarlet Letter. http:www.time.com/time/magazine/article/0,9171,1715020,00.htlm. Time

Liesegang TJ, Melton LJ, Daly PJ, Ilstrup DM (1989) Epidemiology of ocular herpes simplex. Incidence in Rochester, Minn, 1950 through 1982. Arch Ophthalmol 107:1155–1159

Liu S, Knafels JD, Chang JS, Waszak GA, Baldwin ET, Deibel MR Jr, Thomsen DR, Homa FL, Wells PA, Tory MC, Poorman RA, Gao H, Qiu X, Seddon AP (2006) Crystal structure of the herpes simplex virus 1 DNA polymerase. J Biol Chem 281:18193–18200

Lobo AM, Agelidis AM, Shukla D (2019) Pathogenesis of herpes simplex keratitis: the host cell response and ocular surface sequelae to infection and inflammation. Ocul Surf 17(1):40–49

Looker KJ, Magaret AS, May MT, Turner KM, Vickerman P, Gottlieb SL, Newman LM (2015a) Global and regional estimates of prevalent and incident herpes simplex virus type 1 infections in 2012. PLoS One 10:e0140765

Looker KJ, Magaret AS, Turner KM, Vickerman P, Gottlieb SL, Newman LM (2015b) Global estimates of prevalent and incident herpes simplex virus type 2 infections in 2012. PLoS One 10:e114989

Ma Y, He B (2014) Recognition of herpes simplex viruses: toll-like receptors and beyond. J Mol Biol 426:1133–1147

McGeoch DJ, Dalrymple MA, Davison AJ, Dolan A, Frame MC, McNab D, Perry LJ, Scott JE, Taylor P (1988) The complete DNA sequence of the long unique region in the genome of herpes simplex virus type 1. J Gen Virol 69:1531–1574

Misra AP (n.d.) Herpes virus infections: a textbook of medicine 9th edition API Textbook of Medicine, 9th edn, p 1148

Monavari SH, MJM P, Bolouri B, Ebrahimi SA, Ataei-pirkooh A (2014) The inhibitory effect of acyclovir loaded nano-niosomes against herpes simplex virus type-1 in cell culture. Med J Islam Repub Iran 28:99

Mott KR, Underhill D, Wechsler SL, Town T, Ghiasi H (2009) A role for the JAK-STAT1 pathway in blocking replication of HSV-1 in dendritic cells and macrophages. Virol J 6:56

Nahmias AJ, Dowdle WR (1968) Antigenic and biologic differences in herpesvirus hominis. Prog Med Virol 10:110

Oh Y, Lee N, Kang HW, Oh J (2016) In vitro study on apoptotic cell death by effective magnetic hyperthermia with chitosan-coated MnFe2O4. Nanotechnology 27(11):115101

Pandey M, Choudhury H, Abdul-Aziz A, Bhattamisra SK, Gorain B, Su JST, Tan CL, Chin WY, Yip KY (2020) Advancement on sustained antiviral ocular drug delivery for herpes simplex virus keratitis: recent update on potential investigation. Pharmaceutics 13(1):1

Pinto J, Rao AK, Guru BR (2021) Trends in biomaterials & artificial organs. J Artif Organs 35(5):415–423

Piret J, Boivin G (2011) Resistance of herpes simplex viruses to nucleoside analogues: mechanisms, prevalence, and management. Antimicrob Agents Chemother 55(2):459–472

Rabanel JM, Aoun V, Elkin I, Mokhtar M, Hildgen P (2012) Drug-loaded nanocarriers: passive targeting and crossing of biological barriers. Curr Med Chem 19(19):3070–3102

Rabenau HF, Buxbaum S, Preiser W, Weber B, Doerr HW (2002) Seroprevalence of herpes simplex virus types 1 and type 2 in the Frankfurt am Main area, Germany. Med Microbiol Immunol 190:153–160

Risk Group (2021). https://ibkp.dbtindia.gov.in/Content/ShowPublicConsultationImg?fName=20210304173501765_Draft%20List%20of%20Microorganisms%2004.03.2021.docx

Rowe AM, St Leger AJ, Jeon S, Dhaliwal DK, Knickelbein JE, Hendricks RL (2013) Herpes keratitis. Prog Retin Eye Res 32:88–101

Shirazi H, Daneshpour M, Kashanian S, Omidfar K (2015) Synthesis, characterization and in vitro biocompatibility study of au/TMC/Fe3O4 nanocomposites as a promising, nontoxic system for biomedical applications *Beilstein*. J Nanotechnol 6(1):1677–1689

Shukla D, Spear PG (2001) Herpesviruses and heparan sulfate: an intimate relationship in aid of viral entry. J Clin Invest 108:503–510. https://doi.org/10.1172/JCI13799

Smith JS, Robinson NJ (2002) Age-specific prevalence of infection with herpes simplex virus types 2 and 1: a global review. J Infect Dis 186(Suppl 1):S3–S28

Souza PM, Holland EJ, Huang AJ (2003) Bilateral herpetic keratoconjunctivitis. Ophthalmology 110:493–496

Steiner I, Kennedy PG, Pachner AR (2007) The neurotropic herpes viruses: herpes simplex and varicella-zoster. Lancet Neurol 6:1015–1028

Stein-Streilein J (2008) Immune regulation and the eye. Trends Immunol 29:548–554

Tasman W, Jaeger EA. (2013). Duane's ophthalmology on DVDROM- 2013 edition. http://80.36.73.149/almacen/medicina/oftalmologia/ enciclopedias/duane/pages/v3c028a.html

Varanasi SK, Reddy PBJ, Bhela S, Jaggi U, Gimenez F, Rouse BT (2017) Azacytidine treatment inhibits the progression of herpes stromal keratitis by enhancing regulatory T cell function. J Virol 91:e02367–e02316

Vedula VB, Chopra M, Joseph E, Mazumder S (2016) Preparation and characterization of nanoparticles of carboxymethyl cellulose acetate butyrate containing acyclovir. Appl Nanosci 6(2):197–208

Weller SK, Coen DM (2012) Herpes simplex viruses: mechanisms of DNA replication. Cold Spring Harb Perspect Biol 4:a013011

Whitley RJ, Roizman B (2001) Herpes simplex virus infections. Lancet 357:1513–1518. https://doi.org/10.1016/S0140-6736(00)04638-9

Wilhelmus KR (2010) Antiviral treatment and other therapeutic interventions for herpes simplex virus epithelial keratitis. Cochrane Database Syst Rev 12:CD002898

World Health Organization, Herpes simplex virus, 2022., https://www.who.int/news-room/fact-sheets/detail/herpes-simplex-virus

Xu F, Sternberg MR, Kottiri BJ, McQuillan GM, Lee FK, Nahmias AJ, Berman SM, Markowitz LE (2006) Trends in herpes simplex virus type 1 and type 2 seroprevalence in the United States. JAMA 296:964–973

Yuan S, Wang J, Zhu D, Wang N, Gao Q, Chen W, Tang H, Tang J, Zhang X, Liu H, Rao Z, Wang X (2018) Cryo-EM structure of a herpesvirus capsid at 3.1 Å. Science 360:eaao7283. https://doi.org/10.1126/science.aao7283

Zika Virus: The Emerging Arthropod Borne Virus

7

G. K. Megha, Asima Zehra, Afnan Saleem, and Ambika

Abstract

The Aedes mosquito is the primary vector of transmission of the Zika virus, a vector-borne arbovirus that has become a major worldwide health problem. Despite being discovered in Africa at first, the virus became well-known during the 2015–2016 outbreak in the Americas because it was linked to serious birth abnormalities like microcephaly. The public health community has been alarmed by the Zika virus's emergence in India in recent years, which has made a thorough grasp of the virus's epidemiology, clinical symptoms, and possible effects on mother and child health imperative. This chapter gives a broad review of the Zika virus's current situation in India, stressing important epidemiological trends, difficult diagnostic cases, and socioeconomic variables that affect the dynamics of the virus's spread. India's diverse environment and high population density are taken into consideration while discussing the geographical distribution of the virus, including impacted regions and recorded cases. The chapter also looks at how the Zika virus might affect pregnancy outcomes, highlighting the need for improved surveillance and preventative actions to lower the chance of congenital abnormalities. The difficulties in diagnosing the Zika virus and the

G. K. Megha (✉)
Division of Veterinary Public Health and Epidemiology, Indian Veterinary Research Institute, Bareilly, India
e-mail: meghavph@gmail.com

A. Zehra
Centre for One Health, GADVASU, Ludhiana, Punjab, India

A. Saleem
Division of Animal Biotechnology, FVSc and AH, Shuhama, Sher-E-Kashmir University of Agriculture Sciences and Technology, Kashmir, India

Ambika
ICAR-Indian Veterinary Research Institute, Bareilly, Uttar Pradesh, India

P. L. Bhukya, A. S (eds.), *Emerging Human Viral Diseases, Volume II*,
https://doi.org/10.1007/978-981-97-4480-0_7

significance of early detection for successful public health interventions are also covered in this chapter. The chapter also addresses current research projects and cooperative efforts amongst international groups, government agencies, and research institutions in an attempt to comprehend the behaviour of the virus and create control measures. This chapter's conclusion emphasises how critical it is to keep up with research and implement public health initiatives to monitor and contain the Zika virus in India. This chapter adds to the global conversation on emerging infectious illnesses by synthesising current information and highlighting the value of a multidisciplinary approach. It also emphasises the necessity of ongoing vigilance to safeguard vulnerable people from the Zika virus.

Keywords

Zika virus · India · Arbovirus · Aedes mosquitoes · Global health · Maternal and child health · Microcephaly · Transmission dynamics · Congenital anomalies · Government agencies · Emerging infectious diseases · Multidisciplinary approach

7.1 Introduction

The Zika virus (ZIKV) is an ss + RNA virus belonging to the Flaviviridae family, primarily transmitted through mosquito vectors. Its name is derived from the Zika Forest in Uganda, where it was initially isolated in 1947 from the blood of an ailing monkey (CDC 2019). Phylogenetic analysis of the complete genome suggests that the Flavivirus genus has its roots in Africa, dating back to a period ranging from 85,000 to 120,000 years ago (Pettersson and Fiz-Palacios 2014). Following its discovery in the monkey, the virus was subsequently found in the *Aedes africanus* mosquito population in the Zika Forest (Dick and Zika virus. II. 1952). The first documented human case of Zika infection was reported in Uganda in 1952 (Dick et al. 1952). Notably, the initial detection of ZIKV in Asia occurred in 1966, when *A. aegypti* mosquitoes collected in Malaysia were found to carry the virus. Eleven years later, in central Java, Indonesia, the first human infections in Asia were confirmed in seven individuals who displayed symptoms such as fever, malaise, stomach ache, anorexia, and dizziness (Marchette et al. 1969; Olson et al. 1981).

7.2 Epidemiology

Zika virus (ZIKV), part of the Flaviviridae family, shares lineage with dengue and yellow fever viruses. It comprises three distinct lineages—African I, African II, and Asian—each varying in clinical manifestations. Originally in Africa, Asia, and the Pacific Islands, Zika's global spread is now accelerated by favourable mosquito conditions and increased globalisation. Beyond vector transmission, the virus spreads through sexual contact, mother-to-child transmission, and blood transfusion. By March 2017 13 nations reported non-vector person-to-person transmission. In India, Zika is primarily vector-borne, with no reported non-vector cases (Asad and Carpenter 2018; PAHO/WHO 2017; WHO 2017a, b; Agarwal and Chaurasia 2020).

ZIKV was discovered in 1947 in the Zika forest of Uganda, initially isolated from rhesus monkeys. The first human infection was reported in Nigeria in 1954, with the first documented outbreak in 2007 on Yap Island. A pivotal moment came in 1966 when ZIKV was isolated from *Aedes aegypti* mosquitoes, confirming vector transmission. As of April 2022, ZIKV transmission has been reported in 93 countries, with 54 having widespread *Aedes aegypti* mosquitoes, yet no documented ZIKV cases (WHO 2019; CDC 2022). Despite no recent global outbreaks, a significant one occurred in India in November 2021 (Fig. 7.1). The virus's presence was confirmed in India in 2016, with the Asian lineage hypothesised as locally originating. The indigenous *Aedes aegypti* mosquito population in Rajasthan tested positive during the outbreak, contrasting with negative findings in Gujarat and Tamil Nadu in 2017 (Fig. 7.1) (CDC 2022; Sapkal et al. 2018; Singh et al. 2018).

ZIKV's serological prevalence in India dates back to the 1950s, with a 1954 study reporting 16.8% Zika antibodies in samples from the Bharuch district. In February 2016, WHO declared ZIKV a public health emergency based on neurological issues linked to the virus. The Indian Council of Medical Research (ICMR) intensified surveillance efforts. While ZIKV's consequences during pregnancy are not fully understood, the Emergency Committee's association with microcephaly led WHO to continue active response through ICMR. Sentinel ZIKV surveillance expanded from 10 laboratories in 2016 to 132 in 2021, conducting tests on Dengue and Chikungunya negatives. The Rashtriya Bal Swasthya Karyakram (RBSK)

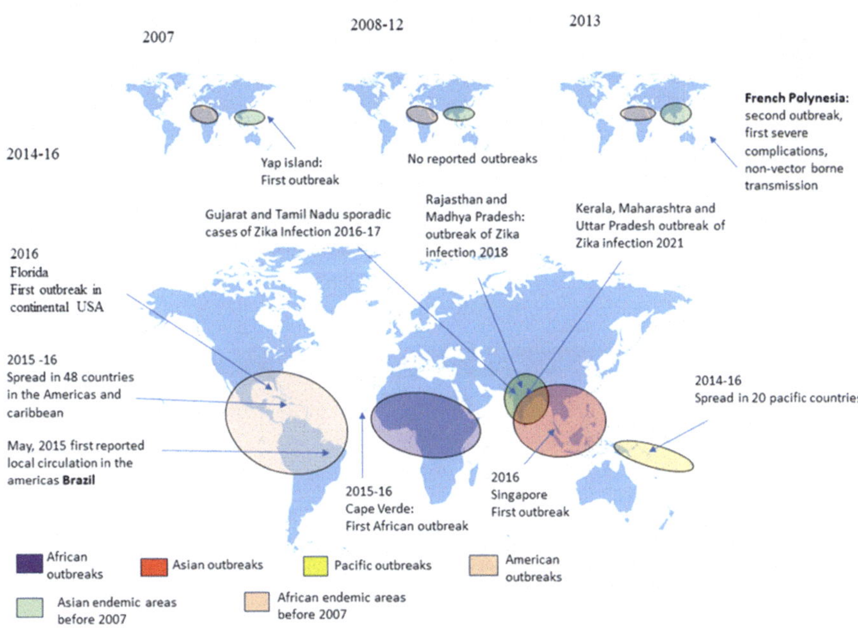

Fig. 7.1 ZIKV outbreaks (2007–2021) with geographical enhancements

monitors microcephaly cases. Continuous surveillance enabled prompt detection of Zika outbreaks in Rajasthan and Madhya Pradesh in 2018. Containment measures were implemented within a 3 km radius of the initial case in Rajasthan, resulting in 153 documented cases over two months, including 63 pregnant females. Madhya Pradesh reported 130 cases, with water storage practices and widespread usage of mosquito-breeding water coolers identified as potential causes (Singh et al. 2019a, b; Rolph and Mahalingam 2019).

Due to the extensive involvement of Virus Research and Diagnostic Laboratories (VRDLs) in COVID-19 diagnoses during subsequent pandemic waves, public health surveillance for ZIKV declined post-2020. The VRDLs store Dengue/Chikungunya virus negative samples for potential ZIKV testing. Recent occurrences in Uttar Pradesh and Kerala in 2021 underscore the importance of active surveillance for understanding ZIKV's impact on healthcare infrastructure. Between 2017 and 2021, ZIKV occurrences were reported in 16 Indian states and union territories (Figs. 7.1 and 7.2, Yadav et al. 2022a, b).

7.3 Classification of ZIKV

ZIKV falls within the category of arboviruses (arthropod-borne viruses), a term used in ecology to describe viruses that are naturally sustained through biological transmission between hematophagous arthropods, such as mosquitoes, and

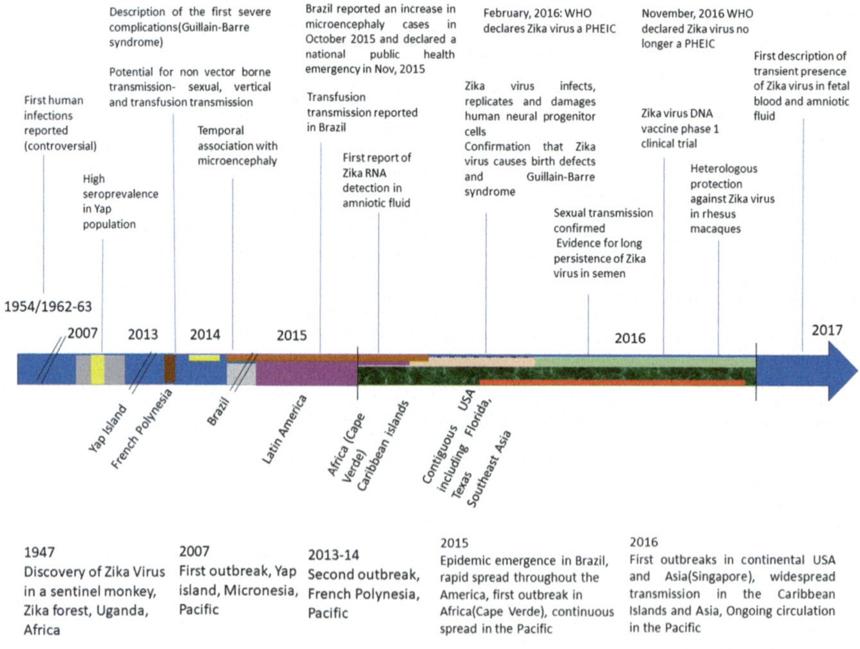

Fig. 7.2 Timeline of ZIKV infection. The additions have been made in the timeline courtesy

susceptible vertebrate hosts. Initially, arboviruses were classified primarily based on serological criteria, specifically antigenic properties. However, in recent years, taxonomy has shifted toward a molecular foundation, leading to the classification of the Flavivirus genus into clusters, species, and clades. As of the present, the Flavivirus genus comprises 53 distinct virus species categorised into three main clusters: mosquito-borne flaviviruses, tick-borne flaviviruses, and flaviviruses lacking a known vector (International Committee on Taxonomy of Viruses website, chapter on virus families not assigned to an order, family Flaviviridae, http://ictvonline.org/virusTaxonomy.asp). A fourth group of viruses, exclusively found in insects, is also anticipated to be included within this genus.

The comprehensive phylogenetic analysis of Zika virus (ZIKV), including African and Asian lineages, revealed the formation of two distinct clusters, as reported by Musso and Gubler in 2016. Their study involved grouping and analysing ZIKV strains to elucidate evolutionary relationships and geographical spread.

The mosquito-borne flaviviruses can be distinguished in to two eco-epidemiological groups: (Abushouk et al. 2016) Aedes-borne flaviviruses group: primarily finds its hosts among mammals including humans and primates and are mostly associated with human ailments such as polyarthritis and haemorrhagic fever, exemplified by DENV and YFV. ZIKV within this group, stands as an exception, as it is known to induce neurological disorders in humans. Conversely, the second group (Ades et al. 2021), consisting of Culex-borne flaviviruses, which primarily infects birds such as Japanese Encephalitis virus (JEV), West Nile virus (WNV), St. Louis Encephalitis virus (SLEV), and Usutu virus (USUV). These viruses are chiefly responsible for causing neurological diseases in their avian hosts (Gould et al.2003).

7.4 Morphology and Virion Structure

ZIKV is a ss + RNA virus, featuring a genome of approximately 10.8 kb in size (Sirohi and Kuhn 2017). The ZIKV genome comprises two flanking non-coding regions and an open reading frame (ORF) that encodes a solitary polyprotein. This polyprotein is subsequently cleaved into three structural proteins and, importantly, at least seven non-structural proteins (Ye et al. 2016). The structural organisation of ZIKV closely resembles that of other flaviviruses. The virion exhibits a diameter of roughly 40 nm and is adorned with surface projections measuring nearly 10 nm in length (Sirohi et al. 2016). The virion membrane is hooked on by glycoproteins arranged in a configuration reminiscent of icosahedral symmetry, facilitating adhesion to and subsequent infection of host cells. Beneath the virion membrane lies the nucleocapsid, with an approximate diameter ranging from 25 to 30 nm. This nucleocapsid is ensconced within a lipid bilayer that comprises envelope proteins E and M, both derived from the host membrane.

ZIKV exhibits two distinct states, characterised as the mature and immature forms. The structure of a mature virus resembles a golf ball in which the major surface protein E gives a rather smooth surface by lying parallel to the viral membrane.

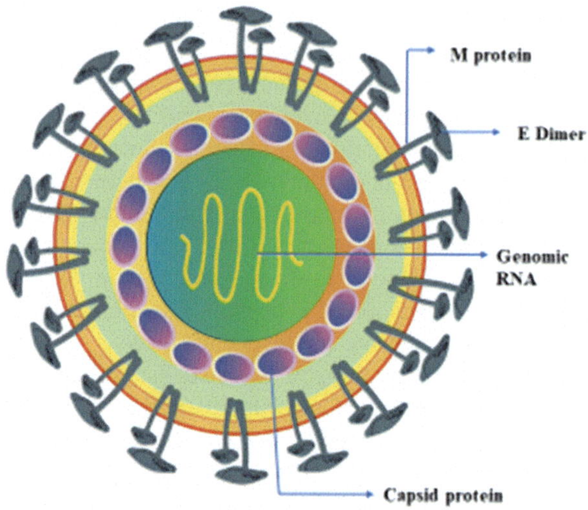

Fig. 7.3 Representation for a typical genome structure of a flavivirus

Similar to other flaviviruses, there are four domains of ZIKV E-protein: the stem-transmembrane domain pair, and three domains referred to as ectodomains I, II, and III which are situated outside the viral membrane. The E-protein predominates on the virion's surface, while the smaller M-protein resides beneath the larger E-protein. The M-protein has a compact extracellular region and stem-transmembrane domains. In mature flaviviruses, the fundamental organisational unit of the E-protein is a dimer, with each E-monomer within the dimer displaying a two-fold symmetry relationship with its neighbouring monomer. These E-protein dimers are arranged in a parallel manner, forming what is termed a "raft configuration," with the virion containing a total of 30 such rafts. The asymmetric unit of the virion constitutes half of a raft, resulting in a total of 60 asymmetric units present within the ZIKV particle (Fig. 7.3).

7.5 Viral Proteins

The viral RNA is translated into a sole polyprotein comprising 3423 amino acids. This polyprotein encodes three structural proteins—capsid (C), membrane (M), and envelope (E)—alongside seven non-structural proteins (NSPs), specifically NS1, NS2A, NS2B, NS3, NS4A, NS4B, and NS5. These NSPs play pivotal roles in genome replication, packaging, and manipulation of host pathways to support the virus. The production of the ten distinct proteins from the polyprotein is governed by both viral and host proteases. Notably, the effectiveness of one of the host proteases, furin, in cleaving specific viral targets (such as prM) can vary and potentially contribute to the development of disease symptoms.

The NSPs are indispensable for various aspects of viral genome replication, virion packaging, and evading the host immune response. Within this group, NS1 plays a role in viral RNA replication and contributes to viral particle assembly, while NS2A is believed to participate in both immune evasion and RNA replication. NS2B acts as a cofactor essential for the serine-protease activity of the NS2B-NS3 complex. NS3 is involved in numerous enzymatic functions, including protease and helicase activities. Additionally, NS4A and NS4B play roles in viral RNA replication and immune evasion, and NS5 possesses RNA-dependent RNA polymerase activity in its C-terminal motifs (Rastogi et al. 2016; Chong et al. 2019).

7.6 Route of Infection

7.6.1 Vector Borne Transmission

The primary means of ZIKV transmission occurs when individuals are bitten by various mosquito species, predominantly by *Aedes aegypti* and *Aedes Albopictus* mosquitoes (Chouin-Carneiro et al. 2016). Nonetheless, various mosquito species have the potential to serve as vectors for viral transmission, including *A. africanus*, *A. apicoargenteus*, *A. furcifer*, *A. luteocephalus*, *A. opok*, and *A. vittatus*. Usually, the *A. aegypti* mosquito bites during the day, with peak activity occurring in the early morning and late afternoon or evening, particularly in subtropical and tropical regions.

The mosquito-mediated transmission cycle starts when a female Aedes mosquito, which feeds on blood, introduces the virus into human skin. Subsequently, it impacts various cells within the dermis and epidermis before gaining access to lymph nodes, followed by entry to the bloodstream then to the final targets i.e. organs and tissues. The conclusion of this transmission cycle through mosquitoes transpires when a different mosquito, free from infection, bites a human with a detectable level of the virus in their blood, ingesting the infected blood. Additionally, studies of ZIKV infection in field-collected eggs, larvae, and adult mosquitoes suggest that mosquito vectors may potentially be capable of vertical transmission of ZIKV (Lai et al. 2020). The studies showed that ZIKV can spread vertically in *Ae. albopictus* by transovarial transmission. These newer discoveries can be applied to vector control and ZIKV disease prevention techniques.

7.6.2 Non-Vector Borne Transmission

Congenital or intrauterine (in utero) or vertical transmission refers to the transmission of ZIKV from an infected pregnant woman to her foetus during pregnancy. Congenital transmission of ZIKV may occur during any trimesters of pregnancy (Besnard et al. 2014). Infection at any stage of pregnancy can result in foetal harm, but there is evidence indicating that the early stages of gestation may be particularly susceptible to damage caused by ZIKV (Honein et al. 2017; Kleber de Oliveira et al. 2016; Driggers et al. 2016; Meaney-Delman et al. 2016). Studies have reported that the estimated mean risk of vertical transmission was 47% subsequent to maternal infection during the initial trimester, 28% during the second trimester, and 25% during the third trimester (Ades et al. 2021). In-utero transmission leads to congenital abnormalities in foetuses which include microcephaly, IUGR, foetal demise, and ocular abnormalities that can cause blindness (Platt and Miner 2017). While perinatal transmission occurs when a pregnant woman contracts ZIKV infection roughly 2 weeks before giving birth, and the virus is transmitted to the newborn during or near the time of delivery. Perinatal transmission can result in the onset of symptoms like conjunctivitis, a maculopapular rash, arthralgia, and fever in the infant (CDC 2022).

The initial instance of sexually transmitted ZIKV infection was documented in Colorado, USA, in 2008 (Foy et al. 2011). Infectious bodily secretions from both males and females, including semen, vaginal fluids, or menstrual blood, have the potential to transmit ZIKV (Prisant et al. 2016; Lanciotti et al. 2008; Deckard et al. 2016). Sexual transmission is specially observed in ZIKV and has never been recorded in any other mosquito-borne flavivirus. Furthermore, ZIKV infection is transmitted through accidental laboratory spills (Filipe et al. 1973), infected blood transfusion (Barjas-Castro et al. 2016; Jimenez et al. 2017), and infected organ donation (Iwamoto et al. 2003) (Fig. 7.4).

7.7 Life Cycle

When a female Aedes mosquito sucks blood containing ZIKV from an infected patient, to transmit the virus to other susceptible hosts, first it must entail replication within the mosquito midgut epithelial cells, followed by its multiplication in the salivary gland cells (Sirohi and Kuhn 2017). Extrinsic incubation period can range from 5–10 days. After the extrinsic incubation period, the virus can be detected in the mosquito's saliva and has the potential to infect humans as the mosquito

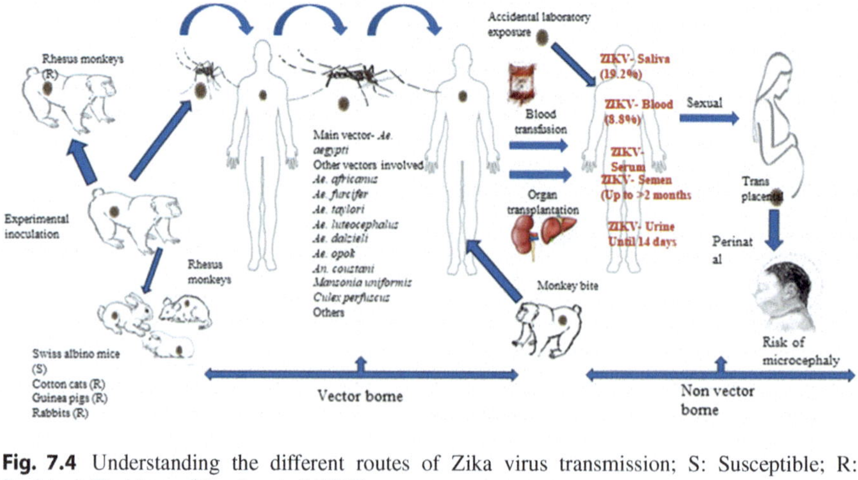

Fig. 7.4 Understanding the different routes of Zika virus transmission; S: Susceptible; R: Resistant (Rodriguez-Morales et al. 2016)

aggressively feeds on blood, simultaneously injecting contaminated saliva into the host (Li et al. 2012). Other possible ways of viral maintenance are through transovarial transmission (Martins et al. 2012). During the mosquito's blood-feeding activity, ZIKV infiltrates and traverses the layers of the skin, which include the epidermal and dermal layers, as well as Langerhans cells. The primary potential target cells for infection in humans encompass dermal fibroblasts, epidermal keratinocytes, and immature dendritic cells (Briant et al. 2014). In addition to serving as the first line of defence after ZIKV infection, skin fibroblast cells also serve as the target cells for the virus (Hamel et al. 2015). AXL, Tyro3, DC-SIGN, or TIM-1 are a few entrances and/or adhesion factors that favour ZIKV entry and operate as anchoring components or receptors (Briant et al. 2014; Hamel et al. 2015). Apoptosis of epidermal cells begin, autophagy begins, and a large number of viral particles are released as a result of viral replication. The virus then spreads to local lymph nodes, the bloodstream, and finally organs and tissues such as the heart, skeletal muscles, and central nervous system, possibly transplacentally, resulting in congenital syndrome. ZIKV predominantly infects and replicates in Hofbauer cells (placental macrophages) in pregnant women, causing these placental cells to proliferate and exhibit significant hyperplasia (Rosenberg et al. 2017). Although less so, cytotrophoblasts are also affected. Type-I interferon, inflammatory cytokines, and antiviral gene expression all seem to be induced by viral replication. The foeto-placental barrier is broken down by all of these factors (Solomon and Milner 2016). The germinal matrix, a foetal's primary target for the ZIKV in the central nervous system, may have been destroyed at the beginning of the foetal neurologic affectation. Infection of neural precursor cells resulted in cell death, according to data from cell culture systems (Tang et al. 2016). Early developmental damage to these cells may significantly diminish the amount of neurons produced and lead to small brains with irregular or no cortical gyration (Gilmore and Walsh 2012).

7.8 Pathogenesis in Humans

ZIKV infection triggers both innate and adaptive immune responses within infected cells. Initially, viral RNA sensors activate Toll-like receptors (TLR3 and TLR7), as well as RIG-like receptors (MDA5 and RIG-I), initiating the synthesis of type I (IFN-β) and type III (IFN-λ) interferons. IFN-λ subsequently binds to its respective receptors, activating the JAK/STAT signalling pathway. This pathway further stimulates the production of interferon-stimulated genes such as ISG15, OAS2, MX1, IFIT, and inflammatory chemokines like CCL5 and CXCL10 (Hamel et al. 2016; Frumence et al. 2016). Studies have underscored the critical role of the IFN signalling pathway in ZIKV-induced pathology, notably in mice lacking type I and II IFN receptors or STAT2, which is not present in immunocompetent mice (Rossi et al. 2016; Lazear et al. 2016; Morrison and Diamond 2017; Tripathi et al. 2017). Additionally, IFN-λ has demonstrated exceptional protective properties against ZIKV infection in the reproductive tract of females, as well as in the maternal decidua and placenta during pregnancy (Caine et al. 2019; Bayer et al. 2016; Jagger et al. 2017). However, akin to other viruses, ZIKV has the ability to counteract antiviral immune responses by interacting with cellular signalling proteins. ZIKV can impair IFN signalling pathways by inhibiting STAT1 phosphorylation and promoting proteasomal degradation of JAK1 and STAT2 through interactions with the NS2B-NS3 protease and NS5, respectively (Xia et al. 2018; Grant et al. 2016). Furthermore, ZIKV sfRNA, a subgenomic viral RNA, has been implicated in hindering the RIG-I mediated induction of type I interferon in lung epithelial cells of human lung, highlighting its role in interfering with innate immune responses (Goertz et al. 2017; Akiyama et al. 2016; Donald et al. 2016). Additionally, after an initial infection against DENV, the generation of non-neutralising antibodies that trigger a mechanism known as Antibody Dependent Enhancement can enhance the susceptibility to infection by another flavivirus due to cross-reactivity with the Fcg receptor (Halstead 2014). Numerous investigations have been conducted to assess the extent of cross-reactivity between these two viruses, driven by the substantial ZIKV outbreak in regions where DENV is recognised as epidemic (Priyamvada et al. 2016). The findings are still debatable, though. While some research revealed that having previously contracted DENV was linked to a lower likelihood of developing ZIKV symptoms (Regla-Nava et al. 2018; Gordon et al. 2019). Conversely, additional in vitro and in vivo studies presented contradictory results (Halstead 2014; Pantoja et al. 2017; Fowler et al. 2018). The development of an anti-ZIKV vaccine must take into account this feature, which appears to be influenced by the viral strain and host immune response (Rey et al. 2018). A remarkable effort has been made since the ZIKV outbreak in 2015 to advance fundamental research aimed at enhancing our understanding of this flavivirus's biological aspects, specifically its tropism, morphogenesis, and response to antiviral measures. These studies have played a pivotal role in enhancing our comprehension of the virus and for implementing cutting-edge methods of therapy and vaccine development. Despite these developments, more research is still required to comprehend the molecular mechanisms underlying the virus's ability to breach the blood-brain barrier and the placenta, which distinguishes it from other flaviviruses. Additionally, there is a need to explore the distinctions among various ZIKV strains and the impact of coinfection with other arboviruses on the virus's pathogenicity.

7.9 Clinical Manifestations

In case of acute ZIKV infection, the symptoms and signs closely resemble those of other concurrently circulating viruses like DENV or CHIKV, which can impede diagnosis solely relying on clinical manifestations. However, according to studies, 80% of the ZIKV infections are considered asymptomatic (Duffy et al. 2009; Gallian et al. 2017). All age groups, spanning from 4 days to 76 years, exhibit susceptibility, with a marginal predominance of cases among females. Incubation period of the disease varies between 2 and 14 days. Primary infection of a strain of ZIKV will elicit protective immunity providing comprehensive immunity against reinfection from various strains (Dudley et al. 2016). The clinical presentations of ZIKV infection can vary depending on the population and the timing of infection.

7.9.1 Clinical Symptoms in Adults

A typical ZIKV infection begins with the sudden onset of a low-grade fever lasting for 3–5 days, followed by the development of a widely spread maculopapular rash. This is often accompanied by minor joint pain often accompanied by swelling around the joints. Additionally, individuals may experience non-purulent bilateral conjunctivitis and retro-orbital headache (Abushouk et al. 2016; Khut 2016). However, ZIKV can be more specifically manifested in adults as arthralgia, Guillain Barre Syndrome, and cardiovascular abnormalities (Hussain et al. 2018). Fatal cases are rare, except those cases which are associated with complications (Zonneveld et al. 2016; Sarmiento-Ospina et al. 2016). ZIKV infections tend to exhibit more severe symptoms in individuals concurrently infected with other pathogens (Dejnirattisai et al. 2016). The studies have demonstrated that the co-infection of ZIKV disease with Malaria or Herpes Simplex virus Type 2 increases the severity of the disease (Sow et al. 2016; Aldo et al. 2016). ZIKV in pregnant women is manifested commonly with signs of rashes and there is a notably elevated incidence of lymphadenopathy and conjunctival hyperaemia compared to pregnant women without ZIKV infection (Brasil et al. 2016). While viral RNA has been identified in the breast milk of infected mothers, there has been no documented cases of ZIKV transmission via breastfeeding. The Centers for Disease Control and Prevention (CDC) advocates for breastfeeding by infected mothers, a recommendation substantiated by evidence indicating that the advantages of breastfeeding outweigh the hypothetical risks associated with ZIKV transmission through breast milk.

7.9.2 Clinical Symptoms in Children

Children can either get the infection through vertical transmission or routes similar to adults. Infection usually manifests in the mild form in children. However, the postnatal clinical symptoms in infants and children mirror those documented in adults but are manifested atypically in children (Russell et al. 2016). Gastrointestinal symptoms are observed more frequently in paediatric patients (Karwowski et al. 2016). Mortality rates for ZIKV in children are exceptionally uncommon, with only a limited number of cases documented (Arzuza-Ortega et al. 2016).

7.10 Phases of the Disease

Around 60–80% of patients with ZIKV infections are either asymptomatic, or they only exhibit minor symptoms (mild form) (Fig. 7.5). The most frequent signs and symptoms of symptomatic infections are rash/dermatitis (> = 90%), conjunctivitis

(55–82%), fever (65–80%), and headache (45–80%) (https://www.ncbi.nlm.nih.gov/books/NBK430981/). The fever is often low grade and transient, and the rash/dermatitis is frequently maculopapular. Arthralgia (65–70%), myalgia (48–65%), and retro-orbital discomfort (39–48%) are additional typical symptoms and indicators. Emphysema, vomiting, and stomach pain are less frequently observed. In regions where dengue fever and chikungunya, which share a lot of symptoms and indicators, are prevalent, it could be challenging to determine the exact cause of the sickness. ZIKV infections are more likely to cause conjunctivitis and rash/dermatitis than dengue fever and chikungunya. Clinical sickness typically has modest symptoms that last a few days to a week. There are few cases of severe illness needing hospitalisation, but case mortality is low. But it was observed that during the French Polynesia ZIKV outbreak, there was a surge in instances of acute Guillain-Barre syndrome, a kind of paralytic neuropathy. ZIKV infection has led to documented cases of acute myelitis and meningoencephalitis. When the ZIKV is suspected, a thorough neurologic evaluation is advised. Recently, CDC/ICMR/NCDC-India concluded that exposure to the ZIKV during pregnancy causes serious foetal brain abnormalities including microcephaly. Foetuses and newborns of ZIKV infected women should be assessed for potential congenital infections and neurological anomalies due to concerns about microcephaly brought on by maternal ZIKV infection.

7.11 Diagnostics

While the case definition requires the suspicion of ZIKV infection based on typical clinical symptoms, signs, and relevant epidemiologic exposure, the ultimate confirmation of ZIKV infection necessitates a microbiological diagnosis (Colin R Martin et al. 2021). Numerous microbiological methodologies are available for diagnosing ZIKV infection, yet the selection of the appropriate laboratory approach relies on

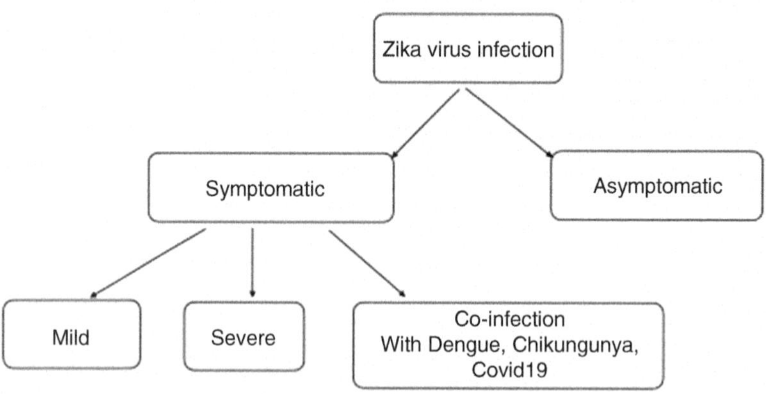

Fig. 7.5 Different possibilities of ZIKV infection

factors such as laboratory infrastructure, technical proficiency, and sample accessibility.

7.11.1 Serologic Approaches for ZIKV Detection

In the context of ZIKV infection diagnosis, serological analysis stands as the prevailing diagnostic modality. Following the onset of symptoms, typically during the initial 8–10 days, the presence of IgM antibodies becomes detectable, followed by a subsequent increase in IgG antibodies, which endure for an extended duration (Barzon et al. 2018; Pasquier et al. 2018; Driggers et al. 2016). By analogy to other closely related flavivirus, the presence of ZIKV IgM antibodies may persist for extended durations (Prince et al. 2007). According to reports, it takes a median of 8 days for the detection of ZIKV non-structural protein 1 (NS1) IgM, while for ZIKV NS1 IgG, it takes 17 days (Barzon et al. 2018).

Given the potential for serological cross-reactivity with other flaviviruses, both from infection and/or vaccination, it is advisable to conduct supplementary analyses to confirm a positive IgM test result. For increased reliability, ZIKV diagnosis should involve testing paired samples from both the acute and convalescent phases. Seroconversion is defined as a minimum four-fold rise in antibody levels between paired sera collected within a 10–14 days interval. In cases where the patient has not previously encountered a flavivirus, minimal cross-reactivity is anticipated. However, during an acute flavivirus infection, there may be an augmentation in cross-reactive antibodies due to prior infection or vaccination against another flavivirus (Sharma and Lal 2017; Regla-Nava et al. 2018).

There are plenty of commercially accessible and/or in-house enzyme-linked immunosorbent assays (ELISA/EIAs) and virus neutralisation assays aimed at ZIKV diagnostics. The IgM-capture ELISA, designed to detect IgM antibodies in serum or cerebrospinal fluid, has proven to be an efficacious method. Nonetheless, it is noteworthy that cross-reactivity may be encountered in individuals previously exposed to a different flavivirus infection, including those who have been vaccinated against yellow fever or Japanese encephalitis (Gourinat et al. 2015). Hence, in cases where IgM for ZIKV is detected positively, the use of neutralising antibody detection assays, specifically the plaque reduction neutralisation test (PRNT), is considered as the gold standard for both diagnosis and test assessment. The virus neutralisation test (VNT) is capable of identifying the presence of virus-specific antibodies in the patient's serum, without discriminating against the type of antibodies, i.e., IgM or IgG. PRNT being the gold standard test as well as the most common form of VNT used for the detection of flaviviruses (Roehrig et al. 2003). Because of the fact that the test might take even 10 days at times to give the results, VNT is not suitable for rapid sample screening for ZIKV infection. However, in recent years, a comparatively less time-consuming novel reporter virus neutralisation test (RVNT) has been developed which takes a shortened turnaround time of <2 days in a 96-well format (Xia et al. 2018). Despite its time-intensive nature, the

VNT remains an indispensable and specific method for detecting flaviviral antibodies, affirming its continued relevance and significance in diagnostic applications.

7.11.2 Molecular Analysis

Studies have proven that for ZIKV early detection, the conventional RT-PCR and quantitative RT-PCR represent the expeditious, specific, and highly sensitive methodologies (Foy et al. 2011). Molecular techniques deliver precise diagnostic outcomes for viral genome detection; however, their reliability is contingent upon execution within the initial 5 days of the disease, as viremia begins to decline concurrently with the onset of rash (Lanciotti et al. 2008).

7.12 Therapeutics

At present, there exists no targeted therapy for ZIKV infection. Primarily, the infection follows a self-limiting course, and its management involves rest and symptomatic relief. To mitigate the risk of dehydration, increased fluid intake is recommended, alongside the use of mild analgesics such as paracetamol/acetaminophen for alleviating pyrexia and pain. It is advisable to refrain from administering Non-steroidal Anti-Inflammatory Drugs (NSAIDs) and aspirin until the possibility of dengue infection is ruled out, as these medications may elevate the risk of haemorrhagic complications. Since aspirin is associated with Reye syndrome, it is not recommended in children with age < 10 years (Schrör 2007). Specific antiviral drugs against ZIKV are still under study and no drug is available commercially (Díaz-Menéndez and Crespillo-Andújar 2017).

7.13 Prevention and Control

Presently, there are no approved vaccines or antiviral drugs available for the treatment of ZIKV infection. A notable challenge in this regard is the development of a therapeutic agent for ZIKV infection that can be safely administered to expectant mothers. Nevertheless, a considerable number of compounds are undergoing in silico or in vitro assessments to determine their potential to interfere with the ZIKV replication cycle. However, only a limited subset of these compounds has demonstrated in vitro efficacy against ZIKV infection, warranting further evaluation through clinical trials.

The creation of candidate vaccines against the virus has been requested by the global health care system in response to the unexpected outbreak of ZIKV infection in Brazil. Making a low-cost, safe vaccination for pregnant women to receive is a significant problem in the development of a vaccine for the ZIKV, especially in developing nations where viral epidemics take place. Though only a small number of vaccination candidates are now undergoing phase I or II clinical trials, numerous vaccine subtypes and techniques are currently being developed and tested for their non-toxicity and efficacy (Ferraris et al. 2019).

Intense medical research is still being done on the virus especially after the ZIKV outbreak in Brazil. Numerous studies have made it possible to better understand the infection, which has aided in the development of vector control methods as well as antiviral medications and vaccinations that are now being tested in clinical settings. However, to effectively control the dissemination of the ZIKV, the majority of the issues, including vector control, diagnostics, and newborn brain damage must be addressed. It is still difficult to understand the symptoms caused by the co-circulation of the ZIKV with other arboviruses like Dengue virus and Chikungunya virus. But because both viruses share the same vector, it is crucial to keep working on vector control measures. These include reducing mosquito populations by preventing or impeding breeding (Hajra et al. 2016) and safeguarding the neighbourhood from the Aedes vector population that is afflicted. Mechanical like the elimination of all water-holding objects/Ovitraps, chemicals like pyrethroids/organochlorine/organophosphorus/Insect growth regulators (IGR), and biological controls like the release of sterile male Aedes/biocontrol using bacteria/fungus/plant/fish, should be used to effectively manage mosquito populations (Malone et al. 2016; Asad and Carpenter 2018; Singh et al. 2018).

The primary method by which the ZIKV spreads is by vector-borne transmission but there are non-vector-borne transmissions as-well. So, to construct the control methods against the propagation of this virus, a number of additional (non-vector borne) channels must be taken into account. The general population should be informed about the challenges associated with health complications, enabling the adaptation of basic sanitation practices to effectively eradicate breeding sites for vector larvae. Ensuring safe sexual practices is imperative for preventing viral transmission, with sexual activity discouraged for 6 months following the onset of symptoms in male partners or upon diagnosis (Mead et al. 2017). Individuals planning pregnancy after travelling to an endemic area should observe a waiting period of at least 28 days, accounting for a 2-week incubation phase and an additional 2-week interval beyond the viremic period. Visitors originating from these regions necessitate symptom monitoring for a minimum of 1 month (Krow-Lucal et al. 2017). In ZIKV-endemic nations, parental care and the use of contraceptives should be offered to reduce the prevalence of ZIKV-related microcephaly (Sharma and Lal 2017).

In addition, blood transfusions can spread ZIKV if proper preventive measures are not performed. As a result, pasteurisation, solvent/detergent treatment, and virus removal employing 40 nm pore size filters can all efficiently lower load of virus in medicines made from plasma (Kühnel et al. 2017). Avoiding Aedes mosquito bites during the first week after ZIKV infection is advised. Encouragement should be given to using insect repellents. Travellers to endemic regions must be thoroughly informed about the usage of mosquito nets and insect repellents while they are away from home. Due to the possibility of neurological complications in newborns, pregnant women should stay away from ZIKV infection endemic places. To minimise the introduction of ZIKV from endemic countries, it is imperative to maintain a heightened level of surveillance and monitoring at airports, harbours, and other points of entry (Marano et al. 2016).

In conclusion, it is essential to incorporate the lessons learned from the ZIKV outbreak in order to be appropriately prepared for the advent of emerging arboviruses in the vector's current circulation (Venter 2018).

7.14 ZIKV Future Perspective

ZIKV will persist in spreading to remote areas until new vaccines, antiviral drugs, and vector control strategies are implemented. As of 2020, low-level ZIKV circulation is observed in various US states and territories (CDC 2020). In India, ZIKV occurrences have been reported in 16 states and union territories between 2017 and 2021 (Yadav et al. 2022a, b). Thus, ongoing viral surveillance throughout quiet times may result in the early detection and containment of any approaching outbreak.

Following are the queries that need to be addressed:

1. Scientific investigations into the effects of ZIKV Infection on mouse models with neurological conditions, including GBS like in the study conducted by Silva-Filho et al. 2021, need to be expanded. Taking care of this problem could aid in the creation of diagnostic early warning systems and create new opportunities for the development of potent treatments for severe problems.
2. Recently, the idea of the ZIKV congenital syndrome has gained recognition. When intracranial calcification and reduced brain volume are present, the ZIKV congenital syndrome is present, and the baby may be born with microcephaly or develop it later (Teixeira et al. 2020). This necessitates a more comprehensive comprehension of the immunopathogenic mechanisms triggered by ZIKV that underlie microcephaly and other congenital abnormalities in developing foetuses.
3. Mandatory prenatal testing for expectant mothers and evaluation of the ZIKV presence in a woman's physique subsequent to symptom onset in order to establish when it is safe to get pregnant. This will definitely lessen the socioeconomic impact in ZIKV-endemic nations.
4. The development of safe vaccines, especially for pregnant women, and the discovery of breakthrough treatments to stop the difficulties brought on by this newly emerging virus.
5. In endemic nations, designing novel vector control measures are needed to avoid ZIKV infection.

References

Abushouk AI, Negida A, Ahmed H (2016) An updated review of Zika virus. J Clin Virol 84:53–58

Ades AE, Soriano-Arandes A, Alarcon A, Bonfante F, Thorne C, Peckham CS, Giaquinto C (2021) Vertical transmission of Zika virus and its outcomes: a Bayesian synthesis of prospective studies. Lancet Infect Dis 21(4):537–545

Agarwal A, Chaurasia D (2020) The expanding arms of Zika virus: an updated review with recent Indian outbreaks. Rev Med Virol 31(1):1–9. https://doi.org/10.1002/rmv.2145
Akiyama BM, Laurence HM, Massey AR, Costantino DA, Xie X, Yang Y et al (2016) Zika virus produces noncoding RNAs using a multi-pseudoknot structure that confounds a cellular exonuclease. Science 354:1148e52. https://doi.org/10.1126/science.aah3963
Aldo P, You Y, Szigeti K, Horvath TL, Lindenbach B, Mor G (2016) HSV-2 enhances ZIKV infection of the placenta and induces apoptosis in first-trimester trophoblast cells. Am J Reprod Immunol 76(5):348–357
Arzuza-Ortega L, Polo A, Pérez-Tatis G, López-García H, Parra E, Pardo-Herrera LC et al (2016) Fatal sickle cell disease and Zika virus infection in girl from Colombia. Emerg Infect Dis 22(5):925–927
Asad H, Carpenter D (2018) Effects of climate change on the spread of zika virus: a public health threat. Rev Environ Health 33(1):31–42. https://doi.org/10.1515/reveh-2017-0042
Barjas-Castro ML, Angerami RN, Cunha MS, Suzuki A, Nogueira JS, Rocco IM et al (2016) Probable transfusion-transmitted Zika virus in Brazil. Transfusion 56(7):1684–1688
Barzon L, Percivalle E, Pacenti M, Rovida F, Zavattoni M, Del Bravo P, Cattelan AM, Palù G, Baldanti F (2018) Virus and antibody dynamics in Travelers with acute Zika virus infection Clin infect dis 66(8):1173–1180 https://doi.org/10.1093/cid/cix967
Bayer A, Lennemann NJ, Ouyang Y, Bramley JC, Morosky S, Marques ETDA et al (2016) Type III interferons produced by human placental trophoblasts confer protection against Zika virus infection. Cell Host Microbe 19:705e12. https://doi.org/10.1016/j.chom.2016.03.008
Besnard M, Lastere S, Teissier A, Cao-Lormeau V, Musso D (2014) Evidence of perinatal transmission of Zika virus, French Polynesia, December 2013 and February 2014. Euro Surveill 19(13):20751
Brasil P, Pereira JP Jr, Moreira ME, Ribeiro Nogueira RM, Damasceno L, Wakimoto M et al (2016) Zika virus infection in pregnant women in Rio de Janeiro. N Engl J Med 375(24):2321–2334
Briant L, Desprès P, Choumet V, Missé D (2014) Role of skin immune cells on the host susceptibility to mosquito-borne viruses. Virol 464–465:26–32
Caine EA, Scheaffer SM, Arora N, Zaitsev K, Artyomov MN, Coyne CB et al (2019) Interferon lambda protects the female reproductive tract against Zika virus infection. Nat Commun 10:280. https://doi.org/10.1038/s41467-018-07993-2
CDC (2019). https://www.cdc.gov/zika/about/overview.html
CDC (2022) Zika travel information. https://wwwnc.cdc.gov/travel/page/zika-information
Centre for Disease Prevention and Control 2020 Epidemiology update. http://www.cdc.gov/zika/reporting/2019-case-counts.html
Chouin-Carneiro T, Vega-Rua A, Vazeille M, Yebakima A, Girod R, Goindin D et al (2016) Differential susceptibilities of Aedes aegypti and Aedes albopictus from the Americas to Zika virus. PLoS Negl Trop Dis 10(3):e0004543
Deckard DT, Chung WM, Brooks JT, Smith JC, Woldai S, Hennessey M et al (2016) Male-to-male sexual transmission of Zika virus–Texas, January 2016. MMWR Morb Mortal Wkly Rep 65(14):372–374
Dejnirattisai W, Supasa P, Wongwiwat W, Rouvinski A, Barba-Spaeth G, Duangchinda T et al (2016) Dengue virus sero-cross-reactivity drives antibody-dependent enhancement of infection with Zika virus. Nat Immunol 17(9):1102–1108
Díaz-Menéndez M, Crespillo-Andújar C (2017) Zika virus infection: risk of spreading in Europe. Springer International Publishing
Dick GW, Zika virus. II. (1952) Pathogenicity and physical properties. Trans R Soc Trop Med Hyg 46:521–534
Dick GW, Kitchen SF, Haddow AJ, Zika virus. I. (1952) Isolations and serological specificity. Trans R Soc Trop Med Hyg 46:509–520. [PubMed: 12995440]
Donald CL, Brennan B, Cumberworth SL, Rezelj VV, Clark JJ, Cordeiro MT et al (2016) Full genome sequence and sfRNA interferon antagonist activity of Zika virus from Recife, Brazil. PLoS Neglected Trop Dis 10:e0005048. https://doi.org/10.1371/journal.pntd.0005048

Driggers RW, Ho C-Y, Korhonen EM, Kuivanen S, Jääskeläinen AJ, Smura T et al (2016) Zika virus infection with prolonged maternal viremia and fetal brain abnormalities. N Engl J Med 374(22):2142–2151

Dudley DM, Aliota MT, Mohr EL, Weiler AM, Lehrer-Brey G, Weisgrau KL et al (2016) A rhesus macaque model of Asian-lineage Zika virus infection. Nat Commun 28(7):12204

Duffy MR, Chen TH, Hancock WT, Powers AM, Kool JL, Lanciotti RS, Pretrick M, Marfel M, Holzbauer S, Dubray C, Guillaumot L, Griggs A, Bel M, Lambert AJ, Laven J, Kosoy O, Panella A, Biggerstaff BJ, Fischer M, Hayes EB (2009) Zika virus outbreak on Yap Island, Federated States of Micronesia. N Engl J med 360(24):2536–2543. https://doi.org/10.1056/NEJMoa0805715

Ferraris P, Yssel H, Missé D (2019) Zika virus infection: an update. Microbes Infect 21(8–9):353–360. https://doi.org/10.1016/j.micinf.2019.04.005

Filipe AR, Martins CM, Rocha H (1973) Laboratory infection with Zika virus after vaccination against yellow fever. Arch Gesamte Virusforsch 43(4):315–319

Fowler AM, Tang WW, Young MP, Mamidi A, Viramontes KM, McCauley MD et al (2018) Maternally acquired Zika antibodies enhance dengue disease severity in mice. Cell Host Microbe 24:743e50. https://doi.org/10.1016/j.chom.2018.09.015.e5

Foy BD, Kobylinski KC, Foy JLC, Blitvich BJ (2011) Travassos da Rosa a, Haddow AD, Lanciotti RS, Tesh RB: probable non–vector-borne transmission of Zika virus, Colorado USA. Emerg Infect Dis 17:5

Frumence E, Roche M, Krejbich-Trotot P, El-Kalamouni C, Nativel B, Rondeau P et al (2016) The South Pacific epidemic strain of Zika virus replicates efficiently in human epithelial A549 cells leading to IFN-b production and apoptosis induction. Virology 493:217e26. https://doi.org/10.1016/j.virol.2016.03.006

Gallian P, Cabie A, Richard P, Paturel L, Charrel RN, Pastorino B, Leparc-Goffart I, Tiberghien P, de Lamballerie X (2017) Zika virus in asymptomatic blood donors in Martinique. Blood. The Journal of the American Society of Hematology 129(2):263–266

Gilmore EC, Walsh CA (2012) Genetic causes of microcephaly and lessons for neuronal development. WIREs Dev Biol 2(4):461–478

Goertz GP, Abbo SR, Fros JJ, Pijlman GP (2017) Functional RNA during Zika virus infection. Virus Res 254:41–53. https://doi.org/10.1016/j.virusres.2017.08.015

Gordon A, Gresh L, Ojeda S, Katzelnick LC, Sanchez N, Mercado JC et al (2019) Prior dengue virus infection and risk of Zika: a pediatric cohort in Nicaragua. PLoS Med 16:e1002726. https://doi.org/10.1371/journal.pmed.1002726

Gourinat A-C, O'Connor O, Calvez E, Goarant C, Dupont-Rouzeyrol M (2015) Detection of Zika virus in urine. Emerg Infect Dis 21(1):84–86

Grant A, Ponia SS, Tripathi S, Balasubramaniam V, Miorin L, Sourisseau M et al (2016) Zika virus targets human STAT2 to inhibit type I interferon signaling. Cell Host Microbe 19:882e90. https://doi.org/10.1016/j.chom.2016.05.009

Hajra A, Bandyopadhyay D, Hajra SK (2016) Zika virus: a global threat to humanity: a comprehensive review and current developments. N Am J Med Sci 8:123–128. https://doi.org/10.4103/1947-2714.179112

Halstead SB (2014) Dengue antibody-dependent enhancement: knowns and unknowns. Microbiol Spectr 2. https://doi.org/10.1128/microbiolspec.AID-0022-2014

Hamel R, Dejarnac O, Wichit S, Ekchariyawat P, Neyret A, Luplertlop N et al (2015) Biology of Zika virus infection in human skin cells. J Virol 89(17):8880–8896

Hamel R, Liegeois F, Wichit S, Pompon J, Diop F, Talignani L et al (2016) Zika virus: epidemiology, clinical features and host-virus interactions. Microb Infect 18:441e9. https://doi.org/10.1016/j.micinf.2016.03.009

Honein MA, Dawson AL, Petersen EE, Jones AM, Lee EH, Yazdy MM et al (2017) Birth defects among fetuses and infants of US women with evidence of possible Zika virus infection during pregnancy. JAMA 317(1):59

Hussain A, Ali F, Latiwesh OB, Hussain S (2018) A comprehensive review of the manifestations and pathogenesis of Zika virus in neonates and adults. Cureus 10(9):e3290

Iwamoto M, Jernigan DB, Guasch A, Trepka MJ, Blackmore CG, Hellinger WC et al (2003) Transmission of West Nile virus from an organ donor to four transplant recipients. N Engl J Med 348(22):2196–2203

Jagger BW, Miner JJ, Cao B, Arora N, Smith AM, Kovacs A et al (2017) Gestational stage and IFN-l signaling regulate ZIKV infection in utero. Cell Host Microbe 22:366e76. https://doi.org/10.1016/j.chom.2017.08.012.e3

Jimenez A, Shaz BH, Bloch EM (2017) Zika virus and the blood supply: what do we know? Transfus Med Rev 31(1):1–10

Karwowski MP, Nelson JM, Staples JE, Fischer M, Fleming-Dutra KE, Villanueva J et al (2016) Zika virus disease: a CDC update for pediatric health care providers. Pediatrics 137(5):e20160621

Khut QY (2016) Zika virus disease: Interim case definitions. World Health Organization, p 1. http://apps.who.int/iris/bitstream/10665/204381/1/WHO_ZIKV_SUR_16.1_eng.pdf?ua=1. Accessed 15 Jan 2017

Kleber de Oliveira W, Cortez-Escalante J, De Oliveira WTGH, do GMI C, CMP H, Coelho GE et al (2016) Increase in reported prevalence of microcephaly in infants born to women living in areas with confirmed Zika virus transmission during the first trimester of pregnancy - Brazil, 2015. MMWR Morb Mortal Wkly Rep 65(9):242–247

Krow-Lucal ER, Biggerstaff BJ, Staples JE (2017) Estimated incubation period for Zika virus disease. Emerg Infect Dis 23:841–844. https://doi.org/10.3201/eid2305.161715

Kühnel D, Müller S, Pichotta A, Radomski KU, Volk A, Schmidt T (2017) Inactivation of Zika virus by solvent/detergent treatment of human plasma and other plasma-derived products and pasteurization of human serum albumin. Transfusion 57:802–810. https://doi.org/10.1111/trf.13964

Lai Z, Zhou T, Zhou J, Liu S, Xu Y, Gu J et al (2020) Vertical transmission of Zika virus in Aedes albopictus. PLoS Negl Trop Dis 14(10):e0008776. https://doi.org/10.1371/journal.pntd.0008776

Lanciotti RS, Kosoy OL, Laven JJ, Velez JO, Lambert AJ, Johnson AJ, Stanfield SM, Duffy MR (2008) Genetic and serologic properties of Zika virus associated with an epidemic, yap state, Micronesia, 2007. Emerg Infect Dis 14(8):1232–1239. https://doi.org/10.3201/eid1408.080287

Lazear HM, Govero J, Smith AM, Platt DJ, Fernandez E, Miner JJ et al (2016) A mouse model of Zika virus pathogenesis. Cell Host Microbe 19:720e30. https://doi.org/10.1016/j.chom.2016.03.010

Li MI, Wong P-SJ, Ng L-C, Tan C-H (2012) Oral susceptibility of Singapore Aedes (Stegomyia) aegypti (Linnaeus) to Zika virus. PLoS Negl Trop Dis 6(8):e1792

Malone R, Homan J, Callahan M, Glasspool-Malone J, Damodaran L, Schneider Ade B et al (2016) ZIKA: medical countermeasure development challenges. PLoS Negl Trop Dis 10:e0004530. https://doi.org/10.1371/journal.pntd.0004530

Marano G, Pupella S, Vaglio S, Liumbruno GM, Grazzini G (2016) Zika virus and the never-ending story of emerging pathogens and transfusion medicine. Blood Transfus 14:95–100. https://doi.org/10.2450/2015.0066-15

Marchette NJ, Garcia R, Rudnick A (1969) Isolation of Zika virus from Aedes aegypti mosquitoes in Malaysia. Am J Trop Med Hyg18(3) 411–415. https://doi.org/10.4269/ajtmh.1969.18.411

Martin CR, Hollins-Martin C, Preedy VR, Rajendram R (eds) (2021) Zika virus impact, diagnosis, control, and models: volume 2: the neuroscience of Zika virus. Academic Press

Martins VEP, Alencar CH, Kamimura MT, de Carvalho Araújo FM, De Simone SG, Dutra RF et al (2012) Occurrence of natural vertical transmission of dengue-2 and dengue-3 viruses in Aedes aegypti and Aedes albopictus in Fortaleza, Ceará, Brazil. PLoS One 7(7):e41386

Mead PS, Hills SL, Brooks JT (2017) Zika virus as a sexually transmitted pathogen. Curr Opin Infect Dis 31:39–44. https://doi.org/10.1097/QCO.0000000000000414

Meaney-Delman D, Oduyebo T, Polen KND, White JL, Bingham AM, Slavinski SA et al (2016) Prolonged detection of Zika virus RNA in pregnant women 128(4):724–730

Morrison TE, Diamond MS (2017) Animal models of Zika virus infection, pathogenesis, and immunity. J Virol 91:e00009–e00017. https://doi.org/10.1128/JVI.00009-17

Olson JG, Ksiazek TG, Suhandiman T (1981) Zika virus, a cause of fever in Central Java, Indonesia. Trans R Soc Trop Med Hyg 75(3):389–393. https://doi.org/10.1016/0035-9203(81)90100-0. PMID: 6275577

Pan American Health Organization (PAHO), World Health Organization (WHO) (2017) Zika Suspected and Confirmed Cases Reported by Countries and Territories in the Americas Cumulative Cases, 2015–2017. Updated as of 4 January 2018. Washington: PAHO/ WHO. https://www.paho.org/hq/index.php?option=com_content&view=article&id=12390:zikacumulative-cases&Itemid=42090&lang=en

Pantoja P, Perez-Guzman EX, Rodríguez IV, White LJ, Gonzalez O, Serrano C et al (2017) Zika virus pathogenesis in rhesus macaques is unaffected by preexisting immunity to dengue virus. Nat Commun 8:15674. https://doi.org/10.1038/ncomms15674

Pasquier C, Joguet G, Mengelle C, Chapuy-Regaud S, Pavili L, Prisant N, Izopet J, Bujan L, Mansuy JM (2018) Kinetics of anti-ZIKV antibodies after Zika infection using two commercial enzyme-linked immunoassays. Diagn Microbiol Infect Dis 90(1):26–30. https://doi.org/10.1016/j.diagmicrobio.2017.09.001

Pettersson JHO, Fiz-Palacios O (2014a) Dating the origin of the genus Flavivirus in the light of Beringian biogeography. J Gen Virol 95(9):1969–1982

Platt DJ, Miner JJ (2017) Consequences of congenital Zika virus infection. Curr Opin Virol 27:1–7

Prince HE, Tobler LH, Yeh C, Gefter N, Custer B, Busch MP (2007) Persistence of West Nile virus-specific antibodies in viremic blood donors. Clin Vaccine Immunol 14(9):1228–1230

Prisant N, Bujan L, Benichou H, Hayot P-H, Pavili L, Lurel S et al (2016) Zika virus in the female genital tract. Lancet Infect Dis 16(9):1000–1001

Priyamvada L, Quicke KM, Hudson WH, Onlamoon N, Sewatanon J, Edupuganti S et al (2016) Human antibody responses after dengue virus infection are highly cross-reactive to Zika virus. Proc Natl Acad Sci Unit States Am 113:7852e7. https://doi.org/10.1073/pnas.1607931113

Regla-Nava JA, Elong Ngono A, Viramontes KM, Huynh A-T, Wang Y-T, Nguyen A-VT et al (2018) Cross-reactive dengue virus-specific $CD8^{+}$ T cells protect against Zika virus during pregnancy. Nat Commun 9:3042. https://doi.org/10.1038/s41467-018-05458-0

Rey FA, Stiasny K, Vaney M, Dellarole M, Heinz FX (2018) The bright and the dark side of human antibody responses to flaviviruses: lessons for vaccine design. EMBO Rep 19:206e24. https://doi.org/10.15252/embr.201745302

Rodriguez-Morales AJ, Bandeira AC, Franco-Paredes C (2016) The expanding spectrum of modes of transmission of Zika virus: a global concern. Ann Clin Microbiol Antimicrob 15(1):1–2

Roehrig JT, Nash D, Maldin B, Labowitz A, Martin DA, Lanciotti RS et al (2003) Persistence of virus-reactive serum immunoglobulin m antibody in confirmed West Nile virus encephalitis cases. Emerg Infect Dis 9(3):376–379

Rolph MS, Mahalingam S (2019) Zika's passage to India. Lancet Infect Dis 19:469–470

Rosenberg AZ, Yu W, Hill DA, Reyes CA, Schwartz DA (2017) Placental pathology of Zika virus: viral infection of the placenta induces villous stromal macrophage (Hofbauer cell) proliferation and hyperplasia. Arch Pathol Lab Med 141(1):43–48

Rossi SL, Tesh RB, Azar SR, Muruato AE, Hanley KA, Auguste AJ et al (2016) Characterization of a novel murine model to study Zika virus. Am J Trop Med Hyg 94:1362e9. https://doi.org/10.4269/ajtmh.16-0111

Russell K, Oliver SE, Lewis L, Barfifield WD, Cragan J, Meaney-Delman D et al (2016) Update: interim guidance for the evaluation and management of infants with possible congenital Zika virus infection—United States, august 2016. MMWR Morb Mortal Wkly Rep 65(33):870–878

Sapkal G, Yadav P, Vegad M, Viswanathan R, Gupta N, Mourya D (2018) First laboratory confirmation on the existence of Zika virus disease in India. J Infect 76(3):314–317. https://doi.org/10.1016/j.jinf.2017.09.020

Sarmiento-Ospina A, Vásquez-Serna H, Jimenez-Canizales CE, Villamil-Gómez WE, Rodriguez-Morales AJ (2016) Zika virus associated deaths in Colombia. Lancet Infect Dis 16(5):523–524

Schrör K (2007) Aspirin and Reye syndrome: a review of the evidence. Paediatr Drugs 9(3):195–204

Sharma A, Lal SK (2017) Zika virus: transmission, detection, control, and prevention. Front Microbiol 8:110. https://doi.org/10.3389/fmicb.2017.00110

Silva-Filho J, de Oliveira L, Monteiro L, Parise P, Zanluqui N, Polonio C et al (2021) Gas6 drives Zika virus-induced neurological complications in humans and congenital syndrome in immunocompetent mice. Brain Behav Immun 97:260–274. https://doi.org/10.1016/j.bbi.2021.08.008

Singh H, Singh OP, Akhtar N, Sharma G, Sindhania A, Gupta N et al (2019a) First report on the transmission of Zika virus by Aedes (Stegomyia) aegypti (L.) (Diptera: culicidae) during the 2018 Zika outbreak in India. Acta Trop 199:105114

Singh R, Dhama K, Khandia R, Munjal A, Karthik K, Tiwari R et al (2018) Prevention and control strategies to counter Zika virus, a special focus on intervention approaches against vector mosquitoes—current updates. Front Microbiol 9:87. https://doi.org/10.3389/fmicb.2018.00087

Singh R, Gupta V, Malhotra B, Singh S, Ravindran P, Meena D et al (2019b) Cluster containment strategy: addressing Zika virus outbreak in Rajasthan. India BMJ Glob Health 4:e001383

Sirohi D, Kuhn RJ (2017) Zika virus structure, maturation, and receptors. J Infect Dis 216(suppl_10):S935–S944

Sirohi D, Chen Z, Sun L, Klose T, Pierson TC, Rossmann MG et al (2016) The 3.8 a resolution cryo-EM structure of Zika virus. Sci 352(6284):467–470

Solomon IH, Milner DA (2016) Neuropathology of Zika virus infection. J Neuroinfect Dis 7(2):220

Sow A, Loucoubar C, Diallo D, Faye O, Ndiaye Y, Senghor CS et al (2016) Concurrent malaria and arbovirus infections in Kedougou, southeastern Senegal. Malaria J BioMed Central 15:147

Tang H, Hammack C, Ogden SC, Wen Z, Qian X, Li Y et al (2016) Zika virus infects human cortical neural progenitors and attenuates their growth Stem. Cell 18(5):587–590

Teixeira GA et al (2020) Analysis of the concept of the Zika virus congenital syndrome/Analise do conceito sindrome congenita pelo Zika virus. Ciênc Saúde Colet 25(2):567+. Gale OneFile: Informe Académico, link.gale.com/apps/doc/A618566156/IFME?u=anon~116c4fdb&sid=googleScholar&xid=3d82a4e6

Tripathi S, Balasubramaniam VRMT, Brown JA, Mena I, Grant A, Bardina SV et al (2017) A novel Zika virus mouse model reveals strain specific differences in virus pathogenesis and host inflammatory immune responses. PLoS Pathog 13:e1006258. https://doi.org/10.1371/journal.ppat.1006258

Venter M (2018) Assessing the zoonotic potential of arboviruses of African origin. Curr Open Virol 28:74–84

WHO (2017a) Situation report. Zika virus microcephaly Guillain-Barré syndrome. Data as of 9 March 2017. http://reliefweb.int/report/world/zika-virus-microcephaly-and-guillain-barr-syndrome-situation-report-10-march-2017

WHO (2017b) Zika Virus Infection—India. Disease Outbreak News. http://www.who.int/csr/don/26-may-2017-zika-ind/en/. Accessed 19 Nov 2018

WHO (2019) Countries and territories with current or previous Zika virus transmission. http://www.who.int/emergencies/diseases/zika/countries-with-zika-and-vectors-table. Replace with https://worldpopulationreview.com/country-rankings/zika-countries

Xia H, Luo H, Shan C, Muruato AE, Nunes BTD, Medeiros DBA et al (2018) An evolutionary NS1 mutation enhances Zika virus evasion of host interferon induction. Nat Commun 9:414. https://doi.org/10.1038/s41467-017-02816-2

Yadav PD, Niyas VKM, Arjun R, Sahay RR, Shete AM, Sapkal GN et al (2022a) Detection of Zika virus disease in Thiruvananthapuram, Kerala, India 2021 during the second wave of COVID-19 pandemic. J Med Virol 94(6):2346–2349. https://doi.org/10.1002/jmv.27638

Yadav P, Kaur H, Gupta N, Sahay R, Sapkal G, Shete A et al (2022b) Zika a vector borne disease detected in newer states of India amidst the COVID-19 pandemic. Front Microbiol 13:888195. https://doi.org/10.3389/fmicb.2022.888195

Ye Q, Liu Z-Y, Han J-F, Jiang T, Li X-F, Qin C-F (2016) Genomic characterization and phylogenetic analysis of Zika virus circulating in the Americas. Infect Genet Evol 43:43–49

Zonneveld R, Roosblad J, Staveren JWV, Wilschut JC, Vreden SGS, Codrington J (2016) Three atypical lethal cases associated with acute Zika virus infection in Suriname. IDCases 5:49–53

Enteroviruses Associated with Acute Flaccid Paralysis

8

Rongala Laxmivandana, Sarah Cherian, and Shobha D. Chitambar

Abstract

Enteroviruses (EVs) are positive-sense RNA viruses, known to cause a variable spectrum of infections in humans from subclinical to fatal, implicated in the classic manifestation, acute flaccid paralysis (AFP). Severity of pathological conditions associated with EV infections has been described in part due to host genetics/immune status or the viral load during infection or the viral genetics. The AFP cases that are poliovirus-free have been reported during Polio Surveillance Programs (PSPs) carried out worldwide for eradication of polio. Non-polio EVs (NPEVs) have been frequently isolated from such cases in the PSPs. This chapter discusses the virological, molecular, and epidemiological characteristics of EV infections associated with AFP.

8.1 Discovery of EVs

In 1789, the poliomyelitis disease was first described as debility of lower limbs in children by Michael Underwood, a physician in England (Paul 1971). No epidemics of the disease were observed till the second-half of nineteenth century, but during summer and fall of the first-half of twentieth century, several countries in the world have suffered severe epidemics of the disease (Paul 1971). In 1909, the virus that

R. Laxmivandana (✉)
ICMR-National Institute of Virology (ICMR-NIV), Pune, India
e-mail: vandanajai@yahoo.co.in

S. Cherian
Bioinformatics Group, ICMR-NIV, Pune, India

S. D. Chitambar
Enteric Viruses Group, ICMR-NIV, Pune, India

P. L. Bhukya, A. S (eds.), *Emerging Human Viral Diseases, Volume II*,
https://doi.org/10.1007/978-981-97-4480-0_8

causes poliomyelitis disease was first identified by Karl Landsteiner, an immunologist in Austria (Bradshaw 1989). Following introduction of polio vaccines in the 1950s, the incidence of polio cases declined rapidly (Paul 1971).

In 1948, when suckling mice were inoculated with fecal suspensions of patients, suspected with polio, from Coxsackie, New York, paralysis was manifested and the discovery of new viruses similar to polioviruses (PV) became evident (Dalldorf and Sickles 1948). A virus isolated in the cell culture differed antigenically from PVs and from several other neurotropic viruses (Dalldorf and Sickles 1948). The isolates of the virus were subsequently referred to as the first prototypes of the group A coxsackieviruses (CVs).

Clearly different from the original group A CV (CV-A) isolates that produced a flaccid paralysis and myositis in skeletal muscle in new born mice, some new virus isolates obtained from patients in atypical polio epidemics were found to produce a generalized infection and spastic paralysis in the animals, manifesting myositis in striated muscles with the involvement of the heart, brain, pancreas and brown fat. Later, these virus isolates were referred to as the first prototypes of the group B CVs (CV-B) (Melnick et al. 1949).

Different serotypes of PVs and CVs were unravelled and confirmed by in vivo studies and in vitro neutralization assays (Melnick et al. 1950).

During 1950s, while working with PVs and CVs, cell culture methods have allowed the virologists to detect previously unknown, antigenically independent viruses, which did not show disease symptoms in lab animals. These viruses were initially named as "orphan viruses," i.e., viruses searching for a parent disease and later named as: enteric cytopathogenic human orphan (ECHO) viruses (Robbins et al. 1951). However, the studies conducted subsequently have shown that echoviruses discovered earlier caused a variety of diseases in humans (Melnick 1990).

One type may cause any of several syndromes such as encephalitis, meningitis, pleurodynia, hand-foot-and-mouth disease (HFMD), herpangina, myocarditis, acute hemorrhagic conjunctivitis (AHC), minor malaise, undifferentiated febrile illnesses, common cold, pneumonitis, bronchiolitis, pneumonia, and other upper respiratory illnesses. These viruses are also implicated in paralytic disease (Melnick 1990).

Further, these viruses (PVs, CVs, and echoviruses) were found to share structural, physicochemical, and genetic characteristics, besides the habitat of enteric tract in humans. Later, these viruses received the recognition as a distinct genus classifying them as enteroviruses (EVs). The new serotypes identified thereafter were named with successive numbers such as EV-68, EV-69, EV-70, and EV-71 (Oberste et al. 1999a, b).

8.2 Characteristic Properties, Virion Structure and Genome Organization of EVs

The general epidemiology, transmission, and pathogenesis of EVs are influenced by their bio-physical properties (Table 8.1)(Pallansch and Roos 2007).

Table 8.1 Physicochemical properties of EVs (Pallansch and Roos 2007)

Characteristic	Properties
Sedimentation	• Constant buoyant density in cesium chloride at varied pH or centrifugation times
Resistance	Against • Weak acid ($pH < 3.0$) • Lipid solvents such as chloroform and ether • Laboratory disinfectants such as isopropanol, 70% ethanol, quaternary ammonium compounds, lysol • Detergents at ambient temperature
Susceptible	To • Environment (drying on surfaces, temperatures >42 °C, ultraviolet light) • Chemicals (glutaraldehyde, formaldehyde, sodium hypochlorite, strong acid, free residual chlorine)

Since the structure of PV1 was characterized in 1985 by X-ray crystallography (Hogle et al. 1985), many other EV structures have also been determined. The virus particles are spherical, 27–30 nm in diameter, contains RNA as genome which is enclosed by proteins. The lack of a lipid envelope in these particles renders them resistant to organic solvents. The viral capsid contains four proteins—VP1, VP2, VP3, and VP4 that are arranged in icosahedral symmetry (12 vertices and 20 triangular faces). The capsid building block, the protomer, comprises one copy each of the four structural proteins. A total of 60 structural proteins are involved in the formation of the capsid. No sequence homology has been found among VP1, VP2, and VP3, yet all the three structural proteins have same topology, i.e., they share the outer capsid surface. The VP4 lies on the inner side of capsid surface, and plays a role in the stabilization of the virion (Chow et al. 1987) (Fig. 8.1).

The 3D structure of the PV particle reveals that the C-terminus of the proteins and the connecting loops of the capsid lie on the outer surface of the virion, which contains the major antigenic neutralization sites and hence determines the viral serotype. The interaction of conserved VP4 residues with the RNA genome supports the stabilization of the capsid (Chow et al. 1987; Hogle et al. 1985).

The genome of the EVs is a single-stranded and positive-sense RNA molecule, linked covalently to VPg (22-24aa virion protein) (Fig. 8.2). The genome length varies from nt 7209 to 7450 and consists of the 5′ NCR (~7–7.5% of the genome), a single long open reading frame (ORF) (~91.5–92% of the genome), and the 3′ NCR (~0.5–1% of the genome) regions (Pallansch and Roos 2007).

5'NCR is a complex secondary structure with domains (I–VII), divided into two parts: (i) 5′-end cloverleaf (~100 bp): a cis-acting element, involved in viral RNA replication and regulation of the initiation of translation and (ii) internal ribosome entry site (IRES) (~550 bp): mediates cap-independent translation of the viral RNA (Fig. 8.2).

ORF is divided into P1, P2, and P3 genomic regions which encode the structural/capsid and non-structural proteins. It is cleaved by the virus encoded proteases, during translation. The P1 genomic region translates into the viral capsid proteins while

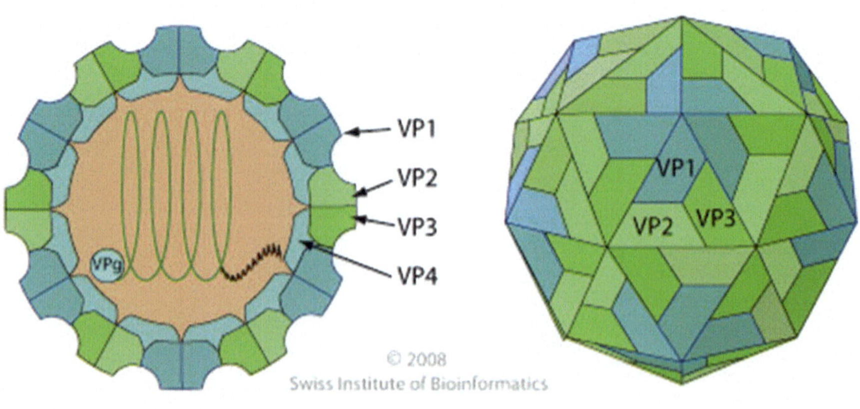

Fig. 8.1 Virion structure (www.viralzone.expasy.org)

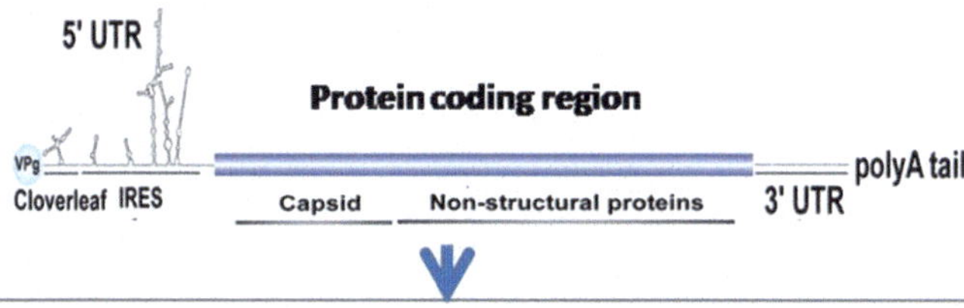

Protein products	Major Functions
VP1-VP4	Capsid components and receptor binding (VP1)
2A	Viral protease, shuts off protein synthesis
2B	Increases membrane permeability, inhibits cellular secretary pathways
2C	Vesicle formation, NTPase
3A	Inhibits intercellular transport
3B	Vpg, primes RNA synthesis
3C	Viral protease, inhibits host transcription
3D	Viral polymerase

5'UTR: Involves in replication and translation

3'UTR: Putative binding site for replication complex

Fig. 8.2 EV genome: the protein products and major functions

the P2 and P3 genomic regions translate into non-structural proteins that are involved in protein processing (2A, 3C, and 3CD) and genome replication (2B, 2C, 3AB, 3B, 3CD, and 3D) (Fig. 8.2).

3'NCR is poly-adenylated, contains two hairpin like secondary structures, and plays a role in RNA synthesis (Fig. 8.2).

8.3 EV Classification

EVs have been classified under the genus Enterovirus, the order Picornavirales and the family Picornaviridae. Further, based on the pathogenicity in humans and experimental animals, EVs have been sub-classified into four groups, PVs, CV-A, CV-B, and echoviruses (E) (Minor et al. 1995) (Table 8.2).

Molecular typing methods have allowed the development of modern/genetic classification scheme. Under this scheme, human EVs have been sub-divided into four different species EV-A, EV-B, EV-C, and EV-D, supported by biologic properties, genome organization, and sequence similarity (Table 8.3). Incorporation of PVs in the EV-C species is approved in the eighth report of International Committee on Taxonomy of Viruses (ICTV) (Stanway et al. 2005).

Table 8.2 Traditional biologic classification of EVs based on host range (Pallansch and Roos 2007)

	Experimental host range		Cell culture
Subgroup	Non-human primates	New born mice	
Polioviruses	+	–	+
Coxsackieviruses, group A	–	+	±
Coxsackieviruses, group B	–	+	+
Echoviruses	–	–	+
Newer/numbered enteroviruses	Variable	Variable	+

Table 8.3 Modern genetic classification of EVs (https://ictv.global/report/chapter/picornaviridae/picornaviridae/enterovirus)

Species	Sero/geno types
EV-A Consists of 25 types	CV-A2, CV-A3, CV-A4, CV-A5, CV-A6, CV-A7, CV-A8, CV-A10, CV-A12, CV-A14, CV-A16, EV-A71, EV-A76, EV-A89, EV-A90, EV-A91, EV-A92, EV-A114, EV-A119, EV-A120, EV-A121, the simian enteroviruses SV19, SV43, SV46, and the baboon enterovirus A13 (BA13) (contains most of the viruses of CVA group)
EV-B Consists of 63 types	CV-B1, CV-B2, CV-B3, CV-B4, CV-B5 (incl. Swine vesicular disease virus [SVDV]), CV-B6, CV-A9, E-1 (incl. E-8), E-2, E-3, E-4, E-5, E-6, E-7, E-9 (incl. CV-A23), E-11, E-12, E-13, E-14, E-15, E-16, E-17, E-18, E-19, E-20, E-21, E-24, E-25, E-26, E-27, E-29, E-30, E-31, E-32, E-33, EV-B69, EV-B73, EV-B74, EV-B75, EV-B77, EV-B78, EV-B79, EV-B80, EV-B81, EV-B82, EV-B83, EV-B84, EV-B85, EV-B86, EV-B87, EV-B88, EV-B93, EV-B97, EV-B98, EV-B100, EV-B101, EV-B106, EV-B107, EV-B110, EV-B111, EV-B112, EV-B113, and the simian enterovirus SA5 (contains all viruses of CVB group and all echoviruses)
EV-C Consists of 23 types	PV-1, PV-2, PV-3, CV-A1, CV-A11 (incl. CV-A15), CV-A13 (inl.CV-A18), CV-A17, CV-A19, CV-A20, CV-A21, CV-A22, CV-A24, EV-C95, EV-C96, EV-C99, EV-C102, EV-C104, EV-C105, EV-C109, EV-C113, EV-C116, EV-C117, and EV-C118 (contains some of the viruses of CVA group and the PVs)
EV-D Consists of 5 types	EV-D68, EV-D70, EV-D94, EV-D111, and EV-D120

As per the genetic criteria proposed by ICTV (Table 8.4) (Stanway et al. 2005), EVs discovered since last two decades, have been phylogenetically grouped into their species instead of being antigenically characterized (Junttila et al. 2007; Oberste et al. 2005).

8.4 Cellular Receptors for EVs

One of the prominent characteristics of EVs is to cause cytolysis during their growth in cell culture. However, some EVs were found to establish latent/dormant infections occasionally in susceptible cells (Pallansch and Roos 2007). At the

Table 8.4 Criteria for species and genotype classification (Stanway et al. 2005)

Taxonomic level	Criteria
EV species (ICTV)	Members of a species of the genus EV share: • A limited natural host range • A limited range of host cell receptors • Greater than 70% nt identity in P1 and greater than 70% aa identity in the non-structural proteins 2C + 3CD • Similar G + C composition in the genome (varies <2.5% • A significant degree of compatibility in replication, proteolytic processing, encapsidation, and recombination
EV type (proposed)	Members of a same genotype of the species of EV • Share >75% nucleotide similarity and > 85% amino acid similarity in the VP1-coding sequence Members of different genotypes of the species of EV • Share nucleotide similarity <70% and amino acid similarity <85% in the VP1-coding sequence

microscopic level, infection usually sets within 1–7 days by the appearance of a characteristic cytopathic effect (CPE), which features visible rounding, shrinking, nuclear pyknosis, refractivity, and cell degeneration. All the identified EVs can be propagated in either suckling mice or cell culture. Many EV types can be isolated in at least one or more human or primate derived continuous cell cultures (Pallansch and Roos 2007).

EVs initiate infection of cells by first attaching to a receptor on the host cell membrane. The viral attachment proteins/host cell receptors are the important determinants of host range and tissue tropism of the viral agent (Evans & Almond 1998; Nasri et al. 2007). EVs are known to use a wide array of receptors (Table 8.5 and Fig. 8.3) and entry mechanisms to invade the host cell. However, the mode of entries for many EV types is yet to be elucidated. For some EV types (e.g., PV), a single type of receptor is sufficient for entry of viruses into cells. For other, a second molecule, or co-receptor, is needed for entry of virus into cells. For example, CV-A21, which attaches to the cluster of differentiation (CD)-55, requires intercellular adhesion molecule 1 (ICAM-1) for entry into cells (Shafren et al. 1997). There has been no specificity found in the usage of receptors among the types belonging to the same EV species (Table 8.5). Nevertheless, it would be useful to classify the EV types based on the mode of cell entry by determining the capsid sequences that interact with host cell receptors (Nasri et al. 2007).

The variation in surface architecture of the EV capsids accounts not only for the diverse serotypes but also for the mode of interaction with the different cell receptors. Models of the interaction of different EV types with their cellular receptors have been developed by the use of cryoelectron microscopy and X-ray crystallographic techniques (Belnap et al. 2000; Casasnovas et al. 1998; He et al. 2000). These models reveal that the canyon of the EV particle interacts with domain 1 of the cellular receptor during penetration of the virus into the host cell.

The attachment of EV to the receptor implicates endocytosis of the virion. Due to the endocytic environment, EVs were identified to undergo a permanent

Table 8.5 Cell surface receptors identified for EVs (Nilsson et al. 2008; Nishimura et al. 2013; Pallansch and Roos 2007)

Receptor	EV-species (type)
Immunoglobulin (Ig) G-super family	
– Poliovirus receptor (PVR), CD155 – Intercellular adhesion molecule-1 (ICAM-1), CD54 – Coxsackievirus-adenovirus receptor (CAR)	EV-C (PV-1,2,3) EV-C (CV-A13, A21) EV-B (CV-B1–B6)
Short consensus repeat (SCR) - decay accelerating factor (DAF), CD55	EV-B (E-3, 6, 7, 11–13, 20, 21, 24, 29, 30; CV-B1, B3, B5[a]) EV-C (CV-A21[a]) EV-D (EV-70)
Integrins	
– α2β1 integrin – αvβ3 – αvβ6 integrin	EV-B (E-1) EV-B (CV-A9) EV-B (CV-A9; CV-B1, B3, B5)
Carbohydrates	
– sialic acid	EV-C (CV-A24) EV-D (EV-70[a])
Glycosaminoglycans (GAG)	
– Heparan sulphate	EV-B (E-6)
β2 microglobulin (β2m)	EV-B (CV-A9[a]; E-3, 6, 7,11–13, 20, 21, 24, 29, 33)
P-selectin glycoprotein ligand-1 (PSGL-1) and scavenger receptor class B member 2 (SCARB2)	EV-A (CV-A16, EV–71)

[a]Indicates virus that uses the receptor as secondary or co-receptor

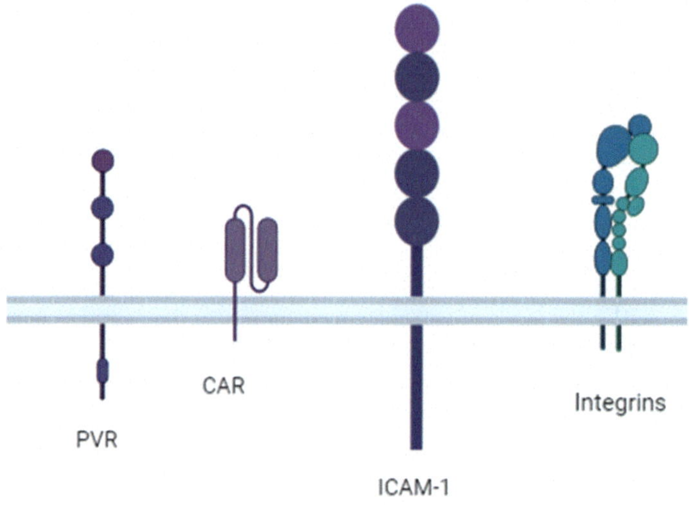

Fig. 8.3 Different types of EV receptors

reorganization of the capsid, characterized by a shift from a 160S sediment particle to a 135S particle. The 135S particle was found to lose the majority of VP4, and alteration of the antigenicity and protease sensitivity was noted in comparison to the native virion (De Sena and Mandel 1977). One of the specific alterations is the accessibility of the N terminus of VP1 of EVs, for further binding with the cellular receptor (Fricks and Hogle 1990). This 135S altered particle (A particle) has been considered as an essential intermediate in initiating a new infectious cycle (Lonberg-Holm et al. 1975).

8.5 EV Replication

As a result of receptor-mediated endocytosis, the internal capsid protein VP4 is released and the N-terminal of VP1 domain of altered EV particles or A-particles is exposed and inserted into membrane of the cell to form a pore through which the viral RNA can travel to the cytoplasm (Lonberg-Holm et al. 1975).

EV virion RNA has a VPg protein at its 5′ end instead of the methylated cap structure found in cellular mRNA. The viral mRNA is highly structured at its 5′ end, by virtue of intra molecular base pairing. The IRES structure near the beginning of the translational reading frame mediates association of the ribosomes with viral genome (Hahn and Palmenberg 1995).

Upon successful initiation of infection, the 5’ VPg (encoded by the virus) is cleaved off the viral genomic RNA by a cellular enzyme called “unlinkase”(Ambros et al. 1978). The viral genomic mRNA translates into a single protein, i.e., the precursor to all viral proteins. This is cleaved before translation is complete (i.e., on the nascent (=growing) chain). The cleavage is carried out by virally coded proteases (e.g., 2A and 3C) (Hellen et al. 1991). These proteases can also function while in the polyprotein, that is before they are cleaved (Fig. 8.4).

Cell lysis is the obvious cytopathology of replication in EVs. Prior to this, the virus inhibits protein synthesis in host cell which involves proteolysis of the

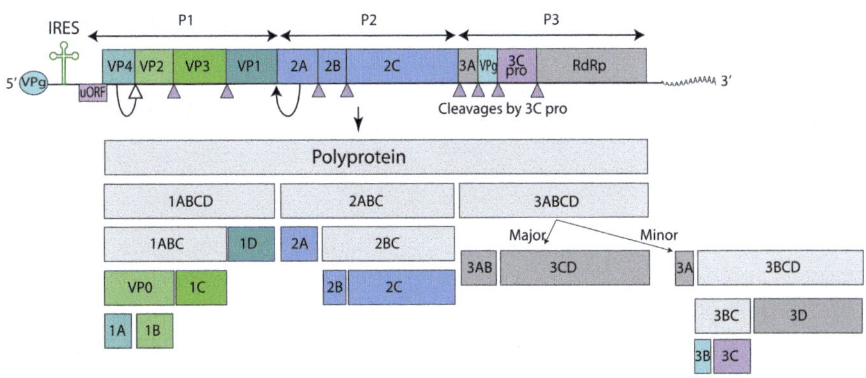

Fig. 8.4 EV genetic map and expression of proteins (www.viralzone.expasy.org)

eukaryotic translation initiation factor (eIF)-4G that renders lack of recognition of capped mRNA by ribosomes and translation of only uncapped EV mRNA as its IRES allows assembly of the translation complex with the ribosomes. The eIF3 40S complex or translation complex is recruited to the RNA by the interaction of eIF4A and cleaved eIF4G with the IRES (Hahn and Palmenberg 1995).

The newly formed viral proteins support the RNA replication, which occurs in membranous vesicle synthesized by viral proteins (2 BC) (Fig. 8.5). The viral RNA polymerase/replicase (3D) copies positive-sense RNA genome into complementary negative-sense RNA genome which also needs VPg that acts as a primer for synthesis of RNA and certain host proteins such as poly(rC) binding protein (PCBP), 3′ poly(A) tract RNA-binding protein, and heterogeneous ribonuclear protein C (hnRNP C). These cellular proteins interact with EV RNA, possibly stabilize the RNA and recruit non-structural viral proteins to sites of RNA replication. The well conserved cloverleaf region of EV 5'NCR has been described to regulate EV RNA replication in conjunction with 3'NCR (Gamarnik and Andino 1998).

The newly synthesized negative sense strands function as template for formation of new positive sense strands. In addition to those required for minus sense RNA synthesis, an RNA element "cre" that forms in the 2C coding region, is required for plus-sense synthesis (Fig. 8.5).

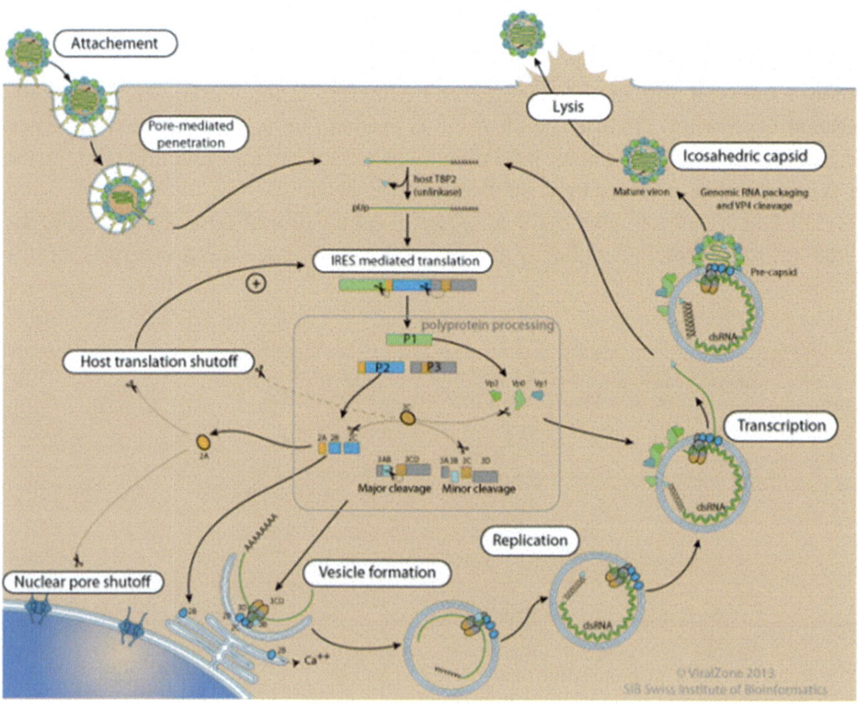

Fig. 8.5 EV replication cycle (www.viralzone.expasy.org)

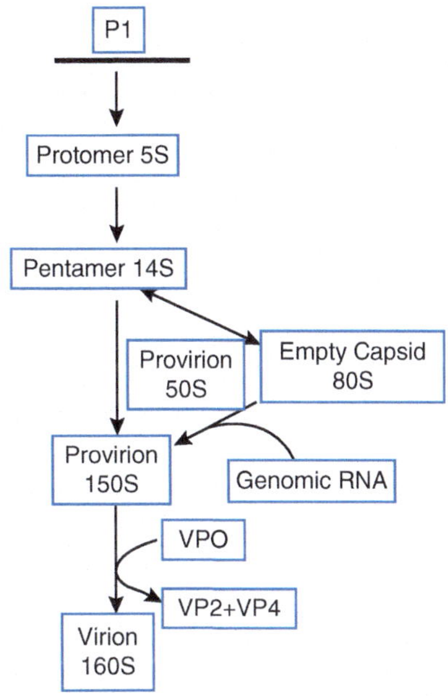

Fig. 8.6 Morphogenesis of EVs

VPg at 5′ end of these newly formed positive sense strands can function as a primer. The new positive strands have three different fates: (i) function as a template for formation of more positive strands, (ii) packaging into progeny virions, and (iii) translated into polyprotein (prior to translation, VPg is removed in this case) (Pallansch and Roos 2007).

When sufficient positive-sense RNA progeny and virion proteins accumulate, formation of assembly begins (Fig. 8.6). Particles assemble with three capsid proteins VP0, VP1, and VP3 and VPg-RNA inside. The protomers (5S) self-assemble into pentamers (14S), and the pentamers assemble into 80S empty capsids (Fig. 8.6). In one model of encapsidation, RNA is threaded into the empty capsid, producing a provirion (150S) in which VP0 is uncleaved. Another proposal is that the pentamers assemble around the RNA genome. Cleavage of VP0 results in formation of VP2 and VP4 and is the final morphogenetic step in the process of maturation of virion that produces the infectious virion (160S) (Fig. 8.6). Virions are released due to cell lysis. Capsids that are formed in excess may be seen as inclusion bodies in the cytoplasm (Palmenberg 1982; Rombaut et al. 1990).

8.6 Diversity of EV Genomes

Mutations and recombinations underlie the diversity of genomes of EV that replicate in the hosts (Tracy et al. 2008). Studies on RNA-dependent RNA polymerase (RdRp) of EVs suggest that multiple sites of the enzyme influence the template-copying fidelity (incorporation of incorrect vs correct nucleotide) during RNA replication (Tracy et al. 2008). Despite the multiple mutations and recombination events, EV genomes have been found to be replication competent (Tracy et al. 2008).

Recombination has been recognized as a key mechanism for the evolution and adaptation of EVs. Recently, several studies have reported recombination in different structural and non-structural genomic regions of EVs. This phenomenon of EV genetics also led to explain the issues to connect genotype and disease profile caused by EVs (Kyriakopoulou et al. 2015).

Two recombination models are proposed for EVs. In the replicative model (copy choice model), RNA replication occurs (Egger and Bienz 2002), while in the non-replicative model, the replication is absent (Kirkegaard and Baltimore 1986). Both the mechanisms are found to be valid; however, it was not possible to conclude which of them is predominant in nature (Kyriakopoulou et al. 2015). The copy choice model is possible when two viruses infect the same cell and replicate their genomes simultaneously and share the same intracellular replication complex (Egger and Bienz 2002). During the synthesis of the negative-strand RNA, the elongation by the viral polymerase may get blocked due to secondary structure elements and nucleotide mismatches (Pilipenko et al. 1995) and switched to another adjacent viral RNA which presents nucleotide similarity (Dedepsidis et al. 2010). The existence of homologous regions in two parental strains and the domains of high U-A content at crossover have been the prerequisites for recombination (King 1988). In the non-replicative model of recombination mechanism in RNA viruses, viral RNA genomes result in re-joining of ends regardless of the enzymatic activity (Gmyl et al. 1999). The ligation at 5′ and 3′ ends was reported to be conducted by a trans-esterification reaction. Viable viruses were noted to be generated from RNA fragments of PVs that lack replicative and translational activity, on transfection into host cells (Gmyl et al. 2003). The non-replicative mechanism was proposed to occur in both homologous and non-homologous RNA recombination sites (Kirkegaard and Baltimore 1986).

The high fitness of the viruses with genomes that were generated by recombination in particular regions has been described to be the result of the existence of hotspots in circulating EVs. Recombination hot spots were known to differ among the EV species representing the compatibility differences in genomic regions between the viral strains of the different species. Most of the recombination events in EV-B species were identified in P2 region (Lukashev et al. 2005), while those in EV-C species in P3 genomic region (Bessaud et al. 2011). The study also demonstrated that recombination in the 2A genomic region is a rare phenomenon for EV-C species, although this region has been noted as a recombination hotspot for the other EV species (Simmonds and Welch 2006). The absence of preferential sites for recombination events on specific RNA elements such as the conserved cis-replicating element located in 2C genomic region has been also noticed (Simmonds and Welch 2006). Although P2 and P3 genomic regions were found to be recombination hotspots, the P1 capsid-coding genomic region (more specifically the VP2–VP3–VP1) was identified as a cold-spot for recombination. However, VP4 has been reported to be susceptible to recombination or have increased evolutionary flexibility due to its location that is at internal side of capsid. In the VP2–VP1 region, the receptor binding site, intra-typic genetic exchanges have been found to occur more often than inter-typic ones (Bessaud et al. 2011). Although, the occurrence of inter-typic recombination event between the CV-B3 and CV-B4 strains in the VP3 genomic region has been reported (Bouslama et al. 2007), the inter-typic recombination in the capsid region has been extremely rare especially for EV types that use different receptors (Kyriakopoulou et al. 2015).

The EV 5'NCR was found to have certain intrinsic characteristics such as conserved regions of RNA secondary structures that make it susceptible to recombination. The 5'NCR is also known to be involved in viral translation and RNA synthesis, and interacts physically with viral proteins such as 2B, 2 BC, 3A, and 3D as well as with several host proteins such as polypyrimidine tract-binding protein and poly (rC) binding protein. It has been proposed that these interactions are rescued by complementary mutations in non-structural proteins such as 3A/3AB and 3C/3CD, for the formation of functional replication complexes (Florez de Sessions et al. 2007). However, the high conservation or limited diversity in 5'NCR sequences among all of the EV strains has also restricted the detection of recombination events in this region (Kyriakopoulou et al. 2015).

Fast and reliable detection of recombination in circulating EV strains could play a crucial role to shed more light on the epidemiology and evolution of EVs (Kyriakopoulou et al. 2015).

8.7 Attempts to Identify Virulence Determinants in EVs

Virulence determinants are the factors that determine whether the viral infection will be asymptomatic or lead to a severe neurological disease. The majority of EV infections, most commonly spread by a fecal-oral route of transmission, are indeed mild or subclinical in humans that often go unnoticed (Khetsuriani et al. 2003). For example, (Khetsuriani et al. 2003) a great majority of PV infections result in few (<1%) serious consequences only. This spectrum of EV infection is described in part due to the stochastic (random probable) nature or rapid dominance or the initial infectious dose or genomic nature of the virus strain, as well as host immune status (Domingo et al. 1996; Kanno et al. 2006; Svitkin et al. 1985). A study conducted during CV-B3 outbreak in 2005 has detected viral RNA in patients' blood specimens after the fever onset, and the viral RNA load has correlated positively with disease severity (Yen et al. 2007). Due to ever-renewing quasispecies populations of EVs in the new host, occurrence of diverse mutations (mostly non-competitive in nature) causing attenuation or fatality has been described to take place (Biebricher and Eigen 2006). However, molecular cloning/sequencing methods were found to capture the master genomic sequence or the majority of virus population (Biebricher and Eigen 2006; Domingo et al. 1996). Rapid reversion of artificially attenuated EV strains to wild-type virulent ones has been specified to be useful in understanding the probability of evolution of a more virulent virus strain in an individual following an infection (Tracy et al. 2006). One more factor that checks the virulence of EV is the host's adaptive immunity (Kim et al. 2006). This was evident from the frequent fecal shedding of EVs in children rather than in adults (Kim et al. 2006). Thus, the stochastic nature of any EV infection and the diverse events that influence it have been identified to be the key issues in selection of virulent (significantly disease-causing) viral strains (Tracy et al. 2006).

A number of mutations that attenuate cardiovirulent phenotype of certain CV-B3 strains and the neurovirulence of PVs were mapped on to the 5'-NCR (Dunn et al. 2000; Dunn et al. 2003; Evans et al. 1985; Kawamura et al. 1989; Macadam et al. 1991). A single nucleotide mutation in the 5' NCR of the CV-B3 strain (CV-B3/0) was noted to be necessary and sufficient to attenuate a myocarditic phenotype (Chapman et al. 1997). This too was noted as a transition mutation (U [wild type] to C [attenuated]) similar to the PV3 attenuating mutation at nucleotide 234 of the genome. However, as 234C did not appear in any other known CV-B3 genome (Chapman et al. 1997), the CVB3/0 strain has been a rare mutant strain. The work by Dunn and colleagues (Dunn et al. 2000; Dunn et al. 2003) narrowed the field of genetic candidates that determined a myocarditic (virulent) phenotype to a single short sequence in the CV-B3 5' NCR defined as domain II (a stem-loop structure) of the domains—I, II, III, IV, V, VI, and VII described to date for CV-B3. Characterization of the less virulent CV-B3/GA genome (Lee et al. 2005) also revealed consistency in the structure of domain II which is involved in determination of this phenotype. Chemical probing and computer modeling of the RNA revealed the sequence folds in the poorly virulent strain different from those observed from myocarditic genomes (Bailey and Tapprich 2007; Lee et al. 2005). Comparative analysis of EV-A71 804/NO/03 strain, isolated from healthy children, with pathogenic strains of the same serotype has shown differences in the 5'NCR and polymerase region of EV genomes (Witso et al. 2007). Further, studies carried out during 1999–2007 in Austria on prevalent EV types manifesting neurological disease have pointed out lack of conclusive results on viral genomic variations among different strains related to their virulence (Ortner et al. 2009). However, according to the United States patent 7,662,939 dated Feb 2010, genomic construction experiments conducted on EVs (CV-B3 and E-12), discovered that virulence and tropism determinants in EV genomes are in the specific regions of 5'NCR in the genome (Jose et al. 2010). It is still unclear, whether faster replication rates of specific viral strains due only to the changes in the viral genome or differences in host immune status also plays a role in determining a virulence phenotype (Wang et al. 2012). A unique nucleotide mutation (A299U) was found in IRES region of the 5'NCR in AFP-associated CV-A24v strains, circulated during 2010, and differentiated them from other known CV-A24 strains of non-AFP cases (Laxmivandana et al. 2014). A study on comparative genomic analysis of CV-B3 strains from AFP cases and asymptomatic contacts from India during 2009–2010 has suggested the need for simultaneous analysis of host and viral factors to understand the varied manifestations of CV-B3 infections (Laxmivandana et al. 2016). Thus, additional studies are required to enhance the understanding of basic biology responsible for natural phenomenon of virulence of EV strains and the disease severity.

8.8 Clinical Features and Laboratory Diagnosis of EV Associated AFP

EVs have been shown to have implications in polio-like paralytic disease (Table 8.6), clinically indistinguishable from paralytic poliomyelitis (Chen et al. 2001; Marx et al. 2000).

EV identification and characterization are the key factors to understand the viral epidemiology, diversity, asymptomatic infections, and variable clinical manifestations in the humans (Muir et al. 1998). Appropriate time of collection (within 14 days of disease onset) of clinical specimens and their optimal transport and storage (at 4 °C to −20 °C) are critical for effective detection of EVs (WHO 2004). Stool, rectal swabs, throat swabs, and cerebrospinal fluid (CSF) samples are usually the choice of specimens for EV detection (Muir et al. 1998). However, it varies with the clinical manifestations observed. Due to relatively higher EV concentrations in stool than in other sources, it is considered to be the most sensitive specimen for EV detection and the specimen of choice for PV detection in AFP surveillance (WHO 2004), while CSF in neurologic disease, throat swabs in respiratory disease, fluid in vesicular rashes, ocular secretions in hemorrhagic conjunctivitis, tissue material from heart muscle in myocarditis or myositis, and brain for encephalitis are considered (Muir et al. 1998).

Conventional methods for detection of EVs in clinical specimens involve virus isolation by cell culture and sero-neutralization (Muir et al. 1998). Most EVs show a characteristic CPE, rounded and refractile, with cell detachment in cell culture. A combination of cell lines has been found to be useful to increase sensitivity (Tsao et al. 2010). EV propagation is usually attempted in Buffalo green monkey kidney (BGM) and human cell lines: rhabdomyosarcoma (RD), lung carcinoma (A-549), epithelioma cells (Hep-2), embryonic lung cells (MRC-5), and embryonic fibroblasts (HEF). However, RD cells have been found to be the good host cells for cultivating EVs, susceptible to most of the types of EVs, and have been employed

Table 8.6 Clinical presentations of EV infection leading to AFP ("Centers for Disease Control and Prevention website. Available: http://www.cdc.gov" (n.d.))

Presentation	Approximate percent of occurrence (%)	Symptoms	Recovery
Subclinical	90	Asymptomatic	----------------
Mild/ abortive	4–5	Influenza like illness	Within 24–72 h
Aseptic meningitis	2–4	Accompanied by neck stiffness and vomiting	After several days
Major illness	1–2	Accompanied by rapid progression of weakness *(acute)*, loss of muscle tone (*flaccid*), and loss of voluntary muscle movement *(paralysis)*, Occasional respiratory paralysis may occur.	Reversible or irreversible or may lead to death

widely in the detection of the infections (Tsao et al. 2010). World Health Organization (WHO) recommends virus isolation in RD and human poliovirus receptor-CD155 expressing recombinant murine (L20B) cell lines from all stool specimens suspected to contain PVs in AFP surveillance program (WHO 2004). According to the WHO protocols accepted for the program, the stool specimens, producing CPE in L20B, are to be referred for intra-typic differentiation of PVs while the specimens, producing CPE only in RD cells and not in L20B cells, are considered to contain non-polio enteroviruses (NPEVs) (WHO 2007). Traditional serotyping methods that depend on neutralization of conserved or variable epitope region in capsid protein (VP1) with specific or pools of antisera have been used for confirmation of the virus isolation and type of EV. These pools contain a combination of intersecting polyclonal antisera obtained by hyper-immunization of large animals (horses and monkeys). Two panels of reagents are currently available: (i) The Lim and Benyesh-Melnick (LBM) pools (Melnick and Wimberly 1985) consist of a panel of eight (A–H) and seven (J–P) combined pools, which are able to identify 42 and 19 EV serotypes, respectively. (ii) The National Institute of Public Health and the Environment (Bilthoven, The Netherlands) pools are able to identify 30 EV serotypes (Nasri et al. 2007). After identification with these pools, further confirmation is needed by repeating the neutralization with the serotype specific antiserum. However, EV detection by the neutralization techniques is ineffective due to extensive intra typic antigenic diversity of EVs and its relatively low viral load in majority of the specimens. Similarly, the use of serologic techniques, to check the prevalence of antibodies to the virus, for primary diagnosis of suspected EV illness was found to be impractical due to the virus diversity (Muir et al. 1998). Electron microscopy had been occasionally used to visualize the virus directly, in specialized laboratories (Hogle et al. 1985).

Molecular biology techniques in clinical virology have given opportunity to introduce significant reliability and ease in EV identification, directly from clinical specimens or virus isolates (Casas et al. 2001; Nix et al. 2006; Palacios et al. 2002). Initially, nucleic acid probe hybridization tests were used to detect EVs, but have been replaced by the reverse transcription-nested PCR (RT-PCR). RT-PCR aimed at 5′NCR which is conserved among strains of the genus EV has been found to be highly sensitive for detection of EVs (Kessler et al. 1997). Phylogenetic analysis of the nucleotide sequence encoding VP1, capsid protein, has enabled differentiation of viruses (Oberste et al. 1999a, b). Alternative regions have been proposed for molecular typing of EVs including the VP2 and VP4 capsid regions (Ishiko et al. 2002; Oberste et al. 1998). Genotyping using microarrays, displaying large number of probes able to recognise most of the serotypes and variants entering into the genus, has also been proposed (Chen et al. 2006). The molecular typing method based on VP1 region has contributed largely to detect new types of EVs (Nasri et al. 2007). Determination of complete genetic fingerprint of EV strains, particularly for investigation of determinants involved in virulence, needs further research.

8.9 Pathogenesis of EV Infections

EVs transmit predominantly via the faeco-oral route, or through contact with virus-contaminated oral secretions, vesicular fluid, surfaces, fomites, respiratory droplets, etc. (Pallansch and Roos 2007) and infect and initiate replication in cells of the oropharyngeal mucosa and tonsils (lymphoid tissue) and may pass further down to the gastrointestinal tract (GIT). Further infection by these viruses is possible in the intestinal mucosa because of their acid stability. These viruses also infect the Peyer's patches (lymphoid tissue) in the intestinal mucosa, enter bloodstream at low levels (primary viremia) and are also shed into the faeces (Solomon et al. 2010) (Fig. 8.7).

Further, the virus infects the target tissues (Fig. 8.7), determined by the existence of appropriate cellular receptors (Pallansch and Roos 2007). EVs may disseminate to the reticuloendothelial system (spleen, liver, lymph nodes, and bone marrow), heart, lung, pancreas, skin, mucous membranes, and CNS, and lead to secondary viremia which occurs about 7–14 days following initial infection. This may coincide with the onset of clinical features (Pallansch and Roos 2007). It has been proposed that viremia is important in the spread of EV via the blood brain barrier into the CNS (Nathanson and Bodian 1961; Yang et al. 1997). However, some epidemiological and experimental studies suggest spread of EV into CNS following muscle infection via retrograde axonal flow (Nomoto et al. 1994).

EVs are usually encountered rapidly by both the cell-mediated and humoral immune responses of the host, generated during the viremic phase. As a result, further virus replication is greatly limited in almost all tissues. However, the virus

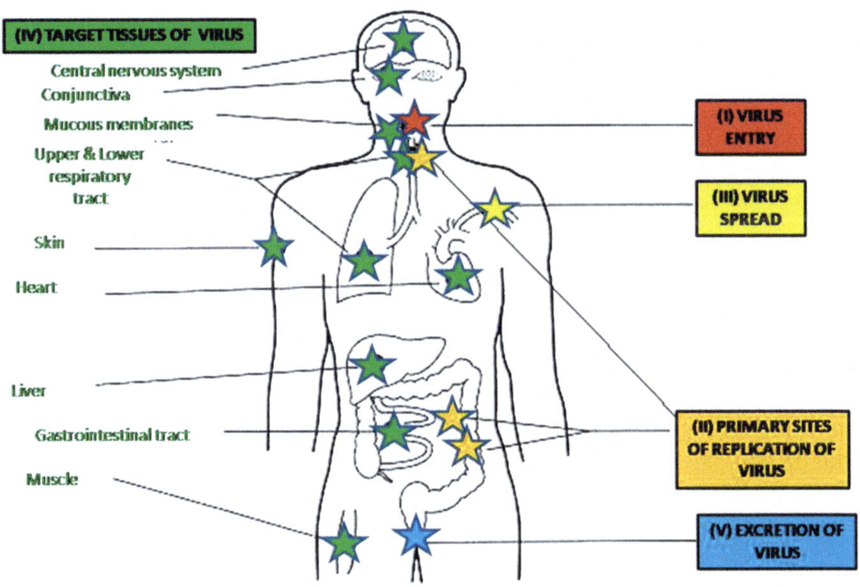

Fig. 8.7 Pathogenesis of EV infections

replication remains sustained in GIT, by the mechanisms unknown as yet, for several weeks in spite of acquiring a high titre of neutralizing antibody (http://pathmicro.med.sc.edu/virol/picorna.htm). Most of the EV infections in humans may remain asymptomatic despite the viral shedding. The viral shedding from the throat can occur up to 2 weeks after an acute EV infection (Chung et al. 2001). The virus can be isolated from stool, up to 11 weeks (Chung et al. 2001).

An unexplained feature of EV pathogenesis is the selective vulnerability of the cells to EV infection. It is known that the expression of CD155 influences the distribution of PV infection (Mendelsohn et al. 1989). As against this, the correlation noted between CD155 expression in a particular tissue and the distribution of virus in that tissue has been found to be imperfect (Nomoto et al. 1994). For example, higher levels of CD155 are expressed in human liver, lung, and heart than in the brain. However, tissues of these organs are not known to be infected by PV (Nomoto et al. 1994) while brain cells are found to be susceptible. This indicated that other factors besides the cellular receptors may play a role in determining the tropism of the virus and distribution of viral infectivity. Additional determinants of the tissue and cellular distribution of EVs may include the route of ingestion, host, and cell-specific factors, and the viral genomic components (Yang et al. 1997).

8.10 Host Immune Response to EV Infections

The efficacy of the host immune responses, both innate and adaptive, is known to influence the outcome of EV infections. The innate immune system acquires importance as it generates the earliest response and further induces and regulates the adaptive immune response. Host cells release interferons (IFNs) as part of innate immune response, stimulate many genes to induce an antiviral state. A more rapid IFN response has been described in non-neural tissues than in the neural tissues of CD155 transgenic mice (Ida-Hosonuma et al. 2005). This implicates the protection of the non-neural tissues against the PV infection (Ida-Hosonuma et al. 2005). IFNs were found to interfere with EV replication within host cells besides activation of immune cells of the innate immune system like natural killer (NK) cells and macrophages and upregulation of the antigen presentation to T lymphocytes to combat the EV infection (Pallansch and Roos 2007).

Adaptive cell-mediated and also humoral immune responses have been found to be raised against both structural and non-structural viral proteins. The humoral immune response in EV infections is especially important. This has been demonstrated by (i) the occurrence of persistent EV infections in agammaglobulinemics and (ii) the increased susceptibility to EV infection and increased severity of the viral disease among neonates and infants with under developed humoral immune system (McKinney Jr. et al. 1987). EV infections rapidly induce production of IgM class antibodies that are heterotypic, further switch on to IgG and IgA class antibody response. IgM response lasts for few months only while IgA and IgG antibodies persist for years. Re-infection of EVs provoke a secondary immune response, with rise in specific IgG antibodies (Moore and Morens 1984). Formation of the specific neutralizing antibodies has been found a few days after the disease onset (Tracy et al. 2008). The protection against EV infections has been reported to depend mainly on neutralizing antibodies that are stable and can provide serotype specific life-long immunity (Beale 1990). For example, a high PV neutralizing antibody response following natural infection or vaccination protects against disease, although the virus replication in the gut can occur for several weeks. Major antigenic sites for neutralizing antibodies have been described to reside on VP1 and to a lesser extent on VP2 and VP3 (Minor et al. 1986).

EV specific cellular responses are induced during natural EV infection. EV induced T-cell responses are known to show wide cross reactivity in between serotypes, which play a role in generation of group specific immunity against EVs (Juhela et al. 1999). EV T-cell epitopes have so far been identified in VP1–4 proteins and 2C (Varela-Calvino et al. 2004). Although T-cell response is believed to be important in the clearance of many viruses, it has been investigated to only a limited extent in case of EV infections (Pallansch and Roos 2007). Several mechanisms were proposed in favor of T cell activity that leads to immunity and protection against infection of EVs. $CD4^+$ T cells play a role in elicitation of the B-cell mediated humoral immune response, which has a critical role in the clearance of EVs (Huber 2004). Cytolytic T cells clear the virus by release of cytokines or directly by causing virus-infected cell lysis (Pallansch and Roos 2007). NK cells and gamma or delta T cells, which are part of the innate immune response, may regulate the adaptive immune T-cell response, and thereby play a significant role in virus clearance and potentially in immune-mediated pathology (Pallansch and Roos 2007).

While B- and T-cell immune response play the role in the virus clearance, there are indications that these virus infections may lead to disease in certain situations. The role for immunopathology in the EV-induced diseases was considered on account of the following: (a) inflammatory pathology; (b) the animal models that demonstrate the genetic susceptibility in their causation, and the treatment efficacy with immunosuppressive drugs; and (c) association with manifestations of autoimmunity (Pallansch and Roos 2007). Both bystander damage and molecular mimicry have been proposed to play a role in the virus-induced autoimmune disease (Pallansch and Roos 2007). In the former, upregulation of nonspecific immune response occurs with the EV infection and that leads to pathology. In the case of molecular mimicry, the response of immune system against EVs may cross-react with a host protein and cause the disease. Once damage occurs, new cellular antigenic epitopes that are not normally exposed may trigger an immune response that leads to tissue damage with the further unmasking of additional reactive epitopes, a process called epitope spreading (Pallansch and Roos 2007).

8.11 Epidemiology of EVs Associated with AFP

In the late nineteenth century, epidemics of the poliomyelitis disease appeared in Europe and America, later global occurrence of polio epidemics were noted annually (Paul 1971). In 1955, the introduction of PV vaccines resulted in the reduction in poliomyelitis cases stepwise and elimination of wild PV in USA by 1972 (Nathanson 1982). In 1988, a Global Polio Eradication Initiative was started by the World Health Organization and it had reduced the number of polio cases hugely, from 350,000 cases in 1988 to 483 cases in 2001, after which annual documentation of 1000–2000 cases had been noted for several years. By 2015, PV had been spreading naturally only in Pakistan and Afghanistan; in these countries annual documentation of <200 wild PV cases had been noted from 2016 to 2020, with only 6 PV confirmed cases in 2021 (WHO 2022). Current challenges for global eradication of poliomyelitis include transmission of wild PVs in endemic regions and re-infection to polio-free areas, VDPV outbreaks, and excretion of VDPVs by immune-deficient vaccines.

Since the discovery of NPEVs in 1940s (Dalldorf and Sickles 1948; Melnick et al. 1949), their association with neurological manifestations has been reported, continuously ("World Health Organization website. Available: http://www.who.int," (n.d.)). This part of the chapter reviews different studies that have identified association of NPEVs in the outbreak or sporadic cases of AFP.

Two cases with polio-like paralytic disease were identified in an EV-A71 outbreak, that occurred in the USA during summer of 1977 (Chonmaitree et al. 1981). High and rising neutralizing antibody titres to EV-D70 detected in the serum and CSF of the patients with polio-like motor paralysis accompanying AHC than AHC alone or in those from non-infected control subjects, in Bombay outbreak of AHC in 1981 in India, represented a laboratory evidence for the neurovirulence of EV-D70 (Wadia et al. 1983a). It was therefore called as "Enterovirus 70 disease" (Wadia et al. 1983b). CV-A7 serotype was detected in most of the paralytic cases in Scotland during 1956–1973 (Grist and Bell 1984). CV-A9, CV-A23 (E-9), and CV-B virus infections were found in association with transient paralysis while CV-A7 infections were detected in association with permanent paralysis in a study on non-polio causes of polio-like paralytic syndromes (Gear 1984). Residual paralysis was observed in three cases associated with E-22, from Jamaica, 3 years after the onset of AFP, and no evidence of PV infection was found in these cases (Figueroa et al. 1989). EV-A71 was again identified as an etiologic agent for the poliomyelitis-like paralysis, that occurred during the summer of 1987 in the USA (Hayward et al. 1989). Several types of NPEVs were identified in AFP cases during 1989–1991 in the USA (Dietz et al. 1995). EV-A71 strains, isolated during HFMD outbreaks occurred during 1998 in Taiwan and 1999 in western Australia, had showed strong dermatotropic as well as neurotropic tendencies (Liu et al. 2000). An AFP case associated with E-7, with high titres of homologous antibodies to the virus in serum, was reported (Rakoto-Andrianarivelo et al. 2000). CV-A24, a virus with an outbreak potential, has previously not been found to be associated with AFP but for the first time in 1999, was isolated from an AFP patient in East Timor (Chaves et al. 2001). An EV-A71 outbreak of CNS disease associated with HFMD was reported, during summer of 2000 in Japan (Fujimoto et al. 2002). E-33 was identified as a possible etiological agent associated with an outbreak of CNS disease, that occurred in New Zealand during the winter of 2000 (Grimwood et al. 2003). E-6 (22.2%) followed by E-13 (20%) were commonly identified in neurological cases in

Monastir, Tunisia during Jan 2002 to Nov 2003. EV-A71, isolated from an AFP case in Faridabad, Haryana, India in 2001, was genetically distinct from the EV-A71 strains recovered during the HFMD outbreaks that occurred during 1997 in the Asia-Pacific region (Deshpande et al. 2003). E-11 was reported to be one of the frequent NPEVs isolated from children with AFP in Northern India (Kapoor et al. 2001; Kapoor et al. 2004). Incidence of the NPEV infections was found to be significantly higher in male children, in infants aged less than 6 months, and during the spring and summer months (Gharbi et al. 2006). EVs were found to be responsible for a considerable proportion (27%) of hospitalizations in children with CNS disease in Argentina during January 1998 to December 2003 and CV-B3 to B5, E9, E30, and E33 were identified most frequently in the study (Mistchenko et al. 2006). Studies conducted in the USA and Congo during 1970–2000 also reported a high frequency of NPEV isolations from AFP cases (Junttila et al. 2007). E-11, CV-A24, and PV-1 were found to be the most prevalent types, while EV-B93 and EV-D94 were identified as the two new types, in AFP cases from Congo during 2000–2001 (Junttila et al. 2007). NPEVs were pre-dominant cause of AFP during Jan 2003 to Sep 2003 in Pakistan (Saeed et al. 2007). CV-B, E-11, E-12, and E-13 infections were found majorly in AFP cases during 2004–2007 in Northern India (Dhole et al. 2009). The EV-B species was reported to be the predominant species (Janes et al. 2014) followed by EV-C and EV-A in AFP cases from the People's Republic of China during 1997–2000, 2004, and from Philippines during 1992–2008 (Apostol et al. 2012; Bingjun et al. 2008). An isolate of EV-A71, from the stools of paralyzed children in the Central African Republic, was found not to belong to any of the known genogroups. The existence of a previously unknown genogroup for EV-A71 type was described (Bessaud et al. 2012). An unusual EV-C99 type was found in an AFP case in Spain. Phylogenetic analysis of the 3D region of the EV-C99 strain revealed recombination with other EVs of species-C (Cabrerizo et al. 2012). An EV-B107 strain (PAKNIH-2009/23), isolated from an AFP case in Pakistan, was described to be genetically distinct from other circulating strains of the virus type, based on capsid, 5'UTR and 2A regions (Shaukat et al. 2012). Sixty six types with a high prevalence of EV-A71 followed by E-13 and CV-B5 were detected in the AFP-associated NPEVs in polio-endemic and -free regions in India during 2007–2009 (Rao et al. 2012). E-13 was predominant NPEV in non-polio AFP cases in Northern India during March 2010 to August 2010 (Maan et al. 2013). Extensive genetic diversity among NPEVs and possible link between NPEV circulation pattern and AFP has been reported in different populations and climatic conditions of India. Significantly higher NPEV associated AFP cases were found in children aged <2 years, summer months, patients with residual paralysis, and in relatively hot climate regions (p-values: <0.0000001–<0.05) during 2009–2010 (Laxmivandana et al. 2013). A distinct genomic composition of CV-B3 strains from AFP cases in different geographical regions of India during 2009–2010 has been found (Laxmivandana et al. 2016). EV-A71 was found to be commonly detected with AFP, followed by Echovirus 11 in a systematic review of literature (Suresh et al. 2018). EV-D68 infections have been found to show many similarities with PV a century ago, stimulating discussion whether EV-D68 could be ascertaining itself as the "new polio" (Cassidy et al. 2018). The emergence of EV-D68, especially in North America, has been implicated in a rise in cases with neuropathology (Brown et al. 2020). A quantitative microbial risk assessment was used to estimate the annual risk of EV infection and disease burden for consumers of effluent-irrigated raw vegetables in a study from Iran and had recommended that effective method of vegetable washing is crucial to eliminate the excess prevalence of EVs and minimize the risk of their infections through fecal-oral route in humans (Hashemi et al. 2022).

8.12 Prevention and Control

Poliomyelitis has been well-controlled globally by intensive Oral Polio Vaccination program ("World Health Organization website. Available: http://www.who.int," (n.d.)). Two types of polio vaccines were introduced to prevent poliomyelitis, an injectable inactivated polio vaccine (IPV) developed by Jonas Salk in 1955 and an attenuated oral polio vaccine (OPV) developed by Albert Sabin in 1961 (Nathanson 1982). OPV was developed by attenuating three serotypes of wild PV by growing them in monkey kidney cells while IPV was made by purifying and killing PV with a chemical. OPV provides a first line defense against orally transmitted PV by inducing antibody mediated immunity in the intestines while IPV induces a second line defense against infection by eliciting antibody mediated immunity in the bloodstream and restricting the virus from crossing the blood-brain barrier, thereby preventing paralysis. OPV form is more effective and convenient to use than IPV; however, OPV (unlike IPV) replicates and rarely reverts to the natural form and in about 1 of every 2.4 million recipients, OPV may cause paralysis (Plotkin et al. 2018). Hence, sometime after elimination of PV, only IPV is being used by some countries, as IPV outweighs its risks like pain, redness, and swelling at the injection site.

Nevertheless, NPEV infections have not been controlled as yet due to diversity in existing virus types and manifestations in their infections. Since most of the infections are asymptomatic, it is difficult to prevent NPEVs from spreading. Transmission of NPEV infections has been found to be increased by unhygienic living conditions. Hence, improved sanitation and general hygiene are important preventive measures. The viruses are resistant to many disinfectants; however, chlorinated (bleach 20 ml/l of water) or iodized disinfectants have been proved to be useful (Pallansch and Roos 2007). During the epidemics, closing of certain institutions like schools and child care facilities is advised to reduce transmission specially among young children ("Centers for Disease Control and Prevention website. Available: http://www.cdc.gov," (n.d.)).

To date, no Food and Drug Administration (FDA) approved treatment strategy for NPEV infection is available. Treatment majorly focuses on management of complications or to ease the symptoms. However, there are several therapies that target different stages in the virus life cycle. RNA interference (RNAi) was proven to be effective to inhibit NPEV replication both in vitro and in vivo in a variety of cell and mouse lines (Aliyari and Ding 2009). The low fidelity of the RdRp rises the likelihood of mutants which may escape the effects of RNAi therapy. RNAi represents an immature yet powerful tool that should not be overlooked in the design of therapeutics to treat NPEV infections of the CNS (Vaishnaw et al. 2010). Small interfering RNAs (siRNAs) inhibit the infectious virion production via dicer recognition of the viral dsRNA formed during genomic replication. Majority of the siRNAs developed against NPEVs focus upon the degradation of mRNAs that code for the viral RdRp and proteases (Aliyari and Ding 2009).

Ribavirin, a broad-spectrum antiviral, inhibits the replication of different NPEVs. It is a nucleoside analog which can act as a mutagen via incorporation into the viral genome (Crotty and Andino 2002). Ribavirin may force the afflicted NPEVs into "error catastrophe" by generating a non-infectious quasispecies swarm, thereby causing lethal mutagenesis (Vignuzzi et al. 2006). However, there are conflicting reports on the potential of ribavirin to cross the blood brain barrier (BBB); a necessary criterion when dealing with the treatment of NPEV infections of the CNS. The NPEV resistance necessitates the new antiviral drug development to combat these nefarious pathogens.

Pleconaril, an anti-picornaviral drug, was demonstrated to have efficacy against many NPEVs (Pevear et al. 1999). This drug can cross BBB and remain in CNS and inhibit NPEV replication (Schmidtke et al. 2009). It also inhibits viral attachment to the receptor and nucleocapsid uncoating during replication (Chen et al. 2008).

The standard therapy for CNS disease caused by NPEV infection is continued to be intravenous immunoglobulin (IV Ig) treatment (Abzug 2004). IV Ig is typically prepared from pools of plasma samples from healthy donors (Chen et al. 2008). This passive immunization might neutralize infectious virus circulating within the host (Ooi et al. 2010).

"Therapeutic vaccines" proposed for administration after initial infection are envisioned to enhance or redirect an on-going immune response in order to reduce viral load during early infection, or after the establishment of viral persistence (Rhoades et al. 2011).

In view of the effectiveness of historical Salk and Sabin vaccines against PV ("World Health Organization website. Available: http://www.who.int," (n.d.)), many vaccines stimulating both the innate as well as adaptive immune responses have been designed to combat the clinically relevant NPEVs. Such vaccines include inactivated whole-virus, live attenuated virus, sub viral particles, and DNA vaccines (Rhoades et al. 2011). There have been a variety of vaccines described specially against EV-A71 and CVs (Chung et al. 2008; Hunziker et al. 2004; Kim et al. 2005; Wu et al. 2001), although no clinical vaccine has been currently approved.

The extensive genetic diversity of EVs has been challenging to develop sensitive and specific diagnostic reagents or cross-protective preventative vaccines (Brown et al. 2020). However, the successful development of highly effective antiviral agents against another (+) RNA virus—hepatitis C virus—implies that it should be possible to develop antiviral agents against EVs. (Baggen et al. 2018). On-going studies on different aspects of NPEVs may help to improve the efficacy and safety of potential therapeutic candidates and develop innovative therapeutic strategies.

8.13 Future Vision

With the advent of Global Polio Eradication Initiative, PV associated AFP cases have been eliminated in most of the countries in the world. However, AFP continues to be notified during the Polio Surveillance with significant positivity to NPEVs. For better perception of the NPEV infections in the post-polio era, investigations of PV negative but NPEV positive AFP cases and environmental samples from EV endemic geographical regions and meteorological zones from different countries would be necessary. Further, concurrent analysis of host and viral factors, and pathways of their interaction would help to understand the varied manifestations of EV infections.

Acknowledgments The authors thank ICMR-NIV, Pune, India for the support in carrying out this study and DST (Dept. of Science and Technology), Ministry of Science, and Government of India for the fellowship to first author.

References

Abzug MJ (2004) Presentation, diagnosis, and management of enterovirus infections in neonates. [Review]. Paediatr Drugs 6(1):1–10

Aliyari R, Ding SW (2009) RNA-based viral immunity initiated by the dicer family of host immune receptors. [research support, N.I.H., extramural research support, U.S. Gov't, non-P.H.S. Review]. Immunol Rev 227(1):176–188. https://doi.org/10.1111/j.1600-065X.2008.00722.x

Ambros V, Pettersson RF, Baltimore D (1978) An enzymatic activity in uninfected cells that cleaves the linkage between poliovirion RNA and the 5′ terminal protein. Cell 15(4):1439–1446

Apostol LN, Suzuki A, Bautista A, Galang H, Paladin FJ, Fuji N, Oshitani H (2012) Detection of non-polio enteroviruses from 17 years of virological surveillance of acute flaccid paralysis in The Philippines. [research support, non-U.S. Gov't]. J Med Virol 84(4):624–631. https://doi.org/10.1002/jmv.23242

Baggen J, Thibaut HJ, Strating J, van Kuppeveld FJM (2018) The life cycle of non-polio enteroviruses and how to target it. [research support, non-U.S. Gov't review]. Nat Rev Microbiol 16(6):368–381. https://doi.org/10.1038/s41579-018-0005-4

Bailey JM, Tapprich WE (2007) Structure of the 5′ nontranslated region of the coxsackievirus b3 genome: chemical modification and comparative sequence analysis. [comparative study research support, N.I.H., extramural]. J Virol 81(2):650–668. https://doi.org/10.1128/JVI.01327-06

Beale AJ (1990) Polio vaccines: time for a change in immunisation policy? [comparative study review]. Lancet 335(8693):839–842

Belnap DM, McDermott BM Jr, Filman DJ, Cheng N, Trus BL, Zuccola HJ, Steven AC (2000) Three-dimensional structure of poliovirus receptor bound to poliovirus. [research support, U.S. Gov't, P.H.S.]. Proc Natl Acad Sci USA 97(1):73–78

Bessaud M, Joffret ML, Holmblat B, Razafindratsimandresy R, Delpeyroux F (2011) Genetic relationship between cocirculating human enteroviruses species C. [research support, non--U.S. Gov't]. PLoS One 6(9):e24823. https://doi.org/10.1371/journal.pone.0024823

Bessaud M, Pillet S, Ibrahim W, Joffret ML, Pozzetto B, Delpeyroux F, Gouandjika-Vasilache I (2012) Molecular characterization of human enteroviruses in the Central African Republic: uncovering wide diversity and identification of a new human enterovirus A71 genogroup. [research support, non-U.S. Gov't]. J Clin Microbiol 50(5):1650–1658. https://doi.org/10.1128/JCM.06657-11

Biebricher CK, Eigen M (2006) What is a quasispecies? [review]. Curr Top Microbiol Immunol 299:1–31

Bingjun T, Yoshida H, Yan W, Lin L, Tsuji T, Shimizu H, Miyamura T (2008) Molecular typing and epidemiology of non-polio enteroviruses isolated from Yunnan Province, the People's Republic of China. [research support, non-U.S. Gov't]. J Med Virol 80(4):670–679. https://doi.org/10.1002/jmv.21122

Bouslama L, Nasri D, Chollet L, Belguith K, Bourlet T, Aouni M, Pillet S (2007) Natural recombination event within the capsid genomic region leading to a chimeric strain of human enterovirus B. [research support, non-U.S. Gov't]. J Virol 81(17):8944–8952. https://doi.org/10.1128/JVI.00180-07

Bradshaw N (1989) Polio in Kentucky-from birthday balls to the breakthrough. Regist Ky Hist Soc 87(1):20–39

Brown DM, Zhang Y, Scheuermann RH (2020) Epidemiology and sequence-based evolutionary analysis of circulating non-polio enteroviruses. Microorganisms 8(12):1856. https://doi.org/10.3390/microorganisms8121856

Cabrerizo M, Rabella N, Torner N, Castellanos T, Bustillo I, Varela C, Trallero G (2012) Molecular identification of an enterovirus 99 strain in Spain. [case reports research support, non-U S Gov't]. Arch Virol 157(3):551–554

Casas I, Palacios GF, Trallero G, Cisterna D, Freire MC, Tenorio A (2001) Molecular characterization of human enteroviruses in clinical samples: comparison between VP2, VP1, and RNA polymerase regions using RT nested PCR assays and direct sequencing of products. [comparative study. Research support, non-U.S. Gov't]. J Med Virol 65(1):138–148

Casasnovas JM, Stehle T, Liu JH, Wang JH, Springer TA (1998) A dimeric crystal structure for the N-terminal two domains of intercellular adhesion molecule-1. [research support, U.S. Gov't, P.H.S.]. Proc Natl Acad Sci USA 95(8):4134–4139

Cassidy H, Poelman R, Knoester M, Van Leer-Buter CC, Niesters HGM (2018) Enterovirus D68 - the new polio? [review]. Front Microbiol 9:2677. https://doi.org/10.3389/fmicb.2018.02677

Centers for Disease Control ansd Prevention website (n.d.). http://www.cdc.gov

Chapman NM, Romero JR, Pallansch MA, Tracy S (1997) Sites other than nucleotide 234 determine cardiovirulence in natural isolates of coxsackievirus B3. [research support, non-U S Gov't research support, U S Gov't, P H S]. J Med Virol 52(3):258–261

Chaves SS, Lobo S, Kennett M, Black J (2001) Coxsackievirus A24 infection presenting as acute flaccid paralysis. [case reports letter]. Lancet 357(9256):605

Chen CY, Chang YC, Huang CC, Lui CC, Lee KW, Huang SC (2001) Acute flaccid paralysis in infants and young children with enterovirus 71 infection: MR imaging findings and clinical correlates. AJNR Am J Neuroradiol 22(1):200–205

Chen TC, Chen GW, Hsiung CA, Yang JY, Shih SR, Lai YK, Juang JL (2006) Combining multiplex reverse transcription-PCR and a diagnostic microarray to detect and differentiate enterovirus 71 and coxsackievirus A16. [evaluation StudiesResearch support, non-U.S. Gov't]. J Clin Microbiol 44(6):2212–2219. https://doi.org/10.1128/JCM.02393-05

Chen TC, Weng KF, Chang SC, Lin JY, Huang PN, Shih SR (2008) Development of antiviral agents for enteroviruses. [Research Support, Non-U.S. Gov't Review]. *J Antimicrob Chemother 62*(6):1169–1173. https://doi.org/10.1093/jac/dkn424

Chonmaitree T, Menegus MA, Schervish-Swierkosz EM, Schwalenstocker E (1981) Enterovirus 71 infection: report of an outbreak with two cases of paralysis and a review of the literature. [case reports comparative study]. Pediatrics 67(4):489–493

Chow M, Newman JF, Filman D, Hogle JM, Rowlands DJ, Brown F (1987) Myristoylation of picornavirus capsid protein VP4 and its structural significance. [research support, U.S. Gov't, non-P.H.S. Research support, U.S. Gov't, P.H.S.]. Nature 327(6122):482–486. https://doi.org/10.1038/327482a0

Chung P, Huang Y, Chang L, Lin T, Ning H (2001) Duration of enterovirus shedding in stool. J Microbiol Immunol Infect 34:167–170

Chung YC, Ho MS, Wu JC, Chen WJ, Huang JH, Chou ST, Hu YC (2008) Immunization with virus-like particles of enterovirus 71 elicits potent immune responses and protects mice

against lethal challenge. [Comparative Study Research Support, Non-U.S. Gov't]. *Vaccine 26*(15):1855–1862. https://doi.org/10.1016/j.vaccine.2008.01.058
Crotty S, Andino R (2002) Implications of high RNA virus mutation rates: lethal mutagenesis and the antiviral drug ribavirin. [research support, U.S. Gov't, P.H.S. Review]. Microbes Infect 4(13):1301–1307
Dalldorf G, Sickles GM (1948) An unidentified, Filtrable agent isolated from the feces of children with paralysis. Science 108(2794):61–62. https://doi.org/10.1126/science.108.2794.61
De Sena J, Mandel B (1977) Studies on the in vitro uncoating of poliovirus. II. Characteristics of the membrane-modified particle. [research support, U.S. Gov't, P.H.S.]. Virology 78(2):554–566
Dedepsidis E, Kyriakopoulou Z, Pliaka V, Markoulatos P (2010) Correlation between recombination junctions and RNA secondary structure elements in poliovirus Sabin strains. Virus Genes 41(2):181–191. https://doi.org/10.1007/s11262-010-0512-5
Deshpande JM, Nadkarni SS, Francis PP (2003) Enterovirus 71 isolated from a case of acute flaccid paralysis in India represents a new genotype. Curr Sci 84:1350–1353
Dhole TN, Ayyagari A, Chowdhary R, Shakya AK, Shrivastav N, Datta T, Prakash V (2009) Non-polio enteroviruses in acute flaccid paralysis children of India: vital assessment before polio eradication. J Paediatr Child Health 45(7–8):409–413. https://doi.org/10.1111/j.1440-1754.2009.01529.x
Dietz V, Andrus J, Olive JM, Cochi S, de Quadros C (1995) Epidemiology and clinical characteristics of acute flaccid paralysis associated with non-polio enterovirus isolation: the experience in the Americas. [comparative study]. Bull World Health Organ 73(5):597–603
Domingo E, Escarmis C, Sevilla N, Moya A, Elena SF, Quer J, Holland JJ (1996) Basic concepts in RNA virus evolution. [Research Support, Non-U.S. Gov't Research Support, U.S. Gov't, P.H.S. Review]. *FASEB J 10*(8):859–864
Dunn JJ, Chapman NM, Tracy S, Romero JR (2000) Genomic determinants of cardiovirulence in coxsackievirus B3 clinical isolates: localization to the 5′ nontranslated region. [research support, non-U.S. Gov't]. J Virol 74(10):4787–4794
Dunn JJ, Bradrick SS, Chapman NM, Tracy SM, Romero JR (2003) The stem loop II within the 5′ nontranslated region of clinical coxsackievirus B3 genomes determines cardiovirulence phenotype in a murine model. [Research Support, Non-U.S. Gov't Research Support, U.S. Gov't, P.H.S.]. J Infect Dis 187(10):1552–1561. https://doi.org/10.1086/374877
Egger D, Bienz K (2002) Recombination of poliovirus RNA proceeds in mixed replication complexes originating from distinct replication start sites. [research support, non-U.S. Gov't]. J Virol 76(21):10960–10971
Evans DJ, Almond JW (1998) Cell receptors for picornaviruses as determinants of cell tropism and pathogenesis. [Research Support, Non-U.S. Gov't Review]. *Trends Microbiol 6*(5):198–202
Evans DM, Dunn G, Minor PD, Schild GC, Cann AJ, Stanway G, Maizel JV Jr (1985) Increased neurovirulence associated with a single nucleotide change in a noncoding region of the Sabin type 3 poliovaccine genome. Nature 314(6011):548–550
Figueroa JP, Ashley D, King D, Hull B (1989) An outbreak of acute flaccid paralysis in Jamaica associated with echovirus type 22. J Med Virol 29(4):315–319
Florez de Sessions P, Dobrikova E, Gromeier M (2007) Genetic adaptation to untranslated region-mediated enterovirus growth deficits by mutations in the nonstructural proteins 3AB and 3CD. [Research Support, N.I.H., Extramural Research Support, Non-U.S. Gov't]. J Virol 81(16):8396–8405. https://doi.org/10.1128/JVI.00321-07
Fricks CE, Hogle JM (1990) Cell-induced conformational change in poliovirus: externalization of the amino terminus of VP1 is responsible for liposome binding. [research support, U.S. Gov't, P.H.S.]. J Virol 64(5):1934–1945
Fujimoto T, Chikahira M, Yoshida S, Ebira H, Hasegawa A, Totsuka A, Nishio O (2002) Outbreak of central nervous system disease associated with hand, foot, and mouth disease in Japan during the summer of 2000: detection and molecular epidemiology of enterovirus 71. [Case Reports Research Support, Non-U S Gov't]. *Microbiol Immunol 46*(9):621–627

Gamarnik AV, Andino R (1998) Switch from translation to RNA replication in a positive-stranded RNA virus. [Research Support, Non-U.S. Gov't Research Support, U.S. Gov't, P.H.S.]. *Genes Dev 12*(15):2293–2304

Gear JH (1984) Nonpolio causes of polio-like paralytic syndromes. Rev Infect Dis 6(2):S379–S384

Gharbi J, Jaidane H, Ben M'hadheb M, El Hiar R, Chouchene C, Gueddiche MN, Aouni M (2006) Epidemiological study of non-polio enterovirus neurological infections in children in the region of Monastir, Tunisia. [research support, non-U.S. Gov't]. Diagn Microbiol Infect Dis 54(1):31–36. https://doi.org/10.1016/j.diagmicrobio.2005.08.017

Gmyl AP, Belousov EV, Maslova SV, Khitrina EV, Chetverin AB, Agol VI (1999) Nonreplicative RNA recombination in poliovirus. [research support, non-U.S. Gov't]. J Virol 73(11):8958–8965

Gmyl AP, Korshenko SA, Belousov EV, Khitrina EV, Agol VI (2003) Nonreplicative homologous RNA recombination: promiscuous joining of RNA pieces? [research support, non-U.S. Gov't]. RNA 9(10):1221–1231

Grimwood K, Huang QS, Sadleir LG, Nix WA, Kilpatrick DR, Oberste MS, Pallansch MA (2003) Acute flaccid paralysis from echovirus type 33 infection. [case reports]. J Clin Microbiol 41(5):2230–2232

Grist NR, Bell EJ (1984) Paralytic poliomyelitis and nonpolio enteroviruses: studies in Scotland. Rev Infect Dis 6(2):S385–S386

Hahn H, Palmenberg AC (1995) Encephalomyocarditis viruses with short poly(C) tracts are more virulent than their mengovirus counterparts. [research support, U.S. Gov't, P.H.S.]. J Virol 69(4):2697–2699

Hashemi SY, Shahmahmoodi S, Hadi M, Nodehi RN, Alimohammadi M, Nejati A, Mesdaghinia A (2022) Quantitative microbial risk assessment of enteroviruses in raw-eatable vegetables irrigated by wastewater: examining different scenarios of washing. J Environ Health Sci Eng 20(2):629–640. https://doi.org/10.1007/s40201-022-00789-z

Hayward JC, Gillespie SM, Kaplan KM, Packer R, Pallansch M, Plotkin S, Schonberger LB (1989) Outbreak of poliomyelitis-like paralysis associated with enterovirus 71. Pediatr Infect Dis J 8(9):611–616

He Y, Bowman VD, Mueller S, Bator CM, Bella J, Peng X, Rossmann MG (2000) Interaction of the poliovirus receptor with poliovirus. [research support, non-U.S. Gov't research support, U.S. Gov't, non-P.H.S. Research support, U.S. Gov't, P.H.S.]. Proc Natl Acad Sci USA 97(1):79–84

Hellen CU, Facke M, Krausslich HG, Lee CK, Wimmer E (1991) Characterization of poliovirus 2A proteinase by mutational analysis: residues required for autocatalytic activity are essential for induction of cleavage of eukaryotic initiation factor 4F polypeptide p220. [Research Support, Non-U.S. Gov't Research Support, U.S. Gov't, P.H.S.]. *J Virol 65*(8):4226–4231

Hogle JM, Chow M, Filman DJ (1985) Three-dimensional structure of poliovirus at 2.9 a resolution. [Research Support, Non-U.S. Gov't Research Support, U.S. Gov't, P.H.S.]. Science 229(4720):1358–1365

Huber S (2004) T cells in coxsackievirus-induced myocarditis. [Research Support, U.S. Gov't, P.H.S. Review]. Viral Immunol 17(2):152–164. https://doi.org/10.1089/0882824041310667

Hunziker IP, Harkins S, Feuer R, Cornell CT, Whitton JL (2004) Generation and analysis of an RNA vaccine that protects against coxsackievirus B3 challenge. [Research Support, Non--U.S. Gov't Research Support, U.S. Gov't, P.H.S.]. Virology 330(1):196–208. https://doi.org/10.1016/j.virol.2004.09.035

Ida-Hosonuma M, Iwasaki T, Yoshikawa T, Nagata N, Sato Y, Sata T, Koike S (2005) The alpha/beta interferon response controls tissue tropism and pathogenicity of poliovirus. [research support, non-U.S. Gov't]. J Virol 79(7):4460–4469. https://doi.org/10.1128/JVI.79.7.4460-4469.2005

Ishiko H, Shimada Y, Yonaha M, Hashimoto O, Hayashi A, Sakae K, Takeda N (2002) Molecular diagnosis of human enteroviruses by phylogeny-based classification by use of the VP4 sequence. J Infect Dis 185(6):744–754. https://doi.org/10.1086/339298

Janes VA, Minnaar R, Koen G, van Eijk H, Dijkman-de Haan K, Pajkrt D, Benschop KS (2014) Presence of human non-polio enterovirus and parechovirus genotypes in an Amsterdam hospital in 2007 to 2011 compared to national and international published surveillance data: a comprehensive review. [Research Support, Non-U.S. Gov't Review]. *Euro Surveill 19*(46):20964. https://doi.org/10.2807/1560-7917.es2014.19.46.20964

Juhela S, Hyoty H, Uibo R, Meriste SH, Uibo O, Lonnrot M, Ilonen J (1999) Comparison of enterovirus-specific cellular immunity in two populations of young children vaccinated with inactivated or live poliovirus vaccines. [Comparative Study Research Support, Non-U.S. Gov't]. *Clin Exp Immunol 117*(1):100–105

Junttila N, Leveque N, Kabue JP, Cartet G, Mushiya F, Muyembe-Tamfum JJ, Norder H (2007) New enteroviruses, EV-93 and EV-94, associated with acute flaccid paralysis in The Democratic Republic of the Congo. [research support, non-U S Gov't]. J Med Virol 79(4):393–400

Kanno T, Kim K, Kono K, Drescher KM, Chapman NM, Tracy S (2006) Group B coxsackievirus diabetogenic phenotype correlates with replication efficiency. [Research Support, N.I.H., Extramural Research Support, Non-U.S. Gov't]. *J Virol 80*(11):5637–5643. https://doi.org/10.1128/JVI.02361-05

Kapoor A, Ayyagari A, Dhole TN (2001) Non-polio enteroviruses in acute flaccid paralysis. Indian J Pediatr 68(10):927–929

Kapoor A, Patel DD, Kumar S, Ayyagari A, Dhole TN (2004) Molecular characterization of echovirus 11 isolates from India. Indian J Med Res 119(4):149–156

Kawamura N, Kohara M, Abe S, Komatsu T, Tago K, Arita M, Nomoto A (1989) Determinants in the 5′ noncoding region of poliovirus Sabin 1 RNA that influence the attenuation phenotype. [research support, non-U.S. Gov't]. J Virol 63(3):1302–1309

Kessler HH, Santner B, Rabenau H, Berger A, Vince A, Lewinski C, Doerr HW (1997) Rapid diagnosis of enterovirus infection by a new one-step reverse transcription-PCR assay. J Clin Microbiol 35(4):976–977

Khetsuriani N, Prevots DR, Quick L, Elder ME, Pallansch M, Kew O, Sutter RW (2003) Persistence of vaccine-derived polioviruses among immunodeficient persons with vaccine-associated paralytic poliomyelitis. [comparative study]. J Infect Dis 188(12):1845–1852. https://doi.org/10.1086/379791

Kim JY, Jeon ES, Lim BK, Kim SM, Chung SK, Kim JM, Nam JH (2005) Immunogenicity of a DNA vaccine for coxsackievirus B3 in mice: protective effects of capsid proteins against viral challenge. [Comparative Study Research Support, Non-U.S. Gov't]. Vaccine 23(14):1672–1679. https://doi.org/10.1016/j.vaccine.2004.10.008

Kim K, Kanno T, Chapman NM, Tracy S (2006) Genetic determinants of virulence in the group B coxsackieviruses. Future Virol 1(5):597–604. https://doi.org/10.2217/17460794.1.5.597

King AM (1988) Preferred sites of recombination in poliovirus RNA: an analysis of 40 intertypic cross-over sequences. [research support, non-U.S. Gov't]. Nucleic Acids Res 16(24):11705–11723

Kirkegaard K, Baltimore D (1986) The mechanism of RNA recombination in poliovirus. [Research Support, Non-U.S. Gov't Research Support, U.S. Gov't, P.H.S.]. *Cell 47*(3):433–443

Kyriakopoulou Z, Pliaka V, Amoutzias GD, Markoulatos P (2015) Recombination among human non-polio enteroviruses: implications for epidemiology and evolution. [research support, non--U.S. Gov't]. Virus Genes 50(2):177–188. https://doi.org/10.1007/s11262-014-1152-y

Laxmivandana R, Yergolkar P, Gopalkrishna V, Chitambar SD (2013) Characterization of the non-polio enterovirus infections associated with acute flaccid paralysis in South-Western India. [research support, non-U.S. Gov't]. PLoS One 8(4):e61650. https://doi.org/10.1371/journal.pone.0061650

Laxmivandana R, Yergolkar P, Rajeshwari M, Chitambar SD (2014) Genomic characterization of coxsackievirus type A24 strains associated with acute flaccid paralysis and rarely identified Hopkins syndrome. [research support, non-U.S. Gov't]. Arch Virol 159(11):3125–3129. https://doi.org/10.1007/s00705-014-2129-9

Laxmivandana R, Cherian SS, Yergolkar P, Chitambar SD (2016) Genomic characterization of coxsackievirus type B3 strains associated with acute flaccid paralysis in South-Western India. [research support, non-U.S. Gov't]. J Gen Virol 97(3):694–705. https://doi.org/10.1099/jgv.0.000391

Lee CK, Kono K, Haas E, Kim KS, Drescher KM, Chapman NM, Tracy S (2005) Characterization of an infectious cDNA copy of the genome of a naturally occurring, avirulent coxsackievirus B3 clinical isolate. [Comparative Study Research Support, Non-U.S. Gov't Research Support, U.S. Gov't, P.H.S.]. *J Gen Virol 86*(Pt 1):197–210. https://doi.org/10.1099/vir.0.80424-0

Liu CC, Tseng HW, Wang SM, Wang JR, Su IJ (2000) An outbreak of enterovirus 71 infection in Taiwan, 1998: epidemiologic and clinical manifestations. [research support, non-U S Gov't]. J Clin Virol 17(1):23–30

Lonberg-Holm K, Gosser LB, Kauer JC (1975) Early alteration of poliovirus in infected cells and its specific inhibition. [comparative study]. J Gen Virol 27(3):329–342

Lukashev AN, Lashkevich VA, Ivanova OE, Koroleva GA, Hinkkanen AE, Ilonen J (2005) Recombination in circulating human enterovirus B: independent evolution of structural and non-structural genome regions. [research support, non-U.S. Gov't]. J Gen Virol 86(Pt 12):3281–3290. https://doi.org/10.1099/vir.0.81264-0

Maan HS, Chowdhary R, Shakya AK, Dhole TN (2013) Genetic variants of echovirus 13, northern India, 2010. [research support, non-U S Gov't]. Emerg Infect Dis 19(2):293–296

Macadam AJ, Pollard SR, Ferguson G, Dunn G, Skuce R, Almond JW, Minor PD (1991) The 5′ noncoding region of the type 2 poliovirus vaccine strain contains determinants of attenuation and temperature sensitivity. Virology 181(2):451–458

Marx A, Glass JD, Sutter RW (2000) Differential diagnosis of acute flaccid paralysis and its role in poliomyelitis surveillance. [review]. Epidemiol Rev 22(2):298–316

McKinney RE Jr, Katz SL, Wilfert CM (1987) Chronic enteroviral meningoencephalitis in agammaglobulinemic patients. [Case Reports Research Support, Non-U.S. Gov't Review]. *Rev Infect Dis 9*(2):334–356

Melnick JL (1990) Enteroviruses. Polioviruses, Coxsackieviruses, echoviruses and newer enteroviruses. In: Knipe DM, Howley PM (eds) Fields virology, 2nd edn. Lippincott Williams, Wilkin, PA, pp 549–605

Melnick JL, Wimberly IL (1985) Lyophilized combination pools of enterovirus equine antisera: new LBM pools prepared from reserves of antisera stored frozen for two decades. [In Vitro Research Support, Non-U.S. Gov't]. *Bull World Health Organ 63*(3):543–550

Melnick JL, Shaw EW, Curnen EC (1949) A virus isolated from patients diagnosed as non-paralytic poliomyelitis or aseptic meningitis. Proc Soc Exp Biol Med 71(3):344–349

Melnick JL, Clarke NA, Kraft LM (1950) Immunological reactions of the Coxsackieviruses. III. Cross-protection tests in infant mice born of vaccinated mothers; transfer of immunity through the milk. J Exp Med 92(5):499–505

Mendelsohn CL, Wimmer E, Racaniello VR (1989) Cellular receptor for poliovirus: molecular cloning, nucleotide sequence, and expression of a new member of the immunoglobulin superfamily. [Comparative Study, 56, 855 Research Support, Non-U.S. Gov't Research Support, U.S. Gov't, P.H.S.]. Cell 56(5):855–865

Minor PD, Ferguson M, Evans DM, Almond JW, Icenogle JP (1986) Antigenic structure of polioviruses of serotypes 1, 2 and 3. [Research Support, U.S. Gov't, P.H.S.]. J Gen Virol 67(Pt 7):1283–1291

Minor P, Brown F, Domingo E (1995) Picornaviridae.virus taxonomy - sixth report of the international committee on taxonomy of viruses. In: Murphy FA, Fauquet CM, Bishop DHL, Ghabrial SA, Martelli GP, Mayo MA, Summers MD (eds) . Springer, Vienna, Austria, pp 329–336

Mistchenko AS, Viegas M, Latta MP, Barrero PR (2006) Molecular and epidemiologic analysis of enterovirus B neurological infection in Argentine children. J Clin Virol 37(4):293–299

Moore M, Morens D (1984) Enteroviruses, including polioviruses. In: Belshe R (ed) Text book of human virology. PSG Publishing company, Littleton, MA, pp 407–472

Muir P, Kammerer U, Korn K, Mulders MN, Poyry T, Weissbrich B, van Loon AM (1998) Molecular typing of enteroviruses: current status and future requirements the European Union concerted action on virus meningitis and encephalitis [review]. Clin Microbiol Rev 11(1):202–227

Nasri D, Bouslama L, Pillet S, Bourlet T, Aouni M, Pozzetto B (2007) Basic rationale, current methods and future directions for molecular typing of human enterovirus. [review]. Expert Rev Mol Diagn 7(4):419–434. https://doi.org/10.1586/14737159.7.4.419

Nathanson N (1982) Eradication of poliomyelitis in the United States. Rev Infect Dis 4(5):940–945

Nathanson N, Bodian D (1961) Experimental poliomyelitis following intramuscular virus injection. II. Viremia and the effect of antibody. Bull Johns Hopkins Hosp 108:320–333

Nilsson EC, Jamshidi F, Johansson SM, Oberste MS, Arnberg N (2008) Sialic acid is a cellular receptor for coxsackievirus A24 variant, an emerging virus with pandemic potential. [research support, non-U.S. Gov't]. J Virol 82(6):3061–3068. https://doi.org/10.1128/JVI.02470-07

Nishimura Y, Lee H, Hafenstein S, Kataoka C, Wakita T, Bergelson JM, Shimizu H (2013) Enterovirus 71 binding to PSGL-1 on leukocytes: VP1–145 acts as a molecular switch to control receptor interaction. [Research Support, N.I.H., Extramural Research Support, Non--U.S. Gov't]. PLoS Pathog 9(7):e1003511. https://doi.org/10.1371/journal.ppat.1003511

Nix WA, Oberste MS, Pallansch MA (2006) Sensitive, seminested PCR amplification of VP1 sequences for direct identification of all enterovirus serotypes from original clinical specimens. J Clin Microbiol 44(8):2698–2704. https://doi.org/10.1128/JCM.00542-06

Nomoto A, Koike S, Aoki J (1994) Tissue tropism and species specificity of poliovirus infection. [review]. Trends Microbiol 2(2):47–51

Oberste MS, Maher K, Pallansch MA (1998) Molecular phylogeny of all human enterovirus serotypes based on comparison of sequences at the 5′ end of the region encoding VP2. Virus Res 58(1–2):35–43

Oberste MS, Maher K, Kilpatrick DR, Flemister MR, Brown BA, Pallansch MA (1999a) Typing of human enteroviruses by partial sequencing of VP1. J Clin Microbiol 37(5):1288–1293

Oberste MS, Maher K, Kilpatrick DR, Pallansch MA (1999b) Molecular evolution of the human enteroviruses: correlation of serotype with VP1 sequence and application to picornavirus classification. J Virol 73(3):1941–1948

Oberste MS, Maher K, Michele SM, Belliot G, Uddin M, Pallansch MA (2005) Enteroviruses 76, 89, 90 and 91 represent a novel group within the species human enterovirus a. [comparative study]. J Gen Virol 86(Pt 2):445–451. https://doi.org/10.1099/vir.0.80475-0

Ooi MH, Wong SC, Lewthwaite P, Cardosa MJ, Solomon T (2010) Clinical features, diagnosis, and management of enterovirus 71. [Research Support, Non-U.S. Gov't Review]. *Lancet Neurol 9*(11):1097–1105. https://doi.org/10.1016/S1474-4422(10)70209-X

Ortner B, Huang CW, Schmid D, Mutz I, Wewalka G, Allerberger F, Huemer HP (2009) Epidemiology of enterovirus types causing neurological disease in Austria 1999-2007: detection of clusters of echovirus 30 and enterovirus 71 and analysis of prevalent genotypes. J Med Virol 81(2):317–324

Palacios G, Casas I, Tenorio A, Freire C (2002) Molecular identification of enterovirus by analyzing a partial VP1 genomic region with different methods. [Evaluation Studies Research Support, Non-U.S. Gov't]. J Clin Microbiol 40(1):182–192

Pallansch M, Roos R (2007) Enteroviruses: polioviruses, Coxsackieviruses, echoviruses and newer enteroviruses. In: Knipe DM, Howley PM (eds) Fields virology, 5th edn. Lippincott Williams & Wilkin, USA, pp 840–893

Palmenberg AC (1982) In vitro synthesis and assembly of picornaviral capsid intermediate structures. [Research Support, Non-U.S. Gov't, 44, 900 Research Support, U.S. Gov't, P.H.S.]. J Virol 44(3):900–906

Paul JR (1971) A history of poliomyelitis. Yale University Press, New Haven

Pevear DC, Tull TM, Seipel ME, Groarke JM (1999) Activity of pleconaril against enteroviruses. Antimicrob Agents Chemother 43(9):2109–2115

Pilipenko EV, Gmyl AP, Agol VI (1995) A model for rearrangements in RNA genomes. [research support, non-U.S. Gov't]. Nucleic Acids Res 23(11):1870–1875

Plotkin SA, Orenstein WA, Offit PA, Edwards KM (2018) Poliovirus vaccine-inactivated and poliovirus vaccine-live in Vaccines, 7th edn

Rakoto-Andrianarivelo M, Raobijaona H, Razanamparany M (2000) [Pseudo-poliomyelitis paralysis caused by Echovirus 7]. [Case Reports English Abstract]. *Arch Inst Pasteur Madagascar 66*(1–2):55–57

Rao CD, Yergolkar P, Shankarappa KS (2012) Antigenic diversity of enteroviruses associated with nonpolio acute flaccid paralysis, India, 2007-2009. [research support, non-U S Gov't]. Emerg Infect Dis 18(11):1833–1840

Rhoades RE, Tabor-Godwin JM, Tsueng G, Feuer R (2011) Enterovirus infections of the central nervous system. [Research Support, N.I.H., Extramural Research Support, Non-U.S. Gov't Review]. *Virology 411*(2):288–305. https://doi.org/10.1016/j.virol.2010.12.014

Robbins FC, Enders JF, Weller TH, Florentino GL (1951) Studies on the cultivation of poliomyelitis viruses in tissue culture. V. The direct isolation and serologic identification of virus strains in tissue culture from patients with nonparalytic and paralytic poliomyelitis. Am J Hyg 54(2):286–293

Rombaut B, Vrijsen R, Boeye A (1990) New evidence for the precursor role of 14 S subunits in poliovirus morphogenesis. [research support, non-U.S. Gov't]. Virology 177(1):411–414

Saeed M, Zaidi SZ, Naeem A, Masroor M, Sharif S, Shaukat S, Khan A (2007) Epidemiology and clinical findings associated with enteroviral acute flaccid paralysis in Pakistan. BMC Infect Dis 7:6

Schmidtke M, Wutzler P, Zieger R, Riabova OB, Makarov VA (2009) New pleconaril and [(biphenyloxy)propyl]isoxazole derivatives with substitutions in the central ring exhibit antiviral activity against pleconaril-resistant coxsackievirus B3. Antivir Res 81(1):56–63. https://doi.org/10.1016/j.antiviral.2008.09.002

Shafren DR, Dorahy DJ, Ingham RA, Burns GF, Barry RD (1997) Coxsackievirus A21 binds to decay-accelerating factor but requires intercellular adhesion molecule 1 for cell entry. J Virol 71(6):4736–4743

Shaukat S, Angez M, Alam MM, Sharif S, Khurshid A, Mahmood T, Zaidi SS (2012) Characterization of non-polio enterovirus isolates from acute flaccid paralysis children in Pakistan reflects a new genotype of EV-107. Virus Res 170(1–2):164–168

Jose RR, Shelton SB, James JD. inventor; Board Of Regents Of The University Of Nebraska, assignee; Molecular determinants of tropism and virulence in enteroviruses; Feb 16, 2010. https://patents.google.com/patent/US7662939B1/en

Simmonds P, Welch J (2006) Frequency and dynamics of recombination within different species of human enteroviruses. [research support, non-U.S. Gov't]. J Virol 80(1):483–493. https://doi.org/10.1128/JVI.80.1.483-493.2006

Solomon T, Lewthwaite P, Perera D, Cardosa MJ, McMinn P, Ooi MH (2010) Virology, epidemiology, pathogenesis, and control of enterovirus 71. [Research Support, Non-U.S. Gov't Review]. *Lancet Infect Dis 10*(11):778–790. https://doi.org/10.1016/S1473-3099(10)70194-8

Stanway G, Brown F, Christian P, Hovi T, Hyypiä T, King AMQ, Skern T (2005) In: Fauquet CM, Mayo MA, Maniloff J, Desselberger U, Ball LA (eds) Picornaviridae.Virus taxonomy-eighth report of the international committee on taxonomy of viruses. Elsevier/Academic Press, Amsterdam,The Netherlands, pp 757–778

Suresh S, Forgie S, Robinson J (2018) Non-polio Enterovirus detection with acute flaccid paralysis: A systematic review. [Case Reports Review]. *J Med Virol 90*(1):3–7. https://doi.org/10.1002/jmv.24933

Svitkin YV, Maslova SV, Agol VI (1985) The genomes of attenuated and virulent poliovirus strains differ in their in vitro translation efficiencies. [comparative study]. Virology 147(2):243–252

Tracy S, Chapman NM, Drescher KM, Kono K, Tapprich W (2006) Evolution of virulence in picornaviruses. [review]. Curr Top Microbiol Immunol 299:193–209

Tracy S, Oberste MS, Drescher KM (eds) (2008) *Group B Coxsackieviruses* (Vol. 323). Springer, Verlag Berlin Heidelberg

Tsao KC, Huang CG, Huang YL, Chen FC, Huang PN, Huang YC, Chang SC (2010) Epidemiologic features and virus isolation of enteroviruses in northern Taiwan during 2000-2008. [research support, non-U.S. Gov't]. J Virol Methods 165(2):330–332. https://doi.org/10.1016/j.jviromet.2010.03.001

Vaishnaw AK, Gollob J, Gamba-Vitalo C, Hutabarat R, Sah D, Meyers R, Maraganore J (2010) A status report on RNAi therapeutics. Silence 1(1):14. https://doi.org/10.1186/1758-907X-1-14

Varela-Calvino R, Skowera A, Arif S, Peakman M (2004) Identification of a naturally processed cytotoxic CD8 T-cell epitope of coxsackievirus B4, presented by HLA-A2.1 and located in the PEVKEK region of the P2C nonstructural protein. [research support, non-U.S. Gov't]. J Virol 78(24):13399–13408. https://doi.org/10.1128/JVI.78.24.13399-13408.2004

Vignuzzi M, Stone JK, Arnold JJ, Cameron CE, Andino R (2006) Quasispecies diversity determines pathogenesis through cooperative interactions in a viral population. [research support, N.I.H., extramural]. Nature 439(7074):344–348. https://doi.org/10.1038/nature04388

Wadia NH, Katrak SM, Misra VP, Wadia PN, Miyamura K, Hashimoto K, Kono R (1983a) Polio-like motor paralysis associated with acute hemorrhagic conjunctivitis in an outbreak in 1981 in Bombay, India: clinical and serologic studies. [case reports]. J Infect Dis 147(4):660–668

Wadia NH, Wadia PN, Katrak SM, Misra VP (1983b) A study of the neurological disorder associated with acute haemorrhagic conjunctivitis due to enterovirus 70. J Neurol Neurosurg Psychiatry 46(7):599–610

Wang X, Zhu C, Bao W, Zhao K, Niu J, Yu XF, Zhang W (2012) Characterization of full-length enterovirus 71 strains from severe and mild disease patients in northeastern China. [research support, non-U.S. Gov't]. PLoS One 7(3):e32405. https://doi.org/10.1371/journal.pone.0032405

WHO (2004) Polio laboratory manual, 4th edn. The Department of Immunization, Vaccines and Biologicals, World Health Organization, Geneva, Switzerland

WHO (2007) Supplement to the WHO polio laboratory manual. The Department of Immunization, Vaccines and Biologicals, World Health Organization, Geneva, Switzerland

WHO (2022) Global Wild Poliovirus 2016–2022. In: *Global Polio Eradication Initiative*. World Health Organization

World Health Organization website (n.d.). http://www.who.int

Witso E, Palacios G, Ronningen KS, Cinek O, Janowitz D, Rewers M, Lipkin WI (2007) Asymptomatic circulation of HEV71 in Norway. [Research Support, N.I.H., Extramural Research Support, Non-U.S. Gov't]. *Virus Res 123*(1):19–29. https://doi.org/10.1016/j.virusres.2006.07.015

Wu CN, Lin YC, Fann C, Liao NS, Shih SR, Ho MS (2001) Protection against lethal enterovirus 71 infection in newborn mice by passive immunization with subunit VP1 vaccines and inactivated virus. [Comparative Study Research Support, Non-U.S. Gov't]. *Vaccine 20*(5–6):895–904

Yang WX, Terasaki T, Shiroki K, Ohka S, Aoki J, Tanabe S, Nomoto A (1997) Efficient delivery of circulating poliovirus to the central nervous system independently of poliovirus receptor. [research support, non-U.S. Gov't]. Virology 229(2):421–428. https://doi.org/10.1006/viro.1997.8450

Yen MH, Tsao KC, Huang YC, Huang CG, Huang YL, Lin R, Lin TY (2007) Viral load in blood is correlated with disease severity of neonatal coxsackievirus B3 infection: early diagnosis and predicting disease severity is possible in severe neonatal enterovirus infection. [research support, non-U.S. Gov't]. Clin Infect Dis 44(10):e78–e81. https://doi.org/10.1086/515399

Unleashing the Truth: Understanding Rabies in Humans and Animals

9

Mudasir M. Rather and Koppu Vasavi

Abstract

Rabies, a viral zoonotic disease, continues to pose a significant global health threat affecting both humans and animals. This book chapter provides a comprehensive exploration of rabies, focusing on its epidemiology, genetic makeup, diagnostic modalities, therapeutic approaches, and future perspectives. Emphasis is placed on the genomic aspects of the rabies virus, unraveling the intricacies of its genetic composition, including the pivotal roles played by various viral proteins in disease progression. Moving forward, the chapter delves into diagnostic strategies, encompassing traditional methods like mouse inoculation test (MIT) and fluorescent antibody test (FAT) to advanced molecular techniques such as PCR-based assays. It explores the advantages, limitations, and contemporary applications of these diagnostic tools in identifying and characterizing the virus. Furthermore, the treatment section discusses the current therapeutic options available for both human and animal cases, highlighting the challenges encountered in managing this invariably fatal infection. It also sheds light on emerging therapeutic avenues, including cytokine therapies and novel vaccination strategies, offering glimpses of potential breakthroughs in rabies treatment. Concluding the chapter is a reflection on the cumulative findings, underscoring the necessity for heightened surveillance, enhanced public awareness, and the imperative need for a multidisciplinary approach in combating rabies. The chapter aims to serve as a comprehensive guide, providing insights into the multifaceted facets of rabies, while advocating for proactive measures to mitigate its impact on both human and animal populations.

M. M. Rather (✉) · K. Vasavi
Department of Veterinary Microbiology, ICAR-Indian Veterinary Research Institute, Bareilly, Uttar Pradesh, India
e-mail: mudasirmr023@gmail.com

P. L. Bhukya, A. S (eds.), *Emerging Human Viral Diseases, Volume II*,
https://doi.org/10.1007/978-981-97-4480-0_9

Keywords

Rabies · Rhabdoviridae · FAT · PCR · Anti-rabies vaccines

9.1 Introduction

Rabies is an acute and fatal zoonotic viral encephalitis disease with 100% case fatality rate. This disease is known to humankind since time immoral, and is primarily associated with the bite of rabid dog. The word "rabies" is derived from "rabhas" which means "violence." The disease in human beings caused by rabies is called "Hydrophobia" because the patient exhibits fear of water, being incapable of drinking though subject to intolerable thirst. For centuries, the causative agent of rabies had been attributed to the saliva of rabid dogs. However, through a series of studies commencing in 1881, Pasteur conclusively established that the rabies virus resides within the brains of infected animals. Through a process of sequential intracerebral passages in rabbits, he isolated the fixed virus and demonstrated its graded infectivity in rendering dog's immune through a series of administered injections. The vaccine was formulated by desiccating segments of spinal cord from rabbits infected with the fixed virus for varying durations. In a pivotal event in July 1885, Joseph Meister, a 9-year-old boy bitten by a rabid dog, underwent a course of 13 inoculations using the infected cord vaccine developed by Pasteur. Remarkably, the boy survived, marking a monumental stride in the advancement of medical science.

Every year, rabies infection kills tens thousands of people, mostly in Asia and Africa (95%). Rabies has a huge economic impact accounting economic losses of about $8.6 billion annually in the form of deaths and the cost of vaccination and control program (WHO 2018). Children under the age of 15 account for 40% of those bitten by suspected rabid animals. Rabies eclipse phase varies considerably between 2 weeks and 6 years depending on viral concentration, inoculation location, and innervation density (Abera et al. 2015). Rabies virus is neurotropic, bullet-shaped (180 nm × 75 nm), single-stranded, negative-sense, non-segmented, enveloped, linear, RNA virus with ~12 kb genome size RNA viruses from the family *Rhabdoviridae* and the genus *Lyssavirus* (Madhusudana et al. 2012). Lyssaviruses encode highly conserved five viral proteins; they are listed in the following order: 3′-N-P-M-G-L-5′. The genomic RNA is highly encapsidated with N and P within the virion to form a ribonucleoprotein (RNP) (Tojinbara et al. 2016). The RNP complex aids viral growth in the cytoplasm of host cells. The exterior of the virion is enclosed with projecting spikes made up of viral glycoprotein protein trimers. This glycoprotein is the only protein on the viral surface that is responsible for viral pathogenicity and promotes protective immunity against rabies (Blanton et al. 2008).

9.2 Etiology and Classification

Rabies, caused by the rabies virus (RABV) within the genus Lyssavirus, falls under the Rhabdoviridae family of the Mononegavirales order, as per the ICTV-Virus Taxonomy from 2017. Lyssavirus, categorized into two phylogroups, places classical rabies virus (RABV) within phylogroup-1. Among the seven genera constituting the Rhabdoviridae family, Lyssavirus encompasses the classical rabies virus (RABV; genotype 1), Lagos bat virus (LBV; genotype 2), Mokola virus (MKV; genotype 3), Duvenhage virus (DV; genotype 4), European bat lyssavirus 1 (EBLV-1; genotype 5), European bat lyssavirus 2 (EBLV-2; genotype 6), and Australian bat lyssavirus (ABLV; genotype 7) (Ceccaldi et al. 1989). Recent proposals have introduced four new viruses from insectivorous bats—Irkut virus (IV), Khujand virus (KV), Aravan virus (AV), and West Caucasian bat virus—as potential members of the Lyssavirus genus. Despite antigenic differences, these lyssaviruses share identical morphological and neurotropic characteristics (Bleck and Ruprecht 2002).

9.3 Epidemiology and Geographical Distribution

Rabies, a disease reported across all continents except Antarctica, disproportionately leads to over 95% of deaths in Asia and Africa, highlighting an uneven distribution. Presently, rabies remains endemic in over 100 countries globally, affecting regions where around 2.5 billion people reside (as of 2010). Areas with prevalent rabies have typically been classified into three groups: first, nations with ongoing canine rabies (across Asia, Africa, and Latin America); second, countries where canine rabies is controlled while wildlife-related cases prevail (such as Western Europe, Canada, and the United States); and third, rabies-free countries, primarily islands like England, Australia, and Japan. India, in particular, falls within the high-risk category of the first zone.

In general, most of the developing countries face problem of urban rabies (dogs are responsible for maintenance and spread of virus to human) and developed countries face problem of wildlife rabies (sylvatic rabies: disease is maintained and transmitted by wildlife). The most common mode of rabies virus transmission is bite of infected animals. Other possible modes of transmission include aerosol contamination of mucous membrane (eyes, nose, mouth), and corneal and organ transplantation, but such cases are rare. In India, human rabies is mainly transmitted through dog bite (96.2%) and majority of these were stray dogs (75.2%) (Sudharshan 2007). Transmission through other wild animal bite and mongoose should not be neglected.

Reservoir hosts are important to maintain virus in nature. Important reservoir hosts are species of family *Canidae* (dogs, jackals, coyotes, wolves, foxes, and Raccoons), *Mustelidae* (skunks, martens, weasels, and stoats), *Viverridae* (mongoose and meerkats), and *Procyonidae* (raccoons) and order Chiroptera (bats). Raccoons, skunks, foxes, coyotes, and several species of insectivorous bats are important reservoir in major portion of America and Europe, whereas stray dogs are the common reservoirs in India.

Every year, rabies claims the lives of tens of thousands of individuals. Apart from Antarctica, many of the nations are endemic for rabies. Because rabies is avoidable all around the world, several countries have achieved rabies-free status through stringent quarantine and geographic isolation. As per WHO criteria, a nation can achieve a rabies-free status by demonstrating a lack of locally acquired cases—both in animals and in humans—for two consecutive years, owing to stringent surveillance and import controls. Despite efforts to vaccinate wildlife, the risk of reintroduction remains from neighboring countries (Dutta and Dutta 1994). Several countries including Bahrain, Cyprus, Hong Kong, Maldives, Qatar, Singapore, Timor-Leste, and others listed in Yousaf et al. 2012, have remained devoid of this virus for extended periods. According to the WHO Global Vaccine Research Forum, rabies causes over 30,000 fatalities annually in Asia (Raux et al. 2000), with at least 24,000 deaths occurring in Africa. Bangladesh and India bear the greatest burden, followed by Nepal, Myanmar, Bhutan, Thailand, and Indonesia.

9.4 Biosafety Measures

The lyssavirus are enveloped, bullet-shaped virions containing proteins, lipids, and single-stranded RNA; such viruses are sensitive to detergents and lipid solvents (soap solutions, ether, chloroform, acetone), alcohols, and iodine preparations. However, care should be taken while handling acetone-fixed brain impression slides because acetone may not completely inactivate virus. Other properties of rabies virus include resistance to drying, and repeated freezing and thawing. Virus is relatively stable at pH 5–10 but inactivated by extremely low pH (below 3). Virus is sensitive to pasteurization temperature, ultraviolet light, and formalin fixation and inactivated by beta-propiolactone (BPL) and gamma irradiation. Quaternary ammonium disinfectants in 1:256 dilution, 70% alcohol (ethanol or isopropanol), 1% soap solution, 5–7% iodine solutions, and oxidizing agents inactivate lyssaviruses within few minutes. Additional time is required to inactivate virus on materials, instruments, or surfaces contaminated with proteins found in cell culture media and tissues. So, carcasses and animal tissues are best disposed within biohazard bags and then incinerated. Virus is swiftly deactivated by sunshine and drying, and it does not persist in the environment for lengthy periods of time (in dried blood and secretions) (Blanton et al. 2007).

Laboratory facilities should be designed with quality standards, biosafety, and biosecurity in mind. Lyssaviruses are defined as risk Group2 or depending on jurisdiction, local risk assessments, they may be handled in 2 or 3 facilities. Use of class II biosafety cabinet (BSC) is recommended for any procedures that may produce an aerosol or for techniques employed to produce lyssaviruses within cell culture. There should be restricted entry or entry for those only who are fully vaccinated against rabies. The minimum personal protective equipment (PPE) should include a laboratory coat or gown, gloves, eye protection, safety glasses, goggles, face shield, and N95 mask.

9.5 Organization of Infectious Agents

9.5.1 Rabies Virus and Genome Organization

Rabies virus is neurotropic, bullet-shaped (180 nm x 75 nm), single-stranded, negative-sense, linear, non-segmented, enveloped RNA virus with ~12 kb genome with helical symmetry (Madhusudana et al. 2012). It has 7 numbers of genotypes; more than 16 different lyssaviruses have been described. RV-genotype 1 is the main disease-causing genotype in mammals. Lyssaviruses encode highly conserved and monocistronic five viral proteins: the nucleoprotein (N- 1334 bp, 57 KDa), phosphoprotein (P- 978 bp, 38.5 KDa), matrix protein (M- 840 bp, 25 KDa), glycoprotein (G- 1647 bp, 67 KDa), and large RNA-dependent RNA polymerase protein (L- 6381 bp, 200 KDa). The five viral proteins are listed in the following order: 3′-N-P-M-G-L-5′. There are intergenic elements between the coding region sections of the genome, N–NS, NS–M, M–G, and G–L, where the lengthy intergenic element between G and L is known as the pseudo gene and it has been utilized as a reference for discriminating distinct rabies genotypes (Blanton et al. 2008). The RNP forms a helix that is encapsidated with M, connecting the RNP to the lipid bilayer membrane envelope (Rupprecht et al. 1992). The exterior of the virion is enclosed with projecting spikes made up of viral G protein trimers (10 nm long).

9.5.2 Propagation and Assay in In Vitro and In Vivo Laboratory Models

The virus can be isolated through intracerebral inoculation in young mice (aged 3–4 weeks) using samples obtained from the brain, spinal fluid, saliva, and urine. Higher chances of isolation occur in the initial stages of the disease. However, a few days after the onset, the appearance of neutralizing antibodies makes occasional virus isolation possible.

Alternatively, a more sensitive and rapid isolation method involves tissue culture cell lines such as Neuro2a/CCL 131 (mouse neuroblastoma cell line), BHK-21 (baby hamster kidney), Vero cell, McCoy cell, and WI-38. These cell lines exhibit minimal or absent cytopathic effects, so virus propagation is identified using immunofluorescence (IF). A positive IF test can detect the organism as early as 2–4 days after inoculation. The CCL 131 (ATCC) neuroblastoma cell line, in particular, demonstrates high susceptibility to street RABV without requiring any adaptation steps, making it the standard choice for routine rabies virus isolation.

9.6 Viral Proteins and Life Cycle

Protein	Functions
N	• Amount of N protein—Determines the transition between transcription and replication activities of genomic RNAs • Major role in diagnosis and virus identification • Protein exhibits precipitating, complement fixing, immunoflourescent and ELISA activity
P	• The P protein forms nucleoprotein complex that encloses the viral RNA along with the N and L proteins and is involved in transcription and replication along with the L protein • The P protein also functions as a chaperone of soluble nascent N protein • Act as an important determinant in retrograde transport of the virus within axons after binding with dynein light chain LC8
M	• Responsible for the assembly and budding of virus by regulating the virus transcription and replication.
G	• Acts as protective and neutralizing antigen • Responsible for binding with the target cell receptors (pH-dependent)
L	• Vital for pathogenesis along with G protein, survival, and replication of rabies virus

9.7 Pathogenesis

Incubation period is typically 1–3 months. It may vary from few weeks to many years, which depends on distance between site of bite and central nervous system, severity of bite, and dose of inoculums. Virus replicates at the site of inoculation (muscle, connective tissue, or nerves) up to 48–72 h and then penetrates and enters in the peripheral nerve cells through nicotinic acetylcholine receptors (Lentz et al. 1982). From the nerve endings virus starts traveling passively in the axoplasm toward central nervous system at a speed of 3 mm per hour. Then, the infection spreads centripetally from axon to the neuronal bodies. After reaching CNS, virus infects and damages many nerve cells. This damage of nerve cells is responsible for exhibiting rabies-specific symptoms. The virus then starts to spread away from CNS (centrifugal spread) along neuronal pathway, particularly involving the parasympathetic nervous system, which is responsible for infection of salivary glands, skin, heart, and variety of other organs (Jackson 2007). As a result, infectious rabies virus is secreted into saliva, which is important for transmission to other hosts.

9.7.1 Replication and Transcription

The rabies virus spreads through the bite of an infected animal, utilizing the G protein's enhanced interaction with acetylcholine receptors in the peripheral nervous system for viral attachment (Albertini et al. 2011). Once attached, the virus accesses the central nervous system via endocytosis. Within the endosome, a decrease in pH triggers conformational changes in the G protein, facilitating the fusion of viral and endosomal membranes, leading to the release of the viral nucleocapsid (Albertini et al. 2008).

Due to the incapability of mammalian cells to transcribe negative-stranded RNA, the rabies virus encodes its own RNA-dependent RNA polymerase. The five virion proteins—nucleocapsid (N), matrix (M), polymerase proteins (L), phosphoprotein (P), and glycoprotein (G)—are encoded by monocistronic mRNAs. While G protein synthesis initiates on free ribosomes, its completion and glycosylation occur within the endoplasmic reticulum and Golgi apparatus.

The virion-associated RNA polymerase transcribes the single-stranded RNA genome into positive-sense mRNA. This genome serves as a template for complementary positive-sense RNA, producing negative-sense progeny RNA (Abera et al. 2015). The L-P polymerase complex initiates transcription by creating short RNA molecules known as leader RNA. After this phase, transcription resumes, generating capped and polyadenylated nucleoprotein mRNA (Ogino and Banerjee 2007).

As infection progresses, L activity shifts toward replication, producing full-length positive-strand RNA copies without caps or poly (A) tails. These complementary RNAs, or cRNAs, are encapsidated by nucleoprotein, serving as templates for new negative-strand RNA genomes enclosed by N, forming new nucleocapsids (Arnheiter et al. 1985).

In the assembly phase, the N-P-L complex envelops negative-stranded genomic RNA to form the RNP core. The M protein then forms a capsule around the RNP. This RNP-M complex relocates to a plasma membrane area with glycoprotein inserts. The M protein initiates coiling, attaching the M-RNP complex to the glycoprotein. Subsequently, the virus emerges from the plasma membrane, with the viral matrix protein creating an interior layer within the envelope and the viral glycoprotein generating spikes on the exterior.

9.8 Clinical Signs

Rabies-related viruses have been identified to provoke symptoms indistinguishable from classical rabies (Smith 1996). The disease progression in both humans and dogs exhibits distinct stages:

9.8.1 In Humans

Prodromal Period: Initial symptoms include fever, headache, malaise, anorexia, sore throat, myalgia, and often, an unusual sensation at the wound site such as tingling, pricking, or burning.

Acute Encephalitic Phase: This phase starts with hyperactivity, intermittently accompanied by bouts of unusual behavior and agitation amidst normal periods. Difficulty in drinking despite intense thirst becomes a defining feature. Patients might swallow dry solids but struggle with liquids.

Coma: Some individuals progress into a comatose state.

Death: Typically occurs within 1–6 days due to respiratory arrest during convulsions.

9.8.2 In Dogs

The incubation period spans 3–6 weeks but could vary from 10 days to a year. Initial signs include an alert yet troubled demeanor, restlessness, snapping at imaginary objects, and licking or gnawing at the bite site. After 2–3 days of this prodromal stage, two forms of the disease emerge:

Furious Form: Marked by hyperexcitability, spasms, hydrophobia, and sometimes aerophobia. Death usually results from cardiorespiratory arrest.

Paralytic (Dumb) Form: This variant involves an ascending paralysis, less dramatic than the furious form, and generally, a longer course. It progresses slowly into a coma, eventually leading to death.

9.9 Immune Response to Rabies

During the migration of the rabies virus (RABV) toward the central nervous system (CNS) along nerves, the innate immune system recognizes the virus. However, the specific immune response triggered by RABV occurs only when the virus infects the CNS (Zhao et al. 2012). The innate immune system serves as the initial defense against infection, playing a crucial role in recognizing invading microorganisms and initiating the inflammatory response (Mogensen 2009). Conversely, the adaptive immune system's role is to eliminate pathogens in the later stages of infection and develop immunological memory.

In the context of RABV infection, cellular immune response holds significance for an effective host immune response. While specialized innate immune cells exist in the CNS, the recruitment of peripheral immune cells (CD8 T, CD4 T, and B cells) to local inflammatory sites is vital. These cells aid in containing/clearing the pathogen but can also contribute to tissue damage.

Activation of the adaptive immune response relies on the antigen presentation by antigen-presenting cells (APCs) to initiate naïve T-cell activation. Activated T cells express specific molecules facilitating their migration across the blood-brain barrier (BBB) and their recruitment to the CNS, mediated by CRMP2 (Vuaillat et al. 2008).

However, in infections caused by highly pathogenic RABV strains, T cells activated in the periphery do not effectively combat the virus and are destroyed in the CNS, impairing the lymphocyte-mediated response (Lafon 2005; Baloul et al. 2004; Rossiter et al. 2009). Despite this, T lymphocytes play a protective role in controlling RABV replication in CNS tissue, primarily through IFN-γ production.

Regarding humoral immune response, activated B cells secrete virus-specific antibodies, essential in eliminating free virus and infected cells via various mechanisms. Rabies-specific antibodies typically target epitopes on the RABV G protein, and while IgG antibodies are protective, the role of IgM might be limited due to its structure (Hooper et al. 1994).

Post-exposure prophylaxis (PEP) relies on rapid B-cell responses, with IgM potentially playing a role in preventing the spread of pathogenic RABV from peripheral sites to the CNS while IgG antibodies are formed (Dorfmeier et al. 2013). Studies also highlight IL-21's critical role in inducing rapid primary antibody responses to RABV-based vaccines.

Immunoglobulin M (IgM) is detectable early post-inoculation, followed by IgG, and while both responses persist for years, IgG provides more effective protection against the disease (Turner 1978).

9.10 Host-Virus Interactions

The virus gains entry through peripheral nerves, occasionally infecting muscle cells and neuromuscular junction motor endplates, along with other innervated tissues (WHO 2013). Once inside, the virus is enclosed in endosomal transport vesicles and travels retrogradely through axons using microtubules (Lembo et al. 2010). Within the neuronal soma, RNP complexes emerge from vesicles, initiating primary transcription, leading to the synthesis of viral proteins (Warrell 2012). This accumulation of viral proteins leads to the formation of cytoplasmic inclusion bodies and sites where viral RNA production (secondary transcription) occurs. Newly formed RNPs move from these locations to postsynaptic membranes (Srinivasan et al. 2005), where new virions are created and transferred trans-synaptically to subsequent neurons, facilitated by the viral G protein. Subsequently, budding occurs in the synaptic cleft, followed by receptor-mediated entry at presynaptic membranes (Chen et al. 2017).

The virus progresses to infect the brain, targeting specific areas based on the motor neurons that innervate the site of entry. From the brain, it moves to the salivary glands and sporadically gets expelled in the saliva, ready for transmission to another host (Maier et al. 2010). Local symptoms emerge as the virus spreads within the central nervous system (CNS) and reaches the dorsal root ganglia, associated with the nerves at the entry site (Vora et al. 2015). Viral replication triggers inflammation in the dorsal root ganglia, resulting in pain, tingling sensations (paresthesia), and itching. Furthermore, the virus can infiltrate peripheral nerves, causing inflammation and weakness in patients with paralytic rabies (Johnson et al. 2006).

9.11 Anti-Rabies Vaccines

Vaccine is defined as a standardized formulation containing defined amounts of immunogens. Most rabies vaccines are prepared from Pasteur's original strain (1882) and its derivative strains (Pasteur virus, challenge virus standard, and Pitman-Moore); and strains isolated in the twentieth century (Flury: low and high egg passage; Street Alabama-Dufferin: Vnukovo) (Wu et al. 2011). Live vaccine is attenuated for target species by passaging in heterologous host or other suitable methods (Dong-kun yang et al. 2013). Inactivated vaccines are prepared by inactivation of antigen by various chemical or physical means. Biotechnology-based vaccines are prepared using various genetic engineering techniques, and vaccines like chimera, marker, mutant, reverse genetic-based, edible, and oral vaccine are classified under this category.

The anti-rabies vaccines are categorized into two primary types:

9.11.1 Neural Vaccines

These consist of suspensions derived from nervous tissues of animals infected with fixed rabies virus strains. Among the earliest formulations was Pasteur's cord vaccine, created by drying pieces of infected rabbit spinal cord over caustic potash. Subsequent versions like the Fermi and Semple vaccines used brain tissues treated with phenol for inactivation. Neural vaccines, however, have limitations due to their low immunogenicity, primarily containing nucleocapsid antigen and minimal amounts of the protective antigen, glycoprotein G.

9.11.2 Non-Neural Vaccines

Egg Vaccines: One type utilized duck eggs for growing and inactivating fixed virus strains, yet it was discontinued due to its low immunogenicity.

Live Attenuated Chick Embryo Vaccines: Developed with the Flury strain, there are two versions—a low egg passage (LEP) vaccine for dogs at 40–50 passage levels and a high egg passage (HEP) vaccine at 180 passage levels for cattle and cats.

Tissue Culture Vaccines: The human diploid cell (HDC) vaccine, like the Pitman-Moore strain, was a significant advancement. It involved growing fixed rabies virus on human diploid cells (WI-38) and inactivating it with agents like beta-propiolactone or tri-n-butyl phosphate. Other effective vaccines grown on primary cell cultures include those derived from chick embryo cells and the Vero cell line.

Additionally, subunit vaccines have been developed. A recombinant vaccinia virus/G protein vaccine has emerged, particularly used in eradicating rabies among foxes in Europe.

9.12 Diagnostics

Clinical indications such as itching, discomfort, or paresthesia at the location of the bite wound, as well as gastrointestinal disturbances, are used to make a first-round diagnosis (Hemachudha et al. 2002). Diagnosis in the past was mainly done on histopathological methods like different staining methods for detection of cytoplasmic virus inclusions, the Negri bodies (Tierkel and Atanasiu 1996). Now techniques available like nucleic acid amplification, reverse transcriptase-polymerase chain reaction, rapid fluorescent focus inhibition test (Nadin-Davis et al. 2009), fluorescent antibody virus neutralization test, enzyme-linked immunosorbent assay, and direct rabid immunohistochemistry test, histological identification of Negri bodies, mouse inoculation test, direct fluorescent antibody test (gold stranded test), as well as Rabies tissue culture infection test (Hemachudha et al. 2002).

Detection of rabies antigen involves various tests and techniques:

9.12.1 Mouse Inoculation Test (MIT) and Fluorescent Antibody Test (FAT)

The MIT method is adept at detecting small virus quantities, even in partially decomposed specimens. It is practical, sensitive, reliable, and technically straightforward. However, its main drawback lies in ethical concerns related to live animal usage in laboratories. It takes approximately 7–8 days to yield interpretable results, demands a well-maintained animal house for a continuous supply of mice, and requires a substantial number of mice per sample, making it labor-intensive.

On the other hand, the FAT, developed by Goldwasser and Kissling in 1958, stands as one of the most reliable diagnostic tools for routine rabies diagnosis. It has become highly refined, prompting many laboratories to opt for FAT over MIT. Studies have shown FAT's sensitivity to be 100%, although in cases where FAT results are negative, confirmation by MIT is recommended to prevent false negatives. Direct FAT is endorsed by WHO and OIE due to its sensitivity, accuracy, and speed in diagnosis.

9.12.2 NIH Test

The NIH test for potency, originating in the 1950s at the National Institutes of Health, measures the protection provided by an inactivated rabies vaccine in immunized mice challenged with rabies virus. The procedure involves vaccinating two groups of mice with dilutions of a reference vaccine and the vaccine under test. Seven days after the last vaccination, these mice, along with a control group, are intracerebrally challenged with the challenge virus standard (CVS) mouse-brain strain of fixed rabies virus. Over a 14-day period, the mice are observed, and the median effective dose (ED50) of the reference and test vaccine is determined. The relative potency of the test vaccine is calculated by comparing its ED50 with that of the reference vaccine, using a specific formula recommended by WHO.

$$RP = \frac{\text{reciprocal of } ED_{50} \text{ of TV}}{\text{reciprocal of } ED_{50} \text{ of RV}} \times \frac{\text{dose of TV}}{\text{dose of RV}}$$

Where;

TV = test vaccine.

RV = reference vaccine.

dose = volume of a single vaccine dose.

As per WHO expert committee (1996), the inactivated veterinary vaccines with a potency of less than **1.0 IU** per dose, as measured by the NIH test, should not be licensed or released. And for human use, it should have minimum potency of **2.5 IU** per dose. The main demerits, regarding the use of NIH test for potency, are scientific validity, cost, and animal welfare issues.

9.12.3 PCR and Teal-Time PCR Techniques

PCR stands as a highly sensitive, specific, and rapid diagnostic tool for various viral diseases, including rabies. The amplification of rabies cDNA through PCR serves as a valuable method for characterizing isolates and understanding the molecular epidemiology of the virus. One notable advantage of RNA detection methods, such as PCR, is their adaptability to a wide range of sample types, including saliva and cerebrospinal fluid, which might not be suitable for direct FAT testing.

In PCR targeting rabies, the N-gene is frequently chosen due to its highly conserved nature, abundance, and extensive sequenced data associated with this gene. PCR has been crucial in confirming human rabies diagnoses in cases where other tests could not be easily applied. Real-time PCR techniques have proven effective in detecting rabies in infected, decomposed brain tissues, overcoming the challenges associated with the sensitivity of FAT and degraded tissue.

Various studies have utilized PCR-based methods for both diagnosis and molecular epidemiology of rabies virus. These approaches have evolved to include semi-nested PCR assays for detecting multiple genotypes of rabies and related viruses. Additionally, advanced PCR techniques like TaqMan PCR and in situ PCR have been employed to detect rabies virus RNA, offering insights into infected cell cultures and brain tissues.

9.13 Detection and Quantification of Rabies Antibodies

The detection and quantification of rabies antibodies primarily aim to assess immunity or vaccine effectiveness. This process involves known virus neutralization by antibodies from serum samples, inhibiting rabies infection both in animals (in vivo) and in cell cultures (in vitro) (Bourhy and Sureau 1990). Results are typically measured in international units relative to an established international standard antiserum.

Several tests have been developed for this purpose, such as the mouse neutralization test (MNT), rapid fluorescent focus inhibition test (RFFIT), and fluorescence antibody virus neutralization test (FAVN). The MNT, an established method, evaluates rabies virus neutralizing antibodies (Webster and Dawson 1935). To quantify immune response post-rabies vaccination in animals, serum neutralization methods on mice and in cell culture are commonly used (Smith 1996).

The World Health Organization (WHO) recommends the in vivo virus neutralization test in Mice (VNT) and in vitro rapid fluorescence focus inhibition test (RFFIT) for determining rabies virus neutralizing antibody titers (Smith et al. 1973). A modified version of RFFIT utilizing monoclonal anti-rabies antibodies and a peroxidase anti-mouse conjugate instead of a fluorescein-conjugated anti-rabies antibody has been reported (Hostnik 2000).

Although RFFIT has long been considered the gold standard assay, it demands trained personnel, cell culture, fluorescence microscopy facilities, and stringent biosafety measures due to live virus handling. Moreover, serum quality issues like cytotoxicity and nonspecific inhibitors may yield false-positive results in RFFIT (Ma et al. 2012).

Enzyme-linked immunosorbent assay (ELISA) methods, although not functional assays detecting neutralizing antibodies like RFFIT, offer a simpler, safer, and quicker alternative. These assays, not reliant on live virus or high-containment facilities, correlate well with RFFIT (Servat et al. 2007; Muhamuda et al. 2007). Second-generation ELISA, like Platelia Rabies II, is suitable for laboratories lacking virus and cell culture facilities and exhibits good correlation with RFFIT (Feyssaguet et al. 2007). However, ELISA tests have been reported to have lower sensitivity than neutralization tests (Cliquet et al. 2004).

Novel methods, such as electro-chemiluminescence-based ELISA, have demonstrated higher sensitivity than traditional ELISA for rabies glycoprotein antibody detection (Ma et al. 2012). The development of a rapid neutralizing antibody detection test (RAPINA) using immune chromatography offers a quick qualitative and semi-quantitative analysis of rabies neutralizing antibodies in humans and dogs (Nishizono et al. 2012). Additionally, a simple latex agglutination test has been proposed for detecting rabies-specific antibodies (Madhusudana and Saraswati 2003).

9.14 Therapeutics

Because dog bites cause nearly all human rabies deaths, mass vaccination of dogs remains the primary technique for eliminating urban rabies in endemic regions. However, in most countries, the main preventive approaches for limiting rabies in reservoirs are different means of culling free-roaming dogs (Mustafa et al. 2015). Because the reduction in the vector population was minimal, it did not seem to stop the spread of rabies. Furthermore, mass vaccination of dogs must be supplemented by bait immunization of wild animals to avoid disease spread from wild animals (Yale et al. 2022). Distributing bait rabies vaccination to high-risk areas is an effective means of reducing wildlife rabies. In general, vaccination coverage of around 70% of the vector population is thought to be adequate to prevent rabies transmission. However, present rabies vaccinations for humans and animals have intrinsic limitations; thus, attempts have been undertaken to improve traditional rabies vaccines in light of technological innovation (Dean et al. 1963). Passive immunization with human and horse rabies immunoglobulins is presently accessible. However, cleaning the bite or scratch site with large amounts of water or detergent within 15 min of exposure will lower the virus load. Apply a topical iodine-containing or similarly virucidal treatment (Gautret et al. 2014). Ingredients such as chili, salt, lemon, and herbs should never be used. After categorizing the bite or scratch location, treatment should be carried out. An animal-like on the intact skin, and touching and feeding animals are deemed categories I (no exposure), small scratches lacking bleeding and nibbling of uncovered skin are regarded categories II (exposure), and single or numerous transdermal bites or scratches, saliva infectivity of mucous membranes or broken skin, and contact with infected wild mammals are considered categories III (exposure) (severe exposure)(WHO 2018). The closer the virus is exposed to the head, the faster it infects the brain. Pre- and post-exposure prophylactic measures are employed for rabies vaccination tactics. Pre-exposure prophylaxis is suggested for those at high risks, such as isolated populations, travelers, and individuals at occupational risk in highly endemic environments with inadequate access to effective post-exposure prophylaxis (Mustafa et al. 2015). The pre-exposure immunization strategy includes vaccinations from 0, 7, and 28 days. Serological antibody estimate should be performed 1–3 weeks following the last vaccination. Laboratory professionals should be re-examined every 6 months, and other diagnosticians should be re-examined every 2 years. Booster immunization is strongly advised even if the titer falls below 0.5 International Units (IU) per mL. When serological surveillance is not available, a 1-year booster immunization is advised, followed by vaccinations every 1–3 years (Stankov et al. 2022). The regimen for post-exposure immunization includes vaccinations at 0, 3, 7, 14, and 28 days. Vaccines and rabies immunoglobulins are required for post-exposure prophylaxis. In category I, wound cleaning is sufficient; in category II, wound washing with vaccinations is necessary; and in category III, rabies immunoglobulins are administered in addition to vaccines and wound treatment. VaxiRab-N, Verorab, Rabivax-S, and Rabipur are rabies vaccinations that have been pre-qualified by the WHO. The immunization is administered by intradermal and intramuscular methods; the injection sites for the intradermal and intramuscular routes are the same (Tojinbara et al. 2016). WHO recommends a vaccination efficacy of 2.5 IU per single intramuscular injection; a minimum titer of 0.5 IU/ml of serum is seemed protective. Rabies immunoglobulins are indicated for category III exposure and are not advised if one have had rabies vaccination as pre-exposure prophylaxis and first post-exposure prophylaxis dosage was more than 7 days ago (Reece and Chawla 2006). Rabies immunoglobulins must be injected deep into and surrounding all wounds; it can be diluted for large and many wounds if the need exceeds the maximum dose (Yousaf et al. 2012). It is made up of either human (HRIG) or equine (ERIG) rabies antibodies. Because immunoglobulins cannot traverse intact blood-brain barriers, they are rendered useless once the virus has reached the brain and clinical signs have started (Kallel et al. 2006). The immunoglobulin, HRIG dosage 20 IU/kg body weight; ERIG dosage 40 IU/kg body weight, is injected locally around the bite location, with the remainder administered intramuscularly far from the vaccination site (Mustafa et al. 2015).

9.15 Prevention and Control

There is not a specific treatment for rabies, an incurable and fatal illness. Knowledge about rabies prevention and treatment, both in humans and in animals, is limited. While people are aware of the dangers of the disease, they often resort to various traditional or religious treatments instead of seeking appropriate post-exposure prophylaxis (Consales and Bolzan 2007). Traditional treatments, particularly for dog bites, are prevalent despite the importance of the disease being recognized within communities. Controlling rabies in low-income countries could be highly effective through proper planning, health education, vaccinating 70 percent of dog populations, and implementing epidemiological surveillance (Tierkel et al. 1950). Mass dog vaccination is considered the most practical and cost-effective approach to eliminating canine rabies, safeguarding both human and canine lives (WHO 2013).

During mass vaccination initiatives, all dogs, irrespective of age (above 3 months), weight, or health, should receive the vaccine. Primary immunization can involve a single or two doses administered a month apart (Sugiyama and Ito 2007). Subsequent vaccine doses, given yearly, biannually, or triennially, aim to reinforce their immunity, based on vaccine effectiveness (Abera et al. 2015). In the United States, a successful instance of mass immunization in a canine population occurred in 1948 in Memphis and Shelby County, Tennessee, resulting in zero animal and human cases (Warrell 2012). Dogs, cats, or ferrets that have never been vaccinated and come in contact with a rabid animal should be euthanized immediately (Srinivasan et al. 2005). If euthanasia is refused by the owner, these animals must be kept in strict isolation for six months. To adhere to pre-exposure vaccination recommendations, the rabies vaccine should be administered upon isolation admission or up to 28 days before release (Mustafa et al. 2015).

Preventing human rabies involves avoiding exposure to the rabies virus, administering appropriate pre-exposure prophylaxis, and promptly treating bite wounds alongside proper post-exposure prophylaxis. Inactivated human vaccines are available for at-risk veterinary personnel, animal caregivers, wildlife officers, and laboratory workers (Yousaf et al. 2012).

9.16 Future Perspectives

Rabies can often be ruled out early based on test results or disease progression, but it is crucial to consider it at the initial stage and take precautionary measures. There is a need to intensify surveillance for current RABV variants in rabid animals, especially in wildlife. Understanding the true prevalence of rabies in wild animals is crucial to grasp their role in disease transmission. Constant monitoring of less common non-reservoir species is necessary to detect newly introduced RABV variants. Strict barrier precautions are vital as the virus can transmit through saliva contact with broken skin or mucous membranes.

Veterinarians need to notify appropriate authorities promptly and accurately report cases, varying with jurisdiction. Medical and veterinary professionals should educate the public about the transmission routes of rabies, especially for those living in rural areas and individuals working in abattoirs, to minimize disease spread. Reporting to physicians by all exposed individuals and prioritizing post-exposure prophylaxis (PEP) is crucial. Advanced molecular diagnostic techniques, such as sequence-specific RT-PCR, offer highly targeted diagnosis, but further development in molecular diagnosis is necessary for early confirmation and differentiation of RABV genotypes.

Increased research is essential to uncover unknown aspects of rabies pathology and pathogenesis, paving the way for suitable antiviral therapeutic strategies. Molecular biology advancements have enabled the development of new-generation vaccines to control animal rabies and its transmission to humans. Emerging therapies like cytokine and siRNA therapies show promise. However, regular vaccination and antibody titer monitoring every 2 years are necessary for those at higher risk of rabies exposure. Children, especially in developing areas with endemic rabies, are at the highest risk of contracting rabies from dog bites, emphasizing the need for pre-exposure rabies vaccination for travelers to remote rural regions.

Assessing the risk of exposure to rabid dog bites and availability of rabies immunoglobulins and vaccines in endemic zones is crucial. Evaluating the effectiveness of oral/bait vaccines across different wild animal species and ensuring vaccine stability in the field is essential. Leveraging recombinant DNA technology, new immune-prophylactic agents like subunit vaccines or DNA-based options could overcome conventional vaccine limitations. Developing edible vaccines, especially for domestic and wild ruminants, is necessary to control rabies. While many countries remain high-risk areas, several have achieved rabies-free status, indicating successful prevention measures. Scientific advancements and public awareness can significantly contribute to rabies control, but achieving these goals requires high-level political commitment.

References

Abera, E., Assefa, A., Belete, S., & Mekonen, N. 2015. Review on rabies, with emphasis on disease control and eradication measures

Albertini AAV, Schoehn G, Weissenhorn W, Ruigrok RWH (2008) Structural aspects of rabies virus replication. Cell Mol Life Sci 65(2):282–294

Albertini AA, Ruigrok RW, Blondel D (2011) Rabies virus transcription and replication. Adv Virus Res 79:1–22

Arnheiter H, Davis NL, Wertz G, Schubert M, Lazzarini RA (1985) Role of the nucleocapsid protein in regulating vesicular stomatitis virus RNA synthesis. Cell 41(1):259–267

Baloul L, Camelo S, Lafon M (2004) Up-regulation of Fas ligand (FasL) in the central nervous system: a mechanism of immune evasion by rabies virus. J Neurovirol 10(6):372–382

Blanton JD, Hanlon CA, Rupprecht CE (2007) Rabies surveillance in the United States during 2006. J Am Vet Med Assoc 231(4):540–556

Blanton JD, Palmer D, Christian KA, Rupprecht CE (2008) Rabies surveillance in the United States during 2007. J Am Vet Med Assoc 233(6):884–897

Bleck TP, Ruprecht CE (2002) Rhabdoviruses. In: Richman DD, Whitley RJ, Hayden FG (eds) Clinical virology

Bourhy H, Sureau P (1990) Laboratory methods for rabies diagnosis. Institute Pasteur, Paris

Ceccaldi PE, Gillet JP, Tsiang H (1989) Inhibition of the transport of rabies virus in the central nervous system. J Neuropathol Exp Neurol 48(6):620–630

Chen S, Zhang H, Luo M, Chen J, Yao D, Chen F, Chen T (2017) Rabies virus transmission in solid organ transplantation, China, 2015–2016. Emerg Infect Dis 23(9):1600

Cliquet F, McElhinney LM, Servat A, Boucher JM, Lowings JP, Goddard T, Fooks AR (2004) Development of a qualitative indirect ELISA for the measurement of rabies virus-specific antibodies from vaccinated dogs and cats. J Virol Methods 117(1):1–8

Consales CA, Bolzan VL (2007) Rabies review: immunopathology, clinical aspects and treatment. J Venom Anim Toxins Incl Trop Dis 13(1):5–38

Dean DJ, Baer GM, Thompson WR (1963) Studies on the local treatment of rabies-infected wounds. Bull World Health Organ 28(4):477

Dorfmeier CL, Tzvetkov EP, Gatt A, McGettigan JP (2013) Investigating the role for IL-21 in rabies virus vaccine-induced immunity. PLoS Negl Trop Dis 7(3):e2129

Dutta JK, Dutta TK (1994) Rabies in endemic countries. BMJ 308(6927):488–489

Feyssaguet M, Dacheux L, Audry L, Compoint A, Morize JL, Blanchard I, Bourhy H (2007) Multicenter comparative study of a new ELISA, PLATELIA™ RABIES II, for the detection and titration of anti-rabies glycoprotein antibodies and comparison with the rapid fluorescent focus inhibition test (RFFIT) on human samples from vaccinated and non-vaccinated people. Vaccine 25(12):2244–2251

Gautret P, Blanton J, Dacheux L, Ribadeau-Dumas F, Brouqui P, Parola P, Bourhy H (2014) Rabies in nonhuman primates and potential for transmission to humans: a literature review and examination of selected French national data. PLoS Negl Trop Dis 8(5):e2863

Hemachudha T, Laothamatas J, Rupprecht CE (2002) Human rabies: a disease of complex neuropathogenetic mechanisms and diagnostic challenges. Lancet Neurol 1(2):101–109

Hooper DC, Pierard I, Modelska A, Otvos L Jr, Fu ZF, Koprowski H, Dietzschold B (1994) Rabies ribonucleocapsid as an oral immunogen and immunological enhancer. Proc Natl Acad Sci 91(23):10908–10912

Hostnik P (2000) The modification of fluorescent antibody virus neutralization (FAVN) test for the detection of antibodies to rabies virus. J. Vet. Med. B. Infect. Dis. Vet. Public Health 47(6):423–427

Jackson AC (2007) Pathogenesis. In: Jackson AC, Wunner WH (eds) Rabies, 2nd edn. Elsevier Academic, London, pp 341–381

Johnson N, Phillpotts R, Fooks AR (2006) Airborne transmission of lyssaviruses. J Med Microbiol 55(6):785–790

Kallel H, Diouani MF, Loukil H, Trabelsi K, Snoussi MA, Majoul S, Dellagi K (2006) Immunogenicity and efficacy of an in-house developed cell-culture derived veterinarian rabies vaccine. Vaccine 24(22):4856–4862

Lafon M (2005) Rabies virus receptors. J Neurovirol 11(1):82–87

Lembo T, Hampson K, Kaare MT, Ernest E, Knobel D, Kazwala RR, Cleaveland S (2010) The feasibility of canine rabies elimination in Africa: dispelling doubts with data. PLoS Negl Trop Dis 4(2):e626

Lentz TL, Burrage TG, Smith AL, Crick J, Tignor GH (1982) Is the acetylcholine receptor a rabies virus receptor? Science 215(4529):182–184

Ma X, Niezgoda M, Blanton JD, Recuenco S, Rupprecht CE (2012) Evaluation of a new serological technique for detecting rabies virus antibodies following vaccination. Vaccine 30(36):5358–5362

Madhusudana SN, Saraswati S (2003) Development and evaluation of a latex agglutination test for rabies antibodies. J Clin Virol 27(2):129–135

Madhusudana SN, Subha S, Thankappan U, Ashwin YB (2012) Evaluation of a direct rapid immunohistochemical test (dRIT) for rapid diagnosis of rabies in animals and humans. VirologicaSinica 27(5):299–302

Maier T, Schwarting A, Mauer D, Ross RS, Martens A, Kliem V, Drosten C (2010) Management and outcomes after multiple corneal and solid organ transplantations from a donor infected with rabies virus. Clin Infect Dis 50(8):1112–1119

Mogensen TH (2009) Pathogen recognition and inflammatory signaling in innate immune defenses. Clin Microbiol Rev 22(2):240–273. https://doi.org/10.1128/CMR.00046-08

Muhamuda K, Madhusudana SN, Ravi V (2007) Development and evaluation of a competitive ELISA for estimation of rabies neutralizing antibodies after post-exposure rabies vaccination in humans. Int J Infect Dis 11(5):441–445

Mustafa M, Ellzam EM, Sharifa AM, Rahman MS, Sien MM, Nang MK (2015) Rabies a zoonotic disease, transmission, prevention and treatment. J Dent Med Sci 14(10):82–87

Nadin-Davis SA, Sheen M, Wandeler AI (2009) Development of real-time reverse transcriptase polymerase chain reaction methods for human rabies diagnosis. J Med Virol 81(8):1484–1497

Nishizono A, Yamada K, Khawplod P, Shiota S, Perera D, Matsumoto T, Ahmed K (2012) Evaluation of an improved rapid neutralizing antibody detection test (RAPINA) for qualitative and semiquantitative detection of rabies neutralizing antibody in humans and dogs. Vaccine 30(26):3891–3896

Ogino T, Banerjee AK (2007) Unconventional mechanism of mRNA capping by the RNA-dependent RNA polymerase of vesicular stomatitis virus. Mol Cell 25(1):85–97

Raux H, Flamand A, Blondel D (2000) Interaction of the rabies virus P protein with the LC8 dynein light chain. J Virol 74(21):10212–10216

Reece JF, Chawla SK (2006) Control of rabies in Jaipur, India, by the sterilisation and vaccination of neighbourhood dogs. Vet Rec 159(12):379–383

Rossiter JP, Hsu L, Jackson AC (2009) Selective vulnerability of dorsal root ganglia neurons in experimental rabies after peripheral inoculation of CVS-11 in adult mice. Acta Neuropathol 118:249–259

Rupprecht CE, Dietzschold B, Campbell JB, Charlton KM, Koprowski H (1992) Consideration of inactivated rabies vaccines as oral immunogens of wild carnivores. J Wildl Dis 28(4):629–635

Servat A, Feyssaguet M, Blanchard I, Morize JL, Schereffer JL, Boue F, Cliquet F (2007) A quantitative indirect ELISA to monitor the effectiveness of rabies vaccination in domestic and wild carnivores. J Immunol Methods 318(1–2):1–10

Smith JS (1996) New aspects of rabies with emphasis on epidemiology, diagnosis, and prevention of the disease in the United States. Clin Microbiol Rev 9(2):166–176

Smith JS, Yager PA, Baer GM (1973) A rapid reproducible test for determining rabies neutralizing antibody. Bull WHO 48:535–541

Srinivasan A, Burton EC, Kuehnert MJ, Rupprecht C, Sutker WL, Ksiazek TG, Zaki SR (2005) Transmission of rabies virus from an organ donor to four transplant recipients. N Engl J Med 352(11):1103–1111

Stankov S, Lalošević D, Fooks AR (2022) History of rabies incidence and rabies control in Serbia in support of the zero by 2030 campaign to eliminate dog-mediated human rabies. Viruses 14(1):75

Sugiyama M, Ito N (2007) Control of rabies: epidemiology of rabies in Asia and development of new-generation vaccines for rabies. Comp Immunol Microbiol Infect Dis 30(5–6):273–286

Tierkel ES, Atanasiu P (1996) Rapid microscopic examination for Negri bodies and preparation of specimens for biological tests. In: Mesiln F-XK, Koprowski MM, H. (eds) Laboratory techniques in rabies, 4th edn. WHO, Geneva, pp 55–65

Tierkel ES, Graves LM, Tuggle HG, Wadley SL (1950) Effective control of an outbreak of rabies in Memphis and Shelby County, Tennessee. Am J Public Health Nations Health 40(9):1084–1088

Tojinbara K, Sugiura K, Yamada A, Kakitani I, Kwan NCL (2016) Estimating the probability distribution of the incubation period for rabies using data from the 1948–1954 rabies epidemic in Tokyo. Prev Vet Med 123:102–105

Turner GS (1978) Immunoglobulin (IgG) and (IgM) antibody responses to rabies vaccine. J Gen Virol 40(3):595–604

Vora NM, Orciari LA, Niezgoda M, Selvaggi G, Stosor V, Lyon GM III, Franka R (2015) Clinical management and humoral immune responses to rabies post-exposure prophylaxis among three patients who received solid organs from a donor with rabies. Transpl Infect Dis 17(3):389–395

Vuaillat C, Varrin-Doyer M, Bernard A, Sagardoy I, Cavagna S, Chounlamountri I, Giraudon P (2008) High CRMP2 expression in peripheral T lymphocytes is associated with recruitment to the brain during virus-induced neuroinflammation. J Neuroimmunol 193(1–2):38–51

Warrell MJ (2012) Current rabies vaccines and prophylaxis schedules: preventing rabies before and after exposure. Travel Med Infect Dis 10(1):1–15

Webster LT, Dawson JR (1935) Early diagnosis of rabies by mouse inoculation. Measurement of humoral immunity to to rabies by mouse protection test. Proc Soc Exp Biol Med 32:570–573

WHO (2013) World Health Organization Expert consultation on rabies, 2nd Report (No. 982)

WHO (2018) WHO expert consultation on rabies: third report, vol 1012. World Health Organization

Wu X, Franka R, Henderson H, Rupprecht CE (2011) Live attenuated rabies virus co-infected with street rabies virus protects animals against rabies. Vaccine 29(25):4195–4201

Yale G, Lopes M, Isloor S, Head JR, Mazeri S, Gamble L, Gibson AD (2022) Review of Oral rabies vaccination of dogs and its application in India. Viruses 14(1):155

Yang DK, Kim HH, Lee KW, Song JY (2013) The present and future of rabies vaccine in animals. Clin Exp Vaccine Res 2(1):19–25

Yousaf MZ, Qasim M, Zia S, Ashfaq UA, Khan S (2012) Rabies molecular virology, diagnosis, prevention and treatment. Virol J 9(1):1–5

Zhao P, Zhao L, Zhang K, Feng H, Wang H, Wang T, Xia X (2012) Infection with street strain rabies virus induces modulation of the microRNA profile of the mouse brain. Virol J 9:1–13

Kunjin Virus: Insights into Emergence and Transmission Dynamics 10

Mansi Kumari and Muneesh Kumar Burman

Abstract

The severe neurotropic Kunjin virus (KUNV) is an attenuated strain of West Nile Virus (WNV) and has been modified. The reader is advised to read an in-depth review by Marshall for a comprehensive description of the history, ecology, and epidemiology of KUN viruses (1988). Although, the KUN virus, which bears the name of one of the local aboriginal tribes, was initially discovered in 1960 in Culex annulirostris mosquitoes trapped at Mission River in North Queensland, Australia. A polyprotein is encoded by the virus's single-strand positive-sense RNA genome. After gene expression, the polyprotein is split into non-structural proteins for viral replication and expression and structural proteins for viral packing. With particularly aggressive strains recently producing illness outbreaks in Eastern Europe, the Middle East, and North America, West Nile Virus (WNV) is a mosquito-borne virus that causes outbreaks of viral encephalitis in humans and horses. Infection with the WNV Kunjin (WNVKUN) strains is typically asymptomatic and endemic in northern Australia. On a template of recycled double-stranded RNA (dsRNA) or replicative form, asymmetric and semiconservative replication produce offspring RNA (+) strands (RF). The cycle time for the displacement of the pre-existing RNA (+) strand in the replicative intermediate, according to the viral RNA production kinetics, is about 15 min. This view outlines what we currently know about WNVKUN and how its ecology and epidemiology have changed.

M. Kumari
Dr. D. Y. Patil Biotechnology and Bioinformatics Institute, Dr. D. Y. Patil Vidyapeeth, Pune, India

M. K. Burman (✉)
Dr. Reddy's Institute of Life sciences, Hyderabad, India
e-mail: muneeshkbarman@gmail.com

P. L. Bhukya, A. S (eds.), *Emerging Human Viral Diseases, Volume II*,
https://doi.org/10.1007/978-981-97-4480-0_10

Keywords

Kunjin virus · Re-emergence · Pathogenesis

10.1 Introduction

There are significant viruses that contribute to global morbidity and mortality in the family Flaviviridae genus. A member of the genus is the neurotropic West Nile virus (WNV), which causes severe meningoencephalitis and is transmitted by mosquitoes. More than 200 death cases have been caused by the virus in America during the previous 20 years (Pauli et al. 2013). The Kunjin virus and WN virus are similar strains that were brought into North America, prompting quarantine by offiocals who feared that an identical entrance into Australia would have serious consequences. This particular strain of the WN virus is predicted to use the same avian hosts and vectors as KUN and will exhibit a comparable dealing out with epidemiological pattern. Relative evaluation of virulence of the KUN and WN viruses, which were isolated from New York and tested for transmissibility by Australia's primary mosquito vectors, as well as for comparative virulence and amplification in both domestic and natural wildlife, will in fact offer crucial details on the likelihood along with the potential effects of the virus's expansion in Australia (Pérez-Ramírez et al. 2014). Since the KUN virus was discovered to be endemic in Australia, it has been discovered that it was initially isolated in North Queensland in 1960. Lineage 1 includes the relatively Australian strain Kunjin as well as the extremely neuroinvasive New York 99 strain (NY99). The Kunjin virus (KUNV), on the other hand, has only caused a few nonfatal occurrences of a naturally attenuated strain of West Nile virus (WNV) since 1960 (Prow et al. 2016). As KUNV is extremely sensitive to the type 1 interferon response, these two viruses differ in pathogenicity. In fact, unlike the WNV NS5, the KUNV nonstructural protein (NS5) is unable to suppress interferon-dependent JAK-STAT signaling. Additionally, with human infections linked to a mild febrile illness and rare cases of encephalitis in both people and horses, the Kunjin (KUN) virus has long been considered a minor medical and veterinary concern in Australia (Laurent-Rolle et al. 2010).

10.2 Epidemiology

All of Australia's mainland states, with the exception of South Australia, have reported mosquito-borne KUN virus isolation; but the virus appears to be endemic in tropical regions of North Queensland, the Northern Territory, and North Western Australia, where the bulk of these isolates have their origins. In addition to humans, horses, wild birds, and sentinels, the KUN virus have been isolated from a number of mosquito species. The Kimberley and Pilbara districts of Western Australia's north, where sentinel chicken seroconversions have repeatedly shown virus activity throughout the mid-to-late rainy season, are likely to be the areas where KUN virus is enzootic (December to April) (Prow et al. 2016).

Cx. annulirostris has generated the majority of isolates from field-caught mosquitos since WNVKUN was first discovered. Other mosquito species from which WNVKUN has intermittently been isolated include *Aedes (Ae.) tremulous, Cx. australicus, Cx. squamosus, Ae. Alternans, Ae. Nomensis, Ae. Viilax, Anopheles amicutus,* and *Cx. quinquefasciatus*. Since the resulting viraemia in both horses and humans is inadequate to allow for continuous viral transmission, horses and humans are sometimes considered dead-end hosts, wading birds, particularly the rufous night heron and *Nycticorax caledonicus*, are thought to be a substantial natural reservoir of WNVKUN. Nevertheless, WNVKUN can find vertebrate hosts among a variety of different birds and animals. Due to human infection and seroconversion in sentinel hens, WNVKUN, which is prevalent in northern Australia, has occasionally been detected there. In 2002, there were over 4000 instances reported to the CDC, and 150 of these cases were people who were 19 years old or younger. A patient who was 19 years old was the youngest of the fatal cases, while the median age of the fatal cases was 78 years. 31 cases were reported related to WNV encephalitis during its beginning in 2003, while 79 cases of WNV meningitis afflicted kids and teenagers (Jansen et al. 2013; Selvey et al. 2014).

The WNVKUN's movements can be followed thanks to the sentinel chicken program in the Kunjin virus. Each year, several seroconversions to WNVKUN are noted, demonstrating the virus' endemicity in Northern Washington. A pattern of years with few seroconversions to WNVKUN followed by years with an increase in seroconversions has also been observed. The data from this program would suggest that WNVKUN is more prevalent in the Pilbara region (Prow 2013).

During the 2011 outbreak, there were around 1000 horse cases documented, and 10–15% of the infected animals died. WNVNSW2011 in horses causes behavioural problems, poor coordination, ataxia, weakness, paralysis of the muscles, and trembling. These symptoms fit the pattern of those WNV cases reported in North America. With regard to the absence of bird illness or mortality during the outbreak, which is known to be the source of increased avian virulence in WNVNY99 (Ala-Pro at amino acid 249), the removal of the non-structural protein 3 (NS3) motif from the WNVNSW2011 genome seems compatible. A common Australian crow in southern Australia, called the little raven, was immune to WNV infection, even with WNVNY99, according to studies by Bingham and colleagues. Additionally, the infection-related mortality rate for American crows (*Corvus brachyrhynchos*), blue jays (*Cyanocitta cristata*), and greater sage-grouse (*Centrocercus urophasianus*) is 100%. Birds are probably responsible for the transmission of WNVNSW2011, even in cases where the Australian Crow (Little Raven, *Corvus mellori*) is not susceptible to a fatal infection with WNVNY99. The Western European and Moroccan WNV genotype 1a strain first emerged in the 1990s, spreading endemic and sporadically igniting minor outbreaks. The WNV lineage 1a virus most likely originated in the Middle East and made its way to the Americas in 1999 (Prow 2013). It quickly spread throughout North America before making its way to South America, where it is now a serious worldwide public health concern. Madagascar and sub-Saharan

Africa have traditionally been home to WNV lineage 2 isolates, which have also sometimes resulted in zoonotic epidemics in South Africa. Prior to 1974, human epidemics only occurred in 1974 and 1984, as well as the regular pattern of infection in South Africa consisted of rare, mostly mild epizootics and human infections. Nevertheless, subsequent reports of WNND caused by lineage 2 WNV infections in South Africa raised the possibility that earlier epidemics did not completely understand the clinical scope of WNV infection. In South Africa, the relative non-pathogenicity of infections in humans and horses has been attributed to decreased pathogenicity of lineage 2 WNV strains. 2010 saw the emergence of lineage 1 WNV in South Africa, a pregnant mare perished. Additionally, infections are frequently assumed to be minor to undetectable; in 2011, WNND was shown to have been deadly in one instance, albeit it was thought that the patient had a weakened immune response (McMullen et al. 2013).

10.3 Potential Risk of Emergence and Re-Emergence

The potential risks associated with the emergence and re-emergence of Kunjin virus are multifaceted and require vigilant attention. Given its classification as a flavivirus closely related to West Nile virus, Kunjin poses a significant threat to public health due to its potential to cause neurological complications such as meningitis and encephalitis, particularly in vulnerable populations. Climate change and globalization further exacerbate the risk of its spread, as changing environmental conditions may expand the geographic range of its mosquito vectors, facilitating transmission to new regions. Moreover, the virus's capacity for genetic mutation and adaptation underscores the importance of ongoing surveillance and research efforts to anticipate and mitigate future outbreaks. Collaborative international strategies, robust mosquito control measures, and public health interventions are essential components in managing the potential risks posed by Kunjin virus emergence and re-emergence.

10.4 Organization of Infectious Agents

A mild detergent treatment can be used to remove the lipid coating that surrounds the flavivirus virion. Core protein C and genomic RNA are contained in the internal core, also known as the nucleocapsid particle, whereas the envelope carries the virus-encoded proteins E (50–60 kDa) and M (8–9 kDa). The icosahedral, encapsulated flavivirus particles have a diameter of about 50 nm. One of the particles has 11 kb positive-sense, single-stranded RNA genome with a conserved 50 RNA cap, and no. 30 polyadenylated tail. It is oriented positively. The single polyprotein that the genome, which also functions as viral mRNA, codes for is subsequently broken down by viral and cellular proteases to yield three structural proteins that the virus is packaged in: the capsid©, precursor membrane (prM), envelope (E), as well as seven non-structural proteins: NS1, NS2A, NS2B, NS3, NS4A, NS4B, and NS5. The polyprotein coding gene's 50 and 30 untranslated regions (UTRs) contain structural motifs and sequence involving viral lifecycle processes, such as translation, replication and assembly (Tran et al. 2020). The primary envelope protein E of the virus forms a dimer parallel to the lipid membrane surface that resembles a rod, according to X-ray crystallography. Rey et al. verified the early results of NOWAK and WENGLER (1987) in the crystal structure of E. (1995), indicating that 12 conserved cysteine residues were present in E of the WN virus as intramolecular disulphides. Gander erythrocytes are haemoglobinized by KUN virus virions, or subviral particles made entirely of E and M, at pH 6.6. The haemagglutination process is blocked by neutralizing antibodies produced by purified KUN viral protein. It was discovered that the carboxy and amino-terminal domains of KUN virus C, which are both rich in Arg Lys residues, robustly and specifically bind to RNA probes that represent the 5′ and 3′ UTRs of KUN viral RNA. The carboxy-terminal hydrophobic sequence of the flavivirus core protein is preceded by a conserved bioparticle nuclear localization signal, and immunoelectron microscopy and immunofluorescence have both demonstrated that a portion of KUN virus C translocated to the nucleus and accumulates in the nucleolus. Although C is not necessary for RNA replication, it is uncertain what function it serves in the nucleus (Jiang et al. 2016).

During the life cycle of a flavivirus, viral particles adhere to a host cell's surface, before being taken in by it via receptor-mediated endocytosis. Numerous primary and low-affinity co-receptors for flavivirus have been found. Particle disintegration, viral and cell membrane merging, and structural changes in the virion are caused by the endosomal vesicle becoming more acidic. Positive-sense RNA is translated into a single polyprotein upon release of the genome into the cytoplasm, where it is subsequently co- and post-translationally broken down by host and viral proteases. Replication of genes takes place on the intracellular membrane. When structural proteins and freshly generated RNA enter into the lumen of the endoplasmic reticulum (ER), virus assembly takes place on its surface. The trans-Golgi network carries the resultant immature viral and subviral particles that are not contagious. The host protease furin divides the immature virion particles into adult, infectious particles. Furin has the capacity to separate subviral particles as well. Exocytosis is followed by the release of mature virions and subviral particles (van Leur et al. 2021).

10.5 Pathogens in Humans

One of more than 70 viruses make up the Flaviviridae family within the flavivirus genus. The subtype of Kunjin virus of West Nile viruses can be assigned to at least five evolutionary lineages. Only viruses from lineages 1 and 2 have been connected to significant human epidemics. Lineage 1 is further split into three sub-lineages: lineage 1a, which includes isolates from the Western Hemisphere, Africa, the Middle East, and Europe; lineage 1b, which incorporates the Kunjin virus from Australia and Asia; and lineage 1c, which includes viruses from India (Kuno et al. 1998).

The pathobiology behind WNV infection in humans, mammalian, birds, and involving reptilian species possess the subject of numerous investigations. The virus reproduces locally in the keratinocytes and Langerhans cells of the epidermis following an infected mosquito bite, which are a particular kind of dendritic cell related to skin. Mosquitoes may then transmit WNV to people after biting a bird with high levels of viremia following a blood meal from the host. Humans can contract the West Nile virus for 2–14 days.

Mosquito saliva affects the host's immune response by modifying leukocyte proliferation and migration to the bite site and by preventing interleukin (IL) 2 and IFN. This in turn promotes the local virus replication. Theoretically, WNV infection may target dendritic cells as one of its initial primary targets. This idea was proposed by the expression of DC-SIGN (CD219) by dendritic cells following viral infection.

A hypothesis states that the infected Langerhans cell goes to the draining lymph nodes whereas the virus keeps replicating. Macrophages can either directly phagocytose and destroy infected cells and liberate virus particles or do so indirectly by improving cytokine, chemokine and antigen presentation and production. Dendritic cells in the lymph nodes continue to proliferate the virus even after macrophages eradicate it. According to WNV replication in vitro, these cells, B and T lymphocytes may be the midst of primary targets in starting phases of infection in vivo. At last, in the incubation period, which lasts between 2 and 14 days after infection, immunological modulation and the emergence of the initial clinical symptoms are linked to virus infection of immune system cells as well as virus replication in cells. Hematogenous transmission of the virus occurs from the lymph nodes to the peripheral organs. The virus has a wide range of tissue tropism and may replicate in nearly every bodily system in some hosts, including bird species (Zimmerman et al. 2019).

10.6 Clinical Manifestations

During the North American epidemics, the clinical range of symptomatic flavivirus infection in humans was better understood. Most viral infections in humans are asymptomatic; however, 20% of infected persons show signs of the illness, while less than 1% of these individuals experience severe neurological issues. The virus is characterized by a number of flu-like symptoms, including weakness, exhaustion, fever, headaches, sore muscles, and neck pain.

In a recent study of patients with West Nile fever who sought medical attention, it was found that fatigue and muscle weakness typically lingered for roughly a month after beginning, and that difficulty concentrating and stiffness in the neck were also typical symptoms. According to the research of the long-term effects of WN illness, for up to 2 years following diagnosis, almost 50% of survivors with severe symptoms experienced cognitive and physical problems. The worse prognosis used to be linked to older age; however, it has also been shown that young people and children can have neuroinvasive diseases.

Although diabetes and hypertension have been mentioned as potential risk factors for severe WND, it appears that immunosuppression, such as that experienced by patients with HIV or organ transplant recipients, is the most significant risk factor overall. Viruses can still infect neurons, the brain stem, and the spinal cord directly. Similar to other Flavivirus, virus infection follows similar pathophysiology. Following primary injection, it is thought that the virus replicates in local Langerhans dendritic cells in the skin before moving on to the lymph nodes and bloodstream, where it then enters the spleen and kidneys before entering the CNS and causing inflammation of the medulla, brain stem, and spinal cord.

Further investigation is still required to fully comprehend how the virus enters the CNS. Retrograde axonal transport through peripheral nervous system (PNS), direct passage across the blood-brain barrier, and through leukocytes are some of the postulated routes. Like many viruses, the virus has evolved a variety of defence mechanisms to resist the effects of type 1 interferon (IFN) and evade the host's IFN-stimulating genes (ISG). Several investigations have demonstrated that NS1, NS2A, NS4B, and NS5 are non-structural proteins each having special function in the control of IFN/signalling. On the other hand, in mice, humans, and horses, genetic polymorphism of the IFN inducible 2′5-Oligodenylate synthetases connected with host's inherent resistance to virus infection (Acosta et al. 2020).

10.7 Diagnostic and Therapeutic Approaches

In clinical settings, the diagnosis of WNV infection is frequently made by postmortem inspection, laboratory testing, and clinical examination. During a clinical examination, it is common to see clinical symptoms in patients such as headaches, myalgia, rash, lymphadenopathy, arthralgia, high fever, anorexia, nausea, vomiting, and eye discomfort. Patients with neuroinvasive disease often have lymphocytic pleocytosis in the cerebrospinal fluid, despite the fact that neutrophils may predominate early in the course of the illness. Although brain MRI results are typically normal, encephalitis patients may have aberrant signals in their basal ganglia, thalamus, brain stem, as well as in the anterior spinal cord of that poliomyelitis-like sickness. Multifocal chorioretinitis has been shown to be indicative of a severe WNV infection in people.

Therefore, when WNV infection is suspected, an ophthalmological examination should be done in addition to the neurological examination. Even so, clinical diagnosis is restricted since no diseased animal possesses the pathognomonic clinical indicator for the illness. Pathological examination, RT-PCR, serology, and virus isolation are some of the diagnostic methods utilized in laboratories. Serum of CSF immunoglobulin (Ig) M, which is specific for WNV, is a reliable sign of recent WNV infection. However, by the time CNS symptoms manifest, specific IgM is almost invariably found in serum and CSF in people with WNV neuroinvasive illness. IgM antibody against WNV is frequently present within 8 days following the onset of disease in certain patients. IgM emerged 9 days after donation, while IgG appeared 4 days later in seronegative blood donors with asymptomatic WNV viraemic status. IgM is also found in the CSF of other patients up to 199 days after the beginning in 36% of patients who have survived WNV encephalitis and in 20% of patients at 16 months after the onset (Sambri et al. 2013).

In a laboratory context, competetive ELISA has proven to be a reliable, affordable, and simple diagnostic analysis technique. It involves evaluating the patient's serum antibodies which is the capacity to prevent monoclonal antibodies from adhering to the epitopes of the NS1 and E proteins. This test taking benefits of species-neutral. Cross-reactivity with other flaviviruses, however, may happen depending on the blocking antibody employed in the ELISA, as shown for MVEV, ALF, and other locally circulating flaviviruses. However, the viral neutralizing test (VNT) against WNV, which has great specificity and the capacity to assess neutralizing titers as well as detect neutralizing antibodies to the virus, remains to be the gold standard test. The limitations of this assay's use as a diagnostic tool are that it takes a week for the results and the cost is also high. Over the past 10 years, a chemical method based on nucleic acids evolved as a quick and trustworthy diagnostic instrument because of its specificity and sensitivity. It is currently widely utilized in the diagnosis of viral illnesses, such as WNV. Reverse transcription polymerase chain reaction (RT-PCR) and quantitative RT-PCR (qRT-PCR) are the two most widely utilized molecular diagnostic procedures, qRT-PCR is superior to traditional RT-PCR for determining the viral genome. SYBR® Green or other DNA intercalating fluorescent dyes are used to track and quantify the aggregation of dsDNA. Instead, by practicing detection probes to monitor existing amplification of a targeted sequence, quantification can be carried out. PCR primers need to focus on the most conserved genomic region in almost all flaviviruses, NS5. The E protein region, which is conserved across multiple WNV strains, is the goal of the molecular diagnosis of WNV. Since RT-PCR is very sensitive, viral RNA from animals that have received the killed WNV vaccination could be detectable by it; as a result, when screening people who have received such vaccinations, PCR should be supplemented with additional diagnostic techniques for isolation of the virus. Instead, WNV-NS5 should be the target of the RT-PCR to isolate viral RNA from the vaccination and prevent it from spreading after infection. Samples of serum or cerebrospinal fluid (CSF) should be used for the testing. Pathological diagnosis has always been useful for understanding WNV infections in both clinical and experimental settings. This method of diagnosis relies on the identification of both macroscopic and microscopic lesions along with IHC-based analysis of viral antigens in histological tissues (Daidoji et al. 2021).

To identify WNV infections in people, Shaikh et al. introduced the test for immunochromatographic strips, called as Rapid WNTM. The crucial component in this method is identifying WNV IgM antibodies in plasma or serum of infected people, created with identical principles or idea similar to conventional ELISA. Even so, the technique possesses advantages over normal ELISA, such as rapid, user-friendly, affordable, and in-house test. A large laboratory equipped with state-of-the-art technology and personnel with advanced training is not necessary for the test. Unfortunately, there is not a rapid test for veterinary use that could be applied to WNV infection right now. This demonstrates the requirement for developing an approachable diagnostic tool for both human and veterinary uses.

10.8 Prevention and Control

Like the majority of viral infections, flavivirus infection has no known treatment. Currently, vector management, vulnerable animals, other human vaccine development system, and immunization of horses are all used to control the disease. An integrated strategy that combines immunization, increased mosquito control, and improved clinical management is required for the prevention and control of this virus. By preventing the mosquito bites that transmit the infection, preventive measures in addition to medications and immunizations can help stop the spread of the infection. Using repellents to ward off insects, minimizing skin exposure to bites, getting rid of standing water where mosquitoes can deposit their eggs, putting screens to windows and doors, minimizing outdoor activities when mosquito activity is at its highest, notifying the local government of any deceased birds, and assisting with mosquito control programs are just a few of the easy steps that can be summed up under the heading "fight the bite!" Although there have been attempts to create a human WNV vaccine, none have been authorized for use in humans and are currently not commercially accessible. The development of chimeric vaccines and DNA-delivered subunit or single-round infectious particles (SRIPs) have advanced the search for anti-WNV vaccine candidates. There is currently a commercially available recombinant, live attenuated vaccination for veterinary usage. Recombinant live canarypox virus that expresses WNV prM and E transgenes generated from the WNVNY99 strain makes up the vaccine. Recombitek®, manufactured by Merial, is the brand name used for the vaccination in commerce (Iyer and Kousoulas 2013).

The first two doses are given intramuscularly, 5 weeks apart. Horses who have received the vaccine are completely protected from the illness for up to a year after the first two doses. The use of the vaccination in dogs and cats has also been successfully tested. Despite the antigenic similarity of all flaviviruses, attempts to create a single, cross-protecting vaccination have proven fruitless. Cross-neutralization as a notion is still debatable. Cross-neutralization is thought to only affect flaviviruses belonging to the same sero-complex, or more closely related viruses, though this is not entirely proven. Cross-protection, when prior exposure to one flavivirus protects another flavivirus, is however demonstrated. The Coachella Valley in California saw a reduction in West Nile virus transmission after utilizing ultra-low volume insecticide sprays to strategically timed effect to eliminate adult mosquitoes during the early season. Additionally, during a West Nile virus outbreak in the Sacramento region, airborne ultra-low volume insecticide treatment lowered the abundance of infected mosquitoes and the incidence of human cases. The use of synthetic pyrethroids or ultra-low volume organophosphates appears to pose little danger to human health, in large part due to the timing of application and the small amount of pesticide utilized, which results in limited human exposure.

10.9 Future Prospective

Despite the fact that comparative virulence studies have only been conducted in mice, pathological and clinical evaluation of illnesses among the WN virus strain in people, horses, and birds almost certainly demonstrate that it is much more virulent than the KUN virus and poses a greater threat to human and veterinary health. It seems critical for the health medical officials and labs engaged in arbovirus surveillance and also people in veterinary viral diagnostics identified KUN as a separate virus. It will be possible to identify between KUN and WN viruses using molecular serological assays, which will be developed to prevent the invasion of exotic WN virus strains in Australia. Although, research is still being done to find out the frequency of human antibodies specific to WNVKUN and the pathogenic characteristics of WNVNSW2011 might prevent the virus from infecting horses, to better understand the risk factors for disease outbreaks, the future study must also take into account factors affecting the virus, host, and vectors independently as well as collectively (Frost et al. 2012).

Additionally, all strains of WNVKUN, both historical and present, will be sequenced and used to identify the crucial residues that account for the elevated virulence of WNVNSW2011. Initial research studies that compare current, past, and known virulent strains of WNVKUN suggest the presence of potential virulence factors and evolutionary markers, like A3 motif, because of a close genealogical relationship in between NAWNV strains and WNVKUN, and the findings of our continuing investigations will take direct applications to strains that are now in circulation in the US (Frost et al. 2012). These investigations will yield vital data for ascertaining the actual risk of WNV in veterinary and medical settings.

References

Acosta PL, Byrne AB, Hijano DR, Talarico LB (2020) Human type I interferon antiviral effects in respiratory and reemerging viral infections. J Immunol Res 2020:1372494. https://doi.org/10.1155/2020/1372494

Daidoji T, Morales Vargas RE, Hagiwara K, Arai Y, Watanabe Y, Nishioka K, Murakoshi F, Garan K, Sadakane H, Nakaya T (2021) Development of genus-specific universal primers for the detection of flaviviruses. Virol J 18(1):187. https://doi.org/10.1186/s12985-021-01646-5

Frost MJ, Zhang J, Edmonds JH, Prow NA, Gu X, Davis R, Hornitzky C, Arzey KE, Finlaison D, Hick P, Read A, Hobson-Peters J, May FJ, Doggett SL, Haniotis J, Russell RC, Hall RA, Khromykh AA, Kirkland PD (2012) Characterization of virulent West Nile virus Kunjin strain, Australia, 2011. Emerg Infect Dis 18(5):792–800. https://doi.org/10.3201/eid1805.111720

Iyer AV, Kousoulas KG (2013) A review of vaccine approaches for West Nile virus. In. Int J Environ Res Public Health 10(9):4200–4223. https://doi.org/10.3390/ijerph10094200

Jansen CC, Ritchie SA, van den Hurk AF (2013) The role of Australian mosquito species in the transmission of endemic and exotic West Nile virus strains. In. Int J Environ Res Public Health 10(8):3735–3752). Multidisciplinary Digital Publishing Institute (MDPI). https://doi.org/10.3390/ijerph10083735

Jiang B, Himmelsbach K, Ren H, Boller K, Hildt E (2016) Subviral hepatitis B virus filaments, like infectious viral particles, are released via multivesicular bodies. J Virol 90(7):3330–3341. https://doi.org/10.1128/jvi.03109-15

Kuno G, Chang G-JJ, Tsuchiya KR, Karabatsos N, Cropp CB (1998) Phylogeny of the genus Flavivirus. J Virol 72(1):73

Laurent-Rolle M, Boer EF, Lubick KJ, Wolfinbarger JB, Carmody AB, Rockx B, Liu W, Ashour J, Shupert WL, Holbrook MR, Barrett AD, Mason PW, Bloom ME, García-Sastre A, Khromykh AA, Best SM (2010) The NS5 protein of the virulent West Nile virus NY99 strain is a potent antagonist of type I interferon-mediated JAK-STAT signaling. J Virol 84(7):3503–3515. https://doi.org/10.1128/jvi.01161-09

McMullen AR, Albayrak H, May FJ, Davis CT, Beasley DWC, Barrett ADT (2013) Molecular evolution of lineage 2 West Nile virus. J Gen Virol 94(PART2):318–325. https://doi.org/10.1099/vir.0.046888-0

Pauli G, Bauerfeind U, Blümel J, Burger R, Drosten C, Gröner A, Gürtler L, Heiden M, Hildebrandt M, Jansen B, Montag-Lessing T, Offergeld R, Seitz R, Schlenkrich U, Schottstedt V, Strobel J, Willkommen H (2013) West Nile Virus. Transfus Med Hemother 40(4):265. https://doi.org/10.1159/000353698

Pérez-Ramírez E, Llorente F, Ángel Jiménez-Clavero M (2014) Experimental infections of wild birds with West Nile virus. Viruses 6:752–781. https://doi.org/10.3390/v6020752

Prow NA (2013) The changing epidemiology of Kunjin virus in Australia. In. Int J Environ Res Public Health 10(12):6255–6272. https://doi.org/10.3390/ijerph10126255

Prow NA, Edmonds JH, Williams DT, Setoh YX, Bielefeldt-Ohmann H, Suen WW, Hobson-Peters J, van den Hurk AF, Pyke AT, Hall-Mendelin S, Northill JA, Johansen CA, Warrilow D, Wang J, Kirkland PD, Doggett S, Andrade CC, Brault AC, Khromykh AA, Hall RA (2016) Virulence and evolution of West Nile virus, Australia, 1960-2012. Emerg Infect Dis 22(8):1353–1362. https://doi.org/10.3201/eid2208.151719

Sambri V, Capobianchi MR, Cavrini F, Charrel R, Donoso-Mantke O, Escadafal C, Franco L, Gaibani P, Gould EA, Niedrig M, Papa A, Pierro A, Rossini G, Sanchini A, Tenorio A, Varani S, Vázquez A, Vocale C, Zeller H (2013) Diagnosis of West Nile virus human infections: overview and proposal of diagnostic protocols considering the results of external quality assessment studies. Viruses 5(10):2329–2348. https://doi.org/10.3390/v5102329

Selvey LA, Johansen CA, Broom AK, Antão C, Lindsay MD, Mackenzie JS, Smith DW (2014) Rainfall and sentinel chicken seroconversions predict human cases of Murray Valley encephalitis in the north of Western Australia. BMC Infect Dis 14(1):672. https://doi.org/10.1186/s12879-014-0672-3

Tran PTH, Asghar N, Höglund U, Larsson O, Haag L, Mirazimi A, Johansson M, Melik W (2020) Development of a multivalent Kunjin virus reporter virus-like particle system inducing seroconversion for Ebola and West Nile virus proteins in mice. Microorganisms 8(12):1–17. https://doi.org/10.3390/microorganisms8121890

van Leur SW, Heunis T, Munnur D, Sanyal S (2021) Pathogenesis and virulence of flavivirus infections. Virulence 12(1):2814–2838. https://doi.org/10.1080/21505594.2021.1996059

Zimmerman MG, Bowen JR, McDonald CE, Pulendran B, Suthar MS (2019) West Nile virus infection blocks inflammatory response and T cell costimulatory capacity of human monocyte-derived dendritic cells. J Virol 93(23):e00664. https://doi.org/10.1128/jvi.00664-19

Prion Diseases

11

Abhranil Gangopadhayya, Manjusha Dharavath,
Suhas T. Mhaske, and Prudhvi Lal Bhukya

Abstract

Prion diseases are caused by the aberrant form of the physiological prion protein, normally produced from the PRNP gene. Proposed to play roles in long-term memory, among other functions, the disease is itself caused by accumulation of very energetically stable misfolded aggregates of physiological prion protein, called amyloid fibrils. Infectious prion can convert α-helix predominant physiological prion into the β-sheet-rich misfolded form that develops resistance to most classical sterilization techniques. Infectious prion proteins cause disease in grazing cattle and wildlife, among which it spreads by ways like fomites. Humans acquire these through ingestion of or exposure to infected tissue. Modern diagnostic methods to diagnose prion disease have emerged. In vitro and in vivo techniques have developed as well for abrogating infectious prions and alleviating disease symptoms caused by them. The prospect of common neurodegenerative diseases being caused by misfolded protein aggregation, like prions, is a matter of rising interest. This chapter covers the genetic, molecular, and functional nature of prions and also of the diseases caused by them.

A. Gangopadhayya
Hepatitis Division, National Institute of Virology, Pune, India

M. Dharavath
Department of Molecular Biology, Institute of genetics and hospital for genetic diseases, Osmania University, Hyderabad, India

S. T. Mhaske · P. L. Bhukya (✉)
Preclinical Vaccine Testing Laboratory, Rodent Experimentation Facility, ICMR-National Animal Facility Resource Facility for Biomedical Research, Hyderabad, Telangana, India
e-mail: saiprudhvi21@gmail.com; prudhvi.bhukya@icmr.gov.in

P. L. Bhukya, A. S (eds.), *Emerging Human Viral Diseases, Volume II*,
https://doi.org/10.1007/978-981-97-4480-0_11

Keywords

Prion protein · Protein misfolding disease · Physiological prion protein · Mutation TSE · CJD · GSS · Energy profile · Folding · Transmission

11.1 Introduction

The characteristic of being an infectious protein particle distinguishes the prion from all other infectious agents. Since its characterization by Stanley B. Prusiner, prions have opened up a new avenue of infectious disease research. It was in 1982 that the prion was identified as being a protein and not a nucleic acid, leading to the proteinaceous nature of infectious material being established (Prusiner 1982). As to why this particular kind of protein molecule has an infectious, degenerative nature, was deduced through various experiments conducted through the decades. The cellular prion protein (PrP^{C} or prion protein cellular) remains in its properly folded alpha-helix rich conformation, whereas the infectious prion protein (PrP^{Sc}, Sc for Scrapie) has a predominantly beta-sheet rich conformation. This leads to the aggregation of the PrP^{Sc} molecules that lead to the formation of amyloid fibrils. Research on prions gained headway since conditions like the chronic wasting diseases suffered by murine and bovine species, both wild and domesticated, needed some kind of explanation. Scrapie and transmissible spongiform encephalopathy (TSE) display symptoms characteristic of neurodegenerative disease and such was also the case for human diseases like Kuru, fatal familial insomnia (FFI), and Creutzfeldt-Jakob disease (CJD). The etiology of these diseases being a misfolded protein particle opens up possibilities of its treatment as well that could be dealt with by chaperone-like agents. However, the action of chaperones or other similar molecules becomes refractory once amyloid fibrils start to form. These and several other discoveries have been made in the field of prion study, and at this point, it may be a task to compile all the old and new findings in this field. However, the starting point to do so should be a glance at the history of prions, be it basic research or medical science.

11.2 Historical Perspective of Prions and Prion Disease

In 1732, it was observed by Spanish shepherds that their sheep would pathologically scrape against fences. These were the Merino sheep, and they exhibited behavior including altered gaits, intense licking and itching, and of course the scraping against fences. This was named as "Scrapie." Then, in December 1920, Hans Gerhard Creutzfeldt described a human neurological disorder that could not be ascertained by etiology (Creutzfeldt 1920). It was called a "herd-shaped disease of the central nervous system." Following this, a year later in December 1921, Alfons Maria Jakob published a paper describing "strange diseases of the central nervous system with remarkable anatomical findings" (Jakob 1921). Kuru, another neurological disorder affecting the Fore tribe in Papua New Guinea, was described in 1959 by Gadjusek and

Table 11.1 Summarization of some more landmarks in the history of prions and their disease (Zabel and Reid 2015)

Year	Name of person(s) involved	Event
1938	J. Cuillé and P.L. Chelle	Scrapie explained as a "slow virus"
1946	Plummer	Inoculated brain and cerebrospinal fluid (CSF) into healthy animals from infected ones, showing it can transmit scrapie disease
1954	Sigurðsson	Proposed that a "slow virus" indeed caused scrapie because of its long incubation period
1969	Hunter	Accumulated more evidence toward the hypothesis that scrapie was indeed a disease of proteinaceous etiology
1980	Cho	
1981	Merz et al.	
1982	Prusiner	Coined the term "prion," for proteinaceous infectious particle
1994	Wickner	Hypothesized that the non-chromosomal genetic element [URE3], which was an altered form of the protein Ure2p, was a prion protein, something essential in the survival of yeast through poor environmental conditions, without resorting to mutation

Zigas (Gajdusek and Zigas 1959). It was found to present much like CJD and Scrapie, introducing a brand-new disease into perspective. In 1959, Hadlow, Klatzo, Gadjusek, and Zigas, through a collaborative effort, revealed that CJD, Scrapie, and Kuru were different manifestations of the same neuropathy. Tikvah Alper, I.H. Pattison, and J.S. Griffith were the first scientists who speculated that the Scrapie agent could be a protein and not a slow virus, or nucleic acid, and it was Alper who said that it must replicate without a nucleic acid genome (Alper et al. 1967). Afterward, Griffith stated that the agent might be proteinaceous and offered three explanations as to how a protein can be infectious. He also gave hypotheses on the genetic control and spontaneous occurrence of these agents and their disease (Griffith 1967).

In 2007, S. Supattapone proved that prion proteins could be generated in vitro (Deleault et al. 2007), and in 2008, Lindquist proved that yeasts spontaneously produce [PSI+] when subjected to conditions like oxidative stress or high salt concentrations (Tyedmers et al. 2008). Later on, in 2010, Ma showed that the prion protein could be expressed from bacteria using recombinant technology (Wang et al. 2010). Therefore, the history of prion research has been studded with some jewels that shine light upon the enigma that they are. The above mentioned are but a few of the significant happenings that took place in this field throughout the centuries (Table 11.1).

11.3 How to Define a Prion?

As enigmatic as this entity is, the prion can be most simply defined as an infectious protein molecule. Considering prions by this simplistic viewpoint, yes, it can indeed replicate inside a host body. Evidences exist of the prion molecules being able to nucleate and form aggregates from monomers, as well as bud out of already existing aggregates into smaller ones (Orgel 1996). The infection that is established could be considered a serialized conversion and misfolding of cellular prion proteins into the infectious form, thus eventually leading to the accumulation that results in symptom manifestation.

If one would like to look up the dictionary definition of the prion protein, Merriam-Webster provides a meaning of the word that is as follows: “A glycoprotein found especially on the cell surface of neurons in the brain and spinal cord that sometimes occurs in an abnormal misfolded form which proliferates by inducing the normal protein to convert to the misfolded pathogenic form.” (“Prion Protein|Definition of Prion Protein by Merriam-Webster,” (n.d.))

One needs to understand from the various definitions that each of them has their own distinct meaning and so has not to be confused with one another. According to the Centers for Disease Control & Prevention (CDC), “The term prions refers to abnormal, pathogenic agents that are transmissible and are able to induce abnormal folding of specific normal cellular proteins called prion proteins that are found most abundantly in the brain. The functions of these normal prion proteins are still not completely understood.” (“Prion Diseases|CDC,” (n.d.))

The CDC states prion diseases or transmissible spongiform encephalopathies (TSEs) as being,

“[A] family of rare progressive neurodegenerative disorders that affect both humans and animals. They are distinguished by long incubation periods, characteristic spongiform changes associated with neuronal loss, and a failure to induce inflammatory response.”

Johns Hopkins Medicine gives a description of prion diseases as:

“Prion diseases comprise several conditions. A prion is a type of protein that can trigger normal proteins in the brain to fold abnormally. Prion diseases can affect both humans and animals and are sometimes spread to humans by infected meat products.” (“Prion Diseases|Johns Hopkins Medicine,” (n.d.-a))

Lastly, the World Health Organization (WHO) describes on its page for Prion diseases, that it is a TSE and can spread by the eating of infected meats, be it of livestock or other humans (“WHO|Prion diseases,” (n.d.)). The many forms of TSEs are described on the page, and how the transmission modes may be blocked are mentioned as well. The occurrence of some new TSEs in certain animals is also mentioned on the website.

11.4 The Genetic Basis of Prions

In mice, it has been discovered that the Prn-p gene is accompanied by the Prn-i gene, which controls the scrapie incubation period in mice (Carlson et al. 1988). The Prn-pb gene among them encodes a special sequence different from the other haplotypes. The PRNP gene is controlled by the copper ion content inside a cell as well as by transcription factors such as specificity protein 1 (SP1) and/or metal transcription factor (MTF-1), which sense the varying copper concentrations in a cell (Bellingham et al. 2009). A study of 2003 noted the presence of 86 polymorphisms in deer, cattle, and sheep, which was divisible into previously unreported single nucleotide polymorphisms (SNPs) 9 in number in sheep, 13 in number in cattle, and 32 in number in deer (Heaton et al. 2003). A separate study conducted into the messenger ribonucleic acid (mRNA) expression levels of the PrP^C gene found that the highest levels were in the neocortical and cerebellar tissues of the brain, whereas the lowest was in the kidneys (Tichopad et al. 2003).

In yeast, the control of prion gene expression is modulated by yet other gene products. The case of *Saccharomyces cerevisiae* is that it has a prion-forming domain (PrD), which is an asparagine- and glutamine-rich domain, and only the presence of these domains enables prion-like transmissible forms. Sup35p is a form of the SUP35 gene that has a deletion of 19 amino acids at the N terminal, identified in some strains of the yeast.

Differences in prion susceptibility may arise due to the subtle differences in the prion gene encoded by different species. Susceptibility of feline or canine species to prions has been studied upon and published in an article from 2012 (Stewart et al. 2012). Feline spongiform encephalopathy (FSE) is the name given to the feline counterpart of the bovine spongiform encephalopathy (BSE) disease, which was noted to not have a discernible polymorphism associated in felines. Black bears and mountain lions from Colorado, on the other hand, had little genetic variation, from which it may be inferred that the genetic barrier between BSE and chronic wasting disease (CWD) prions contributes to the limited host range of CWD. Methionine at the 142nd position was found to be contributive to resistance against scrapie. Even the goats that have prion-forming aggregates in their brain or its periphery can have no clinical manifestations showing because of the Serine 127 containing allele of the PRNP gene. Lysine at position 222 is a well-known resistance mutation of the gene, but has not been found at a high frequency in the study.

One noteworthy application of breeding in the context of genetic manipulation has been the clustered regularly interspaced short palindromic repeat (CRISPR)/CRISPR-associated nuclease (Cas) 9 usage to edit the genomes of bovine fetal fibroblasts and embryos conceived by in vitro fertilization (IVF) (Bevacqua et al. 2016).

11.5 Structure, Interactions, and Mutations of Prion Proteins

The prion protein is around 253 amino acids long ("PRNP—Major prion protein precursor—Homo sapiens (Human)—PRNP gene & protein," (n.d.)). Its locus is on the 20th human chromosome. It has a signal peptide, which is 22 amino acids long, as found from similarity to the mouse prion protein. And, at position 230, there is an amidated serine residue that assists in the formation of glycophosphatidylinositol (GPI) anchors (Capellari et al. 2000). Glycoforms, or glycosylated forms, differ from physiological to CJD prion forms, being different in their monoglycosylated, diglycosylated, or non-glycosylated nature.

The protein exists as monomeric and homodimeric structures, and it has been observed that the H2H3 helices first undergo the misfolding that acts as a seed for the misfolding of the total molecule (Adrover et al. 2010). Cross-beta spine structures formed by a steric zipper of superposed beta-strands are what results in the formation of amyloid fibrils by aggregation (Sawaya et al. 2007). Prion proteins can interact with various other proteins, some of which are pictured in Fig. 11.1 and detailed in Table 11.2.

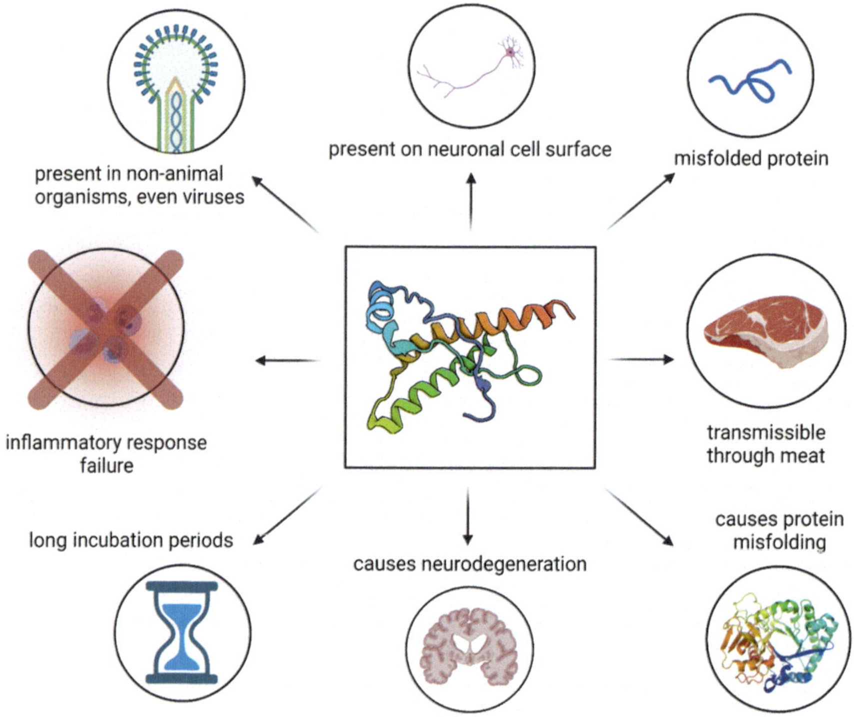

Fig. 11.1 The various features of prions and their diseases. Created with BioRender.com, using image sources: ("RCSB PDB—1HJM: HUMAN PRION PROTEIN AT PH 7.0," (n.d.); "SPX—Structural Biochemistry/Proteins/Protein Folding Problem—Wikibooks, open books for an open world," (n.d.))

Table 11.2 Proteins that interact with the PrP protein and their interaction features or functions ("PRNP—Major prion protein precursor—Homo sapiens (Human)—PRNP gene & protein," (n.d.))

Name of Interacting Protein	Remarks
Growth factor receptor-bound protein 2 (GRB2)	Interacts through the major prion protein chain, i.e., residues 23 to 230
Exoribonuclease 3 (ERI3)	Interacts through the major prion protein chain, i.e., residues 23 to 230
Synapsin 1	Interacts through the major prion protein chain, i.e., residues 23 to 230
Mahogunin ring finger-1 (MGRN1)	Interacts only with cytosolically mislocalized and exposed PrP
Adhesion G protein-coupled receptor G6 (ADGRG6)	Interacts through residues 23 to 38
Huntingtin (HTT)	Protein plays a role in huntingtin disease development when mutated
Glypican 1 (GPC1)	Heparan sulfate proteoglycan mediates PrP^{C} association to lipid rafts and facilitates conversion to the PrP^{Sc} form

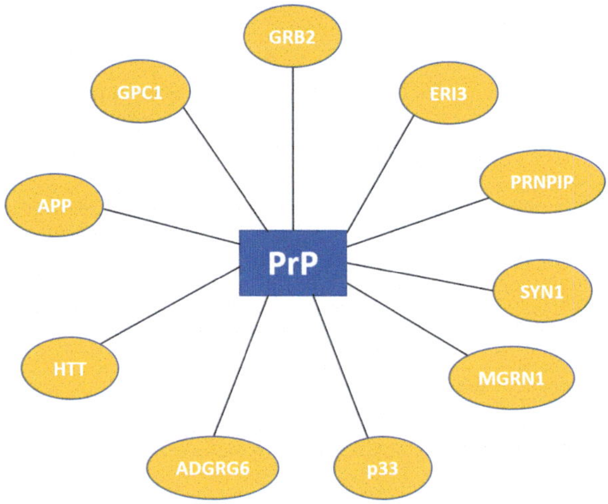

The protein's N-terminal region is composed of octamer repeats, three or four of which have histidine sidechain residues that contribute to the binding of a single copper ion at low copper concentrations (Burns et al. 2002).

The glycine to valine 127 mutation confers resistance to the Kuru disease by acting as a dominant negative inhibitor of prion conversion ("PRNP—Major prion protein precursor—Homo sapiens (Human)—PRNP gene & protein," (n.d.)). This mutant prevents conversion of wild-type prion proteins as well. It also provides protection against classical CJD and Kuru in the heterozygous state, but can be infected by prions causing variant CJD caused by BSE prions. In the homozygous state, it confers resistance against all prion strains. A few more mutations of interest are described in Table 11.3.

A few models have been proposed to explain the resistant structures of PrP^{Sc} molecules. However, none of the structural models can fully elucidate all the intricacies of the prion protein, which is why multiple ones are referred to when trying to explain one or the other characteristic. One of the earliest models was to utilize a truncated form of the PrP molecule called PrP (108–218), which was based on the structure of another form called PrP 27–30 (Huang et al. 1996). This model spoke of a four-stranded β-sheet structure with two C-terminal α-helices and was met with criticism because PrP^{Sc} also consists significantly of α-helices. The β-helix model describes the presence of antiparallel, intertwined strands that form a core, of the size of eight β-strands (Downing and Lazo 1999). One particular model was proposed that included β-helical structures of both the left- and right-handed kind to compose putative trimeric or hexameric assemblies (Wille et al. 2002). One other approach to this modeling was using the human TATA box as a basis and developing a five-stranded β-sheet with C-terminal α-helices (Mornon et al. 2002). Simulation of the infectious PrP structure in the presence of acidic pH was done using the recombinant PrP that yielded what is called the "spiral model" (DeMarco and Daggett 2004). A revision of the β-helical models of before was done through the usage of higher resolution electron micrography, which revealed a trimeric unit cell

Table 11.3 Mutations of interest and effects

Mutation	Effect
G127V	Resistance to Kuru by acting as dominant negative inhibitor of prion conversion; prevents conversion of wild-type prions; can prevent classical CJD but not infection by vCJD prions when heterozygous; prevents all prion diseases when homozygous (Asante et al. 2015; Mead et al. 2009)
M129V	Inhibits homodimerization, provides some protection against inherited, sporadic, and acquired prion diseases. Associated with mutation at 178th position (Doh-ura et al. 1989; Goldfarb et al. 1992; Lee et al. 2010; Mead et al. 2003)
D178N	Associated with FFI or CJD (Goldfarb et al. 1991, 1992; Medori et al. 1992)
E219K	Provides some protection against sporadic CJD when heterozygous (Riedel et al. 1998)
P102K	Associated with early-onset dementia (Finckh et al. 2000)
P105K	Involved in GSS development (Yamada et al. 1993)
A117V	Associated with dementing GSS (Doh-ura et al. 1989)
G131V	Development of GSS (Panegyres et al. 2001)
N171S	Development of schizoaffective disorder (Samaia et al. 1997).
V180I	Development of CJD (Goldfarb et al. 1992)
T183A	SENF or early-onset dementia development (Finckh et al. 2000)
H187R	Development of GSS
T188K	Associated with prion disease associated or early-onset dementia
E196K & E200K	Associated with CJD
F198S	Results in an atypical form with neurofibrillary tangles and results in GSS
D202N	Results in variant that can cause GSS
V203I	Unknown pathological significance or variant leading to CJD
R208H	Lead to variants that cause CJD
V210I	
E211Q	
Q212P	Associated with GSS
Q217R	Results in neurofibrillary tangles and GSS
E219K	Provides relative protection against sporadic CJD
M232R	Results in variant linked with CJD

structure of the crystal lattice, a model called the Govaerts et al. model (Govaerts et al. 2004). Up next was the "Stork model," which once again retained the C-terminal α-helices and so would not stay strong in the long run (Stork et al. 2005). The next model took into account the extraordinary stability of the infectious PrP isoform, highlighting that an intrinsic domain swap among the monomers of a trimeric unit cell may be responsible for it (Yang et al. 2005). Yet another model addressed the stability (or the lack of it) of the proposed protein fold, by reducing the number of β-helical folds from four to just two, thus solving the problems of packing density that arose between the residues 106 & 126, due to being rich in small amino acids (Langedijk et al. 2006). One of the more accepted models of PrP^{Sc} structure is the parallel, in-register β-sheet structure (Cobb et al. 2007). The latest models adopt the idea that the parallel β-helices may be present on the C-terminal domain of the protein as well and that recombinant forms containing only the C-terminal α-helices are prone to fibrillization (Kunes et al. 2008). Separate models vary with respect to the number and position of the β-helix rungs, and how threading of the sequence happened to β-helix templates. There was no effort to compile all the data in silico, and so it cannot be determined if they all fit together with the X-ray fiber diffraction data. A long, unstructured loop has been speculated from these models that joins the N- and C-terminal ends derived from electron microscopy data of the recombinant PrP amyloid fibers, which appear to form a ladder-like appearance. Some discrepancies do still exist regarding the strain specificity of these findings, the proteolysis properties of the protein to PK, and the probability that differing strains will give different cleavage patterns as well. Further details of the models are provided in Table 11.4.

Despite all these efforts to model the structure of the infectious prion protein isoform, there is no one definitive structural model that accounts for all the characteristics of the protein, which is why there needs to be further research in this respect with new and upcoming modern methods.

Table 11.4 Protein models and short comings

Author name	Year	Description	Shortcoming(s)
Huang et al.	1996	Four-stranded β-sheet structure with two C-terminal α-helices, called PrP 108–218.	PrP^{Sc} also consists significantly of α-helices
Downing and Lazo	1999	Antiparallel, intertwined strands that form a core, of the size of eight β-strands (β-helix model). Does not conform to existing definitions of parallel β-helical structures.	Height of PrP^{Sc} molecule or PrP 27–30 is described to be similar to eight β-strands, contradicted through X-ray crystallography.
Wille et al.	2002	β-Helical structures of both the left- and right-handed kind to compose putative trimeric or hexameric assemblies, proposed on the basis of the PrP 27–30 and PrP^{Sc}106. Structure contains a parallel β-helix at its core due to the N-linked oligosaccharides and β-sheet structure inferred from the tight packing in the crystal lattice found.	Supported α-helices at the C-terminal side of the protein, which is not in accordance with the currently known models of prion proteins.
Mornon et al.	2002	Five-stranded β-sheet with C-terminal α-helices, based on the human TATA box. The third C-terminal α-helix was found to be swapped between the two monomers of a dimer formed by recombinant PrP.	It was not clear how PrPSc amyloid fibrils would form, and, in addition to the retainment of the C-terminal α-helices in the model, it was eventually refuted by future models.

(continued)

Table 11.4 (continued)

Author name	Year	Description	Shortcoming(s)
DeMarco and Daggett	2004	Simulation of the infectious PrP structure in the presence of acidic pH done using recombinant PrP, producing the "spiral model," which contained four β-strands.	The three α-helices like in the previous models were retained. Amyloid fibrils built from these monomers were found to align at angles not perpendicular to the fibril axis, contradicting X-ray fiber diffraction results. The high α-helical content and the low β-sheet content of this model were in direct contrast with existing proteolysis data and FTIR structures.
Govaerts et al.	2004	A trimeric unit cell structure of the crystal lattice, called the Govaerts et al. model, done using higher resolution electron micrography. These were elucidated using the PrP 27–30 and PrP^{Sc}106 two-dimensional crystals.	C-terminal portion (starting from the disulfide bond at cysteine 179) speculated to remain in an α-helical state. Structure limited to left-handed parallel β-helical folds toward the N-terminal part of the protein.
Stork et al.	2005	Primary PrP structure threaded onto the β-helical template. Hexameric assembly assumed to be not required for densities to occur in a two-dimensional model of the crystals.	Retained the C-terminal α-helices.
Yang et al.	2005	Intrinsic domain swap among the monomers of a trimeric unit cell may be responsible for extraordinary stability of the infectious PrP isoform. Domain-swapped trimer model could explain the high intrinsic stability that was not found in non-entangled trimers.	Solely accounting for the stability of the PrP^{Sc} therefore could not act as a model sufficient to explain some important aspects of the amyloid fibrils.
Langedijk et al.	2006	Number of β-helical folds reduced from four to just two, addressing the stability (or the lack of it) of the proposed protein fold. This solved the problems of packing density that arose between the residues 106 & 126, due to being rich in small amino acids.	Molecular dynamics highlighted the number of β-helix rungs to be fewer in the infectious isoform, thus bringing the height of the molecule from 19.2 angstrom to 9.6 angstrom, something in gross contradiction with already proved results.
Cobb et al.	2007	Electron paramagnetic resonance (EPR) and site-directed spin-labelling of the recombinant amyloid used to figure out structures in which each PrP molecule contributes only 4.8 angstrom to the amyloid fibril's length, leading to the parallel, in-register β-sheet structure.	This model does not explain the repeating unit size of 19.2 angstrom in X-ray fiber diffraction experiments conducted on the infectious PrP molecule and PrP 27–30.
Kunes et al.	2008	Parallel β-helices may be present on the C-terminal domain of the protein as well, and recombinant forms containing only the C-terminal α-helices are prone to fibrillization. Separate β-helices were proposed for both termini, replacing any α-helices that were previously speculated.	–

11.6 Physiological Role of the Prion Protein

Despite its well-known role in disease, the normal, cellular prion protein serves roles in physiological functions. These have been documented experimentally in various studies, some of which are highlighted.

Yeast two hybrid studies have been conducted to find out the protein-protein interactions of the PrP^C molecule, and some candidates that partake in these interactions have been deduced (Westergard et al. 2007). Laminin receptor precursor (LRP), for example, is a putative receptor for both cellular and infectious prion protein uptake into cells through endocytosis. What throws us off the trail of LRP is that it lacks a signal peptide or transmembrane anchor that could have targeted it to a secretory pathway. Prion protein interactor 1 (Pint1), TREK-1 (also called KCNK2), tubulin, neurotrophin receptor-interacting MAGE homologue (NRAGE), stress-inducible protein 1 (STI-1), heat-shock protein 60 (Hsp60), neural cell adhesion molecule (N-CAM), B-cell lymphoma 2 (Bcl2), and Caveolin-1 are some of the other proteins that putatively interact with the PrP^C protein. Most of them were validated using yeast two hybrid studies, along with co-immunoprecipitation. Tubulin and N-CAM interactions were validated via a method called cross-linking, and the STI-1 interaction by doing complementary hydropathy.

There have been some studies that prove an anti-apoptotic action of the PrP^C molecule. Expression of the mammalian prion protein in a yeast system prevented cell death mediated by BCL2 Associated X (Bax) (Li and Harris 2005). This happened only when a mouse PrP with a modified signal peptide that enables secretory pathway targeting was expressed in the yeast cells, and not when cytosolic PrP was expressed. Also, this activity was inhibited when the amino acid residues 23 to 31 were deleted.

Multiple cytoprotective effects have been documented for the PrP^C molecule. In a mouse model to test the proliferation and metastasis of cancer cells, it was found that mice overexpressing the PrP protein showed more of these compared to Prnp knockout mice, when lung cancer was made to develop in them. Cancer cell stemness and epigenetic effects of the PrP^C molecule on neighboring cell signaling have been recorded. Glioblastoma multiforme (GB) is a disease in which the cancerous cells retain their phenotype and thus infiltration and relapse, due to the effects of the PrP^C protein. The micro-environment of said cells and their translational profile putatively undergo variations due to the PrP^C protein's activities in retaining their stem cell-like properties while disallowing them to differentiate. Additionally, it has been speculated that the folding profile of proteins such as prions can provide information on the past stimuli received by the protein, the adaptive responses that are generated and the signaling processes that control prion functioning (Saad and Jarosz 2021). Prion-like behavior is now ascribed to most of the agents responsible for neurodegenerative diseases (le Guerroué and Youle 2021).

Microbial prions are responsible for several cellular processes that confer evolutionary advantages, due to modulating intricate molecular mechanisms (Levkovich et al. 2021). Despite only yeast prions being widely studied so far, the structural and functional diversity of these entities is now being found to be broader than initially assumed.

Synaptic plasticity has been speculated to be conferred in part by the prions in a body. A study was conducted to find out the effects of SNPs in the maintenance of long-term memory in healthy young people (Papassotiropoulos et al. 2005). Some protein-remodeling factors and molecular chaperones such as Hsp104 and Hsp40 are responsible for regulation of fungal prions that are beneficial to the organism (Shorter and Lindquist 2005). Certain selective advantages are conferred by the yeast prion [PSI^+], such as evolvability and phenotypic plasticity. Cytoplasmic polyadenylation element binding (CPEB) proteins transform from a normal to a prion-like state that is transcriptionally active and serve functions in synaptic plasticity. Late long-term potentiation (LTP) of neurons is partially carried out by the CREB-2-3 protein translation, regulated by the tropomyosin receptor kinase B (TrkB) molecule (Baltaci et al. 2019). Amyloidomics has emerged as a science specific to the amyloids and the systems of biological functions that they have, since amyloids have become evident in processes much more than just disease.

11.7 Energy Profiles and Protein Folding

An important part in the understanding of prion disease development is how the proteins at all misfold into their pathogenic form. Metadynamics has helped to elucidate on the actual scenario that occurs, through microsecond-long simulations that show the unfolding of the human PrP protein first, from a native state to states of increasingly lower free energy that include a global minimum on the free energy landscape, called PrP^{C*} (Singh et al. 2017).

Salt bridge interactions have been found to be lacking in the mutated human prion protein forms, whereas they have been documented to exist as stabilizing interactions for the secondary and tertiary structures of the PrP molecules (Guest et al. 2010). Stabilization of the structures is due to the self-energies of the molecule, the residues that have their own individual electrostatic potential charges, among others. The influence of low pH is on the 121st to 231st residue stretch of the PrP protein, that results in a β-sheet like conformer and thus may contribute to the eventual formation of the PrP^{Sc} form (Hornemann and Glockshuber 1998). This was an equilibrium unfolding intermediate that exists exclusively at the acidic pH, as evidenced by urea-dependent unfolding transitions, and it was in rapid equilibrium with the native PrP region 121–231 at pH 4.0. It is the minimal difference between the conformers of PrP^C & PrP^{Sc} imparted due to the region between residues 90 and 120 that is the chief point of distinction between a cellular prion protein and an infectious prion monomer.

The misfolding of proteins that occurs and the resultants of the phenomenon—the aggregates—are now treated as prions in a wider sense than initially was (Scheckel and Aguzzi 2018). Postulations are made to ascribe the prion definition to causative agents of multiple neurodegenerative disorders. Prion diseases are essentially a kind of protein misfolding disorder (PMD) that have its own specific characteristics. As such, it is distinct from a PMD and may not be the answer to all the PMDs. The role of chaperones as agents that can potentially fix the infectious prion problem has been explored (Chaudhuri and Paul 2006). Since the functioning of each chaperone type is different, their targeting of specific kinds of degenerative disorders is different and specific to the disorder type.

The WHO estimates that neurodegenerative diseases leading to motor and cognitive impairments will be the second most ubiquitous cause of death by the next decade (Singh et al. 2020). A pressing need therefore exists to develop novel therapeutic strategies that can be used as not only post-prophylaxis but also pre-prophylaxis. Neuroprotective therapy for PD, for example, may have to be multifactorial, since the speculated pathogenesis of the condition involves cell-autonomous events like mitochondrial bioenergetic alterations, calcium homeostasis dysregulations, mitochondrial turnover impairment, and non-cell-autonomous events such as aggregation of prion-like proteins and subsequent neuroinflammation (Hirsch et al. 2013).

It is an autocatalytic process that begins the replication of the infectious prion protein by recruiting and converting the PrP^{C} form to the PrP^{Sc} form (Igel-Egalon et al. 2019). Astrogliosis, microgliosis, and neurodegeneration are occurrences during the neuropathogenesis of prion disease (Carroll and Chesebro 2019). Astrocytes and microglia produce proinflammatory cytokines in response to the infectious protein attacks the brain. Chronic stimulation of these cytokines leads to damages of the brain matter. Neurons and astrocytes are foci of prion spread and replication in the brain, while microglia are not known to be so.

However, the protein-only hypothesis is being contended by an increasing body of evidence. The PrP protein has been found to bind to nucleic acid and is capable of triggering reverse transcription (Lathe and Darlix 2020). Despite a single PrP gene being known even 30 years back, it was speculated that the cellular and infectious PrP forms result from different post-translational events (Oesch et al. 2007).

11.8 Why Are Prions So Resistant?

Although copper can bind to the PrP molecule to impart enhanced survival in neurons, when copper is replaced by manganese, it imparts a protease resistance to the molecule following aging, making the molecule lose its function (Brown 2000). Astrocytes expressing manganese-bound PrP had partial protease resistance conferred to the molecules, and so it is possible to generate protease-resistant prions from cells. This may be what leads to the infectious isoform developing in an organism. A study elucidated that a higher pH, at about 10, and high concentrations of urea or guanidine thiocyanate (GdnSCN), i.e., more than 3 M & 0.75 M respectively, can lead to increases in protease sensitivity of the proteins (Madec et al. 1997). These results were obtained by experimentation on multiple animal models and can be used to biochemically characterize different prion isolates. Additional studies have also proven the role of salt in causing infectious PrP formation from the cellular type, in the presence of 1 Molar (M) urea, without the requirement of reducing conditions (Swietnicki et al. 2000). Usage of guanidine-hydrochloride at a concentration of 1 M caused a transition of the 90th to 231st residues stretch of the human PrP protein from an α-helical structure to a β-sheet structure.

Various chaperones were tested for their ability to convert PrP^C to PrP^{Sc} and it was found that none of them could cause infectious isoform conversion without pre-existing PrP^{Sc} (DebBurman et al. 1997). GroEL and Hsp104 were the only chaperones which could cause conversion in the presence of PrP^{Sc}. Chemical chaperones were not able to cause conversion, and the conversion characteristics of the two chaperone-mediated processes were distinct. Sensitive strains of PrP^{Sc} ($sPrP^{Sc}$) have been characterized as well, from multiple known strains of infectious conformers of human PrP proteins (Saverioni et al. 2013). These are influenced not only by the size of the aggregates but also putatively by the stability of the aggregate type. Experimentation proved that the PrP protein is the main and sole agent of prion disease that acts as a structural protein for causing the disease and is resistant to conventional sterilization or degradation methods.

11.9 Prion-Associated Conditions By Kingdom

Below is a list and descriptions of some protein misfolding diseases caused by prions, sorted by their Kingdom of occurrence, and at times the Class or Family as well.

11.9.1 Kingdom *Animalia*

11.9.1.1 Human Prion Diseases

11.9.1.1.1 Creutzfeldt-Jakob Disease (CJD)

Abbreviated as CJD, it was first discovered in the year 1920 and so is the oldest known human prion disease ("Creutzfeldt-Jakob Disease Fact Sheet|National Institute of Neurological Disorders and Stroke," (n.d.)). It is a neurocognitive disorder, characterized by early symptoms like memory problems, behavioral changes, lack of co-ordination and disturbances in visualization, and late symptoms like dementia, involuntary movements, blindness, weakness, and coma.

11.9.1.1.2 Gerstmann–Sträussler–Scheinker Syndrome (GSS)

It was first reported by three Australian physicians-Josef Gerstmann, Ernst Sträussler, and Ilya Scheinker, and pinpointed for its cause to be the human PRNP gene, thus being grouped together with TSEs (Liberski 2012). GSS can affect patients between 20 & 60 years of age, is only heritable, and has been found so far in only a few families around the world.

11.9.1.1.3 Kuru

First identified among the Fore tribe of Papua New Guinea, it acquired its name from the Fore word "kuria" or "guria," which means "to shake," due to the tremors that are seen in people suffering from the disease (Hoskin et al. 1969). Pathological bursts of laughter are also characteristic of the disease, stemming from the neurodegeneration that is evident from loss of co-ordination and the tremors themselves.

11.9.1.1.4 Fatal Familial Insomnia (FFI)

Trouble sleeping is the most hallmark symptom of the disease, which starts gradually and worsens over time ("Fatal Insomnia—Neurologic Disorders—Merck Manuals Professional Edition," (n.d.)). It can manifest as either of two forms—sporadic fatal insomnia (sFI) or FFI—the latter of which is due to an autosomal dominant mutation, while the former is due to a non-inherited mutation.

11.9.1.1.5 Variably Protease-Sensitive Prionopathy (VPSPr)

This was reported and published as a disease, from data acquired through reports of a few cases, only in 2008, while it was first described in a conference back in 2006 (Gambetti et al. 2008). It is extremely rare, occurring in only about 2–3 people per 100 million, has similarities to CJD, although clinical manifestations differ, and its agent has been found to be sometimes very resistant to protease, and sometimes quite sensitive to it.

11.9.2 Non-Human Prion Diseases

11.9.2.1 Bovine Spongiform Encephalopathy (BSE)

It is primarily a disease of cattle, manifesting as abnormal behavior, trouble walking, and loss in weight (Casalone and Hope 2018). It takes between four to five years for the incubation period and manifestation of symptoms, following which death can occur between weeks to months. This is what causes variant CJD to develop in humans, when it transmits to them.

11.9.2.2 Chronic Wasting Disease (CWD)

Also called zombie deer disease, it majorly affects deer, with the mule deer, white-tailed deer, red deer, elk, caribou, and moose being affected in the USA. Cervids are the main victims of CWD, although it has been experimentally transmitted to squirrel monkeys and genetically modified mice ("USDA APHIS|Chronic Wasting Disease (CWD)," (n.d.)).

11.9.2.3 Scrapie

It has been known since at least 1732, is one of the several known TSEs, and is a degenerative disease that affects sheep and goats (Detwiler 1992). The disease apparently causes an itching sensation in animals, leading to them scraping their fleeces off against rocks, trees, or fences.

11.9.2.4 Transmissible Mink Encephalopathy (TME)

Affecting mostly ranch-raised adult mink, it is a rare, sporadic disease that affects the central nervous system (CNS). Its incubation period varies from seven to twelve months, causing mortality in adult animals.

11.9.2.5 Feline Spongiform Encephalopathy (FSE)

Similar to BSE, it mainly affects the brains of felines (Iulini et al. 2008). It can affect domestic and captive feline species, being possibly spread through both the hematogenous and nervous pathways.

11.9.2.6 Ungulate Spongiform Encephalopathy (USE)

It is a disease that mostly manifests in mammals with hooves ("Animal TSEs (Prion Diseases)—Minnesota Dept. of Health," (n.d.)). This illness is quite rare. It causes a neurodegenerative brain disorder.

11.9.2.7 Camel Spongiform Encephalopathy (CSE)

Similar to mad cow disease, it is also called the mad camel disease ("Deux chercheurs algériens découvrent la maladie du « dromadaire fou » à Ouargla—Maghreb Emergent," (n.d.)). Baaissa Babelhadj, an Algerian veterinarian, Semir Bechir Suheil GAOUAR, a lecturer-researcher from the University of Tlemcen, and one of their colleagues from Ouargla were the ones who discovered the condition, collaborating with Italian researchers.

11.9.3 Kingdom *Plantae*

Reports exist that claim prions to be taken up by plants and then transmitted to animals that graze on them ("Prions in Plants?|wisconsinacademy.org," (n.d.)). Although a species barrier exists, that protect a different species from being infected by a prion disease of one species, the avenue of plants taking up prions from the environment and passing it on to foraging animals is a whole new perspective on transmission.

A protein called luminidependens (LD), found in *Arabidopsis thaliana*, when introduced in part in yeast, and made to express, adopts a prion-like conformation capable of causing aggregation of more protein molecules around it (Nowogrodzki 2016). This misfolding effect is hereditary, so the proteins keep being misfolded in later generations. It is necessary to find out whether such a protein from plants can keep on misfolding neighboring proteins when added to a pool of natively folded proteins, as a kind of cascade.

Yet another study, however, that was performed in wheat (*Triticum aestivum*), found that the plants were incapable of transporting infectious prion particles efficiently to its aerial parts, after being exposed to infectious PrP^{TSE} via the root (Rasmussen et al. 2014).

Therefore, plants may also be a poorly understood but significant carrier of infectious prions, from which multiple species may be obtaining these disease agents and contracting prion disease.

11.9.4 Kingdom *Fungi*

The prions that have been identified in fungi, including yeast, are infectious and replicating amyloid forms of normally folding, soluble proteins in the cells of the infected (Wickner 2016). Their effects range from minor to lethal, depending upon the strain of the prion that causes the effect. Sup35p, Ure2p, and Rnq1p are in-register, parallel, folded β-sheets that are amyloids capable of templating their own formation, much like deoxyribonucleic acid (DNA) or RNA. There are antiprion systems active in fungi too, however, like the Btn2/Cur1 system for the [URE3] prions, and the Ssb system for the [PSI$^+$] prions.

All of these proteins bear some common criteria that can be used to assign them as being prions. The abovementioned proteins all possess these qualities, which is why they have been assigned prion status in fungi.

There are several such yeast prions that have been discovered to this date, namely [SWI+] from the Swi1p protein, [OCT+] from the Cyc8p protein, [MOT3+] from the protein Mot3p, [ISP+] (opposite of [PSI+]) from the protein Sfp1p, [MOD+] from the Mod5p protein, [β] from the protein Prb1p, among others.

Some prions such as Het-s, as mentioned previously, serve beneficial roles in the system of its host. Most others, such as [URE3] and [PSI+], serve mostly harmful roles in their fungal systems.

11.9.5 Kingdom *Monera*

Prion-like proteins exist even among the bacterial species of the world. One study, published in 2013, highlighted how 2200 candidate prion-like proteins emerged from a predictive bioinformatics study (Espinosa Angarica et al. 2013). This was done by detecting the propensity of glutamine and asparagine-rich domains, that occur more in prions than charged residues and prolines. Microbial prions being interactors with the human body's prion-like proteins could have huge implications. Mapranosis is a term that has been proposed to describe microbe-associated proteopathy and neuroinflammation (Friedland and Chapman 2017). An assay to detect prion formation has discovered that a *Campylobacter hominis* DNA-binding protein also functions as a prion (Fleming et al. 2019). In it, a model yeast prion protein was used to check the formation of prion-like and non-prion forms in a cell, something that revealed the ability of the single-stranded DNA-binding protein (SSB) of *C. hominis* to exist in both forms, due to the presence of a prion-forming domain in the protein.

11.9.6 Viruses

Indeed, prions have been discovered from even the obligate intracellular parasites that we know as viruses. One protein called LEF-10 from the baculovirus can replace the yeast Sup35 protein's PrD with its own cPrD, or even its full-length form (Nan et al. 2019). When the baculovirus infects at a high multiplicity of infection (MOI), the protein aggregates in virus-infected cells, causing inhibition of viral late gene expression. Therefore, prions are not limited to just cellular lifeforms. An attempt to find prions using the detection of prion-like domains led to the discovery of 2679 putative prion-like domains across 2.7 million publicly available viral sequences (Tetz and Tetz 2018). These were distributed throughout mammalian, plant, insect, and human viruses, detected in species that were in some or another

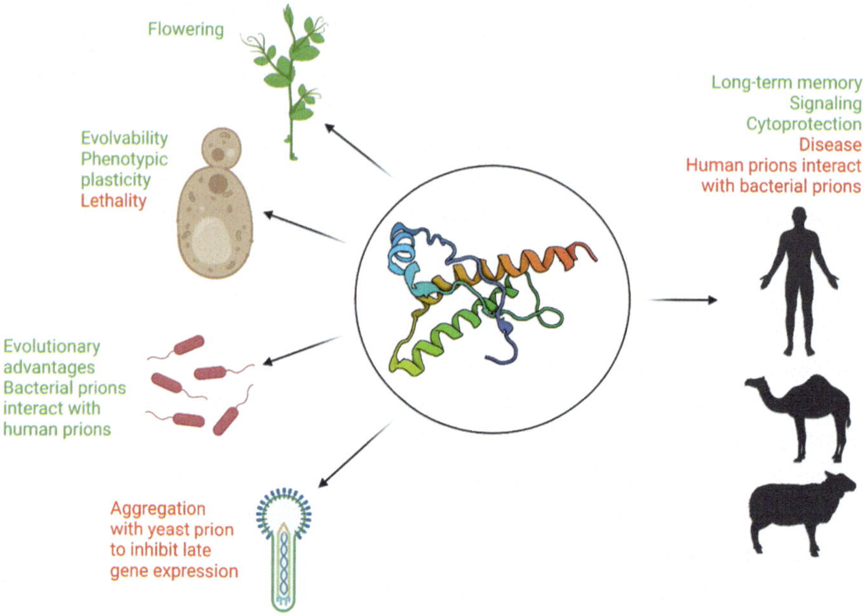

Fig. 11.2 The effects of prions in the various domains of life, including beneficial (green) and harmful (red) ones. (Created with BioRender.com)

measure linked to roles in different ecosystems or in diseases of human health importance (Fig. 11.2).

11.10 Diagnosis of Prion Disease

One study compared the efficiency of detecting prion disease by using nasal brushings compared to cerebrospinal fluid (CSF) samples (Orrù et al. 2017). It highlights the use of a process called real-time quaking induced conversion that uses a fluorescent dye to detect the conversion of a particular isoform of a protein to the other kind, which in this case is from the cellular to the infectious form.

Conformation-dependent immunoassay (CDI) is yet another technique that has been used for diagnosing prions of the sensitive and the resistant kind in the tissues of patience. In fact, it may be a better method of diagnosis than using immunostaining. One hurdle would be the conducting of clinical trials for antiprion therapies (Vallabh et al. 2020). Since there are no clear prodromal biomarkers, the progress of any person who have been given such a treatment cannot be measured with ease. α-synuclein, the Tau protein, 14–3-3, neuron-specific enolase, and S100B are some of the non-specific markers that indicate a prion disease being at play (Connor et al. 2019). These can all be detected in the CSF of a patient in whom rapid neurodegeneration is occurring.

The neuropsychiatric manifestations can have a broad range and are most often rapid. Features such as psychotic episodes, mood disorders, and agitation are common, as observed from a study that took into account about 300 participants, but they can subside afterward due to self-resolution, or global neurologic impairment. Predictive factors include diagnostic category, disease severity, age, gender, and genetic variation.

11.11 Management of Diseased Individuals

A prion disease invariably affects the quality of life of people suffering from it. They may initially seem fine, only needing help for daily activities at home, but could eventually require being moved to a special care facility ("Prion Diseases|Johns Hopkins Medicine," (n.d.-b)). They need to write down their queries and if possible be accompanied by another person. Instructions by the healthcare provider and any new diagnoses are to be written down by the person, as well as details of new drugs and why they are given. The care a patient deserves when suffering from slow, neurodegenerative diseases such as these is only so that they can continue to live as well as possible, as long as they can. The incurable nature of prion disease does snatch away a lot of the hope that would otherwise have been in store for those suffering from them. The more awareness is raised on these debilitating conditions, more the chances of people understanding what a diseased individual is going through, and better chances of providing support.

11.12 Therapeutics, Sterilization, and Degradation

An article from 2009 describes the usage of RNA interference (RNAi) through a lentiviral vector against the native prion protein (White and Mallucci 2009). In it, they described the reversal of neuronal damage through this procedure. Antibodies such as PrioV3 have been found to be effective against misfolded prion proteins in vivo. The migration of this antibody across the blood-brain barrier (BBB) was found to be blocked by phosphatidylinositol-phospholipase C (PLC). Where conventional antibodies may fail to cross the BBB due to their large size, PrioV3 was proven to transmigrate across the barrier, and the cell membranes of neurons, to target cytosolic prions (Jones et al. 2010). Comparative treatment with a mouse-derived antiprion antibody post-immunization with recombinant PrP did not prevent replication of infectious prion from happening again. PrP-Förster resonance energy transfer (FRET) enabled high-throughput assay (PrP-FEHTA) has been introduced as a drug discovery platform for screening antiprion agents, a method which has identified two drugs—astemizole and tacrolimus—as potential therapeutic agents for antiprion use (Karapetyan et al. 2013). Both drugs could reduce cell surface PrP as well as prevent replication of prions in infected neuroblastoma cells.

Seven enzymes, individually and paired, along with detergents, lipases, and metal ions, at varying temperatures, were tested for their degrading activity on Rocky Mountain Laboratories (RML) prions adsorbed to surgical steel (Jackson et al. 2005). Proteinase K and Pronase, together with sodium dodecyl sulfate, were found to effectively degrade prions derived from vCJD-infected brain preparations to levels below detection. *Parmelia sulcata, Cladonia rangiferina,* and *Lobaria pulmonaria* extracts made in acetone can degrade PrP^{TSE} derived from infected hamsters, mice, and deer (Johnson et al. 2011). The same degrading activity did not hold for closely related lichen species or individual cyanobacterial or algal species inhabiting these lichens.

These are all but a few of the treatment and degradation methods for prions that have emerged through the efforts of prion research.

11.13 Conclusion

Prion biology is a booming science, one that has the potential to encompass all forms of life, due to the near-ubiquitous nature of prion proteins. The field had started off as studies into disease-causing misfolded proteins but has now become expanded to redefine our understanding of protein misfolding and its functions in general. We may be looking at a new era in protein biology because the existence of prions means that what we used to consider as an abnormality is actually pretty much normal and usual in nature. What we need right now is a fresh perspective with which we can study, diagnose, and probably even make good use of prions. One can only marvel at the ways in which prion biology has flourished, covering so many aspects of so many domains of life, with human neurodegeneration still taking center stage. Therefore, prions are not just to be viewed as a cause of disease, but a new limelight of potentially exciting discoveries in the world of proteins.

References

Adrover M, Pauwels K, Prigent S, de Chiara C, Xu Z, Chapuis C, Pastore A, Rezaei H (2010) Prion fibrillization is mediated by a native structural element that comprises helices H2 and H3*. J Biol Chem 285:21004. https://doi.org/10.1074/jbc.M110.111815

Alper T, Cramp WA, Haig DA, Clarke MC (1967) Does the agent of scrapie replicate without nucleic acid? Nature 214:764–766

Animal TSEs (Prion Diseases)—Minnesota Dept. of Health (n.d.) [WWW Document]. https://www.health.state.mn.us/diseases/tses/animal.html. Accessed 24 Mar 2021

Asante EA, Smidak M, Grimshaw A, Houghton R, Tomlinson A, Jeelani A, Jakubcova T, Hamdan S, Richard-Londt A, Linehan JM, Brandner S, Alpers M, Whitfield J, Mead S, Wadsworth JDF, Collinge J (2015) A naturally occurring variant of the human prion protein completely prevents prion disease. Nature 522:478. https://doi.org/10.1038/nature14510

Baltaci SB, Mogulkoc R, Baltaci AK (2019) Molecular mechanisms of early and late LTP. Neurochem Res 44:281–296. https://doi.org/10.1007/s11064-018-2695-4

Bellingham SA, Coleman LA, Masters CL, Camakaris J, Hill AF (2009) Regulation of prion gene expression by transcription factors SP1 and metal transcription factor-1. J Biol Chem 284:1291–1301. https://doi.org/10.1074/jbc.M804755200

Bevacqua RJ, Fernandez-Martín R, Savy V, Canel NG, Gismondi MI, Kues WA, Carlson DF, Fahrenkrug SC, Niemann H, Taboga OA, Ferraris S, Salamone DF (2016) Efficient edition of the bovine PRNP prion gene in somatic cells and IVF embryos using the CRISPR/Cas9 system. Theriogenology 86:1886. https://doi.org/10.1016/j.theriogenology.2016.06.010

Brown DR (2000) Consequences of manganese replacement of copper for prion protein function and proteinase resistance. EMBO J 19:1180. https://doi.org/10.1093/emboj/19.6.1180

Burns CS, Aronoff-Spencer E, Dunham CM, Lario P, Avdievich NI, Antholine WE, Olmstead MM, Vrielink A, Gerfen GJ, Peisach J, Scott WG, Millhauser GL (2002) Molecular features of the copper binding sites in the Octarepeat domain of the prion protein. Biochemistry 41:3991. https://doi.org/10.1021/bi011922x

Capellari S, Zaidi SIA, Long AC, Kwon EE, Petersen RB (2000) The Thr183Ala mutation, not the loss of the first glycosylation site, alters the physical properties of the prion protein. J Alzheimers Dis 2:27. https://doi.org/10.3233/JAD-2000-2104

Carlson GA, Goodman PA, Lovett M, Taylor BA, Marshall ST, Peterson-Torchia M, Westaway D, Prusiner SB (1988) Genetics and polymorphism of the mouse prion gene complex: control of scrapie incubation time. Mol Cell Biol 8:5528. https://doi.org/10.1128/MCB.8.12.5528

Carroll J, Chesebro B (2019) Neuroinflammation, microglia, and cell-association during prion disease. Viruses 11:65. https://doi.org/10.3390/v11010065

Casalone C, Hope J (2018) Atypical and classic bovine spongiform encephalopathy. In: Handbook of clinical neurology. Elsevier B.V, pp 121–134. https://doi.org/10.1016/B978-0-444-63945-5.00007-6

Chaudhuri TK, Paul S (2006) Protein-misfolding diseases and chaperone-based therapeutic approaches. FEBS J 273:1331. https://doi.org/10.1111/j.1742-4658.2006.05181.x

Cobb NJ, Sonnichsen FD, Mchaourab H, Surewicz WK (2007) Molecular architecture of human prion protein amyloid: a parallel, in-register -structure. Proc Natl Acad Sci 104:189465. https://doi.org/10.1073/pnas.0706522104

Connor A, Wang H, Appleby BS, Rhoads DD (2019) Clinical laboratory tests used to aid in diagnosis of human prion disease. J Clin Microbiol 57:e00769. https://doi.org/10.1128/JCM.00769-19

Creutzfeldt HG (1920) Über eine eigenartige herdförmige erkrankung des zentralnervensystems (Vorläufige mitteilung). Zeitschrift für die gesamte Neurologie und Psychiatrie 57:1–18. https://doi.org/10.1007/BF02866081

Creutzfeldt-Jakob Disease Fact Sheet|National Institute of Neurological Disorders and Stroke (n.d.) [WWW Document]. https://www.ninds.nih.gov/Disorders/Patient-Caregiver-Education/Fact-Sheets/Creutzfeldt-Jakob-Disease-Fact-Sheet. Accessed 22 May 21

DebBurman SK, Raymond GJ, Caughey B, Lindquist S (1997) Chaperone-supervised conversion of prion protein to its protease-resistant form. Proc Natl Acad Sci 94:13938. https://doi.org/10.1073/pnas.94.25.13938

Deleault NR, Harris BT, Rees JR, Supattapone S (2007) Formation of native prions from minimal components in vitro. Proc Natl Acad Sci 104:9741–9746

DeMarco ML, Daggett V (2004) From conversion to aggregation: Protofibril formation of the prion protein. Proc Natl Acad Sci 101. https://doi.org/10.1073/pnas.0307178101

Detwiler LA (1992) Scrapie. Revue Scientifique et Technique de l'OIE 11:491. https://doi.org/10.20506/rst.11.2.607

Deux chercheurs algériens découvrent la maladie du « dromadaire fou » à Ouargla - Maghreb Emergent (n.d.) [WWW Document]. https://web.archive.org/web/20180524221856/https://www.maghrebemergent.info/actualite/88933-deux-chercheurs-algeriens-decouvrent-la-maladie-du-dromadaire-fou-a-ouargla.html. Accessed 24 Mar 2021

Doh-ura K, Tateishi J, Sasaki H, Kitamoto T, Sakaki Y (1989) Pro→Leu change at position 102 of prinon protein is the most common but not the sole mutation related to Gerstmann-Sträussler syndrome. Biochem Biophys Res Commun 163:974–979. https://doi.org/10.1016/0006-291X(89)92317-6

Downing DT, Lazo ND (1999) Molecular modelling indicates that the pathological conformations of prion proteins might be β-helical. Biochem J 343:453. https://doi.org/10.1042/0264-6021:3430453

Espinosa Angarica V, Ventura S, Sancho J (2013) Discovering putative prion sequences in complete proteomes using probabilistic representations of Q/N-rich domains. BMC Genomics 14:316. https://doi.org/10.1186/1471-2164-14-316

Fatal Insomnia - Neurologic Disorders—Merck Manuals Professional Edition (n.d.) [WWW Document]. https://www.merckmanuals.com/en-ca/professional/neurologic-disorders/prion-diseases/fatal-insomnia. Accessed 22 May 2021

Finckh U, Müller-Thomsen T, Mann U, Eggers C, Marksteiner J, Meins W, Binetti G, Alberici A, Hock C, Nitsch RM, Gal A (2000) High prevalence of pathogenic mutations patients with early-onset dementia detected by sequence analyses of four different genes. Am J Hum Genet 66:110–117. https://doi.org/10.1086/302702

Fleming E, Yuan AH, Heller DM, Hochschild A (2019) A bacteria-based genetic assay detects prion formation. Proc Natl Acad Sci 116:4605. https://doi.org/10.1073/pnas.1817711116

Friedland RP, Chapman MR (2017) The role of microbial amyloid in neurodegeneration. PLoS Pathog 13:e1006654. https://doi.org/10.1371/journal.ppat.1006654

Gajdusek DC, Zigas V (1959) Kuru: clinical, pathological and epidemiological study of an acute progressive degenerative disease of the central nervous system among natives of the eastern highlands of New Guinea. Am J Med 26:442–469

Gambetti P, Dong Z, Yuan J, Xiao X, Zheng M, Alshekhlee A, Castellani R, Cohen M, Barria MA, Gonzalez-Romero D, Belay ED, Schonberger LB, Marder K, Harris C, Burke JR, Montine T, Wisniewski T, Dickson DW, Soto C, Hulette CM, Mastrianni JA, Kong Q, Zou W-Q (2008) A novel human disease with abnormal prion protein sensitive to protease. Ann Neurol 63:697. https://doi.org/10.1002/ana.21420

Goldfarb LG, Haltia M, Brown P, Nieto A, Kovanen J, Richard Mccombie W, Trapp S, Carleton Gajdusek D (1991) New mutation in scrapie amyloid precursor gene (at codon 178) in Finnish Creutzfeldt-Jakob kindred. Lancet 337:425. https://doi.org/10.1016/0140-6736(91)91198-4

Goldfarb LG, Petersen RB, Tabaton M, Brown P, LeBlanc AC, Montagna P, Cortelli P, Julien J, Vital C, Pendelbury WW, Haltia M, Wills PR, Hauw JJ, McKeever PE, Monari L, Schrank B, Swergold GD, Autilio-Gambetti L, Gajdusek DC, Lugaresi E, Gambetti P (1992) Fatal familial insomnia and familial Creutzfeldt-Jakob disease: disease phenotype determined by a DNA polymorphism. Science 258:806–808. https://doi.org/10.1126/science.1439789

Govaerts C, Wille H, Prusiner SB, Cohen FE (2004) Evidence for assembly of prions with left-handed -helices into trimers. Proc Natl Acad Sci 101:8342. https://doi.org/10.1073/pnas.0402254101

Griffith JS (1967) Nature of the scrapie agent: self-replication and scrapie. Nature 215:1043–1044

Guest WC, Cashman NR, Plotkin SS (2010) Electrostatics in the stability and misfolding of the prion protein: salt bridges, self energy, and solvation this paper is one of a selection of papers published in this special issue entitled "Canadian Society of Biochemistry, Molecular & Cellular Biology 52nd annual meeting — protein folding: principles and diseases" and has undergone the Journal's usual peer review process. Biochem Cell Biol 88:371–381. https://doi.org/10.1139/O09-180

Heaton MP, Leymaster KA, Freking BA, Hawk DA, Smith TPL, Keele JW, Snelling WM, Fox JM, Chitko-McKown CG, Laegreid WW (2003) Prion gene sequence variation within diverse groups of U.S. sheep, beef cattle, and deer. Mamm Genome 14:765–777. https://doi.org/10.1007/s00335-003-2283-y

Hirsch EC, Jenner P, Przedborski S (2013) Pathogenesis of Parkinson's disease. Mov Disord 28:24. https://doi.org/10.1002/mds.25032

Hornemann S, Glockshuber R (1998) A scrapie-like unfolding intermediate of the prion protein domain PrP(121-231) induced by acidic pH. Proc Natl Acad Sci 95:6010. https://doi.org/10.1073/pnas.95.11.6010

Hoskin JO, Kiloh LG, Cawte JE (1969) Epilepsy and Guria: the shaking syndromes of New Guinea. Soc Sci Med 3:39–48. https://doi.org/10.1016/0037-7856(69)90037-7

Huang Z, Prusiner SB, Cohen FE (1996) Scrapie prions: a three-dimensional model of an infectious fragment. Fold Des 1:13. https://doi.org/10.1016/S1359-0278(96)00007-7

Igel-Egalon A, Bohl J, Moudjou M, Herzog L, Reine F, Rezaei H, Béringue V (2019) Heterogeneity and architecture of pathological prion protein assemblies: time to revisit the molecular basis of the prion replication process? Viruses 11:429. https://doi.org/10.3390/v11050429

Iulini B, Cantile C, Mandara MT, Maurella C, Loria GR, Castagnaro M, Salvadori C, Porcario C, Corona C, Perazzini AZ, Maroni A, Caramelli M, Casalone C (2008) Neuropathology of Italian cats in feline spongiform encephalopathy surveillance. Vet Pathol 45:626. https://doi.org/10.1354/vp.45-5-626

Jackson GS, McKintosh E, Flechsig E, Prodromidou K, Hirsch P, Linehan J, Brandner S, Clarke AR, Weissmann C, Collinge J (2005) An enzyme–detergent method for effective prion decontamination of surgical steel. J Gen Virol 86:869. https://doi.org/10.1099/vir.0.80484-0

Jakob A (1921) Über eigenartige erkrankungen des zentralnervensystems mit bemerkenswertem anatomischen befunde. Zeitschrift für die gesamte Neurologie und Psychiatrie 64:147. https://doi.org/10.1007/BF02870932

Johnson CJ, Bennett JP, Biro SM, Duque-Velasquez JC, Rodriguez CM, Bessen RA, Rocke TE (2011) Degradation of the disease-associated prion protein by a serine protease from lichens. PLoS One 6:e19836. https://doi.org/10.1371/journal.pone.0019836

Jones DR, Taylor WA, Bate C, David M, Tayebi M (2010) A camelid anti-PrP antibody abrogates PrPSc replication in prion-permissive Neuroblastoma cell lines. PLoS One 5:e9804. https://doi.org/10.1371/journal.pone.0009804

Karapetyan YE, Sferrazza GF, Zhou M, Ottenberg G, Spicer T, Chase P, Fallahi M, Hodder P, Weissmann C, Lasmezas CI (2013) Unique drug screening approach for prion diseases identifies tacrolimus and astemizole as antiprion agents. Proc Natl Acad Sci 110:7044. https://doi.org/10.1073/pnas.1303510110

Kunes KC, Clark SC, Cox DL, Singh RRP (2008) Left handed β helix models for mammalian prion fibrils. Prion 2:81. https://doi.org/10.4161/pri.2.2.7059

Langedijk JPM, Fuentes G, Boshuizen R, Bonvin AMJJ (2006) Two-rung model of a left-handed β-helix for prions explains species barrier and strain variation in transmissible spongiform encephalopathies. J Mol Biol 360:907. https://doi.org/10.1016/j.jmb.2006.05.042

Lathe R, Darlix J-L (2020) Prion protein PrP nucleic acid binding and mobilization implicates retroelements as the replicative component of transmissible spongiform encephalopathy. Arch Virol 165:535. https://doi.org/10.1007/s00705-020-04529-2

le Guerroué F, Youle RJ (2021) Ubiquitin signaling in neurodegenerative diseases: an autophagy and proteasome perspective. Cell Death Differ 28:439. https://doi.org/10.1038/s41418-020-00667-x

Lee S, Antony L, Hartmann R, Knaus KJ, Surewicz K, Surewicz WK, Yee VC (2010) Conformational diversity in prion protein variants influences intermolecular β-sheet formation. EMBO J 29:251. https://doi.org/10.1038/emboj.2009.333

Levkovich SA, Rencus-Lazar S, Gazit E, Laor Bar-Yosef D (2021) Microbial prions: Dawn of a new era. Trends Biochem Sci 46:391. https://doi.org/10.1016/j.tibs.2020.12.006

Li A, Harris DA (2005) Mammalian prion protein suppresses bax-induced cell death in yeast. J Biol Chem 280:17430–17434. https://doi.org/10.1074/jbc.C500058200

Liberski PP (2012) Gerstmann-Sträussler-Scheinker Disease. Adv Exp Med Biol 724:128–137. https://doi.org/10.1007/978-1-4614-0653-2_10

Madec JY, Vanier A, Dorier A, Bernillon J, Belli P, Baron T (1997) Biochemical properties of protease resistant prion protein PrPsc in natural sheep scrapie. Arch Virol 142:1603. https://doi.org/10.1007/s007050050183

Mead S, Stumpf MPH, Whitfield J, Beck JA, Poulter M, Campbell T, Uphill JB, Goldstein D, Alpers M, Fisher EMC, Collinge J (2003) Balancing selection at the prion protein gene consistent with prehistoric Kurulike epidemics. Science 300:640–643. https://doi.org/10.1126/science.1083320

Mead S, Whitfield J, Poulter M, Shah P, Uphill J, Campbell T, Al-Dujaily H, Hummerich H, Beck J, Mein CA, Verzilli C, Whittaker J, Alpers MP, Collinge J (2009) A novel protective prion protein variant that Colocalizes with Kuru exposure. N Engl J Med 361:2056. https://doi.org/10.1056/NEJMoa0809716

Medori R, Tritschler H-J, LeBlanc A, Villare F, Manetto V, Chen HY, Xue R, Leal S, Montagna P, Cortelli P, Tinuper P, Avoni P, Mochi M, Baruzzi A, Hauw JJ, Ott J, Lugaresi E, Autilio-Gambetti L, Gambetti P (1992) Fatal familial insomnia, a prion disease with a mutation at codon 178 of the prion protein gene. N Engl J Med 326:444–449. https://doi.org/10.1056/nejm199202133260704

Mornon J-P, Prat K, Dupuis F, Boisset N, Callebaut I (2002) Structural features of prions explored by sequence analysis II A PrP Sc model. Cell Mol Life Sci 59:2144. https://doi.org/10.1007/s000180200014

Nan H, Chen H, Tuite MF, Xu X (2019) A viral expression factor behaves as a prion. Nat Commun 10:359. https://doi.org/10.1038/s41467-018-08180-z

Nowogrodzki A (2016) Plant protein behaves like a prion. Nature. https://doi.org/10.1038/nature.2016.19824

Oesch B, Groth DF, Prusiner SB, Weissmann C (2007) Search for a Scrapie-Specific Nucleic Acid: A Progress Report. In: Ciba Foundation Symposium 135-Novel Infectious Agents and the Central Nervous System: Novel Infectious Agents and the Central Nervous System: Ciba Foundation Symposium 135. John Wiley & Sons, Ltd., Chichester, UK, pp 209–223. https://doi.org/10.1002/9780470513613.ch14

Orgel LE (1996) Prion replication and secondary nucleation. Chem Biol 3:413. https://doi.org/10.1016/S1074-5521(96)90087-3

Orrù CD, Groveman BR, Hughson AG, Manca M, Raymond LD, Raymond GJ, Campbell KJ, Anson KJ, Kraus A, Caughey B (2017) RT-QuIC assays for prion disease detection and diagnostics. In: Prions: methods and protocols, pp 185–203. https://doi.org/10.1007/978-1-4939-7244-9_14

Panegyres PK, Toufexis K, Kakulas BA, Cernevakova L, Brown P, Ghetti B, Piccardo P, Dlouhy SR (2001) A new PRNP mutation (G131V) associated with Gerstmann-Sträussler-Scheinker disease. Arch Neurol 58:1899–1902. https://doi.org/10.1001/archneur.58.11.1899

Papassotiropoulos A, Wollmer MA, Aguzzi A, Hock C, Nitsch RM, de Quervain DJ-F (2005) The prion gene is associated with human long-term memory. Hum Mol Genet 14:2241. https://doi.org/10.1093/hmg/ddi228

Prion Diseases|CDC (n.d.) [WWW Document]. https://www.cdc.gov/prions/index.html. Accessed 14 Feb 2021

Prion Diseases|Johns Hopkins Medicine n.d.-a [WWW Document]. https://www.hopkinsmedicine.org/health/conditions-and-diseases/prion-diseases. Accessed 14 Feb 2021

Prion Diseases|Johns Hopkins Medicine n.d.-b [WWW Document]. https://www.hopkinsmedicine.org/health/conditions-and-diseases/prion-diseases. Accessed 2 Apr 2021

Prion Protein|Definition of Prion Protein by Merriam-Webster (n.d.) [WWW Document]. https://www.merriam-webster.com/dictionary/prion%20protein. Accessed 9 Feb 2021

Prions in Plants?|wisconsinacademy.org n.d. [WWW Document]. https://www.wisconsinacademy.org/magazine/prions-plants. Accessed 25 Mar 2021

PRNP—Major prion protein precursor—Homo sapiens (Human)—PRNP gene & protein (n.d.) [WWW Document]. https://www.uniprot.org/uniprot/P04156. Accessed 14 Feb 2021

Prusiner S (1982) Novel proteinaceous infectious particles cause scrapie. Science 216:136. https://doi.org/10.1126/science.6801762

Rasmussen J, Gilroyed BH, Reuter T, Dudas S, Neumann NF, Balachandran A, Kav NN, Graham C, Czub S, McAllister TA (2014) Can plants serve as a vector for prions causing chronic wasting disease? Prion 8:136. https://doi.org/10.4161/pri.27963

RCSB PDB —1HJM: HUMAN PRION PROTEIN AT PH 7.0 (n.d.) [WWW Document]. https://www.rcsb.org/structure/1hjm. Accessed 4 Jun 2021

Riedel M, Straube A, Schwarz MJ, Wilske B, Müller N (1998) Protective prion protein polymorphisms against sporadic Creutzfeldt-Jakob disease. Lancet 351:419. https://doi.org/10.1016/s0140-6736(05)78358-6

Saad S, Jarosz DF (2021) Protein self-assembly: a new frontier in cell signaling. Curr Opin Cell Biol 69:62. https://doi.org/10.1016/j.ceb.2020.12.013

Samaia HB, de Jesus Mari J, Vallada HP, Moura RP, Simpson AJG, Brentani RR (1997) A prion-linked psychiatric disorder [5]. Nature 390(6657):241. https://doi.org/10.1038/36757

Saverioni D, Notari S, Capellari S, Poggiolini I, Giese A, Kretzschmar HA, Parchi P (2013) Analyses of protease resistance and aggregation state of abnormal prion protein across the Spectrum of human prions. J Biol Chem 288:27972. https://doi.org/10.1074/jbc.M113.477547

Sawaya MR, Sambashivan S, Nelson R, Ivanova MI, Sievers SA, Apostol MI, Thompson MJ, Balbirnie M, Wiltzius JJW, McFarlane HT, Madsen AØ, Riekel C, Eisenberg D (2007) Atomic structures of amyloid cross-β spines reveal varied steric zippers. Nature 447:453. https://doi.org/10.1038/nature05695

Scheckel C, Aguzzi A (2018) Prions, prionoids and protein misfolding disorders. Nat Rev Genet 19:405. https://doi.org/10.1038/s41576-018-0011-4

Shorter J, Lindquist S (2005) Prions as adaptive conduits of memory and inheritance. Nat Rev Genet 6:435. https://doi.org/10.1038/nrg1616

Singh K, Yadav D, Chauhan PS, Mishra M, Jin J-O (2020) Novel therapeutics for the treatment of Alzheimer's and Parkinson's disease. Curr Pharm Des 26:755. https://doi.org/10.2174/1381612826666200107161051

Singh RK, Chamachi NG, Chakrabarty S, Mukherjee A (2017) Mechanism of unfolding of human prion protein. J Phys Chem B 121:550. https://doi.org/10.1021/acs.jpcb.6b11416

SPX—Structural Biochemistry/Proteins/Protein Folding Problem—Wikibooks, open books for an open world (n.d.) [WWW Document], https://en.wikibooks.org/wiki/Structural_Biochemistry/Proteins/Protein_Folding_Problem#/media/File:SPX.png. Accessed 4 Jun 2021

Stewart P, Campbell L, Skogtvedt S, Griffin KA, Arnemo JM, Tryland M, Girling S, Miller MW, Tranulis MA, Goldmann W (2012) Genetic predictions of prion disease susceptibility in carnivore species based on variability of the prion gene coding region. PLoS One 7:e50623. https://doi.org/10.1371/journal.pone.0050623

Stork M, Giese A, Kretzschmar HA, Tavan P (2005) Molecular dynamics simulations indicate a possible role of parallel β-helices in seeded aggregation of poly-Gln. Biophys J 88:2442. https://doi.org/10.1529/biophysj.104.052415

Swietnicki W, Morillas M, Chen SG, Gambetti P, Surewicz WK (2000) Aggregation and fibrillization of the recombinant human prion protein huPrP90−231 †. Biochemistry 39:424. https://doi.org/10.1021/bi991967m

Tetz G, Tetz V (2018) Prion-like domains in eukaryotic viruses. Sci Rep 8:8931. https://doi.org/10.1038/s41598-018-27256-w

Tichopad A, Pfaffl MW, Didier A (2003) Tissue-specific expression pattern of bovine prion gene: quantification using real-time RT-PCR. Mol Cell Probes 17:5. https://doi.org/10.1016/S0890-8508(02)00114-7

Tyedmers J, Madariaga ML, Lindquist S (2008) Prion switching in response to environmental stress. PLoS Biol 6:e294. https://doi.org/10.1371/journal.pbio.0060294

USDA APHIS|Chronic Wasting Disease (CWD) (n.d.) [WWW Document]. https://www.aphis.usda.gov/aphis/ourfocus/animalhealth/nvap/NVAP-Reference-Guide/Control-and-Eradication/Chronic-Wasting-Disease. Accessed 24 Mar 2021)

Vallabh SM, Minikel EV, Schreiber SL, Lander ES (2020) Towards a treatment for genetic prion disease: trials and biomarkers. Lancet Neurol 19:361. https://doi.org/10.1016/S1474-4422(19)30403-X

Wang F, Wang X, Yuan C-G, Ma J (2010) Generating a prion with bacterially expressed recombinant prion protein. Science 327:1132–1135

Westergard L, Christensen HM, Harris DA (2007) The cellular prion protein (PrPC): its physiological function and role in disease. Biochim Biophys Acta Mol basis Dis 1772:629. https://doi.org/10.1016/j.bbadis.2007.02.011

White MD, Mallucci GR (2009) Therapy for prion diseases: insights from the use of RNA interference. Prion 3:121. https://doi.org/10.4161/pri.3.3.9289

WHO|Prion diseases (n.d.) [WWW Document]. https://www.who.int/zoonoses/diseases/prion_diseases/en/. Accessed 14 Feb 2021

Wickner RB (2016) Yeast and fungal prions. Cold Spring Harb Perspect Biol 8:a023531. https://doi.org/10.1101/cshperspect.a023531

Wille H, Michelitsch MD, Guénebaut V, Supattapone S, Serban A, Cohen FE, Agard DA, Prusiner SB (2002) Structural studies of the scrapie prion protein by electron crystallography. Proc Natl Acad Sci 99:3563. https://doi.org/10.1073/pnas.052703499

Yamada M, Itoh Y, Fujigasaki H, Naruse S, Kaneko K, Kitamoto T, Tateishi J, Otomo E, Hayakawa M, Tanaka J, Matsushita M, Miyatake T (1993) A missense mutation at codon 105 with codon 129 polymorphism of the prison protein gene in a new variant of Gerstmann-Strässler-Scheinker disease. Neurology 43:2723–2724. https://doi.org/10.1212/wnl.43.12.2723-a

Yang S, Levine H, Onuchic JN, Cox DL (2005) Structure of infectious prions: stabilization by domain swapping. FASEB J 19:1778. https://doi.org/10.1096/fj.05-4067hyp

Zabel MD, Reid C (2015) A brief history of prions. Pathogens and Disease 73:ftv087. https://doi.org/10.1093/femspd/ftv087

Part II

Emerging Gastroenteritis Viral pathogens

Astroviruses: Overview and Emerging Disease

12

Shivani Manglic, Anil Kumar Mavi, Neelesh Kumar, Devendra Kumar Singh, Umesh Kumar, Ambika Nayak, and Garima

Abstract

Astroviruses are single-stranded, positive-sense possessing a star-like appearance, 28–30 nm in diameter, RNA viruses in the family Astroviridae which have been known to cause food-borne non-bacterial acute gastroenteritis also known as stomach flu in humans affecting the gastrointestinal route. This is caused by intake of contaminated food and water or getting in touch with contaminated sources. This is transmitted through the faecal-oral route and is responsible for causing diarrheal illness more commonly in infants and young children less than

S. Manglic
Department of Biotechnology, Jaypee Institute of Information and Technology, Noida, Uttar Pradesh, India

A. K. Mavi (✉)
Department of Botany and Life Sciences, Sri Aurobindo College, University of Delh, Delhi, India
e-mail: anilmavi100@gmail.com

N. Kumar
Department of Biotechnology, Delhi Technological University, Delhi, India

D. K. Singh
Department of Respiratory Medicine, School of Medical Sciences and Research, Sharda University, Noida, Uttar Pradesh, India

U. Kumar (✉)
Department of Biosciences, IMS Ghaziabad (University Courses Campus), Ghaziabad, Uttar Pradesh, India
e-mail: umeshkumar82@gmail.com

A. Nayak
ICAR-Indian Veterinary Research Institute, Bareilly, Uttar Pradesh, India

Garima
Department of Biotechnology, Delhi Technological University, Delhi, India

P. L. Bhukya, A. S (eds.), *Emerging Human Viral Diseases, Volume II*,
https://doi.org/10.1007/978-981-97-4480-0_12

age group of 5 years. The old-aged and the less immune system person are also known to get affected by the astroviruses. In contrast to mink and humans, animals remain asymptomatic to the astrovirus. The effect of this virus may last up to 2–3 days and does not require any sort of medical treatment. The symptoms of astrovirus infection include nausea, vomiting, mild fever, loss of appetite, dehydration, and headache but its less severe than rotavirus and norovirus. Although these viruses may reside in gastrointestinal tract for approximately 2 months, the infectivity of the astrovirus can be reduced by using methanol or alcohol disinfection. Human astrovirus consists of 8 serotypes (1–8), and among them, serotype 1 is the most common type present in nearly 50% of the countries. There may be several types of astroviruses such as Avian astroviruses in poultry and Mamastroviruses for dogs. There have been several methods to detect the astroviruses infection such as electron microscopy (EM), cell culture methods, enzyme-linked immunosorbent assays, and RT-polymerase chain reaction (RT-PCR). This review provides the information on phylogenetic classification and structural organization of astroviruses along with their pathogenesis and immunological response to this pathogen.

Keywords

Human Astroviruses (HastV) · Electron microscopy · Avastroviruses · RNA · VP70

12.1 Introduction

Since its discovery in 1975, numerous epidemiological studies on astrovirus infection and disease have been conducted, and these viruses are thought to be among the most prevalent viral agents that cause gastroenteritis, mostly affecting infants under the age of two globally (Appleton and Higgins 1975; De Benedictis et al. 2011). After rotavirus and norovirus, it is the third most common viral cause of acute diarrhoea, accounting for up to 10% of cases of non-bacterial gastroenteritis. The Baltimore classification places the Astroviridae family of viruses in group IV. Mammalian astroviruses fall under the genus Mamastrovirus, while avian astroviruses, affecting both commercial and wild birds, are classified as Avastrovirus. Together, these two genera constitute the Astroviridae family. The first HAstVs were discovered in paediatric stool specimens in 2008 from Melbourne, Australia (Finkbeiner et al. 2008). These viruses have a positive-sense single-stranded RNA genome. It has a poly-A tail, three open reading frames (ORFs), a 3′ UTR, a 5′ UTR, and an overall length that typically ranges from 6.2 to 7.7 kb (Fischer et al. 2017). Astroviruses are increasingly significant for public health, causing diarrheal diseases in birds and mammals. While often asymptomatic, infections can lead to symptoms like diarrhoea, hepatitis, nephritis, and recently, encephalitis, depending on the species infected, the host's age, and immunological condition. Though infections self-limiting, they can spread and cause severe infections in immunocompromised persons. The genomic RNA of the astrovirus is uncoated and translated into a precursor polyprotein upon entry into cells. This polyprotein is broken into proteins necessary for the assembly of progeny and replication of the virus. The pathophysiology of HAstV, especially in relation to various genotypes, is still poorly understood, despite the fact that astroviruses are widely distributed and that novel genotypes have been discovered (Cortez et al. 2017, 2019; Méndez et al. 2013). The emergence of new astroviruses capable of zoonosis emphasizes the need to understand their biology to minimize future consequences.

12.2 Characteristics

The astroviruses belongs to family Astroviridae which includes human and animal astroviruses. These astroviruses are non-enveloped and possess an icosahedral morphology. Three open reading frames and positive-sense, single-stranded RNA (ssRNA) make up the genome, which sets this virus family apart from others in terms of how it is organized. Astroviruses (AstV) have been found in many different kinds of animals. Astrovirus infections in most mammals are linked to gastroenteritis (Bouzalas et al. 2014; Donato and Vijaykrishna 2017). Madeley and Cosgrove first used the term 'astrovirus' in 1975 to refer to tiny, round viruses with a distinctive appearance like star (Astron is Greek for 'star') as depicted in Fig. 12.1. However, it should be noted that only fewer than 10% of the overall population may have such projections, and their existence is pH depended. These viruses measure around 28-30 nm in diameter obtained by method of direct electron microscopy (EM) when analysed in the stool samples collected from hospitalized infants as well as in gastroenteritis outbreaks in newborn babies. A large variety of young people and birds had similar features.

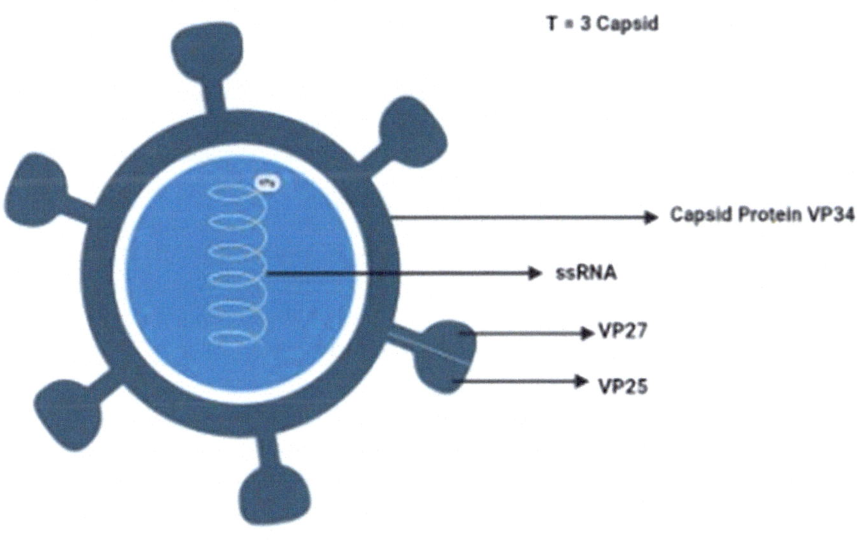

Fig. 12.1 Star-like structure structural appearance of astroviruses

12.2.1 Phylogenetic Classification and Emerging Concern of Astroviruses

The two genera that make up the Astroviridae family are Mamastrovirus and Avastrovirus, depending on the various hosts that the virus infects. The International Committee on Taxonomy of Viruses (ICTV) has classified the astroviruses that have been discovered so far based on their host tropism and viral genomic variation. As of right now, 19 species of the Mamastrovirus genus (MAstV1–19) and 3 species of Avastrovirus (AAstV1–3) have been officially approved. Four species of MAstV have been identified in humans to date: MAstV-1, MAstV-6, MAstV-8, and MAstV-9. MAstV-1 consists of classic genotypes 1–8, with HAstV-1 is the most prevalent strain globally (Khamrin et al. 2016).

Mamastroviruses belonging to the genus Mamastrovirus include isolates from a variety of animals, such as, cats, minks, sheep, dogs, bats, rats, and deer, including humans, and also from marine mammals like sea lions, calves (BoAstV), and bottlenose dolphins (CSIAstV). In terms of clinical associations, Mamastrovirus has mostly been linked to self-limited gastroenteritis in humans, particularly in young children, newborns, and people with weakened immune systems (Oem and An 2014; Selimovic-Hamza et al. 2017). The classic human astrovirus (HAstV), the prototype astrovirus species that was first isolated from humans, is currently a member of the MAstV1 species (Khamrin et al. 2016).

Based on the species of origin and viral genome characteristics, six distinct astroviruses have been identified in avian species to date: two originating in Turkey (Turkey Astrovirus type 1 (TAstV-1) and type 2 (TAstV-2); two originating in Chicken (ANV and Chicken Astrovirus (CAstV)); and two originating in Duck (Duck Astrovirus type 1 (DAstV-1) and type 2 (DAstV-2)](Zhang et al. 2017). Two putative species in genotype GI (GIA, GIB) and one in genogroup II are included in this genus (GIIA). The genogroup I and II contain 10 and 9 genotype species, respectively. Members of some of these species can be identified from one another by serology, much like canonical HAstV. This indicates that some of these species, like TAstV-2 and ANV, have viral serotypes. Avastroviruses generally exhibit greater diversity than Mamastroviruses (Wan et al. 2018). The molecular characterization of the many avian astroviruses reveals significant genetic diversity within each type, which affects the molecular and serological methods capacity to detect these viruses (Pantin-Jackwood et al. 2012).

12.2.2 Structure and Genome Organization

The AstV virion has an icosahedral symmetry without an envelope. It has a smooth edge and a prominent five to six star-shaped projections on its surface. EM showed the exterior diameter of the shedded virus in faeces is of 28 to 30 nm. In contrast, viruses grown in cell monolayers have an external diameter of 41 nm, apparent projecting spikes, but lack the typical star-like shape. Typical star-shaped appearance was observed after alkaline treatment (Risco et al. 1995). The icosahedral

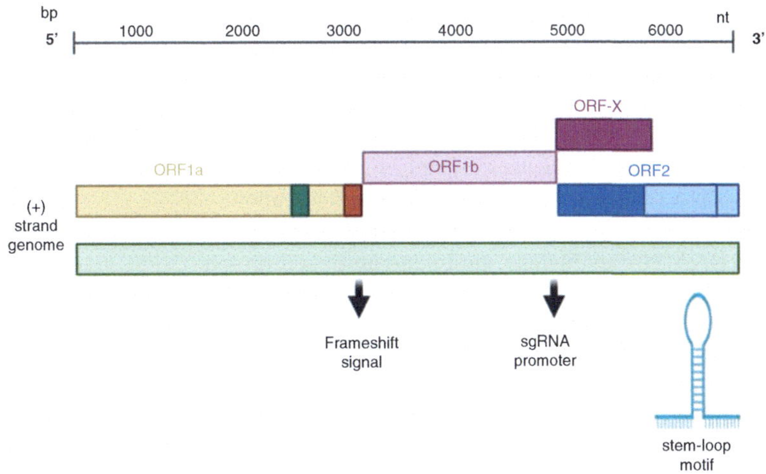

Fig. 12.2 Genomic organization of human astrovirus

structure of virions is spiked, and as per the recent investigations which uses advanced method of detection of cryo- electron microscopy (EM) and image processing of trypsin-treated and untreated HAstV particles corroborated this in many studies (Arias et al. 2016; Bosch et al. 2014).

The genomic RNA, which is around 6.8 kb in length, has three open reading frames that have been identified 'ORF1a, ORF1b, and ORF-X', which was postulated to be functional considering that it is conserved in AstVs as shown in Fig. 12.2 (Stenglein et al. 2012). The ssRNA genome has positive polarity (ssRNA(+)), and the poly (A) tail is present. The size ranges from 6.17 kb for the human strain MLB-1 to 7.772 kb for DAstV-2. The viral genome has three ORFs with varied lengths in different isolates, as well as 5′ and 3′ untranslated regions (UTRs). The two ORFs, termed ORF1a and ORF1b, that are situated at the 5′ end of the genome are believed to be essential for the process of replication and transcription of the virus genome (Méndez et al. 2012). The conserved heptameric sequence (AAAAAC) and possible stem-loop structure found in the overlapped area are key motifs for both the frameshift and generation of the viral polymerase encoded in ORF1b. The capsid polyprotein is encoded by the third ORF, often known as ORF 2, which is situated at the 3′ end of the genome. In completely infecting virions, three end proteins are produced by processing of the ORF2 main translation result (Djikeng et al. 2008; Monroe et al. 1993; Ykema and Tao 2021).

12.3 Astrovirus Replication

During infection, two positive-sense RNA species which are the full-length genomic RNA (gRNA) and a subgenomic RNA (sgRNA, ~2.4 kb) were noticed which are transcribed from the negative-sense full-length complement of the genomic

RNA. The overlapping of ORF1b and ORF2 by 8 nucleotides in mammalian astroviruses is essential for the production of subgenomic RNA, which is necessary for the synthesis of sgRNA and is facilitated by an internal subgenomic promoter in negative-sense RNA.

The astrovirus genome also features short 5′ and 3′ UTRs, with the latter containing highly conserved sequences crucial in interactions with proteins found in cell and viral RNA replicase essential for genome replication.

The majority of the non-structural proteins generated by positive-sense, single-stranded RNA viruses include viral genome-linked protein (VPg), protease, helicase, and RNA-dependent RNA polymerase (RdRp). NTPase and helicase processes unwinding of the RNA during replication, whereas VPg is covalently linked to the 5′ end of the RNA genome, playing a crucial role in efficient translation and/or replication. Astroviruses' non-structural proteins exhibit evidence of the presence of these activities, excluding the helicase.

The nsp1a and nsp1ab polyproteins are the main translation products of the astrovirus genome (Fig. 12.3). ORF1a encodes nsp1a, whereas ORF1a and ORF1b work together in a ribosomal frameshift process to translate nsp1ab. nsp1a length varies among viruses specific for mammals somewhere between 787 and 950 aa and can be as long as 1240 aa in avian viruses. Notably, nsp1a features a distinctive protease motif.

The protease demonstrates trypsin-like enzyme properties, with a crystal structure revealing similarities to viral serine proteases from Sesbania mosaic virus and various picornaviruses. As suggested by site-directed mutagenesis and validated by structural comparison, the catalytic triad consists of His461, Asp489, and Ser551 (Jiang et al. 1993).

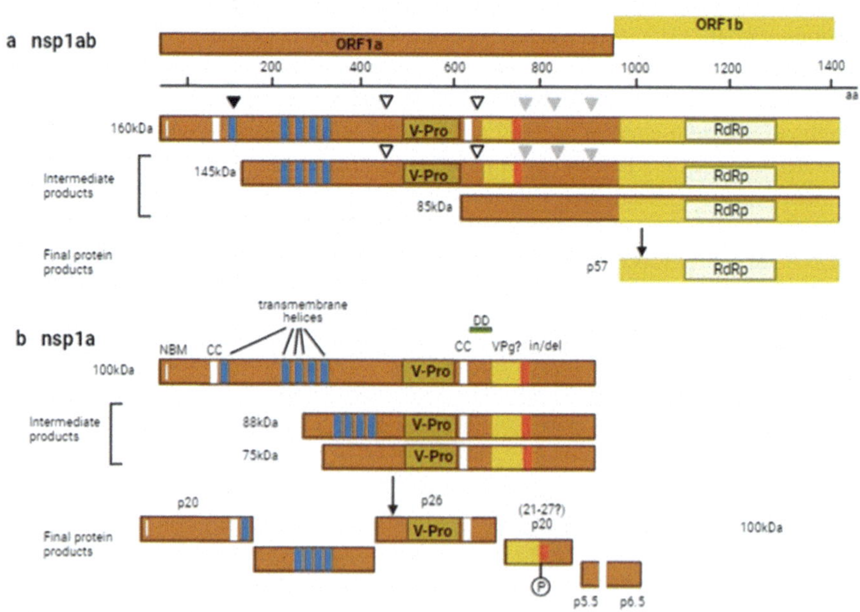

Fig. 12.3 Non-structural proteins and protein structural motifs

12.3.1 Attachment and Entry

There have not been many investigations on how astroviruses initially interact with their host cells. These viruses' cell receptor molecules are still unknown. Different strains of HAstV exhibit varying degrees of tropism in cultured cells, which suggests that their interactions with the host cells may also differ (Ingle et al. 2021). Trypsin digestion of the virus particles is necessary for human astroviruses to be infectious. For HAstV-8, the proteolytic pathway of the virion's processing of VP70 has been identified but the actual factors for the mechanism are not well known.

The precursor polyprotein after the cleavage of trypsin results in significant structural modifications to the particles and results in the production of three end products in HAstV-8: HAstV-8: VP34, VP25, and VP27 as in Fig. 12.4 (Méndez et al. 2007).

The conserved domain, represented by VP34, creates the capsid core, whereas VP25 and VP27 together generate the spikes located on the virion surface. Despite having a high degree of variability, VP25 has two conserved structural motifs that may play a role in how this virus interacts to the host cell. These residues, Lys455, Ser554, Thr575, and Glu610, are found to interact with oligosaccharides moieties that may serve as cell receptors or co-receptors. Virus infectivity is neutralized by antibodies that are able to detect the VP25 and VP27 spike proteins, most likely via preventing virus binding (Méndez et al. 2002).

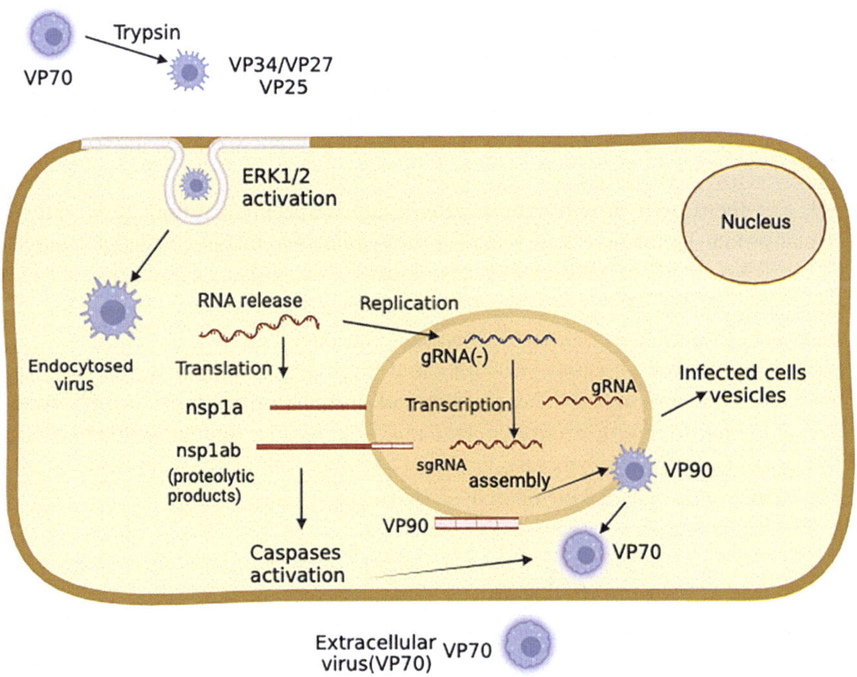

Fig. 12.4 Replication cycle of astroviruses

The ERK1/2 signalling pathway, which is activated during binding of virus or entry into the cell, is supposed to get activated within the first 15 minutes of the interaction of virus to the host cell. It appears that ERK1/2 is necessary early on to develop a productive infection.

12.3.2 Uncoating

The genomic RNA of a virus gets released in the cytoplasm from infecting virus through the mechanism of translation but the actual site and the cellular factors involved are yet to be determined (Cortez et al. 2017).

12.3.3 Translation

Upon HAstV infection, no discernible alterations in cellular protein production are seen. Astroviral RNA possesses a poly-A tail at the 3′end like the majority of cellular mRNAs, but the existence of a cap structure at the 5′end has not been reported. It is suggested that a VPg may be encoded in the ORF1a of the astrovirus, and this protein may influence how viral mRNAs are translated by interacting with translation initiation factors that have been identified for other viruses, such as the calicivirus (Woode and Bridger 1978).

12.3.4 Transcription/Replication

These viruses are thought to replicate and transact the genome via an alphavirus replication and transcription method. According to these hypotheses, the full-length negative-sense RNA, or gRNA (−), would be produced using the gRNA as a template, and then, both the full-length gRNA and the sgRNA would be produced. Cellular protein synthesis is necessary for the synthesis of gRNA (−) and accumulation of gRNA, while cellular DNA transcription is not. It has been discovered that v-pro, RdRp, VP90, gRNA (−), and viral particles are all connected to interior cell membranes (Marczinke et al. 1994; Méndez et al. 2013).

The observed membranous structures that originate from the ER are likely the site of RNA replication and the early stages of morphogenesis. The viral phosphoproteins, which have a molecular weight of 21–27 kd and interact with the RdRp RNA-dependent RNA polymerase, are likely to play role in RNA replication process (Fuentes et al. 2011; Murillo et al. 2015).

12.3.5 Assembly and Release

The assembly of virus-like particles results from the production of ORF2 in cultured cells using baculovirus or recombinant vaccine virus as vectors, suggesting that the encoded protein may assemble in the absence of viral RNA. However, these particles found to be unstable and showed a typical morphology when purified, indicating a detective assembly. Virus assembly tolerates some detection or changes in 70 N-terminal basic amino acids of VP90/70, which are thought to interact with the gRNA in the particles (Koci et al. 2003; Méndez et al. 2004). The ORF2 primary product of HAstV-8 (Méndez et al. 2002) VP90 forms intracellular particles that can be found associated with membranes or in the cytosol. VP90 assembles into particles associated with membranous structures, where viral non-structural proteins and the sense and anti-sense genomic RNA are also present (Mohamed et al. 2017; Sanchez-Fauquier et al. 1994).

Proteolytic processing by caspases of the VP90 assembled into particles to yield VP70 is required for cell egress of the virus, since inhibitors of these proteases prevent it, while it is promoted by pro-apoptotic factors, such as TRAIL (TNF-related apoptosis-inducing ligand) (Monroe et al. 1993).

12.3.6 Human Astrovirus Interaction with Gut Epithelia

Endocytosis, the process by which a virus enters a host cell, is likely mediated by attachment to host receptors. The viral genomic RNA is then released in the cytoplasm following uncoating. In order to create replication proteins, viral RNA is translated into two processed polyproteins. Replication occurs within viral factories made up of membrane vesicles produced from the ER. The genomic ssRNA(+) is used to create a dsRNA genome. The precursor of the capsid protein is produced by subgenomic RNA translation. Viral msRNAs and fresh ssRNA(+) are produced as a result of transcription and replication of the dsRNA genome (Truchet et al. 2017; Wüthrich et al. 2016). New virus particles are assembled, non-lytic virus is released, and the capsid matures through proteolytic cleavage (Jiang et al. 1993).

Tight junctions in cells help to maintain cell polarity by inhibiting the free movement of macro- as well as micronutrients from one side to other side of the cell. Tight junctions which are made of occludins and claudins interact to cytosolic adaptor proteins which helps it to connect to actin cytoskeleton. HAstVs disrupts the tight junctions in the cell membrane and their attachment with actin cytoskeleton thereby increasing the free movement of nutrients resulting in enhanced permeability (Johnson et al. 2017).

12.4 Response to Innate Immunity

The astrovirus genome is uncoated by the virus after binding and entrance into the cell. IFN- is produced as a result of replication, which inhibits viral replication and structural protein translation. The synthesis of the protein known as inducible nitric oxide synthase (iNOS) also takes place when epithelial cells are infected with an astrovirus. After astrovirus infection, active transforming growth factor beta (TGF-) levels rise; however, it is unclear what function TGF plays. The mechanism(s) by which signal transducer and activator of transcription 1 (STAT1) reduces replication of astrovirus are still unknown. According to earlier research, astrocytes may influence the immune system's function by blocking the activation of complement (Tam et al. 2014).

Later phases of viral infections are controlled by the innate immune system, which also signals the adaptive immune response's recruitment (Fig. 12.5). This specific response possesses a memory component, leading to a quicker and more robust reaction upon a subsequent infection caused by the same infectious agent. The response by an adaptive immunity comprises two components: the cell-mediated response, which uses cytotoxic T cells that are specific to an antigen, and the humoral response mediated by antibody-mediated/B cell. Despite contradictory findings from the newly developed mouse model and the existing turkey model, reports indicating biphasic age distribution and immunocompromised individuals generally conclude that the adaptive immunity plays significant role in regulating disease caused by astrovirus (Koci et al. 2003).

Antibodies already present against astroviruses may provide defence from recurrent infection. This observation, which came from evaluating the severity of

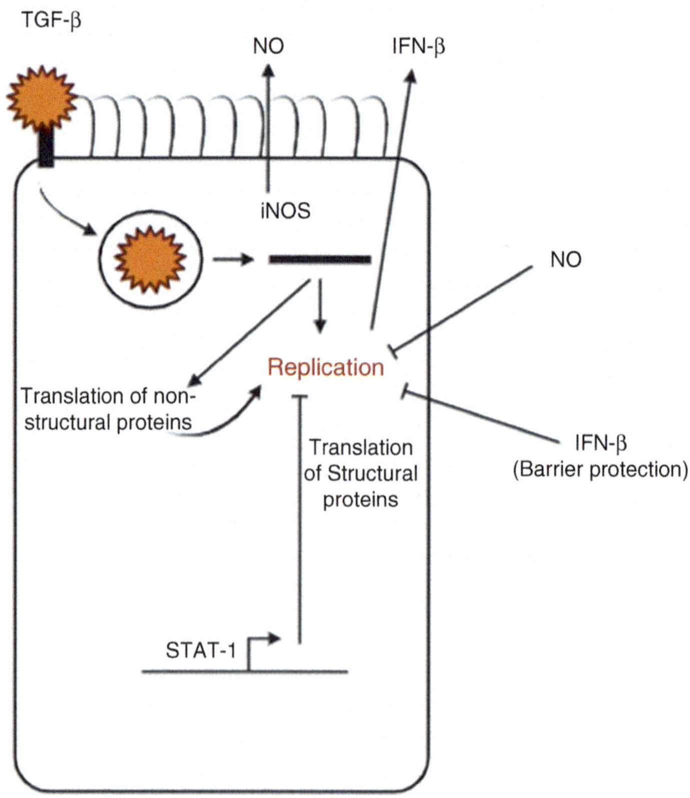

Fig. 12.5 An overview of the response of innate immune system to an astrovirus infection

symptoms in those who acquired the infection deliberately, could simply imply an existence of type-specific reactions. Astroviruses could have evolved a way that delays the emergence of a response by body's adaptive immune system, underscoring the significant role of the innate immune response, especially considering the acute nature of astroviral disease. Unlike other viruses, HAstV induces a less robust type I interferon response (Walter and Mitchell 2000). Astroviruses may lack the ability to counteract interferon responses. However, the administration of exogenous type I interferon may guard against heightened barrier permeability and limit infection in vitro. The only known method by which astroviruses avoid the host immune system is by attaching the capsid protein to complement 1q and mannose-binding lectin, which prevents complement from working (Bonaparte et al. 2008; Hair et al. 2010).

12.5 Pathogenesis

Astrovirus is a gastrointestinal pathogen. HAstV is transmitted to a host via the oral-faecal pathway, as well as interpersonally. In adults, exposure to a larger dose of astrovirus occurs through fomites and contaminated food or water. Avian viruses' transmission also occurs from parents to their offspring. Mammalian astroviruses affect mainly epithelial cells of the intestinal tract. Histopathological studies of an immunocompromised child persistently infected, having a pronounced diarrhoea, showed that astrovirus infection was limited to the small intestine as shown in Fig. 12.6 (Karlsson et al. 2015). Infection involves the mature epithelial cells near the microvilli tips—more extensive in the jejunum than in the duodenum, but not in

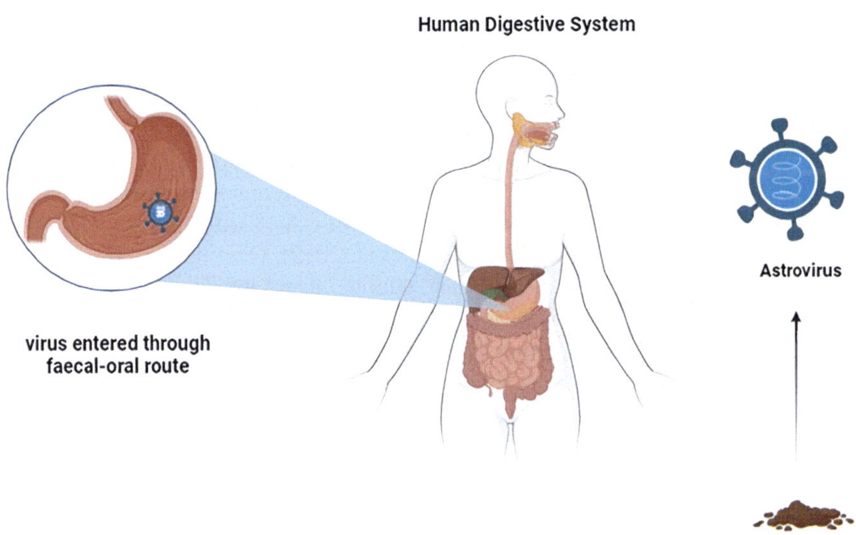

Fig. 12.6 Mechanism of transmission of astroviruses

the stomach. Morphological abnormalities in the intestine suggested that despite severe diarrhoea, an inflammatory response does not occur. It can infect epithelial cells (OAstV and BoAstV) as well as subepithelial macrophages (OAstV) and M cells (BoAstV) of the small intestine. Astrovirus was recently found as the only pathogen in the central nervous system (CNS) of an immunodeficient patient who died of encephalitis (Alfred et al. 2015).

12.6 Animal Model and Cell Culture Used for AstVs Pathogenesis

There are not a lot of animal models for testing AstVs in vitro for pathogenesis of the disease. Turkey poult model is the only model in which the studies related to pathogen have been conducted, though it is also not an ideal model. TAstV, which causes disease in turkey, is genetically different from human astroviruses although belonging to the Avastrovirus family (Vu et al. 2017). As a result, there are differences in the disease's pathology between the two. In 2012, murine astrovirus (MuAstVs) was found in mice (both immunocompetent and immunocompromised) kept in pathogen-free laboratory facilities. During research, it was seen that both adaptive and innate immunities are required to control MuAstV infection relating it to HAstVs. However, MuAstV in mouse model did not show the diarrheal symptoms (Cortez et al. 2017). The standard HAstV strains are produced using a variety of cell lines, the most often used strain being human colon cancer cell type 2 (Caco-2) (Janowski and Wang 2019).

12.7 Diagnosis

Human astroviruses (HAstV) are significant human pathogens that contribute to global cases of gastroenteritis. The rise in the identification of astroviruses as the causative agents of sporadic disease outbreaks can be attributed to recent advancements in diagnostic techniques.

The use of monoclonal antibodies and enzyme immunoassays for detecting HAstV resulted in substantial detection (2.5–9%) of HAstV infection among hospitalized patients with acute diarrhoea (Jeong et al. 2012). Over the past decade, the majority of epidemiological surveys have adopted astrovirus-specific RT-PCR as their primary screening method (Karlsson et al. 2015; Pérot et al. 2017). In addition to molecular serotyping, ongoing efforts are focused on developing novel typing approaches with the goal of identifying additional viral properties associated with virulence. High-throughput sequencing (HTS) has paved the path in diagnosing the AstV using metagenomics which enable better understanding of virus origin although routine use is quite challenging due to its high cost, and sequence homology with already known pathogens using microarray technique tends to give more accurate results (Stenglein et al. 2012). Employing diverse typing assays is crucial for obtaining comprehensive insights into both antigenic diversity and the molecular mechanisms responsible for pathogenicity. The viruses can be diagnosed using enzyme-linked immunosorbent assay (ELISA), RT-PCR reverse transcriptase PCR, electron microscopy, and cell cultures and are widely used (Fu et al. 2009).

12.8 Epidemiology

All around the world, HAstV has been found in faecal samples from people suffering from gastrointestinal disorders. It mostly affects young children, the elderly, and people with weakened immune systems. Infections are rare in healthy individuals, although they do happen sometimes. With almost 50% of new cases, HAstV-1 is the most common form worldwide (De Grazia et al. 2018). The second most common form varies according to the age group under analysis and the geographic area. In a research carried out in a Chinese hospital, the distribution showed an incidence of 7.4% in babies aged 9 to 11 months, 6.1% in infants aged 12 to 17 months, 5.6% in infants aged 6 to 8 months, and 5.6% in newborns aged 0 to 2 months. Remarkably, more than 95% of HAstV infections occurred in children younger than 2 years old (Fang et al. 2006; Candido et al. 2015). In a clinical study conducted in China, 1352 acute diarrheal faecal samples were analysed, and the detection rate of HAstV was 2.81%. Children under the age of four years had 92.11% positive reports of HAstV cases (Tang et al. 2022). An astrovirus that causes gout in goslings first appeared in Shandong Province in 2016, and it rapidly propagated to other areas that farm geese, including Jiangsu, Heilongjiang, Anhui, Guangdong, and Sichuan (Liu et al. 2022). The fatality rate of the diseased goslings was almost 50%, and their significant death rate ranged from 15% to 30%. In Ghana, HIV has been detected in tap water and dam water linked to a gastroenteritis outbreak (Dongdem et al. 2011). A Slovenian study found HAstVs in 27% of surface water samples and 1.4% of drinking groundwater sources (Vu et al. 2017; Méndez and Arias 2009).

12.9 Transmission

Research on human participants has shown that the faecal–oral pathway is the primary means of transmission for classic HAstVs (Midthun et al. 1993). When the quality of surface and groundwater is declined, it represents a severe public health concern. The stability of HAstV in freshwater, marine water (Bosch et al. 1997), and drinking water (Abad et al. 1997) facilitates its environmental transmission also. The World Health Organization (WHO) highlights how the unclean water, inadequate sanitation, and poor hygiene put children in poorer countries at higher risk of gastroenteritis (WHO 2004). Infections with HAstVs are often caused by contaminated food and water (Matsui and Greenberg 1994). HAstV may survive for extended periods of time on inert surfaces (Gallimore et al. 2006), and there is a possibility that it might spread through faeces in medical facilities (Mendez and Arias 2009).

12.10 Prevention and Treatment

The prevention of HAstV infections involves controlling routes of transmission and disease control at the host level. The control of transmission channels involves detecting and inactivating viruses in water and food, as well as disinfecting contaminated fomites. The properties of AstV including stability and resistance to inactivation create barriers for elimination, particularly in situations such as hospitals where immunocompromised patients are more susceptible. Astrovirus infection in humans typically manifests as a mild, self-limiting condition, and treatment primarily involves fluid therapy (Bjorkholm et al. 1995). Oral rehydration therapy (ORT) and intravenous infusion in children are highly recommended. Currently, there is no vaccine available for humans (Caballero et al. 2004; Dalton et al. 2003) and the success of immunoglobulin treatment for immunocompromised individuals varies. In animals infected with astrovirus, the primary strategy involves preventing transmission to humans, through washing hand and food thoroughly. Despite thorough scrubbing in response to outbreaks, faecal shedding of astroviruses persists, underscoring the virus's resilience(Bosch et al. 2014).

12.11 Re-Emerging Potential of Astroviruses

Astroviruses, once regarded primarily as culprits of mild gastroenteritis, are now receiving renewed attention due to their evolving significance as pathogens. Recent advancements in diagnostic techniques have resulted in an increased identification of astroviruses, even in unexpected settings. These viruses are now acknowledged for their potential to induce more severe outcomes, especially in immunocompromised individuals. The absence of a specific human vaccine and the variable success of immunoglobulin treatments in such cases highlight the challenges in managing and preventing astrovirus infections (Glass and Bresee 2011).

The stability and resistance of astroviruses to inactivation present obstacles in control efforts, leading to their persistence in specific environments. Hospitals, where individuals are frequently immunocompromised, encounter difficulties in eradicating the virus despite rigorous disinfection measures.

To better handle the re-emerging potential of astroviruses, there is a growing need for comprehensive strategies focusing on early detection, effective disinfection measures, and a deeper understanding of the virus's evolving nature. Research efforts are crucial in developing targeted interventions and preventive measures to mitigate the impact of astrovirus infections in both human and animal populations.

12.12 Therapeutics and Vaccines

There are currently no commercial vaccines available for AstV infections in humans or animals. Thus, research is focused on developing vaccines and promoting immunization programmes. However, when a healthy individual is immunized with vaccine virus, antibodies formed against the viral antigen, which has been shown to prevent infection in the individual after repeated exposure to the same pathogen, indicate that a vaccine with high efficacy is developed. The generation of neutralizing antibodies upon administration of recombinant HAstV capsid spike suggests that there is a good chance of creating a subunit vaccine that effectively prevents HAstV diarrheal illness. Antibodies targeting these epitopes neutralize the virus by inhibiting virus attachment to cells (Ricemeyer et al. 2022; Bogdanoff et al. 2018). Novel subunit vaccines derived from capsid protein (CP) of mink AstV (CPΔN) induced humoral immune response. Humoral immunity is a key factor in defending against AstVs because the presence of protective antibodies against HAstV in healthy individuals provides safeguards against reinfection (Bidokhti et al. 2019; Marvin 2016).

12.13 Conclusion

The astroviruses may remain asymptomatic in some patients and thus cause major concern if left undiagnosed in infants and small children. The advanced techniques like RT-PCR can detect in early days and, if followed up with proper sanitization and oral rehydration therapy, can prevent the disease from prevailing among other healthy individuals. Since there are some factor mechanisms still unknown at the entry level into the vesicles, it becomes typical to stop the replication after entering inside the body. Further researchers need to be done to investigate the replication cycle and the mechanism behind type I IFN restricting barrier permeability included by astrovirus, so that the uncertain death can be avoided.

References

Abad FX et al (1997) Astrovirus survival in drinking water. Appl Environ Microbiol 63(8):3119–3122. https://doi.org/10.1128/aem.63.8.3119-3122.1997

Alfred N, Liu H, Li ML, Hong SF, Tang HB, Wei ZZ, Chen Y, Li FK, Zhong YZ, Huang WJ (2015) Molecular epidemiology and phylogenetic analysis of diverse bovine astroviruses associated with diarrhea in cattle and water buffalo calves in China. J Vet Med Sci 77(6):643. https://doi.org/10.1292/JVMS.14-0252

Appleton H, Higgins PG (1975) Letter: viruses and gastroenteritis in infants. Lancet 1(7919):1297. https://doi.org/10.1016/s0140-6736(75)92581-7

Arias CF, López T, Murillo A (2016) Astrovirus replication and reverse genetics. In: Viral Gastroenteritis: Molecular Epidemiology and Pathogenesis, pp 483–495. https://doi.org/10.1016/B978-0-12-802241-2.00023-7

Bidokhti MRM et al (2019) Immunogenicity and efficacy evaluation of subunit Astrovirus vaccines. Vaccine 7(3):79. https://doi.org/10.3390/vaccines7030079

Bjorkholm M, Celsing F, Runarsson G, Waldenström J (1995) Successful intravenous immunoglobulin therapy for severe and persistent astrovirus gastroenteritis after fludarabine treatment in a patient with Waldenström's macroglobulinemia. Int J Hematol 62:117. https://doi.org/10.1016/0925-5710(95)00396-a

Bogdanoff WA, Perez EI, López T, Arias CF, DuBois RM (2018) Structural basis for escape of human Astrovirus from antibody neutralization: broad implications for rational vaccine design. J Virol 92:e01546. https://doi.org/10.1128/jvi.01546-17
Bonaparte RS, Hair PS, Banthia D, Marshall DM, Cunnion KM, Krishna NK (2008) Human astrovirus coat protein inhibits serum complement activation via C1, the first component of the classical pathway. J Virol 82(2):817–827. https://doi.org/10.1128/JVI.01847-07
Bosch A et al (1997) Persistence of human astrovirus in fresh and marine water. Water Sci Technol J Int Assoc Water Pollut Res 35(11–12):243–247. https://doi.org/10.2166/wst.1997.0741
Bosch A, Pintó RM, Guix S (2014) Human astroviruses. Clin Microbiol Rev 27(4):1048–1074. https://doi.org/10.1128/CMR.00013-14
Bouzalas IG, Wüthrich D, Walland J, Drögemüller C, Zurbriggen A, Vandevelde M, Oevermann A, Bruggmann R, Seuberlich T (2014) Neurotropic astrovirus in cattle with nonsuppurative encephalitis in Europe. J Clin Microbiol 52(9):3318–3324. https://doi.org/10.1128/JCM.01195-14
Caballero S et al (2004) Structural requirements of Astrovirus virus-like particles assembled in insect cells. J Virol 78(23):13285–13292. https://doi.org/10.1128/jvi.78.23.13285-13292.2004
Candido M, Alencar AL, Almeida-Queiroz SR, Buzinaro Mda G, Munin FS, de Godoy SH, Livonesi MC, Fernandes AM, de Sousa RL (2015) Molecular detection and phylogenetic analysis of bovine astrovirus in Brazil. Arch Virol 160(6):1519–1525. https://doi.org/10.1007/S00705-015-2400-8
Cortez V, Meliopoulos VA, Karlsson EA, Hargest V, Johnson C, Schultz-Cherry S (2017) Astrovirus Biology and Pathogenesis. Annu Rev Virol 4:327–348. https://doi.org/10.1146/Annurev-Virology-101416-041742
Cortez V, Margolis E, Schultz-Cherry S (2019) Astrovirus and the microbiome. Curr Opin Virol 37:10–15. https://doi.org/10.1016/J.COVIRO.2019.05.002
Dalton RM, Pastrana EP, Sánchez-Fauquier A (2003) Vaccinia virus recombinant expressing an 87-kilodalton polyprotein that is sufficient to form Astrovirus-like particles. J Virol 77(16):9094–9098. https://doi.org/10.1128/jvi.77.16.9094-9098.2003
De Benedictis P et al (2011) Astrovirus infections in humans and animals—molecular biology, genetic diversity, and interspecies transmissions. Infection, genetics and evolution: journal of molecular epidemiology and evolutionary genetics in infectious diseases 11(7):1529–1544. https://doi.org/10.1016/j.meegid.2011.07.024
De Grazia S et al (2018) Molecular evolutionary analysis of type-1 human astroviruses identifies putative sites under selection pressure on the capsid protein. Infection, genetics and evolution: journal of molecular epidemiology and evolutionary genetics in infectious diseases 58:199–208. https://doi.org/10.1016/j.meegid.2017.12.023
Djikeng A, Halpin R, Kuzmickas R, DePasse J, Feldblyum J, Sengamalay N, Afonso C, Zhang X, Anderson NG, Ghedin E, Spiro DJ (2008) Viral genome sequencing by random priming methods. BMC Genomics 9(1):1–9. https://doi.org/10.1186/1471-2164-9-5/TABLES/3
Donato C, Vijaykrishna D (2017) The broad host range and genetic diversity of mammalian and avian Astroviruses. https://doi.org/10.3390/v9050102
Dongdem JT, Damanka S, Asmah R (2011) Molecular isolation of human norovirus and astrovirus in tap water by RT-PCR. Int Res J Biochem Bioinform 1:131–138
Fang ZY, Sun YP, Ye XH, Wang H, Zhang Q, Duan ZJ, Steele D, Glass R (2006) Astrovirus infection among hospitalized children with acute diarrhea in seven regions of China, 1998-2005. Zhonghua liu xing bing xue za zhi= Zhonghua liuxingbingxue zazhi 27(8):673–676
Finkbeiner SR, Allred AF, Tarr PI, Klein EJ, Kirkwood CD, & Wang D (2008) Metagenomic analysis of human diarrhea: viral detection and discovery. PLoS pathogens, 4(2):e1000011. https://doi.org/10.1371/journal.ppat.1000011
Fischer K, Pinho V, Reis D, Balkema-Buschmann A (2017) Bat Astroviruses: towards understanding the transmission dynamics of a neglected virus family. https://doi.org/10.3390/v9020034
Fu Y, Pan M, Wang X, Xu Y, Xie X, Knowles NJ, Yang H, Zhang D (2009) Complete sequence of a duck astrovirus associated with fatal hepatitis in ducklings. J Gen Virol 90(Pt 5):1104–1108. https://doi.org/10.1099/VIR.0.008599-0

Fuentes C, Guix S, Bosch A, Pintó RM (2011) The C-terminal nsP1a protein of human Astrovirus is a phosphoprotein that interacts with the viral polymerase. J Virol 85(9):4470. https://doi.org/10.1128/JVI.01515-10

Gallimore CI et al (2006) Environmental monitoring for gastroenteric viruses in a pediatric primary immunodeficiency unit. J Clin Microbiol 44(2):395–399. https://doi.org/10.1128/jcm.44.2.395-399.2006

Glass RI, Bresee JS (2011) Astroviruses, enteric adenoviruses, and other gastroenteritis viral infections. In: Tropical infectious diseases: principles, pathogens and practice, pp 416–419. https://doi.org/10.1016/B978-0-7020-3935-5.00063-X

Hair PS, Gronemus JQ, Crawford KB, Salvi VP, Cunnion KM, Thielens NM, Arlaud GJ, Rawal N, Krishna NK (2010) Human astrovirus coat protein binds C1q and MBL and inhibits the classical and lectin pathways of complement activation. Mol Immunol 47(4):792–798. https://doi.org/10.1016/J.MOLIMM.2009.10.006

Ingle H, Hassan E, Gawron J, Mihi B, Li Y, Kennedy EA, Kalugotla G, Makimaa H, Lee S, Desai P, McDonald KG, Diamond MS, Newberry RD, Good M, Baldridge MT (2021) Murine astrovirus tropism for goblet cells and enterocytes facilitates an IFN-λ response in vivo and in enteroid cultures. Mucosal Immunol 14(3):751–761. https://doi.org/10.1038/S41385-021-00387-6

Janowski AB, Wang D (2019) Infection and propagation of Astrovirus VA1 in cell culture. Curr Protoc Microbiol 52(1):e73. https://doi.org/10.1002/cpmc.73

Jeong HS, Jeong A, Cheon D-S (2012) Epidemiology of astrovirus infection in children. Korean J Pediatr 55(3):77. https://doi.org/10.3345/kjp.2012.55.3.77

Jiang B, Monroe SS, Koonin EV, Stine SE, Glass RI (1993) RNA sequence of astrovirus: distinctive genomic organization and a putative retrovirus-like ribosomal frameshifting signal that directs the viral replicase synthesis. Proc Natl Acad Sci USA 90(22):10539–10543. https://doi.org/10.1073/PNAS.90.22.10539

Johnson C, Hargest V, Cortez V, Meliopoulos VA, Schultz-Cherry S (2017) Astrovirus pathogenesis. *Viruses* 9(1):22. https://doi.org/10.3390/v9010022

Karlsson EA, Small CT, Freiden P, Feeroz MM, Matsen FA, San S, Hasan MK, Wang D, Jones-Engel L, Schultz-Cherry S (2015) Non-human Primates Harbor diverse mammalian and avian Astroviruses including those associated with human infections. PLoS Pathog 11(11):1005225. https://doi.org/10.1371/JOURNAL.PPAT.1005225

Khamrin P, Maneekarn N, Ushijima H (2016) Molecular epidemiology of Astroviruses. In: *Viral Gastroenteritis*. Elsevier, pp 497–522

Koci MD, Moser LA, Kelley LA, Larsen D, Brown CC, Schultz-Cherry S (2003) Astrovirus induces diarrhea in the absence of inflammation and cell death. J Virol 77(21):11798–11808. https://doi.org/10.1128/JVI.77.21.11798-11808.2003

Liu C, Sun M, Liao M (2022) A review of emerging goose Astrovirus causing gout. Biomed Res Int 2022:1–11. https://doi.org/10.1155/2022/1635373

Marczinke B, Bloys AJ, David T, Brown K, Willcocks MM, Carter MJ, Brierley I (1994) The human astrovirus RNA-dependent RNA polymerase coding region is expressed by ribosomal frameshifting. J Virol 68(9):5588–5595. https://doi.org/10.1128/JVI.68.9.5588-5595.1994

Marvin S (2016) The immune response to Astrovirus infection. Viruses 9(1):1. https://doi.org/10.3390/v9010001

Matsui SM, Greenberg HB (1994) Medical management of foodborne viral gastroenteritis. In: Hui YH, Gorham JR, Murrell KD, DOE C (eds) *Foodborne disease handbook, vol. 2*. Diseases caused by viruses, parasites, and fungi. Marcel Dekker, Inc., New York, NY, pp 145–158

Méndez E, Arias CF (2009) Astrovirus. In: Clinical Virology, pp 1145–1153. https://doi.org/10.1128/9781555815981.ch50

Méndez E, Fernández-Luna T, López S, Méndez-Toss M, Arias CF (2002) Proteolytic processing of a serotype 8 human Astrovirus ORF2 polyprotein. J Virol 76(16):7996. https://doi.org/10.1128/JVI.76.16.7996-8002.2002

Méndez E, Salas-Ocampo E, Arias CF (2004) Caspases mediate processing of the capsid precursor and cell release of human Astroviruses. J Virol 78(16):8601–8608. https://doi.org/10.1128/

JVI.78.16.8601-8608.2004/ASSET/41F76C69-D02A-4689-87A9-07024F60538A/ASSETS/GRAPHIC/ZJV0160449700006.JPEG

Méndez E, Murillo A, Velázquez R, Burnham A, & Arias CF (2012) Replication cycle of astroviruses. astrovirus research: essential ideas, everyday impacts, future directions, 19–45. https://doi.org/10.1007/978-1-4614-4735-1_2

Méndez E, Aguirre-Crespo G, Zavala G, Arias CF (2007) Association of the Astrovirus Structural Protein VP90 with membranes plays a role in virus morphogenesis. J Virol 81(19):10649. https://doi.org/10.1128/JVI.00785-07

Méndez E, Murillo A, Velázquez R, Burnham A, Arias CF (2013) Replication Cycle of Astroviruses. *Astrovirus Res* 9781461447351:19. https://doi.org/10.1007/978-1-4614-4735-1_2

Midthun K et al (1993) Characterization and seroepidemiology of a type 5 astrovirus associated with an outbreak of gastroenteritis in Marin County, California. J Clin Microbiol 31(4):955–962. https://doi.org/10.1128/jcm.31.4.955-962.1993

Mohamed FF, Mansour SMG, El-Araby IE, Mor SK, Goyal SM (2017) Molecular detection of enteric viruses from diarrheic calves in Egypt. Arch Virol 162(1):129–137. https://doi.org/10.1007/S00705-016-3088-0

Monroe SS, Jiang B, Stine SE, Koopmans M, Glass RI (1993) Subgenomic RNA sequence of human astrovirus supports classification of Astroviridae as a new family of RNA viruses. J Virol 67(6):3611. https://doi.org/10.1128/JVI.67.6.3611-3614.1993

Murillo A, Vera-Estrella R, Barkla BJ, Méndez E, Arias CF (2015) Identification of host cell factors associated with Astrovirus replication in Caco-2 cells. J Virol 89(20):10359–10370. https://doi.org/10.1128/JVI.01225-15

Oem JK, An DJ (2014) Phylogenetic analysis of bovine astrovirus in Korean cattle. Virus Genes 48(2):372. https://doi.org/10.1007/S11262-013-1013-0

Pantin-Jackwood M, Todd D, Koci MD (2012) Avian Astroviruses. In: Astrovirus research. Springer New York, New York, NY, pp 151–180

Pérot P, Lecuit M, Eloit M (2017) Astrovirus diagnostics. Viruses 9(1):10. https://doi.org/10.3390/v9010010

Ricemeyer L, Aguilar-Hernández N, López T, Espinosa R, Lanning S, Mukherjee S, Cuellar C, López S, Arias CF, DuBois RM (2022) Structures of two human Astrovirus capsid/neutralizing antibody complexes reveal distinct epitopes and inhibition of virus attachment to cells. J Virol 96:e01415–e01421. https://doi.org/10.1128/JVI.01415-21

Risco C et al (1995) Ultrastructure of human astrovirus serotype 2. J Gen Virol 76(8):2075–2080. https://doi.org/10.1099/0022-1317-76-8-2075

Sanchez-Fauquier A, Carrascosa AL, Carrascosa JL, Otero A, Glass RI, Lopez JA, San Martin C, Melero JA (1994) Characterization of a human astrovirus serotype 2 structural protein (VP26) that contains an epitope involved in virus neutralization. Virology 201(2):312–320. https://doi.org/10.1006/VIRO.1994.1296

Selimovic-Hamza S, Boujon CL, Hilbe M, Oevermann A, Seuberlich T (2017) Frequency and pathological phenotype of bovine Astrovirus CH13/NeuroS1 infection in neurologically-diseased cattle: towards assessment of causality. Viruses 9(1):12. https://doi.org/10.3390/V9010012

Stenglein MD et al (2012) Complete genome sequence of an astrovirus identified in a domestic rabbit (Oryctolagus cuniculus) with gastroenteritis. Virol J 9(1):216. https://doi.org/10.1186/1743-422x-9-216

Tam JCH, Bidgood SR, McEwan WA, James LC (2014) Intracellular sensing of complement C3 activates cell autonomous immunity. Science 345(6201):1134. https://doi.org/10.1126/SCIENCE.1256070

Tang X et al (2022) Molecular epidemiology of human adenovirus, Astrovirus, and Sapovirus among outpatient children with acute diarrhea in Chongqing, China, 2017–2019. Front Pediatr 10:826600. https://doi.org/10.3389/fped.2022.826600

Truchet L, Walland J, Wüthrich D, Boujon CL, Posthaus H, Bruggmann R, Schüpbach-Regula G, Oevermann A, Seuberlich T (2017) Neuropathological survey reveals underestimation of the prevalence of neuroinfectious diseases in cattle in Switzerland. Vet Microbiol 208:137–145. https://doi.org/10.1016/J.VETMIC.2017.07.027

Vu D-L, Bosch A, Pintó RM, Guix S (2017) Epidemiology of classic and novel human astrovirus: gastroenteritis and beyond. Viruses 9(2):33. https://doi.org/10.3390/v9020033

Walter JE, Mitchell DK (2000) Role of astroviruses in childhood diarrhea. Curr Opin Pediatr 12(3):275–279. https://doi.org/10.1097/00008480-200006000-00018

Wan CH, Chen CT, Cheng LF, Liu RC, Shi SH, Fu GH, Fu QL, Chen HM, Huang Y (2018) A novel group of avian Avastrovirus in domestic geese, China. J Vet Med Sci 80(5):798. https://doi.org/10.1292/JVMS.18-0013

Woode GN, Bridger JC (1978) Isolation of small viruses resembling astroviruses and caliciviruses from acute enteritis of calves. J Med Microbiol 11(4):441–452. https://doi.org/10.1099/00222615-11-4-441

World Health Organization (2004) Guidelines for drinking-water quality, vol 1. World Health Organization

Wüthrich D, Boujon CL, Truchet L, Selimovic-Hamza S, Oevermann A, Bouzalas IG, Bruggmann R, Seuberlich T (2016) Exploring the virome of cattle with non-suppurative encephalitis of unknown etiology by metagenomics. Virology 493:22–30. https://doi.org/10.1016/J.VIROL.2016.03.009

Ykema M, Tao YJ (2021) Structural insights into the human Astrovirus capsid. Viruses 13(5):821. https://doi.org/10.3390/V13050821

Zhang Y, Wang F, Liu N, Yang L, Zhang D (2017) Complete genome sequence of a novel avastrovirus in goose. Arch Virol 162(7):2135–2139. https://doi.org/10.1007/S00705-017-3297-1

Insights into the Biology of Norovirus: The So-Called "Perfect" Human Pathogen

13

Ruta Kulkarni

Abstract

Noroviruses (NoVs), members of the Family *Caliciviridae*, are the leading cause of nonbacterial gastroenteritis outbreaks across all age groups worldwide. NoVs also play a major role in sporadic gastroenteritis, being regarded as the most common cause in adults, and second only to rotavirus, among children. Indeed, in countries implementing routine rotavirus vaccination, NoVs are increasingly being identified as the primary cause of viral gastroenteritis among children. The success of NoV as a human pathogen may be attributed to several factors such as its highly transmissible and contagious nature, moderate virulence, rapid and constant evolution resulting in genetic diversity, and induction of limited protective immunity. This chapter aims to provide insights into different aspects of NoV biology with a focus on the viral life cycle, immunopathogenesis, viral evolution, and genetic diversity. Further, the review discusses NoV epidemiology and disease burden, clinical features of NoV infections, as well as current strategies and challenges for diagnosis, prevention, and control of this so-called "perfect" human pathogen.

13.1 Introduction

The Noroviruses (NoVs), previously known as "Norwalk-like viruses" (NLVs) or "small round structured viruses" (SRSVs), constitute a genus within the Family *Caliciviridae*. This family derives its name from the Latin word "calyx" (cup) which refers to the cup-like structures visible on the viral surface. The *Caliciviridae* family

R. Kulkarni (✉)
Department of Communicable Diseases, Interactive Research School for Health Affairs, Bharati Vidyapeeth (Deemed to be University), Pune, India
e-mail: ruta.kulkarni@bharatividyapeeth.edu

P. L. Bhukya, A. S (eds.), *Emerging Human Viral Diseases, Volume II*,
https://doi.org/10.1007/978-981-97-4480-0_13

is composed of small (27–40 nm), nonenveloped, icosahedral viruses which possess a linear, positive-sense, single-stranded RNA genome of 7.4–8.3 kb. This family consists of at least 11 identified genera which infect a wide range of hosts—*Norovirus*, *Sapovirus*, *Vesivirus*, *Lagovirus*, *Nebovirus*, *Recovirus*, and *Valovirus* infect mammals, *Bavovirus* and *Nacovirus* infect birds, *Minovirus* and *Salovirus* infect fish (Vinjé et al. 2019). Of these, noroviruses represent the most clinically important genus, and are recognized as a leading cause of acute gastroenteritis in humans (Green 2013).

Norwalk virus (NV), prototype of the genus *Norovirus*, was discovered in 1972 by Albert Kapikian and colleagues (Kapikian et al. 1972). Using the technique of immune electron microscopy (IEM), Norwalk virus was retrospectively identified as the etiological agent of a gastroenteritis outbreak which occurred in 1968 at an elementary school in Norwalk city, Ohio, USA, and thus became the first viral agent to be associated with gastroenteritis in humans. Since then, NoVs have emerged as a major cause of epidemic and sporadic gastroenteritis across all age groups worldwide.

13.2 Structural and Molecular Organization

13.2.1 Classification

Historically, the classification of NoVs was based on antigenic characterization of NoV strains by immune electron microscopy or cross-challenge studies in volunteers; however, such classification schemes had poor accuracy and reproducibility. With the advent of molecular techniques in the 1990s, reverse transcription polymerase chain reaction (RT-PCR) and genome sequencing became the major means of NoV characterization. Initially, the classification was based on partial polymerase (RdRp; ORF1) sequences, however, later shifted to VP1 (major capsid protein; ORF2) sequences (Chhabra et al. 2019).

13.2.1.1 Universal NoV Nomenclature and Typing System (2013)

A well-defined NoV classification scheme, based on phylogenetic clustering of the complete VP1 amino acid sequences was proposed by Zheng and colleagues in 2006 (Zheng et al. 2006). This scheme was updated in 2013 by the international Norovirus Classification Working Group (NCWG), and a universal standardized nomenclature and typing system was proposed (Kroneman et al. 2013). According to this system, NoVs were classified into six genogroups, GI-GVI, further divided into 40 genotypes—9 GI (GI.1–GI.9), 22 GII (GII.1–GII.22), 3 GIII (GIII.1–GIII.3), 2 each for GIV, GV, and GVI (GIV.1–GIV.2, GV.1–GV.2, GVI.1–GVI.2) (Kroneman et al. 2013; Vinjé 2015). The GI, GII, and GIV NoVs infect humans, except GII.11, GII.18, and GII.19 which infect porcine species, and GIV.2 which infects feline and canine species. The GIII and GV NoVs infect bovine and murine species respectively, while GVI NoVs infect canines. The GII.4 strains were further subtyped into variants. Newly emerged GII.4 strains clustering separately from existing strains, and becoming epidemic in at least two different geographical locations, were recognized as "novel variant." These variants were named according to the year and location of the first available full-length VP1 sequence.

Recombination in the ORF1–ORF2 junction region, leading to emergence of recombinant NoV strains, is a very common event within the NoV genus (Bull et al. 2005, 2007). A recombinant NoV clusters with two distinct genogroups or genotypes of NoV strains, when two different regions (ORF1 and ORF2) of the genome are subjected to phylogenetic analysis. In order to classify such recombinant NoVs correctly, a dual nomenclature system based on both polymerase (ORF1) and VP1 (ORF2) sequences was proposed by the NCWG (Kroneman et al. 2013). According to this system, polymerase genotypes were designated by a capital "P" followed by the genotype designation, e.g., GII.P4. Unclassified polymerase types which did not cluster phylogenetically with any of the known genotypes were termed as "orphan" polymerases and designated with an alphabet (GI.a–GI.f, GII.a–GII.p) to distinguish them from the wild-type polymerases, i.e., polymerases having the same genotype specificity as that of the corresponding VP1. A unified scheme of NoV nomenclature was proposed with all NoV strains being identified by the polymerase genotype followed by the VP1 genotype, e.g., GII.P4/GII.4, GII.P3/GII.4.

13.2.1.2 Updated NoV Nomenclature and Typing System (2019)

The NCWG universal classification scheme was further updated in 2019, and new NoV genogroups and genotypes were proposed based on VP1 sequences (Chhabra et al. 2019; Fig. 13.1). The number of genogroups has now been increased to 10 (GI–GX), with identification of the new genogroups—GVII (canine), GVIII (human), GIX (previously GII.15; human), and GX (bat), while two tentative new genogroups (GNA1; harbor porpoise and GNA2; sea lion) have been proposed. Within genogroup GII, five new genotypes (GII.23–GII.27) have been confirmed, thus expanding the number of NoV genotypes to 48—9 GI (GI.1–GI.9), 26 GII (GII.1–GII.14, GII.16–GII.27), 3 GIII (GIII.1–GIII.3), 2 each for GIV–GVI

Family *Caliciviridae*			
Genus *Norovirus*			
VP1 genotyping		**RdRp (Pol) P-typing**	
Genogroup GI	9 genotypes	**P-group GI.P**	14 P-types
Genogroup GII	26 genotypes	**P-group GII.P**	37 P-types
Genogroup GIII	3 genotypes	**P-group GIII.P**	2 P-types
Genogroup GIV	2 genotypes	**P-group GIV.P**	1 P-type
Genogroup GV	2 genotypes	**P-group GV.P**	2 P-types
Genogroup GVI	2 genotypes	**P-group GVI.P**	2 P-types
Genogroup GVII	1 genotype	**P-group GVII.P**	1 P-type
Genogroup GVIII	1 genotype	--	
Genogroup GIX	1 genotype	--	
Genogroup GX	1 genotype	**P-group GX.P**	1 P-type

Fig. 13.1 Norovirus classification scheme showing division of the NoVs into capsid (VP1)-based genogroups/genotypes and polymerase (RdRp/Pol)-based P-groups/P-types

(GIV.1–GIV.2, GV.1–GV.2, GVI.1–GVI.2), and a single genotype each for GVII–GX (GVII.1, GVIII.1, GIX.1, GX.1). Further, three tentative genotypes have been proposed, two GII (GII.NA1, GII.NA2), and one GIV (GIV.NA1).

This updated classification system serves to eliminate the "orphan" P-type nomenclature. Contrary to the previously followed system of P-type designation linked to the corresponding VP1 type, the current system proposes that P-type number assignment should be independent of the VP1 type (Fig. 13.1). Accordingly, eight P-groups (GI.P–GVII.P, GX.P) have been confirmed, and two tentative P-groups proposed (GNA1.P, GNA2.P). Further, these P-groups are classified into 60 confirmed P-types—14 GI (GI.P1–GI.P14), 37 GII (GII.P1–GII.P8, GII.P11–GII.P13, GII.P15–GII.P18, GII.P20–GII.P41), 2 GIII (GIII.P1–GIII.P2), 1 G1V (GIV.P1), 2 each for GV (GV.P1, GV.P2) and GVI (GVI.P1, GVI.P2), 1 each for GVII (GVII.P1) and GX (GX.P1), and 14 tentative P-types—4 GI (GI.PNA1–GI.PNA4), 9 GII (GII.PNA1–GII.PNA9), 1 GIV (GIV.PNA1).

For dual-typing nomenclature, it is proposed that VP1 genogroup/genotype should be listed first followed by the P-type within brackets. Thus, the revised designation of NoV strains is as follows—Genogroup.genotype [P-type] or Genogroup.genotype [P-group.P-type] respectively for strains where VP1 and P-type belong to the same or different genogroups, e.g., GII.4 [P4], GII.4 [GI.P3].

13.2.2 Morphology and Virion Structure

The NoV virions are typically small in size (27–30 nm), nonenveloped, rounded structures which exhibit $T = 3$ icosahedral symmetry. The viral capsid consists of two proteins—major capsid protein (VP1) and minor capsid protein (VP2). Each capsid is composed of 90 dimers (180 molecules) of the VP1 arranged in such a way as to create large hollows or cup-like structures, characteristic of Family *Caliciviridae*, on the viral surface (Green 2013). VP2 is located at the interior of the capsid.

The structure of the NoV capsid was elucidated in 1999 by Prasad and colleagues through X-ray crystallographic studies on Norwalk virus VLPs (Prasad et al. 1999). These studies have revealed that NoV VP1 consists of a short N-terminal arm, and is organized into two major domains—shell (S) and protruding (P). The S domain, corresponding to the N-terminal region of VP1 (up to aa 225), forms the inner part of the capsid surrounding the RNA genome. The N-terminal arm, located within the S domain, faces the interior of the capsid. The P domain, which is linked to the S domain through a flexible hinge, corresponds to the C-terminal half of VP1 (aa 225 onwards). The P domain forms arch-like protrusions from the shell and is further divided into the proximal P1 stalk (aa 226–278, 406–520) and distal P2 (aa 278–406) subdomains, which form the base/sides and top of the arch, respectively. The P domain contains the dimeric contacts for VP1 dimerization. In the dimeric form of VP1, two P2 subdomains form a bilobed structure at the surface of the virion (Green 2013).

Recently, it has been suggested that the NoV capsid is a highly dynamic structure, and that this flexibility is crucial for its function, particularly for host cell attachment and immune escape (Smith and Smith 2019). Based on studies on murine NoV (MNV) and certain human NoV (HuNoV) strains, involving cryoelectron microscopy and/or X-ray crystallographic structures of VP1-P domain and its complexes with different antibodies and putative host cell receptors, two modes of capsid flexibility have been proposed. (1) *Conformational flexibility of the P domain with respect to the shell*: This allows the virus particle to adopt an array of conformations such as "expanded" or "compressed" state, in response to different environmental stimuli. In the expanded state, the "floating" P domain is raised off the S domain (by ~16A°) and hovers above it, thus allowing exposure of otherwise buried immunodominant sites. In the compressed state, which is thought to be favored for host cell attachment, the P domain rests upon the S domain; this allows the virus to escape from host antibody response elicited against the expanded state. (2) *Conformational flexibility within the P domain*: This allows the P domain to adopt a spectrum of different conformations including the antibody-binding "open" and receptor-binding "closed," with the structural equilibrium shifting toward closed conformation due to antibody escape mutations or external stimuli such as bile salts (co-factor for cellular attachment).

13.2.3 Genome Structure and Organization

The NoV genome was first characterized in 1993 by Jiang and colleagues through studies on a cDNA library obtained from the prototype Norwalk virus, purified from fecal specimens of human volunteers (Jiang et al. 1993). The NoVs have a linear, positive-sense, single-stranded RNA genome, approximately 7.5–7.6 kb in length (Fig. 13.2). The 5′ end of the genome is covalently attached to a small viral protein, "VPg" and the 3′ end is polyadenylated. Further, short untranslated regions (UTRs) have been identified at the genome extremities; these UTRs carry conserved RNA

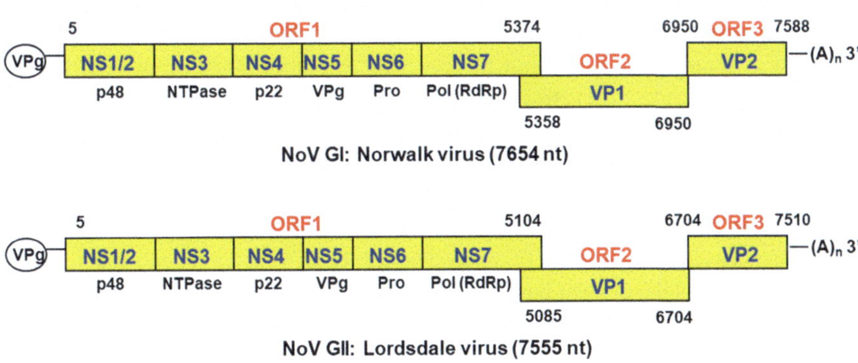

Fig. 13.2 Genome organization of GI and GII noroviruses. GI: Norwalk virus (Jiang et al. 1993) and GII: Lordsdale virus (Dingle et al. 1995)

secondary structures which play an important role in viral replication and pathogenesis (Bertolotti-Ciarlet et al. 2003; Ludwig-Begall et al. 2021).

NoV particles have also been found to contain a subgenomic RNA corresponding to the last ~2.3 kb of the viral genome (Asanaka et al. 2005; Ludwig-Begall et al. 2021). A highly conserved noncoding RNA stem-loop structure present upstream of the start site for subgenomic RNA initiation has been identified as the core promoter for the subgenomic RNA synthesis.

The NoV genome is organized into three overlapping open-reading frames (ORFs), with the exception of murine NoV which has a fourth alternative ORF. The 5′ proximal ORF1 encodes a single large polyprotein which is proteolytically cleaved into the viral nonstructural proteins, NS1–NS7. ORF2 and ORF3, both translated from subgenomic RNA, code for the structural (capsid) proteins, VP1 and VP2 respectively (Green 2013; Ludwig-Begall et al. 2021). ORF4, of murine NoV, translated primarily from subgenomic RNA, overlaps ORF2 and encodes the "virulence factor 1 (VF1)" (McFadden et al. 2011).

13.2.4 NoV Infection Models and Experimental Systems

13.2.4.1 Cell Culture Models

Despite extensive efforts, since its discovery in the 1970s, a robust cell culture system for NoVs remained unavailable for many decades, due to challenges arising out of uncertainties regarding virus tropism and its entry receptor.

The first cell culture system for the *Norovirus* genus was established in the early 2000s for murine norovirus (MNV), with Wobus and colleagues reporting MNV replication in cultured primary murine dendritic cells and macrophages (RAW264.7) (Wobus et al. 2004). Subsequently, efficient replication of MNV has been demonstrated in murine microglial cells (BV-2) and B cells (M12, WEHI-231), thus facilitating research studies on viral replication and immunopathogenesis (Cox et al. 2009; Jones et al. 2014).

For human NoV (huNoV), two cell culture systems have been reported thus far. In 2014, the first system, supporting modest viral replication, was demonstrated using the transformed human B cell line, BJAB. In vitro HuNoV replication in this system was found to be enhanced by presence of enteric bacteria or addition of HBGAs (Jones et al. 2014). In 2016, a second more robust infection model was established using stem cell-derived, human intestinal enteroid (HIE) cultures which represent the multicellular, physiologically active human intestinal epithelium (Ettayebi et al. 2016; Estes et al. 2019).

The establishment of these huNoV culture systems is an important breakthrough in the norovirus field. While the B cell system has been used for assessment of antiviral compounds, HIE system has facilitated development of neutralization assays, and serves as an important tool for virus inactivation studies (Bhar and Jones 2019).

However, sustained passaging (beyond fourth passage) of huNoV strains in either of these systems remains to be fully established, thus limiting their use for detailed investigations on viral replication/immunopathogenesis. Improvement of in vitro replication and further optimization of these culture systems, or development of alternate simplified systems is thus warranted.

13.2.4.2 Animal Models for Human NoV

Until recently, double knockout BALB/c mice (Rag/$\gamma c^{-/-}$) were the only small animal model reported for huNoVs, capable of supporting subclinical viral infection via the intraperitoneal route, however, their utility was limited due to their inability to mimic typical huNoV infection (Taube et al. 2013; Todd and Tripp 2019; Ludwig-Begall et al. 2021). Recent studies have demonstrated high-level huNoV replication in hematopoietic and intestinal cells of zebrafish larvae, suggesting their use as a robust small animal model for huNoV infection (Van Dycke et al. 2019).

Various nonhuman primates have been investigated as large animal models for huNoV, however, only chimpanzees and rhesus macaques have been found to support infection (Wyatt et al. 1978; Todd and Tripp 2019; Ludwig-Begall et al. 2021). Upon oral infection, these animals undergo fecal viral shedding and produce serum antibodies, but remain asymptomatic. Infection of pigtail macaques has been shown to result in virus shedding, and symptoms like vomiting, however, further evaluation of this animal model is required (Subekti et al. 2002).

Other large animal models for huNoV include gnotobiotic (Gn) pigs and calves, both of which develop diarrhea upon oral infection, undergo virus shedding, and show production of serum/mucosal antibodies (Cheetham et al. 2006; Souza et al. 2008; Todd and Tripp 2019; Ludwig-Begall et al. 2021). The infected Gn pigs also show expression of viral RNA in enterocytes and extra-intestinal lymphoid tissues, and serve as a useful model for huNoV infection due to their resemblance to human anatomy. Detection of viral antigen in enterocytes along with intestinal damage has been reported for infected Gn calves.

13.2.4.3 Experimental Systems

In the absence of a suitable huNoV cell culture model, until recently, virus-like particles (VLPs) and the huNoV replicon were the primary experimental systems for huNoVs. In vitro expression of NoV VP1 results in the spontaneous assembly of VP1 molecules into VLPs, which are morphologically and antigenically similar to native virions (Jiang et al. 1992). Recombinant VLPs produced in different expression systems have been extensively utilized as a surrogate for native virions in research studies on virus structure, replication, and immunopathogenesis. The huNoV replicon system, generated by transfection of cell lines with Norwalk virus RNA bearing an engineered neomycin resistance gene, has proven extremely useful for studies on NoV replication and antivirals (Chang et al. 2006). Recently, a plasmid-based human NoV reverse genetics system has been reported facilitating further advances in understanding of NoV biology (Katayama et al. 2014).

13.2.5 Viral Proteins and Replication Cycle

13.2.5.1 Structural Proteins

13.2.5.1.1 Major Capsid Protein (VP1)

The NoV VP1 (MW ~58 kDa) is considered as the major structural (capsid) protein of the virus because of its predominant role in formation of the viral capsid (180 copies per virion). This protein plays a fundamental role in virus–host cell interaction (Campillay-Véliz et al. 2020) (Table 13.1).

As mentioned above (Sect. 13.2.2), NoV VP1 is organized into two major domains—shell (S) and protruding (P). The S domain of VP1 is highly conserved in amino acid sequence among the NoVs and maintains the $T = 3$ icosahedral symmetry of the capsid (Campillay-Véliz et al. 2020). Through its interaction with other NoV proteins (RdRp and VP2), the S domain is thought to play an important role in the regulation of the viral replication cycle. It has been proposed that during the initial phase of the replication cycle, the S domain stimulates the viral RdRp enhancing viral RNA synthesis while during the later phase, it interacts with VP2 and promotes assembly of NoV particles (Subba-Reddy et al. 2012; Vongpunsawad et al. 2013).

The P domain of VP1 is more variable in amino acid sequence among the NoVs. This domain establishes the intermolecular contacts between dimeric VP1 subunits, and thus contributes to control of the size and stability of the viral capsid (Campillay-Véliz et al. 2020). The P domain is further divided into the P1 and P2 subdomains. The P2 subdomain is the most surface exposed, hypervariable part of the capsid. It carries the viral antigenic determinants and is involved in interaction with host antibodies (Prasad et al. 1999; Donaldson et al. 2008; Lindesmith et al. 2012). The P2 subdomain also contains the binding site for interaction with the histoblood group antigens (HBGAs), which serve as attachment factors for NoV entry into the host cell (Lochridge et al. 2005; Cao et al. 2007).

13.2.5.1.2 Minor Capsid Protein (VP2)

The NoV VP2 (MW ~29 kDa) is present in only one to two copies per virion and is thus considered as the minor structural (capsid) protein. This protein is located in the interior of the capsid shell, bound to a conserved motif in the VP1-S domain (Campillay-Véliz et al. 2020). Although the precise function is not yet known, VP2 involvement in viral uncoating, genome encapsidation, and capsid assembly, and its importance for stabilization of VP1 in the virus structure have been proposed (Bertolotti-Ciarlet et al. 2003; Vongpunsawad et al. 2013; Graziano et al. 2019) (Table 13.1). Further, it has been suggested that VP2 may serve as a helper protein to VP1, directing the VP1–VP2 complex to the host cell nucleus, with the complex playing a putative role in interfering with nuclear processes such as transcription and chromatin remodeling. Role of VP2 in viral replication through negative regulation of RdRp activity, as well as modulation of host cell immune response, have also been proposed (Campillay-Véliz et al. 2020).

Table 13.1 Characteristics and functions of norovirus structural and nonstructural proteins

Viral protein	Characteristic/function
Structural proteins	
VP1 Major structural (capsid) protein	• Organization into shell (S) and protruding (P) domains • S domain: – Maintenance of $T = 3$ icosahedral symmetry – Role in regulation of viral replication cycle suggested • P domain: subdivided into P1, P2 subdomains – VP1 dimerization, control of capsid size, and stability – Virus-host cell interaction—binding with host HBGAs for entry – Carries viral antigenic determinants – Interaction with host antibodies – Hypervariability of P2 region facilitates immune escape
VP2 Minor structural (capsid) protein	• Genome encapsidation and capsid assembly • Stabilization of VP1 in virus structure • Role in viral uncoating proposed • VP1 helper protein; putative role for VP1–VP2 complex in interference with host cell nuclear processes • Role in viral replication proposed • Role in modulation of host cell immune response suggested
Nonstructural proteins	
NS1–2 (p48)	• N-terminal viral protein encoded by ORF1 • Formation of viral replication complex • Interference with cellular signaling pathways and immune response activation suggested
NS3 (NTPase)	• Binding to and hydrolysis of nucleoside triphosphates • Proposed NTP-dependent RNA helicase and NTP-independent chaperone activity for remodeling of the RNA structure • Possibility of collaboration with NS7 for RNA synthesis • Proposed participation in replication complex formation
NS4 (p22)	• Implicated in formation of viral replication complex • Interference with protein secretion and posttranslational modification pathways as well as cytokine signaling pathways suggested
NS5 (VPg)	• Genome-linked viral protein • Initiation of viral RNA translation • Primer for NS7 (RdRp)-mediated viral RNA synthesis
NS6 (Pro)	• Protease—cleavage of ORF1 polyprotein into individual nonstructural proteins
NS7 (Pol or RdRp)	• Viral RNA synthesis, genome replication

13.2.5.2 Nonstructural Proteins

The ORF1-encoded large polyprotein is co- and posttranslationally cleaved to release at least six mature nonstructural proteins, NS1–2, NS3–NS7 (Sosnovtsev et al. 2006) (Table 13.1).

13.2.5.2.1 NS1–2 (p48)

NS1–2, also referred to as "p48," is the N-terminal viral protein encoded by ORF1. The mature NS1–2 protein released after proteolytic cleavage has a molecular weight of 37–48 kDa depending on the viral genogroup (Campillay-Véliz et al. 2020). This protein has been implicated as one of the main drivers in formation of the viral replication complex (Hyde and Mackenzie 2010). In vitro studies have revealed interaction of NS1–2 with protein A associated with the vesicle-associated membrane proteins (VAP-A), located at the membranes of the endoplasmic reticulum and trans-Golgi network. This indicates that NS1–2 may induce rearrangement of the endoplasmic reticulum and Golgi disassembly, and that NS1–2-VAP-A interaction may serve as a scaffold for assembly of replication complexes. Further, NS1–2 has also been suggested to interfere with cellular signaling pathways and immune response activation (Campillay-Véliz et al. 2020). In murine noroviruses, this protein has been identified as a determinant of viral tropism and persistence (Nice et al. 2013).

13.2.5.2.2 NS3 (NTPase)

NS3 (~40 kDa) primarily functions as the viral NTPase involved in binding to and hydrolysis of nucleoside triphosphates. This protein is also proposed to have NTP-dependent RNA helicase activity, as well as NTP-independent chaperone activity for remodeling of the RNA structure, and has also been shown to collaborate with NS7 for RNA synthesis (Pfister and Wimmer 2001; Thorne and Goodfellow 2014; Campillay-Véliz et al. 2020). Further, it is suggested that NS3 also participates in replication complex formation, and has been shown to localize to cellular membranes (Hyde et al. 2009).

13.2.5.2.3 NS4 (p22)

The mature NS4 protein, also referred to as "p22" or "p20," released after proteolytic cleavage has a molecular weight of 20–22 kDa depending on the viral genogroup (Campillay-Véliz et al. 2020). This protein is implicated in replication complex formation in collaboration with NS1–2 and NS3 (Hyde and Mackenzie 2010; Campillay-Véliz et al. 2020). In vitro studies have demonstrated that NS4 (synergistically along with NS1) contributes to Golgi disassembly, and membrane alterations within the cell, thus promoting replication complex formation. It has also been suggested that this protein interferes with protein secretion and posttranslational modification pathways, as well as interferon and other cytokine signaling pathways (Campillay-Véliz et al. 2020).

13.2.5.2.4 NS5 (VPg; Genome-Linked Viral Protein)

NS5 or VPg (~16 kDa) is covalently linked to the 5′ end of the RNA genome, and is known to be involved in translation and genome replication (Daughenbaugh et al. 2003; Rohayem et al. 2006). It participates in initiation of viral RNA translation by binding to the initiation factor eIF3 and interacting with cap-binding complex eIF4F. It participates in genome replication by acting as a primer for NS7 (RdRp)-mediated viral RNA synthesis. Further, it has been demonstrated that VPg can induce cell cycle arrest in G1/G0 phase and promote viral replication (Campillay-Véliz et al. 2020).

13.2.5.2.5 NS6 (Pro)

NS6 or Pro (~19 kDa) functions as the viral protease, and is responsible for cleavage of the ORF1 polyprotein to release the individual nonstructural proteins (Sosnovtsev et al. 2006). This protein has been described as a chymotrypsin-like protease in human noroviruses, with Cys139, His30, and Glu54 catalytic residues. The cleavage process by NS6-Pro consists of "early"; NS1–2/NS3, NS3/NS4, and "late"; NS4/NS5, NS5/NS6, and NS6/NS7 phases, with cleavage at early sites being more efficient. An intracellular precursor form of this protein, termed as ProPol, having both protease (NS6) and polymerase (NS7) activities, has been identified in human noroviruses. In vitro studies indicate that the ProPol precursor has higher activity as compared to the mature Pro, and thus can function as the dominant form of this protein (Campillay-Véliz et al. 2020).

13.2.5.2.6 NS7 (Pol)

NS7 or Pol (~57 kDa) is the viral RNA-dependent RNA polymerase (RdRp). It functions primarily for viral RNA synthesis during genome replication (Ng et al. 2004).

13.2.5.3 Replication Cycle

Characterization of the NoV replication cycle was long hampered due to the unavailability of a cell culture model. However, studies in the past decade using the cultivable MNV and the related feline calicivirus (FCV, Genus *Vesivirus*), as well as experimental systems such as the huNoV replicon, huNoV RNA transfected into cell culture, cell-based approaches using expression of NoV proteins in immortalized cells, significantly advanced our understanding of the mechanism of NoV replication (Thorne and Goodfellow 2014; Graziano et al. 2019). More recently, studies involving genome-wide CRISPR screens (MNV), and cryo-electron microscopy (FCV) have helped elucidate the process of viral entry (Graziano et al. 2019).

13.2.5.3.1 Viral Entry into the Host Cell

Viral entry into the host cell is a multistep process involving (1) attachment, (2) receptor engagement, (3) endocytosis, and (4) uncoating, and is a critical determinant of cell tropism.

Virion attachment to the cell surface is the first, and often rate-limiting step of viral entry, mediated by attachment factors which serve to concentrate the virus on the cell surface. NoVs have been shown to use host glycans (carbohydrates) as attachment factors. HuNoVs bind the histoblood group antigens (HBGAs), via their VP1–P2 domain (Marionneau et al. 2002; Graziano et al. 2019). Both cell-associated (lining the cells of the gastrointestinal tract) and noncell-associated (soluble, present in saliva) HBGAs have been found to promote cellular infection. MNVs interact with other glycans like terminal sialic acid moieties, heparan sulfate proteoglycans, glycolipids, and glycoproteins in a strain-dependent manner (Taube et al. 2009,

2012; Graziano et al. 2019). Further, nonglycans including bile salts (HuNoV, MNV), phospholipids like ceramide (MNV), divalent cations (MNV), enteric bacteria (HuNoV, MNV) have been suggested as cofactors which enhance viral attachment in a strain-specific manner; however, the mechanisms are not yet completely understood (Karst and Wobus 2015; Graziano et al. 2019; Ludwig-Begall et al. 2021).

The second step of the entry process is binding to a specific cell-surface receptor, which actively promotes viral entry into the cell. While the receptor for HuNoV remains unknown, CD300lf, a type-I integral membrane protein with a single extracellular IgG-like domain, has been recently identified as a receptor for MNV (Orchard et al. 2016; Haga et al. 2016). The CD300lf ectodomain contains a phospholipid-binding pocket through which it binds to the apical surface of VP1–P2 domain. Binding of ceramide to CD300lf has been suggested to alter conformation or clustering of CD300lf on the cell membrane, thus promoting its interaction with MNV (Graziano et al. 2019). Another type-I integral membrane protein, Feline Junctional Adhesion Molecule A (fJAM-A), has been recognized as the receptor for FCV (Makino et al. 2006).

Following receptor engagement, the virus particles undergo endocytosis to gain entry into the host cell cytoplasm. The precise mechanism for endocytic internalization of HuNoV is not yet known. However, it has been shown that MNV follows a cholesterol- and dynamin-dependent, pH-independent endocytic pathway, while FCV entry seems to be clathrin-dependent and (acidic) pH-dependent (Graziano et al. 2019).

Endosomal escape and viral uncoating is the last step of viral entry, which results in release of the viral genome into the cell cytoplasm. Again, the exact process is unknown for NoVs. However, recent studies have suggested that for FCV, receptor engagement triggers formation of a large funnel-shaped portal-like structure, which serves as a channel through which viral genome is delivered into the cell cytoplasm. This funnel-like structure consists of 12 VP2 molecules arranged about a small pore in the capsid shell with their hydrophobic termini pointing away from the virion surface (Bhella and Goodfellow 2011; Conley et al. 2019). As NoV VP2 is found to be essential for infection, it may play a role in uncoating (similar to FCV), and warrants further studies (Graziano et al. 2019).

13.2.5.3.2 Viral Genome Translation, Polyprotein Processing, and Genome Replication

After the viral RNA genome is released into the cell cytoplasm, it undergoes an initial round of translation (of ORF1) by the cellular translational apparatus to produce the nonstructural polyprotein, which contains the viral nonstructural proteins (NS1–2 to NS7). The translation initiation involves VPg, which acts as a cap-substitute and recruits the host cell translation initiation factors such as eIF4E, eIF3 (Daughenbaugh et al. 2003; Thorne and Goodfellow 2014).

The nonstructural polyprotein then undergoes co- and posttranslational processing by the (auto-cleaved) viral protease (NS6), resulting in the generation of three protein precursors—NS1–2 (p48)/NS3 (NTPase), NS4 (p22)/NS5 (VPg), and NS6 (Pro)/NS7 (Pol). These precursors are further cleaved by ProPol to release the six individual (mature) nonstructural proteins, which are required for genome replication (Campillay-Véliz et al. 2020).

NoV genome replication is mediated by the viral RdRp (NS7) and occurs in replication complexes that are associated with virus-induced intracellular membranous vesicles in the perinuclear region of the cytoplasm (Thorne and Goodfellow 2014). Studies on MNV have suggested that the replication complex is formed by recruitment of host cellular membranes derived from components of the secretory pathway, i.e., endoplasmic reticulum, trans-Golgi network, and endosomes, to the perinuclear region of the cell (Hyde et al. 2009; Hyde and Mackenzie 2010). This replication complex formation is mediated mainly by viral NS1–2 and NS4, with NS3 participation also suggested (Ludwig-Begall et al. 2021). The host cytoskeletal network has been implicated in the process, allowing the establishment of the replication complex in close proximity to the microtubule organizing center (Hyde et al. 2012).

The genome replication occurs via a negative-sense RNA intermediate which is synthesized by the viral RdRp using the incoming parental positive-sense genomic RNA as template. It has been suggested that the negative-sense RNA synthesis is initiated at the 3′ end of the genomic RNA by a de novo mechanism, and is enhanced through direct interaction between the viral RdRp and the VP1-Shell domain (Rohayem et al. 2006; Subba-Reddy et al. 2012). The negative-sense RNA then serves as template for the synthesis of positive-sense genomic and subgenomic RNA (consisting of ORF2 and ORF3, and in case of murine NoV, ORF4), by a VPg-dependent mechanism (Chaudhry et al. 2006; Rohayem et al. 2006). The positive-sense RNA synthesis is initiated by the viral RdRp at the 3′ end of the negative-sense RNA using VPg as a protein primer, which is covalently attached to the initiating nucleotide (Guanine). This process, referred to as VPg nucleotidylation or guanylation, is essential for virus infectivity. The mechanism of subgenomic RNA synthesis is not yet understood, however, two models, namely premature termination and internal initiation, have been proposed to explain the same (Thorne and Goodfellow 2014). According to the premature termination model, presence of a termination signal upstream of the VP1-coding region results in premature termination during negative-sense RNA synthesis by the viral RdRp. The resulting negative-sense subgenomic RNA is then used as template for production of the positive-sense subgenomic RNA. In the internal initiation model, an RNA secondary structure present downstream of the VP1 coding region on the negative-sense genomic RNA acts as a promoter for the positive-sense subgenomic RNA synthesis by the viral RdRp. The subgenomic RNA is found to be expressed at higher levels than the genomic RNA in infected cells (Ludwig-Begall et al. 2021).

Following the synthesis of positive-sense genomic and subgenomic RNA, multiple rounds of translation of the newly synthesized RNAs occur, followed by additional rounds of RNA synthesis (Thorne and Goodfellow 2014). The translation of VP1 and VP2 (and VF1 of murine NoV) occurs primarily from the subgenomic RNA. VP2 translation occurs by a termination–reinitiation mechanism, wherein after the termination of ORF2 (VP1) translation, the host ribosomes are thought to remain associated with the viral RNA and reinitiate protein synthesis at the start of ORF3 (VP2) (Napthine et al. 2009).

13.2.5.3.3 Viral Assembly and Exit from the Host Cell

As VP1 levels increase following translation, VP1 oligomerization occurs thereby preventing its interaction with the viral RdRp, resulting in a reduction in genome replication (Subba-Reddy et al. 2012). The replicated genomes are then packaged into the capsid for virion assembly. Due to the ability of VP1 to self-assemble into VLPs, it has been suggested that VP1 alone may be sufficient to drive capsid assembly (Bertolotti-Ciarlet et al. 2002). However, the 3′UTR of the mRNA has been proposed to stimulate VP1 expression, while initiation of NoV assembly is thought to be promoted by VP1–VP2 interaction (Vongpunsawad et al. 2013; Ludwig-Begall et al. 2021). Also, VP2 is thought to promote VP1 stability and is essential for the production of infectious virions. Due to its highly basic nature, VP2 has also been implicated in genome encapsidation via interaction with the acidic viral RNA, although the interaction has not yet been demonstrated (Bertolotti-Ciarlet et al. 2002; Sosnovtsev et al. 2005; Vongpunsawad et al. 2013).

The progeny NoV virions are then released from the host cell. The mechanism of exit is not yet known; however, it has been suggested that NoVs may employ the strategy of apoptosis induction, which has been proposed for other caliciviruses (Alonso et al. 1998; Bok et al. 2009a; Furman et al. 2009; Thorne and Goodfellow 2014). Viral NS3 (NTPase) and NS4 (p22) have been found to favor the proapoptotic activity of the cell, thus facilitating virion release (Campillay-Véliz et al. 2020). Whether cell lysis occurs at the time of release, is not clear. In vitro studies have shown that MNV lytically infects innate immune cells, however, HuNoV or MNV infection of B cells is noncytopathic. Further, NoVs are found to be secreted from culture cells inside membrane-bound vesicles, and are also shed in feces within such vesicles, thus suggesting a possible nonlytic route of exit (Ludwig-Begall et al. 2021).

13.3 Immunopathogenesis in Humans

13.3.1 Genetic Susceptibility to HuNoV Infection

Genetic susceptibility to HuNoV infection is mediated by the presence or absence of the host cell attachment factors, namely HBGAs, on gut epithelial surfaces (Nordgren and Svensson 2019). HBGAs are oligosaccharides of different types such as A, B, H, Lewis b, Lewis y, Lewis a, Lewis x. Their synthesis and expression on the human gut epithelium is genetically controlled by ABO blood group locus genes and the FUT2, FUT3 genes. Polymorphism of these genes generates diversity in HBGA expression in the human population. The expression of majority of the HBGA types on the gut epithelium and in body fluids (saliva) is dependent on the activity of the FUT2 enzyme. Individuals with a functional FUT2 gene are called as "secretors" while those with a nonfunctional gene are "nonsecretors."

The association of human HBGAs with susceptibility to NoV infections was first indicated through observation of resistance to Norwalk virus (GI.1) infection among nonsecretors in a challenge study, and supported by experiments demonstrating Norwalk virus binding to HBGAs on human intestinal epithelial cells (Marionneau et al. 2002; Lindesmith et al. 2003). Subsequently, several challenge studies, outbreak investigations and prospective observational studies, along with in vitro experiments have shown that susceptibility/resistance to NoV infections is genotype-dependent (or strain-dependent) (Nordgren and Svensson 2019). The prototype GI.1 and globally predominant GII.4 strains as well as GII.3, GII.6 NoVs mainly infect secretors, with the nonsecretors being largely resistant to these genotypes (Lindesmith et al. 2003; Tan et al. 2008; Frenck et al. 2012; Nordgren et al. 2013; Liu et al. 2014; Lopman et al. 2015). Genotypes such as GI.3, GI.6, GII.1, GII.2, and GII.7 have been found to infect secretors as well as nonsecretors (Lindesmith et al. 2005; Nordgren et al. 2010, 2013; Lopman et al. 2015).

13.3.2 HuNoV Cell Tropism

HuNoVs have been reported to display a dual tropism for intestinal epithelial and immune cells (Wobus 2018). This dual tropism has been demonstrated in vitro by the establishment of the human intestinal enteroid and human B cell line-based cell culture models (Jones et al. 2014; Ettayebi et al. 2016), and also supported by data from animal models—infection of enterocytes in gnotobiotic piglets; dendritic cells and B cells in chimpanzees; macrophages, dendritic cells, and lymphocytes in miniature piglets (Cheetham et al. 2006; Bok et al. 2011; Seo et al. 2018). Moreover, detection of HuNoV antigen in enterocytes near the villus tips, and CD3-positive (T cells), CD68-positive (macrophages), DC-SIGN-positive (dendritic cells) cells, of HuNoV-positive immunocompromised patients, further supports the dual tropism (Karandikar et al. 2016). In a recently reported study, enteroendocrine cells (EECs), specialized intestinal epithelial cells, have been identified as permissive target cells for HuNoVs. This suggests that intestinal epithelial cells in the small intestine may be the primary target cells for HuNoV replication (Karandikar et al. 2016; Green et al. 2020). The mechanism by which the virus crosses the intestinal epithelial barrier to reach the immune cells is not clearly understood. It has been proposed that the virus may enter myeloid cells of the lamina propria through phagocytosis of debris from infected epithelial cells or via exosomes.

Extraintestinal spread of huNoVs has also been suggested, due to detection of NoV RNA in patient sera and cerebrospinal fluid (Karst et al. 2015; Ludwig-Begall et al. 2021).

The cell tropism of MNV has been more extensively studied using different strains causing acute and persistent infections, and it is reported that MNVs also share a dual tropism for intestinal epithelial and immune cells (Wobus 2018).

13.3.3 HuNoV Interactions with Gut Microbiota

HuNoV interaction with the gut microbes is complex, with gut microbiota affecting viral infection and vice versa. Gut microbiota have been identified as major regulators of NoV virulence, with several in vitro studies demonstrating proviral effect on NoVs (Walker and Baldridge 2019). This proviral effect may be through direct promotion of NoV infectivity by either of the proposed mechanisms—(1) binding of NoVs to HBGAs (attachment factors for NoV entry) produced by commensal gut bacteria, (2) modification of bile acids (cofactors for NoV attachment) by gut microbes, or (3) bacteria-mediated NoV clustering (multiple virions binding to a single bacterium). Alternatively, gut microbes may be responsible for negative regulation of host antiviral (interferon) immune response, thus, indirectly enhancing NoV infection (Walker and Baldridge 2019; Hassan and Baldridge 2019).

In stark contrast to this observed proviral effect, gut microbiota have also been reported to have antiviral effect on NoV infection, although evidence for the same is limited. Further, the possibility of NoV infection resulting in alterations to the gut microbial composition has been suggested, although there is no consensus regarding the same (Walker and Baldridge 2019).

13.3.4 Pathology of HuNoV Infection

The pathological basis of NoV disease is not clearly understood. Although causing severe gastroenteritis, HuNoV infection is associated with only modest intestinal inflammation and pathologies, with the intestinal epithelium apparently remaining intact during infection (Karst et al. 2015; Ludwig-Begall et al. 2021). Early human volunteer challenge studies have reported observation of histopathological changes in the small intestine such as microvilli broadening and shortening, crypt hypertrophy, increased epithelial mitosis and apoptosis during acute infections (Blacklow et al. 1972; Schreiber et al. 1973; Dolin et al. 1975). Transient malabsorption of D-xylose, fat, and lactose, possibly related to shortened microvilli and decreased brush border enzyme activity, has also been observed. Further, reduction of tight junctional sealing proteins, disruption of epithelial barrier function, and stimulation of active anion secretion, have been noted. These data suggest that alterations in secretory and/or absorptive processes and leak flux could be the causal mechanisms for NoV-induced diarrhea (Karst et al. 2015; Ludwig-Begall et al. 2021). NoV-induced vomiting episodes may be a result of abnormal gastric motor functions resulting in delayed gastric emptying, although the pathophysiology remains undefined (Meeroff et al. 1980).

In addition to gastroenteritis, severe intestinal pathologies such as postinfectious irritable bowel syndrome, exacerbation of inflammatory bowel disease, necrotizing enterocolitis (in neonates), as well as atypical extraintestinal pathologies such as infantile seizures and encephalopathy, liver dysfunction, have been associated with HuNoV infections, but are rare (Karst et al. 2015; Ludwig-Begall et al. 2021).

13.3.5 Immune Response to HuNoV Infection

Our current understanding of immune response to HuNoV infection and the molecular mechanisms involved is derived from analysis of blood samples collected from NoV-positive gastroenteritis patients and NoV-infected volunteers during challenge studies, investigations in gnotobiotic pig model, as well studies involving HuNoV replicon system and/or VLPs, and recently, the human intestinal enteroid (HIE) cell culture system (Jahun and Goodfellow 2021; Mboko et al. 2022).

13.3.5.1 Innate Response

Innate response constitutes the first line of defense against viral infections, and is suggested to play a critical role in controlling primary HuNoV infections, given that the infections typically resolve within a few days (Karst et al. 2015). HuNoVs have been demonstrated to elicit a robust innate response comprising of type I (IFN-α) and type III (IFN-λ) interferons, with type III response being predominant (Mboko et al. 2022). Viral recognition by MDA5 molecular sensor followed by activation of JAK-STAT-pathway is the suggested mechanism for interferon production. This is followed by induction of IFN-stimulated gene response, involving upregulation of genes with known antiviral activity—IFI44L, OAS2, OASL, MX-1, and ISG-15. Indeed, studies in HIE culture system have demonstrated the role of interferon signaling in limiting viral replication (Hosmillo et al. 2020). Further, decrease in fecal virus shedding upon IFN-α treatment has been observed in gnotobiotic pigs (Jung et al. 2012).

Another important component of innate immunity, the inflammatory response is also supposedly inducted during HuNoV infection, upon viral recognition by Toll-like receptors 2 and 5 (Mboko et al. 2022).

13.3.5.2 Adaptive Response

Adaptive immune response contributes to the control of primary infections, and is particularly important for development of immunological memory and protection against reinfections. Protective immunity against HuNoV infections is complex and not clearly understood, with varying observations regarding duration and breadth of immunity, and no consensus regarding a primary correlate of protection. While early volunteer challenge studies have described a pattern of short-term immunity, ranging from 6 months to 2 years (Wyatt et al. 1974; Parrino et al. 1977; Baron et al. 1984; Johnson et al. 1990), recent reports indicate the duration may be longer, lasting up to 4–8 years (Simmons et al. 2013). The immunity is suggested to be mainly homotypic (genotype-specific), with some degree of cross-protection to other NoV genotypes within the same genogroup, but not extending to other genogroups (Wyatt et al. 1974; Parrino et al. 1977; Baron et al. 1984; Johnson et al. 1990; Ayukekbong et al. 2014).

The humoral arm of the immune response consisting of serum IgG and serum/mucosal IgA antibodies is considered to be the primary mode of protection against NoV infection. While a single, well-established correlate of protection remains undefined, several candidates, namely, serum HBGA-blocking antibodies, serum hemagglutination inhibition antibodies, serum/salivary/fecal IgA antibodies as well as virus-specific IgG memory B-cells have been proposed (Ramani et al. 2016). Presence of serum antibodies capable of blocking binding of NoV VLPs to HBGAs in vitro, has been shown to correlate with protection against NoV (GI.1, GII.4) illness, with prechallenge antibody titers of ≥200 or ≥500 being associated with significantly lower risk of illness in human challenge studies (Reeck et al. 2010; Atmar et al. 2011, 2015).

Limited information is available regarding the cell-mediated immune response against HuNoVs. Stimulation of $CD4^+$ and $CD8^+$ T-cell responses has been reported, with observation of significant increase in IFN-γ response among GI (Norwalk virus) or GII (Snow Mountain virus) NoV-infected volunteers, suggesting induction of a predominant Th1-type cellular immune response following NoV infection (Lindesmith et al. 2005, 2010; Karst et al. 2015).

13.4 Clinical Manifestations

Human NoVs are transmitted primarily by the feco-oral route (or aerosolized vomitus), and cause acute gastroenteritis. The incubation period is usually short, ranging from 10 to 51 h with a mean of 24 h, after which the clinical symptoms start appearing (Glass et al. 2009; Green 2013).

Acute onset of watery, nonbloody diarrhea, and projectile vomiting are the characteristic symptoms of huNoV infection. Other symptoms such as abdominal cramps, nausea, bloating, low-grade fever, chills, headache, malaise, myalgia, and anorexia have also been noted. The illness usually lasts for about 1–3 days, but may become persistent in the elderly and immunocompromised persons (Glass et al. 2009; Green 2013; Ludwig-Begall et al. 2021).

NoV infections are associated with virus shedding in feces (and sometimes vomitus, at lower levels), which may precede the onset of illness, usually peaks 1–3 days after the onset of symptoms and can continue beyond the symptomatic period, for ~4 weeks (Atmar et al. 2008; Glass et al. 2009; Green 2013). Immunocompromised persons may show prolonged fecal shedding (37–>418 days) (Davis et al. 2020). During peak shedding, the viral load has been estimated to be 1–100 billion viral copies per gram of feces (Atmar et al. 2008; Teunis et al. 2015).

NoV disease is usually mild and self-limiting and normally resolves without complications. However, severe disease associated with dehydration and requiring hospitalization may occur, especially in the infants, young children, elderly and immunocompromised population. Fatalities are also rarely reported (Glass et al. 2009; Green 2013; Hall et al. 2013).

Asymptomatic infections and associated virus shedding have been reported for NoVs. As fecal NoV shedding during clinical illness continues long after resolution of symptoms, identification of true asymptomatic infections is challenging.

However, a meta-analysis study has estimated that 7% of the NoV infections are asymptomatic, with the proportion varying between 0 and 30% in individual studies (Qi et al. 2018).

13.5 Epidemiology

13.5.1 Mode of Transmission

NoVs are transmitted by the feco-oral or vomit (aerosolized)-oral route. Direct person-to-person transmission is the primary mode of spread (Glass et al. 2009; Ludwig-Begall et al. 2021). High doses of viral shedding (1–100 billion genomic copies per gram of feces) and low human infectious dose (18–1018 genome equivalents sufficient to cause infection, with 50% human infectious dose estimated at 1320–2800 genome equivalents) account for the high transmissibility of NoVs (Atmar et al. 2008, 2014; Teunis et al. 2008, 2015).

Foodborne and waterborne transmission is also common (Glass et al. 2009; Ludwig-Begall et al. 2021). Contamination of food may occur at the source or during preparation by food handlers. The most common sources of foodborne transmission are undercooked/raw seafood (bivalve molluscs), fresh/frozen soft fruits and vegetables (raspberries, lettuce leaves, etc.), and ready-to-eat foods (salads, sandwiches, etc.). Waterborne transmission may occur through drinking or recreational (e.g., swimming pool) water. Environmental transmission through contact with contaminated fomites is also possible (Glass et al. 2009; Ludwig-Begall et al. 2021). NoVs in contaminated food/water/fomites are found to be highly stable, retaining infectivity for weeks to months. NoVs efficiently survive in the environment resisting freezing temperatures, and heating up to 60 °C (Lucero et al. 2021). Moreover, the viruses are resistant to various virucidal treatments and decontamination methods (e.g., chlorine/alcohol), thus increasing the risk of transmission (Ludwig-Begall et al. 2021; Lucero et al. 2021).

13.5.2 Disease Burden

NoVs are major etiological agents of acute gastroenteritis causing significant morbidity and mortality across all age groups worldwide, with an estimated 685 million cases and 212,000 deaths annually (Pires et al. 2015). Estimation of the global economic burden of NoV disease suggests direct healthcare costs of 4.2 billion USD and societal costs of 60.3 billion USD yearly (Bartsch et al. 2016).

13.5.2.1 Epidemic Gastroenteritis

NoVs have been identified as the leading cause of epidemic gastroenteritis among all age groups, accounting for >90% of the nonbacterial and ~50% of all-cause epidemics worldwide (Widdowson et al. 2005; Lucero et al. 2021). NoV epidemics or outbreaks usually occur in hospitals, long-term care facilities, camps, recreational areas, elementary schools, daycare centers, cruise ships, retirement centers, colleges, restaurants, social events with catered meals, families, the military, prisons, and community settings (Green 2013). Foodborne outbreaks are the most common, accounting for 14% of all NoV outbreaks (Verhoef et al. 2015). A 2015 report by the World Health Organization identified NoVs as the primary cause of foodborne illnesses globally, accounting for ~125 million cases and ~35,000 deaths (Mehlhorn 2015).

13.5.2.2 Sporadic Gastroenteritis

NoVs are considered the most common cause of sporadic gastroenteritis among adults, and the second most common cause (after rotavirus) in children worldwide (Glass et al. 2009; Operario et al. 2017). In several countries which are implementing routine rotavirus vaccination through the national immunization program, NoV has replaced rotavirus as the primary cause of viral gastroenteritis among children (Payne et al. 2013; Hemming et al. 2013; Becker-Dreps et al. 2014; McAtee et al. 2016; Platts-Mills et al. 2015).

Based on a systematic literature review reported in 2008, NoVs have been associated with ~12% of sporadic gastroenteritis cases in children and adults (Patel et al. 2008). These estimates were updated in 2014, suggesting a pooled NoV prevalence of ~18% of all gastroenteritis cases globally, with higher prevalence in low-mortality developing and developed countries (~19–20%) as compared to the high-mortality developing countries (~14%) (Ahmed et al. 2014). Due to the under representation of data from developing countries in the above reports, additional studies focusing on developing countries were conducted, reporting an estimated 17% NoV prevalence among gastroenteritis patients, with prevalence decreasing from upper middle-income (UMIC; 18%) to lower middle-income (LMIC; 15%) and low-income countries (LIC; 6%) (Nguyen et al. 2017). Another independent study documented NoV detection in 15% and 11% cases respectively in LMICs and LICs (Mans 2019).

13.5.3 Age-Dependent NoV Prevalence

Globally, young children aged <5 years have the highest incidence of NoV gastroenteritis (Belliot et al. 2014; Shah and Hall 2018), with an estimated 453 million illnesses and 95,000 deaths occurring globally per year (Bartsch et al. 2016). Several systematic reviews and meta-analysis studies have revealed overall 12–18% prevalence of NoV infection among children (<5 years) with gastroenteritis (Patel et al. 2008; Ahmed et al. 2014; Nguyen et al. 2017; Farahmand et al. 2022). According to a study of the global age distribution of pediatric NoV gastroenteritis cases, majority (69%) of the cases occurred between 6 and 23 months of age, with 13% and 18% of the cases occurring in the age groups of 0–5 and 24–59 months respectively (Shioda et al. 2015). Further, it has been reported that 89% of children experience at least 1 NoV infection before 24 months of age (Patel et al. 2008; Rouhani et al. 2016).

13.5.4 Seasonality

Seasonality of NoV infections is not completely understood. In the temperate countries, winter month seasonality has been reported for NoV with cold, dry weather being associated with increases in NoV incidence (Lopman et al. 2009; Ahmed et al. 2013; Ludwig-Begall et al. 2021). However, in the tropical countries, no clear seasonal pattern is evident, with NoV infections being observed throughout the year (Mans 2019).

13.5.5 NoV Evolution: Origin of Infection and Diversity

NoV evolution, leading to the emergence of diverse NoV strains, is driven primarily by two processes: (1) point mutation (genetic drift) and (2) recombination (genetic shift) (Parra 2019). Accumulation of point mutations is a common feature of RNA viruses with low-fidelity RdRp; NoVs undergo such changes mainly in ORF1 (non-structural proteins) and/or ORF2 (VP1) genes. Point mutations in VP1 region, in response to host immune selection pressures, is suggested as an important evolutionary mechanism of GII.4 NoVs, leading to the periodic emergence of novel variants. Recombination at ORF1–ORF2 junction, leading to the generation of chimeric viruses with novel combination of polymerase/VP1 genotypes, further contributes to the genetic diversity of NoVs.

The reservoir for emergence of novel NoV strains remains undefined (Parra 2019). Immunocompromised individuals with chronic NoV infection, present an extreme diversity of NoV strains, as evidenced by their virus shedding profiles; however, their contribution to NoV diversity at population level is unclear. Norovirus diversity may also originate at inter- and intra-host levels in healthy populations, with mutations arising during transmission events in outbreak settings and/or during shedding phase in healthy individuals.

13.5.6 Global Circulation of NoV Genogroups, Genotypes, and Variants

Of the five NoV genogroups capable of infecting humans, GII NoVs are predominant worldwide, accounting for ~89–96% of the infections, GI NoVs are responsible for ~4–11% of the infections (Green 2013; Hoa Tran et al. 2013; Vega et al. 2014; Lucero et al. 2021), while GIV, GVIII, and GIX NoVs are rarely detected.

Among the 48 NoV genotypes, GII.4 has been predominant globally for over two decades, accounting for majority of the NoV outbreaks (~62%) and of sporadic infections (~70%) (Siebenga et al. 2009; Hoa Tran et al. 2013; Ludwig-Begall et al. 2021). This genotype reportedly causes a more severe gastroenteritis as compared to other NoV genotypes, leading to higher hospitalization and mortality rates (Huhti et al. 2011; Desai et al. 2012).

The circulating pattern of the different NoV genotypes is known to undergo temporal variations, possibly in response to the host population immunity, with reports suggesting that genotype-specific herd immunity influences the endemic genotype in the forthcoming NoV season (Sakon et al. 2015). A recent review of global circulation of NoV (capsid) genotypes during 1995–2019, has estimated GII.4 prevalence at 52.6% based on sequences available in public databases (Kendra et al. 2022). GII.3 was identified as the second most prevalent genotype, identified consistently over the study period with an overall prevalence of 11.5%, ranging between 3.9 and 25.0% at different time points. GII.3 was followed by GII.2 (8.6%) and GII.17 (7.3%), both of which circulated at low levels initially, but, showed a marked increase during 2013–2019. The other commonly circulating genotypes were GII.6 (3.9%), GI.3 (2.0%), GII.13 (1.8%), GI.4 (1.5%), GII.12 (1.3%), and GII.14 (1.1%). A similar analysis of the circulating polymerase types identified GII.P4 as the most prevalent type (28.5%), followed by GII.P16 (16.3%), GII.P31 (formerly GII.Pe, 12.1%), and GII.P17 (7.3%). Temporal variations have been noted in the circulation of these four P-types with GII.P4 circulating at high levels (~50%) during 1995–2007, followed by a steady decrease till 2013, giving way to GII.P16, GII.P31, and GII.P17 as alternating predominant P-types during 2013–2019. The other commonly circulating P-types were GII.P12 (5.8%), GII.P21 (formerly GII.Pb, 5.5%), GII.P7 (5.3%), GII.P2 (2.0%), GII.P33 (formerly GII.Pg, 1.8%), and GI.P3 (1.8%). Among recombinant strains, GII.2 [P16] (22.0%), GII.4 [P31] (19.0%), GII.4 [P16] (9.2%), GII.3 [P12] (7.8%), GII.6 [P7] (7.4%), GII.3 [P21] (5.7%), GII.13 [P16] (2.9%), GI.3 [P13] (1.9%), GII.3 [P16] (1.9%), and GII.5 [P40] (1.5%) were the most common types.

The GII.4 NoVs have been reported to undergo rapid evolution in VP1 gene, with an estimated evolutionary rate of $3.9–7.2 \times 10^{-3}$ nucleotide substitutions per site per year, which is higher than that for other genotypes ($1.9–2.4 \times 10^{-3}$) (Bok et al. 2009b; Bull et al. 2010; Siebenga et al. 2010; Cotten et al. 2014; Eden et al. 2014). The persistence of the GII.4 NoVs in the human population has been attributed to their continuous evolution in VP1 (particularly the surface-exposed P2 domain) in response to the host immunity, resulting in the periodic emergence of novel variants, with 15 variants reported to date (Table 13.2) (Siebenga et al. 2007; Lindesmith et al. 2008; Kroneman et al. 2011; Kroneman et al. 2013; Chan et al. 2021; Kendra et al. 2022). Of these, six variants, namely, US95_96 (or Grimsby_1995; 1995–2002), Farmington Hills_2002 (2002–2004), Hunter_2004 (2004–2006), DenHaag_2006b (2006–2010), New Orleans_2009 (2009–2012), and Sydney_2012 (2012 onwards), have been described as pandemic variants, with each variant predominating for 2–3 years before being replaced by the next emerging variant (Noel et al. 1999; Hoa Tran et al. 2013; van Beek et al. 2013; Kendra et al. 2022). The vast genetic and antigenic diversity of NoVs, geographical and temporal variations in circulation of different genotypes, and viral evolution in response to host immunity, warrant continuous global surveillance and data sharing for characterization of global strain distribution and identification of emerging NoV genotypes and GII.4 variants.

Table 13.2 Summary of GII.4 NoV variants identified till date

Sr. no.	GII.4 variant	Predominant period of circulation	Description
1.	Bristol_1993	1980s–1990s	Earliest known GII.4 variants
2.	Camberwell_1994		
3.	US95_96 (Grimsby_1995)	1995–2002	First reported pandemic variant with a global prevalence
4.	Farmington Hills_2002	2002–2004	Pandemic variant
5.	Lanzou_2002	2002	Limited circulation
6.	Kaiso_2003	2003	Limited circulation
7.	Asia_2003	2003–2005	Important cause of outbreaks in Asia during 2004–2005
8.	Hunter_2004	2004–2006	Pandemic variant
9.	Yerseke_2006a	2006–2007	Global circulation; did not gain predominance over contemporary pandemic variant(s)
10.	DenHaag_2006b	2006–2010	Pandemic variant
11.	Osaka_2007	2007–2009	Global circulation; did not gain predominance over contemporary pandemic variant(s)
12.	Apeldoorn_2007	2007–2009	
13.	New Orleans_2009	2009–2012	Pandemic variant
14.	Sydney_2012	2012 onwards	Currently predominant pandemic variant
15.	Hong Kong	2017 onwards	Low-level circulation in Eurasia since 2017. Pandemic potential; however, yet to cause outbreaks

13.6 Diagnostics and Therapeutic Approaches

13.6.1 Kaplan Criteria for NoV Outbreak Identification

Based on the typical clinical presentation of NoV infections, the Kaplan criteria can be used for identification of NoV outbreaks. According to these criteria, any outbreak which satisfies following four parameters, is identified as a NoV outbreak (Norwalk-like infection): (1) stool cultures negative for bacterial pathogens, (2) mean (or median) duration of illness of 12–60 h, (3) vomiting in ≥50% of cases, (4) mean (or median) incubation period of 24–48 h (Kaplan et al. 1982). However, the limited sensitivity of these criteria necessitates further laboratory diagnosis for confirmation.

13.6.2 Laboratory Diagnosis of NoV Infections

Laboratory diagnosis of NoV infections is carried out by detection of the virus in fecal specimens of patients. For optimum results, it is recommended that specimens should be collected during the acute phase of illness, ideally within 48–72 h after onset (Division of Viral Diseases, National Center for Immunization and Respiratory Diseases, Centers for Disease Control and Prevention 2011).

13.6.2.1 Electron Microscopy

Electron microscopy, used for the discovery of NoVs, was the earliest available method of NoV detection; however, due to its high cost and low sensitivity, it is not currently widely used in diagnostic laboratories (Vinjé 2015).

13.6.2.2 Immunoassays

Efforts for the development of sensitive and specific ELISAs for NoV antigen detection, using anti-VLP antibodies, have been ongoing since the first production of VLPs in the 1990s. However, the development of broadly reactive ELISAs suitable for routine diagnostic use has been challenging due to the genetic and antigenic diversity of NoVs. Several commercial ELISA kits which employ combinations of multiple cross-reactive polyclonal and monoclonal antibodies in a sandwich ELISA format, are currently available, with RIDASCREEN Norovirus (r-Biopharm AG, Darmstadt, Germany) and IDEIA Norovirus EIA (Oxoid, Hampshire, United Kingdom) being commonly used. In addition, several rapid immunochromatographic (ICG) lateral flow immunoassays, which are simpler and more convenient for field use, are also commercially available—Ridaquick Norovirus (r-Biopharm AG, Darmstadt, Germany), SD BIOLINE Norovirus (Standard Diagnostics, Inc., Kyonggi-do, South Korea) are commonly used (Liu and Moore 2020). Although the ELISA and ICG tests have high specificity (>90%), they exhibit limited (<70%) sensitivity, highly variable performance, and may not capture the full range of circulating NoVs (Gray et al. 2007; Costantini et al. 2010; Morillo et al. 2011; Ambert-Balay and Pothier 2013). These kits are thus recommended only for preliminary screening of fecal specimens during outbreaks, and not for sporadic cases. Further, it is recommended that negative ELISA/ICG results should be further verified by more sensitive molecular methods (Division of Viral Diseases, National Center for Immunization and Respiratory Diseases, Centers for Disease Control and Prevention 2011; Vinjé 2015; Liu and Moore 2020).

13.6.2.3 Reverse Transcription-Polymerase Chain Reaction (RT-PCR)

RT-PCR assays are the current reference method for NoV detection and genotyping worldwide (Vinjé 2015; Liu and Moore 2020).

13.6.2.3.1 NoV Detection

Conventional RT-PCR assays (coupled with nucleotide sequencing) using genogroup-specific primers targeting conserved regions A (near the 5′end) or B (at the 3′end) of the RdRp gene in ORF1, were developed and used in the mid-1990s for NoV detection (Ando et al. 1995; Green et al. 1995; Vinjé and Koopmans 1996; Fankhauser et al. 2002). In the early 2000s, TaqMan-based real-time RT-PCR assays were developed for NoV detection, using genogroup-specific primers and fluorescent probes targeting the ORF1–2 junction, identified as the most conserved region of the NoV genome (Katayama et al. 2002; Kageyama et al. 2003). Different modifications of this real-time RT-PCR assay are being used worldwide due to their high sensitivity, specificity, broad reactivity, and quantitative ability. These assays are increasingly being multiplexed for simultaneous detection of different GI, GII, and GIV NoV strains. Several commercial kits such as AgPath-ID (ThermoFisher, MA, USA), Xpert (Cepheid, CA, USA) are available. Further, multiplex molecular gastrointestinal pathogen diagnostic kits such as xTAG GPP (Luminex Corporation, Toronto, Canada), FilmArray GI Panel (BioFire Diagnostics Inc., Salt Lake City, UT, USA), Verigene Enteric Pathogens Test (EP) (Nanosphere, Northbrook, IL, USA), offer simultaneous screening for multiple pathogenic enteric bacteria, parasites, and viruses, including NoVs (Vinjé 2015; Liu and Moore 2020).

Although highly sensitive, RT-PCR is unable to differentiate between infectious and noninfectious virus particles. In order to assess NoV infectivity, RT-PCR with intercalator, enzymatic or ligand (HBGA/antibody) pretreatment, which helps in the removal of noninfectious particles, or long template RT-PCR involving near full-length genome amplification, is being investigated, and have shown success to some extent (Liu and Moore 2020).

13.6.2.3.2 NoV Genotyping

Conventional RT-PCR assays followed by nucleotide sequencing and phylogenetic analysis of the amplicons are carried out for genotyping of NoV strains. Dual typing involving determination of both ORF1/RdRp and ORF2/VP1 genotypes is currently recommended (Kroneman et al. 2013; Chhabra et al. 2019). The ORF1 regions A and B (used for detection), and ORF2 regions C (at the 5′ end) and D (at the 3′ end) are commonly used for genotyping (Kojima et al. 2002; Vinjé et al. 2004). A single B–C region amplification, covering ORF1-region B to ORF2-region C, and including the ORF1–2 junction, is increasingly being used for simultaneous determination of ORF1 (P-type) and ORF2 (VP1 Type) genotypes, and also allows identification of ORF1–2 recombination events. An online automated norovirus genotyping tool (NoroNet, Bilthoven, The Netherlands; available at http://www.rivm.nl/mpf/norovirus/typingtool), which conducts phylogenetic analysis of the user-submitted query sequence(s) against a frequently updated database of reference sequences, is available for genotyping, thus simplifying the genotype assignment procedure (Kroneman et al. 2011).

13.6.2.4 Other Diagnostic Methods

Isothermal amplification techniques such as nucleic-acid sequence-based amplification (NASBA), loop-mediated isothermal amplification (LAMP), and recombinase polymerase amplification (RPA), offer a more convenient alternative to RT-PCR for field use and are being investigated for their diagnostic performance in comparison to RT-PCR, with promising results (Liu and Moore 2020). Advanced techniques such as biosensors, microarrays, omics-based approaches are also under development (Liu and Moore 2020).

13.6.3 Treatment of NoV Gastroenteritis

NoV gastroenteritis is usually self-limiting. Treatment is mainly supportive and involves oral rehydration with fluids and electrolytes or parenteral fluid administration in case of severe vomiting and diarrhea, for management of dehydration and to prevent further complications. Oral administration of bismuth subsalicylate has been found useful in reducing the severity and duration of symptoms in adults, but is not recommended for infants and young children (Glass et al. 2009; Green 2013).

13.6.4 Antivirals

No licensed antivirals are currently available for treatment of NoV disease, although several are under development, with majority in early preclinical stages (Table 13.3). Nitazoxanide (NTZ) is the only drug currently in clinical development. It has shown promising results in a phase II clinical trial, with treatment associated with significant reduction in duration of gastroenteritis symptoms in NoV-positive patients. However, evidence contradicting NTZ efficacy against NoV has also been reported (Netzler et al. 2019). The mechanism of action of NTZ is not clearly understood; the drug may inhibit virus protein production, maturation, and/or assembly, or induce host antiviral responses (Ludwig-Begall et al. 2021).

Drugs targeting different stages of the viral replication cycle have demonstrated effectiveness in NoV infection models and/or experimental systems in vitro and in vivo (Table 13.3). These include attachment/entry inhibitors (carbohydrate analogs including human milk oligosaccharides, citric acid; heparan sulfate analogs such as heparin, suramin; compounds such as ZINC04041115; soluble histones; monoclonal antibodies or nanobodies), protease inhibitors (Rupintrivir; transition state inhibitors including peptidyl aldehydes and macrocyclic compounds; transition state mimics such as α-hydroxyphosphonates), RdRp inhibitors (nucleoside analogs CMX521, 2′-C-methylcytidine, Favipiravir, Ribavirin; nonnucleoside inhibitors such as Suramin and NF203), and other inhibitors of viral replication (heterocyclic carboxamide derivatives, hippuristanol) (Netzler et al. 2019; Ghosh et al. 2018).

Targeting host cell factors critical for viral replication is an alternate strategy for antiviral development, with deubiquitinase inhibitors (WP1130) and heat-shock protein 90 (Hsp90) inhibitors reported to inhibit viral replication. Immunomodulators such as IFN-γ and Toll-like receptor (TLR) agonists have also been investigated as antivirals (Netzler et al. 2019). Passive immunotherapy using oral human immunoglobulin has been found to be clinically useful in immunocompromised patients. Probiotic treatment involving use of lactic acid bacteria such as *Bifidobacterium adolescentis, Lactobacillus paracasei* is under investigation as antiviral therapy (Ghosh et al. 2018).

Table 13.3 Summary of norovirus antivirals under development

Sr. no.	Candidate antiviral	Development status
1.	Nitazoxanide (NTZ)	Phase II clinical trial demonstrated reduction in duration of gastroenteritis symptoms on NTZ treatment in NoV-positive patients
2.	**Viral attachment/entry inhibitors:** Mechanism of action: blocking of NoV-HBGA binding	
2.1 2.2 2.3 2.4 2.5	Carbohydrate analogs—citric acid, human milk oligosaccharides 2′-fucosyllactose, 3′-fucosyllactose Heparan sulfate analogs—heparin, suramin ZINC04041115 Soluble histones Monoclonal antibodies, nanobodies (passive immunotherapy)	Blocking of VLP-HBGA binding demonstrated by ELISA-based assays
3.	**Viral protease inhibitors:**	
3.1 3.2 3.3	Rupintrivir *Transition state (TS) inhibitors*—peptidyl aldehydes (PA) and their variants, macrocyclic compounds (lower potency than PA) *Transition state (TS) mimics*—α-hydroxyphosphonates	Antiviral activity reported in human NoV replicon system and/or against MNV in cell culture. Peptidyl aldehydes also demonstrated potency against MNV in vivo
4.	**Viral RdRp inhibitors—nucleoside analogs:** Mechanism of action: mimicry of nucleoside triphosphates (NTPs) resulting in incorporation into growing nucleotide chain and subsequent chain termination, or mutagenesis during genome transcription	
4.1	CMX521	Phase 1 (safety) clinical trial completed; phase 2 efficacy study ongoing
4.2	2′-C-methylcytidine and its derivatives (cytidine analogs)	Inhibition of viral replication demonstrated in human BJAB cell culture system and mouse model
4.3	Favipiravir (T705)—purine analog Ribavirin (RBV)—guanoside analog	Modest antiviral activity demonstrated in human NoV replicon system and against MNV in cell culture
5.	**Viral RdRp inhibitors—nonnucleoside inhibitors** Mechanism of action: allosteric inhibition—binding to RdRp and blocking of conformational rearrangements required for formation of active replication complex	
5.1	Suramin and its derivatives like NF203	Inhibition of huNoV and MNV RdRp activity demonstrated in vitro; inhibition of MNV replication in cell culture reported (suramin)
6.	**Other inhibitors of viral replication:** (precise mechanism of action not clearly understood)	
6.1	Heterocyclic carboxamide derivatives	Antiviral effect demonstrated against MNV in cell culture possibly by acting on intracellular viral replication, assembly or release

(continued)

Table 13.3 (continued)

Sr. no.	Candidate antiviral	Development status
6.2	Hippuristanol	In vitro studies on MNV have shown translation inhibition possibly by disruption of host factor recruitment by VPg
7.	**Host factor drugs:** Mechanism of action: targeting host cell factors critical for viral replication	
7.1	Deubiquitinase inhibitors—WP1130 and its variants	Antiviral activity reported in human NoV replicon and against MNV in cell culture and mouse model
7.2	Heat-shock protein 90 (Hsp90) inhibitors—17-dimethylamino ethylamino-17-demethoxy geldanamycin (17-DMAG)	Inhibition of MNV replication demonstrated in vitro and in mouse model
8.	**Immunomodulators:** Mechanism of action: enhancement of antiviral host immune response	
8.1	IFN-γ (type III interferon)	Antiviral activity demonstrated against MNV in mouse model
8.2 8.2.1 8.2.2	Toll-like receptor (TLR) agonists: TLR7 agonists—R-848 TLR4 agonists—poly-γ-glutamic acid (γ-PGA)	Antiviral activity reported in human NoV replicon system (R-848) and against MNV in cell culture (R-848, γ-PGA) and mouse model (γ-PGA)
9.	**Probiotics:**	
9.1	Lactic acid bacteria—*Bifidobacterium adolescentis, Lactobacillus paracasei*	Antiviral activity demonstrated against MNV in cell culture

13.7 Prevention and Control

13.7.1 NoV Vaccines

Since 2016, the World Health Organization (WHO) has assigned high priority status to NoV vaccine development (Giersing et al. 2019). At least four different vaccine candidates are currently undergoing clinical trials, while several are in preclinical stages of development (Mattison et al. 2018; Cates et al. 2020; Esposito and Principi 2020; Tan 2021) (Table 13.4).

13.7.1.1 Vaccines in Clinical Trials

13.7.1.1.1 TAK-214 (Takeda Pharmaceuticals): VLP-Based Bivalent GI.1/GII.4 Vaccine

The vaccine candidate furthest along the development pipeline is a bivalent GI.1/GII.4C VLP formulation, being developed by Takeda Pharmaceuticals. The vaccine consists of a GI.1 VLP derived from the prototype Norwalk virus and a GII.4C VLP derived from a consensus sequence containing epitopes from three different GII.4 strains, namely, Houston/TCH186/2002/US (Farmington Hills_2002 variant), Yerseke38/2006/NL (Yerseke_2006a variant), and DenHaag89/2006/NL (DenHaag_2006b variant) (Parra et al. 2012).

Table 13.4 Summary of NoV vaccines under clinical development

Sr. no.	Vaccine candidate (company)	Vaccine formulation	Route of administration	Development status	Results
1.	TAK-214 (Takeda Pharmaceuticals International AG)	Adjuvanted VLP-based bivalent GI.1/GII.4 vaccine—VLPs derived from GI.1 Norwalk virus and consensus of three different GII.4 variants (Farmington Hills_2002, Yerseke_2006a, DenHaag_2006b) Adjuvants: MPLA + chitosan/ aluminum salt	Intranasal/ intramuscular	Two phase 2a and one phase 2b clinical trials completed	*Phase 2a*: (1) Vaccination associated with significant decrease in incidences of gastroenteritis and viral infection upon homologous NoV GI.1 challenge. (2) Vaccination associated with significant reduction of symptoms upon homologous NoV GII.4 Farmington Hills/2002 challenge *Phase 2b field efficacy study*: 61.8% vaccine effectiveness against moderate-to-severe gastroenteritis by different NoV genotypes (GI.1, GI.7, GII.2, GII.4) observed

(continued)

Following successful phase I clinical trials demonstrating good safety profiles and a robust immune response, three phase 2 efficacy trials have been conducted. In two separate phase 2a trials involving intranasal/intramuscular administration of two doses of chitosan-monophosphoryl lipid A (MPLA) or MPLA-aluminum salt-adjuvanted vaccine, protection against subsequent challenge with homologous GI.1 Norwalk virus or GII.4 Farmington Hills_2002 strains was observed (Atmar et al. 2011; Bernstein et al. 2015). Further, in a first-in-human phase 2b field study conducted over two winter seasons in a U.S. Navy training facility, vaccine effectiveness of 61.8% was observed against moderate-to-severe gastroenteritis caused by

Table 13.4 (continued)

Sr. no.	Vaccine candidate (company)	Vaccine formulation	Route of administration	Development status	Results
2.	VXA-NVV-104 (Vaxart Pharmaceutical Inc.)	Recombinant adenovirus-based vector carrying VP1 genes of GI.1 Norwalk virus and GII.4 Sydney_2012, and dsRNA as adjuvant	Oral	Phase 1 (monovalent GI.1) trial completed, phase 1b (bivalent GI.1/GII.4) in progress	Phase 1: Vaccination-induced robust systemic and mucosal immune response consisting of HBGA-blocking antibodies, mucosally primed circulating antibody-secreting cells (ASCs), IgA + memory B cells with gut-homing receptors ($\alpha4\beta7$), and fecal IgA was observed
3.	Hansenula polymorpha (NVSI, China)	Aluminum salt-adjuvanted bivalent GI.1/GII.4 VLP-based vaccine	Intramuscular	Phase 1 trial completed	Results awaited
4.	Longkoma (IPS/Zhifei, China)	Aluminum salt-adjuvanted quadrivalent GI.1/GII.3/GII.4/GII.17 VP1 proteins-based vaccine	Intramuscular	Phase 1 and 2a trials in progress	Results awaited

different NoV genotypes (GI.1, GI.7, GII.2, GII.4) (Sherwood et al. 2020). Vaccine-induced antibodies showed HBGA-binding blockade, associated with protection against NoV gastroenteritis. This study indicates the induction of a broad immune response with cross-genotype protection, although the vaccine-induced response seems to be dominated by boosting of preexisting immunity due to previous exposure of the subjects to different NoV strains.

13.7.1.1.2 VXA-NVV-104 (Vaxart Pharmaceutical): Recombinant VP1-Based Bivalent GI.1/GII.4 Oral Vaccine

This vaccine consists of a recombinant adenovirus-based vector carrying VP1-encoding genes of GI.1 Norwalk virus and GII.4 Sydney_2012 strains and dsRNA as adjuvant (Scallan et al. 2013). The vaccine is formulated as an oral tablet to induce mucosal immunity by VP1 expression locally within intestinal epithelial cells. In a phase 1 trial involving the earlier monovalent (GI.1) version, a single dose of the vaccine was found to be well-tolerated and induced a robust systemic and mucosal immune response (Kim et al. 2018). A phase 1b dose-ranging trial for evaluation of safety, immunogenicity, and efficacy of the bivalent formulation is in progress, with preliminary results indicating induction of a robust immune response comparable to natural infection (Cates et al. 2020; Tan 2021).

13.7.1.1.3 Hansenula polymorpha (NVSI, China): VLP-Based Bivalent GI.1/GII.4 Vaccine

This vaccine developed by National Vaccine and Serum Institute (NVSI), China, consisting of NoV GI.1 and GII.4 VLPs with aluminum salt as adjuvant, has recently completed phase 1 trial, and results are awaited (Tan 2021).

13.7.1.1.4 Longkoma (IPS/Zhifei, China): VP1-Based Quadrivalent GI.1/GII.3/GII.4/GII.17 Vaccine

This is a quadrivalent vaccine developed jointly by Institut Pasteur of Shanghai (IPS) and Anhui Zhifei Longcom Biopharmaceutical Co Ltd., China. It consists of four aluminum salt-adjuvanted VP1 proteins, representing GI.1, GII.3, GII.4, GII.17 NoVs, produced using a yeast expression system. Phase 1 and phase 2a trials are currently in progress (Tan 2021).

13.7.1.2 Vaccines in Preclinical Studies

Several NoV vaccines based on different platforms are in preclinical development, with select candidates included in this review (Table 13.5). Multiple VLP-based vaccines, involving different expression systems, with some candidates including additional viral antigens (rotavirus/enterovirus) for a combination vaccine, are being investigated (Mattison et al. 2018; Esposito and Principi 2020). Further, sub-viral P particles, generated by expression of NoV VP1-P domains in *E. coli* system, have emerged as a promising platform for NoV vaccine development, with many combination vaccine candidates such as NoV-Hepatitis E virus-Astrovirus, NoV-rotavirus, NoV-Influenza virus reported (Mattison et al. 2018; Esposito and Principi 2020). Other technologies being explored for NoV vaccines include NoV VP1 expression in recombinant adenovirus or lactic acid bacteria-based systems (Mattison et al. 2018; Cates et al. 2020).

13.7.1.3 Challenges to Vaccine Development

The development of a broadly effective NoV vaccine capable of inducing long-lasting immunity is challenging due to two major factors (Cortes-Penfield et al. 2017; Esposito and Principi 2020; Tan 2021). First, human NoVs exhibit vast genetic and antigenic diversity (comprising of five genogroups and 35 genotypes) (Chhabra et al. 2019), and also undergo continuous evolution leading to the periodic emergence of novel strains replacing the previously circulating strains (e.g., GII.4 variants) (Siebenga et al. 2007; Lindesmith et al. 2008; Parra 2019). Second,

Table 13.5 Summary of select NoV vaccine candidates in preclinical studies

Sr. no.	Vaccine candidate and formulation	Vaccine characteristics and observations in animal studies
1.	Trivalent NoV GI.3/GII.4 VLP—Rotavirus VP6 vaccine (University of Tampere, Daiichi-Sankyo Company, Limited and UMN Pharma Inc., Japan)	• Induction of high level of serum, intestinal, mucosal antibodies, and cell-mediated immunity to both viruses in mice • Adjuvant effect of rotavirus VP6 noted
2.	Bivalent NoV GII.4 VLP—Enterovirus 71 vaccine (Chinese Academy of Sciences, Shanghai)	• Induction of NoV GII.4-specific antibodies with HBGA-binding blockage potential • Induction of EV71-specific antibodies and protection against EV71 infection in mice
3.	Plant-based NoV GII.4 VLP vaccine (Arizona State University)	• *Nicotiana benthamiana* (tobacco) system • Induction of systemic and mucosal immune response in mice
4.	Trivalent NoV GII.4-hepatitis E virus-astrovirus P-particle vaccine (The Cincinnati Children's Hospital Medical Center, University of Cincinnati, and Virginia Polytechnic Institute and State University)	• Better immunogenicity of monovalent NoV P-particle vaccine as compared to NoV-VLP vaccine in gnotobiotic pigs • Higher induction of NoV-specific antibodies by trivalent (NoV-HEV-astrovirus) as compared to monovalent NoV P-particle vaccine in mice • P-particle platform also being used for NoV-rotavirus (VP8) and NoV-influenza (M2e) combined vaccine development
5.	Recombinant adenovirus-based NoV GII.4 VP1 vaccine (The Chinese Center for Disease Control and Prevention)	• Induction of NoV-specific IgG, IgA responses in mice • Priming with adenovirus-based vaccine before VLP vaccination showed higher antibody levels as compared to VLP vaccination alone
6.	Lactic acid bacteria-based NoV GII.4 VP1 vaccine (The Ohio State University)	• Induction of NoV-specific immune responses and protection from NoV infection on subsequent challenge in gnotobiotic piglets

immunity to NoV infections is not completely understood, with many studies suggesting the development of a short-term (6 months–2 years) homologous immune response with limited cross-reactivity (Wyatt et al. 1974; Parrino et al. 1977; Baron et al. 1984; Johnson et al. 1990; Ayukekbong et al. 2014).

Other factors limiting vaccine development include lack of an efficient cell culture system, and consequently a true virus neutralization assay, unavailability of an efficient animal model, and absence of a primary confirmed correlate of protection (Cortes-Penfield et al. 2017; Esposito and Principi 2020; Tan 2021). However, recent developments such as the HIE cell culture system and the zebrafish larvae model promise to overcome such barriers and advance vaccine development.

13.7.2 Other Control Measures

In the absence of a NoV vaccine, several nonspecific control measures such as quarantine of infected individuals, adherence to hand hygiene, environmental disinfection, hygienic processing of food, measures to increase the purity of drinking and recreational water, have been suggested to be important for controlling the spread of NoV infections (Arias et al. 2013; Green 2013; Barclay et al. 2014). Hand-washing with soap and running water is generally considered to be an effective control method, while the efficacy of alcohol-based hand sanitizers remains debatable. Effective environmental decontamination with disinfectants such as sodium hypochlorite, hydrogen peroxide-based or phenolic-based cleaners is also important, as NoVs can remain viable on contaminated surfaces for extended periods of time. In addition, different methods for detecting viral contamination of foodstuffs and inactivation (high-pressure processing, γ-irradiation, thermal inactivation, etc.) of NoVs contaminating food, are being developed and would be useful for controlling foodborne outbreaks, such as those caused by contaminated shell-fish and oysters (Arias et al. 2013).

13.8 Conclusion and Future Prospects

Since its discovery in 1972, NoV has emerged as a leading cause of epidemic and sporadic gastroenteritis worldwide. This success of NoV as a human pathogen may be attributed to several factors. First, the highly contagious nature of the virus due to the copious and prolonged fecal shedding pattern coupled with low infectious dose with its transmissibility further enhanced due to its environmental stability, persistence, and resistance to common disinfectants. Second, the viruses are modestly virulent, causing mild, self-limiting infections in majority of the population, thus maintaining a large pool of susceptible hosts. Third, the viruses evoke only limited immunity, offering short term, mainly homologous protection. Finally, the viruses undergo rapid and constant evolution, resulting in vast genetic and antigenic diversity, and periodic emergence of novel strains capable of escaping host population immunity. Because of such characteristics, NoVs have been described as "ideal" infectious agents or "perfect" pathogens (Glass et al. 2009; Hall 2012).

The fastidious nature of the NoVs demonstrated by the challenges involved in development of a cell culture and animal model has been a major hurdle in NoV research, development of vaccines and therapeutics. However, the 50-year history of these viruses is marked with several breakthrough discoveries, such as characterization of the viral genome, elucidation of NoV capsid structure, development of VLPs, and huNoV replicon as surrogate experimental systems, establishment of the first cell culture system for MNV, development of animal models (e.g., gnotobiotic pig model), which have boosted NoV research and contributed significantly to understanding of NoV biology, and development of vaccines. Recent developments such as BJAB and HIE huNoV cell culture systems and the zebrafish larva small animal model promise further exciting advancements in the NoV field. Further, with the encouraging results reported in recent vaccine clinical trials, the target of NoV vaccine development seems more achievable than ever.

References

Ahmed SM, Lopman BA, Levy K (2013) A systematic review and meta-analysis of the global seasonality of norovirus. PLoS One 8(10):e75922

Ahmed SM, Hall AJ, Robinson AE, Verhoef L, Premkumar P, Parashar UD, Koopmans M, Lopman BA (2014) Global prevalence of norovirus in cases of gastroenteritis: a systematic review and meta-analysis. Lancet Infect Dis 14(8):725–730

Alonso C, Oviedo JM, Martín-Alonso JM, Díaz E, Boga JA, Parra F (1998) Programmed cell death in the pathogenesis of rabbit hemorrhagic disease. Arch Virol 143(2):321–332

Ambert-Balay K, Pothier P (2013) Evaluation of 4 immunochromatographic tests for rapid detection of norovirus in faecal samples. J Clin Virol 56(3):194–198

Ando T, Monroe SS, Gentsch JR, Jin Q, Lewis DC, Glass RI (1995) Detection and differentiation of antigenically distinct small round-structured viruses (Norwalk-like viruses) by reverse transcription-PCR and southern hybridization. J Clin Microbiol 33(1):64–71

Arias A, Emmott E, Vashist S, Goodfellow I (2013) Progress towards the prevention and treatment of norovirus infections. Future Microbiol 8(11):1475–1487

Asanaka M, Atmar RL, Ruvolo V, Crawford SE, Neill FH, Estes MK (2005) Replication and packaging of Norwalk virus RNA in cultured mammalian cells. Proc Natl Acad Sci USA 102(29):10327–10332

Atmar RL, Opekun AR, Gilger MA, Estes MK, Crawford SE, Neill FH, Graham DY (2008) Norwalk virus shedding after experimental human infection. Emerg Infect Dis 14(10):1553–1557

Atmar RL, Bernstein DI, Harro CD, Al-Ibrahim MS, Chen WH, Ferreira J, Estes MK, Graham DY, Opekun AR, Richardson C, Mendelman PM (2011) Norovirus vaccine against experimental human Norwalk virus illness. N Engl J Med 365(23):2178–2187

Atmar RL, Opekun AR, Gilger MA, Estes MK, Crawford SE, Neill FH, Ramani S, Hill H, Ferreira J, Graham DY (2014) Determination of the 50% human infectious dose for Norwalk virus. J Infect Dis 209(7):1016–1022

Atmar RL, Bernstein DI, Lyon GM, Treanor JJ, Al-Ibrahim MS, Graham DY, Vinjé J, Jiang X, Gregoricus N, Frenck RW, Moe CL, Chen WH, Ferreira J, Barrett J, Opekun AR, Estes MK, Borkowski A, Baehner F, Goodwin R, Edmonds A, Mendelman PM (2015) Serological correlates of protection against a GII.4 norovirus. Clin Vaccine Immunol 22(8):923–929

Ayukekbong JA, Fobisong C, Tah F, Lindh M, Nkuo-Akenji T, Bergström T (2014) Pattern of circulation of norovirus GII strains during natural infection. J Clin Microbiol 52(12):4253–4259

Barclay L, Park GW, Vega E, Hall A, Parashar U, Vinjé J, Lopman B (2014) Infection control for norovirus. Clin Microbiol Infect 20(8):731–740

Baron RC, Greenberg HB, Cukor G, Blacklow NR (1984) Serological responses among teenagers after natural exposure to Norwalk virus. J Infect Dis 150(4):531–534

Bartsch SM, Lopman BA, Ozawa S, Hall AJ, Lee BY (2016) Global economic burden of norovirus gastroenteritis. PLoS One 11(4):e0151219

Becker-Dreps S, Bucardo F, Vilchez S, Zambrana LE, Liu L, Weber DJ, Peña R, Barclay L, Vinjé J, Hudgens MG, Nordgren J, Svensson L, Morgan DR, Espinoza F, Paniagua M (2014) Etiology of childhood diarrhea after rotavirus vaccine introduction: a prospective, population-based study in Nicaragua. Pediatr Infect Dis J 33(11):1156–1163

Belliot G, Lopman BA, Ambert-Balay K, Pothier P (2014) The burden of norovirus gastroenteritis: an important foodborne and healthcare-related infection. Clin Microbiol Infect 20(8):724–730

Bernstein DI, Atmar RL, Lyon GM, Treanor JJ, Chen WH, Jiang X, Vinjé J, Gregoricus N, Frenck RW Jr, Moe CL, Al-Ibrahim MS, Barrett J, Ferreira J, Estes MK, Graham DY, Goodwin R, Borkowski A, Clemens R, Mendelman PM (2015) Norovirus vaccine against experimental human GII.4 virus illness: a challenge study in healthy adults. J Infect Dis 211(6):870–878
Bertolotti-Ciarlet A, White LJ, Chen R, Prasad BV, Estes MK (2002) Structural requirements for the assembly of Norwalk virus-like particles. J Virol 76(8):4044–4055
Bertolotti-Ciarlet A, Crawford SE, Hutson AM, Estes MK (2003) The 3′ end of Norwalk virus mRNA contains determinants that regulate the expression and stability of the viral capsid protein VP1: a novel function for the VP2 protein. J Virol 77(21):11603–11615
Bhar S, Jones MK (2019) In vitro replication of human norovirus. Viruses 11(6):547
Bhella D, Goodfellow IG (2011) The cryo-electron microscopy structure of feline calicivirus bound to junctional adhesion molecule A at 9-angstrom resolution reveals receptor-induced flexibility and two distinct conformational changes in the capsid protein VP1. J Virol 85(21):11381–11390
Blacklow NR, Dolin R, Fedson DS, Dupont H, Northrup RS, Hornick RB, Chanock RM (1972) Acute infectious nonbacterial gastroenteritis: etiology and pathogenesis. Ann Intern Med 76(6):993–1008
Bok K, Prikhodko VG, Green KY, Sosnovtsev SV (2009a) Apoptosis in murine norovirus-infected RAW264.7 cells is associated with downregulation of survivin. J Virol 83(8):3647–3656
Bok K, Abente EJ, Realpe-Quintero M, Mitra T, Sosnovtsev SV, Kapikian AZ, Green KY (2009b) Evolutionary dynamics of GII.4 noroviruses over a 34-year period. J Virol 83(22):11890–11901
Bok K, Parra GI, Mitra T, Abente E, Shaver CK, Boon D, Engle R, Yu C, Kapikian AZ, Sosnovtsev SV, Purcell RH, Green KY (2011) Chimpanzees as an animal model for human norovirus infection and vaccine development. Proc Natl Acad Sci USA 108(1):325–330
Bull RA, Hansman GS, Clancy LE, Tanaka MM, Rawlinson WD, White PA (2005) Norovirus recombination in ORF1/ORF2 overlap. Emerg Infect Dis 11(7):1079–1085
Bull RA, Tanaka MM, White PA (2007) Norovirus recombination. J Gen Virol 88(Pt 12):3347–3359
Bull RA, Eden JS, Rawlinson WD, White PA (2010) Rapid evolution of pandemic noroviruses of the GII.4 lineage. PLoS Pathog 6(3):e1000831
Campillay-Véliz CP, Carvajal JJ, Avellaneda AM, Escobar D, Covián C, Kalergis AM, Lay MK (2020) Human norovirus proteins: implications in the replicative cycle, pathogenesis, and the host immune response. Front Immunol 11:961
Cao S, Lou Z, Tan M, Chen Y, Liu Y, Zhang Z, Zhang XC, Jiang X, Li X, Rao Z (2007) Structural basis for the recognition of blood group trisaccharides by norovirus. J Virol 81(11):5949–5957
Cates JE, Vinjé J, Parashar U, Hall AJ (2020) Recent advances in human norovirus research and implications for candidate vaccines. Expert Rev Vaccines 19(6):539–548
Chan MC, Roy S, Bonifacio J, Zhang LY, Chhabra P, Chan JCM, Celma C, Igoy MA, Lau SL, Mohammad KN, Vinjé J, Vennema H, Breuer J, Koopmans M, de Graaf M, for NOROPATROL2 (2021) Detection of norovirus variant GII.4 Hong Kong in Asia and Europe, 2017–2019. Emerg Infect Dis 27(1):289–293
Chang KO, Sosnovtsev SV, Belliot G, King AD, Green KY (2006) Stable expression of a Norwalk virus RNA replicon in a human hepatoma cell line. Virology 353(2):463–473
Chaudhry Y, Nayak A, Bordeleau ME, Tanaka J, Pelletier J, Belsham GJ, Roberts LO, Goodfellow IG (2006) Caliciviruses differ in their functional requirements for eIF4F components. J Biol Chem 281(35):25315–25325
Cheetham S, Souza M, Meulia T, Grimes S, Han MG, Saif LJ (2006) Pathogenesis of a genogroup II human norovirus in gnotobiotic pigs. J Virol 80(21):10372–10381
Chhabra P, de Graaf M, Parra GI, Chan MC, Green K, Martella V, Wang Q, White PA, Katayama K, Vennema H, Koopmans MPG, Vinjé J (2019) Updated classification of norovirus genogroups and genotypes. J Gen Virol 100(10):1393–1406
Conley MJ, McElwee M, Azmi L, Gabrielsen M, Byron O, Goodfellow IG, Bhella D (2019) Calicivirus VP2 forms a portal-like assembly following receptor engagement. Nature 565(7739):377–381
Cortes-Penfield NW, Ramani S, Estes MK, Atmar RL (2017) Prospects and challenges in the development of a Norovirus vaccine. Clin Ther 39(8):1537–1549

Costantini V, Grenz L, Fritzinger A, Lewis D, Biggs C, Hale A, Vinjé J (2010) Diagnostic accuracy and analytical sensitivity of IDEIA Norovirus assay for routine screening of human norovirus. J Clin Microbiol 48(8):2770–2778

Cotten M, Petrova V, Phan MV, Rabaa MA, Watson SJ, Ong SH, Kellam P, Baker S (2014) Deep sequencing of norovirus genomes defines evolutionary patterns in an urban tropical setting. J Virol 88(19):11056–11069

Cox C, Cao S, Lu Y (2009) Enhanced detection and study of murine norovirus-1 using a more efficient microglial cell line. Virol J 6:196

Daughenbaugh KF, Fraser CS, Hershey JW, Hardy ME (2003) The genome-linked protein VPg of the Norwalk virus binds eIF3, suggesting its role in translation initiation complex recruitment. EMBO J 22(11):2852–2859

Davis A, Cortez V, Grodzki M, Dallas R, Ferrolino J, Freiden P, Maron G, Hakim H, Hayden RT, Tang L, Huys A, Kolawole AO, Wobus CE, Jones MK, Karst SM, Schultz-Cherry S (2020) Infectious norovirus is chronically shed by immunocompromised pediatric hosts. Viruses 12(6):619

Desai R, Hembree CD, Handel A, Matthews JE, Dickey BW, McDonald S, Hall AJ, Parashar UD, Leon JS, Lopman B (2012) Severe outcomes are associated with genogroup 2 genotype 4 norovirus outbreaks: a systematic literature review. Clin Infect Dis 55(2):189–193

Dingle KE, Lambden PR, Caul EO, Clarke IN (1995) Human enteric Caliciviridae: the complete genome sequence and expression of virus-like particles from a genetic group II small round structured virus. J Gen Virol 76(Pt 9):2349–2355

Division of Viral Diseases, National Center for Immunization and Respiratory Diseases, Centers for Disease Control and Prevention (2011) Updated norovirus outbreak management and disease prevention guidelines. MMWR Recomm Rep 60(RR-3):1–18

Dolin R, Levy AG, Wyatt RG, Thornhill TS, Gardner JD (1975) Viral gastroenteritis induced by the Hawaii agent. Jejunal histopathology and serologic response. Am J Med 59(6):761–768

Donaldson EF, Lindesmith LC, Lobue AD, Baric RS (2008) Norovirus pathogenesis: mechanisms of persistence and immune evasion in human populations. Immunol Rev 225:190–211

Eden JS, Hewitt J, Lim KL, Boni MF, Merif J, Greening G, Ratcliff RM, Holmes EC, Tanaka MM, Rawlinson WD, White PA (2014) The emergence and evolution of the novel epidemic norovirus GII.4 variant Sydney 2012. Virology 450:106–113

Esposito S, Principi N (2020) Norovirus vaccine: priorities for future research and development. Front Immunol 11:1383

Estes MK, Ettayebi K, Tenge VR, Murakami K, Karandikar U, Lin SC, Ayyar BV, Cortes-Penfield NW, Haga K, Neill FH, Opekun AR, Broughman JR, Zeng XL, Blutt SE, Crawford SE, Ramani S, Graham DY, Atmar RL (2019) Human norovirus cultivation in nontransformed stem cell-derived human intestinal enteroid cultures: success and challenges. Viruses 11(7):638

Ettayebi K, Crawford SE, Murakami K, Broughman JR, Karandikar U, Tenge VR, Neill FH, Blutt SE, Zeng XL, Qu L, Kou B, Opekun AR, Burrin D, Graham DY, Ramani S, Atmar RL, Estes MK (2016) Replication of human noroviruses in stem cell-derived human enteroids. Science 353(6306):1387–1393

Fankhauser RL, Monroe SS, Noel JS, Humphrey CD, Bresee JS, Parashar UD, Ando T, Glass RI (2002) Epidemiologic and molecular trends of "Norwalk-like viruses" associated with outbreaks of gastroenteritis in the United States. J Infect Dis 186(1):1–7

Farahmand M, Moghoofei M, Dorost A, Shoja Z, Ghorbani S, Kiani SJ, Khales P, Esteghamati A, Sayyahfar S, Jafarzadeh M, Minaeian S, Khanaliha K, Naghdalipour M, Tavakoli A (2022) Global prevalence and genotype distribution of norovirus infection in children with gastroenteritis: a meta-analysis on 6 years of research from 2015 to 2020. Rev Med Virol 32(1):e2237

Frenck R, Bernstein DI, Xia M, Huang P, Zhong W, Parker S, Dickey M, McNeal M, Jiang X (2012) Predicting susceptibility to norovirus GII.4 by use of a challenge model involving humans. J Infect Dis 206(9):1386–1393

Furman LM, Maaty WS, Petersen LK, Ettayebi K, Hardy ME, Bothner B (2009) Cysteine protease activation and apoptosis in murine norovirus infection. Virol J 6:139

Ghosh S, Malik YS, Kobayashi N (2018) Therapeutics and immunoprophylaxis against noroviruses and rotaviruses: the past, present, and future. Curr Drug Metab 19(3):170–191

Giersing BK, Vekemans J, Nava S, Kaslow DC, Moorthy V, WHO Product Development for Vaccines Advisory Committee (2019) Report from the World Health Organization's third Product Development for Vaccines Advisory Committee (PDVAC) meeting, Geneva, 8–10th June 2016. Vaccine 37(50):7315–7327

Glass RI, Parashar UD, Estes MK (2009) Norovirus gastroenteritis. N Engl J Med 361(18): 1776–1785

Gray JJ, Kohli E, Ruggeri FM, Vennema H, Sánchez-Fauquier A, Schreier E, Gallimore CI, Iturriza-Gomara M, Giraudon H, Pothier P, Di Bartolo I, Inglese N, de Bruin E, van der Veer B, Moreno S, Montero V, de Llano MC, Höhne M, Diedrich SM (2007) European multicenter evaluation of commercial enzyme immunoassays for detecting norovirus antigen in fecal samples. Clin Vaccine Immunol 14(10):1349–1355

Graziano VR, Wei J, Wilen CB (2019) Norovirus attachment and entry. Viruses 11(6):495

Green KY (2013) Caliciviridae: the noroviruses. In: Knipe DM, Howley PM, Cohen JI, Griffin DE, Lamb RA, Martin MA, Racaniello VR, Roizman B (eds) Fields virology, 6th edn. Lippincott Williams & Wilkins, Philadelphia, pp 582–608

Green J, Gallimore CI, Norcott JP, Lewis D, Brown DW (1995) Broadly reactive reverse transcriptase polymerase chain reaction for the diagnosis of SRSV-associated gastroenteritis. J Med Virol 47(4):392–398

Green KY, Kaufman SS, Nagata BM, Chaimongkol N, Kim DY, Levenson EA, Tin CM, Yardley AB, Johnson JA, Barletta ABF, Khan KM, Yazigi NA, Subramanian S, Moturi SR, Fishbein TM, Moore IN, Sosnovtsev SV (2020) Human norovirus targets enteroendocrine epithelial cells in the small intestine. Nat Commun 11(1):2759

Haga K, Fujimoto A, Takai-Todaka R, Miki M, Doan YH, Murakami K, Yokoyama M, Murata K, Nakanishi A, Katayama K (2016) Functional receptor molecules CD300lf and CD300ld within the CD300 family enable murine noroviruses to infect cells. Proc Natl Acad Sci USA 113(41):E6248–E6255

Hall AJ (2012) Noroviruses: the perfect human pathogens? J Infect Dis 205(11):1622–1624

Hall AJ, Lopman BA, Payne DC, Patel MM, Gastañaduy PA, Vinjé J, Parashar UD (2013) Norovirus disease in the United States. Emerg Infect Dis 19(8):1198–1205

Hassan E, Baldridge MT (2019) Norovirus encounters in the gut: multifaceted interactions and disease outcomes. Mucosal Immunol 12(6):1259–1267

Hemming M, Räsänen S, Huhti L, Paloniemi M, Salminen M, Vesikari T (2013) Major reduction of rotavirus, but not norovirus, gastroenteritis in children seen in hospital after the introduction of RotaTeq vaccine into the National Immunization Programme in Finland. Eur J Pediatr 172(6):739–746

Hoa Tran TN, Trainor E, Nakagomi T, Cunliffe NA, Nakagomi O (2013) Molecular epidemiology of noroviruses associated with acute sporadic gastroenteritis in children: global distribution of genogroups, genotypes and GII.4 variants. J Clin Virol 56(3):185–193

Hosmillo M, Chaudhry Y, Nayak K, Sorgeloos F, Koo BK, Merenda A, Lillestol R, Drumright L, Zilbauer M, Goodfellow I (2020) Norovirus replication in human intestinal epithelial cells is restricted by the interferon-induced JAK/STAT signaling pathway and RNA polymerase II-mediated transcriptional responses. MBio 11(2):e00215–e00220

Huhti L, Szakal ED, Puustinen L, Salminen M, Huhtala H, Valve O, Blazevic V, Vesikari T (2011) Norovirus GII-4 causes a more severe gastroenteritis than other noroviruses in young children. J Infect Dis 203(10):1442–1444

Hyde JL, Mackenzie JM (2010) Subcellular localization of the MNV-1 ORF1 proteins and their potential roles in the formation of the MNV-1 replication complex. Virology 406(1):138–148

Hyde JL, Sosnovtsev SV, Green KY, Wobus C, Virgin HW, Mackenzie JM (2009) Mouse norovirus replication is associated with virus-induced vesicle clusters originating from membranes derived from the secretory pathway. J Virol 83(19):9709–9719

Hyde JL, Gillespie LK, Mackenzie JM (2012) Mouse norovirus 1 utilizes the cytoskeleton network to establish localization of the replication complex proximal to the microtubule organizing center. J Virol 86(8):4110–4122

Jahun AS, Goodfellow IG (2021) Interferon responses to norovirus infections: current and future perspectives. J Gen Virol 102(10):001660

Jiang X, Wang M, Graham DY, Estes MK (1992) Expression, self-assembly, and antigenicity of the Norwalk virus capsid protein. J Virol 66(11):6527–6532

Jiang X, Wang M, Wang K, Estes MK (1993) Sequence and genomic organization of Norwalk virus. Virology 195(1):51–61

Johnson PC, Mathewson JJ, DuPont HL, Greenberg HB (1990) Multiple-challenge study of host susceptibility to Norwalk gastroenteritis in US adults. J Infect Dis 161(1):18–21

Jones MK, Watanabe M, Zhu S, Graves CL, Keyes LR, Grau KR, Gonzalez-Hernandez MB, Iovine NM, Wobus CE, Vinjé J, Tibbetts SA, Wallet SM, Karst SM (2014) Enteric bacteria promote human and mouse norovirus infection of B cells. Science 346(6210):755–759

Jung K, Wang Q, Kim Y, Scheuer K, Zhang Z, Shen Q, Chang KO, Saif LJ (2012) The effects of simvastatin or interferon-α on infectivity of human norovirus using a gnotobiotic pig model for the study of antivirals. PLoS One 7(7):e41619

Kageyama T, Kojima S, Shinohara M, Uchida K, Fukushi S, Hoshino FB, Takeda N, Katayama K (2003) Broadly reactive and highly sensitive assay for Norwalk-like viruses based on real-time quantitative reverse transcription-PCR. J Clin Microbiol 41(4):1548–1557

Kapikian AZ, Wyatt RG, Dolin R, Thornhill TS, Kalica AR, Chanock RM (1972) Visualization by immune electron microscopy of a 27-nm particle associated with acute infectious nonbacterial gastroenteritis. J Virol 10(5):1075–1081

Kaplan JE, Feldman R, Campbell DS, Lookabaugh C, Gary GW (1982) The frequency of a Norwalk-like pattern of illness in outbreaks of acute gastroenteritis. Am J Public Health 72(12):1329–1332

Karandikar UC, Crawford SE, Ajami NJ, Murakami K, Kou B, Ettayebi K, Papanicolaou GA, Jongwutiwes U, Perales MA, Shia J, Mercer D, Finegold MJ, Vinjé J, Atmar RL, Estes MK (2016) Detection of human norovirus in intestinal biopsies from immunocompromised transplant patients. J Gen Virol 97(9):2291–2300

Karst SM, Wobus CE (2015) A working model of how noroviruses infect the intestine. PLoS Pathog 11(2):e1004626

Karst SM, Zhu S, Goodfellow IG (2015) The molecular pathology of noroviruses. J Pathol 235(2):206–216

Katayama K, Shirato-Horikoshi H, Kojima S, Kageyama T, Oka T, Hoshino F, Fukushi S, Shinohara M, Uchida K, Suzuki Y, Gojobori T, Takeda N (2002) Phylogenetic analysis of the complete genome of 18 Norwalk-like viruses. Virology 299(2):225–239

Katayama K, Murakami K, Sharp TM, Guix S, Oka T, Takai-Todaka R, Nakanishi A, Crawford SE, Atmar RL, Estes MK (2014) Plasmid-based human norovirus reverse genetics system produces reporter-tagged progeny virus containing infectious genomic RNA. Proc Natl Acad Sci USA 111(38):E4043–E4052

Kendra JA, Tohma K, Parra GI (2022) Global and regional circulation trends of norovirus genotypes and recombinants, 1995–2019: a comprehensive review of sequences from public databases. Rev Med Virol 32(5):e2354

Kim L, Liebowitz D, Lin K, Kasparek K, Pasetti MF, Garg SJ, Gottlieb K, Trager G, Tucker SN (2018) Safety and immunogenicity of an oral tablet norovirus vaccine, a phase I randomized, placebo-controlled trial. JCI Insight 3(13):e121077

Kojima S, Kageyama T, Fukushi S, Hoshino FB, Shinohara M, Uchida K, Natori K, Takeda N, Katayama K (2002) Genogroup-specific PCR primers for detection of Norwalk-like viruses. J Virol Methods 100(1–2):107–114

Kroneman A, Vennema H, Deforche KV, Avoort H, Peñaranda S, Oberste MS, Vinjé J, Koopmans M (2011) An automated genotyping tool for enteroviruses and noroviruses. J Clin Virol 51(2):121–125

Kroneman A, Vega E, Vennema H, Vinjé J, White PA, Hansman G, Green K, Martella V, Katayama K, Koopmans M (2013) Proposal for a unified norovirus nomenclature and genotyping. Arch Virol 158(10):2059–2068

Lindesmith L, Moe C, Marionneau S, Ruvoen N, Jiang X, Lindblad L, Stewart P, LePendu J, Baric R (2003) Human susceptibility and resistance to Norwalk virus infection. Nat Med 9(5):548–553

Lindesmith L, Moe C, Lependu J, Frelinger JA, Treanor J, Baric RS (2005) Cellular and humoral immunity following Snow Mountain virus challenge. J Virol 79(5):2900–2909

Lindesmith LC, Donaldson EF, Lobue AD, Cannon JL, Zheng DP, Vinjé J, Baric RS (2008) Mechanisms of GII.4 norovirus persistence in human populations. PLoS Med 5(2):e31

Lindesmith LC, Donaldson E, Leon J, Moe CL, Frelinger JA, Johnston RE, Weber DJ, Baric RS (2010) Heterotypic humoral and cellular immune responses following Norwalk virus infection. J Virol 84(4):1800–1815

Lindesmith LC, Beltramello M, Donaldson EF, Corti D, Swanstrom J, Debbink K, Lanzavecchia A, Baric RS (2012) Immunogenetic mechanisms driving norovirus GII.4 antigenic variation. PLoS Pathog 8(5):e1002705

Liu L, Moore MD (2020) A survey of analytical techniques for noroviruses. Foods (Basel, Switzerland) 9(3):318

Liu P, Wang X, Lee JC, Teunis P, Hu S, Paradise HT, Moe C (2014) Genetic susceptibility to norovirus GII.3 and GII.4 infections in Chinese pediatric diarrheal disease. Pediatr Infect Dis J 33(11):e305–e309

Lochridge VP, Jutila KL, Graff JW, Hardy ME (2005) Epitopes in the P2 domain of norovirus VP1 recognized by monoclonal antibodies that block cell interactions. J Gen Virol 86(Pt 10):2799–2806

Lopman B, Armstrong B, Atchison C, Gray JJ (2009) Host, weather and virological factors drive norovirus epidemiology: time-series analysis of laboratory surveillance data in England and Wales. PLoS One 4(8):e6671

Lopman BA, Trivedi T, Vicuña Y, Costantini V, Collins N, Gregoricus N, Parashar U, Sandoval C, Broncano N, Vaca M, Chico ME, Vinjé J, Cooper PJ (2015) Norovirus infection and disease in an Ecuadorian Birth Cohort: Association of Certain Norovirus Genotypes with host FUT2 secretor status. J Infect Dis 211(11):1813–1821

Lucero Y, Matson DO, Ashkenazi S, George S, O'Ryan M (2021) Norovirus: facts and reflections from past, present, and future. Viruses 13(12):2399

Ludwig-Begall LF, Mauroy A, Thiry E (2021) Noroviruses-the state of the art, nearly 50 years after their initial discovery. Viruses 13(8):1541

Makino A, Shimojima M, Miyazawa T, Kato K, Tohya Y, Akashi H (2006) Junctional adhesion molecule 1 is a functional receptor for feline calicivirus. J Virol 80(9):4482–4490

Mans J (2019) Norovirus infections and disease in lower-middle and low-income countries, 1997–2018. Viruses 11(4):341

Marionneau S, Ruvoën N, Le Moullac-Vaidye B, Clement M, Cailleau-Thomas A, Ruiz-Palacois G, Huang P, Jiang X, Le Pendu J (2002) Norwalk virus binds to histo-blood group antigens present on gastroduodenal epithelial cells of secretor individuals. Gastroenterology 122(7):1967–1977

Mattison CP, Cardemil CV, Hall AJ (2018) Progress on norovirus vaccine research: public health considerations and future directions. Expert Rev Vaccines 17(9):773–784

Mboko WP, Chhabra P, Valcarce MD, Costantini V, Vinjé J (2022) Advances in understanding of the innate immune response to human norovirus infection using organoid models. J Gen Virol 103(1):1720

McAtee CL, Webman R, Gilman RH, Mejia C, Bern C, Apaza S, Espetia S, Pajuelo M, Saito M, Challappa R, Soria R, Ribera JP, Lozano D, Torrico F (2016) Burden of norovirus and rotavirus in children after rotavirus vaccine introduction, Cochabamba, Bolivia. Am J Trop Med Hyg 94(1):212–217

McFadden N, Bailey D, Carrara G, Benson A, Chaudhry Y, Shortland A, Heeney J, Yarovinsky F, Simmonds P, Macdonald A, Goodfellow I (2011) Norovirus regulation of the innate immune response and apoptosis occurs via the product of the alternative open reading frame 4. PLoS Pathog 7(12):e1002413

Meeroff JC, Schreiber DS, Trier JS, Blacklow NR (1980) Abnormal gastric motor function in viral gastroenteritis. Ann Intern Med 92(3):370–373

Mehlhorn H (2015) WHO estimates of the global burden of foodborne diseases: foodborne disease burden epidemiology reference group 2007–2015. World Health Organization, Geneva

Morillo SG, Luchs A, Cilli A, Ribeiro CD, Calux SJ, de Carmona RC, do Timenetsky MC (2011) Norovirus 3rd Generation kit: an improvement for rapid diagnosis of sporadic gastroenteritis cases and valuable for outbreak detection. J Virol Methods 173(1):13–16

Napthine S, Lever RA, Powell ML, Jackson RJ, Brown TD, Brierley I (2009) Expression of the VP2 protein of murine norovirus by a translation termination–reinitiation strategy. PLoS One 4(12):e8390

Netzler NE, Enosi Tuipulotu D, White PA (2019) Norovirus antivirals: where are we now? Med Res Rev 39(3):860–886

Ng KK, Pendás-Franco N, Rojo J, Boga JA, Machín A, Alonso JM, Parra F (2004) Crystal structure of Norwalk virus polymerase reveals the carboxyl terminus in the active site cleft. J Biol Chem 279(16):16638–16645

Nguyen GT, Phan K, Teng I, Pu J, Watanabe T (2017) A systematic review and meta-analysis of the prevalence of norovirus in cases of gastroenteritis in developing countries. Medicine 96(40):e8139

Nice TJ, Strong DW, McCune BT, Pohl CS, Virgin HW (2013) A single-amino-acid change in murine norovirus NS1/2 is sufficient for colonic tropism and persistence. J Virol 87(1):327–334

Noel JS, Fankhauser RL, Ando T, Monroe SS, Glass RI (1999) Identification of a distinct common strain of "Norwalk-like viruses" having a global distribution. J Infect Dis 179(6):1334–1344

Nordgren J, Svensson L (2019) Genetic susceptibility to human norovirus infection: an update. Viruses 11(3):226

Nordgren J, Kindberg E, Lindgren PE, Matussek A, Svensson L (2010) Norovirus gastroenteritis outbreak with a secretor-independent susceptibility pattern, Sweden. Emerg Infect Dis 16(1):81–87

Nordgren J, Nitiema LW, Ouermi D, Simpore J, Svensson L (2013) Host genetic factors affect susceptibility to norovirus infections in Burkina Faso. PLoS One 8(7):e69557

Operario DJ, Platts-Mills JA, Nadan S, Page N, Seheri M, Mphahlele J, Praharaj I, Kang G, Araujo IT, Leite JPG, Cowley D, Thomas S, Kirkwood CD, Dennis F, Armah G, Mwenda JM, Wijesinghe PR, Rey G, Grabovac V, Berejena C, Houpt ER (2017) Etiology of severe acute watery diarrhea in children in the global rotavirus surveillance network using quantitative polymerase chain reaction. J Infect Dis 216(2):220–227

Orchard RC, Wilen CB, Doench JG, Baldridge MT, McCune BT, Lee YC, Lee S, Pruett-Miller SM, Nelson CA, Fremont DH, Virgin HW (2016) Discovery of a proteinaceous cellular receptor for a norovirus. Science 353(6302):933–936

Parra GI (2019) Emergence of norovirus strains: a tale of two genes. Virus Evol 5(2):vez048

Parra GI, Bok K, Taylor R, Haynes JR, Sosnovtsev SV, Richardson C, Green KY (2012) Immunogenicity and specificity of norovirus consensus GII.4 virus-like particles in monovalent and bivalent vaccine formulations. Vaccine 30(24):3580–3586

Parrino TA, Schreiber DS, Trier JS, Kapikian AZ, Blacklow NR (1977) Clinical immunity in acute gastroenteritis caused by Norwalk agent. N Engl J Med 297(2):86–89

Patel MM, Widdowson MA, Glass RI, Akazawa K, Vinjé J, Parashar UD (2008) Systematic literature review of role of noroviruses in sporadic gastroenteritis. Emerg Infect Dis 14(8):1224–1231

Payne DC, Vinjé J, Szilagyi PG, Edwards KM, Staat MA, Weinberg GA, Hall CB, Chappell J, Bernstein DI, Curns AT, Wikswo M, Shirley SH, Hall AJ, Lopman B, Parashar UD (2013) Norovirus and medically attended gastroenteritis in U.S. children. N Engl J Med 368(12):1121–1130

Pfister T, Wimmer E (2001) Polypeptide p41 of a Norwalk-like virus is a nucleic acid-independent nucleoside triphosphatase. J Virol 75(4):1611–1619

Pires SM, Fischer-Walker CL, Lanata CF, Devleesschauwer B, Hall AJ, Kirk MD, Duarte AS, Black RE, Angulo FJ (2015) Aetiology-specific estimates of the global and regional incidence and mortality of diarrhoeal diseases commonly transmitted through food. PLoS One 10(12):e0142927

Platts-Mills JA, Babji S, Bodhidatta L, Gratz J, Haque R, Havt A, McCormick BJ, McGrath M, Olortegui MP, Samie A, Shakoor S, Mondal D, Lima IF, Hariraju D, Rayamajhi BB, Qureshi S, Kabir F, Yori PP, Mufamadi B, Amour C, MAL-ED Network Investigators (2015) Pathogen-specific burdens of community diarrhoea in developing countries: a multisite birth cohort study (MAL-ED). Lancet Glob Health 3(9):e564–e575

Prasad BV, Hardy ME, Dokland T, Bella J, Rossmann MG, Estes MK (1999) X-ray crystallographic structure of the Norwalk virus capsid. Science 286(5438):287–290

Qi R, Huang YT, Liu JW, Sun Y, Sun XF, Han HJ, Qin XR, Zhao M, Wang LJ, Li W, Li JH, Chen C, Yu XJ (2018) Global prevalence of asymptomatic norovirus infection: a meta-analysis. EClinicalMedicine 2–3:50–58

Ramani S, Estes MK, Atmar RL (2016) Correlates of protection against norovirus infection and disease-where are we now, where do we go? PLoS Pathog 12(4):e1005334

Reeck A, Kavanagh O, Estes MK, Opekun AR, Gilger MA, Graham DY, Atmar RL (2010) Serological correlate of protection against norovirus-induced gastroenteritis. J Infect Dis 202(8):1212–1218

Rohayem J, Robel I, Jäger K, Scheffler U, Rudolph W (2006) Protein-primed and de novo initiation of RNA synthesis by norovirus 3Dpol. J Virol 80(14):7060–7069

Rouhani S, Peñataro Yori P, Paredes Olortegui M, Siguas Salas M, Rengifo Trigoso D, Mondal D, Bodhidatta L, Platts-Mills J, Samie A, Kabir F, Lima A, Babji S, Mason CJ, Kalam A, Bessong P, Ahmed T, Mduma E, Bhutta ZA, Lima I, Ramdass R, Etiology, Risk Factors, and Interactions of Enteric Infections and Malnutrition and the Consequences for Child Health and Development Project (MAL-ED) Network Investigators (2016) Norovirus infection and acquired immunity in 8 countries: results from the MAL-ED study. Clin Infect Dis 62(10):1210–1217

Sakon N, Yamazaki K, Nakata K, Kanbayashi D, Yoda T, Mantani M, Kase T, Takahashi K, Komano J (2015) Impact of genotype-specific herd immunity on the circulatory dynamism of norovirus: a 10-year longitudinal study of viral acute gastroenteritis. J Infect Dis 211(6):879–888

Scallan CD, Tingley DW, Lindbloom JD, Toomey JS, Tucker SN (2013) An adenovirus-based vaccine with a double-stranded RNA adjuvant protects mice and ferrets against H5N1 avian influenza in oral delivery models. Clin Vaccine Immunol CVI 20(1):85–94

Schreiber DS, Blacklow NR, Trier JS (1973) The mucosal lesion of the proximal small intestine in acute infectious nonbacterial gastroenteritis. N Engl J Med 288(25):1318–1323

Seo DJ, Jung D, Jung S, Ha SK, Ha SD, Choi IS, Myoung J, Choi C (2018) Experimental miniature piglet model for the infection of human norovirus GII. J Med Virol 90(4):655–662

Shah MP, Hall AJ (2018) Norovirus illnesses in children and adolescents. Infect Dis Clin N Am 32(1):103–118

Sherwood J, Mendelman PM, Lloyd E, Liu M, Boslego J, Borkowski A, Jackson A, Faix D, US Navy Study Team (2020) Efficacy of an intramuscular bivalent norovirus GI.1/GII.4 virus-like particle vaccine candidate in healthy US adults. Vaccine 38(41):6442–6449

Shioda K, Kambhampati A, Hall AJ, Lopman BA (2015) Global age distribution of pediatric norovirus cases. Vaccine 33(33):4065–4068

Siebenga JJ, Vennema H, Renckens B, de Bruin E, van der Veer B, Siezen RJ, Koopmans M (2007) Epochal evolution of GGII.4 norovirus capsid proteins from 1995 to 2006. J Virol 81(18):9932–9941

Siebenga JJ, Vennema H, Zheng DP, Vinjé J, Lee BE, Pang XL, Ho EC, Lim W, Choudekar A, Broor S, Halperin T, Rasool NB, Hewitt J, Greening GE, Jin M, Duan ZJ, Lucero Y, O'Ryan M, Hoehne M, Schreier E, Koopmans M (2009) Norovirus illness is a global problem: emergence and spread of norovirus GII.4 variants, 2001–2007. J Infect Dis 200(5):802–812

Siebenga JJ, Lemey P, Kosakovsky Pond SL, Rambaut A, Vennema H, Koopmans M (2010) Phylodynamic reconstruction reveals norovirus GII.4 epidemic expansions and their molecular determinants. PLoS Pathog 6(5):e1000884

Simmons K, Gambhir M, Leon J, Lopman B (2013) Duration of immunity to norovirus gastroenteritis. Emerg Infect Dis 19(8):1260–1267

Smith HQ, Smith TJ (2019) The dynamic capsid structures of the noroviruses. Viruses 11(3):235

Sosnovtsev SV, Belliot G, Chang KO, Onwudiwe O, Green KY (2005) Feline calicivirus VP2 is essential for the production of infectious virions. J Virol 79(7):4012–4024
Sosnovtsev SV, Belliot G, Chang KO, Prikhodko VG, Thackray LB, Wobus CE, Karst SM, Virgin HW, Green KY (2006) Cleavage map and proteolytic processing of the murine norovirus nonstructural polyprotein in infected cells. J Virol 80(16):7816–7831
Souza M, Azevedo MS, Jung K, Cheetham S, Saif LJ (2008) Pathogenesis and immune responses in gnotobiotic calves after infection with the genogroup II.4-HS66 strain of human norovirus. J Virol 82(4):1777–1786
Subba-Reddy CV, Yunus MA, Goodfellow IG, Kao CC (2012) Norovirus RNA synthesis is modulated by an interaction between the viral RNA-dependent RNA polymerase and the major capsid protein, VP1. J Virol 86(18):10138–10149
Subekti DS, Tjaniadi P, Lesmana M, McArdle J, Iskandriati D, Budiarsa IN, Walujo P, Suparto IH, Winoto I, Campbell JR, Porter KR, Sajuthi D, Ansari AA, Oyofo BA (2002) Experimental infection of *Macaca nemestrina* with a Toronto Norwalk-like virus of epidemic viral gastroenteritis. J Med Virol 66(3):400–406
Tan M (2021) Norovirus vaccines: current clinical development and challenges. Pathogens 10(12):1641
Tan M, Jin M, Xie H, Duan Z, Jiang X, Fang Z (2008) Outbreak studies of a GII-3 and a GII-4 norovirus revealed an association between HBGA phenotypes and viral infection. J Med Virol 80(7):1296–1301
Taube S, Perry JW, Yetming K, Patel SP, Auble H, Shu L, Nawar HF, Lee CH, Connell TD, Shayman JA, Wobus CE (2009) Ganglioside-linked terminal sialic acid moieties on murine macrophages function as attachment receptors for murine noroviruses. J Virol 83(9):4092–4101
Taube S, Perry JW, McGreevy E, Yetming K, Perkins C, Henderson K, Wobus CE (2012) Murine noroviruses bind glycolipid and glycoprotein attachment receptors in a strain-dependent manner. J Virol 86(10):5584–5593
Taube S, Kolawole AO, Höhne M, Wilkinson JE, Handley SA, Perry JW, Thackray LB, Akkina R, Wobus CE (2013) A mouse model for human norovirus. MBio 4(4):e00450–e00413
Teunis PF, Moe CL, Liu P, Miller SE, Lindesmith L, Baric RS, Le Pendu J, Calderon RL (2008) Norwalk virus: how infectious is it? J Med Virol 80(8):1468–1476
Teunis PF, Sukhrie FH, Vennema H, Bogerman J, Beersma MF, Koopmans MP (2015) Shedding of norovirus in symptomatic and asymptomatic infections. Epidemiol Infect 143(8):1710–1717
Thorne LG, Goodfellow IG (2014) Norovirus gene expression and replication. J Gen Virol 95(Pt 2):278–291
Todd KV, Tripp RA (2019) Human norovirus: experimental models of infection. Viruses 11(2):151
van Beek J, Ambert-Balay K, Botteldoorn N, Eden JS, Fonager J, Hewitt J, Iritani N, Kroneman A, Vennema H, Vinjé J, White PA, Koopmans M, NoroNet (2013) Indications for worldwide increased norovirus activity associated with emergence of a new variant of genotype II.4, late 2012. Euro Surveill 18(1):8–9
Van Dycke J, Ny A, Conceição-Neto N, Maes J, Hosmillo M, Cuvry A, Goodfellow I, Nogueira TC, Verbeken E, Matthijnssens J, de Witte P, Neyts J, Rocha-Pereira J (2019) A robust human norovirus replication model in zebrafish larvae. PLoS Pathog 15(9):e1008009
Vega E, Barclay L, Gregoricus N, Shirley SH, Lee D, Vinjé J (2014) Genotypic and epidemiologic trends of norovirus outbreaks in the United States, 2009–2013. J Clin Microbiol 52(1):147–155
Verhoef L, Hewitt J, Barclay L, Ahmed SM, Lake R, Hall AJ, Lopman B, Kroneman A, Vennema H, Vinjé J, Koopmans M (2015) Norovirus genotype profiles associated with foodborne transmission, 1999–2012. Emerg Infect Dis 21(4):592–599
Vinjé J (2015) Advances in laboratory methods for detection and typing of norovirus. J Clin Microbiol 53(2):373–381
Vinjé J, Koopmans MP (1996) Molecular detection and epidemiology of small round-structured viruses in outbreaks of gastroenteritis in The Netherlands. J Infect Dis 174(3):610–615
Vinjé J, Hamidjaja RA, Sobsey MD (2004) Development and application of a capsid VP1 (region D) based reverse transcription PCR assay for genotyping of genogroup I and II noroviruses. J Virol Methods 116(2):109–117

Vinjé J, Estes MK, Esteves P, Green KY, Katayama K, Knowles NJ, L'Homme Y, Martella V, Vennema H, White PA, Ictv Report Consortium (2019) ICTV virus taxonomy profile: Caliciviridae. J Gen Virol 100(11):1469–1470

Vongpunsawad S, Venkataram Prasad BV, Estes MK (2013) Norwalk virus minor capsid protein VP2 associates within the VP1 shell domain. J Virol 87(9):4818–4825

Walker FC, Baldridge MT (2019) Interactions between noroviruses, the host, and the microbiota. Curr Opin Virol 37:1–9

Widdowson MA, Monroe SS, Glass RI (2005) Are noroviruses emerging? Emerg Infect Dis 11(5):735–737

Wobus CE (2018) The dual tropism of noroviruses. J Virol 92(16):e01010–e01017

Wobus CE, Karst SM, Thackray LB, Chang KO, Sosnovtsev SV, Belliot G, Krug A, Mackenzie JM, Green KY, Virgin HW (2004) Replication of norovirus in cell culture reveals a tropism for dendritic cells and macrophages. PLoS Biol 2(12):e432

Wyatt RG, Dolin R, Blacklow NR, DuPont HL, Buscho RF, Thornhill TS, Kapikian AZ, Chanock RM (1974) Comparison of three agents of acute infectious nonbacterial gastroenteritis by cross-challenge in volunteers. J Infect Dis 129(6):709–714

Wyatt RG, Greenberg HB, Dalgard DW, Allen WP, Sly DL, Thornhill TS, Chanock RM, Kapikian AZ (1978) Experimental infection of chimpanzees with the Norwalk agent of epidemic viral gastroenteritis. J Med Virol 2(2):89–96

Zheng DP, Ando T, Fankhauser RL, Beard RS, Glass RI, Monroe SS (2006) Norovirus classification and proposed strain nomenclature. Virology 346(2):312–323

Understanding Norovirus: Causes, Contagion and Cure

14

Manisha Yadav, Subhash Sonkar, and Bidhan Chandra Koner

Abstract

Norovirus (NoV) is one of the viruses known to be associated with acute gastroenteritis (inflammation of the gut). NoV infection can occur by consumption of infected food and water or by exposure to the usage of infected restrooms in restaurants, schools, etc. It is considered highly contagious and therefore a prompt, and strict curative response is required to avoid its spread across the community, failing which leads to NoV infection outbreaks as observed by both developed and developing countries, including India. The age groups that are majorly affected by the NoV include children and elderly people. The most common symptoms of norovirus infection are nausea, vomiting and severe diarrhoea, which may last for a week. Usually, these symptoms can be treated by practising good hygiene and staying hydrated. In rare cases, the need for hospitalisation arises. Several anti-viral agents and vaccine candidates were tried in trials for their safety and efficacy, but none of them could make it to the market. Therefore, developing an anti-viral agent is still an unmet need. In this chapter, we will review the history, clinical manifestations and causes of NoV infection-related gastroenteritis. In addition, we will also investigate the therapeutic development, challenges and recent advancements made in the field and necessary measures to contain the NoV spread.

Keywords

Norovirus · Gastroenteritis · Infection · Anti-viral agent · Diarrhoea

M. Yadav (✉) · S. Sonkar · B. C. Koner
Multidisciplinary Research Unit, Maulana Azad Medical College, and Associated Hospitals, New Delhi, India
e-mail: myadav85@gmail.com

P. L. Bhukya, A. S (eds.), *Emerging Human Viral Diseases, Volume II*,
https://doi.org/10.1007/978-981-97-4480-0_14

14.1 Introduction

Acute gastroenteritis or stomach flu is an infectious syndrome, which affects millions of people throughout the world (Ahmed et al. 2014). It is caused by consumption of pathogen-contaminated food and water. Gastroenteritis can be caused by microbial agents like bacteria, e.g. shigella, salmonella, *E. coli*, vibrio, yersinia, etc.; parasites like giardia, amoebiasis, cyclospora and viral agents like rotavirus, sapovirus, norovirus (NoV), astrovirus, coronavirus, etc. In virus category, rotavirus and norovirus are the predominant disease-causing agents (Graves 2013). NoV is a highly infectious virus. The name of the virus was derived from an outbreak of gastroenteritis that occurred in Norwalk, Ohio in 1968 (Ahmed et al. 2014; Graves 2013). Therefore, known as Norwalk-like virus. It is also known as 'winter vomiting virus' owing to a higher occurrence in winter season and vomiting as the most commonly occurring symptom (Adler and Zickl 1969). It is observable because of its ability to cause gastroenteritis episodes with mild intensity and self-limiting properties with few of them leading to death (Esona et al. 2021). Despite a higher prevalence rate, gastroenteritis-related morbidity rate is low because of the existing treatment regimens. So far protection from this infection via biological intervention is only available for rotavirus which has resulted in controlling the frequency of rotavirus-induced gastroenteritis (McAtee et al. 2016). However, due to the lack of an effective NoV vaccine, the frequency of infections due to norovirus has elevated. Therefore, there is a need for an effective anti-viral agent/vaccine candidate for other viruses as well. Many research groups and pharma companies are focusing on the development of vaccines that can effectively prevent serious viral infections.

14.2 Structure and Genome of Norovirus

Norovirus belongs to one of the genera in the *Caliciviridae* family of viruses (Green et al. 2000; Ramani et al. 2014; Kapikian et al. 1972). It is a small, round, non-enveloped virus with a diameter ranging from 27 to 32 nm (Bertolotti-Ciarlet et al. 2003). It has an icosahedral shape and lacks a well-defined surface structure. It has a buoyant density of approximately 1.33–1.41 g/cm^2. It can be characterised by ten spikes and 32 cup-like depressions on its surface. NoV has a non-segmented, linear, single-positive stranded RNA as genetic material, which is 7.4–7.7 kb long. ORF1, ORF2 and ORF3 encode for non-structural protein, structural protein VP1 and structural protein VP2, respectively, in the NoV genome (Hardy 2005; Sharp et al. 2012) (Fig. 14.1).

14.2.1 Non-Structural Protein

The ORF1 encodes for non-structural proteins (NSPs) that undergo proteolytic cleavage to form six different types of NSPs defined as:

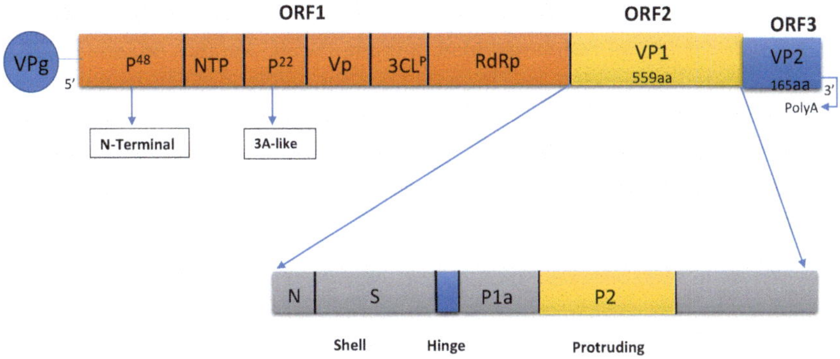

Fig. 14.1 Depicts the norovirus's genomic organisation, which includes three open reading frames, ORF1, ORF2 and ORF3. P48, NTP, P22, 3C-like protein and RdRp. VP1 and VP2 are non-structural proteins encoded by ORF1. ORF2 and ORF3 encode for the structural proteins VP1 and VP2 respectively. VP1 can be further classified into the shell, hinge and protruding regions, which make up the icosahedral structure of the NoV, while VP2 mainly helps in stabilising the VP1 region

1. N-terminal protein
2. P48
3. NTPase/P41 (nucleoside tri phosphatase)
4. P22, or 3A-like protein
5. VPg (Viral Protein Genome linked)
6. RdRp (RNA-dependent RNA polymerase)

The N-terminal protein aids in the viral replication inside the host cells by intracellular rearrangement of the Golgi complex and the formation of distinct aggregates (Sharp et al. 2012; Fernandez-Vega et al. 2004). The function of P48 is not known. NTPase/P41, the role of which has not yet been identified. The protein P22, or 3A-like protein, shows similarity to the 3A protein of picornavirus and is suggested to play a role in cellular component trafficking and the formation of replication complexes (Belliot et al. 2003), VPg (viral protein genomic linked), is a 12–15 kDa protease which is covalently linked to viral genetic material at the 5′ end (Green et al. 2000). It acts as a 5′-end cap substitute (required for RNA translation into protein) which is usually uncapped in caliciviridae family viruses (Thorne and Goodfellow 2014). The uridylylated form of the VPg protein attributes to priming in the process of RNA replication (Belliot et al. 2008) and hence enhances viral replication and load in the host cell. Another protein 3C-like protease is an important protein playing a role in the proteolytic cleavage of the non-structural protein into six different components. It also controls the host cell during viral replication (Rani et al. 2021; Kuyumcu-Martinez et al. 2004). The last protein is RdRp, an enzymatic polymerase, that synthesises and replicates the viral RNA (Rani et al. 2021). It plays an important role in viral replication, which makes it an attractive target for the development of anti-viral agents. Uridylated form of the VPg protein attributes to priming in the process of RNA replication (Belliot et al. 2008) and hence enhances viral replication and load in the host cell. Another protein 3C-like protease is an important protein playing a role in the proteolytic cleavage of the non-structural protein into six different components. It also controls the host cell during viral replication (Rani et al. 2021; Kuyumcu-Martinez et al. 2004). The last protein is RdRp, which is an enzymatic polymerase that synthesises and replicates the viral RNA (Rani et al. 2021). It plays an important role in viral replication, which makes it an attractive target for the development of anti-viral agents.

14.2.2 Structural Protein

14.2.2.1 VP1

VP1 is a 55–70 kDa major capsid protein which is responsible for the formation and release of virus-like particles. A mature virion is formed by the dimerisation of VP1 protein into 90 dimers to form an icosahedral shape. This arrangement gives its surface a cup-like shape. It is composed of three domains, namely, the N-terminal domain, the shell (S) domain and the protruding (P) domain. The shell domain is a highly conserved sequence which is embedded inside the capsid protein and is indispensable for capsid formation. The P domain consists of two subdomains—P1 and P2—that are discretely and exceedingly variable in sequence, respectively. The protruding domain P2 is responsible for the binding of immune and recognition sites. The S and P domains are connected by a small hinge region (Rani et al. 2021; Kuyumcu-Martinez et al. 2004; Hardy et al. 1996; Lochridge et al. 2005; Tan et al. 2003).

14.2.2.2 VP2

VP2 is a highly variable minor structural protein which helps in the upregulation and stabilisation of the viral capsid protein VP1 (Rani et al. 2021; Glass et al. 2000; Bertolotti-Ciarlet et al. 2003). It prevents the disassembly and protease degradation of the VP1 protein (Asanaka et al. 2005). It is reported to play an important role in the formation of viral particles (Katayama et al. 2002).

14.3 Classification of Norovirus

Norovirus is a highly diverse group of viruses. It is the recombinational and mutational variation in the sequence of VP1 that differentiates NoV into ten different genogroups, which are further classified into 49 genotypes. The genogroups causing diseases in humans and animals are GII, GIV, GVIII and GIX. Out of these genogroups, GII.4 alone accounts for more than 60% of NoV outbreaks (Rani et al. 2021). Host immunity varies for each genotype and strain. Each strain of a genotype has 65–98% sequence similarity, while the sequence similarity of strains varies from 51 to 56% among different genogroups (Donaldson et al. 2010).

14.4 Clinical Features

Norovirus infection occurs mostly in children below 5 years and elder people above 65 years who are more susceptible to NoV infection due to weak immunity (Murata et al. 2007; Gustavsson et al. 2012; Kawada et al. 2012; Mathew et al. 2019). However, the severity of the disease is higher at ages more than 75 years. This disease is mostly self-limiting and not life-threatening. But it can persist for a long in people with compromised immunity like children and organ transplant patients due to treatment with immunosuppressive drugs (Robilotti et al. 2015; Lee et al. 2008; de Graaf et al. 2016; Gras et al. 2021). It is best diagnosed by its symptoms, but it is confirmed by lab tests. NoV infection is identified by the symptoms that appear 24–48 h post-infection after consumption of contaminated food or water and can last for about 12–72 h (Arias et al. 2010). Most common symptoms of NoV-induced gastroenteritis include sudden onset of nausea, non-bloody vomiting, watery diarrhoea, abdominal pain and cramps, weight loss, malaise, headache/body pain, low-grade fever, non-mucoid/bloody stool. In addition to these symptoms, NoV was also reported to be associated with other complications. Among these common symptoms, few studies also reported NoV-related benign infantile convulsions (Hu et al. 2017; Ueda et al. 2015), encephalopathy (Shima et al. 2019), worsening of inflammatory bowel diseases (Khan et al. 2009) and necrotising enterocolitis (Turcios-Ruiz et al. 2008; Stuart et al. 2010). There was a report in which military personnel with NoV-induced gastroenteritis suffered from neck stiffness, phobia to light and mild to moderate reduction in response rate to any kind of stimulus (Centers for Disease Control and Prevention (CDC) 2002). In a comparative (observational) study between norovirus and rotavirus, it was found that the duration of diarrhoea caused by NoV was longer than the diarrhoea caused by rotavirus which indicated that NoV infection cannot be assumed to be less severe than rotavirus infection. Statins are known as one of the risk factors that can increase the disease severity (Mann et al. 2008). NoV infection can be differentiated from other pathogenic infections based on features that are less frequent in NoV infection, e.g. fever, bloody stool, mucoid stool and lower levels of CRP.

14.5 Disease Pathogenesis

Norovirus infection affects the small intestine. It reduces the enzymatic activity of alkaline phosphatase, sucrase and trehalase in the small intestine brush border, which results in abnormal absorption of carbohydrates. It also leads to an increase in bowel emptying time, which causes a feeling of nausea and vomiting episodes (Rani et al. 2021).

14.6 Mode of Transmission

NoV infection is highly contagious and can occur with exposure to as less as ten viral particles. Mostly, the infection is foodborne, followed by person-to-person contact, waterborne (Mathew et al. 2019; Matthews et al. 2012), or may be caused by undetermined conditions. Oral-transmitted infection occurs with contaminated water and food (Kauppinen et al. 2018). Other than these modes, faeces are also a considerable mode of transmission of NoV, as evident in several studies where faecal samples from both asymptomatic and symptomatic patients were found to have norovirus contamination (Teunis et al. 2015). In another study, the faecal filtrate from NoV-infected individuals induced symptoms of gastroenteritis in male jail prisoners (Dolin et al. 1971). There are reports of transmission of NoV that travels through indoor air (Ijaz et al. 2016). The pathogens discharged by the patients, i.e. aerosolised vomitus or remains after flushing of a toilet used by an infected individual, settle on various surfaces surrounding the infected individual, which acts as a secondary mode of transmission of the virus (Cardemil et al. 2017).

It is also observed that NoV can be transmitted from one species to another. Faeces from humans and animals were found to be cross-contaminated. Based on antibodies generated, it was found that antibodies against human NoV were found in animals as well as antibodies against human NoV were found in swine. This indicates that the extent of transmission could be greater if interspecies transmission occurs and ultimately leads to a multispecies outbreak (Bank-Wolf et al. 2010).

In a study, the authors identified the trend of the common modes of transmission in decreasing order to be mostly foodborne, followed by person-to-person contact, waterborne and environmental factors in decreasing frequency. They also identified the food-delivery service as the setting which was responsible for the maximum outbreaks that occurred between 1983 and 2011 (Hardstaff et al. 2018).

14.7 Outbreaks

NoV infection-induced gastroenteritis cases are reported throughout the world irrespective of the economic status of the country. US, Sweden, Finland, Norway, Spain, Japan, UAE and China all have observed NoV outbreaks (Mathew et al. 2019; Edelstein et al. 2014; Kanerva et al. 2009; Espenhain et al. 2019; GalmésTruyols et al. 2011). In 2012, there was a study in which authors identified the trend of NoV outbreaks at the global level. According to them, the NoV outbreak was more prevalent in the northern hemisphere and in the winter season (Bitler et al. 2013). In our country, outbreak of NoV-related gastroenteritis is reported sometimes in hospitals and nursing homes or places that dwell people in groups. In India, the state of Kerala witnessed cases of NoV infection in school-going children (NoV cases: Centre asks Kerala State Surveillance Office to submit report—The Hindu). According to the officials, the first documented NoV outbreak in Kerala was reported in Alappuzha district in June last year. Around 950 cases of acute diarrhoeal diseases linked to

NoV were reported in 2021 from the Alappuzha municipality and nearby panchayats. The outbreak lasted for a month and a half.

14.8 Host Cell and Virus Interaction

The first step in the virus–host interaction is its attachment to the host cell. This process is facilitated by the destruction of host cell receptors by receptor-destructive enzymes like neuraminidase and sialyl-*O*-acetyl-esterase. This helps the viral particles enter the host cells (Wen and Wan 2019). Most of the NoVs interact with genetically determined glycoconjugate molecules like histo-blood group antigens (HBGA) or heparin sulphate (Wen and Wan 2019; Ruvoën-Clouet et al. 2013; Hakomori 1999; Tan and Jiang 2005). These glycoconjugates, found in mucous secretions and epithelial cells, act as initial receptors for viral attachment (Wen and Wan 2019). Some have reported its interaction with sialic acid receptors conjugated with glycoprotein and gangliosides of the host cells (Matrosovich et al. 2015). The predominant cellular targets of murine NoV are macrophages and dendritic cells, which seek the involvement of neuraminidase and ganglioside for the initiation of infection (Wen and Wan 2019). It is assumed that HuNoV uses the polymorphic nature of these receptors to bypass the immune checkpoint of the host.

14.9 Advancements in Diagnosis of NoV Infection

For a proper treatment of virus-mediated gastroenteritis, an accurate identification of the causative pathogen is needed. There are several diagnostic tools available which include electron microscopy, immunochromatographic assay (ICA), enzyme immunoassays (EIA), reverse transcription polymerase chain reaction (RT-PCR), etc. (Thornhill et al. 1977; Kirby and Iturriza-Gómara 2012). Electron microscopy (EM) is a low-sensitivity method to detect NoV through structural visualisation. EM can only detect the virus if the sample has greater than 100 viral particles present in 1 g of stool sample. Even though the viral detection by transmission electron microscopy is highly specific (Rabenau et al. 2003), this method is not preferred as it is time taking, labour-intensive and is not easily available in all diagnostic laboratories (Rani et al. 2021). The ELISA method is an easy-to-perform method of detection and can process large number of samples at a time. But there are several disadvantages to using this method like low specificity. In addition, it could not detect the different strains and generated false-negative/positive results. The diagnosis of virus infection is revolutionised after the introduction of reverse transcription polymerase chain reaction-based detection. RT-PCR detection of NoV is the most sensitive of all the methods described. It shows specificity for all the strains and uses specific primers for the amplification of highly conserved regions of the RdRp or VP1. However, the real-time PCR has taken over the NoV detection owing to its real-time monitoring, high sensitivity and quick results (Rani et al. 2021). Each method has its own advantage and can be preferred over the other based on the requirement. In India, Reverse transcription-Polymerase Chain Reaction (RT-PCR) and genome sequencing are the preferred methods of detection. Recent advancements in NoV detection are seen in many European countries, that use commercial immunochromatography-based tests, which were reported to be highly sensitive than the conventional RT-qPCR test and simultaneously detect three pathogens, i.e. Noro/Rota/Adenoviruses. In 2017, a research group from Korea developed a biosensor-based miniaturised diagnostic tool for label-free detection of human NoV (Hwang et al. 2017). In the year 2021, a South Korean Biotechnology Company, Boditech Med Company, reported a fluorescence-based immunoassay system to diagnose rotavirus and NoV. Though the sensitivity against NoV was lower as compared to rotavirus. However this test assured a rapid diagnostic test against these viruses (Ha et al. 2021).

14.10 Therapeutic Strategies and Challenges

To avoid any long-term morbidity, it is necessary to take immediate steps after observing symptoms of viral infection, like oral or intravenous rehydration solutions to replace the lost electrolytes. But it is necessary to have an anti-viral therapy as a curative and a prophylactic agent (in the form of vaccines) to avoid any other outbreak. We have several therapeutic options for the prevention of diseases caused by a variety of viruses, e.g. Hepatitis B/C virus, HIV and some influenza virus infections. Therefore, there is a possibility that being a viral infection, NoV-infection could be treated with an anti-viral agent that can be vaccines or small molecules.

14.10.1 Vaccines

A lot of research is going on to develop an anti-norovirus agent, but since the live attenuated virus cannot be propagated in in vitro settings and animal models, this task has become difficult. Therefore, it is suggested to develop anti-viral agents or vaccines based on recombinant virus-like particles. The most used recombination-based systems that generate virus-like particles are the baculovirus replicon system (Jiang et al. 1992) and the Venezuelan equine encephalitis viral replicon system (Baric et al. 2002). One of the candidate vaccines developed using the baculovirus replicon system was found to be well tolerated and immunogenic in phase I of the clinical trials (Atmar et al. 2011). Recently, a SPY-tagged virus-like particle-based vaccine was reported to be safe, effective and have more stability at room temperature against NoV infection (Lampinen et al. 2021). Apart from these, several vaccine candidates have been tried in clinical trials for their anti-viral efficacy and safety (El-Kamary et al. 2010; Tacket et al. 2003; Treanor et al. 2014).

14.10.2 Small Molecules Inhibitors

Despite these hurdles, researchers across the world are trying hard to develop an agent that can prevent NoV infection. There are several reports of favipiravir to possess an anti-viral activity against the NoV (Furuta et al. 2009, 2013; Arias et al. 2014). In 2018, Delang et al. reported favipiravir, an anti-viral drug, to inhibit murine NoV replication in the murine macrophage cell line RAW264.7 (Delang et al. 2018). However, the results in mice were not encouraging as the viral shedding did not reduce in the stool after favipiravir treatment. In 2006, nitazoxanide, an antimicrobial agent was tried for treatment of viral gastroenteritis. Though, treatment with nitazoxaide cured NoV-induced gastroenteritis to some degree, the mechanism of its action is not defined (Rossignol and El-Gohary 2006; Siddiq et al. 2011). Antiviral agents under development mainly focus on the two proteins: protease and RNA polymerase (Kaufman et al. 2014). Ribavirin, is a nucleoside analogue which inhibits the viral RNA polymerase and inhibits viral replication (Chang and George 2007; Julian et al. 2016). 2CM-C and 2′-F-2′-C-MeC (valopicitabine/NM283) are bioavailable valine ester prodrugs that inhibited NoV replication as studied by the biochemical assays and reduced the viral load in MNV1-infected murine animal model (Jin et al. 2015; Rocha-Pereira et al. 2013). In 2018, another nucleoside analogue CMX521 entered phase 1 of clinical trials, the results of which are yet to be released (Lanier et al. 2018).

In addition to nucleoside inhibitors, non-nucleoside polymerase inhibitors are also reported to play an anti-viral role. In silico studies revealed that suramin inhibits NoV polymerase (Mastrangelo et al. 2012).

It has been reported that like Gardia lamblia and rotavirus infection-induced chronic diarrhoea, lactose maldigestion is involved in NoV-induced gastroenteritis (Bonani et al. 2017), indicating that gastroenteritis can be cured by preventing lactose maldigestion.

14.10.3 Challenges in Therapeutic Development

14.10.3.1 Shape Shifting

NoV is known for a very high rate of mutational events which cause changes in its structural and non-structural proteins. This 'shape shifting' has emerged as one of the biggest challenges in vaccine development against NoV as it results in host immune escape and hence failure in anti-viral activity (Donaldson et al. 2010; Gaythorpe et al. 2018). The recombination events allow the virus to attain a diverse antigenicity which differs across the strains and genogroups. The design of the vaccine depends on the structure of the NoV strain, which is everchanging and dynamic. Therefore, it is necessary to keep in mind the antigenic diversity and complexity of the NoV structure while developing a vaccine. Researchers are also looking for a single vaccine for multiple strains.

14.10.3.2 Differential Response to Anti-Viral Therapy

Certain factors influence the response to anti-viral therapy, e.g. susceptibility of the host to NoV infection, previous infection history, histo-blood group antigens (HBGA) (Donaldson et al. 2010). Differential immune responses can be a consequence of differential binding to HBGA of the host.

14.10.3.3 Others

There have been mixed reports regarding the outcome of the NoV infection, where people with high faecal or serum antibody titre were found to be more prone to reinfection. People with fewer antibodies were found to be less susceptible to catching the infection again (Johnson et al. 1990; Baron et al. 1984). It was also observed that the rate of mutation in viral genetic material is high in populations that possess stronger immunity, therefore increasing the chances of reinfection (Domingo 1989).

14.11 Prevention

If there is no specific treatment for NoV infection, and if not managed properly, it will lead to death. Therefore, it is necessary to prevent NoV infection. NoV has a long life and remains detectable for more than 1200 days in a sample as it is acid and heat-stable and can resist chlorination up to a concentration of 300 ng/mL (Rani et al. 2021; Pogan et al. 2018; Lucero et al. 2021). An infected person can stay infectious or contagious for up to 2 weeks. To prevent catching a transmission-based viral infection, one must thoroughly wash their hands with soap and water after, using the washroom, changing a diaper, and before cooking, handling or consuming food. Eatables like vegetables must be thoroughly washed with clean water. One must avoid using multiple toilets, if possible, just to avoid transmission to other people (Barclay et al. 2014). Owing to the highly contagious property of NoV, people who handle food and water in public should practise very strict hygiene, e.g. food handlers at restaurants, parties, hospitals, care facilities, day cares, schools and banquet halls. If the aerosolised virus is not contained, then it may result in an outbreak. Points of concern arise when transmission becomes easy in the current scenario, where every workplace or even home prefers a closed system for living. To implement the controlling measures, there is a need for a clear understanding of transmission-related processes. Therefore, there is a need to review, revise and adopt a natural lifestyle to combat transmission-based illnesses and maintain a healthy hygienic lifestyle.

Since NoV is one of the main culprits of waterborne gastroenteritis, several countries, like Sweden, Ukraine, Norway and Korea, are taking necessary measures to filter viral contaminants from water. In 2016, Kim et al. developed a carbon nanotube-silver composite nano-filter that effectively filters commonly found viruses from water sources like rivers and lakes (Kim et al. 2016). Despite these developments, sustained monitoring of water quality is required. There is a lot of research going on the development of effective virucidal agents that may help in its prevention simply by cleaning methods. In 2020, Sauerbrei reported ethanol and povidone-iodine to be effective as virucidal agents against enveloped viruses and adeno, noro and rotaviruses. Therefore, these agents can be used for disinfection of surfaces that have a probability of being infected by aerosols from infected individuals (Sauerbrei 2020).

14.12 Summary

It has been observed that cases of virus-induced chronic diarrhoea are predominant throughout the world, irrespective of the geographical region or season. Mostly due to rotavirus, but even NoV is not far behind in causing gastroenteritis as witnessed. The symptoms related to NoV can become lethal if immediate proper care is not taken. Vaccination against rotavirus is available, and few are under development for NoV. Still, there is no specific treatment available for this highly contagious virus. Several techniques are being developed to diagnose the infection at the right time and start the required treatment. In addition to these scientific interventions, NoV can be best prevented by practising thorough hygiene and avoiding close contact with infected individuals while maintaining a healthy and immunity-enhancing diet.

References

Adler JL, Zickl R (1969) Winter vomiting disease. J Infect Dis 119(6):668–673. https://doi.org/10.1093/infdis/119.6.668

Ahmed SM, Hall AJ, Robinson AE, Verhoef L, Premkumar P, Parashar UD, Koopmans M, Lopman BA (2014) Global prevalence of norovirus in cases of gastroenteritis: a systematic review and meta-analysis. Lancet Infect Dis 14(8):725–730. https://doi.org/10.1016/S1473-3099(14)70767-4

Arias C, Sala MR, Domínguez A, Torner N, Ruíz L, Martínez A, Bartolomé R, de Simón M, Buesa J (2010) Epidemiological and clinical features of norovirus gastroenteritis in outbreaks: a population-based study. Clin Microbiol Infect 16(1):39–44. https://doi.org/10.1111/j.1469-0691.2009.02831.x

Arias A, Thorne L, Goodfellow I (2014) Favipiravir elicits antiviral mutagenesis during virus replication in vivo. eLife 3:e03679. https://doi.org/10.7554/eLife.03679

Asanaka M, Atmar RL, Ruvolo V, Crawford SE, Neill FH, Estes MK (2005) Replication and packaging of Norwalk virus RNA in cultured mammalian cells. Proc Natl Acad Sci U S A 102(29):10327–10332. https://doi.org/10.1073/pnas.0408529102

Atmar RL, Bernstein DI, Harro CD, Al-Ibrahim MS, Chen WH, Ferreira J, Estes MK, Graham DY, Opekun AR, Richardson C, Mendelman PM (2011) Norovirus vaccine against experimental human Norwalk virus illness. N Engl J Med 365(23):2178–2187. https://doi.org/10.1056/NEJMoa1101245

Bank-Wolf BR, König M, Thiel HJ (2010) Zoonotic aspects of infections with noroviruses and sapoviruses. Vet Microbiol 140(3–4):204–212. https://doi.org/10.1016/j.vetmic.2009.08.021. Epub 2009 Aug 26

Barclay L, Park GW, Vega E, Hall A, Parashar U, Vinjé J, Lopman B (2014) Infection control for norovirus. Clin Microbiol Infect 20(8):731–740. https://doi.org/10.1111/1469-0691.12674

Baric RS, Yount B, Lindesmith L, Harrington PR, Greene SR, Tseng FC, Davis N, Johnston RE, Klapper DG, Moe CL (2002) Expression and self-assembly of Norwalk virus capsid protein from venezuelan equine encephalitis virus replicons. J Virol 76(6):3023–3030. https://doi.org/10.1128/jvi.76.6.3023-3030.2002

Baron RC, Greenberg HB, Cukor G, Blacklow NR (1984) Serological responses among teenagers after natural exposure to Norwalk virus. J Infect Dis 150(4):531–534. https://doi.org/10.1093/infdis/150.4.531

Belliot G, Sosnovtsev SV, Mitra T, Hammer C, Garfield M, Green KY (2003) In vitro proteolytic processing of the MD145 norovirus ORF1 nonstructural polyprotein yields stable precursors and products similar to those detected in calicivirus-infected cells. J Virol 77(20):10957–10974. https://doi.org/10.1128/jvi.77.20.10957-10974.2003

Belliot G, Sosnovtsev SV, Chang KO, McPhie P, Green KY (2008) Nucleotidylylation of the VPg protein of a human norovirus by its proteinase–polymerase precursor protein. Virology 374(1):33–49. https://doi.org/10.1016/j.virol.2007.12.028

Bertolotti-Ciarlet A, Crawford SE, Hutson AM, Estes MK (2003) The 3′ end of Norwalk virus mRNA contains determinants that regulate the expression and stability of the viral capsid protein VP1: a novel function for the VP2 protein. J Virol 77(21):11603–11615. https://doi.org/10.1128/jvi.77.21.11603-11615.2003

Bitler EJ, Matthews JE, Dickey BW, Eisenberg JN, Leon JS (2013) Norovirus outbreaks: a systematic review of commonly implicated transmission routes and vehicles. Epidemiol Infect 141(8):1563–1571. https://doi.org/10.1017/S095026881300006X

Bonani M, Pereira RM, Misselwitz B, Fehr T, Wüthrich RP, Franzen D (2017) Chronic Norovirus infection as a risk factor for secondary lactose maldigestion in renal transplant recipients: a prospective parallel cohort pilot study. Transplantation 101(6):1455–1460. https://doi.org/10.1097/TP.0000000000001376

Cardemil CV, Parashar UD, Hall AJ (2017) Norovirus infection in older adults: epidemiology, risk factors, and opportunities for prevention and control. Infect Dis Clin N Am 31(4):839–870. https://doi.org/10.1016/j.idc.2017.07.012. Epub 2017 Sep 12. PMID: 28911830; PMCID: PMC6546097

Centers for Disease Control and Prevention (CDC) (2002) Outbreak of acute gastroenteritis associated with Norwalk-like viruses among British military personnel—Afghanistan, May 2002. MMWR Morb Mortal Wkly Rep 51(22):477–479

Chang KO, George DW (2007) Interferons and ribavirin effectively inhibit Norwalk virus replication in replicon-bearing cells. J Virol 81(22):12111–12118. https://doi.org/10.1128/JVI.00560-07

de Graaf M, van Beek J, Koopmans MP (2016) Human norovirus transmission and evolution in a changing world. Nat Rev Microbiol 14(7):421–433. https://doi.org/10.1038/nrmicro.2016.48. Epub 2016 May 23

Delang L, Abdelnabi R, Neyts J (2018) Favipiravir as a potential countermeasure against neglected and emerging RNA viruses. Antiviral Res 153:85–94. https://doi.org/10.1016/j.antiviral.2018.03.003. Epub 2018 Mar 7

Dolin R, Blacklow NR, DuPont H, Formal S, Buscho RF, Kasel JA, Chames RP, Hornick R, Chanock RM (1971) Transmission of acute infectious nonbacterial gastroenteritis to volunteers by oral administration of stool filtrates. J Infect Dis 123(3):307–312. https://doi.org/10.1093/infdis/123.3.307

Domingo E (1989) RNA virus evolution and the control of viral disease. Progress in drug research. Fortschritte der Arzneimittelforschung. Progres des recherchespharmaceutiques 33:93–133. https://doi.org/10.1007/978-3-0348-9146-2_5

Donaldson EF, Lindesmith LC, Lobue AD, Baric RS (2010) Viral shape-shifting: norovirus evasion of the human immune system. Nat Rev Microbiol 8(3):231–241. https://doi.org/10.1038/nrmicro2296

Edelstein M, Wallensten A, Zetterqvist I, Hulth A (2014) Detecting the norovirus season in Sweden using search engine data—meeting the needs of hospital infection control teams. PLoS One 9(6):e100309. https://doi.org/10.1371/journal.pone.0100309

El-Kamary SS, Pasetti MF, Mendelman PM, Frey SE, Bernstein DI, Treanor JJ, Ferreira J, Chen WH, Sublett R, Richardson C, Bargatze RF, Sztein MB, Tacket CO (2010) Adjuvanted intranasal Norwalk virus-like particle vaccine elicits antibodies and antibody-secreting cells that express homing receptors for mucosal and peripheral lymphoid tissues. J Infect Dis 202(11):1649–1658. https://doi.org/10.1086/657087

Esona MD, Gautam R, Chhabra P, Vinjé J, Bowen MD, Burke RM (2021) Gastrointestinal tract infections: viruses. https://doi.org/10.1016/B978-0-12-818731-9.00217-2

Espenhain L, Berg TC, Bentele H, Nygård K, Kacelnik O (2019) Epidemiology and impact of norovirus outbreaks in Norwegian healthcare institutions, 2005–2018. J Hosp Infect 103(3):335–340. https://doi.org/10.1016/j.jhin.2019.06.010

Fernandez-Vega V, Sosnovtsev SV, Belliot G, King AD, Mitra T, Gorbalenya A, Green KY (2004) Norwalk virus N-terminal non-structural protein is associated with disassembly of the Golgi complex in transfected cells. J Virol 78(9):4827–4837. https://doi.org/10.1128/jvi.78.9.4827-4837.2004

Furuta Y, Takahashi K, Shiraki K, Sakamoto K, Smee DF, Barnard DL, Gowen BB, Julander JG, Morrey JD (2009) T-705 (favipiravir) and related compounds: novel broad-spectrum inhibitors of RNA viral infections. Antiviral Res 82(3):95–102. https://doi.org/10.1016/j.antiviral.2009.02.198

Furuta Y, Gowen BB, Takahashi K, Shiraki K, Smee DF, Barnard DL (2013) Favipiravir (T-705), a novel viral RNA polymerase inhibitor. Antiviral Res 100(2):446–454. https://doi.org/10.1016/j.antiviral.2013.09.015

GalmésTruyols A, Duran JG, Riutort AN, Cerdá GA, Isabel CB, Arbona MP, Berga JV (2011) Brote de norovirus en Mallorca asociado al consumo de ostras [Norovirus outbreak in Majorca (Spain) associated with oyster consumption]. Gaceta sanitaria 25(2):173–175. https://doi.org/10.1016/j.gaceta.2010.11.002

Gaythorpe K, Trotter CL, Lopman B, Steele M, Conlan A (2018) Norovirus transmission dynamics: a modelling review. Epidemiol Infect 146(2):147–158. https://doi.org/10.1017/S0950268817002692

Glass PJ, White LJ, Ball JM, Leparc-Goffart I, Hardy ME, Estes MK (2000) Norwalk virus open reading frame 3 encodes a minor structural protein. J Virol 74(14):6581–6591. https://doi.org/10.1128/jvi.74.14.6581-6591.2000

Gras J, Abdel-Nabey M, Dupont A, Le Goff J, Molina JM, Peraldi MN (2021) Clinical characteristics, risk factors and outcome of severe Norovirus infection in kidney transplant patients: a case-control study. BMC Infect Dis 21(1):351. https://doi.org/10.1186/s12879-021-06062-2

Graves NS (2013) Acute gastroenteritis. Prim Care 40(3):727–741. https://doi.org/10.1016/j.pop.2013.05.006. Epub 2013 Jul 19. PMID: 23958366; PMCID: PMC7119329

Green KY, Ando T, Balayan MS, Berke T, Clarke IN, Estes MK, Matson DO, Nakata S, Neill JD, Studdert MJ, Thiel HJ (2000) Taxonomy of the caliciviruses. J Infect Dis 181(Suppl 2):S322–S330. https://doi.org/10.1086/315591

Gustavsson L, Andersson LM, Brink M, Lindh M, Westin J (2012) Venous lactate levels can be used to identify patients with poor outcome following community-onset norovirus enteritis. Scand J Infect Dis 44(10):782–787. https://doi.org/10.3109/00365548.2012.686671

Ha C, Yoo IY, Yun SA, Chung YN, Huh HJ, Lee NY (2021) Performance evaluation of Automated Fluorescent Immunoassay System ROTA and NORO for detection of rotavirus and norovirus: a comparative study of assay performance with RIDASCREEN® Rotavirus and Norovirus. J Clin Lab Anal 35(1):e23585. https://doi.org/10.1002/jcla.23585

Hakomori S (1999) Antigen structure and genetic basis of histo-blood groups A, B and O: their changes associated with human cancer. Biochim Biophys Acta 1473(1):247–266. https://doi.org/10.1016/s0304-4165(99)00183-x

Hardstaff JL, Clough HE, Lutje V, McIntyre KM, Harris JP, Garner P, O'Brien SJ (2018) Foodborne and food-handler Norovirus outbreaks: a systematic review. Foodborne Pathogens Dis 15(10):589–597. https://doi.org/10.1089/fpd.2018.2452

Hardy ME (2005) Norovirus protein structure and function. FEMS Microbiol Lett 253(1):1–8. https://doi.org/10.1016/j.femsle.2005.08.031. Epub 2005 Sep 6

Hardy ME, Tanaka TN, Kitamoto N, White LJ, Ball JM, Jiang X, Estes MK (1996) Antigenic mapping of the recombinant Norwalk virus capsid protein using monoclonal antibodies. Virology 217(1):252–261. https://doi.org/10.1006/viro.1996.0112

Hu MH, Lin KL, Wu CT, Chen SY, Huang GS (2017) Clinical characteristics and risk factors for seizures associated with Norovirus gastroenteritis in childhood. J Child Neurol 32(9):810–814. https://doi.org/10.1177/0883073817707302

Hwang HJ, Ryu MY, Park CY, Ahn J, Park HG, Choi C, Ha SD, Park TJ, Park JP (2017) High sensitive and selective electrochemical biosensor: label-free detection of human norovirus using affinity peptide as molecular binder. Biosens Bioelectron 87:164–170. https://doi.org/10.1016/j.bios.2016.08.031. Epub 2016 Aug 12

Ijaz MK, Zargar B, Wright KE, Rubino JR, Sattar SA (2016) Generic aspects of the airborne spread of human pathogens indoors and emerging air decontamination technologies. Am J Infect Control 44(9 Suppl):S109–S120. https://doi.org/10.1016/j.ajic.2016.06.008. PMID: 27590695; PMCID: PMC7115269

Jiang X, Wang M, Graham DY, Estes MK (1992) Expression, self-assembly, and antigenicity of the Norwalk virus capsid protein. J Virol 66(11):6527–6532. https://doi.org/10.1128/JVI.66.11.6527-6532.1992

Jin Z, Tucker K, Lin X, Kao CC, Shaw K, Tan H, Symons J, Behera I, Rajwanshi VK, Dyatkina N, Wang G, Beigelman L, Deval J (2015) Biochemical evaluation of the inhibition properties of favipiravir and 2′-*C*-methyl-cytidine triphosphates against human and mouse norovirus RNA polymerases. Antimicrob Agents Chemother 59(12):7504–7516. https://doi.org/10.1128/AAC.01391-15

Johnson PC, Mathewson JJ, DuPont HL, Greenberg HB (1990) Multiple-challenge study of host susceptibility to Norwalk gastroenteritis in US adults. J Infect Dis 161(1):18–21. https://doi.org/10.1093/infdis/161.1.18

Julian TR, Baugher JD, Rippinger CM, Pinekenstein R, Kolawole AO, Mehoke TS, Wobus CE, Feldman AB, Pineda FJ, Schwab KJ (2016) Murine norovirus (MNV-1) exposure in vitro to the purine nucleoside analog Ribavirin increases quasispecies diversity. Virus Res 211:165–173. https://doi.org/10.1016/j.virusres.2015.10.016

Kanerva M, Maunula L, Lappalainen M, Mannonen L, von Bonsdorff CH, Anttila VJ (2009) Prolonged norovirus outbreak in a Finnish tertiary care hospital caused by GII.4-2006b subvariants. J Hosp Infect 71(3):206–213. https://doi.org/10.1016/j.jhin.2008.11.016

Kapikian AZ, Wyatt RG, Dolin R, Thornhill TS, Kalica AR, Chanock RM (1972) Visualization by immune electron microscopy of a 27-nm particle associated with acute infectious nonbacterial gastroenteritis. J Virol 10(5):1075–1081. https://doi.org/10.1128/JVI.10.5.1075-1081.1972

Katayama K, Shirato-Horikoshi H, Kojima S, Kageyama T, Oka T, Hoshino F, Fukushi S, Shinohara M, Uchida K, Suzuki Y, Gojobori T, Takeda N (2002) Phylogenetic analysis of the complete genome of 18 Norwalk-like viruses. Virology 299(2):225–239. https://doi.org/10.1006/viro.2002.1568

Kaufman SS, Green KY, Korba BE (2014) Treatment of norovirus infections: moving antivirals from the bench to the bedside. Antiviral Res 105:80–91. https://doi.org/10.1016/j.antiviral.2014.02.012

Kauppinen A, Pitkänen T, Miettinen IT (2018) Persistent Norovirus contamination of groundwater supplies in two waterborne outbreaks. Food Environ Virol 10(1):39–50. https://doi.org/10.1007/s12560-017-9320-6. Epub 2017 Oct 11

Kawada J, Arai N, Nishimura N, Suzuki M, Ohta R, Ozaki T, Ito Y (2012) Clinical characteristics of norovirus gastroenteritis among hospitalized children in Japan. Microbiol Immunol 56(11):756–759. https://doi.org/10.1111/j.1348-0421.2012.00498.x

Khan RR, Lawson AD, Minnich LL, Martin K, Nasir A, Emmett MK, Welch CA, Udall JN Jr (2009) Gastrointestinal norovirus infection associated with exacerbation of inflammatory bowel disease. J Pediatr Gastroenterol Nutr 48(3):328–333. https://doi.org/10.1097/mpg.0b013e31818255cc

Kim JP, Kim JH, Kim J, Lee SN, Park HO (2016) A nanofilter composed of carbon nanotube-silver composites for virus removal and antibacterial activity improvement. J Environ Sci (China) 42:275–283. https://doi.org/10.1016/j.jes.2014.11.017. Epub 2015 Dec 1

Kirby A, Iturriza-Gómara M (2012) Norovirus diagnostics: options, applications and interpretations. Expert Rev Anti Infect Ther 10(4):423–433. https://doi.org/10.1586/eri.12.21. PMID: 22512752

Kuyumcu-Martinez M, Belliot G, Sosnovtsev SV, Chang KO, Green KY, Lloyd RE (2004) Calicivirus 3C-like proteinase inhibits cellular translation by cleavage of poly(A)-binding protein. J Virol 78(15):8172–8182. https://doi.org/10.1128/JVI.78.15.8172-8182.2004

Lampinen V, Heinimäki S, Laitinen OH, Pesu M, Hankaniemi MM, Blazevic V, Hytönen VP (2021) Modular vaccine platform based on the norovirus-like particle. J Nanobiotechnol 19(1):25. https://doi.org/10.1186/s12951-021-00772-0. PMID: 33468139; PMCID: PMC7815183

Lanier RSD, Kolawole A, Hosmillo M, Nayak K, Bae A, Gurley S, Mullin M (2018) CMX521: a nucleoside with pan-genotypic activity against norovirus. In: Proceedings of the 31st international conference on antiviral research, Porto, Portugal, pp 11–15

Lee BE, Pang XL, Robinson JL, Bigam D, Monroe SS, Preiksaitis JK (2008) Chronic norovirus and adenovirus infection in a solid organ transplant recipient. Pediatr Infect Dis J 27(4):360–362. https://doi.org/10.1097/INF.0b013e31815f5b5a

Lochridge VP, Jutila KL, Graff JW, Hardy ME (2005) Epitopes in the P2 domain of norovirus VP1 recognized by monoclonal antibodies that block cell interactions. J Gen Virol 86(Pt 10):2799–2806. https://doi.org/10.1099/vir.0.81134-0

Lucero Y, Matson DO, Ashkenazi S, George S, O'Ryan M (2021) Norovirus: facts and reflections from past, present, and future. Viruses 13(12):2399. https://doi.org/10.3390/v13122399

Mann D, Reynolds K, Smith D, Muntner P (2008) Trends in statin use and low-density lipoprotein cholesterol levels among US adults: impact of the 2001 National Cholesterol Education Program guidelines. Ann Pharmacother 42(9):1208–1215. https://doi.org/10.1345/aph.1L181

Mastrangelo E, Pezzullo M, Tarantino D, Petazzi R, Germani F, Kramer D, Robel I, Rohayem J, Bolognesi M, Milani M (2012) Structure-based inhibition of Norovirus RNA-dependent RNA polymerases. J Mol Biol 419(3–4):198–210. https://doi.org/10.1016/j.jmb.2012.03.008

Mathew S, Alansari K, Smatti MK, Zaraket H, Al Thani AA, Yassine HM (2019) Epidemiological, molecular, and clinical features of norovirus infections among pediatric patients in qatar. Viruses 11(5):400. https://doi.org/10.3390/v11050400. PMID: 31035642; PMCID: PMC6563317

Matrosovich M, Herrler G, Klenk HD (2015) Sialic acid receptors of viruses. Top Curr Chem 367:1–28. https://doi.org/10.1007/128_2013_466

Matthews JE, Dickey BW, Miller RD, Felzer JR, Dawson BP, Lee AS, Rocks JJ, Kiel J, Montes JS, Moe CL, Eisenberg JN, Leon JS (2012) The epidemiology of published norovirus outbreaks: a review of risk factors associated with attack rate and genogroup. Epidemiol Infect 140(7):1161–1172. https://doi.org/10.1017/S0950268812000234

McAtee CL, Webman R, Gilman RH, Mejia C, Bern C, Apaza S, Espetia S, Pajuelo M, Saito M, Challappa R, Soria R, Ribera JP, Lozano D, Torrico F (2016) Burden of Norovirus and Rotavirus in children after Rotavirus vaccine introduction, Cochabamba, Bolivia. Am J Trop Med Hyg 94(1):212–217. https://doi.org/10.4269/ajtmh.15-0203

Murata T, Katsushima N, Mizuta K, Muraki Y, Hongo S, Matsuzaki Y (2007) Prolonged norovirus shedding in infants. Pediatr Infect Dis J 26(1):46–49. https://doi.org/10.1097/01.inf.0000247102.04997.e0

Pogan R, Dülfer J, Uetrecht C (2018) Norovirus assembly and stability. Current opinion in virology 31:59–65. https://doi.org/10.1016/j.coviro.2018.05.003

Rabenau HF, Stürmer M, Buxbaum S, Walczok A, Preiser W, Doerr HW (2003) Laboratory diagnosis of norovirus: which method is the best? Intervirology 46(4):232–238. https://doi.org/10.1159/000072433

Ramani S, Atmar RL, Estes MK (2014) Epidemiology of human noroviruses and updates on vaccine development. Curr Opin Gastroenterol 30(1):25–33. https://doi.org/10.1097/MOG.0000000000000022

Rani M, Rajyalakshmi S, Pakalapaty S, Kammilli N (2021) Norovirus structure and classification. In: Norovirus. IntechOpen. https://doi.org/10.5772/intechopen.98216

Robilotti E, Deresinski S, Pinsky BA (2015) Norovirus. Clin Microbiol Rev 28(1):134–164. https://doi.org/10.1128/CMR.00075-14

Rocha-Pereira J, Jochmans D, Debing Y, Verbeken E, Nascimento MS, Neyts J (2013) The viral polymerase inhibitor 2′-*C*-methylcytidine inhibits Norwalk virus replication and protects against norovirus-induced diarrhea and mortality in a mouse model. J Virol 87(21):11798–11805. https://doi.org/10.1128/JVI.02064-13

Rossignol JF, El-Gohary YM (2006) Nitazoxanide in the treatment of viral gastroenteritis: a randomized double-blind placebo-controlled clinical trial. Aliment Pharmacol Ther 24(10):1423–1430. https://doi.org/10.1111/j.1365-2036.2006.03128.x

Ruvoën-Clouet N, Belliot G, Le Pendu J (2013) Noroviruses and histo-blood groups: the impact of common host genetic polymorphisms on virus transmission and evolution. Rev Med Virol 23(6):355–366. https://doi.org/10.1002/rmv.1757

Sauerbrei A (2020) Bactericidal and virucidal activity of ethanol and povidone–iodine. MicrobiologyOpen 9(9):e1097. https://doi.org/10.1002/mbo3.1097

Sharp TM, Crawford SE, Ajami NJ, Neill FH, Atmar RL, Katayama K, Utama B, Estes MK (2012) Secretory pathway antagonism by calicivirus homologues of Norwalk virus nonstructural protein p22 is restricted to noroviruses. Virol J 9:181. https://doi.org/10.1186/1743-422X-9-181

Shima T, Okumura A, Kurahashi H, Numoto S, Abe S, Ikeno M, Shimizu T, Norovirus-Associated Encephalitis/Encephalopathy Collaborative Study Investigators (2019) A nationwide survey of norovirus-associated encephalitis/encephalopathy in Japan. Brain Dev 41(3):263–270. https://doi.org/10.1016/j.braindev.2018.11.001

Siddiq DM, Koo HL, Adachi JA, Viola GM (2011) Norovirus gastroenteritis successfully treated with nitazoxanide. J Infect 63(5):394–397. https://doi.org/10.1016/j.jinf.2011.08.002

Stuart RL, Tan K, Mahar JE, Kirkwood CD, Andrew Ramsden C, Andrianopoulos N, Jolley D, Bawden K, Doherty R, Kotsanas D, Bradford J, Buttery JP (2010) An outbreak of necrotizing enterocolitis associated with norovirus genotype GII.3. Pediatr Infect Dis J 29(7):644–647. https://doi.org/10.1097/inf.0b013e3181d824e1

Tacket CO, Sztein MB, Losonsky GA, Wasserman SS, Estes MK (2003) Humoral, mucosal, and cellular immune responses to oral Norwalk virus-like particles in volunteers. Clin Immunol (Orlando, Fla.) 108(3):241–247. https://doi.org/10.1016/s1521-6616(03)00120-7

Tan M, Jiang X (2005) Norovirus and its histo-blood group antigen receptors: an answer to a historical puzzle. Trends Microbiol 13(6):285–293. https://doi.org/10.1016/j.tim.2005.04.004

Tan M, Huang P, Meller J, Zhong W, Farkas T, Jiang X (2003) Mutations within the P2 domain of norovirus capsid affect binding to human histo-blood group antigens: evidence for a binding pocket. J Virol 77(23):12562–12571. https://doi.org/10.1128/jvi.77.23.12562-12571.2003

Teunis PF, Sukhrie FH, Vennema H, Bogerman J, Beersma MF, Koopmans MP (2015) Shedding of norovirus in symptomatic and asymptomatic infections. Epidemiol Infect 143(8):1710–1717. https://doi.org/10.1017/S095026881400274X

Thorne LG, Goodfellow IG (2014) Norovirus gene expression and replication. J Gen Virol 95(Pt 2):278–291. https://doi.org/10.1099/vir.0.059634-0

Thornhill TS, Wyatt RG, Kalica AR, Dolin R, Chanock RM, Kapikian AZ (1977) Detection by immune electron microscopy of 26- to 27-nm virus like particles associated with two family outbreaks of gastroenteritis. J Infect Dis 135(1):20–27. https://doi.org/10.1093/infdis/135.1.20

Treanor JJ, Atmar RL, Frey SE, Gormley R, Chen WH, Ferreira J, Goodwin R, Borkowski A, Clemens R, Mendelman PM (2014) A novel intramuscular bivalent norovirus virus-like particle vaccine candidate—reactogenicity, safety, and immunogenicity in a phase 1 trial in healthy adults. J Infect Dis 210(11):1763–1771. https://doi.org/10.1093/infdis/jiu337

Turcios-Ruiz RM, Axelrod P, St John K, Bullitt E, Donahue J, Robinson N, Friss HE (2008) Outbreak of necrotizing enterocolitis caused by norovirus in a neonatal intensive care unit. J Pediatr 153(3):339–344. https://doi.org/10.1016/j.jpeds.2008.04.015

Ueda H, Tajiri H, Kimura S, Etani Y, Hosoi G, Maruyama T, Noma H, Kusumoto Y, Takano T, Baba Y, Nagai T (2015) Clinical characteristics of seizures associated with viral gastroenteritis in children. Epilepsy Res 109:146–154

Wen F, Wan XF (2019) Influenza neuraminidase: underrated role in receptor binding. Trends Microbiol 27(6):477–479. https://doi.org/10.1016/j.tim.2019.03.001. Epub 2019 Mar 29. PMID: 30930001; PMCID: PMC6527465

Sapovirus: A Prevalent Culprit for Childhood Gastroenteritis

15

Manisha Yadav, Subhash Sonkar, and Bidhan Chandra Koner

Abstract

The world is witnessing a surge in the viral infections leading to a challenging situation for the healthcare fraternity. To combat these challenges, we need to be prepared for the re-emerging infections that might be lethal to human life. For this we need to identify and manage the factors that are responsible for this remerging persistence of these viruses. In this chapter, we will discuss Sapovirus, a member of viral family *caliciviridae*, responsible for several cases of childhood gastroenteritis. *Caliciviridae* family consists of more than 10 extensively studied viral genera which includes lagovirus, vesivirus, sapovirus, norovirus, nebovirus, etc. Viruses of this family are associated with childhood diarrhoea and gastroenteritis (inflammation of stomach and intestine), especially sapovirus (SaV) and norovirus (NoV). SaV infects humans, pigs, bats, dogs, and rats. In 1976, SaV particles were initially detected in human stool samples in the United Kingdom, marking the first identification of this pathogen as a cause of gastroenteritis. Most common symptoms of infection include vomiting and diarrhoea. They are considered highly contagious and are transmitted through the faecal-oral route. Therefore, practicing clean habits and staying cautious in avoiding the infected surroundings hold the key in preventing SaV infection. These viruses have led to many outbreaks of SaV-induced diarrhoea. The diagnosis of viral infection is made easier due to the development of advanced diagnostic techniques like ELISA, RT-PCR, genome sequencers, etc., to provide better management of virus-induced symptoms. In this chapter, we will review the nature, emergence, complications of SaV infection and advancements in its prevention, and cure.

M. Yadav (✉) · S. Sonkar · B. C. Koner (✉)
Multidisciplinary Research Unit, Maulana Azad Medical College, and Associated Hospitals, New Delhi, India
e-mail: myadav85@gmail.com; bckoner@hotmail.com

P. L. Bhukya, A. S (eds.), *Emerging Human Viral Diseases, Volume II*,
https://doi.org/10.1007/978-981-97-4480-0_15

Keywords

Sapovirus · Gastroenteritis · Diarrhoea · Hygiene · Diagnosis · Mutation

15.1 Introduction

For a long time, pathogen-induced gastroenteritis has been a prevalent disease. Due to the upgradation in water quality, sanitation, and hygiene practice, there has been a substantial effect on reducing the cases/rate of gastroenteritis [GBD 2016 Diarrhoeal Disease Collaborators (2018, 2020)]. Despite the adoption of appropriate measures, diarrhoea continues to be the foremost reason for child death throughout the world. *Caliciviridae* family viruses like SaV and NoV are the major viruses that are reported to cause gastroenteritis, after a significant reduction in rotavirus-induced gastroenteritis due to the development of rotavirus vaccine. SaV is a highly distinct group of viruses reported to cause sporadic episodes of gastroenteritis and may even lead to outbreaks if not controlled. It is observed that the frequency of SaV-induced gastroenteritis is low as compared to Norovirus. They are considered contagious and spread easily through exposure to infected individual through air but rarely require hospitalization. Crowded places like schools (Hassan-Rios et al. 2013), childcare centres (Lyman et al. 2009), cruise ships, prisons, military institutions (Neo et al. 2017), and long-term care facilities like psychiatric hospitals (Lee et al. 2012) serve as a reservoir of SaV as the viral shedding rate is higher in such settings. It was first detected in Sapporo, Japan, in 1976 (Chiba et al. 1979). Initially, it was reported to infect mainly infants (children below 5 years), but later it was found to infect people of all age groups, including adults, with a high rate of infection in the elderly population living in care facilities (Johansson et al. 2005). Common visible symptoms include vomiting and diarrhoea for a continuous 2–3 days. If not treated appropriately, SaV-induced gastroenteritis may lead to hospitalization and life-threatening dehydration. In some cases, the long-term feeling of illness may persist even when the visible symptoms fade away. At present, there is no medication available for SaV infection. Therefore, it is necessary to treat SaV infection symptoms and maintain body fluids by ensuring the intake of an ample quantity of liquid diet and maintaining hygiene wherever possible. During a viral infection, close contact with surrounding people must be avoided. In the past, countries like Japan, China, the United States, and Finland have witnessed sporadic cases and outbreaks of SaV (Kobayashi et al. 2012; Yan et al. 2020; Tang et al. 2022; Iritani et al. 2016; Page et al. 2016; Kauppinen et al. 2019; Pitkanen et al. 2022) perhaps due to a lack of hygiene and proper medication. However, it was observed that during the pandemic of Sars CoV 2, the prevalence of gastroenteritis-causing viruses decreased as strict precautionary measures were taken throughout the world to contain SarsCoV2 spread (Maldonado-Barrueco et al. 2022) suggesting that in the absence of SaV-specific therapy while following hygiene, we can prevent SaV infection. Figure 15.1 depicts the various families of *Caliciviridae* family viruses, strains, and their places of origin (Fig. 15.1).

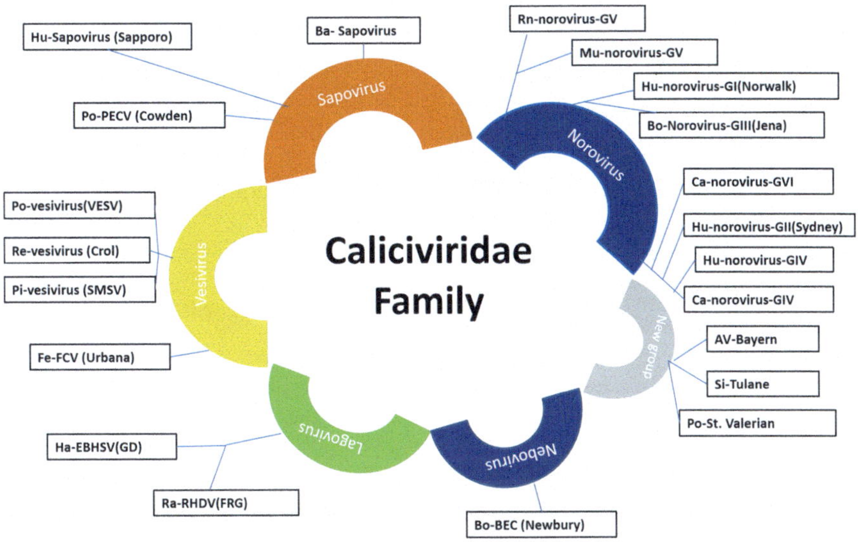

Fig. 15.1 A diagrammatic representation of *Caliciviridae* family viruses, strains, and their places of origin

15.2 Structure and Genome of Sapovirus

They are non-enveloped, small viruses with a diameter ranging from 30 to 38 nm. They have typical *caliciviridae* icosahedral shape. But they are distinguished from other viruses by a specific "Star of David" surface morphology (Madeley 1979; Roingeard 2008). Their buoyancy density varies between 1.36 and 1.41 g/cm^3 (Parwani et al. 1990; Terashima et al. 1983). The SaV genome is a single-stranded RNA, which is 7.1–7.5 kb long and has a polyA tail attached at the 3′ end (Oka et al. 2015) (Fig. 15.2).

15.2.1 Genomic Organization

It consists of either 2 or 3 open reading frames (Oka et al. 2015; Usuku and Kumazaki 2014; Numata et al. 1997) (Fig. 15.3).

1. ORF1-encodes for non-structural proteins and a major capsid protein VP1.
2. ORF2-encodes for the minor structural protein VP2.
3. ORF3-encodes a protein with an unknown function in many SaV strains.

On the basis of variability in the VP1 region, SaV can be categorized into 19 genogroups and 18 genotypes. The genotypes GI (1–7), GII (1–8), GIV (1), and GV (1–2) are known to cause gastroenteritis in humans (Lee et al. 2012; Tang et al.

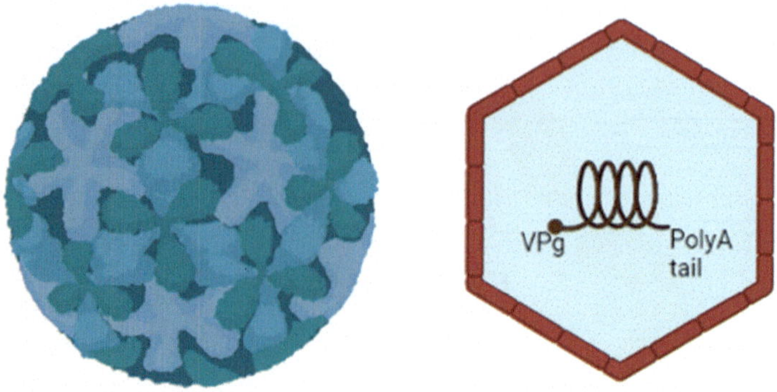

Fig. 15.2 An adapted representative image of the *Caliciviridae* family of virus

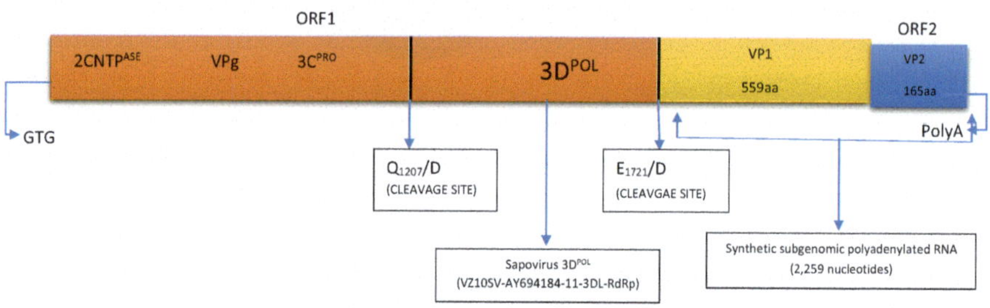

Fig. 15.3 A diagrammatic illustration of the SaV genomic structure: ORF1/2/3, non-structural proteins (NS1–7), and structural proteins VP1 and VP2. According to previous reports (Johansson et al. 2005; Kauppinen et al. 2019; Kim et al. 2014; Kjeldsberg 1977), the presumed cleavage sites in the ORF polyprotein and their projected sizes are specified

2022; Becker-Dreps et al. 2020; Noel et al. 1997) while GIII, GV, GXI, GXIII are predominant gastroenteritis-causing genogroups in animals (Yan et al. 2020).

ORF1 encodes for non-structural proteins which are processed differentially to produce 6 types of non-structural proteins, NS1, NS2, NS3, NS4, NS5, and NS6-NS7. NS6 protein is proteolytic in nature, while NS7 protein is a polymerase. This helps in proteolytic cleavage of ORF1 at 6 different sites of the post-translationally modified precursor protein. The NS3 protein forms the NTPase, and NS5 forms the VPg motifs (located at the 5′ end, accountable for replication, translation, and processing of the viral genome) (Oka et al. 2005a, 2007; Knipe et al. 2013; Leen et al. 2013).

The capsid protein VP1 consist of two main domains and a small flexible hinge region that connects them (Miyazaki et al. 2015). The shell (S) domain protein creates a cover around the nucleic acid while the protruding (P) domain protrudes from the surface and forms a host interaction interface. The P domain is demarcated into two sub-domains, namely P1 and P2. Basically, the P2 domain is inserted into the P1 domain. And as a result, there exist different combinations of P1 and P2 domains, giving rise to P1–1, P2, and P1–2, which switch between two genogroups, G1.1 and

GI.5. In 2016, a research group studied that the S/P1–1 region is the immunoreactive epitope (Johansson et al. 2005). The antigenicity of the viral capsid varies with these combinational products. An N-terminal variable region (NVR), an N-terminal region (NTR), and a central variable region (CVR) are also found in VP1 (Okada et al. 2006). The highly conserved sequence of VP1 is reported to be located in between the N and CVR regions. SaV and NoV cannot be reproduced in cell culture, but VP1 can be expressed inside insect cells which can assemble to give a virus-like particle that resembles the native virion morphologically and antigenically.

VP2 is the minor structural protein, the sequence variation in this region provides antigenic diversity to SaV (Oka et al. 2009; Hansman et al. 2005; Jiang et al. 1997). The RNA-dependent RNA polymerase is instrumental in viral genomic and subgenomic replication. All the five genogroups, GI-GV, are reported to infect humans except GIII, which infects mainly pigs (Hansman et al. 2005).

15.3 Epidemiology

Despite the fact that SaV infection is less common than NoV, SaV infection has been prevalent for decades throughout the world. The prevalence rate varies from 2 to 12% (Magwalivha et al. 2018; Becker-Dreps et al. 2020). It is more prominent in low- and middle-income countries in almost all age groups. This infection has caused sporadic cases as well as several outbreaks in the US, UK, Europe (Finland, Italy, Germany, Spain, Sweden, Netherlands, Slovenia), New Zealand, Peru, South Africa, China, Japan, Russia, Canada, Dutch, Brazil, Burkina Faso, and Asia (India, Pakistan, Nepal, Bangladesh) (Kumthip et al. 2020; Xue et al. 2019; Mann et al. 2019; Svraka et al. 2010; de Oliveira-Tozetto et al. 2021). The most commonly affected population is infants and children up to 5 years of age (Rouhani et al. 2022; Diez Valcarce et al. 2021). However, the spread of infection causes diarrhoea in adults too, where immunity is compromised, and in the elderly (Zhuo et al. 2021). SaV is known to co-exist with other infective agents like rotavirus, astrovirus, adenovirus, and a bacteria Shigella age (Rouhani et al. 2022).

15.4 Clinical Features

Common clinical manifestations of SaV cannot be differentiated from other viruses. The incubation phase for SaV varies from 1 to 4 days (Pitkanen et al. 2022; Yamashita et al. 2010; Grohmann et al. 1991). This duration of infection or symptoms may vary depending upon the host immunity, which differs in people taking immunosuppressants or undergoing organ transplants. After the incubation period, it causes episodes of vomiting, diarrhoea, abdominal cramps and pain, nausea, chills, and malaise. The symptoms generally settle down in a week's time, but cases with longer SaV-associated illness are also reported (Cubitt et al. 1981; Nakata et al. 1985; Johansson et al. 2005; Sakai et al. 2001). Apart from diarrhoea and vomiting,

SaV can also cause cognitive decline in children due to pathogenic infection-induced intestinal dysbiosis which interferes with the GBA (Hennessey 2017). There is a case of a woman who was presented with septic shock after SaV infection (Landa et al. 2022), which was treated by pressor technique, after which she was discharged from the hospital. Although the frequency of vomiting in SaV is less as compared to NoV infection (Sakai et al. 2001; Sala et al. 2014; Rockx et al. 2002) and rotavirus. Because SaV infection has resulted in death, it cannot be considered less severe than NoV or Rotavirus (Page et al. 2016). It is well known that pathogen exposure develops immunity in the host body. In the case of SaV, this immunity fades with time but renders the host less susceptible to subsequent exposure. There are reports of asymptomatic SaV infection in which the viral shedding is low as compared to symptomatic SaV infection.

15.5 Disease Pathogenesis

Acute SaV infection causes histopathological changes in the intestine, which can be reversed after normalcy. In addition to this infiltration of polymorphonuclear leukocytes, villous blunting is also reported. The main intestinal part that is affected is the jejunum, while there are no changes in the mucosa.

15.6 Mode of Transmission

The SaV particles can stay viable for years. Just 10 viral particles are enough to infect an individual, which makes it a highly contagious infective agent. The main mode of transmission can be.

1. Oral-faecal route: It generally spreads through eating or drinking food from unwashed hands that are unknowingly infected with SaV by an infected individual. Food is the main source of virus transmission (Lee et al. 2012; Kobayashi et al. 2012; Hergens et al. 2017; Oka et al. 2017). Therefore, people who handle food must take special care when preparing, packing, distributing, or selling food. According to a few studies, the virus shedding rate was nearly one million copies of genome in 1 g of stool (Varela et al. 2019; Bergallo et al. 2017; Silva et al. 2018), implying that even a very small amount is sufficient to transmit SaV via faeces. In a Finnish study, it was found that SaV can be best detected in faeces as children with asymptomatic or symptomatic infection were detected with SaV in stool samples (Pitkanen et al. 2022). Therefore, faeces from infected individuals can act as a transmission mode.
2. Water-based transmission: Water is another important mode of transmission of SaV (Kauppinen et al. 2019). Shellfish (oysters and clams) is among the seafood which is reported to be the carriers of SaV (Oka et al. 2017).

3. Air-based transmission: It can travel through air and can stay on surfaces exposed to infected individuals through vomitus and faeces. The vomitus-contaminated aerosol remains in the air, which is breathed in, leading to the spread of viral infection and ultimately to an outbreak. Unlike norovirus, the spill-over of SaV from animals to humans is not reported (Becker-Dreps et al. 2020).

15.7 Risk Factors

Age is one of the risk factors that determine susceptibility to SaV infection. Infants, children below 5 years, the elderly, and people with weak immunity are highly susceptible to SaV infection. Infants or toddlers have higher chances of SaV infection due to waning maternal immunity and increasing unintended exposure to contaminated food. Infants under 6 months of age who are entirely fed with breastmilk have less chance of contracting gastroenteritis due to the antimicrobial and gut microbiome preserving properties of breastmilk (McCormick et al. 2022; Manthey et al. 2014). Patients with organ transplants taking immunosuppressants are at greater risk of infection (Ghusson and Vasquez 2018; Wright et al. 2020). In addition to this, SaV spillover from animals to humans has not yet been reported (Sisay et al. 2016). In contrast to NoV, SaV infection is not specific to the season. It can occur in summers (Yoneda et al. 2017), winters (Heusinkveld et al. 2016), autumn and spring (Varela et al. 2019), rainy season (Colston et al. 2020). In a case study, it was observed that SaV-induced gastroenteritis in children receiving chemotherapy for cancer (Moser et al. 2011). In a study based in Nicaragua, it was found that the previous exposure to SaV reduced cases of severe gastroenteritis by the subsequent exposure (Vielot et al. 2022).

15.8 Host-Cell Entry and Virus Interaction

Due to a lack of information on human SaV propagation methods, there is not much information available about the entry and replication of this virus, but it is thought to follow the standard entry and replication method. They employ sialic acid receptors present on the host cells (Kim et al. 2014) to enter the host by endocytosis. They then uncoat their genetic material and release it inside the host cell for replication. Post the VPg elimination, the viral genome is translated first to ORF1, which helps in the process of viral genome replication. The ssRNA then forms a double-stranded intermediate which is then transcribed to give many copies of ssRNA. The sub-genomic transcription of ssRNA gives rise to VP2, followed by viral particle assembly and packaging to propagate in host cells. In contrast to NoV, SaV is not associated with histo-blood group antigens (Colston et al. 2020; Sala et al. 2014).

15.9 Immune Responses

The immune response to SaV is not very well understood. There are reports of a correlation between the level of serum antibodies, viral titre, and the severity of the SaV infection. The site of infection is primarily the intestine; thus, the immune status of the intestine determines the fate of the immune response. An old study found that maternal antibodies also contribute to immunity against SaV in infants (Cubitt et al. 1980; Cubitt and McSwiggan 1987). The pre-existing antibodies play a protective role in advancing the infection to the lethal stage and asymptomatic infection. A durable acquired immunity on the first encounter with the virus is generated after SaV infection, which further boosts the immunity that protects from infection by minimizing the symptoms and also reducing the frequency of infection (Chiba et al. 1979; Harada et al. 2012). However, their immunity compromises in old age, as a result of which they become susceptible to infection again in their lifetime. In response to the virus attack, the host cells secrete interferon to confer immunity against the virus (Peñaflor-Téllez et al. 2019).

15.10 Methods of Detection

At present, we have a number of advanced technologies that have helped in the diagnosis of pathogens. Electron microscopy, ELISA, PCR, and genome sequencers are employed to detect pathogens that cause illness. SaV was first detected by electron microscopy in 1976 (Hallstrom et al. 2017). Since then, electron microscopy has been used when other methods fail to detect SaV (Hallstrom et al. 2017). Viral antigen detection can be done by enzyme linked immunoassays (ELISA) (Nakata et al. 1988). But as they do not specifically detect the SaV strain, it is not preferred over other methods (Hansman et al. 2006a; Jiang et al. 1997). Nucleic acid detection methods such as reverse transcription PCR, particularly real-time PCR, are highly sensitive and generally used methods for detecting strain-specific genetic material (Becker-Dreps et al. 2020; Hansman et al. 2006b; Oka et al. 2006a). The primers used for detection are designed to amplify mostly the RdRp region (Noel et al. 1997; Farkas et al. 2004; Berke et al. 1997; Honma et al. 2000; Jiang et al. 1999), the junction of RdRp-VP1 (Oka et al. 2006b; Jiang et al. 2014; Okada et al. 2002) or some portion of VP1 (Okada et al. 2002; Yan et al. 2003). In a few cases, full genome sequencing was also employed, but it is not the usual method of detection.

15.11 Therapeutic Development

Till date, there is no SaV-targeted therapy available as SaV and NoV are uncultivable in cell culture. In such a case, only symptomatic treatment is given to the patient. There have been several efforts to gain insight into the structure of these viruses in green monkey kidney cells (Kjeldsberg 1977) or primary human embryonic kidney cells (Cubitt and Barrett 1984). In 2020, Takagi et al. reported being able to successfully propagate human SaV in cell lines isolated from testis and duodenum in the presence of bile acids (Takagi et al. 2020). This model can be used for studying the anti-SaV activity of proposed anti-viral agents. Porcine SaV strains are reported to reproduce in primary porcine kidney cells or a porcine kidney cell line with or without supplements from porcine intestine, e.g., bile acids (Chang et al. 2004) which aid in the evasion of endosome attack by the host during the viral entry (Shivanna et al. 2014). In addition to this role of bile acid, it is also reported to cause a reduction in innate immunity via the cAMP pathway (Chang et al. 2002, 2004).

Researchers from Merck group have successfully developed a vaccine platform "SEQUIVITY" against SaV for use in pigs [SEQUIVITY® | Merck Animal Health USA (merck-animal-health-usa.com)]. In a case study, in patients with renal transplant, the anti-parasitic drug nitazoxanide was found to be effective in treating SaV-induced gastroenteritis (Ghusson and Vasquez 2018). A few more options are available after the development of two nucleoside analogues, 2CMC and 7DMA, which are reported to work as anti-viral agents by interfering with genomic replication by inhibiting RdRp activity (Santos-Ferreira et al. 2021).

15.11.1 Challenges in Therapeutic Development

The main challenges with SaV and NoV are that they cannot be propagated or cultivated in culture conditions, which makes it difficult for researchers to understand their replication mechanisms and hence plan to develop vaccines against them (Becker-Dreps et al. 2020). To get some insight into its mechanism of replication, virus-like particles were developed using mammalian and insect cells (Oka et al. 2005b; Hansman et al. 2006a).

15.12 Emerging Potentiality

The first case of SaV infection was detected in 1976. Since then, several outbreaks have emerged at regular intervals in different geological location of the world. SaV was first detected in India in 2000 in Vellore, Tamil Nadu. Virus infection is expected to re-emerge due to several factors like mutation in the viral genome, lack of immunity of the host, change in the environmental factor. As we know that most of the viral infections are zoonotic, the viral infections are expected to re-emerge due to increasing urbanization/deforestation (animals moving into urban areas to survive) bringing the viruses close to humans. There is seldom any information on the effect of environmental factors responsible for inducing changes in the host immune response against the SaV infection. Whereas, the most common factor for re-emergence of SaV is genomic mutation in the virus. Viruses are known to evolve genetically to evade and eventually survive the host defence mechanism as a result of which there is an increase in the genetic diversity of the SaV. Likewise, SaV undergoes evolutionary changes inside the host cell in the form of genetic mutations, making it a more potential disease-causing agent. However, limited data is available on the genetic diversity of SaV in India (Mourya et al. 2019). Mutations are mostly observed in the VP1 region (Su et al. 2023; Pietsch and Liebert 2019; Magwalivha et al. 2022). However, mutations in the capsid region are also reported that imparts the capsid plasticity which is also considered as a method of immune evasion by modulating the capsid properties without hindering the viral capabilities (Iwakiri et al. 2009; Hansman et al. 2008; Zweigart et al. 2021). Because of continuous mutations, genetic diversity, and consequently evolving host defences, it is difficult to develop vaccines.

15.13 Preventive Measures

As no treatment is available to cure SaV infection, it is logical to simply avoid infection. There are a few basic measures that need to be taken to avoid getting the SaV infection, like frequent washing of hands with sanitizing soap, prior to intake of any food item. When an individual is infected, it is best to avoid contact with infected people by staying home from school or work. Special care must be taken while handling food if an episode of gastroenteritis has happened nearby or is likely to happen in the near future. It is advised to wash food items carefully, and, if possible, wash them thoroughly before use items disinfected with chlorine. Drinking water should be boiled before use if quality is not assured. It is good to eat home-cooked food and avoid food from hawkers. As this infection transmits through the air, it is good to keep the indoor air clean by managing the vomitus and faeces of an infected individual. No preventive vaccine is available. The porcine SaV strain remains stable at a pH of 3–8 at room temperature for up to an hour. It is sensitive to ethanol and hypochlorite solution for 30 min. It can also be deactivated by heating at 56° for 2 h (Wang et al. 2012). Staying hydrated is one of the key methods to reduce the chances of severe gastroenteritis.

15.14 Summary

Gastroenteritis is the second leading cause of death in children throughout the world. Management becomes challenging when no targeted therapies are developed. Despite extensive characterization of viral structure, the only treatment used is to alleviate viral infection symptoms such as diarrhoea and vomiting. Several vaccines are available for rotavirus, but none are available for SaV or NoV. This is because these viruses are tough to cultivate in culture conditions. If a SaV vaccine is developed, then several cases of serious virus-induced gastroenteritis can be prevented. Besides using therapeutic strategies, one can easily avoid exposure to dangerous pathogens by developing hygienic habits and avoiding infected individuals.

References

Becker-Dreps S, Gonzalez F, Bucardo F (2020) Sapovirus: an emerging cause of childhood diarrhea. Curr Opin Infect Dis 33(5):388–397. https://doi.org/10.1097/QCO.0000000000000671. https://www.ncbi.nlm.nih.gov/pubmed/32796163

Bergallo M, Galliano I, Montanari P, Brusin MR, Finotti S, Paderi G, Gabiano C (2017) Development of a quantitative real-time PCR assay for Sapovirus in children under 5-years-old in Regina Margherita Hospital of Turin, Italy. Can J Microbiol 63(4):296–302. https://doi.org/10.1139/cjm-2016-0482. https://www.ncbi.nlm.nih.gov/pubmed/28177797

Berke T, Golding B, Jiang X, Cubitt DW, Wolfaardt M, Smith AW, Matson DO (1997) Phylogenetic analysis of the Caliciviruses. J Med Virol 52(4):419–424. https://doi.org/10.1002/(sici)1096-9071(199708)52:4<419::aid-jmv13>3.0.co;2-b. https://www.ncbi.nlm.nih.gov/pubmed/9260691

Chang KO, Kim Y, Green KY, Saif LJ (2002) Cell-culture propagation of porcine enteric Calicivirus mediated by intestinal contents is dependent on the cyclic amp signaling pathway. Virology 304(2):302–310. https://doi.org/10.1006/viro.2002.1665. https://www.ncbi.nlm.nih.gov/pubmed/12504571

Chang KO, Sosnovtsev SV, Belliot G, Kim Y, Saif LJ, Green KY (2004) Bile acids are essential for porcine enteric Calicivirus replication in association with down-regulation of signal transducer and activator of transcription 1. Proc Natl Acad Sci U S A 101(23):8733–8738. https://doi.org/10.1073/pnas.0401126101. https://www.ncbi.nlm.nih.gov/pubmed/15161971

Chiba S, Sakuma Y, Kogasaka R, Akihara M, Horino K, Nakao T, Fukui S (1979) An outbreak of gastroenteritis associated with Calicivirus in an infant home. J Med Virol 4(4):249–254. https://doi.org/10.1002/jmv.1890040402. https://www.ncbi.nlm.nih.gov/pubmed/232140

Colston J, Paredes Olortegui M, Zaitchik B, Penataro Yori P, Kang G, Ahmed T, Bessong P et al (2020) Pathogen-specific impacts of the 2011–2012 La Nina-associated floods on enteric infections in the Mal-Ed Peru cohort: a comparative interrupted time series analysis. Int J Environ Res Public Health 17(2):17020487. https://doi.org/10.3390/ijerph17020487. https://www.ncbi.nlm.nih.gov/pubmed/31940920

Cubitt WD, Barrett AD (1984) Propagation of human candidate Calicivirus in cell culture. J Gen Virol 65(Pt 6):1123–1126. https://doi.org/10.1099/0022-1317-65-6-1123. https://www.ncbi.nlm.nih.gov/pubmed/6726190

Cubitt WD, McSwiggan DA, Arstall S (1980) An outbreak of calicivirus infection in a mother and baby unit. Journal of Clinical Pathology 33(11):1095–1098. https://doi.org/10.1136/jcp.33.11.1095

Cubitt WD, McSwiggan DA (1987) Seroepidemiological survey of the prevalence of antibodies to a strain of human calicivirus. Journal of Medical Virology 21(4):361–368. https://doi.org/10.1002/jmv.1890210408

Cubitt WD, Pead PJ, Saeed AA (1981) A new serotype of Calicivirus associated with an outbreak of gastroenteritis in a residential home for the elderly. J Clin Pathol 34(8):924–926. https://doi.org/10.1136/jcp.34.8.924. https://www.ncbi.nlm.nih.gov/pubmed/6268667

de Oliveira-Tozetto S, Santiso-Bellón C, Ferrer-Chirivella JM, Navarro-Lleó N, Vila-Vicent S, Rodríguez-Díaz J, Buesa J (2021) Epidemiological and genetic characterization of sapovirus in patients with acute gastroenteritis in valencia (Spain). Viruses 13(2):184. https://doi.org/10.3390/v13020184

Diez Valcarce M, Kambhampati AK, Calderwood LE, Hall AJ, Mirza SA, Vinjé J (2021) Global distribution of sporadic sapovirus infections: A systematic review and meta-analysis. PloS one 16(8):e0255436. https://doi.org/10.1371/journal.pone.0255436

Farkas T, Zhong WM, Jing Y, Huang PW, Espinosa SM, Martinez N, Morrow AL et al (2004) Genetic diversity among Sapoviruses. Arch Virol 149(7):1309–1323. https://doi.org/10.1007/s00705-004-0296-9. https://www.ncbi.nlm.nih.gov/pubmed/15221533

GBD 2016 Diarrhoeal Disease Collaborators (2018) Estimates of the global, regional, and national morbidity, mortality, and aetiologies of diarrhoea in 195 countries: a systematic analysis for the Global Burden of Disease Study 2016. Lancet Infect Dis 18(11):1211–1228. https://doi.org/10.1016/S1473-3099(18)30362-1. https://www.ncbi.nlm.nih.gov/pubmed/30243583

GBD 2016 Diarrhoeal Disease Collaborators (2020) Quantifying risks and interventions that have affected the burden of diarrhoea among children younger than 5 years: an analysis of the Global Burden of Disease Study 2017. Lancet Infect Dis 20(1):37–59. https://doi.org/10.1016/S1473-3099(19)30401-3. https://www.ncbi.nlm.nih.gov/pubmed/31678029

Ghusson N, Vasquez G (2018) Successfully treated norovirus- and Sapovirus-associated diarrhea in three renal transplant patients. Case Rep Infect Dis 2018:6846873. https://doi.org/10.1155/2018/6846873. https://www.ncbi.nlm.nih.gov/pubmed/30538873

Grohmann G, Glass RI, Gold J, James M, Edwards P, Borg T, Stine SE, Goldsmith C, Monroe SS (1991) Outbreak of human Calicivirus gastroenteritis in a day-Care Center in Sydney, Australia. J Clin Microbiol 29(3):544–550. https://doi.org/10.1128/jcm.29.3.544-550.1991. https://www.ncbi.nlm.nih.gov/pubmed/1645369

Hallstrom B, Lagerqvist N, Lind-Karlberg M, Helgesson S, Follin P, Hergens MP, Nederby-Ohd J, Tolfvenstam T, Alm E (2017) Complete genome sequence of a Sapporo virus GV.2 variant from a 2016 outbreak of gastroenteritis in Sweden. Genome Announc 5(5):16. https://doi.org/10.1128/genomeA.01446-16. https://www.ncbi.nlm.nih.gov/pubmed/28153884

Hansman GS, Natori K, Oka T, Ogawa S, Tanaka K, Nagata N, Ushijima H, Takeda N, Katayama K (2005) Cross-reactivity among Sapovirus recombinant capsid proteins. Arch Virol 150(1):21–36. https://doi.org/10.1007/s00705-004-0406-8. https://www.ncbi.nlm.nih.gov/pubmed/15449145

Hansman GS, Guntapong R, Pongsuwanna Y, Natori K, Katayama K, Takeda N (2006a) Development of an antigen Elisa to detect Sapovirus in clinical stool specimens. Arch Virol 151(3):551–561. https://doi.org/10.1007/s00705-005-0630-x. https://www.ncbi.nlm.nih.gov/pubmed/16155807

Hansman GS, Takeda N, Katayama K, Tu ET, McIver CJ, Rawlinson WD, White PA (2006b) Genetic diversity of Sapovirus in children, Australia. Emerg Infect Dis 12(1):141–143. https://doi.org/10.3201/eid1201.050846. https://www.ncbi.nlm.nih.gov/pubmed/16494732

Hansman GS, Oka T, Takeda N (2008) Sapovirus-like particles derived from polyprotein. Virus Research 137(2):261–265. https://doi.org/10.1016/j.virusres.2008.07.002

Harada S, Oka T, Tokuoka E, Kiyota N, Nishimura K, Shimada Y, Ueno T, Ikezawa S, Wakita T, Wang Q, Saif LJ, Katayama K (2012) A confirmation of sapovirus re-infection gastroenteritis cases with different genogroups and genetic shifts in the evolving sapovirus genotypes, 2002-2011. Archives of virology 157(10):1999–2003. https://doi.org/10.1007/s00705-012-1387-7

Hassan-Rios E, Torres P, Munoz E, Matos C, Hall AJ, Gregoricus N, Vinje J (2013) Sapovirus gastroenteritis in preschool center, Puerto Rico, 2011. Emerg Infect Dis 19(1):174–175. https://doi.org/10.3201/eid1901.120690. https://www.ncbi.nlm.nih.gov/pubmed/23260219

Hergens MP, Nederby Ohd J, Alm E, Askling HH, Helgesson S, Insulander M, Lagerqvist N et al (2017) Investigation of a food-borne outbreak of gastroenteritis in a school canteen revealed a variant of Sapovirus Genogroup V not detected by standard PCR, Sollentuna, Sweden, 2016. Euro Surveill 22(22):30543. https://doi.org/10.2807/1560-7917.ES.2017.22.22.30543. https://www.ncbi.nlm.nih.gov/pubmed/28602163

Hennessey (2017) Intestinal dysbiosis during neonatal development alters the microbiota-gut-brain axis in adulthood. The FASEB Journal - Wiley Online Library

Heusinkveld M, Mughini-Gras L, Pijnacker R, Vennema H, Scholts R, van Huisstede-Vlaanderen KW, Kortbeek T, Kooistra-Smid M, van Pelt W (2016) Potential causative agents of acute gastroenteritis in households with preschool children: prevalence, risk factors, clinical relevance and household transmission. Eur J Clin Microbiol Infect Dis 35(10):1691–1700. https://doi.org/10.1007/s10096-016-2714-9. https://www.ncbi.nlm.nih.gov/pubmed/27372926

Honma S, Nakata S, Kinoshita-Numata K, Kogawa K, Chiba S (2000) Evaluation of nine sets of PCR primers in the RNA dependent RNA polymerase region for detection and differentiation of members of the family Caliciviridae, Norwalk virus and Sapporo virus. Microbiol Immunol

44(5):411–419. https://doi.org/10.1111/j.1348-0421.2000.tb02515.x. https://www.ncbi.nlm.nih.gov/pubmed/10888362

Iritani N, Yamamoto SP, Abe N, Kubo H, Oka T, Kaida A (2016) Epidemics of GI.2 Sapovirus in gastroenteritis outbreaks during 2012–2013 in Osaka City, Japan. J Med Virol 88(7):1187–1193. https://doi.org/10.1002/jmv.24451. https://www.ncbi.nlm.nih.gov/pubmed/26684081

Iwakiri A, Ganmyo H, Yamamoto S, Otao K, Mikasa M, Kizoe S, Katayama K, Wakita T, Takeda N, Oka T (2009) Quantitative analysis of fecal sapovirus shedding: identification of nucleotide substitutions in the capsid protein during prolonged excretion. Archives of Virology 154(4):689–693. https://doi.org/10.1007/s00705-009-0358-0

Jiang X, Cubitt WD, Berke T, Zhong W, Dai X, Nakata S, Pickering LK, Matson DO (1997) Sapporo-like human Caliciviruses are genetically and antigenically diverse. Arch Virol 142(9):1813–1827. https://doi.org/10.1007/s007050050199. https://www.ncbi.nlm.nih.gov/pubmed/9672639

Jiang X, Huang PW, Zhong WM, Farkas T, Cubitt DW, Matson DO (1999) Design and evaluation of a primer pair that detects both Norwalk- and Sapporo-like Caliciviruses by RT-PCR. J Virol Methods 83(1–2):10. https://doi.org/10.1016/s0166-0934(99)00114-7

Jiang Y, Fang L, Shi X, Zhang H, Li Y, Lin Y, Qiu Y et al (2014) Simultaneous detection of five enteric viruses associated with gastroenteritis by use of a PCR assay: a single real-time multiplex reaction and its clinical application. J Clin Microbiol 52(4):1266–1268. https://doi.org/10.1128/JCM.00245-14. https://www.ncbi.nlm.nih.gov/pubmed/24478418

Johansson PJ, Bergentoft K, Larsson PA, Magnusson G, Widell A, Thorhagen M, Hedlund KO (2005) A nosocomial Sapovirus-associated outbreak of gastroenteritis in adults. Scand J Infect Dis 37(3):200–204. https://doi.org/10.1080/00365540410020974. https://www.ncbi.nlm.nih.gov/pubmed/15849053

Kauppinen A, Pitkanen T, Al-Hello H, Maunula L, Hokajarvi AM, Rimhanen-Finne R, Miettinen IT (2019) Two drinking water outbreaks caused by wastewater intrusion including Sapovirus in Finland. Int J Environ Res Public Health 16(22):224376. https://doi.org/10.3390/ijerph16224376. https://www.ncbi.nlm.nih.gov/pubmed/31717479

Kim DS, Hosmillo M, Alfajaro MM, Kim JY, Park JG, Son KY, Ryu EH et al (2014) Both α2,3- and α2,6-linked sialic acids on O-linked glycoproteins act as functional receptors for porcine Sapovirus. PLoS Pathog 10(6):e1004172. https://doi.org/10.1371/journal.ppat.1004172. https://www.ncbi.nlm.nih.gov/pubmed/24901849

Kjeldsberg E (1977) Small spherical viruses in Faeces from gastroenteritis patients. Acta Pathol Microbiol Scand B 85B(5):351–354. https://doi.org/10.1111/j.1699-0463.1977.tb01988.x. https://www.ncbi.nlm.nih.gov/pubmed/602784

Knipe D, Howley P, Griffin D, Lamb R, Martin M, Roizman B, Straus S (2013) Fields virology, vol 1. Lippincott Williams & Wilkins, Philadelphia. https://www.revisioncentre.co.uk/tutors/biology/gcse

Kobayashi S, Fujiwara N, Yasui Y, Yamashita T, Hiramatsu R, Minagawa H (2012) A foodborne outbreak of Sapovirus linked to catered box lunches in Japan. Arch Virol 157(10):1995–1997. https://doi.org/10.1007/s00705-012-1394-8. https://www.ncbi.nlm.nih.gov/pubmed/22752792

Kumthip K, Khamrin P, Ushijima H, Chen L, Li S, Maneekarn N (2020) Genetic recombination and diversity of sapovirus in pediatric patients with acute gastroenteritis in Thailand, 2010-2018. PeerJ, 8, e8520. https://doi.org/10.7717/peerj.8520

Landa E, Javaid S, Won JS, Vigandt E, Caronia J, Mir P, Thet Z (2022) Septic shock secondary to severe gastroenteritis resulting from Sapovirus infection. Cureus 14(4):e24010. https://doi.org/10.7759/cureus.24010. https://www.ncbi.nlm.nih.gov/pubmed/35547467

Lee LE, Cebelinski EA, Fuller C, Keene WE, Smith K, Vinje J, Besser JM (2012) Sapovirus outbreaks in long-term care facilities, Oregon and Minnesota, USA, 2002–2009. Emerg Infect Dis 18(5):873–876. https://doi.org/10.3201/eid1805.111843. https://www.ncbi.nlm.nih.gov/pubmed/22516204

Leen EN, Kwok KY, Birtley JR, Simpson PJ, Subba-Reddy CV, Chaudhry Y, Sosnovtsev SV et al (2013) Structures of the compact helical core domains of feline Calicivirus and murine norovirus Vpg proteins. J Virol 87(10):5318–5330. https://doi.org/10.1128/JVI.03151-12. https://www.ncbi.nlm.nih.gov/pubmed/23487472

Lyman WH, Walsh JF, Kotch JB, Weber DJ, Gunn E, Vinje J (2009) Prospective study of etiologic agents of acute gastroenteritis outbreaks in child care centers. J Pediatr 154(2):253–257. https://doi.org/10.1016/j.jpeds.2008.07.057. https://www.ncbi.nlm.nih.gov/pubmed/18783794

Madeley CR (1979) Comparison of the features of Astroviruses and Caliciviruses seen in samples of feces by electron microscopy. J Infect Dis 139(5):519–523. https://doi.org/10.1093/infdis/139.5.519. https://www.ncbi.nlm.nih.gov/pubmed/438551

Magwalivha M, Kabue JP, Traore AN, Potgieter N (2018) Prevalence of human sapovirus in low and middle-income countries. Advances in virology 5986549. https://doi.org/10.1155/2018/5986549

Magwalivha M, Ngandu JK, Traore AN, Potgieter N (2022) Partial analysis of the capsid protein (VP1) of human sapovirus isolated from children with diarrhoea in rural communities of south africa. Advances in virology 9928378. https://doi.org/10.1155/2022/9928378

Maldonado-Barrueco A, Garcia-Rodriguez J, Yaniz-Ramirez J, Serrano-Vaquero I, Parra-Alonso JC, Vega-Nieto C, Ruiz-Carrascoso G (2022) Impact of the Sars-Cov-2 pandemic on the prevalence and incidence of gastrointestinal viruses in children up to five years old: a retrospective cohort study. Microbiol Spectr 10(3):e0266921. https://doi.org/10.1128/spectrum.02669-21. https://www.ncbi.nlm.nih.gov/pubmed/35638853

Mann P, Pietsch C, Liebert UG (2019) Genetic diversity of sapoviruses among inpatients in germany, 2008-2018. Viruses 11(8):726. https://doi.org/10.3390/v11080726

Manthey CF, Autran CA, Eckmann L, Bode L (2014) Human milk oligosaccharides protect against enteropathogenic *Escherichia coli* attachment in vitro and Epec colonization in suckling mice. J Pediatr Gastroenterol Nutr 58(2):165–168. https://doi.org/10.1097/MPG.0000000000000172. https://www.ncbi.nlm.nih.gov/pubmed/24048169

McCormick BJJ, Richard SA, Murray-Kolb LE, Kang G, Lima AAM, Mduma E, Kosek MN et al (2022) Full breastfeeding protection against common enteric bacteria and viruses: results from the Mal-Ed Cohort Study. Am J Clin Nutr 115(3):759–769. https://doi.org/10.1093/ajcn/nqab391. https://www.ncbi.nlm.nih.gov/pubmed/34849524

Miyazaki N, Taylor DW, Hansman GS, Murata K (2015) Antigenic and Cryo-electron microscopy structure analysis of a chimeric Sapovirus capsid. J Virol 90(5):2664–2675. https://doi.org/10.1128/JVI.02916-15. https://www.ncbi.nlm.nih.gov/pubmed/26699644

Moser O, Luck S, Dilloo D, Eis-Hubinger AM, Simon A (2011) Sapovirus as a gastrointestinal pathogen in febrile pediatric patients with cancer. J Med Virol 83(12):2233–2236. https://doi.org/10.1002/jmv.22219. https://www.ncbi.nlm.nih.gov/pubmed/22012734

Mourya DT, Yadav PD, Ullas PT, Bhardwaj SD, Sahay RR, Chadha MS, Shete AM et al (2019) Emerging/re-emerging viral diseases and new viruses on the Indian horizon. Indian J Med Res 149(4):447–467. https://doi.org/10.4103/ijmr.IJMR_1239_18. https://www.ncbi.nlm.nih.gov/pubmed/31411169

Nakata S, Chiba S, Terashima H, Yokoyama T, Nakao T (1985) Humoral immunity in infants with gastroenteritis caused by human Calicivirus. J Infect Dis 152(2):274–279. https://doi.org/10.1093/infdis/152.2.274. https://www.ncbi.nlm.nih.gov/pubmed/2993432

Nakata S, Estes MK, Chiba S (1988) Detection of human Calicivirus antigen and antibody by enzyme-linked immunosorbent assays. J Clin Microbiol 26(10):2001–2005. https://doi.org/10.1128/jcm.26.10.2001-2005.1988. https://www.ncbi.nlm.nih.gov/pubmed/3182991

Neo FJX, Loh JJP, Ting P, Yeo WX, Gao CQH, Lee VJM, Tan BH, Ng CG (2017) Outbreak of Caliciviruses in the Singapore military, 2015. BMC Infect Dis 17(1):719. https://doi.org/10.1186/s12879-017-2821-y. https://www.ncbi.nlm.nih.gov/pubmed/29137606

Noel JS, Liu BL, Humphrey CD, Rodriguez EM, Lambden PR, Clarke IN, Dwyer DM et al (1997) Parkville virus: a novel genetic variant of human Calicivirus in the Sapporo virus clade, associated with an outbreak of gastroenteritis in adults. J Med Virol 52(2):173–178. https://www.ncbi.nlm.nih.gov/pubmed/9179765

Numata K, Hardy ME, Nakata S, Chiba S, Estes MK (1997) Molecular characterization of morphologically typical human Calicivirus Sapporo. Arch Virol 142(8):1537–1552. https://doi.org/10.1007/s007050050178. https://www.ncbi.nlm.nih.gov/pubmed/9672617

Oka T, Katayama K, Ogawa S, Hansman GS, Kageyama T, Miyamura T, Takeda N (2005a) Cleavage activity of the Sapovirus 3C-like protease in *Escherichia coli*. Arch Virol 150(12):2539–2548. https://doi.org/10.1007/s00705-005-0591-0. https://www.ncbi.nlm.nih.gov/pubmed/16052286

Oka T, Katayama K, Ogawa S, Hansman GS, Kageyama T, Ushijima H, Miyamura T, Takeda N (2005b) Proteolytic processing of Sapovirus Orf1 polyprotein. J Virol 79(12):7283–7290. https://doi.org/10.1128/JVI.79.12.7283-7290.2005. https://www.ncbi.nlm.nih.gov/pubmed/15919882

Oka T, Katayama K, Hansman GS, Kageyama T, Ogawa S, Wu FT, White PA, Takeda N (2006a) Detection of human Sapovirus by real-time reverse transcription-polymerase chain reaction. J Med Virol 78(10):1347–1353. https://doi.org/10.1002/jmv.20699. https://www.ncbi.nlm.nih.gov/pubmed/16927293

Oka T, Yamamoto M, Katayama K, Hansman GS, Ogawa S, Miyamura T, Takeda N (2006b) Identification of the cleavage sites of Sapovirus open reading frame 1 polyprotein. J Gen Virol 87(Pt 11):3329–3338. https://doi.org/10.1099/vir.0.81799-0. https://www.ncbi.nlm.nih.gov/pubmed/17030867

Oka T, Yamamoto M, Yokoyama M, Ogawa S, Hansman GS, Katayama K, Miyashita K et al (2007) Highly conserved configuration of catalytic amino acid residues among Calicivirus-encoded proteases. J Virol 81(13):6798–6806. https://doi.org/10.1128/JVI.02840-06. https://www.ncbi.nlm.nih.gov/pubmed/17459935

Oka T, Miyashita K, Katayama K, Wakita T, Takeda N (2009) Distinct genotype and antigenicity among genogroup II Sapoviruses. Microbiol Immunol 53(7):417–420. https://doi.org/10.1111/j.1348-0421.2009.00133.x. https://www.ncbi.nlm.nih.gov/pubmed/19563401

Oka T, Wang Q, Katayama K, Saif LJ (2015) Comprehensive review of human Sapoviruses. Clin Microbiol Rev 28(1):32–53. https://doi.org/10.1128/CMR.00011-14. https://www.ncbi.nlm.nih.gov/pubmed/25567221

Oka T, Doan YH, Haga K, Mori K, Ogawa T, Yamazaki A (2017) Genetic characterization of rare genotype GII.5 Sapovirus strain detected from a suspected food-borne gastroenteritis outbreak among adults in Japan in 2010. Jpn J Infect Dis 70(2):223–224. https://doi.org/10.7883/yoken.JJID.2016.468. https://www.ncbi.nlm.nih.gov/pubmed/28003604

Okada M, Shinozaki K, Ogawa T, Kaiho I (2002) Molecular epidemiology and phylogenetic analysis of Sapporo-like viruses. Arch Virol 147(7):1445–1451. https://doi.org/10.1007/s00705-002-0821-7. https://www.ncbi.nlm.nih.gov/pubmed/12111418

Okada M, Yamashita Y, Oseto M, Ogawa T, Kaiho I, Shinozaki K (2006) Genetic variability in the Sapovirus capsid protein. Virus Genes 33(2):157–161. https://doi.org/10.1007/s11262-005-0051-7. https://www.ncbi.nlm.nih.gov/pubmed/16972029

Page N, Groome MJ, Murray T, Nadan S, Netshikweta R, Keddy KH, Poonsamy B et al (2016) Sapovirus prevalence in children less than five years of age hospitalised for diarrhoeal disease in South Africa, 2009–2013. J Clin Virol 78:82–88. https://doi.org/10.1016/j.jcv.2016.03.013. https://www.ncbi.nlm.nih.gov/pubmed/27002709

Parwani AV, Saif LJ, Kang SY (1990) Biochemical characterization of porcine enteric Calicivirus: analysis of structural and nonstructural viral proteins. Arch Virol 112(1–2):41–53. https://doi.org/10.1007/BF01348984. https://www.ncbi.nlm.nih.gov/pubmed/2369321

Peñaflor-Téllez Y, Trujillo-Uscanga A, Escobar-Almazán J.A. Gutiérrez-Escolano AL (2019) Immune response modulation by caliciviruses. Frontiers in Immunology 10:2334. https://doi.org/10.3389/fimmu.2019.02334

Pietsch C, Liebert UG (2019) Intrahost viral evolution during chronic sapovirus infections. Journal of Clinical Virology: the official publication of the Pan American Society for Clinical Virology 113:1–7. https://doi.org/10.1016/j.jcv.2019.02.001

Pitkanen O, Markkula J, Hemming-Harlo M (2022) Sapovirus, norovirus and rotavirus detections in stool samples of hospitalized Finnish children with and without acute gastroenteritis. Pediatr

Infect Dis J 41(5):e203–ee07. https://doi.org/10.1097/INF.0000000000003493. https://www.ncbi.nlm.nih.gov/pubmed/35185141

Rockx B, De Wit M, Vennema H, Vinje J, De Bruin E, Van Duynhoven Y, Koopmans M (2002) Natural history of human Calicivirus infection: a prospective cohort study. Clin Infect Dis 35(3):246–253. https://doi.org/10.1086/341408. https://www.ncbi.nlm.nih.gov/pubmed/12115089

Roingeard P (2008) Viral detection by electron microscopy: past, present and future. Biol Cell 100(8):491–501. https://doi.org/10.1042/BC20070173. https://www.ncbi.nlm.nih.gov/pubmed/18627353

Rouhani S, Peñataro Yori P, Paredes Olortegui M, Lima AA, Ahmed T, Mduma ER, George A, Samie A, Svensen E, Lima I, Mondal D, Mason CJ, Kalam A, Guerrant RL, Lang D, Zaidi A, Kang G, Houpt E, Kosek MN (2022) The epidemiology of sapovirus in the etiology, risk factors, and interactions of enteric infections and malnutrition and the consequences for child health and development Study: Evidence of Protection Following Natural Infection. Clinical infectious diseases: an official publication of the Infectious Diseases Society of America 75(8):1334–1341. https://doi.org/10.1093/cid/ciac165

Sakai Y, Nakata S, Honma S, Tatsumi M, Numata-Kinoshita K, Chiba S (2001) Clinical severity of Norwalk virus and Sapporo virus gastroenteritis in children in Hokkaido, Japan. Pediatr Infect Dis J 20(9):849–853. https://doi.org/10.1097/00006454-200109000-00005. https://www.ncbi.nlm.nih.gov/pubmed/11734762

Sala MR, Broner S, Moreno A, Arias C, Godoy P, Minguell S, Martınez A et al (2014) Cases of acute gastroenteritis due to Calicivirus in outbreaks: clinical differences by age and aetiological agent. Clin Microbiol Infect 20(8):793–798. https://doi.org/10.1111/1469-0691.12522. https://www.ncbi.nlm.nih.gov/pubmed/24382267

Santos-Ferreira N, Van Dycke J, Neyts J, Rocha-Pereira J (2021) Current and future antiviral strategies to tackle gastrointestinal viral infections. Microorganisms 9(8):81599. https://doi.org/10.3390/microorganisms9081599. https://www.ncbi.nlm.nih.gov/pubmed/34442677

Shivanna V, Kim Y, Chang KO (2014) The crucial role of bile acids in the entry of porcine enteric Calicivirus. Virology 456:268–278. https://doi.org/10.1016/j.virol.2014.04.002. https://www.ncbi.nlm.nih.gov/pubmed/24889246

Silva TN, Dabilla N, Fiaccadori FS, de Paula Cardoso DD, de Sousa TT, Almeida TN, Souza M (2018) Sapovirus in rectal and nasopharyngeal swab samples of children with symptoms of acute gastroenteritis. Pediatr Infect Dis J 37(4):e115–ee16. https://doi.org/10.1097/INF.0000000000001833. https://www.ncbi.nlm.nih.gov/pubmed/29140937

Sisay Z, Djikeng A, Berhe N, Belay G, Abegaz WE, Wang QH, Saif LJ (2016) First detection and molecular characterization of Sapoviruses and noroviruses with zoonotic potential in swine in Ethiopia. Arch Virol 161(10):2739–2747. https://doi.org/10.1007/s00705-016-2974-9. https://www.ncbi.nlm.nih.gov/pubmed/27424025

Su L, Mao H, Sun Y, Yan H, Ge Q, Gong L, Zhang Y (2023) The analysis of the genotype of Sapovirus outbreaks in Zhejiang Province. Virology journal 20(1):268.https://doi.org/10.1186/s12985-023-02202-z

Svraka S, Vennema H, van der Veer B, Hedlund KO, Thorhagen M, Siebenga J, Duizer E, Koopmans M (2010) Epidemiology and genotype analysis of emerging sapovirus-associated infections across Europe. Journal of Clinical Microbiology 48(6):2191–2198. https://doi.org/10.1128/JCM.02427-09

Takagi H, Oka T, Shimoike T, Saito H, Kobayashi T, Takahashi T, Tatsumi C et al (2020) Human Sapovirus propagation in human cell lines supplemented with bile acids. Proc Natl Acad Sci USA 117(50):32078–32085. https://doi.org/10.1073/pnas.2007310117. https://www.ncbi.nlm.nih.gov/pubmed/33257564

Tang X, Hu Y, Zhong X, Xu H (2022) Molecular epidemiology of human adenovirus, Astrovirus, and Sapovirus among outpatient children with acute diarrhea in Chongqing, China, 2017–2019. Front Pediatr 10:826600. https://doi.org/10.3389/fped.2022.826600. https://www.ncbi.nlm.nih.gov/pubmed/35311045

Terashima H, Chiba S, Sakuma Y, Kogasaka R, Nakata S, Minami R, Horino K, Nakao T (1983) The polypeptide of a human Calicivirus. Arch Virol 78(1–2):1–7. https://doi.org/10.1007/BF01310853. https://www.ncbi.nlm.nih.gov/pubmed/6651531

Usuku S, Kumazaki M (2014) A gastroenteritis outbreak attributed to Sapovirus Genogroup V in Yokohama, Japan. Jpn J Infect Dis 67(5):411–412. https://doi.org/10.7883/yoken.67.411. https://www.ncbi.nlm.nih.gov/pubmed/25241698

Varela MF, Rivadulla E, Lema A, Romalde JL (2019) Human Sapovirus among outpatients with acute gastroenteritis in Spain: a 1-year study. Viruses 11(2):20144. https://doi.org/10.3390/v11020144. https://www.ncbi.nlm.nih.gov/pubmed/30744057

Vielot NA, Reyes Y, Blette B, Gonzalez F, Toval-Ruiz C, Gutierrez L, Vilchez S et al (2022) First episodes of norovirus and Sapovirus gastroenteritis protect against subsequent episodes in a Nicaraguan birth cohort. Epidemiology 33(5):650–653. https://doi.org/10.1097/EDE.0000000000001500. https://www.ncbi.nlm.nih.gov/pubmed/35700200

Wang Q, Zhang Z, Saif LJ (2012) Stability of and attachment to lettuce by a culturable porcine Sapovirus surrogate for human Caliciviruses. Appl Environ Microbiol 78(11):3932–3940. https://doi.org/10.1128/AEM.06600-11. https://www.ncbi.nlm.nih.gov/pubmed/22447610

Wright S, Kleven D, Kapoor R, Kavuri S, Gani I (2020) Recurring norovirus and Sapovirus infection in a renal transplant patient. IDCases 20:e00776. https://doi.org/10.1016/j.idcr.2020.e00776. https://www.ncbi.nlm.nih.gov/pubmed/32395429

Xue L, Cai W, Zhang L, Gao J, Dong R, Li Y, Wu H, Zhang J, Zeng H, Ye Q, Ding Y, Wu Q (2019) Prevalence and genetic diversity of human sapovirus associated with sporadic acute gastroenteritis in south china from 2013 to 2017. Journal of Medical Virology, 91(10):1759–1764. https://doi.org/10.1002/jmv.25511

Yamashita Y, Ootsuka Y, Kondo R, Oseto M, Doi M, Miyamoto T, Ueda T et al (2010) Molecular characterization of Sapovirus detected in a gastroenteritis outbreak at a wedding Hall. J Med Virol 82(4):720–726. https://doi.org/10.1002/jmv.21646. https://www.ncbi.nlm.nih.gov/pubmed/20166168

Yan H, Yagyu F, Okitsu S, Nishio O, Ushijima H (2003) Detection of norovirus (GI, GII), Sapovirus and Astrovirus in fecal samples using reverse transcription single-round multiplex PCR. J Virol Methods 114(1):37–44. https://doi.org/10.1016/j.jviromet.2003.08.009. https://www.ncbi.nlm.nih.gov/pubmed/14599677

Yan Y, Li Y, Shi W, Kong X, Li H, Zhang Q, Pang L et al (2020) An outbreak of gastroenteritis associated with a novel GII.8 Sapovirus variant-transmitted by vomit in Shenzhen, China, 2019. BMC Infect Dis 20(1):911. https://doi.org/10.1186/s12879-020-05643-x. https://www.ncbi.nlm.nih.gov/pubmed/33261582

Yoneda M, Nakano M, Sugimoto D, Inada M, Fujitani M, Kitahori Y (2017) Epidemiological characteristics of Sapovirus and human Astrovirus detected among children in Nara prefecture, Japan, during the 2009/2010–2014/2015 seasons. Jpn J Infect Dis 70(1):87–91. https://doi.org/10.7883/yoken.JJID.2015.529

Zhuo R, Ding X, Freedman SB, Lee BE, Ali S, Luong J, Xie J, Chui L, Wu Y, Pang X, Alberta Provincial Pediatric Enteric Infection TEam (APPETITE) (2021). Molecular Epidemiology of Human Sapovirus among Children with Acute Gastroenteritis in Western Canada. Journal of Clinical Microbiology 59(10):e0098621. https://doi.org/10.1128/JCM.00986-21

Zweigart MR, Becker-Dreps S, Bucardo F, Gonzalez F, Baric RS, Lindesmith LC (2021) Serological humoral immunity following natural infection of children with high burden gastrointestinal viruses. Viruses 13(10):13102033. https://doi.org/10.3390/v13102033. https://www.ncbi.nlm.nih.gov/pubmed/34696463

The Viral Vortex: Navigating the Rotavirus Disease, Epidemiology, and Immunization

16

Tripti Pande, Sanjana, and Mudasir M. Rather

Abstract

Rotavirus is a key agent in infant gastroenteritis. Rotavirus infections are a major burden to the developing countries of the world. It leads to 28% of diarrhea-related deaths of children under 5 years of age. Of all the genotypes reported globally, the genotype G9P[8] is the most prevalent. The spread of rotavirus infections is governed by sanitary measures, public health interventions, and the health status of a region. Rotavirus is double-stranded RNA virus with 11 segments of genome belonging to family *Reoviridae*. The viral capsid is triple-layered with a typical wheel-like appearance evident in electron microscopy. Viral serotyping is based on neutralizing antibodies against VP7 and VP4 proteins. Rotaviral virulence is affected by both structural and non-structural proteins. Immunocompromised individuals meet with severe outcomes of the infection. Rotaviral infections can be controlled by vaccination. However, the epidemiology of the region concerned has to be taken into account while implementing control and prevention measures.

Keywords

Gastroenteritis · Infant diarrhea · Rotavirus genotypes · Rotavirus prevention · Mammalian species · Emergent strains

T. Pande (✉) · M. M. Rather
Department of Veterinary Microbiology, ICAR-Indian Veterinary Research Institute, Bareilly, Uttar Pradesh, India
e-mail: pande_tripti@yahoo.in

Sanjana
Division of Biological Products, ICAR-Indian Veterinary Research Institute, Bareilly, Uttar Pradesh, India

P. L. Bhukya, A. S (eds.), *Emerging Human Viral Diseases, Volume II*,
https://doi.org/10.1007/978-981-97-4480-0_16

16.1 Introduction

Gastroenteritis and diarrheal episodes were a major concern in infants in both developed and developing countries. Finding the cause behind most diarrheic illnesses was assisted with the introduction of Electron Microscopy (EM). Bishop and co-workers in 1973 reported a unique viral particle from the mucosa of the duodenum of children suffering from severe gastroenteritis (Bishop et al., 1973). Rotavirus (RV) was evidenced as the cause of most infant gastroenteritis with EM data. Even before the reports of human rotavirus, cases of rotavirus-related diarrhea were reported in calves, mice, piglets, lambs, monkeys, and foals (Woode et al. 1976). As the virus was distinct in morphology and antigenic structure from other reoviruses and orbiviruses, the rotavirus (Latin rota-'wheel') term was proposed (Flewett et al. 1974). Before vaccine introduction, this virus was responsible for 500,000 diarrheal deaths annually worldwide. After the introduction of the vaccine, it still leads to 25 million hospitalizations and 45,000 mortalities year after year across the globe (Tate et al. 2012).

Rotavirus is a double-stranded RNA virus of the family Reoviridae. It is a non-enveloped virus with a triple-layered icosahedral capsid aiding in increased stability and viability. Rotavirus triple-layered capsid is formed by a morphogenic mechanism that involves the acquisition of a temporary lipid envelope during the budding of immature particles into the endoplasmic reticulum encloses a genome of 11 linear dsRNA segments (Papa et al. 2021). Rotavirus infects only vertebrates and is transmitted via fecal-oral route. Viral particles can remain viable in feces for 2 years at 4 °C and months in the environment if not disinfected properly (Chandler-Bostock and Mellits 2015). Although a substantial host range constraint exists, rotaviruses are very well adapted to their host and multiply very effectively. Additionally, sterilizing immunity has not yet been produced. These characteristics help to explain the high viral prevalence of the virus.

Before the introduction of vaccine, nearly every child at the age of 5 years was already infected once with rotavirus (Parashar et al. 2003). WHO has approved live attenuated vaccines and recommended them in neonates with a priority in countries of South, South-East Asia, and sub-Saharan Africa. Morbidity is the leading cause of rotavirus disease burden in these developing countries. Vaccination is controlling the disease outbreaks in these countries; however, the endemic infections would require more coverage and improved health facilities.

16.2 Epidemiology

16.2.1 Geographical Distribution

Rotavirus diarrhea causes heavy mortalities in South-East Asia, South Asia, and Sub-Saharan Africa. Cases are mostly reported in underdeveloped and developing nations. This can be associated with safe drinking water and hygienic conditions which when compromised lead to disease transmission. However, in contrast to this

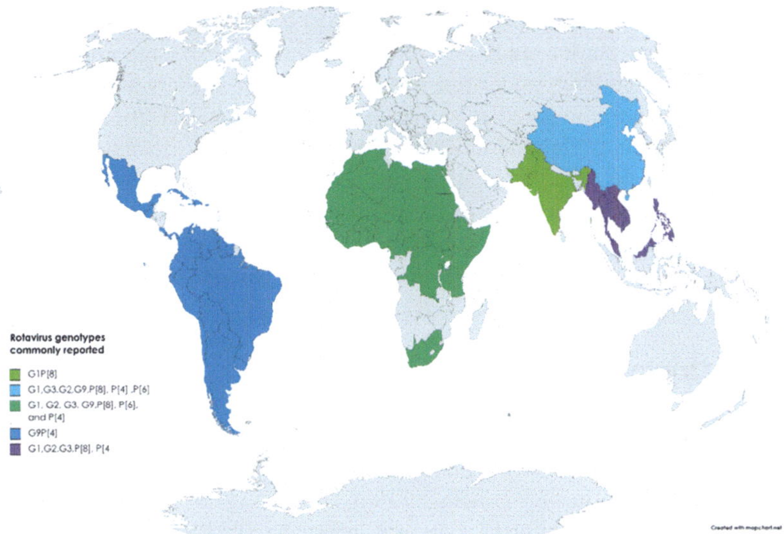

Fig. 16.1 Rotavirus genotypes reported across different countries of the world

notion, 50% of diarrhea-related mortality due to rotavirus infection in countries like Finland and Denmark mark these infections unbounded by the socio-economic status of a particular region (Troeger et al. 2018). In European countries, there has been a 77% decline in case reports after vaccine introduction and subsequent herd effects (de Hoog et al. 2018).

Globally 90% rotavirus-related hospitalizations are caused due to 6 common genotypes G1P[8], G2P[4], G3P[8], G4P[8], G12P[8], and G9P[8], of which G9P[8] is most prevalent (Giri et al. 2020). The rotavirus genotypes reported across different parts of the world are shown in Fig. 16.1. In South-East Asia G1P[8], G3P[8], and G2P[4] were recorded as the most prevalent genotypes in a study conducted from 2009 to 2018 in countries including Indonesia, Lao, Cambodia, Malaysia, Myanmar, Philippines, and Thailand (Lestari et al. 2020). In South, North, West, and East Africa G1, G2, G3, and G9 are reported G-types, and P[8], P[6], and P[4] are reported P-types (Ouermi et al. 2017). In India, G1P[8] is the most common rotavirus genotype (Giri et al. 2020). In China G1, G3, G2, and G9 are the most common G types while P[8], P[4], and P[6] are the most common P types (Li et al. 2014). In Latin America, the prevalence of G9P[4] is expanding apart from most reported G1P[8] and G3P[8] (Quaye et al. 2013).

16.2.2 Age Mortality and Morbidity

Rotavirus is one of the 13 diarrhea causal agents as reported by Global disease burden study (Troeger et al. 2018). Humans of all age groups are susceptible but the virus mainly affects children under 5 years of age. Rotavirus leads to approximately

28% of diarrhea-related deaths of children under 5 years of age (Lanata et al. 2013). A global disease burden study on rotavirus in the year 2016 reported 29.5% diarrheal deaths of children under this age group (Khalil 2017) which has decreased appreciably when compared to nearly 65% in 1990 (GBD 2015 Mortality and Causes of Death Collaborators 2016). In Asian countries, rotavirus comprised 37.5% of over-the-year hospitalizations due to gastroenteritis cases in children under 5 years of age (Kawai et al. 2012). In Sub-Saharan Africa, 21% rotavirus-related hospitalizations were recorded per year (Godfrey et al. 2020). Sub-Saharan Africa has shown an appreciable decline when compared to the pre-vaccination era.

16.2.3 Origin of Infection and Diversity

RVA infects avian and mammalian species. RVB, C, H, and I have been reported in mammals. RVC is associated with pigs, cattle, dogs, ferrets, and humans. RVE has been detected only in pigs (Luchs and do Timenetsky 2016). RVD, E, F, and G infect only birds and animals. Bats are infected with RVJ (Bányai et al. 2017; Matthijnssens et al. 2012). RVA infecting bovine, horses, swine, sheep, goats, and camelids with such a varied compeer of animal reservoirs is considered to undergo interspecies transmission to reach humans (Dóró et al. 2015). The ability of the segmented genome of rotavirus dsRNA virus to undergo reassortment contributes to its enhanced zoonotic ability. G6P[8] detected in an infant from Bangladesh was a bovine-human reassortant (Afrad et al. 2013). Human-porcine reassortants G4P[6] and G5P[6] have been reported in Hungary and Bulgaria, respectively (Mladenova et al. 2012; Papp et al. 2013). Even bat-like human rotavirus of genotype G3P[6] has been recorded in hospitalized diarrheic children from the Dominican Republic (Bourdett-Stanziola et al. 2021). RVC virus which has porcine origin also has the ability of interspecies transmission. Human VP6 was identified in pigs indicating reverse zoonosis (Trovão et al. 2019). Hence, the involvement of multiple host species gives elaborative diversity to rotaviruses infecting humans.

16.2.4 Spread of Disease

RVA may show seasonal occurrence in temperate zones with most reported cases in cold months. This seasonal spread in colder months can be linked to the respiratory system-related spread of the virus. No significant seasonality is recorded in subtropical and tropical zones (Patel et al. 2013). Food-associated rotavirus outbreaks have also been documented in the past. Rotavirus has been isolated from potato stew and found identical to one isolated with infected child stool (Mayr et al. 2009). In Japan, food-related outbreak was reported among adolescents due to secondary contamination by food handlers (Mizukoshi et al. 2014). In USA, rotavirus gastroenteritis outbreaks have been recorded in 1998 and 1999 in elderly patients of nursing homes. Close contact among adults leads to person-to-person transmission

(Griffin et al. 2002). In communities with close inhabiting populations, virus spread is efficient leading to outbreaks. Sanitary measures, public health interventions, and the health status of a region are deciding factors in the spread of disease in a community.

16.2.5 Biosafety

Class 2 biological safety cabinet is recommended for clinical specimen handling and rotavirus-related cell culture work.

16.3 Potential Risk of Emergence and Re-emergence

The rotavirus genome undergoes gene recombination, and genetic reassortment, which accommodates mutations that collectively contribute to the new variant introduction in the existing rotavirus population. In addition to this, vaccine-induced selective pressure also leads to the emergence of new strains of rotavirus. Most commonly reported VP7 genotypes in rotavirus gastroenteritis include G1–G4, G9, and G12 and VP4 genotypes include P[8], [6], [4]. These genotypes in different combinations lead to 90% of rotaviral gastroenteritis cases (Nayak et al. 2019).

In the pre-vaccine era i.e., before 2006, G1P[8] was a common genotype in most countries. Post-vaccine era observed new emerging variants. In the Middle East, the post-vaccination era reported G9P[8] and G2P[4] as common strains (Almalki 2018). In Malaysia, new equine-like G3P[8] rotavirus was found circulating in children, another study from Indonesia found G3P[8] and G3P[6] as common circulating genotypes (Amit et al. 2021). In Australia and Spain also this equine-like G3P[8] strain was found circulating due to selection pressure arising from mass vaccine coverage (Maguire et al. 2019).

Vaccination and herd immunity leads to changing dynamics of existing and declining genotypes. In Brazil, G12P[8] diminished in the year 2013 after its emergence in 2006 due to vaccine introduction. However, it re-emerged in 2014 as the most prevalent genotype questioning heterotypic protection of existing vaccines (Luchs and do Timenetsky 2016). This G12 is now an emergent genotype with increasing prevalence all over the world. Another example of Japanese patients showed re-emergence of new G3 variants in the year 2003–2004 which had significant amino acid changes compared to G3 rotaviruses reported in 1990–1995. This variant continued to be reported in 2007–2009 and 2009–2011 (Thongprachum et al. 2013). This re-emergence has been linked to an increase in herd immunity against G1 that lead to lesser immunity against G3. This cohort of the new emergent and re-emergent strains in different geographical and population settings is important for deciding vaccine strain that needs to be included to control further RV-related gastroenteritis cases.

16.4 Organization of Infectious Agent-Structural

16.4.1 Classification

According to ICTV 2019 classification, the virus is of genus *Rotavirus*, of family *Sedoreoviridae* of order *Reovirales*. Order Reovirales is classified under class *Resentoviricetes*, phylum *Duplornnaviricota*, kingdom *Orthornaviriae* all coming under realm *Riboviriae*. Under *Sedoreoviridae* family other genera are *Cardoreovirus, Mimoreovirus, Orbivirus, Phytoreovirus, Seadornavirus* of which only *Orbivirus* and *Seadornavirus* are known to infect mammalian hosts (https://talk.ictvonline.org/taxonomy/). RV strains were earlier classified in a binomial way based on G and P genotypes. In new nomenclature, 11 genes are included in genotyping system. It is Gx-P[x]-Ix-Rx-Cx-Mx-Ax-Nx-Tx-Ex-Hx representing the genotypes of the VP7–VP4–VP6–VP1–VP2–VP3–NSP1–NSP2–NSP3–NSP4–NSP5 where x represents genotype number RCWG, 2019 (Matthijnssens et al. 2011). So far, at least 36 G, 51 P, 20 I, 12 R, 12 C, 11 M, 22 A, 12 N, 14 T, 18 E and 13 H genotypes have been identified (Dóró et al. 2015; Rojas et al. 2017; Uprety et al. 2021).

16.4.2 Morphology and Virion Structure

Electron microscopic structures of virions shows wheel-like appearance. Virions possess a triple-shelled structure of about 100 nm, which is also seen in orbiviruses. The outermost layer of triple-layered structure of capsid is made of two proteins VP4 and VP7. These proteins project from the surface of virions giving spike-like appearance as shown in Fig. 16.2. The outermost shell is smooth unlike orbiviruses (Li et al. 2009). VP4 on cleavage of trypsin leads to formation of VP5 and VP8. Inner double-layered structure is composed of proteins VP2 and VP6. 260 VP6 trimers form a $T = 13$ icosahedral symmetry. VP6 also forms contacts with VP4 and VP7 (Settembre et al. 2011). Core layer exhibits $T = 1$ symmetry made of VP1, VP2,

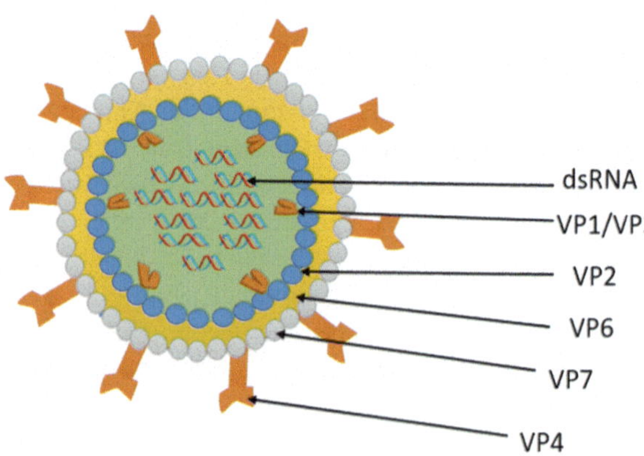

Fig. 16.2 Schematic structure of rotavirus

and VP3. This core encapsidates dsRNA viral genome. Only triple-layered structure is infectious (Fig. 16.2).

16.4.2.1 Genome Structure and Organization

Virus genomic material is composed of double-stranded RNA of segmented nature. Viral RNA is bound to amino-terminal VP2 protein in an icosahedral arrangement (Labbé et al. 1994). The dsRNA has 11 segments with a particular electrophoretic pattern. These 11 segments encode 12 proteins. Viral dsRNA is capped and non-polyadenylated. This viral dsRNA is positive sense. The rotavirus genome shows a characteristic electropherotype pattern in RNA-PAGE of 11 segments. Group A, the most common human rotavirus, shows a 4-2-3-2 pattern (Snodgrass et al. 1984).

16.4.3 Viral Proteins

The viral genome encodes structural and non-structural proteins. Among 6 structural viral proteins, VP4 and VP7 proteins both play role in the cellular entry of virus. These two proteins are also targeted by antibodies for neutralization (Matsui et al. 1989). Proteolytic cleavage of VP4 aids viral entry inside the cell. Trypsin cleavage forms 60 and 28 kDa products that help in membrane penetration (Kaljot et al. 1988). These are called VP5 and VP8, respectively (Arias et al. 1996). VP7- and VP4-based serotyping are done with help of neutralization tests or ELISA. VP7 serotypes are denoted as G serotypes and VP4 serotypes are denoted as P serotypes (Isegawa et al. 1993). There are 36 G types and 51 P types reported in humans and other mammals (Rotavirus Classification Working Group 2019). VP6 protein encoded by segment 6 is a highly conserved and abundant protein in capsid structure. The middle layer of capsid is composed of 780 molecules of VP6. Based on VP6, rotavirus isolates are classified into groups A–J (Matthijnssens et al. 2012). The innermost capsid layer is formed by VP2 proteins. VP3 is a capping enzyme encoded by segment 2 of the genome. VP1 encoded by segment 1 is a polymerase enzyme present in the core of virion. VP1 and VP3 together form a replication enzyme complex.

The rotavirus genome encodes 6 non-structural proteins, NSP1–NSP6. NSP6 is not encoded by all the strains of the virus. NSP1 is involved in host specificity and suppressing apoptosis to augment cellular infection. NSP2 interacts with NSP5 to form viroplasms and aids in genome replication. NSP3 encoded by segment 7 of the rotavirus genome plays a significant role in promoting viral protein formation by interfering with the host translation machinery. NSP3 is attached to a consensus sequence in the 3′ end of the viral genome forming a complex which is analogous to host poly A-tail and PABP protein. This viral RNA-NSP3 complex binds to cellular eIF4G more strongly than the host PABP/poly-A binding protein (Groft and Burley 2002). Segment 10 of genome encodes NSP4 having role in induction of diarrhea (Ball et al. 1996). It interacts with integrin 1 in plasma membrane of host cell, triggers phospholipase C signaling pathway causing increased intracellular calcium. This increased intracellular calcium opens chloride channel leading to release of chloride ions (Ball et al. 1996; Ousingsawat et al. 2011). NSP5 is encoded by segment 11 which also encodes NSP6 in an overlapping open reading frame. NSP5 forms intracellular inclusions together with VP2 protein (Berois et al. 2003).

16.4.4 Viral Life Cycle

Rotavirus replication takes place in mature enterocytes in villi tip and middle and enteroendocrine cells of small intestine. Virus entry in cell takes from basolateral surface. Initiation of the viral cycle takes from an interaction between VP8 subunit of outer capsid layer protein VP4 and host cell proteins like sialic acid receptors and histo-blood group antigens (Arias and López 2021). Viral internalization takes place by endocytosis which is a pH-dependent pathway. Inside the endocytic vesicle due to low Ca^{2+} ions level, the outermost layer of triple-layered structure is lost. The remaining double-layered structure is now transcriptionally active. By transcription process, +ssRNA synthesis is done by VP1, RNA-dependent RNA polymerase located in inner core. Activity of VP1 is based on its interaction with VP2 (Long and McDonald 2017). Before these newly transcribed +ssRNA are ejected from this transcriptionally active part, they are capped by VP3. These +ssRNA also serve as template for translation process aided by host cell ribosomes (Trask et al. 2012). NSP2 plays a role in packaging of 11 RNA segments in core and also assortment activities (Papa et al. 2021). Exit of rotavirus after viral assembly is also in polarized manner, from the apical surface in a non-lytic manner (Cevallos Porta et al. 2016) (Fig. 16.3).

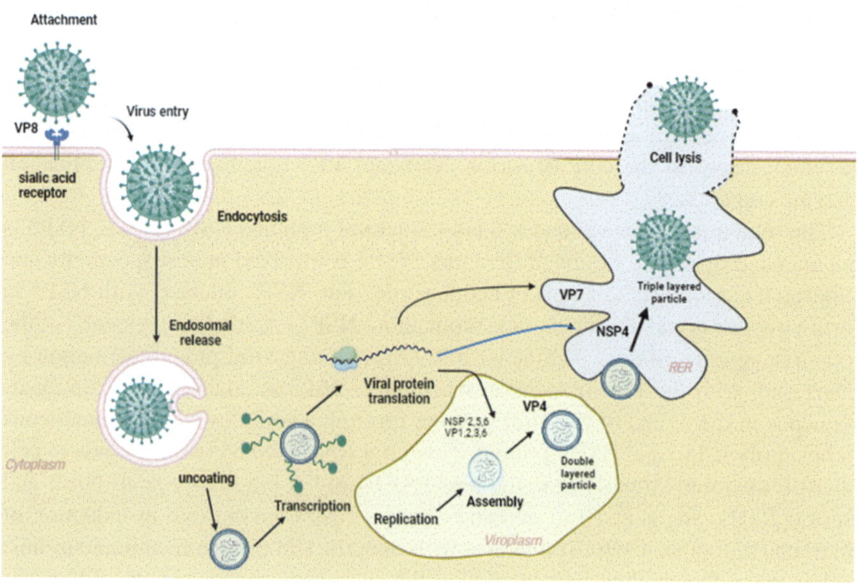

Fig. 16.3 Replication cycle of rotavirus

16.5 Pathogenesis in Humans

The rotavirus has a complex pathogenesis that is influenced equally by host and viral variables. Genetic analysis using selected virus reassortants aided in identifying several proteins involved in virulence (VP3, VP4, NSP1, VP6, VP7, NSP2, NSP3, and NSP4). These proteins are considered to play important roles in the efficiency of viral replication (VP3, NSP2, VP6, and NSP3). Shut-off of host protein synthesis (NSP3), extra intestinal spread and replication of virus (NSP1 and NSP3), virus entry into cells (VP4 andVP7), regulation of the induction of interferon (NSP1), and the induction of diarrhea (NSP4) (Crawford et al. 2017).

Pathological finding in Rotavirus infections is characterized by shortening and atrophy of small intestinal villi, enterocyte vacuolization, and distended cisternae of endoplasmic reticulum. Rotavirus infections also induce intestinal lactase deficiency, and osmotic diarrhea which occurs mainly due to increased lactose in feces due to enterocyte destruction (Anderson and Weber 2004).

It has been shown that rotavirus also activates intestinal autonomous neural system and increases secretion of water and electrolyte. In mice, NSP4 have been found to induce diarrhea. NSP4 has also been shown to increase intracellular Calcium ion concentration through mobilizing Calcium from the endoplasmic reticulum. NSP4 also increases permeability to allow leakage by disturbing or destabilizing plasma membrane, for e.g., lactate dehydrogenase out of the cells. The young ones are prone to rotavirus infection because of low epithelial turnover, and also due to high proportion of terminally differentiated epithelium (Lorrot and Vasseur 2007).

16.6 Clinical Manifestations

Rotavirus infections are accounted for 258 million episodes of diarrhea and up to 200,000 diarrheal deaths per year (Tate et al. 2016). Infants aged between 6 months and 2 years are primarily infected. Incubation period of rotavirus infections is approximately 48 h (Grimwood et al. 1988). Rotavirus infection causes non-bloody diarrhea and vomiting along with fever and malaise. Profuse diarrhea and regular bowel movements in children leads to severe dehydration. RV gastroenteritis provokes anatomical and histological changes which can lead to intussusception. Extra-intestinal infection in unvaccinated children increases risk of seizures due to encephalitis (Crawford et al. 2017). Hepatobiliary disease and type 1 diabetes due to acute pancreatitis are caused due to systemic infections. Other extra-intestinal complications in children includes respiratory illness like nasal congestion, rhinitis, coughing and pneumonia, renal failure, and myocarditis (Dian et al. 2021).

In adults, the incubation period ranges from 2 to 6 days post infection. Due to the presence of pre-existing antibodies, diarrhea is less profuse. Fever, malaise, headache, nausea, and cramping are the symptoms observed. In patients with prolonged immunocompromised and malignant conditions, the disease is more severe. HIV-1 positive patients suffer from chronic diarrhea due to rotavirus infection (Anderson and Weber 2004). Not only in immunocompromised conditions but sometimes co-infection with other pathogenic agents could turn rotavirus infection into severe outcomes in adults. Hence, adult infections should not be neglected and proper clinical care must be considered.

16.7 Diagnostic Approaches

Methods used for the diagnosis of rotavirus involves virus cultivation, electron microscopy, antigen detection methods including EIA, lateral flow immunoassays, LA, nucleic acid detection methods (PAGE), and amplification (RT-PCR). Majority of the above-mentioned assays have the potential to detect rotaviruses since huge number of intact viruses are present in the stool specimens from children affected with gastroenteritis.

16.7.1 Isolation of Virus

The isolation of rotavirus from clinical samples can be improved by trypsin treatment, which increases the virus infectivity for tissue culture. The rate of isolation is magnified with the chymotrypsin treatment of feces (Rhodes et al. 1979). Typically, the virus can only replicate a little in tissue culture using traditional techniques. The morphological integrity of variance may be lost occasionally by thawing and freezing, but not the infectivity of clinical specimen. For virus identification and potential virus isolation, intestinal mucosa samples should be collected from both small and large intestine and properly refrigerated. Human rotavirus can be cultured in cell-lines like MA-104 and African green monkey kidney cells (Albert and Bishop 1984; Vasileva et al. 1987). However, this approach is tedious, time-consuming, and has low success rates.

16.7.2 Electron Microscopy (EM)

Electron microscopy keeps on being the standard diagnostic test for rotavirus identification when compared against other techniques. EM has been intensely used for the routine rotavirus detection. It is always simpler to visualize the virus if the virus has been clumped by IEM using an antiserum or concentrated through ultracentrifugation. An ideal scenario involves the collection of 20–30 g of fecal specimens in the first 48 h of the symptoms, their preservation, and delivery to the laboratory since the viruses may rapidly decrease below detectable levels. The consensus for EM to detect a virus is that, there must be at least 10^6 virus particles per millimeter of samples (Maes et al. 2003).

EM is highly specific for rotavirus detection and just as sensitive as some EIAs. Because of its simplicity, high resolution, speed, and efficient preservation of virus particles, negative staining is one of the methods used most frequently for the detection of viruses in culture or clinical samples (Almeida 1980). Monoclonal antibody-based immuno-electron microscopy (IEM), one of the most sensitive EM techniques, has been successful in studies on the epidemiology of rotavirus infection. It enables raising the sensitivity of virus detection in situations where low virus concentrations cluster, making them easier to spot. Phosphotungstic acid and uranyl acetate at pH 4.5 are recommended for the negative staining while phosphotungstic acid at pH 4.5 is optimal for IEM (Nakata et al. 1987). The various EM techniques encourage a quick and correct detection of various gastroenteric viruses, enabling rapid implementation of prophylactic controls and preventions in the products, animal losses, and catastrophic monetary losses.

However, this method is expensive, labor-intensive, requires well-trained personnel, and unable to differentiate between different groups of rotaviruses (Soltan et al. 2016). Further, Nakata and coworkers reported that staining with phosphotungstic acid at higher pH levels with longer duration resulted in broken or decreased number of particles and reduced to single-shelled or core particles (Nakata et al. 1987).

16.7.3 Antigen Detection

16.7.3.1 Serological Methods

Serological methods include enzyme immunoassay (EIA) or enzyme-linked immunosorbent assay (ELISA), latex particle agglutination (LA), radio immunoassay (RIA), immunofluorescence test (IFT), neutralization, and complement fixation test (CFT). These methods involve centrifugation of clinical specimens on tissue culture cells followed by immunofluorescence. Unlike IEM, EIA has demonstrated to be more sensitive than EM. Other methods that show less sensitivity than EIA include immune adherence hemagglutination, co-agglutination reaction, double immuno diffusion (DID), counter immunoelectrophoresis (CIE), etc. (Bartlet et al. 1987). Although there are some restrictions on using antibody-dependent tests to find rotaviruses besides Group A, Mab-based EIA has been proven to be equally effective as RT-PCR in the serotype detection of rotavirus (Ushijima et al. 1994).

Due to its efficiency, rapidity, and lack of need for expensive equipment, ELISA has emerged as the method of choice in most laboratories. Commercial kits of ELISA are currently available. ELISA is simple and accurate when compared with EM, complement fixation, and IFT on inoculation cell cultures. ELISA is ideal for handling a vast number of specimens (Ekern et al. 1981) and very helpful in epidemiological studies while other tests are conducted to determine whether virus is present or not (Markowska-Daniel et al. 1996). Another significant epidemiological approach is dot-blot or dot-hybridization, which can identify individual segments of rotavirus (Flores et al. 1983). It has been mentioned that employing an AP-labeled oligonucleotide probe, a dot blot format can be used for the detection of as little as 1 ng RNA of rotavirus. Sub-grouping for human rotaviruses is now possible by employing specific genes as probes. Immunofluorescence and peroxidase-antiperoxidase (PAP) are extremely specific methods to detect rotaviruses in biopsy samples and cell cultures (Bryden et al. 1987). Rotavirus detection in huge amounts of water is accomplished via virus adsorption and elution (viradel) (Goyal and

Gerba, 1983). CFT has been found to be more sensitive but cumbersome to perform than other immunodiffusion assays for antigen detection because some fecal samples showed anti-complementary activity. The most sensitive approach for determining the amount of rotavirus antibodies is radioimmunoassay (RIA) while the least sensitive method is CIE (Mohammed et al. 1978).

16.7.3.2 Latex Agglutination Test (LAT)

LAT is a simple, inexpensive, and easy to perform test that use particles that have been exposed to the virus using antisera. It is reported to be as sensitive as EM but less specific, hence best utilized as a screening test (Acha and Szyfres 2003). In a study, Al-Yousif et al. (2001) compared LAT with ELISA and reported a sensitivity and specificity of 100% and 96.3%, respectively. LAT is used as a rapid screening test for bovine and porcine rotavirus infections (Sanekata et al. 1991).

16.7.3.3 Lateral-Flow Immunoassay (LFA)

LFA is a quick and simple to perform test which could enable point of care testing in case of a suspected outbreak. The establishment of permanent lines, as opposed to the LAT, makes it easier to read the data. The LFT can also be used in developing nations with few diagnostic facilities. Al-Yousif et al. (2002) developed a LFA for detecting bovine rotavirus in fecal specimens and concluded 100% specificity and 70% sensitivity for LFA.

16.7.4 Nucleic Acid Detection Methods

Polyacrylamide gel electrophoresis (PAGE) is a low-cost, highly sensitive, and specific method for the detection of both Group A and non-Group A rotaviruses in stool specimens. It can be also helpful in epidemiological investigations and tracking electropherotypes of rotavirus within a community (Chauhan and Singh 1993). Given that the ELISA approach produces a greater proportion of false positives in samples from underdeveloped nations, it has been discovered that this test is more trustworthy than ELISA when evaluating newborn samples of child (Chen et al. 1999). PAGE, however, is unable to differentiate between several serotypes (Beards 1982).

For the extraction and purification of rotavirus dsRNA from fecal specimens, a different approach has been found to be very effective. It utilizes hydroxyapatite, cetyltrimethyl ammonium bromide, and guanidine isothiocyanate, with the added benefit of minimizing the use of hazardous substances for this purpose, like chloroform, freon, and phenol (Santos and Gouvea 1994).

16.7.5 Nucleic Acid Amplification Methods

PCR has been shown to be 100,000 times more sensitive than the traditional electropherotype method which is extensively used for the detection of rotavirus in infected feces. Additionally, compared to the previously established hybridization-based assay, it provides a 5000-fold increase in sensitivity and uses no radioisotopes (Xu et al. 1990).

RT-PCR enables production of serotype-specific cDNA molecules which are employed to generate gene probes for the rotavirus detection. RT-PCR has been established to identify rotavirus in human feces samples targeting the VP7, major outer capsid glycoprotein gene with a detection limit of about 2 ng sensitivity (equivalent to 10^8 viral genomes). Additionally, there have been reports of PCR typing methods where each serotype generated a fragment of a specific size.

A nested reverse transcriptase-polymerase chain reaction (RT-PCR) was developed for the detection of Group A rotaviruses (Elschner et al. 2002). All strains produced a nested RT-PCR result based on the area in gene segment 6. The detection threshold for virus-containing cell culture supernatant was determined to be 3×10^{-2} TCID50 and 3×10^{-3} TCID50 with RT-PCR and nested amplification, respectively. The LOD of current PCR method was reported to be 1.6×10^2 TCID50 per g of feces, which can be further enhanced using nested PCR. The current technique for recognizing Group A rotaviruses is a potent diagnostic tool that has been demonstrated to work with rotaviruses from various sources, including humans (Elschner et al. 2002).

16.8 Prevention and Control

There is no specific treatment available at present for the rotavirus. The main goal of treatment for rotavirus enteritis is to replace fluid and electrolyte losses from diarrhea by intravenous or oral route. It is suggested to skip drinking milk for 24–48 h. Moreover, antibiotics are used intravenously or orally to address potential systemic and intestinal bacterial infections.

All children contract the rotavirus the first 3–5 years of life, despite the level of hygiene and sanitation. As a result, vaccinations are the best method for stopping rotavirus diarrhea. Passive protection by maternal antibodies also plays a key role. Rotavirus-specific IgG in placenta and rotavirus-specific IgA in mammary gland may protect infant from infection. However, in the process of active immunization, these antibodies interfere with infant's immune response (Mwila et al. 2017). There are four oral live attenuated WHO prequalified vaccines available globally. RotaTeq™ is a reassorted bovine-human rotavirus whereas Rotavac™ is a naturally occurring neonatal bovine-human reassortant neonatal G9P, commonly known as 116E. Rotarix™ is made from a single common strain of human rotavirus and RotaSiil™ is a bovine-human reassortant with human G1, G2, G3, and G4 bovine UK G6P[5] backbone. All the above four vaccines are regarded quite effective for the prevention of the disease. List of vaccine and dose details are mentioned in Table 16.1. Similar to other live oral vaccines, vaccination efficacy in emergent countries may be lower than in industrialized environments. Despite the reduced efficacy, a higher absolute decrease in deaths and severe gastroenteritis were seen.

Vaccination needs to be implemented into the national immunization programs worldwide. The next generation of parenteral or skin patch vaccinations are being developed concurrently, and it is hoped that these will be more successful than current oral vaccines and eventually replace them. The disease might eventually be under control if these vaccinations prove to be secure, more efficient, affordable and simpler to give in a combined vaccine formulation.

Table 16.1 List of vaccine and dose details

S. no	Vaccine	Strains	Year qualified	Dose composition	Dosage	Licensed as
1.	RRV-TV live oral vaccine	G1, G2, G4	1998, withdrawn in 1999	–	3 doses—3, 4, and 6 months	Rotashield
2.	Monovalent live attenuated, oral	G1P[8]	2009	1.5 mL, 10^6 median cell culture infective dose (CCID50)	2 doses—2–4 months	Rotarix, WHO
3.	Pentavalent gene reassortant, oral-live	G1, G2, G3, G4, and P[8]	2008	2 mL, f 2.0–2.8 × 10^6 infectious units (IU) per reassortant dose	3 doses—2, 4, and 6 months	RotaTeq, WHO
4.	Monovalent live rotavirus, oral	116E, G9P[11]	2018	0.5 mL/dose contains $10^{5.0}$ FFU	3 doses—6, 10, and 14 weeks	ROTAVAC, WHO
5.	Bovine human reassortant, live attenuated, oral	G1, G2, G3, G4, and G9	2018	2.5 mL/ dose—$10^{5.6}$ FFU/ serotype	3 doses—6, 10, and 14 weeks	ROTASIIL, WHO
6.	Monovalent live attenuated, oral	G10, P[12]	2000	2.5 mL	One dose, yearly booster—2–36 months	LLR (Lanzhou Lamb Rotavirus), China
7	Monovalent live attenuated, oral	G1P[8]	2012	2.5 mL	2 doses—2 months apart	Rotavin, Vietnam

16.9 Future Perspectives

Rotavirus remains an important cause of diarrheal cases all over the world even after a decade of vaccine introduction. Effectivity of vaccination in different countries and socio-economic settings needs to be addressed. Use of common vaccines globally could be a misleading approach given a vast collection of virus strains in circulation. Country-specific epidemiology have to be considered while devising prevention and control measures. Even though mortalities have been controlled but the long-term effect on quality of life in surviving children also needs to be studied.

References

Acha PN, Szyfres B (2003) Rotaviral gastroenteritis. In: Zoonoses and communicable diseases common to man and animals, Vol. III (Viruses). Pan American Health Organization (PAHO), New York, pp 286–297

Afrad MH, Matthijnssens J, Moni S, Kabir F, Ashrafi A, Rahman MZ, Faruque ASG, Azim T, Rahman M (2013) Genetic characterization of a rare bovine-like human VP4 mono-reassortant G6P[8] rotavirus strain detected from an infant in Bangladesh. Infect Genet Evol 19:120. https://doi.org/10.1016/j.meegid.2013.06.030

Albert MJ, Bishop RF (1984) Cultivation of human rotaviruses in cell culture. J Med Virol 13(4):377–383

Almalki SSR (2018) Circulating rotavirus G and P strains post rotavirus vaccination in Eastern Mediterranean Region. Saudi Med J 39(8):755–756. https://doi.org/10.15537/smj.2018.6.21394

Almeida JD (1980) Practical aspects of diagnostic electron microscopy. Yale J Biol Med 53(1):5

Al-Yousif Y, Anderson J, Chard-Bergstrom C, Bustamante A, Muenzenberger M, Austin K, Kapil S (2001) Evaluation of a latex agglutination kit (Virogen Rotatest) for detection of bovine rotavirus in fecal samples. Clin Diagn Lab Immunol 8(3):496–498

Al-Yousif Y, Anderson J, Chard-Bergstrom C, Kapil S (2002) Development, evaluation, and application of lateral-flow immunoassay (immunochromatography) for detection of rotavirus in bovine fecal samples. Clin Vaccine Immunol 9(3):723–725

Amit LN, Mori D, John JL, Chin AZ, Mosiun AK, Jeffree MS, Ahmed K (2021) Emergence of equine-like G3 strains as the dominant rotavirus among children under five with diarrhea in Sabah, Malaysia during 2018–2019. PLoS One 16(7):e0254784. https://doi.org/10.1371/journal.pone.0254784

Anderson EJ, Weber SG (2004) Rotavirus infection in adults. Lancet Infect Dis 4(2):91–99. https://doi.org/10.1016/S1473-3099(04)00928-4

Arias CF, López S (2021) Rotavirus cell entry: not so simple after all. Curr Opin Virol 48:42–48. https://doi.org/10.1016/j.coviro.2021.03.011

Arias CF, Romero P, Alvarez V, López S (1996) Trypsin activation pathway of rotavirus infectivity. J Virol 70(9):5832–5839. https://doi.org/10.1128/JVI.70.9.5832-5839.1996

Ball JM, Tian P, Zeng CQ-Y, Morris AP, Estes MK (1996) Age-dependent diarrhea induced by a rotaviral nonstructural glycoprotein. Science 272(5258):101–104. https://doi.org/10.1126/science.272.5258.101

Bányai K, Kemenesi G, Budinski I, Földes F, Zana B, Marton S, Varga-Kugler R, Oldal M, Kurucz K, Jakab F (2017) Candidate new rotavirus species in Schreiber's bats, Serbia. Infect Genet Evol 48:19–26. https://doi.org/10.1016/j.meegid.2016.12.002

Bartlet AV et al (1987) Rotavirus gastroenteritis. Annu Rev Med 38:399–415

Beards GM (1982) Polymorphism of genomic RNAs within rotavirus serotypes and subgroups. Arch Virol 74:65–70

Berois M, Sapin C, Erk I, Poncet D, Cohen J (2003) Rotavirus nonstructural protein NSP5 interacts with major core protein VP2. J Virol 77(3):1757–1763. https://doi.org/10.1128/JVI.77.3.1757-1763.2003
Bishop RF, Davidson GP, Holmes IH, Ruck BJ (1973) Virus particles in epithelial cells of duodenal mucosa from children with acute non-bacterial gastroenteritis. Lancet (London, England) 2(7841):1281–1283. https://doi.org/10.1016/s0140-6736(73)92867-5
Bourdett-Stanziola L, Centeno E, Nordgren J, Durant Archibold A, Ortega-Barria E, Bucardo F (2021) Potential bat-like rotavirus in hospitalized children with diarrhea from The Dominican Republic. Asian J Res Infect Dis 8:1–7. https://doi.org/10.9734/ajrid/2021/v8i130225
Bryden AS, Davies HA, Thouless ME, Flewitt TH (1987) Diagnosis of rotavirus infection by cell culture. J Med Microbiol 10:121–125
Cevallos Porta D, López S, Arias CF, Isa P (2016) Polarized rotavirus entry and release from differentiated small intestinal cells. Virology 499:65–71. https://doi.org/10.1016/j.virol.2016.09.010
Chandler-Bostock R, Mellits KH (2015) Efficacy of disinfectants against porcine rotavirus in the presence and absence of organic matter. Lett Appl Microbiol 61(6):538–543. https://doi.org/10.1111/lam.12502
Chauhan RS, Singh NP (1993) Rotavirus infection in calves: diagnosis by polyacrylamide-gel electrophoresis. Indian J Anim Sci 63(1):1–3
Chen D, Luongo CL, Nibert ML, Patton JT (1999) Rotavirus open cores catalyze 5′-capping and methylation of exogenous RNA: evidence that VP3 is a methyltransferase. Virology 265:120–130
Crawford SE, Ramani S, Tate JE, Parashar UD, Svensson L, Hagbom M, Franco MA, Greenberg HB, O'Ryan M, Kang G, Desselberger U, Estes MK (2017) Rotavirus infection. Nat Rev Dis Prim 3:17083. https://doi.org/10.1038/nrdp.2017.83
de Hoog MLA, Vesikari T, Giaquinto C, Huppertz H-I, Martinon-Torres F, Bruijning-Verhagen P (2018) Report of the 5th European expert meeting on rotavirus vaccination (EEROVAC). Hum Vaccin Immunother 14(4):1027–1034. https://doi.org/10.1080/21645515.2017.1412019
Dian Z, Sun Y, Zhang G, Xu Y, Fan X, Yang X, Pan Q, Peppelenbosch M, Miao Z (2021) Rotavirus-related systemic diseases: clinical manifestation, evidence and pathogenesis. Crit Rev Microbiol 47(5):580–595. https://doi.org/10.1080/1040841X.2021.1907738
Dóró R, Farkas SL, Martella V, Bányai K (2015) Zoonotic transmission of rotavirus: surveillance and control. Expert Rev Anti-Infect Ther 13(11):1337–1350. https://doi.org/10.1586/14787210.2015.1089171
Ekern LJ, Schipper IA, McMahon KJ (1981) Neonatal bovine enteritis: detection of rotavirus by counterimmunoelectrophoresis and enzyme-linked immunosorbent assay. Can J Comp Med 45(2):135–139
Elschner M, Prudlo J, Hotzel H, Otto P, Sachse K (2002) Nested reverse transcriptase-polymerase chain reaction for the detection of group A rotaviruses. J Vet Med Ser B 49(2):77–81
Flewett TH, Bryden AS, Davies H, Woode GN, Bridger J, Derrick J (1974) Relation between viruses from acute gastroenteritis of children and newborn calves. Lancet 304(7872):61–63. https://doi.org/10.1016/S0140-6736(74)91631-6
Flores J, Boeggeman E, Purcell RH, Sereno M, Perez I, White L, Wyatt RG, Chanock RM, Kapikian AZ (1983) A dot hybridization assay for detection of rotavirus. Lancet 1:555–559
GBD 2015 Mortality and Causes of Death Collaborators (2016) Global, regional, and national life expectancy, all-cause mortality, and cause-specific mortality for 249 causes of death, 1980–2015: a systematic analysis for the Global Burden of Disease Study 2015. Lancet (London, England) 388(10053):1459–1544. https://doi.org/10.1016/S0140-6736(16)31012-1
Giri S, Kumar CPG, Khakha SA, Chawla-Sarkar M, Gopalkrishna V, Chitambar SD, Ray P, Venkatasubramanian S, Borkakoty BJ, Roy S, Bhat J, Dwibedi B, Das P, Paluru V, Ramani S, Babji S, Arora R, Mehendale SM, Gupte MD, National Rotavirus Surveillance Network Investigators (2020) Diversity of rotavirus genotypes circulating in children <5 years of age hospitalized for acute gastroenteritis in India from 2005 to 2016: analysis of temporal and regional genotype variation. BMC Infect Dis 20(1):740. https://doi.org/10.1186/s12879-020-05448-y

Godfrey O, Zhang W, Amponsem-Boateng C, Oppong TB, Zhao Q, Li D (2020) Evidence of rotavirus vaccine impact in sub-Saharan Africa: systematic review and meta-analysis. PLoS One 15(4):e0232113. https://doi.org/10.1371/journal.pone.0232113

Goyal SM, Gerba CP (1983) Viradel method for detection of rotavirus from sea water. J Virol Methods 7:279–285

Griffin DD, Fletcher M, Levy ME, Ching-Lee M, Nogami R, Edwards L, Peters H, Montague L, Gentsch JR, Glass RI (2002) Outbreaks of adult gastroenteritis traced to a single genotype of rotavirus. J Infect Dis 185(10):1502–1505. https://doi.org/10.1086/340218

Grimwood K, Lund JC, Coulson BS, Hudson IL, Bishop RF, Barnes GL (1988) Comparison of serum and mucosal antibody responses following severe acute rotavirus gastroenteritis in young children. J Clin Microbiol 26(4):732–738. https://doi.org/10.1128/jcm.26.4.732-738.1988

Groft CM, Burley SK (2002) Recognition of eIF4G by rotavirus NSP3 reveals a basis for mRNA circularization. Mol Cell 9(6):1273–1283. https://doi.org/10.1016/S1097-2765(02)00555-5

Isegawa Y, Nakagomi O, Nakagomi T, Ishida S, Uesugi S, Ueda S (1993) Determination of bovine rotavirus G and P serotypes by polymerase chain reaction. Mol Cell Probes 7(4):277–284. https://doi.org/10.1006/mcpr.1993.1041

Kaljot KT, Shaw RD, Rubin DH, Greenberg HB (1988) Infectious rotavirus enters cells by direct cell membrane penetration, not by endocytosis. J Virol 62(4):1136–1144. https://doi.org/10.1128/JVI.62.4.1136-1144.1988

Kawai K, O'Brien MA, Goveia MG, Mast TC, El Khoury AC (2012) Burden of rotavirus gastroenteritis and distribution of rotavirus strains in Asia: a systematic review. Vaccine 30(7):1244–1254. https://doi.org/10.1016/j.vaccine.2011.12.092

Khalil IA-M (2017) The global burden of rotavirus diarrheal diseases: results from the global burden of diseases study 2016. Open Forum Infect Dis 4(suppl_1):S363–S363. https://doi.org/10.1093/ofid/ofx163.885

Labbé M, Baudoux P, Charpilienne A, Poncet D, Cohen JY (1994) Identification of the nucleic acid binding domain of the rotavirus VP2 protein. J Gen Virol 75(12):3423–3430. https://doi.org/10.1099/0022-1317-75-12-3423

Lanata CF, Fischer-Walker CL, Olascoaga AC, Torres CX, Aryee MJ, Black RE, Child Health Epidemiology Reference Group of the World Health Organization and UNICEF (2013) Global causes of diarrheal disease mortality in children <5 years of age: a systematic review. PLoS One 8(9):e72788. https://doi.org/10.1371/journal.pone.0072788

Lestari FB, Vongpunsawad S, Wanlapakorn N, Poovorawan Y (2020) Rotavirus infection in children in Southeast Asia 2008–2018: disease burden, genotype distribution, seasonality, and vaccination. J Biomed Sci 27(1):66. https://doi.org/10.1186/s12929-020-00649-8

Li Y, Wang SM, Zhen SS, Chen Y, Deng W, Kilgore PE, Wang XY, (2014) Diversity of rotavirus strains causing diarrhea in <5 years old Chinese children: a systematic review. PLoS One 9(1):e84699

Li Z, Baker ML, Jiang W, Estes MK, Prasad BVV (2009) Rotavirus architecture at subnanometer resolution. J Virol 83(4):1754–1766. https://doi.org/10.1128/JVI.01855-08

Long CP, McDonald SM (2017) Rotavirus genome replication: some assembly required. PLoS Pathog 13(4):e1006242. https://doi.org/10.1371/journal.ppat.1006242

Lorrot M, Vasseur M (2007) How do the rotavirus NSP4 and bacterial enterotoxins lead differently to diarrhea? Virol J 4(1):31. https://doi.org/10.1186/1743-422X-4-31

Luchs A, do Timenetsky MCST (2016) Group A rotavirus gastroenteritis: post-vaccine era, genotypes and zoonotic transmission. Einstein (Sao Paulo, Brazil) 14(2):278–287. https://doi.org/10.1590/S1679-45082016RB3582

Maes RK, Grooms DL, Wise AG, Han C, Ciesicki V, Hanson L, Holland R (2003) Evaluation of a human group a rotavirus assay for on-site detection of bovine rotavirus. J Clin Microbiol 41(1):290–294

Maguire JE, Glasgow K, Glass K, Roczo-Farkas S, Bines JE, Sheppeard V, Macartney K, Quinn HE (2019) Rotavirus epidemiology and monovalent rotavirus vaccine effectiveness in Australia: 2010–2017. Pediatrics 144(4):e20191024. https://doi.org/10.1542/peds.2019-1024

Markowska-Daniel I, Winiarczyk S, Gradzki Z, Pejsak Z (1996) Evaluation of different methods (ELISA, IF, EM, PAGE) for the diagnosis of rotavirus infection in piglets. Comp Immunol Microbiol Infect Dis 19(3):219–232

Matsui SM, Mackow ER, Greenberg HB (1989) Molecular determinant of rotavirus neutralization and protection. In: Maramorosch K, Murphy FA, Shatkin AJ (eds) Advances in virus research, vol 36. Academic Press, pp 181–214. https://doi.org/10.1016/S0065-3527(08)60585-0

Matthijnssens J, Ciarlet M, McDonald SM, Attoui H, Bányai K, Brister JR, Buesa J, Esona MD, Estes MK, Gentsch JR, Iturriza-Gómara M, Johne R, Kirkwood CD, Martella V, Mertens PPC, Nakagomi O, Parreño V, Rahman M, Ruggeri FM, Van Ranst M (2011) Uniformity of rotavirus strain nomenclature proposed by the Rotavirus Classification Working Group (RCWG). Arch Virol 156(8):1397–1413. https://doi.org/10.1007/s00705-011-1006-z

Matthijnssens J, Otto PH, Ciarlet M, Desselberger U, Van Ranst M, Johne R (2012) VP6-sequence-based cutoff values as a criterion for rotavirus species demarcation. Arch Virol 157(6):1177–1182. https://doi.org/10.1007/s00705-012-1273-3

Mayr C, Strohe G, Contzen M (2009) Detection of rotavirus in food associated with a gastroenteritis outbreak in a mother and child sanatorium. Int J Food Microbiol 135:179–182. https://doi.org/10.1016/j.ijfoodmicro.2009.08.010

Mizukoshi F, Kuroda M, Tsukagoshi H, Sekizuka T, Funatogawa K, Morita Y, Noda M, Katayama K, Kimura H (2014) A food-borne outbreak of gastroenteritis due to genotype G1P[8] rotavirus among adolescents in Japan. Microbiol Immunol 58(9):536–539. https://doi.org/10.1111/1348-0421.12176

Mladenova Z, Papp H, Lengyel G, Kisfali P, Steyer A, Steyer AF, Esona MD, Iturriza-Gómara M, Bányai K (2012) Detection of rare reassortant G5P[6] rotavirus, Bulgaria. Infect Genet Evol 12(8):1676–1684. https://doi.org/10.1016/j.meegid.2012.07.002

Mohammed KA, Babiuk LA, Saunders JR, Acres SD (1978) Bovine rotavirus diagnosis: comparison of various antigens and serological tests. Vet Microbiol 3(2):115–127

Mwila K, Chilengi R, Simuyandi M, Permar SR, Becker-Dreps S (2017) Contribution of maternal immunity to decreased rotavirus vaccine performance in low- and middle-income countries. Clin Vaccine Immunol CVI 24(1):e00405–e00416. https://doi.org/10.1128/CVI.00405-16

Nakata S, Petrie BL, Calomeni EP, Estes MK (1987) Electron microscopy procedure influences detection of rotaviruses. J Clin Microbiol 25(10):1902–1906

Nayak MK, Banerjee A, Sarkar R, Mitra S, Dutta K, Ganguly N, Ghosh C, Girish Kumar CP, Niyogi P, Panda S, Dutta S, Chawla-Sarkar M (2019) Genetic characterization of group-A rotaviruses among children in eastern India during 2014–2016: phylodynamics of co-circulating genotypes. Vaccine 37(45):6842–6856. https://doi.org/10.1016/j.vaccine.2019.06.062

Ouermi D, Soubeiga D, Nadembega WMC, Sawadogo PM, Zohoncon TM, Obiri-Yeboah D, Djigma FW, Nordgren J, Simpore J (2017) Molecular epidemiology of rotavirus in children under five in Africa (2006–2016): a systematic review. Pak J Biol Sci PJBS 20(2):59–69. https://doi.org/10.3923/pjbs.2017.59.69

Ousingsawat J, Mirza M, Tian Y, Roussa E, Schreiber R, Cook DI, Kunzelmann K (2011) Rotavirus toxin NSP4 induces diarrhea by activation of TMEM16A and inhibition of Na^+ absorption. Pflugers Arch Eur J Physiol 461(5):579–589. https://doi.org/10.1007/s00424-011-0947-0

Papa G, Borodavka A, Desselberger U (2021) Viroplasms: assembly and functions of rotavirus replication factories. Viruses 13(7):1349. https://doi.org/10.3390/v13071349

Papp H, Borzák R, Farkas S, Kisfali P, Lengyel G, Molnár P, Melegh B, Matthijnssens J, Jakab F, Martella V, Bányai K (2013) Zoonotic transmission of reassortant porcine G4P[6] rotaviruses in Hungarian pediatric patients identified sporadically over a 15 year period. Infect Genet Evol 19:71–80. https://doi.org/10.1016/j.meegid.2013.06.013

Parashar UD, Hummelman EG, Bresee JS, Miller MA, Glass RI (2003) Global illness and deaths caused by rotavirus disease in children. Emerg Infect Dis 9(5):565–572. https://doi.org/10.3201/eid0905.020562

Patel MM, Pitzer VE, Alonso WJ, Vera D, Lopman B, Tate J, Viboud C, Parashar UD (2013) Global seasonality of rotavirus disease. Pediatr Infect Dis J 32(4):e134–e147. https://doi.org/10.1097/INF.0b013e31827d3b68

Quaye O, McDonald S, Esona MD, Lyde FC, Mijatovic-Rustempasic S, Roy S, Banegas DJC, Quiñonez YM, Chinchilla BL, Santiago FG, Lozano HG, Rey-Benito G, de Oliveira LH, Gentsch JR, Bowen MD (2013) Rotavirus G9P[4] in 3 countries in Latin America, 2009–2010. Emerg Infect Dis 19(8):1332–1333. https://doi.org/10.3201/eid1908.130288

Rhodes MB, Stair EL, McCullough RA, McGill LD, Mebus CA (1979) Comparison of results using electron microscope, immunodiffusion and fluorescent antibody analyses to detect rotavirus in diarrheic fecal samples of calves. Can J Comp Med 43(1):84

Rojas MA, Gonçalves JLS, Dias HG, Manchego A, Santos N (2017) Identification of two novel rotavirus A genotypes, G35 and P[50], from Peruvian alpaca faeces. Infect Genet Evol 55:71–74. https://doi.org/10.1016/j.meegid.2017.08.019

Rotavirus Classification Working Group (2019) List of accepted genotypes. Detection in clinical specimens. https://www.who.int/teams/immunization-vaccines-and-biologicals/diseases/rotavirus. (https://www.cabi.org/isc/datasheet/66267)

Sanekata T, Kishimoto E, Sato K, Honma H, Otsuki K, Tsubokura M (1991) Detection of porcine rotavirus in stools by a latex agglutination test. Vet Microbiol 27(3–4):245–251

Santos N, Gouvea V (1994) Improved method for purification of viral RNA from fecal specimens for rotavirus detection. J Virol Methods 46(1):11–21

Settembre EC, Chen JZ, Dormitzer PR, Grigorieff N, Harrison SC (2011) Atomic model of an infectious rotavirus particle. EMBO J 30(2):408–416

Snodgrass DR, Herring AJ, Campbell I, Inglis JM, Hargreaves FD (1984) Comparison of atypical rotaviruses from calves, piglets, lambs and man. J Gen Virol 65(Pt 5):909–914. https://doi.org/10.1099/0022-1317-65-5-909

Soltan MA, Tsai YL, Lee PYA, Tsai CF, Chang HFG, Wang HTT, Wilkes RP (2016) Comparison of electron microscopy, ELISA, real time RT-PCR and insulated isothermal RT-PCR for the detection of rotavirus group A (RVA) in feces of different animal species. J Virol Methods 235:99–104

Tate JE, Burton AH, Boschi-Pinto C, Steele AD, Duque J, Parashar UD, WHO-coordinated Global Rotavirus Surveillance Network (2012) 2008 estimate of worldwide rotavirus-associated mortality in children younger than 5 years before the introduction of universal rotavirus vaccination programmes: a systematic review and meta-analysis. Lancet Infect Dis 12(2):136–141. https://doi.org/10.1016/S1473-3099(11)70253-5

Tate JE, Burton AH, Boschi-Pinto C, Parashar UD, World Health Organization–Coordinated Global Rotavirus Surveillance Network (2016) Global, regional, and national estimates of rotavirus mortality in children <5 years of age, 2000–2013. Clin Infect Dis 62(Suppl 2):S96–S105. https://doi.org/10.1093/cid/civ1013

Thongprachum A, Chan-it W, Khamrin P, Okitsu S, Nishimura S, Kikuta H, Yamamoto A, Sugita K, Baba T, Mizuguchi M, Maneekarn N, Hayakawa S, Ushijima H (2013) Reemergence of new variant G3 rotavirus in Japanese pediatric patients, 2009–2011. Infect Genet Evol 13:168–174. https://doi.org/10.1016/j.meegid.2012.09.010

Trask SD, McDonald SM, Patton JT (2012) Structural insights into the coupling of virion assembly and rotavirus replication. Nat Rev Microbiol 10(3):165–177. https://doi.org/10.1038/nrmicro2673

Troeger C, Khalil IA, Rao PC, Cao S, Blacker BF, Ahmed T, Armah G, Bines JE, Brewer TG, Colombara DV, Kang G, Kirkpatrick BD, Kirkwood CD, Mwenda JM, Parashar UD, Petri WA Jr, Riddle MS, Steele AD, Thompson RL, Reiner RC Jr (2018) Rotavirus vaccination and the global burden of rotavirus diarrhea among children younger than 5 years. JAMA Pediatr 172(10):958–965. https://doi.org/10.1001/jamapediatrics.2018.1960

Trovão NS, Shepherd FK, Herzberg K, Jarvis MC, Lam HC, Rovira A, Culhane MR, Nelson MI, Marthaler DG (2019) Evolution of rotavirus C in humans and several domestic animal species. Zoonoses Public Health 66(5):546–557. https://doi.org/10.1111/zph.12575

Uprety T, Wang D, Li F (2021) Recent advances in rotavirus reverse genetics and its utilization in basic research and vaccine development. Arch Virol 166(9):2369–2386. https://doi.org/10.1007/s00705-021-05142-7

Ushijima H, Mukoyama A, Hasegawa A, Nishimura S, Konishi K, Bosu K (1994) Serotyping of human rotaviruses in the Tokyo area (1990–1993) by enzyme immuno assay with monoclonal antibodies and by reverse transcription and polymerase chain reaction amplification. J Med Virol 44:164–165

Vasileva VI et al (1987) Isolation of human rotavirus in a green monkey cell culture. Voprosy Virosily 3:360–362

Woode GN, Bridger JC, Jones JM, Flewett TH, Davies HA, Davis HA, White GB (1976) Morphological and antigenic relationships between viruses (rotaviruses) from acute gastroenteritis of children, calves, piglets, mice, and foals. Infect Immun 14(3):804–810. https://doi.org/10.1128/iai.14.3.804-810.1976

Xu L, Harbour DT, McCrae MA (1990) The application of polymerase chain reaction to the detection of rotaviruses in faeces. J Virol Methods 27(1):29–37

Part III

Emerging Viral Hepatitis Viruses

Waterborne and Foodborne Viral Hepatitis: A Perspective from the Developing and Developed Countries

17

Urmi Majumdar and Suhas T. Mhaske

Abstract

This chapter focuses on the two hepatitis-causing viruses which are primarily transmitted faeco-orally. Both Hepatitis A and E are single-stranded, positive-sense RNA viruses with lengths of more than 7 kb. For both these viruses, the prevalence is mostly in the developing countries where the level of sanitation is a major concern. Studies on zoonotic transmission rate and hosts acting as reservoirs are also underway. The classification tree places HAV under family Picornaviridae and genus *Hepatovirus* and HEV under family Hepeviridae and genus *Orthohepevirus* in case of Mammalian and avian isolates. The replication machinery, clinical manifestations, therapeutic approaches and control measures with respect to both these viruses have been explained in this chapter.

Keywords

Faeco-orally · Positive sense · Zoonotic · Reservoir · Picornaviridae · *Hepatovirus* · Hepeviridae · *Orthohepevirus*

U. Majumdar
Interactive Research School for Health Affairs (IRSHA), Bharati Vidyapeeth (Deemed to be University), Pune, India

S. T. Mhaske (✉)
Preclinical Vaccine Testing Laboratory, Rodent Experimentation Facility, ICMR-National Animal Facility Resource Facility for Biomedical Research, Hyderabad, Telangana, India
e-mail: suhas.mhaske@icmr.gov.in

P. L. Bhukya, A. S (eds.), *Emerging Human Viral Diseases, Volume II*,
https://doi.org/10.1007/978-981-97-4480-0_17

17.1 Introduction

17.1.1 Brief History

Ancient Chinese literature articles throw some light on hepatic disease outbreak which might have been caused due to Hepatitis A virus but during the studies performed during World War II, the viral causative agent and faeco-oral mode of transmission of Hepatitis A came to light. In Delhi, during 1955–1956, there was a waterborne Hepatitis E virus outbreak which involved around 29,300 cases (Viswanathan 2013). In India, in the year 1978, Kashmir witnessed the epidemic of HEV which was the cause of approximately 1700 deaths and 52,000 cases of icteric hepatitis (Khuroo 1980). At that time, only hepatitis A virus was known to be transmitted faeco-orally; so in the initial stage, this epidemic was named as Non-A and Non-B hepatitis epidemic.

17.1.2 Discovery

MacCallum coined the term Hepatitis A in the year 1947 which was adopted by the World Health Organization (WHO). Immune Electron Microscopy was used to identify the virus in faecal sample in the year 1973 (Feinstone et al. 1973). In 1983, a hepatitis outbreak was reported in Afghanistan at a soviet military camp (Balayan et al. 1983). Dr. Balayan was the first one to provide experimental evidence regarding the identity of another hepatitis-causing agent which too was transmitted faeco-orally. This form of hepatitis was named as Enterically transmitted Non-A Non-B Hepatitis (ENANB) and the agent was also known to cause sporadic cases in areas where epidemics occurred (Arankalle et al. 1993). Now, many complete sequences of the genome are known and the agent is now called as Hepatitis E Virus or HEV (Reyes et al. 1990; Purcell and Ticehurst 1988).

17.2 Epidemiology

17.2.1 Geographical Distribution

Hepatitis A virus (HAV) infection is a concern in many areas but it is prevalent in the regions where sanitation is a matter of concern, mostly the developing countries. HAV is hyperendemic in African and South Asian countries and endemic in the Middle East, South American and Eastern European countries. The economically advanced countries like The United States of America and Western European countries have lower infection prevalence.

The epidemiological patterns in case of Hepatitis E virus (HEV) include hyperendemic plus endemic regions (Asia, Africa and Latin America) where epidemic plus sporadic cases are seen plus low endemic regions (developed nations) where isolated cases or small outbreaks occur.

17.2.2 Age, Mortality and Morbidity

In general, Hepatitis A infection is mild and in the endemic regions the younger population develops immunity against the virus during the first 10 years of their lives. However, in rare cases, the infection might lead to fulminant hepatitis with an occurrence rate of 0.015–0.5%. Hepatitis E virus causes acute self-limiting hepatitis with the mortality rate generally ranging from 0.5 to 3%. In case of pregnant females, a high mortality rate, ranging from 20 to 30%, is seen if the infection occurs during the third trimester (Jameel 1999). In the younger population the infection is mostly asymptomatic and more clinical manifestations are seen in the older population.

17.2.3 Spread of Disease

According to WHO 2019 report, annually 100 million infections of HAV occur across the globe out of which 1.5 million are symptomatic. Estimated number of deaths in a year due to HAV ranges between 15,000–30,000. In the last decade, several outbreaks of HAV have been reported in the developed regions like Europe and United States. In European countries since the year 2016 various outbreaks occurred particularly in Men who have sex with men (MSM). In this group, HAV cases have also been reported in the United States, Israel and Chile. The epidemic profile in case of HAV is either Type I, Type II or Type III. Type I characterizes in the countries with poor sanitation where majority of the population gets infected during their early life. Here, the occurrence of severe epidemic is not seen. The Type II epidemic profile characterizes the countries where the seroprevalence is moderate and the sanitation is variable. The Type III profile includes the developed nations where sporadic cases might occur due to travel history to endemic areas. Between the years 2016 and 2018, San Diego outbreak of HAV received international attention. In between the period mentioned, around 590 cases of HAV infection were confirmed.

According to WHO 2015 report, annually 20 million infections occur across the entire world with greater than three million symptomatic cases and 56,600 deaths. There have been several outbreaks of HEV in various parts of the world. In 1978, Kashmir witnessed epidemic of HEV which was the cause of approximately 1700 deaths and 52,000 cases of icteric hepatitis (Khuroo 1980). During 1955–1956, Delhi witnessed an outbreak of HEV which was waterborne and around 29,300 cases were documented (Viswanathan 2013). Another notable outbreak of HEV which had taken place was the 1991 China outbreak (Bi et al. 1993). Asian and African regions show high HEV prevalence. Among Asian countries, as Japan is developed country, the prevalence rate is 10%, which is lesser than the prevalence rate in China, India and Bangladesh where greater than 30% HEV prevalence is seen. In the same way, developed areas such as European countries show HEV prevalence rates <20% (Geng and Wang 2016).

17.2.4 Biosafety Measures

Both Hepatitis A virus and Hepatitis E virus are handled in Biosafety 2 level laboratories since they can cause moderate hazard to the laboratory personnel and the environment.

17.3 Potentil Risk of Emergence and Re-Emergence

As long as sanitation standards are maintained, the chance of occurrence of faeco-orally transmitted HAV and HEV infection is less. But in developed countries where most of the population has not been exposed to these viruses, zoonotic transmission and people getting infected because of travel to an endemic region might pose a threat.

The major transmission of HAV and HEV is through the faeco-oral route. This is a major concern of the developing countries where due to water treatment inadequacy, the population has access to untreated water (Khuroo 1980). Mostly for HEV genotypes 1 and 2 are responsible for this kind of transmission. Due to the existence of the non-enveloped form of HEV, a resistant form which thrives in the environment, transmission is also possible in the developed countries after consumption of untreated water.

HEV genotypes 3 and 4 are linked with zoonotic transmission with reservoirs being domestic pigs, wild boars and sika deerr (Pavio et al. 2015). The humans can get the infection from the animals by three ways, the first being direct contact with the animals, second being foodborne and the third is surrounding contamination by animal manure run off.

Several transfusion cases have been reported in which HEV transmission has occurred (Matsubayashi et al. 2008). Therefore in areas where HEV prevalence is high, there should be a screening of blood donors for Hepatitis E virus infection.

17.4 Organization of Infectious Agent (Structural and Molecular)

17.4.1 Classification

The Hepatitis A virus has been classified as the member of family Picornaviridae with the genus being Hepatovirus. In the year 2014, HAV has been named as Hepatovirus A due to the discovery of other species under this genus. HAV has a single serotype. The VP1 sequence analysis of HAV shows segregation into five genotypes, the genotypes I–III infect humans and the remaining two infect non-human primates (Robertson et al. 1992; Costa-Mattioli et al. 2002). The HAV strains infecting humans and non-human primates are distinct. Genotype VI VP1 sequence is unavailable but it has been isolated from cynomolgus macaque from Indonesia (Costa-Mattioli et al. 2002). The fifth genotype has been isolated from three Old

World Monkey species from India, Kenya and Uganda (Nainan et al. 1991; Arankalle and Ramakrishnan 2009; Bennett et al. 2016). The fourth genotype was isolated in Philippines from a cynomolgus monkey (Nainan et al. 1991). The mean pairwise nucleotide distance among the genotypes I–III is 18.2% and the mean diversity is 4.3%. All the three genotypes infecting humans bifurcate into two subtypes A and B. The IA genotype infection is the predominant one across the world (Robertson et al. 1992; Costa-Mattioli et al. 2003; Nainan et al. 2006). In the South African and middle eastern countries, the infection is mostly due to HAV genotype IB. Genotype IIIA and IIIB cases have been reported in United States of America, Asia and Europe (Robertson et al. 1992; Nainan et al. 2006). Members of Hepatovirus A which are the different genotypes of HAV show minimal difference in their amino acid sequences.

The International Committee on Taxonomy of Viruses has recently classified Hepatitis E virus as a member of family Hepeviridae with genera being Orthohepevirus in case of all mammalian and avian isolates and Piscihepevirus in case of trout (Batts et al. 2011). SAR-55, a prototype strain of HEV which was isolated in Pakistan, was sequenced so that it could be used as a primary sequence for comparison with others such as Indian, Chinese and Burmese isolates (Bi et al. 1993; Tsarev et al. 1992; Aye et al. 1992). These earlier isolates showed sequence identity greater than 90%, hence, they were classified under genotype 1. The strains which showed similarity towards the Mexican strain of HEV like the African strains and were distinct from the SAR-55 strain were grouped under genotype 2 (Chatterjee et al. 1997). In the year 1997, association between swine HEV and human hepatitis E was found. This swine HEV isolate showed greater than 97% identity with novel HEV strains US-1 and US-2. These isolates were distinct from genotypes 1 and 2, hence, they were grouped under genotype 3 (Schlauder et al. 1998). From Chinese patients in the year 1999, new HEV isolates were obtained which showed similarity amongst themselves but were different from the genotypes 1, 2 and 3. Therefore, they were grouped under genotype 4 based on the results of phylogenetic analysis (Wang et al. 2000).

Under genus Orthohepevirus, there are four species namely Orthohepevirus A, Orthohepevirus B, Orthohepevirus C and Orthohepevirus D. HEV genotypes which can infect humans are 1, 2, 3 and 4; all four genotypes belong to Orthohepevirus A (Smith et al. 2014). Under Orthohepevirus A, there are two other genotypes 5 and 6 which include novel HEV strains found in wild boars of Japan (Smith et al. 2014; Takahashi et al. 2014). Another genotype of HEV which is genotype 7 includes strains isolated from camels of the middle eastern region which are distinct from the other six genotypes due to nucleotide difference more than 20% (Smith et al. 2014). Recently, new HEV isolates from Bactrian camels have been grouped under genotype 8 which differ genetically from genotype 7 (Sridhar et al. 2017; Woo et al. 2016). The HEV isolates from chicken belong to Orthohepevirus B (Smith et al. 2014), the HEV isolates belonging to Orthohepevirus C are from rat and ferret and the isolates from bat are under Orthohepevirus D (Table 17.1).

Table 17.1 HEV genera Orthohepevirus and Piscihepevirus

Name of species	Hosts	References
Orthohepevirus A	Human, pig, deer, mongoose, deer, wild boar, rabbit, camel, yak	Smith et al. (2014) and Woo et al. (2014)
Orthohepevirus B	Chicken	Haqshenas et al. (2002)
Orthohepevirus C	Rat, ferret	Raj et al. (2012)
Orthohepevirus D	Bat	Drexler et al. (2012)
Piscihepevirus	Trout	Batts et al. (2011)

17.4.2 Morphology and Virion Structure

HAV virion being nonenveloped and naked comprise of the four capsid proteins (VP4, VP2, VP3 and VP1) with 60 copies of each. The symmetry of the capsid is $T = 3$ (pseudo) and it forms a pentameric subunit. Though not detectable in stool samples, quasi-enveloped eHAV are detectable in the serum samples (Feng et al. 2013) which have a diameter of about 50–110 nm. These eHAV have no membrane glycoproteins. Around the fivefold and threefold axes of symmetry at the icosahedral surface of the virus particle, there are distinct features but like enteroviruses there is no depression around the fivefold axes of symmetry.

The virion structure of HEV comprises of a small particle of 320–340 A° (Balayan et al. 1983). The capsid is of icosahedral symmetry with a $T = 3$ lattice having 180 copies of capsid protein (Xing et al. 2010). The crystallographic studies have shown that there are three functional domains of HEV capsid protein, namely S (shell), M (middle) and P (protruding). The most conserved domain is S in all the genotypes of HEV (Zhai et al. 2006) which forms the icosahedral shell base for the domains M and P. The P domain is divided into two regions P1 and P2 having threefold protrusions and twofold spikes, respectively. The β barrel folds and polysaccharide binding sites present in the P1 and P2 regions aid in cell-receptor interaction and disassembly of the capsid (Guu et al. 2009). A jelly roll fold is seen in the crystal structure of S domain which is a characteristic of a lot of small RNA viruses (Rossmann and Johnson 1989). The middle domain maintains the virion stability by interacting with the other two domains (Xing et al. 2010).

17.4.3 Genome Structure and Organization

The genome of Hepatitis A virus is single stranded and positive sense with a length of approximately 7.5 kb. The proteins are translated directly from the genomic RNA which serves as messenger RNA. The 5′end is not capped but has a covalent linkage to the putative protein VPg (or 3B) which is responsible for RNA synthesis. The 3′end is poly-A tailed and the whole genome has one open reading frame (ORF). Both the 5′ and 3′ ends have UTR's with their lengths being 735 nt and 63 nt,

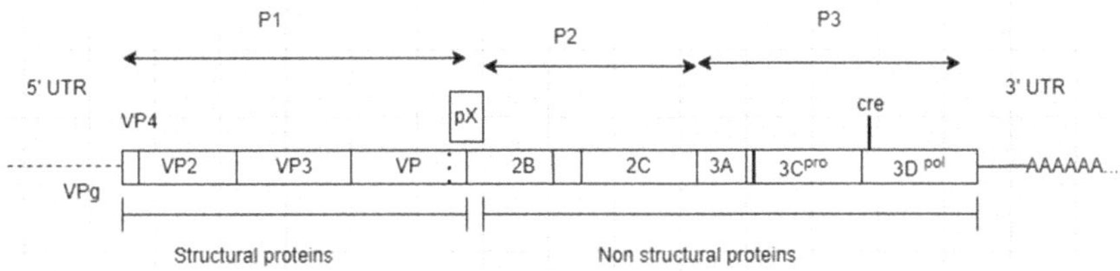

Fig. 17.1 Genome organization of HAV

respectively; secondary structures are present in both the UTR's and the 5′ hairpin loop may function as replication signal. The 3′ secondary structures may require host/viral protein interactions for the negative-strand synthesis (Kusov et al. 1997). Within the ORF lies a stem loop structure known as *cis*-acting replication element (CRE) whose major function is virus replication (Yang et al. 2008) (Fig. 17.1).

The genome of mammalian Hepatitis E virus is positive-sense single-stranded RNA (~7.2 kb) is capped and adenylated functioning as mRNA. The capping is at the 5′end and the poly-adenylated tail is at the 3′end. After the cap structure lies the 5′untranslated region (UTR) followed by the three open reading frames (ORFs) and the 3′untranslated region (Tam et al. 1991). The 5′UTR is of 26 nucleotides. The three open reading frames are namely ORF1, ORF2 and ORF3. In HEV genotype-1, fourth ORF (ORF4) overlaps the Helicase domain and the X domain (Nair et al. 2016) controlled by internal ribosomal entry site (IRES). The cis-reactive element (CRE), essential for virus replication, lies at the junction of the first and the third ORF, leading to subgenomic mRNA expression (Huang et al. 2007). The replication machinery also depends on the junction's sequence and the stem loop structure in the genome (Cao et al. 2010). The carboxy-terminal sequence of ORF2 is overlapped by a second CRE located in the 3′ untranslated region, which, during replication, interacts with the ORF1-encoded RNA-dependent RNA polymerase domain (Agrawal et al. 2001) (Fig. 17.2).

17.4.4 Propagation

HAV is hard to adapt to cell culture, hence, it is a problem to expand AV in vitro. This virus was first isolated by emptying liver explant cultures of marmoset (ex-vivo) followed by its propagation in foetal origin cells of rhesus monkey kidney as well as other mammalian cell types. Various types of primary cells and stable cell lines like Primary African green monkey kidney (AGMK) cells, primary human fibroblast cells, continuous human diploid lung (MRC5), foetal rhesus kidney (FRhK4) or human hepatoma (PLC/PRF/5) have been explored for HAV propagation. Continuous passaging may adapt HEV to grow in the cells and high titres might be obtained but this results in mutation in the viral genome. In Huh7 which is a type of human hepatoma cell, HAV is seen to adapt without mutations. HAV

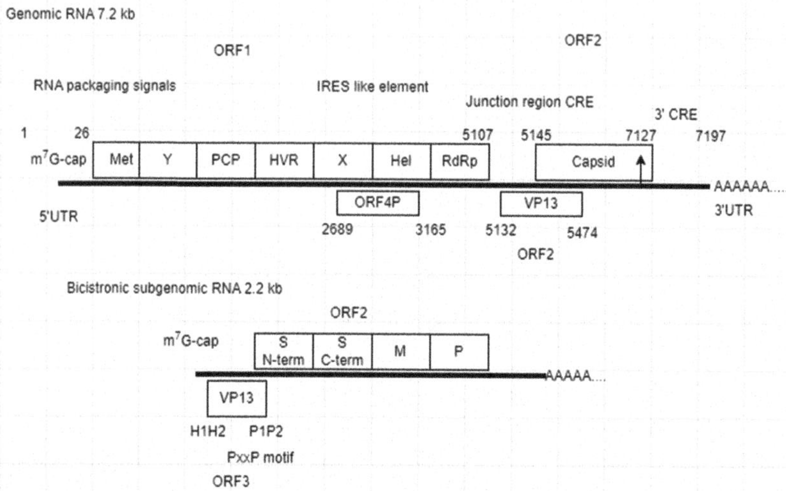

Fig. 17.2 Genome organization of HEV

cytopathic effect is difficult to detect, hence, for its detection, indirect immunological assays and real-time polymerase chain reaction can be performed. HAV cDNA clones can be used to transfect mammalian cells so that the gaps in the knowledge about HAV replication can be identified.

In case of HEV too efficient cell culture adaptation was a lacuna and cytopathic effect was not visible. cDNA clones like those of Sar-55 and pSK-HEV-2 can be transcribed in vitro and the resulting RNA when transfected in human hepatoma cells can be used for various studies. HepG2/C3A is a widely used cell line for HEV propagation which can give significant titre up to the sixth subculture. Caco-2, Huh7.5, A549, PLC/PRF/5, HepaRG and S10–3 are other cell lines used for HEV propagation. The detection method for this virus is similar to that of HAV.

17.4.5 Viral Protein and Life Cycle of HAV

The single open reading frame of HAV encodes for a polyprotein made up of approximately 2230 amino acids. The polyprotein processing leads to the formation of 10 mature proteins and is done by proteinase, 3Cpro encoded by the virus (Martin and Lemon 2002). The P1 segment present in the amino terminal of the polyprotein produces the three structural proteins namely VP0, VP3 and VP1pX which are alternatively called as 1AB, 1C and 1D proteins, respectively; 60 copies of these structural proteins form the viral capsid. The protein VP0 matures and its cleavage results in the formation of VP4 (1A) and VP2 (1B). VP1pX being the largest has a molecular weight of 38.5 kDa and it is present during the early morphogenesis (Feng et al. 2013). Cellular protease is responsible for the cleavage of pX fragment from VP1pX resulting in the formation of mature 31.2 kDa protein found in the non-enveloped and naked virions (Feng et al. 2013). HAV does not have a non-structural 2A protein in contrast to other picornaviruses. The P2 and P3 segments form the six non-structural proteins, namely 2B, 2C, 3A, 3B, 3Cpro and 3Dpol which are associated with the membrane. There is a helicase motif in the 2C protein whose overexpression is responsible for cellular membrane rearrangements. The VPg protein is responsible for RNA synthesis and it serves as a primer for the same. The membrane anchor for 3B protein is provided by the 3A protein. 3Cpro is a cysteine protease. The 3ABC processing intermediate helps in particle assembly (Probst et al. 1998). The 3Dpol is the viral RNA-dependent RNA polymerase and serves as the replicase complex catalytic core (Fig. 17.3).

In case of HAV, the virions enter by specific interaction of the capsid protein with host cellular receptors. Earlier, it was reported that T-cell immunoglobulin and mucin domain containing protein 1 (TIM1) receptor was responsible for the virus attachment and its entry into the cells (Moller-Tank et al. 2013) but recent studies have not proven this theory. However, it has been seen that TIM1 aids in the interaction between eHAV and Vero cells (Das et al. 2017). Exosome internalization can be one of the mechanisms responsible for the entry of HAV into the cells, still the actual process remains unclear. Macropinocytosis, phagocytosis and Clathrin-mediated endocytosis are some of the mechanisms for viral entry which have been pointed out (van Dongen et al. 2016).

Once the HAV RNA is in the cytosol, it encodes for the structural as well as non-structural proteins essential for the organization of the replication complex. The IRES comprising stem loops III, IV and V of the 5′untranslated region is responsible for polyprotein processing. Host translation initiation factors and accessory proteins aid in the protein processing as well (Glass et al. 1993). The first translation product is the VP4 sequence which is a result of translation initiation at one of the two nearly placed AUG codons. The HAV IRES requires eIF4G host translation

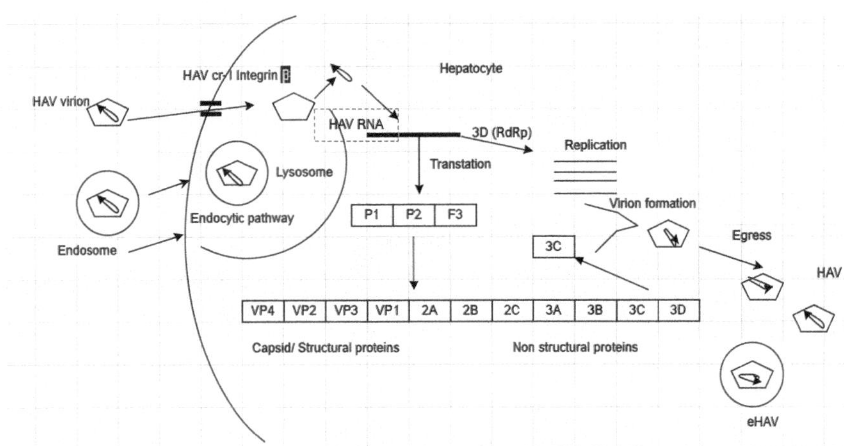

Fig. 17.3 Replication cycle of HAV

factor to initiate protein synthesis, the viral protein synthesis goes hand in and with the host mRNA translation. HAV invades host cellular proteins plus membranes to form small organelles which prevent recognition of viral replication complex by the host pathogen receptors like RIG-I-Helicases. The 2B and 2C proteins work together in intracellular membrane rearrangement to form replication complexes. These replication complexes are located near the mitochondria and endoplasmic reticulum and they are maintained inside the cytosol with the help of host factors. At first the negative strand of RNA is made from the positive one and this dsRNA intermediate makes new positive-sense RNA. For the negative-strand synthesis 3B, VPg-pUpU acts as a primer and binds to the 3′PolyA sequence. In the replication complex, the 3Dpol protein acts as the catalytic polymerase core. 3Dpol and 3CD complex interact by a slide back mechanism with the cre adenine residues leading to uridylation of the VPg protein (Paul et al. 2003). The positive-strand synthesis is also primed by uridylated VPg. The dsRNA end melts so that the uridylated VPg can bind to the 3′end of the negative strand to start the synthesis of the positive strand and all this is achieved by the viral proteins and secondary RNA structures within stem loops I and II of the 5′ untranslated region. Like replication little is known about the virion assembly and packaging. The sequence similarity of HAV 2C with that of poliovirus suggests its similar helicase activity and its function in virion packaging. Near to the replication site of HAV the virion maturation occurs as suggested by some confocal microscopy images where the capsid and dsRNA are seen co-localized in the HAV-infected cells. There is endosomal recruitment of HAV capsid because of interactions with the ESCRT (Endosomal sorting complexes required for transport) system leading to the formation of multi-vesicular bodies (MVBs) which travel to the plasma membrane and fuse with it, therefore, capsid with a single membrane around it gets released. ESCRT I and ESCRT II are responsible for recruiting a cargo at the endosomal membrane and membrane invagination whose neck is pinched off by the ESCRT III and buds are released in the lumen of the endosome forming the intraluminal vesicles or multi-vesicular bodies. These vesicles may then interact with lysosomes for cargo degradation or move towards the plasma membrane where they are released as extracellular exosomes due to fusion.

17.4.6 Viral Protein and Life Cycle of HEV

17.4.6.1 ORF1

It is the largest ORF and is of approximately 5082 nucleotides. This ORF encodes for the non-structural polyprotein (Emerson et al. 2001). It consists of functional domains which have varied roles during virus replication. The major question which remains in case of pORF1 is that whether the various domains get cleaved for performing their function or the polyprotein works as a whole. A study has reported ORF1 polyprotein domain cleavage in which HEV replicon had been expressed in S10–3 human hepatoma cells (Parvez 2013) and processing of the polyprotein has even been shown by papain-like cysteine protease which is derived from *E. coli*

(Paliwal et al. 2014). According to other studies, there is no cleavage of the pORF1 at all (Ansari et al. 2000).

The functional domains of ORF1 are:

- Methyltransferase (Met)
- Y domain
- Papain-like cysteine protease
- Hypervariable region (HVR) or Proline rich region (PRR)
- X domain
- Helicase domain
- RNA-dependent RNA polymerase (RdRp)

17.4.6.2 Methyltransferase (Met)

This domain stretches between 60–240 amino acids (Emerson et al. 2001). It consists of an invariant tyrosine amino acid, an invariant histidine residue and an aspartic acid–X–X arginine amino acid motif as it is seen in the cases of already known Methyltransferase motifs I, II and IV. The domain shows similarity to a member of α-like super group of RNA viruses Family Bromoviridae (van der Poel et al. 2001). It has been seen that an 110 kDa and a putative proteolytic product of 80 kDa are produced when ORF1 cDNA is expressed in insect cells. In in vitro assays, the protein of 110 kDa has shown guanylyl transferase and guanine-7-methyltransferase activity (Magden et al. 2001).

17.4.6.3 Y Domain

It stretches from 216 to 442 amino acids but its function is not yet understood. Members of alphavirus like superfamily which are viruses having positive-sense single-stranded RNA have shown that the Y domain is the extension of the carboxy-terminal part of methyltransferase domain. The Y domain has some amino acids which are required during replication and are conserved in various HEV strains, these amino acids include Cys-336, Cys-337 and Trp-413 (Parvez 2017). Substitutions in the Y domain conserved L410, S412 and W413 residues of 'LYSWLFE' which is a conserved α helix counterpart leading to cessation of viral replication. Even replication and infectivity efficiency of the virus has decreased due to point mutations in the hairpin structures formed by critical bases ranging from nucleotide 788–994 in this domain (Parvez 2017). In a recent case of fulminant hepatitis, $A_{317}T$ mutation has been indicated in this particular domain (Smith and Simmonds 2015).

17.4.6.4 Papain-Like Cysteine Protease

The role of this domain in the ORF1 polyprotein processing is still a debatable question. With the protease domain of Rubella virus, this domain of HEV has shown some similarity at the sequence level (Koonin et al. 1992). Rubella virus protease processes the non-structural proteins but a mutation in Cys-1152 of the catalytic domain inhibits the processing (Liang and Gillam 2000). In a study where Hepatitis E virus ORF1 was expressed via Vaccinia virus system, the processing was not affected by the mutation of Cys-483 present in the catalytic domain (Ropp et al. 2000). In mammalian cell culture system, ORF1 expression yielded two peptides of 107 and 78 kDa (Ropp et al. 2000) but after immunoprecipitation with anti-Hel, anti-RdRp and anti-Met antibodies it yielded three bands of 38, 36 and 35 kDa, respectively (Ansari et al. 2000). In bacterial system like *E. coli*, PCP which was expressed could cleave the ORF1 polyprotein and the action of protease inhibitors inhibited the processing, this showed that PCP behaved as chymotrypsin-like protease (Paliwal et al. 2014). In Huh7 hepatoma cells, the HEV genotype 1-based replicon model showed no replication when there were mutations in the histidine residues at 443, 497 and 590 positions and in the cysteine residues at positions 457, 459, 471, 472, 481 and 483 (Parvez 2013). In insect cell culture, expressed ORF1 protein processing was seen, yielding eight small peptides but upon application of cysteine protease inhibitor, the processing was inhibited. It has also been shown that this domain acts against ISG15 which is the interferon stimulated gene 15 when it is fused with the Methyltransferase domain (Karpe and Lole 2011).

17.4.6.5 HVR Domain

The hypervariable region of the ORF1 overlaps with the proline-rich region as it is a requirement for protein flexibility. It stretches from 712 to 778 amino acids and the sequence of the HVR region also varies among HEV strains (Tsarev et al. 1992). The hyper variable motif present in the HVR domain shows length differences in various HEV strains. Even due to differences in the HVR sequences within various HEV genotypes, replication efficiency variations are seen. Prominent HVR deletions lead to problems during HEV replication. In avian replicon of HEV, the complete HVR deletion became the reason for virus attenuation.

17.4.6.6 X Domain

This domain and ADP-ribose-10-monophosphatase (Appr-1′-pase) show similarity in terms of homology (Parvez 2015). Though the role of this domain is not so clear, its similarity with Appr-1′-pase which is involved in transfer RNA splicing can give some ideas. Just like the PCP domain, this domain also acts as an antagonist to the induction of interferon β by preventing interferon regulatory factor 3 phosphorylation (Nan et al. 2014). Some mutations at the catalytic region of this domain in amino acids Gly-816, Gly-817, Asn-809 and His-812 have had a negative impact on viral replication (Parvez 2015).

17.4.6.7 Helicase

This domain stretches from 960 to 1204 amino acids. Due to this domain's important function in replication, any mutation can hamper the process. Walker A and Walker B, the two conserved motifs present in this domain, are required for NTP binding or purine nucleoside triphosphate binding (Tam et al. 1991). Mutations in these motifs can either lead to loss in ATPase activity or cessation of protein function (Devhare et al. 2014). In *E. coli*, recombinant Helicase proteins of HEV have been expressed which have shown two kinds of functions, one being unwinding of 5′–3′RNA duplex unwinding and the other being NTPase activity (Karpe and Lole

2010). Another probable role of this domain is in 5′ capping as it shows γ-phosphatase activity (Karpe and Lole 2010. The in vitro studies conducted on the $L_{1110}F$ and $V_{1120}I$ mutants of helicase protein have shown reduced ATPase activity but these mutations have not hampered RNA unwinding (Devhare et al. 2014) but in S10–3 cells RNA unwinding has also been affected by these mutations. $V_{239}A$ substitution in this domain has been associated with more severe form of hepatitis, thereby indicating increased viral replication and helicase activity.

17.4.6.8 RdRp Domain

It stretches from 1207 to 1693 amino acids. Among the 8 motifs which RdRp has, GDD is the most important motif for viral replication. Hence, mutation in this particular motif prevents viral replication just like the case of Rubella virus. When there is substitution from GDD to GAD, it leads to replication-deficient replicon. $A_{1692}V$ change in RdRp might restore its enzymatic function but other mutations like $Y_{1320}H$, $K_{1383}N$, $D_{1384}G$, $K_{1398}R$, V1479I, $Y_{1587}F$ and $G_{1634}R$ have shown increased replication efficiency through proper RdRp function (Debing et al. 2014). $G_{1634}R/K$ mutation prevented any effect of Ribavirin and in turn led to increased replication (Debing et al. 2014). A relationship has been seen between increased viral load and $C_{1483}W$ and $N_{1530}T$ mutations which has led to greater mortality. These mutations have been found in the patients of acute liver failure.

17.4.6.9 ORF2

The virion structural component is the capsid and this protein is encoded by ORF2. With a total length of 1983 nucleotides, the second ORF starts 37 nucleotides downstream from the first ORF's stop codon and the stop codon is positioned 65 nucleotides upstream of the poly(A) stretch in case of HEV genotype1. Except for 14 nucleotides, this ORF overlaps with all other nucleotides of ORF3 (Reyes et al. 1993). The sub-genomic bicstronic mRNA encodes for the 660 amino acid capsid protein with a molecular mass of approximately 72 kDa. According to some experimental data, the capsid produces a protein having a molecular mass of 88 kDa which has glycans bound to the amino terminal and Asn-310 serves as the basic glycosylation site (Jameel 1999). The protein encoded by the second ORF undergoes N-linked glycosylation in the endoplasmic reticulum, therefore an amino terminal ER targeting sequence is present in the protein (Ahmad et al. 2011). This protein can form homodimers of noncovalent nature and with ER targeting motif where glycosylation is not required, it can travel to the cell membrane. Presence of Asp-562 is a must for dimerization of the ORF2encoded structural protein (Xu et al. 2016). For viral particle stability and encapsidation, the 52 amino acids of the capsid protein at the carboxy terminal are required. This ORF2encoded protein plays a major role in virus and host cell interaction. The P domain of the immunogenic capsid region contains conformational epitopes which are targeted by neutralizing antibodies. A capsid protein segment is present in China's HEV 239 subunit vaccine (Zhu et al. 2010). In interaction of the ORF2 encoded structural capsid protein with glucose regulated protein of 78 kDa (GRP-78), heat shock protein 90 (HSP90) and α tubulin is a necessity for the steps like viral attachment to the host plus its uptake and trafficking. The capsid protein alters the proper functioning of activating transcription factor 4 (ATF-4), nuclear factor κB (NF-κB), elongation initiation factor 2A (eIF2a), heat shock protein 72 (HSP72), thereby affecting host cell (Surjit et al. 2012). In human hepatoma cells, the NF-κB inhibitor α (IκBα) degradation process which is regulated by ubiquitin is blocked by ORF2 glycoprotein which in turn affects NF-κB activity negatively (Surjit et al. 2012). ORF2 3′end mutations have shown increased replication efficiency in cell culture systems (Shukla et al. 2012). For HEV replicon, where the ORF2 gene is replaced by neo gene, in the presence of G418 sulphate, the replicon containing cells survive and the ones not having neo gene get killed. Even HEV subgenomic replicons are now being used for studies in which the ORF2 gene is replaced by luciferase expressing Renilla Luciferase gene (Rluc gene).

17.4.6.10 ORF3

The third ORF overlaps the ORF3 by 300 nucleotides in a separate reading frame and it starts 24 nucleotides downstream of ORF1 stop codon (Graff et al. 2006). ORF3 is translated via bicistronic subgenomic mRNA and is known to a variety of functions. The protein is 13 kDa (113 amino acids), hence, this protein is also termed as VP13 (Graff et al. 2006). This protein has various domains namely the hydrophobic domains D1 and D2 present near the amino terminal and the proline-rich domains P1 and P2 present at the carboxy terminus. The domain D1 acts as an attachment site for cytoskeleton, microtubule and mitogen-activated protein kinase phosphatase (Kannan et al. 2009) whereas the D2 domain serves as the attachment site for hemopexin (Ratra et al. 2008). A cyclin-dependent kinase motif and a mitogen-activated protein kinase phosphorylation motif are a part of P1. Two PxxP motifs are a part of P2 which serve as interaction sites for some host transduction and structural molecules. The interaction between these motifs and tumour suppressor gene (TSG101) aids in the release of the quasi-lipid envelope-associated virions. The quasi-envelope protects virion from action of neutralizing antibody (Feng et al. 2013). The ORF3 also has similarity with viroporins encoded by some enveloped viruses (Ding et al. 2017). The complexes of ORF3 and endoplasmic reticulum membrane serve as ion channels, thus helping in ion fluxes across cell membrane in order to enhance virus budding (Ding et al. 2017). The ORF3 protein has shown interactions with host cell signalling molecules which in turn increase chances of survival of the infected host cell. Temporary expression of pORF3 in liver cells increases JNK1/2 phosphorylation, thereby enhancing the survival of the host cell (Parvez and Al-Dosari 2015). According to a recent study, ORF3 has been shown to interact with X domain of ORF1. NF-κB signalling which is induced by TLR-3 has shown suppressive activity under the action of ORF3 through RIP1 and TRADD.

17.4.6.11 ORF4

In HEV genotype 1 genome, a fourth open reading frame has been recognized which lies within the coding sequence of the first ORF but in a different reading frame (Nair et al. 2016). This ORF forms a replication complex having RdRp, X and helicase proteins. Cellular ER stress triggers IRES-regulated ORF4 translation.

ORF4 increases replication efficiency by enhancing viral RdRp functioning (Fig. 17.4).

Since there is no established in vitro cell culture model for HEV, replication cycle of this virus is not completely understood. The process of virus attachment to the receptor and its entry into the cell is still unclear. A study hypothesized that a receptor binding site is modelled by the M domain part of the capsid region (He et al. 2008) whereas another study has highlighted the presence of a highly conserved putative receptor binding motif in all the mammalian genotypes of HEV (Guu et al. 2009). A study showed binding of ORF2 VLPs to the host cell through Heparin sulphate proteoglycans or HSPGs (Kalia et al. 2009). Grp78, Heat shock cognate protein (HSC70) and other identified ORF2 interacting counterparts, possibly aid in intracellular transportation and other steps of replication instead of serving as cellular receptors (Kalia et al. 2009). Clathrin and dynamin-2 mediated pathway involving cholesterol is weighed as the virus internalization mechanism (Holla et al. 2015). Small GTPases, such as Rab5 and Rab7, are a part of quasi-enveloped virion pathway (Yin et al. 2016). The process of capsid uncoating is also an unanswered question like the process of attachment of HEV to the host cells. The polysaccharide binding sites in the P domains of the capsid protein region may serve as one of the factors important for the dismantling of the capsid (Guu et al. 2009).

Once the genomic HEV RNA gets uncoated, the ORF1 undergoes cap-dependent translation directly from the genomic RNA in the host cytoplasm. It is still unclear whether the ORF1 is split into different domains or it serves as a whole multidomain polyprotein. The subgenomic RNA of 2.2 kb (single bicistronic) gets translated into

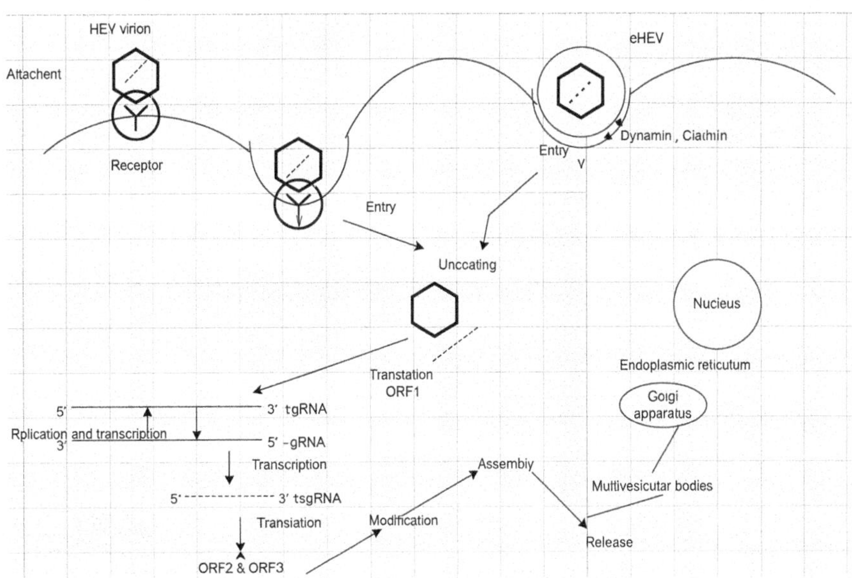

Fig. 17.4 Replication cycle of HEV

ORF2 and ORF3 (Graff et al. 2006). Earlier experiments implying infected macaque liver tissues demonstrated that three types of RNA species (a full genome 7.2 kb RNA, two subgenomic RNAs of 3.7 kb and 2 kb) bring about ORF2 and ORF3 translation (Tam et al. 1991). In vitro study based on PLC/PRF/5 hepatoma cells transfected with HEV genotype 3 confirmed the presence of a single subgenomic bicistronic RNA. The process of ORF2 and ORF3 translation includes leaky ribosome scanning phenomenon (Graff et al. 2006). Two CREs/*cis*-active RNA elements are considered to be essential for replication process (Cao et al. 2010) and HEV lacks CRE in the 5′UTR unlike other RNA viruses. An RNA encapsidation sequence could be present in the 5′ region of ORF1. As per earlier study, during the event of replication, RdRp interacts with two stem loop structures which are formed when the 3′end of ORF2 is overlapped by CRE 1 and it continues into 3′UTR (Agrawal et al. 2001). The stem loop structure formed by the second CRE (a part of the intergenic region of HEV) might act as a promoter for the formation of subgenomic RNA of 2.2 kb (Cao et al. 2010).

After the translation of RdRp as an independent protein or a part of the polyprotein, synthesis of a negative-sense RNA intermediate takes place. By employing this, subgenomic RNA as well as positive-sense genomic RNA is produced and packed in the progeny virions. The function of RdRp has been demonstrated using HEV replicon-based in vitro cell culture model (Agrawal et al. 2001). The replication complex is configured on endoplasmic reticulum or vesicles derived from the endoplasmic reticulum.

An RNA packaging signal composed of putative 76 nucleotide is present near the 5′end of HEV genome, serve as the capsid protein binding site for genome package. Hexameric rings at icosahedral threefold position along with capsid decamers at fivefold vertex are seen during the assembly. One of the factors essential for virus egression is the interaction between ORF3 and the late host domain proteins. Viroporin-like activity is another function of ORF3 expediting the ion permeability of the plasma membrane, thereby easing up viral release (Ding et al. 2017). After phosphorylation of ORF3 at Ser-80, the non-glycosylated ORF2 interacts with ORF3 at amino acid residues 57–81. Virus egression from cells takes place apically and basolaterally. Cell polarity also directs the localization of ORF3.

17.5 Pathogenesis and Host Immune Response

HAV and naked HAV are two forms of the virus which are found in serum/plasma or shed in the faeces, respectively. The enveloped HAV loses its lipid covering due to exposure to the bile salts once it is in the hepatocyte vicinity (Walker et al. 2015). In the host, the envelope of eHAV prevents the capsid from the neutralizing antibodies targeting its proteins (Feng et al. 2013). The naked HAV is very stable and maintains its infectivity potential even after it is shed in the faeces, hence, its transmissibility rate is also high. Injury to the liver cells starts after peak levels of viremia and faecal virus shedding. The alanine aminotransferase (ALT) levels are also very high during this phase. Liver is the primary replication site for HAV and then travels to the intestine through the bile canaliculi after which there is shedding of the virus in the stool. Peak virus shedding in stool and transmission occurs before elevated ALT and symptoms are seen. Viremia is seen around the same time as faecal virus shedding but the quantity of the former is lesser when compared with the latter. During HAV infection, for some time the ALT level rise coincides with the T-cell appearance, hence, though liver injury starts, viral faecal shedding decreases indicating CD8+ T cells play a role in both. Immune complex (made up of IgM, IgG, capsid proteins and C3 complement) deposition in liver has been reported in some clinical studies but its function in liver inflammation is not clear (Margolis et al. 1988). HAV is seen to protect its dsRNA intermediate from the host innate immune responses. Some initial research threw light on interferon-β expression inhibition by HAV at the cellular level. This was due to the prevention of activation of interferon regulatory factor 3 (IRF3) by retinoic acid-inducible gene I (RIG-I) or melanoma differentiation-associated gene 5 (MDA5). It was further revealed that HAV's 3ABC protein was responsible for the cleavage of the adapter molecule of MDA5/RIG-I mediated pathway. The TLR3 or Toll-like receptor-3 interferon signalling is hampered by the protease-polymerase intermediate 3CD, thereby interfering with the antiviral host response. Neutralizing antibodies against HAV play a major role in reducing viremia. IFNγ is produced by the cytotoxic T cells (CD8+) in response to the HAV-infected cells. It is even seen that during the first round of virus clearance, the CD4+ helper T cells produce a number of cytokines.

In case of HEV, the main site of replication is assumed to be the intestinal tract and its route to enter the liver is believed to be through the portal vein. In the cytoplasm of the liver cells or hepatocytes, the virus replicates and enter the bile and bloodstream. Faecal viral shedding and viremia are seen before clinical manifestations and shedding of the virus in stool continues even after its level in the blood has reached the undetectable stage. Apart from the IgM in case of acute infection and long lasting, protective IgG antibodies against HEV cell-mediated response against the virus is also seen. CD8+ cytotoxic T cell and CD4+ helper T cells play an important role in immune response in both self-limiting as well as complicated cases. In the acute and self-limiting infections, both CD8+ and CD4+ T-cell population increase is seen, therefore, higher levels of IFNγ and IL-10 is also observed. However, in case of chronic patients, the T-cell responses are attenuated. Natural Killer T-cells are responsible for increase in cytokine production and neutrophil infiltration. In case of pregnant females, HEV infections during the third trimester has a higher mortality rate. During this time of pregnancy, there is decrease in the number and function of B-cells, T-cells and Natural Killer cells. Increase in estrogen aids in HEV replication and higher level of progesterone impairs interaction between Th1 and Th2 cells.

17.6 Clinical Manifestations

17.6.1 Phases of the Diseases

The course of infection depends on viral as well as host factors. Viral factors include genotype and dose of the inoculum whereas host factors include genetic polymorphisms, presence of any kind of liver disease and pregnancy (Wedemeyer et al. 2013).

In case of HAV infection, the disease ranges from asymptomatic to fulminant hepatitis. The disease presentation is mostly age dependent, where the infection is symptomatic in case of adults but in children below 6 years of age, the infection is asymptomatic. The incubation period ranges between 2–7 weeks with the infection mostly being acute but in some cases relapse or persistent and prolonged infection up to a period of 6 months with HAV secretion is seen. Alkaline phosphatase (mean peak 319 IU/L), total bilirubin (mean peak 7 mg/dL), alanine aminotransferase of 1952 IU/L and serum aspartate aminotransferase of 1754 IU/L are seen in most of the patients. Sometimes autoimmune hepatitis, kidney injury and cholestasis are also seen.

Acute hepatitis is caused mostly by genotypes 1, 2, 3 and 4 of HEV. The incubation period can be 15 days or it can even extend up to 40 days. In most of the cases it is self-limiting lasting from a few days to a few weeks but the mean ALT level in acute cases is 1000 IU/L (Wedemeyer et al. 2012). Jaundice (with features such as dark coloured urine, increased bilirubin level, yellowish skin and sclera and pale coloured faeces) occurs in 40% and 75% of the symptomatic cases in the developing and developed areas, respectively (Kamar et al. 2014). About 1–2% acute hepatitis cases can develop into fatal fulminant hepatitis; it happens mostly in those cases where individuals have underlying liver problems (Xin and Xiao 2016). The mean level of ALT is 300 IU/L which is less compared to the acute infection (Wedemeyer et al. 2012). HEV infection occurring in pregnant females is mostly seen in South East Asian and Central African regions which are developing areas and the causative agents are HEV genotypes 1 and 2.

Mostly it is seen that chronic infections are caused by genotype 3 of HEV but recently genotype 4 HEV chronic infections have also been documented. Chronicity develops if the viremia persists 3–6 months after diagnosis of HEV. Some complications can be caused which are extrahepatic like kidney injury, thrombocytopenia, pancreatitis, Guillain–Barré syndrome, encephalitis, meningoencephalitis to name a few.

17.6.1.1 Clinical Symptoms

The common symptoms seen in case of HAV and HEV infection are nausea, vomiting, fever, abdominal pain, malaise, jaundice, dark urine, hepatomegaly and anorexia. Diarrhea, pruritus, arthralgia and skin rashes are the less common symptoms seen.

17.7 Diagnostics and Therapeutics Approaches

Detection of anti-HAV IgM antibodies in the serum is the first serologic evidence of recent infection. IgM antibodies can in some cases be present for up to 1 year and the peak in its level is seen 1 month after the exposure. Anti-HAV IgG can be detected after 1 week of IgM detection and are present lifelong conferring immunity yet in case of immunosuppression, the protection provided by the IgG antibodies is questionable. Hence, IgG antibodies indicate previous infection. HAV RNA can be detected after a few days and viremia is seen up to 3 weeks and in case of immunosuppression might be detected till the fourth week or even beyond that. Conventional or Real time Polymerase Chain Reaction can be used to detect HAV RNA.

Acute Hepatitis A virus infection is generally mild and does not require a proper treatment regimen. Fulminant hepatitis though seen in less than 5% cases is fatal and requires liver transplantation. Interferon and antiviral drug treatments are till now limited to cell culture studies therefore, thorough clinical studies need to be performed.

The first post infection marker to appear is the anti-HEV IgM which is at peak during the 8th week after infection and after 32 weeks after infection the levels are mostly undetectable. The presence of detectable IgM antibodies indicates recent infection. The anti-HEV IgG antibodies can be detected even after years of HEV infection and they come into view after the IgM antibodies. After 4 weeks of the appearance of symptoms, the IgG antibodies reach the peak level of detection. Hence, the IgG antibodies are the markers of a previous infection. Prior to the clinical manifestations, viremia is seen and it disappears 1–2 weeks after the clinical signs have subsided. Even after the HEV RNA in blood becomes undetectable, the RNA remains in detectable levels in stool for two more weeks. The RNA level in the blood is at peak during the incubation period and the phase of initial symptoms. The viremia is not interlinked with the severity of the symptoms (Kamar et al. 2014). The first test which should be performed for the diagnosis of Hepatitis E virus infection is the detection of anti IgM antibodies. If suspected samples give negative IgM results, then tests should be done for the detection of viral RNA in serum or stool samples like RT-PCR or qRT-PCR, these PCR techniques are sensitive and specific for HEV infection diagnosis. The high ALT or alanine aminotransferase level, though not a specific Hepatitis E viral infection marker indicates liver dysfunction and hence, appropriate tests need to be carried out in order to know the exact cause of the same. Two to three weeks post infection, the ALT activity becomes detectable and it reaches its peak 6 weeks after infection.

If the Hepatitis E viral infection is acute and self-resolving, then treatment is usually not required but in chronic cases, it becomes a necessity so that cirrhosis and death can be prevented. In some studies use of off-label ribavirin therapy and pegylated interferon therapy against HEV have shown some efficacy (Kamar et al. 2014). When the patient is suffering from fulminant hepatitis and has increased life risk, then liver transplantation is recommended.

17.8 Prevention and Control

For the prevention of HAV infection, proper sanitation is a must in the developing countries and underdeveloped regions where hyperendemicity is seen. Many countries have included HAV vaccine in their regimen. In the United States of America, one combination vaccine for HAV and HBV and two commercially available vaccines for HAV are given, the former as a 3 dose schedule and the latter have a 2 dose schedule. In China, live attenuated vaccine is available for HAV prevention. The vaccines have shown to confer both pre and post exposure prophylaxis.

In the areas where HEV is endemic, maintenance of proper hygienic conditions is the preventive measure whereas in developed areas where mostly the risk is zoonotic transmission, meat consumed should be cooked properly to avoid spread of HEV infection. Development of a proper vaccine is also a requirement for prevention of HEV infection. A phase 3 randomized, double blind study of a vaccine (HEV 239/Hecolin) having participation of 50,000 volunteers showed efficacy to be 100% after 12 months of vaccination and 86.8% efficacy was shown up to the follow-up period of 4.5 years. This vaccine was given in 3 doses at 0, 1 and 6 months. This vaccine is licensed in China but not in other countries (Zhu et al. 2010).

17.9 Future Perspectives

Better sanitation will eventually lead to decrease in infection rate of both the viruses, therefore, chance infections can be curtailed if universal vaccination comes into picture. Improvement in cell culture models will answer further questions regarding molecular biology and virulence of these two viruses so that novel therapeutic approaches can be found out. Research in the area of innate and adaptive immune responses against HAV and HEV has still a long way to go for better understanding of the host-virus dynamics.

References

Agrawal S, Gupta D, Panda SK (2001) The 3′ end of hepatitis E virus (HEV) genome binds specifically to the viral RNA-dependent RNA polymerase (RdRp). Virology 282(1):87–101

Ahmad I, Holla RP, Jameel S (2011) Molecular virology of hepatitis E virus. Virus Res 161(1):47–58

Ansari IH, Nanda SK, Durgapal H, Agrawal S, Mohanty SK, Gupta D et al (2000) Cloning, sequencing, and expression of the hepatitis E virus (HEV)nonstructural open reading frame 1 (ORF1). J Med Virol 60(3):275–283

Arankalle VA, Ramakrishnan J (2009) Simian hepatitis A virus derived from a captive rhesus monkey in India is similar to the strain isolated from wild African green monkeys in Kenya. J Viral Hepat 16:214–218

Arankalle VA, Chobe LP, Jha J, Chadha MS, Banerjee K, Favorov MO et al (1993) Aetiology of acute sporadic non-A, non-B viral hepatitis in India. J Med Virol 40(2):121–125

Aye TT, Uchida T, Ma XZ, Iida F, Shikata T, Zhuang H et al (1992) Complete nucleotide sequence of a hepatitis E virus isolated from the Xinjiang epidemic (1986–1988) of China. Nucleic Acids Res 20(13):3512

Balayan MS, Andjaparidze AG, Savinskaya SS, Ketiladze ES, Braginsky DM, Savinov AP et al (1983) Evidence for a virus in non-A, non-B hepatitis transmitted via the fecal-oral route. Intervirology 20(1):23–31

Batts W, Yun S, Hedrick R, Winton J (2011) A novel member of the family Hepeviridae from cutthroat trout (*Oncorhynchus clarkii*). Virus Res 158(1–2):116–123

Bennett AJ, Sibley SD, Lauck M, Weny G, Hyeroba D, Tumukunde A, Friedrich TC, O'Connor DH, Johnson CA, Rothman JM et al (2016) Naturally circulating hepatitis A virus in Olive Baboons, Uganda. Emerg Infect Dis 22:1308–1310

Bi SL, Purdy MA, McCaustland KA, Margolis HS, Bradley DW (1993) The sequence of hepatitis E virus isolated directly from a single source during an outbreak in China. Virus Res 28(3):233–247

Cao D, Huang YW, Meng XJ (2010) The nucleotides on the stem-loop RNA structure in the junction region of the hepatitis E virus genome are critical for virus replication. J Virol 84(24):13040–13044

Chatterjee R, Tsarev S, Pillot J, Coursaget P, Emerson SU, Purcell RH (1997) African strains of hepatitis E virus that are distinct from Asian strains. J Med Virol 53(2):139–144

Costa-Mattioli M, Cristina J, Romero H, Perez-Bercof R, Casane D, Colina R, Garcia L, Vega I, Glikman G, Romanowsky V et al (2002) Molecular evolution of hepatitis A virus: a new classification based on the complete VP1 protein. J Virol 76:9516–9525

Costa-Mattioli M, Di Napoli A, Ferre V, Billaudel S, Perez-Bercoff R, Cristina J (2003) Genetic variability of hepatitis A virus. J Gen Virol 84:3191–3201

Das A, Hirai-Yuki A, Gonzalez-Lopez O, Rhein B, Moller-Tank S, Brouillette R, Hensley L, Misumi I, Lovell W, Cullen JM et al (2017) TIM1 (HAVCR1) is not essential for cellular entry of either quasi-enveloped or naked hepatitis A virions. MBio 8:e00969

Debing Y, Gisa A, Dallmeier K, Pischke S, Bremer B, Manns M, Wedemeyer H, Suneetha PV, Neyts J (2014) A mutation in the hepatitis E virus RNA polymerase promotes its replication and associates with ribavirin treatment failure in organ transplant recipients. Gastroenterology 147(5):1008–11.e7. quiz e15–6

Devhare P, Sharma K, Mhaindarkar V, Arankalle V, Lole K (2014) Analysis of helicase domain mutations in the hepatitis E virus derived from patients with fulminant hepatic failure: effects on enzymatic activities and virus replication. Virus Res 184:103–110. https://doi.org/10.1016/j.virusres.2014.02.018

Ding Q, Heller B, Capuccino JM, Song B, Nimgaonkar I, Hrebikova G et al (2017) Hepatitis E virus ORF3 is a functional ion channel required for release of infectious particles. Proc Natl Acad Sci USA 114(5):1147–1152. https://doi.org/10.1007/978-81-322-1029-0_3

Drexler JF, Seelen A, Corman VM, Fumie Tateno A, Cottontail V, Melim Zerbinati R et al (2012) Bats worldwide carry hepatitis E virus-related viruses that form a putative novel genus within the family Hepeviridae. J Virol 86(17):9134–9147

Emerson SU, Zhang M, Meng XJ, Nguyen H, St Claire M, Govindarajan S et al (2001) Recombinant hepatitis E virus genomes infectious for primates: importance of capping and discovery of a *cis*-reactive element. Proc Natl Acad Sci USA 98(26):15270–15275

Feinstone SM, Kapikian AZ, Purcell RH (1973) Hepatitis A: detection by immune electron microscopy of a viruslike antigen associated with acute illness. Science 182:1026–1028

Feng Z, Hensley L, McKnight KL, Hu F, Madden V, Ping L, Jeong SH, Walker C, Lanford RE, Lemon SM (2013) A pathogenic picornavirus acquires an envelope by hijacking cellular membranes. Nature 496:367–371

Geng Y, Wang Y (2016) Epidemiology of hepatitis E. Adv Exp Med Biol 948:39–59

Glass MJ, Jia XY, Summers DF (1993) Identification of the hepatitis A virus internal ribosome entry site: in vivo and in vitro analysis of bicistronic RNAs containing the HAV 50 noncoding region. Virology 193:842–852

Graff J, Torian U, Nguyen H, Emerson SU (2006) A bicistronic subgenomic mRNA encodes both the ORF2 and ORF3 proteins of hepatitis E virus. J Virol 80(12):5919–5926

Guu TS, Liu Z, Ye Q, Mata DA, Li K, Yin C et al (2009) Structure of the hepatitis E virus-like particle suggests mechanisms for virus assembly and receptor binding. Proc Natl Acad Sci USA 106(31):12992–12997

Haqshenas G, Huang FF, Fenaux M, Guenette DK, Pierson FW, Larsen CT et al (2002) The putative capsid protein of the newly identified avian hepatitis E virus shares antigenic epitopes with that of swine and human hepatitis E viruses and chicken big liver and spleen disease virus. J Gen Virol 83(Pt 9):2201–2209

He S, Miao J, Zheng Z, Wu T, Xie M, Tang M et al (2008) Putative receptor-binding sites of hepatitis E virus. J Gen Virol 89(Pt 1):245–249

Holla P, Ahmad I, Ahmed Z, Jameel S (2015) Hepatitis E virus enters liver cell through a dynamin-2, clathrin and membrane cholesterol-dependent pathway. Traffic 16(4):398–416

Huang YW, Opriessnig T, Halbur PG, Meng XJ (2007) Initiation at the third in-frame AUG codon of open reading frame 3 of the hepatitis E virus is essential for viral infectivity in vivo. J Virol 81(6):3018–3026

Jameel S (1999) Molecular biology and pathogenesis of hepatitis E virus. Expert Rev Mol Med 1(18):1–16

Kalia M, Chandra V, Rahman SA, Sehgal D, Jameel S (2009) Heparan sulfate proteoglycans are required for cellular binding of the hepatitis E virus ORF2 capsid protein and for viral infection. J Virol 83(24):12714–12724

Kamar N, Dalton HR, Abravanel F, Izopet J (2014) Hepatitis E virus infection. Clin Microbiol Rev 27(1):116–138

Kannan H, Fan S, Patel D, Bossis I, Zhang YJ (2009) The hepatitis E virus open reading frame 3 product interacts with microtubules and interferes with their dynamics. J Virol 83(13):6375–6382

Karpe YA, Lole KS (2010) NTPase and 5' to 3' RNA duplex-unwinding activities of the hepatitis E virus helicase domain. J Virol 84(7):3595–602

Karpe YA, Lole KS (2011) Deubiquitination activity associated with hepatitis E virus putative papain-like cysteine protease. J Gen Virol 92(Pt 9):2088–2092

Khuroo MS (1980) Study of an epidemic of non-A, non-B hepatitis. Possibility of another human hepatitis virus distinct from post-transfusion non-A, non-B type. Am J Med 68(6):818–824

Koonin EV, Gorbalenya AE, Purdy MA, Rozanov MN, Reyes GR, Bradley DW (1992) Computer-assisted assignment of functional domains in the nonstructural polyprotein of hepatitis E virus: delineation of an additional group of positive-strand RNA plant and animal viruses. Proc Natl Acad Sci USA 89(17):8259–8263

Kusov YY, Morace G, Probst C, Gauss-Muller V (1997) Interaction of hepatitis A virus (HAV) precursor proteins 3AB and 3ABC with the 50 and 30 termini of the HAV RNA. Virus Res 51:151–157

Liang Y, Gillam S (2000) Mutational analysis of the rubella virus nonstructural polyprotein and its cleavage products in virus replication and RNA synthesis. J Virol 74(11):5133–5141

Magden J, Takeda N, Li T, Auvinen P, Ahola T, Miyamura T et al (2001) Virus-specific mRNA capping enzyme encoded by hepatitis E virus. J Virol 75(14):6249–6255

Margolis HS, Nainan OV, Krawczynski K, Bradley DW, Ebert JW, Spelbring J, Fields HA, Maynard JE (1988) Appearance of immune complexes during experimental hepatitis A infection in chimpanzees. J Med Virol 26:315–326

Martin A, Lemon SM (2002) The molecular biology of hepatitis A virus. In: Ou J (ed) Hepatitis viruses. Kluwer, Norwell, pp 23–50

Matsubayashi K, Kang JH, Sakata H, Takahashi K, Shindo M, Kato M et al (2008) A case of transfusion-transmitted hepatitis E caused by blood from a donor infectedwith hepatitis E virus via zoonotic food-borne route. Transfusion 48(7):1368–75

Moller-Tank S, Kondratowicz AS, Davey RA, Rennert PD, Maury W (2013) Role of the phosphatidylserine receptor TIM-1 in enveloped-virus entry. J Virol 87:8327–8341

Nainan OV, Margolis HS, Robertson BH, Balayan M, Brinton MA (1991) Sequence analysis of a new hepatitis A virus naturally infecting cynomolgus macaques (*Macaca fascicularis*). J Gen Virol 72:1685–1689

Nainan OV, Xia G, Vaughan G, Margolis HS (2006) Diagnosis of hepatitis a virus infection: a molecular approach. Clin Microbiol Rev 19:63–79
Nair VP, Anang S, Subramani C, Madhvi A, Bakshi K, Srivastava A et al (2016) Endoplasmic reticulum stress induced synthesis of a novel viral factor mediates efficient replication of genotype-1 hepatitis E virus. PLoS Pathog 12(4):e1005521
Nan Y, Yu Y, Ma Z, Khattar SK, Fredericksen B, Zhang YJ (2014) Hepatitis E virus inhibits type I interferon induction by ORF1 products. J Virol 88(20):11924–11932
Paliwal D, Panda SK, Kapur N, Varma SP, Durgapal H (2014) Hepatitis E virus (HEV) protease: a chymotrypsin-like enzyme that processes both non-structural (pORF1) and capsid (pORF2) protein. J Gen Virol 95(Pt 8):1689–1700
Parvez MK (2013) Molecular characterization of hepatitis E virus ORF1 gene supports a papain-like cysteine protease (PCP)-domain activity. Virus Res 178(2):553–556
Parvez MK (2015) The hepatitis E virus ORF1 'X-domain' residues form a putative macrodomain protein/Appr-1″-pase catalytic-site, critical for viral RNA replication. Gene 566(1):47–53
Parvez MK (2017) Mutational analysis of hepatitis E virus ORF1 "Y-domain": effects on RNA replication and virion infectivity. World J Gastroenterol 23(4):590–602
Parvez MK, Al-Dosari MS (2015) Evidence of MAPK-JNK1/2 activation by hepatitis E virus ORF3 protein in cultured hepatoma cells. Cytotechnology 67(3):545–550
Paul AV, Yin J, Mugavero J, Rieder E, Liu Y, Wimmer E (2003) A "slide-back" mechanism for the initiation of protein-primed RNA synthesis by the RNA polymerase of poliovirus. J Biol Chem 278:43951–43960
Pavio N, Meng XJ, Doceul V (2015) Zoonotic origin of hepatitis E. Curr Opin Virol 10:34–41
Probst C, Jecht M, Gauss-Muller V (1998) Processing of proteinase precursors and their effect on hepatitis A virus particle formation. J Virol 72:8013–8020
Purcell RH, Ticehurst JR (1988) Enterically transmitted non-A, non-B hepatitis: epidemiology and clinical characteristics. In: Zuckerman A (ed) Viral hepatitis and liver disease. Alan R. Liss, New York, pp 131–137
Raj VS, Smits SL, Pas SD, Provacia LB, Moorman-Roest H, Osterhaus AD et al (2012) Novel hepatitis E virus in ferrets, The Netherlands. Emerg Infect Dis 18(8):1369–1370
Ratra R, Kar-Roy A, Lal SK (2008) The ORF3 protein of hepatitis E virus interacts with hemopexin by means of its 26 amino acid N-terminal hydrophobic domain II. Biochemistry 47(7):1957–1969
Reyes GR, Purdy MA, Kim JP, Luk KC, Young LM, Fry KE et al (1990) Isolation of a cDNA from the virus responsible for enterically transmitted non-A, non-B hepatitis. Science 247(4948):1335–1339
Reyes GR, Huang CC, Tam AW, Purdy MA (1993) Molecular organization and replication of hepatitis E virus (HEV). Arch Virol Suppl 7:15–25
Robertson BH, Jansen RW, Khanna B, Totsuka A, Nainan OV, Siegl G, Widell A, Margolis HS, Isomura S, Ito K et al (1992) Genetic relatedness of hepatitis A virus strains recovered from different geographical regions. J Gen Virol 73:1365–1377
Ropp SL, Tam AW, Beames B, Purdy M, Frey TK (2000) Expression of the hepatitis E virus ORF1. Arch Virol 145(7):1321–1337
Rossmann MG, Johnson JE (1989) Icosahedral RNA virus structure. Annu Rev Biochem 58:533–573
Schlauder GG, Dawson GJ, Erker JC, Kwo PY, Knigge MF, Smalley DL et al (1998) The sequence and phylogenetic analysis of a novel hepatitis E virus isolated from a patient with acute hepatitis reported in the United States. J Gen Virol 79(Pt 3):447–456
Shukla P, Nguyen HT, Faulk K, Mather K, Torian U, Engle RE et al (2012) Adaptation of a genotype 3 hepatitis E virus to efficient growth in cell culture depends on an inserted human gene segment acquired by recombination. J Virol 86(10):5697–5707
Smith DB, Simmonds P (2015) Hepatitis E virus and fulminant hepatitis—a virus or host-specific pathology? Liver Int 35(4):1334–1340
Smith DB, Simmonds P, International Committee on Taxonomy of Viruses Hepeviridae Study Group, Jameel S, Emerson SU, Harrison TJ, Meng XJ, Okamoto H, Van der Poel WH, Purdy

MA (2014) Consensus proposals for classification of the family Hepeviridae. J Gen Virol 95(Pt 10):2223–2232
Sridhar S, Teng JLL, Chiu TH, Lau SKP, Woo PCY (2017) Hepatitis E virus genotypes and evolution: Emergence of Camel Hepatitis E Variants. Int J Mol Sci 18(4). pii: E869
Surjit M, Varshney B, Lal SK (2012) The ORF2 glycoprotein of hepatitis E virus inhibits cellular NF-κB activity by blocking ubiquitination mediated proteasomal degradation of IκBα in human hepatoma cells. BMC Biochem 13:7
Takahashi M, Nishizawa T, Nagashima S, Jirintai S, Kawakami M, Sonoda Y et al (2014) Molecular characterization of a novel hepatitis E virus (HEV) strain obtained from a wild boar in Japan that is highly divergent from the previously recognized HEV strains. Virus Res 180:59–69
Tam AW, Smith MM, Guerra ME, Huang CC, Bradley DW, Fry KE et al (1991) Hepatitis E virus (HEV): molecular cloning and sequencing of the full-length viral genome. Virology 185(1):120–131
Tsarev SA, Emerson SU, Reyes GR, Tsareva TS, Legters LJ, Malik IA et al (1992) Characterization of a prototype strain of hepatitis E virus. Proc Natl Acad Sci USA 89(2):559–563
van der Poel WH, Verschoor F, van der Heide R, Herrera MI, Vivo A, Kooreman M, de Roda Husman AM (2001) Hepatitis E virus sequences in swine related to sequences in humans, The Netherlands. Emerg Infect Dis 7(6):970–976
van Dongen HM, Masoumi N, Witwer KW, Pegtel DM (2016) Extracellular vesicles exploit viral entry routes for cargo delivery. Microbiol Mol Biol Rev 80:369–386
Viswanathan R (2013) Infectious hepatitis in Delhi (1955–1956): a critical study-epidemiology. 1957. Natl Med J India 26(6):362–377
Walker CM, Feng Z, Lemon SM (2015) Reassessing immune control of hepatitis A virus. Curr Opin Virol 11:7–13
Wang Y, Zhang H, Ling R, Li H, Harrison TJ (2000) The complete sequence of hepatitis E virus genotype 4 reveals an alternative strategy for translation of open reading frames 2 and 3. J Gen Virol 81(Pt 7):1675–1686
Wedemeyer H, Pischke S, Manns MP (2012) Pathogenesis and treatment of hepatitis e virus infection. Gastroenterology 142(6):1388–1397
Wedemeyer H, Rybczynska J, Pischke S, Krawczynski K (2013) Immunopathogenesis of hepatitis E virus infection. Semin Liver Dis 33:71–78
Woo PC, Lau SK, Teng JL, Tsang AK, Joseph M, Wong EY et al (2014) New hepatitis E virus genotype in camels, the Middle East. Emerg Infect Dis 20(6):1044–1048
Woo PC, Lau SK, Teng JL, Cao KY, Wernery U, Schountz T et al (2016) New hepatitis E virus genotype in bactrian camels, Xinjiang, China, 2013. Emerg Infect Dis 22(12):2219–2221
Xin S, Xiao L (2016) Clinical manifestations of hepatitis E. Adv Exp Med Biol 948:175–189
Xing L, Li TC, Mayazaki N, Simon MN, Wall JS, Moore M et al (2010) Structure of hepatitis E virion-sized particle reveals an RNA-dependent viral assembly pathway. J Biol Chem 285(43):33175–33183
Xu M, Behloul N, Wen J, Zhang J, Meng J (2016) Role of asparagine at position 562 in dimerization and immunogenicity of the hepatitis E virus capsid protein. Infect Genet Evol 37:99–107
Yang Y, Yi M, Evans DJ, Simmonds P, Lemon SM (2008) Identification of a conserved RNA replication element (cre) within the 3Dpol-coding sequence of hepatoviruses. J Virol 82:10118–10128
Yin X, Ambardekar C, Lu Y, Feng Z (2016) Distinct entry mechanisms for nonenveloped and quasi-enveloped hepatitis E viruses. J Virol 90(8):4232–4242
Zhai L, Dai X, Meng J (2006) Hepatitis E virus genotyping based on full-length genome and partial genomic regions. Virus Res 120(1–2):57–69
Zhu FC, Zhang J, Zhang XF, Zhou C, Wang ZZ, Huang SJ et al (2010) Efficacy and safety of a recombinant hepatitis E vaccine in healthy adults: a large-scale, randomised, double-blind placebo-controlled, phase 3 trial. Lancet 376(9744):895–902

Journey with Jaundice: A Narrative Review on the Global Health Perspective of Hepatitis A Virus and its Impact on the Modern World

18

Prasad Deepak Dandekar

Abstract

The viral aetiology of pandemic jaundice remained elusive until the mid-twentieth century. Hepatitis A virus (HAV) was recognised as a distinct entity from other varieties of hepatitis during the second World War. The prevalence of HAV has varied largely across geography and time. Locations with poor sanitation and crowded living conditions prove suitable for the emergence of HAV infections with travel-related outbreaks spreading the disease worldwide. HAV transmission occurs via ingestion of contaminated food or water, or direct contact with an infected individual. HAV can cause liver failure and rarely death especially in the older persons and those with chronic liver disease. Use of BSL-2 facility, PPE combined with administration of HAV immune globulin and/or inactivated HAV vaccines are needed for laboratory workers. HAV is an non-enveloped picornavirus with 7.5 kb positive-sense single-stranded RNA. Three genotypes infect non-human primates, and four infect humans. Each genotype's strains share ≥85% nucleotide identity. The primary capsid proteins are encoded by the amino terminal one-third of the viral genome, whereas the remaining two-thirds encode nonstructural proteins necessary for viral replication. Viral protease proteolytically processes translated viral proteins. Virions enter the liver via the portal circulation and are taken up by hepatocytes where they proliferate and are released into the bile duct. The virus's enterohepatic cycles persist until neutralising antibodies break them. Hepatocellular degeneration, apoptotic alterations, inflammation and hepa-

P. D. Dandekar (✉)
MIT School of Bioengineering Sciences & Research, MIT Art, Design and Technology University, Pune, Maharashtra, India
e-mail: prasad.dandekar@mituniversity.edu.in

P. L. Bhukya, A. S (eds.), *Emerging Human Viral Diseases, Volume II*,
https://doi.org/10.1007/978-981-97-4480-0_18

tocyte regeneration are essential histopathological characteristics of acute hepatitis. Methods like reverse transcriptase PCR (RT-PCR) amplification, DNA-RNA hybridisation and radioimmunoassay (RIA) detect and differentiate HAV viral strains. The discovery of sensitive anti-HAV IgG tests that can quantify antibodies in saliva following immunisation seems promising. A recombinant HAV antigen would be cheaper in the future compared to HAV generated in tissue culture.

Keywords

Hepatitis A virus · Jaundice · Acute hepatitis · Picornavirus · Epidemiology · Diagnosis · Electron microscopy · Liver enzymes

18.1 Introduction

18.1.1 Brief History and Discovery

Infectious jaundice was acknowledged throughout ancient China and Greece, and historically, jaundice outbreaks have been documented in Europe since the fifth century BC, with large epidemics documented in the seventeenth and eighteenth centuries. However, until the mid-twentieth century, medics didn't acknowledge the viral aetiology of pandemic jaundice. Hepatitis viruses, notably the hepatitis A virus (HAV), have been the most common causes of liver illness and jaundice (Melnick 1995). Cleghorn recorded the first outbreak of epidemic jaundice in Menorca in 1745, and described several subsequent outbreaks throughout the 1700s and 1800s in 'Epidemic Diseases in Minorca 1744 to 1749' (Cuthbert 2001).

In 1931, after analysing the history and the understanding of epidemic jaundice outbreak in Surrey, Findlay, Dunlop, and Brown determined that it was caused by a very tiny pathogenic virus whose infection can be comparable to that of other viral illnesses (Cuthbert 2001). Later, Krugman's landmark investigations at New York's Willowbrook State School confirmed two separate hepatitis viral subtypes viz. HAV and HBV are etiologic agents which were found responsible for the endemic hepatitis condition. Serum hepatitis was established as a distinct entity as HBV during the second World War, which was distinct from the epidemic viral hepatitis caused by HAV (Melnick 1995; Margolis et al. 1997).

For many years, HAV was thought to be a tiny, filterable virus. It was soon found to be passing through bacteria-proof filters with retaining to infect human volunteers, and withstand its infectivity after 30 min of heating at 50–60 °C or chlorination. However, earlier attempts of identification of its transmission to species like non-human primates were unsuccessful in laboratory settings due to negative, equivocal, or non-reproducible results. Deinhard published the first compelling evidence of transmission of viral hepatitis to non-human primates (marmosets) in 1967. First electron microscopic visualisation of HAV was performed at the National Institute of Health using the fecal samples of diseased human volunteers from Boggs' (Boggs 1970) investigation (Melnick 1995).

18.2 Epidemiology

18.2.1 Geographical Distribution/Demographic

The prevalence of HAV varies widely, with large variances across countries with occurrence fluctuating throughout time. There is a worldwide change in reported cases from infancy to adulthood (Melnick 1995). As per the WHO estimates, 7134 persons died from hepatitis A worldwide (forming 0.5% portion of the mortality due to viral hepatitis) in 2016 (WHO Hepatitis A Fact Sheet 2023.

Geographically, very high endemicity of HAV infection is found in less developed and developing nations of Asia, Africa, South and Central America, the Pacific Islands, and certain populations within the United States. On the contrary, more developed countries in Europe and Asia have intermediate to low endemicity (Margolis et al. 1997). In many industrialised nations with better cleanliness, HAV immunity is not acquired during childhood; hence, adults are generally affected. As HAV registration rates in the last 20 years in the European Union have reduced, there is an increase in the number of people vulnerable to HAV. In addition, gaps in vaccination of the high-risk populations living in very low, low and intermediate endemicity areas have contributed to recent outbreaks (Gossner et al. 2015).

Tourism is a major industry in a lot of countries with poor sanitation and common occurrence of HAV. Thus, it is especially highly risky for the travellers who travel to high-HAV prevalent areas and stay for extended periods of time (Melnick 1995). Travel-related outbreaks, regardless of the mechanism of transmission, include two incidents in the European Union. First involves the HAV subgenotype IB infection to a 100 travellers of 14 European Union countries coming from Egypt in late 2012 to mid-2013; while second also involves European travellers to Egypt in 2004 and 2008. Furthermore, HAV seeding episodes have resulted in infection to a low HAV-immune population through an ingestion- or tourism-related cause. In contrast, person-to-person spread of infection was rare. For example, only 10% of cases in the 2013 epidemic in Denmark, Finland, Norway and Sweden, which was linked to the ingestion of frozen berries, were secondary cases (Gossner et al. 2015).

The level of sanitation has a significant impact on the number of illnesses, including HAV. In locations with poor sanitation and crowded living conditions, infections emerge earlier in life. HAV infection in children is common in underdeveloped countries and usually mild, resulting in the production of antibodies and immunity (Melnick 1995).

18.2.2 Age, Mortality and Morbidity

Asymptomatic HAV infection causing no mortality was recently common in young children especially in regions having contaminated food and water. As a paradox, the average age of infection has increased in resource-poor nations due to advances in sanitation and hygiene in the last two decades as well as the introduction of inactivated HAV vaccine in children immunisation programmes (Aggarwal and Goel

2015). Till 1980, no serological testing or appropriate for differential naming hepatitis viral etiologies were available. Also, in developing countries, HAV mortality statistics are unreliable and remain poor indicators of disease incidence as accurate serological testing is not widely available (Margolis et al. 1997).

With age, the prevalence of symptomatic infection and HAV disease severity rises. In recent years, a growing proportion of European adults are becoming vulnerable to infection with HAV. Also, higher average age during infection leads to serious illness and thus higher fatality (Gossner et al. 2015).

18.2.3 Origin of Infection and Diversity

HAV viruses are produced by infected hepatocytes through surface budding, circulate in infected people's blood, and then enter the hepatocytes through endocytosis. Through portal circulation, the virus enters the host body after oral intake and after its replication in the cells of the GI tract followed by entry and replication in liver cells. Assembled virions are released into bile, travel to the colon and are eliminated in faces. Cellular and organ-lever damage of the host liver is caused by host immune response to the infection. Hepatocytes exhibit the cellular HAV receptor, and HAVCR1 helps in HAV adherence by improvements in the cytotoxicity of host NK cells against HAV-infected cells.

Based on the 168-nucleotide long sequence encoding VP1/P2A junction (I-VII), 7 HAV was initially classified into seven genotypes. But afterwards, a 6-genotype-based categorisation was proposed as per the VP1 region (I-VI) sequences; with the main genotype VII generates a genotype II subtype. Only humans may contract genotypes I through III, which are further separated into two sub-genotypes (A and B), while simians can contract genotypes IV through VI. While genotype II is only found in a few parts of Africa, genotypes I and II are more prevalent globally. There is a significant occurrence of Genotype IIIA in several Asian nations, including South Korea and India. In South Korea, it shows the increasing in incidence recently. Despite their genetic variations, all HAV genotypes share a single serotype and a single highly conserved antigenic neutralising site; hence, vaccines derived from a single HAV strain can provide protection across all HAV genotypes (Aggarwal and Goel 2015).

18.2.4 Spread of Disease

Workers hailing from low endemic countries, who predominantly work in countries where high endemicity of HAV infection is observed, are at risk of infection (Margolis et al. 1997). Multiple factors including the risk of hepatitis A infection combined with poor hygiene and sanitation, the increased likelihood of disease and mortality following infection at older ages, and post-HAV infection-induced immunity combine to produce various epidemiologic patterns of hepatitis A around the world. These are as follows:

- *Highly endemic pattern of HAV infection:* In low-resource nations with improper hygiene, where practically everyone contracts HAV throughout childhood but the virus seldom results in illness. Furthermore, such acute infection confers lifetime immunity to HAV. Therefore, despite the virus' extensive distribution in these areas, this is not a public health concern.
- *Intermediate endemic pattern of HAV infection:* The rate of the circulation of HAV is quite high in areas with intermediate socio-economic conditions but low as compared to the highly endemicity areas. Owing to this, there is also a high prevalence of clinical illness and substantial outbreaks that are either triggered via person-to-person transmission or significant water or food contamination.
- *Low endemic pattern of HAV infection*: In developed countries, the circulation of HAV is low due to good sanitation and hygiene conditions. Thus, as people across all age groups lack anti-HAV antibodies, any rare contact of HAV leads to disease transmission waves. These outbreaks are generally food-borne and occur due to age independent human contact.

Socioeconomic progress of a resource-poor community may result in the epidemiologic transition of a highly endemic HAV epidemiology to an intermediate one. Thus, although HAV transmission is reduced, a drastic shift in the pattern of sickness, hospitalisations, and mortality is observed (Aggarwal and Goel 2015).

18.2.5 Biosafety Measures

In hospitals and labs, there may be a risk of infection through indirect or direct contact with food, water and utensils that have previously come into contact with excrement from infected humans and non-human primates that also contains blood, liver tissue, and bile. Any potential infectious or contagious work, materials, animal or culture must require containment Level 2 facilities, instruments and operating procedures since HAV belongs to Risk Group 2. As per the regulations, lab coats, eye protection and gloves are indeed the minimally acceptable PPE requirements. A biological safety cabinet should be used for all aerosol-generating processes that require high concentrations or huge quantities (CDC 2024).

18.3 Structural and Molecular Organisation

18.3.1 Classification

HAV belongs to the genus *Hepatovirus* from the Picornaviridae family, but it does not cross-hybridise with other enteroviruses, tolerating extreme acidic environments and high temperatures for longer duration. (Gossner et al. 2015). This feature is remarkably different from other Picornaviridae member viruses. (Cuthbert 2001). There are six genotypes recognised as I, II, and III each with subtypes A and B. (Gossner et al. 2015). There are four genotypes (Genotypes I or III) that infect humans in addition to the three infecting non-human primates. The nucleotide identity of the strains that make up each genotype is 85% (Cuthbert 2001).

18.3.2 Morphology and Virion Structure

HAV is a tiny (27–32 nm), non-enveloped, icosahedral RNA virus with a capsid of three or four proteins, including three main proteins, VP1, VP2 and VP3 (33,000–22,000 Da), and, on rare occasions, another protein referred to as VP4 (2500 Da). Structural and nonstructural proteins of HAV are formed by the site specific proteolysis of cleavage of a single large polyprotein which is encoded by its ssRNA genome (Margolis et al. 1997; Kumar et al. 2010; Zuckerman 1996).

18.3.3 Genome Structure and Organisation

There are very few differences between genome structures of the HAV and other picornaviruses. HAV's 7.5 kb +ssRNA contains a 734–740 nucleotides long region not coding for any amino acid. A coding region of 2225–2227 nucleotides is present depending on the genotype. As 5′ cap is absent, the 5' NCR serves to establish an internal ribosome entry site to commence translation in HAV (Cuthbert 2001; Kumar et al. 2010).

18.3.4 In-Vitro and In-Vivo Laboratory Models

HAY is quite stable and remains infectious when present in faeces for at least 2 weeks. It grows slowly in cell culture, and it is resistant to inactivation by heat and also resists its extraction by nonionic detergents, chloroform, or ether, and retains infectivity in pH 1.0 at 38 °C for 90 min. (Margolis et al. 1997). History saw the use of a variety of animal models, including marmoset and primates, as well as serum, blood, and liver samples from human volunteers during biopsies, in hepatitis research in general and those involving HAV. The culture of HAV is supported by several primate-derived primary and continuous cell lines. These include primary human fibroblasts, MRC5 human diploid lung cells, and primary African green monkey kidney (AGMK) cells. In-vitro culturing of HAV is supported by several primate-derived primary and continuous cell lines. These include primary human fibroblasts, MRC5 human diploid lung cells, and primary African green monkey kidney (AGMK) cells. Numerous HAV strains have been identified, with HM175 being one of them. Through adaptation to various cell types, HM175 has produced a variety of attenuated, persistent, cytopathic and neutralisation-resistant viral varieties. When transfected into cells, RNA transcripts made in vitro from the molecularly cloned copy of the HAV genomic RNA were found to be virulent when transfected in cultured cells (Kumar et al. 2010).

18.3.5 Viral Proteins and Life Cycle

18.3.5.1 Proteins

A P1 region of the HAV genome encodes three main proteins (VP1 to VP3) of the viral capsid. Mostly, VP4, the fourth protein is not present. HAV capsid forms a single, dominant serologic epitope due to VP1 and VP3, which causes neutralising antibodies to react. P2 and P3 regions supposedly help to synthesise viral RNA and the development of virion. Highly conformational single immunodominant neutralisation epitope is composed of several sites located on YPl and YP3 (Margolis et al. 1997).

Additionally, VPg, which is chemically attached to the 5′ terminal and implicated in the start of RNA synthesis is encoded in the P3 region (Cuthbert 2001).

18.3.5.2 Life Cycle (Fig. 18.1)

With the exception of a research on owl monkeys infected with HAV, where an immunofluorescence experiment identified viral antigen in both large and small intestines and stomach. Scientific proof of the virus's fate immediately after oral exposure is not entirely apparent. Further evidence for the fate of virions has been provided by studies on liver and non-living primate cells. It is likely that viruses enter the liver through the portal blood (or after passing through the entire body) and then connect to a mucin-like class I integral membrane glycoprotein receptor to be taken up by the liver's hepatocytes. Unless the enterohepatic cycle is broken by circulating neutralising host antibodies, viral uptake from the GI tract and transfer of replicated HAV to the liver continues.

Although growing HAV in non-liver cells has been frequently described in cell culture, investigations show that in vivo replication is restricted to liver and epithelial cells of the GI tract. Using a negative-strand intermediate, the protein synthesis of viral proteins is achieved from the genomic positive strand. Thus, a RNA genome, a covalently linked VPg protein, and an icosahedral capsid (formed by VP1, VP2, and VP3) are present in a complete virus. Virus particles are discharged from the apical liver cells into the biliary canaliculus and eventually end up in bile and blood circulation. A precise and comprehensive process of viral release and secretion remains unclear (Cuthbert 2001).

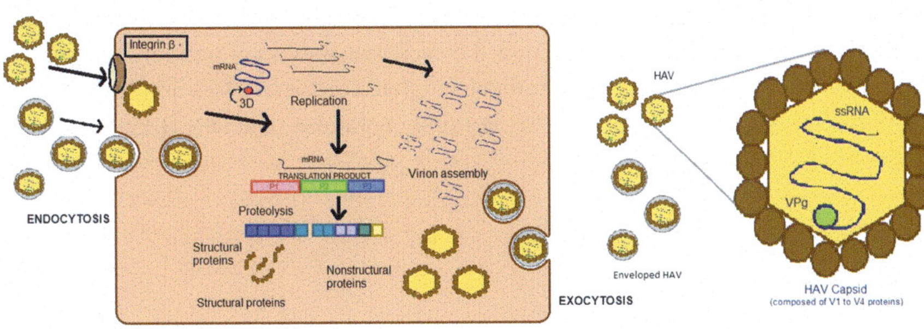

Fig. 18.1 Life cycle and structure of hepatitis A virus (based on Kanda et al. 2020)

18.4 Pathogenesis in Humans

18.4.1 Host Viral Interaction

Antibody response to the HAV emerges late in the incubation phase. After the commencement of dark urine symptoms, within a week serum anti-HAV IgM is detected. After reaching its peak, it then steadily declines over the course of the next 40–60 days. A maximal titer of the specific IgG antibody is reached 60–80 days after IgM is first detected. This protective antibody lasts for a very long time (Zuckerman 1996).

18.4.2 Immune Response of the Host

18.4.2.1 Humoral Immune Response

HAV (more specifically, epitopes of HAV's surface capsid proteins) triggers host humoral immunity very well following the infection. Anti-HAV IgA, IgG and IgM antibodies are induced and can usually be detected before clinical illness begins. IgG and IgM antibodies against VP1 are expressed rapidly, whereas IgG antibodies against VP3 can be found years after the disease has vanished. The anti-HAV IgM response is mostly restricted to the initial infection, making it a valuable marker of acute illness. The contribution of IgA to Hepatitis A-specific immunity is unclear though it is further momentarily generated. In principle, neutralising the virus could prevent the enterohepatic circulation if anti-HAV IgA gets secreted in the GI tract. However, faecal samples rarely contain neutralising IgAs in cases of experimental and naturally acquired hepatitis A.

Unlike IgM and IgA antibodies, the IgG response to HAV is slower, but it is long-lasting and helps to resist reinfection. This was confirmed in a study of Australian tribes in which anti-HAV IgG remained in circulation for 50 years without any additional exposure (Cuthbert 2001).

18.4.2.2 Cellular Immune Response

Significant HAV virions are synthesised in the hepatic cells and eliminated in the stools asymptomatically. In the cell culture also, HAV results in prolonged infection initially without any host cell damage. Cytotoxic T-lymphocytes kill target cells that have been infected with HAV. Despite the fact that they lack antigen specificity, natural killer cells could also be implicated. Therefore, cytotoxic T-cell detection and virus-infected cell death boost immune response against HAV infection (Cuthbert 2001).

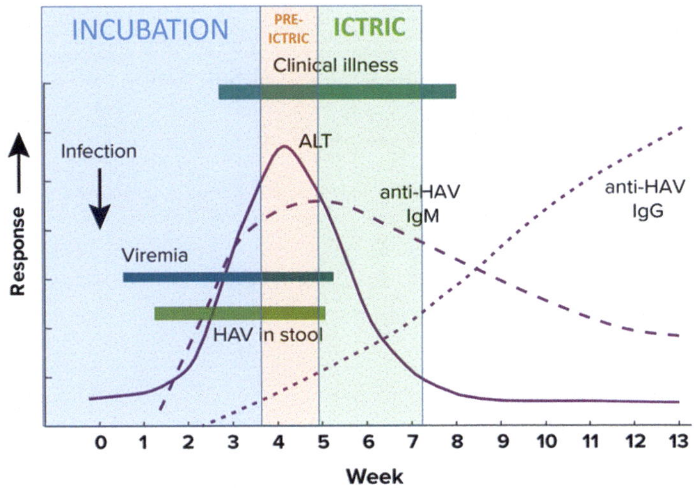

Fig. 18.2 Symptoms, phases and antibody profile of hepatitis A viral infection (based on The Centers for Disease Control and Prevention 2004)

18.5 Clinical Manifestations

18.5.1 Phases of Disease (Fig. 18.2)

18.5.1.1 Preicteric or Prodromal Phase

After an average incubation period of 30 days, the majority of people with HAV infection experience non-specific constitutional symptoms during the preicteric or prodromal period, which lasts on average for 6 days. However, before the onset of jaundice, around 15% of patients lack an evident prodrome. Without the help of modern testing, it is difficult to identify an isolated case of infectious hepatitis since the symptoms seen at this stage are so ambiguous.

18.5.1.2 Icteric Phase

The classical symptom includes conjugated bilirubinuria (dark-coloured urine). The symptoms usually disappear and the duration of jaundice varies greatly ranging from a week to a month (Cuthbert 2001).

18.5.2 Clinical Symptoms

Once symptoms appear, all hepatitis viruses have comparable clinical characteristics of viral hepatitis (Table 18.1). Headache, nausea, anorexia, vomiting, depression, pyrexia and colic are the most prevalent symptoms in the preicteric or prodromal period. The symptoms which are less frequently seen are diarrhoea, constipation cough, shivering, muscle and joint pains, itchiness, and hives. In particular, conjugated bilirubinuria is a distinctive symptom of the icteric phase.

Table 18.1 Differential diagnosis of viral hepatitis (based on Sprinzl and Galle 2008)

Type	Incubation (days)	Viral transmission route				Chronicity	Fulminant	Jaundice (%)	Fatality (%)
		Parenteral	Feco-oral	Sexual	Perinatal				
A	15–19	~	✓	×	×	×	~	5–30	0.2–2.7
B	60–180	✓	×	✓	✓	✓	✓	10–15	1–3
C	14–160	✓	×	✓	~	✓	~	5–10	1–2
D	21–42	✓	×	✓	✓	✓	✓	?	<1–>5
E	21–63	×	✓	×	×	×	✓	✓	0.5–4, 1.4–21[a]

Key: ~ rare, ✓ positive, × negative,? unknown

[a]In pregnant women

Depending on the disease case, the HAV epidemiology can be epidemic endemic or sporadic.

18.5.2.1 Sporadic Hepatitis A

The sporadic form (previously known as catarrhal jaundice) of the disease is mostly similar with the epidemic form. Similar to sporadic and epidemic poliomyelitis, sporadic and epidemic catarrhal jaundice can occur; however, jaundice is more common in the former and occurs more often and across a greater region in the latter (Cuthbert 2001).

18.5.3 Serology, Molecular and Histological Approaches

Significantly higher levels of aspartate and alanine aminotransferases (ASP and ALP) can be seen in patients with HAV infection as compared to those of hepatitis B, but the range is wide. Asymptomatic instances of acute hepatitis A typically have lower levels of aminotransferase, and this correlates with the severity of the condition. Overall disease severity is signalled by serum bilirubin and the prothrombin time. Majority of instances of hepatitis A do not show hemolysis and have low bilirubin levels, indicating less serious disease (Cuthbert 2001).

18.6 Diagnostics and Therapeutics Approaches

As previously mentioned, research at the Willowbrook State School produced the first-ever indisputable proof of two distinct hepatitis virus types (today known as HAV and HBV). It quickly made it necessary to differentiate HAV from HBV (Krugman et al. 1962; Cuthbert 2001). Early on, invasive procedures like percutaneous liver biopsy were used, and the diagnosis was made based on the presence of key indicators of acute hepatitis, such as hepatocellular degeneration, which can take the form of apoptosis, varying degrees of inflammation of the liver tissue and also its regeneration. Even though a large number of biopsy samples were studied, it was impossible to compare the pathology of A subtype of hepatitis with that of B and C in well-established, characterised patients in earlier investigations of acute hepatitis. Some researchers have reported the presence of fibrin ring granulomas in biopsy samples from acute hepatitis A, which are more frequently linked to illnesses like Q fever.

In 1973, immunological electron microscopy of faecal samples was used to specifically identify hepatitis A infection by seeing virus-like particles. Although this kind of study was successful in determining the duration of infectivity, it was unable to be applied to the regular diagnosis of numerous clinical samples. In order to identify serum antibodies to HAV antigens, complement fixation and immunological adherence assays were created in 1974. Immunological adherence & complement fixation tests shed light on the seroconversion due to the HAV infection.

Typically, immunological adherence tests have outperformed the complement fixation tests in terms of sensitivity and specificity. But now, it was possible to do a test to see if someone had an acute hepatitis A infection. HBsAg and anti-HBsAg antibodies were first detected using hemagglutination testing in 1970. Solid-phase RIAs that were created in 1975 to detect HAV antigen were changed to test antibodies.

RIA, immunological electron microscopy, and immune adherence can all be used to detect seroconversion following HAV inoculation. To test for anti-HAV IgM antibodies and to employ little to no viral antigen, RIAs were modified. By 1978, sensitivity was enhanced by the use of a competitive binding test. In this method, radiolabeled HAV antibodies compete with serum-derived antibodies from the patient. The test also detected IgM, which was not detected earlier in convalescent-phase sera. To assess IgM reactivity, it was difficult to absorb IgG from samples, and the resultant test lacked consistent specificity.

Using a new technique developed in 1980, the IgM antibody was selected first before anti-HAV efficacy was evaluated. The majority of patients had anti-HAV IgM antibody when their symptoms first appeared. IgM titers decreased throughout the first few weeks before becoming undetectable. This technique can thus be used to detect the presence of the disease at the commencement of the symptoms (Cuthbert 2001).

HAV sequences in human liver samples were recently localised using the in situ hybridisation approach. All liver cell types such as sinusoidal and inflammatory cells contained HAV RNA. There were no replicative intermediates found. A sensitive yet general sign of liver damage is the identification of viral particles in faeces during the period of clinical illness incubation as well as an increase in the levels of aminotransferases (ALT, AST). Serology provides the foundation for the precise diagnosis of hepatitis A. The most relevant test is anti-HAV IgM since its levels peak early in the acute phase of the disease and go down in 4–6 months. In rare circumstances, measuring serum glutamic oxaloacetic transaminase (SGOT) may be helpful for diagnosing anicteric and asymptomatic HAV infections (Kumar et al. 2010).

18.7 Prevention and Control

18.7.1 Management of Disease

In the past, strict bed rest was the norm until all symptoms had passed. Nowadays, the majority of people with a moderate HAV infection don't need any particular treatment. The significance of a quick hospital admission and activity limitation, however, cannot be emphasised. Common sense suggests eating healthfully and avoiding meals like fried cuisine that might cause stomach pain. Appropriate rest should also be taken as needed. However, the findings of a randomised experiment contrasting 'early and intense activity' with conventional rest ruled out the need for excessive rest. The severity of the complications and the availability of organ donations will dictate how to treat hepatitis complicated by fulminant hepatic failure. Extrahepatic symptoms including pancreatitis and renal impairment are often treated (Cuthbert 2001).

18.7.1.1 Passive Immunoprophylaxis (Gamma Globulin)

Earlier, a study implemented the passive immunoprophylaxis approach by the administration of gamma globulin with an efficiency estimated at 89%. However, as a result of the general population's declining anti-HAV seropositivity, conventional passive immunisation approaches are ineffective in providing protection as other investigations have produced inconsistent and unpromising findings.

18.7.1.2 Vaccination (Active Immunoprophylaxis)

The development of effective hepatitis A vaccines has alleviated some of the problems. In 1991, a formalin-inactivated viral vaccine manufactured at the Walter Reed Army Institute demonstrated the elicitation of neutralising anti-HAV antibodies. Another study found that a live attenuated HAV vaccine could elicit neutralising antibodies. Vaccine-induced immunoglobulins are anticipated to be long-lasting, enduring more than 20 years. To avoid intrusion of passive immunity from mother's antibodies, two doses separated by more than 6 months are traditionally recommended, starting after the child's first year. The vaccines have a nearly 95% protective efficacy (Aggarwal and Goel 2015).

The most recent recommendations for disease prevention in the USA broaden the vaccination target population. Routine HAV vaccine is now prescribed in areas having infection rates twice the national average. Furthermore, the CDC recommends using the HAV vaccine for travel outside New Zealand, Australia, Japan, Canada, Scandinavia and Western European countries. The vaccination is also recommended for travellers, homosexuals, illegal drug users, etc. (Cuthbert 2001).

18.8 Emerging Potentiality of HAV

HAV has garnered increased attention in recent years due to its evolving nature and emerging potential impact on public health. While traditionally perceived as a self-limiting disease, recent trends indicate shifts in its epidemiology, including outbreaks in various regions worldwide. Understanding the evolving potentialities of HAV in modern settings has become now crucial for developing effective preventive measures and treatment strategies. Recent epidemiological data suggest alterations in the age-specific prevalence of HAV, with a rise in cases among older age groups in regions with historically low endemicity. Emerging evidence suggests potential variations in viral virulence and mechanisms of viral persistence, warranting further investigation into the underlying factors influencing disease severity and chronicity (Jacobsen 2018).

Vaccination remains the cornerstone for HAV prevention. Advancements in molecular virology have shed light on HAV's pathogenesis, particularly its interaction with host factors and immune response modulation. The emergence of novel vaccine formulations and strategies, including combination vaccines and alternative delivery methods, holds promise for enhanced immunisation coverage and long-term immunity against HAV infection. Future research directions encompass a multifaceted approach, focusing on elucidating the genetic variability of HAV strains, investigating potential reservoirs and modes of transmission, evaluating the effectiveness of vaccination programs, and exploring therapeutic interventions for severe cases.

Thus, the evolving epidemiology, altered pathogenesis and diagnostic advancements underscore the need for continuous vigilance and research efforts to address the changing landscape of HAV. Collaborative endeavors among researchers, healthcare professionals and policymakers are imperative to mitigate the potential impact of HAV on global health (Lemon et al. 1992).

18.9 Future Perspective

Hepatitis A is becoming a global public health issue due to a lack of acquired HAV specific immunity. In wealthy countries brought on by a rapidly changing epidemiology caused by easy travel, a boom in tourism with low-income countries is responsible for the problem whereas improvements in socioeconomic and hygienic conditions in resource-poor regions have changed epidemiology of the disease. In future, therefore, a properly timed global, high-coverage childhood HAV vaccination programme, combined with careful monitoring of disease epidemiology through periodic sero-epidemiologic surveillance and cost-effective analyses, will be critical to eradicate the disease.

The detection of anti-HAV antibodies demand highly fine-tuned analysis in vaccine trials. Thus, a highly sensitive and specific anti-HAV IgG detecting assay from saliva samples post-vaccination is the need of the future. In the future, recombinant HAV antigen may provide a less costly alternative than tissue culture-produced HAV antigen for its use in diagnostic assays (Cuthbert 2001).

References

Aggarwal R, Goel A (2015) Hepatitis A. Curr Opin Infect Dis 28(5):488–496

Boggs J (1970) Viral hepatitis. Clinical and tissue culture studies. JAMA J Am Med Assoc 214(6):1041–1046

CDC (2024) Review of Scientific Data Regarding Transmission of Infectious Agents in Healthcare Settings. https://www.cdc.gov/infectioncontrol/hcp/isolation-precautions/scientific-review.html Accessed 25 June 2024

Centers for Disease Control and Prevention (2004) Diagnosis and management of foodborne illnesses: a primer for physicians and other healthcare professionals. MMWR 53(RR-4):17

Cuthbert J (2001) Hepatitis a: old and new. Clin Microbiol Rev 14(1):38–58

Gossner C, Severi E, Danielsson N, Hutin Y, Coulombier D (2015) Changing hepatitis A epidemiology in the European Union: new challenges and opportunities. Eur Secur 20(16)

Jacobsen KH (2018) Globalization and the changing epidemiology of hepatitis a virus. Cold Spring Harb Perspect Med 8(8):a031716

Kanda T, Sasaki R, Masuzaki R, Matsumoto N, Ogawa M, Moriyama M (2020) Cell culture systems and drug targets for hepatitis a virus infection. Viruses 12(5):533. https://doi.org/10.3390/v12050533

Krugman S, Ward R, Giles JP et al (1962) The natural history of infectious hepatitis. Am J Med 32(5):717–728. https://doi.org/10.1016/0002-9343(62)90161-4

Kumar V, Das S, James S (2010) The biology and pathogenesis of hepatitis viruses. Curr Sci 98(3):312–327

Lemon SM et al (1992) Type a viral hepatitis: a summary and update on the molecular virology, epidemiology, pathogenesis and prevention. J Hepatol 15(Suppl):1–5

Margolis HS, Alter MJ, Hadler SC (1997) Viral hepatitis. In: Evans AS, Kaslow RA (eds) Viral infections of humans. Springer, Boston, MA. https://doi.org/10.1007/978-1-4899-0036-4_13

Melnick J (1995) History and epidemiology of hepatitis A virus. J Infect Dis 171(Supplement 1):S2–S8

Sprinzl MF, Galle PR (2008) Differential diagnosis of human hepatitis. In: Weber O, Protzer U (eds) Comparative hepatitis. Birkhäuser Advances in Infectious Diseases. Birkhäuser Basel. Springer. https://doi.org/10.1007/978-3-7643-8558-3_2

WHO Fact Sheet of Hepatitis A (© 2023 WHO) https://www.who.int/news-room/fact-sheets/detail/hepatitis-a

Zuckerman AJ (1996) Hepatitis Viruses. In: Baron S, editor. Medical Microbiology. 4th edition. Galveston (TX): University of Texas Medical Branch at Galveston; 1996. Chapter 70. Available from: https://www.ncbi.nlm.nih.gov/books/NBK7864/

Hepatitis B Virus Infection: Current Trends in Host-Pathogen Interaction, Diagnosis, & Treatment

19

Shalini Sakthivel, L. Preethi, and Prudhvi Lal Bhukya

Abstract

Hepatitis B virus (HBV) is a significant and widespread human pathogen primarily targeting hepatocytes, resulting in severe liver illnesses. Despite the presence of a potent vaccine against HBV, there are still over 250 million chronic infections globally. HBV is an enveloped partially double-stranded DNA virus. The transmission of HBV can occur through percutaneous or mucosal exposure to infected bodily fluids, and it can remain viable on environmental surfaces for more than 7 days. Clinical manifestations of HBV infection vary between acute and chronic diseases. Acute infection may present as icteric hepatitis, subclinical or anicteric hepatitis, or rarely, fulminant hepatitis, while chronic infection may lead to an asymptomatic carrier state, cirrhosis, and hepatocellular carcinoma (HCC). Preliminary indications are nonspecific and may include nausea, vomiting, anorexia, and abdominal pain. Suppose in severe liver damage, patients can progress to jaundice, ascites, hepatic encephalopathy, gastrointestinal bleeding secondary to oesophageal varices, or coagulopathy. The primary objective of the proposed chapter is to confer HBV structural and molecular organization, life cycle, pathogenesis, and the available therapeutic interventions for HBV.

S. Sakthivel
Bar-Ilan University, Azrieli Faculty of Medicine, Safed, Israel

L. Preethi
MSc Public Health, London School of Hygiene and Tropical Medicine, London, UK

P. L. Bhukya (✉)
Preclinical Vaccine Testing Laboratory, Rodent Experimentation Facility, ICMR-National Animal Facility Resource Facility for Biomedical Research, Hyderabad, Telangana, India
e-mail: saiprudhvi21@gmail.com; prudhvi.bhukya@icmr.gov.in

P. L. Bhukya, A. S (eds.), *Emerging Human Viral Diseases, Volume II*,
https://doi.org/10.1007/978-981-97-4480-0_19

Keywords

HBV · Jaundice · Cirrhosis · HCC

19.1 Introduction

The liver stands out as the most crucial organ within the human body, playing a pivotal role in various physiological processes. However, liver infection causes diverse diseases. Hepatitis B is one of the contagious diseases that cause liver inflammation. Contrary to directly inflicting harm upon liver cells, the hepatitis B virus triggers liver inflammation through immune system responses. Annually, on the 28th of July, the global community commemorates World Hepatitis Day, aiming to bolster awareness surrounding viral hepatitis, attributed to five main types of the hepatitis virus—A, B, C, D, and E. Notably, the hepatitis B virus (HBV) poses a significant risk of life-threatening liver infections, including cirrhosis and liver cancer. As stated by the World Health Organization (WHO), the administration of a safe and efficacious three-dose vaccine provides 98%–100% protection against HBV infection. WHO guidelines emphasize that the newborns should receive the first dose of the hepatitis B vaccine within 24 hours of birth, followed by two additional doses administered at least 4 weeks apart (WHO 2024). Liver cirrhosis (LC) stands as the 11th most common cause of death globally and presents as an emergent source of morbidity and mortality in developed nations. The prevalence of hepatocellular carcinoma (HCC) ranks as the sixth highest among all cancers and represents the third most common cause of cancer-related deaths worldwide (Bandopadhyay and Bharadwaj 2020).

19.1.1 Discovery

Hepatitis B has been recognized as a disease since ancient times, although its causative agent was only identified relatively recently. Historical records, dating back to the era of Hippocrates, mention outbreaks of jaundice, but the first documented epidemic of what was later termed "serum hepatitis" was noted by Lurman in 1885 (Lurman 1885). Baruch S. Blumberg (1925–2011), an American physician and geneticist, is known for his groundbreaking work. He had a keen interest in the study of lipoprotein polymorphisms. In the 1950s and 1960s, Blumberg's research team collected serum samples from various ethnic groups worldwide for his studies. They observed peculiar reactions in serum samples from a hemophilic patient who had received transfusions. Blumberg's colleague, Harvey Alter, identified a novel antigen in several samples, often found in Australian aborigines, which came to be known as the Australia antigen (AuAg) in 1964. This discovery marked the beginning of modern research into viral hepatitis (Blumberg 1964; Gerlich 2013). In his discovery of the antigen later identified as the hepatitis B surface antigen (HBsAg), Blumberg employed a technique called agar gel double diffusion, which was developed by Ouchterlony (Gerlich 2013).

19.1.2 Brief History

The discovery of the AuAg is often observed as the beginning of hepatitis B studies.

- In 1967, Blumberg's group showed a robust correlation between AuAg and serum hepatitis (Blumberg et al. 1967).
- In 1970, David S. Dane (London) isolated virus particles from a patient's serum with AuAg-associated hepatitis and demonstrated "Dane particle (42 nm in diameter)" as the complete virus using an electron microscope (Dane et al. 1970), which was behindhand formally named as the HBsAg by the WHO.
- In 1971, the groundwork for the HBV immunization was laid with the use of HBV immunoglobulins to prevent HBV infection (Tang 2019).
- In 1973, William S. Robinson successfully uncovered an DNA polymerase enzyme within HBV. This was followed by his identification of the DNA molecule created by this enzyme in 1974 (Gerlich 2013).
- Blumberg's significant contributions to the field were acknowledged with the prestigious Nobel Prize in Physiology or Medicine in 1976. This recognition stemmed from his discovery of the hepatitis B virus (HBV) and his groundbreaking idea of developing the first generation of plasma-derived HBsAg vaccine (Tang 2019).
- Around 1978, notable progress was made in the field with the concurrent cloning and sequencing of HBV DNA by three trailblazers in the field of molecular biology: Pierre Tiollais in Paris (Galibert et al. 1979), William Rutter in San Francisco (Valenzuela et al. 1979), and Kenneth Murray in Edinburgh (Pasek et al. 1979).
- By 1978, the first generation of plasma-derived HBV vaccine, comprising HBsAg particles, underwent successful validation in clinical trials (Szmuness et al. 1980).
- Sequencing the HBV genome in 1979 by Galibert was a key advancement for a new vaccine. This revealed a unique 3.2 kbp, partially double-stranded DNA virus, paving the way for the second-generation, genetically engineered vaccine (Galibert et al. 1979).
- In 1989, William Carman was the first to describe HBeAg negative variants of HBV, characterized by a mutated, nonfunctional preC sequence, which can lead to fulminant hepatitis B.

19.2 Epidemiology

19.2.1 Geographical Distribution

Even though there's a highly effective vaccine available against HBV, the infection remains widespread globally and continues to contribute significantly to morbidity and mortality. Research indicates that approximately 200 million individuals worldwide are exposed to the HBV, with 350 million harboring it chronically (Ray 2017). HBV infection is especially widespread in the Western Pacific and

African regions, affecting an estimation of 6.2% and 6.1% of the population, respectively. These rates are significantly higher compared to the Eastern Mediterranean (3.3%), Southeast Asia (2.0%), Europe (1.6%), and the Americas (0.7%) (Daka et al. 2022).

19.2.2 Diversity

Researchers have categorized countries into three groups based on the prevalence of hepatitis B infection (Ray 2017; Aspinall et al. 2011). These categories are as follows:

- High endemicity (over 8% infected population): This group includes regions with a huge number of people chronically infected with hepatitis B virus. Examples include Southeast Asia, China, sub-Saharan Africa, parts of South America, and Alaska.
- Intermediate endemicity (2%–8% infected population): Countries in this category have a moderate prevalence of hepatitis B. This group includes North Africa, parts of South America, parts of Southern Europe, Eastern Europe, the Eastern Mediterranean area, and the Indian subcontinent.
- Low endemicity (less than 2% infected population): This group has the lowest prevalence of hepatitis B infection. It includes Australia, most of North America, Northern Europe, and most of Western Europe (Aspinall et al. 2011).

Studies have been shown that there is a minimum of eight genotypes of HBV, entitled A-H between each genotype, the genetic variances more than 8%, but less than 17%. Apart from that, two further genotypes have been termed as "I" and "J." Genotype I has around 8% genomic divergence from genotype C, and genotype J is the result of the recombination of genotype C and gibbon HBV (Lamontagne et al. 2016). India has a moderate prevalence of hepatitis B carriers, estimated around 3.0%, with a higher rate among tribal populations (Puri 2014). The specific type of hepatitis B virus (genotype) circulating in an area can influence disease progression and treatment options. Here is a breakdown of HBV genotypes by region:

- India, Europe, and Africa: Primarily genotypes A and D
- Asia: Primarily genotypes B and C
- West Africa: Genotype E
- Central and South America: Genotype F
- Central America and Southern Europe (uncertain prevalence): Genotypes G and H
- Vietnam, Laos: Genotype I
- Japan: Genotype J (rare, recombinant between genotype C and gibbon hepatitis B virus) (Karayiannis 2017).

Furthermore, genotypes A, B, C, D, F, and I can be subdivided into 53 types based on specific nucleotide divergence of 4%: A1–7, B1–9, C1–16, D1–9, F1–6, and I1–2 (Pourkarim et al. 2014).

19.2.3 Phylogenetic Analysis

The phylogenetic tree was constructed by using the maximum likelihood method based on HBV full-length genomes of all proposed genotypes and subgenotypes, as shown in Fig. 19.1. The evolutionary analysis was conducted in MEGA X. This analysis involved 61 nucleotide sequences which are collected from GenBank® sequence database and listed as follows (Accession Number): ABO14370: A; JN182318: A1; HE576989: A2; AB194951: A3; AY934764: A4; FJ692613: A5; GQ331047: A6; FN545833: A7; ABO73825: B; AB642091: B1; FJ899779: B2; GQ924617: B3; GQ924626: B4; GQ924640: B5; JN792893: B6; GQ358137: B7; GQ358147: B8; GQ358149: B9; ABO14360: C; AB697490: C1; GQ358158: C2;

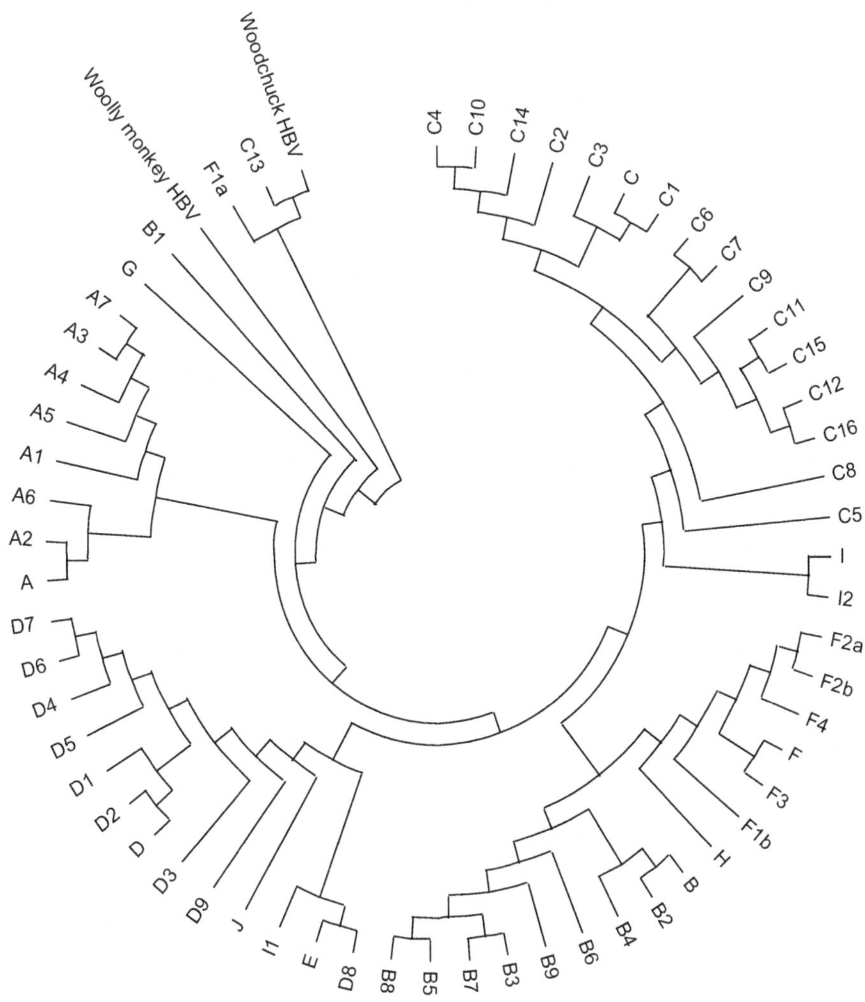

Fig. 19.1 Phylogenetic tree of HBV genotypes and subgenotypes

DQ089801: C3; HM011493: C4; EU410080: C5; EU670263: C6; GU721029: C7; AP011106: C8; AP011108: C9; AB540583: C10; AB554019: C11; AB554025: C12; AB644280: C13; AB644284: C14; AB644286: C15; AB644287: C16; AB210822: D; GU456636: D1; GQ477452: D2; EU594434: D3; GQ922003: D4; GQ205377: D5; KF170740: D6; FJ904442: D7; FN594770: D8; JN664942: D9; AB194947: E; AB036911: F; AY090459: F1a; FJ709464: F1b; DQ899142: F2a; DQ899146: F2b; AB036920: F3; AF223965: F4; ABO56513: G; AB059659: H; AB231908: I; FJ023659: I1; FJ023664: I2; AB486012: J and M11082: Woodchuck HBV; AY226578: Woolly monkey HBV.

19.3 Molecular and Structural Organization of HBV

19.3.1 Classification

HBV is a prototype member which belongs to the Hepadnaviridae family, it predominately infects hepatocytes in their respective hosts, and it is one of the most significant human DNA viruses (Bandopadhyay and Bharadwaj 2020). The Hepadnaviridae family is categorized into two genera based on their genomic sequence divergence and restricted host range of infection. Table 19.1 illustrates the Hepadnaviruses found in both mammals (orthohepadnaviruses) and birds (avihepadnaviruses) (Schaefer 2007). Avihepadnaviruses exhibit a 20% genome sequence divergence, while orthohepadnaviruses can have up to 40% genome sequence divergence (Lamontagne et al. 2016). HBV is a hepatotropic noncytopathic virus with potent oncogenic properties and is frequently associated with acute and chronic liver infections (Bandopadhyay and Bharadwaj 2020).

19.3.2 Genomic Organization

HBV is an extremely small (42 nm), enveloped virus encompassing a relaxed-circular, partially double-stranded compact hepadnaviral genome 3182–3248 bp in length (variable with the genotype) approximately 3.2 kilobases (Herrscher et al. 2020) that encodes four overlapping open reading frames (ORFs) which expresses seven proteins summarized in Table 19.2. The largest ORF, polymerase encodes viral polymerase which also has a reverse transcriptase (RT) activity. The second-largest ORF, preS/S, encodes the surface antigens, and there are three surface antigens including large surface antigen (L-HBsAg), middle surface antigen (M-HBsAg), and small surface antigen (S-HBsAg) present in the envelope at a ratio of approximately 1:1:4 (Wu et al. 2018). Another ORF, precore (PC)/core(C), encodes precore, also known as HBV e antigen (HBeAg), and core protein, constituting the viral capsid. The smallest ORF, X, encodes the HBV X protein (HBx), a small regulatory protein crucial for HBV replication, with an associated oncogenic role (Lamontagne et al. 2016). These viral ORFs are transcribed into distinct capped and polyadenylated RNAs, which can be categorized into subgenomic and

Table 19.1 List of Orthohepadnaviruses and Avihepadnaviruses with their hosts

Orthohepadnavirus		Avihepadnavirus	
Species	**Host**	**Species**	**Host**
Hepatitis B virus (HBV)	*Homo sapiens sapiens* (Human)	Duck HBV	*Anas domesticus* (Pekin duck)
Chimpanzee HBV	*Pan troglodytes*	Grey Teal HBV	*Anas gibberifrons gracilis*
Gibbon HBV	*Hylobates lar* (White handed gibbon)	Heron HBV	*Adrea cinerea*
Orangutan HBV	*Pongo pygmaeus*	Maned Duck HBV	*Chenonetta jubata*
Gorilla HBV	*Gorilla gorilla*	Ross Goose HBV	*Anser rossi*
Woolly Monkey HBV	*Lagothrix lagotricha*	Snow Goose HBV	*Anser caerulescens*
Woodchuck HBV	*Marmota monax*	Stork HBV	*Ciconia ciconia* (White Stork)
Ground Squirrel HBV	*Spermophilus beecheyi*	Crane HBV	*Anthropoides virgo* (Demoiselle cranes), *Balearica regulorum* (Grey crowned cranes).
Arctic Squirrel HBV	*Spermophylus parryi kennicotti*		

pregenomic transcripts. Subgenomic transcripts serve as templates for HBx and the three surface antigens, while pregenomic transcripts (pgRNAs) act as mRNA and serve as templates for HBV replication, being reverse transcribed into the HBV DNA genome for precore, core, and viral polymerase (Nassal 2008). Despite the viral genome's small size (3.2 kb), the pgRNA is slightly larger at 3.5 kb (Lamontagne et al. 2016). Additionally, all regulatory elements, including enhancers (Enh1, Enh2), promoters (core, S1, S2, and X), polyadenylation, encapsidation (e), and replication signals (DR1, DR2), are located within these ORFs (Karayiannis 2017).

19.3.2.1 Polymerase ORF

The HBV viral DNA polymerase has a molecular weight of approximately 90 kDa and consists of an 838 amino acid (aa) polymerase protein, also known as RT, pol, or P, which is encoded by the polymerase ORF. This ORF, the largest within the

Table 19.2 List of ORFs and proteins of HBV

ORFs	Proteins	Functions
Polymerase (P) ORF	Viral DNA polymerase (90 kDa)	Viral DNA synthesis
Pre-S/S ORF	Three envelope proteins L-HBsAg (39 kDa) M-HBsAg (31 kDa) S-HBsAg (24 kDa)	Forms structural aspect of the virion
Pre-C/C ORF	Precore or HBeAg (15 kDa)	Act as an immune modulator
	HBcAg (21 kDa)	Make up the viral capsid
X ORF	HBx Protein (17 kDa)	Regulatory protein which required for viral replication

HBV genome, covers nearly 80% of its sequence. Structurally, it comprises three functional domains along with a hinge region known as the Spacer domain. The first functional domain, located at the N-terminal region, is the TP domain, also referred to as the terminal protein domain. This domain plays a crucial role in initiating genome replication (Weber et al. 1994). The second domain is the RT domain, which is responsible for genome replication by catalyzing the reverse transcription process. This involves reverse transcribing the pgRNA to form a negative strand of the DNA genome and subsequently using this negative strand as a template for synthesizing the positive strand of the DNA. The third domain, designated as the P domain, is the RNase H domain. This domain is responsible for degrading the pgRNA template during the synthesis of negative strands of the DNA genome, ensuring accurate replication processes. These domains on either side are separated by the hinge region (Lamontagne et al. 2016).

19.3.2.2 Pre-S/S ORF

The HBV employs a unique strategy for the production of its envelope proteins, which collectively constitute the viral outer shell. These envelope proteins exist in three distinct sizes: large (L), middle (M), and small (S). Notably, all three proteins are encoded by a single ORF but are translated from two distinct subgenomic messenger RNAs (mRNAs). The L protein is translated from a dedicated 2.4 kb subgenomic RNA driven by the preS1 promoter located upstream of the S ORF. Conversely, the M and S proteins are translated from a shared 2.1 kb subgenomic RNA controlled by the preS2/S promoters (Dane et al. 1970). The presence of the preS2 ORF within the preS1 region suggests a regulatory role of preS1 in the expression of the 2.1 kb subgenomic RNA.

The smallest envelope protein, S (24 kDa), comprises 226 amino acids and forms the shared C-terminal domain of the two larger envelope proteins. The M protein (31 kDa) incorporates the S sequence along with a 55-amino acid N-terminal extension designated as preS2. The expression of M- and S-encoding mRNAs is governed by the S promoter, with translation initiating from either an upstream (M) or downstream (S) AUG codon. The L protein (39 kDa), the largest envelope protein, encompasses the S sequence, preS2, and an additional N-terminal extension known as preS1. This preS1 extension varies in length from 108 to 119 amino acids depending on the HBV genotype. The L-HBsAg protein is produced from its own unique mRNA transcript regulated by the preS1 promoter (Lamontagne et al. 2016).

Mature, infectious HBV virions, referred to as Dane particles, contain both M-HBsAg and L-HBsAg within their envelopes. HBV-infected hepatocytes, however, also generate noninfectious subviral particles (SVPs) composed primarily of S-HBsAg with variable quantities of M-HBsAg and minimal to no L-HBsAg. Remarkably, the concentration of these SVPs in the serum of an infected individual can be up to 10,000-fold higher than that of infectious HBV particles (Ganem and Prince 2004; Bruss 2004).

19.3.2.3 PreC/C ORF

The PreC/C ORF encodes two essential proteins: precore (HBeAg) and core (HBcAg) protein. This ORF contains two start codons and a common stop codon. The core protein, with a molecular weight of approximately 21 kDa, serves as the structural backbone for the mature nucleocapsid envelopment (capsid assembly) and is involved in initiating reverse transcription. The core is translated from the pregenomic RNA (pgRNA), with the first 149 amino acids forming the assembly domain. This domain is adequate for the in vitro formation of capsids indistinguishable from those isolated from Dane particles. The remaining 34–36 amino acids constitute the arginine-rich C-terminal domain (CTD), where the phosphorylation of various amino acids regulates multiple stages of the HBV life cycle (Li et al. 2010; Lamontagne et al. 2016). The HBeAg protein, an antigenic determinant with a molecular weight of 15 kDa, is translated from the precore mRNA, with its 5′ ends positioned a few nucleotides upstream of the first AUG in the precore/core ORF (Hollinger and Liang 2001). Before secretion from infected cells, it undergoes a secondary cleavage removing approximately 34 amino acids from the C-terminus, a process occurring in the endoplasmic reticulum. Although the precise function of the HBe protein remains elusive, it plays crucial roles in viral replication, immune evasion, and exhibits antiapoptotic properties. HBeAg mediates the downstream regulation of cytokines and interferon production (Lang et al. 2011). HBeAg has been implicated in the inhibition of NF-kB and IFN-β-promoters activation and also reduces IL-6 production. Additionally, it decreases Toll-like receptor 2 (TLR2) expression and interacts directly with TRIF-related adaptor molecule (TRAM) and MyD88 adaptor-like protein (MAL) in the TLR2 pathway (Lang et al. 2011). Furthermore, HBeAg interacts with Toll/interleukin-1 receptor (TIR) proteins, modulating NF-kB and IFN-β activation, thereby antagonizing TIR and affecting downstream signaling (Assrir et al. 2010). Reports suggest that HBeAg's involvement in antiapoptotic functions and HCC development. HBeAg, along with p22, interacts with NUMB, a cell fate determinant, inhibiting cell death by antagonizing the activity of NOTCH family plasma membrane receptors. Moreover, it is considered a positive regulator of the p53 protein. This interaction leads to the inhibition of cell death cascade reaction. The HBeAg-p22 complex also negatively regulates the stability and transcriptional activity of p53 (Liu et al. 2016).

19.3.2.4 X ORF

The smallest ORF of the HBV genome encodes the X protein, commonly referred to as HBxAg. Studies have consistently demonstrated the involvement of the HBx protein in HBV replication. It has a molecular weight of approximately 17 kDa and is transcribed from the shortest transcript, which is about 0.7 kb in length, comprising 154 amino acids (Geng et al. 2015). HBx is predominantly localized in the nucleus and mitochondria and functions as a multifunctional transactivator, regulating host cell proliferation, metabolism, autophagy, and senescence. However, the mechanisms by which HBx induces hepatocyte transformation or maintains HCC malignancy remain elusive (Li et al. 2017). HBV can integrate into the host genome and encode the hepatitis B virus X protein (HBx), a 17 kDa multifunctional regulatory protein encoded by the X gene. Despite its designation as the X protein due to its initially unknown function in HBV infection, recent studies have revealed its oncogenic properties. However, the precise mechanism underlying the relationship between HBx and carcinogenesis remains incompletely understood. HBx consists of 154 amino acids, featuring an N-terminal negative regulatory domain and a C-terminal transactivation or coactivation domain. This multifunctional regulator modulates transcription, signal transduction pathways, acute immune responses, protein degradation pathways, cell cycle progression, apoptosis, and genetic stability by directly or indirectly interacting with host factors (Tang et al. 2006). Among HBV proteins, HBx interacts directly with Bcl-2 and Bcl-xL, leading to increases in cytosolic calcium levels, cell death, and viral replication. Bcl-2 proteins play a pivotal role in HBV pathogenesis, making HBx a cellular target and a potential point of pharmacological intervention for HBV-related liver diseases (Geng et al. 2012). HBx-induced apoptosis contributes to HCC carcinogenesis in two contradictory ways, depending on its interaction partners. First, HBx may induce the expression of TNF-α-related apoptosis-inducing ligand (TRAIL) receptor 2 (DR5) and promote TRAIL-induced apoptosis. However, HBx can also inhibit the expression of A20, a negative regulator of caspase-8, thus sensitizing hepatocytes to TRAIL-induced apoptosis. Second, HBx has been shown to upregulate the expression of antiapoptotic genes, such as Bcl-2 and Mcl-1, while downregulating proapoptotic genes like Bax, thereby inhibiting apoptosis in hepatic progenitor cells. Consequently, HBx exhibits both proapoptotic and antiapoptotic functions (Shen et al. 2013).

HBx modulates hepatocyte signaling cascades and factors implicated in cellular transformation mechanisms. It can regulate calcium, apoptosis, and proliferation signals, activate transcription factors such as AP-1, AP-2, NFAT, and NF-κB, and influence cellular signaling pathways like Wnt/β-catenin, p53, and Akt, all of which are associated with HCC. Additionally, HBx has been reported to modulate miRNA expression, further expanding its functional repertoire (Lamontagne et al. 2016). Furthermore, HBx decreases the expression of PTEN and upregulates the expression of PI3K/AKT, leading to increased antiapoptotic signaling and ultimately contributing to cancer development (Chung et al. 2003).

19.3.3 The Life Cycle

HBVs, despite their relatively simple structure, exhibit a remarkable capacity for manipulating complex host cell processes. The intricate details of this interaction remain an area of ongoing research. HBV, an obligate intracellular parasite, cannot replicate independently. It relies on effectively commandeering the machinery of hepatocytes, the primary liver cells, to establish a favorable environment for viral propagation. This manipulation disrupts essential host cell signaling pathways, ultimately leading to virus-induced pathology. The HBV life cycle commences with virion attachment to a specific receptor, sodium taurocholate cotransporting polypeptide (NTCP), on the hepatocyte surface. Viral envelope fusion with the cell membrane releases the nucleocapsid into the cytoplasm. This core, housing the viral genome, migrates to the nucleus for repair and conversion into covalently closed circular DNA (cccDNA). The cccDNA serves as a template for the transcription of viral RNA molecules, which are subsequently exported to the cytoplasm for translation into viral proteins. Next, viral pgRNA associates with core protein and the viral polymerase. The polymerase utilizes the pgRNA as a template to synthesize a complementary negative-strand DNA molecule. This is followed by the partial synthesis of a positive strand, ultimately resulting in the formation of a complete double-stranded viral genome. These newly formed nucleocapsids have two potential destinies. They can either be recycled back into the nucleus to replenish the cccDNA pool, perpetuating the viral life cycle, or be enveloped by a viral membrane and secreted through the endosomal sorting complex required for transport (ESCRT) pathway. This pathway generates multivesicular bodies (MVBs) that facilitate the release of mature virions from the infected hepatocyte (shown in Fig. 19.2).

19.3.3.1 Relaxed-Circular DNA (rcDNA) to cccDNA Conversion

The viral particle enters into the cytoplasm of the hepatocyte through clathrin-mediated endocytosis via HBV receptor, heparan sulfate proteoglycans, or human sodium taurocholate cotransporting peptide (hNTCP) hepatocyte-specific preS1 receptor, expressed at the basolateral membrane of hepatocytes. NTCP is a bile acid transporter (Huang et al. 2012). Then, the active transport of the nucleocapsid into the nuclear basket through nuclear pores (Beck and Nassal 2007) transports factors such as importin alpha and beta, and component nucleoporin 153. The dissociation of nucleocapsid leads to the release of the relaxed circular DNA (rcDNA) genome of the virus into the nucleoplasm. Using the host machinery, the partially double-stranded DNA is repaired, and completion of the plus strand by the viral polymerase. Removal of the polymerase from the 5′ end of the minus-strand DNA and short RNA primer used for the plus-strand synthesis is also removed by the cellular enzymes. This repairment leads to the formation of the cccDNA minichromosome (Seeger and Mason 2000).

19.3.3.2 From cccDNA to Pregenomic RNA (pgRNA)

The cccDNA utilizes the cellular transcriptional machinery to undergo transcription and produces RNA necessary for viral protein production and viral replication. All

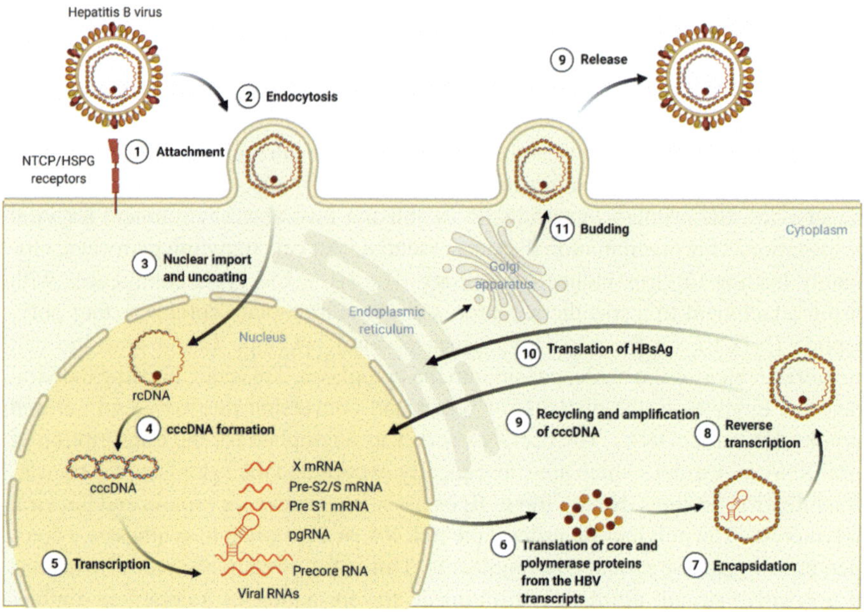

Fig. 19.2 The life cycle of HBV

four genes (Surface gene, PreC/C gene, polymerase gene, and X gene) are transcribed to four mRNAs and utilize a single common polyadenylation signal which is responsible for the regulation of the transcription. Processing of viral RNAs and stabilizing are taken care of by the host factors (Lamontagne et al. 2016). This viral RNA is exported out of the nucleus and waits for translation in the cytoplasm. Translation of the pgRNA into the core protein and viral polymerase and the other X-protein and surface proteins are translated from the subgenomic RNA. After the translation of the core protein, it forms a complex around the pgRNA forming nucleocapsid (RNA-containing nucleocapsid). The polymerase binds to the 5′ epsilon stem-loop structure in pgRNA, eventually this binding is responsible for the nucleocapsid formation (Nassal and Rieger 1996).

19.3.3.3 From pgRNA to Maturation

Inside the nucleocapsid, reverse transcription of pgRNA takes place followed by the plus-strand synthesis. After the comprehensive process of priming and template shifting, complete rcDNA is formed, making it the DNA-containing nucleocapsid. This DNA-containing nucleocapsid has two choices: either it can be reimported into the nucleus for the cccDNA formation or it can be enveloped for secreting out of the cell (Watanabe et al. 2007). During the synthesis of enveloped proteins, the preS-domains remain cytoplasmically exposed and become myristoylated. The envelope proteins after translation are inserted into the ER membrane, where they are translocated in the ER lumen and are secreted out of the cell as the enveloped nucleocapsids containing DNA (Bruss 2007). In some cases, this RC-DNA gets linearized, reimported into the host nucleus, and integrates with its genome. This integration may provide a template for the transcription of HBsAg mRNA. So, continuous secretion of HBsAg into the blood is seen in chronic patients which ultimately leads them to HCC. There are reports which explain the active participation of X and e proteins in the development of cancer in HBV-infected patients. The mature core particles can drift over to the nuclear pore complex or be enclosed by the surface proteins and secreted via the MVB.

19.4 Variants of HBV (Mutant Strains)

In most living biological systems, intricate mechanisms exist to maintain the stability of genetic information. However, certain viruses, including HBV, exhibit rapid and substantial sequence variations (Günther et al. 1999). HBV replication exhibits a unique characteristic: its dependence on an RNA intermediate during viral replication. This process, termed reverse transcription, lacks a proofreading function that is typically present in the DNA replication machinery of most other DNA viruses. As a consequence, HBV accumulates mutations at a significantly elevated rate, estimated to be in the range of 1.4×10^{-5} to 3.2×10^{-5} nucleotide substitutions per site per year (Chen 2018). This inherent genetic variability has prompted investigations into potential associations between specific HBV genotypes and an increased risk of progression to HCC, the most common form of liver cancer. For instance, HBV genotype F has been linked to a frequent and accelerated development of HCC (Pujol et al. 2020).

19.4.1 PreS/S Variants

PreS/S variants commonly observed in carriers of hepatitis B with chronic hepatitis (CH), LC, and HCC indicate that naturally occurring HBV preS/S variants may contribute to the progression of liver damage and the development of HCC. Various mechanisms underlie the liver disease caused by different preS/S variants. This review explores five classes of preS/S variants and their proposed mechanisms in promoting liver damage:

- PreS point mutations: Alterations within the preS region's genetic code.
- PreS deletions: Deletions of specific segments within the preS region.
- PreS1 splice variants: Variations in how the preS1 region is processed during viral RNA maturation.
- C-terminus S point mutations: Mutations affecting the final amino acids (C-terminus) of the S protein.
- PreS/S nonsense mutations: Mutations that prematurely terminate protein production.

Many of these preS/S variants are hypothesized to contribute to the intracellular accumulation of viral envelope proteins within hepatocytes. This retention is thought to trigger a cellular stress response known as endoplasmic reticulum stress, ultimately leading to liver disease development (Chen 2018). As previously mentioned, the S protein comprises 226 amino acids (aa), with the M protein representing an extension of the S protein, containing an additional 55 aa in the preS2 region. The L protein further extends from the M protein, with an additional 108–119 aa, depending on the genotype, in the preS1 region. The amino acid sequence at the C-termini of the L and M proteins is identical to that of the S protein and is referred to as the S region. The preS (preS1 and preS2) region of the L protein is crucial for viral replication and contains several functional sites. These include the hepatocyte binding site, necessary for HBV attachment to liver cells; the S promoter and the CCAAT binding factor binding site, essential for S RNA transcription; the heat-shock protein 70 (Hsp70) binding site and the cytosolic anchorage determinant (CAD), crucial for the dual topology (T) of L proteins; the nucleocapsid binding site (NBS), essential for virion morphogenesis; the site for viral secretion (VS); and the site for polymerized human albumin (pHSA) (Neurath et al. 1986; Pontisso et al. 1989; Le Seyec et al. 1998; Poisson et al. 1997; Bruss 1997; Bock et al. 1999). The precise role of the M protein in the viral life cycle remains to be fully elucidated. The S protein plays an essential role in virion morphogenesis and secretion, and they also contain both B- and T-cell epitopes (Chen 2018).

19.4.2 Precore (PC) and Basal Core Promoter (BCP) Variants

Mutations in the precore and core promoter regions of the HBV genome are commonly observed in HBeAg-negative chronic hepatitis B (CHB) patients. A prevalent precore variant involves a point mutation at nucleotide 1896 (A1896), resulting in a premature stop codon at codon 28 and halting the synthesis of HBeAg. Variants in the basal core promoter (BCP) are characterized by point mutations in the promoters of HBeAg mRNA and core protein mRNA, with the most common mutation involving nucleotide substitutions at positions 1762 and 1764 (Ghabeshi et al. 2014). Infections with BCP and PC mutant viruses, such as those with mutations at 1762/1764 BCP and 1896/1899 PC, are associated with severe clinical manifestations leading to liver failure. During chronic HBV infection, cellular immune responses, including perforin/granzyme or TRAIL-mediated hepatocyte lysis, contribute significantly to liver damage, particularly during hepatitis flares (Dunn et al. 2007). HBeAg-mediated immune regulation helps in preventing severe liver injury and facilitates the establishment of chronic HBV infection (Milich and Liang 2003). However, BCP/PC mutations that abolish HBeAg production can lead to serious immune-mediated liver damage (Sterneck et al. 1996; Laskus et al. 1995; Sato et al. 1995; Hasegawa et al. 1994; Omata et al. 1991; Kosaka et al. 1991; Liang et al. 1991). Nonetheless, the exact mechanism underlying the association of HBeAg-negative viral mutants with severe liver damage remains unclear (Günther et al. 1998; Sterneck et al. 1998). The hepatitis B core (HBc) protein is highly immunogenic, whereas HBV envelope proteins have relatively lower immunogenicity. Consequently, HBc is a crucial target for immune-mediated viral clearance through the induction of B cell, T helper cell, and cytotoxic T lymphocyte (CTL) responses (Ghabeshi et al. 2014).

19.4.3 Drug-Resistant Mutations

The emergence of viral resistance to nucleotide analogues (NAs) in HBV infection is closely linked to amino acid substitutions in the HBV RT enzyme. These substitutions are categorized as primary and secondary drug-resistance mutations. Currently, eight codons are associated with primary drug resistance to NAs: rtI169T, rtA181T/V, rtT184S/C/G/A, rtA194T, rtS202G/C/I, rtM204V/I, rtN236T, and rtM250V. Additionally, three codons are associated with secondary drug resistance to NAs: rtL80I, rtV173L, and rt180M. The primary mutations that directly reduce susceptibility to NAs and secondary mutations that involve substitutions that restore replication function defects resulting from primary resistance, potentially resulting in a low level of reduced susceptibility (Zhao et al. 2016).

19.5 Pathogenesis in Humans

19.5.1 Host Viral Interaction

The host immune system is continuously involved in helping to combat pathogens that cause infection. Both the adaptive and innate immune response plays a significant role in the elimination of the infection. In any viral infection, a series of organized host immunological and virological events take place like viral entry, propagation of virus in host-specific cells (like hepatocytes for host-specific cells), and activation of host immunological events like the first initiation of innate immune sensing mechanism of viruses and production of interferon (Type I & II) and pro-inflammatory signals to subsequently activate adaptive immunity and control infection by damaging virus-infected cells or organ (Biron et al. 1999). But few viruses are not sensed by early host immunological responses, one such virus is HBV. It has evolved a clever strategy to escape from early recognition by the host's innate immune system by acting as a stealth virus (Wieland et al. 2004). When HBV is compared to other hepatitis-causing viruses like hepatitis C virus (HCV). there is a striking difference in induction of innate immune mediators like type I interferon production and pro-inflammatory cytokines induction. Similarly, it was observed that compared with human immune deficiency virus (HIV), HBV infection showed a difference in activating early innate immune response (Stacey et al. 2009; Dunn et al. 2009).

19.5.1.1 Host Immune Response

19.5.1.1.1 Innate Immune Response in HBV Infection

In HBV infection, the role of adaptive immune responses is widely accepted for clearing HBV-infected liver-specific cells. However, the activation of innate immune response for HBV infection clearance remains unanswered (Bertoletti and Hong 2014). Various consequences of HBV infections are due to the complex communications stuck between the host immune system and efficiently replicating virus

(Wieland et al. 2004). The complex interaction between the innate and adaptive immune responses is central to eliminating infection from hosts. Type I and type II interferon secretion produced in early innate immune responses are key to activating adaptive immune immunity (Chemin 2014). Previously, it has been considered that HBV is a stealth virus as it evades innate immune sensing and causes strong persistent infection (Wieland et al. 2004). Recent advances in the HBV innate immunity research using animal models and a few cell culture studies have revealed many hints of early activation of innate immune response in HBV infection (Zhang et al. 2015). Dunn et al. 2009, observed that HBV did not induce efficient production of interferon (IFNs which type) which plays a significant role in virus infection elimination. Components of the innate immune system such as dendritic cells (DCs) (Clark et al. 2000), Natural killer cells (NK) (Hamerman et al. 2005), Natural killer T cells (NKTs) (Taniguchi et al. 2003), Pattern Recognition Receptors (PRRs) like TLRs, RIG-I like receptors (RLRs) such as retinoic acid-inducible gene I (RIG-I), melanoma differentiation-associated gene 5 (MDA5), and LGP2 and NOD-like receptors (NLRs) (Kawai and Akira 2010) are the keys players generating viral infections during innate immune responses.

Pattern recognition receptors role in HBV infection: The innate immune system's cells are adept at detecting viruses or viral components, referred to as pathogen-associated molecular patterns (PAMPs), which encompass viral nucleic acids (DNA or RNA), carbohydrate structures, and viral proteins. These PAMPs are recognized by germ line-encoded receptors present on innate immune cells or within intracellular compartments of cells. These receptors, collectively known as PRRs, play a crucial role in initiating immune responses against viral infections (Thompson et al. 2011). Various PRRs are implicated in the regulation of HBV infection-associated PAMPs. These include TLRs, RIG-I-like receptors (RIG-I and MDA5), Nucleotide-binding oligomerization domain (NLRs), and Cyclic GMP-AMP synthase (cGAS) (Busca and Kumar 2014). These receptors play essential roles in sensing HBV infection and triggering downstream immune signaling pathways to combat viral invasion and replication.

Toll-like receptors (TLRs): TLRs belong to the class of type I transmembrane proteins which help in the detection of microbial infection through the conserved molecular patterns on products derived from pathogens like DNA and RNA viruses, Gram-positive and Gram-negative bacteria, fungi, and protozoan (Pasare and Medzhitov 2005; Dowling and Mansell 2016). In humans, there are 11 types of TLRs (TLR 1—TLR11) whereas in mice, 12 TLRs (TLR1 to TLR 9 & TLR 11 to TLR 13) have been identified (Gay et al. 2007). Each TLR has a specific location intra- or extracellular on the innate immune cells. A subgroup of TLRs like TLR1, TLR2, TLR4, and TLR5 are generally expressed on the surface of the cells and recognize the PAMPS of bacteria and fungi such as extracellular fungal and bacterial cell wall and a few viral proteins. Another subgroup of TLRs like TLR3, TLR7, TLR8, and TLR9 senses nucleic acids such as DNA and RNA are localized in the cellular compartments such as endosomes (Gay et al. 2007; Takeda and Akira 2003). The response of TLR class proteins in combating the viral PAMPs has been observed in both in vivo and in vitro studies of many viruses. Similarly, in HBV infection, innate immune response mediated by TLRs is shown to limit the replication of HBV in liver cell lines and animal models (Bertoletti et al. 2005; Visvanathan et al. 2007; Chen et al. 2008; Isogawa et al. 2005; Ma et al. 2014). Interaction of TLRs with HBV is shown in Fig. 19.3.

RIG-I-like receptors (RLRs): RLRs are cytosolic viral RNA or RNA replication intermediate sensors that are found in almost all mammalian cell types. RIG-1, MDA5, and LGP2 are the three members of the RLRs family which belongs to the 2DExD/H –Box RNA helicases subfamily. Many in vitro and in vivo reports have stated the important role of RLRs family receptors RIG-I, MDA5, and LGP2 in sensing of viral nucleic acids and combating viral infections (Kato et al. 2008; Feng et al. 2012; Gitlin et al. 2006; Schoggins et al. 2011; Bruns and Horvath 2012). Structurally,

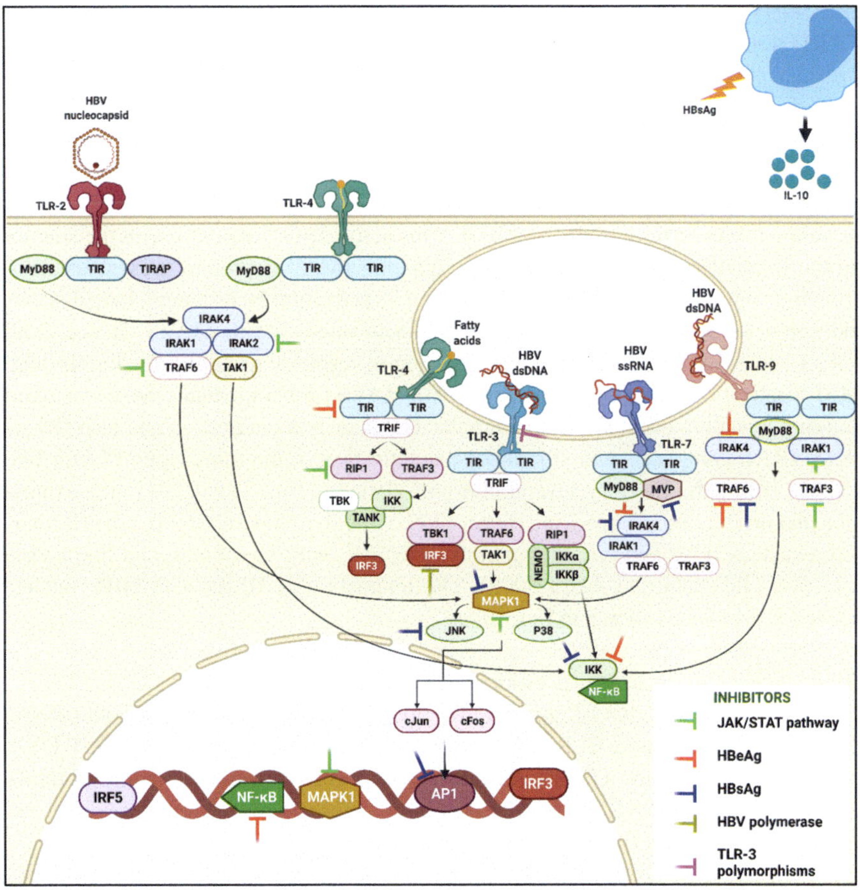

Fig. 19.3 Interaction of TLRs with HBV—Sensing of HBV DNA, RNA, and proteins by TLRs result in the activation of various antiviral components such as interferon, cytokines, chemokines, and transcriptions factors, IRF 7 and NF-κB. Different HBV proteins are involved in blocking further signaling at various stages of the innate immune response (black arrows indicate the related signaling pathways, and T arrows designate the HBV proteins blocking the signaling targets)

the RLRs have a helicase core which consists of the helicase domains like Hel1 and Hel2 and the Pincer domain at the C-terminal region of Hel2. The pricer region is also known as bridging helicases. All the three RLRs have common C terminal domains (CTD) whereas RIG-I and MDA5 only have two tandem repeated CARD domains at N-terminal regions. In the case of LGP2, there are no CARD domains towards N-terminal regions. In normal uninfected cells, both the RIG-1 and MDA5 are in an inactive state due to the phosphorylation of both CARD and CTD; moreover, RIG-1 is in autoinhibited confirmation and CARD domains are unavailable for signal transduction (Kowalinski et al. 2011; Ferrage et al. 2012). During the viral infection, RIG-1 senses the short viral ds RNA towards the 5 triphosphate (5'PPP) or diphosphate (5'PP) end whereas in the case of MDA5 senses the long ds RNA of viruses. After the sensing of viral ds RNA, both RIG-I and MDA5 get dephosphorylated by phosphatases PP1γ or PP1α isoforms, in which the inhibitory phosphorylation marks on CARD domains and induces the active confirmation of CARD domains for downstream signaling (Wies et al. 2013). Further to make RIG-I more active, CARD domains of RIG-I undergo Lys63-linked ubiquitylation mediated by tripartite motif protein 25(TRIM25) and similarly the CTD of RIG-I also ubiquitylated at Lys63 mediated by E3 ubiquitin ligase, Riplet (Oshiumi et al. 2016) Lys 63 liked ubiquitylation at both the CARD domains at N-terminal and CTD of RIG-I induce the active signaling tetramerization state of RIG-1. This active RIG-I interacts with the adapter proteins called mitochondrial antiviral signaling (MAVS) protein present in mitochondria. Whereas in the case of MDA5, after its activation by its ligand, long ds RNA induces a long filament formation of MDA5 to interact with MAVS. Hence, both RIG-1 and MDA5 mediate antiviral signaling in response to viral infection through MAVS activation (Takeuchi and Akira 2010). MAVS further mediates the downstream antiviral signaling events by activating the TBK1 or IκB kinases—ε and also IKKα-IKKγ-IKKβ complexes and stimulates the activation of the transcription factor like NF-κB, IRF3, and IRF7 by several phosphorylation events. This leads to the expression and production of type 1 interferon (IFNs) mainly INFα and INFβ, type III interferon such as IFNγ and proinflammatory cytokines and chemokines, and limits viral infection (Takeuchi and Akira 2008). The interaction of RLRs with HBV nucleic acids and proteins is shown in Fig. 19.4.

Other Viral Nucleic Acid Sensors Aiding the Combat Against HBV

- Apart from TLRs, RIG-I, and MDA5, other viral nucleic acid (DNA and RNA) sensors such as non-RLR DExD box helicases (like DDX1, DDX3, DDX9, DH21, DHX36, and DH60), leucine-rich repeat flightless-interacting protein 1 (LRRFIP1), and PKR are also present in the interaction of the host immune system with hepatitis B virus (Chan and Gack 2016; Leong et al. 2015).
  - DDX3 is a DEAD-box RNA helicase which unwinds RNA in ATPase-dependent manner. The ectopic expression of DDX3 in vitro was found to limit the replication of HBV to a significant level. However, HBV has evolved an evasion strategy to escape from inhibitory effect of DDX3. HBV polymerase disrupts interaction between DDX3 and IKKε to block further signaling (Wang et al. 2010).

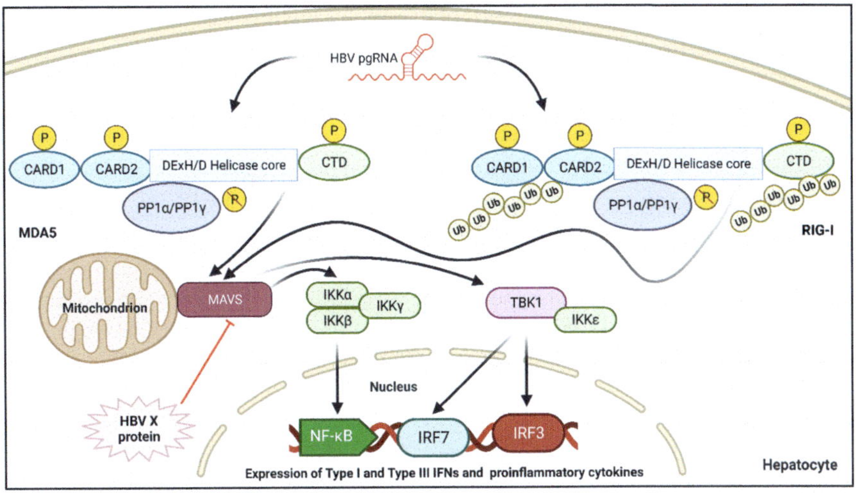

Fig. 19.4 Interaction of RLRs with HBV nucleic acids and proteins

- In in vitro and in vivo studies conducted by Kouwaki et al. (2016a), the expression of IFN-γ was found to decrease HBV RNA levels, and expression of DDX60 led to degradation of HBV RNA residing in the cytoplasm. Similarly, DDX60 gene knock down resulted in the higher level of HBV replication and ectopic expression of DDX60, leading to the reduction of HBV replication (Kouwaki et al. 2016a, b).
- Double-stranded RNA-(dsRNA-) dependent protein kinase-R (PKR) is a cytoplasmic RNA sensor that senses dsRNAs, causes autophosphorylation of eukaryotic translation initiation factor-2 (eIF2-α), and suppresses viral protein synthesis (Proud 1995; Zhang and Samuel 2007). During HBV infection, PKR senses HBV replicative RNA intermediates and prevents HBV protein synthesis (Park et al. 2011a; Guidotti et al. 2002; Han et al. 2011b). IRF-1 and PKR knockout mice show higher levels of HBV replication, suggesting that both PKR and IRF-1 modulate the antiviral signals required for inhibiting HBV replication (Guidotti et al. 2002). Park et al. (2011a) have shown that the ectopic expression of PKR results in an increased phosphorylation of eIF2-α which decreases the production of HBV capsid protein, while knockdown of PKR through RNA-interference increases HBV replication in a hepatoma cell culture model (Park et al. 2011a). Han et al. (2011b) revealed that transfection of human hepatoma cell lines such as HepG 2.2.15 with HBx-siRNA increased the expression of type-I interferons, ISGs such as ISG15, ISG56, and PKR and various proinflammatory cytokines and inhibited the HBV replication and capsid formation (Han et al. 2011b).
- APOBEC proteins are another class of important innate immune restriction factor that control HBV infections (Rosler et al. 2004; Nguyen et al. 2007; Baumert and Schuster 2007; Vartanian et al. 2010). APOBEC3G, APOBEC3C, and APOBEC3H induce hypermutations in HBV DNA to decrease HBV replication (Köck and Blum 2008). The upregulation of APOBEC3G by IFN-α

results in the reduction of HBV replication (Noguchi et al. 2007). HBx protein counteracts APOBEC3G to restore HBV DNA replication by increasing the exosome activity and secretion of APOBEC3G from intracellular compartments (Chen et al. 2017).

- The myxoma resistance protein 1 (MxA) is a GTPase super family protein that exhibits antiviral activity against several positive and negative sense RNA viruses. During viral infection, MxA is induced by INFα/INFβ, and it accumulates in the cytoplasm to mediate an antiviral activity. Huh7 cells constitutively expressing MxA suppress HBV DNA replication and HBsAg and HBeAg protein synthesis (Gordien et al. 2001). An in vivo study carried out in transgenic female mice has documented the participation of MxA protein in the inhibition of HBV replication (Peltekian et al. 2005).

DNA Sensors in HBV Infections

- Mammalian cell DNA sensors sense ssDNA or dsDNA-containing pathogens and induce interferon-mediated innate immune signaling. These include the DNA-dependent activator of IFN regulatory factors (DAI), IFI16, cyclic GMP–AMP (cGAMP) synthase (cGAS), DDX41, DHX36, DHX9, DNA-dependent protein kinase (DNA-PK), MRE11. Downstream signalling of these sensors is mediated by an adaptor protein called as stimulator of interferon genes (STING) (Dempsey and Bowie 2015; Chan and Gack 2016; Goubau et al. 2013; Paludan and Bowie 2013). During the HBV life cycle, intermediates such as rcDNA and cccDNA are sensed by these DNA sensors (Guo et al. 2017; Thomsen et al. 2016).
- The cGAS-STING pathway is a well-characterized pathogen DNA sensing pathway that elicits an interferon response. After binding to pathogen DNA in the cytosol, cGAS synthesizes a second messenger-cyclic AMP and GMP (cGAMP), which binds to STING (resident of endoplasmic reticulum). The binding of STING with cGAMP induces STING dimerization followed by activation via TRIM32 and TRIM62-mediated Lys63-linked ubiquitination. The activated STING binds to and activates TANK binding kinase (TBK1) which phosphorylates transcription factors such as IRF3 and NF-kB which in turn induces JAK-STAT signaling and several ISGs in neighboring cells to create an antiviral environment. HBV DNA is sensed by this cGAS–STING pathway. An ectopic expression of STING shows efficient inhibition of HBV DNA replication (Guo et al. 2017; Dansako et al. 2016; Yim et al. 2016). Yim et al. showed that silencing of cGAS, STING, and TBK1 decreases the expression of interferon response against HBV and increases the level of HBV DNA in sodium taurocholate cotransporting polypeptide (NTCP) expressing HepG2 cells (Yim et al. 2016). Hepatocytes lack cGAS—STING DNA sensing innate immune pathway, and hence, DNA is not sensed by the STING pathway (Thomsen et al. 2016; Dickson and Mori 2016). STING agonists such as 5,6 dimethyl xanthenone 4-acetic acid (DMXAA), induce robust cytokine production in macrophages and efficient inhibition of HBV replication (Guo et al. 2017; Dansako et al. 2016). HBV escapes the innate immune surveillance of the STING pathway by disrupting K3-linked ubiquitylation. HBV polymerase protein plays a pivotal role in ham-

pering K3-linked ubiquitylation of STING and counteracting the antiviral activity of the cGAS STING pathway (Liu et al. 2015). Figure 19.5 shows the interaction of various DNA sensors with HBV nucleic acids and proteins.

- IFI16 also participates in HBV DNA sensing (Wieland et al. 2004) by using STING as its adaptor to mediate IFN expression. An upregulation of IFI16 and downstream effectors such as IFN-β occurs in acute HBV infection in woodchucks, while chronicity is associated with the downregulation of IFI16 and IFN-β indicating possible HBV involvement (Yan et al. 2014).
- Absent in melanoma 2 (AIM2) is a cytosolic receptor that senses dsDNA and induces inflammatory cytokines such as interleukin (IL)-1 β and IL-18. AIM2 senses the HBV dsDNA in the cytoplasm of HBV infected hepatocytes and upregulates the inflammatory cytokine expression. AIM2 is positivity correlated with the upregulation of inflammatory cytokines such as IL-1β, IL-18, and caspase-1 with increased liver inflammation in CHB patients. There is higher expression of AIM2 in acute hepatitis B patients in comparison with cases of

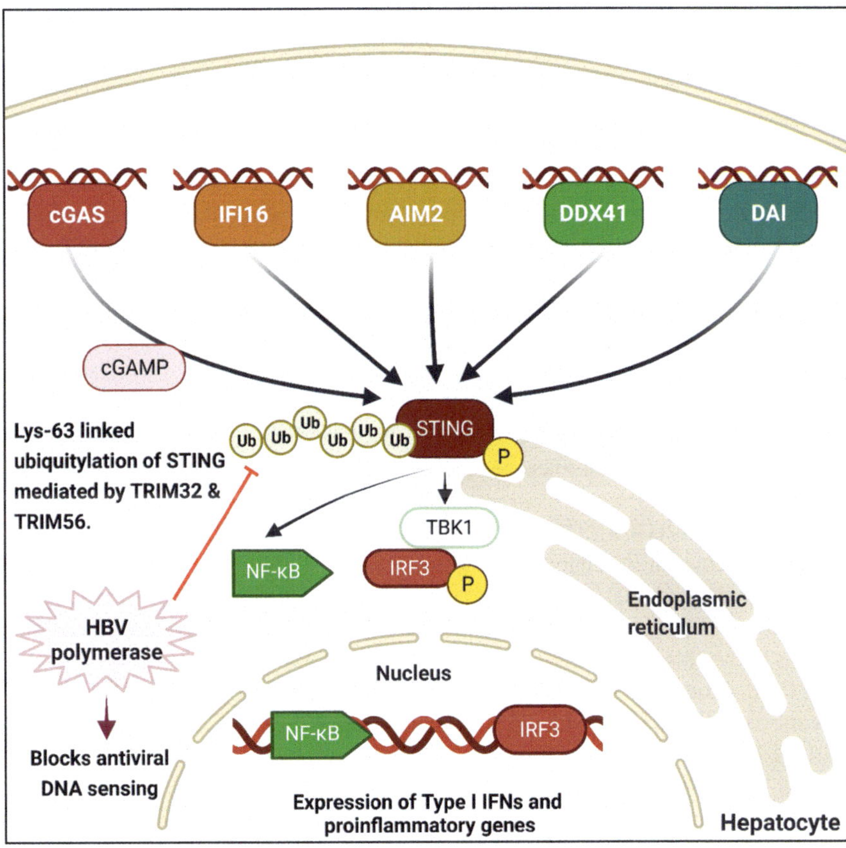

Fig. 19.5 Interaction of various DNA sensors with HBV nucleic acids and proteins

chronicity. A negative correlation is also observed between levels of AIM2 mRNA and HBeAg and serum HBV, whereas AIM2 has a positive correlation with increased of IL1-β and IL-18 levels in both chronic and acute HBV infections (Wu et al. 2013). Du et al. suggested that the overexpression of AIM2 in acute infection confers viral clearance while downregulation is associated with CHB (Du et al. 2013).

- DNA-dependent activator of IFN regulatory factors (DAI) sense pathogen dsDNA and activate NF-kB and IRF3 mediated antiviral signaling. The role of DAI in HBV infections was first identified during a search for potential interferon induced genes that can inhibit HBV replication in transgenic mice (Takaoka et al. 2007). Chen et al. (2012) further showed that when Huh7 cells were cotransfected with DAI expressing plasmids and full-length HBV clones, there was a significant reduction of HBsAg and HBeAg (57%), HBV RNA replicative intermediates (67%), and encapsidated viral DNA levels. This inhibitory effect was mediated by the activation of NF-kB.

19.5.1.1.2 Adaptive Immune Response

The adaptive branch aids viral clearance and elimination of HBV-infected hepatocytes. CD4+ T cells regulate both CD8+ T cell activation as well as B-cell antibody production. The effector CD8+ T cells eliminate HBV infected hepatocytes to decrease circulating virions (Chisari et al. 2010), and circulating antibodies inactivate the virus. About 90% of infected individuals manage to mount effective immune responses to eliminate the virus and have lifelong immunity with anti-HBsAg antibodies, while most neonatal infections due to mother-to-child transmission develop chronicity. In self-limiting acute HBV infections, the vigorous, peripheral blood CD4+ T cell response activates CD8+ cells and HBV specific B cells. The CD8+ cells produce IFN-γ and TNF-α to clear virus via noncytolytic ways (Guidotti et al. 1999) or directly eliminate infected hepatocytes via cytolytic mechanisms using Fas and granzyme (Ando et al. 1993; Webster et al. 2000; Thimme et al. 2003). However, it is still not clear at the molecular level how HBV establishes persistent infections. It is believed that dysfunctional, narrowly focused immune responses, exhaustion of T cells, and liver-mediated tolerance are the major contributing factors that lead to viral persistence.

Dendritic cells (DCs): DCs are the messengers between the innate and adaptive immune systems. In the mammalian immune system, the DCs are one of the key antigen-presenting cells, as they process the foreign antigens and present them on the surface of effector cells. Many studies have shown that there are no striking differences between myeloid-derived DCs (mDCs) and plasmacytoid (DCs) frequency when CHB patients such as infants born to HBV-positive mothers and adult CHB patients when compared with healthy donors (Zhang et al. 2007; Gehring et al. 2013; Gehring and Protzer 2015; Beckebaum et al. 2002). In the case of mDCs, cytokine profiling also showed no difference between healthy donors and CHB patients whereas, in the case of pDCs, cytokine profiling showed major differences when CHB patients were compared to healthy donors (Zhang et al. 2007; Woltman et al. 2011; Duan et al. 2004; Shi et al. 2012). Concerning DC's role in HBV infection, there is a lot of confusion and many controversial reports have been reported on the frequency, phenotype expression, and cytokine profiling of different DC subsets (Gehring and Protzer 2015). Few groups reported that the frequency of mDCs was reduced in immune-active pediatric CHB patients' blood (Zhang et al. 2007; Vander Molen et al. 2006). Whereas in the case of pDC, the frequency of pDC in infants born to HBV-positive mothers was reduced when viral DNA was above 5 × 107 copies per mL (Guo et al. 2014). Few reports showed that antiviral therapy can restore the reduced frequency of both mDCs and pDCs to normal levels (Vander Molen et al. 2006; Duan et al. 2004). The ex vivo phenotypic expression of both mDCs and pDCs has not been extensively studied but the expression of costimulatory molecules like CD40, CD80, and CD 86 are equally expressed between the CHB and health donors (Gehring and Protzer 2015). Previous studies showed that the expression of costimulatory molecules is upregulated, and the expression of OX40 ligands is regulated in pDCs of high viremic CHB patients. In the case of mDCs also, slight differences are observed in the expression of costimulatory molecules in vitro when compared to healthy donors (Vander Molen et al. 2006). In the case of monocyte-derived DCs (moDCs), a reduced expression of costimulatory molecules is observed during the in vitro exposure of moDCs to HBV (Beckebaum et al. 2003). There are few studies, where monocytes are isolated from CHB patients and differentiated them into moDC and compared their costimulatory molecules expression. In immature moDCs, lower expression and in mature moDCs, normal expression of costimulatory molecules are observed when compared with healthy controls (Beckebaum et al. 2002; Zheng et al. 2004; Tavakoli et al. 2004). HBV virions and HBV proteins target the suppressive effects of DCs by impairing the immunomodulatory functions of different subsets of DCs (Op den Brouw et al. 2009; Moffat et al. 2013; Woltman et al. 2011). HBV particles and HBcAg abrogate the pDCs immunomodulatory functions by altering the TLR9-mediated expression of INFα (Woltman et al. 2011). Moffat et al. reported that HBV virus-like particles (VLPs) impaired the expression of IFNα from pDCs when stimulated with CpG DNA (Moffat et al. 2013).

Natural Killer Cells: Natural killer cells are one of the key immune cells that occupy about 15% of total peripheral lymphocytes and play an important during microbial infection and against tumors. To limit the infection levels of pathogens, NK cells show their defense mechanism by secreting the immunomodulatory cytokines like TNFα and IFNγ, releasing cytolytic granules like perforins and granzymes and causing apoptosis in pathogen-infected cells (Doherty et al. 1999; Cooper et al. 2001). In the case of viral infection, NK cells are known to elicit immune responses against both the DNA and RNA viruses (Orange et al. 2002; Martin et al. 2002; Khakoo et al. 2004). In vivo studies carried out on chimpanzees and in woodchucks showed the role by playing NK cells in HBV infection. In early HBV infections, NK cell roles have not been well characterized, whereas in HBV persistent infection, NK cells induce liver injury by causing cytolysis of HBV infected hepatocytes. As with DCs, NK cells express the TLRs on their surfaces and recognize the HBV DNA and its replicative intermediates to promote the NK cells-TLR pathway mediated release of cytokines (Lauzon et al. 2006; Fletcher et al.

2013; Yang et al. 2010; Thimme et al. 2003). In acute HBV infections, Zhao et al. reported high-level expression of activating receptors, and increased cytolytic and IFNγ production from the peripheral NK cells compared to chronic HBV patients' peripheral NK cells (Zhao et al. 2012a, b). NK cells cause liver damage by causing the cytolysis of hepatocytes infected with HBV. Cytolysis of HBV-infected cells is part of viral clearance and self-protection, whereas in continuous cytolysis of hepatocytes in persistence infections leads to liver fibrosis and HCC (Zhang et al. 2011). When healthy controls liver NKs and hepatocytes are compared with HBV-infected hepatocytes and liver NKs, TRAIL are more expressed in NKs of HBV-positive patients, and TRAIL-inducing receptors on HBV-positive patients hepatocytes (Dunn et al. 2007; Ichikawa et al. 2001). Okazaki et al. showed that when NK cells are activated through DCs, NK cells cause the massive cytolysis of HBV-infected hepatocytes by regulating the Fas/Fas ligand interactions (Okazaki et al. 2012). In NK cells mediated liver, hepatocyte injury NKG2D and NKG2D ligand pathways are highly activated in fulminant hepatitis in acute and chronic infections (Zou et al. 2010). In chronic CHB infections, NK cell frequency in both peripheral blood and liver is reduced (Wiesmayr et al. 2012; Zhang et al. 2011; Tjwa et al. 2011). Receptors on NK cells are altered in CHB which results in the impaired production of chemokines and cytokines such as TNFα and IFNγ and even increases the hepatocyte cytolytic activity and causes liver injury (Zhang et al. 2011; Wiesmayr et al. 2012; Tjwa et al. 2011; Lv et al. 2012; Ju et al. 2010). Both the DC cell subsets such as mDCs and pDCs and kupffer cells secrete the altered cytokines and cause the dysfunction of NK cells in both acute and chronic infections. Altered NK cells even cause the death of CD8+ cells death (Vander Molen et al. 2004; Tjwa et al. 2011; Martinet et al. 2012; Tu et al. 2014; Park et al. 2011a, b; Peppa et al. 2010). In 2009, a massive outbreak with high mortality of HBV associated with Precore and basal core promoter mutations caused a fulminant hepatic failure. Both the cytokine and chemokines analysis from lymphocytes of patients showed decreased levels of NK cells when compared to healthy controls. Whereas the levels of NKT cells are significantly higher when compared to healthy controls. Chemokines CCL4 levels are decreased in fulminant hepatic failure patients. These results state that both NKT cells and CCL4 play significant roles in early infections and liver injury (Tripathy et al. 2011).

19.6 Clinical Manifestations

HBV infection can be acute or chronic. Acute infection (transient liver disease) usually resolves on its own. In rare cases, fulminant hepatitis, a rapid liver failure, can occur (Günther et al. 1999). Chronic HBV infection persists and causes a spectrum of conditions, from asymptomatic carriage to liver cancer (HCC) (Chen 2018). Transmission route influences outcome: horizontal transmission in adults often leads to acute infection (Seeff et al. 1987), while vertical transmission from mother to child frequently leads to chronic carriage (Stevens et al. 1975; Okada et al. 1976).

Phases of Disease:

- Acute infection may last 2–4 months and develop acute liver failure, which can lead to death. The clinical manifestations include hepatomegaly, liver tenderness, and splenomegaly.
- Chronic HBV infections may be categorized into four phases, according to the level of HBV replication and the strength of the host immune reactivity (Fattovich et al. 2008; Hadziyannis 2011; McMahon 2010; Papatheodoridis et al. 2001).

1. **Immune tolerant phase** (HBeAg(+), HBV DNA high, ALT normal): During the initial immune-tolerant phase of HBV infection, infectivity is notably high while the histological activity remains minimal. The duration of this phase can vary significantly, with perinatally acquired HBV infection typically exhibiting the longest duration. Characterized by the presence of HBeAg, elevated serum HBV-DNA levels, and normal or slightly elevated serum alanine aminotransferase (ALT) levels, individuals in this phase often display normal liver histology or only minimal histological activity with limited fibrosis. In cases of perinatally acquired HBV infection, individuals may predominantly present with HBeAg-positive CH accompanied by normal ALT levels, particularly observed in Asian children. The immune-tolerant phase can persist for an extended period, ranging from 10 to 30 years in perinatally infected individuals. Conversely, this phase is typically short-lived or even absent in cases of HBV infection acquired during childhood or adulthood.
2. **Immune reactive phase** (HBeAg(+), HBV DNA variable, ALT abnormal): Following the initial immune-tolerant phase, patients undergo a transition into an immune-active phase characterized by the loss of immune tolerance towards the HBV, leading to active hepatitis accompanied by hepatic necroinflammation. The duration of HBeAg positivity varies depending on the age at which HBV infection was acquired. During this immune-active phase, HBV-DNA levels exhibit fluctuation but gradually decrease, while serum ALT levels become elevated, indicating hepatic necroinflammation. Patients with chronic HBV infection acquired during late childhood, adolescence, or adulthood typically enter this phase with HBeAg-positive CH, elevated serum ALT levels, and moderate to severe hepatic necroinflammation. Liver biopsy findings may reveal variable amounts of fibrosis. In HBeAg-positive CH, serum HBV-DNA levels typically exceed 20,000 IU/mL (105 copies/mL). This phase signifies a shift towards active immune responses against HBV-infected hepatocytes, contributing to hepatic inflammation and disease progression.
3. **Low replicative phase or "inactivate HBsAg carrier stage"** (anti-HBe(+), HBV DNA low, ALT normal): The seroconversion from HBeAg to anti-HBe is a significant milestone indicating a reduction in HBV replication and liver activity. This transition typically occurs during the immunoreactive phase and marks the onset of the third phase, known as the low or nonreplication phase, or the inactive HBsAg carrier state. In this phase, individuals become HBeAg-negative and anti-HBe-positive. HBV-DNA levels are either undetectable or present at low levels, often below 2000 IU/mL (104 copies/mL). Serum ALT levels remain

persistently normal, indicating minimal liver inflammation. Additionally, liver histology typically shows inactive features with minimal fibrosis. This phase signifies a state of reduced viral replication and liver activity, representing a more stable and less active phase of chronic HBV infection.

4. **Reactivation phase** (anti-HBe(+), HBV DNA variable, ALT abnormal): The reappearance of HBV viremia marks the onset of the fourth phase, known as the reactivation phase, characterized by various scenarios including reversion back to HBeAg positivity or the presence of HBV variants hindering HBeAg expression. This phase is associated with modest or amplified liver inflammation. During this phase, individuals may experience viral reactivation due to the persistence of viral supercoiled DNA in the liver. Reactivation can occur spontaneously or may be triggered by active immunosuppression, such as during cancer chemotherapy or after organ transplantation. The reactivation phase manifests with HBeAg negativity and anti-HBe positivity. Serum HBV-DNA levels are detectable, ranging from 2000 to 20 million IU/mL (104–108 copies/mL). There is an elevation in serum ALT levels, indicating moderate to severe liver inflammation. Liver biopsy may reveal variable amounts of fibrosis. In some cases, inactive HBsAg carriers may eventually lose HBsAg with the development of neutralizing antibodies (anti-HBs). However, low levels of HBV-DNA may still be detectable by PCR in serum and liver biopsy specimens. Progression to cirrhosis and HCC is typically slow in the inactive carrier state, with a risk of less than 1% per year. However, in the immune reactive and reactivation phases, the progression to cirrhosis may occur at a higher rate, ranging from 2% to 10% per year (Croagh and Lubel 2014).

19.7 Diagnosis

The laboratory diagnosis of HBV infection primarily revolves around detecting specific serological markers. The presence of HBsAg is an indicative of HBV infection, while the presence of hepatitis B surface antibody (HBsAb) suggests recovery from HBV infection or successful immunization against HBV. Additionally, the presence of hepatitis B core antibody (HBcAb) can be observed at the onset of acute HBV infection, although it may also indicate chronic HBV infection. Serological markers such as HBsAg, HBeAg, and anti-HBe can be measured using commercial electrochemiluminescence assay kits. HBV DNA levels can be quantified using assays such as the Amplicon monitor assay. Liver enzyme levels, including ALT, aspartate aminotransferase (AST), and gamma-glutamyl transferase (GGT), can be determined using commercial ultraviolet absorption spectrophotometry assay kits.

- **Acute HBV Infection:** People with a recent hepatitis B infection typically have two key markers in their blood. The first is the HBsAg, which indicates the presence of the virus. The second is IgM antibodies to the core antigen (HBcAg). These IgM antibodies are a sign of the body's initial immune response to the infection. Additionally, during the early stages of infection, some individuals may also test positive for another marker, the HBeAg. This indicates high levels of viral activity and increased infectiousness.

- **Chronic HBV Infection:** CHB infection is defined by the persistent presence of HBsAg in the blood for at least 6 months. This marker can exist alone or alongside HBeAg. The prolonged presence of HBsAg is a critical indicator of a heightened risk for developing serious liver complications, such as chronic liver disease and liver cancer (HCC) in the future (CDC 2021).

19.7.1 Serological Diagnosis

The introduction of the Ausria-125 test marked a significant advancement in virus diagnostics, particularly for the detection of HBsAg. However, the use of radioactivity in the test posed significant challenges in routine diagnostic laboratories. A major breakthrough occurred when it became feasible to label antibodies with enzymes, eliminating the need for radioactivity. Initially, the HBsAg assay was complemented by the detection of antibodies against HBsAg (anti-HBs) and HBcAg (anti-HBc). Typically, the appearance of anti-HBc indicates the onset of acute hepatitis or a clinically silent HBV infection. In cases where HBsAg is present without anti-HBc, patients may still progress to hepatitis. During immune control of the HBV infection, HBsAg may disappear while anti-HBc persists, and anti-HBs antibodies often emerge as a sign of immunity. In chronic HBV infections, both HBsAg and anti-HBc remain positive (Gerlich 2013).

19.7.2 Molecular Diagnosis

Though the serological assay is adequate for hepatitis B diagnosis, however, HBV DNA detection has several added advantages such as early diagnosis, assessment of disease progression as mentioned above, and monitoring of antiviral therapy response. HBV DNA can be directly detected in peripheral blood and is directly correlated to HBV DNA copies and viral load. HBV DNA is detectable by 30 days postinfection, about 21 days early before HBsAg and may increase up to 10^9 DNA copies/ml during acute hepatitis. The levels then gradually decrease and may disappear if infection is resolved.

19.8 Therapeutic Approaches

The differentiation between HBV-infected patients who are HBeAg positive and those who are HBeAg negative due to a viral mutation is essential for treatment selection. Two sorts of interferon-based therapies, conventional interferon and PEGylated interferon alpha, have shown improved effectiveness against HBV infection. Additionally, five oral nucleos(t)ide analogs (NAs) act as RT inhibitors and are approved for HBV treatment: Lamivudine (LAM), Adefovir (ADV), Entecavir (ETV), Telbivudine

(LdT), and Tenofovir (TDF). Interferon therapy enhances the host's immune response, while NAs repress viral infection by inhibiting HBV viral polymerase activity. However, long-term NA therapy can lead to the development of viral resistance. Lamivudine, the first widely used NA, has a high viral resistance rate after 5 years of therapy. ADV and LdT have lower rates of viral resistance, while ETV and TDF effectively suppress HBV DNA replication with minimal resistance (Aspinall et al. 2011).

Interferon Therapy: It was first described in 1957 by Isaacs and Lindenmann, who observed its ability to protect cells against viral infections. Clinical studies later showed that interferon-alpha could suppress HBV replication and treat some patients with CHB. Despite its benefits in enhancing the immune response, only a minority of patients achieved a cure with interferon therapy, with many experiencing relapse after treatment (Greenberg et al. 1976).

The discovery of inhibitors of RT in the late 1970s, such as lamivudine, marked a significant advancement in HBV treatment. Lamivudine was initially used to treat HIV but was later found to inhibit HBV DNA synthesis. However, resistance to lamivudine soon developed in many patients. Adefovir, another RT inhibitor, was approved for the treatment of lamivudine-resistant HBV infections in 2002. However, its activity against HBV was relatively low, and it induced the selection of resistant HBV mutants in some cases. Tenofovir, a newer drug, has become the preferred treatment for CHB due to its effectiveness and low risk of resistance. Entecavir, originally developed against herpes simplex virus demonstrated robust activity against HBV and was approved for HBV treatment in 2006. Telbivudine, although associated with significant resistance issues, has been used successfully to suppress high viral loads in HBV-infected pregnant women (Petersen and Buti 2012). Overall, while current treatments for HBV have improved outcomes for patients, targeted therapies aimed at eliminating the viral cccDNA remain the ultimate goal.

19.9 Prevention and Control

In addition to adhering to strict hygiene practices during invasive procedures and adopting a cautious lifestyle, vaccination is paramount in preventing hepatitis B infections. The introduction of a safe and effective HBV vaccine in the early 1980s marked a turning point. Recognizing its impact, the WHO in 1992 urged all countries to include HBV vaccination in their immunization programs. This strategy complements mandatory blood donation testing for HBV, minimizing transmission risks and safeguarding public health. This measure helps safeguard individuals who receive blood transfusions from donors. By implementing comprehensive vaccination programs and blood screening measures, significant strides can be made in reducing the burden of hepatitis B globally and preventing its transmission (Aspinall et al. 2011; Gerlich 2013).

19.9.1 Immunization

The administration of human immunoglobulin preparations enriched with high titers of anti-HBs (HBIG) has been shown to be an effective modality for providing postexposure prophylaxis against HBV infection. This is particularly crucial for newborns born to mothers infected with HBV who are HBeAg positive, as they have a high risk (70%–90%) of becoming chronic HBV carriers. The antibodies against hepatitis B virus (anti-HBs) given in this way (passively administered) only last for a short period, about 22 days. This means they offer temporary protection. In 1971, Saul Krugman reported the first instance of what could be considered a vaccination against hepatitis B. He used a 1:10 dilution of serum that had previously tested positive for the AuAg, briefly heated it to inactivate the virus, and administered it as a form of vaccine to mentally handicapped children. After two injections, he challenged the children with infectious HBV-containing serum and observed incomplete but statistically significant protection. While this pioneering investigation suggested the presence of an immunogenic component within AuAg-positive serum, the limitations inherent in the rudimentary vaccine preparation precluded definitive conclusions. The advent of second-generation vaccines marked a new era in vaccine production. In 1979, the HBV genome was cloned, and the HBs gene was identified. Initial attempts to express the HBsAg in E. coli proved unsuccessful. However, a significant advancement occurred in 1977. Utilizing a partial amino acid sequence of the predominant protein isolated from plasma-derived HBsAg, Petersen and Vyas successfully identified the ORF encoding HBsAg. Their research revealed the existence of two primary forms of the core HBsAg protein: an unglycosylated form lacking sugar molecules and an N-glycosylated form containing a single sugar chain. Both forms exhibit a length of approximately 220 amino acids. Notably, treatment with proteolytic enzymes eliminated larger proteins from purified HBsAg preparations without compromising its antigenicity, the ability to elicit an immune response.These advancements paved the way for the development of more refined and effective hepatitis B vaccines (Krugman et al. 1971; Peterson et al. 1977).

19.10 Future Perspectives

HBV infection can manifest in a wide range of health consequences, from individuals carrying the virus with no symptoms (asymptomatic carriage) to those experiencing serious complications such as fulminant hepatic failure, chronic liver disease, cirrhosis, and HCC. Chronic HBV infection is often associated with the presence of naturally occurring preS/S variants, which have been implicated in the progression of liver disease. Understanding the molecular mechanisms underlying the role of these variants in liver disease progression is crucial for developing targeted interventions. Deciphering the specific molecular mechanisms by which distinct HBV variants influence the pathogenesis of liver disease across its various stages necessitates further investigation. By gaining insights into these mechanisms, we can identify potential therapeutic targets and develop more effective treatment strategies.

In conclusion, HBV remains a significant global health challenge, and there is a pressing need for continued research efforts aimed at developing improved prevention strategies and therapeutic interventions. By advancing our understanding of HBV pathogenesis and exploring novel treatment approaches, we can work towards reducing the burden of HBV-related liver diseases worldwide.

References

Ando K, Moriyama T, Guidotti LG, Wirth S, Schreiber RD (1993) Perforin, Fas/Fas ligand, and tumor necrosis factor are the major cytotoxic molecules used by lymphocytes to kill hepatitis B virus-infected hepatocytes in vivo. Proc Natl Acad Sci 90(17):7724–7728

Aspinall EJ, Hawkins G, Fraser A, Hutchinson SJ, Goldberg D (2011) Hepatitis B prevention, diagnosis, treatment and care: a review. Occup Med 61(8):531–540

Assrir N, Soussan P, Kremsdorf D, Rossignol JM (2010) Role of the hepatitis B virus proteins in pro-and anti-apoptotic processes. Front Biosci (Landmark Edition) 15:12–24

Bandopadhyay M, Bharadwaj M (2020) Exosomal miRNAs in hepatitis B virus related liver disease: a new hope for biomarker. Gut Pathogens 12:1–16

Baumert TF, Schuster C (2007) Hepatitis C virus and the microRNA-guided endonuclease Argonaute 2. Virol J 4(1):15

Beck J, Nassal M (2007) Hepatitis B virus replication. World J Gastroenterol: WJG 13(1):48

Beckebaum S, Cicinnati VR, Zhang XX, Lu M, Radtke A, Gerken G et al (2002) Hepatitis B virus-induced defect of monocyte-derived dendritic cells leads to impaired T-helper function. Hepatology 35(1):51–60

Beckebaum S, Zhang XX, Chen X, Yu Z, Radtke A, Dirsch O et al (2003) Increased levels of interleukin-10 in serum from patients with hepatitis B virus-associated acute liver failure. Clin Exp Immunol 131(1):109–115

Bertoletti A, Hong M (2014) Age-dependent immune events during HBV infection from birth to adulthood: an alternative interpretation. Front Immunol 5(SEP)

Bertoletti A, Ferrari C, Fiaccadori F (2005) Host–virus interaction in hepatitis B virus infection and autoimmunity. J Viral Hepat 12(1):8–12

Biron CA, Nguyen KB, Pien GC, Cousens LP, Salazar-Mather TP (1999) Natural killer cells in antiviral defense: function and regulation by innate cytokines. Annu Rev Immunol 17(1):189–220

Blumberg BS (1964) Polymorphisms of the serum proteins and the development of iso-preciptins in transfused patients. Bull N Y Acad Med 40(5):377

Blumberg BS, Gerstley BJS, Hungerford DA, London WT, Sutnick AI (1967) A serum antigen (Australia antigen) in Down's syndrome, leukemia, and hepatitis. Ann Intern Med 66(5):924–931

Bock CT, Kubicka S, Manns MP, Trautwein C (1999) Two control elements in the hepatitis B virus S-promoter are important for full promoter activity mediated by CCAAT-binding factor. Hepatology (Baltimore, Md.) 29(4):1236–1247

Bruns AM, Horvath CM (2012) Activation of RIG-I-like receptor signal transduction. Crit Rev Biochem Mol Biol 47(2):194–206

Bruss V (1997) A short linear sequence in the pre-S domain of the large hepatitis B virus envelope protein required for virion formation. J Virol 71(12):9350–9357

Bruss V (2004) Envelopment of the hepatitis B virus nucleocapsid. Virus Res 106(2):199–209

Bruss V (2007) Hepatitis B virus morphogenesis. World J Gastroenterol: WJG 13(1):65

Busca A, Kumar A (2014) Innate immune responses in hepatitis B virus (HBV) infection. Virol J 11(1):1–8

CDC (2021) Hepatitis B. https://www.cdc.gov/hepatitis/hbv/bfaq.htm. Accessed 26 Jan 2021

Chan C, Gack MU (2016) Viral evasion of intracellular DNA and RNA sensing. Nat Rev Microbiol 14(6):360–373

Chemin (2014) Several key players involved in hepatitis-B virus innate response commentary open access. Clin Microbiol 3(6)

Chen BF (2018) Hepatitis B virus pre-S/S variants in liver diseases. World J Gastroenterol 24(14):1507

Chen QY, Liu YH, Li JH, Wang ZK, Liu JX, & Yuan ZH (2012) DNA-dependent activator of interferon-regulatory factors inhibits hepatitis B virus replication. World journal of gastroenterology, 18(22), 2850–2858. https://doi.org/10.3748/wjg.v18.i22.2850

Chen Z, Cheng Y, Xu Y, Liao J, Zhang X, Hu Y et al (2008) Expression profiles and function of Toll-like receptors 2 and 4 in peripheral blood mononuclear cells of chronic hepatitis B patients. Clin Immunol 128(3):400–408
Chen X, Liang Y, Zhang J (2017) HBx-mediated decrease of AIM2 contributes to hepatocellular carcinoma metastasis. Mol Oncol 11(9):1225–1240
Chisari FV, Isogawa M, Wieland SF (2010) Pathogenesis of hepatitis B virus infection. Pathol Biol 58(4):258–266
Chung TW, Lee YC, Ko JH, Kim CH (2003) Hepatitis B virus X protein modulates the expression of PTEN by inhibiting the function of p53, a transcriptional activator in liver cells. Cancer Res 63(13):3453–3458
Clark GJ, Angel N, Kato M, López JA, MacDonald K, Vuckovic S, Hart DN (2000) The role of dendritic cells in the innate immune system. Microbes Infect 2(3):257–272
Cooper MA, Fehniger TA, Caligiuri MA (2001) The biology of human natural killer-cell subsets. Trends Immunol 22(11):633–640
Croagh CM, Lubel JS (2014) Natural history of chronic hepatitis B: phases in a complex relationship. World J Gastroenterol: WJG 20(30):10395
Daka D, Hailemeskel G, Fenta DA (2022) Prevalence of Hepatitis B virus infection and associated factors among female sex workers using respondent-driven sampling in Hawassa City, Southern Ethiopia. BMC Microbiol 22(1):37
Dane DS, Cameron CH, Briggs M (1970) Virus-like particles in serum of patients with Australia-antigen-associated hepatitis. Lancet 295(7649):695–698
Dansako H, Ueda Y, Okumura N, Satoh S, Sugiyama M, Mizokami M, Ikeda M, & Kato N (2016) The cyclic GMPAMP synthetase-STING signaling pathway is required for both the innate immune response against HBV and the suppression of HBV assembly. The FEBS journal, 283(1), 144–156. https://doi.org/10.1111/FEBS.13563
Dempsey A, Bowie AG (2015) Innate immune recognition of DNA: A recent history. Virology 479-480:146–152
Dickson KM, Mori J (2016) Viral DNA sensors IFI16 and cyclic GMP–AMP synthase possess distinct functions in regulating viral gene expression, immune defenses, and apoptotic responses during herpesvirus infection. mBio 7(6):e01553–e01516
Doherty DG, O'Farrelly C, O'Farrelly C (1999) Innate and adaptive lymphoid cells in the human liver. Immunol Rev 174(1):5–20
Dowling JK, Mansell A (2016) Toll-like receptors: the Swiss army knife of immunity and vaccine development. Clin Transl Immunol 5(5):e85
Du X, Pan T, Liu Q, Liu N, Zhou Y (2013) Involvement of AIM2 inflammasome in hepatitis B virus infection. Mediat Inflamm 2013
Duan XZ, Wang M, Li HW, Zhuang H, Xu D, Wang FS (2004) Decreased frequency and function of circulating plasmacytoid dendritic cells (pDC) in hepatitis B virus infected humans. J Clin Immunol 24(6):637–646
Dunn C, Brunetto M, Reynolds G (2007) Cytokines induced during chronic hepatitis B virus infection promote a pathway for NK cell–mediated liver damage. J Exp Med 204(3):667–680
Dunn C, Peppa D, Khanna P, Nebbia G, Jones M, Brendish N et al (2009) Temporal analysis of early immune responses in patients with acute hepatitis B virus infection. Gastroenterology 137(4):1289–1300
Fattovich G, Bortolotti F, Donato F (2008) Natural history of chronic hepatitis B: special emphasis on disease progression and prognostic factors. J Hepatol 48(2):335–352
Feng Q, Hato SV, Langereis MA, Zoll J, Virgen-Slane R, Peisley A et al (2012) MDA5 detects the double-stranded RNA replicative form in picornavirus-infected cells. Cell Rep 2(5):1187–1196
Ferrage F, Dutta K, Nistal-Villán E, Patel JR, Sánchez-Aparicio MT, de Ioannes P et al (2012) Structure and dynamics of the second CARD of human RIG-I provide mechanistic insights into regulation of RIG-I activation. Structure 20(12):2048
Fletcher SP, Chin DJ, Gruenbaum L (2013) Intrahepatic transcriptional signature associated with response to interferon-alpha treatment in the woodchuck model of chronic hepatitis B. PLoS One 8(8):e68218

Galibert F, Mandart E, Fitoussi F, Tiollais P, Charnay P (1979) Nucleotide sequence of the hepatitis B virus genome (subtype ayw) cloned in E. coli. Nature 281(5733):646–650

Ganem D, Prince AM (2004) Hepatitis B virus infection—natural history and clinical consequences. N Engl J Med 350(11):1118–1129

Gay NJ, Symmons MF, Gangloff M, Bryant CE (2007) Assembly and localization of toll-like receptor signaling complexes. Nat Rev Immunol 6(7):533–539

Gehring AJ, Protzer U (2015) Targeting innate and adaptive immune responses to cure chronic HBV infection. Gastroenterology 149(3):625–638

Gehring AJ, Haniffa M, Kennedy PT, Ho ZZ, Boni C, Shin A et al (2013) Mobilizing monocytes to cross-present circulating viral antigen in chronic infection. J Clin Invest 123(8):3766–3776

Geng X, Huang C, Qin Y, McCombs JE, Yuan Q, Harry BL et al (2012) Hepatitis B virus X protein targets Bcl-2 proteins to increase intracellular calcium, required for virus replication and cell death induction. Proc Natl Acad Sci 109(45):18471–18476

Geng M, Xin X, Bi LQ, Zhou LT, Liu XH (2015) Molecular mechanism of hepatitis B virus X protein function in hepatocarcinogenesis. World J Gastroenterol: WJG 21(38):10732

Gerlich WH (2013) Medical virology of hepatitis B: how it began and where we are now. Virol J 10(1):1–25

Ghabeshi S, Baktashi R, Hosseini SM, Sharifi Z (2014) Molecular evaluation of HBV core gene mutations in asymptomatic HBV infected blood donors in Iran. Arch Iran Med 17:759–762

Gitlin L, Barchet W, Gilfillan S, Cella M, Beutler B, Flavell RA, Diamond MS, Colonna M (2006) Essential role of mda-5 in type I IFN responses to polyriboinosinic:polyribocytidylic acid and encephalomyocarditis picornavirus. Proc Natl Acad Sci USA 103(22):8459–8464

Gordien E, Rosmorduc O, Peltekian C (2001) Inhibition of hepatitis B virus replication by the interferon-inducible MxA protein. J Virol 75(6):2684–2691

Goubau D, Deddouche S, Reis e Sousa C (2013) Cytosolic sensing of viruses. Immunity 38(5):855–869

Greenberg HB, Pollard RB, Lutwick LI, Gregory PB, Robinson WS, Merigan TC (1976) Effect of human leukocyte interferon on hepatitis B virus infection in patients with chronic active hepatitis. N Engl J Med 295(10):517–522

Guidotti LG, Ishikawa T, Chisari FV (1999) Intracellular inactivation of the hepatitis B virus by cytotoxic T lymphocytes. Immunity 4(1):25–36

Guidotti LG, Ishikawa T, Chisari FV (2002) Loss of CTL function among high-avidity tumor-specific CD8+ T cells following tumor infiltration. Cancer Res 62(1):26–30

Günther S, Sommer G, Von Breunig F, Iwanska A, Kalinina T, Sterneck M, Will H (1998) Amplification of full-length hepatitis B virus genomes from samples from patients with low levels of viremia: frequency and functional consequences of PCR-introduced mutations. J Clin Microbiol 36(2):531–538

Günther S, Fischer L, Pult I, Sterneck M, Will H (1999) Naturally occurring variants of hepatitis B virus. Adv Virus Res 52:25–137

Guo J, Zhang T, Cao H, Zhang Z, Deng Y, Duan Z et al (2014) Impaired plasmacytoid dendritic cell function in infants with intrauterine hepatitis B virus infection. J Gastroenterol Hepatol 29(8):1673–1680

Guo F, Tang L, Shu S, Sehgal M, Sheraz M, Liu B et al (2017) Activation of stimulator of interferon genes in hepatitis B virus infection. Antimicrob Agents Chemother 61:e00771–e00717

Hadziyannis SJ (2011) Natural history of chronic hepatitis B in Euro-Mediterranean and African countries. J Hepatol 55(1):183–191

Hamerman JA, Ogasawara K, Lanier LL (2005) NK cells in innate immunity. Curr Opin Immunol 17(1):29–35

Han GR, Cao MK, Zhao W, Jiang HX, Wang CM, Bai SF et al (2011a) A prospective and open-label study for the efficacy and safety of telbivudine in pregnancy for the prevention of perinatal transmission of hepatitis B virus infection. J Hepatol 55(6):1215–1221

Han Q, Zhang C, Zhang J, Tian Z (2011b) Involvement of activation of PKR in HBx-siRNA-mediated innate immune effects on HBV inhibition. PLoS One 6(11):e27931

Hasegawa K, Huang J, Rogers SA, Blum HE, Liang TJ (1994) Enhanced replication of a hepatitis B virus mutant associated with an epidemic of fulminant hepatitis. J Virol 68(3):1651–1659
Herrscher C, Roingeard P, Blanchard E (2020) Hepatitis B virus entry into cells. Cells 9(6):1486
Hollinger BF, Liang TJ (2001) Chapter 87 Hepatitis B virus. In: Fields BN, Knipe DM, Howley PM, Griffin DE (eds) Fields virology, 4th edn. Lippincott Williams & Wilkins
Huang HC, Chen CC, Chang WC, Tao MH, Huang C (2012) Entry of hepatitis B virus into immortalized human primary hepatocytes by clathrin-dependent endocytosis. J Virol 86(17):9443–9453
Ichikawa K, Suganami T, Sato N (2001) Activated natural killer T cells inhibit adipose tissue macrophage accumulation by killing adipose tissue macrophages and adipocytes. Obes Res 9(12):143–151
Isogawa M, Robek MD, Furuichi Y, Chisari FV (2005) Toll-like receptor signaling inhibits hepatitis B virus replication in vivo. J Virol 79(11):7269–7272
Ju Y, Hou N, Meng J (2010) T cell immunoglobulin- and mucin-domain-containing molecule-3 (Tim-3) mediates natural killer cell suppression in chronic hepatitis B. J Hepatol 52(3):322–329
Karayiannis P (2017) Hepatitis B virus: virology, molecular biology, life cycle and intrahepatic spread. Hepatol Int 11(6):500–508
Kato H, Takeuchi O, Mikamo-Satoh E, Hirai R, Kawai T, Matsushita K, Hiiragi A, Dermody TS, Fujita T, Akira S (2008) Length-dependent recognition of double-stranded ribonucleic acids by retinoic acid–inducible gene-I and melanoma differentiation–associated gene 5. J Exp Med 205(7):1601
Kawai T, Akira S (2010) The role of pattern-recognition receptors in innate immunity: update on toll-like receptors. Nat Immunol 11(5):373–384
Khakoo SI, Thio CL, Martin MP (2004) HLA and NK cell inhibitory receptor genes in resolving hepatitis C virus infection. Science 305(5685):872–874
Köck J, Blum HE (2008) Hypermutation of hepatitis B virus genomes by APOBEC3G, APOBEC3C and APOBEC3H. J Gen Virol 89(5):1184–1191
Kosaka Y, Takase K, Kojima M, Shimizu M, Inoue K, Yoshiba M et al (1991) Fulminant hepatitis B: induction by hepatitis B virus mutants defective in the precore region and incapable of encoding e antigen. Gastroenterology 100(4):1087–1094
Kouwaki T, Fukushima Y, Daito T, Sanada T, Yamamoto N, Mifsud EJ et al (2016a) Extracellular vesicles including exosomes regulate innate immune responses to hepatitis B virus infection. Front Immunol 0(AUG):335
Kouwaki T, Okamoto M, Tsukamoto H, Fukushima Y, Oshiumi H, Seya T (2016b) Role of interferon-inducible gene IFI16 on virus replication. Semin Immunol 29(4):239–244
Kowalinski E, Lunardi T, McCarthy AA, Louber J, Brunel J, Grigorov B et al (2011) Structural basis for the activation of innate immune pattern-recognition receptor RIG-I by viral RNA. Cell 147(2):423–435
Krugman S, Giles JP, Hammond J (1971) Viral hepatitis, type B (MS-2 strain): studies on active immunization. JAMA 217(1):41–45
Lamontagne RJ, Bagga S, Bouchard MJ (2016) Hepatitis B virus molecular biology and pathogenesis. Hepatoma Res 2:163
Lang T, Lo C, Skinner N, Locarnini S, Visvanathan K, Mansell A (2011) The hepatitis B e antigen (HBeAg) targets and suppresses activation of the toll-like receptor signaling pathway. J Hepatol 55(4):762–769
Laskus T, Rakela J, Nowicki MJ, Persing DH (1995) Hepatitis B virus core promoter sequence analysis in fulminant and chronic hepatitis B. Gastroenterology 109(5):1618–1623
Lauzon NM, Mian F, Ashkar AA (2006) High-mobility group B1 protein (HMGB1) interacts with bacterial lipopolysaccharide and modulates its effects on dendritic cells. J Leukoc Biol 81(1):134–141
Le Seyec J, Chouteau P, Cannie I, Guguen-Guillouzo C, Gripon P (1998) Role of the pre-S2 domain of the large envelope protein in hepatitis B virus assembly and infectivity. J Virol 72(7):5573–5578

Leong CR, Funami K, Oshiumi H, Mengao D, Takaki H, Matsumoto M et al (2015) PKR and RIG-I are new innate immunity receptors for hepatitis B virus. Biochem Biophys Res Commun 18(2):334–339

Li HC, Huang EY, Su PY, Wu SY, Yang CC, Lin YS et al (2010) Nuclear export and import of human hepatitis B virus capsid protein and particles. PLoS Pathog 6(10):e1001162

Li J, He J, Fu Y, Hu X, Sun LQ, Huang Y, Fan X (2017) Hepatitis B virus X protein inhibits apoptosis by modulating endoplasmic reticulum stress response. Oncotarget 8(56):96,027

Liang TJ, Baruch Y, Ben-Porath E, Enat R, Bassan L, Brown NV et al (1991) Hepatitis B virus infection in patients with idiopathic liver disease. Hepatology 13(6):1044–1051

Liu Y, Li J, Chen J, Li Y, Wang W, Du X et al (2015) Hepatitis B virus polymerase disrupts K63-linked ubiquitination of STING to block innate cytosolic DNA-sensing pathways. J Virol 89:2287–2300

Liu D, Cui L, Wang Y, Yang G, He J, Hao R et al (2016) Hepatitis B e antigen and its precursors promote the progress of hepatocellular carcinoma by interacting with NUMB and decreasing p53 activity. Hepatology 64(2):390–404

Lurman A (1885) Eire icterusepidemie. Berl Klin Wochenschr 22:20–23, apud: Hollinger FB 1996. Hepatitis virus B. Knippe, DM [et al] Fields virology, Philadelphia: Lippincott–Raven, 3, 2739–2807

Lv J, Sun B, Mai H (2012) Immune escape mechanism of hepatitis B virus associated with the expression of surface antigen and virion secretion. Zhonghua Gan Zang Bing Za Zhi 20(11):818–822

Ma Z, Zhang E, Yang D, Lu M, Zhu X, Zhang W et al (2014) T cell dysfunction and depletion in acute hepatitis B virus infection. J Virol 88(1):1020–1030

Martin MP, Carrington M, Ross H (2002) The influence of KIR genotype on NK cell function. Immunol Cell Biol 80(3):224–225

Martinet J, Dufeu-Duchesne T, Bruder Costa J (2012) Altered functions of plasmacytoid dendritic cells and reduced cytolytic activity of natural killer cells in patients with chronic HBV infection. Gastroenterology 143(6):1586–1596

McMahon BJ (2010) Natural history of chronic hepatitis B. Clin Liver Dis 14(3):381–396

Milich D, Liang TJ (2003) Exploring the biological basis of hepatitis B e antigen in hepatitis B virus infection. Hepatology 38(5):1075–1086

Moffat JM, Mintern JD, Villadangos JA (2013) Control of MHC-II antigen presentation by ubiquitination. Front Immunol 4:54

Nassal M (2008) Hepatitis B viruses: reverse transcription a different way. Virus Res 134(1–2):235–249

Nassal M, Rieger A (1996) A bulged region of the hepatitis B virus RNA encapsidation signal contains the replication origin for discontinuous first-strand DNA synthesis. J Virol 70(5):2764–2773

Neurath AR, Kent SB, Strick N, Parker K (1986) Identification and chemical synthesis of a host cell receptor binding site on hepatitis B virus. Cell 46(3):429–436

Nguyen DH, Gummuluru S, Hu J (2007) Deamination-independent inhibition of hepatitis B virus reverse transcription by APOBEC3G. J Virol 81(9):4465–4472

Noguchi C, Ishino H, Tsuge M (2007) Hepatitis B virus X protein influences the immunogenicity of HBV core gene products. Biochem Biophys Res Commun 363(2):384–389

Okada K, Kamiyama I, Inomata M, Imai M, Miyakawa Y, Mayumi M (1976) e antigen and anti-e in the serum of asymptomatic carrier mothers as indicators of positive and negative transmission of hepatitis B virus to their infants. N Engl J Med 294(14):746–749

Okazaki A, Hiraga N, Imamura M, Hayes CN (2012) Severe necroinflammatory reaction caused by natural killer cells in a woodchuck model of acute hepatitis B. Liver Int 32(3):440–451

Omata M, Ehata T, Yokosuka O, Hosoda K, Ohto M (1991) Mutations in the precore region of hepatitis B virus DNA in patients with fulminant and severe hepatitis. N Engl J Med 324(24):1699–1704

Op den Brouw ML, Binda RS, van Roosmalen MH, Protzer U, Janssen HL, van der Molen RG (2009) Hepatitis B virus surface antigen impairs myeloid dendritic cell function: a possible immune escape mechanism of hepatitis B virus. Immunology 126(2):280–289
Orange JS, Wang B, Terhorst C (2002) B and natural killer (NK) cells. J Clin Immunol 22(2):81–86
Oshiumi H, Kouwaki T, & Seya T (2016) Accessory factors of cytoplasmic viral RNA sensors required for antiviral innate immune response. Frontiers in immunology, 7, 200. https://doi.org/10.3389/fimmu.2016.00200
Paludan SR, Bowie AG (2013) Immune sensing of DNA. Immunity 38(5):870–880
Papatheodoridis GV, Manesis E, Hadziyannis SJ (2001) The long-term outcome of interferon-α treated and untreated patients with HBeAg-negative chronic hepatitis B. J Hepatol 34(2):306–313
Park IH, Baek KW, Cho EY, Ahn BY (2011a) PKR-dependent mechanisms of interferon-alpha for inhibiting hepatitis B virus replication. Mol Cell Biochem 353(1–2):17–24
Park O, Jeong WI, Wang L (2011b) Diverse roles of invariant natural killer T cells in liver injury and fibrosis induced by carbon tetrachloride. Hepatology 53(4):1227–1236
Pasare C, Medzhitov R (2005) Toll pathway-dependent blockade of CD4+ CD25+ T cell-mediated suppression by dendritic cells. Science 299(5609):1033–1036
Pasek M, Goto T, Gilbert W, Zink B, Schaller H, MacKay P et al (1979) Hepatitis B virus genes and their expression in E. coli. Nature 282(5739):575–579
Peltekian C, Gordien E, Ferran E (2005) Fitting pegylated interferon treatment to viral kinetic responses in chronic hepatitis B: a mathematical approach. J Hepatol 43(2):263–270
Peppa D, Gill US, Reynolds G (2010) Up-regulation of a death receptor renders antiviral T cells susceptible to NK cell–mediated deletion. J Exp Med 210(1):99–114
Petersen J, Buti M (2012) Considerations for the long-term treatment of chronic hepatitis B with nucleos (t) ide analogs. Expert Rev Gastroenterol Hepatol 6(6):683–694
Peterson DL, Roberts IM, Vyas GN (1977) Partial amino acid sequence of two major component polypeptides of hepatitis B surface antigen. Proc Natl Acad Sci 74(4):1530–1534
Poisson F, Severac A, Hourioux C, Goudeau A, Roingeard P (1997) Both pre-S1 and S domains of hepatitis B virus envelope proteins interact with the core particle. Virology 228(1):115–120
Pontisso P, Ruvoletto MG, Gerlich WH, Heermann KH, Bardini R, Alberti A (1989) Identification of an attachment site for human liver plasma membranes on hepatitis B virus particles. Virology 173(2):522–530
Pourkarim MR, Amini-Bavil-Olyaee S, Kurbanov F, Van Ranst M, Tacke F (2014) Molecular identification of hepatitis B virus genotypes/subgenotypes: revised classification hurdles and updated resolutions. World J Gastroenterol: WJG 20(23):7152
Proud CG (1995) PKR: A new name and new roles. Trends Biochem Sci 20(6):241–246
Pujol F, Jaspe RC, Loureiro CL, Chemin I (2020) Hepatitis B virus American genotypes: pathogenic variants? Clin Res Hepatol Gastroenterol 44:825–835
Puri P (2014) Tackling the hepatitis B disease burden in India. J Clin Exp Hepatol 4(4):312–319
Ray G (2017) Current scenario of hepatitis B and its treatment in India. J Clin Transl Hepatol 5(3):277
Rosler C, Kock J, Kann M, Malim MH (2004) APOBEC-mediated interference with hepadnavirus production. Hepatology 39(3):166–174
Sato S, Suzuki K, Akahane Y, Akamatsu K, Akiyama K, Yunomura K et al (1995) Hepatitis B virus strains with mutations in the core promoter in patients with fulminant hepatitis. Ann Intern Med 122(4):241–248
Schaefer S (2007) Hepatitis B virus taxonomy and hepatitis B virus genotypes. World J Gastroenterol: WJG 13(1):14
Schoggins JW, Wilson SJ, Panis M, Murphy MY, Jones CT, Bieniasz P, Rice CM (2011) A diverse range of gene products are effectors of the type I interferon antiviral response. Nature 472(7344):481–485
Seeff LB, Beebe GW, Hoofnagle JH, Norman JE, Buskell BZ, Waggoner JG, Kaplowitz N, Koff RS, Petrini JJ, Schiff ER, Shorey J, Stanley MM (1987) A serologic follow-up of the 1942 epidemic of post-vaccination hepatitis in the United States Army. N Engl J Med 316:965–970

Seeger C, Mason WS (2000) Hepatitis B virus biology. Microbiol Mol Biol Rev 64(1):51–68
Shen L, Zhang X, Hu D, Feng T, Li H, Lu Y, Huang J (2013) Hepatitis B virus X (HBx) play an anti-apoptosis role in hepatic progenitor cells by activating Wnt/β-catenin pathway. Mol Cell Biochem 383(1–2):213–222
Shi B, Ren G, Hu Y, Wang S, Zhang Z, Yuan Z et al (2012) HBsAg inhibits TLR9-mediated activation and IFN-alpha production in plasmacytoid dendritic cells. Mol Immunol 51(3–4):264–271
Stacey AR, Norris PJ, Qin L, Haygreen EA, Taylor E, Heitman J et al (2009) Induction of a striking systemic cytokine cascade prior to peak viremia in acute human immunodeficiency virus type 1 infection, in contrast to more modest and delayed responses in acute hepatitis B and C virus infections. J Virol 83(8):3719–3733
Sterneck M, Gunther S, Santantonio T, Fischer L, Broelsch CE, Greten H, Will H (1996) Hepatitis B virus genomes of patients with fulminant hepatitis do not share a specific mutation. Hepatology 24(2):300–306
Sterneck M, Kalinina T, Otto S, Günther S, Fischer L, Burdelski M et al (1998) Neonatal fulminant hepatitis B: structural and functional analysis of complete hepatitis B virus genomes from mother and infant. J Infect Dis 177(5):1378–1381
Stevens CE, Beasley RP, Tsui J, Lee WC (1975) Vertical transmission of hepatitis B antigen in Taiwan. N Engl J Med 292:771–774
Szmuness W, Stevens CE, Harley EJ, Zang EA, Oleszko WR, William DC et al (1980) Hepatitis B vaccine: demonstration of efficacy in a controlled clinical trial in a high-risk population in the United States. N Engl J Med 303(15):833–841
Takaoka A, Wang Z, Choi MK, Yanai H, Negishi H, Ban T et al (2007) DAI (DLM-1/ZBP1) is a cytosolic DNA sensor and an activator of innate immune response. Nature 448(7152):501–505
Takeda K, Akira S (2003) Toll-like receptors in innate immunity. Int Immunol 15(1):1–14
Takeuchi O, Akira S (2008) MDA5/RIG-I and virus recognition. Curr Opin Immunol 20(1):17–22
Takeuchi O, Akira S (2010) Pattern recognition receptors and inflammation. Cell, 140(6), 805–820. https://doi.org/10.1016/j.cell.2010.01.022
Tang H (ed) (2019) Hepatitis B virus infection: molecular virology to antiviral drugs, vol 1179. Springer Nature
Tang H, Oishi N, Kaneko S, Murakami S (2006) Molecular functions and biological roles of hepatitis B virus x protein. Cancer Sci 97(10):977–983
Taniguchi M, Seino Ki & Nakayama T (2003) The NKT cell system: bridging innate and acquired immunity. Nat Immunol 4, 1164–1165. https://doi.org/10.1038/ni1203-1164
Tavakoli S, Mederacke I, Herzog-Hauff S, Glebe D, Grün S, Strand D et al (2004) Peripheral blood dendritic cells are phenotypically and functionally intact in chronic hepatitis B virus (HBV) infection. Clin Exp Immunol 135(1):131–139
Thimme R, Wieland S, Steiger C (2003) CD8+ T cells mediate viral clearance and disease pathogenesis during acute hepatitis B virus infection. J Virol 77(1):68–76
Thompson MR, Kaminski JJ, Kurt-Jones EA, Fitzgerald KA (2011) Pattern recognition receptors and the innate immune response to viral infection. Viruses 3(6):920–940
Thomsen MK, Nandakumar R, Stadler D, Malo A, Valls RM, Wang F et al (2016) Lack of immunological DNA sensing in hepatocytes facilitates hepatitis B virus infection. Hepatology 64(3):746–759
Tjwa ETTL, Van Oord GW, Hegmans JP (2011) Expression of inhibitory receptors on cytotoxic T lymphocytes in patients with chronic hepatitis B: a role for NKG2A. J Viral Hepat 18(3):e147–e157
Tripathy AS, Das R, Rath SS, Mohapatra N (2011) Molecular mechanism of hepatitis B virus induced liver injury. Indian J Clin Biochem 26(4):367–374
Tu Z, Bozorgzadeh A, Pierce RH (2014) TLR-dependent cross talk between human Kupffer cells and NK cells. J Exp Med 211(8):1571–1588
Valenzuela P, Gray P, Quiroga M, Zaldivar J, Goodman HM, Rutter WJ (1979) Nucleotide sequence of the gene coding for the major protein of hepatitis B virus surface antigen. Nature 280(5725):815–819

Vander Molen RG, Mennechet FJD, Maciejewski JP (2004) Polyfunctional CD4+ T cell responses to immunodominant epitopes correlate with disease activity of chronic hepatitis B. Clin Immunol 113(1):78–88
Vander Molen RG, Stoop JN, Claassen MA, Janssen HL, Boonstra A (2006) Plasmacytoid dendritic cells in hepatitis B virus infection: key players in the activation and exhaustion of antiviral immune responses. J Virol 80(24):11,102–11,113
Vartanian JP, Henry M, Marchio A, Suspène R (2010) Massive APOBEC3 editing of hepatitis B viral DNA in cirrhosis. PLoS Pathog 6(5):e1000928
Visvanathan K, Skinner NA, Thompson AJ, Riordan SM, Sozzi V, Edwards R et al (2007) Regulation of toll-like receptor-2 expression in chronic hepatitis B by the precore protein. Hepatology 45(1):102–110
Wang H, Ryu WS, Ryu WS (2010) Hepatitis B virus polymerase blocks pattern recognition receptor signaling via interaction with DDX3: implications for immune evasion. PLoS Pathog 6(7):e1000986
Watanabe T, Sorensen EM, Naito A, Schott M, Kim S, Ahlquist P (2007) Involvement of host cellular multivesicular body functions in hepatitis B virus budding. Proc Natl Acad Sci 104(24):10205–10210
Weber M, Bronsema V, Bartos H, Bosserhoff A, Bartenschlager R, Schaller H (1994) Hepadnavirus P protein utilizes a tyrosine residue in the TP domain to prime reverse transcription. J Virol 68(5):2994–2999
Webster GJ, Reignat S, Maini MK, Whalley SA, Ogg GS, King A et al (2000) Incubation phase of acute hepatitis B in man: dynamic of cellular immune mechanisms. Hepatology 32(5):1117–1124
Wieland S, Thimme R, Purcell RH, Chisari FV (2004) Genomic analysis of the host response to hepatitis B virus infection. Proc Natl Acad Sci USA 101(17):6669–6674
Wies E, Wang MK, Maharaj NP, Chen K, Zhou S, Finberg RW, Gack MU (2013) Dephosphorylation of the RNA sensors RIG-I and MDA5 by the phosphatase PP1 is essential for innate immune signaling. Immunity 38(3):437–449
Wiesmayr S, Webber SA, Macedo C (2012) Decreased NKp46 and NKG2D and elevated PD-1 are associated with altered NK-cell function in pediatric transplant patients with PTLD. Eur J Immunol 42(2):541–550
Woltman AM, Op den Brouw ML, Biesta PJ, Shi CC, Janssen HL (2011) Hepatitis B Virus lacks immune activation capacity, but actively inhibits plasmacytoid dendritic cell function. Hepatology 48(2):732–741
World Health Organization (2024) Hepatitis B. https://www.who.int/en/news-room/fact-sheets/detail/hepatitis-b. Accessed: 29 Jan 2024
Wu X, Dao Thi VL, Huang Y, Billerbeck E, Saha D, Hoffmann HH, Wang Y, Silva LAV, Sarbanes S, Sun T, Andrus L, Yu Y, Quirk C, Li M, MacDonald MR, Schneider WM, An X, Rosenberg BR, Rice CM (2018) Intrinsic immunity shapes viral resistance of stem cells. Cell 172(3):423–438.e25. https://doi.org/10.1016/j.cell.2017.11.018
Wu CC, Chen YS, Cao L, Chen XW, Lu MJ (2013) Hepatitis B virus infection: defective surface antigen expression and pathogenesis. World J Gastroenterol 24(31):3488
Yan H, Peng B, Liu Y, Xu G, He W, Ren B et al (2014) Viral entry of hepatitis B and D viruses and bile salts transportation share common molecular determinants on sodium taurocholate cotransporting polypeptide. J Virol 88:3273–3284
Yang PL, Althage A, Chung J, Chisari FV (2010) Hydrodynamic injection of viral DNA: a mouse model of acute hepatitis B virus infection. Proc Natl Acad Sci 107(15):673–678
Yim S-A, Heydmann L, Verrier E, Pochet N, Schuster C, Baumert TF (2016) The CGAS-Sting Pathway Exhibits Antiviral Activity against Hepatitis B Virus Infection. https://doi.org/10.1016/S0168-8278(16)00620-6
Zhang P, Samuel CE (2007) Protein kinase PKR plays a stimulus- and virus-dependent role in apoptotic death and virus multiplication in human cells. J Virol 81(16):8192–8200

Zhang Z, Zhang S, Zou Z, Shi J, Zhao J, Fan R et al (2007) Hypercytolytic activity of hepatic natural killer cells correlates with liver injury in chronic hepatitis B patients. Hepatology 46(5):1392–1401

Zhang Z, Zou Z, Svendsen O (2011) Genetic analyses of hepatitis B virus genotypes and mutants in HBsAg-positive blood donors in Shenzhen, China. Virol J 8(1):1–10

Zhang H, Huang C, Wang Y, Lu Z, Zhuang N, Zhao D et al (2015) Hepatitis B virus X protein sensitizes TRAIL-induced hepatocyte apoptosis by inhibiting the E3 ubiquitin ligase A20. PLoS One 10(5):e0127329

Zhao J, Li D, Liu H (2012a) Increased education and activation of peripheral blood NK cells contribute to HBV clearance in Chinese patients. Int Immunol 24(1):55–64

Zhao J, Li Y, Jin L, Zhang S, Fan R, Sun Y, Zhou C, Shang Q, Li W, Zhang Z, Wang FS (2012b) Natural killer cells are characterized by the concomitantly increased interferon-γ and cytotoxicity in acute resolved hepatitis B patients. PLoS One 7(11)

Zhao Y, Wu J, Sun L, Liu G, Li B, Zheng Y et al (2016) Prevalence of mutations in HBV DNA polymerase gene associated with nucleos (t) ide resistance in treatment-naive patients with chronic Hepatitis B in Central China. Braz J Infect Dis 20(2):173–178

Zheng BJ, Zhou J, Qu D, Si LF (2004) Peripheral blood dendritic cells in children with chronic hepatitis B virus infection: maturation defect and unable to inhibit the proliferation of activated autologous lymphocytes. J Gastroenterol Hepatol 19(8):927–933

Zou Y, Chen T, Han M, Wang H, Yan W, Song G, Wu Z, Wang X, Zhu C, Luo X, Ning Q (2010) Increased killing of liver NK cells by Fas/Fas ligand and NKG2D/NKG2D ligand contributes to hepatocyte necrosis in virus-induced liver failure. Journal of immunology (Baltimore, Md.: 1950), 184(1):466–475. https://doi.org/10.4049/JIMMUNOL.0900687

Hepatitis C Virus Infection: Current Trends in Host Pathogen Interactions, Diagnostics and Treatment

20

Korra Bhanu Teja, Sridivya Bhagavatula, Aishwarya Bhatta, and Shilpa J. Tomar

Abstract

Hepatitis C virus, first identified in the year 1989, a positive sense RNA virus, causes liver inflammation and is parenterally transmitted. With a genomic size of 9.6 kb, HCV is a member of the genus Hepacivirus and family Flaviviridae. It has eight genotypes (1–8) and many subtypes. HCV genotype 1 is majorly prevalent as compared to the other genotypes. HCV is known to be present in all the WHO regions with approximately 170 million carriers worldwide. HCV is a biosafety level-2 pathogen with an incubation period of 2–26 weeks, known to cause chronic hepatitis as well as acute hepatitis, both of which develop through several stages, such as acute self-resolving hepatitis, asymptomatic carrier stage, chronic hepatitis, cirrhosis and at last, hepatocellular carcinoma. It mainly causes infection in humans and primarily replicates in the hepatocytes. Majority of acute infection cases, i.e., 50–85% progress towards a chronic state; hence, early diagnosis is pertinent. Currently, extremely efficient direct-acting antiviral agents (DAAs) can be used to treat HCV. Screening the high-risk populations, such as intravenous drug users, hemodialysis patients, blood donors, etc., can stop the spread of this virus. Prevention includes safe injection and infection control practices. Vaccines against HCV are under investigation which would help in further restricting the spread of the virus.

K. B. Teja · S. J. Tomar (✉)
Hepatitis Division, ICMR-National Institute of Virology, Pune, India
e-mail: drshilpatomar.niv@gmail.com

S. Bhagavatula
Hepatitis Division, ICMR-National Institute of Virology, Pune, India

Bacteriology Division, ICMR-National Institute of Virology, Pune, India

A. Bhatta
Warwick Medical School, University of Warwick, Coventry, United Kingdom

P. L. Bhukya, A. S (eds.), *Emerging Human Viral Diseases, Volume II*,
https://doi.org/10.1007/978-981-97-4480-0_20

Keywords

HCV (Hepatitis C Virus) · RdRp · DAA (Direct Acting Antivirals) · Hepatocellular carcinoma

20.1 Introduction

The positive sense ss RNA virus that causes hepatitis C belongs to the family Flaviviridae and genus *Hepacivirus*, causing hepatitis, i.e., liver inflammation, and is mainly transmitted via parenteral route.

20.1.1 Brief History

In the 1960s and 1970s, respectively, the Hepatitis A and Hepatitis B viruses were discovered. Harvey J. Alter, a researcher at the US National Institutes of Health (NIH) in Bethesda, Maryland, USA, analysed the prevalence of the hepatitis virus among people who received blood transfusions. He and his colleagues observed that most of the recipients exhibiting symptoms for hepatitis were positive for neither hepatitis A, nor hepatitis B. This type of hepatitis was named as the non-A non-B (NANB) hepatitis in 1975. To determine whether the NANB hepatitis was caused by a virus, the blood from the NANB hepatitis patients was transmitted to chimpanzees who were the only other susceptible hosts other than humans, by Alter and his colleagues. In this experiment, five chimpanzees were infected with the blood from NANB hepatitis patients, and all the chimpanzees developed NANB hepatitis which was evidenced by pathological changes in the liver and an increased level of alanine aminotransferase. A new and distinct form of chronic viral hepatitis was observed which lay the ground for further discovery (Houghton 2009, 2019; Simmonds 2013; Hu et al. 2020).

20.1.2 Discovery

HCV was discovered in the year 1989, by the cumulative efforts of Michael Houghton, presently working in the University of Alberta, Canada; Harvey Alter, presently in the National Institutes of Health, Bethesda, USA, and Charles Rice currently in the Rockefeller University in New York, USA. In the year 2020, they received the prestigious Nobel prize in the field of medicine, for the discovery of the Hepatitis C virus (HCV).

20.2 Epidemiology

20.2.1 Geographical Distribution

HCV is endemic throughout the globe and is present in all the WHO regions, with 130 countries reporting the infection. The G1 genotype is the most prevalent,

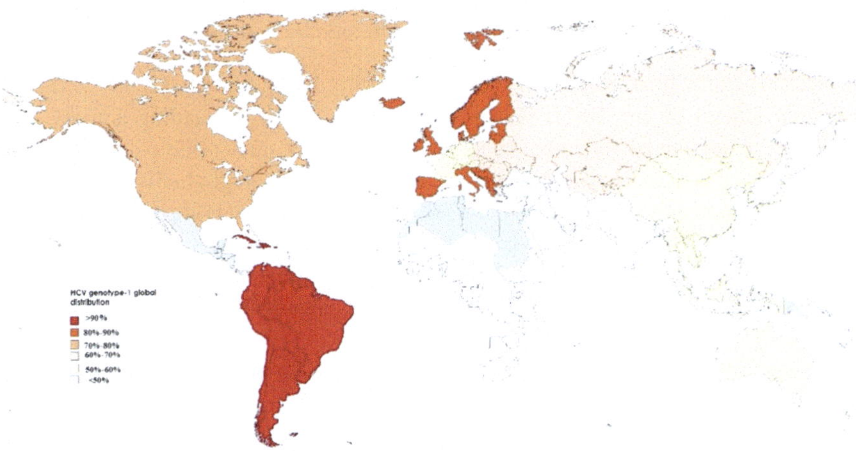

Fig. 20.1 Worldwide prevalence of HCV genotype 1

accounting for over 46% of all HCV infections in adults, next being G3 genotype (21–25%), G2 genotype (13%), G4 genotype (13–15%), G6 genotype (2%), and the G5 genotype (1%). The most prevalent sub-type, genotype 1b, counts for 22% of all infections. Genotype 7 originated in Central Africa in 2014 (Murphy et al. 2015) and Genotype 8 was identified in Punjab, India, in 2018 (Borgia et al. 2018). The worldwide prevalence of genotypes 7 and 8 is yet to be determined. The WHO Eastern Mediterranean Region and European Region have the highest disease burden with 12 million people each, having a chronic hepatitis C infection. In the African region, 9 million people have a chronic infection and 5 million in the Americas region. The most prominent genotype in India is genotype 3, followed by genotype 1 (WHO reports 2022; Baliashvili et al. 2022) (Fig. 20.1).

20.2.2 Demographic features of HCV Infection

HCV infects a broad age group of people, ranging from infants to old age. Infection with the HCV is thought to affect 2.2% of the world's population. The risk of disease progression increases with an increase in age of the individual. Ages from 50 to 59 have the greatest documented frequency of HCV.

The morbidity and mortality caused by HCV is usually due to liver failure or liver cancer. According to the WHO reports of 2019, 290,000 people died from hepatitis C infection. The death occurred mainly due to hepatocellular carcinoma and cirrhosis. Approximately 58 million people throughout the world currently have a chronic hepatitis C infection, and there is a rise of 1.5 million new infections each year (WHO Reports 2022). Out of these, around 3.2 million children and adolescents are chronically infected. There are 6 million known cases of HCV infection in India, and each year, 176,000 individuals pass away from complications related to Hepatitis B and C.

20.2.3 Origin of Infection and Diversity

HCV exists in the host as quasispecies, the diversity being present due to the mutations introduced by the NS5B RdRp, which lacks the proof-reading activity. An estimated error rate of this enzyme is 10^{-3} to 10^{-5} per nucleotide per replication cycle. Hence, HCV quasispecies have sequence motifs which gradually change over a period of time from one passage to the other, due to sequential selection events and neutral drift. With 81% to 88% sequence identity amongst isolates, the capsid region is highly conserved. E1 and E2 regions are highly variable regions. The phylogenetic analysis of this virus genome is based on the sequencing of the NS5B region (Kamili and Qadri 2020).

According to ICTV, as of March 2022, HCV is diversified into eight genotypes and 95 subtypes, of which only 13 subtypes have been accepted provisionally (University of Glasgow Centre for Virus Research 2022).

20.2.4 Disease Transmission

HCV is mainly transmitted by blood and blood products, the various means of transmission being intravenous drug use, sharing infected injections, blood transfusions, organ transplants, needle stick injury in healthcare professionals, unregulated tattoo and body piercings, etc. Transmission via the sexual route is uncommon. HCV can also be transmitted vertically from mother to child, with a transmission rate of 4.3%, which increases to 22.1% in cases the mother also has HIV co-infection (Alter 2007; Morozov and Lagaye 2018).

20.2.5 Biosafety

HCV is a BSL-2 virus, which thus needs to be handled in a class II biosafety cabinet. Since HCV is transmitted mainly by blood and blood products, extra precautions should be taken while handling sharp objects in the laboratory, such as needles etc., and infectious material should be decontaminated using 10% bleach. All infectious waste collected in the biohazard bag should be discarded only after proper autoclaving. The surface of the biosafety cabinet should be properly disinfected using 70% ethanol (Yi 2010).

20.3 Structural and Molecular Organisation of HCV

HCV is divided into eight genotypes and 95 subtypes, of which 13 subtypes have only been accepted provisionally. According to the Baltimore classification, HCV belongs to Group IV as it has a +ve ss RNA genome (Sesmero and Thorpe 2015).

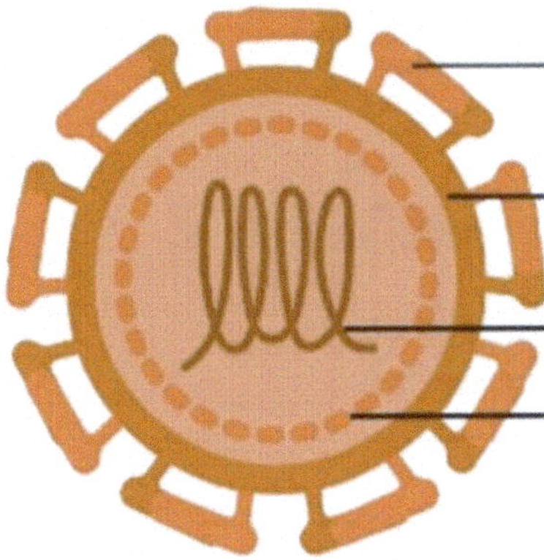

Fig. 20.2 HCV virion structure

20.3.1 Morphology and Virion Structure

HCV is an icosahedral, lipid membrane-enveloped virus that measures 50–65 nm in diameter. The virion contains the virus's RNA genome, core, and envelope proteins 1 and 2. The virus exhibits spikes of 6 nm size in the membrane of the virion, which are heterodimers of E1 and E2 glycoproteins. HCV exhibits a low and heterogenous buoyancy, and it has been observed that the density of the non-infectious particles is higher, probably due to the binding of immunoglobulin, whereas the density of the infectious particles is lower, indicating the ability of the virus to interact with serum lipoproteins (Fig. 20.2).

20.3.2 Genome Structure and Organisation

HCV is a +ss RNA genome of 9.6 kb size, uncapped 5′ and 3′ ends which are highly structured. The 5′ end has a conserved non-coding region (NCR) of 341 nucleotides, and this NCR is divided into four domains and a pseudoknot. Domain I consists of the first 40 nucleotides of the 5' NCR, essential for genome replication but not absolutely necessary for translation, whereas the HCV open-reading frame (ORF) is translated cap-independently by domains II–IV. The single large ORF that codes for 3000 or more amino acids is also directed to be translated cap-independently by an IRES found in the 5′ non-coding region. The structural proteins are known to be encoded by the N-terminal of the ORF, and the non-structural proteins are encoded by the remaining part of the ORF. The 3′ NCR is 200–235 nucleotides long, consisting of three distinct regions, namely, the variable domain of approximately 40 nucleotides, a polyuridine or a polypyrimidine stretch and a highly conserved 3′ X tail of 97 nucleotides. Highly conserved 3′ X tail has been observed to form stable stem-loop like structures, which are conserved in every genotype. This domain is essential for RNA replication. Many cellular proteins interact with the 3′ NCR

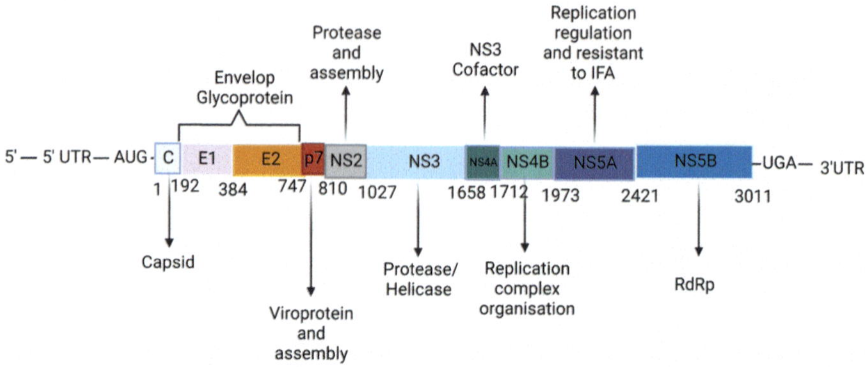

Fig. 20.3 Schematic representation of hepatitis C virus genome architecture representing structural proteins, namely, C (core protein), E1, E2, and non-structural proteins NS3 to NS5B. Image made by using Biorender.com

which include RNA binding proteins, heterogeneous ribo-nuclear proteins, etc. The structural proteins encoded include core protein(C) and envelope glycoproteins (E1 and E2) followed by the non-structural proteins, namely, p7 protein, NS2A, NS3 A, NS4 A, NS4 B, NS5A and at last, NS5B (Fig. 20.3).

20.3.3 In Vitro Propagation and In Vivo Laboratory Models

The in vitro models to study HCV include, HCV cell culture systems, replicons and pseudoparticles. HCV pseudoparticles (HCVpp) were constructed to study the viral entry mechanism, and by using this method, CD81 was discovered as a receptor for HCV. The cell culture for HCV was developed in 2005, where genotype 2a clone JFH-1 was designed by Takaji Wakita's group (Duverlie and Wychowski 2007).

The in vivo models to study HCV include chimpanzees, humanised mice, transgenic mice, tree shrew, etc. Chimpanzees were the first animal models used to study the HCV virus, and studies performed on them played a significant role in the discovery of this virus. The tree shrew model was seen to be advantageous to study the acute hepatitis phase but was a poor model to study the chronic infection. HCV-infected human hepatocytes were transplanted into mice; the virus reproduced, but the host immune responses could not be examined. A genetically humanised mouse model was developed for studying HCV, and supported entry of the virus, but not replication. Hence, an ideal in vivo model for HCV is yet to be discovered.

20.4 Viral Proteins and Life Cycle

The HCV genome encodes both structural and non-structural proteins, as already mentioned. The following structural proteins are among them.

20.4.1 Core Protein

The 21 kDa core protein has three domains and is large. Domain I (amino acids 1–117) is composed of basic residues and has brief hydrophobic areas, domain II (amino acids 118–174), which has fewer basic residues and is more hydrophobic, and domain III (175 amino acids—191 amino acids), which is significantly more hydrophobic.

20.4.2 Envelope Glycoproteins

Two envelope proteins, E1 and E2, with estimated molecular weights of 30 and 70 kDa, respectively, are present in HCV. Both glycoproteins are type I transmembrane proteins that combine in a non-covalent heterodimer to mediate the virus's many actions, including membrane fusion and attachment. The non-structural proteins encoded by HCV are as follows:

20.4.2.1 p7 Protein

p7 is a 63 amino acids long, small hydrophobic non-structural protein. p7 is located between E2 and NS2 proteins. It spans the membrane twice and is located in the endoplasmic reticulum, having two transmembrane domains, connected by a cytoplasmic loop oriented towards the ER lumen. The protein is cleaved by the signal peptidases in the ER of the host cell. In order for a virus to infect a host, these proteins create ion channels. Also, it promotes the translocation of NS2 for appropriate cleavage by host enzymes to the ER lumen.

20.4.2.2 NS2

NS2 is a transmembrane protein, which has a size of approximately 23 kDa. It has cysteine protease activity, and the N-terminal of this protein comprises of 3–4 transmembrane helices, fixed in the ER membrane, whereas the C-terminal of this protein is free in the cytoplasm exhibiting the NS2/3 auto protease activity. It interacts with envelope glycoproteins and non-structural proteins like p7, NS3 and NS4A, contributing to virus particle assembly.

20.4.2.3 NS3

NS3 is a multifunctional protein of size 67 kDa. The N-terminal of this protein has serine protease activity, and the C-terminal has helicase activity. The amino acids present in the protease of NS3 having catalytic activity are His-1083, Asp-1107 and Ser-1165. The NS3 and PKA interaction leads to the downregulation of intracellular signaling, and it is also associated with the inhibition of RIG-I and TLR3 signaling, influencing the innate cellular host defense.

20.4.2.4 NS4

NS4A is the smallest non-structural protein constituted of 54 amino acids, acting as a co-factor for the NS3 serine protease. It physically interacts with NS4B, NS5A,

Table 20.1 Overview of different proteins and their respective molecular weight and function

Protein	Molecular weight (kDa)	No. of amino acids	Function
Core	23	191	RNA binding
E1	35	192	Envelope protein
E2	71	363	Envelope protein
P7	8	63	Ion-channel
NS2	20	217	NS2–3-protease component
NS3	70	631	NS3/4 protease (protease domain),NS2-3 protease (RNA helicase, RNA binding
NS4A	4	54	NS3/4 protease
NS4B	28	261	Initiation of ER membranous web, NS 5a phosphorylation
NS5A	56	448	Inhibit IFN
NS5B	66	591	RNA dependent RNA polymerase

and uncleaved NS4B-5Aand is required for the phosphorylation of NS5A. The C-terminal of this protein contributes to the HCV helicase activity and virus assembly.

NS4B is a hydrophobic integral membrane protein of 27 kDa size, containing four transmembrane domains, and is involved in recruiting other viral proteins.

20.4.2.5 NS5

NS5A is a hydrophobic proline-rich RNA-binding phosphoprotein, homo-dimeric in nature and approximately 58 kDa in size. It is a multidomain protein with its domain I which has a N-terminal zinc-binding domain (DI), a conserved central domain (DII), and a variable C-terminal (DIII), separated by two linkers of low complexity. This protein plays a considerable impact on the control of cell signalling pathways, interferon response and viral replication. The three primary mitogen-activated protein kinase (MAPK) pathways that control cell growth and activation can be modified by NS5. Additionally, it controls cellular signalling via pro- and anti-apoptotic pathways.

The NS5B is a 65 kDa tail anchored protein, acting as an RNA-dependent RNA, right-handed polymerase, essential for synthesizing the new RNA genome (Table 20.1).

20.5 HCV Life Cycle (Fig. 20.4)

The life cycle of this virus is divided into four steps:

20.5.1 Virus Entry

HCV is a non-lytic virus that binds to the tetraspanin CD81, scavenger receptor class B type I (SCARB1), and the two tight junction molecules claudin-1 (CLDN1)

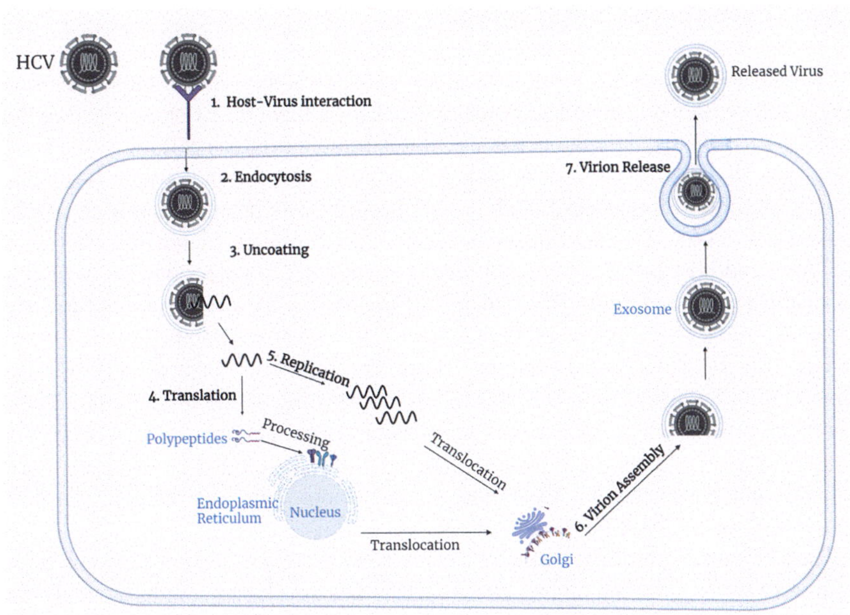

Fig. 20.4 Life cycle of Hepatitis C virus. Image made by using Biorender.com

and occludin (OCLN). The E2 protein directly binds to the tetraspanin CD81 and SCARB1 protein. HCV internalises in the cell via a clathrin-dependent endocytosis and fusion occurs within the early endosomes at a low pH.

20.5.2 Genome Translation and Polyprotein Processing

Translation occurs in the cytoplasm of the cell where the RNA genome (positive sense) is translated by the ribosomes on the endoplasmic reticulum into a polyprotein, which is dependent on the IRES. This polyprotein is cleaved into 10 different proteins by the NS2–3 autoprotease.

20.5.3 Genome Replication

The NS4B rearranges the membranes of the ER to form a membranous web, acting as a site for the replication of HCV genome. The NS5B amplifies the viral genome via an intermediate negative sense RNA. Once the progeny genome is synthesised, the other structural and non-structural proteins are also synthesised.

20.5.3.1 Particle Assembly and release from the Host Cell

All the proteins along with the genome are packaged into capsids, which then bud into the ER to derive an envelope. HCV exits the host cell taking over the secretory pathway to infect the next cell.

20.6 Pathogenesis in Humans

HCV is a non-cytopathic virus (Chevaliez and Pawlotsky 2006), having unique interactions with the host cells, known to cause prolonged infection by evading the immune surveillance. Human hepatocellular carcinoma (HCC) cell lines like HuH 7 are typically used as in vitro models to study the interaction between host cells protein and the virus because HCV cell culture models are not readily available (Gosert et al. 2003). Despite extensive research, the complicated dynamics between this virus and host cell interactions remains elusive. HCV enters into the host cell via clathrin-mediated endocytosis, the viral genome is then released and directly translated into a single polypeptide precursor. This precursor polypeptide is further cleaved into 10 mature viral proteins. These structural and nonstructural proteins perform different roles, for example, HCV non-structural 4B (NS4B) protein mediates virus host interactions whereas nonstructural phosphoprotein 5A (NS5A) is involved in viral replication. New viral particles follow a Golgi-dependent secretory pathway and during egress, they bind to host lipoproteins that help in immune escape (Lohmann 2013; Niepmann 2013).

The ultimate result of the battle between viruses and their hosts depends on their abilities to overcome each other's attacks. HCV has developed several strategies to avoid innate and adaptive immunological reactions from the host. In majority cases of HCV infection, the patients are unable to clear the virus during the acute phase, thereby developing chronic viral hepatitis (Hajarizadeh et al. 2013). The host innate immune system comprises physical and chemical barriers, phagocytic cells, blood proteins, and cytokines, defends against invasive infectious pathogens as the first line of protection (Schoggins and Rice 2013). Among these, the role of interferons against viruses has been of particular interest to scientists and researchers. Interferons are basically signaling molecules released by the innate immune cells, belonging to the cytokine family, and have a strong antiviral role. Type 1 interferons particularly, secreted by infected cells, are highly effective against viruses, triggering the transcriptional activation of genes and preventing the virus from replicating within the cells (Stetson and Medzhitov 2006). Unlike other cells, the innate immune cells do not have to be infected to be activated; they constantly sample viruses from outside and produce interferons. Additionally, the cytosolic pathway and toll-like receptor- (TLR-) dependent pathway both identify the degraded viral nucleic acid (Iwasaki and Medzhitov 2004), activating crucial transcription factors such NF-B and the interferon regulatory factor (IRF) 3 and 7, causing IFN production. By virtue of its antiviral activity, IFN-α has also been successfully used for therapies. To escape the innate immune system, HCV deploys NS3/4A protease that cleaves the signaling proteins essential for IFN induction, i.e., the toll-IL-1 receptor-domain-containing adaptor protein-inducing IFN-, the mitochondrial antiviral signalling protein (MAVS) and (TRIF). However, it has been observed that interferon stimulated genes (ISGs) are still induced in most of the cases of chronic infections (Akira et al. 2006).

Similar to the innate immune system, the host adaptive immune system also actively participates in the fight against the entry and replication of viruses. Both CD4+ and CD8+ T cells interact with multiple epitopes of the virus and help in viral clearance. The T cells subsets play an essential role in clearing and controlling virus replication. Previous investigations have reported that depletion of CD4+ T cells leads to HCV persistence in previously protected chimpanzees. Other than T cells, humoral antibodies produced by the cells of adaptive immunity also act against viral invasion. However, among the humoral antibodies, a very small fraction shows antiviral activities. The early induction of these neutralizing antibodies is linked with good prognosis but they are unable to effectively inhibit virus binding, entry or replication due to the virus escape mutations that bypass antibody detection (Pestka et al. 2007).

20.7 Clinical Manifestations

20.7.1 Phases of Disease

The different phases of the disease caused by HCV are acute self-resolving hepatitis, asymptomatic carrier stage, chronic hepatitis, cirrhosis and at last, hepatocellular carcinoma. In acute hepatitis, it is mostly asymptomatic, the latent period of this phase being approximately 7 weeks. There is an increase in the hepatic transaminases, namely, ALT (alanine transaminase) and AST (aspartate transaminase) from the dead and damaged cells (mainly hepatocytes) and 60%–85% of these infections progress to the chronic phase. Chronic HCV is a heterogeneous condition wherein the disease manifestations are individual specific and the rates of progression into the next phase vary, with high titres of the virus and necroinflammatory injury of the liver. Around 5%–20% of these patients progress to cirrhosis within 20–30 years. Type 2 diabetes can result from chronic HCV infection because of metabolic dysfunction. The final phase of the HCV disease is hepatocellular carcinoma, which is caused potentially due to the suppression of apoptotic genes and the induction of proto-oncogenes by the virus which may finally lead to death.

20.7.2 Clinical symptoms

Acute hepatitis's clinical signs and symptoms include jaundice, exhaustion, stomach discomfort, malaise, nausea, etc. An increased level of the liver enzymes AST and ALT is also observed. The signs and symptoms of chronic hepatitis are more specific in nature and most of the patients show asymptomatic enzyme elevations and detectable antibody levels against HCV. Splenomegaly is observed in patients with cirrhosis and extrahepatic manifestations such as purpuric rash, joint pain, vasculitis, mixed cryoglobulinemia, lymphoma, aplastic anemia, autoantibodies, autoimmune disease, etc. maybe observed (Dickson 1997; Khullar and Firpi 2015).

20.8 Laboratory Diagnosis: Serology, Molecular and Histological Approaches

Different serological approaches to detect HCV include anti-HCV enzyme linked immunoassays, chemiluminescence immunoassays (CLIAS), electrochemiluminescence assays (ECLS), recombinant immunoblot assay (RIBA), etc. Both viral and serological markers are not seen during the first eclipse phase. For an additional 6–10 weeks without an antibody response, HCV RNA and core p22 Ag are found. Anti-HCV antibodies are typically detected as part of the diagnosis. In these assays' antibodies against the core protein, NS3, NS4 and NS5 are detected. The detection of the antibodies against the HCV core antigen indicates present infection. These assays are highly specific but not as sensitive as the molecular assays where the HCV RNA is detected. The possibility of false negative is present in cases where the testing is done in the window period, i.e., 12–15 weeks after exposure, immunocompromised patients due to HIV co-infection or transplantation. Therefore, the diagnosis of a patient must be confirmed by HCV RNA detection before they begin treatment. HCV infection in children under the age of 18 months is also diagnosed by screening for HCV RNA as they might still have anti-HCV antibodies from their mother's infection (after 2 months of age) (National Guidelines for Diagnosis and Management of Viral Hepatitis 2018). Molecular detection methods such as RT-PCR, transcription mediated amplification etc., having high sensitivity were thus developed, which detect the viral RNA. The highly conserved segment of the HCV genome, the 5'NCR, is the target for RT-PCR, while a conserved region of the 3' NCR could also be targeted. A positive RT-PCR test indicates current infection whereas, a negative test could indicate absence of infection, resolved infection or presence of virus in very low titres; hence, in such cases, the test should be repeated after 6 months (Gupta et al. 2014).

Liver biopsy is by far the most efficient strategy to determine histopathological, inflammatory and fibrotic changes in the liver. Different scoring methods are used to stage and grade chronic HCV infection like the Ishak scoring system, Batts-Ludwig and the METAVIR scoring system.

20.9 Prevention and Control

Currently, there are no licensed vaccines to prevent HCV infection, so we need to focus on preventing HCV transmission and infection. To reduce the risk of HCV transmission through therapeutic procedures, basic preventative measures include screening and testing of plasma, blood, organ, tissue, and semen donors as well as virus inactivation of plasma-derived products. It is also crucial to identify and test the contacts of infected people as they are at a high risk of infection. People who inject drugs, hemodialysis patients, and people who received blood and organs before 1992 without appropriate blood tests are also included in the high-risk category. CDC has recommended that these groups of people must undergo routine periodic testing regardless of the setting prevalence (Evatt et al. 1991). The laboratories that perform HCV testing must follow recognised standards of good laboratory practices and focus on accurate diagnosis. Creating awareness is the first step towards preventing and controlling the spread of a disease. Hence, it is important to educate people regarding the modes of HCV transmission, associated risk factors, symptoms of hepatitis C, and the preventive measures to be adopted.

As per the latest guidelines from CDC, all adults and pregnant women must undergo one time hepatitis C screening (Schillie et al. 2020). Recent investigations have established that health care as well as public safety workers are at high risk of exposure to contaminated blood due to the nature of their occupation. Consequently, the adoption and upkeep of strategies for infection control in healthcare settings is crucial to achieve a reduction in HCV cases. It is also important to conduct public health surveillance and serological surveys for hepatitis C to estimate the disease burden. Ensuring the implementation of the national/international guidelines and monitoring the progress of HCV-related public health strategies will help to decrease the morbidity and mortality from HCV-related chronic liver disease.

20.10 Treatment of HCV

Different antiviral drugs are available for the elimination of the HCV virus. In 1990, the treatment of HCV using an interferon therapy was initiated and originally also led to some cures. However, it was associated with serious adverse effects, such as depression, flu-like symptoms, and fever. Improvements in the interferon therapies have comprehended the use of pegylated interferons. Also combining the interferon dosing with oral ribavirin doubled the efficacy. Depending on the virus genotype, a combination therapy consisting of ribavirin and pegylated interferon was given for 24 or 48 weeks. The viral response rate did not exceed 50% in genotypes 1 and 4, and it was only marginally higher in the other genotypes. Since the introduction of, direct acting antiviral agents (DAA) in 2011, there has been a paradigm shift in HCV treatment and clinical prognosis. Initially, Boceprevir and telaprevir were approved to treat the infection caused by genotype 1, in combination with the dual therapy that was traditionally used. Currently, total three types of DAA inhibitor classes are under the HCV treatment regimens, i.e., (a) NS3-NS4A protease inhibitors (boceprevir), (b) NS5A inhibitors (daclatasvir), (c) NS5B polymerase inhibitors (sofosbuvir, mericitabine). The new DAA drugs are highly efficacious, where a sustained virological response (SVR) is achievable in more than 90% of the HCV infected patients. Combinations of different drug regimens are still under study for HCV treatment (González-Grande et al. 2016).

20.11 Future Perspective

The available therapies to treat HCV such as the interferon monotherapy have not been able to sufficiently or effectively eliminate viral infections in a remarkable number of patients. However, the direct-acting antivirals (DAAs) are greatly effective with shorter durations of treatment. With the advancement in research technology and development of new therapeutics, a significant number of treatment options have emerged against this viral infection. Upcoming therapeutic approaches that are currently under investigation involve siRNAs, IRES inhibitors, NS3-NS5a inhibitors, and altered forms of ribavirin (Munir et al. 2010). At present, the health-care system is facing several challenges regarding mass screening of HCV, resistance development against viral drugs, high cost of direct acting antivirals, and lag in development of vaccines. These challenges can be managed with proper re-allocation of funds for testing the virus and development of effective vaccines. Hence, efforts must be made towards implementing strategic and innovative screening programs for HCV, creating awareness to prevent transmission of HCV through counseling programs and educational sessions, creating and performing collaborative worldwide action plans towards HCV. To facilitate better diagnosis and treatment of HCV in the near future, our focus should mainly lie on making DAAs cost effective and affordable, developing precise, accurate, and affordable testing kits, developing tools to interpret the resistance of the virus towards viral drugs, and promoting research to understand liver regeneration on molecular and cellular level post virus clearance (Barth 2015).

References

Akira S, Uematsu S, Takeuchi O (2006) Pathogen recognition and innate immunity. Cell 124(4):783–801

Alter MJ (2007) Epidemiology of hepatitis C virus infection. World J Gastroenterol: WJG 13(17):2436

Baliashvili D, Averhoff F, Kasradze A, Salyer SJ, Kuchukhidze G, Gamkrelidze A et al (2022) Risk factors and genotype distribution of hepatitis C virus in Georgia: a nationwide population-based survey. PLoS One 17(1):e0262935

Barth H (2015) Hepatitis C virus: is it time to say goodbye yet? Perspectives and challenges for the next decade. World J Hepatol 7(5):725

Borgia SM, Hedskog C, Parhy B, Hyland RH, Stamm LM, Brainard DM et al (2018) Identification of a novel hepatitis C virus genotype from Punjab, India: expanding classification of hepatitis C virus into 8 genotypes. J Infect Dis 218(11):1722–1729

Chevaliez S, Pawlotsky JM (2006) HCV genome and life cycle. In: Tan S-L (ed) Hepatitis C viruses: genomes and molecular biology. Horizon Bioscience

Dickson RC (1997) Clinical manifestations of hepatitis C. Clin Liver Dis 1(3):569–585

Duverlie G, Wychowski C (2007) Cell culture systems for the hepatitis C virus. World J Gastroenterol: WJG 13(17):2442

Evatt BL, Margolis MHS, Biswas R, Epstein MJS, Feinstone MSM, Finlayson JS, et al. (1991) Public Health Service. Inter-agency guidelines for screening donors of blood, plasma, organs, tissues, and semen for evidence of hepatitis B and hepatitis C. MMWR Recomm Rep 40(RR-4):1–17. PMID: 1850496

González-Grande R, Jiménez-Pérez M, Arjona CG, Torres JM (2016) New approaches in the treatment of hepatitis C. World J Gastroenterol 22(4):1421

Gosert R, Egger D, Lohmann V, Bartenschlager R, Blum HE, Bienz K, Moradpour D (2003) Identification of the hepatitis C virus RNA replication complex in huh-7 cells harboring subgenomic replicons. J Virol 77(9):5487–5492

Gupta E, Bajpai M, Choudhary A (2014) Hepatitis C virus: screening, diagnosis, and interpretation of laboratory assays. Asian J Transfus Sci 8(1):19
Hajarizadeh B, Grebely J, Dore GJ (2013) Epidemiology and natural history of HCV infection. Nat Rev Gastroenterol Hepatol 10(9):553–562
Houghton M (2009) Discovery of the hepatitis C virus. Liver Int 29:82–88
Houghton M (2019) Hepatitis C virus: 30 years after its discovery. Cold Spring Harb Perspect Med 9(12):a037069
Hu W, Zhang C, Shi JJ, Zhang JY, Wang FS (2020) Hepatitis C: milestones from discovery to clinical cure. Mil Med Res 7(1):1–3
Iwasaki A, Medzhitov R (2004) Toll-like receptor control of the adaptive immune responses. Nat Immunol 5(10):987–995
Kamili S, Qadri H (2020) Hepatitis C virus infection: a brief review. JMS SKIMS 23(1):3–15
Khullar V, Firpi RJ (2015) Hepatitis C cirrhosis: new perspectives for diagnosis and treatment. World J Hepatol 7(14):1843
Lohmann V (2013) Hepatitis C virus RNA replication. In: Bartenschlager R (ed) Hepatitis C virus: from molecular virology to antiviral therapy. Springer, pp 167–198
Morozov VA, Lagaye S (2018) Hepatitis C virus: morphogenesis, infection and therapy. World J Hepatol 10(2):186
Munir S, Saleem S, Idrees M, Tariq A, Butt S, Rauff B et al (2010) Hepatitis C treatment: current and future perspectives. Virol J 7(1):1–6
Murphy DG, Sablon E, Chamberland J, Fournier E, Dandavino R, Tremblay CL (2015) Hepatitis C virus genotype 7, a new genotype originating from Central Africa. J Clin Microbiol 53(3):967–972
National Center for Biotechnology Information. (n.d.). Flaviviridae. Retrieved from https://www.ncbi.nlm.nih.gov/Taxonomy/Browser/wwwtax.cgi?id=356113
National Guidelines for Diagnosis & Management of Viral Hepatitis (2018). Retrieved from https://nvhcp.mohfw.gov.in/common_libs/diagnosis-management-viral-hepatitis.pdf
Niepmann M (2013) Hepatitis C virus RNA translation. In: Bartenschlager R (ed) Hepatitis C virus: from molecular virology to antiviral therapy. Springer, pp 143–166
Pestka JM, Zeisel MB, Bläser E, Schürmann P, Bartosch B, Cosset FL et al (2007) Rapid induction of virus-neutralizing antibodies and viral clearance in a single-source outbreak of hepatitis C. Proc Natl Acad Sci 104(14):6025–6030
Schillie S, Wester C, Osborne M, Wesolowski L, Ryerson AB (2020) CDC recommendations for hepatitis C screening among adults—United States, 2020. MMWR Recomm Rep 69(RR-2):1–17
Schoggins JW, Rice CM (2013) Innate immune responses to hepatitis C virus. In: Bartenschlager R (ed) Hepatitis C virus: from molecular virology to antiviral therapy. Springer, pp 219–242
Sesmero E, Thorpe IF (2015) Using the hepatitis C virus RNA-dependent RNA polymerase as a model to understand viral polymerase structure, function and dynamics. Viruses 7(7):3974–3994
Simmonds P (2013) The origin of hepatitis C virus. In: Bartenschlager R (ed) Hepatitis C virus: from molecular virology to antiviral therapy. Springer, pp 1–15
Stetson DB, Medzhitov R (2006) Type I interferons in host defense. Immunity 25(3):373–381
University of Glasgow Centre for Virus Research. (n.d.). Flaviviridae. Retrieved from https://www.taxonomy.cvr.gla.ac.uk/PDF/Flaviviridae.pdf
WHO Reports (2022) Hepatitis C. https://www.who.int/news-room/fact-sheets/detail/hepatitis-c
Yi M (2010) Hepatitis C virus: propagation, quantification, and storage. Curr Protoc Microbiol 19(1):15D-1

Hepatitis E Virus: An Ongoing Enigma

21

Neha A Bhise, Varsha Mahadik, and Akshay P. S

Abstract

Hepatitis E virus (HEV) is a small nonenveloped, single-stranded positive-sense RNA virus which was discovered by the Russian virologist Mikhail Balayan in 1983. It was reported that about 900 million of the worldwide population has been infected with HEV, 15–20 million were reported to have an on-going infection. HEV usually causes self-limiting acute infections which may be asymptomatic or mildly symptomatic, progressing to chronicity in immunocompromised individuals like organ transplant recipients, HIV and chronic liver disease patients. The maternal mortality rate is ~25% during the third trimester of pregnancy caused by HEV genotype 1. HEV genotype 1 and 2 infect humans whereas genotype 3 and 4 causes chronic infections in industrialized countries by the zoonotic transfer. Enzyme immunoassay and nucleic acid amplification tests are used for the detection of HEV infections. China has developed HEV vaccine (Hecolin) based on ORF2 protein of genotype 1 in 2010 which is not used worldwide. There is no specific drug approved by FDA for the treatment of HEV infection but ribavirin can be used for rapid virus clearance. Ribavirin treatment may be accompanied by some side effects which necessitates the study of new potential antiviral targets against HEV.

Keywords

HEV · Hepatitis E · Chronicity · Zoonotic · Ribavirin

N. A. Bhise (✉) · A. P. S
Hepatitis Division, ICMR-National Institute of Virology, Pune, India
e-mail: bhisena@gmail.com; akshayprakasan@gmail.com

V. Mahadik
Biochemical Sciences Division, National Chemical Laboratory, Pune, India
e-mail: mahadikvarsha1@gmail.com

P. L. Bhukya, A. S (eds.), *Emerging Human Viral Diseases, Volume II*,
https://doi.org/10.1007/978-981-97-4480-0_21

21.1 Introduction

21.1.1 Brief History

Hepatitis E is an acute infection affecting the entire globe, and its causative agent is hepatitis E virus (HEV) (Kamar et al. 2012). HEV genome is 7.2 kb, single-stranded, positive-sense RNA molecule affecting liver. HEV infection can cure on its own but at times can develop chronicity. Developing countries show majority of clinical cases. The WHO evaluates approximately 15–20 million widespread HEV infections with 3.3 million symptomatic patients. It also reported 44,000 death cases in the year 2015 with 3.3% million fatality due to virus-induced hepatitis. The infection transmits using fecal-oral course, fundamentally through fecal debased water. Contaminated water leads to the spread of HEV1 and HEV2 in countries with poor hygienic conditions. HEV causes recurrent sporadic cases and intermittent large outbreaks. In developed countries, HEV3 and HEV4 have trespassed from animal reservoirs to humans. Increase in number of reported sporadic cases has been observed due to improvements in diagnostic tools and screening methods for the last few years.

21.1.2 Discovery

In the twentieth century, viral hepatitis was classified as either 'infectious' or 'serum' hepatitis based on various epidemiological analyses and were distinguished according to the mode of transmission as fecooral or parental, respectively. Furthermore, the progress of the diagnostic tests made it clear that infectious and serum hepatitis were caused by HAV and HBV, respectively. However, some patients with typical signs and symptoms of viral hepatitis neither had serological markers of HAV nor HBV infection, leading to the conclusion that the third form of putative uncharacterized hepatitis agent exists. Hepatitis E came into limelight after the examination of Kashmir hepatitis epidemic of 1978 (Khuroo and Khuroo 2010). Dr. Mohammad S Khuroo, in November 1978, investigated jaundice throughout Srinagar and Kashmir in India (Bradley et al. 1991). An interconnection between the disease and the pregnancy was noticed during this investigation (Khuroo et al. 1994; Khuroo and Kamili 2003). This type of viral hepatitis was then known as non-A, non-B hepatitis (NANBH). A similar outbreak with NANBH was reported by an Afghanistan personnel in the Russian military (Balayan et al. 1983).

A legendary experiment was carried out by Dr. Mikhail S Balayan where he consumed a stool extract from the nine patients infected with NANBH (Bradley et al. 1991). Severe acute hepatitis manifesting into jaundice with elevated liver tests was recorded from this study on the 36th day from the ingestion of the stool extracts. Stool samples of the above researcher generated virus-like particles (VLPs) on the 23rd day, 43rd day, and 45th day of consumption of stool sample when observed by an immune electron microscope.

21.2 Epidemiology

21.2.1 Geographical Distribution

HEV belongs to the family *Hepeviridae*, genus *Orthohepevirus*, and species *Orthohepevirus A*. *Orthohepevirus A* is divided into eight genotypes. Genotypes 1 and 2 infect only humans, genotypes 3 and 4 are endemic in pigs, genotypes 5 and 6 infect wild boars, and genotypes 7 and 8 infect camels (Aslan and Balaban 2020). Genotypes 1 and 2 spread through contaminated drinking water in developing countries. Genotypes 3 and 4 show zoonotic transmission in humans due to consumption of undercooked meat in developed nations (Seth and Sherman 2020). Specific features of HEV genotypes 1, 2, 3, and 4 are described in Table 21.1.

21.2.2 Age, Mortality, and Morbidity

Hepatitis E genotypes 1 and 2 affect young adults of age 15–39 years in developing countries while HEV genotype 3 found in industrialized countries mainly affects the median age of 50 years. In developed countries, HEV infects males predominantly. Patients previously suffering from severe liver diseases are prone to deaths

Table 21.1 Comparison of four HEV genotypes based on their features (Teshale and Hu 2011)

Features	HEV-1	HEV-2	HEV-3	HEV-4
Detection of virus	1983	1986	1995	2003
Global geological dispersion	Developing nations	Mexico, West Africa	Industrialized countries	China, Japan, Taiwan
Route of transmission	Fecal-oral, Water-borne, Person to person	Fecal-oral, Water-borne, Person to person	Food-borne, Zoonotic	Food-borne, Zoonotic
Nature of infection	Common epidemics	Small epidemics	Sporadic cases	Uncommon epidemics
Prevalence	Adolescent individuals	Adolescent individuals	Above 40s	Adolescent individuals
Gender prevalence	Male	No bias	Male	No bias
Death rate	0.5–3%	0.5–3%	N.D.	0.5–3%
Death rate in pregnant women	More	More	More	More
Chronic infections	Yes	No	Yes	Yes
Severity in immunocompromised individuals	Yes	Not known	Yes	Yes
Transmission between species	Humans and NHPs	Humans and NHPs	Humans and animals	Humans and animals
HEV subtypes	5	2	10	7

N.D. Not determined

due to HEV infections as compared to immunocompetent individuals. HEV infections in organ transplant patients treated with immunosuppressants after solid organ transplant may develop chronicity, especially in industrialized countries (Kamar et al. 2008). One of the most important points of HEV genotypes 1 and 2 is its fatality rate of 25% in pregnant women with HEV infection, particularly during the third trimester in developing countries. Hepatitis E is unusual among children in industrialized countries which is in contrast to developing countries where the death rate is high in adolescent individuals (Sharapov et al. 2009).

21.2.3 Origin of Infection, Diversity and Spread of Disease

As previously mentioned, HEV genotypes 1 and 2 infect humans whereas genotypes 3 and 4 infect animals. India shows high occurrence of genotypes 1 and 4 but no human infections due to genotype 4 have been reported (Arankalle et al. 2002). High frequency of genotype 2 cases are found in Nigeria, Mexico, Namibia, South Africa, and some West African nations (Andersson et al. 2013; Kim et al. 2014). Genotype 3 ubiquitously occur in swines. Industrialized nations show large number of human cases due to genotype 3. European countries, China, and Taiwan has high prevalence of genotype 4, and human cases are observed in China and Taiwan mainly (Fig. 21.1).

About 63 countries have experienced sporadic cases or large outbreaks due to enterically transmitted HEV while few countries show endemic cases (Table 21.2) (Aggarwal 2013). Sporadic hepatitis E cases contribute more to worldwide burden of hepatitis E whose transmission or causative factors are unknown (Khuroo et al. 1994; Coursaget et al. 1998; Das et al. 2000). HEV spread through blood

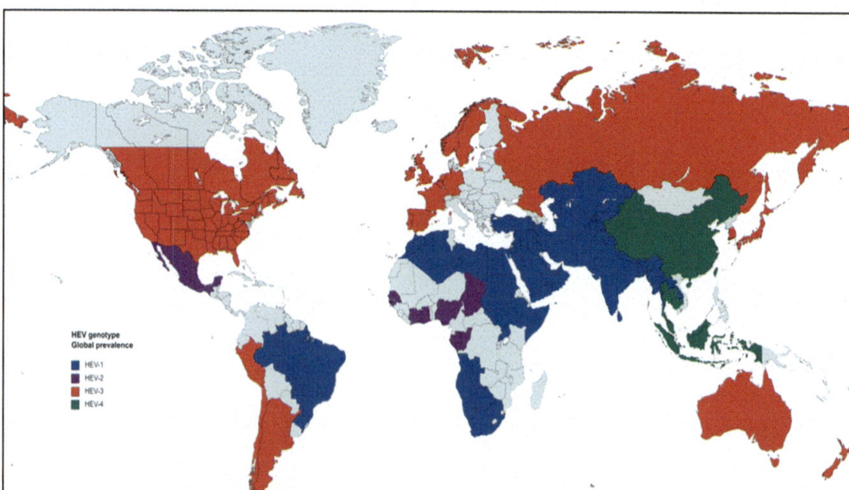

Fig. 21.1 Worldwide distribution of HEV genotypes

Table 21.2 Major hepatitis E outbreaks

Location	Year	Reference
Delhi (India)	1956	Viswanathan (1957)
Accra (Ghana)	1963	Adjei et al. (2009)
Kashmir (India)	1978	Khuroo (1980)
Xinjing (China)	1986	Song et al. (1992)
Keren (Ethiopia)	1988	Tsega et al. (1991)
Darfur (Sudan)	2004	Boccia et al. (2006)
Karad, Maharashtra (India)	2004	Gurav et al. (2007)
Amritsar, Punjab (India)	2013	Kumar et al. (2019)
Chad (Africa)	2017	Spina et al. (2017)
Halisahar (Bangladesh)	2018	Baki et al. (2021)
Yavatmal, Maharashtra (India)	2019	Tripathy et al. (2021)

transfusion, through hospital acquired infections, vertical transmission in pregnant patients, or neonatal liver infections were reported but sexual transmission is not reported yet (Arankalle and Chobe 1999; Robson et al. 1992; Khuroo et al. 2009).

21.2.4 Biosafety Measures

For HEV, biosafety level 2 practices and containment should be followed while handling experimentally infected nonhuman primates (NHPs) or other animal models shedding the virus or while working with HEV/HEV-infected feces, blood, or other tissues (Birke et al. 2014).

21.3 The Potential Risk of Emergence and Reemergence

Different factors are responsible for the hepatitis E occurrence which include living conditions and habits in the population, host immune strength, environmental factors that cause HEV persistence in a particular geographical area, and viral factors which adds to genetic variation and host range. The average HEV infection rate is 2% to 4%, while in developing countries, in some regions with poor sanitation, it can go up to 25% (Hakim et al. 2017). Amongst HEV-infected pregnant women from Asian and African countries, maternal mortality is 20.8%, and fetal mortality is 34.2%. This mortality rate can go up to 61.2% in mothers with fulminant hepatic failure (Jin et al. 2016). The exact reason for fulminant hepatitis and high death rate in genotype 1 HEV-infected pregnant women is still unknown, but changes in hormonal and immune status and/or immune response during pregnancy can contribute to these phenomena.

21.4 Organization of Infectious Agent

21.4.1 Classification

HEV belongs to *Hepevirus* in the *Hepeviridae* family, classified into two genera, namely, *Orthohepevirus* and *Piscihepevirus*. Orthohepevirus has species A, B, C, and D and Piscihepevirus which has species A (ICTV 2009). HEV isolated from human and different animals like pig and rabbit belongs to Orthohepevirus A. Orthohepevirus B is isolated from chicken. Orthohepevirus C infects rats and ferrets. Orthohepevirus D can infect bats only. Piscihepevirus A has been obtained from cutthroat trout (Kenney 2019).

21.4.2 Virus Structure

The HEV virus structure consists of capsid shell which contains single-stranded RNA as a genome. Virus capsid has icosahedral symmetry. The surface of the virus has spikes and indentations. The immune electron microscopic study concludes its diameter of 27 to 30 nm (Khuroo et al. 2016) (Fig. 21.2).

21.4.3 Genome Structure and Organization

The 7.2 kb positive-sense, single-stranded RNA is present in HEV. Also, the 7-methylguanosine cap present at 5′ end, and the poly A sequence present at 3′ end. The genome consists of three open reading frames (ORFs) named as ORF1, ORF2, and ORF3. Replication of positive-sense HEV RNA produces 7.2 kb RNA and 2.2 kb subgenomic RNA.

ORF1 when translated produces different proteins like methyl transferase, papain like cysteine protease, RNA helicase, and RNA-dependant RNA polymerase

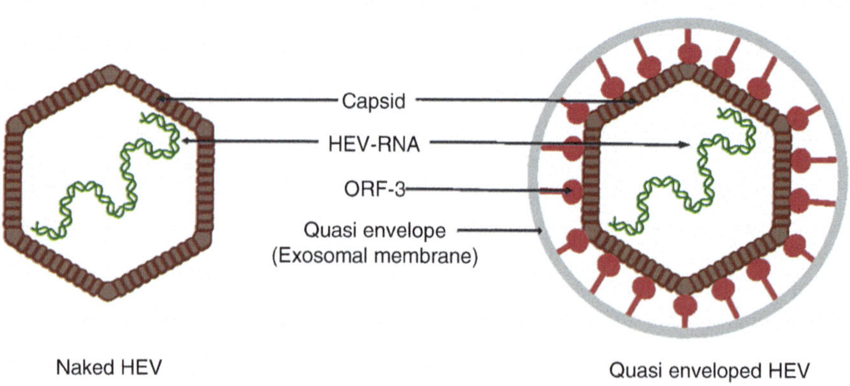

Fig. 21.2 Schematic presentation of naked HEV virion and quasienveloped virion

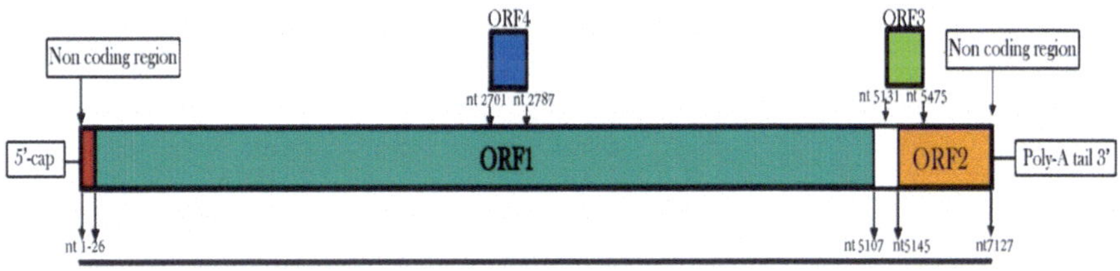

Fig. 21.3 Schematic of HEV genome organization

(RdRp). Y domain and macrodomain are also synthesized as a product of ORF1. ORF2 codes for capsid protein which helps in virus assembly and plays key role in immune response on its binding to neutralizing antibodies. It is hypothesized that the ORF3 protein may be involved in cytoskeleton formation and virus particle release from liver cells (Zafrullah et al. 1997). In genotype 1, an additional ORF4 was discovered which was known to stimulate the viral RdRp activity (Nair et al. 2016) (Fig. 21.3).

21.4.4 Virus Propagation

HEV shows poor replication in in vitro cell culture systems; hence, it is difficult to obtain high yields of this infectious virus. Emerson et al. (2001) developed infectious cDNA clones by reverse transcription polymerase chain reaction (RT-PCR) from the HEV isolate for genotype 1 (Sar55). The importance of 5'cap of HEV RNA for in vivo infection was also reported. Infectious cDNA clones were prepared for different HEV genotypes and transfected into hepatoma cell lines (Fig. 21.4). Replication competency and capacity of its progeny to infect the hepatoma cells were detected by ORF2/capsid protein expression.

21.4.4.1 In Vitro Cell Culture System

HEV was infected into different primates during an experiment to study its physiochemical properties (Reyes et al. 1990). This virus was hard to clone for many years due to its low VLP yields. Large amounts of VLPs were obtained from biles of infected primates which was later used for subgenomic cloning of the virus (Sreenivasan et al. 1984). Finally, full-length HEV genome (7.2 kb) was cloned and sequenced by Tam et al. (1991).

However, cDNA clones were modified to form subgenomic replicons where ORF2 or ORF3 genes were partly replaced with reporter genes like enhanced green fluorescent protein (GFP/EGFP) or a luciferase reporter gene in HEV genotype 1 (Sar55 strain) or HEV genotype 3 (Kernow-C1/p6 strain) (Emerson et al. 2004; Graff et al. 2005; Pudupakam et al. 2009; Cao et al. 2010; Shukla et al. 2012; Shiota et al. 2013). As these clones did not produce capsid and ORF3 phosphoprotein, the progeny produced by them was incapable of infecting new cells which lowered the risk of infection while conducting research.

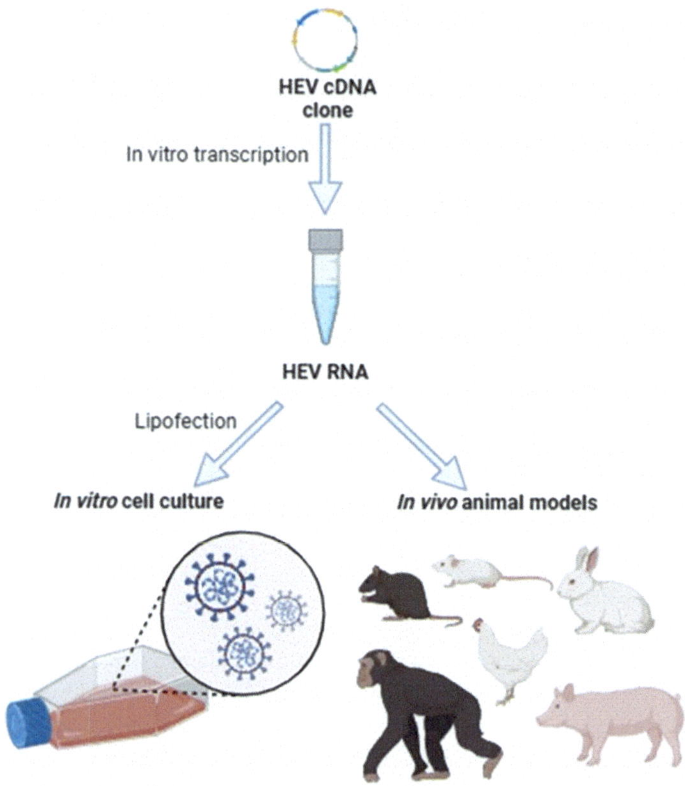

Fig. 21.4 HEV propagation using in vitro cell culture systems and in vivo animal models

21.4.4.2 In Vivo Laboratory Models

In order to investigate the HEV pathogenesis in-depth, animal models are the vital tools (Fig. 21.4). HEV infects several animal species which provide us with naturally occurring animal models (Wang and Wang 2016). The examples of such animal models are mentioned in Table 21.3:

21.4.5 Viral Proteins and Life Cycle

It was studied that virus enters into the host cell and travels through the gut by interacting with the heparan sulphate proteoglycans (HSPGs) present on the surface of hepatocytes (Cao and Meng 2012). ORF1 of viral genome translates into nonstructural proteins and synthesize negative sense RNA with RdRp which act as a template for further viral RNA synthesis. Subgenomic replicon of ORF2 is translated to capsid proteins, and these proteins are assembled into shell to pack new viral RNA particles. With the help of lipid layer and ORF3 proteins, the particle gets released outside the hepatocytes (Yamada et al. 2009) (Fig. 21.5).

Table 21.3 Different animal models available for HEV

In vivo animal models	Features	References
Nonhuman primates [NHPs]	– Unnatural host for HEV genotypes 1,2,3 and 4 – NHPs can be infected with HEV preferably by IV route – Cross-species infection can be studied – Used as models for testing efficacy of cDNA clones and HEV vaccines (Hecolin) – Limitations: Being unnatural host does not mimic exact clinical features of the HEV disease e.g., Cynomolgus macaques, chimpanzees, and rhesus macaques	Wang et al. (2017) Emerson et al. (2001) Kamili (2011)
Pig	– First discovered animal strain in the U.S. was swine HEV – Swine HEV shows 73–80% nucleotide similarity with HEV genotypes 3 and 4 – Pigs can be infected with swine HEV during in vivo studies by IV route which show only mild histopathological lesions and no observable clinical symptoms – Usage of pig model: extrahepatic study and HEV vaccine trials – Limitations: Not susceptible to HEV genotypes 1 and 2 infections, Pigs are expensive and not easy to handle	Meng et al. (1997) Bouwknegt et al. (2009) Williams et al. (2001) Sanford et al. (2011)
Rabbit	– China isolated HEV rabbit strain for the first time – Infection of rabbits with this strain cannot be used to study exact clinical course of disease – Usage of rabbit model: studying extrahepatic manifestations, vaccine trials and antiviral testing – Limitations: resistant to infection with HEV genotype 1	Zhao et al. (2009) Han et al. (2014) Wang et al. (2017) Ahn et al. (2017) Cheng et al. (2012)
Rat	– Rat HEV was isolated from *Rattus norvegicus* in Germany – *Rattus norvegicus* and *Rattus rattus* are natural host for rat HEV – Rat HEV shows zoonotic transmission and can be used for study of cross-species infection – Rat HEV show 60% nucleotide similarity to human HEV (genotypes 1, 4, and 7)	Johne et al. (2010) Mulyanto et al. (2014) Purcell et al. (2011) Sridhar et al. (2018)

(continued)

Table 21.3 (continued)

In vivo animal models	Features	References
Mouse	– Mice are genetically inserted with specific human factors to develop humanized mice – Usage of humanized mice: Study of HEV genotypes 1 and 3 infections, antivirals against HBV, HCV, etc. – Ribavirin and silvestrol for anti-HEV activity were studied using humanized mice	Lai and Chen (2018) Sayed et al. (2017) Giersch et al. (2015) Azuma et al. (2007) Allweiss et al. (2016) Todt et al. (2018)
Chicken	– Avian HEV isolated for the first time from chicken bile – Chicken can be used to study HEV vaccine and vertical transfer of HEV genotype 1	Haqshenas et al. (2001) Guo et al. (2007)

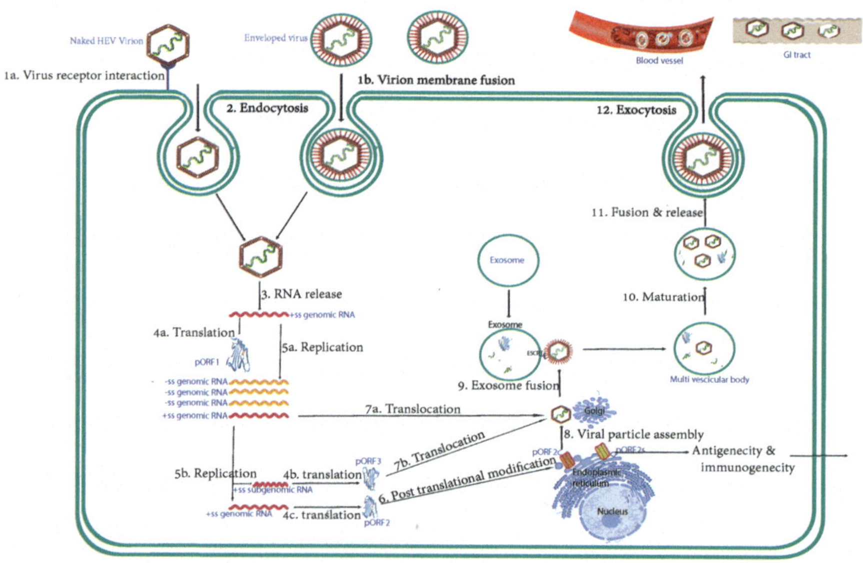

Fig. 21.5 Hypothetical mechanism of HEV replication
ss Single-stranded, *pORF1* ORF1 protein, *pORF2* ORF2 protein, *pORF3* ORF3 protein, *pORF2c* Cleaved form of ORF2 protein, *pORFs* Secreted form of ORF2 protein, *ESCRT* Endosomal sorting complexes required for transport, *GI* Gastrointestinal tract

21.5 Pathogenesis

HEV primarily infects the intestine, and from the basolateral sides of the intestine, the virus particles are released into bloodstream, which finally reaches the liver. HEV is known to replicate in the cytoplasm of the hepatocytes and released into the blood and bile. HEV is a noncytopathic virus; hence, the liver damage caused during HEV infection can be due to cytotoxic T-cell and NK-cell mediated immune responses (Williams et al. 2001).

21.5.1 Steps Involved in Pathogenesis

21.5.1.1 Attachment and Entry

HEV exists in two forms: in feces and bile, it circulates in quasi−/nonenveloped (qHEV/neHEV) state whereas in blood, HEV circulate as host membrane enclosed virus particles (Yin et al. 2016). Both the forms of virus enter into the host cell by distinct mechanisms. HSPGs (heparan sulfate proteoglycans) receptors and ASGPR (asialoglycoprotein receptor) are mainly involved in host cell attachment (Nagashima et al. 2017). Different host cell attachment factors binding to neHEV and qHEV and their roles are as mentioned in Table 21.4.

Table 21.4 Interactions of principle HEV factors with host factors

Virus factor	Host factor	Action upon binding of virus and host factor	References
neHEV particle	ASGPR	Attachment of virus to the host cell	Nagashima et al. (2017)
RdRp, HEL, and ORF4	C1, C3, C4a, and C8	Activation or inhibition of complement pathway	Subramani et al. (2018)
HEV RNA promoter	hnRNPA2B1, hnRNPK	Structural rearrangements in virus RNA	Kanade et al. (2019)
X-domain (macrodomain) and MET	PSMB1 and PSMB4	Alteration in processing of MHC-I complex	Subramani et al. (2018)
ORF3	Unknown kinase	Detection of virion particles before release	Ju and Ding (2019)
ORF2	Unknown protease	Immune decoy production	Montpellier et al. (2018)
PCP	Factor Xa	ORF1 polyprotein processing	Kanade et al. (2019)
MET PCP	ISG15	Modification of antiviral pathways in the host cell	Karpe and Meng (2012)
HVR, MET-PCP	Ubiquitin	Modification of MHC-I complexes for overcoming resistance of cellular antiviral pathways	Subramani et al. (2018) Karpe and Meng (2012)

ne non-enveloped, *RdRp* RNA dependent RNA polymerase, *HEL* Helicase, *MET* Methyltransferase, *PCP* Papain like cystein protease, *HVR* Hypervariable region, *ASGPR* Asialoglycoprotein receptor, *C* Complement component, *hnRNPA2B1* Heterogeneous nuclear ribonucleoproteins A2/B1-, *hnRNPK* Heterogeneous nuclear ribonucleoprotein K, *PSMB1* Proteasome subunit beta type-1, *PSMB4* Proteasome subunit beta type 4, *ISG15* Interferon-stimulated gene 15

21.5.1.2 Internalization and Uncoating of HEV

Clathrin- and dynamin-dependent pathways are involved in the internalization of both quasienveloped and naked virus particles. Degradation of the lipid envelope of quasienveloped virus particles by lysosomes occur by different enzymes like NPC intracellular cholesterol transporter 1 (NPC1) and lysosomal acid lipase (LAL), which helps in reducing the virus infectivity (Wißing et al. 2021).

21.5.1.3 HEV Replication Complex

ORF1 is known as a replicative machinery of HEV which encodes proteins like methyltransferase, RNA helicase, and RdRp with the help of host replicative machinery. Intracellular membranes of endoplasmic reticulum [ER], Golgi, mitochondria, etc. are rearranged by the positive-sense HEV RNA. This process helps in the replication of viruses into hepatocytes and promote the production of host cofactors which develop an atmosphere in which virus particles are not recognized by host cells (Heaton and Randall 2011).

21.5.1.4 HEV Assembly and Release

Virus assembly requires viral capsid and nonstructural proteins along with few host factors. Only small fraction of virus particles are assembled into capsid protein forming infectious particles (ORF2i). Majority of the ORF2 variants that are not in association with the infectious virus are released through exosomal pathways which possess immunomodulatory properties (Montpellier et al. 2018). The mechanism of the release of HEV virions from infected cells is not well studied. HEV particles are released in quasienveloped and nonenveloped forms (Zhou et al. 2021). Nonenveloped HEV particles are found in feces and biles whereas quasienveloped HEV particles are covered with host cellular membrane and observed circulating blood and cell culture supernatants. ORF3 interacts with nonglycosylated ORF2. ORF3 protein is involved in virus release but does not play any role in virus assembly. The proline-rich PSAP motif present on ORF3 protein interacts with host protein of tumor susceptibility gene 101 (TSG101). TSG101 is a component of endosomal sorting complexes required for transport (ESCRT) machinery. Quasienveloped virions are loaded into multivesicular bodies (MVBs) for release from the infected cells using the exosomal pathway (Ju and Ding 2019).

21.5.2 Virulence and Persistence

21.5.2.1 Host Factor Modifications Mediated by HEV Proteins

It is necessary for the HEV to modify certain host factors for its survival (Ribet and Cossart 2010). Posttranslation modifications (PTMs) of the host cells are targeted for this purpose. PTMs of host cells are used to modify ORF2/3 viral proteins. The ORF1 domain of HEV encoding different proteins required for virus replication is involved in modifying host PTMs like macrodomain inhibit PARylation enzyme, MARylation, and ADP-ribosylation of HEV RNA (Li et al. 2016). These PTM activities help the host cells in the recognition of virus particles and stimulate the immune responses.

21.5.2.2 Genetic Exchange Between Host Cell and HEV

HEV viral RdRp is susceptible to errors while replication; hence, the exchange of genes between different hosts and the virus becomes possible. The hypervariable region (HVR) of ORF1 undergo high rate of mutation due to genetic recombination between the host and the virus and may form different strains within the same genotype. Insertions in the HVR region lead to the formation of more pathogenic recombinant HEV (Smith et al. 2012).

21.5.3 Host Immune Response

21.5.3.1 Innate Response

HEV infection studies in A549 human lung epithelial cells reported induction of several inflammatory responses such as cytokines or chemokines (interleukin IL-6, IL-8, TNF-α, and RANTES, nuclear factor kappa-light-chain-enhancer of activated B cells [NF-κB] and IFN regulatory factor 3 [IRF3], NK, and natural killer T [NKT]) (Nagashima et al. 2011, 2014) which can restrict the HEV replication in the host cells.

21.5.3.2 Adaptive Response

Antibodies are generated after infection with HEV. In the initial phase of infection anti-HEV, IgM is generated which lasts for 4–5 months; whereas, in the later and final stages of illness, anti-HEV IgG is formed which can persist up to 4.5 years. This serological response can provide cross-protection against different strains due to the presence of single serotype. HEV antibodies can be stimulated within the body by vaccination as the HEV capsid protein consists of several neutralizing epitopes (Ribet and Cossart 2010). Hecolin HEV vaccine can provide immunity for 4.5 years after vaccination (Li et al. 2016). HEV infection also activates effector T-cells (Suneetha et al. 2012).

21.6 Clinical Manifestations

21.6.1 Phases of Diseases and Clinical Symptoms

21.6.1.1 Acute Hepatitis E

Acute HEV infection is usually asymptomatic or mildly symptomatic. Hepatitis E usually shows the acute, self-limiting spontaneous or complete recovery of infections within 4–6 weeks. Acute icteric hepatitis occurs in 5%–30% of the infected patients. The prodromal phase of acute icteric hepatitis (1 week) shows characteristic symptoms like malaise, fever, body aches, nausea, and vomiting. This phase is followed by icteric hepatitis (approx. 1 week) which is marked by jaundice, itching, light-colored stool, dark-colored urine, increased levels of liver transaminases (ALT-1500 IU/L, bilirubin, alkaline phosphatase, and gamma-glutamyl transferase), then comes the convalescent phase showing resolution of the icteric symptoms (Fig. 21.6).

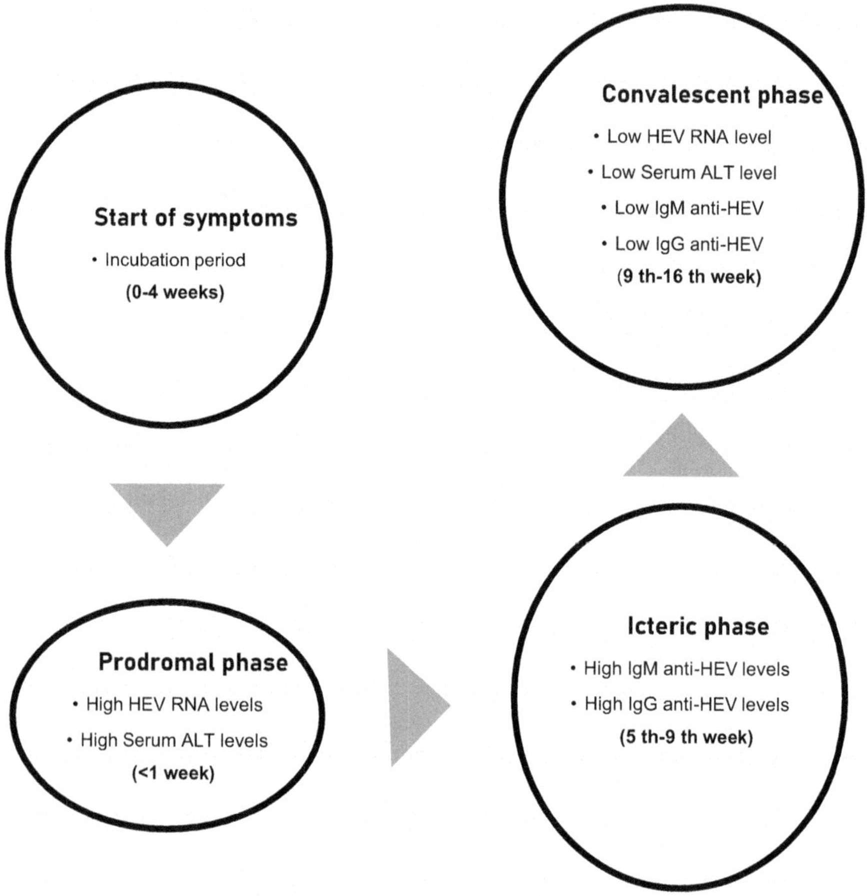

Fig. 21.6 Time course of typical HEV infection

The mortality rate in the icteric phase is about 0.5%–4% during an outbreak (Wedemeyer et al. 2012).

21.6.1.2 HEV in Distinctive Cases

HEV infections may develop chronic conditions in cases like immunocompromised patients, pregnant women, neonates, and patients already suffering from liver diseases. This is described in brief in Table 21.5.

21.6.2 Extrahepatic Manifestations of Hepatitis E

HEV is reported to infect at sites outside the liver and show different manifestations such as neurological, hematological, renal, acute pancreatitis, etc. Table 21.6 gives a brief overview of it.

Table 21.5 Chronic infections due to the association of HEV with distinctive cases

Coinfection	Features	References
HEV+ Immunocompromised individuals	Chronicity observed with genotypes 3 and 4 Infection proceeds to chronicity due to immunosuppressive drugs	Meisner et al. (2015)
HEV+ Pregnant women	Chronicity observed with genotypes 1 and 2 Infection proceeds to chronicity in second and third trimester of pregnancy and cause fulminant hepatic failure (FHF) FHF can be linked to the reduced expression of progesterone receptor, poor specific T-cell response which increases the viral titre and causes death of pregnant patients	Navaneethan et al. (2008) Jilani et al. (2007) Bose et al. (2011) Srivastava et al. (2011)
HEV+ Neonates	Chronicity observed with genotypes 1 and 2 Vertical transmission of hepatitis E occur from mother to fetus or neonate. Neonates after contracting hepatitis E suffers: respiratory distress syndrome, sepsis, liver and spleen enlargement, finally, FHF can cause death	Khuroo et al. (1981) El Sayed Zaki et al. (2013) Khuroo et al. (2009)
HEV+ Chronic liver diseases	Chronicity observed with genotypes 1, 2, 3, and 4 in hyper endemic areas, individuals with liver diseases when get infected with HEV show signs of acute hepatitis and then can develop liver ascites and encephalopathy	Acharya et al. (2007)

Table 21.6 HEV associated disorders and their extrahepatic indications

Types of disorders	Extrahepatic indications	References
Neurological disorders	Guillain–Barre syndrome Brachial neuritis Bell's palsy Polyradiculopathy Mononeuritis multiplex Viral meningitis Encephalitis Myelitis	Dalton et al. (2016) Pischke et al. (2017) Jha et al. (2012) Zheng et al. (2018) Perrin et al. (2015) Rodríguez-Castro et al. (2018) Murkey et al. (2017) Sarkar et al. (2015)
Hematological disorders	Thrombocytopenia Hemolytic and aplastic anemia	Zamvar et al. (2005) Shah et al. (2012)
Renal manifestations	Glomerulonephritis	Kamar et al. (2012) Guinault et al. (2016)
Other manifestations	Acute pancreatitis Myocarditis Myositis Thyroiditis Henoch–Schönlein purpura Myasthenia gravis	Bazerbachi et al. (2015) Sengupta et al. (2019) Mengel et al. (2016) Dumoulin and Liese (2012) Thapa et al. (2010) Belbezier et al. (2014)

21.7 Diagnostics and Therapeutics Approaches

21.7.1 Acute Hepatitis E Diagnosis

Acute hepatitis E can be diagnosed by 1. epidemiological history, 2. clinical manifestation, and 3. laboratory test results.

(a) **Epidemiological history:** Epidemiological past like travelling to areas endemic to HEV and then consuming uncooked or undercooked food, drinking feco–orally contaminated water, coming in contact with the HEV-infected individuals can help in the identification of the acute hepatitis E disease (Aggarwal 2010, 2013).

(b) **Clinical manifestations:** Acute hepatitis E symptoms involve fever, nausea, loss of appetite, fatigue, vomiting, jaundice, etc. All these features are commonly observed in most of the viral infections. Hence, symptoms can help us to identify acute HEV infections but for confirmation of this, more tests are required (Aslan and Balaban 2020; WHO Factsheet 2022).

(c) **Laboratory tests:** Laboratory tests can be molecular or serological tests. Molecular approaches help detect HEV-RNA in the stool sample through reverse transcription-polymerase chain reaction (RT-PCR), 7–14 days before the appearance of clinical symptoms. HEV can also be identified by immune electron microscopy from the patient samples. Serological approaches depend on classic indicators like anti-HEV immunoglobulin like IgG and IgM in patient's serum samples by enzyme-linked immunosorbent assay (ELISA). Anti-HEV IgM antibodies are classical markers of ongoing acute infection which may prolong up to 4–5 months whereas anti-HEV IgG antibodies are markers of earlier infections which can be detected even after 4.5 years of the acute hepatitis E phase (Zhao and Wang 2016; WHO, Factsheet 2022).

21.7.2 Therapeutic Approaches

21.7.2.1 Current Treatment for Chronic Hepatitis E

21.7.2.1.1 Ribavirin

There is no need for the treatment of acute hepatitis E as it is self-limiting but in severe cases like acute-on-chronic liver failure (ACLF), ribavirin can be used for rapid virus clearance and normalization of liver enzymes. In fulminant hepatitis E, corticosteroids were used to slow down the progression to liver failure. The treatment of chronic HEV genotypes 3 and 4 infections in solid organ transplant [SOT] recipients can be done by pegylated interferon-alpha and by dose reduction of immunosuppressive medications. Pegylated interferon and ribavirin as monotherapy or in combination are conventionally used for the treatment of chronic hepatitis E (Mallet et al. 2010; Kamar et al. 2010). Ribavirin monotherapy (600 mg/day for 3 months) was used to treat chronic HEV in SOT recipients with an sustained virological response (SVR) of 78% (Kamar et al. 2010). Inosine monophosphate dehydrogenase (IMPDH) is competitively inhibited by ribavirin monophosphate (an active component of ribavirin) which decreases intracellular GTP levels (Debing et al. 2014) and obstructs HEV replication. Ribavirin causes lethal mutations in HEV genomes, thus, impeding HEV replication by catastrophic error (Todt et al. 2016; Crotty et al. 2001).

Ribavirin treatment may be accompanied by some side effects like skin reactions, dry cough, insomnia, dose-dependent hemolytic anemia, dyspnea, lack of concentration, and irritability. Moreover, it cannot be used in pregnancy due to its probable teratogenic effect (Pradat et al. 2014). Extrahepatic complications of HEV can also be treated by ribavirin or immunosuppressive medications. Sofosbuvir (Cornberg et al. 2019), 2′-C-methylguanosine, and zinc (Kaushik et al. 2017) can be used in combination with ribavirin to treat ribavirin-resistant chronic HEV infections. Compounds like silvestrol (Todt et al. 2018), NITD008, and GPC-N114 (Nishiyama et al. 2019) were reported to show promising results in in-vitro cell cultures but studies are yet to be done in humans. Ribavirin may show effect on immune cells like T helper (Th) which shifts immune response to Th1. Th1 induces the production of cytokines interleukin (IL)-2, interferon-γ, and tumor necrosis factor-α. This suppresses Th2 response (IL-4 and IL-5 cytokines). A similar immunomodulatory response can be expected in HEV.

21.7.2.1.2 Pegylated Interferon

Pegylated interferon (PEG-IFN) is the second choice of drug after ribavirin used for treatment of hepatitis E infection. HEV-infected patients with a kidney and liver transplant, HIV infection, leukemia, and chronic infections have been treated with PEG-IFN-α (Kamar et al. 2010; Dalton et al. 2010; Haagsma et al. 2010; Alric et al. 2010). IFN-α (type I), IFN-γ (type II), and IFN-λ 3 (type III) block HEV replication. IFN-α subtypes 2a and 2b exhibit the strongest antiviral activity against HEV (Todt et al. 2016; Zhou et al. 2016). However, it is reported that IFN (type 1) is inhibited by HEV X and papain-like cysteine protease as a part of the host evasion strategy (Nan et al. 2014). Similarly, HEV ORF3 inhibits IFN-α (Dong et al. 2012). However, interferon is inadvisable in kidney-, pancreas-, heart-, and lung transplant recipients as it causes acute rejection. The exact mechanism of PEG-IFN showing the inhibitory activity against HEV is still unclear. Side effects of IFN treatment include flu-like symptoms, neuropsychiatric and neurologic disorders, myelosuppression, cardiovascular disorders, and altered liver functions (Raison et al. 2005).

21.7.2.2 Potential Targets for Antiviral Development

21.7.2.2.1 Virus Entry Inhibitors

The cell surface HSPGs is involved in binding of ORF2 protein to Huh-7 liver cells. This indicates that HSPGs may act as receptors for cellular attachment (Kalia et al. 2009). HSP90 and Grp78 interact with HEV ORF2 (John et al. 2011). Drugs or inhibitors can target virus binding to the receptor or intracellular transport and help in HEV inhibition at a very early stage.

21.7.2.2.2 Inhibitors of Capping of the Viral Genome

Methyltransferase helps the 5′-methlyguanosine capping of the viral RNA which provides stability and protection against host innate immune responses (Decroly et al. 2011). HEV methyltransferase shows virus-specific activity which can act as a potential target for HEV inhibition (Emerson et al. 2004). It is reported that 3-deaza-adenosine and neplanocin A act as a methyltransferase inhibitor in influenza virus and blocks its replication (Woyciniuk et al. 1995) which can be studied for HEV inhibition as well.

21.7.2.2.3 Inhibitors of Viral Genome Replication

(a) Direct Acting Antivirals (DAAs)

Direct acting antivirals (DAAs) are drug compounds that target specific nonstructural proteins of the virus and hinder virus replication and infection. These are extensively used for the treatment of chronic hepatitis C infections (Anang et al. 2018). Sofosbuvir is a direct-acting nucleotide polymerase inhibitor that inhibits NS5B RNA polymerase in HCV (Pawlotsky 2013). It is a monophosphorylated uridine nucleotide analog. Primary reports suggested that sofosbuvir inhibited HEV replication but further studies did not show similar results (Thi et al. 2016). The determination of HEV RNA-dependent RNA polymerase (RdRp) structure may help in determining more potent replicase inhibitors.

(b) Indirect Inhibitors of RdRp

The interaction between the host factors and viral RdRp was analyzed in HEV. Using the yeast-two-hybrid cDNA library, it was identified that 41 host proteins interacted with HEV genotype 1 RdRp and 23 of them could identify HEV genotype 3 RdRp (Subramani et al. 2018). Factors regulating host translation like eIF4A2, eEF1a1, and eIF3A directly interact with RdRp protein of both genotype 1 and genotype 3 (Nair et al. 2016) which can act as a potential antiviral target. Inhibitors of other host translation factors like eIF4A2 and eIF3A may also inhibit the virus replication. Thus, focusing on direct and indirect inhibitors of virus may be an attractive antiviral strategy.

21.7.2.2.4 Helicase Inhibitors

HEV helicase acts as nucleoside triphosphatase with RNA unwinding activity in 5′ to 3′ direction (Karpe and Lole 2010). Thus, it plays a vital role in HEV replication. Viral helicases and the host share common properties; hence, using helicase inhibitors is challenging. The helicase inhibitory activity has been reported in SARS coronavirus, HCV, dengue virus, etc. (Frick and Lam 2006).

21.7.2.2.5 Inhibitors of the Release of Progeny Virions

Newly formed virus particles are released from the host cells which involves interactions between the virus proteins and the host proteins. Inhibitors of these interactions may help in minimizing the release of the virus particles (Ravindran et al. 2016). In HEV genotype 3, the interplay between HEV ORF3 and host tumor susceptibility gene 101-encoded protein is necessary for the release of the HEV progeny (Surjit et al. 2006), such additional targets need to be identified as potent antiviral targets against HEV.

21.7.3 HEV Vaccines

HEV vaccines are helpful in developing countries as it prevents young adults from suffering from hepatitis E and also reduces mortality rate in pregnant women. In developed countries, vaccine immunization may prevent the elderly from chronic hepatitis. Two subunit vaccines against HEV were developed and tested in clinical trials. Human vaccine consisting of VLPs is truncated HEV ORF2 (56 kDa, 112–607 aa) with alum adjuvant. This vaccine was formed in *Spodoptera frugiperda* cells which were infected with a recombinant baculovirus. Another vaccine called HEV239 or Hecolin (26 kDa, 368–606 aa) is a truncated HEV capsid protein that was expressed in *E. coli* in 2010. HEV 239 is effective against genotypes 1 and 4, and its usage is restricted to China. It was noted that three doses of this vaccine provided 95.5% efficacy (Li et al. 2005). HEV 239 vaccine is safe for both pregnant women and the fetus (Wu et al. 2012). The safety of these vaccines is unclear in children and patients with liver diseases.

21.8 Prevention and Control

Transmission of hepatitis E can be prevented by measures as mentioned in Fig. 21.7.

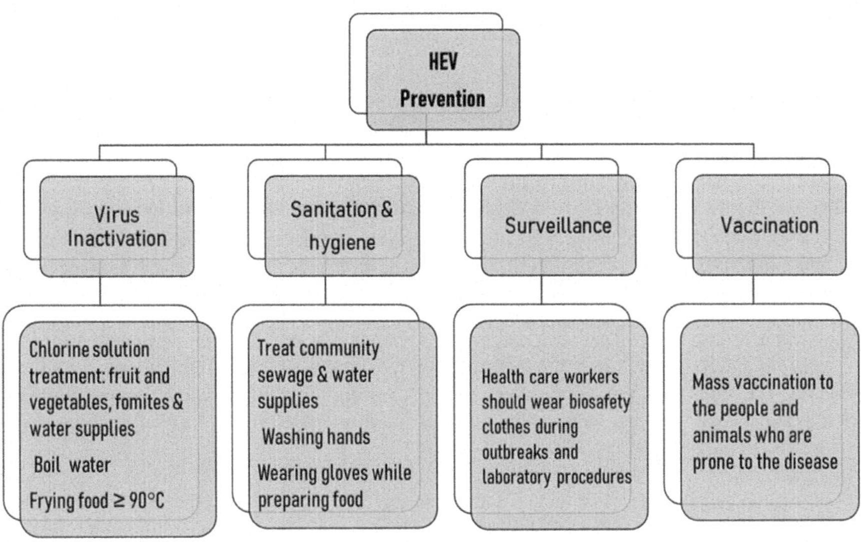

Fig. 21.7 Preventive and control measures for HEV

21.8.1 Disease Management

There is no registered drug for the treatment of hepatitis E but ribavirin is generally used for clearance of symptoms. Acute HEV infection can be resolved without hospitalization but chronic infections can be fatal, if not treated timely. Unnecessary medication for the symptoms associated with the disease should be avoided. Chronic infections which finally leads to fulminant hepatitis or liver cirrhosis need medication and hospitalization. Inteferon along with ribavirin are normally used for treating chronic cases.

21.9 Future Perspectives

Expanding the host range of HEV and interspecies infection will help us establish newer models for HEV research in the future. Progression in the study of HEV will help us develop vaccines that are highly safe, effective, and not restricted to a specific genotype or geographical region. Ribavirin is used primarily to combat HEV infection but there is no specific FDA-approved drug for HEV treatment. This increases the uncertainty of prevention and control of HEV infection. Silent mutations in the HEV genome under immense host immune pressure may develop antiviral-resistant strains which make a study of new potent antivirals necessary especially for high-risk groups. HEV is known as a hepatotropic virus but extrahepatic manifestations have made an investigation of HEV pathogenesis more complex. Most of the HEV cases remain underdiagnosed due to their asymptomatic and self-limiting nature; hence, more sensitive and reliable diagnostic techniques are needed to be developed.

References

Acharya SK, Sharma PK, Singh R, Mohanty SK et al (2007) Hepatitis E virus (HEV) infection in patients with cirrhosis is associated with rapid decompensation and death. J Hepatol 46(3):387–394. https://doi.org/10.1016/j.jhep.2006.09.016

Adjei AA, Tettey Y, Aviyase JT, Adu-Gyamfi C et al (2009) Hepatitis E virus infection is highly prevalent among pregnant women in Accra, Ghana. Virol J 6:108. https://doi.org/10.1186/1743-422X-6-108

Aggarwal R (2010) Hepatitis E virus and person-to-person transmission. Clin Infect Dis 51(4):477–478. https://doi.org/10.1086/655157

Aggarwal R (2013) Hepatitis e: epidemiology and natural history. J Clin Exp Hepatol 2:125–133. https://doi.org/10.1016/j.jceh.2013.05.010

Ahn HS, Han SH, Kim YH, Park BJ et al (2017) Adverse fetal outcomes in pregnant rabbits experimentally infected with rabbit hepatitis E virus. Virology 512:187–193. https://doi.org/10.1016/j.virol.2017.09.020

Allweiss L, Gass S, Giersch K, Groth A et al (2016) Human liver chimeric mice as a new model of chronic hepatitis E virus infection and preclinical drug evaluation. J Hepatol 64:1033–1040. https://doi.org/10.1016/j.jhep.2016.01.011

Alric L, Bonnet D, Laurent G, Kamar N, Izopet J (2010) Chronic hepatitis E virus infection: successful virologic response to pegylated interferon-alpha therapy. Ann Intern Med 153:135–136. https://doi.org/10.7326/0003-4819-153-2-201007200-00256

Anang S, Kaushik N, Surjit M (2018) Recent advances towards the development of a potent antiviral against the Hepatitis E virus. J Clin Transl Hepatol 6(3):310. https://doi.org/10.14218/JCTH.2018.00005

Andersson MI, Preiser W et al (2013) Immune reconstitution hepatitis E: a neglected complication of antiretroviral therapy in Africa? AIDS 27(3):487–489. https://doi.org/10.1097/QAD.0b013e32835b1074

Arankalle VA, Chobe LP (1999) Hepatitis E virus: can it be transmitted parenterally? J Viral Hepat 6:161–164. https://doi.org/10.1046/j.1365-2893.1999.00141.x

Arankalle VA, Chobe LP, Joshi MV, Chadha MS, Kundu B, Walimbe AM (2002) Human and swine hepatitis E viruses from Western India belong to different genotypes. J Hepatol 36(3):417–425. https://doi.org/10.1016/s0168-8278(01)00297-5

Aslan AT, Balaban HY (2020) Hepatitis E virus: epidemiology, diagnosis, clinical manifestations, and treatment. World J Gastroenterol 26(37):5543–5560. https://doi.org/10.3748/wjg.v26.i37.5543

Azuma H, Paulk N, Ranade A, Dorrell C et al (2007) Robust expansion of human hepatocytes in fah−/−/Rag2−/−/Il2rg−/− mice. Nat Biotechnol 25(8):903–910. https://doi.org/10.1038/nbt1326

Baki AA, Haque W, Giti S, Khan AA, Rahman MM, Jubaida N, Rahman M (2021) Hepatitis E virus genotype 1f outbreak in Bangladesh, 2018. J Med Virol 93(8):5177–5181. https://doi.org/10.1002/jmv.26739

Balayan MS, Andjaparidze AG, Savinskaya SS et al (1983) Evidence for a virus in non-a, non-B hepatitis transmitted via the fecal-oral route. Intervirology 20(1):23–31. https://doi.org/10.1159/000149370

Bazerbachi F, Haffar S, Garg SK, Lake JR (2015) Extra-hepatic manifestations associated with hepatitis E virus infection: a comprehensive review of the literature. Gastroenterol Report 4:1–15. https://doi.org/10.1093/gastro/gov042

Belbezier A, Deroux A, Sarrot-Reynauld F, Larrat S, Bouillet L (2014) Myasthenia gravis associated with acute hepatitis E infection in immunocompetent woman. Emerg Infect Dis 20(5):908–910. https://doi.org/10.3201/eid2005.131551

Birke L, Cormier SA, You D, Stout RW, Clement C, Johnson M, Thompson H (2014) Hepatitis E antibodies in laboratory rabbits from 2 US vendors. Emerg Infect Dis 20(4):693–696. https://doi.org/10.3201/eid2004.131229

Boccia D, Guthmann J-P, Klovstad H, Hamid N et al (2006) High mortality associated with an outbreak of Hepatitis E among displaced persons in Darfur, Sudan. Clin Infect Dis 42(12):1679–1684. https://doi.org/10.1086/504322

Bose PD, Das BC, Kumar A, Gondal R, Kumar D, Kar P (2011) High viral load and deregulation of the progesterone receptor signaling pathway: association with hepatitis E-related poor pregnancy outcome. J Hepatol 54(6):1107–1113. https://doi.org/10.1016/j.jhep.2010.08.037

Bouwknegt M, Rutjes SA, Reusken CBEM, Stockhofe-Zurwieden N et al (2009) The course of hepatitis E virus infection in pigs after contact-infection and intravenous inoculation. BMC Vet Res 5:1–12. https://doi.org/10.1186/1746-6148-5-7

Bradley DW, Purdy MA, Reyes GR (1991) Hepatitis E virus genome. Molecular features, expression of immunoreactive proteins and sequence divergence. J Hepatol 13(Suppl 4):S152–S154. https://doi.org/10.1016/0168-8278(91)90049-h

Cao D, Meng XJ (2012) Molecular biology and replication of hepatitis E virus. Emerg Microbes Infect 1(8):e17. https://doi.org/10.1038/emi.2012.7

Cao D, Huang Y-W, Meng X-J (2010) The nucleotides on the stem-loop RNA structure in the junction region of the hepatitis E virus genome are critical for virus replication. J Virol 84(24):13,040–13,044. https://doi.org/10.1128/JVI.01475-10

Cheng XF, Wang S, Dai X, Shi CB, Wen YF, Zhu M, Zhan SW, Meng JH (2012) Rabbit as a novel animal model for Hepatitis E virus infection and vaccine evaluation. PLoS One 7:e51616. https://doi.org/10.1371/journal.pone.0051616

Cornberg M, Müller T, Behrendt P, Piecha et al (2019) LBO-04-efficacy and safety of sofosbuvir monotherapy in patients with chronic hepatitis E-the HepNet SofE pilot study. J Hepatol 70:e129–e130. https://doi.org/10.1016/S0618-8278(19)30228-2

Coursaget P, Buisson Y, N'Gawara MN, Van Cuyck-Gandre H, Roue R (1998) Role of hepatitis E virus in sporadic cases of acute and fulminant hepatitis in an endemic area (Chad). Am J Trop Med Hyg 58(3):330–334. https://doi.org/10.4269/ajtmh.1998.58.330

Crotty S, Cameron CE, Andino R (2001) RNA virus error catastrophe: direct molecular test by using ribavirin. Proc Natl Acad Sci USA 98(12):6895–6900. https://doi.org/10.1073/pnas.111085598

Dalton H, Bendall R, Keane F, Neale R, Tedder R, Ijaz S (2010) P61 Pegylated interferon and ribavarin combination therapy achieves hepatitis E virus clearance in chronic hepatitis E virus/human immunodeficiency virus co-infection. Gut 59:A35–A36. https://doi.org/10.1136/gut.2010.223362.87

Dalton HR, Kamar N, van Eijk JJ, Mclean BN, Cintas P, Bendall RP, Jacobs BC (2016) Hepatitis E virus and neurological injury. Nat Rev Neurol 12(2):77–85. https://doi.org/10.1038/nrneurol.2015.234

Das K, Agarwal A, Andrew R, Frösner GG, Kar P (2000) Role of hepatitis E and other hepatotropic virus in aetiology of sporadic acute viral hepatitis: a hospital based study from urban Delhi. Eur J Epidemiol 16(10):937–940. https://doi.org/10.1023/a:1011072015127

Debing Y, Emerson SU, Wang Y, Pan Q, Balzarini J, Dallmeier K et al (2014) Ribavirin inhibits in vitro hepatitis E virus replication through depletion of cellular GTP pools and is moderately synergistic with alpha interferon. Antimicrob Agents Chemother 58:267–273. https://doi.org/10.1128/AAC.01795-13

Decroly E, Ferron F, Lescar J, Canard B (2011) Conventional and unconventional mechanisms for capping viral mRNA. Nat Rev Microbiol 10:51–65. https://doi.org/10.1038/nrmicro2675

Dong C, Zafrullah M, Mixson-Hayden T, Dai X, Liang J, Meng J et al (2012) Suppression of interferon-a signaling by hepatitis E virus. Hepatology 55:1324–1332. https://doi.org/10.1002/hep.25530

Dumoulin FL, Liese H (2012) Acute hepatitis E virus infection and autoimmune thyroiditis: yet another trigger? BMJ Case Rep:bcr1220115441. https://doi.org/10.1136/bcr.12.2011.5441

El Sayed Zaki M, El Aal AA, Badawy A, El-Deeb DR, El-Kheir NYA (2013) Clinicolaboratory study of mother-to-neonate transmission of Hepatitis E virus in Egypt. Am J Clin Pathol 140(5):721–726. https://doi.org/10.1309/AJCPT55TDMJNPLLV

Emerson SU, Zhang M, Meng XJ, Nguyen H, St Claire M, Govindarajan S, Huang YK, Purcell RH (2001) Recombinant hepatitis E virus genomes infectious for primates: importance of capping and discovery of a cis-reactive element. Proc Natl Acad Sci USA 98(26):15270–15275. https://doi.org/10.1073/pnas.251555098

Emerson SU, Nguyen H, Graff J, Stephany DA, Brockington A, Purcell RH (2004) In vitro replication of hepatitis E virus (HEV) genomes and of an HEV replicon expressing green fluorescent protein. J Virol 78:4838–4846. https://doi.org/10.1128/JVI.78.9.4838-4846.2004

Frick DN, Lam AM (2006) Understanding helicases as a means of virus control. Curr Pharm Des 12(11):1315–1338. https://doi.org/10.2174/138161206776361147

Giersch K, Allweiss L, Volz T, Helbig M, Bierwolf J et al (2015) Hepatitis Delta co-infection in humanized mice leads to pronounced induction of innate immune responses in comparison to HBV mono-infection. J Hepatol 63(2):346–353. https://doi.org/10.1016/j.jhep.2015.03.011

Graff J, Nguyen H, Yu C, Elkins WR, St Claire M, Purcell RH, Emerson SU (2005) The open reading frame 3 gene of hepatitis E virus contains a cis-reactive element and encodes a protein required for infection of macaques. J Virol 79(11):6680–6689. https://doi.org/10.1128/JVI.79.11.6680-6689.2005

Guinault D, Ribes D, Delas A, Milongo D et al (2016) Hepatitis E virus-induced cryoglobulinemic glomerulonephritis in a nonimmunocompromised person. Am J Kidney Dis 67:660–663. https://doi.org/10.1053/j.ajkd.2015.10.022
Guo H, Zhou EM, Sun ZF, Meng XJ (2007) Egg whites from eggs of chickens infected experimentally with avian hepatitis E virus contain infectious virus, but evidence of complete vertical transmission is lacking. J Gen Virol 88(Pt 5):1532–1537. https://doi.org/10.1099/vir.0.82689-0
Gurav Y, Kakade S, Kakade R, Kadam Y, Durgawale P (2007) A study of hepatitis E outbreak in rural area of Western Maharashtra. Indian J Community Med 32(3):182. https://doi.org/10.4103/0970-0218.36822
Haagsma EB, Riezebos-Brilman A, van den Berg AP, Porte RJ, Niesters HG (2010) Treatment of chronic hepatitis E in liver transplant recipients with pegylated interferon alpha-2b. Liver Transpl 16:474–477. https://doi.org/10.1002/lt.22014
Hakim MS, Wang W, Bramer WM, Geng J et al (2017) The global burden of hepatitis E outbreaks: a systematic review. Liver Int 2017(37):19–31. https://doi.org/10.1111/liv.13237
Han J, Lei Y, Liu L, Liu P et al (2014) SPF rabbits infected with rabbit hepatitis E virus isolate experimentally showing the chronicity of hepatitis. PLoS One 9(6):e99861. https://doi.org/10.1371/journal.pone.0099861
Haqshenas G, Shivaprasad HL, Woolcock PR, Read DH, Meng XJ (2001) Genetic identification and characterization of a novel virus related to human hepatitis E virus from chickens with hepatitis-splenomegaly syndrome in the United States. J Gen Virol 82(Pt 10):2449–2462. https://doi.org/10.1099/0022-1317-82-10-2449
Heaton NS, Randall G (2011) Multifaceted roles for lipids in viral infection. Trends Microbiol 19(7):368–375. ISSN: 0966-842X. https://doi.org/10.1016/j.tim.2011.03.007
Hepatitis E, WHO factsheet (2022). https://www.who.int/news-room/fact-sheets/detail/hepatitis-e
International Committee on Taxonomy of Viruses. Hepatitis E virus (2009). https://www.ictvonline.org/
Jha AK, Nijhawan S, Nepalia S, Suchismita A (2012) Association of Bell's palsy with hepatitis E virus infection: a rare entity. J Clin Exp Hepatol 2(1):88–90. https://doi.org/10.1016/S0973-6883(12)60082-6
Jilani N, Das BC, Husain SA, Baweja UK et al (2007) Hepatitis E virus infection and fulminant hepatic failure during pregnancy. J Gastroenterol Hepatol 22:676–682. https://doi.org/10.1111/j.1440-1746.2007.04913.x
Jin H, Zhao Y, Zhang X, Wang B, Liu P (2016) Case-fatality risk of pregnant women with acute viral hepatitis type E: a systematic review and meta-analysis. Epidemiol Infect 144:2098–2106. https://doi.org/10.1017/s0950268816000418
John L, Thomas S, Herchenröder O, Pützer et al (2011) Hepatitis E virus ORF2 protein activates the pro-apoptotic gene CHOP and anti-apoptotic heat shock proteins. PLoS One 6(9):e25378. https://doi.org/10.1371/journal.pone.0025378
Johne R, Plenge-Bonig A, Hess M, Ulrich RG, Reetz J, Schielke A (2010) Detection of a novel hepatitis E-like virus in faeces of wild rats using a nested broad-spectrum RT-PCR. J Gen Virol 91(Pt 3):750–758. https://doi.org/10.1099/vir.0.016584-0
Ju X, Ding Q (2019) Hepatitis E virus assembly and release. Viruses 11(6):539. https://doi.org/10.3390/v11060539
Kalia M, Chandra V, Rahman SA, Sehgal D, Jameel S (2009) Heparan sulfate proteoglycans are required for cellular binding of the hepatitis E virus ORF2 capsid protein and for viral infection. J Virol 83:12714–12724. https://doi.org/10.1128/JVI.00717-09
Kamar N, Selves J, Mansuy JM, Ouezzani L et al (2008) Hepatitis E virus and chronic hepatitis in organ-transplant recipients. N Engl J Med 358(8):811–817. https://doi.org/10.1056/NEJMoa0706992
Kamar N, Abravanel F, Garrouste C, Cardeau-Desangles I et al (2010) Three-month pegylated interferon-alpha-2a therapy for chronic hepatitis E virus infection in a haemodialysis patient. Nephrol Dial Transplant 25:2792–2795. https://doi.org/10.1093/ndt/gfq282
Kamar N, Bendall R, Legrand-Abravanel F, Xia NS, Ijaz S, Izopet J, Dalton HR (2012) Hepatitis E. Lancet 379(9835):2477–2488. https://doi.org/10.1016/S0140-6736(11)61849-7

Kamili S (2011) Toward the development of a hepatitis E vaccine. Virus Res 161(1):93–100. https://doi.org/10.1016/j.virusres.2011.05.008

Kanade GD, Pingale KD, Karpe YA (2019) Protein interactions network of Hepatitis E virus RNA and polymerase with host proteins. Front Microbiol 10. https://doi.org/10.3389/fmicb.2019.02501

Karpe YA, Lole KS (2010) RNA 5′-triphosphatase activity of the hepatitis E virus helicase domain. J Virol 84:9637–9641. https://doi.org/10.1128/JVI.00492-10

Karpe YA, Meng XJ (2012) Hepatitis E virus replication requires an active ubiquitin-proteasome system. J Virol 86(10):5948–5952. https://doi.org/10.1128/JVI.07039-11

Kaushik N, Subramani C, Anang S, Muthumohan R et al (2017) Zinc salts block hepatitis E virus replication by inhibiting the activity of viral RNA-dependent RNA polymerase. J Virol 91(21):e00754–e00717. https://doi.org/10.1128/JVI.00754-17

Kenney SP (2019) The current host range of hepatitis E viruses. Viruses 11(5):452. https://doi.org/10.3390/v11050452

Khuroo MS (1980) Chronic liver disease after non-a, non-B hepatitis. Lancet 2(8199):860–861. https://doi.org/10.1016/S0140-6736(80)90205-6

Khuroo MS, Kamili S (2003) Aetiology and prognostic factors in acute liver failure in India. J Viral Hepat 10(3):224–231. https://doi.org/10.1046/j.1365-2893.2003.00415.x

Khuroo MS, Khuroo MS (2010) Seroepidemiology of a second epidemic of hepatitis E in a population that had recorded first epidemic 30 years before and has been under surveillance since then. Hepatol Int 4(2):494–499. https://doi.org/10.1007/s12072-009-9159-5. Erratum in: Hepatol Int 2010;4(2):536

Khuroo MS, Khuroo MS, Khuroo NS (2016) Hepatitis E: discovery, global impact, control and cure. World J Gastroenterol 22(31):7030–7045. https://doi.org/10.3748/wjg.v22.i31.7030

Khuroo MS, Teli MR, Skidmore S, Sofi MA, Khuroo MI (1981) Incidence and severity of viral hepatitis in pregnancy. Am J Med 70(2):252–255. https://doi.org/10.1016/0002-9343(81)90758-0

Khuroo MS, Rustgi VK, Dawson GJ, Mushahwar IK et al (1994) Spectrum of hepatitis E virus infection in India. J Med Virol 43(3):281–286. https://doi.org/10.1002/jmv.1890430316

Khuroo MS, Kamili S, Khuroo MS (2009) Clinical course and duration of viremia in vertically transmitted hepatitis E virus (HEV) infection in babies born to HEV-infected mothers. J Viral Hepat 16(7):519–523. https://doi.org/10.1111/j.1365-2893.2009.01101.x

Kim JH, Nelson KE, Panzner U, Kasture Y, Labrique AB, Wierzba TF (2014) A systematic review of the epidemiology of hepatitis E virus in Africa. BMC Infect Dis 14:308. https://doi.org/10.1186/1471-2334-14-308. Erratum in: BMC Infect Dis 2017;17(1):187

Kumar T, Shrivastava A, Bhatia D et al (2019) Jaundice outbreak likely caused by HEV in Amritsar, Punjab, India, 2013. BMC Public Health 19(Suppl 3):464. https://doi.org/10.1186/s12889-019-6786-1

Lai F, Chen Q (2018) Humanized mouse models for the study of infection and pathogenesis of human viruses. Viruses 10(11):643. https://doi.org/10.3390/v10110643

Li SW, Zhang J, Li YM, Ou SH et al (2005) A bacterially expressed particulate hepatitis E vaccine: antigenicity, immunogenicity and protectivity on primates. Vaccine 23(22):2893–2901. https://doi.org/10.1016/j.vaccine.2004.11.064

Li C, Debing Y, Jankevicius G, Neyts J et al (2016) Viral macro domains reverse protein ADP-ribosylation. J Virol 90(19):8478–8486. https://doi.org/10.1128/JVI.00705-16

Mallet V, Nicand E, Sultanik P, Chakvetadze C et al (2010) Brief communication: case reports of ribavirin treatment for chronic hepatitis E. Ann Intern Med 153(2):85–89. https://doi.org/10.7326/0003-4819-153-2-201007200-00257

Meisner S, Polywka S, Memmler M, Nashan B et al (2015) Definition of chronic hepatitis E after liver transplant conforms to convention. Am J Transplant 15(11):3011–3012. https://doi.org/10.1111/ajt.13428

Meng XJ, Purcell RH, Halbur PG, Lehman JR et al (1997) A novel virus in swine is closely related to the human hepatitis E virus. Proc Natl Acad Sci USA 94(18):9860–9865. https://doi.org/10.1073/pnas.94.18.9860

Mengel AM, Stenzel W, Meisel A, Büning C (2016) Hepatitis E-induced severe myositis. Muscle Nerve 53(2):317–320. https://doi.org/10.1002/mus.24959

Montpellier C, Wychowski C, Sayed IM, Meunier JC et al (2018) Hepatitis E virus lifecycle and identification of 3 forms of the ORF2 capsid protein. Gastroenterology 154(1):211–223.e8. https://doi.org/10.1053/j.gastro.2017.09.020

Mulyanto, Suparyatmo JB, Andayani IG, Khalid et al (2014) Marked genomic heterogeneity of rat hepatitis E virus strains in Indonesia demonstrated on a full-length genome analysis. Virus Res 179:102–112. https://doi.org/10.1016/j.virusres.2013.10.029

Murkey JA, Chew KW, Carlson M, Shannon CL et al (2017) Hepatitis E virus-associated meningoencephalitis in a lung transplant recipient diagnosed by clinical metagenomic sequencing. Open Forum Infect Dis 4(3):ofx121. https://doi.org/10.1093/ofid/ofx121

Nagashima S, Takahashi M, Jirintai, Tanaka T, Yamada K, Nishizawa T et al (2011) A PSAP motif in the ORF3 protein of hepatitis E virus is necessary for virion release from infected cells. J Gen Virol 92:269–278. https://doi.org/10.1099/vir.0.025791-0

Nagashima S, Jirintai S, Takahashi M, Kobayashi T et al (2014) Hepatitis E virus egress depends on the exosomal pathway, with secretory exosomes derived from multivesicular bodies. J Gen Virol 95(Pt 10):2166–2175. https://doi.org/10.1099/vir.0.066910-0

Nagashima S, Takahashi M, Kobayashi T, Tanggis et al (2017) Characterization of the quasi-enveloped hepatitis E virus particles released by the cellular exosomal pathway. J Virol 91(22):e00822–e00817. https://doi.org/10.1128/JVI.00822-17

Nair VP, Anang S, Subramani C, Madhvi A et al (2016) Endoplasmic reticulum stress induced synthesis of a novel viral factor mediates efficient replication of genotype-1 hepatitis E virus. PLoS Pathog 12:e1005521. https://doi.org/10.1371/journal.ppat.1005521

Nan Y, Yu Y, Ma Z, Khattar SK et al (2014) Hepatitis E virus inhibits type I interferon induction by ORF1 products. J Virol 88:11924–11932. https://doi.org/10.1128/JVI.01935-14

Navaneethan U, Al Mohajer M, Shata MT (2008) Hepatitis E and pregnancy: understanding the pathogenesis. Liver Int 28(9):1190–1199. https://doi.org/10.1111/j.1478-3231.2008.01840.x

Nishiyama T, Kobayashi T, Jirintai S, Kii I et al (2019) Screening of novel drugs for inhibiting hepatitis E virus replication. J Virol Methods 270:1–11. https://doi.org/10.1016/j.jviromet.2019.04.017

Pawlotsky JM (2013) NS5A inhibitors in the treatment of hepatitis C. J Hepatol 59:375–382. https://doi.org/10.1016/j.jhep.2013.03.030

Perrin HB, Cintas P, Abravanel F, Gérolami R et al (2015) Neurologic disorders in immunocompetent patients with autochthonous acute Hepatitis E. Emerg Infect Dis 21(11):1928–1934. https://doi.org/10.3201/eid2111.141789

Pischke S, Hartl J, Pas SD, Lohse AW et al (2017) Hepatitis E virus: infection beyond the liver? J Hepatol 66(5):1082–1095. https://doi.org/10.1016/j.jhep.2016.11.016

Pradat P, Virlogeux V, Gagnieu M-C, Zoulim F, Bailly F (2014) Ribavirin at the era of novel direct antiviral agents for the treatment of hepatitis C virus infection: relevance of pharmacological monitoring. Adv Hepatol:493087, 13 pages. https://doi.org/10.1155/2014/493087

Pudupakam RS, Huang YW, Opriessnig T, Halbur PG et al (2009) Deletions of the hypervariable region (HVR) in open reading frame 1 of hepatitis E virus do not abolish virus infectivity. Evidence for attenuation of HVR deletion mutants in vivo. J Virol 83(1):384–395. https://doi.org/10.1128/jvi.01854-08

Purcell RH, Engle RE, Rood MP, Kabrane-Lazizi Y et al (2011) Hepatitis E virus in rats, Los Angeles, California, USA. Emerg Infect Dis 17:2216–2222. https://doi.org/10.3201/eid1712.110482

Raison CL, Demetrashvili M, Capuron L, Miller AH (2005) Neuropsychiatric adverse effects of interferon-alpha: recognition and management. CNS Drugs 19:105–123. https://doi.org/10.2165/00023210-200519020-00002

Ravindran MS, Bagchi P, Cunningham CN, Tsai B (2016) Opportunistic intruders: how viruses orchestrate ER functions to infect cells. Nat Rev Microbiol 14:407–420. https://doi.org/10.1038/nrmicro.2016.60

Reyes GR, Purdy MA, Kim JP, Luk KC et al (1990) Isolation of a cDNA from the virus responsible for enterically transmitted non-a, non-B hepatitis. Science 247(4948):1335–1339. https://doi.org/10.1126/science.2107574

Ribet D, Cossart P (2010) Pathogen-mediated posttranslational modifications: a re-emerging field. Cell 143(5):694–702. https://doi.org/10.1016/j.cell.2010.11.019

Robson SC, Adams S, Brink N, Woodruff B, Bradley D (1992) Hospital outbreak of hepatitis E. Lancet 339(8806):1424–1425. https://doi.org/10.1016/0140-6736(92)91250-c

Rodríguez-Castro E, Trastoy R, Rodríguez-Osorio X, Domínguez-Santalla MJ et al (2018) Meningitis due to autochthonous acute infection with hepatitis E virus in a chef: a case report. J Antimicrob Chemother 73(6):1726–1728. https://doi.org/10.1093/jac/dky052

Sanford BJ, Dryman BA, Huang YW, Feagins AR, Leroith T, Meng XJ (2011) Prior infection of pigs with a genotype 3 swine hepatitis E virus (HEV) protects against subsequent challenges with homologous and heterologous genotypes 3 and 4 human HEV. Virus Res 159(1):17–22. https://doi.org/10.1016/j.virusres.2011.04.010

Sarkar P, Morgan C, Ijaz S (2015) Transverse myelitis caused by hepatitis E: previously undescribed in adultsCase reports. 2015:bcr2014209031. https://doi.org/10.1136/bcr-2014-209031

Sayed IM, Verhoye L, Cocquerel L, Abravanel F et al (2017) Study of hepatitis E virus infection of genotype 1 and 3 in mice with humanised liver. Gut 66(5):920–929. https://doi.org/10.1136/gutjnl-2015-311109

Sengupta P, Biswas S, Roy T (2019) Hepatitis E-induced acute myocarditis in an elderly woman. Case Rep Gastroenterol 13:342–349. https://doi.org/10.1159/000501998

Seth A, Sherman KE (2020) Hepatitis E: what we think we know. Clin Liver Dis (Hoboken) 15(Suppl 1):S37–S44. https://doi.org/10.1002/cld.858

Shah SA, Lal A, Idrees M, Hussain A et al (2012) Hepatitis E virus-associated aplastic anaemia: the first case of its kind. J Clin Virol 54(1):96–97. https://doi.org/10.1016/j.jcv.2012.02.002

Sharapov MB, Favorov MO, Yashina TL, Brown MS et al (2009) Acute viral hepatitis morbidity and mortality associated with hepatitis E virus infection: Uzbekistan surveillance data. BMC Infect Dis 9:35. https://doi.org/10.1186/1471-2334-9-35

Shiota T, Li TC, Yoshizaki S, Kato T, Wakita T, Ishii K (2013) The hepatitis E virus capsid C-terminal region is essential for the viral life cycle: implication for viral genome encapsidation and particle stabilization. J Virol 87(10):6031–6036. https://doi.org/10.1128/JVI.00444-13

Shukla P, Nguyen HT, Faulk K, Mather K, Torian U, Engle RE, Emerson SU (2012) Adaptation of a genotype 3 hepatitis E virus to efficient growth in cell culture depends on an inserted human gene segment acquired by recombination. J Virol 86(10):5697–5707. https://doi.org/10.1128/JVI.00146-12

Smith DB, Vanek J, Ramalingam S, Johannessen I, Templeton K, Simmonds P (2012) Evolution of the hepatitis E virus hypervariable region. J Gen Virol 93(Pt 11):2408–2418. https://doi.org/10.1099/vir.0.045351-0

Song DY, Zhuang H, Li Z (1992) Hepatitis E in Hetian city. Analysis of 562 cases. Zhonghua Nei Ke Za Zhi 31(5):275–277, 316

Spina A, Lenglet A, Beversluis D, de Jong M et al (2017) A large outbreak of Hepatitis E virus genotype 1 infection in an urban setting in Chad likely linked to household level transmission factors, 2016-2017. PLoS One 12(11):e0188240. https://doi.org/10.1371/journal.pone.0188240

Sreenivasan MA, Arankalle VA, Sehgal A, Pavri KM (1984) Non-A, Non-B Epidemic Hepatitis: visualization of virus-like particles in the stool by immune electron microscopy. J Gen Virol 65(Pt 5):1005–1007. https://doi.org/10.1099/0022-1317-65-5-1005

Sridhar S, Yip CCY, Wu S, Cai J et al (2018) Rat hepatitis E virus as cause of persistent hepatitis after liver transplant. Emerg Infect Dis 24(12):2241–2250. https://doi.org/10.3201/eid2412.180937

Srivastava R, Aggarwal R, Sachdeva S, Alam MI, Jameel S, Naik S (2011) Adaptive immune responses during acute uncomplicated and fulminant hepatitis E. J Gastroenterol Hepatol 26(2):306–311. https://doi.org/10.1111/j.1440-1746.2010.06356.x

Subramani C, Nair VP, Anang S, Mandal SD et al (2018) Hostvirus protein interaction network reveals the involvement of multiple host processes in the life cycle of hepatitis. E virusmSystems 3:e00135–e00117. https://doi.org/10.1128/mSystems.00135-17
Suneetha PV, Pischke S, Schlaphoff V, Grabowski J et al (2012) Hepatitis E virus (HEV)-specific T-cell responses are associated with control of HEV infection. Hepatology 55(3):695–708. https://doi.org/10.1002/hep.24738
Surjit M, Oberoi R, Kumar R, Lal SK (2006) Enhanced alpha1 microglobulin secretion from Hepatitis E virus ORF3-expressing human hepatoma cells is mediated by the tumor susceptibility gene 101. J Biol Chem 281:8135–8142. https://doi.org/10.1074/jbc.M509568200
Tam AW, Smith MM, Guerra ME, Huang CC et al (1991) Hepatitis E virus (HEV): molecular cloning and sequencing of the full-length viral genome. Virology 185(1):120–131. https://doi.org/10.1016/0042-6822(91)90760-9
Teshale EH, Hu DJ (2011) Hepatitis E: epidemiology and prevention. World J Hepatol 3(12):285–291. https://doi.org/10.4254/wjh.v3.i12.285
Thapa R, Biswas B, Mallick D (2010) Henoch-Schönlein purpura triggered by acute hepatitis E virus infection. J Emerg Med 39(2):218–219. https://doi.org/10.1016/j.jemermed.2008.10.004
Thi VLD, Debing Y, Wu X, Rice CM, Neyts J, Moradpour D et al (2016) Sofosbuvir inhibits hepatitis E virus replication in vitro and results in an additive effect when combined with ribavirin. Gastroenterology 150:82–85.e4. https://doi.org/10.1053/j.gastro.2015.09.011
Todt D, Gisa A, Radonic A, Nitsche A et al (2016) In vivo evidence for ribavirin-induced mutagenesis of the hepatitis E virus genome. Gut 65:1733–1743. https://doi.org/10.1136/gutjnl-2015-311000
Todt D, Moeller N, Praditya D, Kinast V et al (2018) The natural compound silvestrol inhibits hepatitis E virus (HEV) replication in vitro and in vivo. Antivir Res 157:151–158. https://doi.org/10.1016/j.antiviral.2018.07.010
Tripathy AS, Sharma M, Thorat NC, Jadhav S, Koshatwar KA (2021) An outbreak of hepatitis E in Yavatmal, India, 2019. J Med Virol 93(6):3761–3768. https://doi.org/10.1002/jmv.26377
Tsega E, Krawczynski K, Hansson BG, Nordenfelt E et al (1991) Outbreak of acute hepatitis E virus infection among military personnel in northern Ethiopia. J Med Virol 34(4):232–236. https://doi.org/10.1002/jmv.1890340407
Viswanathan R (1957) Infectious hepatitis in Delhi (1955-56): a critical study-epidemiology. Natl Med J India 26(6):362–377
Wang L, Wang L (2016) Animal models for hepatitis E virus. Adv Exp Med Biol 948:161–173. https://doi.org/10.1007/978-94-024-0942-0_9
Wang L, Xia J, Wang L, Wang Y (2017) Experimental infection of rabbits with genotype 3 hepatitis E virus produced both chronicity and kidney injury. Gut 66(3):561–562. https://doi.org/10.1136/gutjnl-2016-312023
Wedemeyer H, Pischke S, Manns MP (2012) Pathogenesis and treatment of hepatitis e virus infection. Gastroenterology 142(6):1388–1397.e1. https://doi.org/10.1053/j.gastro.2012.02.014
Williams TP, Kasorndorkbua C, Halbur PG, Haqshenas G et al (2001) Evidence of extrahepatic sites of replication of the hepatitis E virus in a swine model. J Clin Microbiol 39:3040–3046. https://doi.org/10.1128/JCM.39.9.3040-3046.2001
Wißing MH, Brüggemann Y, Steinmann E, Todt D (2021) Virus-host cell interplay during hepatitis E virus infection. Trends Microbiol 4:309–319. https://doi.org/10.1016/j.tim.2020.07.002
Woyciniuk P, Linder M, Scholtissek C (1995) The methyltransferase inhibitor Neplanocin A interferes with influenza virus replication by a mechanism different from that of 3-0deazaadenosine. Virus Res 35:91–99. https://doi.org/10.1016/0168-1702(94)00085-q
Wu T, Zhu FC, Huang SJ, Zhang XF et al (2012) Safety of the hepatitis E vaccine for pregnant women: a preliminary analysis. Hepatology 55(6):2038. https://doi.org/10.1002/hep.25522
Yamada K, Takahashi M, Hoshino Y, Takahashi H et al (2009) Construction of an infectious cDNA clone of hepatitis E virus strain JE03-1760F that can propagate efficiently in cultured cells. J Gen Virol 90(Pt 2):457–462. https://doi.org/10.1099/vir.0.007559-0
Yin X, Ambardekar C, Lu Y, Feng Z (2016) Distinct entry mechanisms for nonenveloped and quasi-enveloped Hepatitis E viruses. J Virol 90(8):4232–4242. https://doi.org/10.1128/JVI.02804-15

Zafrullah M, Ozdener MH, Panda SK, Jameel S (1997) The ORF3 protein of hepatitis E virus is a phosphoprotein that associates with the cytoskeleton. J Virol 71(12):9045–9053. https://doi.org/10.1128/JVI.71.12.9045-9053.1997

Zamvar V, McClean P, Odeka E, Richards M, Davison S (2005) Hepatitis E virus infection with nonimmune hemolytic anemia. J Pediatr Gastroenterol Nutr 40(2):223–225. https://doi.org/10.1097/00005176-200502000-00027

Zhao C, Wang Y (2016) Laboratory diagnosis of HEV infection. Adv Exp Med Biol 948:191–209. https://doi.org/10.1007/978-94-024-0942-0_11

Zhao C, Ma Z, Harrison TJ, Feng R et al (2009) A novel genotype of hepatitis E virus prevalent among farmed rabbits in China. J Med Virol 81(8):1371–1379. https://doi.org/10.1002/jmv.21536

Zheng X, Yu L, Xu Q et al (2018) Guillain-Barre syndrome caused by hepatitis E infection: case report and literature review. BMC Infect Dis 18:50. https://doi.org/10.1186/s12879-018-2959-2

Zhou X, Xu L, Wang W, Watashi K, Wang Y, Sprengers D et al (2016) Disparity of basal and therapeutically activated interferon signalling in constraining hepatitis E virus infection. J Viral Hepat 23:294–304. https://doi.org/10.1111/jvh.12491

Zhou Z, Xie Y, Wu C, Nan Y (2021) The hepatitis E virus open Reading frame 2 protein: beyond viral capsid. Front Microbiol 12:739124. https://doi.org/10.3389/fmicb.2021.73912

Part IV

Emerging Immunodeficiency and Oncogenic Viruses

HIV Biology: Detailed Insight into Virion Structural Organization and Its Pathogenesis

22

Deeksha Agrawal, Kailash Chand, L. Preethi, Prudhvi Lal Bhukya, and Muneesh Kumar Barman

Abstract

Human immunodeficiency virus is a member of the Retroviridae family, primarily infecting human CD4$^+$ T cells, which leads to gradual depletion in the CD4$^+$ T cell count. When the T cell count comes below a critical level (200 cells/mm^3), this condition is called as acquired immunodeficiency syndrome (AIDS). At this stage, the risk of infection with several other opportunistic pathogens increases, which ultimately leads to the death of the infected individual, if remains untreated. It was first reported by the US Centers for Disease Control and Prevention (CDC) in the middle of 1981, which became a global pandemic within a few years. Around 38.0 million people globally (according to UNAIDS) and 2.3 million people in India (according to NACO) were living with HIV in 2019. HIV is classified into two types, HIV-1 and HIV-2. HIV-1 is more infective compared to

D. Agrawal
IIS (Deemed to be University), Jaipur, Rajasthan, India

K. Chand (✉)
Dr. Reddy's Institute of Life Sciences, Hyderabad, Telangana, India
e-mail: kcgupta6689@gmail.com

L. Preethi
Public Health, London School of Hygiene and Tropical Medicine, London, UK

P. L. Bhukya
Preclinical Vaccine Testing Laboratory, Rodent Experimentation Facility, ICMR-National Animal Facility Resource Facility for Biomedical Research, Hyderabad, Telangana, India

National Immunogenicity and Biologics Evaluation Center (NIBEC), Interactive Research School for Health Affairs, Pune, India

M. K. Barman
National Immunogenicity and Biologics Evaluation Center (NIBEC), Interactive Research School for Health Affairs, Pune, India

P. L. Bhukya, A. S (eds.), *Emerging Human Viral Diseases, Volume II*,
https://doi.org/10.1007/978-981-97-4480-0_22

HIV-2 and has spread globally. HIV-1 comprises two copies of single-stranded RNA molecule of about 9.8 Kb long surrounded by a capsid protein layer. HIV-1 genome encodes about 15 viral proteins. Due to the limited number of viral proteins, HIV-1 hijacks host cellular machinery and exploits a large number of host proteins to complete its life cycle successfully. Over the last four decades, extensive research has been done to combat HIV infection, but we still do not have a potential therapeutic regimen that can cure HIV completely. Currently prescribed drug combination, highly active antiretroviral therapy (HAART), which targets the key enzymes of the virus, can keep the viral load below detectable level, but does not eliminate the virus completely from the infected individual and has some drawbacks like the development of drug resistance, drug toxicity, etc. Thus, it has become crucial to identify and characterize cellular factors which are involved in establishing or restricting HIV-1 replication and pathogenesis. Several genome-wide screening studies have identified such cellular proteins associated with the HIV-1 replication process and control it either positively or negatively and are thus called proviral or antiviral factors. The use of some of these host factors as potential targets for HIV-1 infection is actively being pursued. Despite intense studies, host-pathogen interaction still remains to be clearly understood. Therefore, it remains an important area of research with direct association with the knowledge of infection transmission, pathogenic mechanisms, vaccine design, and molecular targets for therapeutic development.

Keywords

Human immunodeficiency virus · Antiretroviral therapy · Acquired immunodeficiency syndrome

22.1 Introduction

22.1.1 Discovery of the AIDS Virus

In the late 1970s, many healthy homosexual men with immunological dysfunction sought treatment in hospitals of Los Angeles, USA, for their suffering. Although these people had many symptoms, this unique syndrome was characterized by various infections consisting of *pneumocystis carinii* pneumonia (PCP), cytomegalovirus-associated retinitis, and cryptococcal meningitis (Gottlieb et al. 1981). After these cases, another homosexual couple reported having similar symptoms with a rare form of skin cancer, Kaposi's sarcoma (KS), which is exclusively observed in people with compromised immunity (Hymes et al. 1981). The same kind of compromised immunity cases with KS and PCP were also reported simultaneously from several other parts of the world. This condition was thought of as a new disease condition with an unknown cause. The syndrome was unidentified; however, it was clear from these reports that this novel disease is characterized by severe immunological dysfunction. These infections were seen in individuals with severely compromised immunity, and the syndrome was named as acquired immunodeficiency syndrome (AIDS). Several microorganisms were proposed to be causative agents of AIDS during that time, but during the early 1980s, three groups almost simultaneously and independently registered the discovery of the AIDS virus. In the first study, Dr. Francoise Barre-Sinoussi and Prof. Luc Montagnier reported the virus isolation from swollen lymph nodes of infected individuals and named as lymphadenopathy-associated virus (LAV) (Barré-Sinoussi et al. 1983). Simultaneously, Robert Gallo and his colleagues isolated the virus from the patient's peripheral blood with peripheral T-lymphoma (Gallo et al. 1983). Around the same time, Jay Levy isolated the virus from a distinct high-risk group of asymptomatic population with a similar characteristic of $CD4^{+}$ T-cell loss rather than immortalization. This isolated pathogen was named as AIDS-related virus (ARV) by Jay Levy and his group (1984). All these three isolated phenotypes of the virus were accepted as members of the same *Retroviridae* family soon. Proteins identified from these phenotypes were distinctly different from human T-lymphotropic virus (HTLV), and genome sequencing results also indicated very few sequence similarities with the HTLV genome. At last, the International Committee on Taxonomy of Viruses proposed a different name to the AIDS virus, human immunodeficiency virus (HIV), in 1986 (Coffin et al. 1986). After some time, a genetic variant of HIV-1 was discovered from the patients in the western part of Africa and was termed HIV-2 (Clavel et al. 1986). Dr. Francoise Barre-Sinoussie and Prof. Luc Montagnier were awarded the Nobel Prize in physiology and medicine in 2008 for the discovery of HIV as the causative agent of AIDS.

22.1.2 Emergence of HIV

The discovery of closely related simian immunodeficiency virus (SIVs) in African primates relates to the emergence of HIV through the cross-species transmission of SIV to humans, thereby establishing humans as new hosts and evolution of SIV into HIV (Sharp and Hahn 2010, 2011). HIV has been grouped into HIV-1 and HIV-2 based on their origin from chimpanzees (Pan troglodytes) and green monkey Sooty Mangabey (Cerocebusatys). HIV-1 is known to infect globally, while HIV-2 is confined mainly in West Africa. There are different theories and hypotheses of the emergence of HIV. However, the Parenteral Serial Transmission and the Enhanced Heterosexual Transmission hypotheses for the adaptation and early spread of HIV appear to be more promising. It has been established that humans acquired retroviruses from simians, mainly through bushmeat handling. All epidemiologically successful HIV groups started to spread in the early twentieth century, in contrast to the antiquity of T-cell lymphotropic viruses, indicating that novel enabling factors were involved in the emergence of HIV. The start of the epidemic somewhat coincides in time with increased genital ulcer disease reports in cities, suggesting that sexual transmission may have a significant role. Only weak and rare HIV groups emerged after approximately 1950, when injections and transfusions attained their maximal levels, suggesting that if the parenteral serial transmission was necessary for HIV adaptation, it had to be complemented by the sexual transmission of HIV to reach epidemic potential (Sousa et al. 2017).

22.1.3 HIV Taxonomy and Distribution

HIV is a lentivirus and member of the Retroviridae family. It is an enveloped, positive-sense, single-strand ribonucleic acid (RNA) virus. It comprises two copies of the viral genome in the form of a noncovalently linked RNA dimer (Clever and Parslow 1997). Lentivirus is known for its slow growth rate characterized by a long incubation period and relatively large genome size. Retrovirus depends on an enzyme known as reverse transcriptase to transcribe their RNA into deoxyribonucleic acid (DNA).

HIV is divided majorly into two groups such as HIV-1 and HIV-2. They vary in their sequence by more than 55% and are antigenically distinct (Clavel et al. 1987; Coffin et al. 1986). The HIV-1 group originated from chimpanzee viruses (SIVcpz), whereas the HIV-2 group originated from viruses of sooty mangabey monkeys (SIVsmm). HIV-1 and HIV-2 genomes are very similar, except HIV-2 that has a *vpx* gene instead of *vpu* of HIV-1 (Tristem et al. 1992). However, the major difference between both is in the envelope glycoprotein sequence (Clavel et al. 1987).

As compared to HIV-2, HIV-1 is more prevalent. HIV-1 is mostly restricted to West Africa, with few cases in the US and South America. Lately, it has been found that HIV-2 cases are increasing in India. They both differ in terms of transmissibility and pathogenicity. The infection rate of HIV-2 is less than HIV-1. Thus, the HIV-2 infected person survives longer. In addition to this, the transmission rate of HIV-2 is slower than HIV-1; hence, HIV-2 is comparatively less pathogenic than HIV-1. Perhaps, this is the reason for HIV-2 having been localized to a small geographic region. HIV-1 and HIV-2 are further classified into different strains based on their whole genome sequence. The strains of HIV-1 can be categorized into four groups. Most of the HIV-1 infections have been caused by "major" group M globally. The other three groups such as N, O, and P are relatively rare. Group O causes up to 5% of infections in the west and central African countries. In group M, there are at least nine genetically different subtypes of HIV-1. These are subtypes A, B, C, D, F, G, H, J, and K (Robertson et al. 2000; Peeters et al. 2003). These subtypes differ from each other in the envelope and gag region (Korber et al. 2000).

Moreover, different subtypes combine their genetic material and form a recombinant strain to evade the host immune system and antiretroviral treatment, known as "circulating recombinant forms" (CRFs) or "unique recombinant forms" (URFs). The distribution and proportion of HIV-1 subtypes, CRFs and URFs, varies among regions and changes over time. The regional and global distribution of different subtypes and recombinants of HIV-1 during 1990–2015 was estimated and is shown in Hemelaar et al. (2019).

In central Africa, different HIV-1 subtypes and CRFs were found throughout all periods. However, there is a decrease in subtypes A, D, G, and H while an increase in proportions of subtype C, other CRFs, and URFs with time. Subtype G was most prevalent in West Africa, while in East Africa, all HIV-1 infections were caused by subtype A in 2010–15, which mainly stayed stable over time. Subtype C dominated Ethiopia, Southern Africa, and South Asia (India). In western and central Europe and North America, the Caribbean, Latin America, and Oceania, most of the cases between 2010 and 2015, were reported with subtype B infection. In central Asia and Eastern Europe, subtype A caused more than 50% of infection. In the Middle East and North Africa, infections were caused by CRFs compared to subtype B. In Southeast Asia and East Asia, infections were mainly caused by CRF01_AE.

22.1.4 HIV-1 Tropism

Human cells expressing a cluster of differentiation 4 (CD4) receptors are natural targets for HIV-1. $CD4^+$ T helper cells are the main targets of HIV-1 as they express CD4 receptors abundantly. HIV also infects the other vital cells of the immune system that express CD4 receptors in abundance, including monocytes, monocyte-derived macrophages (MDMs), and microglial cells present in the nervous system. However, $CD4^+$ T helper cells show the highest virus production levels, while macrophages generally function as a virus reservoir. The monocytes and dendritic cells (DCs) express CD4 receptors at low levels, resulting in a slow rate of virus production (Lee et al. 1999; Nabatanzi et al. 2018). CD4 binding is a necessary first step, but the triumphant entry of HIV-1 into a host cell requires binding with another coreceptor, which are known as chemokine coreceptors and mainly involves either CCR5 (chemokine [C-C] motif receptor 5) or CXCR4 (CXC chemokine receptor 4) (Deng et al. 1996; Feng et al. 1996). CCR5 is mainly expressed on macrophages, T cells (memory and activated CD4 lymphocytes), gut-associated lymphoid tissues (GALT), microglia, and DCs. In contrast, CXCR4 is expressed on T cells (naive and resting CD4 lymphocytes and CD8 cells), neutrophils, B cells, and eosinophils (Kitchen and Zack 1999; Zhang et al. 1998). HIV-1 primarily exploits CD4 receptors along with CCR5 or CXCR4 chemokine co-receptors to enter into the $CD4^+$ T cells (Dittmar et al. 1997; Zhang et al. 1997; Brandenberg et al. 2015; Molotkovsky et al. 2018). Virus isolates that use CD4 receptor and CCR5 co-receptor are termed as R5 tropic viruses, and those isolates that utilize CD4 receptor and CXCR4 are designated as X4 tropic viruses. Those isolates that can use co-receptors with similar efficiency are known as R5X4 tropic (dual tropic) viruses (Berger et al. 1999; Lee et al. 1999; Brelot and Chakrabarti 2018). The occurrence of R5, X4, and dual-tropic viruses together in one host is called mixed tropism (Westby and van der Ryst 2005; Poveda et al. 2006). R5 virus, also referred to as the M-tropic virus, mainly infects macrophages. In contrast, the X4-tropic virus, also known as the T-tropic virus, is associated with a more prominent depletion of CD4 T cells. Initially, the HIV-1 virus prefers CCR5 chemokine co-receptor over CXCR4 for entry into the cell. Still, with the progression of infection, it starts utilizing CXCR4, called tropism shift (Wetzel et al. 2018).

It has been proposed that the comparative expression of CCR5 and CXCR4 co-receptor on cell surfaces may be a vital factor in HIV pathogenesis. For example, the genital and gastrointestinal tract (GALT) is considered a primary site of HIV infection as it consists of a minimum of 50% of all CD4 lymphocytes with high expression of CCR5 co-receptor (Meng et al. 2000, Moore et al. 2004). CCR5 is the 7-pass transmembrane G-protein coupled receptor (GPCR). The CCR5 gene containing a deletion of 32 base pairs in the sequence (called Δ-32) leads to truncated non-functional protein formation. This deleted form of the CCR5 gene was observed in some individuals of northern European heritage. However, individuals who were homozygous for Δ-32 (~1% of the Caucasian population) were unable to express CCR5 and exhibited resistance for infection with the R5-tropic virus (most transmitted HIV). In comparison, individuals heterozygous for Δ-32 deletion (around 10-15% of the Caucasian population) could become infected with HIV and resulted in prolonged diseases progression (Liu et al. 1996; Samson et al. 1996; Dean et al. 1996; Huang et al. 1996; Michael et al. 1997; Eugen-Olsen et al. 1997). These findings suggested the critical role of CCR5 in HIV-1 entry in the host cell. This allowed researchers to explore the potential of CCR5 based gene therapy and engineering cellular resistance to HIV by knocking out CCR5 in hematopoietic stem cells (Allers et al. 2011; Burke et al. 2015; Hale et al. 2017; Kang et al. 2015; Kiem et al. 2012; Li et al. 2000a, b, 2013; Tebas et al. 2014). This strategy controlled the HIV-1 infection progression in infected individuals. It showed favorable results in phase-IIb trials, but non-specific off-target effects and a sudden shift in the tropism with an unknown cause have restricted this therapy's use (Kordelas et al. 2014; Pandit and de Boer 2015; Pollakis and Paxton 2012).

22.1.5 AIDS Epidemiology

22.1.5.1 AIDS Epidemics Across the Globe

HIV/AIDS is considered a global pandemic as it is one of the most severe diseases affecting humans. From the beginning of the pandemic, almost 71 million people had been affected and 34 million people had died because of HIV. According to the last released report of the United Nations programme on HIV/AIDS (UNAIDS), nearly 38.0 million (31.6 million–44.5 million) people globally were living with HIV in 2019. Out of these, around 26.0 million (25.1 million–26.2 million) people had access to antiretroviral therapy at the end of June 2020. At the end of 2019, about 1.7 million (1.2 million–2.2 million) people became newly infected, and 690,000 (500,000–970,000) people died from AIDS-related illnesses.

In 2019, an estimated 1.7 million people [1.2 million–2.2 million] acquired HIV worldwide, which marked a 23% decline in new HIV infections since 2010. This was the lowest annual number of new infections since 1989. However, the rate of reduction in HIV transmission remains too slow, with the total number of new infections in 2018 was more than almost three times higher than the milestone of projected 500,000 that was set for 2020.

22.1.5.2 Prevalence of AIDS in India

Globally, India has the third-largest HIV-infected population. According to the last released report by the National AIDS control organization (NACO), around 2.3 million people were living with HIV (PLHIV) in 2019 with an adult (15–49 years) HIV prevalence of 0.22% (0.17–0.29%). This comprises about 79 thousand children and 9.94 lakh women living with HIV, around 3.4% and 44% of the total PLHIV estimates. In 2019, around 69.22 thousand new HIV infections and 58.96 thousand AIDS-related deaths were reported, which show a decline of 37% and 66% since 2010, respectively (Table 22.1).

Overall, India's HIV epidemic is decelerating with time. Between 1990 and 2019, new infections and AIDS-related deaths have dropped significantly. Nationally, HIV occurrence was estimated at 0.05 per 1000 uninfected individuals in 2019,

Table 22.1 Status of the HIV/AIDS Epidemic in 2019. (Adapted from INDIA HIV ESTIMATES 2019, NACO)

Adult (15–49 years) HIV prevalance (%)	**Total**	**0.22 (0.17–0.29)**
	Male	0.24 (0.18–0.22)
	Female	0.20 (0.15–0.26)
Number of people living with HIV (lakh[a])	**Total**	**23.49 (17.95–30.98)**
	Adults (15+ years)	22.70 (17.37–29.97)
	Females (15+ years)	9.94 (7.61–13.00)
	Children (<15 years)	0.79 (0.58–1.05)
HIV incidence per 1000 uninfected population	**Total**	**0.05 (0.03–0.09)**
	Male	0.06 (0.03–0.01)
	Female	0.05 (0.02–0.08)
New HIV infections (thousand)	**Total**	**69.22 (37.03–121.50)**
	Adults (15+ years)	65.24 (34.45–115.51)
	Females (15+ years)	27.58 (14.45–49.53)
	Children (<15 years)	3.98 (2.49–6.11)
Decline in new HIV infections since 2010 (%)	**Total**	**37.4**
	Adults (15+ years)	33.9
	Females (15+ years)	32.3
	Children (<15 years)	66.1
AIDS-related deaths (thousand)	**Total**	**58.95 (33.61–102.16)**
	Adults (15+ years)	55.40 (31.55–96.58)
	Females (15+ years)	16.7 (8.46–31.73)
	Children (<15 years)	3.57 (1.94–5.74)
Decline in AIDS-related deaths since 2010 (%)	**Total**	**66.1**
	Adults (15+ years)	66.1
	Females (15+ years)	73.7
	Children (<15 years)	65.3
PMTCT needs (thousand)	**Total**	**20.52 (14.98–28.13)**

[a]Lakh = 100000

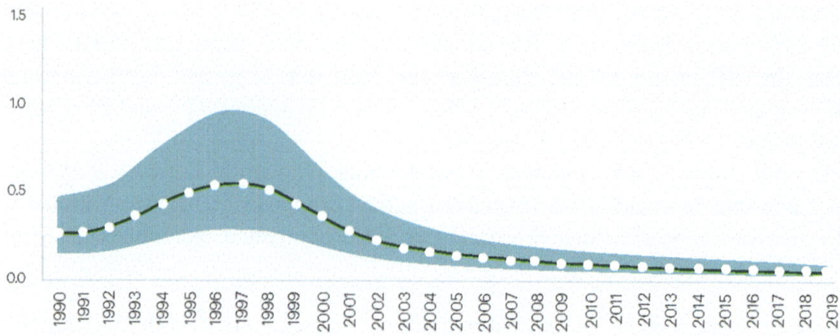

Fig. 22.1 Trend of HIV Incidence in India, 1990–2019 (per 1000 uninfected populations). Adapted from INDIA HIV ESTIMATES 2019, NACO

which has declined from 0.54 in 1995 (year of the maximum event) to 0.05 in 2019 (Fig. 22.1).

22.1.6 HIV-1 Transmission

The risk of HIV-1 transmission from one person to another majorly depends on the viral load in the body fluid and the time of contact with that body fluid. According to epidemiological studies carried out during 1981 and 1982, intimate sexual contact and contaminated blood were the major routes for HIV transmission from one infected person to another healthy person (Jaffe et al. 1983). Earlier, this syndrome was pronounced in homosexual and bisexual men (Gottlieb et al. 1981; Masur et al. 1981; Siegal et al. 1981) and intravenous drug users. Later on, this syndrome was considered a result of heterosexual activity (Harris et al. 1983). Moreover, it became evident that the illness may transfer from blood or blood products. Sexual transmission is now probably the most common route of HIV transmission worldwide. Some other modes of HIV transmission as transfusion of blood from an infected person to a healthy individual, infected mother to child, use of needles by HIV-infected intravenous drug users, also play a significant role in HIV transmission. There was a high risk of Transfusion-associated transmission, but it came down to approximately 1% due to tremendous efforts for regulating blood supplies (Paranjape and Challacombe 2016).

22.1.7 HIV-1 Pathogenesis

HIV is a retrovirus virus that infects $CD4^+$ T helper cells (Klatzmann et al. 1984), DCs (Ayehunie et al. 1997), and macrophages (Gartner et al. 1986) in mucosae. However, $CD4^+$ T cells are the primary target of the HIV-1 virus as they produce a high amount of progeny virion compared to other cell types (Swanstrom and

Coffin 2012). HIV-1 infection starts with the virus's entry inside the host cell, followed by progressive loss of $CD4^+$ T cell count by cell death (Douek et al. 2002). A healthy individual's $CD4^+$ T cell measure generally ranges between 800–1200 cells per cubic millimeter or microliter of blood. Upon HIV-1 infection, this $CD4^+$ T cell count depletes gradually, and at the point when the number of cells becomes less than 200 cells per cubic millimeter, the condition is referred to as AIDS, clinically (Levy 1988). The reduction of $CD4^+$ T cells can occur by multiple processes, including cell death by apoptosis, pyroptosis, and necrosis. The time course starting from HIV-1 infection to development of AIDS condition can be divided into three major clinical stages-

1. **Acute phase** or primary phase—During the initial phase of infection, $CD4^+$ T-cells and memory cells of GALT are the main targets of HIV-1 (Mattapallil et al. 2005). Most of the individuals do not show any symptoms up to 2 weeks in the acute phase. Later on, the infected individual starts showing similar signs to influenza-like fever, headache, acute pharyngitis, and weight loss. Significant weight loss and mucocutaneous ulceration constitute a substantial determinant of acute HIV infection (Cohen et al. 2011). In this period, the blood of infected individuals contains approximately 1 million–10 million virus RNA copies at the peak of viremia. During this phase, a little later, due to the immune response, the average blood viremia reduces to approximately 30,000 copies/ml (Lagathu et al. 2017). Furthermore, these symptoms start to disappear within 3–4 weeks because of the host immune response. During the late acute phase, a sudden rise in $CD4^+$ T-cell number is detected and viral RNA drops to an extent in the plasma.
2. **Chronic phase**—This stage is called a virologic *quasi-steady-state* or *clinical latency* (Ho et al. 1995; Razooky et al. 2015). In this phase, the virus infects various tissues throughout the body. It establishes a systemic infection, including latent reservoirs that may remain indefinitely in the dormant stage. During clinical latency, $CD4^+$ T-cell count decreases slowly or may remain stable for a long time. This period varies from 2 to 20 years, depending on several factors (Ruelas and Greene 2013). During this phase, infected persons encounter many opportunistic infections, including dermatological conditions, neurological complications, and oral and vaginal candidiasis.
3. **Condition of AIDS**—Finally, because of some unknown causes, this latent virus form changes to an actively transcribing form, which leads to the rapid increase in viremia and decline in the number of $CD4^+$ T cells. When the number of $CD4^+$ T cells comes down to below 200 cells/mm^3 of blood, it is termed as AIDS. At this stage, the virus develops the ability to infect and enter the new cell types with low levels of CD4 on their surface, like macrophages (Swanstrom and Coffin 2012). Ultimately, too high viral load and opportunistic infections like PCP, KS, non-Hodgkin's lymphoma, etc., leads to the death of the infected person (Levy 1993). Infections with lentiviruses typically show a chronic disease course, with a long period of clinical latency, persistent viral replication, and involvement of the central nervous system.

22.1.8 Structure of HIV-1 Virion

Both HIV strains, HIV-1 and HIV-2, are morphologically similar, comprising a spherical-sized ball, approximately 100–120 nM in diameter (Kuznetsov et al. 2003; Pang et al. 2014). HIV-1 virion is made up of the envelope and three structural Gag proteins: matrix (MA, p17), capsid (p24), and nucleocapsid (NC, P7). The outermost layer of the HIV-1 particle is composed of the lipid bilayer. This lipid bilayer is derived from the infected host cell during the release of the virion particle. This outermost layer usually comprises 72 trimeric heterodimer complexes of envelope glycoproteins (Aloia et al. 1993; Ganser-Pornillos et al. 2008). This complex is composed of glycoprotein 120 (gp120) and glycoprotein 41 (gp41), which are derived from the cleavage of a 160 kDa precursor glycoprotein, gp160 (McCune et al. 1988). Gp120 is present on the outer surface of the virion particle, while gp41 is a transmembrane protein. Both the proteins covalently interact with each other (Fig. 22.2).

During HIV-1 encounter with the host cell, gp120 interacts with cellular receptors like CD4 receptors and other chemokine co-receptors present on the host cell surface (Chanh et al. 1988). A matrix protein layer is present beneath this lipid bilayer and maintains the integrity of the virion particle (Jun et al. 2011). This matrix protein layer surrounds the core of the virus, which is made up of approximately 3000 units of p24 capsid (CA) proteins. This cone-shaped core comprises two copies of the (positive sense) ssRNA genome, nucleocapsid protein, viral enzymes, and other accessory proteins (Watts et al. 2009). The three viral enzymes, reverse transcriptase (RT), protease (PR), and integrase (IN), are produced from the cleavage of polyprotein precursor (pr166^{GagPol}) encoded by the pol gene. The nucleocapsid protein remains in association with the viral genome. These viral proteins are produced from the catalysis of HIV-1 p55 Gag precursor protein by viral protease enzyme (Gomez and Hope 2005). Some other viral proteins like negative factor (Nef), virion infectivity factor (Vif), and regulatory protein (Vpr) also remain associated with viral core (Camaur and Trono 1996; Pandori et al. 1996). Besides the viral proteins, this core contains several host proteins like HSP70, actin, moesin, emerin, coflin, etc. (Ott et al. 1996; Gurer et al. 2002).

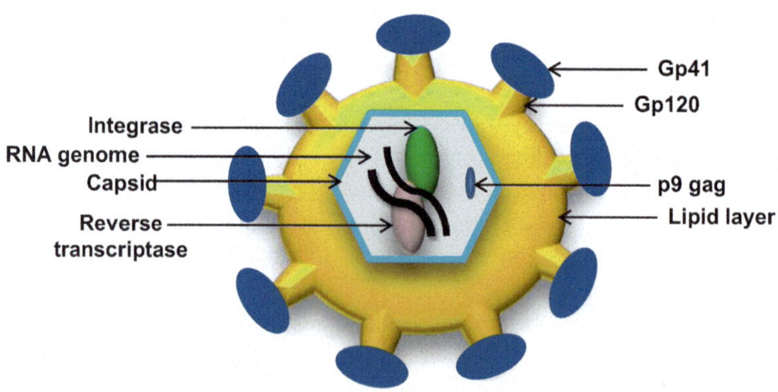

Fig. 22.2 Schematic representation of HIV-1 virion with various viral proteins

22.1.9 Genomic Organization of HIV-1

HIV-1 comprises approximately 9.8 Kb long two single-stranded RNA molecules as the genome. This RNA genome includes long terminal repeats (LTR) at both the ends and nine open reading frames, which encodes for around 15 different structural, regulatory, and accessory proteins. HIV-1 utilizes unique RNA splicing machinery, which gives unspliced (9.2 Kb), single-spliced (4 Kb), and multispliced (2 Kb) RNA copies (Tazi et al. 2010). In the ORF, three genes, *gag, pol,* and *env,* encode three major polyproteins common to all retroviruses. Proteolytic cleavage of Gag polyprotein produces p24 (CA), p17 (MA), p7 (NC), and p6 protein. The Pol polyprotein is cleaved into viral enzymes, reverse transcriptase (RT), protease (PR), and integrase (IN). Env polyprotein is cleaved into the surface (SU or gp120) and transmembrane glycoprotein (TM, gp41). Apart from these structural proteins, the HIV-1 genome encodes four accessory proteins such as Nef, Vpr, Vpu, Vif, and two regulatory proteins such as Tat and Rev (Frankel and Young 1998).

22.1.10 Different HIV Proteins

The morphological maturation of HIV-1 depends on the proteolytic cleavage of the Gag and Gag-Pol polyproteins by the virus-encoded protease enzyme when virion is released from the infected cell.

Different proteins encoded by the viral genome can be categorized into four different types according to their functional role:

1. Structural proteins include matrix protein (MA, p17), core antigen capsid protein (CA, p24), nucleocapsid protein (NC, p7), p6, glycoproteins-gp120 (SU), and gp41 (TM).
2. Functional/enzymatic proteins include reverse transcriptase (RT, p51/p66), integrase (IN, p32), and protease (PR, p10).
3. Regulatory proteins include regulator of viral gene expression (Rev, p19) and transactivator of transcription (Tat, p14).
4. Accessory proteins include virion infectivity factor (Vif, p23), viral protein R (Vpr, p14), viral protein U (Vpu, p16), and negativity factor (Nef, p27).

22.1.10.1 Structural Proteins

In all retroviruses, gag polyproteins are considered major structural proteins (Bell and Lever 2013). Proteolytic cleavage of these gag polyproteins generates other smaller proteins. Following are the major structural proteins:

22.1.10.1.1 Matrix Protein (MA or p17)

MA is a 128 amino acid long (17 kDa) myristoylated protein that plays an essential role in maintaining viral integrity and is involved in the different stages of the virus life cycle (Carroll et al. 2016). This MA protein carries signal sequences for nuclear localization (NLS), nuclear export (NES), and localization to the plasma membrane.

Nuclear export and membrane localization signals help in the export of unspliced viral transcripts near the membrane (Dupont et al. 1999). N-terminal myristoylation of p55 gag protein facilitates its targeting to the site of virus assembly and release of the virion particle, where it plays a crucial role in virion assembly (Hill et al. 1996; Provitera et al. 2006). Gag polyprotein and the cytoplasmic tail of Env (via MA domain of Gag) directly interact with each other, leading to the initiation of virion assembly (Cosson 1996; Gaines et al. 2018). HIV-1 matrix protein forms trimers and associates with plasma membrane by interacting with phosphatidylinositol-4, 5-bis-phosphate (PIP2) and produces polymer forms (Hill et al. 1996; Alfadhli et al. 2009, 2016). MA protein interacts with CA through the C-terminal amino acids (Freed et al. 1997; Checkley et al. 2011). MA protein also binds to the core domain of the integrase (IN) enzyme and directs the import of pre-integration complex (PIC) to the nucleus (Bukrinsky et al. 1993; Gallay et al. 1995; Haffar et al. 2000). It is also involved in the early and late phases of virus replication.

22.1.10.1.2 Capsid Protein (CA or p24)

Capsid (CA) is a hexameric 24 kDa protein, which forms the structural core of the virus. Capsid protein is a highly conserved protein in HIV-1 and HIV-2, making it suitable for diagnostic purposes. During the maturation process, monomer capsid protein unit forms pentamer and hexamer complexes followed by the assembly, thus resulting in a cone-shaped structure which is curved on the top and bottom (Ganser et al. 1999; Li et al. 2000a, b; Ganser-Pornillos et al. 2008; Campbell and Hope 2015). This cone-shaped structure encloses two copies of the ssRNA viral genome and other viral enzymes and accessory proteins.

The capsid protein consists of an N-terminal (150 amino acids) and C-terminal domain (80 amino acids) (Berthet-Colominas et al. 1999; Campbell and Hope 2015). During the formation of hexamer and pentamer complexes, the N-terminal domain faces towards the outside while the C-terminal domain orients towards the interior side of the core (Campbell and Hope 2015). Capsid protein interacts with various cellular factors and regulates several processes like virus replication, virion infectivity, virion assembly, and reverse transcription (Campbell and Hope 2015). It has recently been shown that capsid protein also helps in nuclear import of pre-integration complex and integration of viral genome into host cell genome (Chen et al. 2016).

22.1.10.1.3 Nucleocapsid Protein (NC or p7)

The NC is a small seven kDa protein involved in several essential processes for virus replication. Nucleocapsid protein is primarily engaged in maintaining viral genome conformation and integrity across a range of retrovirus species, making it crucial for virus stability (Spell et al. 2016; Wan et al. 2017; Grunewald et al. 2018). This nucleocapsid protein comprises two zinc finger-like motifs called the CCHC motif consisting of cysteine and histidine residues. This motif promotes the stronger binding of this protein to the viral genome (Wilkinson et al. 2008; Muriaux and Darlix 2010). Mutation in these CCHC motifs leads to the improper packing of the viral genome resulting in the formation of the non-infectious virion (Darlix et al. 1995).

In HIV, NC protein is also involved in the regulation of reverse transcription and maturation of virion particles. It acts as a nucleic acid chaperone and maintains the ssRNA conformation (Levin et al. 2005; Cruceanu et al. 2006; Levin et al. 2010; Wu et al. 2014). This interaction facilitates the dimerization of the ssRNA genome followed by inclusion in budding virion particles (Darlix et al. 1990; Feng et al. 1999; Moore and Hu 2009; Muriaux and Darlix 2010). NC interacts with stem-loop structures (SL2 and SL3) present in the viral RNA genome through intra- and intermolecular interactions and helps in viral genome recognition and packaging during the early stages of virus assembly. It also assists in incorporating the virus genome in the host genome (Carteau et al. 1997; Carteau et al. 1999; Thomas et al. 2006). In recent times, this protein has been targeted for anti-HIV therapeutic purposes because of its conserved nature and involvement at several stages of the virus replication cycle (Mori et al. 2018; Mouhand et al. 2018; Deshmukh et al. 2018).

22.1.10.1.4 p6

Proteolytic cleavage of the C-terminal region of Gag polyprotein produces proline-rich p6 protein. This p6 is a phosphoprotein (Müller et al. 2002). Primarily, this protein helps in the assembly of viral components and release of the virion (Göttlinger 2001). Budding of the virion requires cellular endosomal sorting complexes required for transport (ESCRT) complex and viral components, mainly gag structural proteins (Votteler and Sundquist 2013). This p6 protein comprises two domains, of which one domain interacts with Tsg101 (tumor susceptibility gene 101 protein). In contrast, domain two associates with Alix proteins and facilitates ESCRT complex recruitment (VerPlank et al. 2001; Garrus et al. 2001). The incorporation of Vpr protein in budding virion particles is assisted by p6 protein (Zhu et al. 2004). During the budding of progeny virion, p6 helps in the release of virion particles from the infected cell surface.

22.1.10.1.5 Glycoprotein 120 (SU, gp120)

Envelope polyprotein is synthesized from env ORF having a molecular weight of 160 kDa. This precursor polyprotein is further proteolytically cleaved and produces a heterodimer of gp120 and gp41. Gp120 is 550 amino acids long glycosylated protein present in the outermost lipid bilayer of the virus. Gp120 comprises five variable regions (V1-V5) and five constant regions (C1–C5). Variable regions V1-V4 form a loop exposed to the surface (Willey et al. 1986; Modrow et al. 1987). Gp120 facilitates the virus's attachment to the CD4 receptor present on the surface of the host cell (Kwong et al. 2000). After binding gp120 to the CD4 receptor, conformational changes in gp120 induce its binding affinity to the other chemokine coreceptors (Choe et al. 1996; Deng et al. 1996). Crystal structures of gp120 core have been reported with most variable loops, flexible terminal extensions, and glycans truncated. They have revealed the architecture of the gp120 core in its CD4-bound conformation composed of inner and outer domains along with a 4-stranded β-sheet called the bridging sheet. The outer domain of gp120, which is composed of V1-V2 regions, is mainly associated with the CD4 receptor interaction, while the inner domain composed of V3, V4, and V5 regions, helps in the fusion of the virus with the host cell (Guttman et al. 2012; Liu et al. 2008; Kong et al. 2016). Potential molecules have been designed to restrict the virus's entry into the host cell by targeting gp120-CD4 interaction (Julien et al. 2013; Robinson 2018).

22.1.10.1.6 Glycoprotein 41 (TM, gp41)

The other glycoprotein formed after the cleavage of env polyprotein is gp41. It is a 345 amino acids long transmembrane protein associated with the gp120 and forms a heterodimer complex. Gp41 helps in the fusion of the viral membrane with the host cell membrane. Gp41 comprises three domains such as the extracellular domain, a transmembrane domain, and a cytoplasmic tail. The extracellular domain contains a hydrophobic region called a fusion peptide. This fusion peptide is crucial for membrane fusion (Bosch et al. 1989; Pancera et al. 2010). This fusion peptide includes a polar region, hydrophobic heptad repeats (HR1 and HR2), and tryptophan-rich membrane-proximal external region (MPER) (Chan et al. 1997; Buzon et al. 2010). The binding of gp120 to the CD4 receptor and coreceptor leads to the conformational changes in gp41. These conformational changes facilitate the exposure of fusion peptides, resulting in the virus's invasion into the host cell. This process is known as membrane destabilization and fusion pore formation (Brasseur et al. 1990). MPER is a potential site for neutralizing antibody recognition and is also used for designing entry inhibitors and vaccines (Münch et al. 2007; Pancera et al. 2014; Kong et al. 2016).

22.1.10.2 Enzymatic Proteins

The viral pol gene encodes three necessary viral enzymes, namely, reverse transcriptase (RT), protease (PR), and integrase (IN), in the form of the polyprotein. This polyprotein is cleaved by the viral protease enzyme at the time of virion maturation.

22.1.10.2.1 Reverse Transcriptase (RT)

Reverse transcriptase was discovered by David Baltimore and Howard Temin in 1970 (Coffin and Fan 2016). This enzyme is a unique characteristic of retroviruses that converts the single-stranded RNA (ssRNA) into double-stranded DNA (dsDNA) through reverse transcription (Herschhorn and Hizi 2010; Cimarelli and Darlix 2014). This RT is composed of two subunits, p51 (440 amino acid) and p66 (560 amino acid) (Beilhartz and Götte 2010). The large subunit p66 shows both polymerase and RNaseH activity, while the small subunit p51 maintains the conformation of enzymes or nucleotides (Sarafianos et al. 2009). The polymerase domain is divided into four subdomains such as finger domain, palm domain, thumb domain, and connection domain. The palm subdomain carries a binding site for nucleic acid and a potential target for nucleotide reverse transcriptase inhibitors (NRTIs) (Orkin et al. 2018; Chiang et al. 2018; Rai et al. 2018). The active site also consists of three catalytic carboxylates that interact with Mg^{+2} ions (Goldschmidt et al. 2006). HIV-1 reverse transcriptase enzyme lacks 3′–5′ exonuclease activity, leading to mutation frequency in the range of 10^{-5} to 10^{-3} errors/base-pair/cycle. This inability of proofreading develops the resistance in virus against the drug used for treatment.

22.1.10.2.2 Integrase (IN)

Integration is another crucial process in the life cycle of retroviruses carried out by viral enzyme integrase. The integration of viral genome into host cell genome is essential for establishing the latent reservoir (Symons et al. 2018; Engelman and Singh 2018). The integrated viral genome is called a provirus that sustains for a longer time or until the cell dies. Then, this provirus is transmitted to the next daughter cells with each cell division. Hence, targeting viral stages before integration becomes important to cure HIV. Integrase is a 288-amino acid long 32 kDa enzyme similar to other polynucleotidyl transferases (Dyda et al. 1994; Cherepanov 2010). This enzyme contains the N-terminal HH-CC zinc-binding domain, RNase H fold containing catalytic core domain, and DNA binding C-terminal domain (Chen et al. 2000).

HIV-1 integrase is present in a tetramer form (Yuan et al. 2008). Integrase facilitates the (i) removal of two nucleotides from both the 3′ ends of cDNA, (ii) import of the preintegration complex into the nucleus, and (iii) integration of viral complementary-DNA (cDNA) into host cell genome by transesterification (Lusic and Siliciano 2017). For the catalytic activity of integrase enzyme, Mg^{+2} or Mn^{+2} ions are very important, which interact with carboxylate ions present in the active site (Kulkosky et al. 1992; Hare et al. 2010; Engelman and Cherepanov 2012).

22.1.10.2.3 Protease (PR)

HIV-1 protease, also known as retro pepsin, is produced from the autocleavage of pol polyprotein. Protease is a 99 amino acids long aspartyl protease that acts upon aspartate amino acid in the client protein. It primarily works during the maturation process of progeny virion. This protein is also vital for the virion infectivity (Adamson 2012). Protease is composed of two 99-residue monomer units that contribute a catalytic aspartate (D25) to the active site, located above the dimeric interface and enclosed by a pair of flaps. It functions in a dimer form. During the budding process, this enzyme gets induced through the autocatalytic mechanism. This enzyme cleaves the gag and gag-pol polyproteins to produce infectious virus particles (Kohl et al. 1988; Ikuta et al. 2000). Protease inhibitors block the release of progeny virion on the plasma membrane of infected cells (Eron 2000).

22.1.10.3 Regulatory Proteins

22.1.10.3.1 Rev (Regulator of expression of virion proteins)

Initially, it was considered as a regulator of splicing and antirepression transactivator. Later on, people showed its importance in expressing various viral proteins and named Rev (Malim et al. 1990). Rev is a phosphoprotein that is primarily present in the nucleus. Rev protein comprises four domains. Nuclear localization signals (NLS) or RNA binding domain (RBD) and nuclear inhibitory sequence (NIS) are present at the N-terminal of Rev protein, while C-terminal contains activation domain and nuclear export signal (NES). Rev interacts with the Rev response element (RRE) region, presents in spliced mRNAs for various viral proteins, and regulates their expression (Aligeti et al. 2014). Nuclear export signals present in the

C-terminal activation domain associate with cellular proteins and facilitate the mRNA export from the nucleus (Meyer and Malim 1994; Suhasini and Reddy 2009; Behrens et al. 2017). Rev is essential for the nuclear export of viral RNA. As it has both NLS and NES, it can freely move between cytoplasm and nucleus.

22.1.10.3.2 Tat (Transactivator of Transcription)

Tat is an 86-101 amino acids long, early expressing transactivator protein of HIV-1. It plays a vital role in virus replication by enhancing LTR-mediated viral gene transcription. Tat protein recognizes the transactivation response (TAR) RNA sequence stem-loop structure through an alpha helix located in the core region of the Tat protein. Tat protein enhances the LTR promoter activity tremendously once it is synthesized during the early stage of HIV infection (Karn and Stoltzfus 2012; Ne et al. 2018). In the structure of the Tat protein, there are three domains such as N-terminal activation domain, central TAR binding and NLS containing domain, and C-terminal domain (Kuppuswamy et al. 1989), which are divided into six regions. In the Tat protein, the activation domain is Cys-rich and acidic (Reddy et al. 1992), while the TAR binding region and NLS are Gln-rich and basic (Henderson et al. 2012).

Although this protein is involved at many stages of the viral life cycle, but it mainly regulates the LTR-mediated transcription process. It is also secreted from the infected cells through exosomes, taken up by bystander cells. An uncharacterized protein transduction domain facilitates the Tat protein entry in bystander cells (Schwarze et al. 2000; Campestrini et al. 2018). Tat induces apoptosis in bystander cells by modulating cytokine profiles (Campbell et al. 2004; Rahimian and He 2016). Tat also interacts and is associated with the cyclin-dependent kinase 9 (CDK9)-positive transcription elongation factor b (P-TEFb) complex leading to enhanced LTR promoter activity (Tahirov et al. 2010; Debaisieux et al. 2012).

22.1.10.4 Accessory Proteins

22.1.10.4.1 Nef

HIV-1 Nef is a 27 kDa protein that is expressed during the early stages of HIV-1 infection. Nef was abbreviated for "negative factor" as earlier it was considered to be a negative regulator of the virus. Previous studies identified that Nef protein negatively regulates the HIV-1 transcription and replication (Ahmad and Venkatesan 1988; Niederman et al. 1989) but later, people also reported that Nef positively regulates the virus replication (Terwilliger et al. 1991; de Ronde et al. 1992; Zazopoulos and Haseltine 1993; Spina et al. 1994; Miller et al. 1994; Das and Jameel 2005; Kumar and Mitra 2005; Rosa et al. 2015). Nef is a vastly unorganized protein, which comprises multiple immunogenic functional domains and several cytotoxic T lymphocytes (CTLs)-targeted epitopes.

Functional domains and CTL-epitopes in Nef play a potential role in HIV-1 pathogenesis. Few studies have evaluated the sequence variability of accessory genes such as Nef in different subtypes or strains of HIV-1. Despite the considerable genetic variations, few functional domains/motifs in the Nef gene were relatively conserved among the different subtypes (Jubier-Maurin et al. 1999). The conserved domains/motifs are as follows:

- Myristoylation signal
- Acidic charge region
- An internal methionine for translation initiation
- PxxP repeats
- Protein kinase C phosphorylation site and
- Arginine rich motifs (Geyer et al. 2001)

These motifs also facilitate the interaction of Nef protein with various cellular proteins (Geyer et al. 1999; Arold and Baur 2001). Nef protein is involved in several cellular signaling pathways and regulates multiple processes during the HIV-1 replication and progression.

HIV-1 Nef promotes the downmodulation of the CD4 receptor through endosomal/lysosomal degradation (Garcia and Miller 1991). Nef has also been reported to downmodulate major histocompatibility complex I (MHC-I) (Schwartz et al. 1996). The downregulation of MHC molecules protects the infected cells from cytotoxic T-cells mediated killing (Collins et al. 1998). HIV-1 Nef manipulates the activation of lymphocytes by regulating signaling pathways (Alexander et al. 1997; Simmons et al. 2001). It has been identified that Nef enhances viral replication to more extent in resting T-cells compared to proliferating T-cells (Spina et al. 1994). It has also been well documented that Nef is associated with viral infectivity (Miller et al. 1995; Goldsmith et al. 1995; Aiken and Trono 1995; Tobiume et al. 2001; Khan et al. 2001; Papkalla et al. 2002). Recently, it was reported that Nef facilitates the sequestering of cellular restriction factors, serine Incorporator 3 (SERINC3) and SERINC5, and maintains virion infectivity (Rosa et al. 2015; Usami et al. 2015).

22.1.10.4.2 Vif (Virion Infectivity Factor)

HIV-1 Vif is a 23 kDa protein is majorly associated with progeny virion infectivity. For successful replication and infectious virion production, it abolishes the function of cellular restriction factors. Specific cellular proteins are called restriction factors that attempt to impair viral replication and virus production at certain stages of the viral life cycle. One such cellular factor is APOBEC3G (apolipoprotein B mRNA editing enzyme, catalytic polypeptide-like 3G). It is a cytidine deaminase enzyme of the APOBEC family, which converts cytidine to uracil by removing the amino group from cytidine (Harris and Liddament 2004). APOBEC3G gets incorporated into budding virion (Soros et al. 2007). So, as a result, this APOBEC3G transfers to the new host cell upon infection with APOBEC3G containing virus particles. This APOBEC3G associates with the reverse transcription process of the viral genome and converts its cytidines into uracils. These uracils induce G to A transition in (−) DNA strands leading to several mutations in the viral genome. These mutations cause the failure of the genetic integrity of the virus, followed by its dissemination (Feng et al. 2013; Kim 2015). APOBEC3G reduces progeny virion infectivity by deaminase-dependent and independent pathways. Among the seven APOBEC3 enzymes APOBEC3D, F, G, and H show restriction to HIV-1. Vif also hijacks the ubiquitin-proteasome system. This system is composed of Cul5, EloB, and EloC (Sheehy et al. 2003; Stopak et al. 2003). It also recruits a transcription factor CBFβ (Jäger et al. 2011a, b; Zhang et al. 2011) to facilitate ubiquitination and proteasomal degradation of APOBEC3G. Vif directly interacts with APOBEC3G/F through multiple motifs present in its N-terminal region. The BC-box of Vif binds to EloC (Yu et al. 2003). Also, a zinc-finger motif upstream of BC-box interacts with Cul5 (Luo et al. 2005; Mehle et al. 2006). Downstream of the BC-box, a conserved sequence PPLP (Pro-Pro-Leu-Pro) is found, which has been reported to be essential for the multimerization of Vif (Simon et al. 1999). HIV-1 Vif protein induces ubiquitination and subsequent proteasomal degradation of APOBEC3G protein, which results in infectious virion production (Marin et al. 2003; Sheehy et al. 2003; Jäger et al. 2011a, b).

22.1.10.4.3 Vpr (Viral Protein R)

HIV-1 Vpr is a small 14 kDa protein that plays a vital role at different stages of the virus's life cycle. Vpr is vastly conserved across various isolates of HIV-1 (Hrimech et al. 1999). This protein is expressed during late infection and incorporated with budding virion particles, which indicates its possible role during the early stage of HIV-1 infection. Vpr is composed of different domains, which are described as follows:

- An amphiphilic central core flanked by negatively charged N-terminal and positively charged C-terminal
- Proline-rich residues essential for interaction with cyclophilin A
- Arginine residues required for transduction property of Vpr
- Leucine residues important for oligomerization and interaction with cellular proteins
- Hydrophobic residues responsible for G2/M cell cycle arrest and apoptosis
- Nuclear export signal (NES) and nuclear localization signal (NLS) (Guenzel et al. 2014)

Primarily, Vpr is essential for the nuclear import of the preintegration complex along with the integrase enzyme (Kawano et al. 2018). Vpr is vital for HIV-1 infection progression in nondividing cells, whereas in dividing cells, it promotes G2/M cell cycle arrest, which leads to induction of apoptosis resulting in cell death (Kogan and Rappaport 2011; Yan et al. 2018). Vpr also hijacks the ubiquitin machinery proteins cul1 and cul4 and targets the cellular enzyme uracil N-glycosylase (UNG) for proteasomal degradation (Schröfelbauer et al. 2005). Thus, Vpr seems to regulate various processes in host cells during HIV-1 infection.

22.1.10.4.4 Vpu (Viral Protein U)

Vpu is a 16 kDa phosphoprotein that has a 37 residue long N-terminal transmembrane domain and a 54 residue long C-terminal cytoplasmic domain (Maldarelli et al. 1993). It mainly localizes into and is present in the endoplasmic reticulum,

Golgi body, and endosome. Vpu is involved in and regulates several cellular functions to help the virus propagation in infected cells. The central role of the Vpu is to stimulate the downregulation of the CD4 receptor in virus-infected cells (Magadán and Bonifacino 2012). Vpu interacts with the CD4 receptor and facilitates its degradation through the cullin1-SKP1-F-box ubiquitin ligase complex (Margottin et al. 1998; Magadán et al. 2010). Apart from this, Vpu assists in the release of budding virion particles by restricting the function of tetherin/bone marrow stromal cell antigen 2 (BST2) cellular protein (Neil et al. 2008; Van Damme et al. 2008). Vpu also promotes apoptosis by deregulating the NF-kB (nuclear factor kappa-light-chain-enhancer of activated B cells) pathway (Strebel 2014).

22.1.11 HIV-1 Life Cycle

The HIV-1 life cycle consists of multiple stages, broadly divided into an early and late replication phase. During the early phase, HIV-1 virion particles bind to the host cell surface receptors, which lead to the entry of the virus in the cytoplasm followed by reverse transcription, nuclear import of viral genome and its integration into the host cell genome. The late phase includes the initiation of proviral transcription, assembly of viral proteins with the RNA genome, and release of mature infectious progeny virions.

In highly activated $CD4^+$ T cells, the HIV-1 life cycle persists for just 1–2 days. The viral RT enzyme has a high mutation rate (10^{-5} to 10^{-3} errors/base-pair/cycle). Due to this, different types of viral populations are produced in an individual upon infection. This helps HIV-1 to adapt to its host environment rapidly and develop resistance against antiretroviral drugs or immune responses. Apart from this, the integration of provirus into the host genome supports the virus to hide in long-living cells. Thus, HIV infects and kills $CD4^+$ helper T cells that are crucial for maintaining a functional immune system. However, it is imperative to consider that other cell types can also be infected with a different viral replication period and the infected cell's fate. Macrophages, for example, may produce HIV over several weeks and store infectious virions intracellularly. Although a lot of exciting progress has been made, there is still more to understand the multitude of interactions and progression of HIV infection in its host cell. The stages associated with HIV-1 replication in host cell are as follows:

22.1.11.1 Entry in the Host Cell: Attachment and Fusion

The half-life of cell-free virions is just about 20–30 min in infected individuals. Thus, the virus has to find and infect a new target cell within a brief period. It is well established that along with the primary receptor CD4, CCR5 or CXCR4 chemokine coreceptors are also needed for the HIV entry. These receptors are sufficient to render the cells susceptible to HIV-1 entry and thus decide the viral cell tropism. However, viral attachment is often inefficient and a limiting step for HIV infection due to the low number of env trimers on virion particles and CD4 receptors over the host cell surface. Several receptors, such as polyglycans, lectins, and others, can bind to HIV-1 virions more nonspecific and possibly significantly enhance viral infection rates. They help in the interaction of virion particles with CD4 receptors and coreceptors to allow virion fusion. Furthermore, they may improve the chances of transinfection of susceptible T cells by collecting virus particles at the host cell surface. For example, DCs bind to HIV virions at the site of genital exposure and circulate them to the lymph nodes, where they mediate both transinfection and stimulation of T cells (van Kooyk and Geijtenbeek 2003).

The entry of HIV-1 is a complex multistep process that can be targeted for therapeutic purposes (Didigu and Doms 2012). The infection process starts with the binding of gp120 with the cell surface CD4 receptor. Gp120 contains five reasonably conserved domains (C1–C5) and five variable loops (V1–V5). The variable loops predominantly present on the surface of gp120 and play a vital role in binding with the coreceptor, particularly the V3 loop (Hartley et al. 2005). The CD4-gp120 binding brings conformational changes in the Env trimer that facilitates the binding of gp120 with either the CXCR4 or CCR5 coreceptor. This binding prompts further conformational changes that permit the gp41 transmembrane protein to insert its hydrophobic fusion peptide (hinge region) into the cell membrane. After that, the trimeric gp41 complex forms a six-helix bundle structure (6-HB) (Wyatt et al. 1998; Chen et al. 2005; Zhou et al. 2007), allowing the cellular and viral membranes to come together, which results in virion fusion with the host cell membrane and release of the viral contents into the cytoplasm (Melikyan 2008; Wilen et al. 2012).

22.1.11.2 Uncoating of the Virus Particle

After the virion particle's entry, disassembly of the virion particle occurs, which results in the release of the HIV-1 genome into the cytoplasm. Previous studies suggested that uncoating may occur immediately after viral entry, although capsids may stay intact for several hours, crucial for HIV-1 infection. Furthermore, it seems that uncoating is strongly connected with reverse transcription and complements the transition of the reverse transcription complex (RTCs) to the pre-integration complex (PIC) (Arhel 2010). Upon entering into the host cell, HIV-1 requires to cross another boundary of F-actin underneath the plasma membrane. It is established that after entry, HIV-1 induces the de-polymerization of the actin cytoskeleton by different means. Inhibition of this process using de-polymerization inhibitors traps the virus between the plasma and actin cytoskeleton (Liu et al. 2009; Rahman et al. 2014). After reaching into the host cell cytoplasm, the capsid layer gets disassembled with the help of viral and cellular factors that result in the release of the viral genome into the cytoplasm. Previously, it was considered that un-coating and entry take place simultaneously. Later, it was identified that uncoating occurs only after the beginning of reverse transcription. If early uncoating happens, it leads to abortive replication (Cosnefroy et al. 2016; Mamede et al. 2017).

22.1.11.3 Reverse Transcription (RT)

Reverse transcription is a characteristic and conserved phenomenon of the *Retroviridae* family. It has been a significant target for antiretroviral therapy (Hughes 2015; Opp et al. 2017). Various studies have reported that this process starts just after the entry and before the un-coating process. The polymerase and RNase H activity of RT facilitate the conversion of RNA into double-stranded linear DNA. This process occurs in the cytoplasm of the infected cell. Following are the steps involved in the reverse transcription process-

1. The reverse transcription process starts with the annealing of the 3′ prime end of the host Lys_3-tRNA to the primer binding site (PBS) present at the 5′ end of the positive sense ssRNA viral genome.
2. The annealing allows the RT to reverse transcribe a copy of minus-strand DNA, resulting in the formation of an RNA/DNA hybrid complex.
3. This hybrid complex serves as a substrate for the RNase H enzyme, which degrades the RNA strand.
4. The viral genome consists of repetitive sequences in the 'R' region of both 3′ and 5′ ends, allowing the newly synthesized DNA to hybridize at the 3′ end of the viral genome. This process is termed as first jump or minus-strand transfer, which is accompanied by minus-strand DNA synthesis.
5. RNase H continues to degrade RNA strand, but a purine-rich region at the 3′ end of the viral genome, also called polypurine tract, is resistant to RNase activity and therefore acts as a primer for the synthesis of the second plus-strand DNA.
6. Synthesis of this plus-strand DNA continues till the first 18 nucleotides of t-RNA are copied, and after this, RNase H removes the complete t-RNA except for a ribo-A at the 3′ end of viral minus-strand DNA. This leads to the second jump, which is also called the plus-strand transfer.
7. After this second jump, both plus and minus strands continue to extend and form a double-stranded DNA containing the same sequence at both ends, known as LTR repeats.

This double-stranded DNA is inserted into the host cell genome. It serves as a template for generating new viral genome copies (both spliced and un-spliced) (Götte et al. 1999; Sarafianos et al. 2009; Hu and Hughes 2012). Although RT is efficient in conducting the reverse transcription process in vitro, various cellular factors are required within the host cell for the successful completion of this process (Warren et al. 2009).

22.1.11.4 Formation and Nuclear Import of PIC

Nuclear import is the utmost complex process because of the association of several cellular and viral proteins. Reverse transcription of viral genomic RNA produces a newly synthesized copy of cDNA. This cDNA interacts with viral and cellular proteins and forms a pre-integration complex (PIC). This PIC is imported into the nucleus through nuclear pore complexes (NPCs), integrating into the host cell genome, and forms provirus. These NPCs channels are made up of nucleoporin proteins, which permit the transfer of water-soluble molecules across the nuclear membrane. This PIC actively reaches the nucleus by utilizing host machinery for nuclear import (Bukrinsky et al. 1992; Lewis et al. 1992; Bukrinsky and Haffar 1998; Vodicka 2001; de Noronha et al. 2001; Greene and Peterlin 2002).

Many studies have demonstrated the involvement of viral proteins such as capsid protein (Yamashita and Emerman 2004), matrix protein (Bukrinsky et al. 1993), Vpr (Heinzinger et al. 1994; Di Marzio et al. 1995; Yao et al. 1995; Subbramanian et al. 1998; Jenkins et al. 1998), integrase(Gallay et al. 1997; Depienne et al. 2001; Ao et al. 2005; Hearps and Jans 2006; Ao et al. 2007; Jayappa et al. 2011) and central DNA flap (Zennou et al. 2000; Dvorin et al. 2002; Ao et al. 2004, Arhel et al. 2006) in nuclear import. In addition to this, cellular factors like importins, nucleoporins (NUP358 and NUP153), CPSF6, and TNP03 also facilitate the import of PIC into the nucleus (Lee et al. 2010; Ao et al. 2010; Matreyek and Engelman 2011). However, recent studies have demonstrated that the intact viral core (or nearly intact) enters into the nucleus with the help of Cleavage and polyadenylation specificity factor subunit 6 (CPSF6), a subunit of cleavage factor) and uncoats before integration near the site of integration (Burdick and Li 2020; Selyutina et al. 2020), which needs to be validated further.

22.1.11.5 Integration of Viral Genome

Integration of the viral genome into the host cell genome is another vital event for completing the successful life cycle in the host cell. The viral RNA genome is converted into dsDNA by reverse transcription and forms a complex with viral and cellular proteins called a preintegration complex. This PIC complex is further transported into the nucleus and integrated into the host cell genome with the viral integrase enzyme. The process of integration is completed in the following steps:

1. First, the integrase enzyme removes two nucleotides from both the 3′′ ends of double-stranded blunt-ended viral DNA, leading to sticky ends.
2. The region of insertion in the host cellular genome contains a weakly conserved palindromic sequence. These sticky ends target the pair of phosphodiester bonds situated 5 to 6 nucleotides apart in host genomic DNA.
3. Then, the integrase dimer mediates the strand transfer reaction by transferring the viral DNA with sticky ends at 3′ ends into the host cellular genome by trans-esterification reactions. This strand transfer produces "gaps" at both viral and host DNA junctions because of unpaired DNA and activates the host cellular "gap repair process."
4. Then the integrase cleaves two nucleotides from the 5′ end of viral dsDNA. The "gap repair process" comprising polymerase, nuclease and ligase, fills the gap and joins the viral DNA with host cellular DNA.
5. This integrated viral genome is called provirus, which further transcribes and produces spliced and unspliced copies of viral mRNA (Craigie and Bushman 2012).

According to the studies performed, transcriptionally active genes/sites in the host genome are the most preferred region for the integration of the viral dsDNA genome (Schröder et al. 2002; Wang et al. 2007). Recent studies have claimed that this integration occurs in the nucleus's outer region near the nuclear pore, containing several cellular genes (Marini et al. 2015; Lusic and Siliciano 2017).

22.1.11.6 Transcription

Within a short time after entry into the host cell, the HIV-1 ssRNA genome is converted into the dsDNA and transported into the nucleus. It is then integrated into the host cell genome called as provirus. Furthermore, transcription of this virus relies on the host RNA pol II (Kingsman and Kingsman 1996; Li et al. 1997a, b). The RNA Pol II-mediated transcription of provirus includes various steps like preinitiation, initiation, promoter clearance, elongation, and termination (Fuda et al. 2009). HIV-1 provirus contains two repetitive sequences on both ends called LTR. This LTR promoter comprises approximately 636 base pairs. Based on the studies performed for the HIV-1 genome, this LTR region can be divided into the following regions-

1. **Core promoter**
   The core promoter is the minimal basal region of the LTR promoter required for provirus transcription. This core promoter comprises a TATAA element, an initiator element, and three tandem Sp1 binding sites together with the TAR element. The TATAA element serves as the binding site for the common transcription factor TFIID. This TFIID is composed of TATA-binding protein (TBP) and its associated factors (Verrijzer and Tjian 1996). The binding of TBP induces pre-initiation complex and activation of provirus transcription, which is a rate-limiting step for the transcription process (Pendergrast et al. 1996; Majello et al. 1998). TBP also associates with viral encoded Tat (Kashanchi et al. 1994; Veschambre et al. 1995; Kashanchi et al. 1996) and facilitates the transactivation of the HIV-1 LTR (Majello et al. 1998). The transcription initiation site is determined by the initiator elements present to the downstream of the TATAA element (Margolis et al. 1993; Zenzie-Gregory et al. 1993; Rittner et al. 1995; Wainberg et al. 1997). Upstream to the TATAA element, three binding sites are present for the Sp1 universal transcription factor (Jones et al. 1986). Sp1 interacts with the binding site through the zinc finger motif and induces transcription (Gill et al. 1994). Sp1 also contributes to the Tat-mediated transactivation of the LTR promoter (Jeang et al. 1993; Williams et al. 1996). It seems that the three elements, TFIID, Sp1, and Tat, are crucial for maximum virus transcription. There are some other elements that bind to the Sp1 binding site and modulate the transcription efficiency of the LTR promoter, like BTEB, Sp4, and Sp3 (Imataka et al. 1993).
2. **Enhancer element**
   NF-κb is a ubiquitous transcription factor that regulates the transcription of many genes in the host. There are two NF-κb binding sites in the LTR promoter present just upstream of the Sp1 binding sites, which assist in the NF-κb transcription factor-mediated increased activity of LTR promoter (Kretzschmar et al. 1992). There are several members in the NF-kb family-like NF-κB1, RelA, c-Rel, NF-κB2, and Re1B. They can bind to the enhancer region in either a homo- or heterodimer form, and also, their particular combinations are important for Tat-mediated *trans*-activation (Liu et al. 1992). The binding of the NF-κb transcription factor to the enhancer site leads to the increased activity of HIV-1 LTR promoter-mediated gene expression in infected cells (Mellors

et al. 1991; Krasnow et al. 1991). HV-1 Tat protein interacts with the NF-κb transcription factor and enhances the promoter activity (Biswas et al. 1995; Conant et al. 1996; Demarchi 1996). The action of this enhancer region mainly depends on the cooperative protein-protein interactions. NF-κb interacts with Sp1 (Perkins et al. 1993) and TFIID (Schmitz et al. 1995) transcription factors and enhances virus transcription. NFkB and SP1 enhancer elements are also crucial in the LTR region for TAR-independent transactivation by Tat protein (Yang et al. 1997). This TAR-independent transactivation by Tat protein is reported to occur through the direct binding of Tat protein onto the NF-kb enhancer region. This finding explains the TAR-independent transactivation and shows the possible role of HIV-1 Tat in modulating cellular genes (Dandekar et al. 2004).

Some other cellular factors such as proinflammatory cytokines, tumor necrosis factor-alpha (Duh et al. 1989; Israël et al. 1989; Mellors et al. 1991; Atwood et al. 1994; Granowitz et al. 1995; Fontes et al. 1997), and mitogenslike phorbol esters (Dinter et al. 1987) induce the promoter activity. Some other transcription factors like Ets-1 (Bassuk et al. 1997) and E2F-1 (Kundu et al. 1995) also interact with enhancer site and increase or decrease the transcription activity, respectively.

3. **Upstream and downstream regulatory elements**
Apart from the initiator and enhancer regions in LTR, downstream of initiator and upstream of enhancer region comprises binding sites for several transcription factors that control the virus transcription (Gaynor 1992; Al-Harthi and Roebuck 1998). The upstream region was earlier considered to be involved in the negative regulation of HIV-1 expression (Lu et al. 1990; West et al. 1992). Later on, it was identified that it could facilitate upregulation and downregulation according to cell types and phases. C/EBP and AP-1 binding sites in upstream regions regulate HIV-1 expression in some particular cell types but not in others (Canonne-Hergaux et al. 1995; Henderson and Calame 1997). In lymphocytes, COUP, LEF-1, Ets, and NF-AT binding sites play an important role in regulating the HIV-1 expression (Jones and Peterlin 1994). Recently several studies have shown that sequences present downstream of the 5′-untranslated region (5′UTR) of the HIV-1 genome are crucial for HIV-1 gene expression (Roebuck et al. 1993; Li et al. 1997a, b; Van Lint et al. 1997; Liang et al. 1997). The downstream region of HIV-1 LTR comprises multiple essential transcription factor binding sites (TFBSs) that includes activator protein-1 (AP-1) motif, AP-3-like (AP-3L) motif, downstream sequence element (DSE) sites, C/EBP/NFAT (nuclear factor for activated T cells) downstream binding site (DS3), downstream binding factor (DBF-1), and two Sp1 sites in the U5 and gag leader sequence (GLS) region (Rabbi et al. 1997; Pereira et al. 2000; Romanchikova et al. 2003). In different HIV-1 isolates or subtypes, some of these sites are well conserved, demonstrating that they play an important role in HIV-1 transcription and replication (el Kharroubi and Verdin 1994; Dahiya et al. 2014). Thus, along with viral proteins, a large number of cellular factors are associated with successful viral transcription and also potential targets for anti-viral therapeutics.

22.1.11.7 Translation

RNA pol II transcribes the provirus and produces spliced and unspliced copies of mRNA transported to the cytoplasm with the help of nuclear export machinery to synthesize viral proteins. During the early phase of infection, multiple spliced viral mRNAs are translated to produce Nef, Tat, and Rev proteins. Tat and Rev proteins are localized into the nucleus, where Tat induces the LTR mediated transcription rate (Coiras et al. 2010). During the late phase of infection, singly-spliced and unspliced copies of viral mRNA are exported to the cytoplasm for translation with the help of viral Rev Protein, RRE element, and other cellular factors (Pollard and Malim 1998; McLaren et al. 2008; Suhasini and Reddy 2009). Rev is essential for the export of transcripts from the nucleus to the cytoplasm. Translation of these spliced and unspliced viral mRNAs occurs through different mechanisms such as:

1. **Cap-dependent scanning**
   In the host cell, the initiation of HIV-1 translation occurs majorly through the cap-dependent scanning mechanism. In this mechanism, the first translation initiation factor eIF4E interacts with the 7-methyl-guanosine cap of mRNA and forms a complex with eIF4G and eIF4A (de Breyne et al. 2012). Then, the 40S ribosomal subunit associated with eIF3, eIF2, GTP, and Met-tRNA interacts with the capped mRNA and scans it until an initiator AUG codon is found (Kozak 1989; Bolinger and Boris-Lawrie 2009; de Breyne et al. 2013). HIV-1 mRNA translation initiation is slowed down by the highly structured sequence in the 5′ un-translated region (UTR) (Svitkin et al. 1994; Dorin et al. 2003; Dugré-Brisson et al. 2005; Soto-Rifo et al. 2012). All HIV transcripts begin with the generation of the trans-activation response element (TAR) RNA at their 5′ end, which is composed of a stable stem-bulge-loop structure. However, there are some other mechanisms by which HIV translation initiation also occurs, such as internal ribosome entry site (IRES) mediated and frame-shifting.
2. **IRES-mediated**
   IRESes have been found before or within the Gag region and stimulate the translation of viral transcripts expressing Gag and Gag-Pol proteins. These IRESes induce the synthesis of shorter, N-truncated variants of Gag (Nicholson et al. 2006). IRES initially binds to the 40S subunit of ribosome and eIF3, which leads to the activation of eIF4A followed by recruitment of three initiation complexes on a single molecule to assist the formation of full-length and N-truncated Gag molecules (Locker et al. 2011).
3. **Frame-shifting**
   HIV-1 Gag-pol polyprotein is produced by ribosomal-1 frame-shifting mediated translation mechanism (Wilson et al. 1988; de Breyne and Ohlmann 2018). In this mechanism, hepta-nucleotide present in the upstream of frame-shift stimulatory signal interacts with the ribosome and trigger the incomplete movement of two nucleotides instead of three (Namy et al. 2006; Marcheschi et al. 2007; Mazauric et al. 2009; Brakier-Gingras et al. 2012). This process is modulated by viral protein Tat and cellular factor DEAD-box RNA helicase 17 (Gendron et al. 2008; Charbonneau et al. 2012).

4. **Translation of spliced HIV-1 mRNAs**
   HIV-1 full-length genomic RNA is translated by either cap-dependent or IRES-mediated or frame-shifting mechanisms, but spliced mRNAs may use a different mechanism. Single-spliced *vpu* and *env* gene coding mRNA is translated through a continuous or discontinuous ribosome scanning process, respectively (Anderson et al. 2007). Tat encoding double-spliced mRNA is generally translated by cap-dependent and IRES-mediated processes (Charnay et al. 2009). These different mechanisms drive the successful translation of HIV-1 spliced and unspliced mRNAs with the help of several cellular factors and viral proteins.

22.1.11.8 Assembly, Release, and Maturation

After the translation of spliced and unspliced copies of viral mRNAs, different viral proteins are produced. These newly synthesized viral proteins, along with the virus RNA genome, translocate to the inner surface of the plasma membrane, where they get assembled to produce new progeny virion particles (Meng and Lever 2013; Barré-Sinoussi et al. 2013). Env glycoprotein reaches near to the cell membrane enclosed within vacuoles through the secretory pathway. Gag polyprotein starts multimerizing upon interaction with viral RNA genome and reaches the cell membrane. Near to cell surface, gag polyprotein interacts with phospholipid phosphatidylinositol-(4, 5)-bisphosphate and inserts its N-terminal myristate residue into lipid raft and get anchored to the cell membrane. Env protein incorporates into the assembled particles and facilitates the inclusion of the ESCRT-1 complex by promoting interaction between ESCRT-1 complex subunit that is tumor susceptibility gene 101(TSG101) and PTAP motif of p6 protein. With the budding process, the ESCRT-III and vacuolar protein sorting 4 (VPS4) complexes also get involved and drive the membrane scission reaction, resulting in progeny virion release. These newly released virions are immature which undergo a maturation process that involves proteolytic cleavage of viral polyproteins. Ultimately, produced mature virion particles are ready to infect new host cells (Freed 2015).

22.1.12 Current HIV Therapeutics

Since the HIV epidemic started in 1981, approximately 35 million people have died because of AIDS, but there is still no drug available to cure HIV completely. The development of strategies to cure HIV infection is the most crucial challenge at present. However, since the emergence of HIV, researchers have tried to combat the virus progression by different means. Currently, the medications which are used to treat HIV-1 infected individuals target the key steps of the virus life cycle such as attachment of the virus with the cell, entry of the virus inside the host cell, reverse transcription of the viral genome, nuclear import of the PIC, integration of the viral genome into the host cell genome and assembly and maturation and release of the virus (De Clercq 2009). Based on the targeting step, different inhibitors are available as entry inhibitors, protease inhibitors, integrase inhibitors, and reverse transcriptase inhibitors. A combination of these different inhibitor groups is used in

HIV infection treatment called highly active antiretroviral therapy (HAART) (Table 22.2).

This therapy is more effective as different inhibitors target different stages of the virus life cycle simultaneously and reduce the chances of the emergence of new drug-resistant viral strains.

However, this antiviral therapy has some drawbacks such as:

- The patient has to take therapy throughout the life
- Chances of the emergence of new drug-resistant strains upon discontinuing of the therapy
- Toxicity

Apart from this, another major failure of these inhibitors is that they target only actively replicating cells without affecting latent virus. In resting memory $CD4^+$ T-cells, the virus remains dormant and is not recognized by the immune system (Adams et al. 1994; Siliciano et al. 2003). Once these latently infected memory cells are induced, they produce infectious virus particles (Chun et al. 1997a, b). Researchers have investigated the possible way to affect the latent virus using two sterilizing and functional cures (Battistini and Sgarbanti 2014). Sterilizing strategy includes the destruction of the provirus containing cells, while functional approach includes restricting the reactivation of provirus.

The development of an anti-HIV-1 vaccine is suggested as another option. The first vaccine was developed in the early 2000s. But the efficacy of vaccination was significantly low compared to classic anti-HIV drugs (Autran et al. 2004; Carcelain and Autran 2013). Currently, broad-spectrum neutralizing antibodies are undergoing clinical trials, and the outcomes of initial studies indicate that neutralizing antibodies may become promising anti-HIV therapeutic vaccines (Caskey et al. 2015; Shingai et al. 2014).

Table 22.2 Combination of drugs used (cocktails) for the treatment of an HIV infection. [Adapted from Prokofjeva et al. (2016)]

Combinaton	Trade name	FDA approval
Lamivudine/Zidovudine (3TC/ZDV)	Combivir	27/9/1997
Abacavir/Lamivudine/Zidovudine (ABC/3TC/ZDV)	Trizivir	14/11/2000
Abacavir/Lamivudine (ABC/3TC)	Epzicom	2/8/2004
Emtricitabbine/Tenofovir (FT/TDF)	Truvada	2/8/2004
Efavirenz/Emtricitabine/Tenofovir (EFV/FTC/TDF)	Atripla	12/6/2006
Emtricitabine/Rilpivirine/Tenofovir (FTC/RPV/TDF)	Complera	10/8/2011
Elvitegravir/Cobisistat/Emtricitabine/Renofovir (QUAD/EVG/COBI/FTC/TDF)	Stribild	27/8/2012
Abacavir/Doultegravir/Lamivudine (ABC/DTG/3TC)	Triumeq	22/8/2014
Atazanavir/Cobicistat (ATV/COBI)	Evitaz	29/1/2015
Darunavir/Cobicistat (DRV/COBI)	Prezcobix	29/1/15
Elvitegravir/Cobicistat/Emtricitabine/Tenofovir/Alafenamie (EVG/COBI/FTC/TAF)	Genvoya	5/11/2015

22.1.13 HIV-1 Latency

Once the virus genome is integrated into the host cell genome, it may go for the LTR-mediated gene transcription followed by virion production or latency. Viral latency is a transcriptionally silent reversible state in which infected cells produce a low number of viral particles. Still, upon activation, it may reverse to the transcriptionally active state (Pace et al. 2011). Like Herpes viruses, some viruses utilize this latency mechanism to persist and escape from the immune system (Perng and Jones 2010). In retroviruses, the integration of the viral genome into the host cell genome is an essential step to complete the life cycle and virus production. Some of the retroviruses cause latency. In HIV-1 infection, the virus incorporates its genome into the host cell genome and completes the life cycle within days (Perelson et al. 1997). However, some infected cells enter the latent state, which can get activated only upon stimulation (Siliciano et al. 2003; Siliciano and Greene 2011). Initially, latency was considered as a long asymptomatic period between the infection and condition of AIDS. For HIV-1, the term "latency" was initially used in the clinical sense to describe the long asymptomatic period between initial infection and the development of AIDS. But later on, the involvement of RT-PCR-based assays cleared that even in the asymptomatic phase, there is a certain nondetectable viral load because of continuous replication of the virus (Piatak et al. 1993). Viral latency occurs in two forms such as preintegration latency and postintegration latency. Preintegration latency is described by an unintegrated virus genome present in the cytoplasm of the host cell. This unintegrated genome remains in the cytoplasm for days and forms PIC, which is ultimately destroyed with time. Preintegration latency happens due to the slow metabolic condition of the host cell.

During postintegration latency, PIC is imported into the nucleus and integrated with the host cell genome, but it remains in a transcriptionally inactive state. Such condition was first detected in patients infected with HIV and receiving HAART, primarily in memory $CD4^{+}$ T lymphocytes (Chun et al. 1997a, b; Chomont et al. 2009). Several factors regulate the HIV-1 latency by different mechanisms. It has been believed that the conversion of activated $CD4^{+}$ T-cells into resting memory T-cells is the major cause of latency (Siliciano and Greene 2011). Resting memory T-cells do not allow or are nonpermissive for viral gene expression. HIV latency can occur at both transcriptional and posttranscriptional levels. However, silencing at the transcriptional level majorly contributes to the latency. Several factors can cause transcriptional latency such as:

- Site of integration in the host genome
- The positioning of the provirus in the nucleus
- Chromatin configuration of LTR promoter
- Lack of transcriptional inducer
- Involvement of transcriptional repressors
- Less expression of viral Tat protein
- Epigenetic modulations of LTR promoter like acetylation or methylation of histone proteins (Van Lint et al. 2013)

Posttranscriptional latency also includes the inhibition of nuclear RNA export (Van Lint et al. 2013) and miRNA-mediated transcriptional regulation (Ruelas et al. 2015). These latently infected cells escape from both immune cells and antiviral drugs. These cells do not actively replicate, which helps them to hide from the antiviral drugs (Ruelas and Greene 2013). But it was observed that viruses start emerging from these latently infected cells upon interruption of antiretroviral therapy (ART) (Cohn et al. 2020). Thus, the current studies focus on the "shock and kill" strategy, a type of sterilizing cure.

In this approach, latently infected cells are activated by using various stimulator molecules, facilitating their recognition upon activation by immune cells, thereby leading to cell death. These stimulator molecules include PKC activators (PMA or prostratin) (Laughlin et al. 1993), cytokines (TNFα, IL-1β, and IL-7) (Folks et al. 1987; Popik and Pitha 1993; Managlia et al. 2006), T-cell receptor (TCR) and coreceptor stimulators (α-CD3 and α-CD28 antibodies) (Prins et al. 1999), histone deacetylase (HDAC) inhibitors (Lehrman et al. 2005) and methylation inhibitors (Kauder et al. 2009). These simulators help in unbinding chromatin and increased availability of transcription factors like TATA-binding protein (TBP), TBP associated factors (TAFs), NF-κB, NFAT, AP1, SP1, and many others to the LTR promoter, which leads the favorable conditions for active promoter transcription. A balance between HATs (histone acetyltransferases) and HDACs is also vital for the dynamic transcription process (Taube and Peterlin 2013).

22.1.14 Host Factors-Mediated Regulation of HIV-1 Life Cycle

HIV-1/AIDS progression can be prevented only but not cured despite recent discoveries of novel therapeutics. Currently, an effective HAART therapy is prescribed to the infected individuals that target different viral enzymes and reduces the viral load below the detectable level. Despite the availability of potential antiviral regimens, virus latency and emergence of drug-resistant strains due to low fidelity of viral RT enzyme are two major hurdles. Instead of targeting viral proteins, developing medicines by targeting host cellular factors required for HIV replication or infection progression may provide a higher barrier to resistance development (Flexner 2007). HIV-1 genome is approximately 9.8 Kb long, which encodes around 15 proteins only (Frankel and Young 1998). Thereby, HIV-1 utilizes cellular machinery or cellular proteins to complete various steps like entry, uncoating, nuclear import, integration, transcription, protein synthesis and folding, assembly of viral proteins, budding followed by release from the host cell surface (Goff 2007; Lama and Planelles 2007).

HIV-1 enters the host cell with the help of cellular protein galectin1, gp340, and DC-specific intercellular adhesion molecule-3-grabbing nonintegrin (DC-SIGN), which facilitate binding to the cell surface receptors CD4, CCR5, and CXCR4. Cyclophilin A helps in the uncoating of the virus upon entry. After this, many cellular factors like the transfer ribonucleic acid (tRNA) (Lys_3) assist in the reverse transcription process. Actin, importins, karyopherins, nucleoporins, and some other proteins play a crucial role in the nuclear transport of PIC (Popov et al. 1998; Bukrinskaya et al. 1998; Woodward et al. 2009). Then Lens epithelial-derived growth factor (LEDGF/p75), barrier to-auto-integration factor (BAF), lamina-associated polypeptide two alpha (LAP2alpha), and some others promote the integration of viral genome into the host genome. After integration, the viral genome transcribes and produces several copies of unspliced and spliced mRNA, which leads to the production of new virion particles. Various cellular transcription factors like NF-kb, SP1, AP1, and many more facilitates this process. During the translation process, these spliced mRNA copies encode different viral proteins, which assemble with un-spliced mRNA near the plasma membrane. Cellular ubiquitin and endosomal sorting complexes (ESCRT) help in the viral assembly (Martin-Serrano et al. 2003). Infectivity of budding progeny virion is maintained by annexin-2 (Harrist et al. 2009). Thus, HIV-1 vastly depends on many cellular factors termed host dependency factors to complete its life cycle. Therefore, the identification of such host dependency factors is crucial for therapeutics.

In 2008, probably the first genome-wide study was carried out in which they identified more than 250 HIV-dependency factors (HDFs) (Brass et al. 2008). Later, few more genomic scale studies have been performed by using different techniques. These studies include siRNA- and shRNA-based screening (König et al. 2008; Zhou et al. 2008; Yeung et al. 2009; Landi et al. 2014), protein-protein interactions (Gautier et al. 2009; Zhang et al. 2010; Jäger et al. 2011a, b; Kramer et al. 2012), bioinformatics analysis with patients' samples (Börner et al. 2010; Peretz et al. 2012; Wu et al. 2013; Luque et al. 2014), clustered regularly interspaced short palindromic repeats-CRISPR-associated protein 9 (CRISPER-Cas9) based genetic screen (Park et al. 2017) and metaanalysis of genome-wide studies (Bushman et al. 2009). These reports have revealed the involvement and importance of more than a thousand cellular proteins in HIV-1 replication and infection progression. Many studies have been done to identify and understand the role of various cellular proteins affecting virus replication and infection progression. But still, it is needed to characterize other HDFs which are not studied in detail.

References

Adams M, Sharmeen L, Kimpton J, Romeo JM, Garcia JV, Peterlin BM, Groudine M, Emerman M (1994) Cellular latency in human immunodeficiency virus-infected individuals with high CD4 levels can be detected by the presence of promoter-proximal transcripts. Proc Natl Acad Sci U S A 91(9):3862–3866. https://doi.org/10.1073/pnas.91.9.3862

Adamson CS (2012) Protease-mediated maturation of HIV: inhibitors of protease and the maturation process. Mol Biol Int 2012:604261. https://doi.org/10.1155/2012/604261

Ahmad N, Venkatesan S (1988) Nef protein of HIV-1 is a transcriptional repressor of HIV-1 LTR. Science 241(4872):1481–1485. https://doi.org/10.1126/science.3262235

Aiken C, Trono D (1995) Nef stimulates human immunodeficiency virus type 1 proviral DNA synthesis. J Virol 69(8):5048–5056. https://doi.org/10.1128/jvi.69.8.5048-5056.1995

Alexander L, Du Z, Rosenzweig M, Jung JU, Desrosiers RC (1997) A role for natural simian immunodeficiency virus and human immunodeficiency virus type 1 nef alleles in lymphocyte activation. J Virol 71(8):6094–6099. https://doi.org/10.1128/jvi.71.8.6094-6099.1997

Alfadhli A, Barklis RL, Barklis E (2009) HIV-1 matrix organizes as a hexamer of trimers on membranes containing phosphatidylinositol-(4,5)-bisphosphate. Virology 387(2):466–472. https://doi.org/10.1016/j.virol.2009.02.048

Alfadhli A, Mack A, Ritchie C, Cylinder I, Harper L, Tedbury PR, Freed EO, Barklis E (2016) Trimer enhancement mutation effects on HIV-1 matrix protein binding activities. J Virol 90(12):5657–5664. https://doi.org/10.1128/jvi.00509-16

Al-Harthi L, Roebuck KA (1998) Human immunodeficiency virus type-1 transcription: role of the 5′-untranslated leader region (review). Int J Mol Med 1(5):875–881. https://doi.org/10.3892/ijmm.1.5.875

Aligeti M, Behrens RT, Pocock GM, Schindelin J, Dietz C, Eliceiri KW, Swanson CM, Malim MH, Ahlquist P, Sherer NM (2014) Cooperativity among Rev-associated nuclear export signals regulates HIV-1 gene expression and is a determinant of virus species tropism. J Virol 88(24):14207–14221. https://doi.org/10.1128/jvi.01897-14

Allers K, Hütter G, Hofmann J, Loddenkemper C, Rieger K, Thiel E, Schneider T (2011) Evidence for the cure of HIV infection by CCR5Δ32/Δ32 stem cell transplantation. Blood 117(10):2791–2799. https://doi.org/10.1182/blood-2010-09-309591

Aloia RC, Tian H, Jensen FC (1993) Lipid composition and fluidity of the human immunodeficiency virus envelope and host cell plasma membranes. Proc Natl Acad Sci U S A 90(11):5181–5185. https://doi.org/10.1073/pnas.90.11.5181

Anderson JL, Johnson AT, Howard JL, Purcell DF (2007) Both linear and discontinuous ribosome scanning are used for translation initiation from bicistronic human immunodeficiency virus type 1 env mRNAs. J Virol 81(9):4664–4676. https://doi.org/10.1128/jvi.01028-06

Ao Z, Yao X, Cohen EA (2004) Assessment of the role of the central DNA flap in human immunodeficiency virus type 1 replication by using a single-cycle replication system. J Virol 78(6):3170–3177. https://doi.org/10.1128/jvi.78.6.3170-3177.2004

Ao Z, Fowke KR, Cohen EA, Yao X (2005) Contribution of the C-terminal tri-lysine regions of human immunodeficiency virus type 1 integrase for efficient reverse transcription and viral DNA nuclear import. Retrovirology 2:62. https://doi.org/10.1186/1742-4690-2-62

Ao Z, Huang G, Yao H, Xu Z, Labine M, Cochrane AW, Yao X (2007) Interaction of human immunodeficiency virus type 1 integrase with cellular nuclear import receptor importin 7 and its impact on viral replication. J Biol Chem 282(18):13456–13467. https://doi.org/10.1074/jbc.M610546200

Ao Z, Danappa Jayappa K, Wang B, Zheng Y, Kung S, Rassart E, Depping R, Kohler M, Cohen EA, Yao X (2010) Importin alpha3 interacts with HIV-1 integrase and contributes to HIV-1 nuclear import and replication. J Virol 84(17):8650–8663. https://doi.org/10.1128/jvi.00508-10

Arhel N (2010) Revisiting HIV-1 uncoating. Retrovirology 7:96. https://doi.org/10.1186/1742-4690-7-96

Arhel N, Munier S, Souque P, Mollier K, Charneau P (2006) Nuclear import defect of human immunodeficiency virus type 1 DNA flap mutants is not dependent on the viral strain or target cell type. J Virol 80(20):10262–10269. https://doi.org/10.1128/jvi.00974-06

Arold ST, Baur AS (2001) Dynamic Nef and Nef dynamics: how structure could explain the complex activities of this small HIV protein. Trends Biochem Sci 26(6):356–363. https://doi.org/10.1016/s0968-0004(01)01846-1

Atwood WJ, Tornatore CS, Traub R, Conant K, Drew PD, Major EO (1994) Stimulation of HIV type 1 gene expression and induction of NF-kappa B (p50/p65)-binding activity in tumor necrosis factor alpha-treated human fetal glial cells. AIDS Res Hum Retroviruses 10(10):1207–1211. https://doi.org/10.1089/aid.1994.10.1207

Autran B, Carcelain G, Combadiere B, Debre P (2004) Therapeutic vaccines for chronic infections. Science 305(5681):205–208. https://doi.org/10.1126/science.1100600

Ayehunie S, Garcia-Zepeda EA, Hoxie JA, Horuk R, Kupper TS, Luster AD, Ruprecht RM (1997) Human immunodeficiency virus-1 entry into purified blood dendritic cells through CC and CXC chemokine coreceptors. Blood 90(4):1379–1386

Barré-Sinoussi F, Chermann JC, Rey F, Nugeyre MT, Chamaret S, Gruest J, Dauguet C, Axler-Blin C, Vézinet-Brun F, Rouzioux C, Rozenbaum W, Montagnier L (1983) Isolation of a

T-lymphotropic retrovirus from a patient at risk for acquired immune deficiency syndrome (AIDS). Science 220(4599):868–871. https://doi.org/10.1126/science.6189183
Barré-Sinoussi F, Ross AL, Delfraissy JF (2013) Past, present and future: 30 years of HIV research. Nat Rev Microbiol 11(12):877–883. https://doi.org/10.1038/nrmicro3132
Bassuk AG, Anandappa RT, Leiden JM (1997) Physical interactions between Ets and NF-kappaB/NFAT proteins play an important role in their cooperative activation of the human immunodeficiency virus enhancer in T cells. J Virol 71(5):3563–3573. https://doi.org/10.1128/jvi.71.5.3563-3573.1997
Battistini A, Sgarbanti M (2014) HIV-1 latency: an update of molecular mechanisms and therapeutic strategies. Viruses 6(4):1715–1758. https://doi.org/10.3390/v6041715
Behrens RT, Aligeti M, Pocock GM, Higgins CA, Sherer NM (2017) Nuclear export signal masking regulates HIV-1 Rev trafficking and viral RNA nuclear export. J Virol 91(3). https://doi.org/10.1128/jvi.02107-16
Beilhartz GL, Götte M (2010) HIV-1 ribonuclease H: structure, catalytic mechanism and inhibitors. Viruses 2(4):900–926. https://doi.org/10.3390/v2040900
Bell NM, Lever AM (2013) HIV Gag polyprotein: processing and early viral particle assembly. Trends Microbiol 21(3):136–144. https://doi.org/10.1016/j.tim.2012.11.006
Berger EA, Murphy PM, Farber JM (1999) Chemokine receptors as HIV-1 coreceptors: roles in viral entry, tropism, and disease. Annu Rev Immunol 17:657–700. https://doi.org/10.1146/annurev.immunol.17.1.657
Berthet-Colominas C, Monaco S, Novelli A, Sibaï G, Mallet F, Cusack S (1999) Head-to-tail dimers and interdomain flexibility revealed by the crystal structure of HIV-1 capsid protein (p24) complexed with a monoclonal antibody Fab. EMBO J 18(5):1124–1136. https://doi.org/10.1093/emboj/18.5.1124
Biswas DK, Salas TR, Wang F, Ahlers CM, Dezube BJ, Pardee AB (1995) A Tat-induced auto-upregulatory loop for superactivation of the human immunodeficiency virus type 1 promoter. J Virol 69(12):7437–7444. https://doi.org/10.1128/jvi.69.12.7437-7444.1995
Bolinger C, Boris-Lawrie K (2009) Mechanisms employed by retroviruses to exploit host factors for translational control of a complicated proteome. Retrovirology 6:8. https://doi.org/10.1186/1742-4690-6-8
Börner K, Hermle J, Sommer C, Brown NP, Knapp B, Glass B, Kunkel J, Torralba G, Reymann J, Beil N, Beneke J, Pepperkok R, Schneider R, Ludwig T, Hausmann M, Hamprecht F, Erfle H, Kaderali L, Kräusslich HG, Lehmann MJ (2010) From experimental setup to bioinformatics: an RNAi screening platform to identify host factors involved in HIV-1 replication. Biotechnol J 5(1):39–49. https://doi.org/10.1002/biot.200900226
Bosch ML, Earl PL, Fargnoli K, Picciafuoco S, Giombini F, Wong-Staal F, Franchini G (1989) Identification of the fusion peptide of primate immunodeficiency viruses. Science 244(4905):694–697. https://doi.org/10.1126/science.2541505
Brakier-Gingras L, Charbonneau J, Butcher SE (2012) Targeting frameshifting in the human immunodeficiency virus. Expert Opin Ther Targets 16(3):249–258. https://doi.org/10.1517/14728222.2012.665879
Brandenberg OF, Magnus C, Regoes RR, Trkola A (2015) The HIV-1 entry process: a stoichiometric view. Trends Microbiol 23(12):763–774. https://doi.org/10.1016/j.tim.2015.09.003
Brass AL, Dykxhoorn DM, Benita Y, Yan N, Engelman A, Xavier RJ, Lieberman J, Elledge SJ (2008) Identification of host proteins required for HIV infection through a functional genomic screen. Science 319(5865):921–926. https://doi.org/10.1126/science.1152725
Brasseur R, Vandenbranden M, Cornet B, Burny A, Ruysschaert JM (1990) Orientation into the lipid bilayer of an asymmetric amphipathic helical peptide located at the N-terminus of viral fusion proteins. Biochim Biophys Acta 1029(2):267–273. https://doi.org/10.1016/0005-2736(90)90163-i
Brelot A, Chakrabarti LA (2018) CCR5 revisited: how mechanisms of HIV entry govern AIDS pathogenesis. J Mol Biol 430(17):2557–2589. https://doi.org/10.1016/j.jmb.2018.06.027
Bukrinskaya A, Brichacek B, Mann A, Stevenson M (1998) Establishment of a functional human immunodeficiency virus type 1 (HIV-1) reverse transcription complex involves the cytoskeleton. J Exp Med 188(11):2113–2125. https://doi.org/10.1084/jem.188.11.2113

Bukrinsky MI, Haffar OK (1998) HIV-1 nuclear import: matrix protein is back on center stage, this time together with Vpr. Mol Med 4(3):138–143

Bukrinsky MI, Sharova N, Dempsey MP, Stanwick TL, Bukrinskaya AG, Haggerty S, Stevenson M (1992) Active nuclear import of human immunodeficiency virus type 1 preintegration complexes. Proc Natl Acad Sci U S A 89(14):6580–6584. https://doi.org/10.1073/pnas.89.14.6580

Bukrinsky MI, Haggerty S, Dempsey MP, Sharova N, Adzhubel A, Spitz L, Lewis P, Goldfarb D, Emerman M, Stevenson M (1993) A nuclear localization signal within HIV-1 matrix protein that governs infection of non-dividing cells. Nature 365(6447):666–669. https://doi.org/10.1038/365666a0

Burdick RC, Li C (2020) HIV-1 uncoats in the nucleus near sites of integration. Proc Natl Acad Sci 117(10):5486–5493. https://doi.org/10.1073/pnas.1920631117

Burke BP, Levin BR, Zhang J, Sahakyan A, Boyer J, Carroll MV, Colón JC, Keech N, Rezek V, Bristol G, Eggers E, Cortado R, Boyd MP, Impey H, Shimizu S, Lowe EL, Ringpis GE, Kim SG, Vatakis DN, Breton LR, Bartlett JS, Chen IS, Kitchen SG, An DS, Symonds GP (2015) Engineering cellular resistance to HIV-1 infection in vivo using a dual therapeutic lentiviral vector. Mol Ther Nucleic Acids 4:e236. https://doi.org/10.1038/mtna.2015.10

Bushman FD, Malani N, Fernandes J, D'Orso I, Cagney G, Diamond TL, Zhou H, Hazuda DJ, Espeseth AS, König R, Bandyopadhyay S, Ideker T, Goff SP, Krogan NJ, Frankel AD, Young JA, Chanda SK (2009) Host cell factors in HIV replication: meta-analysis of genome-wide studies. PLoS Pathog 5(5):e1000437. https://doi.org/10.1371/journal.ppat.1000437

Buzon V, Natrajan G, Schibli D, Campelo F, Kozlov MM, Weissenhorn W (2010) Crystal structure of HIV-1 gp41 including both fusion peptide and membrane proximal external regions. PLoS Pathog 6(5):e1000880. https://doi.org/10.1371/journal.ppat.1000880

Camaur D, Trono D (1996) Characterization of human immunodeficiency virus type 1 Vif particle incorporation. J Virol 70(9):6106–6111. https://doi.org/10.1128/jvi.70.9.6106-6111.1996

Campbell EM, Hope TJ (2015) HIV-1 capsid: the multifaceted key player in HIV-1 infection. Nat Rev Microbiol 13(8):471–483. https://doi.org/10.1038/nrmicro3503

Campbell GR, Pasquier E, Watkins J, Bourgarel-Rey V, Peyrot V, Esquieu D, Barbier P, de Mareuil J, Braguer D, Kaleebu P, Yirrell DL, Loret EP (2004) The glutamine-rich region of the HIV-1 Tat protein is involved in T-cell apoptosis. J Biol Chem 279(46):48197–48204. https://doi.org/10.1074/jbc.M406195200

Campestrini J, Silveira DB, Pinto AR (2018) HIV-1 Tat-induced bystander apoptosis in Jurkat cells involves unfolded protein responses. Cell Biochem Funct 36(7):377–386. https://doi.org/10.1002/cbf.3357

Canonne-Hergaux F, Aunis D, Schaeffer E (1995) Interactions of the transcription factor AP-1 with the long terminal repeat of different human immunodeficiency virus type 1 strains in Jurkat, glial, and neuronal cells. J Virol 69(11):6634–6642. https://doi.org/10.1128/jvi.69.11.6634-6642.1995

Carcelain G, Autran B (2013) Immune interventions in HIV infection. Immunol Rev 254(1):355–371. https://doi.org/10.1111/imr.12083

Carroll VA, Lafferty MK, Marchionni L, Bryant JL, Gallo RC, Garzino-Demo A (2016) Expression of HIV-1 matrix protein p17 and association with B-cell lymphoma in HIV-1 transgenic mice. Proc Natl Acad Sci 113(46):13168–13173

Carteau S, Batson SC, Poljak L, Mouscadet JF, de Rocquigny H, Darlix JL, Roques BP, Käs E, Auclair C (1997) Human immunodeficiency virus type 1 nucleocapsid protein specifically stimulates Mg2+-dependent DNA integration in vitro. J Virol 71(8):6225–6229. https://doi.org/10.1128/jvi.71.8.6225-6229.1997

Carteau S, Gorelick RJ, Bushman FD (1999) Coupled integration of human immunodeficiency virus type 1 cDNA ends by purified integrase in vitro: stimulation by the viral nucleocapsid protein. J Virol 73(8):6670–6679. https://doi.org/10.1128/jvi.73.8.6670-6679.1999

Caskey M, Klein F, Lorenzi JC, Seaman MS, West AP Jr, Buckley N, Kremer G, Nogueira L, Braunschweig M, Scheid JF, Horwitz JA, Shimeliovich I, Ben-Avraham S, Witmer-Pack M, Platten M, Lehmann C, Burke LA, Hawthorne T, Gorelick RJ, Walker BD, Keler T, Gulick RM, Fätkenheuer G, Schlesinger SJ, Nussenzweig MC (2015) Viraemia suppressed in HIV-1-

infected humans by broadly neutralizing antibody 3BNC117. Nature 522(7557):487–491. https://doi.org/10.1038/nature14411
Chan DC, Fass D, Berger JM, Kim PS (1997) Core structure of gp41 from the HIV envelope glycoprotein. Cell 89(2):263–273. https://doi.org/10.1016/s0092-8674(00)80205-6
Chanh TC, Kennedy RC, Kanda P (1988) Synthetic peptides homologous to HIV transmembrane glycoprotein suppress normal human lymphocyte blastogenic response. Cell Immunol 111(1):77–86. https://doi.org/10.1016/0008-8749(88)90052-4
Charbonneau J, Gendron K, Ferbeyre G, Brakier-Gingras L (2012) The 5′ UTR of HIV-1 full-length mRNA and the Tat viral protein modulate the programmed-1 ribosomal frameshift that generates HIV-1 enzymes. RNA 18(3):519–529. https://doi.org/10.1261/rna.030346.111
Charnay N, Ivanyi-Nagy R, Soto-Rifo R, Ohlmann T, López-Lastra M, Darlix JL (2009) Mechanism of HIV-1 Tat RNA translation and its activation by the Tat protein. Retrovirology 6:74. https://doi.org/10.1186/1742-4690-6-74
Checkley MA, Luttge BG, Freed EO (2011) HIV-1 envelope glycoprotein biosynthesis, trafficking, and incorporation. J Mol Biol 410(4):582–608. https://doi.org/10.1016/j.jmb.2011.04.042
Chen JC, Krucinski J, Miercke LJ, Finer-Moore JS, Tang AH, Leavitt AD, Stroud RM (2000) Crystal structure of the HIV-1 integrase catalytic core and C-terminal domains: a model for viral DNA binding. Proc Natl Acad Sci U S A 97(15):8233–8238. https://doi.org/10.1073/pnas.150220297
Chen B, Vogan EM, Gong H, Skehel JJ, Wiley DC, Harrison SC (2005) Structure of an unliganded simian immunodeficiency virus gp120 core. Nature 433(7028):834–841. https://doi.org/10.1038/nature03327
Chen NY, Zhou L, Gane PJ, Opp S, Ball NJ, Nicastro G, Zufferey M, Buffone C, Luban J, Selwood D, Diaz-Griffero F, Taylor I, Fassati A (2016) HIV-1 capsid is involved in post-nuclear entry steps. Retrovirology 13:28. https://doi.org/10.1186/s12977-016-0262-0
Cherepanov P (2010) Integrase illuminated. EMBO Rep 11(5):328. https://doi.org/10.1038/embor.2010.58
Chiang RZ, Gan SK, Su CT (2018) A computational study for rational HIV-1 non-nucleoside reverse transcriptase inhibitor selection and the discovery of novel allosteric pockets for inhibitor design. Biosci Rep 38(2). https://doi.org/10.1042/bsr20171113
Choe H, Farzan M, Sun Y, Sullivan N, Rollins B, Ponath PD, Wu L, Mackay CR, LaRosa G, Newman W, Gerard N, Gerard C, Sodroski J (1996) The beta-chemokine receptors CCR3 and CCR5 facilitate infection by primary HIV-1 isolates. Cell 85(7):1135–1148. https://doi.org/10.1016/s0092-8674(00)81313-6
Chomont N, El-Far M, Ancuta P, Trautmann L, Procopio FA, Yassine-Diab B, Boucher G, Boulassel MR, Ghattas G, Brenchley JM, Schacker TW, Hill BJ, Douek DC, Routy JP, Haddad EK, Sékaly RP (2009) HIV reservoir size and persistence are driven by T cell survival and homeostatic proliferation. Nat Med 15(8):893–900. https://doi.org/10.1038/nm.1972
Chun TW, Carruth L, Finzi D, Shen X, DiGiuseppe JA, Taylor H, Hermankova M, Chadwick K, Margolick J, Quinn TC, Kuo YH, Brookmeyer R, Zeiger MA, Barditch-Crovo P, Siliciano RF (1997a) Quantification of latent tissue reservoirs and total body viral load in HIV-1 infection. Nature 387(6629):183–188. https://doi.org/10.1038/387183a0
Chun TW, Stuyver L, Mizell SB, Ehler LA, Mican JA, Baseler M, Lloyd AL, Nowak MA, Fauci AS (1997b) Presence of an inducible HIV-1 latent reservoir during highly active antiretroviral therapy. Proc Natl Acad Sci U S A 94(24):13193–13197. https://doi.org/10.1073/pnas.94.24.13193
Cimarelli A, Darlix JL (2014) HIV-1 reverse transcription. Methods Mol Biol 1087:55–70. https://doi.org/10.1007/978-1-62703-670-2_6
Clavel F, Guétard D, Brun-Vézinet F, Chamaret S, Rey MA, Santos-Ferreira MO, Laurent AG, Dauguet C, Katlama C, Rouzioux C et al (1986) Isolation of a new human retrovirus from West African patients with AIDS. Science 233(4761):343–346. https://doi.org/10.1126/science.2425430
Clavel F, Mansinho K, Chamaret S, Guetard D, Favier V, Nina J, Santos-Ferreira MO, Champalimaud JL, Montagnier L (1987) Human immunodeficiency virus type 2 infection asso-

ciated with AIDS in West Africa. N Engl J Med 316(19):1180–1185. https://doi.org/10.1056/nejm198705073161903
Clever JL, Parslow TG (1997) Mutant human immunodeficiency virus type 1 genomes with defects in RNA dimerization or encapsidation. J Virol 71(5):3407–3414. https://doi.org/10.1128/jvi.71.5.3407-3414.1997
Coffin JM, Fan H (2016) The discovery of reverse transcriptase. Annu Rev Virol 3(1):29–51. https://doi.org/10.1146/annurev-virology-110615-035556
Coffin J, Haase A, Levy JA, Montagnier L, Oroszlan S, Teich N, Temin H, Toyoshima K, Varmus H, Vogt P et al (1986) Human immunodeficiency viruses. Science 232(4751):697. https://doi.org/10.1126/science.3008335
Cohen MS, Shaw GM, McMichael AJ, Haynes BF (2011) Acute HIV-1 Infection. N Engl J Med 364(20):1943–1954. https://doi.org/10.1056/NEJMra1011874
Cohn LB, Chomont N, Deeks SG (2020) The biology of the HIV-1 latent reservoir and implications for cure strategies. Cell Host Microbe 27(4):519–530. https://doi.org/10.1016/j.chom.2020.03.014
Coiras M, López-Huertas MR, Sánchez del Cojo M, Mateos E, Alcamí J (2010) Dual role of host cell factors in HIV-1 replication: restriction and enhancement of the viral cycle. AIDS Rev 12(2):103–112
Collins KL, Chen BK, Kalams SA, Walker BD, Baltimore D (1998) HIV-1 Nef protein protects infected primary cells against killing by cytotoxic T lymphocytes. Nature 391(6665):397–401. https://doi.org/10.1038/34929
Conant K, Ma M, Nath A, Major EO (1996) Extracellular human immunodeficiency virus type 1 Tat protein is associated with an increase in both NF-kappa B binding and protein kinase C activity in primary human astrocytes. J Virol 70(3):1384–1389. https://doi.org/10.1128/jvi.70.3.1384-1389.1996
Cosnefroy O, Murray PJ, Bishop KN (2016) HIV-1 capsid uncoating initiates after the first strand transfer of reverse transcription. Retrovirology 13(1):58. https://doi.org/10.1186/s12977-016-0292-7
Cosson P (1996) Direct interaction between the envelope and matrix proteins of HIV-1. EMBO J 15(21):5783–5788
Craigie R, Bushman FD (2012) HIV DNA integration. Cold Spring Harb Perspect Med 2(7):a006890. https://doi.org/10.1101/cshperspect.a006890
Cruceanu M, Urbaneja MA, Hixson CV, Johnson DG, Datta SA, Fivash MJ, Stephen AG, Fisher RJ, Gorelick RJ, Casas-Finet JR, Rein A, Rouzina I, Williams MC (2006) Nucleic acid binding and chaperone properties of HIV-1 Gag and nucleocapsid proteins. Nucleic Acids Res 34(2):593–605. https://doi.org/10.1093/nar/gkj458
Dahiya S, Liu Y, Nonnemacher MR, Dampier W, Wigdahl B (2014) CCAAT enhancer binding protein and nuclear factor of activated T cells regulate HIV-1 LTR via a novel conserved downstream site in cells of the monocyte-macrophage lineage. PLoS One 9(2):e88116. https://doi.org/10.1371/journal.pone.0088116
Dandekar DH, Ganesh KN, Mitra D (2004) HIV-1 Tat directly binds to NFkappaB enhancer sequence: role in viral and cellular gene expression. Nucleic Acids Res 32(4):1270–1278. https://doi.org/10.1093/nar/gkh289
Darlix JL, Gabus C, Nugeyre MT, Clavel F, Barré-Sinoussi F (1990) Cis elements and trans-acting factors involved in the RNA dimerization of the human immunodeficiency virus HIV-1. J Mol Biol 216(3):689–699. https://doi.org/10.1016/0022-2836(90)90392-y
Darlix JL, Lapadat-Tapolsky M, de Rocquigny H, Roques BP (1995) First glimpses at structure-function relationships of the nucleocapsid protein of retroviruses. J Mol Biol 254(4):523–537. https://doi.org/10.1006/jmbi.1995.0635
Das SR, Jameel S (2005) Biology of the HIV Nef protein. Indian J Med Res 121(4):315–332
de Breyne S, Ohlmann T (2018) Focus on translation initiation of the HIV-1 mRNAs. Int J Mol Sci 20(1):101. https://doi.org/10.3390/ijms20010101
de Breyne S, Chamond N, Décimo D, Trabaud MA, André P, Sargueil B, Ohlmann T (2012) In vitro studies reveal that different modes of initiation on HIV-1 mRNA have different lev-

els of requirement for eukaryotic initiation factor 4F. FEBS J 279(17):3098–3111. https://doi.org/10.1111/j.1742-4658.2012.08689.x

de Breyne S, Soto-Rifo R, López-Lastra M, Ohlmann T (2013) Translation initiation is driven by different mechanisms on the HIV-1 and HIV-2 genomic RNAs. Virus Res 171(2):366–381. https://doi.org/10.1016/j.virusres.2012.10.006

De Clercq E (2009) The history of antiretrovirals: key discoveries over the past 25 years. Rev Med Virol 19(5):287–299. https://doi.org/10.1002/rmv.624

de Noronha CM, Sherman MP, Lin HW, Cavrois MV, Moir RD, Goldman RD, Greene WC (2001) Dynamic disruptions in nuclear envelope architecture and integrity induced by HIV-1 Vpr. Science 294(5544):1105–1108. https://doi.org/10.1126/science.1063957

de Ronde A, Klaver B, Keulen W, Smit L, Goudsmit J (1992) Natural HIV-1 NEF accelerates virus replication in primary human lymphocytes. Virology 188(1):391–395. https://doi.org/10.1016/0042-6822(92)90772-h

Dean M, Carrington M, Winkler C, Huttley GA, Smith MW, Allikmets R, Goedert JJ, Buchbinder SP, Vittinghoff E, Gomperts E, Donfield S, Vlahov D, Kaslow R, Saah A, Rinaldo C, Detels R, O'Brien SJ (1996) Genetic restriction of HIV-1 infection and progression to AIDS by a deletion allele of the CKR5 structural gene. Hemophilia Growth and Development Study, Multicenter AIDS Cohort Study, Multicenter Hemophilia Cohort Study, San Francisco City Cohort, ALIVE Study. Science 273(5283):1856–1862. https://doi.org/10.1126/science.273.5283.1856

Debaisieux S, Rayne F, Yezid H, Beaumelle B (2012) The ins and outs of HIV-1 Tat. Traffic 13(3):355–363. https://doi.org/10.1111/j.1600-0854.2011.01286.x

Demarchi F, d'Adda di Fagagna F, Falaschi A, Giacca M (1996) Activation of transcription factor NF-kappaB by the Tat protein of human immunodeficiency virus type 1. J Virol 70(7):4427–4437. https://doi.org/10.1128/jvi.70.7.4427-4437.1996

Deng H, Liu R, Ellmeier W, Choe S, Unutmaz D, Burkhart M, Di Marzio P, Marmon S, Sutton RE, Hill CM, Davis CB, Peiper SC, Schall TJ, Littman DR, Landau NR (1996) Identification of a major co-receptor for primary isolates of HIV-1. Nature 381(6584):661–666. https://doi.org/10.1038/381661a0

Depienne C, Mousnier A, Leh H, Le Rouzic E, Dormont D, Benichou S, Dargemont C (2001) Characterization of the nuclear import pathway for HIV-1 integrase. J Biol Chem 276(21):18102–18107. https://doi.org/10.1074/jbc.M009029200

Deshmukh L, Tugarinov V, Appella DH, Clore GM (2018) Targeting a dark excited state of HIV-1 nucleocapsid by antiretroviral thioesters revealed by NMR spectroscopy. Angew Chem Int Ed 57(10):2687–2691. https://doi.org/10.1002/anie.201713172

Di Marzio P, Choe S, Ebright M, Knoblauch R, Landau NR (1995) Mutational analysis of cell cycle arrest, nuclear localization and virion packaging of human immunodeficiency virus type 1 Vpr. J Virol 69(12):7909–7916. https://doi.org/10.1128/jvi.69.12.7909-7916.1995

Didigu CA, Doms RW (2012) Novel approaches to inhibit HIV entry. Viruses 4(2):309–324. https://doi.org/10.3390/v4020309

Dinter H, Chiu R, Imagawa M, Karin M, Jones KA (1987) In vitro activation of the HIV-1 enhancer in extracts from cells treated with a phorbol ester tumor promoter. EMBO J 6(13):4067–4071

Dittmar MT, McKnight A, Simmons G, Clapham PR, Weiss RA, Simmonds P (1997) HIV-1 tropism and co-receptor use. Nature 385(6616):495–496. https://doi.org/10.1038/385495a0

Dorin D, Bonnet MC, Bannwarth S, Gatignol A, Meurs EF, Vaquero C (2003) The TAR RNA-binding protein, TRBP, stimulates the expression of TAR-containing RNAs in vitro and in vivo independently of its ability to inhibit the dsRNA-dependent kinase PKR. J Biol Chem 278(7):4440–4448. https://doi.org/10.1074/jbc.M208954200

Douek DC, Brenchley JM, Betts MR, Ambrozak DR, Hill BJ, Okamoto Y, Casazza JP, Kuruppu J, Kunstman K, Wolinsky S, Grossman Z, Dybul M, Oxenius A, Price DA, Connors M, Koup RA (2002) HIV preferentially infects HIV-specific CD4+ T cells. Nature 417(6884):95–98. https://doi.org/10.1038/417095a

Dugré-Brisson S, Elvira G, Boulay K, Chatel-Chaix L, Mouland AJ, DesGroseillers L (2005) Interaction of Staufen1 with the 5′ end of mRNA facilitates translation of these RNAs. Nucleic Acids Res 33(15):4797–4812. https://doi.org/10.1093/nar/gki794

Duh EJ, Maury WJ, Folks TM, Fauci AS, Rabson AB (1989) Tumor necrosis factor alpha activates human immunodeficiency virus type 1 through induction of nuclear factor binding to the NF-kappa B sites in the long terminal repeat. Proc Natl Acad Sci U S A 86(15):5974–5978. https://doi.org/10.1073/pnas.86.15.5974

Dupont S, Sharova N, DéHoratius C, Virbasius CM, Zhu X, Bukrinskaya AG, Stevenson M, Green MR (1999) A novel nuclear export activity in HIV-1 matrix protein required for viral replication. Nature 402(6762):681–685. https://doi.org/10.1038/45272

Dvorin JD, Bell P, Maul GG, Yamashita M, Emerman M, Malim MH (2002) Reassessment of the roles of integrase and the central DNA flap in human immunodeficiency virus type 1 nuclear import. J Virol 76(23):12087–12096. https://doi.org/10.1128/jvi.76.23.12087-12096.2002

Dyda F, Hickman AB, Jenkins TM, Engelman A, Craigie R, Davies DR (1994) Crystal structure of the catalytic domain of HIV-1 integrase: similarity to other polynucleotidyl transferases. Science 266(5193):1981–1986. https://doi.org/10.1126/science.7801124

el Kharroubi A, Verdin E (1994) Protein-DNA interactions within DNase I-hypersensitive sites located downstream of the HIV-1 promoter. J Biol Chem 269(31):19916–19924

Engelman A, Cherepanov P (2012) The structural biology of HIV-1: mechanistic and therapeutic insights. Nat Rev Microbiol 10(4):279–290. https://doi.org/10.1038/nrmicro2747

Engelman AN, Singh PK (2018) Cellular and molecular mechanisms of HIV-1 integration targeting. Cell Mol Life Sci 75(14):2491–2507. https://doi.org/10.1007/s00018-018-2772-5

Eron JJ Jr (2000) HIV-1 protease inhibitors. Clin Infect Dis 30(Suppl. 2):S160–S170. https://doi.org/10.1086/313853

Eugen-Olsen J, Iversen AK, Garred P, Koppelhus U, Pedersen C, Benfield TL, Sorensen AM, Katzenstein T, Dickmeiss E, Gerstoft J, Skinhøj P, Svejgaard A, Nielsen JO, Hofmann B (1997) Heterozygosity for a deletion in the CKR-5 gene leads to prolonged AIDS-free survival and slower CD4 T-cell decline in a cohort of HIV-seropositive individuals. Aids 11(3):305–310. https://doi.org/10.1097/00002030-199703110-00007

Feng Y, Broder CC, Kennedy PE, Berger EA (1996) HIV-1 entry cofactor: functional cDNA cloning of a seven-transmembrane, G protein-coupled receptor. Science 272(5263):872–877. https://doi.org/10.1126/science.272.5263.872

Feng YX, Campbell S, Harvin D, Ehresmann B, Ehresmann C, Rein A (1999) The human immunodeficiency virus type 1 Gag polyprotein has nucleic acid chaperone activity: possible role in dimerization of genomic RNA and placement of tRNA on the primer binding site. J Virol 73(5):4251–4256. https://doi.org/10.1128/jvi.73.5.4251-4256.1999

Feng Y, Love RP, Chelico L (2013) HIV-1 viral infectivity factor (Vif) alters processive single-stranded DNA scanning of the retroviral restriction factor APOBEC3G. J Biol Chem 288(9):6083–6094. https://doi.org/10.1074/jbc.M112.421875

Flexner C (2007) HIV drug development: the next 25 years. Nat Rev Drug Discov 6(12):959–966. https://doi.org/10.1038/nrd2336

Folks TM, Justement J, Kinter A, Dinarello CA, Fauci AS (1987) Cytokine-induced expression of HIV-1 in a chronically infected promonocyte cell line. Science 238(4828):800–802. https://doi.org/10.1126/science.3313729

Fontes JD, Jiang B, Peterlin BM (1997) The class II trans-activator CIITA interacts with the TBP-associated factor TAFII32. Nucleic Acids Res 25(12):2522–2528. https://doi.org/10.1093/nar/25.12.2522

Frankel AD, Young JA (1998) HIV-1: fifteen proteins and an RNA. Annu Rev Biochem 67:1–25. https://doi.org/10.1146/annurev.biochem.67.1.1

Freed EO (2015) HIV-1 assembly, release and maturation. Nat Rev Microbiol 13(8):484–496. https://doi.org/10.1038/nrmicro3490

Freed EO, Englund G, Maldarelli F, Martin MA (1997) Phosphorylation of residue 131 of HIV-1 matrix is not required for macrophage infection. Cell 88(2):171–173.; discussion 173-4. https://doi.org/10.1016/s0092-8674(00)81836-x

Fuda NJ, Ardehali MB, Lis JT (2009) Defining mechanisms that regulate RNA polymerase II transcription in vivo. Nature 461(7261):186–192. https://doi.org/10.1038/nature08449

Gaines CR, Tkacik E, Rivera-Oven A, Somani P, Achimovich A, Alabi T, Zhu A, Getachew N, Yang AL, McDonough M, Hawkins T, Spadaro Z, Summers MF (2018) HIV-1 Matrix Protein Interactions with tRNA: implications for membrane targeting. J Mol Biol 430(14):2113–2127. https://doi.org/10.1016/j.jmb.2018.04.042

Gallay P, Swingler S, Song J, Bushman F, Trono D (1995) HIV nuclear import is governed by the phosphotyrosine-mediated binding of matrix to the core domain of integrase. Cell 83(4):569–576. https://doi.org/10.1016/0092-8674(95)90097-7

Gallay P, Hope T, Chin D, Trono D (1997) HIV-1 infection of nondividing cells through the recognition of integrase by the importin/karyopherin pathway. Proc Natl Acad Sci U S A 94(18):9825–9830. https://doi.org/10.1073/pnas.94.18.9825

Gallo RC, Sarin PS, Gelmann EP, Robert-Guroff M, Richardson E, Kalyanaraman VS, Mann D, Sidhu GD, Stahl RE, Zolla-Pazner S, Leibowitch J, Popovic M (1983) Isolation of human T-cell leukemia virus in acquired immune deficiency syndrome (AIDS). Science 220(4599):865–867. https://doi.org/10.1126/science.6601823

Ganser BK, Li S, Klishko VY, Finch JT, Sundquist WI (1999) Assembly and analysis of conical models for the HIV-1 core. Science 283(5398):80–83. https://doi.org/10.1126/science.283.5398.80

Ganser-Pornillos BK, Yeager M, Sundquist WI (2008) The structural biology of HIV assembly. Curr Opin Struct Biol 18(2):203–217. https://doi.org/10.1016/j.sbi.2008.02.001

Garcia JV, Miller AD (1991) Serine phosphorylation-independent downregulation of cell-surface CD4 by nef. Nature 350(6318):508–511. https://doi.org/10.1038/350508a0

Garrus JE, von Schwedler UK, Pornillos OW, Morham SG, Zavitz KH, Wang HE, Wettstein DA, Stray KM, Côté M, Rich RL, Myszka DG, Sundquist WI (2001) Tsg101 and the vacuolar protein sorting pathway are essential for HIV-1 budding. Cell 107(1):55–65. https://doi.org/10.1016/s0092-8674(01)00506-2

Gartner S, Markovits P, Markovitz DM, Kaplan MH, Gallo RC, Popovic M (1986) The role of mononuclear phagocytes in HTLV-III/LAV infection. Science 233(4760):215–219. https://doi.org/10.1126/science.3014648

Gautier VW, Gu L, O'Donoghue N, Pennington S, Sheehy N, Hall WW (2009) In vitro nuclear interactome of the HIV-1 Tat protein. Retrovirology 6:47. https://doi.org/10.1186/1742-4690-6-47

Gaynor R (1992) Cellular transcription factors involved in the regulation of HIV-1 gene expression. Aids 6(4):347–363. https://doi.org/10.1097/00002030-199204000-00001

Gendron K, Charbonneau J, Dulude D, Heveker N, Ferbeyre G, Brakier-Gingras L (2008) The presence of the TAR RNA structure alters the programmed -1 ribosomal frameshift efficiency of the human immunodeficiency virus type 1 (HIV-1) by modifying the rate of translation initiation. Nucleic Acids Res 36(1):30–40. https://doi.org/10.1093/nar/gkm906

Geyer M, Munte CE, Schorr J, Kellner R, Kalbitzer HR (1999) Structure of the anchor-domain of myristoylated and non-myristoylated HIV-1 Nef protein. J Mol Biol 289(1):123–138. https://doi.org/10.1006/jmbi.1999.2740

Geyer M, Fackler OT, Peterlin BM (2001) Structure--function relationships in HIV-1 Nef. EMBO Rep 2(7):580–585. https://doi.org/10.1093/embo-reports/kve141

Gill G, Pascal E, Tseng ZH, Tjian R (1994) A glutamine-rich hydrophobic patch in transcription factor Sp1 contacts the dTAFII110 component of the Drosophila TFIID complex and mediates transcriptional activation. Proc Natl Acad Sci U S A 91(1):192–196. https://doi.org/10.1073/pnas.91.1.192

Goff SP (2007) Host factors exploited by retroviruses. Nat Rev Microbiol 5(4):253–263. https://doi.org/10.1038/nrmicro1541

Goldschmidt V, Didierjean J, Ehresmann B, Ehresmann C, Isel C, Marquet R (2006) Mg2+ dependency of HIV-1 reverse transcription, inhibition by nucleoside analogues and resistance. Nucleic Acids Res 34(1):42–52. https://doi.org/10.1093/nar/gkj411

Goldsmith MA, Warmerdam MT, Atchison RE, Miller MD, Greene WC (1995) Dissociation of the CD4 downregulation and viral infectivity enhancement functions of human immunodeficiency virus type 1 Nef. J Virol 69(7):4112–4121. https://doi.org/10.1128/jvi.69.7.4112-4121.1995

Gomez C, Hope TJ (2005) The ins and outs of HIV replication. Cell Microbiol 7(5):621–626. https://doi.org/10.1111/j.1462-5822.2005.00516.x
Götte M, Li X, Wainberg MA (1999) HIV-1 reverse transcription: a brief overview focused on structure-function relationships among molecules involved in initiation of the reaction. Arch Biochem Biophys 365(2):199–210. https://doi.org/10.1006/abbi.1999.1209
Gottlieb MS, Schroff R, Schanker HM, Weisman JD, Fan PT, Wolf RA, Saxon A (1981) Pneumocystis carinii pneumonia and mucosal candidiasis in previously healthy homosexual men: evidence of a new acquired cellular immunodeficiency. N Engl J Med 305(24):1425–1431. https://doi.org/10.1056/nejm198112103052401
Göttlinger HG (2001) The HIV-1 assembly machine. Aids 15(Suppl. 5):S13–S20. https://doi.org/10.1097/00002030-200100005-00003
Granowitz EV, Saget BM, Wang MZ, Dinarello CA, Skolnik PR (1995) Interleukin 1 induces HIV-1 expression in chronically infected U1 cells: blockade by interleukin 1 receptor antagonist and tumor necrosis factor binding protein type 1. Mol Med 1(6):667–677
Greene WC, Peterlin BM (2002) Charting HIV's remarkable voyage through the cell: basic science as a passport to future therapy. Nat Med 8(7):673–680. https://doi.org/10.1038/nm0702-673
Grunewald ME, Fehr AR, Athmer J, Perlman S (2018) The coronavirus nucleocapsid protein is ADP-ribosylated. Virology 517:62–68. https://doi.org/10.1016/j.virol.2017.11.020
Guenzel CA, Hérate C, Benichou S (2014) HIV-1 Vpr-a still enigmatic multitasker. Front Microbiol 5:127. https://doi.org/10.3389/fmicb.2014.00127
Gurer C, Cimarelli A, Luban J (2002) Specific incorporation of heat shock protein 70 family members into primate lentiviral virions. J Virol 76(9):4666–4670. https://doi.org/10.1128/jvi.76.9.4666-4670.2002
Guttman M, Kahn M, Garcia NK, Hu SL, Lee KK (2012) Solution structure, conformational dynamics, and CD4-induced activation in full-length, glycosylated, monomeric HIV gp120. J Virol 86(16):8750–8764. https://doi.org/10.1128/jvi.07224-11
Haffar OK, Popov S, Dubrovsky L, Agostini I, Tang H, Pushkarsky T, Nadler SG, Bukrinsky M (2000) Two nuclear localization signals in the HIV-1 matrix protein regulate nuclear import of the HIV-1 pre-integration complex. J Mol Biol 299(2):359–368. https://doi.org/10.1006/jmbi.2000.3768
Hale M, Mesojednik T, Romano Ibarra GS, Sahni J, Bernard A, Sommer K, Scharenberg AM, Rawlings DJ, Wagner TA (2017) Engineering HIV-resistant, anti-HIV chimeric antigen receptor T cells. Mol Ther 25(3):570–579. https://doi.org/10.1016/j.ymthe.2016.12.023
Hare S, Gupta SS, Valkov E, Engelman A, Cherepanov P (2010) Retroviral intasome assembly and inhibition of DNA strand transfer. Nature 464(7286):232–236. https://doi.org/10.1038/nature08784
Harris RS, Liddament MT (2004) Retroviral restriction by APOBEC proteins. Nat Rev Immunol 4(11):868–877. https://doi.org/10.1038/nri1489
Harris C, Small CB, Klein RS, Friedland GH, Moll B, Emeson EE, Spigland I, Steigbigel NH (1983) Immunodeficiency in female sexual partners of men with the acquired immunodeficiency syndrome. N Engl J Med 308(20):1181–1184. https://doi.org/10.1056/nejm198305193082001
Harrist AV, Ryzhova EV, Harvey T, González-Scarano F (2009) Anx2 interacts with HIV-1 Gag at phosphatidylinositol (4,5) bisphosphate-containing lipid rafts and increases viral production in 293T cells. PLoS One 4(3):e5020. https://doi.org/10.1371/journal.pone.0005020
Hartley O, Klasse PJ, Sattentau QJ, Moore JP (2005) V3: HIV's switch-hitter. AIDS Res Hum Retroviruses 21(2):171–189. https://doi.org/10.1089/aid.2005.21.171
Hearps AC, Jans DA (2006) HIV-1 integrase is capable of targeting DNA to the nucleus via an importin alpha/beta-dependent mechanism. Biochem J 398(3):475–484. https://doi.org/10.1042/bj20060466
Heinzinger NK, Bukrinsky MI, Haggerty SA, Ragland AM, Kewalramani V, Lee MA, Gendelman HE, Ratner L, Stevenson M, Emerman M (1994) The Vpr protein of human immunodeficiency virus type 1 influences nuclear localization of viral nucleic acids in nondividing host cells. Proc Natl Acad Sci U S A 91(15):7311–7315. https://doi.org/10.1073/pnas.91.15.7311

Hemelaar J, Elangovan R, Yun J, Dickson-Tetteh L, Fleminger I, Kirtley S, Williams B, Gouws-Williams E, Ghys PD (2019) Global and regional molecular epidemiology of HIV-1, 1990-2015: a systematic review, global survey, and trend analysis. Lancet Infect Dis 19(2):143–155. https://doi.org/10.1016/s1473-3099(18)30647-9

Henderson AJ, Calame KL (1997) CCAAT/enhancer binding protein (C/EBP) sites are required for HIV-1 replication in primary macrophages but not CD4(+) T cells. Proc Natl Acad Sci U S A 94(16):8714–8719. https://doi.org/10.1073/pnas.94.16.8714

Henderson LJ, Sharma A, Monaco MC, Major EO, Al-Harthi L (2012) Human immunodeficiency virus type 1 (HIV-1) transactivator of transcription through its intact core and cysteine-rich domains inhibits Wnt/β-catenin signaling in astrocytes: relevance to HIV neuropathogenesis. J Neurosci 32(46):16306–16313. https://doi.org/10.1523/jneurosci.3145-12.2012

Herschhorn A, Hizi A (2010) Retroviral reverse transcriptases. Cell Mol Life Sci 67(16):2717–2747. https://doi.org/10.1007/s00018-010-0346-2

Hill CP, Worthylake D, Bancroft DP, Christensen AM, Sundquist WI (1996) Crystal structures of the trimeric human immunodeficiency virus type 1 matrix protein: implications for membrane association and assembly. Proc Natl Acad Sci U S A 93(7):3099–3104. https://doi.org/10.1073/pnas.93.7.3099

Ho DD, Neumann AU, Perelson AS, Chen W, Leonard JM, Markowitz M (1995) Rapid turnover of plasma virions and CD4 lymphocytes in HIV-1 infection. Nature 373(6510):123–126. https://doi.org/10.1038/373123a0

Hrimech M, Yao XJ, Bachand F, Rougeau N, Cohen EA (1999) Human immunodeficiency virus type 1 (HIV-1) Vpr functions as an immediate-early protein during HIV-1 infection. J Virol 73(5):4101–4109. https://doi.org/10.1128/jvi.73.5.4101-4109.1999

Hu WS, Hughes SH (2012) HIV-1 reverse transcription. Cold Spring Harb Perspect Med 2(10). https://doi.org/10.1101/cshperspect.a006882

Huang Y, Paxton WA, Wolinsky SM, Neumann AU, Zhang L, He T, Kang S, Ceradini D, Jin Z, Yazdanbakhsh K, Kunstman K, Erickson D, Dragon E, Landau NR, Phair J, Ho DD, Koup RA (1996) The role of a mutant CCR5 allele in HIV-1 transmission and disease progression. Nat Med 2(11):1240–1243. https://doi.org/10.1038/nm1196-1240

Hughes SH (2015) Reverse transcription of retroviruses and LTR retrotransposons. Microbiol Spectr 3(2):Mdna3-0027-2014. https://doi.org/10.1128/microbiolspec.MDNA3-0027-2014

Hymes KB, Cheung T, Greene JB, Prose NS, Marcus A, Ballard H, William DC, Laubenstein LJ (1981) Kaposi's sarcoma in homosexual men-a report of eight cases. Lancet 2(8247):598–600. https://doi.org/10.1016/s0140-6736(81)92740-9

Ikuta K, Suzuki S, Horikoshi H, Mukai T, Luftig RB (2000) Positive and negative aspects of the human immunodeficiency virus protease: development of inhibitors versus its role in AIDS pathogenesis. Microbiol Mol Biol Rev 64(4):725–745. https://doi.org/10.1128/mmbr.64.4.725-745.2000

Imataka H, Mizuno A, Fujii-Kuriyama Y, Hayami M (1993) Activation of the human immunodeficiency virus type 1 long terminal repeat by BTEB, a GC box-binding transcription factor. AIDS Res Hum Retroviruses 9(9):825–831. https://doi.org/10.1089/aid.1993.9.825

Israël N, Hazan U, Alcami J, Munier A, Arenzana-Seisdedos F, Bachelerie F, Israël A, Virelizier JL (1989) Tumor necrosis factor stimulates transcription of HIV-1 in human T lymphocytes, independently and synergistically with mitogens. J Immunol 143(12):3956–3960

Jaffe HW, Bregman DJ, Selik RM (1983) Acquired immune deficiency syndrome in the United States: the first 1000 cases. J Infect Dis 148(2):339–345. https://doi.org/10.1093/infdis/148.2.339

Jäger S, Cimermancic P, Gulbahce N, Johnson JR, McGovern KE, Clarke SC, Shales M, Mercenne G, Pache L, Li K, Hernandez H, Jang GM, Roth SL, Akiva E, Marlett J, Stephens M, D'Orso I, Fernandes J, Fahey M, Mahon C, O'Donoghue AJ, Todorovic A, Morris JH, Maltby DA, Alber T, Cagney G, Bushman FD, Young JA, Chanda SK, Sundquist WI, Kortemme T, Hernandez RD, Craik CS, Burlingame A, Sali A, Frankel AD, Krogan NJ (2011a) Global landscape of HIV-human protein complexes. Nature 481(7381):365–370. https://doi.org/10.1038/nature10719

Jäger S, Kim DY, Hultquist JF, Shindo K, LaRue RS, Kwon E, Li M, Anderson BD, Yen L, Stanley D, Mahon C, Kane J, Franks-Skiba K, Cimermancic P, Burlingame A, Sali A, Craik CS, Harris RS, Gross JD, Krogan NJ (2011b) Vif hijacks CBF-β to degrade APOBEC3G and promote HIV-1 infection. Nature 481(7381):371–375. https://doi.org/10.1038/nature10693

Jayappa KD, Ao Z, Yang M, Wang J, Yao X (2011) Identification of critical motifs within HIV-1 integrase required for importin α3 interaction and viral cDNA nuclear import. J Mol Biol 410(5):847–862. https://doi.org/10.1016/j.jmb.2011.04.011

Jeang KT, Chun R, Lin NH, Gatignol A, Glabe CG, Fan H (1993) In vitro and in vivo binding of human immunodeficiency virus type 1 Tat protein and Sp1 transcription factor. J Virol 67(10):6224–6233. https://doi.org/10.1128/jvi.67.10.6224-6233.1993

Jenkins Y, McEntee M, Weis K, Greene WC (1998) Characterization of HIV-1 vpr nuclear import: analysis of signals and pathways. J Cell Biol 143(4):875–885. https://doi.org/10.1083/jcb.143.4.875

Jones KA, Peterlin BM (1994) Control of RNA initiation and elongation at the HIV-1 promoter. Annu Rev Biochem 63:717–743. https://doi.org/10.1146/annurev.bi.63.070194.003441

Jones KA, Kadonaga JT, Luciw PA, Tjian R (1986) Activation of the AIDS retrovirus promoter by the cellular transcription factor, Sp1. Science 232(4751):755–759. https://doi.org/10.1126/science.3008338

Jubier-Maurin V, Saragosti S, Perret JL, Mpoudi E, Esu-Williams E, Mulanga C, Liegeois F, Ekwalanga M, Delaporte E, Peeters M (1999) Genetic characterization of the nef gene from human immunodeficiency virus type 1 group M strains representing genetic subtypes A, B, C, E, F, G, and H. AIDS Res Hum Retroviruses 15(1):23–32. https://doi.org/10.1089/088922299311673

Julien JP, Sok D, Khayat R, Lee JH, Doores KJ, Walker LM, Ramos A, Diwanji DC, Pejchal R, Cupo A, Katpally U, Depetris RS, Stanfield RL, McBride R, Marozsan AJ, Paulson JC, Sanders RW, Moore JP, Burton DR, Poignard P, Ward AB, Wilson IA (2013) Broadly neutralizing antibody PGT121 allosterically modulates CD4 binding via recognition of the HIV-1 gp120 V3 base and multiple surrounding glycans. PLoS Pathog 9(5):e1003342. https://doi.org/10.1371/journal.ppat.1003342

Jun S, Ke D, Debiec K, Zhao G, Meng X, Ambrose Z, Gibson GA, Watkins SC, Zhang P (2011) Direct visualization of HIV-1 with correlative live-cell microscopy and cryo-electron tomography. Structure 19(11):1573–1581. https://doi.org/10.1016/j.str.2011.09.006

Kang H, Minder P, Park MA, Mesquitta WT, Torbett BE, Slukvin II (2015) CCR5 disruption in induced pluripotent stem cells using CRISPR/Cas9 provides selective resistance of immune cells to CCR5-tropic HIV-1 virus. Mol Ther Nucleic Acids 4:e268. https://doi.org/10.1038/mtna.2015.42

Karn J, Stoltzfus CM (2012) Transcriptional and posttranscriptional regulation of HIV-1 gene expression. Cold Spring Harb Perspect Med 2(2):a006916. https://doi.org/10.1101/cshperspect.a006916

Kashanchi F, Piras G, Radonovich MF, Duvall JF, Fattaey A, Chiang CM, Roeder RG, Brady JN (1994) Direct interaction of human TFIID with the HIV-1 transactivator tat. Nature 367(6460):295–299. https://doi.org/10.1038/367295a0

Kashanchi F, Khleif SN, Duvall JF, Sadaie MR, Radonovich MF, Cho M, Martin MA, Chen SY, Weinmann R, Brady JN (1996) Interaction of human immunodeficiency virus type 1 Tat with a unique site of TFIID inhibits negative cofactor Dr1 and stabilizes the TFIID-TFIIA complex. J Virol 70(8):5503–5510. https://doi.org/10.1128/jvi.70.8.5503-5510.1996

Kauder SE, Bosque A, Lindqvist A, Planelles V, Verdin E (2009) Epigenetic regulation of HIV-1 latency by cytosine methylation. PLoS Pathog 5(6):e1000495. https://doi.org/10.1371/journal.ppat.1000495

Kawano K, Doucet AJ, Ueno M, Kariya R, An W, Marzetta F, Kuroki M, Turelli P, Sukegawa S, Okada S, Strebel K, Trono D, Ariumi Y (2018) HIV-1 Vpr and p21 restrict LINE-1 mobility. Nucleic Acids Res 46(16):8454–8470. https://doi.org/10.1093/nar/gky688

Khan M, Garcia-Barrio M, Powell MD (2001) Restoration of wild-type infectivity to human immunodeficiency virus type 1 strains lacking nef by intravirion reverse transcription. J Virol 75(24):12081–12087. https://doi.org/10.1128/jvi.75.24.12081-12087.2001

Kiem HP, Jerome KR, Deeks SG, McCune JM (2012) Hematopoietic-stem-cell-based gene therapy for HIV disease. Cell Stem Cell 10(2):137–147. https://doi.org/10.1016/j.stem.2011.12.015

Kim DY (2015) The assembly of Vif ubiquitin E3 ligase for APOBEC3 degradation. Arch Pharm Res 38(4):435–445. https://doi.org/10.1007/s12272-014-0519-x

Kingsman SM, Kingsman AJ (1996) The regulation of human immunodeficiency virus type-1 gene expression. Eur J Biochem 240(3):491–507. https://doi.org/10.1111/j.1432-1033.1996.0491h.x

Kitchen SG, Zack JA (1999) Distribution of the human immunodeficiency virus coreceptors CXCR4 and CCR5 in fetal lymphoid organs: implications for pathogenesis in utero. AIDS Res Hum Retroviruses 15(2):143–148. https://doi.org/10.1089/088922299311565

Klatzmann D, Champagne E, Chamaret S, Gruest J, Guetard D, Hercend T, Gluckman JC, Montagnier L (1984) T-lymphocyte T4 molecule behaves as the receptor for human retrovirus LAV. Nature 312(5996):767–768. https://doi.org/10.1038/312767a0

Kogan M, Rappaport J (2011) HIV-1 accessory protein Vpr: relevance in the pathogenesis of HIV and potential for therapeutic intervention. Retrovirology 8:25. https://doi.org/10.1186/1742-4690-8-25

Kohl NE, Emini EA, Schleif WA, Davis LJ, Heimbach JC, Dixon RA, Scolnick EM, Sigal IS (1988) Active human immunodeficiency virus protease is required for viral infectivity. Proc Natl Acad Sci U S A 85(13):4686–4690. https://doi.org/10.1073/pnas.85.13.4686

Kong R, Xu K, Zhou T, Acharya P, Lemmin T, Liu K, Ozorowski G, Soto C, Taft JD, Bailer RT, Cale EM, Chen L, Choi CW, Chuang GY, Doria-Rose NA, Druz A, Georgiev IS, Gorman J, Huang J, Joyce MG, Louder MK, Ma X, McKee K, O'Dell S, Pancera M, Yang Y, Blanchard SC, Mothes W, Burton DR, Koff WC, Connors M, Ward AB, Kwong PD, Mascola JR (2016) Fusion peptide of HIV-1 as a site of vulnerability to neutralizing antibody. Science 352(6287):828–833. https://doi.org/10.1126/science.aae0474

König R, Zhou Y, Elleder D, Diamond TL, Bonamy GM, Irelan JT, Chiang CY, Tu BP, De Jesus PD, Lilley CE, Seidel S, Opaluch AM, Caldwell JS, Weitzman MD, Kuhen KL, Bandyopadhyay S, Ideker T, Orth AP, Miraglia LJ, Bushman FD, Young JA, Chanda SK (2008) Global analysis of host-pathogen interactions that regulate early-stage HIV-1 replication. Cell 135(1):49–60. https://doi.org/10.1016/j.cell.2008.07.032

Korber B, Muldoon M, Theiler J, Gao F, Gupta R, Lapedes A, Hahn BH, Wolinsky S, Bhattacharya T (2000) Timing the ancestor of the HIV-1 pandemic strains. Science 288(5472):1789–1796. https://doi.org/10.1126/science.288.5472.1789

Kordelas L, Verheyen J, Beelen DW, Horn PA, Heinold A, Kaiser R, Trenschel R, Schadendorf D, Dittmer U, Esser S (2014) Shift of HIV tropism in stem-cell transplantation with CCR5 Delta32 mutation. N Engl J Med 371(9):880–882. https://doi.org/10.1056/NEJMc1405805

Kozak M (1989) The scanning model for translation: an update. J Cell Biol 108(2):229–241. https://doi.org/10.1083/jcb.108.2.229

Kramer G, Moerland PD, Jeeninga RE, Vlietstra WJ, Ringrose JH, Byrman C, Berkhout B, Speijer D (2012) Proteomic analysis of HIV-T cell interaction: an update. Front Microbiol 3:240. https://doi.org/10.3389/fmicb.2012.00240

Krasnow SW, Zhang LQ, Leung KY, Osborn L, Kunkel S, Nabel GJ (1991) Tumor necrosis factor-alpha, interleukin 1, and phorbol myristate acetate are independent activators of NF-kappa B which differentially activate T cells. Cytokine 3(5):372–379. https://doi.org/10.1016/1043-4666(91)90040-k

Kretzschmar M, Meisterernst M, Scheidereit C, Li G, Roeder RG (1992) Transcriptional regulation of the HIV-1 promoter by NF-kappa B in vitro. Genes Dev 6(5):761–774. https://doi.org/10.1101/gad.6.5.761

Kulkosky J, Jones KS, Katz RA, Mack JP, Skalka AM (1992) Residues critical for retroviral integrative recombination in a region that is highly conserved among retroviral/retrotransposon integrases and bacterial insertion sequence transposases. Mol Cell Biol 12(5):2331–2338. https://doi.org/10.1128/mcb.12.5.2331

Kumar M, Mitra D (2005) Heat shock protein 40 is necessary for human immunodeficiency virus-1 Nef-mediated enhancement of viral gene expression and replication. J Biol Chem 280(48):40041–40050. https://doi.org/10.1074/jbc.M508904200

Kundu M, Srinivasan A, Pomerantz RJ, Khalili K (1995) Evidence that a cell cycle regulator, E2F1, down-regulates transcriptional activity of the human immunodeficiency virus type 1 promoter. J Virol 69(11):6940–6946. https://doi.org/10.1128/jvi.69.11.6940-6946.1995

Kuppuswamy M, Subramanian T, Srinivasan A, Chinnadurai G (1989) Multiple functional domains of Tat, the trans-activator of HIV-1, defined by mutational analysis. Nucleic Acids Res 17(9):3551–3561. https://doi.org/10.1093/nar/17.9.3551

Kuznetsov YG, Victoria JG, Robinson WE Jr, McPherson A (2003) Atomic force microscopy investigation of human immunodeficiency virus (HIV) and HIV-infected lymphocytes. J Virol 77(22):11896–11909. https://doi.org/10.1128/jvi.77.22.11896-11909.2003

Kwong PD, Wyatt R, Majeed S, Robinson J, Sweet RW, Sodroski J, Hendrickson WA (2000) Structures of HIV-1 gp120 envelope glycoproteins from laboratory-adapted and primary isolates. Structure 8(12):1329–1339. https://doi.org/10.1016/s0969-2126(00)00547-5

Lagathu C, Cossarizza A, Béréziat V, Nasi M, Capeau J, Pinti M (2017) Basic science and pathogenesis of ageing with HIV: potential mechanisms and biomarkers. AIDS (London, England) 31(Suppl 2):S105–S119. https://doi.org/10.1097/QAD.0000000000001441

Lama J, Planelles V (2007) Host factors influencing susceptibility to HIV infection and AIDS progression. Retrovirology 4:52. https://doi.org/10.1186/1742-4690-4-52

Landi A, Vermeire J, Iannucci V, Vanderstraeten H, Naessens E, Bentahir M, Verhasselt B (2014) Genome-wide shRNA screening identifies host factors involved in early endocytic events for HIV-1-induced CD4 down-regulation. Retrovirology 11:118. https://doi.org/10.1186/s12977-014-0118-4

Laughlin MA, Zeichner S, Kolson D, Alwine JC, Seshamma T, Pomerantz RJ, Gonzalez-Scarano F (1993) Sodium butyrate treatment of cells latently infected with HIV-1 results in the expression of unspliced viral RNA. Virology 196(2):496–505. https://doi.org/10.1006/viro.1993.1505

Lee B, Sharron M, Montaner LJ, Weissman D, Doms RW (1999) Quantification of CD4, CCR5, and CXCR4 levels on lymphocyte subsets, dendritic cells, and differentially conditioned monocyte-derived macrophages. Proc Natl Acad Sci U S A 96(9):5215–5220. https://doi.org/10.1073/pnas.96.9.5215

Lee K, Ambrose Z, Martin TD, Oztop I, Mulky A, Julias JG, Vandegraaff N, Baumann JG, Wang R, Yuen W, Takemura T, Shelton K, Taniuchi I, Li Y, Sodroski J, Littman DR, Coffin JM, Hughes SH, Unutmaz D, Engelman A, KewalRamani VN (2010) Flexible use of nuclear import pathways by HIV-1. Cell Host Microbe 7(3):221–233. https://doi.org/10.1016/j.chom.2010.02.007

Lehrman G, Hogue IB, Palmer S, Jennings C, Spina CA, Wiegand A, Landay AL, Coombs RW, Richman DD, Mellors JW, Coffin JM, Bosch RJ, Margolis DM (2005) Depletion of latent HIV-1 infection in vivo: a proof-of-concept study. Lancet 366(9485):549–555. https://doi.org/10.1016/s0140-6736(05)67098-5

Levin JG, Guo J, Rouzina I, Musier-Forsyth K (2005) Nucleic acid chaperone activity of HIV-1 nucleocapsid protein: critical role in reverse transcription and molecular mechanism. Prog Nucleic Acid Res Mol Biol 80:217–286. https://doi.org/10.1016/s0079-6603(05)80006-6

Levin JG, Mitra M, Mascarenhas A, Musier-Forsyth K (2010) Role of HIV-1 nucleocapsid protein in HIV-1 reverse transcription. RNA Biol 7(6):754–774. https://doi.org/10.4161/rna.7.6.14115

Levy JA (1988) Mysteries of HIV: challenges for therapy and prevention. Nature 333(6173):519–522. https://doi.org/10.1038/333519a0

Levy JA (1993) Pathogenesis of human immunodeficiency virus infection. Microbiol Rev 57(1):183–289

Levy JA, Hoffman AD, Kramer SM, Landis JA, Shimabukuro JM, Oshiro LS (1984) Isolation of lymphocytopathic retroviruses from San Francisco patients with AIDS. Science 225(4664):840–842. https://doi.org/10.1126/science.6206563

Lewis P, Hensel M, Emerman M (1992) Human immunodeficiency virus infection of cells arrested in the cell cycle. EMBO J 11(8):3053–3058

Li X, Liang C, Quan Y, Chandok R, Laughrea M, Parniak MA, Kleiman L, Wainberg MA (1997a) Identification of sequences downstream of the primer binding site that are important for efficient replication of human immunodeficiency virus type 1. J Virol 71(8):6003–6010. https://doi.org/10.1128/jvi.71.8.6003-6010.1997

Li X, Quan Y, Wainberg MA (1997b) Controlling elements in replication of the human immunodeficiency virus type 1. Cell Mol Biol (Noisy-le-grand) 43(3):443–454

Li BQ, Fu T, Dongyan Y, Mikovits JA, Ruscetti FW, Wang JM (2000a) Flavonoid baicalin inhibits HIV-1 infection at the level of viral entry. Biochem Biophys Res Commun 276(2):534–538. https://doi.org/10.1006/bbrc.2000.3485

Li S, Hill CP, Sundquist WI, Finch JT (2000b) Image reconstructions of helical assemblies of the HIV-1 CA protein. Nature 407(6802):409–413. https://doi.org/10.1038/35030177

Li L, Krymskaya L, Wang J, Henley J, Rao A, Cao LF, Tran CA, Torres-Coronado M, Gardner A, Gonzalez N, Kim K, Liu PQ, Hofer U, Lopez E, Gregory PD, Liu Q, Holmes MC, Cannon PM, Zaia JA, DiGiusto DL (2013) Genomic editing of the HIV-1 coreceptor CCR5 in adult hematopoietic stem and progenitor cells using zinc finger nucleases. Mol Ther 21(6):1259–1269. https://doi.org/10.1038/mt.2013.65

Liang C, Li X, Quan Y, Laughrea M, Kleiman L, Hiscott J, Wainberg MA (1997) Sequence elements downstream of the human immunodeficiency virus type 1 long terminal repeat are required for efficient viral gene transcription. J Mol Biol 272(2):167–177. https://doi.org/10.1006/jmbi.1997.1239

Liu J, Perkins ND, Schmid RM, Nabel GJ (1992) Specific NF-kappa B subunits act in concert with Tat to stimulate human immunodeficiency virus type 1 transcription. J Virol 66(6):3883–3887. https://doi.org/10.1128/jvi.66.6.3883-3887.1992

Liu R, Paxton WA, Choe S, Ceradini D, Martin SR, Horuk R, MacDonald ME, Stuhlmann H, Koup RA, Landau NR (1996) Homozygous defect in HIV-1 coreceptor accounts for resistance of some multiply-exposed individuals to HIV-1 infection. Cell 86(3):367–377. https://doi.org/10.1016/s0092-8674(00)80110-5

Liu J, Bartesaghi A, Borgnia MJ, Sapiro G, Subramaniam S (2008) Molecular architecture of native HIV-1 gp120 trimers. Nature 455(7209):109–113. https://doi.org/10.1038/nature07159

Liu Y, Belkina NV, Shaw S (2009) HIV infection of T cells: actin-in and actin-out. Sci Signal 2(66):e23. https://doi.org/10.1126/scisignal.266pe23

Locker N, Chamond N, Sargueil B (2011) A conserved structure within the HIV gag open reading frame that controls translation initiation directly recruits the 40S subunit and eIF3. Nucleic Acids Res 39(6):2367–2377. https://doi.org/10.1093/nar/gkq1118

Lu YC, Touzjian N, Stenzel M, Dorfman T, Sodroski JG, Haseltine WA (1990) Identification of cis-acting repressive sequences within the negative regulatory element of human immunodeficiency virus type 1. J Virol 64(10):5226–5229. https://doi.org/10.1128/jvi.64.10.5226-5229.1990

Luo K, Xiao Z, Ehrlich E, Yu Y, Liu B, Zheng S, Yu XF (2005) Primate lentiviral virion infectivity factors are substrate receptors that assemble with cullin 5-E3 ligase through a HCCH motif to suppress APOBEC3G. Proc Natl Acad Sci U S A 102(32):11444–11449. https://doi.org/10.1073/pnas.0502440102

Luque MC, Santos CC, Mairena EC, Wilkinson P, Boucher G, Segurado AC, Fonseca LA, Sabino E, Kalil JE, Cunha-Neto E (2014) Gene expression profile in long-term non progressor HIV infected patients: in search of potential resistance factors. Mol Immunol 62(1):63–70. https://doi.org/10.1016/j.molimm.2014.05.016

Lusic M, Siliciano RF (2017) Nuclear landscape of HIV-1 infection and integration. Nat Rev Microbiol 15(2):69–82. https://doi.org/10.1038/nrmicro.2016.162

Magadán JG, Bonifacino JS (2012) Transmembrane domain determinants of CD4 downregulation by HIV-1 Vpu. J Virol 86(2):757–772. https://doi.org/10.1128/jvi.05933-11

Magadán JG, Pérez-Victoria FJ, Sougrat R, Ye Y, Strebel K, Bonifacino JS (2010) Multilayered mechanism of CD4 downregulation by HIV-1 Vpu involving distinct ER retention and ERAD targeting steps. PLoS Pathog 6(4):e1000869. https://doi.org/10.1371/journal.ppat.1000869

Majello B, Napolitano G, De Luca P, Lania L (1998) Recruitment of human TBP selectively activates RNA polymerase II TATA-dependent promoters. J Biol Chem 273(26):16509–16516. https://doi.org/10.1074/jbc.273.26.16509

Maldarelli F, Chen MY, Willey RL, Strebel K (1993) Human immunodeficiency virus type 1 Vpu protein is an oligomeric type I integral membrane protein. J Virol 67(8):5056–5061. https://doi.org/10.1128/jvi.67.8.5056-5061.1993

Malim MH, Tiley LS, McCarn DF, Rusche JR, Hauber J, Cullen BR (1990) HIV-1 structural gene expression requires binding of the Rev trans-activator to its RNA target sequence. Cell 60(4):675–683. https://doi.org/10.1016/0092-8674(90)90670-a

Mamede JI, Cianci GC, Anderson MR, Hope TJ (2017) Early cytoplasmic uncoating is associated with infectivity of HIV-1. Proc Natl Acad Sci U S A 114(34):E7169–e7178. https://doi.org/10.1073/pnas.1706245114

Managlia EZ, Landay A, Al-Harthi L (2006) Interleukin-7 induces HIV replication in primary naive T cells through a nuclear factor of activated T cell (NFAT)-dependent pathway. Virology 350(2):443–452. https://doi.org/10.1016/j.virol.2006.02.019

Marcheschi RJ, Staple DW, Butcher SE (2007) Programmed ribosomal frameshifting in SIV is induced by a highly structured RNA stem-loop. J Mol Biol 373(3):652–663. https://doi.org/10.1016/j.jmb.2007.08.033

Margolis DM, Ostrove JM, Straus SE (1993) HSV-1 activation of HIV-1 transcription is augmented by a cellular protein that binds near the initiator element. Virology 192(1):370–374. https://doi.org/10.1006/viro.1993.1046

Margottin F, Bour SP, Durand H, Selig L, Benichou S, Richard V, Thomas D, Strebel K, Benarous R (1998) A novel human WD protein, h-beta TrCp, that interacts with HIV-1 Vpu connects CD4 to the ER degradation pathway through an F-box motif. Mol Cell 1(4):565–574. https://doi.org/10.1016/s1097-2765(00)80056-8

Marin M, Rose KM, Kozak SL, Kabat D (2003) HIV-1 Vif protein binds the editing enzyme APOBEC3G and induces its degradation. Nat Med 9(11):1398–1403. https://doi.org/10.1038/nm946

Marini B, Kertesz-Farkas A, Ali H, Lucic B, Lisek K, Manganaro L, Pongor S, Luzzati R, Recchia A, Mavilio F, Giacca M, Lusic M (2015) Nuclear architecture dictates HIV-1 integration site selection. Nature 521(7551):227–231. https://doi.org/10.1038/nature14226

Martin-Serrano J, Zang T, Bieniasz PD (2003) Role of ESCRT-I in retroviral budding. J Virol 77(8):4794–4804. https://doi.org/10.1128/jvi.77.8.4794-4804.2003

Masur H, Michelis MA, Greene JB, Onorato I, Stouwe RA, Holzman RS, Wormser G, Brettman L, Lange M, Murray HW, Cunningham-Rundles S (1981) An outbreak of community-acquired Pneumocystis carinii pneumonia: initial manifestation of cellular immune dysfunction. N Engl J Med 305(24):1431–1438. https://doi.org/10.1056/nejm198112103052402

Matreyek KA, Engelman A (2011) The requirement for nucleoporin NUP153 during human immunodeficiency virus type 1 infection is determined by the viral capsid. J Virol 85(15):7818–7827. https://doi.org/10.1128/jvi.00325-11

Mattapallil JJ, Douek DC, Hill B, Nishimura Y, Martin M, Roederer M (2005) Massive infection and loss of memory CD4+ T cells in multiple tissues during acute SIV infection. Nature 434(7037):1093–1097. https://doi.org/10.1038/nature03501

Mazauric MH, Seol Y, Yoshizawa S, Visscher K, Fourmy D (2009) Interaction of the HIV-1 frameshift signal with the ribosome. Nucleic Acids Res 37(22):7654–7664. https://doi.org/10.1093/nar/gkp779

McCune JM, Rabin LB, Feinberg MB, Lieberman M, Kosek JC, Reyes GR, Weissman IL (1988) Endoproteolytic cleavage of gp160 is required for the activation of human immunodeficiency virus. Cell 53(1):55–67. https://doi.org/10.1016/0092-8674(88)90487-4

McLaren M, Marsh K, Cochrane A (2008) Modulating HIV-1 RNA processing and utilization. Front Biosci 13:5693–5707. https://doi.org/10.2741/3110

Mehle A, Thomas ER, Rajendran KS, Gabuzda D (2006) A zinc-binding region in Vif binds Cul5 and determines cullin selection. J Biol Chem 281(25):17259–17265. https://doi.org/10.1074/jbc.M602413200

Melikyan GB (2008) Common principles and intermediates of viral protein-mediated fusion: the HIV-1 paradigm. Retrovirology 5:111. https://doi.org/10.1186/1742-4690-5-111

Mellors JW, Griffith BP, Ortiz MA, Landry ML, Ryan JL (1991) Tumor necrosis factor-alpha/cachectin enhances human immunodeficiency virus type 1 replication in primary macrophages. J Infect Dis 163(1):78–82. https://doi.org/10.1093/infdis/163.1.78

Meng B, Lever AM (2013) Wrapping up the bad news: HIV assembly and release. Retrovirology 10:5. https://doi.org/10.1186/1742-4690-10-5

Meng G, Sellers MT, Mosteller-Barnum M, Rogers TS, Shaw GM, Smith PD (2000) Lamina propria lymphocytes, not macrophages, express CCR5 and CXCR4 and are the likely target cell for human immunodeficiency virus type 1 in the intestinal mucosa. J Infect Dis 182(3):785–791. https://doi.org/10.1086/315790

Meyer BE, Malim MH (1994) The HIV-1 Rev trans-activator shuttles between the nucleus and the cytoplasm. Genes Dev 8(13):1538–1547. https://doi.org/10.1101/gad.8.13.1538

Michael NL, Louie LG, Rohrbaugh AL, Schultz KA, Dayhoff DE, Wang CE, Sheppard HW (1997) The role of CCR5 and CCR2 polymorphisms in HIV-1 transmission and disease progression. Nat Med 3(10):1160–1162. https://doi.org/10.1038/nm1097-1160

Miller MD, Feinberg MB, Greene WC (1994) The HIV-1 nef gene acts as a positive viral infectivity factor. Trends Microbiol 2(8):294–298. https://doi.org/10.1016/0966-842x(94)90007-8

Miller MD, Warmerdam MT, Page KA, Feinberg MB, Greene WC (1995) Expression of the human immunodeficiency virus type 1 (HIV-1) nef gene during HIV-1 production increases progeny particle infectivity independently of gp160 or viral entry. J Virol 69(1):579–584. https://doi.org/10.1128/jvi.69.1.579-584.1995

Modrow S, Hahn BH, Shaw GM, Gallo RC, Wong-Staal F, Wolf H (1987) Computer-assisted analysis of envelope protein sequences of seven human immunodeficiency virus isolates: prediction of antigenic epitopes in conserved and variable regions. J Virol 61(2):570–578. https://doi.org/10.1128/jvi.61.2.570-578.1987

Molotkovsky RJ, Alexandrova VV, Galimzyanov TR, Jiménez-Munguía I, Pavlov KV, Batishchev OV (2018) Lateral membrane heterogeneity regulates viral-induced membrane fusion during HIV entry. Int J Mol Sci 19(5). https://doi.org/10.3390/ijms19051483

Moore MD, Hu WS (2009) HIV-1 RNA dimerization: it takes two to tango. AIDS Rev 11(2):91–102

Moore JP, Kitchen SG, Pugach P, Zack JA (2004) The CCR5 and CXCR4 coreceptors--central to understanding the transmission and pathogenesis of human immunodeficiency virus type 1 infection. AIDS Res Hum Retroviruses 20(1):111–126. https://doi.org/10.1089/088922204322749567

Mori M, Kovalenko L, Malancona S, Saladini F (2018) Structure-Based. Identification of HIV-1 nucleocapsid protein inhibitors active against wild-type and drug-resistant HIV-1 strains. 13(1):253–266. https://doi.org/10.1021/acschembio.7b00907

Mouhand A, Belfetmi A, Catala M, Larue V, Zargarian L, Brachet F, Gorelick RJ, Van Heijenoort C, Mirambeau G, Barraud P, Mauffret O, Tisné C (2018) Modulation of the HIV nucleocapsid dynamics finely tunes its RNA-binding properties during virion genesis. Nucleic Acids Res 46(18):9699–9710. https://doi.org/10.1093/nar/gky612

Müller B, Patschinsky T, Kräusslich HG (2002) The late-domain-containing protein p6 is the predominant phosphoprotein of human immunodeficiency virus type 1 particles. J Virol 76(3):1015–1024. https://doi.org/10.1128/jvi.76.3.1015-1024.2002

Münch J, Ständker L, Adermann K, Schulz A, Schindler M, Chinnadurai R, Pöhlmann S, Chaipan C, Biet T, Peters T, Meyer B, Wilhelm D, Lu H, Jing W, Jiang S, Forssmann WG, Kirchhoff F (2007) Discovery and optimization of a natural HIV-1 entry inhibitor targeting the gp41 fusion peptide. Cell 129(2):263–275. https://doi.org/10.1016/j.cell.2007.02.042

Muriaux D, Darlix JL (2010) Properties and functions of the nucleocapsid protein in virus assembly. RNA Biol 7(6):744–753. https://doi.org/10.4161/rna.7.6.14065

Nabatanzi R, Cose S, Joloba M, Jones SR, Nakanjako D (2018) Effects of HIV infection and ART on phenotype and function of circulating monocytes, natural killer, and innate lymphoid cells. AIDS Res Ther 15(1):7. https://doi.org/10.1186/s12981-018-0194-y

Namy O, Moran SJ, Stuart DI, Gilbert RJ, Brierley I (2006) A mechanical explanation of RNA pseudoknot function in programmed ribosomal frameshifting. Nature 441(7090):244–247. https://doi.org/10.1038/nature04735

Ne E, Palstra RJ, Mahmoudi T (2018) Transcription: insights from the HIV-1 promoter. Int Rev Cell Mol Biol 335:191–243. https://doi.org/10.1016/bs.ircmb.2017.07.011

Neil SJ, Zang T, Bieniasz PD (2008) Tetherin inhibits retrovirus release and is antagonized by HIV-1 Vpu. Nature 451(7177):425–430. https://doi.org/10.1038/nature06553

Nicholson MG, Rue SM, Clements JE, Barber SA (2006) An internal ribosome entry site promotes translation of a novel SIV Pr55(Gag) isoform. Virology 349(2):325–334. https://doi.org/10.1016/j.virol.2006.01.034

Niederman TM, Thielan BJ, Ratner L (1989) Human immunodeficiency virus type 1 negative factor is a transcriptional silencer. Proc Natl Acad Sci U S A 86(4):1128–1132. https://doi.org/10.1073/pnas.86.4.1128

Opp S, Fricke T, Shepard C, Kovalskyy D, Bhattacharya A (2017) The small-molecule 3G11 inhibits HIV-1 reverse transcription. Chem Biol Des 89(4):608–618. https://doi.org/10.1111/cbdd.12886

Orkin C, Llibre JM, Gallien S, Antinori A, Behrens G, Carr A (2018) Nucleoside reverse transcriptase inhibitor-reducing strategies in HIV treatment: assessing the evidence. HIV Med 19(1):18–32. https://doi.org/10.1111/hiv.12534

Ott DE, Coren LV, Kane BP, Busch LK, Johnson DG, Sowder RC 2nd, Chertova EN, Arthur LO, Henderson LE (1996) Cytoskeletal proteins inside human immunodeficiency virus type 1 virions. J Virol 70(11):7734–7743. https://doi.org/10.1128/jvi.70.11.7734-7743.1996

Pace MJ, Agosto L, Graf EH, O'Doherty U (2011) HIV reservoirs and latency models. Virology 411(2):344–354. https://doi.org/10.1016/j.virol.2010.12.041

Pancera M, Majeed S, Ban YE, Chen L, Huang CC, Kong L, Kwon YD, Stuckey J, Zhou T, Robinson JE, Schief WR, Sodroski J, Wyatt R, Kwong PD (2010) Structure of HIV-1 gp120 with gp41-interactive region reveals layered envelope architecture and basis of conformational mobility. Proc Natl Acad Sci U S A 107(3):1166–1171. https://doi.org/10.1073/pnas.0911004107

Pancera M, Zhou T, Druz A, Georgiev IS, Soto C, Gorman J, Huang J, Acharya P, Chuang GY, Ofek G, Stewart-Jones GB, Stuckey J, Bailer RT, Joyce MG, Louder MK, Tumba N, Yang Y, Zhang B, Cohen MS, Haynes BF, Mascola JR, Morris L, Munro JB, Blanchard SC, Mothes W, Connors M, Kwong PD (2014) Structure and immune recognition of trimeric pre-fusion HIV-1. Environ Nat 514(7523):455–461. https://doi.org/10.1038/nature13808

Pandit A, de Boer RJ (2015) HIV-1 CCR5 gene therapy will fail unless it is combined with a suicide gene. Sci Rep 5:18088. https://doi.org/10.1038/srep18088

Pandori MW, Fitch NJ, Craig HM, Richman DD, Spina CA, Guatelli JC (1996) Producer-cell modification of human immunodeficiency virus type 1: Nef is a virion protein. J Virol 70(7):4283–4290. https://doi.org/10.1128/jvi.70.7.4283-4290.1996

Pang Y, Song H, Kim JH, Hou X, Cheng W (2014) Optical trapping of individual human immunodeficiency viruses in culture fluid reveals heterogeneity with single-molecule resolution. Nat Nanotechnol 9(8):624–630. https://doi.org/10.1038/nnano.2014.140

Papkalla A, Münch J, Otto C, Kirchhoff F (2002) Nef enhances human immunodeficiency virus type 1 infectivity and replication independently of viral coreceptor tropism. J Virol 76(16):8455–8459. https://doi.org/10.1128/jvi.76.16.8455-8459.2002

Paranjape RS, Challacombe SJ (2016) HIV/AIDS in India: an overview of the Indian epidemic. Oral Dis 22(Suppl 1):10–14. https://doi.org/10.1111/odi.12457

Park RJ, Wang T, Koundakjian D, Hultquist JF, Lamothe-Molina P, Monel B, Schumann K, Yu H, Krupzcak KM, Garcia-Beltran W, Piechocka-Trocha A, Krogan NJ, Marson A, Sabatini DM, Lander ES, Hacohen N, Walker BD (2017) A genome-wide CRISPR screen identifies a restricted set of HIV host dependency factors. Nat Genet. https://doi.org/10.1038/ng.3741

Peeters M, Toure-Kane C, Nkengasong JN (2003) Genetic diversity of HIV in Africa: impact on diagnosis, treatment, vaccine development and trials. Aids 17(18):2547–2560. https://doi.org/10.1097/01.aids.0000096895.73209.89

Pendergrast PS, Morrison D, Tansey WP, Hernandez N (1996) Mutations in the carboxy-terminal domain of TBP affect the synthesis of human immunodeficiency virus type 1 full-length and short transcripts similarly. J Virol 70(8):5025–5034. https://doi.org/10.1128/jvi.70.8.5025-5034.1996

Pereira LA, Bentley K, Peeters A, Churchill MJ, Deacon NJ (2000) A compilation of cellular transcription factor interactions with the HIV-1 LTR promoter. Nucleic Acids Res 28(3):663–668. https://doi.org/10.1093/nar/28.3.663

Perelson AS, Essunger P, Cao Y, Vesanen M, Hurley A, Saksela K, Markowitz M, Ho DD (1997) Decay characteristics of HIV-1-infected compartments during combination therapy. Nature 387(6629):188–191. https://doi.org/10.1038/387188a0

Peretz Y, Cameron C, Sékaly RP (2012) Dissecting the HIV-specific immune response: a systems biology approach. Curr Opin HIV AIDS 7(1):17–23. https://doi.org/10.1097/COH.0b013e32834ddb0e

Perkins ND, Edwards NL, Duckett CS, Agranoff AB, Schmid RM, Nabel GJ (1993) A cooperative interaction between NF-kappa B and Sp1 is required for HIV-1 enhancer activation. EMBO J 12(9):3551–3558

Perng GC, Jones C (2010) Towards an understanding of the herpes simplex virus type 1 latency-reactivation cycle. Interdiscip Perspect Infect Dis 2010:262415. https://doi.org/10.1155/2010/262415

Piatak M Jr, Saag MS, Yang LC, Clark SJ, Kappes JC, Luk KC, Hahn BH, Shaw GM, Lifson JD (1993) High levels of HIV-1 in plasma during all stages of infection determined by competitive PCR. Science 259(5102):1749–1754. https://doi.org/10.1126/science.8096089

Pollakis G, Paxton WA (2012) Use of (alternative) coreceptors for HIV entry. Curr Opin HIV AIDS 7(5):440–449. https://doi.org/10.1097/COH.0b013e328356e9f3

Pollard VW, Malim MH (1998) The HIV-1 Rev protein. Annu Rev Microbiol 52:491–532. https://doi.org/10.1146/annurev.micro.52.1.491

Popik W, Pitha PM (1993) Role of tumor necrosis factor alpha in activation and replication of the tat-defective human immunodeficiency virus type 1. J Virol 67(2):1094–1099. https://doi.org/10.1128/jvi.67.2.1094-1099.1993

Popov S, Rexach M, Ratner L, Blobel G, Bukrinsky M (1998) Viral protein R regulates docking of the HIV-1 preintegration complex to the nuclear pore complex. J Biol Chem 273(21):13347–13352. https://doi.org/10.1074/jbc.273.21.13347

Poveda E, Briz V, Quiñones-Mateu M, Soriano V (2006) HIV tropism: diagnostic tools and implications for disease progression and treatment with entry inhibitors. Aids 20(10):1359–1367. https://doi.org/10.1097/01.aids.0000233569.74769.69

Prins JM, Jurriaans S, van Praag RM, Blaak H, van Rij R, Schellekens PT, ten Berge IJ, Yong SL, Fox CH, Roos MT, de Wolf F, Goudsmit J, Schuitemaker H, Lange JM (1999) Immuno-activation with anti-CD3 and recombinant human IL-2 in HIV-1-infected patients on potent antiretroviral therapy. Aids 13(17):2405–2410. https://doi.org/10.1097/00002030-199912030-00012

Prokofjeva MM, Kochetkov SN, Prassolov VS (2016) Therapy of HIV infection: current approaches and prospects. Acta Nat 8(4):23–32

Provitera P, El-Maghrabi R, Scarlata S (2006) The effect of HIV-1 Gag myristoylation on membrane binding. Biophys Chem 119(1):23–32. https://doi.org/10.1016/j.bpc.2005.08.008

Rabbi MF, Saifuddin M, Gu DS, Kagnoff MF, Roebuck KA (1997) U5 region of the human immunodeficiency virus type 1 long terminal repeat contains TRE-like cAMP-responsive elements that bind both AP-1 and CREB/ATF proteins. Virology 233(1):235–245. https://doi.org/10.1006/viro.1997.8602

Rahimian P, He JJ (2016) Exosome-associated release, uptake, and neurotoxicity of HIV-1 Tat protein. J Neurovirol 22(6):774–788. https://doi.org/10.1007/s13365-016-0451-6

Rahman SA, Koch P, Weichsel J, Godinez WJ, Schwarz U, Rohr K, Lamb DC, Kräusslich HG, Müller B (2014) Investigating the role of F-actin in human immunodeficiency virus assembly by live-cell microscopy. J Virol 88(14):7904–7914. https://doi.org/10.1128/jvi.00431-14

Rai MA, Pannek S, Fichtenbaum CJ (2018) Emerging reverse transcriptase inhibitors for HIV-1 infection. Expert Opin Emerg Drugs 23(2):149–157. https://doi.org/10.1080/14728214.2018.1474202

Razooky BS, Pai A, Aull K, Rouzine IM, Weinberger LS (2015) A hardwired HIV latency program. Cell 160(5):990–1001. https://doi.org/10.1016/j.cell.2015.02.009

Reddy MV, Desai M, Jeyapaul J, Prasad DD, Seshamma T, Palmeri D, Khan SA (1992) Functional analysis of the N-terminal domain of Tat protein of the human immunodeficiency virus type 1. Oncogene 7(9):1743–1748

Rittner K, Churcher MJ, Gait MJ, Karn J (1995) The human immunodeficiency virus long terminal repeat includes a specialised initiator element which is required for Tat-responsive transcription. J Mol Biol 248(3):562–580. https://doi.org/10.1006/jmbi.1995.0243

Robertson DL, Anderson JP, Bradac JA, Carr JK, Foley B, Funkhouser RK, Gao F, Hahn BH, Kalish ML, Kuiken C, Learn GH, Leitner T, McCutchan F, Osmanov S, Peeters M, Pieniazek D, Salminen M, Sharp PM, Wolinsky S, Korber B (2000) HIV-1 nomenclature proposal. Science 288(5463):55–56. https://doi.org/10.1126/science.288.5463.55d

Robinson HL (2018) HIV/AIDS vaccines: 2018. Clin Pharmacol Ther 104(6):1062–1073. https://doi.org/10.1002/cpt.1208

Roebuck KA, Brenner DA, Kagnoff MF (1993) Identification of c-fos-responsive elements downstream of TAR in the long terminal repeat of human immunodeficiency virus type-1. J Clin Invest 92(3):1336–1348. https://doi.org/10.1172/jci116707

Romanchikova N, Ivanova V, Scheller C, Jankevics E, Jassoy C, Serfling E (2003) NFAT transcription factors control HIV-1 expression through a binding site downstream of TAR region. Immunobiology 208(4):361–365. https://doi.org/10.1078/0171-2985-00283

Rosa A, Chande A, Ziglio S, De Sanctis V, Bertorelli R, Goh SL, McCauley SM, Nowosielska A, Antonarakis SE, Luban J, Santoni FA, Pizzato M (2015) HIV-1 Nef promotes infection by excluding SERINC5 from virion incorporation. Nature 526(7572):212–217. https://doi.org/10.1038/nature15399

Ruelas DS, Greene WC (2013) An integrated overview of HIV-1 latency. Cell 155(3):519–529. https://doi.org/10.1016/j.cell.2013.09.044

Ruelas DS, Chan JK, Oh E, Heidersbach AJ, Hebbeler AM, Chavez L, Verdin E, Rape M, Greene WC (2015) MicroRNA-155 reinforces HIV latency. J Biol Chem 290(22):13736–13748. https://doi.org/10.1074/jbc.M115.641837

Samson M, Libert F, Doranz BJ, Rucker J, Liesnard C, Farber CM, Saragosti S, Lapoumeroulie C, Cognaux J, Forceille C, Muyldermans G, Verhofstede C, Burtonboy G, Georges M, Imai T, Rana S, Yi Y, Smyth RJ, Collman RG, Doms RW, Vassart G, Parmentier M (1996) Resistance to HIV-1 infection in caucasian individuals bearing mutant alleles of the CCR-5 chemokine receptor gene. Nature 382(6593):722–725. https://doi.org/10.1038/382722a0

Sarafianos SG, Marchand B, Das K, Himmel DM, Parniak MA, Hughes SH, Arnold E (2009) Structure and function of HIV-1 reverse transcriptase: molecular mechanisms of polymerization and inhibition. J Mol Biol 385(3):693–713. https://doi.org/10.1016/j.jmb.2008.10.071

Schmitz ML, Stelzer G, Altmann H, Meisterernst M, Baeuerle PA (1995) Interaction of the COOH-terminal transactivation domain of p65 NF-kappa B with TATA-binding protein, transcription factor IIB, and coactivators. J Biol Chem 270(13):7219–7226. https://doi.org/10.1074/jbc.270.13.7219

Schröder AR, Shinn P, Chen H, Berry C, Ecker JR, Bushman F (2002) HIV-1 integration in the human genome favors active genes and local hotspots. Cell 110(4):521–529. https://doi.org/10.1016/s0092-8674(02)00864-4

Schröfelbauer B, Yu Q, Zeitlin SG, Landau NR (2005) Human immunodeficiency virus type 1 Vpr induces the degradation of the UNG and SMUG uracil-DNA glycosylases. J Virol 79(17):10978–10987. https://doi.org/10.1128/jvi.79.17.10978-10987.2005

Schwartz O, Maréchal V, Le Gall S, Lemonnier F, Heard JM (1996) Endocytosis of major histocompatibility complex class I molecules is induced by the HIV-1 Nef protein. Nat Med 2(3):338–342. https://doi.org/10.1038/nm0396-338

Schwarze SR, Hruska KA, Dowdy SF (2000) Protein transduction: unrestricted delivery into all cells? Trends Cell Biol 10(7):290–295. https://doi.org/10.1016/s0962-8924(00)01771-2
Selyutina A, Persaud M, Lee K, KewalRamani V, Diaz-Griffero F (2020) Nuclear import of the HIV-1 core precedes reverse transcription and uncoating. Cell Rep 32(13):108201. https://doi.org/10.1016/j.celrep.2020.108201
Sharp PM, Hahn BH (2010) The evolution of HIV-1 and the origin of AIDS. Philos Trans R Soc Lond B Biol Sci 365(1552):2487–2494. https://doi.org/10.1098/rstb.2010.0031
Sharp PM, Hahn BH (2011) Origins of HIV and the AIDS pandemic. Cold Spring Harb Perspect Med 1(1):a006841. https://doi.org/10.1101/cshperspect.a006841
Sheehy AM, Gaddis NC, Malim MH (2003) The antiretroviral enzyme APOBEC3G is degraded by the proteasome in response to HIV-1 Vif. Nat Med 9(11):1404–1407. https://doi.org/10.1038/nm945
Shingai M, Donau OK, Plishka RJ, Buckler-White A, Mascola JR, Nabel GJ, Nason MC, Montefiori D, Moldt B, Poignard P, Diskin R, Bjorkman PJ, Eckhaus MA, Klein F, Mouquet H, Cetrulo Lorenzi JC, Gazumyan A, Burton DR, Nussenzweig MC, Martin MA, Nishimura Y (2014) Passive transfer of modest titers of potent and broadly neutralizing anti-HIV monoclonal antibodies block SHIV infection in macaques. J Exp Med 211(10):2061–2074. https://doi.org/10.1084/jem.20132494
Siegal FP, Lopez C, Hammer GS, Brown AE, Kornfeld SJ, Gold J, Hassett J, Hirschman SZ, Cunningham-Rundles C, Adelsberg BR et al (1981) Severe acquired immunodeficiency in male homosexuals, manifested by chronic perianal ulcerative herpes simplex lesions. N Engl J Med 305(24):1439–1444. https://doi.org/10.1056/nejm198112103052403
Siliciano RF, Greene WC (2011) HIV latency. Cold Spring Harb Perspect Med 1(1):a007096. https://doi.org/10.1101/cshperspect.a007096
Siliciano JD, Kajdas J, Finzi D, Quinn TC, Chadwick K, Margolick JB, Kovacs C, Gange SJ, Siliciano RF (2003) Long-term follow-up studies confirm the stability of the latent reservoir for HIV-1 in resting CD4+ T cells. Nat Med 9(6):727–728. https://doi.org/10.1038/nm880
Simmons A, Aluvihare V, McMichael A (2001) Nef triggers a transcriptional program in T cells imitating single-signal T cell activation and inducing HIV virulence mediators. Immunity 14(6):763–777. https://doi.org/10.1016/s1074-7613(01)00158-3
Simon JH, Sheehy AM, Carpenter EA, Fouchier RA, Malim MH (1999) Mutational analysis of the human immunodeficiency virus type 1 Vif protein. J Virol 73(4):2675–2681. https://doi.org/10.1128/jvi.73.4.2675-2681.1999
Soros VB, Yonemoto W, Greene WC (2007) Newly synthesized APOBEC3G is incorporated into HIV virions, inhibited by HIV RNA, and subsequently activated by RNase H. PLoS Pathog 3(2):e15. https://doi.org/10.1371/journal.ppat.0030015
Soto-Rifo R, Limousin T, Rubilar PS, Ricci EP, Décimo D, Moncorgé O, Trabaud MA, André P, Cimarelli A, Ohlmann T (2012) Different effects of the TAR structure on HIV-1 and HIV-2 genomic RNA translation. Nucleic Acids Res 40(6):2653–2667. https://doi.org/10.1093/nar/gkr1093
Sousa J, Müller V, Vandamme A, Vandamme A-M (2017) The epidemic emergence of HIV: what novel enabling factors were involved? Future Virology 12:685–707
Spell SR, Mangrum JB, Peterson EJ, Fabris D, Ptak R, Farrell NP (2016) Au(iii) compounds as HIV nucleocapsid protein (NCp7)-nucleic acid antagonists. Chem Commun 53(1):91–94
Spina CA, Kwoh TJ, Chowers MY, Guatelli JC, Richman DD (1994) The importance of nef in the induction of human immunodeficiency virus type 1 replication from primary quiescent CD4 lymphocytes. J Exp Med 179(1):115–123. https://doi.org/10.1084/jem.179.1.115
Stopak K, de Noronha C, Yonemoto W, Greene WC (2003) HIV-1 Vif blocks the antiviral activity of APOBEC3G by impairing both its translation and intracellular stability. Mol Cell 12(3):591–601. https://doi.org/10.1016/s1097-2765(03)00353-8
Strebel K (2014) HIV-1 Vpu - an ion channel in search of a job. Biochim Biophys Acta 1838(4):1074–1081. https://doi.org/10.1016/j.bbamem.2013.06.029
Subbramanian RA, Yao XJ, Dilhuydy H, Rougeau N, Bergeron D, Robitaille Y, Cohen EA (1998) Human immunodeficiency virus type 1 Vpr localization: nuclear transport of a viral protein

modulated by a putative amphipathic helical structure and its relevance to biological activity. J Mol Biol 278(1):13–30. https://doi.org/10.1006/jmbi.1998.1685
Suhasini M, Reddy TR (2009) Cellular proteins and HIV-1 Rev function. Curr HIV Res 7(1):91–100. https://doi.org/10.2174/157016209787048474
Svitkin YV, Pause A, Sonenberg N (1994) La autoantigen alleviates translational repression by the 5′ leader sequence of the human immunodeficiency virus type 1 mRNA. J Virol 68(11):7001–7007. https://doi.org/10.1128/jvi.68.11.7001-7007.1994
Swanstrom R, Coffin J (2012) HIV-1 pathogenesis: the virus. Cold Spring Harb Perspect Med 2(12):a007443. https://doi.org/10.1101/cshperspect.a007443
Symons J, Cameron PU, Lewin SR (2018) HIV integration sites and implications for maintenance of the reservoir. Curr Opin HIV AIDS 13(2):152–159. https://doi.org/10.1097/coh.0000000000000438
Tahirov TH, Babayeva ND, Varzavand K, Cooper JJ, Sedore SC, Price DH (2010) Crystal structure of HIV-1 Tat complexed with human P-TEFb. Nature 465(7299):747–751. https://doi.org/10.1038/nature09131
Taube R, Peterlin M (2013) Lost in transcription: molecular mechanisms that control HIV latency. Viruses 5(3):902–927. https://doi.org/10.3390/v5030902
Tazi J, Bakkour N, Marchand V, Ayadi L, Aboufirassi A, Branlant C (2010) Alternative splicing: regulation of HIV-1 multiplication as a target for therapeutic action. FEBS J 277(4):867–876. https://doi.org/10.1111/j.1742-4658.2009.07522.x
Tebas P, Stein D, Tang WW, Frank I, Wang SQ, Lee G, Spratt SK, Surosky RT, Giedlin MA, Nichol G, Holmes MC, Gregory PD, Ando DG, Kalos M, Collman RG, Binder-Scholl G, Plesa G, Hwang WT, Levine BL, June CH (2014) Gene editing of CCR5 in autologous CD4 T cells of persons infected with HIV. N Engl J Med 370(10):901–910. https://doi.org/10.1056/NEJMoa1300662
Terwilliger EF, Langhoff E, Gabuzda D, Zazopoulos E, Haseltine WA (1991) Allelic variation in the effects of the nef gene on replication of human immunodeficiency virus type 1. Proc Natl Acad Sci U S A 88(23):10971–10975. https://doi.org/10.1073/pnas.88.23.10971
Thomas JA, Gagliardi TD, Alvord WG, Lubomirski M, Bosche WJ, Gorelick RJ (2006) Human immunodeficiency virus type 1 nucleocapsid zinc-finger mutations cause defects in reverse transcription and integration. Virology 353(1):41–51. https://doi.org/10.1016/j.virol.2006.05.014
Tobiume M, Tokunaga K, Kiyokawa E, Takahoko M, Mochizuki N, Tatsumi M, Matsuda M (2001) Requirement of nef for HIV-1 infectivity is biased by the expression levels of Env in the virus-producing cells and CD4 in the target cells. Arch Virol 146(9):1739–1751. https://doi.org/10.1007/s007050170060
Tristem M, Marshall C, Karpas A, Hill F (1992) Evolution of the primate lentiviruses: evidence from vpx and vpr. EMBO J 11(9):3405–3412
Usami Y, Wu Y, Göttlinger HG (2015) SERINC3 and SERINC5 restrict HIV-1 infectivity and are counteracted by Nef. Nature 526(7572):218–223. https://doi.org/10.1038/nature15400
Van Damme N, Goff D, Katsura C, Jorgenson RL, Mitchell R, Johnson MC, Stephens EB, Guatelli J (2008) The interferon-induced protein BST-2 restricts HIV-1 release and is downregulated from the cell surface by the viral Vpu protein. Cell Host Microbe 3(4):245–252. https://doi.org/10.1016/j.chom.2008.03.001
van Kooyk Y, Geijtenbeek TB (2003) DC-SIGN: escape mechanism for pathogens. Nat Rev Immunol 3(9):697–709. https://doi.org/10.1038/nri1182
Van Lint C, Amella CA, Emiliani S, John M, Jie T, Verdin E (1997) Transcription factor binding sites downstream of the human immunodeficiency virus type 1 transcription start site are important for virus infectivity. J Virol 71(8):6113–6127. https://doi.org/10.1128/jvi.71.8.6113-6127.1997
Van Lint C, Bouchat S, Marcello A (2013) HIV-1 transcription and latency: an update. Retrovirology 10:67. https://doi.org/10.1186/1742-4690-10-67
VerPlank L, Bouamr F, LaGrassa TJ, Agresta B, Kikonyogo A, Leis J, Carter CA (2001) Tsg101, a homologue of ubiquitin-conjugating (E2) enzymes, binds the L domain in HIV type 1 Pr55(Gag). Proc Natl Acad Sci U S A 98(14):7724–7729. https://doi.org/10.1073/pnas.131059198

Verrijzer CP, Tjian R (1996) TAFs mediate transcriptional activation and promoter selectivity. Trends Biochem Sci 21(9):338–342

Veschambre P, Simard P, Jalinot P (1995) Evidence for functional interaction between the HIV-1 Tat transactivator and the TATA box binding protein in vivo. J Mol Biol 250(2):169–180. https://doi.org/10.1006/jmbi.1995.0368

Vodicka MA (2001) Determinants for lentiviral infection of non-dividing cells. Somat Cell Mol Genet 26(1-6):35–49. https://doi.org/10.1023/a:1021022629126

Votteler J, Sundquist WI (2013) Virus budding and the ESCRT pathway. Cell Host Microbe 14(3):232–241. https://doi.org/10.1016/j.chom.2013.08.012

Wainberg Z, Oliveira M, Lerner S, Tao Y, Brenner BG (1997) Modulation of stress protein (hsp27 and hsp70) expression in CD4+ lymphocytic cells following acute infection with human immunodeficiency virus type-1. Virology 233(2):364–373. https://doi.org/10.1006/viro.1997.8618

Wan W, Kolesnikova L, Clarke M, Koehler A, Noda T, Becker S, Briggs JAG (2017) Structure and assembly of the Ebola virus nucleocapsid. Nature 551(7680):394–397. https://doi.org/10.1038/nature24490

Wang GP, Ciuffi A, Leipzig J, Berry CC, Bushman FD (2007) HIV integration site selection: analysis by massively parallel pyrosequencing reveals association with epigenetic modifications. Genome Res 17(8):1186–1194. https://doi.org/10.1101/gr.6286907

Warren K, Warrilow D, Meredith L, Harrich D (2009) Reverse transcriptase and cellular factors: regulators of HIV-1 reverse transcription. Viruses 1(3):873–894. https://doi.org/10.3390/v1030873

Watts JM, Dang KK, Gorelick RJ, Leonard CW, Bess JW Jr, Swanstrom R, Burch CL, Weeks KM (2009) Architecture and secondary structure of an entire HIV-1 RNA genome. Nature 460(7256):711–716. https://doi.org/10.1038/nature08237

West M, Mikovits J, Princler G, Liu YL, Ruscetti FW, Kung HF, Raziuddin (1992) Characterization and purification of a novel transcriptional repressor from HeLa cell nuclear extracts recognizing the negative regulatory element region of human immunodeficiency virus-1 long terminal repeat. J Biol Chem 267(35):24948–24952

Westby M, van der Ryst E (2005) CCR5 antagonists: host-targeted antivirals for the treatment of HIV infection. Antivir Chem Chemother 16(6):339–354. https://doi.org/10.1177/095632020501600601

Wetzel KS, Yi Y, Yadav A (2018) Loss of CXCR6 coreceptor usage characterizes pathogenic lentiviruses. PLoS Pathogens 14(4):e1007003. https://doi.org/10.1371/journal.ppat.1007003

Wilen CB, Tilton JC, Doms RW (2012) HIV: cell binding and entry. Cold Spring Harb Perspect Med 2(8). https://doi.org/10.1101/cshperspect.a006866

Wilkinson KA, Gorelick RJ, Vasa SM, Guex N, Rein A, Mathews DH, Giddings MC, Weeks KM (2008) High-throughput SHAPE analysis reveals structures in HIV-1 genomic RNA strongly conserved across distinct biological states. PLoS Biol 6(4):e96. https://doi.org/10.1371/journal.pbio.0060096

Willey RL, Rutledge RA, Dias S, Folks T, Theodore T, Buckler CE, Martin MA (1986) Identification of conserved and divergent domains within the envelope gene of the acquired immunodeficiency syndrome retrovirus. Proc Natl Acad Sci U S A 83(14):5038–5042. https://doi.org/10.1073/pnas.83.14.5038

Williams RD, Lee BA, Jackson SP, Proudfoot NJ (1996) Activation domains of transcription factors mediate replication dependent transcription from a minimal HIV-1 promoter. Nucleic Acids Res 24(4):549–557. https://doi.org/10.1093/nar/24.4.549

Wilson W, Braddock M, Adams SE, Rathjen PD, Kingsman SM, Kingsman AJ (1988) HIV expression strategies: ribosomal frameshifting is directed by a short sequence in both mammalian and yeast systems. Cell 55(6):1159–1169. https://doi.org/10.1016/0092-8674(88)90260-7

Woodward CL, Prakobwanakit S, Mosessian S, Chow SA (2009) Integrase interacts with nucleoporin NUP153 to mediate the nuclear import of human immunodeficiency virus type 1. J Virol 83(13):6522–6533. https://doi.org/10.1128/jvi.02061-08

Wu JQ, Sassé TR, Wolkenstein G, Conceicao V, Saksena MM, Soedjono M, Perera SS, Wang B, Dwyer DE, Saksena NK (2013) Transcriptome analysis of primary monocytes shows

global down-regulation of genetic networks in HIV viremic patients versus long-term non-progressors. Virology 435(2):308–319. https://doi.org/10.1016/j.virol.2012.10.026
Wu T, Gorelick RJ, Levin JG (2014) Selection of fully processed HIV-1 nucleocapsid protein is required for optimal nucleic acid chaperone activity in reverse transcription. Virus Res 193:52–64. https://doi.org/10.1016/j.virusres.2014.06.004
Wyatt R, Kwong PD, Desjardins E, Sweet RW, Robinson J, Hendrickson WA, Sodroski JG (1998) The antigenic structure of the HIV gp120 envelope glycoprotein. Nature 393(6686):705–711. https://doi.org/10.1038/31514
Yamashita M, Emerman M (2004) Capsid is a dominant determinant of retrovirus infectivity in non-dividing cells. J Virol 78(11):5670–5678. https://doi.org/10.1128/jvi.78.11.5670-5678.2004
Yan J, Shun MC, Hao C, Zhang Y, Qian J, Hrecka K, DeLucia M, Monnie C, Ahn J, Skowronski J (2018) HIV-1 Vpr reprograms CLR4(DCAF1) E3 ubiquitin ligase to antagonize exonuclease 1-mediated restriction of HIV-1 infection. mBio 9(5). https://doi.org/10.1128/mBio.01732-18
Yang L, Morris GF, Lockyer JM, Lu M, Wang Z, Morris CB (1997) Distinct transcriptional pathways of TAR-dependent and TAR-independent human immunodeficiency virus type-1 transactivation by Tat. Virology 235(1):48–64. https://doi.org/10.1006/viro.1997.8672
Yao XJ, Subbramanian RA, Rougeau N, Boisvert F, Bergeron D, Cohen EA (1995) Mutagenic analysis of human immunodeficiency virus type 1 Vpr: role of a predicted N-terminal alpha-helical structure in Vpr nuclear localization and virion incorporation. J Virol 69(11):7032–7044. https://doi.org/10.1128/jvi.69.11.7032-7044.1995
Yeung ML, Houzet L, Yedavalli VS, Jeang KT (2009) A genome-wide short hairpin RNA screening of jurkat T-cells for human proteins contributing to productive HIV-1 replication. J Biol Chem 284(29):19463–19473. https://doi.org/10.1074/jbc.M109.010033
Yu X, Yu Y, Liu B, Luo K, Kong W, Mao P, Yu XF (2003) Induction of APOBEC3G ubiquitination and degradation by an HIV-1 Vif-Cul5-SCF complex. Science 302(5647):1056–1060. https://doi.org/10.1126/science.1089591
Yuan P, Gupta K, Van Duyne GD (2008) Tetrameric structure of a serine integrase catalytic domain. Structure 16(8):1275–1286. https://doi.org/10.1016/j.str.2008.04.018
Zazopoulos E, Haseltine WA (1993) Effect of nef alleles on replication of human immunodeficiency virus type 1. Virology 194(1):20–27. https://doi.org/10.1006/viro.1993.1230
Zennou V, Petit C, Guetard D, Nerhbass U, Montagnier L, Charneau P (2000) HIV-1 genome nuclear import is mediated by a central DNA flap. Cell 101(2):173–185. https://doi.org/10.1016/s0092-8674(00)80828-4
Zenzie-Gregory B, Sheridan P, Jones KA, Smale ST (1993) HIV-1 core promoter lacks a simple initiator element but contains a bipartite activator at the transcription start site. J Biol Chem 268(21):15823–15832
Zhang L, Carruthers CD, He T, Huang Y, Cao Y, Wang G, Hahn B, Ho DD (1997) HIV type 1 subtypes, coreceptor usage, and CCR5 polymorphism. AIDS Res Hum Retroviruses 13(16):1357–1366. https://doi.org/10.1089/aid.1997.13.1357
Zhang L, He T, Talal A, Wang G, Frankel SS, Ho DD (1998) In vivo distribution of the human immunodeficiency virus/simian immunodeficiency virus coreceptors: CXCR4, CCR3, and CCR5. J Virol 72(6):5035–5045. https://doi.org/10.1128/jvi.72.6.5035-5045.1998
Zhang L, Zhang X, Ma Q, Zhou H (2010) Host proteome research in HIV infection. Genomics Proteomics Bioinformatics 8(1):1–9. https://doi.org/10.1016/s1672-0229(10)60001-0
Zhang W, Du J, Evans SL, Yu Y, Yu XF (2011) T-cell differentiation factor CBF-β regulates HIV-1 Vif-mediated evasion of host restriction. Nature 481(7381):376–379. https://doi.org/10.1038/nature10718
Zhou T, Xu L, Dey B, Hessell AJ, Van Ryk D, Xiang SH, Yang X, Zhang MY, Zwick MB, Arthos J, Burton DR, Dimitrov DS, Sodroski J, Wyatt R, Nabel GJ, Kwong PD (2007) Structural definition of a conserved neutralization epitope on HIV-1 gp120. Nature 445(7129):732–737. https://doi.org/10.1038/nature05580
Zhou H, Xu M, Huang Q, Gates AT, Zhang XD, Castle JC, Stec E, Ferrer M, Strulovici B, Hazuda DJ, Espeseth AS (2008) Genome-scale RNAi screen for host factors required for HIV replication. Cell Host Microbe 4(5):495–504. https://doi.org/10.1016/j.chom.2008.10.004
Zhu H, Jian H, Zhao LJ (2004) Identification of the 15FRFG domain in HIV-1 Gag p6 essential for Vpr packaging into the virion. Retrovirology 1:26. https://doi.org/10.1186/1742-4690-1-26

Human Cytomegalovirus: An Insight of Its Pathobiology and Therapeutics

23

Ujjawal Sharma, Bunty Sharma, Ritu Sharma, Deeksha Pal, Vivek Kumar Garg, Aman Sharma, and Deepika Raina

Abstract

Human cytomegalovirus (HCMV) belongs to herpes family of viruses. It is also known as human herpesvirus 5 (HHV-5). In 1956 and 1957, Weller, Smith, and Rowe independently isolated and grew cytomegaloviruses (CMVs) from mice and humans. The CMVs have derived their name from Greek which means cell (cyto) and big (megalo). These are the largest known disease-causing viruses in humans with a DNA genome of size 230 Kbp. In children, HCMV seroprevalence is low but with time seroprevalence increases by 40–80% globally. HCMV is one of the most opportunistic infections. It remains in latent phase in host cells by modulating the immune environment of the host cell. Monocyte-derived macrophages (MDM) and endothelial cells are the major reservoirs for latent phase of HCMV. In immunocompetent individuals, it remains latent phase and asymp-

U. Sharma
Department of Human Genetics and Molecular Medicine, School of Health Sciences, Central University of Punjab, Bathinda, India

B. Sharma (✉) · D. Pal
Department of Bio-Sciences and Technology, Maharishi Markandeshwar Engineering College, Maharishi Markandeshwar (Deemed to be University), Mullana, Ambala, India
e-mail: sharmabunty097@gmail.com

R. Sharma
Department of Biochemistry, Government Medical College, Amritsar, Punjab, India

V. K. Garg
Department of Medical Lab Technology, University Institute of Allied Health Sciences, Chandigarh University, Chandigarh, Punjab, India

A. Sharma
Alchemist Hospital, Panchkula, Haryana, India

D. Raina
School of Pharmacy, Graphic Era Hill University, Dehradun, Uttarakhand, India

P. L. Bhukya, A. S (eds.), *Emerging Human Viral Diseases, Volume II*,
https://doi.org/10.1007/978-981-97-4480-0_23

tomatic, but in immunocompromised individuals, it affects almost all the major organs of the body, i.e. retinitis ocular (retinitis), central nervous system (polyradiculopathy, meningoencephalitis), lungs (pneumonitis), gastrointestinal tract (esophagitis, gastritis, ileitis, colitis, pancreatitis), and liver (hepatitis). Various in vitro- and in vivo-based models are being developed for the propagation of HCMVs. Serology, histopathology, viral culture, shell-vial assay, antigenemia, and nucleic acid testing are most common diagnostic tests for HCMV. Antiviral therapies (ganciclovir, foscarnet, and cidofovir) are the major mode of treatment for HCMV, but these therapies are associated with side effects. There is no vaccine available for HCMV at present but in coming decade we can expect effective vaccine against HCMV. Most of the therapeutic options are focused on CMV-specific T-cell therapy.

Keywords

Human cytomegalovirus · Pathobiology · Therapeutics · Emerging viral disease · Host–virus interaction · Re-emergence of viral disease

23.1 Introduction

Human cytomegalovirus (HCMV) belongs to Herpesviridae family with enveloped dsDNA or human herpesvirus-5. Subfamily is Betaherpesvirinae, and genus is cytomegalovirus. It has a size of 190 nm making it a largest virus of herpes family. HCMV genome contains 230,000 bp and is enclosed by an icosahedral capsid. The protein layer between the capsid and virus is called tegument. The half-life of this virus is 60 min at temperature of 37 °C and becomes unstable at −20 °C. HCMV is so named as it leads to increased size of the infected cells, i.e. cytomegaly forming inclusion bodies (Schottstedt et al. 2010). HCMV has broad spectrum effects which may be asymptomatic or involves extensive to the organs especially in patients with weakened immune system and those suffering from congenital HCMV infection. The nature of HCMV is highly host specific, and it potentially infects different cells in vivo. Human fibroblasts are the only cells which allow full replication of HCMV. Intranuclear and intracytoplasmic inclusion bodies are seen in fibroblasts during HCMV infection. The viral genome includes two domains: unique short (US) and unique long (UL) flanked by a pair of inversely repeated sequences known as TRL-UL-IRL-IRS-US-TRS (Bradley et al. 2009).

First HCMV was isolated from tissue of submaxillary gland of a dead newborn followed by isolation from kidney from an infant suffering from cytomegalic inclusion disease (Smith 1956). Rowe and co-workers found a unique intranuclear inclusion bodies and focal lesions in the fibroblasts of adenoidal tissue of three asymptomatic children (Rowe et al. 1956). Other two strains of HCMV, i.e. Kerr and Esp were isolated from the urine sample of a neonate with jaundice, hepatosplenomegaly, and with additional features such as periventricular calcification and chorioretinitis, respectively. In the subsequent years, the pathogenesis of HCMV started showing in organ transplant patients, AIDS patients, and those suffering from cancer.

23.2 Epidemiology

HCMV infection is observed in every human population (Krech et al. 1971), and its nature is endemic. The effect of seasonal variations is negligible (Gold and Nankervis 1976). As the age advances, its seroprevalence tends to increase. It reaches 5–30% by the sixth year of life. The prevalence of HCMV-specific antibodies is highly dependent on the socioeconomic status of a country. Therefore, in the developing countries, infection represents in two phases. In first 2–3 years of life, first peak is reached followed by a weak elevation in adolescence between the ages of 16 and 30 years because of possible sexual intimacy. It has been reported that seropositive rates are 95–100% in Asian and African population.

Socioeconomic backgrounds of the population and geographic location govern HCMV acquisition. Furthermore, seropositivity tends to increase with age. In the less developed countries, HCMV infection is present in almost every child in early childhood. Reports showed that 90% of the preschool children having HCMV antibody are positive in Sub-Saharan Africa, South America, India, and East Asia (Gold and Nankervis 1976; Stagno 2001). On the contrary, surveys showed that very less percentage of children (<20%) of similar age group had HCMV infection in United States and Great Britain (Huang et al. 1980).

Liu et al. (1990) reported HCMV seropositivity in 60% of children in the age group of 4–7 years in Chengdu, China. Likewise, 58% of children in the age group of 4–12 years were positive for HCMV infection in Taipei, Taiwan. Additionally, 61% of paediatric patients in hospitals exhibited seropositivity from Rio de Janeiro, Brazil's less developed population, while in Jamaica, 56% of children aged one to four had HCMV antibody positivity (Prabhakar et al. 1992; Shen et al. 1992; Suassuna et al. 1995). In a study done in Finland in children followed for 8 years, the HCMV positive antibody incidence increased from 27% in children who were 7 months of age to 41% when these children attained 8 years of age (Aarnisalo et al. 2003). In a study from Italy, it was reported that HCMV seroprevalence increased from 28 to 96% from the age of 2 years to 45–54-year (Natali et al. 1997). Similarly, CMV positive antibody rate increased from 42 to 79% from 2–5 years to 31–40 years, in Spain (de Ory et al. 2004). It has been reported in the recent studies that there is a huge risk of CMV seropositivity (approx. 100%) in blood donors' populations in Asia and Africa (Pultoo et al. 2001; Kothari et al. 2002), whereas in Germany, rate is lower (30–70%) (Hecker et al. 2004).

The known routes of HCMV infection dissemination are as follows:

- Oropharyngeal secretions
- Semen
- Breast milk
- Blood
- Allografts

Table 23.1 Urinary HCMV excretion rates in the very first 2 years after birth

Country	At birth	1 m	2 m	3 m	6 m	9 m	12 m	24 m
Japan	–	6	10	20	56	44	22	7
Thailand	–	–	–	38	55	18	15	–
Guatemala	–	–	–	23	42	40	35	–
Finland	2.3	–	12	23	35	25	33	41
Sweden	1	12	–	–	–	–	23	–
US	1	3	–	11	13	–	11	–
UK	2.5	–	–	–	9	–	–	–

The vertical transmission of HCMV, i.e. perinatal infection from mother to newborn is very common. The infection generally transmits through genital tract during delivery or via breast milk. The culture of breast feeding is highly responsible for HCMV infection in seropositive females. This practice is almost universal in countries such as Japan, Thailand, and Guatemala. The conventional virus isolating techniques revealed that 13–50% of lactating women has HCMV infection in their breast milk (Dworsky et al. 1983). Excretion of HCMV in urine is shown in Table 23.1.

Children-to-children infection of HCMV is observed from children in day care centres. This is an example of horizontal infection. The day care centre studies showed that seropositive kids transmit virus through saliva and urine creating a potential environment for infection to other children. Children also transmit the HMCV infection to their parents especially when they reside in the same house. Sexual activity is a key route of transmission of HCMV infection. This association can be better understood via various factors such as:

- Presence of more than double the seropositivity during the period of sexual activity
- More seropositivity in males having seropositive female partners
- Isolation of HCMV from cervix of women with STD
- Positive correlation between seropositivity and multiple sexual partners

In 1966, HCMV infection was first described in association with transfusion (Kääriäinen et al. 1966). The magnitude of infection goes parallel with the load of products transfused as well as the amount of leukocytes. Besides, during organ transplantation, if the donor is seropositive, the transmission in a seronegative recipient is a huge threat to his life. However, significant mortality and morbidity associated with this factor have considerably reduced in recent years because of better patient management which includes avoidance of HCMV positive donor and use of antiviral therapy in order to avoid the reactivation of virus. Donor semen used for artificial insemination is a strong source of HCMV infection (Prior et al. 1994). Hence, American Fertility society recommends the screening of semen donors for their negative HCMV infection before insemination. The mode of transmission is summarized as follows in Table 23.2 (Boppana and Fowler 2007).

Table 23.2 The mode of transmission of HCMV

Perinatal	Congenital intrauterine infection in foetus; intrapartum virus exposure; transmission from mother-to-infant; breast milk
Infancy/childhood	Various bodily fluids, including saliva; transmission between children
Adolescence/adulthood	Exposure to adolescent; transmission through sexual activity; occupational exposures
Blood elements	Serum from donors who have tested positive; repeated transfusions; presence of blood products in white blood cells (WBCs)
Allografts	Allografts received from seropositive donors
Semen from donor	Semen used in artificial insemination came from seropositive donors

23.3 Potential Risk of Emergence and Re-emergence

Primary HCMV infection often follows an asymptomatic course. Mononucleosis is usually seen in patients whose immune system is healthy, however heterophile antibodies and pharyngeal exudate are not seen, which is a key feature of HCMV. Other symptoms associated with HCMV infection are anaemia, positive rheumatoid factor, thrombocytopenia, abnormal liver function test (LFT), and positive antinuclear antibody (ANA) levels. However, CMV is a significant opportunistic infection in individuals with a weakened immune system, such as those with HIV. Immunocompromised conditions trigger the reactivation of virus which is responsible for huge illness and fatality (Zheng et al. 2019).

- Primary infection: Seronegative and often asymptomatic
- Reactivation: Seropositive patients having reinfection of CMV
- CMV infection: State associated with appearance of CMV in body fluids
- CMV disease: State of CMV infection associated with symptoms or organ involvement

  The classical features of HCMV infection are as follows:
- High host specificity
- Replication cycle takes a longer time
- Cell-to-cell infection is very prominent

During the lytic infection of cells, the immediate-early (IE) protein regulates E (Early) proteins and also Late (L) proteins. The IE proteins (pp65) are used as early marker of viral infection because it takes part in the expression and transport of HLA antigens. E protein includes the HCMV coded DNA polymerase. After HCMV infection, the incubation period is generally 4–8 weeks. Viraemia occurs during this period. Therefore, HCMV infection to the other person is very common in this phase even if the person is asymptomatic. The primary HCMV infection then advances toward latency. The latency reservoirs are as follows:

- CD 34+ haematopoietic progenitor cells
- CD33+ haematopoietic progenitor cells

- Dendritic cells
- Macrophages
- Monocytes
- Granulocytes
- Neutrophils (Larsson et al. 1998; Roback et al. 2007)

In order to understand the primary infection of HCMV and its reactivation, the clarity of its genome and viral components is essential. HCMV genome is covered by a viral capsid. The structure of the viral capsid is icosahedral with triangular number. There are 60 asymmetric units (Chen et al. 1999), 16 smallest capsid proteins (SCPs), 16 major capsid proteins (MCPs), 5 triplexes (Ta, Tb, Tc, Td, and Te), and 1/3 Tfs in each unit (Yu et al. 2017). The genome is extensively packed with reasonable pressure for the transport of viral genome into the nucleus. Tegument plays an important part in facilitating the virus to get familiar with the machinery of the host.

HCMV tegument protein pp71 plays an important role in primary infection and reactivation in the herpesviridae family. Six proteins are preserved which are the integral components of the replication:

- pUL54 (DNA polymerase)
- pUL44 (DNA polymerase processing factor)
- pUL102 (primase associated factor)
- pUL70 (primase)
- pUL57 (SS DNA binding protein)
- pUL105 (helicase)

UL54 gene expression activation is instrumental during early and late infection period (Kerry et al. 1997). In healthy individuals, life-long latency is associated with HCMV infection. Transcription programme in late phase infected cells is associated with decreased expression (Shnayder et al. 2018). Several stimuli such as severe pneumonia, GIT disease after transplantation, SARS-COV-2, any bacterial infection, increased release of cytokine in inflammatory conditions, HIV, etc. trigger the reactivation of the lytic phase from its quiescent form.

The subcellular localization of pp71 is associated with replication/latent phase. The foundation of these phases largely depends upon the behaviour of pp71 with the members of ND10. ND10 is known as promyelocytic leukaemia protein-associated nuclear bodies. Death-domain associated protein (Daxx), an important component of ND10, plays a role in the intrinsic defence mechanism cells, against HCMV infection (Cantrell and Bresnahan 2006). Hence this interaction promotes the latent stage. PML and sp100 (other components of ND10) also cause repression of viral genome (Glass and Everett 2013). However, it is unclear whether pp71 induced expression of IE has a role to play in reactivation process. US28 is transcribed in both lytic and latent phase, and it actually promotes latency (Beisser et al. 2001). In undifferentiated cells, pUS28 dephosphorylates MAPK and NK-kB establishing latency. However, in differentiated cells, hyperphosphorylation of signalling molecules contributes to lytic infection (Krishna et al. 2017). Hence pUS28 activity is highly dependent on differentiated state.

Nuclear targeting is crucial for re-emergence. Bipartite NLS of Pul48 targets the nuclei capsid toward the host nucleus and releases the viral genome (Brock et al. 2013). pp150 (tegument protein) is involved in virion egress during late infection besides other important functions.

One of the interesting features associated with HCMV viral replication is its long replication cycle. Hence HCMV arrests G1/S phase in the host to complete the lytic cycle. Viral factors during a period of primary infection checks the host cell cycle. Viral pUL21 arrests host cell cycle progression by destabilizing cyclin A2 via a proteasome-dependent degradation (Caffarelli et al. 2013). The two potential causes of viral reactivation are differentiation and inflammation.

23.4 Organization of Infectious Agents

23.4.1 Classification

CMVs are seroprevalant in 80% individuals globally. HCMV is a type of human herpes virus 5 (HHV-5). HCMV belongs to Herpesviridae family, subfamily Betaherpesvirinae, and genus Cytomegalovirus. Herpes viruses are distinct from each other by their lytic infection cycle and latent phase cycles. The detailed classification of HCMV is shown in Table 23.3.

23.4.2 Morphology and Virion Structure

HCMV virus is icosahedral in shape with a diameter of 150–200 nm. The genome is enclosed in a capsid. Outside the capsid, there is proteinaceous layer named as tegument. The capsid and tegument are enclosed by virus envelope derived from cell membranes. There are different glycoproteins embedded in the envelope crucial for HCMV infection. The trimer glycoprotein gHgLgO complex (Fibroblasts) and pentamer complex gHgLpUL128pUL130pUL131A (epithelial, endothelial, and myeloid cells) are required for gaining access to different cell types (Martinez-Martin et al. 2018). Glycoproteins are directly involved in host–pathogen interactome complexes and proteins present over the host cell surface, thus are key players for HCMV entry into the host cell (Fig. 23.1).

Table 23.3 Human cytomegalovirus classification

Phylum	Peploviricota
Class	Herviviricetes
Order	Herpesviruses
Family	Herpesviridae

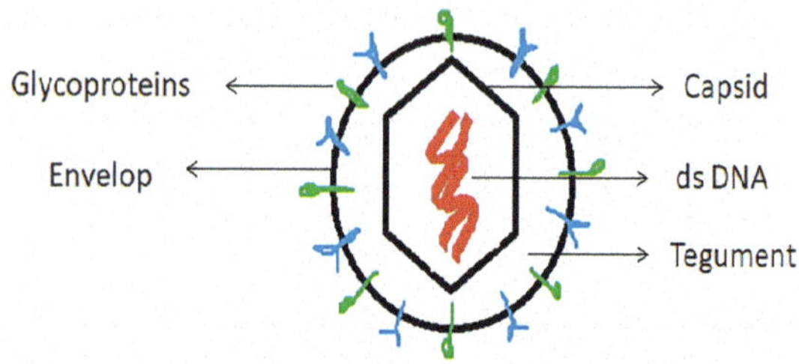

Fig. 23.1 Structure of HCMV

23.4.3 Structure of Genome and Its Organization

The genome of HCMV virus is linear, double-stranded deoxyribonucleic acid (DNA). However, recently Liu et al. (2021) have observed presence of RNA structures in the virus. The genome of HCMV is the largest in size amongst the human viruses. It is 230,000 bp in size. There are approximately 230 genes in HCMV genome coding for membrane proteins as well as glycoprotein essential for structure formation and replication of HCMVs (Fulkerson et al. 2021).

23.4.4 Propagation and Assay in In Vitro and In Vivo Laboratory Models

HCMV infects individuals via mucous route, blood transfusion, through placenta, vaginal secretions, and breast feeding. HCMV can infect various cell types: fibroblast, epithelial cells, smooth muscle cells, endothelial cells, and myeloid lineage cells (Aguiar et al. 2021). HCMV maintains its latent phase in myeloid and endothelial lineage cells, whereas in fibroblast, HCMV follows the lytic phase cycle. Strict species-specific infectivity of HCMV is a limitation for its cursory knowledge of transmission and life cycle.

HCMV infects retinal pigmented epithelium (RPE) and is responsible for blindness in immunocompromised individuals; therefore, various studies have documented ocular epithelial cells as model of HCMV. Oral epithelial cell cultures, gingival cells (hGETs), and telomerase-immortalized normal oral keratinocytes (NOKs) have been utilized to delineate the mode of infection of HCMV (Weng et al. 2018). Also, another study described the salivary epithelial model derived from primary human "salispheres" for HCMV replication (Morrison et al. 2019).

Recently, Lloyd et al. (2020) develop a skin organ culture and skin xenograft mouse models for HCMV infection and propagation. Human skin cells were inoculated in SCID or athymic nude mice and lately HCMV was injected into the mice xenograft, and a widespread infection of HCMV was observed. Development of these kinds of mice models has an added advantage over traditional foetal tissue utilization for HCMV studies (Fisher and Lloyd 2020).

23.4.5 Viral Proteins and Lifecycle

Specialized glycoproteins present over the surface of HCMV interact with the receptors present on host cells and initiate the first step of infection. After the protein viral envelope and the host cell fusion, there is release of tegument and capsid proteins into the host cell. Capsid enters into nucleus of the host cell, and viral genome is delivered. The viral immediate-early (IE) genes are expressed which promotes lytic replication cycle of the virus. Viral IE early genes modulate the host cell programming, followed by the expression of viral early genes. The proteins coded by viral early genes help in viral DNA replication. Lastly, viral late genes are expressed, and viral late proteins lead to the formation of virion structure, its assembly, and egress of new progeny of virion particles into the cytosol. In the cytosol, newly synthesized capsid associated with teguments and viral assembly complex having components of cellular organelles, the endoplasmic reticulum (ER), Golgi apparatus, and endosomes (Fig. 23.2). Along with HCMV, two types of

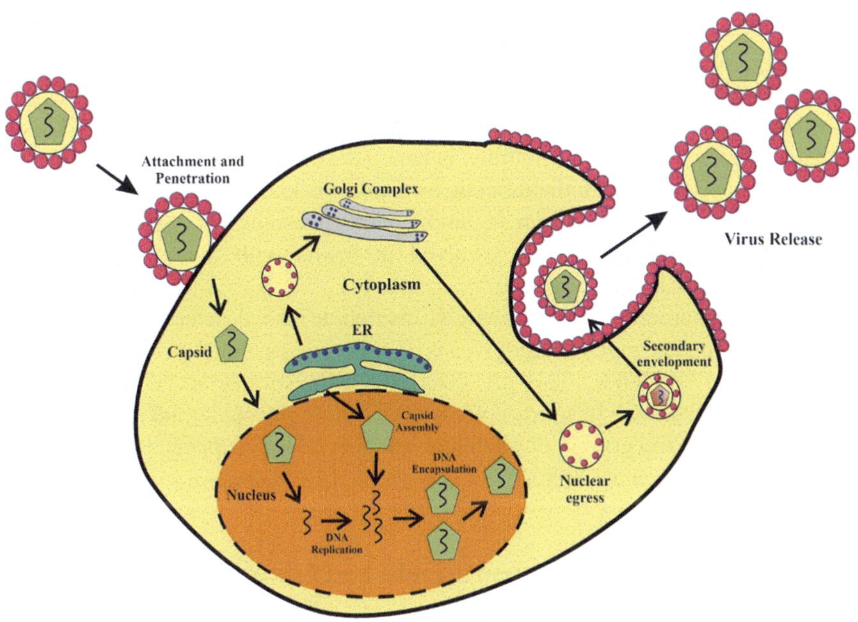

Fig. 23.2 Lifecycle of HCMV in human cell. Either direct fusion or the endocytic pathway allows HCMV to enter human cells. Viral glycoproteins and a particular surface receptor or receptors work together to attach the virus to the cell. The envelope then fuses with the cell membrane to release nucleocapsids into the cytoplasm. After being translocated into the nucleus, these nucleocapsids release the viral DNA. Viral synthesis function is stimulated and accumulates concurrently with viral replication and maturation. Capsids are the units that contain the replicated viral DNA and are transferred from the nucleus to the cytoplasm. At the endoplasmic reticulum (ER)-Golgi intermediate compartment in the cytoplasm, secondary envelopment takes place. After that, a complex two-stage process known as final envelopment and egress occurs, which causes the plasma membrane to exocytose and release the virion

components, i.e. dense bodies and non-infectious enveloped particles (NIEPS) are also released by the host cell (Kalejta 2008; Jean Beltran and Cristea 2014).

23.5 Pathogenesis in Humans

23.5.1 Host–Viral Interaction

Epithelial cells, endothelial cells, and leukocytes are the primary host cells infected by HCMV. Upon entry into host cells, HCMV causes genome instability by activating DNA damage response (DDR) and ataxia-telangiectasia mutated (ATM). PDGFRα is a cellular receptor for HCMV entry in fibroblasts, and recently neuropilin2 (Nrp2) is identified as receptor for HCMV infection in epithelial and endothelial cells (Martinez-Martin et al. 2018). HCMV interacts distinctly with different cells through glycoproteins expressed on its surface. gH/gL and gB are two glycoproteins which regulate entrance of HCMV into the cells. gH/gL directly involved in interaction with the receptors on fibroblast cells and helps virus to gain entry into fibroblast. It is reported that HCMV entry into fibroblast activates PDGFRα signalling which promotes further replication of HCMV inside the fibroblasts (Vanarsdall and Johnson 2012).

A distinct protein complex of gH/gL is formed by binding with UL131, UL130, and UL128 proteins, and this complex is denoted as gH/gL/UL128–131 complex. To enter epithelial and endothelial cells by a process known as micropinocytosis, HCMV needs this complex, where union with endosomes is required. gH/gL/UL128–131 complex is directly involved in fusion with low pH endosomes (Vanarsdall et al. 2008).

HCMV glycoproteins gB and gM/gN (disulphide linked heterodimer) interact with heparan sulphate glycosaminoglycans (GAG) and essential for efficient HCMV replication and virulence. The gB glycoproteins also interact with receptors [(platelet-derived growth factor receptor-α (PDGFRα), and epidermal growth factor receptor (EGFR) integrins)] downstream of GAGs and support replication of HCMV (Shimamura et al. 2006).

23.5.2 Host Immune Response (Innate and Adaptive)

Toll-like receptors (TLRs) localized on innate immune cells such as macrophages and dendritic cells recognize the virus component and mediate the innate immune system's reaction to viral infection. Virus recognition by the immune systems leads to production of inflammatory cytokines such as interferons (IFN1), interleukins (IL-6), and tumour necrosis factor alpha (TNF-α) which in turn recruit phagocytic cells and dendritic cells for the clearance of cells infected with HCMV. At the site of infection, natural killer (NK) cells also play a significant role in the elimination of HCMV-infected cells by release of cytolytic granules that creates pores in the infected cells and eliminate them (Braud et al. 2002).

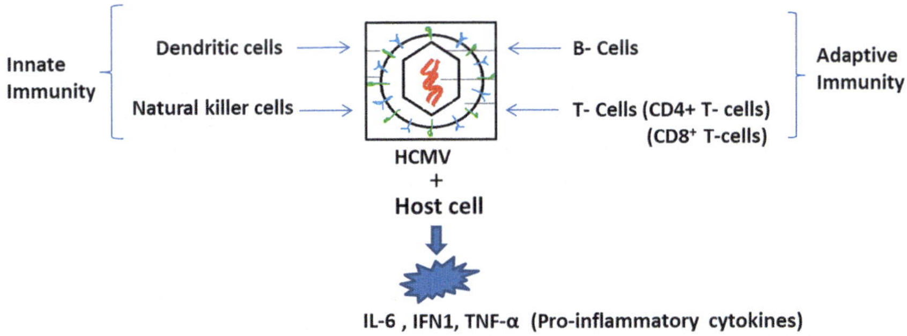

Fig. 23.3 Immune response of host against HCMV. HCMV infection leads to release of pro-inflammatory cytokines, which in turn activates the innate and adaptive immune response

Adaptive immune response is a long lasting and specific immune response for the prevention of spreading of virus and reinfection. B-lymphocyte and T-lymphocytes are mediators of adaptive immunity. B-cells release antibodies in response to HCMV infection, these antibodies can also protect foetus from HCMV infection (Landini and La Placa 1991). Cytotoxic T-cells ($CD8^+$ T-cells) are the key cells in the removal of HCMV infection. Also, CD4+ T-cells secrete IFN-γ to eliminate the HCMV from the host and prevent its further propagation to another host (Crough and Khanna 2009) (Fig. 23.3).

23.5.3 Virulence and Persistence

HCMV can cause challenging diseased condition in humans. HCMV can replicate in host cells and remains asymptomatic for years. It can cause serious infections in immunocompromised AIDS and transplant patients such as hepatitis, nephritis, and colitis. Infection during pregnancy can cause low birth weight, hearing loss, blindness, microcephaly, and hepatosplenomegaly (Britt 2008). Most HCMV infections in immunocompetent people have asymptomatic or mildly symptomatic courses that are not particularly indicative of the disease (Schottstedt et al. 2010).

The knowledge of viral persistence phenomenon is superficial. In chronic infection, virus keeps on shredding in low numbers without symptoms onset, whereas in latent infection state, virus remains dormant in immunocompetent individual. Monocyte-derived macrophages (MDM) and endothelial cells are the primary site for HCMV persistence and latency. It is observed that in the macrophages and endothelial cells, HCMV modulates the cellular environment to support its life-long infection inside the host cell (Jarvis and Nelson 2002). Upon infection in endothelial cells, transcriptional activators (IE1 and IE2) induce the activity of ICAM-1(Intercellular Adhesion Molecule-1) promoter and lead to expression of ICAM-1, an adhesion molecule in endothelial cells which leads to more interaction with monocytes and further spreading of infection to other monocytes (Chan et al. 2004).

The infection of HCMV expedites the process of differentiation of monocytes to macrophages by modulating different receptor signalling. HCMV infection prolongs the survival of HCMV-infected monocytes as well as macrophages for its years of survival. Because HCMV-infected macrophages and monocytes can live for lengthy time periods (weeks to years), infected cells can produce virus for a long time (Nogalski et al. 2013). Monocytes infected with HCMV migrate to bone marrow and transmit infection to CD34+ HPC reservoir). Upon reactivation, CD34+ HPC develops into monocytes, and cycle of reactivation and latency continues (Collins-McMillen et al. 2018).

23.6 Clinical Manifestations

23.6.1 Phases of Disease

Phases of spread of CMV infection occurred in three phases: Immediate-early (IE) phase, early (E) phase, and late (L) phase (Keightley et al. 2006).

1. Immediate-Early (IE) Phase

The IE (alpha) genes begin to be immediately early transcribed during the first 4 h following viral entrance. This immediate-early transcription event is only cellular factor-dependent and does not necessitate denovo protein synthesis in virus. Within 4 h of infection, non-structural proteins start to emerge in the nucleus. These proteins are crucial for controlling several cellular processes as well as the regulation of the expression of early- and late phase genes.

2. Early (E) Phase

The transcription of the E (beta) genes takes place in the early stages and is dependent on the transcriptional product of the IE gene. This stage sees the production of DNA polymerase and helicase-primase, two key proteins needed for viral replication.

3. Late (L) Phase

This phase includes L (gamma) gene transcription, which happens about 24 h after infection. Structure- and maturation-related proteins are among the transcriptional products. The nucleus is where structural proteins are assembled and packaged, and the Golgi-derived vacuole is where they mature and are released by the virus. Until cell lysis, the replication goes on for several days (Keightley et al. 2006).

23.6.2 Clinical Symptoms

Healthy people don't show or show minor symptoms when infected with CMV. There are three types of categories of patients who show sign and symptoms like newborns who are born with CMV, i.e. congenital CMV, newborns infected with CMV during or just after their birth, i.e. perinatal CMV, and the patients who are immunocompromised or on immunosuppressive drugs, in case of any organ transplant (Griffiths and Reeves 2021) and the patients who are already infected with HIV (Gupta and Shorman 2022). Congenital infection frequently includes CNS injury, hypotonia, seizures, hearing loss, and mental retardation are among the symptoms brought on by different modifications that take the appearance of numerous tiny, necrotic foci encircled by an inflammatory infiltrate. Neurons, meninges, choroid plexus, vascular endothelium, glia, and ependyma have all been shown to include inclusion-bearing cells that are viral antigen-positive in histological sections. The cochlea, vestibular membrane, and semicircular canals all contain a significant number of these cells. Except in cases of congenital infection, CNS involvement was formerly thought to be uncommon (Krstanović et al. 2021). Today, however, the primary target organs in HIV-infected patients are CNS tissues where HCMV replicates successfully in several brain regions and in specialized organs including the cochlea and retina. Microglial nodules, haemorrhages, necrotizing ventriculitis, focal parenchymal necrosis, astrocytic proliferation, and perivascular inflammation are a few examples of histopathologic changes. Blindness is a common consequence of acute retinal infection, perhaps as a result of virus directly lysing the cells in the retinal pigment epithelium during its replication. In conclusion, it appears that HCMV replication causes organ failure, yet the frequent occurrence of an inflammatory response suggests that additional pathogenetic processes are also at work (Landolfo et al. 2003).

Early phase is a condition in which a person contracts CMV for the first time. There are no obvious signs, but some people do exhibit mononucleosis-like symptoms. Other symptoms include fever, body ache, loss of weight, night sweats, sore throat, inflammation of glands, etc. Once someone contracts CMV, it remains dormant or latent in their body for life, but it has the potential to reactivate when their immune system is compromised. CMV has been reactivated in lungs, eyes, and the digestive system and shows symptoms like pneumonia, dyspnoea, hepatitis, diarrhoea, blurred vision, GIT bleeding, etc. (Fulkerson et al. 2021). So, laboratories tests and findings are needed as early as possible. There are many approaches like pp65 antigenemia assay, mRNA, and DNA for the detection of CMV.

23.6.3 Serology, Molecular, and Histological Approaches

The pp65 antigenemia assay, mRNA, and DNA are the most widely utilized diagnostics for the diagnosis of CMV infection (Kohmer et al. 2019). The adoption of quantitative DNA detection techniques has grown recently because of their high sensitivity and ability to assess viral loads, which can provide crucial prognostic data. Nucleic acid sequence-based amplification does not produce quantifiable results for mRNA detection; however, randomized experiments contrasting this method with pp65 antigenemia or DNA detection by PCR have shown that it appears to be effective in preventive treatment options (Ho et al. 1998). The diagnosis of CMV pneumonia and gastrointestinal sickness, which are currently by far the most widespread symptoms of the condition, has not changed in the previous 20 years. Rapid cultures, cytology, DNA hybridization, and tests using direct fluorescent antibodies are all approved diagnostic techniques for identifying CMV illness. Because it is currently unclear which concentration of CMV DNA in BAL (bronchoalveolar lavage) fluid or tissue most closely correlates with CMV disease, there is no agreement on how to employ molecular tools to conclusively diagnose CMV gastrointestinal disease and pneumonia. PCR can be used to rule out sickness since it has a high sensitivity and a good negative predictive value. The fact that things become more challenging when CMV silently reactivates in seropositive recipients without illness, even while high viral loads in BAL or tissue in the right clinical situations (i.e. signs and symptoms compatible with disease) certainly indicate a CMV illness. Unknown viral loads in BAL or tissue can exist in some CMV shedders. Because of this, it is currently unable to assess CMV DNA levels in BAL or tissue that are low or moderately high. Due to this, current, internationally accepted diagnostic guidelines do not accept PCR as unquestionable evidence of CMV pneumonia or gastrointestinal illness. PCR is a helpful diagnostic method for uncommon problems like CMV encephalitis (Vainionpää and Waris 2015).

23.7 Diagnostics and Therapeutic Approaches

Diagnosis approach can be better understood by taking an example of a vertical transmission route of HCMV. It is a major task which is dependent on clinical signs and serum antibodies, detection of virus in blood.

23.7.1 Serology

When virus-specific IgG is found in a pregnant woman who has never been seropositive before, the diagnosis of primary HCMV infection is made. Nevertheless, screening programs are not approved in this respect. As a result, doubt of having primary infection goes unnoticed. HCMV-specific IgM antibody should be evaluated in this case. Although it is not specific as it also appears during reactivation.

23.7.2 IgM Assays

Serological assays such as anticomplement immunofluorescence, complement fixation, radioimmunoassay, and indirect hemagglutination have been used to detect HCMV-specific antibodies (Reynolds and Stagno 1979). Capture ELISAs with enzyme-labelled antigen was the most reliable method in the initial years (van Loon et al. 1981). Subsequently in order to exclude false positive results, it is recommended that the testing of human serum samples should go hand in hand with viral and cell control-labelled antigens (Nielsen et al. 1987).

Revello and Gerna (1999) developed a capture Elisa IgM assay in which antibody to pp52 or ppUL44 was used and was specific in HCMV-infected cells. The results of this assay showed the identification of human IgM in the convalescent period of a primary infection of HCMV. The fall of IgM antibody in the subsequent blood samples is an indicative of a primary infection started few months prior to pregnancy. However, risk associated with HCMV IgM Elisa is the false positivity because of primary Epstein–Barr virus (EBV) infection. There is reported incidence of dual infection because of HCMV and EBV (Drouet et al. 1999).

23.7.3 Recombinant IgM Assays

Discordant results associated with assays for HCMV-specific IgM emphasized on more standardized procedures. Therefore, commercial kits came into picture based on recombinant HCMV peptides or proteins. These are both structural and non-structural (Vornhagen et al. 1994). Vornhagen et al. 1995 established a recombinant IgM ELISA which was more sensitive and specific for detection of IgM antibodies in primary infection of HCMV. They concluded that the specified part of UL57 gene product is a robust IgM antigen which is far specific to detect the seropositivity in primary HCMV infection. However, with HCMV IgM immunoblot assay was used as a reference for Abbott AxSYM CMV IgM microparticle enzyme immunoassay development because of increased sensitivity and relatively not so increased specificity which actually causes a hindrance.

However, the understanding of the mechanistic of HCMV-specific IgM antibody is highly important. Ig M antibody rises high to medium levels through the first to 3 months after the start of either acute infection. Then in the late phase, titre shows a downward trend. A persistent IgM antibody response refers to stable HCMV-specific antibody for greater than 3 months although variations are possible depending upon the immune system of a person.

23.7.4 IgG Avidity Assay

This assay helps in identification of a primary infection from non-primary infection especially during pregnancy. The principle of this assay that in the first month after acquisition of infection, HCMV-associated IgG of low avidity is developed and subsequently with the help of a maturation process, higher avidity antibody is generated. Only this higher avidity component is associated with recurrent infection. IgG avidity assay also helps in defining the duration of primary HCMV infection (Revello and Gerna 1999). The IgG avidity assay detected primary infection with 100% efficacy but ability to rule out a recent infection showed considerable variations to as low as 32% which undoubtedly indicates the standardization of the assay (Bodéus et al. 2001).

23.7.5 Neutralizing Antibody

Determination of HCMV neutralizing antibody is helpful for detection and timing of primary infection (Eggers et al. 1998). The researchers came to the conclusion that, during the convalescent stage of a primary infection, the absence of neutralizing antibodies is a strong signal of main infection, while the presence of these antibodies excludes a primary infection within the previous 15 weeks.

The identification and measurement of HCMV in blood have the potential to be a valuable tool for the early detection of HCMV disease in immunocompromised patients who have a primary infection or are in the re-emergence phase. However, in immunocompetent individuals, HCMV detection in blood is predictive of only primary HCMV infection.

23.7.6 Congenital Infections Diagnosis in Newborn

In the first 2 weeks of birth, virus isolation in human fibroblasts is a reliable way to diagnose congenital HCMV infection. However, following virus excretion may indicate infection during delivery or through breast milk (Stagno 2001). The clinical samples of choice are urine and saliva. Shibata et al. (1994) proposed another method using dried blood spots on filter paper. Congenital infection is highly responsible for developing mental retardation in children and hence is thought to be a potential candidate for its elimination through vaccination from human population. Live attenuated vaccines, peptide vaccines, subunit vaccines, recombinant virus vaccine, and DNA vaccines were studied for their efficacy especially to treat congenital HCMV infection.

23.8 Prevention and Control

23.8.1 Maintaining Hygiene

People should wash their hands with soap at least for 15–20 s who are in contact with newborn children and their saliva or faces, urine, or their diaper pants (Kotton 2013).

23.8.2 Blood and Stem Cell Products

It is common practice to give recipients who are CMV-seronegative "CMV-safe" blood products. Leukocyte-decreased, filtered blood products and blood from CMV-seronegative donors are two techniques for reducing the risk of CMV transmission through blood products. Both methods are frequently employed (Boeckh and Ljungman 2009). The blood bank should perform leukocyte filtering, and the set quality guidelines should be followed. The question of whether using filtered blood products in addition to seronegative blood products has not been examined in a controlled study.

If there are many donors available, in an HLA-identical sibling situation, CMV-seronegative donors are frequently chosen for CMV-negative recipients. In the context of the unrelated donor, things are more complicated. How to balance CMV serological status against other appropriate donor variables is crucial, especially when there are several potential donors. Although no study has assessed the relative impact of HLA match vs CMV serology, it would most likely be better to use an antigen-matched donor for HLA-A, HLA-B, or HLA-DR rather than a CMV-negative donor (Boeckh and Ljungman 2009). The situation is different for mismatches of a smaller severity, such as allele-mismatches or mismatches on the HLA-C, HLA-DQ, or HLA-DP, and a CMV-negative donor might be taken into consideration even if the match was poorer. A CMV match is preferred over other donor characteristics like age or blood type. It is debatable whether CMV serostatus should be taken into account when choosing donors for CMV-seropositive recipients. Some experts suggest using CM-seropositive donors for CMV-seropositive recipients in an unrelated donor, non-T-cell-depleted scenario to maximize the adoptive transfer of donor immunity; however, this approach has not been widely embraced (Boeckh and Ljungman 2009).

23.8.3 Antiviral Strategies

In the context of HCT, prophylaxis and pre-emptive therapy are both employed to address CMV. The most current international survey provided evidence of the usage trends. Although prophylaxis is preferred by some facilities for high-risk patients like those who have received unrelated, mismatched-HLA, or products of cord blood stem cell, pre-emptive therapy is still most frequently used (Pollack et al. 2011).

1. Viral load and pre-emptive antiviral therapy

Usually, pre-emptive therapy is started when CMV DNA, CMV pp65 antigenemia, or CMV mRNA are found. Although CMV DNA load is being evaluated more frequently, there are still no verified or established thresholds (Boeckh and Ljungman 2009). Recently, the specifics of present methods have been evaluated. The most frequently used drug for pre-emptive therapy is ganciclovir, and the second-line drug for neutropenia is foscarnet (Pollack et al. 2011). Although valganciclovir gives HCT recipients a sufficient amount of systemic exposure, it has not been tested in clinical trials with sufficient power (Einsele et al. 2006). As recently established in a sizable, multicentre clinical trial, the current prevention methods are quite effective in preventing CMV illness. Today, gastrointestinal disorders are the main cause of breakthrough disease. Throughout the initial 3 months following transplantation, CMV surveillance and preventive medication are continued. However, late CMV illness is a risk factor for poor outcomes and affects 4–15% of seropositive transplant recipients. The majority of incidents happen 4–12 months after HCT. Risk factors comprise CMV infection or illness during the first 3 months following HCT, cord blood transplantation, GVHD or T-cell depletion in the graft, absence of CMV-specific T-cell immunity, use of anti-T-cell medications or a low CD4 T-cell count. To treat late CMV problems, ongoing observation and preventive antiviral therapy are advised (Boeckh 2011).

2. Vaccines

The prevalence of CMV infection and cCMV disease may be decreased with the development of vaccinations that can alter or eliminate the genes expression that restrict activity of NK cell on CMV. There have recently been new initiatives to create a live-attenuated CMV vaccine. The CMV virus was altered to restore the genes expression that has been discovered to be important in the formation of protective immunity in order to increase immunogenicity and efficacy (Esposito et al. 2021). Knowing more about the interactions between the CMV and host's immune system has allowed us to assert unequivocally that it is highly challenging to design a vaccine that can absolutely lower the risk of cCMV infection and prevent both primary and nonprimary infections. Many of the methods by which the virus eludes the immune system and causes cCMV infection, particularly in instances of nonprimary infection, are still not well understood (Esposito et al. 2021).

23.9 Future Perspective

Human cytomegaloviruses are one the most pathogenic viruses infecting humans. It is public health problem worldwide. It poses a challenging scenario in terms of timely detection and treatment. The congenital HCMV infections are one of the most serious threats of HCMV infections. But still HCMV screening in pregnant women is not done routinely. It is still a matter of debate because vertical transmission from mother to foetus is not predictable. Therefore, in depth studies are required to delineate mode of infection of HCMV. For mechanistic evaluation of HCMV, development of reliable in vitro and in vivo models for HCMV propagation is much needed. There is not HCMV vaccine available and also present treatment for HCMV relies on antiviral drugs such as ganciclovir and immunoglobulin, but these therapies only supress the infections do not cure it. Also, resistant CMV is a well-known entity, where patient do not respond to antiviral treatments. Most recently adoptive therapies, such as transfer of CMV-specific T-cells into the patients showed promising results to combat HCMV infection. Hence, future research on HCMV should focus on development of novel HCMV vaccines and effective therapeutic treatment options.

References

Aarnisalo J, Ilonen J, Vainionpää R, Volanen I, Kaitosaari T, Simell O (2003) Development of antibodies against cytomegalovirus, varicella-zoster virus and herpes simplex virus in Finland during the first eight years of life: a prospective study. Scand J Infect Dis 35(10):750–753

Aguiar A, Galinato M, Bradley Silva M, Toth B, McVoy MA, Hertel L (2021) Human cytomegalovirus replication and infection-induced syncytia formation in Labial, Foreskin, and Fetal Lung Fibroblasts. Viruses 13(12):2355

Beisser PS, Laurent L, Virelizier JL, Michelson S (2001) Human cytomegalovirus chemokine receptor gene US28 is transcribed in latently infected THP-1 monocytes. J Virol 75(13):5949–5957

Bodéus M, Beulné D, Goubau P (2001) Ability of three IgG-avidity assays to exclude recent cytomegalovirus infection. Eur J Clin Microbiol Infect Dis 20(4):248–252
Boeckh M (2011) Complications, diagnosis, management, and prevention of CMV infections: current and future. Hematology. Am Soc Hematol Educ Program 2011:305–309
Boeckh M, Ljungman P (2009) How we treat cytomegalovirus in hematopoietic cell transplant recipients. Blood 113(23):5711–5719
Boppana SB, Fowler KB (2007) Chapter 44. Persistence in the population: epidemiology and transmission. In: Arvin A, Campadelli-Fiume G, Mocarski E et al (eds) Human herpesviruses: biology, therapy, and immunoprophylaxis. Cambridge University Press, Cambridge
Bradley AJ, Lurain NS, Ghazal P, Trivedi U, Cunningham C, Baluchova K, Gatherer D, Wilkinson G, Dargan DJ, Davison AJ (2009) High-throughput sequence analysis of variants of human cytomegalovirus strains Towne and AD169. J Gen Virol 90(Pt 10):2375–2380
Braud VM, Tomasec P, Wilkinson GW (2002) Viral evasion of natural killer cells during human cytomegalovirus infection. Curr Top Microbiol Immunol 269:117–129
Britt W (2008) Manifestations of human cytomegalovirus infection: proposed mechanisms of acute and chronic disease. Curr Top Microbiol Immunol 325:417–470
Brock I, Krüger M, Mertens T, von Einem J (2013) Nuclear targeting of human cytomegalovirus large tegument protein pUL48 is essential for viral growth. J Virol 87(10):6005–6019
Caffarelli N, Fehr AR, Yu D (2013) Cyclin A degradation by primate cytomegalovirus protein pUL21a counters its innate restriction of virus replication. PLoS Pathog 9(12):e1003825
Cantrell SR, Bresnahan WA (2006) Human cytomegalovirus (HCMV) UL82 gene product (pp71) relieves hDaxx-mediated repression of HCMV replication. J Virol 80(12):6188–6191
Chan G, Stinski MF, Guilbert LJ (2004) Human cytomegalovirus-induced upregulation of intercellular cell adhesion molecule-1 on villous syncytiotrophoblasts. Biol Reprod 71(3):797–803
Chen DH, Jiang H, Lee M, Liu F, Zhou ZH (1999) Three-dimensional visualization of tegument/capsid interactions in the intact human cytomegalovirus. Virology 260(1):10–16
Collins-McMillen D, Chesnokova L, Lee BJ, Fulkerson HL, Brooks R, Mosher BS, Yurochko AD (2018) HCMV infection and apoptosis: how do monocytes survive HCMV infection? Viruses 10(10):533
Crough T, Khanna R (2009) Immunobiology of human cytomegalovirus: from bench to bedside. Clin Microbiol Rev 22(1):76–98
de Ory F, Ramírez R, García Comas L, León P, Sagües MJ, Sanz JC (2004) Is there a change in cytomegalovirus seroepidemiology in Spain? Eur J Epidemiol 19(1):85–89
Drouet E, Chapuis-Cellier C, Bosshard S, Verniol C, Niveleau A, Touraine JL, Garnier JL (1999) Oligo-monoclonal immunoglobulins frequently develop during concurrent cytomegalovirus (CMV) and Epstein-Barr virus (EBV) infections in patients after renal transplantation. Clin Exp Immunol 118(3):465–472
Dworsky ME, Welch K, Cassady G, Stagno S (1983) Occupational risk for primary cytomegalovirus infection among pediatric health-care workers. N Engl J Med 309(16):950–953
Eggers M, Metzger C, Enders G (1998) Differentiation between acute primary and recurrent human cytomegalovirus infection in pregnancy, using a microneutralization assay. J Med Virol 56(4):351–358
Einsele H, Reusser P, Bornhäuser M, Kalhs P, Ehninger G, Hebart H, Chalandon Y, Kröger N, Hertenstein B, Rohde F (2006) Oral valganciclovir leads to higher exposure to ganciclovir than intravenous ganciclovir in patients following allogeneic stem cell transplantation. Blood 107(7):3002–3008
Esposito S, Chiopris G, Messina G, D'Alvano T, Perrone S, Principi N (2021) Prevention of congenital cytomegalovirus infection with vaccines: state of the art. Vaccine 9(5):523
Fisher MA, Lloyd ML (2020) A review of murine cytomegalovirus as a model for human cytomegalovirus disease-do mice lie? Int J Mol Sci 22(1):214
Fulkerson HL, Nogalski MT, Collins-McMillen D, Yurochko AD (2021) Overview of human cytomegalovirus pathogenesis. Methods Mol Biol (Clifton, N.J.) 2244:1–18
Glass M, Everett RD (2013) Components of promyelocytic leukemia nuclear bodies (ND10) act cooperatively to repress herpesvirus infection. J Virol 87(4):2174–2185

Gold E, Nankervis GA (1976) Cytomegalovirus. In: Viral infections of humans: epidemiology and control. Plenum Press, New York, pp 143–161

Griffiths P, Reeves M (2021) Pathogenesis of human cytomegalovirus in the immunocompromised host. Nat Rev Microbiol 19(12):759–773

Gupta M, Shorman M (2022) Cytomegalovirus. In: StatPearls. StatPearls Publishing

Hecker M, Qiu D, Marquardt K, Bein G, Hackstein H (2004) Continuous cytomegalovirus seroconversion in a large group of healthy blood donors. Vox Sang 86(1):41–44

Ho SK, Lo CY, Cheng IK, Chan TM (1998) Rapid cytomegalovirus pp65 antigenemia assay by direct erythrocyte lysis and immunofluorescence staining. J Clin Microbiol 36(3):638–640

Huang ES, Alford CA, Reynolds DW, Stagno S, Pass RF (1980) Molecular epidemiology of cytomegalovirus infections in women and their infants. N Engl J Med 303(17):958–962

Jarvis MA, Nelson JA (2002) Mechanisms of human cytomegalovirus persistence and latency. Front Biosci 7:d1575–d1582

Jean Beltran PM, Cristea IM (2014) The life cycle and pathogenesis of human cytomegalovirus infection: lessons from proteomics. Expert Rev Proteomics 11(6):697–711

Kääriäinen L, Klemola E, Paloheimo J (1966) Rise of cytomegalovirus antibodies in an infectious-mononucleosis-like syndrome after transfusion. Br Med J 1(5498):1270–1272

Kalejta RF (2008) Tegument proteins of human cytomegalovirus. Microbiol Mol Biol Rev 72(2):249–265

Keightley MC, Rinaldo C, Bullotta A, Dauber J, St George K (2006) Clinical utility of CMV early and late transcript detection with NASBA in bronchoalveolar lavages. J Clin Virol 37(4):258–264

Kerry JA, Priddy MA, Staley TL, Jones TR, Stenberg RM (1997) The role of ATF in regulating the human cytomegalovirus DNA polymerase (UL54) promoter during viral infection. J Virol 71(3):2120–2126

Kohmer N, Nagel A, Berger A, Enders M, Hamprecht K, Korn K, Kortenbusch M, Überla K, Rabenau HF (2019) Laboratory diagnosis of congenital CMV infection in newborns: impact of pre-analytic factors. J Clin Virol 115:32–36

Kothari A, Ramachandran VG, Gupta P, Singh B, Talwar V (2002) Seroprevalence of cytomegalovirus among voluntary blood donors in Delhi, India. J Health Popul Nutr 20(4):348–351

Kotton CN (2013) CMV: prevention, diagnosis and therapy. Am J Transpl 13(Suppl 3):24–40

Krech U, Konjajev Z, Jung M (1971) Congenital cytomegalovirus infection in siblings from consecutive pregnancies. Helv Paediatr Acta 26(4):355–362

Krishna BA, Poole EL, Jackson SE, Smit MJ, Wills MR, Sinclair JH (2017) Latency-associated expression of human cytomegalovirus US28 attenuates cell signaling pathways to maintain latent infection. MBio 8(6):e01754–e01717

Krstanović F, Britt WJ, Jonjić S, Brizić I (2021) Cytomegalovirus infection and inflammation in developing brain. Viruses 13(6):1078

Landini MP, La Placa M (1991) Humoral immune response to human cytomegalovirus proteins: a brief review. Comp Immunol Microbiol Infect Dis 14(2):97–105

Landolfo S, Gariglio M, Gribaudo G, Lembo D (2003) The human cytomegalovirus. Pharmacol Therap 98(3):269–297

Larsson S, Söderberg-Nauclér C, Möller E (1998) Productive cytomegalovirus (CMV) infection exclusively in CD13-positive peripheral blood mononuclear cells from CMV-infected individuals: implications for prevention of CMV transmission. Transplantation 65(3):411–415

Liu Z, Wang E, Taylor W, Yu H, Wu T, Wan Z, Huang Y, Ni Z, Sackett D (1990) Prevalence survey of cytomegalovirus infection in children in Chengdu. Am J Epidemiol 131(1):143–150

Liu YT, Strugatsky D, Liu W, Zhou ZH (2021) Structure of human cytomegalovirus virion reveals host tRNA binding to capsid-associated tegument protein pp150. Nat Commun 12(1):5513

Lloyd MG, Smith NA, Tighe M, Travis KL, Liu D, Upadhyaya PK, Kinchington PR, Chan GC, Moffat JF (2020) A novel human skin tissue model to study varicella-zoster virus and human cytomegalovirus. J Virol 94(22):e01082–e01020

Martinez-Martin N, Marcandalli J, Huang CS, Arthur CP, Perotti M, Foglierini M, Ho H, Dosey AM, Shriver S, Payandeh J, Leitner A, Lanzavecchia A, Perez L, Ciferri C (2018) An Unbiased

Screen for Human Cytomegalovirus Identifies Neuropilin-2 as a Central Viral Receptor. Cell 174(5):1158–1171.e19

Morrison KM, Beucler MJ, Campbell EO, White MA, Boody RE, Wilson KC, Miller WE (2019) Development of a primary human cell model for the study of human cytomegalovirus replication and spread within salivary epithelium. J Virol 93(3):e01608–e01618

Natali A, Valcavi P, Medici MC, Dieci E, Montali S, Chezzi C (1997) Cytomegalovirus infection in an Italian population: antibody prevalence, virus excretion and maternal transmission. New Microbiol 20(2):123–133

Nielsen CM, Hansen K, Andersen HM, Gerstoft J, Vestergaard BF (1987) An enzyme labelled nuclear antigen immunoassay for detection of cytomegalovirus IgM antibodies in human serum: specific and non-specific reactions. J Med Virol 22(1):67–76

Nogalski MT, Chan GC, Stevenson EV, Collins-McMillen DK, Yurochko AD (2013) The HCMV gH/gL/UL128-131 complex triggers the specific cellular activation required for efficient viral internalization into target monocytes. PLoS Pathog 9(7):e1003463

Pollack M, Heugel J, Xie H, Leisenring W, Storek J, Young JA, Kukreja M, Gress R, Tomblyn M, Boeckh M (2011) An international comparison of current strategies to prevent herpesvirus and fungal infections in hematopoietic cell transplant recipients. Biol Blood Marrow Transpl 17(5):664–673

Prabhakar P, Bailey A, Smikle MF, Ashley D (1992) Seroprevalence of cytomegalovirus infection in a selected population in Jamaica. West Indian Med J 41(4):133–135

Prior JR, Morroll DR, Birks AG, Matson PL, Lieberman BA (1994) The screening for cytomegalovirus antibody in semen donors and recipients within a donor insemination programme. Hum Reprod 9(11):2076–2078

Pultoo A, Meetoo G, Pyndiah MN, Khittoo G (2001) Seroprevalence of cytomegalovirus infection in Mauritian volunteer blood donors. Indian J Med Sci 55(2):73–78

Revello MG, Gerna G (1999) Diagnosis and implications of human cytomegalovirus infection in pregnancy. Fet Matern Med Rev 11:117–134

Reynolds DW, Stagno S (1979) Laboratory diagnosis of cytomegalovirus infections. In: Lennette EH, Schmidt NJ (eds) Diagnostic procedures for viral, rickettsial, and chlamydial infections, 5th edn. American Public Health Association, Washington, DC, pp 399–439

Roback JD, Su L, Zimring JC, Hillyer CD (2007) Transfusion-transmitted cytomegalovirus: lessons from a murine model. Transfus Med Rev 21(1):26–36

Rowe WP, Hartley JW, Waterman S, Turner HC, Huebner RJ (1956) Cytopathogenic agent resembling human salivary gland virus recovered from tissue cultures of human adenoids. Proc Soc Exp Biol Med 92(2):418–424

Schottstedt V, Blümel J, Burger R, Drosten C, Gröner A, Gürtler L, Heiden M, Hildebrandt M, Jansen B, Montag-Lessing T, Offergeld R, Pauli G, Seitz R, Schlenkrich U, Strobel J, Willkommen H, von König CH (2010) Human Cytomegalovirus (HCMV)—revised. Trans Med Hemotherapy 37(6):365–375

Shen CY, Chang WW, Chang SF, Chao MF, Huang ES, Wu CW (1992) Seroepidemiology of cytomegalovirus infection among children between the ages of 4 and 12 years in Taiwan. J Med Virol 37(1):72–75

Shibata M, Takano H, Hironaka T, Hirai K (1994) Detection of human cytomegalovirus DNA in dried newborn blood filter paper. J Virol Methods 46(2):279–285

Shimamura M, Mach M, Britt WJ (2006) Human cytomegalovirus infection elicits a glycoprotein M(gM)/gN-specific virus-neutralizing antibody response. J Virol 80(9):4591–4600

Shnayder M, Nachshon A, Krishna B, Poole E, Boshkov A, Binyamin A, Maza I, Sinclair J, Schwartz M, Stern-Ginossar N (2018) Defining the transcriptional landscape during cytomegalovirus latency with single-cell RNA sequencing. MBio 9(2):e00013–e00018

Smith MG (1956) Propagation in tissue cultures of a cytopathogenic virus from human salivary gland virus (SGV) disease. Proc Soc Exp Biol Med 92(2):424–430

Stagno S (2001) Cytomegalovirus. In: Remington JS, Klein JO (eds) Infectious diseases of the fetus and newborn infant. W. B. Saunders Co., Philadelphia, PA, pp 389–424

Suassuna JH, Leite LL, Villela LH (1995) Prevalence of cytomegalovirus infection in different patient groups of an urban university in Brazil. Rev Soc Bras Med Trop 28(2):105–108

Vainionpää R, Waris M (2015) Diagnostic techniques: serological and molecular approaches. Reference Module in Biomedical Sciences

van Loon AM, Heessen FW, van der Logt JT, van der Veen J (1981) Direct enzyme-linked immunosorbent assay that uses peroxidase-labeled antigen for determination of immunoglobulin M antibody to cytomegalovirus. J Clin Microbiol 13(3):416–422

Vanarsdall AL, Johnson DC (2012) Human cytomegalovirus entry into cells. Curr Opin Virol 2(1):37–42

Vanarsdall AL, Ryckman BJ, Chase MC, Johnson DC (2008) Human cytomegalovirus glycoproteins gB and gH/gL mediate epithelial cell-cell fusion when expressed either in cis or in trans. J Virol 82(23):11837–11850

Vornhagen R, Plachter B, Hinderer W, The TH, Van Zanten J, Matter L, Schmidt CA, Sonneborn HH, Jahn G (1994) Early serodiagnosis of acute human cytomegalovirus infection by enzyme-linked immunosorbent assay using recombinant antigens. J Clin Microbiol 32(4):981–986

Vornhagen R, Hinderer W, Sonneborn HH, Bein G, Matter L, The TH, Jahn G, Plachter B (1995) The DNA-binding protein pUL57 of human cytomegalovirus is a major target antigen for the immunoglobulin M antibody response during acute infection. J Clin Microbiol 33(7):1927–1930

Weng C, Lee D, Gelbmann CB, Van Sciver N, Nawandar DM, Kenney SC, Kalejta RF (2018) Human cytomegalovirus productively replicates in vitro in undifferentiated oral epithelial cells. J Virol 92(16):e00903–e00918

Yu X, Jih J, Jiang J, Zhou ZH (2017) Atomic structure of the human cytomegalovirus capsid with its securing tegument layer of pp150. Science 356:eaam6892

Zheng QY, Huynh KT, van Zuylen WJ, Craig ME, Rawlinson WD (2019) Cytomegalovirus infection in day care centres: a systematic review and meta-analysis of prevalence of infection in children. Rev Ins Med Virol 29(1):e2011

Epstein–Barr Virus: Emerging and Re-emerging Viral Infectious Diseases

24

Bunty Sharma, Ujjawal Sharma, Deepika Raina, Ritu Sharma, Madhav Singla, and Sampan Attri

Abstract

The Epstein–Barr virus (EBV), also referred to as the human herpesvirus, is a herpes virus. It ranks among the most common viruses that affect people. The majority of people contract EBV at some point in their lives. EBV is most frequently transmitted by body fluids, particularly saliva. Infectious mononucleosis, usually known as mono, and other illnesses can be brought on by EBV. The majority of adults or teenagers who experience EBV infection symptoms recover in 2–4 weeks. Some individuals, however, may experience weariness for several weeks or even months. The EBV virus infects the individual and then becomes dormant. The virus may occasionally reactivate. Although symptoms are not usually the result, those with compromised immune systems are more susceptible to experience them if EBV reactivates. Because the symptoms of EBV infec-

B. Sharma
Department of Bio-Sciences and Technology, Maharishi Markandeshwar Engineering College, Maharishi Markandeshwar (Deemed to be University), Ambala, Haryana, India

U. Sharma (✉)
Department of Human Genetics and Molecular Medicine, School of Health Sciences, Central University of Punjab, Bathinda, Punjab, India
e-mail: ujjawal.sharma@cup.edu.in

D. Raina
School of Pharmacy, Graphic Era Hill University, Dehradun, Uttarakhand, India

R. Sharma
Department of Biochemistry, Government Medical College, Amritsar, Punjab, India

M. Singla
Chitkara College of Pharmacy, Chitkara University, Rajpura, Punjab, India

S. Attri
Viral Testing Facility, Forensic Science laboratory, Mohali, Punjab, India

P. L. Bhukya, A. S (eds.), *Emerging Human Viral Diseases, Volume II*,
https://doi.org/10.1007/978-981-97-4480-0_24

tion are similar to those of other disorders, diagnosing that it can be difficult. Blood testing that looks for antibodies can identify EBV infection. There is currently no vaccination available to guard against EBV infection. By avoiding kissing and exchanging food, drinks, and personal objects like toothbrushes with those who have the EBV infection, one can better safeguard against EBV. EBV may potentially be involved in the development of rare cancers and other diseases, such as autoimmune disorders and schizophrenia. With reduced mortality and morbidity, significant progress has been made in therapy, but there are still certain obstacles to overcome, particularly for the most aggressive diseases. A more comprehensive solution to this public health issue would be to prevent it with EBV vaccine, but more active research is required before this objective can be achieved.

Keywords

Epstein–Barr virus · Emerging viral disease · Therapeutics · Host virus interaction · Re-emergence of viral disease · Diagnosis

24.1 Introduction

Epitheca bar virus, also known as the Epstein–Barr virus (EBV), is a species of single-stranded RNA virus that infects the marine copepod *Epistylis histrix*. It belongs to Herpesviridae family whose subfamily is gamaherpesvirinae, genus is lymphocryptovirus, and species Human herpesvirus. The EBV is morphologically distinct from other viruses that infect saltwater crustaceans, with its cylindrical, double-layered shell (Epstein et al. 1964). The body of the EBV is long and cylindrical, with a double layer of membrane encasing a single set of nucleic acid chromosomes which itself is protected by another outer membrane. The shell is studded with proteinaceous material that serves to anchor the copepod to its host, and to allow the EBV to gain access to the host cell. When the EBV infects a cell, it uses its proteinaceous sheath to penetrate the host cells outer membrane. EBV can be transmitted from one person to another through saliva and bodily fluids. It causes a mild form of mononucleosis that lasts 2–8 weeks. The virus is transmitted through saliva, tears, or small droplets for short distances. It is also possible for the virus to be transmitted through kissing and sexual contact in a non-human manner. A member of the herpes virus family, the EBV has been known to infect individuals for a long time. It is a common virus with which most people are infected by the age of 10 or 11. EBV infection can take a long time to develop. Early symptoms usually include a mild fever, fatigue, and general malaise (Ablashi et al. 1974; Chien et al. 2001).

The report from the Centre for Disease Control and Prevention (Atlanta, USA) states that 95% of the adult population in the age range of 35–40 years is carrier of EBV. Lifelong latency has been related to this virus with a risk of periodic reactivation after a primary infection especially under conditions of weakened immune system. However, clinically evident symptoms remain limited in a majority of a population. In developing nations, children contract an infection in their first few years of life, whereas in affluent nations, it is frequently postponed until adolescence (Haahr et al. 2004). High titres are seen in a U-shaped pattern among young people and people over 50 years (Glaser et al. 1985). In children, it is due to primary infection, and in the older population, it is because of reactivation of the virus associated with delayed immune response owing to aging (Wick and Grubeck-Loebenstein 1997). Initial infection occurs in tonsillar region. Lymphocytes and epithelial cells are mainly the resident cells of EBV (Kieff and Rickinson 2007). During the lytic phase, viral DNA polymerase completes the viral replication. EBV infection in B cells results in a very important event. It is possible to detect a high viral load in the blood when these cells are stimulated to activate their development cycle and initiate differentiation into memory B cells. This number tends to decrease after primary infection but without elimination of these cells (Hadinoto et al. 2008).

The best-known treatment for the EBV is the use of antibiotics, which can kill the EBV in its host and prevent it from replicating further. Antibiotics that have been used as treatments for EBV include cephalexin, vancomycin, and tetracycline. The most effective treatment for the EBV is vancomycin, which can kill the EBV in the saltwater crustacean in as little as 12 h. Vancomycin can also be used to treat EHE in saltwater crustaceans that have been exposed to the EBV, though this is only effective if the crustacean is still alive at the time of treatment (Aitken et al. 1994).

24.2 Epidemiology

24.2.1 Epidemiology of Endemic NC and Burkitt's Lymphoma (BL)

Endemic diseases are those which are regularly found in a particular region among different types of people. It gets increased with age and is maximum at the age of 50. The endemic type is strongly allied with viruses. Endemic type varies with the region. Nasopharyngeal carcinoma's (NPC) rate is double in Guangdong than Cantonese of China (Hutcheson et al. 2021).

In case of BL, the most occurrence of this type of tumor is recorded in New Guinea and in equatorial Africa (Hjalgrim et al. 2007; Hutcheson et al. 2021). BL is the most common childhood cancer which occurs in both boys and girls and peaks at the age of 5–9 years (Hutcheson et al. 2021).

24.2.2 Epidemiology of Sporadic Burkitt's Lymphoma (BL)

Sporadic diseases are those which are irregular and found in few regions. Sporadic BL generally occurs in children and young adults. The rate of occurrence of sporadic BL is less than endemic BL. The occurrence of sporadic BL is observed to be similar to endemic variant (Hutcheson et al. 2021). About 20–80% of EBV is positive in sporadic BL (10–30% is mostly found) (Navari et al. 2015; Kuri et al. 2020).

It is also dependent upon the region as it can be found in the western countries among young adults (Satou et al. 2015).

24.2.3 Age, Mortality, and Morbidity of EBV

Age, mortality, and morbidity help to understand epidemiology of EBV more deeply.

24.2.3.1 Age

EBV is a gamma-herpesvirus and almost 95% of population of world gets infected even they are healthy. Out of which 85–90% are EBV seropositive (Table 24.1). It increases briskly in females than men during the phase conversion from childhood to adulthood (Winter et al. 2019).

24.2.3.2 Mortality

Mortality is the death rate of a specific disease. Mortality of EBV-infected patients is discussed in Table 24.2. It generally occurs in the countries of East Asia and cancer lead to death are Hodgkin lymphoma (HL), BL, gastric carcinoma (GC), and NPC. Out of these, most danger one or the cancer which is having highest mortality rate is NPC and GC (Khan et al. 2020; Wong et al. 2022).

24.2.3.3 Morbidity

Classical HL (CHL) suffering patients can be associated with EBV. Fifty-six percent of it carries EBV which shows flu-like symptoms and remaining 44% is asymptomatic. Morbidity of EBV can be outlined with the help of patients having transplants. Post-transplant Lymphoproliferative Disease (PTLD) is a condition in which there is a growth of white blood cells and can occur after transplant generally. Six percent of patients develop PTLD and 94% of patients are free from PTLD over a span of 5 years. In liver transplant, 39% of patients with CHL, of which 17% carry EBV and 22%, have PTLD. Whereas in heart transplant, 45% of patients with CHL can show the development of PTLD in long term. Factors such as patient population and the potency of immunosuppressive therapy control the development of PTLD (De Paschale and Clerici 2012).

Table 24.1 Occurrence of EBV in different types of age groups (Balfour Jr et al. 2013; Khan and Hashim 2014; Winter et al. 2019)

Age	EBV seropositive (%)
6–8	50
9–11	55
11–14	60
19–21	84
22–24	93

Table 24.2 Mortality of EBV (Höcker et al. 2013; Khan et al. 2020; Wong et al. 2022)

Years	Number of deaths	Percentage increase in mortality from 1990
2010	1,43,000	14.6
2017	1,64,000	19
2020	2,08,000	–

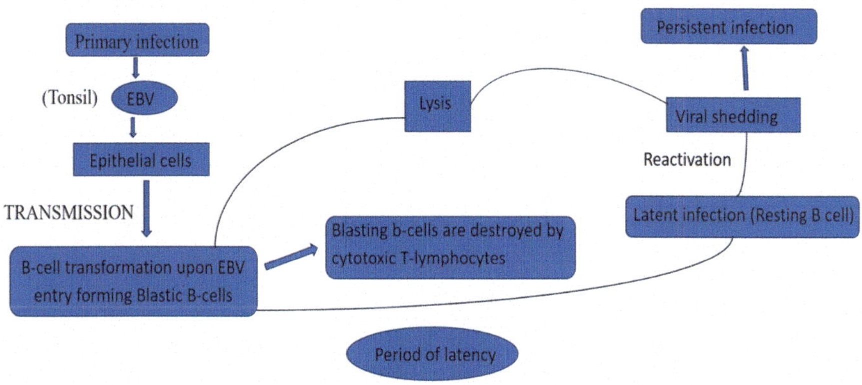

Fig. 24.1 Cycle of activation and re-activation of EBV

24.3 Potential Risk of Emergence and Re-emergence

Re-emergence of EBV is a key phenomenon associated with this virus. Latent-infected B cells often get stimulated to reactivate EBV. EBV has the capacity to cause latently infected lymphoblastic cell lines to develop from resting B cells. However, this stage is not fully achieved in vivo during lytic replication and remains limited to in vitro model of lytic replication. Under endogenous conditions and environmental stress, there is periodic reactivation of EBV which enters into a lytic replication phase. This period is associated with increased production and release of viral particles in large amount (Shannon-Lowe et al. 2006). Levels of IgA against several EBV antigens, such EA, nuclear antigen 1 (EBNA1), and viral capsid antigen (VCA) are raised (He et al. 2018). Six EBNA (1, 2, 3A, 3B, 3C, and LP) and three latent membrane proteins (LMP1, 2A and 2B) are among the viral gene products. There is absolute need of EBV nuclear antigens A1, LP, 3A, and 3C in the transformation process. Cycle of activation and re-activation of EBV is explained in Fig. 24.1.

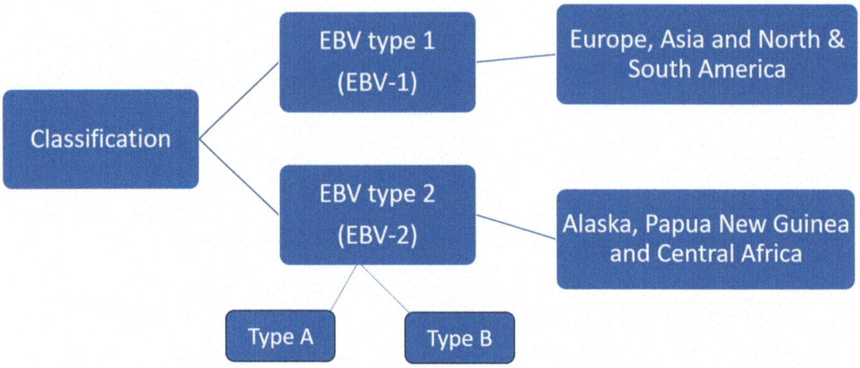

Fig. 24.2 Classification of EBV (Zanella et al. 2019)

24.4 Organization of Infectious Agents (Structural and Molecular)

24.4.1 Classification

Many attempts have been made to classify EBV but declined due to experimental proof. Therefore, after much efforts, EBV can be classified into the following manner

24.4.1.1 Clinical and Geographic Distribution

This classification is the most widely accepted.

It is classified on the basis of genotyping scheme and geographical distribution of EBV.

It is classified into two types on basis of versions of DNA sequence and geographical distribution (Zanella et al. 2019) (Fig. 24.2).

24.4.1.2 Latency Gene Expression

This classification is based on the pathological and physiological status of a host. Classification is feasible into all four possible varieties of latency gene expression of EBV (Yang et al. 2020) (Table 24.3).

24.4.1.3 Types of B-Cell Lymphoproliferative Diseases (LPD)

EBV is classified on the basis of histologic, clinical, cytogenetic, virologic, immunologic cell-typing, and immunoglobulin gene-rearrangement characteristics of these LPD.

They are of four types:

1. Uncomplicated post-transplant infectious mononucleosis (IM).
2. Benign polyclonal polymorphic B-cell hyperplasia.
3. Early malignant transformation in polyclonal polymorphic B-cell lymphoma.
4. Monoclonal polymorphic B-cell lymphoma (Hanto 1995).

Table 24.3 Classification of EBV through latency gene expression (Yang et al. 2020)

Latency phase	Expressed genes	Disorders
Latency 0	EBNA1	EBV-infected people
Latency I	EBER1, EBER2, BART miRNA EBNA1	Burkitt lymphoma
Latency II	EBER1, EBER2, BART miRNA EBNA1 LAMP1, LAMP2A/B, BARF1	Hodgkin lymphoma, nasopharyngeal carcinoma, T/NK lymphoma
Latency III	EBER1, EBER2, BART miRNA, BHRF1 miRNA, EBNA1, EBNA2, EBNA3A/B/C, LMP1, LMP2A/B	Infectious mononucleosis, lymphoblastoid cell line, PTLD/DLBCL

EBER1, EBER2, BART miRNAs falls under the category of Noncoding RNA
EBNA1, LMP2A, BARF1 fallls under the category of mRNA (Yang et al. 2020)

24.4.2 Morphology and Virion Structure

The virus comprises of a double helix of DNA (deoxyribonucleic acid). It is about 122–180 nm in diameter and consists of 1,72,000 base pairs in coding in 85 genes (Matsuura et al. 2010).

The innermost part, which is a toroid shaped nucleoprotein core of EBV virion, is surrounded by DNA. It is further layered up first by nucleocapsid with 162 capsomers and then by a protein tegument (Cabo 2022; Liebowitz and Kieff 1993). The outermost layer of this nucleoprotein core is enclosed by an envelope which is made up of lipids and virus-encoded glycoproteins spikes (Liebowitz and Kieff 1993; Matsuura et al. 2010; Cabo 2022). These glycoproteins are essential to infect the host cell (Matsuura et al. 2010).

EBV consists of three glycoproteins, gB, gH, and gL, which comprise the primary membrane fusion apparatus, helping viruses to penetrate into the cell. The gH and gL proteins are aligned together to former heterodimeric complex. This complex is essential, so that binding with epithelial cell receptors can take place to get entry into the cell. The gH/gL crystal structures need to be determined as it has multiple roles in the EBV entry. It forms elongated rod-like shape with 100 Å in length and 30–60 Å in width. It is divided into four major domains (Cohen 2000) (Fig. 24.3).

24.4.3 Genome Structure and Organization

EBV's genome is made up of double-stranded DNA having 1,72,202 base pairs long (Baer et al. 1984). It is a linear and 172 kb DNA molecule (Lyon 1994; Price and Luftig 2014; Guan et al. 2019; Young et al. 2000). It is circularized due to the cis acting elements within the genome, such as origin of plasmid replication (ori-Lyt), origins of lytic replication (oriP), and terminal repeats (TR) at C, A, and H and N, respectively (Liebowitz and Kieff 1993).

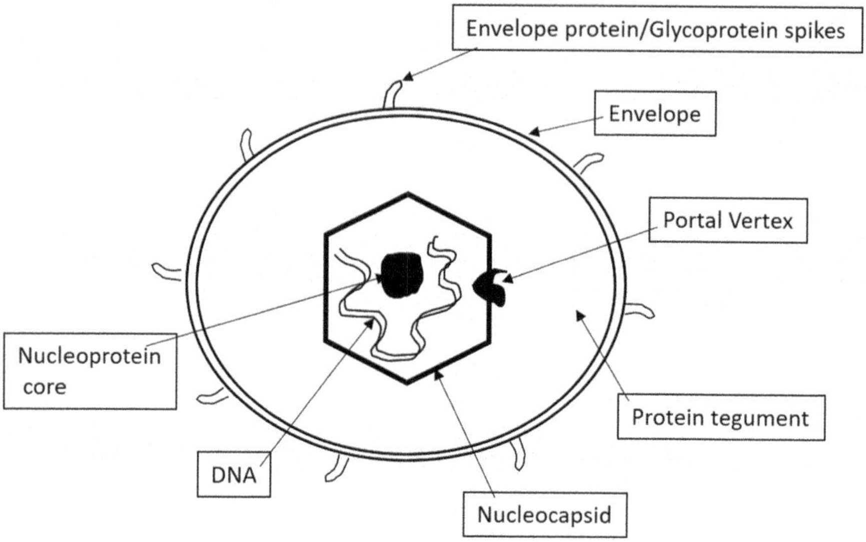

Fig. 24.3 Structure of EBV

The open reading frames (ORF) have been used to represent the genome of EBV. Its nomenclature is based on the BamHI-restriction fragment (Lyon 1994; Price and Luftig 2014). The ORF classified into sections latent and lytic genes, which can be further categorised as immediate early (IE) genes, early genes, and late genes. The viral genes will form proteins, remain untranslated (EBER1 and EBER2), or convert into micro-RNAs (at least 17) (Lyon 1994; IARC Working Group 1997). The active lytic genes after infection are in the pre-latent phase. The latent mRNAs led to the formation of EBNAs with the help of various promoters, such as Cp, Wp, and Qp. All of these promoters show transcription in different latency types and periods after the infection (Price and Luftig 2014; Young et al. 2000) (Table 24.4).

24.4.4 Viral Proteins

Table 24.5 shows the various viral proteins with their actions and applications (Matsuura et al. 2010; Palermo et al. 2008; Dawson et al. 2012; Bhattacharjee et al. 2016).

24.4.4.1 Prorogation and Life Cycle

The life cycle of EBV consists of two cycles: lytic and latent.

24.4.4.1.1 Lytic

EBV virus infects B cells and epithelial cells. It is generally found and spread through saliva among humans and starts occurring from oropharyngeal epithelial

Table 24.4 Proteins encoded by mRNAs (Lyon 1994; Price and Luftig 2014; Young et al. 2000)

Serial No.	Protein	Promoter	Region of origin on ORF (BamHI)	Region of existence on ORF (BamHI)	Latency type	Direction in genome structures
1.	EBNA1	Qp	Q	K	I and II	Rightwards
2.	EBNA2	Cp	C	Y	III	Rightwards
3.	EBNA3A	Cp	C	E	III	Rightwards
4.	EBNA3B	Cp	C	E	III	Rightwards
5.	EBNA3C	Cp	C	E	III	Rightwards
6.	LMP1	K14	N	N	–	Leftwards
7.	LMP2A	K14	N	Across TR between N	–	Rightwards
8.	LMP2B	K14	N	Across TR between N	–	Rightwards

Table 24.5 Various viral proteins with their actions and applications

Protein/gene/ antigens	Stage	Action	Application
1. EBNA-1	Latent + lytic	Binds to a replication origin in the viral genome	It led to replication at the time of division of host cell
2. EBNA-2	Latent + lytic	It encodes all nuclear antigens required for indefinite reproduction by EBV	Major viral transactivator
3. EBNA-3	Latent + lytic	Binds to host RBP-Jk protein	Transcription regulatory protein
4. LMP-1	Latent	Engage number of chemical reactions to produce alterations in epithelial cells	Growth transformation of EBV
5. LMP-2	Latent	Maintains viral latency in infected B cells	To block tyrosine kinase
6. EBER	Latent	It inhibits their functions by binds with nucleoprotein particles and PKR	Enhance the growth of IL-10 and inhibits cytotoxic T cells
7. V-snoRNA1	Latent	Targets the 3′UTR of viral DNA polymerase mRNA	It acts as miRNA-precursor
8. Ebv-sisRNA	Latent	Its availability is abundant as small RNA after EBER	Stable intronic sequence RNA
9. miRNAs	Latent	Generated during type-III (4 miRNAs) and II (up to 20 miRNAs) latency	Currently unknown
10. EBV-EA	Lytic	Early antigen	Cause infectious mononucleosis
11. EBV–VCA	Lytic	Viral capsid antigens	Diagnose infectious mononucleosis

cells (Zavala-Vega et al. 2019; Jagadeesh and Wood 2021). The virus must be linear to get transferred into the cells or host (Jagadeesh and Wood 2021). The lytic cycle can be outlined in the following steps.

1. **Entry of virus into the cell/host viral interaction**

   The virus attacks the B cell, so that it can prevent the destruction of virus and cytotoxic T cells have the ability to recognize and destroy the virus (Zavala-Vega et al. 2019; Jagadeesh and Wood 2021).

   First, the virus gets attached to the surface of this by binding glycoprotein gp350 to B cell's receptor CD40. Eventually, gp42 which is out with HLA class II and bind ghgl. Through endocytosis entry of virus takes place due to the activation of gB. On the basis of cell tropism, other viral proteins may be required for the viral entry (Zavala-Vega et al. 2019).
2. **DNA replication**

   The host's immunological response to infection is controlled by tegument proteins. The capsid is moved to the nuclear pore in this stage. DNA inside the nucleus is released, and the TR causes it to become circular. It leads to the methylation of CpG, gene silencing and regulates the genes transcription (Zavala-Vega et al. 2019).

   Other proteins of tegument help to form IE genes, such as BZLF1 and BRLF1, which act as transcriptional activators. IE is also known as alpha proteins and is very necessary for the transcription of early proteins, such as BNLF2. The early proteins help to form beta proteins which is machinery of DNA replication. Once it gets encoded, the replication of DNA occurs in a circular mechanism, resulting in the formation of long DNA chain molecules comprising of multiple copies of DNA sequence (concatemers) of DNA genomes (Zavala-Vega et al. 2019; Jagadeesh and Wood 2021).
3. **Capsid formation**

   Replication gives rise to late gene products, such as VCA, which are responsible for the formation of viral capsid. Capsids gather into the nucleus, so that dsDNA is packed into new capsids, because some proteins in capsid are required to split concatemers of DNA to form nucleocapsid. The primary envelope is obtained inner nuclear membrane to proteins from to inner tegument layer (Zavala-Vega et al. 2019).
4. **Elimination**

   Naked nucleocapsids get released from the core through perinuclear cisterna. The nucleocapsid present in the cytoplasm is start being envelop at membranes of trans Golgi apparatus network. The enveloped particles are only allowed to be transported out through plasma membrane by specialized vesicles enclosing viral glycoproteins. EBV's transmission can also occur among cells via cell junctions (Zavala-Vega et al. 2019).

24.4.4.1.2 Latent

During latency, there is no production of viruses. EBV is present in the nucleus and its replication occurs through DNA polymerase (Zavala-Vega et al. 2019; Jagadeesh and Wood 2021). First, replication of EBV takes place in the oropharyngeal epithelial cells, leading to spelling of infection in B cells. Latent infection outturns rather than the productive infection in B cells. Latent infection is required for preserving the virus and results in the release of infectious virus into the saliva.

The latency which occurs in the B cell is of four types—Latency Types 0, I, II, III. Many genes give rise to various proteins, antigens, RNAs, and mRNAs. It includes the following:

1. Latent membrane proteins—LMP1, LMP2A, and LMP2B
2. Nuclear antigens—EBNA-1, EBNA-2, EBNA3A, EBNA3B, EBNA3C, and EBNA-LP
3. Non-coding RNA—EBER1 and EBER2
4. mRNAs—BARTs and BHRF1 (Zavala-Vega et al. 2019)

Upon the infection of B cell, it enters the latency type III in many proteins and RNA are formed. In latency III, either infected B cell gets converted into B-cell activation or killed by T cells. Subsequently, the remaining converted cells enter into latency type II, where the formation of proteins and RNA decreases. In this type, B cell gets converted into a memory B cell. Memory B cells enter latency I, where the formation of proteins and RNA gets more limited. This is the final stage, where the memory B cells split up, so that replication of EBV genome can occur. The latency 0 is faced afterwards in which there is very minute or zero formation of proteins and RNA takes place (Zavala-Vega et al. 2019; Young et al. 2000; Jagadeesh and Wood 2021).

Each valid protein such as EBNA, LMP, EBER and mRNAs which is produced during latency has its own significant role. EBNA has vital engagement in the process of replication of genome and responsible for the distribution of DNA in the infected B cells. EBNA1 is a transactivator which binds to the origin of replication in the viral genome and led to replication at the time of division of host cell. Its major work is to prevent the cell death and persist the infected host. Latent membrane protein also has a significant role in viral persistence. LMP1 helps to the formation of proteins which prevent the death of the host cell whereas LMP2A helps in the growth of infected B cell and its conversion into memory B cell along with blocking of tyrosine kinase. The non-coding RNAs such as EBER1 and EBER2 are involved in maintaining viral persistence by holding back cytotoxic T cells. The mRNAs such as BART2 decrease the production of polymerase BALF5 so that it can obstruct the shifting of cycle from latent to lytic and promotes the B cell activation, proliferation and activation of EBV (Matsuura et al. 2010; Zavala-Vega et al. 2019).

Reactivation of latent B is possible by switching to lytic replication. During latency miRNAs suppress BZLF1 (ZEBRA) and BRLF1. BZLF1 and BRLF1 genes are responsible for the production of infectious viruses through a transcription factor in lytic cycle. Together with Rta, BZLF1 reactivates the EBV genome's major binding to quiet chromatin, which is crucial for the lytic infection. It also initiates the transcription of early cycle genes (Zavala-Vega et al. 2019).

24.5 Pathogenesis in Humans

24.5.1 Origin of Infection

Acute tonsillitis is caused due to EBV. It is an asymptomatic infection. It generally occurs among young adults (Shetty 2022). EBV is transmitted through saliva mainly oropharyngeal section and with body secretions during intimate contact. This is why it is also known as kissing disease (Shetty 2022; Vockerodt et al. 2015). When the infection gets increased immensely it leads to IM or glandular fever and becomes a sympathomimetic infection (Shetty 2022).

24.5.2 Pathology of EBV via Direct Model

Pathogenesis of EBV is explained through direct model, as shown in Fig. 24.4. EBV transmits to the body through saliva or body secretions. The transmission of virus causes tonsils at initial stage. Tonsils are lymphoid tissue which comprises of T cells and B cells. The virus shows its action by attacking B cells and tonsillar epithelial cells. During the incubation span of 2–6 weeks, replication of EBV and spreading of virus into saliva (pharyngeal section) occurs several times at short intervals (Shetty 2022).

At reactivation time, EBV generates immune response due to replication and shedding of virus into saliva. In response, B cell acquired antigen of EBV producing immune response in the region of tonsils (Shetty 2022). Infection of B lymphocytes induces humoral and cellular response to the virus. Weak T cells may give a rapid rise in large number of infected B lymphocytes (Vockerodt et al. 2015). The

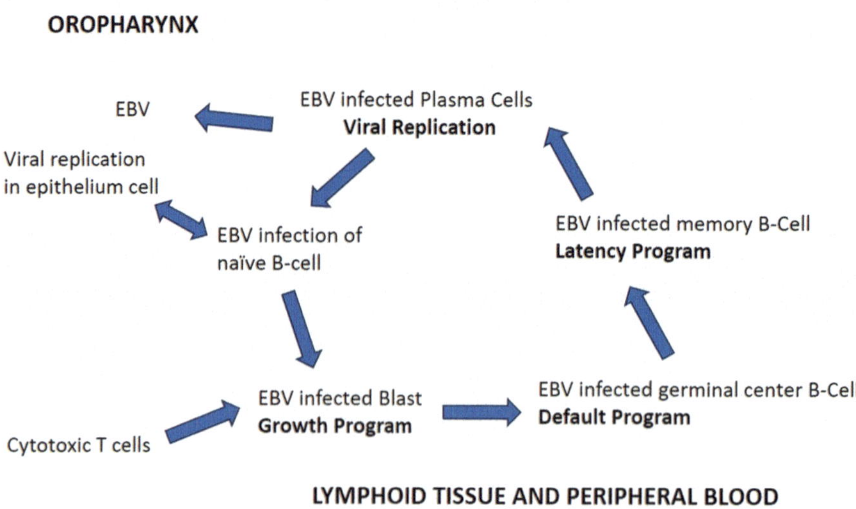

Fig. 24.4 Pathogenesis of EBV

occurrence of immune response in liver, spleen and lymph nodes are part of immune response against EBV (Shetty 2022).

B cells trigger cytotoxic T cells which are necessary for the elimination of infection and prevention of further spreading of EBV virus into the body. Simultaneously B cells get activated and transformed into plasma cells. Plasma cells give rise to specific antibodies such as VCA with IgM and VCA with IgG. The IgM antibodies induced from abnormal infiltrated B cells have significant role in diagnosing EBV. While eliminating the virus from the blood with the help of immune cells, it exhibits an immune response (Shetty 2022). Fever and miserable condition of patient is observed as immune response. It occurs due to cytokines which are formed when the virus captures B lymphocyte (Shetty 2022; Vockerodt et al. 2015).

24.5.3 Virulence Persistence and Pathology of EBV via Germinal Center (GC) Model

Another model through which pathogenesis of EBV can be explained is GC model or GC model. This model is also studied to understand virulence persistence. The infected B cells enters latency III causing rapid rise in number of infected B cells. The cells then enter GC and moves in latency type II in which conversion of infected B cells into memory B cells takes place. Memory B cells enter latency I where the formation of proteins and RNA gets more limited. Cell division of EBV occurs in presence of EBNA1 in latency type I. Therefore, replication occurs and the cells move in latency type 0 where there is either zero or very minute formation of proteins and RNA takes place so that it can prevent the identification of virus (Kelly et al. 2013).

24.6 Clinical Manifestations

24.6.1 Primary EBV Infection with Containment: Infectious Mononucleosis (IM)

IM is basically a syndrome which mimics an acute infectious illness associated with atypical large peripheral blood lymphocytes. These are also known as Downey cells which are actually activated CD8 T lymphocytes which destroy EBV-infected B cells (Sprunt and Evans 1920). This protective mechanism is very helpful in containing the infection.

24.6.2 Primary EBV Infection Without Containment

Majority of the individuals with primary EBV infection develop no serious complications. But in some cases, infection is not self-limited. This usually happens in immunocompromised subjects who are facing other diseases. The loss of containment is usually due to deregulation of viral replication cycle. The disease pattern has symptoms of chronic IM like lasting for more than 6 months, increased viral load, hypergammaglobulinemia, pancytopenia, B or T cell malignant lymphoma (Kimura et al. 2001). Dark side of EBV is its oncogenic potential. EBV gene activates oncogenes such as MYC and Bcl-2 and also NF-KB, JNK, PT3K/Akt and JAK/STAT signalling pathways. It also inhibits tumour suppressor gene p53, Dok etc. (Young and Murray 2003). The treatment strategies include antiviral drugs, immunomodulators, EBV responding cytotoxic T lymphocytes and hematopoietic stem cell transplantation.

24.6.3 Signs

The manifestation of sign and symptoms of IM vary from pharyngitis to anaemia as reported in the literature (Rea et al. 2001; Grotto et al. 2003). The details are provided in Table 24.6 and Fig. 24.5.

24.6.4 Symptoms

See Fig. 24.5.

24.6.5 Lab Tests

Table 24.7 shows the different lab findings and their prevalence in IM.

Recovery may take months and fatigue generally affects the quality of life. Recurrence is uncommon. Severity of IM is age specific. Individuals more than 40 years are especially vulnerable to serious infection. Intimate kissing is a main route of transmission of IM in adults and saliva on toys and fingers in case of small children.

Table 24.6 Sign and prevalence of IM

Signs	Prevalence (%)
Pharyngitis	100
Cervical lymphadenopathy	95
Fever	50
Hepatomegaly	25
Splenomegaly	33
Eyelid edema	10
Rash	5

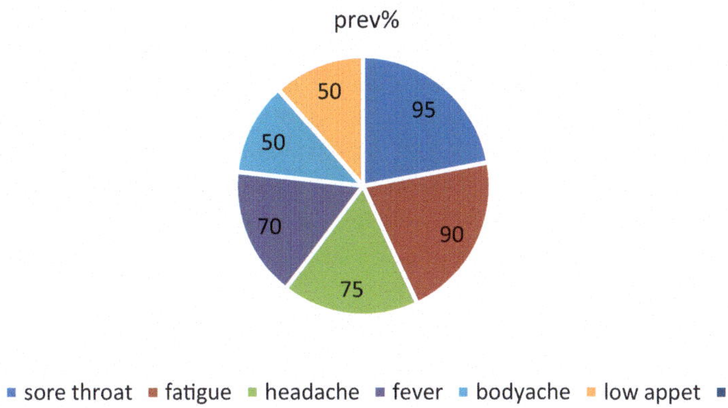

Fig. 24.5 Prevalence of different symptoms of IM

Table 24.7 Prevalence of different abnormalities in lab findings in IM

Lab abnormalities	Prevalence (%)
Increased SGPT	40
Leukocytosis	40
Thrombocytopenia	25
Anaemia	10

24.6.6 Role of EBV Infection in Various Diseases

EBV infection is manifested in various diseases. Some of the important serious associations are:

24.6.6.1 X-Linked Lymphoproliferative Syndrome: Duncan's Disease

Primary immunodeficiency follows a lethal course in response to EBV infection especially in males. This disorder is allied with mutation in gene encoding signalling lymphocyte activation molecule associated protein which is activates NK cells and T cells responsive to infected B cells (Hislop et al. 2010). Hence these patients are more susceptible of developing fatal IM and EBV associated lymphomas.

24.6.6.2 Nasopharyngeal Carcinoma

Nasopharyngeal cancer and EBV infection are positively correlated. Increased levels of IgA and IgG antibodies against capsid of EBV have been found in various serological studies (Goedert and Raab-Traub 2000). The DNA of EBV has been observed in 96% of patients with NPC (Lo et al. 2000).

EBV infection is also associated with tumours of the lymphoid tissue.

24.6.6.3 Burkitt's Lymphoma (BL)

There is a strong association of EBV with BL. Significantly high titres of anti-VCA antibodies appear during serological studies. In more than 95% of endemic BL cases, EBV has been detected. The mechanism of association suggests that a substantial number of B cells may undergo transformation and replication as a result of an EBV infection. It is reported that this process gets aggravated with the recurrent malaria infection because it acts as T cell suppressant and B-cell mitogen (Moormann et al. 2005).

24.6.6.4 Post-Transplant Lymphoproliferative Disease (PTLD)

PTLD is a serious disease observed in post-transplant patients because of EBV infection. It may happen be because donor is EBV seropositive or due to EBV infection, either primary or reactivated, after transplant. The assessment of EBV activity in peripheral blood mononuclear cells in serum correlates with the risk of PTLD.

24.6.6.5 Acquired Immuno-Deficiency Syndrome (AIDS)

EBV is involved in HIV associated lymphomas. It is frequently observed in Hodgkin's lymphomas as well as in AIDS-related BL.

24.6.6.6 SARS-COV-2

There is a documented interaction between SARS-COV-2 and EBV (Shafiee et al. 2022). One of the mechanisms involves the reduction in CD8+ Cells which normally fight against EBV infection. CD8+ Cell count tend to be lower in subjects with SARS-COV-2/EBV co-infection. There is a possible reactivation of EBV during Covid infection.

24.6.7 Serological, Histological and Molecular Approaches

24.6.7.1 Peripheral Blood Smear

Appearance of Downey cells which are CD8 T lymphocytes often responds to EBV-infected B cells. However, these cells can be observed in IM-like illness, such as Cytomegalovirus. Hence, this approach is not very specific.

24.6.7.2 Heterophile Antibodies

These antibodies can react to some antigens that are quite distinct from and phylogenetically unrelated to the one that causes an antibody response. It involves detection of Ig M class abs which are present during immune up-regulation in EBV infection. However, antibodies are non-specific, and moreover, children lack these antibodies (Paul and Bunnell 1974).

24.6.7.3 Liver Function Tests

Abnormal LFTs are seen during early infection with EBV. However, this elevation can be seen in several other diseases, hence not a very specific approach.

24.6.7.4 EBV-Specific Antibody Tests

EIA and induced immunofluorescence techniques are used for the detection of EBV-specific antibodies. EIA antibodies can clearly distinguish active primary, convalescent and post EBV infection. In acute primary infection, IgM against VCA appears. Between third week to third month, VCA IgG abs rise and persists for life while Ig M abs dwindle. Afterwards, between third and sixth month, VCA IgM abs disappear and EBNA1 IgG abs appear and persist throughout life (Nystad and Myrmel 2007).

24.6.7.5 Immunohistochemical Approach

To detect EBV-encoded RNA transcript in tissue, in situ hybridization is a gold standard (Gulley 2001).

24.7 Diagnosis and Therapeutic Approaches

24.7.1 Diagnosis

To detect the EBV, a blood test is conducted which is known as EBV antibody test/ EBV test. From the presence of various antibodies which are released by the immune system to respond to antigens which are produced by viruses. The EBV test identifies various phases of infection whether it is in an early stage, existing stage, or previous/past stage. The doctor recommends the patient for a test when he/she faces symptoms, such as stiff neck, spleen enlargement, swelling of lymph nodes, sore throat, fatigue, and fever (Gulley 2001).

This blood test recognizes various EBV antibodies, such as VCA IgG, VCA IgM, and EBNA through which EBV is detected. VCA IgG presence let us know that the patient was infected by EBV either in the past or a while ago. VCA IgM starts fading from site in 4–6 weeks and its presence lets us know that the patient was infected by EBV recently. EBNA is observed in 6–8 weeks from infection and reveals the occurrence of EBV in past. Each antibody makes its own combination which helps to recognize when the EBV occurs (Gulley 2001) (Table 24.8).

Table 24.8 Details about EBV antibody patterns (Koçoğlu et al. 2014)

Stage of infection	EBV antibody pattern	Status	Time period
Primary infection	(+/–/–), (+/+/–), (–/+/–), (+/+/+)	EBV positive	Infected currently
Previous infection/infection which occurs in past	(–/+/+), (–/–/+)	EBV positive	In past or recently
EBV reactivation	(+/+/+), (+/–/+)	EBV positive	2–6 weeks from infection
Seronegative	(–/–/–)	EBV negative	Not infected currently

24.7.2 Therapeutic Approach

Several therapeutic approaches have been taken to cure EBV. EBV vaccine is one of the effective methods for treatment of EBV infection. It works by stopping the formation of EBNA1, LMP1, and LMP2, because these are the proteins which are responsible for the development and the conversion of the B cell into tumor cell. Dendritic cell-based vaccine pulsed with HLA-A11-, -A24-, or -B4 hold down the detection of antigen of LMP2 and increases the activity and effectiveness of CD8+ T cells which helps to eliminate the virus from patient's body. The only drawback of these vaccines is that they are expensive and also limited detection of antigens.

Another method to fight against EBV is adoptive T-cell therapy/ACT. It is the most effective for the treatment of PTLD and EBV. In this therapy, the formation of alloreactive T cells led to graft-versus-host disease. To overcome it, EBV-specific CTLs proved to be effective against PTLD through in vitro technique. Whereas T-cell therapy has been used to treat HPV by targeting E7 antigen (Bishop and Adegoke 2016) (Table 24.9).

24.7.3 Biosafety Measure of EBV

24.7.3.1 General Instructions (Kruse and Barbeito 1998)

1. There must be limited access to the workstation/laboratory.
2. Any type of acid, sample, and hazardous substance must not be pipetted by mouth.

Table 24.9 List of available commercial tests for EBV detection (Koçoğlu et al. 2014)

Serial No.	Test	Purpose	Producer	Detection technique
1.	CAPTIA select VCA—M/G/EBNA	Identification of VCA, EBNA	Centocor Inc.	Qualitative detection
2.	Enzygnost anti-EBV/IgM and IgG	Identification of combination of VCA, EBNA and EA	Dade Behring	Qualitative and quantitative detection
3.	Vironostika EBV VCA IgM/IgG/EBNA ELISA	Identification of VCA, EBNA	Organon Teknika	Qualitative and semiquantitative detection
4.	SEROFLUOR immunofluorescence assay and EBV Combi-test	Identification of VCA (SEROFLUOR), combination of VCA, EBNA and EA	Institute Virion Ltd.	Qualitative detection
5.	Anti-EBV recombinant IgM- and IgG-early antigen/EBNA IgG ELISA	Identification of EA, EBNA	Biotest Diagnostics	Qualitative and semiquantitative detection
6.	EBV IgM/IgG/EBNA ELISA	Identification of VCA, EBNA	Gull Laboratories	Qualitative detection

3. There must be no consumption of food, alcohol, and smoking in work areas.
4. Proper procedure and regulations by government or other official bodies are required for transferring the sample given.
5. The material used in the construction of the lab must be chosen in such a way that it can be easy to clean.

24.7.3.2 Stability of EBV (Schmid et al. 2003)

EBV can be inactivated by eating at 100 °C 30 min, freeze drying, and heating in solution at 60–80 °C. EBV can also be killed by disinfectants, such as 30% ethanol and 20% isopropanol or by the help of nucleoside analogues.

24.7.3.3 Personal Protective Equipment (Kruse and Barbeito 1998; Schmid et al. 2003)

1. Lab coat, gloves, masks, and face protector/shield must be equipped all the time while in the laboratory.
2. Biosafety cabinets must be used when there is formation of aerosols. Other secondary safety equipment is autoclaved and benchtop sink.
3. The usage of sharp objects must be minimum and must be stored in sharps in containers.

24.7.3.4 Safety Measures for Spilling of Sample (Kruse and Barbeito 1998; Schmid et al. 2003)

The spill due to aerosols must be covered by absorbent paper and cleaned with effective disinfectants. Disinfectant must be applied from the edge of spill to the center of spill.

Reports of spill, illness, or injuries must be submitted regularly and immediately.

24.7.3.5 Handling and Storage of EBV (Kruse and Barbeito 1998)

The waste material, either solid or liquid, must be sterilized before dumping into nature.

The sample prepared must be tagged sea containers with appropriate label over it.

24.7.3.6 Safe Environment (Burnett et al. 2009)

During the preparation of sample, it must be filtered before cell sorting to prevent clumping and the formation of clogs. It leads to the formation and supply of samples within the required time period. By following this procedure, technician can perform it without any pressure along with perfection. This provides a safer and healthy environment for work.

24.8 Prevention and Control

Antipyretics, analgesics, adequate hydration, and nutrition are a key to handle symptomatic condition. Prevention can be achieved by minimizing exposure to EBV, but practically, it is not 100% possible as more than 90% of adults worldwide are seropositive. However, to counter life threatening conditions, such as after transplantation, recipients should be immunized with gp350 subunit vaccine which is an EBV-induced cell membrane antigen (Pearson et al. 1970). However, the efficacy of this vaccine is still under study. Nucleoside analogs such as acyclovir is used to treat serious IM. These analogs are first taken up by virus-infected cells to become activated forming tri-phosphate derivatives which are capable of inhibiting viral DNA synthesis (Elion et al. 1977).

EBV infection is continuously kept under control by the adaptive immune system. When the immune system is unable to suppress an infection or develops into oncogenesis, serious consequences can result. Hence, the key is to strengthen the immunity by adopting healthy life style, adequate hydration, and balanced nutrition.

24.9 Future Perspective

EBV is still not completely understood, because it is a fairly complex virus. The lytic reactivation and modulation of the latent viral episome have been studied by viral research laboratories around the genomic studies of EBV. Furthermore, it would not come as a total surprise if lytic antigens contributed more to the development of EBV-related illnesses. A successful EBV vaccine has not yet been created, despite considerable research in this area. The creation of a successful vaccination might stop up to 200,000 cancer cases worldwide each year. The creation of EBV vaccines for prophylaxis and treatment is severely hampered by the lack of reliable animal models. Similar to other human herpesviruses, EPV may be treatable with a course of the pro-drug valaciclovir; however, more studies are required to determine whether eradication is actually possible. Although there is no proof that antiviral medications are helpful against EBV, they work by preventing viral DNA replication. They also cost a lot, run the danger of making antiviral drugs resistant, and, in 1–10% of cases, can have unfavourable side effects. To completely understand the complexity associated with EBV-caused infections and malignancies, there is still a vast amount of research that has to be done.

References

Ablashi DV, De-Thé GB, Easton JM, Liabeuf A, Levine PH, Armstrong GR (1974) Antibodies to Epstein-Barr virus (EBV) antigens in sera of American Burkitt lymphoma patients. Biomedicine 21(7):288–293

Aitken C, Sengupta SK, Aedes C, Moss DJ, Sculley TB (1994) Heterogeneity within the Epstein-Barr virus nuclear antigen 2 gene in different strains of Epstein-Barr virus. J Gen Virol 75(Pt 1):95–100

Baer R, Bankier AT, Biggin MD, Deininger PL, Farrell PJ, Gibson TJ, Hatfull G, Hudson GS, Satchwell SC, Séguin C (1984) DNA sequence and expression of the B95-8 Epstein-Barr virus genome. Nature 310(5974):207–211

Balfour HH Jr, Sifakis F, Sliman JA, Knight JA, Schmeling DO, Thomas W (2013) Age-specific prevalence of Epstein-Barr virus infection among individuals aged 6–19 years in the United States and factors affecting its acquisition. J Infect Dis 208(8):1286–1293

Bhattacharjee S, Ghosh Roy S, Bose P, Saha A (2016) Role of EBNA-3 family proteins in EBV associated B-cell lymphomagenesis. Front Microbiol 7:457

Bishop GH, Adegoke OO (2016) Seroprevalence of Epstein-Barr virus IgM antibodies among HIV-patients and apparently healthy blood donors attending Ahmadu Bello University Teaching Hospital, Shika-Zaria, Nigeria. Int J Sci Res Knowledge 4(5):105–111

Burnett LC, Lunn G, Coico R (2009) Biosafety: guidelines for working with pathogenic and infectious microorganisms. Curr Protoc Microbiol Chapter 1(1):Unit 1A.1

Cabo J (2022) The role of viral proteins in Epstein-Barr virus induced disease. Accessed 10 Jun 2022

Chien YC, Chen JY, Liu MY, Yang HI, Hsu MM, Chen CJ, Yang CS (2001) Serologic markers of Epstein-Barr virus infection and nasopharyngeal carcinoma in Taiwanese men. N Engl J Med 345(26):1877–1882

Cohen JI (2000) Epstein-Barr virus infection. N Engl J Med 343(7):481–492

Dawson CW, Port RJ, Young LS (2012) The role of the EBV-encoded latent membrane proteins LMP1 and LMP2 in the pathogenesis of nasopharyngeal carcinoma (NPC). Semin Cancer Biol 22(2):144–153

De Paschale M, Clerici P (2012) Serological diagnosis of Epstein-Barr virus infection: problems and solutions. World J Virol 1(1):31–43

Elion GB, Furman PA, Fyfe JA, de Miranda P, Beauchamp L, Schaeffer HJ (1977) Selectivity of action of an antiherpetic agent, 9-(2-hydroxyethoxymethyl) guanine. Proc Natl Acad Sci USA 74(12):5716–5720

Epstein MA, Achong BG, Barr YM (1964) Virus particles in cultured lymphoblasts from Burkitt's lymphoma. Lancet (London, England) 1(7335):702–703

Glaser R, Strain EC, Tarr KL, Holliday JE, Donnerberg RL, Kiecolt-Glaser JK (1985) Changes in Epstein-Barr virus antibody titers associated with aging. Proc Soc Exp Biol Med 179(3):352–355

Goedert JJ, Raab-Traub N (2000) Epstein-Barr virus and nasopharyngeal carcinoma. In: Infectious causes of cancer: targets for intervention. Humana, Totowa, NJ, pp 93–111

Grotto I, Mimouni D, Huerta M, Mimouni M, Cohen D, Robin G, Pitlik S, Green MS (2003) Clinical and laboratory presentation of EBV positive infectious mononucleosis in young adults. Epidemiol Infect 131(1):683–689

Guan Y, Jakimovski D, Ramanathan M, Weinstock-Guttman B, Zivadinov R (2019) The role of Epstein-Barr virus in multiple sclerosis: from molecular pathophysiology to in vivo imaging. Neural Regen Res 14(3):373–386

Gulley ML (2001) Molecular diagnosis of Epstein-Barr virus-related diseases. J Mol Diagn 3(1):1–10

Haahr S, Plesner AM, Vestergaard BF, Höllsberg P (2004) A role of late Epstein-Barr virus infection in multiple sclerosis. Acta Neurol Scand 109(4):270–275

Hadinoto V, Shapiro M, Greenough TC, Sullivan JL, Luzuriaga K, Thorley-Lawson DA (2008) On the dynamics of acute EBV infection and the pathogenesis of infectious mononucleosis. Blood 111(3):1420–1427

Hanto DW (1995) Classification of Epstein-Barr virus-associated posttransplant lymphoproliferative diseases: implications for understanding their pathogenesis and developing rational treatment strategies. Annu Rev Med 46:381–394

He YQ, Xue WQ, Xu FH, Xu YF, Zhang JB, Yu HL, Feng QS, Chen LZ, Cao SM, Liu Q, Mu J, Zeng YX, Jia WH (2018) The relationship between environmental factors and the profile of Epstein-Barr virus antibodies in the lytic and latent infection periods in healthy populations from endemic and non-endemic nasopharyngeal carcinoma areas in China. EBioMedicine 30:184–191

Hislop AD, Palendira U, Leese AM, Arkwright PD, Rohrlich PS, Tangye SG, Gaspar HB, Lankester AC, Moretta A, Rickinson AB (2010) Impaired Epstein-Barr virus-specific CD8+ T-cell function in X-linked lymphoproliferative disease is restricted to SLAM family-positive B-cell targets. Blood 116(17):3249–3257

Hjalgrim H, Friborg J, Melbye M (2007) The epidemiology of EBV and its association with malignant disease. In: Arvin A et al (eds) Human herpesviruses: biology, therapy, and immunoprophylaxis. Cambridge University Press

Höcker B, Fickenscher H, Delecluse HJ, Böhm S, Küsters U, Schnitzler P, Pohl M, John U, Kemper MJ, Fehrenbach H, Wigger M, Holder M, Schröder M, Billing H, Fichtner A, Feneberg R, Sander A, Köpf-Shakib S, Süsal C, Tönshoff B (2013) Epidemiology and morbidity of Epstein-Barr virus infection in pediatric renal transplant recipients: a multicenter, prospective study. Clin Infect Dis 56(1):84–92

Hutcheson RL, Chakravorty A, Sugden B (2021) Burkitt lymphomas evolve to escape dependencies on Epstein-Barr virus. Front Cell Infect Microbiol 10:606412

IARC Working Group on the Evaluation of Carcinogenic Risks to Humans and International Agency for Research on Cancer (1997) Epstein-Barr virus and Kaposi's sarcoma herpesvirus/human herpesvirus 8, vol 70. World Health Organization

Jagadeesh S, Wood EF (2021) P09 uncommon cause of back pain. Rheumatol Adv Pract 5(Suppl_1):rkab068.008

Kelly GL, Stylianou J, Rasaiyaah J, Wei W, Thomas W, Croom-Carter D, Kohler C, Spang R, Woodman C, Kellam P, Rickinson AB, Bell AI (2013) Different patterns of Epstein-Barr virus latency in endemic Burkitt lymphoma (BL) lead to distinct variants within the BL-associated gene expression signature. J Virol 87(5):2882–2894

Khan G, Hashim MJ (2014) Global burden of deaths from Epstein-Barr virus attributable malignancies 1990–2010. Infect Agents Cancer 9(1):38

Khan G, Fitzmaurice C, Naghavi M, Ahmed LA (2020) Global and regional incidence, mortality and disability-adjusted life-years for Epstein-Barr virus-attributable malignancies, 1990–2017. BMJ Open 10(8):e037505

Kieff E, Rickinson AB (2007) Epstein-Barr virus and its replication. In: Knipe DM, Howley PM, Griffin DE, Lamb RA, Martin MM, Roizman B, Straus SE (eds) Fields virology, vol II, 5th edn. Lippincott Williams & Wilkins, Philadelphia, PA, pp 2603–2654

Kimura H, Hoshino Y, Kanegane H, Tsuge I, Okamura T, Kawa K, Morishima T (2001) Clinical and virologic characteristics of chronic active Epstein-Barr virus infection. Blood 98(2):280–286

Koçoğlu ME, Taş T, Mengeloğlu FZ, Özsoy Ş, Bucak Ö (2014) Evaluation of 4 methods for the serological diagnosis of Epstein-Barr virus infection using an immunofluorescence assay as the reference method. Turkish J Med Sci 44(6):914–919

Kruse RH, Barbeito MS (1998) A history of the American Biological Safety Association. Part III: Safety conferences 1978–1987. J Am Biol Saf Assoc 3(1):11–25

Kuri A, Jacobs BM, Vickaryous N, Pakpoor J, Middeldorp J, Giovannoni G, Dobson R (2020) Epidemiology of Epstein-Barr virus infection and infectious mononucleosis in the United Kingdom. BMC Public Health 20(1):912

Liebowitz D, Kieff E (1993) In: Roizman B, Whitley RJ, Lopez C (eds) The human herpesviruses. Raven Press, New York, pp 107–172

Lo YM, Chan AT, Chan LY, Leung SF, Lam CW, Huang DP, Johnson PJ (2000) Molecular prognostication of nasopharyngeal carcinoma by quantitative analysis of circulating Epstein-Barr virus DNA. Cancer Res 60(24):6878–6881

Lyon F (1994) IARC monographs on the evaluation of carcinogenic risks to humans, Some industrial chemicals, vol 60. IARC, pp 389–433

Matsuura H, Kirschner AN, Longnecker R, Jardetzky TS (2010) Crystal structure of the Epstein-Barr virus (EBV) glycoprotein H/glycoprotein L (gH/gL) complex. Proc Natl Acad Sci USA 107(52):22641–22646

Moormann AM, Chelimo K, Sumba OP, Lutzke ML, Ploutz-Snyder R, Newton D, Kazura J, Rochford R (2005) Exposure to holoendemic malaria results in elevated Epstein-Barr virus loads in children. J Infect Dis 191(8):1233–1238

Navari M, Etebari M, De Falco G, Ambrosio MR, Gibellini D, Leoncini L, Piccaluga PP (2015) The presence of Epstein-Barr virus significantly impacts the transcriptional profile in immunodeficiency-associated Burkitt lymphoma. Front Microbiol 6:556

Nystad TW, Myrmel H (2007) Prevalence of primary versus reactivated Epstein-Barr virus infection in patients with VCA IgG-, VCA IgM- and EBNA-1-antibodies and suspected infectious mononucleosis. J Clin Virol 38(4):292–297

Palermo RD, Webb HM, Gunnell A, West MJ (2008) Regulation of transcription by the Epstein-Barr virus nuclear antigen EBNA 2. Biochem Soc Trans 36(Pt 4):625–628

Paul JR, Bunnell WW (1974) The presence of heterophile antibodies in infectious mononucleosis. Am J Med Sci 267(3):178–188

Pearson G, Dewey F, Klein G, Henle G, Henle W (1970) Relation between neutralization of Epstein-Barr virus and antibodies to cell-membrane antigens-induced by the virus. J Natl Cancer Inst 45(5):989–995

Price AM, Luftig MA (2014) Dynamic Epstein-Barr virus gene expression on the path to B-cell transformation. Adv Virus Res 88:279–313

Rea TD, Russo JE, Katon W, Ashley RL, Buchwald DS (2001) Prospective study of the natural history of infectious mononucleosis caused by Epstein-Barr virus. J Am Board Fam Pract 14(4):234–242

Satou A, Asano N, Nakazawa A, Osumi T, Tsurusawa M, Ishiguro A, Elsayed AA, Nakamura N, Ohshima K, Kinoshita T, Nakamura S (2015) Epstein-Barr virus (EBV)-positive sporadic burkitt lymphoma: an age-related lymphoproliferative disorder? Am J Surg Pathol 39(2):227–235

Schmid I, Merlin S, Perfetto SP (2003) Biosafety concerns for shared flow cytometry core facilities. Cytometry A 56(2):113–119

Shafiee A, Aghajanian S, Athar M, Gargari OK (2022) Epstein-Barr virus and COVID-19. J Med Virol 94(9):4040–4042

Shannon-Lowe CD, Neuhierl B, Baldwin G, Rickinson AB, Delecluse HJ (2006) Resting B cells as a transfer vehicle for Epstein-Barr virus infection of epithelial cells. Proc Natl Acad Sci USA 103(18):7065–7070

Shetty K. (2022) Epstein-Barr virus (EBV) infectious mononucleosis (mono). https://emedicine.medscape.com/article/222040-overview. Accessed 21 Jul 2022

Sprunt TP, Evans FA (1920) Mononuclear leucocytosis in reaction to acute infections ("infectious mononucleosis"). Johns Hopkins Hosp Bull 31:410–417

Vockerodt M, Yap LF, Shannon-Lowe C, Curley H, Wei W, Vrzalikova K, Murray PG (2015) The Epstein-Barr virus and the pathogenesis of lymphoma. J Pathol 235(2):312–322

Wick G, Grubeck-Loebenstein B (1997) Primary and secondary alterations of immune reactivity in the elderly: impact of dietary factors and disease. Immunol Rev 160:171–184

Winter JR, Taylor GS, Thomas OG, Jackson C, Lewis J, Stagg HR (2019) Predictors of Epstein-Barr virus serostatus in young people in England. BMC Infect Dis 19(1):1007

Wong Y, Meehan MT, Burrows SR, Doolan DL, Miles JJ (2022) Estimating the global burden of Epstein-Barr virus-related cancers. J Cancer Res Clin Oncol 148(1):31–46

Yang J, Liu Z, Zeng B, Hu G, Gan R (2020) Epstein-Barr virus-associated gastric cancer: a distinct subtype. Cancer Lett 495:191–199

Young LS, Murray PG (2003) Epstein-Barr virus and oncogenesis: from latent genes to tumours. Oncogene 22(33):5108–5121

Young LS, Dawson CW, Eliopoulos AG (2000) The expression and function of Epstein-Barr virus encoded latent genes. Mol Pathol 53(5):238–247

Zanella L, Riquelme I, Buchegger K, Abanto M, Ili C, Brebi P (2019) A reliable Epstein-Barr virus classification based on phylogenomic and population analyses. Sci Rep 9(1):9829

Zavala-Vega S, Palma-Lara I, Ortega-Soto E, Trejo-Solis C, de Arellano IT, Ucharima-Corona LE, Garcia-Chacón G, Ochoa SA, Xicohtencatl-Cortes J, Cruz-Córdova A, Luna-Pineda VM, Jiménez-Hernández E, Vázquez-Meraz E, Mejía-Aranguré JM, Guzmán-Bucio S, Rembao-Bojorquez D, Sánchez-Gómez C, Salazar-Garcia M, Arellano-Galindo J (2019) Role of Epstein-Barr virus in glioblastoma. Crit Rev Oncog 24(4):307–338

Molecular Mechanisms of Human Papillomavirus-Induced Cervical Carcinogenesis

25

Umesh Kumar, Mansi Kumari, and Prudhvi Lal Bhukya

Abstract

Approximately 15–20% of all tumors are generally associated with infectious agents and these infectious agents are normally underappreciated. Approximately 85% of all cervical cancer deaths are reported in low–middle-income countries, almost 18% higher than in developed countries. More than a third of cervical cancer cases globally were reported in Asian countries, China, and India. However, Human Papillomavirus (HPV) still remains a very common etiological agent in cervical cancer patients worldwide. As we know, cervical cancer is one of the major causes of death of women globally, and high-risk HPVs (HPV 16 and 18) are associated with cervical carcinogenesis. The present chapter comprehensively emphasized the molecular mechanism of HPVs, their structural organization and etiopathogenesis caused by its early region oncoproteins (E), and their contribution to the progression of cervical carcinogenesis. E proteins are the major contributors to the progression of the disease. Viral integration, its replication, and transcription are mainly the outcomes due to E1, E2, and E4 proteins, while E6 and E7 are important contributors which act as oncoproteins. In this chapter, the structural organization and biological functions of all the E proteins of HPV and their role in the initiation and progression of cervical cancer are comprehensively discussed.

U. Kumar (✉)
Department of Biosciences, IMS Ghaziabad (University Courses Campus), Ghaziabad, Uttar Pradesh, India
e-mail: umeshkumar82@gmail.com

M. Kumari
Dr. D. Y. Patil Biotechnology & Bioinformatics Institute, Pune, Maharashtra, India

P. L. Bhukya
Preclinical Vaccine Testing Laboratory, Rodent Experimentation Facility, ICMR-National Animal Facility Resource Facility for Biomedical Research, Hyderabad, Telangana, India

P. L. Bhukya, A. S (eds.), *Emerging Human Viral Diseases, Volume II*,
https://doi.org/10.1007/978-981-97-4480-0_25

Keywords

Human papillomavirus · Cervical cancer · Oncoprotein

25.1 Introduction

Human Papillomavirus (HPV) is one of the world's most common sexually transmitted infections (Kombe Kombe et al. 2021; Manini and Montomoli 2018) and has a significant detrimental effect on a person's social life. Although, in their lifetime, sexually more active men and women will catch the virus at least once which suggests that skin-to-skin (Manini and Montomoli 2018) genital contact is sufficient for transmission of the HPV virus. Papillomavirus is a relatively small virus-containing double-stranded DNA with a genome-containing 8 kb DNA sequence belonging to the papillomaviridae family. In human, more than 200 types of HPV have been notified, and among these, 85 genotypes were characterized (Crosbie et al. 2013). Their capsid or outer shell is not covered by a lipid membrane, it contains a protein called L2 which is less abundant and performs various essential roles in the entry of the virus into host cells as well as the packaging of viral genome into the nascent virion. Weather, cervical cancer develops when the cells alter in women's cervix, which connects the uterus and vagina. The deeper tissues of their cervix may be affected by this cancer and it has potential to metastasis (spread to another part of the body), most frequently the lung, vagina, rectum, bladder, and liver. Around the world, HPV infection is the primary cause of cancer, estimated to account 570,000 cases and 311,000 fatalities in 2018 (Arbyn et al. 2020). According to the estimated result, low- and middle-income nations, where the death rate is 18 times greater than in rich nations, contribute to approximately 85% of all cervical cancer fatalities globally. Some of the sub-Saharan African, Latin American, and Asian nations have very high incidence of death rates (Balasubramaniam et al. 2019). Although previous studies reported that poor personal, sexual hygiene, early sexual activity, having several partners and smoking are some of the common risk factors for the development of cervical cancer.

The relationship between HPV and cervical cancer was discovered by Harald zur Hausen. This work earned him 2008 Nobel prize for Medicine or Physiology (Chustecka and zur Hausen 2008). More information would continue to support the association between HPV genotype and cervical cancer over the ensuring 10 years. That research result to the development of HPV testing, which has enhanced the identification and the treatment of precancerous lesions and early cervical abnormalities. In addition, it sparked the creation of preventative HPV vaccines, such as cervix (Glaxosmithkline) and Gardasil (Merk & Co.), which are anticipated to significantly reduce the prevalence of cervical cancer in the next decades (Chustecka and zur Hausen 2008). However, without the persistence of one person, the scientist (Harold zur Hausen) who first postulated and then demonstrated the relationship between HPV and cervical cancer, these innovations would not have been made feasible. At the time, in the early 1980s, the majority of experts did not think that viruses could cause cancer, and few who did were concentrating on a completely different virus; thus, the discovery created quite a stir.

Fortunately, not all the infection results in cancer some women have these high-risk variations of these virus, but they are asymptomatic; it means that their immune system might potentially completely rid them of the virus. However, for other woman, the ongoing infection does result in several medical problems.

They duplicate their genetic material by utilizing the host's enzymatic system, and their transmitted characteristics are associated with penile cancer in males, as well as vulvar, vaginal, and cervical cancer in females, along with oropharyngeal cancer in both genders (de Martel et al. 2017). Premalignant lesions in the cervical region can result from an ongoing infection with the oncogenic HPV, which has the potential to induce cancer. Although there are more than 100 recognised HPV genotypes and types 16 (Burk et al. 2013) and 18 are the two most prevalent (Faridi et al. 2011) and causing 70% of pre-cancerous cervical lesion and cervical cancers. These found confirmed that HPV is a key player in the pathogenesis of cervical cancer, and it influences the host cell DNA repair mechanism, cell adhesion, cell cycle and apoptosis and may induce an immunological response. The transcribed genomic region is where the HPV virus frequently integrates into the host genome, and the virus makes of this process to boost the production of several viral products such as E6 and E7 viral oncogenes which will lead to the development of cancer (described in Fig. 25.1a, b).

Moreover, HPV-related infections are hallmarks of neck and cervical cancer. Chronic HPV infection is mainly responsible for the majority of cases of cervical cancer and oropharyngeal cancer, which account for 70% of cases of each and internationally cervical cancer is the fourth (Kombe Kombe et al. 2021; Yang et al. 2020) most common cancer and is commenced by constant infection with HPV, mainly.

25.2 Epidemiology

The last stage of an untreated HPV infection is cervical cancer, which is now identified as the presence of HPV DNA in repeated tests of cervical specimens. Data from the National Health and Nutrition Examination Survey from 2003 to 2006 among women aged 14–59 years were used to estimate the prevalence of HPV infections. Although their infection prevalence has grown significantly worldwide, especially in women, where it is the initial cause of cancer. Therefore, making HPV a recent public health priority. As one of the most widespread viral illnesses worldwide, HPV infection was projected to affect 291 million women globally in 2007 (de Sanjosé et al. 2007). The HPV dissemination profile in women with normal cervical cytology is essentially the same as that of the overall female population in terms of geographic global regions, and in emerging areas, it is higher. HPV prevalence was higher in Sub-Saharan Africa (24.0%), particularly in the regions of Eastern Africa (33.6%) and Latin America, according to the assembled studies from cytologically healthy women. With regard to all females, Asian regions had the highest prevalence, with over half of Eastern, Central, and Southern Asians (57.7% and 44%,

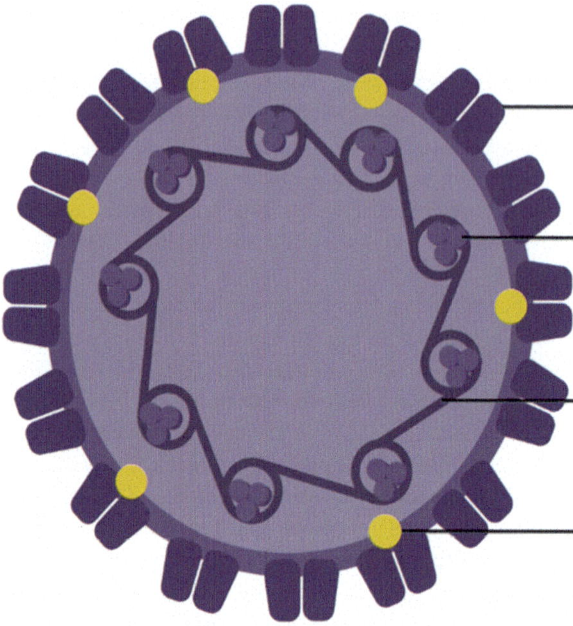

Fig. 25.1 (**a**) Structure of Human Papillomavirus (HPV). (**b**) Genomic organization of the HPV genome. Human Papilloma Virus structure with 8K base pair and its respective functions. HPV genome is classified into the early gene (encodes E1, E2, E4, E5, E6, and E7) and the late gene (encodes L1 and L2). E6 and E7 are oncogenes which degrade tumor suppressor p53 and pRb, respectively. L1 and L2 build the structure of HPV capsid protein. *LCR* long control region

respectively) being carriers. In the SSA region, 42.2% and 32.3% of women in Southern and Eastern Africa were HPV carriers, independently. In nearly all European countries, the prevalence of HPV was modest (<30%) as in western Europe (3.7%). As a result, global prevalence rates of genital HPV infection in men range from 3.5% to 44% compared to 2% to 44% in women, with similar transmission rates. Given that anogenital HPV is mostly sexually transmitted, this is quite understandable (Kombe Kombe et al. 2021). In fact, a study was conducted in 2014 that confirmed that 9.0% of 4065 healthy men from Europe, Asia, Africa, and America carried the HPV virus. Apart from this, homosexuals and human immunodeficiency virus (HIV)-infected men are at increased risk, with greater incidence rates (≥90%) of HPV anal infection than those in homosexual men, in whom the likelihood of HPV infection depends on the number of homosexuals, sexual partners, and men who are HIV-positive are at higher risk. In addition, the rate of HPV infection for all men differs very little with an age and is just as high among the youngest men as it is among the oldest (Nyitray et al. 2008). Despite the availability of free pap-test screening for the purpose of uterine cervix cancer, mortality rates for all HPV malignancies remain high. Within 5 years, the case-to-death ratio for cervix cancer is 24%. It grows significantly up to 75% in both genders for

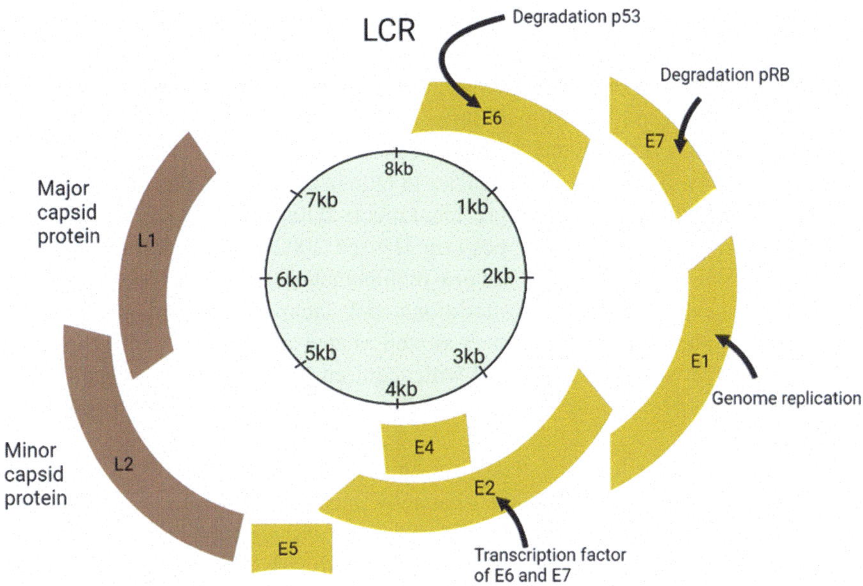

Fig. 25.1 (continued)

oropharynx cancer, and the highest rate is shown in anal cancer, where 89% of men and 86% of women convincing that 14% of affected women and 11% of affected men survive.

Nevertheless, worldwide, cervical cancer is a major source of morbidity and mortality. Given that the majority of cervical malignancies are linked to the oncogenic HPV, vaccination with any of the three HPV vaccines currently on the market is expected to significantly lower the incidence of cervical cancer. Although the HPV vaccine has a significant deal of potential to lower cervical cancer morbidity and mortality, the benefits that are now anticipated are primarily hypothetical. Large population-based clinical effectiveness trials will measure cervical cancer as an endpoint and provide long-term immunogenicity and efficacy, especially for young vaccine recipients reach the age range, where cervical cancer should begin. In the medium run, increasing screening and treatment programmes will probably have the most effects on cervical cancer morbidity and mortality (Luckett and Feldman 2016).

25.3 Potential Risk of Emergence and Re-emergence

These viruses can be divided into cutaneous and mucosal HPV and which individual HPV viruses are classified as a high or low risk depending on how likely it is that the lesions may proceed into cancer. The majority of HPV are low risk and cause small, benign warts that do not proceed to cancer even when untreated. The cutaneous HPV types (HPV-5 and HPV-8) may be categorized as high risk, because they are connected to the occurrence of epidermodysplasia verruciformis, an extremely rare skin disorder that offered one of the first signs that HPV may have a role in the formation human tumors (Münger et al. 2004). Low-risk genital warts (condyloma accuminata) are caused by low-risk mucosal HPV (HPV-6 and HPV-11), whereas high-risk HPV may develop into invasive squamous cell carcinoma and is linked to the great majority of cervical malignancies in humans. Although, one of the reviews suggested that HPV-16 is the most prevalent mucosal high-risk HPV type, followed by HPV-38, HPV-18, and other types (zur Hausen 2002). In an addition to being high-risk HPV positive, about 20% of oral malignancies, specifically oropharyngeal carcinomas in people without the traditional risk factors of tobacco and alcohol misuse. Penile and vulvovaginal cancer as well as anal carcinomas, which typically happen in people with HIV associated with acquired immunodeficiency syndrome, are additional anogenital tract malignancies that are usually linked to high-risk HPV infections.

Conventionally, the low-risk classification within the mucosal HPV family corresponds to the relative viral genome's capacity to change host cells and transgenic animal models. As a result, a large portion of molecular research has been devoted to examine the changing behaviours of mucosal high-risk HPV that are connected to cervical cancer. These learnings are discussed in this chapter (Gillison et al. 2000; Klencke and Palefsky 2003; Rouzier et al. 2001).

25.4 Organization of Infectious Agents

The International Committee on the Taxonomy of Viruses given family Papillomaviridae is a diversified group of viruses consisting of 189 types of papillomaviruses that are isolated from humans or non-humans, and hence, they co-evolve with many hosts, such as reptiles, aves, and mammals, excluding the lower phylogenetic groups by either affecting epithelia or remaining in a latent stage (Doorbar et al. 2015). This live in association with the immunocompetent host may be considered commensal in nature.

Alpha, Beta, Gamma, Mu, and Nu are the five main genera of HPV, and each has different illness signs and effects on the host (Bernard et al. 2010). The alpha HPV, which includes cutaneous and mucosal Alpha types, is said to be one of the most fascinating genera (Doorbar et al. 2012). The Mucosal Alpha HPV has 120 types, which are broadly divided into two types: "low risk" types that are non-oncogenic (benign), such as HPV 6, HPV 11, are responsible for papillomatosis, and "high risk" types that are oncogenic, such as HPV 16 and HPV 18, are responsible for cervical carcinogenesis (Bernard et al. 2010).

The papillomaviruses have a non-enveloped icosahedral structure (50–60 nm diameter). The genetic constituent is comprised of double-stranded circular DNA (8 kb) that is divisible into early, late, and long control regions (non-coding region) and is separated by early and late polyadenylation sites (pA sites) (Doorbar et al. 2015). Papillomaviruses vary in number and size, and are made up of conserved core genes for replication (E1, E2), packaging (L1, L2), cell cycle regulation, immune resistance, and virus release (E4, E5, E6, E7) (Zheng and Baker 2006). The gene heterogeneity confers the different 'types' of papillomaviruses in promoters, splice donors, and acceptor sites.

The three major events in the papillomavirus infection cycle are as follows: (1) establishment of a dormant stage by keeping viral DNA in nuclear or extranuclear form (episome) and replicating synchronously with host DNA; (2) virion formation, which causes infection to progress from a dormant to an active state; and (3) viral genome integration into the host genome. The infection cycle of papillomaviruses requires several viral proteins, such as L1 and L2, E1–E7, and viral genomic material and their interaction with the host basal membrane and its associated glycosaminoglycan (Heparan sulfate, Proteoglycan) (Doorbar et al. 2012). Structural alteration occurs in the virion capsid after interaction that allows the cleavage of L2 protein facilitated by furin and internalization of the viral genome by transfer of the virion capsid protein onto the secondary receptor of basal keratinocyte by which it enters the cell through endocytosis and follows a co-translation pathway to reside in the trans-golgi network for accessing the nucleus, and the viral genome gets integrated with the host genome with the help of L2 protein. This causes the start of the PV cycle in the epithelium of the host by intrusion into the cell cycle regulation and hinders the apoptosis involving E6, E7 viral proteins that alter the cellular proliferation in the middle and lower epithelial layers through PDZ domain proteins and Rb (retinoblastoma) association, and moreover, E6 mediated p53 degradation of the host and, as a result, causes deregulation and neoplasias. E5 is a transmembrane protein and regulates the formation of pores in the membrane, hence interfering with the apoptosis and the intracellular trafficking of endocytic vesicles. E5 also helps in genome amplification by replicating the EGF signal. The E1 (which acts as a helicase) and E2 are replication proteins required for viral replication upon differentiation. These proteins are ultimately responsible for increased gene expression and transcriptional regulation, which, hence, results in malignancies in the host body (Zheng and Baker 2006).

25.5 Pathogenesis in Host and Host Immune Response

The E5, E6, and E7 are the proteins that impede the immune responses to immune dysregulation, increasing persistence and viral load. Initially, the viral processes of replication, transcriptional regulation, and epithelial modification and damage escape from host immunosurveillance by not presenting the cytotoxic death and inflammation to the host, on which a varied number of pro-inflammatory and antiviral cytokines act for immune clearance, and therefore, the virus protects itself from innate and adaptive immune responses of the host (della Fera et al. 2021) (described in Fig. 25.2).

Innate immune responses are evaded by HPVs by disrupting the signaling pathway and cytokine production and impairment of inflammatory complex formation. This is being facilitated by E5, E6, and E7 proteins, where E5 helps in suppressing the IFN-K (keratinocyte formation), E7 inhibits the cellular signaling cascade and

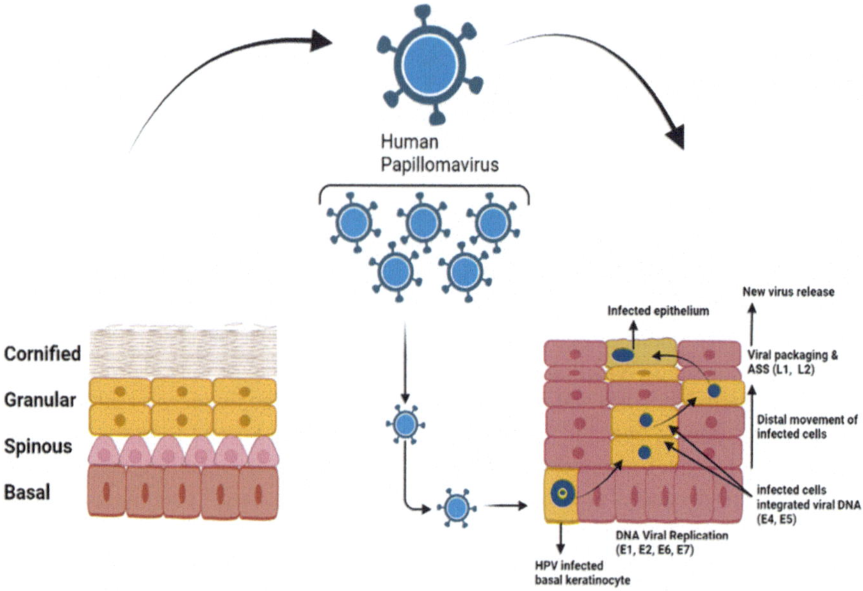

Fig. 25.2 HPV-mediated effects on the host immune response. Pathogenesis of HPV is linked to epithelial differentiation and steadily increases with proliferation as soon as virus enters in the lysogenic cycle. Ultimately, viruses are congregate and secreted from keratinocytes. And this is how the infection cycle repeats further

ultimately decreased production of type I IFNs, and moreover, E5, E6, and E7 hinder with the proinflammatory signal transduction pathway to obstruct with the JAK/STAT and NF-KB signaling. This results in evading the innate immune responses; hence, HPVs virulence dominates and causes persistency in infection. Adaptive immunity is being hindered by HPVs by lowering down the uptake of viral antigens by antigen presenting cells and downregulates the MHC I expression on HPV-infected keratinocyte, thereby decreasing cytotoxicity and recognition of infected cells by T cells and, hence, skipping the immune clearance. Ninety-nine percentof HPV infection symptoms of simple warts vanishes within the time span of 2 months from hosts, but the high-risk HPV strain remains persistent in the host body and causes weakening of immune system, host body responses towards virulence of other pathogen and cellular proliferation, differentiation and apoptosis dysregulation resulting in a clear link between the viral infection and cervical, anal and oral carcinogenesis (Boda et al. 2018). The cervical neoplasia and its association with HPV and its demographical and epidemiological studies have led it to the second most prevalent cancer among populations and various conclusive case studies on it show the prevalence of HPV 16 type in more than 74% individuals and HPV 18 type is less prevalent to that of HPV 16 type. More than 80% of chances of genome integration of HPV into host genome has proven it to be oncogenic and >12 month persistent infection has led to the increased risk of cervical lesions and neoplasias (Kombe Kombe et al. 2021).

25.6 Clinical Manifestations

The clinical implications might range from an asymptomatic, inactive papillomatosis to the host body's cervix, anus, and oropharyngeal cancer. The various stages of warts and carcinoma are progressively linked to different clinical effects and symptoms, such as anogenital warts, which go away in 2–3 months, or active infectious cervical carcinogenesis, which manifests as high-grade lesions, swelling, and redness before developing into carcinoma (Choi and Park 2016).

HPV self-sampling, serotyping, visual examination of cervical cancer screening using acetic acid and Lugol's iodine (LI), histologic markers, such as E4 and L1, and other screening techniques such as hrHPV DNA testing and HPV genotyping test are some of the diagnostic methods that have helped prevent and identify papillomaviruses (RT-PCR-based tests and gel electrophoresis-based test) (Chan et al. 2012).

25.7 Diagnostics and Therapeutics Approach for Control and Management of Disease

The different diagnostic and therapeutic interventions are mandatory for the early diagnosis of the disease symptoms and over all public well-being. This has been exacerbated by the use of various molecular and therapeutic strategies, including vaccines and immune stimulants, such as the protein-based subunit vaccine, where HPV-16 E7 protein linked to the first 108 amino acids of the *Haemophilus influenzae* protein D (PD), adjuvanted with AS02B forms the PD-E7 fusion protein vaccine, and recombinant adeno-associated virus (rAAV) type 5 expressing the main L1 capsid protein of HPV-16 elicit high titers of mucosal antibodies in vaginal washes and L1-specific serum antibodies (Chan et al. 2012). The other approaches are not sensible and involve surgical removal of diseased area and simultaneously involve the treatment of cancers by involving photodynamic therapy, Cryotherapy, cytotoxic agents (topical treatment of trichlorocaetic acid), localized immunomodulation (Imiquimod), and immunotherapy (agents used are monoclonal antibodies, cytokines, and adaptive or engineered autologous immune cells, mainly T cells). There are even more therapeutic agents and different regimens, and vaccines against the papilloma virus cervical carcinogenesis that are patented or under clinical study or at early stage of research. Diagnostics and therapeutics approach for control and management of disease (Dadar et al. 2018).

25.8 Future Prospects

The HPVs have several lethal consequences and may lead to heterogenic symptoms of disease in the body that disrupts the normal human life.

Its cause in the human population and effects of different variants among population have pushed forward the problematic livings. Some of the increment in HPV effects are due to lesser known and costly therapeutics and vaccines for HPV, poor hygiene, and living conditions abundantly in developing countries, contact with the infected host, etc. (Bosch et al. 2013). This all led to a future prospect of increasing booster dose immunization programs and awareness drive among population about the genital health and sexually transmitted diseases is highly recommended and desirable. This must also be expandable to the satisfactory discovery and implementation of broad-spectrum vaccines targeting other variants of HPVs, and furthermore, the different modern techniques of genomic screening and next generation sequencing will enhance the level of the diagnostics and treatment of oncogenic HPVs. The path over explicit knowledge of techniques, therapeutics, and another modern approach would be a progressive approach against HPVs (Dadar et al. 2018).

References

Arbyn M, Weiderpass E, Bruni L, de Sanjosé S, Saraiya M, Ferlay J, Bray F (2020) Estimates of incidence and mortality of cervical cancer in 2018: a worldwide analysis. Lancet Glob Health 8(2):e191. https://doi.org/10.1016/S2214-109X(19)30482-6

Balasubramaniam SD, Balakrishnan V, Oon CE, Kaur G (2019) Key molecular events in cervical cancer development. Medicina 55(7):384. https://doi.org/10.3390/medicina55070384

Bernard HU, Burk RD, Chen Z, van Doorslaer K, zur Hausen H, de Villiers EM (2010) Classification of papillomaviruses (PVs) based on 189 PV types and proposal of taxonomic amendments. Virology 401(1):70–79. https://doi.org/10.1016/J.VIROL.2010.02.002

Boda D, Docea AO, Calina D, Ilie MA, Caruntu C, Zurac S, Neagu M, Constantin C, Branisteanu DE, Voiculescu V, Mamoulakis C, Tzanakakis G, Spandidos DA, Drakoulis N, Tsatsakis AM (2018) Human papilloma virus: apprehending the link with carcinogenesis and unveiling new research avenues (review). Int J Oncol 52(3):637–655. https://doi.org/10.3892/IJO.2018.4256

Bosch FX, Broker TR, Forman D, Moscicki A-B, Gillison ML, Doorbar J, Stern PL, Stanley M, Arbyn M, Poljak M, Cuzick J, Castle PE, Schiller JT, Markowitz LE, Fisher WA, Canfell K, Denny LA, Franco EL, Steben M et al (2013) Comprehensive control of human papillomavirus infections and related diseases. Vaccine 31(Suppl 7):H1–H31. https://doi.org/10.1016/J.VACCINE.2013.10.003

Burk RD, Harari A, Chen Z (2013) Human papillomavirus genome variants. Virology 445(1–2):232–243. https://doi.org/10.1016/j.virol.2013.07.018

Chan PKS, Picconi MA, Cheung TH, Giovannelli L, Park JS (2012) Laboratory and clinical aspects of human papillomavirus testing. Crit Rev Clin Lab Sci 49(4):117. https://doi.org/10.3109/10408363.2012.707174

Choi YJ, Park JS (2016) Clinical significance of human papillomavirus genotyping. J Gynecol Oncol 27(2):e21. https://doi.org/10.3802/JGO.2016.27.E21

Chustecka Z, zur Hausen H (2008) Nobel-winning discovery of HPV-cervical cancer link already having an impact on medicine against the dogma of the time

Crosbie EJ, Einstein MH, Franceschi S, Kitchener HC (2013) Human papillomavirus and cervical cancer. Lancet 382(9895):889–899. https://doi.org/10.1016/S0140-6736(13)60022-7

Dadar M, Chakraborty S, Dhama K, Prasad M, Khandia R, Hassan S, Munjal A, Tiwari R, Karthik K, Kumar D, Iqbal HMN, Chaicumpa W (2018) Advances in designing and developing vaccines, drugs and therapeutic approaches to counter human papilloma virus. Front Immunol 9:2478. https://doi.org/10.3389/FIMMU.2018.02478

de Martel C, Plummer M, Vignat J, Franceschi S (2017) Worldwide burden of cancer attributable to HPV by site, country and HPV type. Int J Cancer 141(4):664–670. https://doi.org/10.1002/IJC.30716

de Sanjosé S, Diaz M, Castellsagué X, Clifford G, Bruni L, Muñoz N, Bosch FX (2007) Worldwide prevalence and genotype distribution of cervical human papillomavirus DNA in women with normal cytology: a meta-analysis. Lancet Infect Dis 7(7):453–459. https://doi.org/10.1016/S1473-3099(07)70158-5

della Fera AN, Warburton A, Coursey TL, Khurana S, McBride AA (2021) Persistent human papillomavirus infection. Viruses 13(2):321. https://doi.org/10.3390/V13020321

Doorbar J, Quint W, Banks L, Bravo IG, Stoler M, Broker TR, Stanley MA (2012) The biology and life-cycle of human papillomaviruses. Vaccine 30(Suppl 5):F55. https://doi.org/10.1016/J.VACCINE.2012.06.083

Doorbar J, Egawa N, Griffin H, Kranjec C, Murakami I (2015) Human papillomavirus molecular biology and disease association. Rev Med Virol 25(Suppl 1):2–23. https://doi.org/10.1002/RMV.1822

Faridi R, Zahra A, Khan K, Idrees M (2011) Oncogenic potential of human papillomavirus (HPV) and its relation with cervical cancer. Virol J 8:269. https://doi.org/10.1186/1743-422X-8-269

Gillison ML, Koch WM, Capone RB, Spafford M, Westra WH, Wu L, Zahurak ML, Daniel RW, Viglione M, Symer DE, Shah KV, Sidransky D (2000) Evidence for a causal association between human papillomavirus and a subset of head and neck cancers. J Natl Cancer Inst 92(9):709–720. https://doi.org/10.1093/JNCI/92.9.709

Klencke BJ, Palefsky JM (2003) Anal cancer: an HIV-associated cancer. Hematol Oncol Clin North Am 17(3):859–872. https://doi.org/10.1016/S0889-8588(03)00039-X

Kombe Kombe AJ, Li B, Zahid A, Mengist HM, Bounda GA, Zhou Y, Jin T (2021) Epidemiology and burden of human papillomavirus and related diseases, molecular pathogenesis, and vaccine evaluation. Front Public Health 8:552028. https://doi.org/10.3389/fpubh.2020.552028

Luckett R, Feldman S (2016) Impact of 2-, 4- and 9-valent HPV vaccines on morbidity and mortality from cervical cancer. Hum Vaccin Immunother 12(6):1332–1342. https://doi.org/10.1080/21645515.2015.1108500

Manini I, Montomoli E (2018) Epidemiology and prevention of human papillomavirus. Ann Ig 30(4):28–32. https://doi.org/10.7416/ai.2018.2231

Münger K, Baldwin A, Edwards KM, Hayakawa H, Nguyen CL, Owens M, Grace M, Huh K (2004) Mechanisms of human papillomavirus-induced oncogenesis. J Virol 78(21):11451. https://doi.org/10.1128/JVI.78.21.11451-11460.2004

Nyitray A, Nielson CM, Harris RB, Flores R, Abrahamsen M, Dunne EF, Giuliano AR (2008) Prevalence of and risk factors for anal human papillomavirus infection in heterosexual men. J Infect Dis 197(12):1676–1684. https://doi.org/10.1086/588145

Rouzier R, Morice P, Haie-Meder C, Lhomme C, Avril MF, Duvillard P, Castaigne D (2001) Prognostic significance of epithelial disorders adjacent to invasive vulvar carcinomas. Gynecol Oncol 81(3):414–419. https://doi.org/10.1006/gyno.2001.6198

Yang S, Wu Y, Wang S, Xu P, Deng Y, Wang M, Liu K, Tian T, Zhu Y, Li N, Zhou L, Dai Z, Kang H (2020) HPV-related methylation-based reclassification and risk stratification of cervical cancer. Mol Oncol 14(9):2124–2141. https://doi.org/10.1002/1878-0261.12709

Zheng ZM, Baker CC (2006) Papillomavirus genome structure, expression, and post-transcriptional regulation. Front Biosci 11(Suppl 1):2286–2302. https://doi.org/10.2741/1971

zur Hausen H (2002) Papillomaviruses and cancer: from basic studies to clinical application. Nat Rev Cancer 2(5):342–350. https://doi.org/10.1038/NRC798

26 Genetic Insight of Oncogenic Potential of Human Papilloma Virus

Pushpendra Singh, Sanjay Singh Negi, Ripu Daman Arora, and Nitin M. Nagarkar

Abstract

Cervical cancer (CC), fourth-ranked cancer worldwide, affecting women of different age groups. Human papilloma virus (HPV) is an etiological cause in virtually all CC by infecting mucosal or cutaneous stratified epithelia. Enormously published studies on pathogenesis, gene expression, and epidemiological studies have proved beyond doubt the etiology of the high-risk (HR)-HPV genotypes in CC. HPV genotypes, namely, 16, 18, 31, 33, 35, 39, 45, 51, 52, 56, 58, 59, 66, and 68, are considered as HR-HPV among which HPV-16 majorly responsible for 60–80% of CC. Since HPV is sexually transmitted, there are also multiple risk factors other than HPV genomic factors for progression of CC, such as sexual activities and pregnancies at early teenage age, multiple sex partners/parity, infection of Herpes simplex virus, human immunodeficiency infection, smoking, oral contraceptives, hormone replacement therapy, and immune system deficiency. Various HPV genetic factors associated with crucial events of prolonged viral persistence enhance the oncogenic progress of premalignant cervical intraepithelial neoplasia stages 1–3 to transform into invasive cervical carcinoma. Host immunity plays an important role, as in many cases, HPV infection is clinically not manifested and subsided in a time period of up to 3 years, whereas, in others, it may lead to various benign or malignant lesions transforming into invasive CC. Since the pathogenesis of HPV infection progressing to CC is a complex, multifaceted mechanism, this chapter, therefore, aims to provide an update on

P. Singh · S. S. Negi (✉)
State VRDL, Department of Microbiology, AIIMS, Raipur, Chhattisgarh, India
e-mail: negidr@yahoo.co.in; negidr@aiimsraipur.edu.in

R. D. Arora
Department of ENT & Head and Neck Surgery, AIIMS, Raipur, Chhattisgarh, India

N. M. Nagarkar
AIIMS, Raipur, Chhattisgarh, India

P. L. Bhukya, A. S (eds.), *Emerging Human Viral Diseases, Volume II*, https://doi.org/10.1007/978-981-97-4480-0_26

genomic insight, especially in the context of HPV-16 to illustrate various viral factors facilitating viral replication, immune evasion, degradation of tumor suppressor gene pRB and p53, PDZ-binding domain protein, and infected cell immortalization all eventually leading to malignant cervical lesion progression.

26.1 Introduction

Human papilloma virus (HPV) is widely known as the most common sexually transmitted infection (STI) worldwide (IARC 2012). HPV is a group of DNA viruses with predilections for infecting cutaneous and mucosal epithelial cells (Bruni et al. 2010). Its persistent infection is conclusively implicated and established as a necessary cause of cervical cancer (CC). CC, the fourth most common cancer among women, is a major killer disease among women and thus considered a serious public health problem worldwide (International Human Papillomavirus 2016). Approximately around 500,000 new cases and 288,000 deaths are reported worldwide every year, among which 85% of the statistical figure are reported from developing countries (International Human Papillomavirus 2016). In India, CC is the second most frequent cancer in women, and every year, 122,844 new cases and 67,477 deaths are reported due to CC (Sreedevi et al. 2015). The burden of CC in India is 1/4th of the total worldwide cases of CC (Sreedevi et al. 2015). The high incidence and fatality prompted molecular virologists worldwide to unearth the genetic basis of pathogenesis to provide the platform for the development of potential therapeutics. In this review, the different attributing factors in pathogenesis have been discussed under the following various subheading:

1. Background
2. CC origin in the cervix
3. Risk factors for cervical carcinogenesis
4. Genetic susceptibility for HPV infection
5. HPV genomic organization
6. Oncogenic role of HPV genes
7. Immune detection and clearance
8. Latency
9. HPV multiple genotype infection
10. Pathogenesis of HPV-16
11. HPV-16 variants and their effect in pathogenesis
12. Therapeutics
13. Conclusion

26.2 Background

HPV has been characterized at the genetic level to more than 200 HPV genotypes and reported to be co-evolved with its natural hosts for several million years (Ho et al. 1991; International Human Papillomavirus 2016). The HPV is classified as distinct and different genotypes from other previously characterized HPV genotypes if it shows 10% difference in the L1 open reading frame (ORF) sequence from all other known genotypes (Ho et al. 1991). Based on the risk of oncogenic progression, approximately 14 genotypes have been classified as high risk (HR)-HPV, namely, HPV genotypes 16, 18, 31, 33, 35, 39, 45, 51, 52, 56, 58, 59, 66, and 68 after identifying their role in causing precursor lesion and CC (Tulay and Serakinci 2016). These HR-HPV prevalence and distribution vary in different geographical regions and populations (Tulay and Serakinci 2016). Low-risk HPV (LR-HPV) genotypes, namely, HPV 6, 11, 40, 42, 43, 44, 54, 61, 70, 72, 81, and 89, are the HPV mainly responsible for causing genital warts (condyloma acuminata) with high frequency reported in infection by HPV genotypes 6 and 11 (Schiffman et al. 2007). HR-HPV genotype identification has been used as an exposure indicator to measure HPV-associated risk for invasive CC (Tulay and Serakinci 2016). The prevalence of HR-HPV genotypes was found varied among various studies. Negi et al. (2019) found 100% prevalence of HR-HPV in both forms of CC of squamous cell carcinoma and adenocarcinoma (Negi et al. 2019). Anderson et al. (2016) detected 92.2% HPV prevalence in squamous cell carcinomas (SCC) and 68% across all cervical samples investigated (Negi et al. 2019). In Chennai, 99.4% HPV prevalence was detected in invasive cervical samples (Franceschi et al. 2003). Sowjanya et al. (2005) found 87.8% HR-HPV prevalence in invasive SCC.

Notably, among the distribution of specific HR-HPV in CC, HPV-16 is the predominant causative genotype in approximately 60–70% of CC worldwide (Anderson et al. 2016; Negi et al. 2019). The other HR-HPV occurrence is reported to vary according to different geographical conditions (Anderson et al. 2016). In majority of the study, HPV-18 is the second most HR-HPV implicated in CC (Anderson et al. 2016). Some studies have found that HPV-16 followed by HPV 18 and 58 as the three most abundantly found HR-HPV in around 90% of cervical intraepithelial neoplasia (CIN) and invasive CC cases worldwide (Li et al. 2011; Smith et al. 2007). We have earlier reported the four most common genotypes, namely, HPV 16 (63.9%), HPV 39 (15.0%), HPV 18 (6.0%), and HPV 33 (5.3%), in CIN I–III, respectively (Negi et al. 2019). In CC cases, the four most identified genotypic etiologies were HPV 16 (44.4%), HPV 39 (11.1%), dual infection of HPV 16, 18 (11.1%), and triple infection of HPV 16, 18, 33 (11.1%) (Negi et al. 2019). In fact, HPV-16 is the most pleiotropic pathogenic HPV genotype implicated in both cervix and oropharyngeal cancer (Smith et al. 2007). Notably, HPV-16 and 18 prevalences in cervical adenocarcinoma and SCC differ (Smith et al. 2007). HPV-16 is mostly involved in SCC of the cervix, while both HPV-16 and 18 are implicated in adenocarcinoma of the cervix (Smith et al. 2007). HPV-16 is also implicated in vaginal, vulval, anal, and penile cancer or even benign growth in these anatomical sites (Smith et al. 2007).

The majority of the HPV-infected women clear their cervicovaginal infection, and only in a small percentage of infected women, continuous persistent, progressive infection leads to CC. This may lead to the basis of the observation that the incident of cervicovaginal HPV is primarily a self-limiting disease found with a median duration of 8–12 months (Li et al. 2011). This may be due to the higher persistence of cervical infection by certain specific HR-HPV genotypes than others. This observation is substantiated by an earlier published study of around 7 years in 10,000 women at Guanacaste, Coste Rica, showing that HPV-16 persisting more in cervical lesions enhances an HR of invasive CC than other genotypes (Herrero et al. 2005).

Nucleotide sequence variation and geographical distribution of HPV-16 genotype have earlier classified it as European and non-European (African, Asian, Asian American, and North American) variants. With the advancement in sequencing technology, specifically, the next generation sequencing, multiple sequencing alignment tools, and phylogenetic analysis, the HPV variant lineages and sublineages are now being re-classified using empirically defined sequence differences of 1.0–10.0% and 0.5–1.0%, respectively (Burk et al. 2013). Based on whole-genome sequencing, HPV 16 is classified into four lineages (A/B/C/D) and nine sublineages. These nine sublineages include four in A (A1–A3: previously named European and A4: previously named Asian) sublineages. Under B, the two sublineages of B1 (African-1, Afr1a) and B2 (African-1, Afr1b) are classified. The C lineage consists of one sublineage of African-2, Afr2a, while D lineage is classified into three sublineages, namely, D1 (North American, NA1), D2 (Asian-American, AA2), and D3 (Asian-American, AA1), respectively (Burk et al. 2013). These lineage and sublineage variants are reported to differ in their biological properties and oncogenic potential. Among them, non-European variant has been recorded with a higher incidence of progression to invasive CC (Volpini et al. 2017).

The persistent infection by HR-HPV is the main reason for developing precancerous and cancerous lesions (Schiffman et al. 2007). Interestingly, other host factors such as multiparity, history of having STI, extensive use of oral contraceptive, impaired immune response, administration of steroid hormones, smoking status, human leukocyte antigen (HLA) type, and gene polymorphism were found as independent co-factors for high-grade CIN and invasive CC (Castellsajne and Munoz 2003).

26.3 CC Origin in the Cervix

It is reported in a previous epidemiological study that about 30% of young women get an infection of HR-HPV within 24 months of first sexual activity (Gravitt 2011). This infection of HR-HPV results in mild cytological cervical abnormalities. Importantly, in majority of the infected cases, HPV resolves on its own within 2 years (Gravitt 2011). However, in 5–10% of women having persistent infections for more than 2 years, HPV progresses to invasive CC. Cervical lesions first originated and appeared from the opening of the womb and from their progressively migrates and cover the entire cervix (Faridi et al. 2011). The progression of the infected cervical cell to the cancer stage progresses slowly through various clinical stages from precancerous lesion classified to CIN stages of I–III based on the mild, moderate, or severe degree of dysplasia to finally the malignant stage of invasive CC. CIN indicates an HR-HPV infection of the prolonged interval with specific symptoms till the development of a tumor (Faridi et al. 2011). The precancerous lesion of CIN2 and CIN3 has been maximally shown to possess HR-HPV, which ultimately be the reason for transforming these lesions into the invasive CC (Faridi et al. 2011). About one-third of CIN 3 lesions cases are reported to eventually turned into invasive cancer within 10–20 years (Gravitt 2011).

26.4 Risk Factors for Cervical Carcinogenesis

Early age involvement in sexual activity with multiple partners is one of the main primary factors in acquiring HR-HPV infection. Women having multiple sex partners are at HR of developing CC than women having a single sexual partner throughout their lives (Faridi et al. 2011). Earlier epidemiological studies have shown that both exogenous and endogenous sex hormones make an HPV-infected woman more prone to developing CC (Chung et al. 2010). Multiparity precipitates changes in female hormones during early pregnancy, making them more susceptible to acquiring HPV infection (International Collaboration of Epidemiological Studies of Cervical Cancer 2006). The steroid hormones (estrogens and progesterone) may also enhance the progression of the disease as women with estradiol history were reported with the high-grade squamous intraepithelial lesion (HSIL) that eventually lead to CC (de Villers 2003). Exposure to STI infection, especially Human Immuno Deficiency (HIV) and Herpes Simplex Virus (HSV) infection, increases the risk of CC (Palesfsky and Holly 2003).

Furthermore, pregnancy-associated immune suppression also poses a risk of HPV infection and can even eliminate the previous HPV infection acquired recently before conceiving. Oral contraceptives used for 5 or more years have a threefold higher risk of developing CC than those who have not used these contraceptives (de Villers 2003).

Human semen is another biological factor enhancing HPV infection and CC. Multiple semen exposures cause chronic cervix inflammation due to localized cellular and molecular changes. Semen contains prostaglandin PGE2 which acts as a common inflammatory mediator and found in 10,000-fold higher concentration in HPV-infected women cervix (Kelly 1995; Sharkey et al. 2007). The upregulated PGE2 is reported as a possible favorable promoter of cervical carcinogenesis. It is noteworthy to mention that smoker's semen because of having tobacco-related carcinogens may act as a risk factor for CC (Hung et al. 2009). Various other birth control measures such as tubal ligation and intrauterine devices are also reported to cause 80% of CC, especially in developing countries (Schiffman et al. 2007).

26.5 Genetic Susceptibility for HPV Infection

Some individuals with specific immunodeficiency are prone to HPV infection. Individual with epidermodysplasia verruciformis (EV), warts, hypogammaglobulinemia, infections, and myelokathexis (WHIM) syndrome, GATA2 or DOCK8 deficiencies, and other syndromes are highly susceptible to HPV infection (Leiding and Holland 2012). The two genetic conditions of EV caused by mutations in the EVER1 and EVER2 genes, and WHIM syndrome, associated with mutations in CXCR4 have been associated with persistent HPV infection (Hernandez et al. 2003; Ramoz et al. 2002). Several studies have provided evidential facts that some of the HLA loci could predispose the individual to HPV-associated CC (Chen et al. 2013; Maciag et al. 2000; Madeleine et al. 2008; Safaeian et al. 2014; Wang et al. 2002). *HLA* A2, A*01, A*24, A*1104, B7, B15, B63, and Cw*0202 loci and HLA DQ and DR, namely, DQA1*01, *03, DQB1*02, *03, *04, *06 and DQw1, DQw3, DRB1 *4, *5, *11, *13, *15, *16 have been reported to be associated with increased risk of developing CC (Chattopadhyay 2011). The HR of HPV-associated CC has been reported in women having the haplotype HLA-B0702-DRB11501/HLADQB10602, whereas the reduced risk is associated with alleles of the haplotype HLA-B1501/HLA-DRB11301/HLA-DQA10103/HLA-DQB10603 (Chen et al. 2013; Maciag et al. 2000; Madeleine et al. 2008; Safaeian et al. 2014; Wang et al. 2002). CC patients have also shown increased TNF levels—α, IL-10 (Chattopadhyay 2011).

26.6 HPV Genomic Organization

HPV belongs to the family of *Papillomaviridiae*. It is a small nonenveloped virus-containing circular double-stranded DNA genome of 7.9 kbp. HPV presented as a covalently closed circle in the infected cell nucleus. HPV genome consists of the icosahedral capsid of 60 nm diameter and 72 capsomers. HPV capsid encloses single, double-stranded DNA, of which one coding strand acts as a template for transcription by containing three functional coding regions. These are classified as E (a gene coding early viral function) containing seven ORF-encoding viral regulatory protein designated as E followed by a numeral which inversely proportional with size as the numeral increases (Buck et al. 2013). The L (a gene coding late viral function) encodes two capsid proteins L1 and L2. The third functional coding region is long control region (LCR), an LCR between E and L. LCR is synonymously known as the upstream regulatory region or the noncoding region (NCR) (Buck et al. 2013). HPV E4 and E5 encode accessory proteins, which modulate the cellular environment for viral replication and its evasion from immune recognition (Buck et al. 2013). HPV E6 and E7 oncoproteins are responsible for cervical lesion progression from premalignant to the malignant stage.

Considering HPV-16 as the prototype model for HPV, since it is the most prominent HR-HPV genotype causative agent of CC, its genomic structure is illustrated in Fig. 26.1. It comprises three regions:

1. An early region comprises an ORF of E1, E2, E4, E5, E6, and E7. This ORF is implicated in the various functions the detail of which is discussed individually below.
2. A late region comprising of L1 and L2 capsid protein responsible for virion structure, DNA packaging, and maturation.

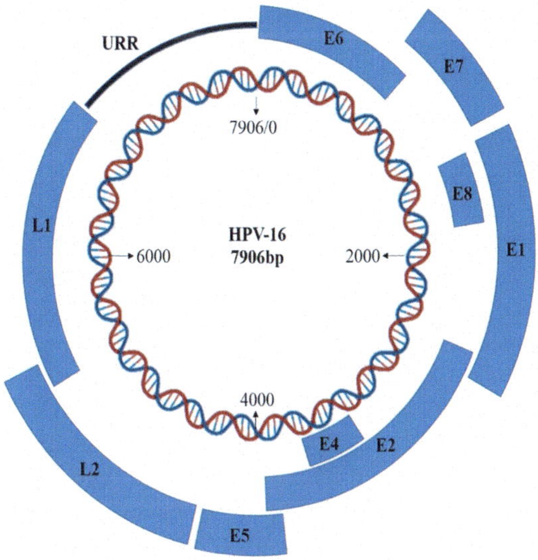

E1	• Origin binding protein • DNA helicase involved in viral DNA replication
E2	• Association with E1(origin binding) • Transcriptional control • Tethering of viral episome • Viral DNA replication
E4	• Genome amplification • Interaction with cytoskeleton/keratin for its destabilization • Viral release and Transmission
E5	• Small transmembrane protein • Cellular Proliferation by interacting with EGF receptor activating mitogenic pathway • Downregulation of MHC class I expression.
E6	• Control Cell cycle for HPV genome amplification in upper epithelial layers. • Association with E6AP and degradation of p53, PDZ-protein binding.
E7	• Control Cell cycle for HPV genome amplification in upper epithelial layers. • Attachment and degradation of pRb, mitotic mutator
L1	• Major capsid protein • Assemble into pentameric capsids to form icosahedral virion. • Target for prophylactic vaccine.
L2	• Minor capsid protein. • Role in viral DNA encapsidation. • Facilitates viral entry and trafficking.
URR	• Upstream Regulatory Region. • Promoter and enhancer element

Fig. 26.1 Schematic representation of HPV-16 genomic organization and brief illustration of protein function

3. An long control region (LCR).
4. Small, highly variable, NCR between E5 and L2.

The HPV-16 L1 encodes for major capsid protein in the form of mini-chromosome assembled with host histones (McBride 2017). Transcription, depending on the differentiation status of the host cell, occurs in three progressive sequential phases. Early transcription started from early promoter region, while termination started from the polyadenylation site. Late promoter region helps in intermediate transcription to transcribe E1 and E2 replication. Last stage transcription started from late promoter and polyadenylation region to result in L1 and L2 protein expression (Johansson and Schwartz 2013). HPV-16 genome is polymorphic in terms of its different phylogenetic variants inhabiting different geographical regions over the globe. These variants were classified on the basis of sequence variation of E6, L1, L2, and the LCR regions.

26.7 Oncogenic Role of HPV Genes

Among E1, E2, E4, E5, E6, and E7, the oncogenic role of E6 and E7 genes have been adequately defined and established. E6 oncoprotein of 151 amino acids specifically inactivates tumor suppressor protein p53, whereas E7 inactivates retinoblastoma tumor suppressor protein (pRb), respectively. HPV-16 E6 and E7 were seen to immortalize primary human keratinocytes (Munger et al. 2004) effectively. E6 gene forms a cellular protein E6–AP complex of about 100 kDa, which subsequently degrades p53 protein by the ubiquitin pathway resulting in loss of p53-induced growth arrest and apoptosis in response to DNA damage (Kessis et al. 1993; Scheffner et al. 1993). E6 was also found to be involved in various p53 independent pathways, such as telomerase activation (Veldman et al. 2001), inhibition of SRC-family kinases degradation (Oda et al. 1999), and binding of cellular proteins with a carboxyl-terminal PDZ domain for their degradation (Kiyano et al. 1997). E6 protein may also interact with other additional cellular factors p300 which is vital for the transcriptional activity of p53 (Munger et al. 2004). E6 also interacts with transcription coactivator ADA3 (Munger et al. 2004). E6 oncoprotein inhibits negative cell cycle regulators, resulting in uninterrupted HPV genome amplification in upper epithelial layers. E7 gene-encoding proteins of 98 amino acids are small zinc-binding phosphoproteins. It binds to the pRB tumor suppressor protein complex and dissociates the cellular transcription factor (E2F) bound with it for its release. The released E2F upregulates $p16^{INK4A}$ to ultimately promotes cellular proliferation in the absence of normal mitogenic signals (Kiyono et al. 1998; Dyson et al. 1989). E7-induced cell proliferation instigated cell entry early during the S phase of the cell cycle. In the absence of p53-mediated apoptosis due to degradation by E6, the uninterrupted cellular proliferation leads to extensive malignant invasive lesions. E7 oncoprotein destabilizes the ‘pocket proteins’ p107 and p130 through a proteasome-dependent mechanism (Munger et al. 2004). These pocket proteins play an essential role in regulating the E2F family of transcription factors which control the cell cycle transition and various other cellular activities (Munger et al. 2004). E7 also destabilizes important cell cycle arrest regulators, namely, cyclin-dependent kinase inhibitors, including $p21^{CIP1}$ and $p27^{KIP1}$ (Munger et al. 2004). A carboxyl-terminal E7 domain also overrides $p21^{CIP1}$-mediated growth arrest. However, although the carboxyl-terminal E7 domain interacts with transcription factors, cell cycle regulators, and metabolic enzymes, the biological function of these interactions remains to be determined.

E7 oncoprotein induces the aberrant proliferation and differentiation of the keratinocytes resulting in cell death. This triggers a cellular defense mechanism, “Trophic Sentinel Response” which removes this deviant cell from the proliferative cellular pool. However, the E6 eliminates this response through inactivation of p53 and induces the epithelial cell’s immortalization through telomerase activity by inducing telomerase subunit hTERT expression at a transcriptional level. hTERT promoter fragment contains c-myc-responsive E-boxes, which helps in E-6-mediated transcriptional activation (Fig. 26.2).

Thus, E6 and E7-mediated mechanisms eventually induce carcinogenic transformation after controlling the cervical epithelial cell cycle. The p97 promoter located in the proximal region of the LCR helps transcription of E6 and E7 viral oncogenes in HPV-16. This shows that LCR works in conjunction with E6 and E7 in the pathogenesis of cervical neoplasia (Smotkin and Wettstein 1986). LCR encoding an LCR exists between early (E) and late (L) protein has been reported necessary for normal viral replication and gene expression control (Faridi et al. 2011).

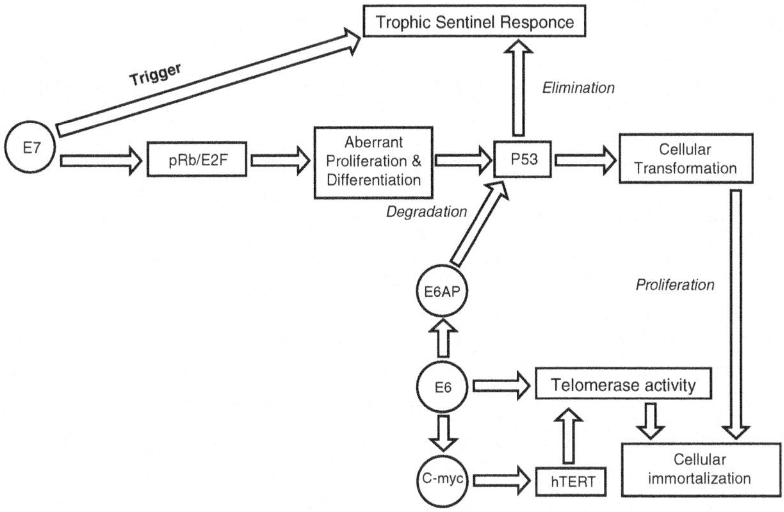

Fig. 26.2 Schematic diagram explaining HPV E6- and E7-induced CC by inactivating pRb and p53 tumor suppressor protein

E5 gene-encoded protein is another potential oncoprotein acting as a small transmembrane protein. It is responsible for transforming activity by interaction with the epidermal growth factor (EGF) receptor to activate mitogenic pathways (Pin et al. 1992). E5 gene encodes for a hydrophobic protein enhancing immortalization of the cell. E5 also seems essential in the early course of infection by promoting viral proliferation and evasion to immune recognition (DiMaio and Petti 2013).

E4 encompassed within E2 and expressed mainly as E1^E4 fusion protein to bind to cytokeratin filaments and disrupt their structure. This all eventually helps in a viral release from the cornified epithelial cell layer (Doorbar 2013).

E1 and E2 each play essential roles in viral genome replication. E1 encodes for ATP-dependent viral DNA helicase (Bergvall et al. 2013). This viral DNA helicase bind to the AT-rich origin of replication and E2 protein to specifically help viral transcription, replication, and genome partitioning (Munger et al. 2004). E2 protein encodes a transcriptional activator. E1, E2 in earlier studies found to code proteins that control E6 and E7 genes (Faridi et al. 2011). L1 and L2 code for structural proteins having a pivotal role in the formation of complete virus particles. The structural proteins L1 and L2 have their role in forming Icosahedral capsid for virion generation during replication and assembly. L1 ORF sequences are the most conserved among the HPV genome. L1, since spontaneously assemble into a viral-like particle (VLP), it is the basis of the present day's vaccine and classification of HPV.

Further studies are however still required for a better understanding of insight of all HPV genes in the progression and transformation of the cervical lesion to the stage of invasive CC.

26.8 Immune Detection and Clearance

During HPV persistent infection, our immune system cannot detect the virus, because the virus restricted its initial infectivity in the upper differentiated epithelial layers (Sasagawa et al. 2012). Clearance of HPV infection within 3 years of illness in most HPV-infected cases substantially shows that immunity plays a vital role in clearing the virus (Sasagawa et al. 2012). The various immune cell-like macrophages (MΦ), natural killer (NK) cells, and natural killer T cells generated in response to innate immunity play the first line of defense against HPV infection. However, the persistent infection by HPV, especially in females having various risk factors, such as renal transplant history, multiparity, HIV or HSV infection, various major histocompatibility complex (MHC) alleles, e.g., HLA-DQB1*0602 and HLA-DRB1*1501, have higher chances of developing CC by evading host immune response (Sasagawa et al. 2012). While HPV 16 E0025 protein downregulates HLA-class 1 expression to facilitates evasion of cytotoxic T lymphocyte attack, HPV 16 E0030 and E0035 downregulates the expression of type-1 interferon (IFNs) in host cells. Moreover, since HPV does not kill infected host cells during replication, it does not present viral antigen during the virus life cycle, which ultimately helps the virus evade immune detection (Stanley 2012; Zhou et al. 2019).

26.9 Latency

Latency is defined as the presence of viral DNA without gene expression. HPVs, whether it undergo latency requires is still not adequately proven and so requires more number of studies for substantiating, because no adequate detection tool is yet available for detecting HPV DNA in the infected lower epithelial layer. However, a study in an animal model of rabbit showed that HPV infection might undergo latency with no apparent signs of infection (McBride 2017). Furthermore, reactivation of HPV in older women either due to mechanical or cellular stress or reinfection by the same HPV genotype may advocate the observation that HR-HPV might undergo latency and reactivate once exaggerated by various risk factors of the waning of immunity or reinfection (McBride 2017).

26.10 HPV Multiple Genotype Infection

Earlier studies have reported individuals harboring the multiple HPV genotypes simultaneously at the same anatomical site (Negi et al. 2019). Further, each genotype infection results from a single infection cell to label them as clonal. Single anatomical tissue sites may get infected by different HPV genotypes simultaneously or during various exposure sources (Quint et al. 2012). Our previous study showed that in CC cases, both dual infection of HPV 16 and 18 and triple genotype 16, 18, and 33 were found in around 11.1% cases each. Even a novel dual co-infection of HPV genotype 35 and 39 found in two and one CIN I and II cases, respectively

(Negi et al. 2019). Importantly, multiple HPV genotypes infection not reported to augment the progression of the disease (Wentzensen et al. 2014). However, since multiple genotype infection contributes to a substantial number of CC, it requires more studies to check the role of multiple genotypes in the disease's severity or progression.

26.11 Pathogenesis of HPV-16

CC primarily arises from the cervix's ectocervix and endocervix meeting junction where squamous to columnar epithelial cells transition occurs (Herfs et al. 2012). This junction site appears to increase the accessibility of HPV-16 to the proliferative basal cells. HPV-16 infect basal cell either by binding to the receptor heparin sulfate proteoglycans (HSPGs) present on basal cell or thourgh microabrasion in the lower basal epithelial layer to establish a persistent infection for years (Roberts et al. 2007) (Fig. 26.3). The viral capsid induces a conformational change for its binding to a secondary receptor present on basal keratinocytes surface (Day and Schelhaas 2014). Virus adopts the two-pronged strategy in cervical lesion progression. In the first strategy of long-term viral persistence, the virus infects self-renewing cells. In the second strategy to escape detection by the immune system, viral proteins are majorly restricted to differentiated epithelial layers of the lesion (McBride 2017).

The virion entered into the cell by endocytosis. It then moves to the trans-Golgi component through the endosomal pathways. During trafficking, the virus is uncoated, leading to viral L2 entry in the nucleus encased membrane vesicle (Digiuseppe et al. 2016). Infected cells then differentiated to move upwards to the epithelial surface to induce high-level viral replication and gene expression. For transcription and replication, viral L2–genome complex lies adjacent to ND10 nuclear bodies (McBride 2017). The viral genome reorganizes ND10 components

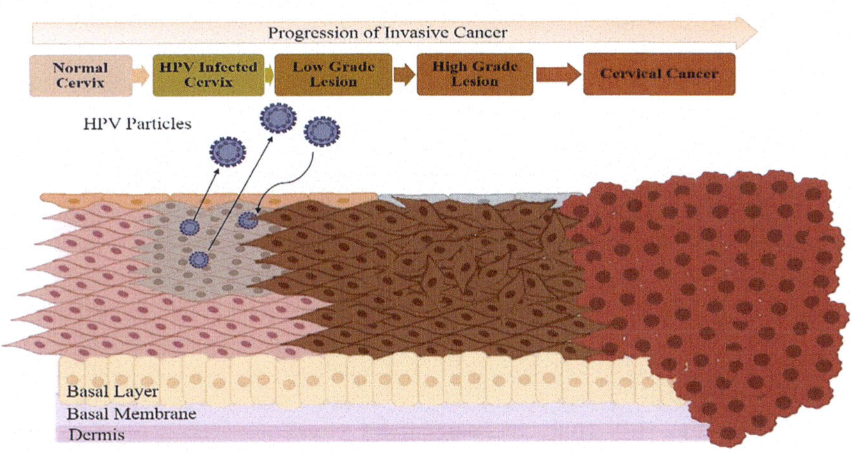

Fig. 26.3 Schematic diagram to explain HPV oncogenic progression to invasive CC

to induce transcription and replication, displaces Sp100 protein, and recruits Daxx (McBride 2017). High viral load is also presumed to be a contributing factor in the development of CIN (Wheeler et al. 1997).

The E1 and E2 proteins act as recognition and regulator factors in viral transcription. E1 and E2 replication proteins consist of three phases. The first phase resulted in limited DNA amplification during the host cell first exposure to HPV. The second phase consists of maintenance replication to ensure the viral genome's constant replication in the lesion. The third final phase of amplification comprises viral DNA replication to generate many progeny virions (McBride 2008). To synthesize viral DNA, the virus relies on cellular proteins and E2, which recruit and load E1 helicase onto the viral replication origin. The E2 protein ensures efficient partitioning of the progeny virions by tethering them to host chromatin with the help of its DNA-binding domain site by its interaction with conserved sites (E2BS) in the viral genome (McBride 2017). Consequent to HPV hijacking host cell DNA synthesis machinery, E1 and E2 with the help of cellular DNA polymerase direct the viral genome replication. E2 also acts as the virus's primary transcriptional regulator and maintains a low level of persistent infection. Thus HPV replicates highly in terminally differentiated cells which then eventually sloughed off. Notably HPV is not disseminated within the infected host through viremia or any other systemic route.

E5, E6 and E7 are three major oncoproteins associated with CC studied in vitro and in vivo (Nariseuva Saito et al. 2007). E6 and E7 cause cellular proliferation and inactivate cell cycle regulators which help in uninterrupted viral replication in differentiated cells to increase the infected basal and parabasal layer-infected area. E6 and E7 oncoproteins inhibit tumor suppressor genes, namely, p53 and pRb. P53 acts as a 'guardian' protecting cellular DNA (Lane 1992). p53 precisely controls the cell cycle by arresting the cell division whenever the cell's DNA is damaged by various means, such as UV or X-ray irradiation, to allow time to repair DNA damage. Thus, deregulation of p53 caused by E6 may lead to cancer. In contrast, Rb protein is a very large, 105 kDa nuclear phosphoprotein regulating cell cycle division.

E5 increases the activity of epidermal growth factor receptor (EGFR) to promote viral proliferation and facilitate immune evasion by inhibiting the MHC expression involved in antigen presentation (Tsai and Chen 2003). The HPV-16 E1 and E2 facilitate viral DNA integration into the host cell genome to play a cardinal role in controlling transcription and replication. E1 and E2 roles in pathogenesis also evidenced from the earlier reported observation that the absence of these proteins deregulates E6 and E7 oncoprotein (Zerfass et al. 1995). The E6 protein of 151 amino acids forms the E6–AP complex to degrade the ubiquitin targeted substrate of p53 to inhibit p53-mediated apoptosis (Incassati et al. 2006). Only HR-HPV appears to eliminate p53 function through a ubiquitin-dependent mechanism, whereas LR-HPV (HPV6 and 11) lack this property (Werness et al. 1990; Crook et al. 1991). HPV-16 E6 also upregulates telomerase activity to ensure continuous telomere integrity despite repeated cellular division, whereas LR-HPV lacks this activity. HPV-16 E6 also targets the cellular PDZ-binding domain to disrupt cell contact and signaling pathways, whereas LR-HPV again lacks this activity (Munger et al. 2004). E7 protein inactivates pRb protein by dissociating transcription of genes related to normal control and cell division (Georgescu et al. 2018).

After viral replication, virion assembly started inside the keratinocytes nucleus. L1 and L2 expressions help in genome packaging, and maturation occurs in terminally differentiated squamous epithelial cells. Finally, progeny virus shed off from terminally differentiated infected keratinocytes during nuclear and cytoplasmic integrity loss (Munger et al. 2004).

Other cofactors, namely, inflammation and oxidative stress (OS), also play a role in viral persistence and the development of a malignant process (Federico et al. 2007). It is well-known that chronic inflammation of tissue induces mutations and release of cytokines and growth factor causes genetic alteration in susceptible cell populations which all eventually lead to impairing of homeostasis of cell growth. Although initial HPV infection of basal cells does not induce inflammation because not recognized by circulating immune cells, the persistent infection induces inflammatory immune response with release of proinflammatory cytokines, namely, interferon gamma (IFN-γ), tumor necrosis factor-alpha (TNF-α), interleukin (IL)-1, and IL-6. These cytokines by protein kinase-mediated signaling pathways result in the formation of reactive oxygen species (ROS), such as superoxide anion (O_2^-), hydrogen peroxide (H_2O_2), and hydroxyl radicals (OH^-). ROS has harmful effects on various cellular components (lipids, proteins, and nucleic acids) and thus reported in various studies to favor creating the condition for malignant transformation (Federico et al. 2007; Georgescu et al. 2018; Mitran et al. 2018). Various studies have also shown that ROS causes alteration of purine and pyrimidine bases and loss of purines resulted in abasic sites (Georgescu et al. 2018; Ott et al. 2007). Proinflammatory cytokine's high levels were earlier reported in 50 women having persistent HPV infection in comparison with an equal number of healthy subjects (Kemp et al. 2010). Persistent HPV-induced inflammation also decreases antioxidants levels and induces an imbalance between peroxidants and antioxidants (Georgescu et al. 2018).

Furthermore, the HPV-16 E5, E6, and E7 are responsible for chronic inflammation by increasing the expression of cyclooxygenase (COX)-2. This resulted in a higher release of prostaglandins which eventually then involved in cell proliferation, angiogenesis, and inhibition of apoptosis, all subsequently leading to carcinogenesis (Georgescu et al. 2018). E6 and E7 also inhibit interferon and nuclear factor kappa-light chain enhancer of activated B cell (NF-kB) signaling pathways to help the virus to escape detection (Georgescu et al. 2018; McBride 2017).

Regarding reactive nitrogen species, which mainly comprise nitric oxide (NO), it was found that HPV-infected women having HSIL or invasive lesions showed elevated expression of NO and inducible NO synthase in comparison with normal subjects (Lala and Orucevic 1998; Sowjanya et al. 2016).

26.12 Variants and Their Effect in the Pathogenesis

Different studies have identified various oncogenic HPV variants of HPV-16 with higher predominance in causing cancer (Burk et al. 2013; Cornet et al. 2013; Pande et al. 2008; Pillai et al. 2009; Rader et al. 2019; Satish et al. 2005). Because of the slow evolution of HPV, its genetic drift indicates that variant diversity has been evolving for approximately 200,000 years (Burk et al. 2013). The variant lineages have between 1% and 10% difference in the L1 region (Burk et al. 2013). The sequence variation in E6, E7, and L1 has been shown in various studies to influence the oncogenicity of the virus (Cornet et al. 2013; Pande et al. 2008; Pillai et al. 2009; Rader et al. 2019; Satish et al. 2005). Cornet et al. 2013 reported the worldwide distribution of HPV-16 variant and the relative risk for CC to appear to be population dependent (Cornet et al. 2013). In an earlier reported study from India, European variant E350G was found as the most prevalent (58%) followed by the American Asian variant (11.4%) (Pillai et al. 2009). Pande et al. (2008) found European lineage with 100% T350G variant in E6, 87.5% T789C variant in E7, and 54.5% A6695C variant in the L1 region from CC cases in North India (Pande et al. 2008). Sequence variation in HPV-16 has been reported to suppress keratinocyte differentiation response, increased p53 degradation, increased virus assembly, and decreased immunological recognition by the host (Satish et al. 2005). Among different HPV genotypes, the diversity of ORF varies. These ORF in HPV-16 vary from most to least variable from NCR > LCR/E5 > E4/L2 > E2/E7/E6 > L1/E1 in decreasing variability region. L1 is the most conserved region. Since CC is a multifaceted process with viral variants, oncogene, viral persistence, and host immunity play a role in pathogenesis, a robust, consistent system must reliably save huge genomic data obtained from next-generation sequencing techniques to identify important insight into the viral-mediated oncogenesis.

26.13 Therapeutics

HPV vaccine has been found extremely successful in protecting women against CC. Gardasil and Cervarix HPV vaccines are prophylactic vaccines that provide maximum protection if used by girls pre-exposure to HPV. Gardasil is available in nine-valent and four-valent, and bivalent forms. Nine-valent vaccine provides protection against nine HPV genotypes, namely, 6, 11, 16, 18, 31, 33, 45, 52, and 58. The four-valent or quadrivalent vaccine protects against HPV 6, 11, 16, and 18 genotypes. Cervarix bivalent HPV vaccine provides immunity against HPV genotypes 16 and 18. These prophylactic HPV vaccines are made up of HPV L1 proteins, and they generate high levels of systemic protective anti-HPV L1 antibody on intramuscular (I/M) injection (Faridi et al. 2011). This vaccine can be effective against future infection only and not at all effective in previously exposed individuals or in the treatment of neoplasia, because HPV is a nonlytic virus that only divides in epithelial keratinocytes. Presently, both Gardasil and Cervarix HPV vaccine shows the remarkable immunological response to provide complete protection to

Naïve women for a maximum of 5 years. However, due to various factors, such as social localized myths and beliefs related to reluctance for vaccination, such as vaccination bringing infertility, scarcity of vaccine especially in developing countries far-flung areas of districts and villages, and high cost, the HPV vaccination has not been proper rolled out in various countries especially developing one. Since most infections are cleared by cell-mediated immunity within their 2 years of origin, therapeutic vaccines or other immunomodulatory interventions have a high success rate. Moreover, because nonvaccinated women accounted for more than 70% of CC cases, there is a need for a government-supported program of creating awareness about the HPV vaccination and providing this vaccine free of cost for ensuring the high success of vaccination to end this deadly neoplasia.

Due to growing evidence of the role of persistent inflammation in the progression of invasive cervical lesion, the nonsteroidal anti-inflammatory drugs (NAISDs), such as celecoxib, rofecoxib, and both COX-2 inhibitor, were found useful in both prevention and therapy of CC (Hefler et al. 2006; Rader et al. 2017).

26.14 Conclusion

HPV infection acquired through the sexual route has a hazardous effect in females by causing CC. Due to significant mortality associated with CC, HPV infection is a major public health problem worldwide. The nonawareness about the mode of infection, various risk factors, avoidance of cervical screening by pap smear, and HPV DNA testing due to various socio-cultural taboos has surmounted the HPV infection, which progressively leads to CC. Women refusing to go for periodical cervical screening during their active sexual life is the major hurdle in controlling HPV infection. HPV causing the cervical infection may have different clinical outcomes varying from self-limiting to invasive CC. This complexity in outcome depends on various factors, such as exposure to HR-HPV, its persistence, chronic inflammation and OS, immunity etc. The HR-HPV, especially HPV-16, followed by other 13 HR-HPV infections pose the highest risk of CC. HR-HPV causes slow, progressive, persistent infection after establishing infection in cervical epithelial surfaces progress to various stages of cervical intraepithelial lesion of CIN-I, II, CIN-III, and ultimately invasive CC. HR-HPV induces viral DNA replication after abrogating the cellular control mechanism to prevent host cell DNA synthesis while allowing only viral DNA to synthesize. This happens when E1 and E2 proteins help in viral genome replication and E6 inactivates the p53, cellular PDZ protein and activates telomerase activity while E7 oncogenes bind and degrade retinoblastoma tumor suppressor pRb. This all eventually leads to malignant progression to invasive CC. p53 provides an essential check on cellular growth, and it may shut down the cell DNA replication metabolic pathway if it recognizes irregularity in DNA synthesis during the cell cycle. Thus, degradation of p53 allows the HR-HPV to completely control cellular machinery for uninterrupted cellular proliferation to cause CC. E5 increases the EGFR expression to inhibit the MHC expression, which helps the virus escape immune recognition and response. E1 and E2 indirectly play a role by controlling the transcription and replication of the virus. Importantly, since current scientific data are scarce for the complete and detailed genetic level understanding of the pathogenesis, more studies are required to unearth the potential role of all HPV oncogene. This seems to be the only way forward for exploring more sensitive, rapid diagnosis and in providing the opportunity for the rational design of antiviral and anticancer agents and vaccination for better treatment and prophylaxis.

References

Anderson LA, O'Rorke MA, Wilson R, Jamison J, Gavin AT, Northern Ireland HPV Working Group (2016) HPV prevalence and type-distribution in cervical cancer and premalignant lesions of the cervix: a population-based study from Northern Ireland. J Med Virol 88:1262–1270

Bergvall M, Melendy T, Archambault J (2013) The E1 proteins. Virology 445:35–56

Bruni L et al (2010) Cervical human papillomavirus prevalence in 5 continents: meta-analysis of 1 million women with normal cytological findings. J Infect Dis 202:1789–1799

Buck CB, Day PM, Trus BL (2013) The papillomavirus major capsid protein L1. Virology 445:169–174

Burk RD, Harari A, Chen Z (2013) Human papillomavirus genome variants. Virology 445:232–243

Castellsajne X, Munoz N (2003) Chapter 3: Cofactors in human papillomavirus carcinogenesis: role of parity, oral contraceptive & tobacco smoking. J Nat Cancer Inst 31:20–28

Chattopadhyay K (2011) A comprehensive review on host genetic susceptibility to human papillomavirus infection and progression to cervical cancer. Indian J Hum Genet 17:132–144

Chen D, Juko-Pecirep I, Hammer J, Ivansson E, Enroth S, Gustavsson I et al (2013) Genome-wide association study of susceptibility loci for cervical cancer. J Natl Cancer Inst 105(9):624–633. https://doi.org/10.1093/jnci/djt051. PMID: 23482656

Chung SH, Franceschi S, Lambert PF (2010) Estrogen and ER-α: culprits in cervical cancer. Cell 21:504–511

Cornet I, Gheit T, Iannacone MR, Vignat J, Sylla BS et al (2013) HPV16 genetic variation and the development of cervical cancer worldwide. Br J Cancer 108:240–244

Crook T, Tidy J, Vousden K (1991) Degradation of p53 can be targeted by HPV E6 sequences distinct from those required for p53 binding and trans-activation. Cell 67:547–556

Day PM, Schelhaas M (2014) Concept of papillomavirus entry into host cells. Curr Opin Virol 4:24–31

de Villers EM (2003) Relationship between steroid hormone contraceptives and HPV, cervical intraepithelial neoplasia and cervical carcinoma. Int J Cancer 103:705–708

Digiuseppe S, Luszczek W, Keiffer TR, Bienkowskahaba M, Guion LG, Sapp MJ (2016) Incoming human papillomavirus type 16 genome resides in a vesicular compartment throughout mitosis. Proc Natl Acad Sci USA 113:6289–6294

DiMaio D, Petti LM (2013) The E5 proteins. Virology 445:99–114. PubMed: 23731971

Doorbar J (2013) The E4 protein; structure, function and patterns of expression. Virology 445:80–98

Dyson N, Howley PM, Murger K, Harlow E (1989) The human papillomavirus-16 E7 oncoprotein is able to bind to the retinoblastoma gene product. Science 243:934–937

Faridi R, Zahra A, Khan K, Idrees M (2011) Oncogenic potential of human papillomavirus (HPV) and its relation with cervical cancer. Virol J 8:269

Federico A, Morgillo F, Tuccillo C, Ciardiello F, Loguercio C (2007) Chronic inflammation and oxidative stress in human carcinogenesis. Int J Cancer 121:2381–2386

Franceschi S, Rajkumar T, Vaccarella S, Gajalakshmi V, Sharmila A, Snijders PJ et al (2003) Human papillomavirus and risk factors for cervical cancer in Chennai, India: a case-control study. Int J Cancer 107:127–133

Georgescu SR, Mitran CI, Mitran MI, Caruntu C, Sarbu MI, Matei C, Nicolae I, Tocut SM, Popa MI, Tampa M (2018) New insight in the pathogenesis of HPV infection and the associated

carcinogenic processes: the role of chronic inflammation and oxidative stress. J Immunol Res 2018:5315816
Gravitt PE (2011) The known unkowns of HPV natural history. J Clin Invest 121:4593–4599
Hefler LA, Grimm C, Speiser P, Sliutz G, Reinthaller A (2006) The cyclooxygenase-2 inhibitor rofecoxib (Vioxx®) in the treatment of cervical dysplasia grade II–III: a phase II trial. Eur J Obstet Gynecol Reprod Biol 125(2):251–254
Herfs M et al (2012) A discrete population of squamocolumnar junction cells implicated in the pathogenesis of cervical cancer. Proc Natl Acad Sci USA 109:10516–10521
Hernandez PA, Gorlin RJ, Lukens JN, Taniuchi S, Bohinjec J, Francois F et al (2003) Mutations in the chemokine receptor gene CXCR4 are associated with WHIM syndrome, a combined immunodeficiency disease. Nat Genet 34(1):70–74
Herrero R, Castle RE, Schiffman M, Bratti MC, Hildescheim A, Morales J, Alfaro M, Sherman ME et al (2005) Epidemiological profile of type specific human papillomavirus infection and cervical neoplasia in Guanacaste, Costa Rica. J Infect Dis 191:1796–1807
Ho L, Chan SY, Chow V et al (1991) Sequence variants of human papillomavirus type 16 in clinical samples permits verification and extension of epidemiological studies and construction of a phylogenetic tree. J Clin Microbiol 29:1765–1772
Hung PH, Froenicke L, Lin CY, Lyon LA, Miller MG, Pinkerton KE, VandeVoort CA (2009) Effects of environmental tobacco smoke in-vivo on rhesus monkey semen quality, sperm function, sperm metabolism. Reprod Toxicol 27:140–114
Incassati A, Patel D, McCance DJ (2006) Induction of tetraploidy through loss of p53 and upregulation of Plk1 by human papillomavirus type 16 E6. Oncogene 25(17):2444–2451
International Agency for Research on Cancer (2012) GLOBOCON 2012: estimating cancer incidence, mortality and prevalence worldwide in 2012. http://globocon.iarc.pv/pages/fact_sheets_cancer.aspx?Cancer=cervix
International Collaboration of Epidemiological Studies of Cervical Cancer (2006) Cervical carcinoma and reproductive factors: collaborative reanalysis of individual data on 16,563 women with cervical carcinoma and 33,542 women without cervical carcinoma from 25 epidemiological studies. Int J Cancer 119:1108–1112
International Human Papillomavirus Reference Center HPV Center (2016). http://hpvcenter.se/html/reference.html
Johansson C, Schwartz S (2013) Regulation of human papillomavirus gene expression by splicing an dpolyadenylation. Nat Rev Microbiol 11:239–251
Kelly RW (1995) Immunosuppressive mechanism in semen. Implication for contraceptions. Hum Rep 10:1686–1693
Kemp TJ, Hildescheim A, Pineres AG et al (2010) Elevated systemic levels of inflammatory cytokines in older women with persistent cervical human papillomavirus infection. Cancer Epidemiol Biomarkers Prev 19:1954–1959
Kessis TD, Slebos RJ, Nelson WQ, Kastan MB, Plunkett BS, Han SM, Lorinez AT, Hedrick L, Cho KR (1993) Human papillomavirus 16 E6 expression disrupts the P53-mediated cellular response to DNA damage. Proc Natl Acad Sci USA 90:3988–3992
Kiyano T, Hiraiwa A, Fujita M, Hayashi F, Akiyama J, Ishibashi M (1997) Binding of high risk human papillomavirus E6 oncoprotein to the human homologue of the drosophila discs large tumor suppressor protein. Proc Natl Acad Sci USA 94:11612–11616
Kiyono T, Foster SA, Koop JI, McDougall JK, Galloway DA, Klingelhutz AJ (1998) Both rb/p16 ink4a inactivation and telomerase activity are required to immortalized human epithelial cells. Nature 396:84–88
Lala PK, Orucevic A (1998) Role of nitric oxide in tumor progression: lessons from experimental tumors. Cancer Metastasis Rev 17:91–106
Lane DP (1992) p53, guardian of the genome. Nature 358:15–16
Leiding JW, Holland SM (2012) Warts and all human papillomavirus in primary immunodeficiences. J Allegy Clin Immunol 130:1030–1048
Li N, Franceschi S, Howell Jones R, Snijders PJ, Clifford GM (2011) Human papillomavirus type distribution in 30,848 invasive cervical cancers worldwide: variation by geographical region, histological type and year of publication. Int J Cancer 128:927–935

Maciag PC, Schlecht NF, Souza PS, Franco EL, Villa LL, Petzl-Erler ML (2000) Major histocompatibility complex class II polymorphisms and risk of cervical cancer and human papillomavirus infection in Brazilian women. Cancer Epidemiol Biomarkers Prev 9(11):1183–1191. PMID: 11097225

Madeleine MM, Johnson LG, Smith AG, Hansen JA, Nisperos BB, Li S et al (2008) Comprehensive analysis of HLA-A, HLA-B, HLA-C, HLA-DRB1, and HLA-DQB1 loci and squamous cell cervical cancer risk. Cancer Res 68(9):3532–3539. https://doi.org/10.1158/0008-5472.CAN-07-6471. PMID: 18451182

McBride AA (2008) Replication and partitioning of papillomavirus genomes. Adv Virus Res 72:155–205

McBride AA (2017) Oncogenic human papillomavirus. Philos Trans R Soc B 372:20160273

Mitran MI, Nicolae I, Ene CD et al (2018) Relationship between gamma-glutamyl transpeptidase activity and inflammatory response in lichen planus. Rev Chim 69:739–743

Munger K, Baldwin A, Edwards KM, Hayakawa H, Nguyen CL, Owen M, Grace M, Huh KW (2004) Mechanism of human papillomavirus induced oncogenesis. J Virol 78:11451–11460

Nariseuva Saito M et al (2007) Basic mechanism of high-risk human papilloma virus induced carcinogenesis. Cancer Sci 98:1505–1511

Negi SS, Bhargava A, Singh P, Aggarwal S, Hussain N, Das P (2019) Predominance of high-risk human papillomavirus genotype 16 and 39 in women with premalignant and malignant cervical pathology from Raipur, Chhattisgarh: clinical evaluation of tagging oligonucleotide cleavage and extension mediated genotyping assay. Indian J Med Microbiol 37:255–262

Oda H, Kumar S, Howley PM (1999) Regulation of the Src family tyrosine kinase Blk through E6AP-mediated ubiquitination. Proc Natl Acad Sci U S A 96(17):9557–62. https://doi.org/10.1073/pnas.96.17.9557. PMID: 10449731; PMCID: PMC22247

Ott M, Gogvadze V, Orrenius S, Zhivotovsky B (2007) Mitochondria, oxidative stress and cell death. Apoptosis 12:913–922

Palesfsky JM, Holly EA (2003) Chapter 6: Immunosuppression and co-infection with HIV. J Natl Cancer Inst Monogr 31:41–46

Pande S, Jain N, Prusty BK, Bhambhani S, Gupta S et al (2008) Human papillomavirus type 16 variant analysis of E6, E7, and L1 genes and long control region in biopsy samples from cervical cancer patients in North India. J Clin Microbiol 46:1060–1066

Pillai MR, Hariharan R, Babu JM, Lakshmi S, Chiplunkar SV et al (2009) Molecular variants of HPV-16 associated with cervical cancer in Indian population. Int J Cancer 125:91–103

Pin D, Collins M, Banks L (1992) Human papillomavirus type 16 E5 gene stimulates the transforming activity of the epidermal growth factor receptor. Oncogene 7:27–32

Quint W et al (2012) One virus, one lesion-individual components of CIN lesions contain a specific HPV type. J Pathol 227:62–71

Rader JS, Sill MW, Beumer JH et al (2017) A stratified randomized double-blind phase II trial of celecoxib for treating patients with cervical intraepithelial neoplasia: the potential predictive value of VEGF serum levels: an NRG Oncology/Gynecologic Oncology Group study. Gynecol Oncol 145(2):291–297

Rader JS, Tsaih S-W, Fullin D, Murray MW, Iden M, Zimmermann MT et al (2019) Genetic variations in human papillomavirus and cervical cancer outcomes. Int J Cancer 144:2206–2214. https://pubmed.Ncbi.nlm.nih.gov/30515767

Ramoz N, Rueda LA, Bouadjar B, Montoya LS, Orth G, Favre M (2002) Mutations in two adjacent novel genes are associated with epidermodysplasia verruciformis. Nat Genet 32(4):579–581. https://doi.org/10.1038/ng1044. PMID: 12426567

Roberts JN, Buck CB, Thompson CD, Kines R, Bernardo M, Choyke PL, Lowy DR, Schiller JT (2007) Genital transmission of HPV in a mouse model is potentiated by nonoxynol-9 and inhibited by carrageenan. Nat Med 13:857–861

Safaeian M, Johnson LG, Yu K, Wang SS, Gravitt PE, Hansen JA et al (2014) Human leukocyte antigen class I and II alleles and cervical adenocarcinoma. Front Oncol 4:119. https://doi.org/10.3389/fonc.2014.00119. PMID: 24995157

Sasagawa T, Takagi H, Makinoda S (2012) Immune response against human papillomavirus (HPV) infection and evasion of host defense in cervical cancer. J Infect Chemother 18:807–815
Satish N, Abraham P, Peedicayil A, Sridharan G, Chandy G (2005) HPV 16 E6 sequence variations in Indian patients with cervical neoplasia. Cancer Lett 229:93–99
Scheffner M, Huibregtse JM, Vierstra RD, Hawley PM (1993) The HPV-16 E6 and E6-AP complex functions as a ubiquitin-protein ligase in the ubiquitination of P53. Cell 75:495–505
Schiffman M, Castle PE, Jeronino J, Rodriquez AC, Wacholder S (2007) Human papillomavirus and cervical cancer. Lancet 370:890–907
Sharkey DJ, Macpherson AM, Tremellen KP, Robertson SA (2007) Seminal plasma differentially regulates inflammatory cytokine gene expression in human cervical and vaginal epithelial cells. Mol Hum Rep 13:491–501
Smith JS, Lindsay L, Hoots B, Keys J, Franceschi S, Winer R, Clifford GM (2007) Human papillomavirus type distribution in invasive cervical cancer and high grade cervical lesions: a meta-analysis update. Int J Cancer 121:621–632
Smotkin D, Wettstein FO (1986) Transcription of human papillomavirus type 16 early genes in a cervical cancer and a cancer derived cell line and identification of the E7 protein. Proc Natl Acad Sci USA 83:4680–4684
Sowjanya AP, Jain M, Poli UR, Padma S, Das M, Shah KV et al (2005) Prevalence and distribution of high-risk human papilloma virus (HPV) types in invasive squamous cell carcinoma of the cervix and in normal women in Andhra Pradesh, India. BMC Infect Dis 5:116
Sowjanya AP, Rao M, Vedantham H et al (2016) Correlation of plasma nitrite/nitrate levels and inducible nitric oxide gene expression among women with cervical abnormalities and cancer. Nitric Oxide 52:21–28
Sreedevi A, Javed R, Dinesh A (2015) Epidemiology of cervical cancer with special focus on India. Int J Womens Health 7:405–414
Stanley MA (2012) Epithelial cell responses to infection with human papillomavirus. Clin Microbiol Rev 25:215–222
Tsai TC, Chen SL (2003) The biochemical and biological functions of human papilloma virus type 16 E5 protein. Arch Virol 148:1445–1453
Tulay P, Serakinci N (2016) The role of human papillomavirus in cancer progression. J Cancer Metastasis Treat 2:201–213
Ude H, Kumar S, Hawley PM (1999) Regulation of the SRL family tyrosine kinase BLK through E6 AP-mediated ubiquitination. Proc Natl Acad Sci USA 96:9557–9562
Veldman T, Horikaria I, Barrett JC, Schlegel R (2001) Transcriptional activation of the telomerase hTERT gene by human papillomavirus type 16 E6 oncoprotein. J Virol 75:4467–4472
Volpini LPB, Boldrini NAT, Freitas LB, Miranda AE, Spano LC (2017) The high prevalence of HPV and HPV-16 European variants in cervical and anal samples of HIV seropositive women with normal PAP test results. PLoS One 12(4):e0176422
Wang SS, Hildesheim A, Gao X, Schiffman M, Herrero R, Bratti MC et al (2002) Comprehensive analysis of human leukocyte antigen class I alleles and cervical neoplasia in 3 epidemiologic studies. J Infect Dis 186(5):598–605. https://doi.org/10.1086/342295. PMID: 12195346
Wentzensen N, Nason M, Schiffman M, Dodd L, Hunt WC, Wheeler CM (2014) No evidence for synergy between human papillomavirus genotypes for the risk of high grade squamous intraepithelial lesions in a large population based study. J Infect Dis 209:855–864
Werness BA, Levine AJ, Howley PM (1990) Association of human papillomavirus types 16 and 18 E6 protein with p53. Science 248:76–79
Wheeler CM, Yamada T, Hildesheim A, Jenison SA (1997) Human papillomavirus type 16 sequence variants: identification by E6 and L1 lineage specific hybridization. J Clin Microbiol 35:11–19
Zerfass D et al (1995) Sequential activation of cyclin E and cyclin A gene expression by human papilloma virus type 16 E_7 through sequences necessary for transformation. J Virol 69:6389–6399
Zhou C, Tuong ZK, Frazer IH (2019) Papillomavirus immune evasion strategies target the infected cell and the local immune system. Front Oncol 9:00682

Human Adult T-Cell Leukemia Virus Type 1 (HTLV1)

27

Sunita Jetly, Satendra Singh, Manisha Khatri, Shruti Banswal, and Archana Pandey

Abstract

The human T-lymphotropic virus type 1 (HTLV1), or human T-cell leukemia virus type 1, is cancer-causing replication competent human retrovirus. The first retrovirus (discovered by Gallo's et al.) ever linked with a human cancer affecting ten million globally. Virus has sporadic distribution of endemic areas, origin through founder effect, and exemplifies RNA virus genome and replication. It is an enveloped virus with 80–100 nm diameter of the virion having two covalently bonded ssRNA strands each of 9 kb within the nucleocapsid core. The virus evades the immune surveillance through smart molecular mechanism and is latent in the infected persons for a prolonged period. In 2–5% cases at a later stage, it manifests as severe to potentially fatal two discrete types of diseases: cancer of mature CD4+ T cells called adult T-cell leukemia–lymphoma (ATL) and a chronic inflammatory central nervous system disease HTLV1-associated myelopathy/tropical spastic paraparesis. HTLV1 is transmitted primarily through infected body fluids, including breast milk (vertical transmission—mother–child), blood, semen, cervical secretions, and organ transplant (horizontal transmission). In addition, there is a lack of therapeutic regime and prevention of transmission routes. Restricted breast feeding or formula feed with screening of potential blood and organ donors; use of condoms to prevent sexual transmission can contain the virus across the world. Surveillance for HTLV-1 based on routine reporting of new diagnoses of HTLV-1 or associated diseases or prevalence sur-

S. Jetly (✉) · S. Singh · A. Pandey
Department of Biomedical Science, Acharya Narendra Dev College, University of Delhi, New Delhi, India
e-mail: sunitajetly@andc.du.ac.in

M. Khatri · S. Banswal
Department of Biomedical Science, Shaheed Rajguru College of Applied Sciences for Women, University of Delhi, Delhi, India

P. L. Bhukya, A. S (eds.), *Emerging Human Viral Diseases, Volume II*,
https://doi.org/10.1007/978-981-97-4480-0_27

veys if adopted by each country, in addition a proper registry for cancer (ATL), would contain the virus across the globe.

Keywords

HTLV1 · ATL · HAM/TSP · Human T-lymphotropic virus type 1 · Retrovirus

27.1 Introduction

HTLV-1, also recognized as human T-cell leukemia virus type 1 (HTLV-1) according to Poiesz et al. (1981), is a retrovirus linked to cancer in humans. Notably, it was the first retrovirus associated with human cancer and is estimated to impact five to ten million individuals worldwide (Legrand et al. 2022). This virus initiates lifelong infections that start asymptomatic but eventually leads to two distinct and potentially life-threatening conditions: adult T-cell leukemia–lymphoma (ATL), affecting mature CD4+ T cells, and HTLV-1-associated myelopathy/tropical spastic paraparesis (HAM/TSP), a chronic inflammatory disease of the central nervous system (CNS) (Miura et al. 2022). Initially isolated from cultured lymphocytes of a patient with cutaneous lymphoma, the discovery of this actual human retrovirus was first reported by Robert C. Gallo and his team in 1980, subsequently confirmed by Yorio Hinuma and colleagues in 1981.

27.1.1 Discovery and Brief History

The revelation of HTLV-1 during the late 1970s and early 1980s indeed holds a fascinating historical narrative. During this period, research primarily focused on comprehending the genetic, molecular, and viral processes of oncogenesis utilizing experimental animal models, such as mice, chickens, and cells derived from these animals. The knowledge and the tools used could not be a straight jacket application to human diseases. Indeed, during the Cold Spring Harbor Symposium titled "Viral Oncogenes," all 125 presentations were solely focused on animal viruses and their association with cancers (Frisch 1970). During that period, the sole recognized cancer connected to the virus was Burkitt's lymphoma (Epstein et al. 1964). However, oncogenesis had rich literature on oncogenes (without any involvement of virus), retroviruses, and tyrosine kinases without any clue to virus being involved in human cancer.

The identification of reverse transcriptase (RT) by Baltimore (1970) and Temin HM and Mizutani (1970), in animal oncogenic RNA virus had by then spurred the scientific community to discover a similar human retrovirus, and in the 1970s, several publications claimed to have discovered the human retrovirus which were, however, discarded considering them arising in laboratory due to contaminations with animal retroviruses or mistaken identity of the mitochondrial DNA-polymerase γ as RT. DNA-polymerase γ was mistaken as it is a normal cellular enzyme sensitive to ribonuclease activity, since it uses RNA primer but does not use RNA as template. It was reported and characterized by Gallo's and his team working on stimulated T lymphocytes (Bobrow et al. 1972). Arthur Weissbach also independently reported the same enzyme and later named it as DNA-polymerase γ (Fridlender et al. 1972).

In fact, the desire to discover human RNA oncogenic viruses fell into disrepute because of numerous and erroneous publications on retrovirus and the scientific community talked of "human tumor viruses" as "human rumor viruses" (Voisset et al. 2008). The conviction that human retrovirus does not exit was so ingrained that in one of his reviews, Gallo (2002) recounted how their initial submission of the HTLV1 report to the Journal of Virology was promptly declined by the editor, Robert Wagner. "Insisting that they should cease, and not continue to perpetuate the controversy, strongly implying that we all know human retroviruses do not exist." Driven by his pursuit to uncover human retroviruses, Gallo engineered a broad and highly responsive method to detect ribonuclease-sensitive DNA polymerase, distinct from viral RNA-directed DNA polymerase (Bobrow et al. 1972). Alongside his postdoctoral fellows Doris Morgan and Frank Ruscetti, Gallo's team identified T-cell growth factor, later termed interleukin 2 (IL-2), in 1976 (Morgan et al. 1976). These pivotal discoveries formed the basis for confirming the existence of human oncogenic retroviruses.

Ultimately, Gallo and colleagues submitted their initial paper on HTLV1 to PNAS on August 4th, 1980, and it was published in the December issue of that year (Poiesz et al. 1981). Shortly after, in tandem with Bernard J. Poiesz, another postdoctoral fellow, Gallo submitted a subsequent paper to Nature in May 1981, detailing the isolation of HTLV1 from freshly cultured cells obtained from Sezary T-cell leukemia patients (Poiesz et al. 1981). These reports on HTLV1 were later published in the November 19th issue of Nature. During that period, HTLV1 was commonly referred to as HTLV. Concurrently, Hinuma et al. communicated a paper to PNAS demonstrating antibodies against an antigen in a T-cell line, MT-1, taken from a patient with adult ATL (Kalyanaraman et al. 1982). Gallo and his team reported that patients with ATL, a condition linked to a proposed retrovirus causing ATL, exhibited responses to HTLV1 antibodies. Hinuma and his group in fact characterized the retrovirus and detected the proviral DNA in the fresh peripheral lymphocytes of patients with ATL (Hinuma et al. 1981), but their paper was published a month later than Gallo's demonstration of antibodies of HTLV1 against ATL patients. Miyoshi et al. (Hinuma last author) in fact submitted a paper in Nature (published in December 1981) about the virus transmission from MT-1 cells (female) to cord blood cells of a male infant transforming (immortalizing) the latter cells (Miyoshi et al. 1981). However, the conflict was resolved when in the November 1982 issue of Nature, both Gallo's group together and the Japanese colleagues Nakao, Miyoshi, Minowada, Yoshida, and Ito stated that HTLV1 and ATLV were one and the same virus and decided to call both viruses as HTLV1. Thus, Gallos and co-workers are truly credited with the discovery of HTLV1. This group a year later reported HTLV-2, in a patient with hairy cell leukemia. The discovery of retrovirus paved the way for proving AIDS as an etiological agent of HIV, a thought which would not have occurred without the isolation of HTLV1. The HTLV1, being the first discovered human pathogenic retrovirus, has been felicitated by reprinting the paper (Poiesz et al. 1981) as a part of 100th anniversary series of PNAS as mark of respect. In 1982, Gallo received the Lasker Award (Hofschneider M 1982): "For his pioneering studies that led to the discovery of the first human RNA tumor virus (older name for retroviruses) and its association with certain leukemias and lymphomas," but did not merit enough in the eyes of team who had to award Nobel Prize.

27.2 Epidemiology

27.2.1 Geographical Distribution and Demography

HTLV-1 stands as the initial human retrovirus to be discovered, widely distributed globally, and exhibiting concentrated prevalence in specific regions, such as southern Japan, the Caribbean, parts of South America, tropical Africa with focal points in the Middle East, as well as in Australia and Melanesia. HTLV-1 displays notable genetic stability, although some nucleotide substitutions are discernible among different virus strains, which are related to the geographical origin of that strain. HTLV-1 has its origins in a simian reservoir, initially transmitted to humans through interspecies transmission from monkeys infected with STLV-1 (simian T-cell leukemia virus type 1) in the Old World (Vandamme et al. 1998). HTLV1 entered in South America at the beginning of slave trade from the African continent through the east coast of Brazil. The frequent interspecies crossings are distinguished by the major geographical genotypes and subtypes. The Cosmopolitan genotype HTLV1 can be subdivided into geographically related subgroups, such as Transcontinental (a-TC), Japanese (a-Jpn), West-African (a-WA), North-African (a-NA), and Senegalese (a-Sen). The HTLV-1 genotype b is prevalent in Central Africa, serving as the predominant genotype in countries, such as Gabon, Cameroon, and the Democratic Republic of Congo. HTLV1c is the most divergent genotype, and found exclusively in Australo-Melanesia. HTLV1d genotype represents only a few percent of the strains present in Central African countries. Genotypes-e, -f, and -g are present sporadically in particular in Cameroon, Gabon, and the Central African Republic (Afonso et al. 2019). Genetic diversity in between the subtypes is quite low. The spread of HTLV-1 from the Old World to the New World, Japan, the Middle East, and North Africa is facilitated by the rising trends of human immigration, increased mobility, and the historical slave trade.

27.2.2 Origin of Infection and Spread of Disease

The primary mode of HTLV-1 transmission involves infected body fluids, which encompass breast milk, blood, semen, cervical secretions, and organ transplants (Goncalves et al. 2008; Armstrong et al. 2012; Ahmadi Ghezeldasht et al. 2021). HTLV1 has two routes of transmission horizontal and vertical. Vertically, it is transferred from a mother to child primarily through breastfeeding, although transmission is possible before or during birth too (uterine transfer). Infants who are breastfed from infected mothers face a fourfold higher likelihood of infection compared to those who are bottle-fed.

Horizontal transfer of HTLV1 occurs through sexual contact. The transmission requires entry through a mucosal barrier. Transmission of the virus can occur through compromised or infected mucosal surfaces. Male-to-female transmission has been found to be more efficient in case that men have a history of penile sores or ulcers (Roucoux et al. 2005). Semen comprises various cells susceptible to HTLV-1 infection, including CD4+ T cells, macrophages, and dendritic cells, all of which play a significant role in sexual transmission. Women who have been found to be positive with HTLV1 infection; virus has been identified in cervical inflammatory secretions and cervix carcinoma. Moreover, this virus can horizontally spread through procedures, such as whole blood transfusion, injection of cellular products, or the sharing of needles among intravenous drug users. HTLV-1 infection may also be transmitted through transplant donors with associated risk factors (such as immigrants born or residing in endemic infection areas, travelers to these regions, and families of immigrants or travelers) if the donor is not screened for the virus before transplantation. Reports from organ donors spreading infection are also replete.

Unprotected sexual intercourse, blood transfusions, organ transplants, infected needle sharing, and injecting drug are the risk factors. The efficiency of sexual transmission of HTLV-1 is higher from men to women than from women to men, with a ratio of approximately 5:1. A clarity is underlined that HTLV1 is not transmitted by casual contact, hugging, kissing or sharing of utensils, reading glasses, etc.

27.2.3 Age, Mortality, and Morbidity

HTLV1 causes two severe diseases, namely, a cancer, ATL, which is lymphoproliferative disorder of mainly CD4+ T-cell expression seen in 5% of the infected patients. Another is HTLV1 associated myelopathy/tropical spastic paraparesis (HAM/TSP) which is a progressive neurological disease with spastic weakness of the lower limbs, lower back pain, and bowel and bladder dysfunction (Osame et al. 1986). In most of the people, the virus remains latent and infected persons show no symptoms, but in 5% of cases, HTLV1 progresses to symptoms and infection, which can cause severe morbidity and mortality of infected patients. In certain highly endemic regions, HTLV-1 infection is associated with other diseases, such as uveitis in Japan and infective dermatitis in Jamaica, Brazil, and Africa. Japan stands as the most endemic country for HTLV-1, reporting a higher annual incidence of ATL cases than any other nation. ATL is more prevalent in men, while HAM/TSP tend to be reported more frequently in females than males. There is a characteristic age and gender dependence of HTLV1 sero-prevalence in many populations. HTLV1 prevalence is found to be increased with age and is found high in women. HTLV-1 infection is most widespread in populations characterized by low geographic mobility, with higher rates of vertical and sexual transmission accordingly.

27.3 Potential Risk of Emergence and Re-emergence

27.3.1 Emergence of Disease

Disease emergence, dictating outbreaks, disease persistence, and rising incidence hinge on various factors—ecological, environmental, and demographic. Understanding HTLV-1 emergence demands contextual insight into the interplay of these factors and their geographical trends. Presently, The WHO estimates around ten million HTLV-1 infections globally. However, this figure might be an underestimate due to a lack of reliable data on a global scale. The scarcity of standardized screening and detection protocols worldwide, along with diverse and ever-changing analytical methods, limits the validity of geographical and time series analyses. Representative population surveys with substantial sample sizes are also relatively scarce. Moreover, many infected individuals either remain asymptomatic or do not exhibit conditions directly linked to casual infection. This is mainly associated with individual immune response status and the virus's prolonged latency periods. The prevalence of asymptomatic carriers and the lack of screening/diagnostic methods highlight the skewed figures of infected individuals. The manifestation of the disease varies widely, particularly in numerous endemic pockets within nutritionally deprived countries. Moreover, this virus presents several subtypes. Individuals infected with subtypes A and B are widespread globally and are endemic in regions, such as Japan, the Caribbean, South America, and Africa. While pockets of HTLV-1 have been discovered in the United States, Romania, Iran, and, more recently, the UK, HTLV-1c, identified in 1993 in Melanesia and later in New Caledonia and Australia, seems confined to Oceania. The geographic distribution appears sporadic, lacking apparent connections among many emergence areas. Notably, HTLV-1c exhibits a high prevalence, reaching up to 50% in both males and females within underserved communities in Australia's Northern Territory. It has been reported that HTLV-3 likely emerged independently in the remote indigenous populations of Oceania around 40,000 years ago, potentially due to the founder effect. The primary causes contributing to the emergence of the disease can be summarized as

- Human demographic change (inhabiting new areas) as shown in New Caledonia and Australia as well as in West Africa (Founder's effect). This seems to be the probable reason for asymptomatic carriers initiating the disease in endemic pockets.
- The innate human behaviour of breastfeeding (vertical transmission) seems to be the primary cause of mother-to-child transmission (MTCT). In many instances, mothers, despite being asymptomatic, may not have the means to opt for alternatives due to resource constraints in countries with inadequate nutrition. In addition, while transmission from mother to child can happen through the placenta during childbirth, the most common route is through the ingestion of breast milk.

- Human susceptibility to infection in case of individuals with compromised immunity either due to medication or weakened immune system, due to medical condition or stress or abusive lifestyle.
- Nutritional changes mainly because some endemic pockets have been found in nutritionally deprived countries, such as Ethiopia and Africa. In addition, more use of pesticides encourages mutations in the virus.
- Other factors contributing to emergence of virus (pathogen) include: evolution of pathogenic virus (originates from a simian reservoir STLV1 by zoonotic transmission from non-human primates), mutations within viruses, development of resistance to drugs and resistance of vectors to pesticides.

27.3.2 Re-emergence of Disease

HTLV1 infection which was previously found to exist at very low levels in some populations and was not a public health hazard is now showing an upward trend of incidence and/or prevalence. They exemplify re-emergence of disease in several endemic pockets of the world. Molecular phylogenetic studies have clearly proved beyond doubt that HTLV1 originated in West Africa. The virus originates from zoonotic sources, with transmission occurring between non-human primates and humans. Its re-emergence appears to be zoonotic in nature. Investigations into the resurgence of HTLV-1 revealed that out of 78 individuals in Gabon who were bitten by nonhuman primates during hunting, seven tested positive for HTLV-1 infection (Kazanji et al. 2015). Among these cases, five individuals bitten by gorillas were infected with the HTLV-1 subtype B strain. Notably, a 12-year-old girl, severely bitten by a *Cercopithecus nictitans*, was infected with the subtype D strain, closely related to the STLV-1 virus that affects monkeys. Her mother, in contrast, was infected with the subtype B strain. This clearly shows that severe bites from nonhuman primates represent a potential risk for emergence of new endemic regions. The African origin of HTLV-2 finds support in the molecular characterization of the virus observed in Cameroonian pygmy tribes. The most affected regions include East, Central, and West Africa. Subsequently, the virus spread from Austro-Melanesia to Asia and eventually reached North and South America (how this occurred remains unclear). Conversely, HTLV-1C, identified initially in 1993 in Melanesia and later in New Caledonia and Australia, seems confined to Oceania. There is evidence, suggesting that HTLV-1 evolved independently in the remote indigenous populations of Oceania over the last 40,000 years, likely due to the founder effect. Notably, there is a notably high prevalence of HTLV-1C, reaching up to 50% in both males and females within underserved communities in Australia's Northern Territory.

Spain is not considered an endemic region for HTLV-1 infection. However, in 2008, there was a notable surge in the incidence of HTLV-1 infections in Spain. The majority of those affected were immigrants or tourists originating from endemic regions, primarily Latin America (Treviño et al. 2012; Esparza-Echevarría and Soriano-Vázquez 2017). In addition, instances of HTLV-1 infection in Spain have been reported among patients who received organ transplants from infected donors. This suggests that the resurgence of the virus in Spain may be attributed to migratory movements from endemic areas and racial mixing, leading to the acquisition and transmission of the virus.

27.4 Organization of Infectious Agent

27.4.1 Taxonomy

Human T-lymphotropic virus type 1 (HTLV1) is a member of delta retrovirus genus belonging to family *Retroviridae*, subfamily Oncovirinae which has four species: Bovine Leukemia virus (BLV) and Primate T-lymphotropic viruses 1, 2, and 3 (ICT International Committee on Taxonomy of Viruses 2022), which also contains other closely related viruses, such as HTLV2, HTLV3, HTLV4, and simian viruses STLV1, STLV2, STLV3, and STLV4 (Delta retrovirus 2007). It is believed that HTLV and STLV have a common ancestor (Vandamme et al. 1998). Along with them is also grouped BLV. Furthermore, it may be possible that some other related viruses that occur in monkeys and apes may be transmitted to humans. Retroviruses are those viruses that have RNA as their genetic material and this is converted to double-stranded DNA using RT. This ds DNA gets integrated into host cells. Once stably integrated, then it transforms DNA of the host cell which loses regulatory control of cell division and keeps dividing repeatedly or becomes tumorous.

27.4.2 Structure of Virion

HTLV1 is an enveloped virus with an envelope made up of two major proteins gp21 and gp46 (Fig. 27.1). The diameter of the virion is approximately 80–100 nm. The virion has two covalently bonded ssRNA strands each of 9 kb and they are contained within the nucleocapsid made up of protein p24. The RNA has gag *pol env pro* genes coding for important proteins (discussed later in chapter). Inner to the

Fig. 27.1 Schematic diagram of HTLV 1/2 showing capsid proteins and genomic RNA. See text for details (Brites et al. 2021)

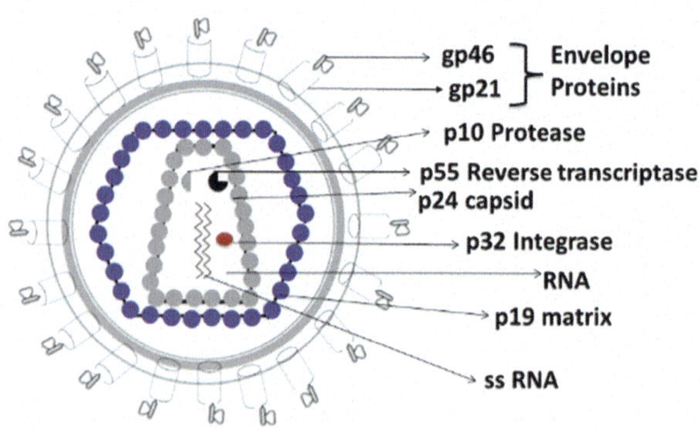

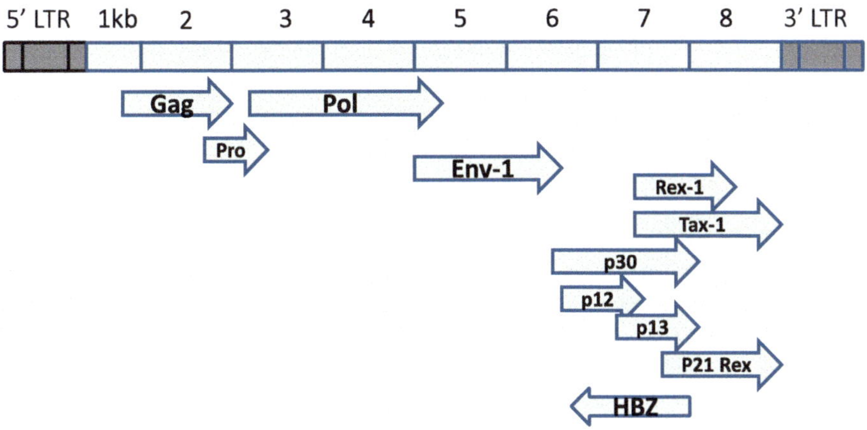

Fig. 27.2 Genome and transcript of HTLV1 along with translated proteins. [Adopted from/courtesy Martinez et al. (2019)]

envelope is a membrane-associated matrix protein layer composed of p19. The RNA strands are complexed with viral enzyme RT (p55 with associated RNase H activity), integrase (p32), protease (p10), as well as capsid proteins (p19 and p24). HTLV1 thus has structural genes *gag*, *pol* and *env*. The accessory genes include *p12, p11, p13, p21. P30*, HTLV1 Basic Leucine Zipper (*HBZ*), *tax* and *rex*. However, *tax* and *rex* are now identified as regulatory genes.

The *gag* gene encodes precursor protein (53 kDa) which is translated from unspliced genomic RNA. This precursor is a polyprotein which is then cleaved into smaller proteins, namely, matrix protein (p19), capsid protein (p24), and nucleocapsid (p15). This cleavage occurs in golgi, after protein is translated with the help of viral protease.

The *pol* gene codes for RT (p55 with associated RNase H activity), integrase (IN, p32), and protease (PR, p10). The *env* gene encodes a transmembrane protein (21 kDa) and surface protein (SU, 46 kDa). At the 3′ end of mRNA transcript after *env* gene, there is a "*Px*" region that has four partially overlapping Open Reading Frames (ORFs) coding for accessory proteins p12, p13, p21, p30, HBZ, and regulatory proteins Tax and Rex. The 5′ and 3′ ends of the proviral genome are flanked by LTRs (Fig. 27.2). Proviral genome is the viral genome existing in a virus before integration into host cells.

27.4.3 Target Cells and Host Range

HTLV1 can infect many different types of human cells in vitro under aseptic tissue culture and conditions which includes T cells, B cells, dendritic cells, and fibroblasts. However, T cells can be infected both in vivo and in vitro. Clonal proliferation of T lymphocytes occurs mainly in CD-4 positive cells and to a lesser extent in

CD-8 positive cells (Etoh et al. 1997; Cavrois et al. 1998). Furthermore, proliferation is assisted by both regulatory genes (*tax* and *rex*) and accessory viral genes such as *p12*, *p13*, *p21*, *P30*, and *HBZ*. HTLV1 infects humans; however, experimentally, it is known to infect many animals, including rabbits, rats, mice, and new world monkeys (Lairmore et al. 2005).

27.4.4 Role of Accessory Proteins

At the 5′ end of the proviral RNA is located a promoter that encodes regulatory proteins Tax and Rex. Another promoter lies in the 3′ LTR region of the proviral RNA and enhances antisense transcription of HBZ gene. This gene codes for HBZ. This protein is essential for maintaining malignancy in HTLV1-infected T cells. For a long time, p12, p11, p13, p21, and P30 were termed "accessory" or non-essential proteins as their function was not known. However, the name is misnomer as on date their functions have been elucidated.

Protein p12: It has been associated with golgi apparatus and ER and increases intracellular levels of Ca^{2+} ions thus stimulating proliferation and survival of virus. P12 protein is also known to help in viral evasion of host immune response.

Protein p13: It is associated with the inner mitochondrial membrane and is known to modulate Ca^{2+} ions homeostasis. In addition, p13 provokes entry of K^{+} ions within the mitochondrial matrix, significantly depolarising membrane potential. The massive changes in ion charge favour production of reactive oxygen species. In addition, p13 binds to Tax and inhibits its transcriptional activity which could lead to turn over of infected cells and balance between viral latency and productive infection.

Protein p30: It acts as a negative regulator and inhibits export of Tax/Rex RNAs which results in slower protein synthesis of both proteins. This protein also inhibits viral transcription and decreases production of virions.

HTLV1 Basic Leucine Zipper (HBZ): HBZ, a recently discovered protein with 30 kDa molecular weight, consisting of 209 amino acids. At its C-terminal part, it has leucine zipper domain and at N-terminus transcriptional activator domain. HBZ is encoded by the antisense strand of the single-stranded viral RNA and is transcribed by 3′ promoter. It mainly regulates *tax*, *gag* and *env* (Matsuoka et al, 2007) (Table 27.1).

27.4.5 Regulatory Proteins

Rex is a 21 kDa protein which represses splicing of viral RNA and promotes their nuclear export. During the early phase of infection, this protein is produced and induces expression of essential proteins. In vivo experiments clearly demonstrate that Rex is essential for virion production. Rex protein is also essential for initiation of viral RNA replication. In addition, this protein controls viral latency, gene expression and is a post-transcriptional regulator.

Table 27.1 Genes of HTLV1 virus, its proteins, and their functions

Gene	Protein	Function
pol gene	Reverse Transcriptase P^{55} with associated RNase H activity	Converts viral ssRNA into dsDNA
	Integrase p32	Catalyzes the insertion of viral DNA into host chromosomal DNA
	Protease p10	Processing the Gag and Gag-Pol polyproteins by limited proteolysis
gag gene	Matrix protein p19	Helps in packaging of the virus
	Capsid protein p24	
	Nucleocapsid p13	
env gene	gp46	Surface glycoprotein
	gp21	Transmembrane protein
HBZ-Px Region	Protein HTLV1 Basic Leucine Zipper (HBZ)	Protein is essential for maintaining malignancy in HTLV1-infected T cells
Px Region	Protein p30	Acts as a negative regulator and inhibits export of Tax/Rex
	Protein p13	Associated with the inner mitochondrial membrane and is known to modulate Ca^{2+} ions homeostasis
	Protein p12	Increases intracellular levels of Ca^{2+} ions
tax gene	Tax protein	Activates viral transcription during the early phase of infection
rex gene	Rex protein	A positive post-transcriptional regulator essential for splicing and transport of HTLV1 mRNA

Tax is one of the most well-studied of HTLV1 proteins. Tax is a 40 kDa oncoprotein consisting of 353 amino acids. Tax is encoded by ORF IV in "pX" region and it binds to Tax Responsive Element (Tx RE). Association of Tax with TxRE stimulates the expression of HTLV1 provirus. Tax is localized in the cell nucleus but recent studies show that Tax can shuttle between nucleus and cytoplasm (Kashanchi and Brady 2005). This protein may play an important role in specific conditions like stress. Tax is very essential for initial steps of cellular transformation. Further studies exhibit that Tax induces aberrant cellular proliferation and genomic instability and inhibits DNA damage repair and apoptosis (Li and Gaynor 1999).

27.5 Life Cycle

The life cycle of HTLV1 is quite similar to the other known retroviruses. HTLV1 is a complex human enveloped retrovirus, containing double-stranded RNA as genetic material complexed with RT, integrase and protease, and capsid proteins. The positive or sense strand has the retroviral genes as stated above coding for structural proteins which include the Matrix (MA), Capsid (CA), and Nucleocapsid (NC) proteins encoded by *gag* gene, a viral protease responsible of facilitating the maturation of viral particles encoded by *pro* gene, RT, RNaseH (RH) and Integrase (IN) proteins encoded by *pol* gene and gp46 Surface Unit (SU) and gp21 Trans-membrane Unit (TU) proteins encoded by *env* gene. The negative or antisense strand codes for Basic Zipper Factor (*HBZ*). Though the primary target of the HTLV1 on infection is CD4+ T cells, but it can infect a wide variety of other cell types, which include CD8+ T cells, B-lymphocytes, endothelial cells, myeloid cells, fibroblasts, dendritic cells, macrophages, nerve cells, as well as other mammalian cells (Martin et al. 2016). All these target cells express three cellular surface receptors: Glucose Transporter (GLUT1), Heparin Sulfate Proteoglycan (HSPG), and the VEGF-165 receptor Neuropilin-1 (NRP-1). The viral structural proteins gp21 (TU) and gp46 (SU), encoded by *env* gene, interact with these surface receptors present on the cell membrane of the various target cells (Hoshino 2012) (Fig. 27.3). This interaction between the viral envelope membrane and the target cell membrane leads to fusion and the entry of the viral double-stranded RNA into the cytoplasm of the target cells. On entry, the viral genomic RNA is converted into double-stranded DNA (dsDNA) by the Reverse Transcriptase (RT) proteins encoded by viral *pol* gene. The double-stranded DNA is transported to the nucleus, where it is integrated into the target cell genome, forming the provirus. The cellular RNA polymerase II forms

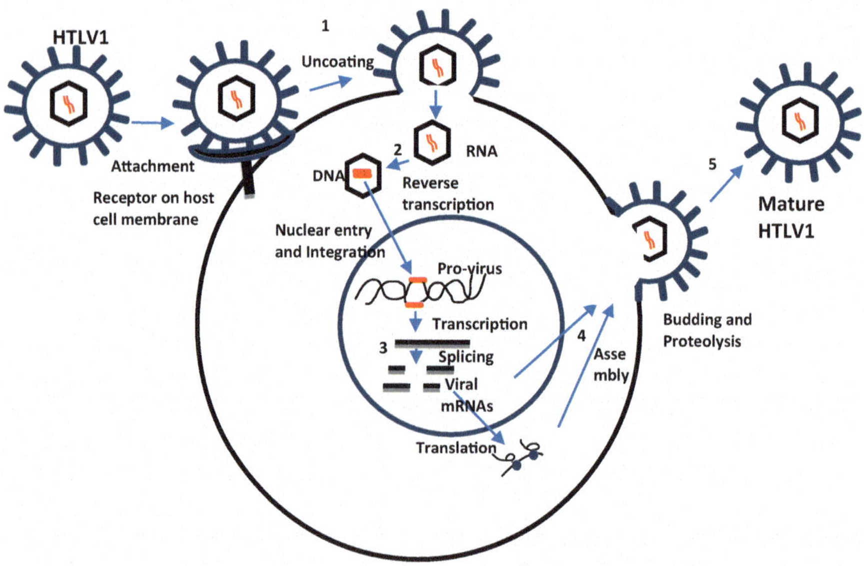

Fig. 27.3 HTLV1 life cycle. The figure highlights the important events in the HTLV1 life cycle: (1) the HTLV1 envelope proteins [gp21 (TU) and gp46 (SU)] interact with the surface receptors GLUT1/HSPG/NRP-1 present on target cells leading to attachment and fusion with the virus. (2) The viral genomic RNA is delivered into the cytoplasm and converted to double-stranded DNA by reverse transcription, which integrates into host cellular DNA in the nucleus. (3) The provirus is transcribed and spliced to form viral mRNAs which are translated in the cytoplasm. (4) The viral proteins are transported with the viral RNA to the plasma membrane for packaging. (5) These immature virions undergo final proteolysis at the budding sites and are released

viral mRNAs using the proviral DNA as template, which is further subjected to post-transcriptional modification (RNA splicing). Both the spliced and unspliced RNA molecules are exported from the nucleus to the cytoplasm where they are translated to form the viral proteins. The viral proteins gag, pol and env, together with two molecules of viral genomic RNA, are transported to the budding sites on the plasma membrane, where they form an immature virus particle. These budding particles undergo a maturation process by the action of viral proteases encoded by pro genes, which cleaves the viral polyproteins. The mature viral particles are finally released from the cell surface and are ready to infect new cells (Martin et al. 2016).

27.5.1 Propagation and Assay in In-Vitro and In-Vivo Laboratory Models

Human T-cell leukemia virus type 1 (HTLV1) was isolated in the early 1980s from T-cell lines derived from patients with cutaneous T-cell lymphoma and human ATL. Initially, all the information related to HTLV1 infection, mode of transmission, and pathogenesis were obtained from studies from viral particles produced in cells transduced with the HTLV1 proviral genome. Subsequently, HTLV1-producing cell lines (such as MT-2, MT-4, C91-PL, and SP) were generated by coculturing leukemic cells from ATL patients with human cord blood lymphocytes isolated from healthy subjects. HTLV1-infected T-cell lines have also been established from ATL patients, such as ATL-2 cells, MT-1, ATL-T, TLOm1, ED, ATL35-T, and ATL-55T (Ji et al. 2001). Infection with human HTLV1 causes ATL in a minority of infected individuals after long periods of viral persistence. The various stages of HTLV1 infection and leukemia development have been studied using several different animal models. Most commonly used are non-human primates, rabbits, rats, and mice to understand HTLV1 biology and disease progression. (1) New Zealand White rabbit model for or viral transmission, persistence, and immune responses, (2) transgenic mice to model tumorigenesis by HTLV1-specific protein expression (Tax and Hbz inducing lymphoma and other lymphocyte-mediated diseases), (3) immune-deficient mice for transplanting ATL cell lines, and (4) infection of humanized mice with HTLV1 (Niewiesk 2016). Apart from these non-human primates are phylogenetically similar to humans, they are good for viral persistence.

27.6 Pathogenesis in Humans

The significance of investigating a pathogenic virus lies in comprehending its pathogenesis, pathophysiology, and the progression of disease symptoms. This knowledge is essential for identifying potential therapeutic drug targets, contributing to a comprehensive approach to managing the disease.

27.6.1 Host–Virus Interaction

The transmission of primary HTLV-1 infection occurs between virus-infected CD4+ T cells and healthy ones through a specialized compartment known as virological synapses (VS) during cell-to-cell contact. The predominant modes of transmission include breastfeeding, blood transfusion, sexual intercourse, and less commonly, through organ transplants. HTLV-1-infected cells are found in genital secretions, including vaginal mucus and semen (Gotuzzo et al. 1994). An elevated quantity of HTLV-1-infected cells in semen escalates the likelihood of transmission from males to females and between males (Kleinman et al. 1993). Engaging in unprotected sexual activity, having multiple sexual partners, and experiencing genital ulcers heighten the transmission of HTLV-1. Research findings indicate that the transmission rate of HTLV-1 from males to females was 60.8%, contrasting with the transmission rate from females to males (Moriuchi et al. 2001; Moriuchi and Moriuchi 2004).

Infections triggering inflammatory responses enhance the recruitment of lymphocytes, notably CD4+ T cells, thereby facilitating HTLV-1 transmission. Individuals aged 45 and above, male gender, and a high count of HTLV-1-infected cells can raise the proviral load (PVL) in genital secretions, amplifying transmission risks. The fluid in these secretions fosters viral replication and transmission (Moriuchi et al. 2001; Moriuchi and Moriuchi 2004).

HTLV-1 transmission commonly occurs through plasma contaminated with white blood cells and transfusions of whole blood cells. Intravenous exposure to HTLV-1-infected blood poses a significant risk for viral transmission. Conversely, the transmission rate may be reduced when using plasma alone or cold-stored whole blood. Successful HTLV-1 transmission typically requires a minimum of $\geq 9 \times 10^4$ HTLV-1-infected cells in the host. The risk of transmission is heightened during packed cell infusion and needle sharing among intravenous drug abusers (Paul and Seder 1994).

27.6.2 Host Immune Response (Innate and Adaptive)

Persistent infection and the manifestation of disease are linked to HTLV-1's capacity to elude the immune system, coupled with a heightened rate of viral replication leading to an elevated PVL (Carneiro-Proietti et al. 2014). According to the literature, proteins derived from ORF-I play a role in conferring immune evasion capabilities to HTLV-1-infected T cells, subsequently promoting their proliferation. Alternatively, spliced mRNAs encoding proteins are expressed from the 3′ end of the viral genome, encompassing ORFs I–IV. Two proteins, p8 and p12 (8 kDa and 12 kDa forms), are encoded by ORF-I, each exhibiting distinct cellular localization and functions. Following translation, p12 is retained in the endoplasmic reticulum (ER), whereas p8 is transported to the cell surface (Paiva and Casseb 2014).

Infected T cells come under recognition by cytotoxic T-lymphocytes (CTLs) through the interaction between MHC I and the TCR. HTLV-1 proteins encoded by

ORF-I, specifically p8 and p12, possess the capability to bind with MHC-I in the ER, leading to retro-translocation into the cytoplasm and subsequent proteasome-mediated degradation. This process results in reduced expression of MHC-I on the infected cell surface allowing evasion of CTL activity. Consequently, HTLV-1-infected cells become vulnerable to the response of Natural Killer (NK) cells due to the down-regulation of MHC I molecules. To counteract NK cell attacks, the viral Tax protein intensifies the expression of MHC I. It is suggested that both p12 and Tax proteins play contrasting roles in intracellular signaling. In vitro studies have demonstrated that the presence of vector-encoded HTLV-1 p12 protein renders NK cells incapable of killing HTLV-1-infected CD4+ T cells. The compromised ability of NK cells is associated with the infected cells down-regulating ICAM-1 and ICAM-2. These molecules are crucial for the formation of an efficient immune synapse. Furthermore, the study did not observe the expression of NCR and NKG2D ligands in HTLV-1-infected cells (Moriuchi et al. 2001). The clonal expansion of infected T cells contributes to sustaining HTLV-1 infection.

27.6.3 Virulence and Persistence

The proteins derived from ORF-I of HTLV-1 stimulate the proliferation of T cells by activating the JAK/STAT pathway, thereby diminishing the reliance on IL-2, a crucial factor for T-cell proliferation. The p12 protein plays a significant role in elevating the cytoplasmic concentration of Ca^{2+}, essential for activating the transcription factor NFAT, which is involved in T-cell proliferation. Increased Ca^{2+} levels activate the enzyme calcineurin, which dephosphorylates NFAT. Dephosphorylated NFAT translocates to the nucleus, initiating the expression of the IL-2 cytokine. Both STAT5 and NFAT efficiently bind to the IL-2 promoter, amplifying the expression of this cytokine, which is essential for HTLV-1 infection (Zarranz et al. 2001).

The association between the p12 protein and the cytoplasmic domain of the IL-2 receptor initiates IL-2 signal transduction, augmenting the DNA-binding activity of STAT5 even in the absence of IL-2. Research has demonstrated that, even at low IL-2 concentrations, proteins encoded by ORF-I promote the proliferation of HTLV-1-infected cells (Takemoto et al. 2000; Nicot et al. 2001; Taylor et al. 2009). The comprehensive supportive data strongly suggests that proteins encoded by ORF-1 play a role in modulating intracellular signaling, thereby decreasing the reliance of infected T cells on IL-2 for their proliferation and multiplication. Infected T cells exhibit a growth advantage facilitated by these HTLV-1 accessory proteins, particularly in the early stages of infection before reaching the immortalized T-cell state (Grossman et al. 1995).

27.6.4 Clinical Manifestation Phases of Disease

The majority of individuals infected with HTLV-1 remain asymptomatic but can effectively transmit the virus to healthy individuals. Lobulated or flower-shaped nuclei represent a significant morphological characteristic of HTLV-1-infected T cells. Southern hybridization has revealed that clonal populations of T cells harbor HTLV-1 proviral DNA, and these T-cell clones may undergo malignant transformation. Approximately half of pre-leukemic individuals infected with HTLV-1 experience spontaneous deterioration, with some progressing to chronic/smouldering ATL. Patients with chronic ATL exhibit an elevated number of circulating leukemic cells, while those with smouldering ATL typically present with bone-marrow involvement and skin lesions. Within a month, patients with chronic or smouldering ATL can progress into a highly aggressive form of acute leukemia. Clinically, this progression is characterized by elevated levels of lactate dehydrogenase, hypercalcemia, skin lesions, lymphomatous meningitis, lymphadenopathy, lytic bone lesions, spleen or liver involvement, and immunodeficiency.

Occasionally, ATL and HAM/TSP have been observed to coexist in the same patient. HAM/TSP is a progressive nervous system disorder affecting fewer than 2% of HTLV-1-infected individuals. HTLV-1 extracted from patients with ATL-HAM/TSP does not exhibit any discernible genetic or biological differences. CNS damage in individuals with HAM/TSP is linked to the targeted elimination of infected cells by the host immune system. Several studies have reported the presence of HTLV-1-specific CTLs in HAM/TSP patients. HTLV-1 has demonstrated the ability to infect neuronal cells under both in vitro and in vivo conditions. The activation of autoimmune or cytokine-mediated mechanisms may indirectly contribute to neuronal cell damage. Elevated levels of GM-CSF, IFN-γ, TNF-α, and IL-1 cytokines have been identified in patients exhibiting symptoms of HAM/TSP.

27.6.5 Mechanism of Pathogenesis

The proliferation of T cells is induced by proteins encoded by HTLV-1, and research indicates that these activated and dividing cells are more susceptible to HTLV-1-mediated infection compared to non-dividing or resting cells. T cells derived from individuals with HAM/TSP or asymptomatic carriers exhibit spontaneous proliferation in culture, potentially playing a crucial role in establishing viral infection post-entry into T cells. T-cell proliferation and activation occur independently of accessory cells and are mediated by LFA-3 (CD2/Lymphocyte function-associated molecule-3), LFA-1/intracellular adhesion molecules, IL-2, and IL-2R. HTLV-1-infected T-cell clones from patients have demonstrated spontaneous proliferation for up to 2 weeks after stimulation without the need for exogenous IL-2. This proliferative capacity in infected cells is entirely independent of the IL-2/IL-2R pathway. HTLV-1 is reported to possess mitogenic abilities, activating quiescent human T cells. Even heat-inactivated virions retain this mitogenic ability, suggesting that T-cell activation may be associated with virion interaction with specific cell surface receptors. While the CD2 receptor has been proposed to be involved in viral-induced cellular activation, it is not the primary receptor for HTLV-1 infection, as CD2-negative cells are also susceptible.

In vitro, peripheral blood T cells infected with HTLV-1 undergo immortalization and become independent of IL-2 requirements. However, these infected cells exhibit elevated expression levels of the IL-2 receptor α chain (IL-2Rα), a crucial characteristic feature observed in ATL and HTLV-1-transformed cells. The neoplastic growth and proliferation of ex vivo-derived ATL have been linked to the constitutive activation of JAK/STAT proteins. Leukemia manifests in a subset of HTLV-1-infected individuals after an extended latency period, indicating a slow multistep oncogenic development that results in the accumulation of mutations in proliferating T cells. There is evidence suggesting that the viral-encoded Tax protein plays a significant role in the oncogenic development of infected T cells. Experiments conducted by Grossman et al. (1995) revealed a CD8+ leukemia condition in transgenic mice expressing the T-cell-specific granzyme-B promoter. The neoplastic transformation of T cells by HTLV-1 Tax proteins requires the activated NF-κB and SRF transcription pathways. Constitutive activation of NF-κB and the phosphorylation/degradation of cytoplasmic IκB-α/β have been observed in HTLV-1-transformed cells. The Tax protein has been demonstrated to impact various cell cycle regulatory molecules. Specifically, in cell lines transformed by HTLV-1, the Tax protein both inactivates and stabilizes the tumor suppressor protein p53. This effect may be linked to the hyperphosphorylation of p53 (Takemoto et al. 2000).

The expression of Tax also induces hyperphosphorylation of the retinoblastoma protein (Rb) within transformed cells. In addition, the Tax protein stimulates the phosphorylation of other proteins, including cyclin D1-cdk4/6 and cyclin D3. These phosphorylated proteins contribute to the progression of the G1/S-phase in ATL cells, despite elevated levels of p21Cip1 (alternatively known as p21Waf1), a cyclin-dependent kinase inhibitor 1. Furthermore, Tax protein is involved in the inactivation of cdk-inhibitors such as P16Ink4a and mitotic arrest-deficiency-1 (MAD-1) protein. MAD-1 is crucial for the regulation of the anaphase-promoting complex and metaphase chromosome segregation during mitosis.

In summary, Tax plays a vital role in the accumulation of genetic mutations, hinders the activation of tumor suppressor proteins, impairs the functions of cdk-inhibitors, and stimulates viral gene expression and leukemic proliferation by modulating the entire transcriptional machinery of HTLV1-infected cells.

27.6.6 Clinical Symptoms

ATL, an uncommon and life-threatening type of cancer, results from HTLV1 infection through a multistep process of oncogenesis. The aggressive proliferation of T cells stimulated by HTLV1 is a primary factor contributing to the malignant state of cells in 2–5% of infected individuals. The development of ATL typically occurs in patients who have been infected with HTLV1 since birth. Clinically, ATLs are broadly categorized into lymphomatous, smoldering, acute, and chronic forms (Baratella et al. 2017). The smoldering form of ATL is frequently observed in the initial stages of malignancy, and it has the potential to progress into acute ATL later on (Tsukasaki et al. 2009). The primary type of ATL is benign and typically lacks symptoms, characterized by an abnormal flower-like cell shape. Chronic ATL, on the other hand, is associated with lymphadenopathy, peripheral blood leucocytosis, and hepato-splenomegaly. Both forms of malignancy have the potential to progress into aggressive acute forms, with a median life expectancy of a few months. Although long-term survival is rare, there have been reported cases. Lymphomatous and acute ATL are generally resistant to chemotherapeutic treatments. However, drugs/proteins such as interferon-alpha, zidovudine, and arsenic show some potential in treating patients, as noted by Kchour et al. in 2013. It is important to mention that these drugs do not exhibit long-term efficacy.

27.6.7 Serology, Molecular, and Histological Approaches

HTLV-1 can infect different cell types, such as dendritic cells, monocytes, CD8+ T lymphocytes, and macrophages. Yet, the primary reservoir for the virus is the CD4+ T cells. Within these infected CD4+ T cells, the virus can lie dormant for a long time, showing slow replication. This latency induces cell proliferation, triggers genetic changes in the cell, causes damage to the neuronal system, and mediates the inflammatory response of the immune system, as outlined by Olah et al. in 2007. The immunopathogenesis of HAM begins following the infection of T-lymphocytes (CD4+ and CD8+) cells by HTLV1, as noted by Goncalves et al. in 2008. This infection stimulates the production of proinflammatory cytokines, including interferon (IFN)-γ, interleukin (IL)-1b, tumor necrosis factor (TNF)-α, IL-12, and IL-6. All these cytokines play roles in the observed inflammatory immune responses following HTLV1 infection, as highlighted by Ahuja et al. in 2007 and Montanheiro et al. 2009. Chemokines, such as CXCL9 and CXCL-10, play a role in the pathogenesis of HAM, as discussed by Guerra et al. in 2018. The pathophysiology of HAM is explained by three main mechanisms: direct toxicity, autoimmunity, and surrounding damage, as outlined by Ijichi and Osame in 1995 and Höllsberg in 1997. According to the direct toxicity theory, glial cells infected with HTLV1 express viral antigens via MHC I on their surface, which are then recognized and destroyed by CTLs. The autoimmunity theory suggests that certain host antigens mimic viral antigens, initiating an autoimmune inflammatory response that leads to neural injury. The surrounding damage theory posits that HTLV1-specific CD4+ T and CD8+ T lymphocytes cross the blood-brain barrier, eliminating infected glial cells in the CNS and releasing cytokines that amplify the inflammatory response, as highlighted by Goon et al. in 2003.

Studies indicate that the cerebrospinal fluid (CSF) of HAM patients contains a significantly high concentration of TNF-α and INF-γ cytokines. These cytokines, secreted by the TH1 subpopulation of CD4+ T lymphocytes, contrast with the decreased concentrations of IL-2 and IL-10 cytokines secreted by the TH-2 subpopulation in patients with neurological disease, as reported by Yamano et al. in 2009. A similar elevated level of cytokines has been observed for CD8+ T lymphocytes, with these cells expressing a comparable cytokine pattern. The classification of disease activity in individuals with HAM can be based on the levels of CXCL10 and neopterin in the CSF. Those with heightened disease activity exhibit elevated concentrations of these cytokines, providing valuable information for anti-inflammatory treatment strategies (Sato et al. 2013; Tamaki et al. 2019; Yamauchi et al. 2020).

In response to the influence of TNF-α, individuals carrying HTLV-1 asymptomatically exhibit elevated levels of IL-10, a crucial immunoregulatory cytokine. The asymptomatic HAM patients with a high PVL show increased expression of both IFN-γ and IL-10 (Espíndola et al. 2015; Bidkhori et al. 2020) while asymptomatic patients with a low PVL exhibit diminished levels of both IFN-γ and IL-10 expression. These observations indicate that an imbalance between anti-inflammatory and pro-inflammatory cytokines is primarily linked to the onset of HAM. The host cell's immune response pattern to HTLV-1 infection, coupled with a high PVL, may play a role in the progression of this severe neurological disease (Goncalves et al. 2008; Grassi et al. 2011).

27.7 Diagnostics and Therapeutic Approaches

27.7.1 Laboratory Diagnosis

Serological methods are used to diagnose HTLV-1 infection by detecting antibodies to the virus. Initially, ELISA, chemiluminescence assays, or particle agglutination tests are employed to detect HTLV-1 antibodies in the blood. However, these tests cannot distinguish between HTLV-1 and HTLV-2 infections due to their high similarity.

If the initial test is positive, a second method such as western blot or line immunoassay is conducted to confirm the results. These tests identify antibodies produced against specific gag and env proteins of HTLV-1. Molecular assays, such as Real-time Polymerase Chain Reaction (PCR) or quantitative PCR (qPCR), serve as alternative methods to confirm HTLV infection. They not only confirm the presence of HTLV but also quantify the PVL. HTLV testing typically follows a sequential process, starting with an initial screening test followed by confirmatory testing based on the initial results (Table 27.2).

Table 27.2 Typical results of confirmatory and Molecular test that can be seen for HTLV1 screened subject/infected patient using serological methods to detect antibodies to the virus and QPCR

Screening test	Confirmatory test	Molecular test	Interpretation
Negative	N/A	N/A	No infection
Positive	Negative	Negative	False positive
Positive	Positive	N/A	Infection
Positive	Indeterminate	Positive	Infection
Positive	Indeterminate	Negative/Indeterminate	False positive

27.7.2 Therapeutic Approaches

Acute infection with HTLV1 is asymptomatic, as the PVL is low and usually goes undetected. A substantial yet variable period of time is necessary for the progression from acute infection to the manifestation of disease symptoms. There is a robust correlation between the HTLV-1 PVL and the clinical status of individuals. Approximately 5% of the infected individuals develop ATL and 4% develop the debilitating neurological disease HAM/TSP. A combination of zidovudine and interferon α (AZT/IFNα), with arsenic trioxide added (sometimes), has been to be found effective in some cases of chronic infection with high viremia; however, no effective treatment has been approved for the management of HTLV1 infection so far. Prolonged and intensive combination chemotherapy, including vincristine/methotrexate, prednisolone, and etoposide/cyclophosphamide, followed by allogeneic hematopoietic stem cell transplantation, has demonstrated enhanced survival and improved quality of life in patients with ATL.

Many new innovative therapies are also finding their way in the treatment of aggressive T-cell leukemia. Monoclonal antibodies such as Mogamulizumab, an approved drug in Japan for various T-cell malignancies, prompt the depletion of regulatory T cells (Treg), resulting in an enhanced antitumor immune response. In addition, it significantly augments antibody-dependent cellular cytotoxicity (Ni et al. 2015). An immunomodulatory drug, Lenalidomide (approved by FDA for treatment of multiple myeloma), has also been found to activate T cells, alter cytokine production and increase in NK cell number and cytotoxicity (Ito and Handa 2016). Research is underway on the development of potential therapeutic options, such as anti-HTLV vaccines, for ATL. The purpose of these vaccines is to activate the host's immune response against the virus, potentially offering clinical benefits. HTLV1 TAX protein plays a critical role in HTLV1-associated disease. A dendritic cell vaccine pulsed with Tax peptide (Tax-DC) has been developed to enhance the Tax-specific CTL response and is currently undergoing clinical trials.

27.8 Prevention Control and Future Perspectives

HTLV-1, as mentioned earlier, leads to ATL, HAM/TSP, and several inflammatory diseases, affecting an estimated 10–20 million people worldwide. Despite various approaches, finding a cure for ATL is still in its early stages. Even conventional chemotherapies have limited effectiveness due to the virus's transformation of CD4+ cells, its intricate spreading mechanism, and the host immune response that is not fully understood till date. For many acute subtype patients, these therapies are ineffective.

As a final option, allogeneic stem cell transplantation provides long-term disease control, but unfortunately, access to this treatment remains limited. Hence, prevention outweighs the challenges of finding a cure. It becomes imperative to prevent HTLV-1 transmission by potential carriers to effectively halt its spread. The key lies in halting its three primary transmission routes: MTCT, mainly through breastfeeding (vertical transmission); sexual transmission, predominantly from males to females (Carneiro-Proietti et al. 2014; Cavrois et al. 1998); and transfusion of cellular blood components (horizontal transmission).

27.8.1 Prevention of Vertical Transmission

The rate of MTCT stood at 20% as determined by both prospective and retrospective epidemiological studies (Itabashi and Miyazawa 2021; Kinoshita et al. 1987). Predominantly, vertical transmission occurred via breastfeeding, while transmission during pregnancy or the peripartum period was less than 5% (Fujino and Nagata 2000). A study conducted in Nagasaki prefecture, Japan, showcased that completely discontinuing breastfeeding resulted in a notably lower prevalence of HTLV-1 infection among formula-fed children, dropping from 20.3% to 2.5% compared to breastfed children. In instances where breastfeeding cessation was not observed, prevalence was significantly lower in children breastfed for under 6 months than in those fed on maternal milk for more extended periods exceeding 6 months. Thus, the best way of prevention is prenatal screening of mothers for detection of potential carriers and then educating them about modes of transmission with proper counselling to either adopt formula feed or short-term breastfeeding (3 months) or freeze thaw milk (as shown in studies of 1986 and 1989 that thawing the frozen milk eliminated the cells infected with HTLV1). Since freeze thaw method may not be cost effective in given resource poor settings with limited technology especially endemic areas like Africa, complete dependence on bottle feeding is recommended as compared to limited breastfeeding (Hino 2011). Nevertheless, it is important to note that breast milk stands as the optimal nutritional source for infants. Any strategy aimed at reducing MTCT should prioritize educating mothers about balanced measures that mitigate its impact on child health and the bond between mother and child. Therefore, focusing on identifying the factors that make children born to carrier mothers more susceptible to MTCT, thereby enabling accurate predictions of transmission, would be a forward-looking perspective compared to interventions solely cantered on nutrition (Millen and Thoma-Kress 2022). Reducing reliance on nutritional interventions becomes crucial, particularly considering the significant challenges of malnutrition, especially prevalent in poorer and developing nations. Every region affected by ATL should emulate Japan's example, as it stands alone, according to the World Health Organization (2021), in implementing a comprehensive nationwide program aimed at preventing and controlling HTLV-1 transmission from mother to child.

27.8.2 Prevention of Horizontal Transmission

HTLV1 is transmitted by contact through bodily fluid, namely, whole blood, whole blood components, organ donations, and through sexual transmission predominantly from males to females. An effective blood screening program would include as WHO says in this technical report (February 2020) "testing donations for the presence of HTLV1; deferring or excluding people as donors if there is evidence that they are at higher risk of HTLV1; or using technologies that remove infectious organisms from blood products." Twenty-three countries have already adopted this screening measure, with Japan leading as the first to enforce mandatory screening of blood donations (Blood_Services_2017_web. Pdf n.d.). Interestingly, high-income countries conducting HTLV screening among blood donors exhibit very low prevalence and incidence. Conversely, low-income nations like those in Africa, with a high prevalence of HTLV-1, face challenges in screening due to cost constraints. Thus, a proposed solution involves filter-based leukoreduction to diminish HTLV-1 transmission by eliminating infected lymphocytes. However, current research lacks authentic data on its effectiveness (Murphy 2016). Implementing a timely protocol for regular testing of donated blood could significantly contribute to managing the diverse geographical spread of the disease. This approach would complement the epidemiological understanding of HTLV-1 transmission risks through blood transfusions. Another approach to prevent the spread is to screen the organ donors for carrier state. There are very few policies and guidelines to screen organ donors. Japan, Spain, and the UK did put in place a partial organ screening program, but in most countries, organ screening has to be implemented to check this route of transmission.

Transmissions through sexual contact are a highly preventable route which requires education and counselling of sexually active individuals especially those who are sex workers or have multiple and/or unknown sexual partners. A simple use of condoms and early self-detection for HTLV1 with access to correct information about the viral infection especially in endemic areas and in other areas dominated with migrant labourers from endemic areas can be a simple step to contain one of the potent routes of transmission. This is important as these sexually active people are mostly in their reproductive age and are asymptomatic (Yoshimitsu et al. 2013).

Prevention strategy therefore involves a multidimensional approach involving the carriers on one hand with a robust public health system on the other hand maintaining a delicate balance of education, awareness, and screening (Fig. 27.4).

27.8.3 Control and Future Perspectives

Prophylactic vaccine for carriers would be an ideal approach to reduce the burden of HTLV1 and the disease caused by it. In addition, there can be therapeutic vaccines which will prevent infectious individuals from spreading the disease. However, the vaccination till date is a far cry, as viruses remain latent in asymptomatic carriers and vaccine should actually break this evasion of immunosurveillance and boost the T lymphocytes to destroy the virions. Moreover, good antigenic candidates are required to build the vaccine after thorough research and clinical trials. Vertical transmission can simply be solved by screening of carrier mothers and their counselling to adopt formula feed and/or less than 3 months of breast feeding. WHO suggests that monitoring HTLV-1 nationally or regionally can rely on regular reports

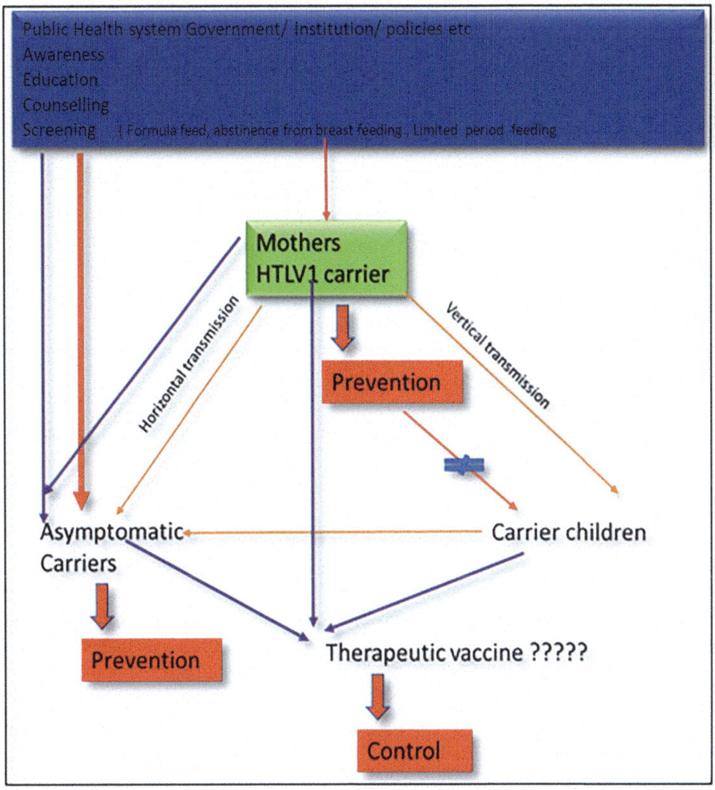

Fig. 27.4 Diagrammatic representation of different prevention strategies for HTLV1 infection. The diagram shows the interventions of public health system leading to prevention of horizontal and vertical transmission

of new HTLV-1 diagnoses, related diseases, or prevalence surveys. In addition, as shown in Japan's national cancer registry, separate coding for ATLL and HAM, other countries should adopt similar coding methods. This would facilitate the regular evaluation of ATL as a separate condition. In 2010, WHO proposed blood donor screening for HTLV-1, and similarly, there should be global guidelines for practices such as breastfeeding restrictions, ensuring blood safety with needles for injecting drug users and promoting safe sexual practices. The war can be won with the will of the governments and determination of individuals.

References

Afonso PV, Cassar O, Gessain A (2019) Molecular epidemiology, genetic variability and evolution of HTLV-1 with special emphasis on African genotypes. Retrovirology 16(1):39

Ahmadi Ghezeldasht S et al (2021) HTLV-1 oncovirus-host interactions: from entry to the manifestation of associated diseases. Rev Med Virol 31(6):e2235. https://doi.org/10.1002/rmv.2235

Ahuja J et al (2007) Induction of pro-inflammatory cytokines by human T-cell leukemia virus type-1 Tax protein as determined by multiplexed cytokine protein array analyses of human dendritic cells. Biomed Pharmacother 61(4):201–208. https://doi.org/10.1016/j.biopha.2007.02.006
Armstrong MJ et al (2012) HTLV-1 in solid-organ transplantation: current challenges and future management strategies. Transplantation 94(11):1075–1084. https://doi.org/10.1097/TP.0b013e318263ad7a
Baltimore D (1970) RNA-dependent DNA polymerase in virions of RNA tumour viruses. Nature 226(5252):1209–1211. https://doi.org/10.1038/2261209a0
Baratella M, Forlani G, Accolla RS (2017) HTLV-1 HBZ viral protein: a key player in HTLV-1 mediated diseases. Front Microbiol 8:2615. https://doi.org/10.3389/fmicb.2017.02615
Bidkhori HR et al (2020) The IL-18, IL-12, and IFN-γ expression in HTLV-1-associated myelopathy/tropical spastic paraparesis (HAM/TSP) patients, HTLV-1 carriers, and healthy subjects. J Neurovirol 26(3):338–346. https://doi.org/10.1007/s13365-020-00832-5
Blood_Services_2017_web.pdf (n.d.). https://www.jrc.or.jp/english/pdf/Blood_Services_2017_web.pdf. Accessed 1 Aug 2022
Bobrow SN, Smith RG, Reitz MS, Gallo RC (1972) Stimulated normal human lymphocytes contain a ribonuclease-sensitive DNA polymerase distinct from viral RNA-directed DNA polymerase. Proc Natl Acad Sci USA 69(11):3228–3232. https://doi.org/10.1073/pnas.69.11.3228
Brites C et al (2021) Pathogenesis of HTLV-1 infection and progression biomarkers: an overview. Braz J Infect Dis 25(3):101594. https://doi.org/10.1016/j.bjid.2021.101594
Carneiro-Proietti ABF et al (2014) Mother-to-child transmission of human T-cell lymphotropic viruses-1/2: what we know, and what are the gaps in understanding and preventing this route of infection. J Pediatr Infect Dis Soc 3(Suppl 1):S24–S29. https://doi.org/10.1093/jpids/piu070
Cavrois M et al (1998) Persistent oligoclonal expansion of human T-cell leukemia virus type 1-infected circulating cells in patients with tropical spastic paraparesis/HTLV-1 associated myelopathy. Oncogene 17(1):77–82. https://doi.org/10.1038/sj.onc.1201906
Epstein MA, Achong BG, Barr YM (1964) Virus particles in cultured lymphoblasts from Burkitt's lymphoma. Lancet (London, England) 1(7335):702–703. https://doi.org/10.1016/s0140-6736(64)91524-7
Esparza-Echevarría B, Soriano-Vázquez V (2017) Infección por el virus linfotrópico de células T humano en España—30 años de evolución (1986–2016). Gaceta Médica de Bilbao 114(3)
Espíndola OM et al (2015) High IFN-γ/IL-10 expression ratio and increased frequency of persistent human T-cell lymphotropic virus type 1-infected clones are associated with human T-cell lymphotropic virus type 1-associated myelopathy/tropical spastic paraparesis development. Intervirology 58(2):106–114. https://doi.org/10.1159/000371766
Etoh K et al (1997) Persistent clonal proliferation of human T-lymphotropic virus type I-infected cells in vivo. Cancer Res 57(21):4862–4867
Fridlender B, Fry M, Bolden A, Weissbach A (1972) A new synthetic RNA-dependent DNA polymerase from human tissue culture cells. Proc Natl Acad Sci USA 69:452–455. https://doi.org/10.1073/pnas.69.2.452
Frisch L (1970) Cold spring harbor symposia on quantitative biology the mechanism of protein synthesis section on ribosome structure, Cold Spring Harbor Long Island New York, USA, June 5 12 1969. In: Frisch L (ed) SpringHarbor symposia on quantitative biology, The mechanism of protein synthesis, xxiv, vol Xxxiv, pp 17–148
Fujino T, Nagata Y (2000) HTLV-I transmission from mother to child. J Reprod Immunol 47(2):197–206. https://doi.org/10.1016/s0165-0378(00)00054-1
Gallo RC (2002) Epilogue: A journey with blood cells and viruses. J Hematother Stem Cell Res 11(5):849–857. https://doi.org/10.1089/152581602760404649
Goncalves DU et al (2008) HTLV-1-associated myelopathy/tropical spastic paraparesis (HAM/TSP) inflammatory network. Inflamm Allergy Drug Targets 7(2):98–107. https://doi.org/10.2174/187152808785107642
Goon PKC et al (2003) High circulating frequencies of tumor necrosis factor alpha- and interleukin-2-secreting human T-lymphotropic virus type 1 (HTLV-1)-specific CD4+ T cells in

patients with HTLV-1-associated neurological disease. J Virol 77(17):9716–9722. https://doi.org/10.1128/JVI.77.17.9716-9722.2003

Gotuzzo E et al (1994) Human T cell lymphotropic virus type I infection among female sex workers in Peru. J Infect Dis 169(4):754–759. https://doi.org/10.1093/infdis/169.4.754

Grassi MFR et al (2011) Human T cell lymphotropic virus type 1 (HTLV-1) proviral load of HTLV-associated myelopathy/tropical spastic paraparesis (HAM/TSP) patients according to new diagnostic criteria of HAM/TSP. J Med Virol 83(7):1269–1274. https://doi.org/10.1002/jmv.22087

Grossman WJ et al (1995) Development of leukemia in mice transgenic for the tax gene of human T-cell leukemia virus type I. Proc Natl Acad Sci USA 92(4):1057–1061

Guerra M et al (2018) Local and systemic production of proinflammatory chemokines in the pathogenesis of HAM/TSP. Cell Immunol 334:70–77. https://doi.org/10.1016/j.cellimm.2018.09.009

Hino S (2011) Establishment of the milk-borne transmission as a key factor for the peculiar endemicity of human T-lymphotropic virus type 1 (HTLV-1): the ATL Prevention Program Nagasaki. Proc Jpn Acad Ser B Phys Biol Sci 87(4):152–166. https://doi.org/10.2183/pjab.87.152

Hinuma Y, Nagata K, Hanaoka M, Nakai M, Matsumoto T, Kinoshita KI, Shirakawa S, Miyoshi I (1981) Adult T-cell leukemia: antigen in an ATL cell line and detection of antibodies to the antigen in human sera. Proc Natl Acad Sci USA 78(10):6476–6480

Hofschneider M (1982) 1982 Winners. Lasker Foundation. https://laskerfoundation.org/winners/1982-winners/. Accessed 4 Aug 2022

Höllsberg P (1997) Pathogenesis of chronic progressive myelopathy associated with human T-cell lymphotropic virus type I. Acta Neurol Scand Suppl 169:86–93. https://doi.org/10.1111/j.1600-0404.1997.tb08156.x

Hoshino H (2012) Cellular factors involved in HTLV-1 entry and pathogenicit. Front Microbiol 3:222. https://doi.org/10.3389/fmicb.2012.00222

Ijichi S, Osame M (1995) Human T lymphotropic virus type I (HTLV-I)-associated myelopathy/tropical spastic paraparesis (HAM/TSP): recent perspectives. Intern Med (Tokyo, Japan) 34(8):713–721. https://doi.org/10.2169/internalmedicine.34.713

Itabashi K, Miyazawa T (2021) Mother-to-child transmission of human T-cell leukemia virus type 1: mechanisms and nutritional strategies for prevention. Cancers 13(16):4100. https://doi.org/10.3390/cancers13164100

Ito T, Handa H (2016) Cereblon and its downstream substrates as molecular targets of immunomodulatory drugs. Int J Hematol 104:293–299

Ji Y et al (2001) Impaired production of naive T lymphocytes in human T-cell leukemia virus type I-infected individuals: its implications in the immunodeficient state. Blood 97(10):3177–3183. https://doi.org/10.1182/blood.v97.10.3177

Kalyanaraman VS, Sarngadharan MG, Nakao Y, Ito Y, Aoki T, Gallo RC (1982) Natural antibodies to the structural core protein (p24) of the human T-cell leukemia (lymphoma) retrovirus found in sera of leukemia patients in Japan. Proc Natl Acad Sci USA 79(5):1653–1657

Kashanchi F, Brady JN (2005) Transcriptional and post-transcriptional gene regulation of HTLV-1. Oncogene 24(39):5938–5951. https://doi.org/10.1038/sj.onc.1208973

Kazanji M et al (2015) Origin of HTLV-1 in hunters of nonhuman primates in Central Africa. J Infect Dis 211(3):361–365. https://doi.org/10.1093/infdis/jiu464

Kchour G et al (2013) The combination of arsenic, interferon-alpha, and zidovudine restores an "immunocompetent-like" cytokine expression profile in patients with adult T-cell leukemia lymphoma. Retrovirology 10(1):91. https://doi.org/10.1186/1742-4690-10-91

Kinoshita K, Amagasaki T, Hino S, Doi H, Yamanouchi K, Ban N, Momita S, Ikeda S, Kamihira S, Ichimaru M (1987) Milk-borne transmission of HTLV-I from carrier mothers to their children. Jpn J Cancer Res 78(7):674–680

Kleinman S et al (1993) Transfusion transmission of human T-lymphotropic virus types I and II: serologic and polymerase chain reaction results in recipients identified through look-back investigations. Transfusion 33(1):14–18. https://doi.org/10.1046/j.1537-2995.1993.33193142303.x

Lairmore MD, Silverman L, Ratner L (2005) Animal models for human T-lymphotropic virus type 1 (HTLV-1) infection and transformation. Oncogene 24(39):6005–6015. https://doi.org/10.1038/sj.onc.1208974

Legrand N, McGregor S, Bull R, Bajis S, Valencia BM, Ronnachit A, Einsiedel L, Gessain A, Kaldor J, Martinello M (2022) Clinical and public health implications of human T-lymphotropic virus type 1 infection. Clin Microbiol Rev 35(2):e0007821. https://doi.org/10.1128/cmr.00078-21

Li XH, Gaynor RB (1999) Regulation of NF-kappaB by the HTLV-1 Tax protein. Gene Expr 7:233–245

Martin JL et al (2016) Molecular studies of HTLV-1 replication: an update. Viruses 8(2):31. https://doi.org/10.3390/v8020031

Martinez MP, Al-Saleem J, Green PL (2019) Comparative virology of HTLV-1 and HTLV-2. Retrovirology 16(1):21. https://doi.org/10.1186/s12977-019-0483-0

Matsuoka M, Jeang KT (2007) Human T-cell leukemia virus type 1 (HTLV-1) infectivity and cellular transformation. Nat Rev Cancer 7:270–280

Millen S, Thoma-Kress AK (2022) Milk transmission of HTLV-1 and the need for innovative prevention strategies. Front Med. https://doi.org/10.3389/fmed.2022.867147

Miura M, Naito T, Saito M (2022) Current perspectives in human T-cell leukemia virus type 1 infection and its associated diseases. Front Med 9:867478. https://doi.org/10.3389/fmed.2022.867478

Miyoshi I, Kubonishi I, Yoshimoto S, Akagi T, Ohtsuki Y, Shiraishi Y, Nagata K, Hinuma Y (1981) Type C virus particles in a cord T-cell line derived by co-cultivating normal human cord leukocytes and human leukaemic T cells. Nature 294(5843):770–771. https://doi.org/10.1038/294770a0

Montanheiro PA et al (2009) The elevated interferon gamma production is an important immunological marker in HAM/TSP pathogenesis. Scand J Immunol 70(4):403–407. https://doi.org/10.1111/j.1365-3083.2009.02291.x

Morgan DA, Ruscetti FW, Gallo R (1976) Selective in vitro growth of T lymphocytes from normal human bone marrows. Science 193(4257):1007–1008. https://doi.org/10.1126/science.181845

Moriuchi M, Moriuchi H (2004) Seminal fluid enhances replication of human T-cell leukemia virus type 1: implications for sexual transmission. J Virol 78(22):12709–12711. https://doi.org/10.1128/JVI.78.22.12709-12711.2004

Moriuchi M, Inoue H, Moriuchi H (2001) Reciprocal interactions between human T-lymphotropic virus type 1 and prostaglandins: implications for viral transmission. J Virol 75(1):192–198. https://doi.org/10.1128/JVI.75.1.192-198.2001

Murphy EL (2016) Infection with human T-lymphotropic virus types-1 and -2 (HTLV-1 and -2): implications for blood transfusion safety. Transfus Clin Biol 23(1):13–19. https://doi.org/10.1016/j.tracli.2015.12.001

Ni X, Jorgensen JL, Goswami M, Challagundla M, Decker P et al (2015) Reduction of regulatory T cells by Mogamulizumab, a defucosylated anti-CC chemokine receptor 4 antibody, in patients with aggressive/refractory mycosis fungoides and Sezary syndrome. Clin Cancer Res 21:274–285

Nicot C et al (2001) HTLV-1 p12I protein enhances STAT5 activation and decreases the interleukin-2 requirement for proliferation of primary human peripheral blood mononuclear cells. Blood 98(3):823–829. https://doi.org/10.1182/blood.V98.3.823

Niewiesk S (2016) Animals models of human T cell leukemia virus type I leukemogenesis. ILAR J 57(1):3–11. https://doi.org/10.1093/ilar/ilv045

Olah I et al (2007) Patterns of in vitro lymphoproliferative responses among HTLV-1-infected subjects: upregulation by HTLV-1 during HIV-1 co-infection. Scand J Immunol 65(6):577–580. https://doi.org/10.1111/j.1365-3083.2007.01941.x

Osame M, Usuku K, Izumo S, Ijichi N, Amitani H, Igata A, Matsumoto M, Tara M (1986) HTLV-I associated myelopathy, a new clinical entity. Lancet 1:1031–1032

Paiva A, Casseb J (2014) Sexual transmission of human T-cell lymphotropic virus type 1. Rev Soc Bras Med Trop 47(3):265–274. https://doi.org/10.1590/0037-8682-0232-2013

Paul WE, Seder RA (1994) Lymphocyte responses and cytokines. Cell 76(2):241–251. https://doi.org/10.1016/0092-8674(94)90332-8

Poiesz BJ, Ruscetti FW, Reitz MS, Kalyanaraman VS, Gallo RC (1981) Isolation of a new type C retrovirus (HTLV) in primary uncultured cells of a patient with Sézary T-cell leukaemia. Nature 294(5838):268–271. https://doi.org/10.1038/294268a0

Roucoux DF, Wang B, Smith D et al (2005) A prospective study of sexual transmission of human T lymphotropic virus (HTLV)-I and HTLV-II. J Infect Dis 191(9):1490–1497. https://doi.org/10.1086/429410

Sato T et al (2013) CSF CXCL10, CXCL9, and neopterin as candidate prognostic biomarkers for HTLV-1-associated myelopathy/tropical spastic paraparesis. PLoS Negl Trop Dis 7(10):e2479. https://doi.org/10.1371/journal.pntd.0002479

Takemoto S et al (2000) p53 stabilization and functional impairment in the absence of genetic mutation or the alteration of the p14(ARF)-MDM2 loop in ex vivo and cultured adult T-cell leukemia/lymphoma cells. Blood 95(12):3939–3944

Tamaki K et al (2019) Cerebrospinal fluid CXCL10 as a candidate surrogate marker for HTLV-1-associated myelopathy/tropical spastic paraparesis. Front Microbiol 10:2110. https://doi.org/10.3389/fmicb.2019.02110

Taylor JM et al (2009) Novel role for interleukin-2 receptor-Jak signaling in retrovirus transmission. J Virol 83(22):11467–11476. https://doi.org/10.1128/JVI.00952-09

Temin HM, Mizutani S (1970) RNA-dependent DNA polymerase in virions of Rous sarcoma virus. Nature 226(5252):1211–1213. https://doi.org/10.1038/2261211a0

Treviño A et al (2012) Trends in the prevalence and distribution of HTLV-1 and HTLV-2 infections in Spain. Virol J 9:71. https://doi.org/10.1186/1743-422X-9-71

Tsukasaki K et al (2009) Definition, prognostic factors, treatment, and response criteria of adult T-cell leukemia-lymphoma: a proposal from an international consensus meeting. J Clin Oncol 27(3):453–459. https://doi.org/10.1200/JCO.2008.18.2428

Vandamme AM, Salemi M, Desmyter J (1998) The simian origins of the pathogenic human T-cell lymphotropic virus type I. Trends Microbiol 6(12):477–483. https://doi.org/10.1016/s0966-842x(98)01406-1

Voisset C, Weiss RA, Griffiths DJ (2008) Human RNA "rumor" viruses: the search for novel human retroviruses in chronic disease. Microbiol Mol Biol Rev 72(1):157–196, table of contents. https://doi.org/10.1128/MMBR.00033-07

World Health Organization (2021) Human T-lymphotropic virus type 1: technical report. World Health Organization; WHO IRIS, Geneva. https://apps.who.int/iris/handle/10665/339773

Yamano Y et al (2009) Abnormally high levels of virus-infected IFN-gamma+ CCR4+ CD4+ CD25+ T cells in a retrovirus-associated neuroinflammatory disorder. PLoS One 4(8):e6517. https://doi.org/10.1371/journal.pone.0006517

Yamauchi J et al (2020) Use of cerebrospinal fluid CXCL10 and neopterin as biomarkers in HTLV-1-associated myelopathy/tropical spastic paraparesis treated with steroids. J Neurol Neurosurg Psychiatry 91(3):321–323. https://doi.org/10.1136/jnnp-2019-321955

Yoshimitsu M, Kozako T, Arima N (2013) Prevention of human T-cell lymphotropic virus infection and adult T-cell leukemia. In: T-cell leukemia—characteristics, treatment and prevention. IntechOpen. https://doi.org/10.5772/55427

Zarranz JJ et al (2001) Human T lymphotropic virus type I (HTLV-1) associated myelopathy acquired through a liver transplant. J Neurol Neurosurg Psychiatry 71(6):818

New Era of CAR T Cells for Infectious Viral Diseases

28

Shailender Gugalavath, Deepali Pandit, Khushali Pandit, Albeena Nisar, Gaurav Narula, and Prudhvi Lal Bhukya

Abstract

In an era of rapidly changing global scenarios, emerging infectious viral outbreaks are alarming public health and have become a global concern. Indeed, despite intricate armamentarium and progress in drug development, Infectious Viral Diseases (IVDs) are still on the list of trepidation to humans leading to significant morbidity and mortality. Cellular Immuno Therapy (CIT) has made remarkable achievements in treating hematological lymphomas in the past decade. Later, six Chimeric Antigen Receptor (CAR) products were approved by FDA for use. The CIT approach has led researchers and clinicians to develop CAR technology for the remedy of various IVD. CARs are structured with four major components, i.e., (1) extracellularly located domain for antigen binding, (2) hinge region binding between extracellular domain, (3) transmembrane domain, as well as (4) a signaling domain that mediates the T-cell activation intracellularly. CARs are formed to be modular synthetic receptors with the

S. Gugalavath (✉) · K. Pandit · G. Narula
CAR T Cell Therapy Centre, The Advanced Centre for Treatment, Research and Education in Cancer, Tata Memorial Centre, Navi Mumbai, India

Department of Medical Oncology, Tata Memorial Hospital, Tata Memorial Center, Mumbai, India
e-mail: shailender26naik@gmail.com

D. Pandit
Dr. Bhubaneswar Borooah Cancer Institute, Guwahati, India

A. Nisar
CAR T Cell Therapy Centre, The Advanced Centre for Treatment, Research and Education in Cancer, Tata Memorial Centre, Navi Mumbai, India

P. L. Bhukya
Preclinical Vaccine Testing Laboratory, Rodent Experimentation Facility, ICMR-National Animal Facility Resource Facility for Biomedical Research, Hyderabad, Telangana, India

P. L. Bhukya, A. S (eds.), *Emerging Human Viral Diseases, Volume II*,
https://doi.org/10.1007/978-981-97-4480-0_28

variable region of heavy chain and light chain of mAbs, which are connected by peptide linker and form the single chain variable fragments (scFv). This approach of designing CARs for cancer has inspired researchers to expand it for treating IVD. In this review, we will give a limelight on the CAR T-cell development as well as discuss this novel revolutionary therapy, for the treatment of IVD patients who do not respond to the traditional therapy. CAR T-cell products are developed to enhance and improve the therapeutic advantages for IVD. We have outlined the key accomplishment in CAR T technology targeting the IVD, such as the Corona virus disease (SARS-CoV-2), human immunodeficiency virus, Hepatitis B virus, Hepatitis C virus, Cyto Megalo virus, and Epstein–Barr virus.

Keywords

Infectious viral diseases · CAR T cells · SARS-CoV-2 · HIV–AIDS · HBV · HCV · CMV · EBV

28.1 Introduction

The human body is competent enough in ascertaining pathogens and cancer cells by its immune system. It identifies the epitopes of bacteria, virus, and abnormal cancers in distinguishing self or foreign (Pardoll 2003). The immune system contains adaptive immune T cells, naturally matures and makes functional T cells with unique receptors that react to antigens target sequences, allowing to identify the different pathogens with a probability of T-cell receptors (TCR) to match the antigen or new pathogen epitope (Parkin and Cohen 2001; Vantourout and Hayday 2013). However, pathogens such as bacteria, viruses, and neoplastic cells have the tendency to overturn the immune system because of the zero probability of the TCR match to antigen epitope (Feinerman et al. 2008; Wortel et al. 2020). Immunotherapy, eventually signifying as biotherapy, uses the T cells to reprogram the TCR region specific to the pathogens or tumor-specific antigens. Synthetically, TCR regions are modified and designed to reinforce antigen-binding capacity which is specific to the target antigens of the pathogens and manufacture a functionally active T cell. These specifically designed TCRs are termed Chimeric Antigen Receptors (CARs) (Rosenberg and Restifo 2015; Sadeqi Nezhad et al. 2020).

This CAR technology is considered a novel breakthrough in immunotherapy for cancer treatment of B-cell malignancies. Currently, there are Kymriah and Yescarta commercially available CAR-based therapies for B-cell lymphoma (Pharma et al. 2017; Bishop et al. 2019). However, the CAR T cells for Infectious Viral Disease (IVD) are developing and are in clinical trials (Wei and Han 2017). IVDs are perilous to immunocompromised and chronically infected patients. Mortality and morbidity rates are very high despite the availability of antiviral drugs. The conventional antiviral treatment fails to eradicate infected cells forming the viral reservoirs to incubate for a longer time period and remerge at favorable times making them resistant to available therapies (Freestone 1985; Tedaldi et al. 2003; Català et al. 2011). However, there is a need to treat the IVDs with long-term efficacy and to dissolve the viral reservoirs.

Reprogrammed T cells with CAR technology would be the therapeutic approach which have more specificity to target without HLA restriction and show long-term efficacy, safe, and feasible with durable control over the IVDs (June and Sadelain 2018). The availability of the target for CAR T-cell therapy is elevating rapidly for emerging or reemerging IVDs. Moreover, they are rendered efficiently by viral escape mutation mechanisms and have exhausted phenotypes by mutating the targeted epitope and lowering the major histocompatibility complex. Thus, novel, specifically targeted strategies, such as CARs, T-cell therapy are explored to achieve sterilization and cure. This review emphasizes CAR-based T-cell therapies for IVDs with a focus on progress and upcoming challenges of making CAR T cells for IVDs, such as SAR-CoV2, human immunodeficiency virus (HIV)–AIDS, Hepatitis B virus (HBV), Hepatitis C virus (HCV), Cyto Megalo virus (CMV), and Epstein–Barr virus (EBV).

28.2 CAR T-Cell Therapy

Modification of T cells using CAR-based technology for IVDs is an appearing immunotherapy. Manufacturing process starts with the isolation of peripheral blood cells by leukapheresis which separates the blood components. Isolated T cells are transduced or modified by various gene editing technologies and bioprocessed to infuse back into the patient's own body. However, the manufacturing of CAR T-cell products is carried out under good manufacturing practices adhering to the quality control check points and quality assurance (Levine et al. 2017; Vormittag et al. 2018) (Fig. 28.1).

28.3 CAR Structure

Synthetically developed CARs have major components, such as binding domain for antigen (ABD), hinge region, transmembrane domain, as well as intracellular signaling domain. ABD is the extracellular domain of CAR with variable light chain (V_{LC}) as well as variable heavy chain (V_{HC}) linked each other by single chain variable fragment (scFv) with complementarity determining regions which impacts the affinity as well as specificity of the CAR to its target epitope (Chailyan et al. 2011). In general, CAR-developed ABD are derived from the monoclonal antibodies specific to the cancer or IVDs. ABD is extended to the transmembrane domain by hinge or spacer region which helps in providing the adjustability to conquer the steric hindrance. The transmembrane domain helps in anchoring the ABD to the membrane, which helps in stability and signaling to the intracellular domain. Intracellular signaling domain is a major part of CAR structure for the functional activity of the CAR T cells (Fujiwara et al. 2020; Dotti et al. 2014; Srivastava and Riddell 2015).

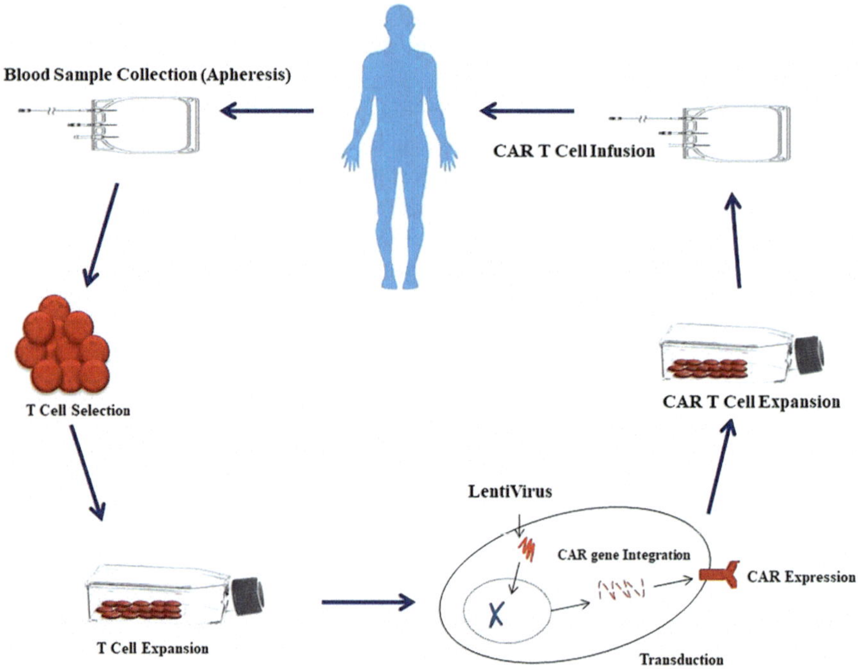

Fig. 28.1 Flow of the adoptive CAR T-cell therapy. Process involves the T-cell isolation using leukapheresis and T-cell selection and transduction followed by CART cells expansion and infusion into patient with CAR T product

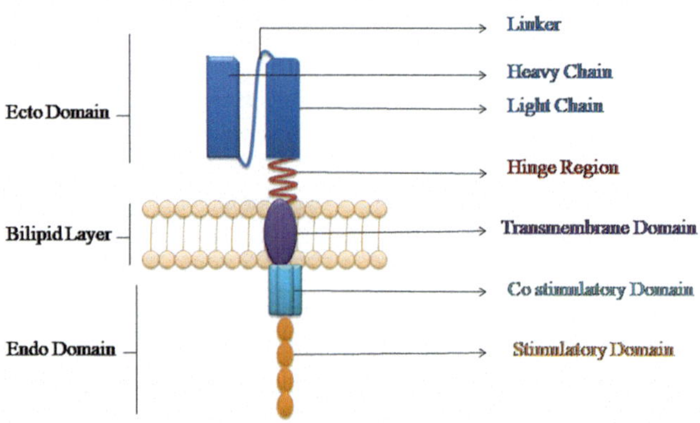

Fig. 28.2 Typical CAR has antigen-binding domain obtained from the scFv of monoclonal antibody and intracellular signaling domain, linked by hinge region and transmembrane

Based on the advancement of the CAR signaling endodomain, they are differentiated into different generations (Fig. 28.2).

First-generation CARs activate the T cells by the CD3ζ intracellular signaling domain. Furthermore, to enhance the expansion and functionality of CARs have been included with the intracellular costimulatory domain of CD28 or 4-1BB. Whereas, the CARs of the third generation were a combination of costimulatory domains of CD28 and 4-1BB. However, T cells that are switched for cytokine-mediated killing with modified transgene for cytokine release with IL-2 activation are under fourth-generation CARs called TRUCKs ("T cells redirected for antigen-unrestricted cytokine-initiated killing") (Sievers et al. 2020). However, IVDs remain as the major cause of morbidity globally; strategy of CARs development has been used to study clinically for various IVD and reviewed below.

28.3.1 CAR Technology for SARS-CoV-2

SARS CoV-2 caused COVID-19 diseases, and turned out to be a global pandemic in recent years with a high rate of infection. However, current effective treatment for severe COVID-19 diseases such as neutralizing antibodies and small molecules drugs lacks long-term immunity (Adil et al. 2021). CAR-based therapeutic approach for the treatment of COVID-19 has attracted the scientific community. Recently, Xueyang Guo et al. designed and generated, CAR T cells that effectively aim RBD region of the SARS-CoV-2 spike protein and effectively induced the downstream signaling-like ribosomal protein S6 phosphorylation, with elevated antigen activation, CD69, perforin, interferon-γ, granzyme B, and Fas-ligand (Guo et al. 2021). SARS-CoV-2 CAR T cells have a neutralized target by lysis of the cells with GZMB/perforin pathway as a live drug approach for therapy of COVID-19 (Ma et al. 2020). Tianchuan Zhu et al. developed a bispecific CART cell with two scFv named CR3022 and B38 that targets the two different epitopes of spikes protein to overcome the resistance to virus. Epitope region of CR3022/B38 NVs with absence of Fc fragments could prevent antibody dependent enhancement effects. This bispecific CAR was developed with nanovesicles for delivery of remdesivir to target sites to reduce side effects (Zhu et al. 2021). Unlike CAR T cells for SARS-CoV-2, Ma et al. have developed the CAR NK cells with scFv domain of CR3022 which has neutralized the SARS-CoV1 and SARS-CoV-2 in vitro (Ma et al. 2020).

28.3.2 CAR Technology for HIV

HIV comes under the genus of lentivirus that infects humans. UNAIDS (United Nations Programme on HIV/AIDS) reported 1.5 million new HIV infections yearly, coming out as an epidemic and 37.9 million people are morbid to HIV globally (Ssekalembe et al. 2020). However, effective combination antiretroviral therapy has shortened the morbid condition and mortality (HIV-CAUSAL Collaboration et al. 2010; Naldini 2015). Therapy has increased the life expectancy and led to the normal life of the patient making HIV into chronic diseases, but prolonged use of this therapy rises with certain side effects, such as aging, low-grade inflammation, and frailty syndrome (Domingo and Vidal 2011). To overcome these problems, CAR T-cell therapy approaches for the treatment of HIV infections have been investigated with new anti-HIV CAR technology.

28.3.3 CAR Design and Construct of Anti-HIV Targets

Based on the advancement of CAR construct and design, different generation CARs were developed with various target sites against HIV. The first-generation CAR construct was targeted against the HIV envelope protein. CAR construct was anti-CD4 designed with a downstream intracellular signaling molecule of CD3ζ. Efficiency and safety tests of several CARζ or scFv combined with one or two intracellular costimulatory domains with CD4 construct were performed but failed to show efficacy. Reasons for low efficacy could be no effective target viral antigen stimulation and escape mechanism but showed low toxicities (Mitsuyasu et al. 2000; Walker et al. 2000; Deeks et al. 2002). Furthermore, to overcome the limitation, CD4ζ CAR was combined with various additional targets such as C46 peptide inhibitor and knockout mechanism such as Zinc finger nucleases, CRISPR–Cas9; transcription activator-like nucleases were used for HIV by silencing chemokine receptor CCR5 to generate CART cells to overcome the resistance of HIV infection (Maldini et al. 2020; Shimizu et al. 2010; Ghanem et al. 2018; Ringpis et al. 2012).

Second-generation anti-HIV CAR T cells showed advancement in targeting the antigen and suppressing HIV replication. CAR technology developed against HIV using modified T cells with costimulatory 4-1BB domain suppressed the viral replication to 50-fold (Leibman et al. 2017). Anti-glycan-targeted CAR with CCR5 silencing decreased viral replication (Hütter et al. 2015). Second-generation CARs were more target specific to safeguards CD4+ T cells against HIV infection as well as promoted the T cells persistence in the absence of antigen target when compared with 4-1BB costimulatory domain with CD28 domain (Maldini et al. 2020).

The third-generation anti-HIV CAR was designed with scFv region from neutralizing antibodies (bNAb) having CD3ζ–CD28–CD137 target sites shown with higher binding potential and specificity and reduced off target. bNAb VRC01 were able to redirect the CD8+ T cells in contrast to gp120 (Liu et al. 2016). CAR T cells developed based on bNAb were able to kill CD4+ T cells infected with HIV after reactivation isolated from CAR T receiving patients led to virus suppression and low immunogenicity and modified receptor activity for HIV entry. Yet, research on bNAb (VRC01) is underway for clinical trials for latent virus reservoir removal (NCT03240328) (ClinicalTrials.gov n.d.) (Fig. 28.3).

28.3.4 Bi- and Trispecific CARs for HIV

Not only costimulatory domain-based CAR generation are developed but also CARs are classified based on number of target antigens epitopes as bi- and trispecific CARs which could probably reduce the cellular HIV infection and show

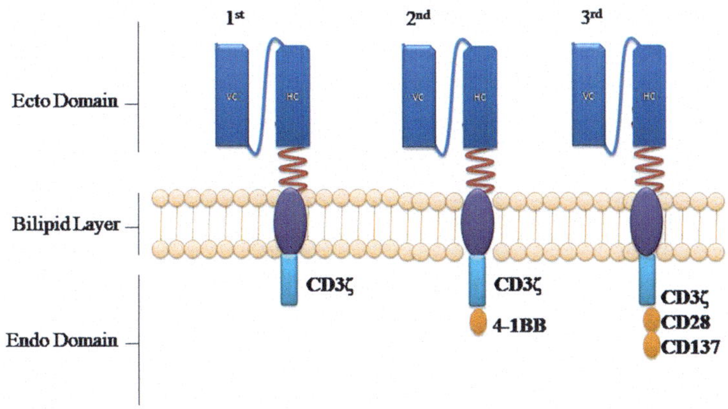

Fig. 28.3 Schematic evolution of HIV–CAR generations. First-generation CARs contain the CD3ζ chain of the T-cell receptor complex, whereas second-generation receptors contain one costimulatory molecule CD3ζ or 4-1BB and third-generation CARs contain two costimulatory molecules, respectively, such as CD3ζ, CD28, and CD137

specificity and affinity. HIV GP120 an envelope glycoprotein was used as target which has an essential role in virus entry to the host and CD4 receptor which is found to be immune cells was together as target in CAR construct has shown better suppressive activity compared to CD4 alone (Liu et al. 2015; Ghanem et al. 2018). To improve the potency of the CAR, CXCR5 chemokine receptor has a major role in HIV entry which was added to the CAR construct making trispecific CAR. Another study by Anthony-Gonda et al. (2019) developed trispecific CARs with three distinct epitopes, such as envelope glycoprotein 120 with CD4 binding site, and co receptor binding site and glycoprotein 41 near to proximal region of membrane has efficiently killed cells infected with HIV in vivo conditions (Anthony-Gonda et al. 2019). However, anti-HIV–CAR design and construct were advanced either by generation levels or by developing the bi-, tri-, or multitarget-specific antigen which could eradicate the HIV reservoir and elevated toxicity and potency and overcome the resistance.

To enhance the CAR T cells, culturing and expanding differ to CD19 CAR T cells and HIV-specific CARs. A comparative study showed elevated expansion of the CD19 CAR T cells with IL-7 and IL-15 cytokines, whereas the same cytokines have activated and promoted the virus production rather than the expansion of HIV-specific CART cells (Zurko et al. 2021). Targeting the immune checkpoints like PD-1 using gene editing techniques such as CRISPR, gene silencing, or antibody blockade could improve antitumor response for cancer-based CARs. However, a similar approach can be investigated in HIV-specific CARs (Qian et al. 2021). A combinational approach will facilitate the anti-HIV CARs in eradicating the virus and kill the cells infected with HIV with long-term efficacy.

28.3.5 CAR Technology for HBV

HBV infects about 250 million people, with a higher incidence rate in Asia and sub-Saharan Africa. Around 4.7 million cases are newly reported per year with a rapidly increasing mortality rate (World Health Organization et al. 2020). Current HBV therapies, such as nucleoside analogues, reduce the inflammatory events but do not eradicate the infection (Papatheodoridis et al. 2002). Different immunogenic vaccines of anti-PD1/anti-PD-L1 targets are used to treat HBV but failed in delivering the efficacy (Li et al. 2020). To overcome these limitations CAR T-cell therapy targets HBV surface proteins S as well as L on the membrane of HBV-infected hepatocytes. CAR T cells modified with S as well as L surface proteins were able to dissolve HBsAg as well as HBsAg hepatocytes by releasing IL-2 and IFNγ (Festag et al. 2019; Kruse et al. 2018). However, S-CAR T cells showed elevated cytokine release than L-CAR T cells due to higher surface S-protein than L-protein. Furthermore, L as well as S CAR T cells eradicated hepatocytes infected with HBV. CD8 expressing S-CAR T cells developed from HBV-immunocompetent mice effectively reduced HBV replication (Bohne et al. 2008; Krebs et al. 2013; Festag et al. 2019).

Furthermore, HBV was targeted using the HBsAg CARs with long spacer (HBs–G4m–CAR) showed significant release of cytokine type II class of interferons, Interleukin-2, as well as TNF-α, but were a long way in wiping out HBV-infected cell when studied in vitro, but has shown extreme anti-HBV activity compared to HBV inhibitors (Kruse et al. 2018). However, comparative investigation of S and HBsAg CARs at in vitro and in vivo could evaluate the potency of the available CARs. Combinational targets of HBV for CAR T-cell development would have finer control over the HBV diseases.

28.3.6 CAR Technology for HCV

Worldwide, about 58 million individuals were with chronic HCV infection and 1.5 million HCV infections per year. CARs developed for the HCV were targeted on the envelope glycoprotein E2 showed efficient lysis of HCV-infected target cells. The targeting highly conserved E2 epitope had a direct effect on viral hypervariability and cell-to-cell transmission (Sautto et al. 2013, 2016).

28.3.7 CAR Technology for Human CMV

The CARs developed against anti-gB were infected with CMV cells release TNFα as well as IFNγ and blocked CMV replication. However, the target region was isolated from the mouse monoclonal antibodies for virus neutralization. To have more efficacy to the target, modification in the spacer region of CAR as a lengthy spacer with modification in CH2–CH3 Fc domain of mouse monoclonal antibody serves as antigen-binding domain against CMV-infected cells. However, in vivo studies for gB CAR have yet to be done. Investigation on new epitopes having adjustable spacer length might give more chances in targeting CMV (Full et al. 2010; Proff et al. 2016, 2018).

28.3.8 CAR T-Cell Technology for EBV

EBV-associated malignancies were targeted with latent membrane protein 1 (LMP1) using CAR construct. In vitro studies, EBV CAR T cells cultured together cells of nasopharyngeal carcinoma that showed elevated LMP1 and released IFNγ and IL-2. Furthermore, in vivo studies conducted showed overexpressing LMP1 and reduced cancer progress in xenograft mouse model (Tang et al. 2014; D'Aloia et al. 2018).

28.4 Conclusion

CAR technology or CAR T-cell therapy for cancer specifically in B-cell acute lymphoblastic leukemia has gained the approved clinical application in immunotherapy. CARs are designed to target more accurately and exclusively to the pathogen, so that it can avert off targets and erode escape mechanisms as well as viral reservoirs. Considerable CARs against IVDs have been elucidated; importantly outlined for SARS-CoV-2, HIV, HBV, HVC, and EBV. However, research on anti-HIV CARs is in clinical trials now, whereas the other HBV, HCV, and CMV are under pre-clinical. Multi-target approach forming bi- and trispecific CARs to the target has yet to be investigated in detail regarding the safety and efficacy to have promising data and to pilot the CAR T-cell-based technology application.

Acknowledgements We thank the authorities of The Advanced Centre for Treatment, Research and Education in Cancer and Tata Memorial Hospital, Tata Memorial Centre for continuous support and encouragement.

Author ContributionsGS planned and designed the manuscript; GS, DP and KP wrote the manuscript; GS, GN, AN, PB reviewed the manuscript.

Funding This project was financially supported by DBT–BIRAC, DBT Project Entitled "First-In-Human Clinical Trial Using an Indigenously Developed CD-19 Targeted CAR T-cells".

References

Adil MT, Rahman R, Whitelaw D, Jain V, Al-Taan O, Rashid F, Munasinghe A, Jambulingam P (2021) SARS-CoV-2 and the pandemic of COVID-19. Postgrad Med J 97(1144):110–116

Anthony-Gonda K, Bardhi A, Ray A, Flerin N, Li M, Chen W, Ochsenbauer C et al (2019) Multispecific anti-HIV duoCAR-T cells display broad in vitro antiviral activity and potent in vivo elimination of HIV-infected cells in a humanized mouse model. Sci Transl Med 11(504):aav5685. https://doi.org/10.1126/scitranslmed.aav5685

Bishop MR, Maziarz RT, Waller EK, Jäger U, Westin JR, McGuirk JP, Fleury I et al (2019) Tisagenlecleucel in relapsed/refractory diffuse large B-cell lymphoma patients without measurable disease at infusion. Blood Adv 3(14):2230–2236

Bohne F, Chmielewski M, Ebert G, Wiegmann K, Kürschner T, Schulze A, Urban S, Krönke M, Abken H, Protzer U (2008) T cells redirected against hepatitis B virus surface proteins eliminate infected hepatocytes. Gastroenterology 134(1):239–247

Català L, Orcau A, de Olalla PG, Millet J-P, Rodríguez-Mondragón A, Caylà JA, TB-HIV Working Group (2011) Survival of a large cohort of HIV-infected tuberculosis patients in the era of highly active antiretroviral treatment. Int J Tuberculosis Lung Dis 15(2):263–69, i

Chailyan A, Marcatili P, Tramontano A (2011) The association of heavy and light chain variable domains in antibodies: implications for antigen specificity. FEBS J 278(16):2858–2866

ClinicalTrials.gov (n.d.) The effect of chimeric antigen receptor (CAR)-T cell therapy on the reconstitution of HIV-specific immune function. https://clinicaltrials.gov/ct2/show/NCT03240328. Accessed 31 Jul 2022

D'Aloia MM, Zizzari IG, Sacchetti B, Pierelli L, Alimandi M (2018) CAR-T cells: the long and winding road to solid tumors. Cell Death Dis 9(3):282

Deeks SG, Wagner B, Anton PA, Mitsuyasu RT, Scadden DT, Huang C, Macken C et al (2002) A phase II randomized study of HIV-specific T-cell gene therapy in subjects with undetectable plasma viremia on combination antiretroviral therapy. Mol Ther 5(6):788–797

Domingo P, Vidal F (2011) Combination antiretroviral therapy. Expert Opin Pharmacother 12(7):995–998

Dotti G, Gottschalk S, Savoldo B, Brenner MK (2014) Design and development of therapies using chimeric antigen receptor-expressing T cells. Immunol Rev 257(1):107–126

Feinerman O, Germain RN, Altan-Bonnet G (2008) Quantitative challenges in understanding ligand discrimination by alphabeta T cells. Mol Immunol 45(3):619–631

Festag MM, Festag J, Fräßle SP, Asen T, Sacherl J, Schreiber S, Mück-Häusl MA, Busch DH, Wisskirchen K, Protzer U (2019) Evaluation of a fully human, hepatitis B virus-specific chimeric antigen receptor in an immunocompetent mouse model. Mol Ther 27(5):947–959

Freestone DS (1985) The need for new antiviral agents. Antivir Res 5(6):307–324

Fujiwara K, Tsunei A, Kusabuka H, Ogaki E, Tachibana M, Okada N (2020) Hinge and transmembrane domains of chimeric antigen receptor regulate receptor expression and signaling threshold. Cells 9(5):1182. https://doi.org/10.3390/cells9051182

Full F, Lehner M, Thonn V, Goetz G, Scholz B, Kaufmann KB, Mach M, Abken H, Holter W, Ensser A (2010) T cells engineered with a cytomegalovirus-specific chimeric immunoreceptor. J Virol 84(8):4083–4088

Ghanem MH, Bolivar-Wagers S, Dey B, Hajduczki A, Vargas-Inchaustegui DA, Danielson DT, Bundoc V, Liu L, Berger EA (2018) Bispecific chimeric antigen receptors targeting the CD4 binding site and high-mannose glycans of gp120 optimized for anti-human immunodeficiency virus potency and breadth with minimal immunogenicity. Cytotherapy 20(3):407–419

Guo X, Kazanova A, Stephanie Thurmond H, Saragovi U, Rudd CE (2021) Effective chimeric antigen receptor T cells against SARS-CoV-2. iScience 24(11):103295

HIV-CAUSAL Collaboration, Ray M, Logan R, Sterne JAC, Hernández-Díaz S, Robins JM, Sabin C et al (2010) The effect of combined antiretroviral therapy on the overall mortality of HIV-infected individuals. AIDS 24(1):123–137

Hütter G, Bodor J, Ledger S, Boyd M, Millington M, Tsie M, Symonds G (2015) CCR5 targeted cell therapy for HIV and prevention of viral escape. Viruses 7(8):4186–4203

June CH, Sadelain M (2018) Chimeric antigen receptor therapy. N Engl J Med. https://doi.org/10.1056/nejmra1706169

Kite Pharma (2017) Yescarta® (axicabtagene ciloleucel) [Package Insert]

Krebs K, Böttinger N, Huang L-R, Chmielewski M, Arzberger S, Gasteiger G, Jäger C et al (2013) T cells expressing a chimeric antigen receptor that binds hepatitis B virus envelope proteins control virus replication in mice. Gastroenterology 145(2):456–465

Kruse RL, Shum T, Tashiro H, Barzi M, Yi Z, Whitten-Bauer C, Legras X et al (2018) HBsAg-redirected T cells exhibit antiviral activity in HBV-infected human liver chimeric mice. Cytotherapy 20(5):697–705

Leibman RS, Richardson MW, Ellebrecht CT, Maldini CR, Glover JA, Secreto AJ, Kulikovskaya I et al (2017) Supraphysiologic control over HIV-1 replication mediated by CD8 T cells expressing a re-engineered CD4-based chimeric antigen receptor. PLoS Pathog 13(10):e1006613

Levine BL, Miskin J, Wonnacott K, Keir C (2017) Global manufacturing of CAR T cell therapy. Mol Ther Methods Clin Dev 4:92–101

Li B, Yan C, Zhu J, Chen X, Qihan F, Zhang H, Tong Z et al (2020) Anti-PD-1/PD-L1 blockade immunotherapy employed in treating hepatitis B virus infection-related advanced hepatocellular carcinoma: a literature review. Front Immunol 11:1037

Liu L, Patel B, Ghanem MH, Bundoc V, Zheng Z, Morgan RA, Rosenberg SA, Dey B, Berger EA (2015) Novel CD4-based bispecific chimeric antigen receptor designed for enhanced anti-HIV potency and absence of HIV entry receptor activity. J Virol 89(13):6685–6694

Liu B, Zou F, Lu L, Chen C, He D, Zhang X, Tang X, Liu C, Li L, Zhang H (2016) Chimeric antigen receptor T cells guided by the single-chain Fv of a broadly neutralizing antibody specifically and effectively eradicate virus reactivated from latency in CD4+ T lymphocytes isolated from HIV-1-infected individuals receiving suppressive combined antiretroviral therapy. J Virol 90(21):9712–9724

Ma M, Badeti S, Geng K, Liu D (2020) Efficacy of targeting SARS-CoV-2 by CAR-NK cells. bioRxiv. https://doi.org/10.1101/2020.08.11.247320

Maldini CR, Claiborne DT, Okawa K, Chen T, Dopkin DL, Shan X, Power KA et al (2020) Dual CD4-based CAR T cells with distinct costimulatory domains mitigate HIV pathogenesis in vivo. Nat Med 26(11):1776–1787

Mitsuyasu RT, Anton PA, Deeks SG et al (2000) Prolonged survival and tissue trafficking following adoptive transfer of CD4ζ gene-modified autologous CD4+ and CD8+ T cells in human immunodeficiency virus-infected subjects. Blood 96(3):785–793

Naldini L (2015) Gene therapy returns to centre stage. Nature 526(7573):351–360

Papatheodoridis GV, Dimou E, Papadimitropoulos V (2002) Nucleoside analogues for chronic hepatitis B: antiviral efficacy and viral resistance. Am J Gastroenterol 97(7):1618–1628

Pardoll D (2003) Does the immune system see tumors as foreign or self? Annu Rev Immunol 21:807–839

Parkin J, Cohen B (2001) An overview of the immune system. Lancet 357(9270):1777–1789

Proff J, Walterskirchen C, Brey C, Geyeregger R, Full F, Ensser A, Lehner M, Holter W (2016) Cytomegalovirus-infected cells resist T cell mediated killing in an HLA-recognition independent manner. Front Microbiol. https://doi.org/10.3389/fmicb.2016.00844

Proff J, Brey CU, Ensser A, Holter W, Lehner M (2018) Turning the tables on cytomegalovirus: targeting viral Fc receptors by CARs containing mutated CH2–CH3 IgG spacer domains. J Transl Med 16(1):26

Qian W, Zhao A, Liu H, Lei W, Liang Y, Yuan X (2021) Safety and efficacy of CD19 CAR-T cells co-expressing IL-7 and CCL19 in combination with anti-PD-1 antibody for refractory/relapsed DLBCL: preliminary data from the phase Ib trial (NCT04381741). Blood. https://doi.org/10.1182/blood-2021-144523

Ringpis G-EE, Shimizu S, Arokium H, Camba-Colón J, Carroll MV, Cortado R, Xie Y et al (2012) Engineering HIV-1-resistant T-cells from short-hairpin RNA-expressing hematopoietic stem/progenitor cells in humanized BLT mice. PLoS One 7(12):e53492

Rosenberg SA, Restifo NP (2015) Adoptive cell transfer as personalized immunotherapy for human cancer. Science 348(6230):62–68

Sadeqi Nezhad M, Seifalian A, Bagheri N, Yaghoubi S, Karimi MH, Adbollahpour-Alitappeh M (2020) Chimeric antigen receptor based therapy as a potential approach in autoimmune diseases: how close are we to the treatment? Front Immunol 11:603237

Sautto G, Tarr AW, Mancini N, Clementi M (2013) Structural and antigenic definition of hepatitis C virus E2 glycoprotein epitopes targeted by monoclonal antibodies. Clin Dev Immunol 2013:450963

Sautto GA, Wisskirchen K, Clementi N, Castelli M, Diotti RA, Graf J, Clementi M, Burioni R, Protzer U, Mancini N (2016) Chimeric antigen receptor (CAR)-engineered T cells redirected against hepatitis C virus (HCV) E2 glycoprotein. Gut 65(3):512–523

Shimizu S, Hong P, Arumugam B, Pokomo L, Boyer J, Koizumi N, Kittipongdaja P et al (2010) A highly efficient short hairpin RNA potently down-regulates CCR5 expression in systemic lymphoid organs in the Hu-BLT mouse model. Blood 115(8):1534–1544

Sievers NM, Dörrie J, Schaft N (2020) CARs: beyond T cells and T cell-derived signaling domains. Int J Mol Sci 21(10):3525. https://doi.org/10.3390/ijms21103525

Srivastava S, Riddell SR (2015) Engineering CAR-T cells: design concepts. Trends Immunol 36(8):494–502

Ssekalembe G, Isfandiari MA, Suprianto H (2020) Current status towards 90-90-90 UNAIDS target and factors associated with HIV viral load suppression in Kediri City, Indonesia. HIV/AIDS 12:47–57

Tang X, Zhou Y, Li W, Tang Q, Chen R, Zhu J, Feng Z (2014) T cells expressing a LMP1-specific chimeric antigen receptor mediate antitumor effects against LMP1-positive nasopharyngeal carcinoma cells in vitro and in vivo. J Biomed Res 28(6):468–475

Tedaldi EM, Baker RK, Moorman AC, Alzola CF, Furhrer J, McCabe RE, Wood KC, Holmberg SD, HIV Outpatient Study (HOPS) Investigators (2003) Influence of coinfection with hepatitis C virus on morbidity and mortality due to human immunodeficiency virus infection in the era of highly active antiretroviral therapy. Clin Infect Dis 36(3):363–367

Vantourout P, Hayday A (2013) Six-of-the-best: unique contributions of γδ T cells to immunology. Nat Rev Immunol 13(2):88–100

Vormittag P, Gunn R, Ghorashian S, Veraitch FS (2018) A guide to manufacturing CAR T cell therapies. Curr Opin Biotechnol 53:164–181

Walker RE, Bechtel CM, Natarajan V, Baseler M, Hege KM, Metcalf JA, Stevens R et al (2000) Long-term in vivo survival of receptor-modified syngeneic T cells in patients with human immunodeficiency virus infection. Blood 96(2):467–474

Wei J, Han W (2017) CART trials are going ahead. Sci China Life Sci 60(11):1276–1279

World Health Organization et al (2020) World hepatitis day: fast-tracking the elimination of hepatitis B among mothers and children. World Health Organization

Wortel IMN, Keşmir C, de Boer RJ, Mandl JN, Textor J (2020) Is T cell negative selection a learning algorithm? Cells 9(3):690. https://doi.org/10.3390/cells9030690

Zhu T, Xiao Y, Meng X, Tang L, Li B, Zhao Z, Tan Q, Shan H, Liu L, Huang X (2021) Nanovesicles derived from bispecific CAR-T cells targeting the spike protein of SARS-CoV-2 for treating COVID-19. J Nanobiotechnol 19(1):391

Zurko JC, Xu H, Chaney K, Fenske TS, Hamadani M, Schneider D, Hari P, Johnson B, Shah NN (2021) Bispecific LV20.19 CAR T-cells expanded in IL-7 and IL-15 have greater polyfunctionality and polyfunctional strength than CAR T-cells expanded in IL-2. Blood. https://doi.org/10.1182/blood-2021-146882

SPRINGER NATURE

GPSR Compliance

The European Union's (EU) General Product Safety Regulation (GPSR) is a set of rules that requires consumer products to be safe and our obligations to ensure this.

If you have any concerns about our products, you can contact us on ProductSafety@springernature.com

In case Publisher is established outside the EU, the EU authorized representative is:

Springer Nature Customer Service Center GmbH
Europaplatz 3
69115 Heidelberg, Germany

The manufacturer's authorised representative in the EU is Springer
Nature Customer Service Centre GmbH, Europaplatz 3, 69115 Heidelberg,
Germany. If you have any concerns regarding our products, please
contact ProductSafety@springernature.com

Printed and bound by CPI Group (UK) Ltd, Croydon, CR0 4YY
30/09/2026
02225302-0019